SLOPE**STABILITY**

in Surface Mining

Good luck, Anthony!

EDITED BY

William A. Hustrulid
Michael K. McCarter
Dirk J.A. Van Zyl

William Hustrulid

Best wishes on a successful career

Published by the
Society for Mining, Metallurgy, and Exploration, Inc.

Society for Mining, Metallurgy, and Exploration, Inc. (SME)
8307 Shaffer Parkway
Littleton, Colorado, USA 80127
(303) 973-9550 / (800) 763-3132
www.smenet.org

SME advances the worldwide minerals community through information exchange and professional development. With more than 16,000 members in 50 countries, SME is the world's largest professional association of mineral professionals.

Disclaimer
The papers contained in this volume are published as supplied by individual authors. Any statement or views presented here are those of individual authors and are not necessarily those of the Society for Mining, Metallurgy, and Exploration, Inc. The mention of trade names for commercial products does not imply the approval or endorsement of SME.

Cover photograph courtesy of Robert Sharon, Director, Geotechnical Engineering, Barrick Goldstrike Mines Inc. It shows the southeast highwall of Barrick Goldstrike's Betze-Post pit, located in northern Nevada. This August 1999 photograph shows final cleanup of an instability that occurred in early 1997 and successful development of the ultimate highwall in this section of the open pit.

ISBN 0-87335-194-0

Library of Congress Cataloging-in-Publication Data

Slope stability in surface mining / edited by William A. Hustrulid, Michael K. McCarter, and Dirk J.A. Van Zyl.
p. cm.
Includes bibliographical references and index.
ISBN 0-87335-194-0
1. Strip mining. 2. Slopes (Soil mechanics) I. Hustrulid, W.A. II. McCarter, M.K. (M. Kim) III. Van Zyl, Dirk J.A.

TN291 .S56 2001
622'.292–dc21

00-052235

Contents

This book is dedicated to Chuck Brawner
for his pioneering efforts in slope stability and his leadership in
establishing symposia and publications to advance the
art and science of slope stability in surface mining.

Preface

The First International Conference on Stability in Open Pit Mining was held in 1970 in Vancouver, British Columbia. The purpose of the conference was to review, in practical terms, what was known about the rock mechanics relating to slope stability. The guiding force for this meeting was C.O. Brawner. Sponsorship came from the Center for Continuing Education, University of British Columbia; the Engineering Institute of Canada (B.C. Section); and the Canadian Institute of Mining and Metallurgy (B.C. Section). The contents of the proceedings published by the Society of Mining Engineers of the American Institute of Mining, Metallurgical, and Petroleum Engineers (AIME) included:

Introduction, *C.O. Brawner*

The Role of Slope Stability in the Economics, Design and Operation of Open Pit Mines, *Richard M. Stewart and Bruce A. Kennedy*

Geologic Factors Controlling Slope Stability in Open Pit Mines, *F.D. Patton and D.U. Deere*

Influence of Rock Structure on the Stability of Rock Slopes, *Evert Hoek*

The Influence of Groundwater on Stability, *Norbert R. Morgenstern*

The Influence and Evaluation of Blasting on Stability, *Alan Bauer and Peter N. Calder*

Influence of Earthquakes on Stability, *Robert V. Whitman*

Methods of Analysis of Stability of Rock Slopes, *David L. Pentz*

Field Instrumentation for Rock Slopes, *K. Barron, D.G.F. Hedley, and D.F. Coates*

The Stabilization of Slopes in Open-Pit Mining, *H.Q. Golder*

Design and Construction of Tailings Dams, *Leo Casagrande and B.N. McIver*

Case Studies of Stability on Mining Projects, *C.O. Brawner*

Panel Discussion, *K. Barron, C.O. Brawner, L. Casagrande, H.Q. Golder, B.A. Kennedy, and D.L. Pentz*

Conference Summary, *Evert Hoek*

This conference generated so much interest that the Second International Conference on Stability in Open Pit Mining was held in 1971 in Vancouver. At this conference, the practical application of the factors influencing stability was stressed, with significant emphasis placed on case studies. Again the guiding force behind the meeting was C.O Brawner. The sponsors for this conference were the Center for Continuing Education, University of British Columbia; the Engineering Institute of Canada (B.C. Section); the Canadian Institute of Mining and Metallurgy (B.C. Section); and AIME. The proceedings covered the requirements for stability in mining; a review of recent stability research in the United States, England, and Canada; the influence of pit slopes on the economics of open pit mining; and the requirements for stability investigations and geological studies for slopes in soil, rock, and tailings dams.

The contents of the proceedings published by the Society of Mining Engineers of the AIME included:

Requirements for Stability in Open Pit Mining, *Richard M. Stewart and B.L. Seegmiller*

The Practical Side of Mining Research at Kennecott Copper Corporation, *C.D. Broadbent*

Recent Rock Slope Stability Research at the Royal School of Mines, London, *E. Hoek*

Recent Research on Rock Slope Stability by the Mining Research Center, *G. Herget*

Pit Slopes—Their Influence on the Design and Economics of Open Pit Mines, *R.B. Moffitt, T.W. Friese-Greene, and R.M. Lillico*

The Stability of Natural and Man-Made Slopes in Soil and Rock, *H.Q. Golder*

Stability of Slopes in Overburden Excavations, *T. Cameron Kenney*

Stability Investigations for Tailings Dams, *J.C. Osler*

Geological Investigations to Evaluate Stability, *Richard E. Goodman*

Redesign and Construction of a Tailings Dam to Resist Earthquakes, *C.O. Brawner*

Tailings Dams in British Columbia, *Earle J. Klohn*

The Control of Water in Tailings Ponds, *A.L. Galpin*

Blasting Effects and Their Control in Open Pit Mining, *L. Oriard*

A Slide in Cretaceous Bedrock at Devon, Alberta, *K.D. Eigenbrod and N.R. Morgenstern*

A Study of the Stability of a Disused Limestone Quarry Face in the Mendip Rills, England, *D. Roberts and E. Hoek*

Rock Mechanics and Slope Stability at Mount Isa, Australia, *K. Rosengren*

The Third International Conference on Stability in Open Pit Mining was held in 1981 in Vancouver. As with the first and second conferences, the guiding force behind the meeting was C.O. Brawner. This time the advisory organizations were the Society of Mining Engineers of the AIME, the B.C. and Yukon Chamber of Mines, the B.C. Mining Association, and the B.C. Department of Mines and Petroleum Resources.

The program was developed in three parts:

1. State of the Art–Rock Slope Stability
2. Investigation, Research, and Design for Stability in Surface Mining
3. Case Examples of Stability in Surface Mining

State of the Art

The Role of Slope Stability in the Economics, Design and Operation of Open Pit Mines–An Update, *Michael R. Richings*

Influence of Rock Structure on Stability, *Carl D. Broadbent and Zavis M. Zavodni*

Influence and Control of Groundwater in Large Slopes, *Adrian Brown*

Influence of Blasting on Slope Stability; State of the Art, *L.L. Oriard*

Influence of Earthquakes on Rock Slope Stability, *Charles E. Glass*

Mechanics of Rock Slope Failure, *Douglas R. Pitea and Dennis C. Martin*

Shear Strength Investigations for Surface Mining, *Nick Barton*

Slope Stability Analysis Techniques Incorporating Uncertainty in Critical Parameters, *D.L. Pentz*

Monitoring Pit Slope Behavior, *Richard D.Call*

Artificial Support of Rock Slopes, *Ben L. Seegmiller*

Stabilization of Rock Slopes, *C.O. Brawner*

Research Requirements in Surface Mine Stability and Planning, *G. Herget and O. Garg*

Investigation, Research, and Design

Sedimentological Control of Mining Conditions in the Permian Measures of the Bowen Basin, Australia, *C.W. Mallett*

Geology and Rock Slope Stability–Application of the "Key Block" Concept for Rock Slopes, *Richard E. Goodman, and Gen-Hua Shi*

Analysis of Slope Stability in Very Heavily Jointed or Weathered Rock Masses, *Evert Hoek*

The Application of Stochastic Medium Theory to the Problem of Surface Movements Due to Open Pit Mining, *Liu Baoshen and Lin Dezhang*

Analytical Estimation of Parabolic Water Table Drawdown to a Slope Face, *Stanley M. Miller*

A Computer Program for Footwall Slope Stability Analysis in Steeply Dipping Bedded Deposits, *Brian Stimpson and Keith E. Robinson*

Analysis of Bolt Reinforcement in Rock Slopes, *Francois E. Heuze*

A Simple Core Orientation Technique, *R.D. Call, J.P. Savely, and R. Pakalnis*

Monitoring the Behavior of High Rock Slopes, *W.B. Tijmann*

Blasting to Achieve Slope Stability in Weak Rock, *G. Harries*

Blasting Practices for Improved Coal Strip Mine Highwall Safety and Cost, *Francis S. Kendorski and Michael F. Dunn*

Production Blasting and the Development of Open Pit Slopes, *John P. Ashby*

Case Examples

Practical Aspects of Wall Stability at Brenda Mines Ltd., Peachland, B.C., *G.H. Blackwell and Peter N. Calder*

Slope Instability at Inspiration's Mine, *James P. Savely and Victor L. Kastner*

Open Pit Slope Stability Investigation of the Hasancelebi Iron Ore Deposit, Turkey, *Caner Zanbak, Kemal A. Erguvanli, Erdogan Yuzer, and Mahir Vardar*

Redesign of the West Wall Kanmantoo Mine, South Australia, *Barry K. McMahon*

Design Examples of Open Pit Slopes Susceptible to Toppling, *Douglas R. Piteau, Alan F. Stewart, and Dennis C. Martin*

Successful Implementation of Steeper Slope Angles in Labrador, Canada, *Om P. Garg*

The Northeast Tripp Slide–11.7 Million Cubic Meter Wedge Failure at Kennecott's Nevada Mine Division, *Victor J. Miller*

Back Analysis of Slope Failure in the Cercado Uranium Mine (Brazil), *C. Dinis da Gama*

Case Examples of Blasting Damage and Its Influence on Slope Stability, *Roger Holmberg and Kenneth Maki*

Waste Dump Stability at Fording Coal Limited in B.C., *Robert S. Nichols*

Evaluation of Surface Coal Mine Spoil Pile Failure, *Peter M. Douglass and Michael J. Baile*

Slope Stability in Reclaimed Contour Stripping, *G. Faulkner, C. Haycocks, M. Karmis, and E. Topuz*

The Impact of the Federal Surface Mining Control and Reclamation Act, *R.W. Thompson and D.A. Ferguson*

It is now the occasion of the Fourth International Conference on Slope Stability in Open Pit Mining. Almost 30 years have passed since the first conference was held and nearly 20 years since the third conference. Some things have changed in the intervening years but many have not. To provide a basis for the reader to judge the changes, it is of interest to consider some of the remarks presented by C.O. Brawner in his introduction to the first international conference. The reader is strongly encouraged to read these comments in their entirety. For those who do not have access to the proceedings of the first conference, the following extracts are presented:

The advent of larger drilling, excavation, and milling equipment is resulting in a tremendous increase in the scale and annual tonnage of open pit mining.

Stability at open pit mining developments must be assessed for tailings dams, waste dumps, open pit slopes in overburden soil, and open pit slopes in rock.

The science of soil mechanics developed by Dr. Karl Terzaghi provides the basis for analyzing:

- *bearing capacity of soil under dams and waste dumps*
- *amount of seepage under and through dams*
- *stability of slopes and determination of safe slope angles for tailings dams, waste dumps, and open pits*
- *influence of earthquakes on stability*
- *influence and cost of different excavation and construction techniques*

If soil mechanics principles had been applied, many major failures involving tailings, tailings dams, or waste dumps, such as those at Aberfan, Mufulira, Luanshya, and El Cobre would not have occurred.

Construction procedures used in the past have paid little attention to the compaction of the shell material or to the influence of earthquakes on stability. This neglect has led to failures. While failures in low dams may not be too serious, for high dams such failures could have catastrophic results.

Another major area of concern is the stability of rock slopes. From the standpoint of long-term safety and economics, rock mechanics problems are frequently more serious than soil mechanics problems in open pit mining. This is partly due to the increasing depths of proposed open pit mines.

General application of rock mechanics to open pit stability was delayed for several reasons:

- *There has been a reluctance to spend money on rock mechanics because of an apparent lack of certainty of economic reward.*
- *Rock strength parameters relating to rock masses are infinitely variable and difficult, if not impossible, to determine precisely.*
- *Generalized models and theories of rock behavior are complex, as are the mathematics involved.*
- *Field conditions are extremely difficult, and often impossible to duplicate in the laboratory.*
- *Field testing is usually complicated, time consuming, and almost always very expensive.*

An important factor to recognize is that while rock mechanics is a new science, we can make use of theory and experience from many other related fields. The following are examples:

- *The theory of elasticity may be used for studies of stress distribution in rock masses.*
- *The techniques of triaxial and uniaxial compression, direct shear and permeability tests, as applied to soils, can also be used to test rock.*
- *The principle of effective stress and the mechanics of the stability analysis, developed many years ago in soil mechanics, are applicable to rock slopes.*
- *The measurement of water pressures and the flow of water through porous media can take advantage of experience used in investigation for the construction of major dams. These studies date back to the 1930s.*
- *Geologic mapping techniques to classify rock and to define the orientation and frequency of discontinuities are essential in rock mechanics studies.*
- *The principles and techniques of stabilization of landslides in soil are usually equally effective for many types of slides in rock.*

The advent of the computer and the finite element technique have made it possible to incorporate the third dimension in stability studies. However, these new techniques are not a cure-all. When used with discretion, in the hands of experienced engineers, they are a powerful tool. In the hands of the inexperienced, they may only provide a quicker way of getting the wrong answer.

Many fallacies regarding stability have developed over the years. It is important to recognize that:

- *Water does not generally act as a lubricant in slides.*
- *Reduction in water pressure is a more important factor in stability than is the amount of water that is intercepted.*
- *The influence of blasting on stability is far more important than most mining engineers recognize.*
- *Placement of tailings by hydraulic means does not provide a high degree of density.*
- *The length of time that the seismic vibrations last during an earthquake can have a major influence on stability.*

If an open pit operation does not have some evidence of instability, money is being wasted. However, if potential instability exists and is unsuspected, lives may be lost. Controlled stability is good mine management. Accordingly, operating mine engineers should have a knowledge of the general techniques of stabilization. This should also include a knowledge of the instrumentation used to monitor slope movement.

The late Dr. R.A.L. Black, former Head of the Dept. of Mines, Royal School of Mines, London, and Dr. E. Hoek, Head of the Rock Mechanics Project, made some very pertinent comments at the 9th US Symposium on Rock Mechanics (1967) Conference held in Golden, Colorado. [Black, R.A.L., and E. Hoek. 1967. Status of rock mechanics as applied to mining. Status of Practical Rock Mechanics. Proceedings of the 9th U.S. Symposium on Rock Mechanics (edited by N.E. Grosvenor and P.W. Paulding, Jr.). AIME., pp 5–27].

- *"Rock mechanics is no longer only an esoteric study for academic initiates: It is the essential basis of practical mine design."*
- *"A body of really useful knowledge in rock slope stability cannot be developed unless the more academic studies are developed and validated in practice in the field. There are no short cuts, no cheap solutions and no substitutes for a proper and complete appreciation of the basic problems involved. The nature of the forces which are at work when ancient equilibrium is disturbed by the creation of mining excavations must be understood before real progress in the application of rock mechanics to mine design can be made."*
- *"The penalty for the practical engineer for attempting to defy the known laws of strata behavior is relentlessly exacted in hard economic terms and sometimes in blood."*

We must apply the knowledge that we already have. We must build the knowledge up from carefully validated experience. The work must be coordinated between research establishments, such as universities, and the mining companies concerned.

We have developed analysis and design techniques that are far beyond our capability to determine the necessary strength and boundary parameters. We are desperately in need of case studies and analysis of field failures to test theoretical concepts. Only then will real practical progress be made.

Apart from the economic advantages of applying rock mechanics principles correctly, there is the certain benefit of better and safer working conditions.

To make or to save money we must be prepared to spend it in commensurate amounts. There are no ready made solutions in the field of mining rock mechanics, but given time and reasonable financial aid and other support from industry, solutions can be found.

The content of this fourth conference, like that of the preceding conferences, is heavily focused toward rock slope stability. This reflects, in some sense, the need for more published information in this area. A number of other publications and conference proceedings are available on the topics of waste rock and tailings embankments. Particularly noteworthy is the Engineering and Design Manual–Coal Refuse Disposal Facilities/ (I69.8:EN3) prepared for the U.S. Department of the Interior, Mining Enforcement and Safety Administration by E. D'Appolonia Consulting Engineers, Inc. This publication appeared in 1975 and provided the first definitive resource for practitioners in the United States subsequent to the Buffalo Creek disaster in 1972. In recognition of the need to expand available resources to include waste rock embankments other than those associated with coal, a workshop was held in conjunction with a meeting of the Society of Mining Engineers, AIME. A proceedings of this workshop, Non-Impounding Mine Waste Dumps, appeared in 1985 published by SME and edited by M.K. McCarter and an organizing committee composed of Bruce C. Vandre, John D. Welsh and Zavis M. Zavodni. This publication contained the following papers:

Classification and Surface Water Controls, *M.J. Taylor and R.J. Greenwood*

Planning Models: Operating and Environmental Implications, *Thomas R. Couzens*

Optimum Dump Planning in Rugged Terrain, *Ernest L. Bohnet*

Geotechnical Site Investigation, *John D. Welsh*

Evaluation of Material Properties, *Richard D. Call*

Simplified Stability Analysis, *Jack A. Caldwell and Allan S.E. Moss*

Limit Equilibrium Slope Analysis, *Steven G. Wright*

Scoping Regulatory Requirements, *Bruce Vandre*

Surface and Groundwater Pollution Potential, *Duane L. Whiting*

Water Movement, *John D. Nelson and David B. McWhorter*

Design of Drainage Systems for Embankments and Other Civil Engineering Works, *Harry R. Cedergren*

Influence of Earthquakes, *Charles E. Glass*

Failure Mode, *Geoff Blight*

Construction and Performance in Mountainous Terrain, *David B. Campbell*

Foundation Investigation and Treatment, *Z.M. Zavodni, B.D. Trexler and J. Pilz*

Stability Monitoring, *M.K. McCarter*

Reclamation in the Intermountain Rocky Mountain Region, *Bland Z. Richardson*

At the same time, efforts were underway by Bruce C. Vandre to produce an engineering guide for the U.S. Department of Agriculture, Forest Service Intermountain Region. This guide is titled Stability of Non Water Impounding Mine Waste Embankments. More recently, a series of manuals was produced for the Province of British Columbia Ministry of Energy, Mines, and Petroleum Resources by the British Columbia Mine Waste Rock Pile Research Committee. These manuals deal with mined rock and overburden piles and include titles such as *Investigation and Design Manual,* 1991; *Operating and Monitoring Manual,* 1991; *Methods of Monitoring,* 1992; *Failure Runout Characteristics,* 1992; and *Review and Evaluation of Failures,* 1992. These manuals provide a wealth of information pertaining to conditions in the mountainous terrain of British Columbia and similar environs.

Major contributions to the technical aspects of waste rock and tailings have been documented in a series of annual conferences held at Colorado State University and published under the title of *Tailings and Mine Waste.* Additional information applicable to tailings disposal structures is also contained in a series of publications by the U.S. Committee on Large Dams and the International Committee on Large Dams (many of these contributions may be found in the paper by Davies et al., which is included in this volume).

Large-scale field applications of heap leaching for the extraction of gold, silver, and copper developed during the last 20 years. Extensive use is made of synthetic materials (including geomembranes and geotextiles) in the construction of these facilities. This has added a new dimension to the stability evaluation of surface mine facilities. In 1987, a two-session symposium dealing with heap leaching was held at the annual meeting of SME. The resulting SME publication, *Geotechnical Aspects of Heap Leach Design,* edited by Dirk van Zyl, includes the following papers:

Optimizing Technology for Leach Pad Liner Selection, *D.R. East, J.R. Haile and R.V. Beck*

Design of Chemically Amended Soil Liners, *M.E. Smith and G.J. Gierzerski*

Compression Testing of Geomembrane Soil Interfaces, *L.A. Hanson and J.D. Deatherage*

Practical Design Considerations for the Installation of Leach Pad Liners, *J.D. Welsh*

Heap Leach Construction Over Tailings, *W.J. Attwooll and C. Gerity*

Slope Stability in Heap Leach Design, *R.T. Tape*

Construction of Leach Pads on Steeply Sloping Ground, *N.C. Shaver and A. Tapp*

Shear Testing of Geomembrane Interfaces, *L.A. Hanse*

Potential for Heap Leach Mass Instability, *R.T. Tape and T.G. Harper*

Engineering Properties of Agglomerated Ore in a Heap Leach Pile, *D.T. Kennard and A.A. Schweizer*

Feasibility Assessment for Increasing Heap Thickness at the Alligator Ridge Mine, *C. Stachan and D. Van Zyl*

Heap Leach Recovery Rates vs. Heap Height–A Case Study at Alligator Ridge Mine, Nevada, *A. Kuzycki and R.A. Womack*

In reviewing the titles contained in the preceding conferences and other volumes mentioned, it is noted that many of the papers are now regarded as classics and the information contained is highly relevant even today. By including these titles in this preface, it is hoped that the reader will become reacquainted with this valuable knowledge and the pioneers in this field. The editors expect that a number of the papers in this volume will enter into the "classic" category as well.

Because of the long time lapse between the third and fourth conferences and the significant changes in the knowledge base affecting all aspects of stability in surface mining, the editors made the decision to invite the papers included in this volume rather than to employ a Call-for-Papers. By doing this, it was hoped to ensure a more complete coverage by *Slope Stability in Surface Mining.* The volume has been divided into the following 4 sections:

1. Rock Slope Design Considerations
2. Case Studies in Rock Slope Stability
3. Stability of Waste Rock Embankments
4. Tailings and Heap Leaching

Immediately following the release of this volume, a symposium was held in conjunction with the 2001 SME Annual Meeting in Denver, Colorado. Participants were encouraged to expand the present volume by bringing their own contributions in electronic format to facilitate distribution.

In this way, we as editors and organizers of the symposium encourage on-going dialog via electronic means for professionals in this developing field. Never has there been such an opportunity as now exists through CD, email, and Web-based means to exchange ideas and advance the art and science of slope stability. We look forward with great optimism to potential benefits of improved safety, resource recovery, environmental stewardship, and economic reward that can result from such a dialog.

As we enter the twenty-first century, mines reaching depths of more than 1,100 m are being planned, waste rock embankments have surpassed 600 m in height, tailings dams have reached heights of 200 m, and heap leach facilities have reached heights of 150 m. There are great challenges to the rock- and geomechanics communities to ensure that these facilities are safe and economic today and in the future. It is only through information collection and exchange that improvements can be made in design concepts, construction methods, monitoring strategies, and reclamation practices. We hope, and expect, that the next state-of-the-art summary as contained in this volume will not require a gestation period of 20 years. This series of volumes and symposia initiated some 30 years ago by C.O. Brawner should be continued on a regular basis.

ACKNOWLEDGMENTS

To the authors of the chapters in this volume and their employers, we express sincere gratitude for the time and effort each of you has expended, along with your willingness to share your expertise. We also gratefully acknowledge the significant support provided by the Itasca Consulting Group in the initial stages of this project and the commitment of the departments of Mining Engineering at the University of Utah and the Mackay School of Mines.

W.A. Hustrulid, M.K. McCarter, and D.J.A. Van Zyl

Introduction

MINE ECONOMIC CONDITIONS

The first of the International Conferences on Stability in Open Pit Mining took place nearly 30 years ago. From the point of view of mineral commodity prices, the period since then has not been a good one for the metals mining industry. Although there have been some ups and downs, the overall price expressed in constant dollars has steadily declined. This trend is shown in Figures 1 and 2 for copper and iron ore fines, respectively. Figure 3 shows the situation for gold, which, contrary to the other two, has experienced a slight increase over this period. The increase is small, however, when compared with the general cost of living in this same period.

In light of this, the surface-mining industry has responded by finding ways to reduce costs. One way has been to turn to ever-larger equipment and a corresponding increase in bench geometry to suit this equipment. The decrease in open-pit mining costs has meant that the break-even depth for the consideration of underground mining methods has steadily increased. Today there are pits with depths of 800 m and plans to go to 1,200 m. Such deep pits have a number of consequences. First, an emphasis on having slopes as steep as possible to minimize stripping exerts considerable pressure on the mine's geotechnical team to define, test, and push limits. Second, there is a growing gap between the equipment fleet being used for production and that most appropriate for constructing these slopes. Third, waste-rock embankments have reached gargantuan heights due to limitations on available space and haulage distances. Again, the geotechnical team is challenged to examine the limits. Finally, as appropriate, lower-grade material present in these high embankments is being leached to recover the minerals they contain. The stability issues involved also place increasing demands upon the geotechnical team. It is both challenging and stimulating to try and provide the required basis and justification for the designs.

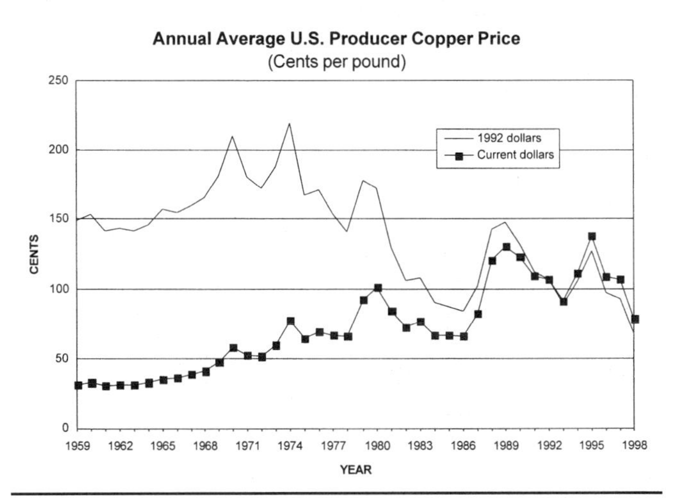

FIGURE 1 The price trend for copper (After Edelstein 2000)

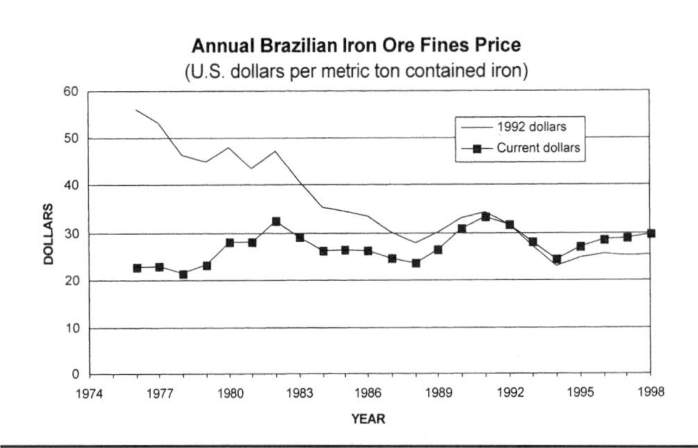

FIGURE 2 The price trend for iron ore fines (After Kirk 2000)

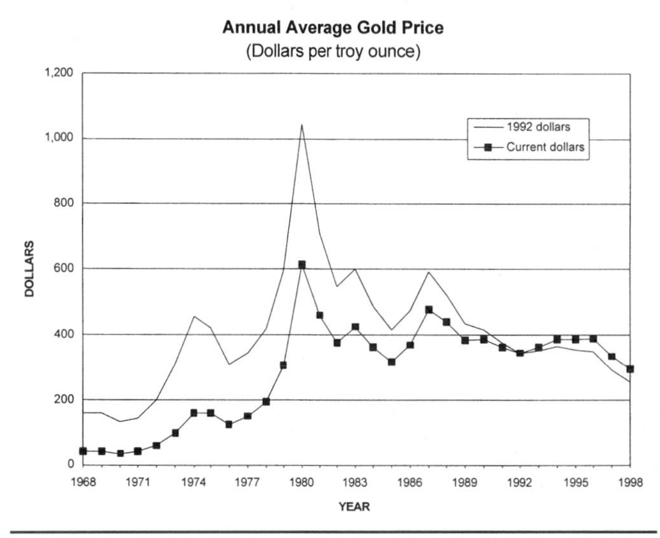

FIGURE 3 The price trend for gold (After Amey 2000)

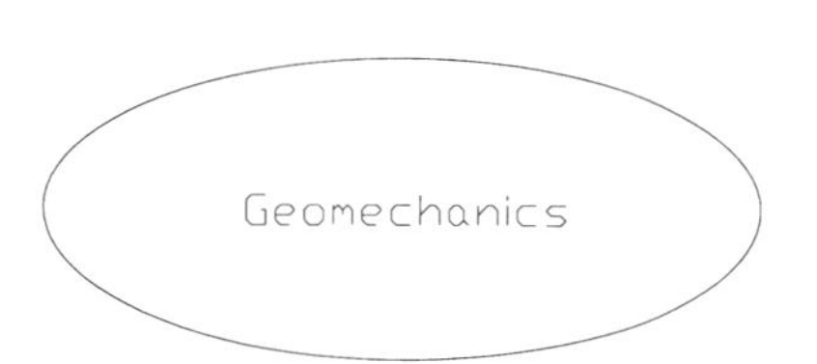

FIGURE 4 Diagrammatic representation of the geomechanics group's role of rock slope design

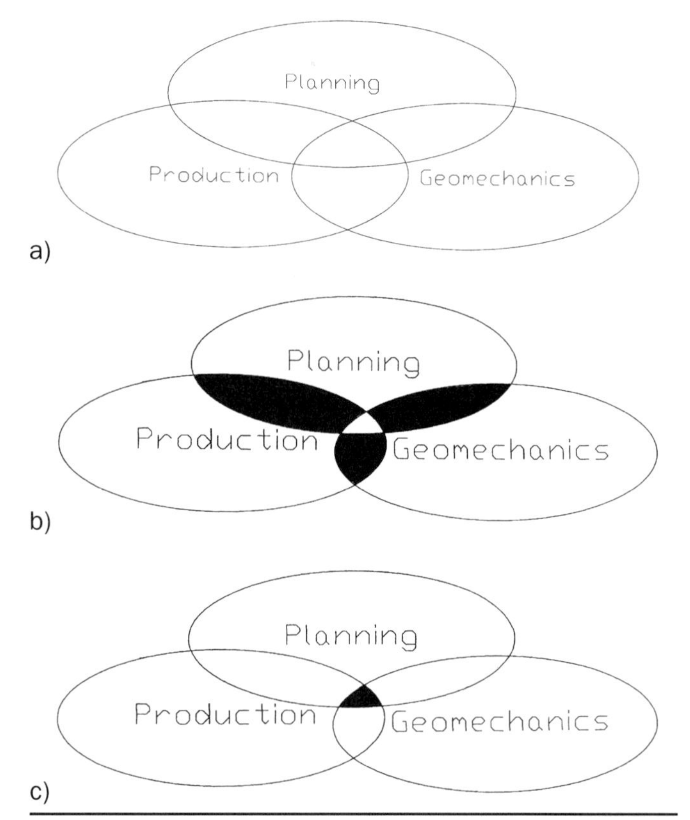

FIGURE 5 Idealized interactions between the geomechanics, planning, and production groups in pit wall creation. 5a. Overall interactions, 5b. Pair-wise interactions, 5c. Three-way interaction

The first two sections of this book deal with the design and the reality of rock slopes. Later sections deal with waste-rock embankments, tailings impoundments, and heap leach facilities.

ROCK SLOPE

The rock slope design role, as performed by the geomechanics group, is represented by the ellipse shown in Figure 4. The total process of creating the slopes (design, layout, and excavation) actually consists of the interaction of three groups: planning, geomechanics, and production. Ideally, the interaction should appear as shown in Figure 5a. As can be seen, there are certain issues addressed by the three groups working in pairs (Figure 5b): geomechanics-planning, production-geomechanics, and planning-production, and there are issues in which all three–geomechanics, planning, and production–interact (Figure 5c).

In Figure 5 all three of the responsibility areas have been drawn to the same size, suggesting that there is an equal weighting of their inputs and their influence in reaching the final action path. In practice, the "kingdoms" involved are not of the same size as measured in terms of personnel, budgets, influence, or power. The interactions in the worst case may appear as shown in Figure 6, where there is no interaction. In a more typical case, the interactions and relative group sizes are as shown in Figure 7. In this case, geomechanics interfaces with planning, which interfaces with production.

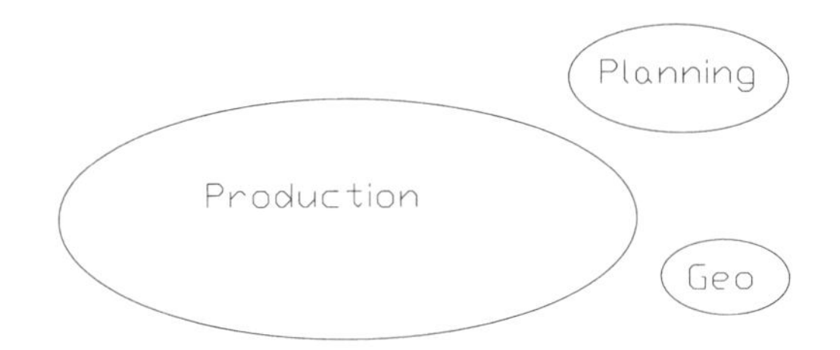

FIGURE 6 A possible interaction between the geomechanics, planning, and production groups in pit wall creation

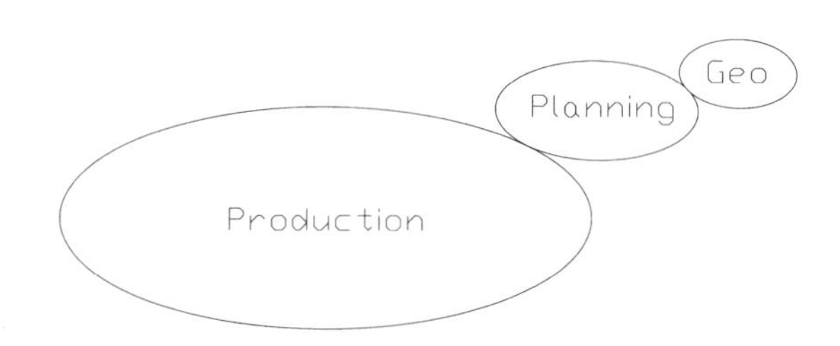

FIGURE 7 A typical interaction between the geomechanics, planning, and production groups in pit wall creation

Good planning and good geomechanics are both necessary for the preparation of good designs, and a good production group is required to ensure that the "as-built" slopes strongly resemble the "as-designed." Otherwise, the overall slope operation will be far from optimum. There is a natural but unfortunate tendency for each group to specialize and as a result each has little knowledge about each other's business. However, such a knowledge is necessary if the designers are to recommend designs that can be built and the builders are to understand the reasons why the designs must be built just so. Lacking this understanding, builders have a natural tendency to invoke a certain poetic license and, due to production or other pressures, add that certain "special touch" that large equipment and techniques can bring to a design when they try first and foremost to get "rock-in-the-box."

Each group–planning, geomechanics, and production–has certain special concerns with respect to the slope mining operation. These interests are represented diagrammatically in Figure 8. Planning (Plan) is largely interested in the pit being located correctly with respect to position in space. Thus, it is the positions of the bench toes that are their primary concern. Geomechanics (Geo) is interested in the bench face angle and the width of the safety bench. Production (Prod) is interested in removing the rock mass at the lowest possible (operating) cost. All three have an interest in the overall slope angle. Although satisfaction of each of their specific demands is possible, it does not happen by chance and it does not happen easily.

During the period 1970 through 2000, the size of the available production equipment has markedly increased. Table 1 summarizes the largest available loading and hauling models supplied by several well-known manufacturers for years 1970, 1985, and 2000.

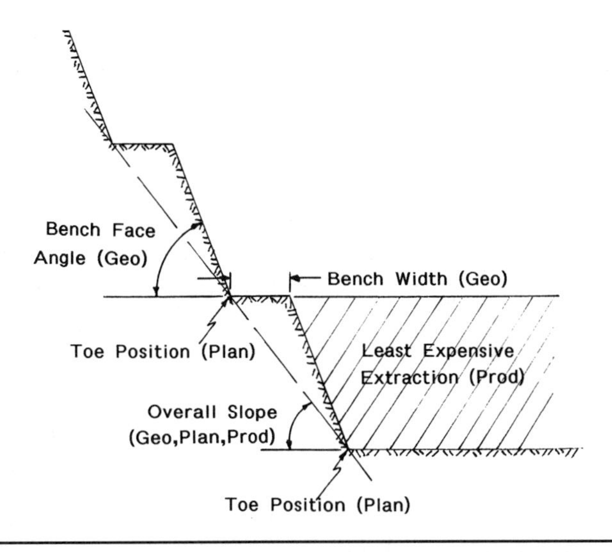

FIGURE 8 The prime areas of concern for the geomechanics (Geo), planning (Plan), and production (Prod) groups

TABLE 1 Generalized development of mining production equipment

	Equipment Capacity	
Year	**Shovels (yd^3)**	**Truck Payload (tons)**
1970	15	50
1985	46	150
2000	67	360

After Caterpillar 1999, Dietz 2000, Jackson 2000, and Reise 2000.

During this period, the shovel size has increased by a factor of more than 4 and the truck size has increased by a factor of 7. To keep these large machines operating at greatest efficiency, a large amount of room in the pit is required and large blasts involving hundreds of holes are needed to provide the daily tons. For the sake of factors such as efficiency, ease of maintenance, and keeping a spare-part inventory, there has been a push for the mine to standardize on a relatively few machine sizes. These high-ball production conditions are not, however, especially conducive to the gentle handling of the slopes. Such gentle handling is required if current slope angles are to be maintained or possibly even steepened with increasing pit depth. Quite another equipment fleet from that used in production is required to gently sculpt the slopes. There is, in short, a production fleet and a geomechanics fleet. To emphasize the differences in the two, the machines shown in Figures 9 through 11 have been drawn to the same scale. Figure 9 shows a 67-yd^3 shovel used in production compared with a 20-yd^3 shovel that is appropriate for creating the bench face. Figure 10 shows the 360-t truck that matches the production shovel and the 100-t capacity truck matched with the bench-shaping shovel. The corresponding drills are shown in Figure 11. This particular production drill, if required, can drill holes up to 22 inches in diameter, whereas the bench-shaping drill creates the 4-inch- or 6½-inch-diameter holes needed for presplitting or smoothwall blasting.

Given the different priorities of the three groups, it is only with a respect for and an understanding of the other groups' requirements and possibilities that an overall optimum result can emerge. The need for understanding, respect, and communication is greatest between the production and the geomechanics groups, which are often the furthest apart in both "kingdom" size and in background.

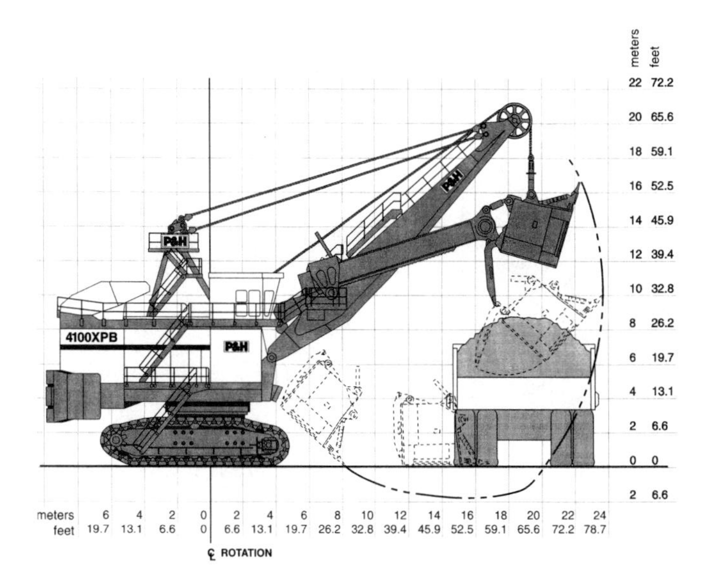

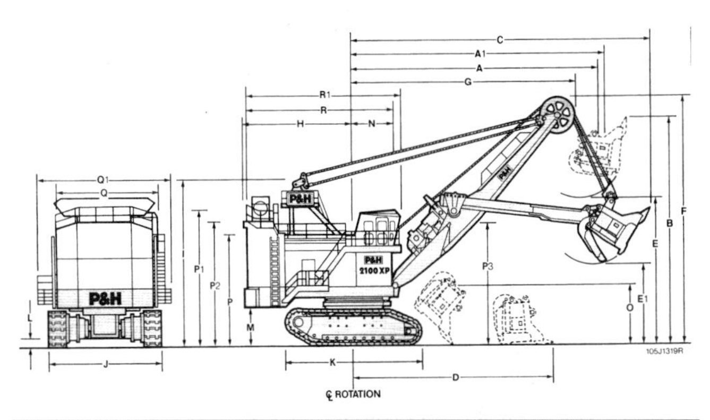

FIGURE 9 a) Cross section of a 67-yd^3 capacity production shovel (Dietz 2000) b) Cross section of a 20-yd^3 capacity trim shovel (Dietz 2000)

FIGURE 10 Comparative cross sections of the 360-t and 100-t capacity trucks (Caterpillar 1999)

The gaps between the groups must be bridged if the most economic pits of the future are to be designed and built. Large-diameter vertical holes completely filled with explosives are not compatible with the creation of the sculpted surfaces required to create the steepest, deepest slopes. The one-size-fits-all fleet consisting of the very latest and largest equipment is not compatible with the digging and hauling of the rock at the periphery of the steepest, deepest slopes. This disconnect must be acknowledged and the hard decisions made on that basis. The justification for operating-cost increases to capture capital-cost benefits (steeper slopes and hence less stripping or greater ore recovery) in the longer term is a hard one to get across.

One of the major problems is that the production group generally has much more confidence in their ability to move rock

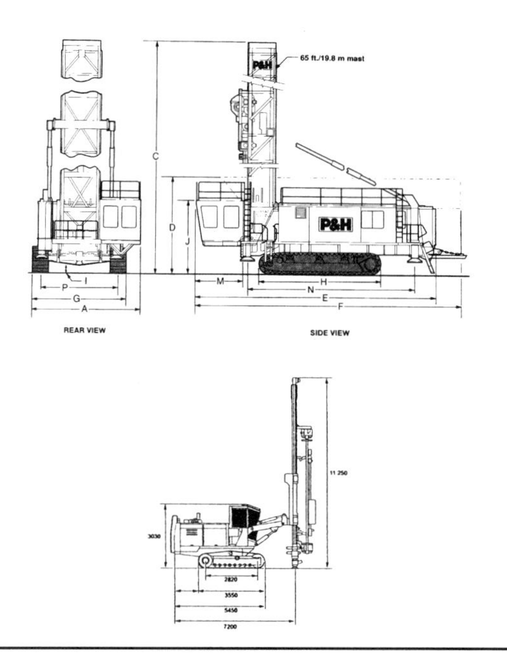

FIGURE 11 a) Cross section of a large production drill (Dietz 2000) b) Cross section of a trim drill (Shellhammer 2000)

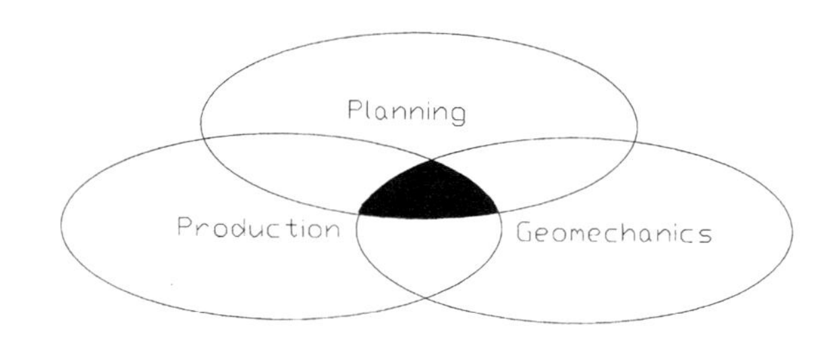

FIGURE 12 Future interaction of the geomechanics, planning, and production groups

than in the basis for the geomechanics designs. Although this attitude is changing as the tools for collecting and presenting data, describing the rock mass, and for analyzing slopes have improved, there is still suspicion regarding the basis for the design and hence the reliability of the design.

An educational process is in order. The first step in the process is to discuss, agree upon, and communicate among ourselves the best, most current, and most appropriate approaches to slope design. The next step is to have presentation materials in a form that encourages meaningful dialog with the other parties. The goal is to make correct decisions based upon the same understanding of the overall opportunity. The ellipses of the future must reflect an equality of understanding of the total slope-creation process. The degree of ellipse overlap (Figure 12) must be increased from that of today.

WASTE-ROCK EMBANKMENTS

The design objectives of waste-rock embankments or "dumps" have evolved over the past 30 years. Even the terms commonly applied prior to the 1980s have taken on new meanings. Sensitivity to these new interpretations has created reticence to using "waste" and "dumps" in describing facilities and operations related to disposing of barren rock. In this book, an attempt has been made to avoid these terms, especially when discussing embankments that are carefully designed to be repositories for materials that are essentially inert or that can be rendered nonreactive by appropriate construction.

In the early part of this century, waste-rock embankments were designed to minimize overall costs of mining. Normally, this required building a repository that would minimize haulage distance, vertical lift, and the possibility of moving the material a second time should pit limits expand. At times, the selected site was not optimally chosen based on stability or for eventual reclamation. Scale-up in truck haulage and power shovels in the 1960s and 1970s resulted in higher production and the need to dispose of larger volumes of barren rock. Larger embankments and higher rates of deposition frequently resulted in increased rates of settlement and occasionally mass instability. The design objective then changed to include not only the economics of materials handling but safety for personnel involved with waste-rock disposal and others downhill from the facility.

Design objectives have continued to evolve and now include reduction or elimination of contaminant transport, surface stabilization, and reclamation. Not only must the waste-rock embankments function in an economic, structurally stable, and environmentally acceptable manner, but there is also considerable time, effort, and financial resources expended to address the "view-scape" of even infrequent observers.

Useful design criteria, methods, and operational considerations can be found in a number of past publications. However, this field is evolving and changing continuously. It is appropriate therefore to periodically summarize current practice. It is for this purpose that the chapters have been incorporated into this volume on surface mine stability.

The information presented includes site selection, characterization, and assessment; evaluation of embankment parameters; mechanisms that control the near-surface stability of angle-of-repose embankments; embankment hydrology; reclamation and surface stabilization; case histories for three noteworthy projects; and a preview of an important emerging issue that of selenium contamination of surface and groundwater associated with phosphate waste-rock embankments.

TAILINGS EMPOUNDMENTS/HEAP LEACH

Tailings dams and heap leach facilities contain chemicals, at different concentrations, used in the recovery of metals. Failure in these facilities can result in releases of such chemicals to the environment and therefore to potential environmental impacts. Consequently, stability evaluations of tailings and heap leach facilities have evolved into a serious environmental issue. While stability may be maintained from a classical circular or other stability failure mode perspective, an overtopping of a tailings dam can result in overall instability of the structure. Evaluation of the stability of tailings dams and heap leach facilities must therefore include a thorough understanding of the operating issues associated with these structures.

Failures of tailings dams often lead to legal proceedings where a number of entities (and sometimes individuals, depending on the jurisdiction) are sued. The outcome of these proceedings is very often a settlement before the case goes to court. In most of these cases, the technical information about the failure is sealed and cannot be released in the open literature. This makes it impossible to learn from the causes of such failures and could lead to mistakes being repeated. While little may be done to change this situation, it clearly adds to the responsibility of tailings design engineers to remain informed on the causes of these failures, and it is often necessary to obtain this information informally.

SUMMARY

In summary, 30 years have passed since the first International Conference on Stability in Open Pit Mining. Much has been accomplished, as is shown by the contributions in this volume. The push for both higher and steeper slopes is challenging our tools and capabilities. Furthermore, the mines' change to larger, more-productive equipment means that new approaches to sculpt the required slopes must be implemented. Mine management must recognize this potential challenge and take the appropriate actions. The rush to bigness should be evaluated with respect to the total opportunity of optimizing the operation. It is vitally important that the geomechanics input be heard with the same tone and amplitude as the production voice. The best way for this to occur is for each group to better understand each other's specialty and then together explore the opportunities to tailor an optimum design based upon that understanding. The collection and presentation of the knowledge base through forums like this are an essential part of creating a common platform for making the correct slope design and construction decisions.

REFERENCES

Amey, E.B. 2000. Gold. U.S. Geological Survey.

Caterpillar. 1999. Caterpillar Performance Handbook. Edition 30.

Dietz, M. 2000. P&H Mining Equipment. Personal communication.

Edelstein, D. 2000. Copper. U.S. Geological Survey.

Jackson, B. 2000. Wheeler Machinery Company. Personal communication.

Kirk, W.S. 2000. Iron Ore. U.S. Geological Survey.

Riese, M. 2000. Bucyrus International, Inc. Personal communication.

Shellhammer, D. 2000. Atlas Copco Rock Drills AB. Personal communication.

SECTION 1

Rock Slope Design Considerations

CHAPTER 1

Large-Scale Slope Designs—A Review of the State of the Art

E. Hoek,* K.H. Rippere,† and P.F. Stacey‡

1.1 INTRODUCTION

As part of an ongoing design effort by the Toquepala and Cuajone open-pit mines, Southern Peru Copper Corporation (SPCC) invited a group of slope-stability specialists to participate in a state-of-the-art review meeting. The meeting was held in Vancouver, Canada, from August 6 to 8, 1998. This meeting was the first stage in a slope-design project undertaken by a consulting team led by Golder Associates Ltd. and including Piteau Associates Engineering Ltd. and the Itasca Consulting Group.

The purpose of the meeting was to assess the state of the art in the design of large slopes. Particular emphasis was placed on slopes higher than 1,000 m that are excavated in rock masses associated with copper porphyry type ore deposits, which represent the majority of pits of that scale.

The three co-authors of this paper are, respectively, the meeting's recorder (E. Hoek), the SPCC project manager (K. Rippere), and the study project manager (P. Stacey). A complete list of meeting participants, all of whom contributed to the paper, is given in Table 1.1.

TABLE 1.1 List of meeting participants

Representatives from Southern Peru Copper Corporation
Robert H. Collier, Toquepala Mine Manager
Clifford T. Smith, Cuajone Mine Manager
Elsiario Antunez de Mayolo, Technical Services Manager
Kenneth H. Rippere, Chief Geotechnical Engineer
Yonglian Sun, Senior Geotechnical Engineer
Meeting Participants
Peter F. Stacey, Principal, Golder Associates
R. John Byrne, Principal, Golder Associates
Al V. Chance, Associate, Golder Associates
Jose L. Carvalho, Senior Rock Mechanics Engineer, Golder Associates
P. Mark Hawley, Vice-President, Piteau Associates
Brent W. Gilmore, Associate, Piteau Associates
Loren J. Lorig, General Manager, Itasca S.A., Santiago, Chile
Evert Hoek, Consulting Engineer, Vancouver, Canada
Oskar Steffen, Steffen Robertson and Kirsten, South Africa
John Read, Consulting Engineer, Brisbane, Australia
John C. Sharp, Geo-Engineering, New Jersey
Antonio Karzulovic, Consulting Engineer, Santiago, Chile
Chen Zuyu, Tsinghua University, Beijing, China
T.K. Krishnan, Consulting Geologist, Vancouver, Canada
Paul R. Pryor, Call & Nicholas Inc., Tucson, Arizona

1.2 STATE OF THE ART AND STATE OF PRACTICE

It became evident early in the meeting that a distinction had to be made between the state of the art and the state of practice in the design of large slopes. Some concepts that were discussed were certainly state-of-the-art concepts, but it will probably be many years before these concepts find their way into routine slope-design practice. Consequently, an attempt is made here to differentiate between the concepts and tools that are currently available and can impact slope designs in the near future and those that will not be used in practice for many years to come.

1.3 THE GEOLOGICAL MODEL

The availability of a comprehensive geological model is fundamental to any slope design. Without such a model, slope designers have to resort to crude empiricism, and the usefulness of the resulting designs, except possibly for very simple prefeasibility evaluations, is highly questionable.

The availability of a number of computer-based geological modeling tools is an important aspect of modern rock-slope design. These tools, such as the Vulcan three-dimensional solid modeling system, permit the visualization and construction of comprehensive models that can include geological and structural geologic information, ore grade distributions, groundwater distributions, and a variety of geotechnical details. Constructing these models is a useful exercise since deficiencies in the database are highlighted and the user is forced to consider the interrelationships between the various types of information displayed by the model. These tools are operational and, with software prices in the range of $50,000 (US), need to be considered by any mining company embarking on major open-pit development.

Many open-pit mines already use some form of three-dimensional modeling. In some cases, information is transferred from these three-dimensional models to limit equilibrium or numerical models for slope-stability analyses. For example, at the Chuquicamata and Escondida mines in Chile, ASCII files generated from Vulcan models are used for input into the program's Universal Distinct Element Code (UDEC) and Fast Langrangian Analysis of Continua (FLAC). In other cases, AutoCAD files are used to transfer such information. Examples of three-dimensional solid models for the Chuquicamata Mine in Chile are given in Figures 1.1 and 1.2.

Under development are interfaces that will allow transfer of geometrical and geotechnical information directly from a three-dimensional geological model into a two-dimensional limit equilibrium or numerical model for slope-stability analysis.

* Independent Consultant, Vancouver, British Columbia, Canada.
† Chief Geotechnical Engineer, Southern Peru Copper Corp, Peru.
‡ Principal, Golder Associates Ltd., Vancouver, British Columbia, Canada.

FIGURE 1.1 A computer-generated three-dimensional solid model of the rock mass in which the Chuquicamata open-pit copper mine in northern Chile is being mined (Prepared by R. Torres of the Chuquicamata Geotechnical Group using the program Vulcan.)

FIGURE 1.2 Chuquicamata open-pit mine in 1998 showing the geological units exposed in the walls of the 750-m-deep pit (Prepared by R. Torres of the Chuquicamata Geotechnical Group using the program Vulcan.)

1.3.1 Collecting Engineering Geological Data

The geological model forms the fundamental basis for all slope designs. Its importance cannot be overstated. The model should form the basis for all geotechnical and hydrogeological studies.

The tools and techniques for collecting engineering and structural geologic data are well developed. Most of these tools have been available for the past 20 years and have been widely used by the mining industry. Consequently, there does not appear to be any need for the development of new tools and techniques. It is likely that the future will see more digital cameras, borehole television cameras, electronic compasses, and other "high-tech" devices being used by geologists and geotechnical engineers to gather data. Such devices should never be regarded as replacements for existing tools, and the need for good practical engineering and structural geologic observations and interpretations remains as strong as ever.

The definitions of what information to collect and the efficient use of the collected data are matters to which significant uncertainties are attached. Currently, there are few standards for data collection, and the type and quantity of information gathered depends very much on the personal opinions and preferences of the individuals or organizations carrying out the work. The disadvantage of this approach is that it is very difficult to maintain continuity when staff or organizational changes occur. It is also difficult to make meaningful comparisons between deposits with similar geology.

While there does not appear to be any immediate prospect for the adoption of international or even industry-wide standards for geological data collection, there are organizations that have established company-wide standards. For example, in 1997 the Codelco mining organization in Chile organized a three-day workshop for geologists and geotechnical engineers from all of its operating divisions. The outcome of this workshop was a manual of procedures for collecting geological data, laboratory testing of rock mechanics, and interpreting and presenting results. Such manuals are considered useful providing they do not impose rigid boundaries on the data collection and interpretation process.

One suggestion that merits serious consideration is to categorize engineering and structural geology data in terms of different levels of confidence. For example, mineral resources are usually described in terms of proven, probable, and possible reserves, and there is no reason why the geological and geotechnical databases and the resulting slope designs should not be defined in the same terms. This type of description would probably help in the communication process between the geology and geotechnical groups and the mine planners (Steffen 1997).

The meeting participants emphasized the need to collect good-quality basic data in the form of structural information, detailed core logs, and photographs that can be interpreted by any good engineering geologist at any time. It was recognized that the current state of practice includes the use of rock-mass classifications such as rock-quality designation (RQD), rock-mass rating (RMR), Q, and geological strength index (GSI). The usefulness of these classifications in rock-slope engineering was seriously questioned. Many of these classifications were developed specifically for the confined conditions that apply in underground engineering, and they have been found to produce unreliable results in rock-slope engineering. Where classification systems are used, the need to calibrate them to local mine site conditions was emphasized.

In spite of their limitations and the care that must be taken when using rock-mass classifications in an open-pit mining environment, several meeting participants felt these classification systems have a definite role in the process of assessing the characteristics of the rock masses in which the slopes are to be excavated. At the very least, they provide a checklist of items that should be considered. In the hands of an experienced engineering or structural geologist, they can play a more significant role. This is because the classification systems or selected components of these systems can be calibrated to provide a site-specific index of one or more rock-mass characteristics. When using the available rock-mass classifications in this way, great care must be taken to ensure that not too much reliance is placed on published correlations that may have been derived for totally different circumstances. In each case, it is necessary to consider what components are included in the classification and how they relate to the specific rock masses being considered (Hawley, Newcomen, and Gilmore 1994). Data collection should focus primarily on fundamental parameters such as rock strength, rock fabric, major structure data, and hydrogeologic information. Indices and classification systems can be derived later if required.

As discussed above, inputting the results from engineering and structural geology data collection into a geological model, preferably by the geologists themselves, is important to understanding the processes that control the rock-mass behavior. Deficiencies and anomalies in the data become obvious during this model construction process and these provide useful guidance to the geologists in developing further site-investigation programs.

1.3.2 The Role of Major Structures

No one questions the role of major structures, such as faults and shear zones, in controlling the stability of large slopes. It is essential that the data-gathering process include the development of regional and local geological models that incorporate these major structures. It is also important that ongoing mapping of these structures, as they are exposed in the pit walls, be performed as a way to continually update the structural geology model.

From a practical point of view, the role of second-order structures, such as joints, is more of a problem. The importance of rock fabric (mass or material) in controlling the rock-mass deformation/strength behavior is clearly recognized. However, because of the large number of such features in high rock slopes, the question that arises is: How much data should be collected and how should they be used in the design of the slopes? Further, the interpretation of structure needs to account for potential mechanisms of deformation and cannot be based entirely on outcrop mapping. The use of computer-generated structural interpretations may prove beneficial with the guidance of an experienced structural geologist and adequate data.

The current state of practice tends to separate slope designs into two distinct categories. The first is for those designs that can be dealt with in terms of kinematically possible, structurally controlled failures. For example, failures that involve wedges sliding along the line of intersection of two intersecting faults can be analyzed using well-established limit-equilibrium models. This approach tends to be used to assess bench stability or interramp slope sections. However, major structural features can also result in overall slope failure, and this possibility should never be ignored.

The second category includes nonstructurally controlled failures in which some or all of the failure surface passes through a rock mass that has been weakened by the presence of joints, bedding planes, and other second-order structural features. It is commonly assumed that these second-order structural features are randomly or chaotically distributed and that the rock-mass strength can be defined by a simple failure criterion in which "average" or "smeared" nondirectional strength properties are assigned to the rock mass. This approach is sometimes used to assess the stability of overall mine slopes.

State-of-the-art slope-stability specialists recognize that it is too simplistic to consider slope-stability problems in terms of these two categories. In reality, most slope failures involve a complex mixture of structural and nonstructural failures, and it is very rare to find a true "wedge" failure or a well-defined "circular" failure.

During the meeting, extensive discussions took place on how the deficiencies of this inadequate approach could be remedied. One of the most promising solutions involves the generation of a statistically equivalent model of the numerous structural features, such as joints, measured in the field. Using principles of fracture mechanics, a Monte Carlo simulation process can be used to generate failure paths through this jointed rock mass. The failure path providing the lowest shearing resistance would be considered the most likely "candidate failure surface" for an overall slope failure.

This type of analysis, with or without the statistical simulation component, has been used for the past 20 years by individual slope-stability specialists and consulting organizations. Numerous attempts at this analysis have been made that include sliding on structural discontinuities and failure of "intact" rock bridges. Chen Zuyu described an application of this technique for the design of very steep slopes at the Three Gorges hydroelectric project in China. A similar technique, used extensively since the early 1980s by Call and Nicholas Inc., was described by Paul Pryor. John Read described a project that is being undertaken by the CSIRO in Australia to utilize this process.

What became clear from the discussion is that few of these techniques have been published in any formal way. Many of the analyses have been performed by consulting organizations that tend to regard their particular approach as a "tool of the trade," and there has been an understandable reluctance to publish details of these approaches. This has the disadvantage that free debate of the various techniques is limited and critical examination of the assumptions inherent in each approach is seldom possible.

The industry-sponsored project currently being undertaken by CSIRO and directed by J. Read should overcome these problems, since all the results will be available to the sponsors and the research will be subjected to extensive review and discussion. Ultimately, the results of this work will be published and available to open-pit mining organizations everywhere.

There is a good possibility that these methods, currently in limited use, will be incorporated into practical rock-slope design tools and will have more widespread use within the next few years. Hence, given a good set of structural geology data, it may be possible to investigate a wide range of potential slope-failure models without having to resort to the simplified "smeared" rock-strength models described earlier. This question is explored further in the next section.

1.4 DETERMINATION OF ROCK-MASS PROPERTIES

The determination of rock-mass strength was clearly identified as a major deficiency in current rock-slope design practice. Even in terms of the state of the art, there are many unanswered questions and many opinions on how this task should be performed in the low-stress environment that is characteristic of rock slopes. In the case of copper porphyry deposits, this problem is further complicated by the influence of alteration, which is discussed later in this chapter.

In terms of state of practice, many slope designers tend to use some form of "smeared" failure criterion to estimate the rock's shear-strength properties. This information is then used to analyze nonstructurally controlled failures of blocks bounded by major structural features such as faults, shear zones, and dykes.

The Hoek–Brown failure criterion is commonly used to estimate the properties of these "homogeneous" rock masses. It is widely recognized that this failure criterion, while it may be the best that is currently available, has many limitations when applied to the low confining stress conditions that prevail in slopes. Some of these limitations can be minimized by adjusting the estimated strengths based on judgment or observations. For example, anisotropic strength can be simulated by using the ubiquitous joint model in the program FLAC.

Until better alternatives are available, the Hoek–Brown or similar rock-mass failure critera will continue to be used by slope designers. However, great care should be taken to ensure that the estimated strength values are not too optimistic and, wherever possible, that more than one method of estimating rock-mass strength is used. At every opportunity, back-analysis of slope behavior should be used to check the rock-mass strength estimates.

An alternative approach is to generate a realistic rock-mass shear-strength criterion from the component parts. Hence, a statistically generated "joint" pattern would be created to reproduce, as accurately as possible, the distribution of structural discontinuities in the rock mass. The shear strength of these structural features can be adequately described by available methods such as the Barton–Bandis shear failure criterion. The failure of "rock bridges" between these discontinuities would be simulated by a process that satisfied the conditions dictated by fracture mechanics principles.

While some models of this type have been constructed and used for specific projects, for example the Three Gorges project in China and the models used by Call & Nicholas, these studies certainly fall in the category of state of the art rather than industry-wide state of practice. Therefore, at this stage, the rock bridge theory, which is based on joint persistence interpretations, should be used with extreme caution and should be considered only as a specific, secondary approach to augment other assessments.

The meeting participants felt that in the future there will be great merit in having some qualified researcher carry out a systematic study of rock-mass strength using existing numerical modeling tools. For example, FLAC, UDEC, UTAH2, and PHASE2 all contain interface elements that can be used to simulate some aspects of this problem. It may also be possible to use other numerical models such as PFC or ABACUS to study rock-mass strength. However, it is not clear where suitably qualified researchers could be found and how such research would be funded. Therefore, it is unlikely that these ideas will produce short-term results.

One of the more promising research projects currently in progress is the one being undertaken by CSIRO with funding from a number of mining companies. This research attempts to answer some, but not all, of the questions discussed above. A Monte Carlo process will be used to find candidate failure surfaces through "jointed" rock in order to better represent rock-mass failure than is possible by using the current simplified models.

1.5 ROLE OF GROUNDWATER

The presence of groundwater in a rock slope is critical in any assessment of that slope's stability. Water pressure, acting within discontinuities in the rock mass, reduces effective stresses with a consequent reduction of shear strength. Depressurization using horizontal or vertical wells or drainage galleries is a powerful tool in controlling slope behavior.

It was agreed that the technology and tools for groundwater pressure and flow evaluation and control are well developed and that no further research into this area is required. It was further agreed that applying these techniques and tools in practice does not always satisfy the requirements for data interpretation. Consequently, any programs for measuring and assessing groundwater effects require careful planning and execution. For example, the difficulty maintaining piezometers and drains in an operating pit is often used as an excuse for not maintaining adequate control or monitoring of groundwater conditions. It is also important that water flowing from horizontal drains be collected and piped or pumped to a disposal area away from active slope-stability problem areas.

As with the geological and geotechnical models discussed earlier, development of a good groundwater model is an important component in the rational design of large open-pit slopes. It is important that resources be provided to ensure that sufficient information is collected to permit the construction of such a model, which must address any geological controls on the hydrogeologic regime.

There was considerable discussion on the issue of drainage versus depressurization. Water pressure creates slope-stability problems; if these water pressures are reduced, it is not necessary for a "drainage" hole or well to produce large water flows. This common misconception leads operators to abandon "drains" that do not appear to be working because they do not produce much water. The judgment should, however, be based on the response of piezometers, which reflect water pressure change, rather than on flow volume.

Drainage galleries have an important function in large slopes, not only because of the depressurization that can result from their construction but because of the valuable geological information that can be collected from locations that are not normally accessible. Drainage-gallery construction costs in South America average around $1,800 (US) per meter, with an additional cost of $45 (US) per meter for 75-mm-diameter drain holes drilled from the gallery. This can sometimes give an overall depressurization scheme that is comparable in cost to one based on horizontal holes and/or vertical pumped wells. It was agreed that more serious consideration be given to galleries for slope depressurization and that a reference database of previous practice in groundwater control using gallery systems be developed.

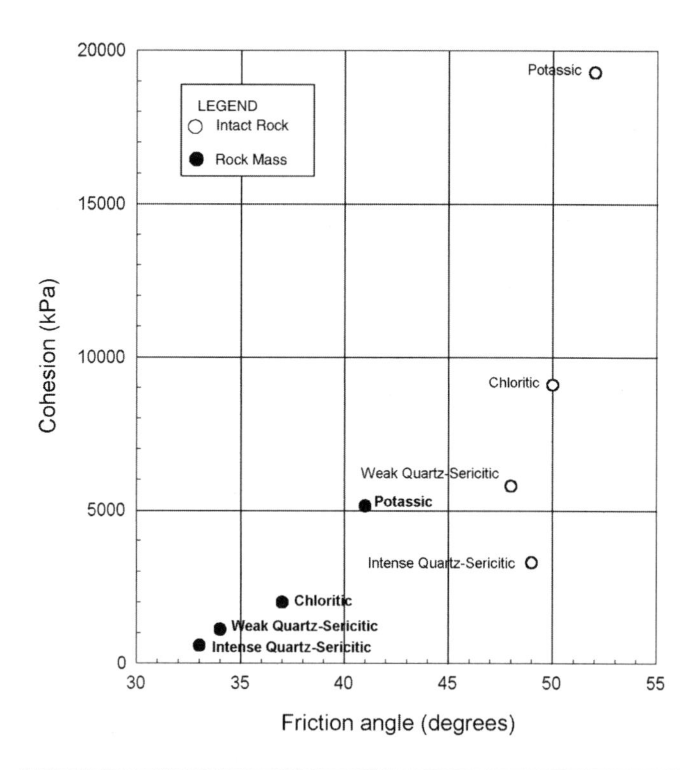

FIGURE 1.3 Influence of alteration on the strength of east porphyry from the Chuquicamata Mine in Chile (Permission from the Chuquicamata Mine to publish these data is acknowledged.)

General experience, including that of Call & Nicholas at Bingham Canyon, indicates that horizontal drain holes up to approximately 300 m in length can be effectively installed, with the effective dewatering depth being approximately half the drain length, or about 150 m. Response to pumping of underground workings at Bingham and Golder's experience at Twin Buttes indicate that drainage galleries can be effectively used to dewater beyond the limits achieved by horizontal drains alone.

1.6 IMPACT OF ALTERATION

For orebodies associated with hydrothermal processes, alteration can significantly impact both the rock material and the discontinuity state compared with the unaltered host rocks. As a result of the ore emplacement process, all copper porphyry deposits are associated with a zone of altered rock. In some areas, such as the copper porphyry deposits of northern British Columbia, this alteration is relatively mild and does not have a major impact on rock-mass strength. In other areas, for example in southern British Columbia, the western United States, and most open-pit mines in the South American Andes, the orebodies are surrounded by a halo of strongly altered rock and the impact on rock-mass strength, and hence slope stability, is significant.

Antonio Karzulovic gave the results of laboratory tests of rock strength for the East Porphyry at the Chuquicamata Mine in Chile. He showed that potassic alteration has the least impact on rock strength, chloritic alteration has a significant impact, and sericitic alteration has a major impact. A plot of estimated cohesive and frictional strengths for altered porphyry is given in Figure 1.3.

Since alteration also has significant implications in terms of copper grades, Karzulovic suggested that a good orebody geology map, one that shows zones and degrees of alteration, is useful to engineering geologists and geotechnical engineers concerned with slope-stability problems. Such maps should include information on the degree and type of alteration. The maps should emphasize the impact on the strength and durability of the rock mass rather than the alteration of specific minerals that may be of interest to geologists.

For any new mine or any mine where no such information exists, it is also worth performing specific laboratory studies on rock strength for different degrees of alteration. These laboratory results then form a basis for estimating the rock-mass strength for input into slope-stability analyses. In some cases, alteration is also associated with intense microfracturing, particularly adjacent to major fault structures. This makes it very difficult to obtain representative samples for laboratory testing; consequently, some laboratory strength values may not be too reliable. In such cases, it may be necessary to resort to the techniques used in soil mechanics to collect and test weak soil samples.

Some of the geophysical techniques used by the oil, gas, and coal industries could be adapted for use in defining alteration grades in copper porphyry deposits.

1.7 IN SITU ROCK STRESS

Rock "noses," or slopes, that are convex in plan are less stable than concave slopes. This is generally because of the lack of confinement in convex slopes and the beneficial effects of confinement in concave slopes. These observations provide practical evidence that lateral stresses in the rock in which slopes are excavated can have an important influence on slope stability.

Stress relief due to removal of the material within the open pit results in "deconfinement" of the near-surface rock mass and the resulting freedom to dilate results in a reduction of rock-mass strength. On the other hand, concentration of the stress at the toe of high slopes could cause rock-mass failure in this region, which could lead to propagation of failure that, in the worst case, could lead to slope failure.

In the current state of practice in rock-slope design, lateral stresses are usually ignored or are dealt with in a very simplistic manner. In fact, all limit-equilibrium models are based on gravity loading only and, hence, lateral stresses are excluded from any slope-stability analysis that uses these models. Numerical models can incorporate lateral stresses, but most analyses using these models are based on a very simple approximation in which the horizontal stress applied to the model is some proportion of the vertical stress.

The in situ stress field near large open-pit mines is seldom measured, in spite of the fact that such measurements are entirely feasible. This is because the in situ stress field is generally considered to be of minor significance. This assumption may be adequate for small open-pit mines but needs to be questioned for the design of very large slopes.

Before taking in situ stress measurements at any site, it is worth performing a parametric study using numerical models to determine whether variations in horizontal in situ stresses significantly impact the stresses induced in the near-surface rock in which slope failures could occur. When lateral stresses in potential failure zones show a large variation, serious consideration should be given to a field program to measure the in situ stress field.

In order for a stress-measurement program to be effective, a sufficient number of measurements must be carried out so that statistically representative results can be obtained. Consequently, these programs are both time consuming and expensive. Because such a stress-measurement program need only be carried out once or twice in the life of an open-pit mine, there is usually no justification for purchasing equipment and training mine staff to carry out this task. It is generally more efficient to bring in a qualified contractor to take the stress measurements.

1.8 BLAST DAMAGE

The damage from large production blasts in open-pit mines can extend many tens of meters into the rock mass behind the slope face. This blast damage is due to rock fracture and joint opening that result from the dynamic stresses induced by the blast. In addition, penetration of gas pressure from the blast can open existing discontinuities for considerable distances from the face. This damage causes loosening of the rock mass with a consequent reduction in strength.

Controlling the amount of explosive detonated per delay is the method commonly used to minimize blast damage; where this is done correctly on a routine basis, it is certainly effective. However, a certain amount of blast damage is inevitable, and this must be considered by the geotechnical engineer when gathering data and carrying out slope-stability analyses.

One consequence of blast damage is that the appearance of the rock mass exposed on the bench faces may not be representative of the undisturbed rock mass through which a potential failure surface may develop. Since most geotechnical mapping is carried out on these bench faces, the shear strength of joints and the overall strength of the rock mass, estimated on the basis of this mapping, may be unrealistically low. Therefore, it is important that observations from diamond drill core and exposures in underground excavations be compared with surface mapping data to differentiate between natural fractures and those induced by blast damage. This is a largely qualitative process since none of the methods currently used to estimate joint shear strength or rock-mass strength incorporate corrections for blast damage.

1.9 DESIGN PROCESS

The proper understanding and presentation of precedence for major slopes in related rock masses, e.g., similar copper porphyries, is critical for future slope design. The back-analysis of representative failures or stable slope configurations through well-documented case studies can be fundamentally important in understanding large-scale failure mechanisms.

The presentation of precedent slope-design data in terms of a slope height/slope angle plot is inevitably simplistic but does allow a valid model of design experience to be simulated and presented. A chart of this type, developed with input from several sources, is presented in Figure 1.4, a review of which suggests that current design practices do not result in stable slopes in a high proportion of the cases.

1.10 LIMIT-EQUILIBRIUM AND NUMERICAL MODELING OF SLOPE INSTABILITY

The design of any slope must involve some form of analysis in which the disturbing forces, due to gravity and water pressure, are compared to the available strength of the rock mass. Traditionally, these analyses have been carried out using limit-equilibrium models, but, more recently, numerical models have been used for this purpose.

Limit-equilibrium models fall into two main categories: models that deal with structurally controlled planar or wedge slides and models that deal with circular or near-circular failure surfaces in "homogeneous" materials. Many of these models have been available for more than 25 years and can be considered reliable slope-design tools if used in appropriate situations.

One type of limit-equilibrium model that is not as reliable is one that involves sliding on a combination of failure planes and failure through "nonstructured" rock masses. Current methods of noncircular failure analysis do not produce reliable results for many of these cases. Chen Zuyu described work that he has done on extending and improving methods of noncircular failure originally developed by Sarma. This work has the potential to significantly improve this class of slope-stability analysis. Work is currently under way to incorporate

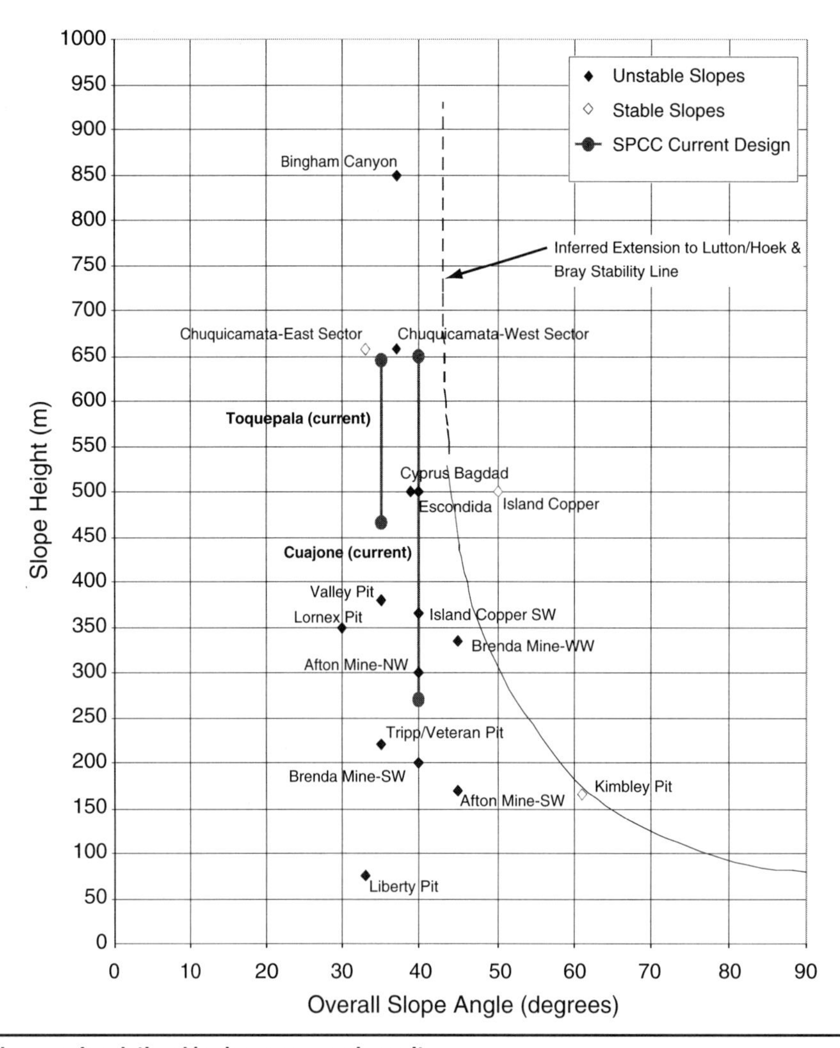

FIGURE 1.4 Slope height:slope angle relationships in copper porphyry pits

Chen's work into the analytical tool kit used by practical slope designers.

Numerical modeling of slope deformation behavior is now routine on many large open-pit mines. Programs such as FLAC and UDEC are often used for such modeling, and these codes do not require any further development to meet the needs of slope modeling. Using these codes correctly is not a trivial process, and mines embarking on a numerical modeling program should anticipate a learning curve of one to two years, even with expert help from consultants. Obtaining realistic input information for these models and interpreting the results produced by the models are the most difficult aspects of numerical modeling in the context of large-scale slopes.

In practice, both limit-equilibrium and numerical modeling tools are used to generate possible solutions for the range of input parameters that exist for a particular site. Because the geotechnical department does not appear to be capable of producing a single definitive design, it may be frustrating for mine management and mine planners. However, it is far more realistic to look at the results of a parametric study than to rely on a single analysis.

1.11 DEFINITION OF SLOPE FAILURE

Slopes that fall down and deposit a pile of rubble on a haulage ramp or on a conveyor belt are failures in every sense of the word. The rock mass has failed and the resulting pile of rubble or the "hole in the road" has resulted in a major disruption to the business of getting ore out of the ground and into the mill.

Many slope failures are more subtle than the simple case described above. The gradual deformation of a slope, even when movement is on the order of 4 m/yr, as in the case of the Chuquicamata west wall, does not represent "failure" but rather is regarded as a slope problem that has to be managed. This is in contrast to civil engineering practice where slope deformations of this type would certainly be considered failures.

One threat in open-pit mining is the potential for the gradual deformation of a large slope to develop into a fast-moving catastrophic slide. This is a very poorly understood process, as there are no reliable documented case histories. However, it must be considered a potential threat where large deforming slopes occur in an open-pit mine. Numerical modeling of these slopes for possible combinations of structural and rock-mass failure and comparison of the results of these models with observations and measurements of actual slope behavior is probably our best hope of understanding this problem.

In any economic open-pit mine, a variety of slope "failures" will be present at various locations in the mine at any time. The successful management of these failures is the art of good open-pit mining. The absence of any failures is a sign of overconservative, hence, inefficient slope design and mine management. It is therefore an absolute requirement that geologists, geotechnical engineers, and mine planners work together to ensure that appropriate data are collected, appropriate analyses are carried out, and slope designs are clearly conveyed to and understood by mine planners. It is also important to draw up contingency plans to deal with the inevitable surprises that occur during the life of the mine.

Defining slope failure, even if only on a crude, qualitative basis, is an important component of risk management in open-pit mining. The definition will vary from mine to mine; engineering geologists, geotechnical engineers, planning engineers, and operators should be involved in its development. Each mine has to develop a set of "acceptability" criteria against which the performance of the slopes can be measured; a process that must involve the mine management. These criteria will vary from design sector to design sector and will depend on the nature of the rock mass and groundwater conditions and on the function of each slope. Temporary slopes being developed as part of a pushback are obviously much less sensitive to slope deformation and instability than are slopes adjacent to haul roads and plant such as in-pit crushers.

1.11.1 Slope Management

A slope design is based on the best possible evaluations of the rock types and characteristics, the structural geology, and the groundwater conditions in the slope. Even the best slope designs require some averaging of all of this information, and local variations in geology or groundwater conditions are not incorporated into the design. These local variations can significantly impact slope stability and, depending on the location of the instability, may have important consequences for the open-pit operation. For example, a relatively small failure in the benches immediately above or below a main ramp can seriously affect production.

Advance warning of these slope instability problems is very important to management and to the planning and operations groups at the mine because this information can be used to adjust the mine plan and operation to compensate for local problems. These adjustments require a significant amount of time, usually several months, which has to be taken into account in short-term mine planning. The most frequently used adjustments for controlling slope instability are to accelerate pushbacks, delay excavation of the toe (thereby reducing the effective overall slope angle), or "unload" the top of the slope by accelerating the excavation of upper benches. These adjustments are made individually or in combination, depending on the slope configuration and the nature of the problem.

Monitoring slope movement has proved to be the most reliable method for detecting slope instability; the more accurate this measurement, the earlier the developing problem can be detected. Tools for slope-displacement monitoring are well developed and used routinely on most large open-pit mines. These tools are generally based on observations of numerous targets placed at carefully selected locations on the benches of the mine. Electro-optical distance measuring (EDM) equipment and, more recently, global positioning by satellite (GPS) systems are used to monitor the relative positions of these targets on a regular basis. High-quality EDM or GPS systems can provide an accuracy of less than 1 cm over measuring distances of 1 km or more. This order of measurement accuracy is generally sufficient to give advance warning of most slope-stability problems.

A typical state-of-practice monitoring system exists at the Chuquicamata Mine in Chile. This system consists of two computer-controlled Leica distance-measuring systems mounted in air-conditioned cabins on the pit crest. These devices scan several hundred prism targets mounted on the slopes and check their locations by triangulation onto remote stable base stations. A 1997 study to determine the horizontal and vertical accuracy of this equipment was carried out over a distance of 2,600 m. The results indicated that an accuracy of ±20 mm can be achieved consistently with this equipment.

The use of downhole inclinometers and extensometers tends to be restricted to local slope-instability problems. Maintaining this equipment for any length of time in a deforming operating slope is very difficult. Consequently, this type of equipment tends not to be used for large-scale slope deformation or failure studies. This situation may change if equipment to be installed in the west wall of the Chuquicamata open-pit mine proves successful. Developed by the CSIRO in Australia, this equipment consists of self-contained, battery-operated sensors that can be located so that they report to three or more surface antennas with positions that are tied in real time to the surface-deformation monitoring system.

The anticipated accuracy of these devices is on the order of millimeters and their design life on the order of five years. Installed in boreholes at depths of up to 150 m, these devices will be able to track the movement of the rock mass behind the slope. This information is very important in developing an understanding of how some large slopes deform.

One slope-observation tool that tends to be ignored is simple visual observation by geologists, geotechnical engineers, and mine staff. The development of tension cracks, the appearance of bench faces, the presence of rocks that have fallen from steep faces are important signs of slope behavior. If these are observed routinely and recorded systematically, a feel for the behavior of the slopes in the mine can gradually be developed. This important information can be taken into account when signs of significant slope instability appear and when remedial measures and contingency plans are discussed. When failure is imminent, it is a useful technique for determining when to remove equipment and close access.

One tool that has been tried unsuccessfully at a number of mines is microseismic monitoring. Background noise from truck and shovel operations, blasting, and regional seismic activity tend to mask any measurements of the microseismic noise generated by moving slopes.

Measurement and interpretation of slope deformation is one of the only means available for calibrating the analytical tools available to the slope designer. It is only by detecting the onset of movement or following the progressive development of a failure that we have any hope of understanding the reliability and predictive capability of the various models used to analyze slope deformation and failure. Consequently, slope deformation measurement is more than just a safety measure—it is an essential component in the development of the art and science of slope design.

1.12 SUMMARY

The meeting served to summarize both the state of the art and the state of practice in the design of large slopes. Discussions confirmed that the technology available for designing large slopes has advanced little since the early 1970s, although the body of experience has increased significantly.

It was generally agreed that a good geological model is the prime prerequisite for designing any large slope. In addition, this model must consider the distribution of alteration, both in type and degree; the rock type distribution; structural elements, both major and minor; and water.

The methodology for assessing rock-mass strength as a function of the interaction between structure and intact rock properties requires detailed study, since the state of practice currently relies heavily on empirical methods such as the Hoek–Brown criterion. State-of-the-art studies in this area are in progress, although it is likely to be several years before full results are obtained and alternative methods become available to the practitioner.

On the other hand, methods for groundwater control are relatively well developed, although modeling in complex geological environments can be difficult. As a result, the application of the methodology to the scale of large slopes must be carefully assessed, since it frequently includes expensive approaches such as deep wells and drainage galleries.

The role of stress conditions, particularly at the toe of large slopes, is not well understood and must certainly be the subject of further general study.

In the design of large slopes, precedent, either from initial expansion slopes in the same pit or from experience in similar geological environments, can be invaluable. However, any comparison must include full consideration of the respective geological models and other factors such as groundwater conditions.

In the current state of practice, heavy reliance is placed on limit-equilibrium analyses, which are often too simplistic, particularly for larger slopes. Numerical modeling is finding increasing application, particularly for defining potential failure modes. However, some degree of calibration is required before the numerical models can be considered predictive in the design sense. This also is an area requiring considerable future investigation.

The current state of practice for large open-pit slopes relies heavily on slope management in which the monitoring of slope performance is used as a basis for design modifications as the slopes are excavated. Tools and techniques for slope monitoring and management are well developed. In the context of slope management, a definition of failure is a critical issue, since many mines live with moving slopes that have very little impact on the efficiency or economics of the operations. On the other hand, even a relatively small failure in the wrong location can seriously impact even a large operation.

Overall, the meeting provided several areas of study to improve the state of the art for large slope design while clarifying the general agreement as to the current state of practice.

1.13 ACKNOWLEDGMENTS

The authors thank Southern Peru Copper Corporation for permission to publish this summary of the meeting and acknowledge the input of all meeting participants.

Dr. Dennis Martin, president of Piteau Associates, Vancouver, Canada, passed away on August 29, 1998. His contribution to rock-slope engineering applied to open-pit mining is appreciated by all the participants at the meeting who wish to record their respect and sympathy for his family.

1.14 REFERENCES

Hawley, P.M., H.W. Newcomen, and B.W. Gilmore. 1994. Application of rock mass classification to open pit slope design. ISRM Symposium, Santiago, Chile, 1994.

Steffen, O. 1997. Planning open pit mines on a risk basis. *Jour. South African Inst. Min. Metall.,* March 1997.

CHAPTER 2

Collecting and Using Geologic Structure Data for Slope Design

David E. Nicholas* and Danny B. Sims*

2.1 INTRODUCTION

It is often said that the three most important factors in evaluating residential and commercial real estate are location, location, and location. Similarly, the three most important factors in the evaluation of open-pit slope stability are structure, structure, and structure. The adverse interaction of geologic structures with the mine walls is the greatest contributing factor to slope instability in open-pit mines, and the success of slope-stability analysis depends upon the level of understanding of the characteristics of geologic structure throughout the deposit.

In order to maintain clarity throughout this chapter, we will first define the terminology used to describe geologic structure. For data collection purposes, the geologic structures are divided into fractures and major structures. Fractures are those geologic structures that are too small and usually too numerous to be mapped and located individually. The major structures are those geologic structures that are long enough to be individually located on a geologic map. There is a continuum between fractures and major structures, and the differentiation between the two is based on a lower-truncation limit for the length of major structures. The minimum length for a major structure is usually equal to the height of one design bench.

For purposes of structure analysis, the geologic structures are divided into rock fabric, intermediate structures, and regional structures. The rock fabric may include both fractures and major structures. The intermediate and regional structures only include the major structures.

The geologic structure terminology used in this chapter is as follows:

- *Geologic structures,* which comprise all fractures and major structures, regardless of their length.
- *Fractures and/or joints,* which are geologic structures that break the intact rock into more or less discrete blocks. They usually comprise the rock fabric and sometimes the intermediate structures. The fractures and joints are too numerous and too short to be mapped individually throughout a deposit.
- *Major structures,* which are geologic structures, such as faults, that are large enough to be mapped and located as individual structures. There is actually a continuum between fractures and major structures, but the differentiation is useful for design purposes. The lower-length truncation limit can vary, but generally it is equal to the height of a single design bench.
- *Rock fabric,* which is defined as geologic structures that are too numerous to be evaluated individually. They are therefore treated statistically in a slope-design analysis. The rock fabric may include both fractures and major structures.
- *Intermediate structures,* which are geologic structures that are too numerous to be evaluated individually. They are treated statistically in a slope-design analysis. The intermediate structures only include major structures that are longer than a given, lower-length truncation limit. The lower-length truncation limit can vary, but it generally is equal to or greater than the height of two design benches.
- *Regional structures,* which are major structures that are of a regional scale. These structures generally have a minimum length of 100 m; when these structures are faults, they are usually assigned unique names on geologic maps.
- *Structural domains,* which are zones in which the distributions of orientation, length, spacing, and shear strength are similar.

The geologic structure attributes that are most critical include orientation, length, spacing, overlaps, and shear strength. Structure length and overlap must be measured from surface exposures. It is best to measure structure orientation and spacing from surface exposures as well, but these data may also be obtained from oriented core. Shear-strength data can be obtained equally well from either surface or core samples.

Although it is best to measure most of the critical geologic structure attributes from surface exposures, there is usually a limited surface area that is exposed and accessible for surface structure data collection. It is therefore necessary to extrapolate the available data and develop an accurate structure model as a basis for the slope-stability analysis.

The rock fabric, intermediate structures, and regional structures are defined to match the pit-slope design. A pit slope has three major components: bench configuration, interramp slope, and overall slope. The bench configuration is defined by bench height, catch bench width, and bench face angle. The interramp slope is formed by a series of uninterrupted benches, and the overall slope is formed by a series of interramp slopes separated by haul roads (Figure 2.1).

The bench height is determined by the size of the mining equipment. The required bench width is based on either expected failure volumes or on relationships developed by Ritchie (1963) and the State of Washington. The bench face angle is determined by the geologic structure, given that there are good blasting and digging practices. The structures controlling the bench face angle are usually the rock fabric because of their high frequency of occurrence, but the bench face angle can also be controlled by the intermediate and regional structures.

Interramp, multiple-bench, slope-stability analysis concerns only those failures that incorporate two or more benches. Structures that affect the interramp stability must therefore be equal to

* Call & Nicholas, Inc., Tucson, Arizona.

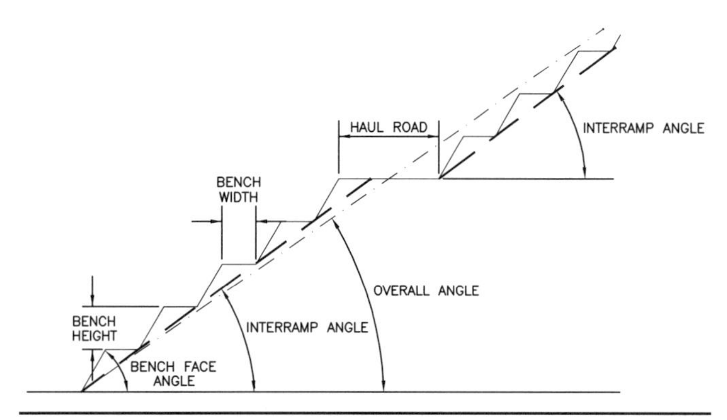

FIGURE 2.1 Typical open-pit design cross section

or greater in length than the height of two design benches. The structures that have a minimum cutoff length equal to the height of two design benches are, by definition, the intermediate structures. The intermediate structures are analyzed statistically to determine the probability of structurally controlled multiple bench failures that affect the interramp slope stability.

For both the bench face angle and the interramp slope analysis, the distribution of the geologic structure characteristics is required for statistical analysis. Collecting the data and interpreting these distributions are a challenge for the geologist and geotechnical engineer.

Overall slope stability concerns only those failures that incorporate most of, if not all, the height of the pit slope. The overall analysis can incorporate all of the geologic structure database. The regional structures are projected to future pit designs, and the interaction of the designed pit walls with each unique regional structure is analyzed. The rock fabric data and intermediate structure data are used to characterize the rock mass to determine whether there is a potential for rock-mass failure, either solely through rock fabric, intermediate structures, and intact rock or in combination with a regional structure.

The greatest challenge in overall stability analysis is probably in determining the rock-mass strength for a pit highwall. An overall weak rock mass can fail through areas that do not contain regional structures. Also, a non-daylighted regional structure may require a rock-mass buttress between the fault and the pit. In this case, the rock-mass strength must be known to determine the required buttress width. The rock-mass failures generally occur through rock-fabric-scale structures, intermediate structures, and intact rock. The rock fabric and intermediate structures that provide the weakest failure path in the rock mass generally occur in regular orientations within a structure domain; therefore, the rock-mass strength is directional dependent.

2.2 MAJOR STRUCTURES

The interramp and overall slope analysis mainly utilizes the intermediate and regional structures, both of which are part of the major-structure database. Major-structure data have a more significant impact on the slope design than that of the rock fabric, and the major-structure data are therefore the most important geologic structure data. The major-structure data are also the most difficult to collect and interpret.

2.2.1 Regional Structures

Regional structures are greater than 100 m in length, but major structures can only be mapped one bench at a time. Accurate survey, mapping, and interpretation methods are essential to determine bench-to-bench continuity and true structure length. (Mapping techniques are discussed in more detail later in this chapter.) Another problem in mapping regional structures is that it is difficult to determine the continuity of flat major structures, dipping 30° to 60°; yet these structures are the most critical to slope design. Therefore, it is important for the geologist to look for these flatter structures and, if uncertain, put them on the map and let the geotechnical engineer justify the drilling to determine the structural continuity. The regional structures have to be correctly projected on to each future pit plan, and, if possible, a structure contour has to be developed of that regional structure for future planning.

The shear strength of the structure should be determined from direct-shear test of the material that defines the fault zone. Fault zones are variable; the strength that usually controls the behavior of the fault is the strength of the material that is the weakest and comprises at least 20% of the fault zone.

2.2.2 Intermediate Structures

The intermediate and regional structure data are collected by the same major structure-mapping technique. Those structure types are differentiated in the interpretation of the data. The intermediate structures are those that have interpreted lengths that are greater than the height of two design benches and are less than 100 m.

Intermediate structures are mapped individually as major structures; however, in the stability analysis, they may be used as either unique structures or as part of a database for statistical analysis. The characteristics required are orientation, spacing, length, and shear strength. Similar to the collection of regional structure data, it is important that the geology staff follow rigorous surveying, mapping, and interpretation methods to ensure that structure lengths are properly represented on maps and in the database.

2.2.3 Mapping Techniques

Several methods are commonly used to map major structure in open pits: (1) face method, (2) mid-bench method, and (3) Anaconda (touch point) method. Call & Nicholas, Inc. (CNI) recommend the face method. The relative advantages and disadvantages of each method are presented in the following discussion.

The discussion of each method is aided by a hypothetical, plan-view geologic map, which would be produced with each different method for the same geology. An oblique view of a hypothetical pit wall is presented in Figure 2.2.

Two rock types are present, and they are cut by faults and single joints. The 3,715-level bench face has a wedge failure, and the 3,710-level bench face has a plane shear on the lithologic contact. Survey marks are represented on each bench.

A plan-view base map for the oblique view map is presented in Figure 2.3. It is important that a toe-crest base map be used regardless of the mapping method. A toe-crest map accurately represents areas that are horizontal benches and areas where the topography is sloping.

Face Method. The face method (Peters 1978) differs from the Anaconda and mid-bench methods in that the face method does not utilize a horizontal datum plane. The face method is a surface trace method that is similar to common techniques used for most surface mapping.

The face method is conducted by first surveying the bench area to be mapped. Survey marks are made in bright paint on the middle of the bench at 5- to 10-m intervals. If mapping an existing highwall, the mapping should begin at the lowest level and progress upward. Following this method, the survey marks on the bench below can be easily used to locate where structures reached the crest on the bench below, even though these structures cannot be seen by looking over the crest from above. The ability to know exactly where a structure reaches the crest is important because the mapper can stand at that crest point and look in the direction of strike across the horizontal bench to see whether the structure continues to the next level.

The survey data are recorded on the same opaque paper on which the toe and crest map is plotted. A gridded Mylar sheet is overlain for collection of the structure data. The mapper

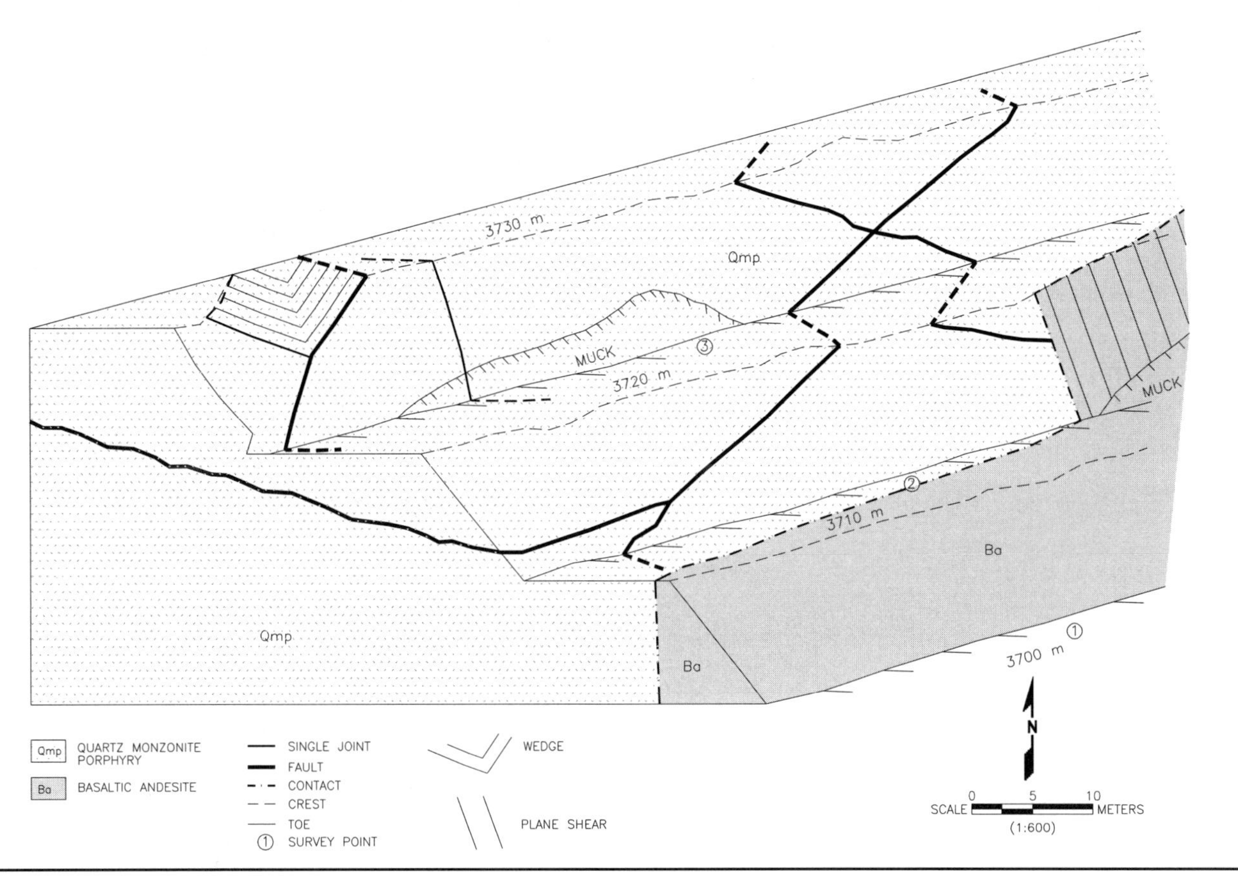

FIGURE 2.2 Oblique view pit geology

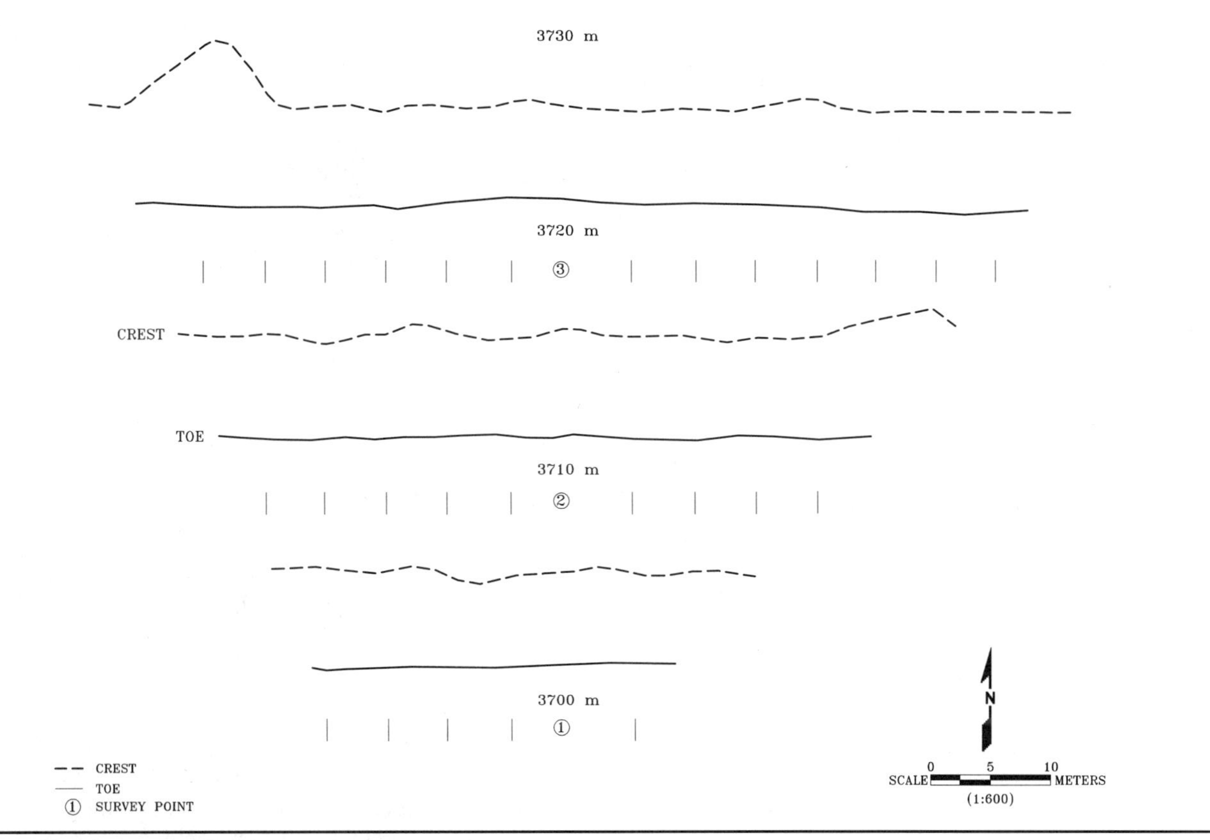

FIGURE 2.3 Plan view toe-crest pit topography map

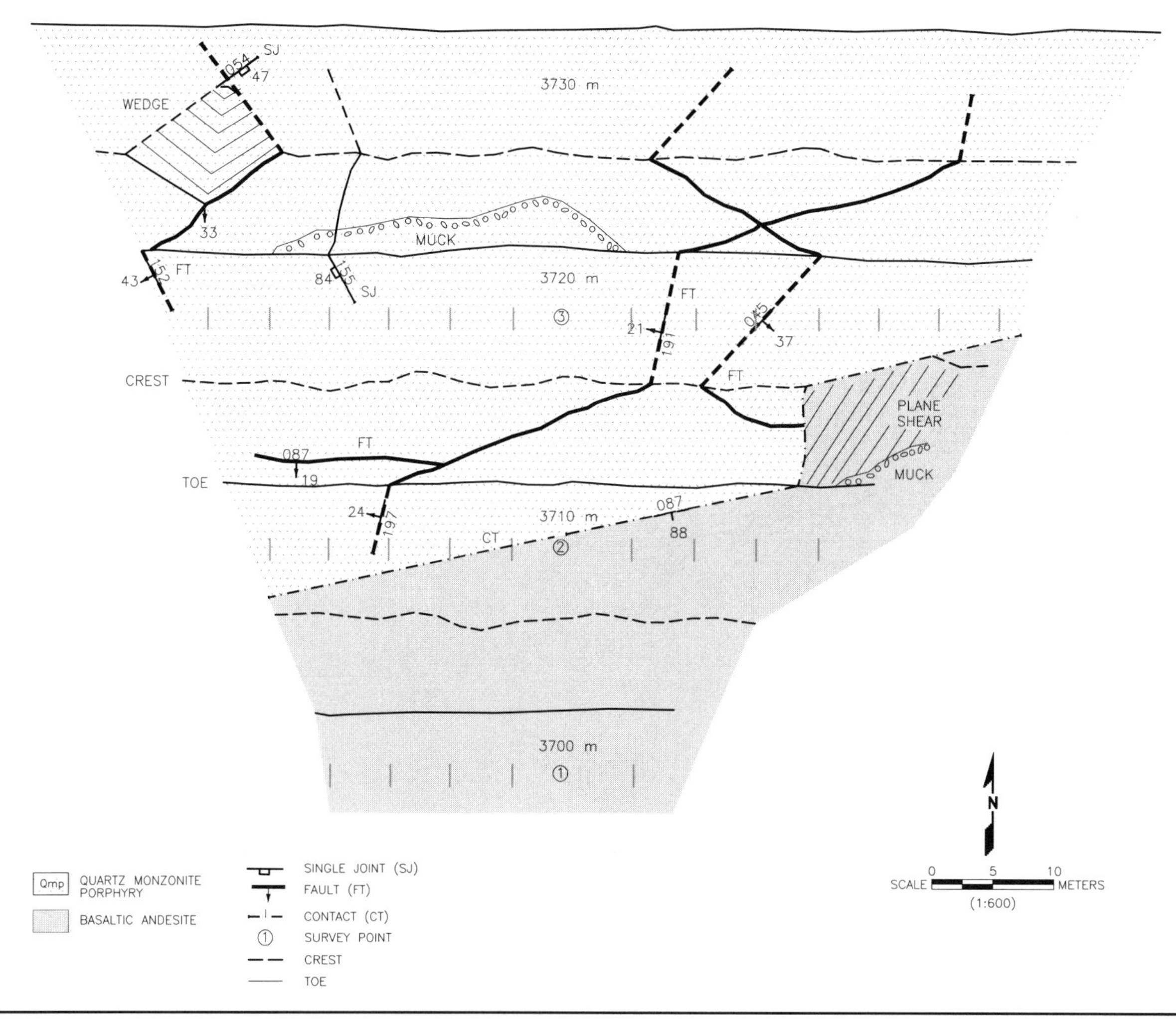

FIGURE 2.4 Face method structure map for geology shown on Figure 2.2

traverses the bench and identifies those structures that are continuous for a length equal to or greater than a single bench height. The structure is traced on the map exactly as it appears on the face. The exact locations of the crest or toe intersections are critical for determining bench-to-bench continuity. The strike and dip of the structure are measured with a compass, and the structure is dashed on the bench above and below in the strike direction. The dip is plotted with conventional dip-direction symbols for different structure types. The structure type, strike, and dip are noted next to the structure for ease in building a structure database from the map. Important features, such as plane shears, wedges, muck piles, tension cracks, and water seeps, should also be drawn onto the structure map.

The structure data should be posted to a master map at the end of each day. The new data should be interpreted in the context of previous data so that structure continuity and potential structure domain boundaries can be detected. This also gives the mapper direction when mapping future bench faces as mine development progresses.

A face-type major-structure map (Figure 2.4) was generated for the geology shown on Figure 2.2. All structures can be represented on the map because there is no horizontal datum plane to which the structures are projected. The relationship of structure to physical condition of the face can be clearly demonstrated.

Advantages to the face method include the following:

- Structures are plotted as surface traces at their actual locations, and actual structure lengths are drawn directly on the map.
- It is possible to accurately represent the relationship of structure to physical conditions of pit walls.
- Notation of the exact location of structure intersections with the toe and crest makes it possible to accurately determine bench-to-bench continuity of structures. This is especially true for the low-angle structures.

Mid-Bench Method. The mid-bench method uses a horizontal datum plane located at the mid-bench elevation of the face. A mid-bench major-structure map (Figure 2.5) was generated for the geology shown on Figure 2.2. Structures are only represented by a strike-and-dip symbol at the point where the structure crosses the mid elevation of the face. One potentially important low-angle major structure does not cross the mid elevation of the face and therefore is not represented on the map.

The advantage of the mid-bench method over the Anaconda method is that structures are less likely to be covered by muck at the mid-bench elevation than they are at waist height. Overall disadvantages to using this method include the following:

- Difficulties arise in plotting structures that do not cross the mid-bench elevation datum plane in the face.
- It is difficult to accurately represent the relationship of structures to the physical conditions of pit walls.

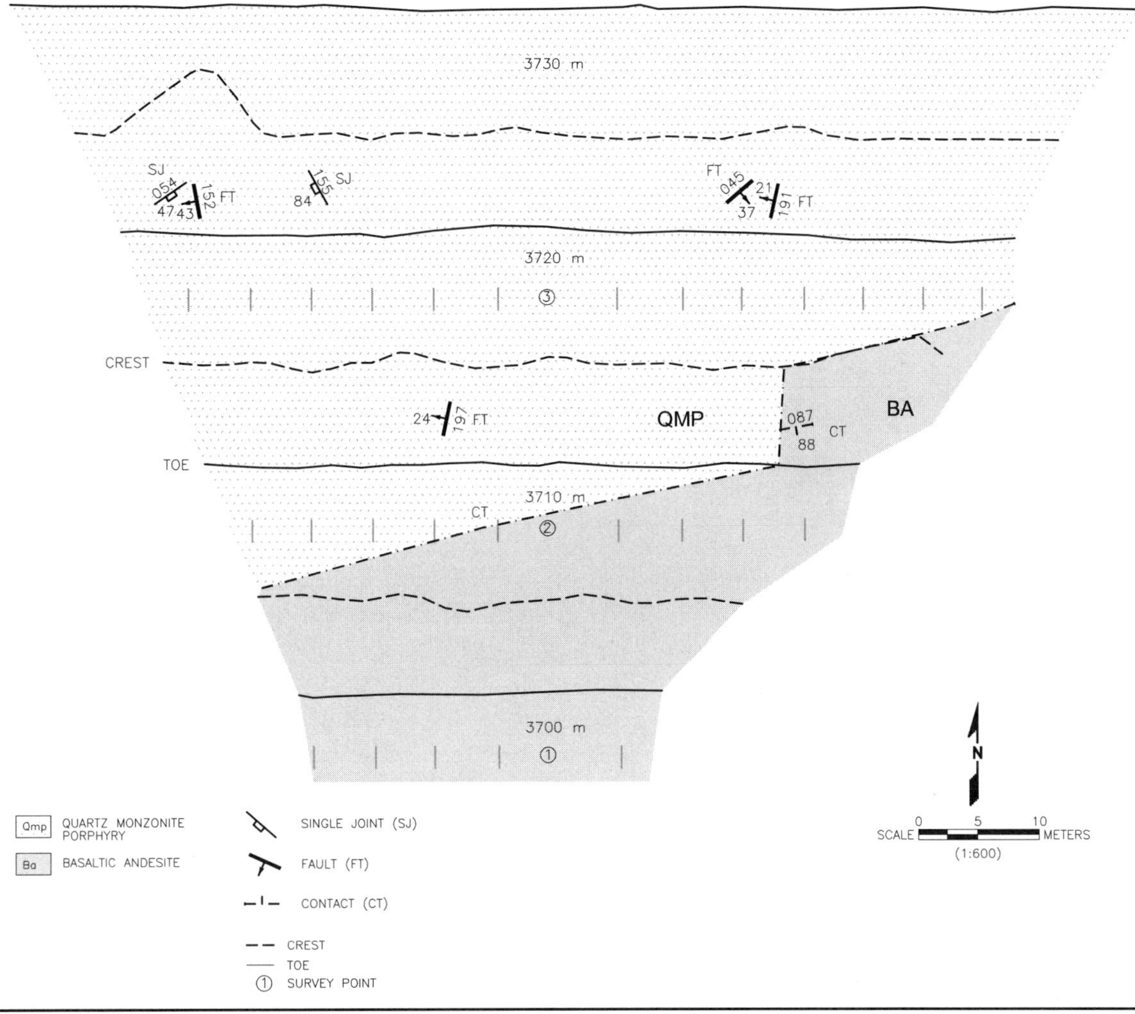

FIGURE 2.5 Mid-bench structure maps for geology shown on Figure 2.2

- It is difficult to determine bench-to-bench continuity of structures because the intersection of a structure with the toe and crest is not specifically noted. This is especially true for the highly important low-angle major structures.

Anaconda Method. The Anaconda method was developed for underground mapping, and it is probably the best method for that situation. The method is essentially the same as the mid-bench method, except the Anaconda method uses a waist-high or shoulder-high datum line to which all structures are projected. It is therefore primarily a method for producing level maps.

In an open pit, a measuring tape is laid out on the flat bench for survey reference. Structures are plotted at the face where they intersect the datum line. A strike-and-dip symbol is placed on the map at the intersection of the structure projection with the tape. The greatest operational difficulty with this method is that the waist-high or shoulder-high datum is often covered with muck.

An Anaconda-type major-structure map (Figure 2.6) was generated for the geology shown in Figure 2.2. Notice that structures are only represented by a strike-and-dip symbol at the point where the strike of the structure crosses the survey line on the bench face. Structure surface traces are not shown because the map is a two-dimensional level map. Two structures do not cross the survey line, and they cannot be displayed on the map. In an underground mapping situation, these structures would be projected to the intersection of the structure plane with the horizontal datum plane. However, on an open-pit topography map, this line of intersection would either be behind the current pit wall or in open air.

The following are disadvantages to using this method in an open pit:

- Low-angle structures that are striking parallel to the face will be projected either into air within the pit or into rock far behind the exposed face that is being mapped.
- It is difficult to represent the relationship of structure to physical conditions of pit walls. For example, the relationship of individual structures to wedge and plane-shear failures is not easily represented.
- It is difficult to determine bench-to-bench continuity of structures.

2.2.4 Major-Structure Data Management

Major-structure data should be compiled graphically on current pit maps and on level maps. Regional structures should also be graphically projected to future pit designs and cross sections. An example of a current pit, face method, major-structure map for a limestone mine in North America is presented as Figure 2.7.

Level maps should be constructed for structure data interpretation. Level maps can be constructed from face method maps at any elevation, including the toe elevation, crest elevation, or at the mid-bench elevation, because the actual surface traces are mapped over topography with this method. The toe elevation is

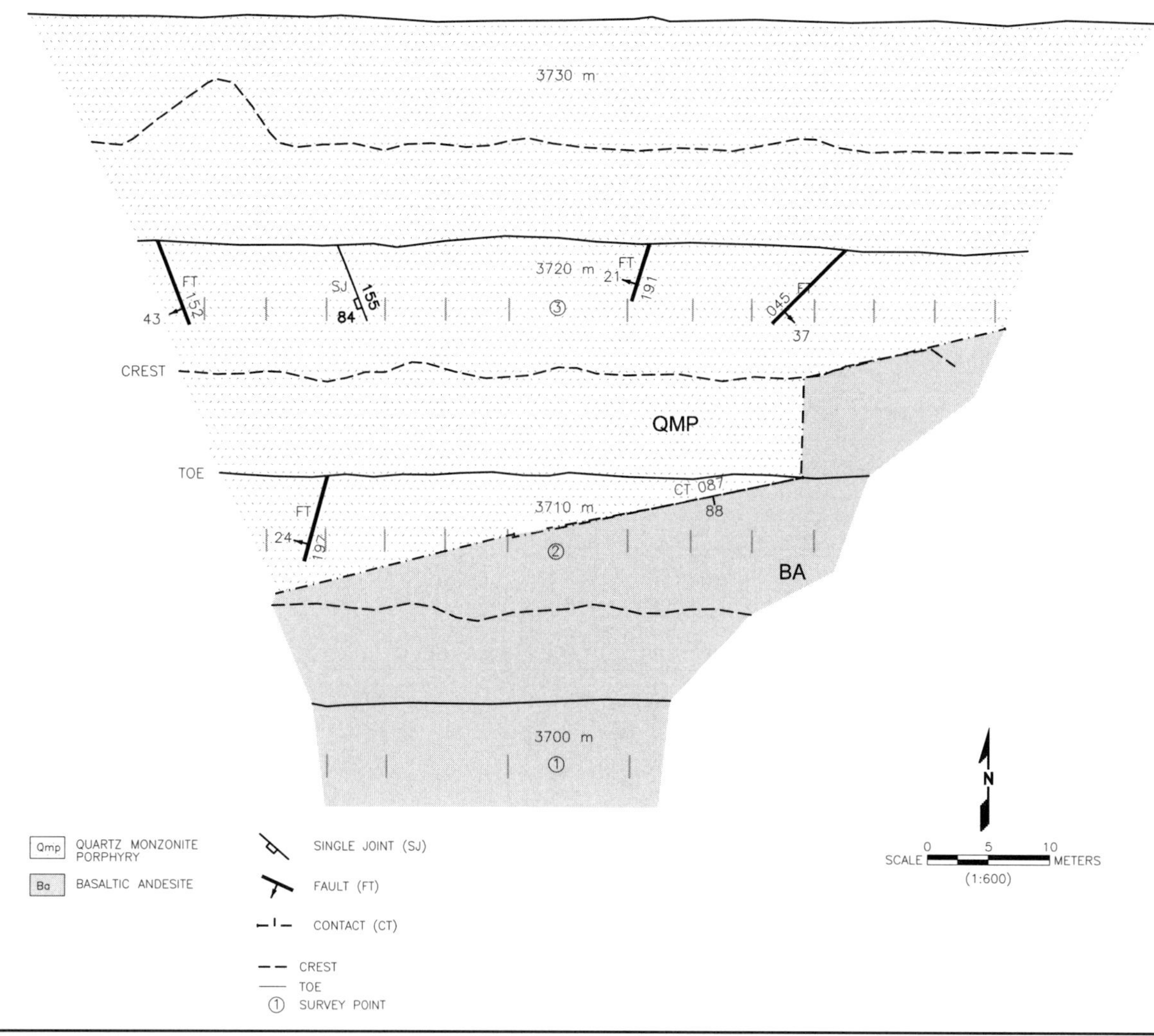

FIGURE 2.6 Anaconda method structure map for geology shown on Figure 2.2

most commonly used for level maps that are constructed from face maps.

Major-structure data should also be entered into a database for stability analysis. The data for each structure should include an ID number, northing, easting, elevation of the structure mid-point, rock type, structure type, dip direction, dip, length, thickness, and filling materials. Other data may be pertinent at different properties. With a spreadsheet, it is possible to sort structures by any variable for analytical purposes.

Any mine employee can identify a major slope failure after it has occurred. It is the job of the geologist and the geotechnical engineer to anticipate and avoid potential problems. To accomplish this, regional structures and rock types should be projected to future pit plans so that potential failure geometries can be recognized. Plan maps and cross sections should be produced for all annual pit plans through the final design. If the potential failure geometries are recognized early, the mine design might be altered to either avoid or control interramp and overall failures.

2.3 FRACTURES

The rock fabric is used in the bench design and in the determination of the rock-mass strengths. Fractures are too numerous to be analyzed as unique geologic structures, and the fracture data are generally analyzed by statistical methods. The fracture data are collected by mapping methods that provide the appropriate input parameters. Most of the fractures that are measured usually have lengths that are less than the bench height, and the minimum cutoff length is usually 0.3 to 1 m. Fracture spacing is usually less than 5 m.

2.3.1 Mapping Techniques

The fracture data are most commonly collected by using the fracture-set mapping method, scan-line method, cell-mapping method, and oriented-core method.

The various fracture data collection methods were summarized by Call (1992) as follows:

- *Fracture-Set Mapping.* Fracture sets are visually identified during the course of regular geologic mapping, and the fracture set orientation, length, and spacing are recorded.
- *Scan-Line Mapping* (also commonly called Detail-Line). The scan-line method is a systematic spot sampling method in which a measuring tape is stretched along the bench face or outcrop to be measured. For all the fractures along the tape, the point of intersection with the tape, orientation, length, roughness, filling type, and thickness are recorded.
- *Cell Mapping.* The bench face or outcrop is divided into cells. Normally, the width of the cell is equal to one to two times the height of the cell. Within each cell, the fracture sets are visually identified, and the orientation, length, and spacing characteristics are recorded along a line that is oriented in any direction.

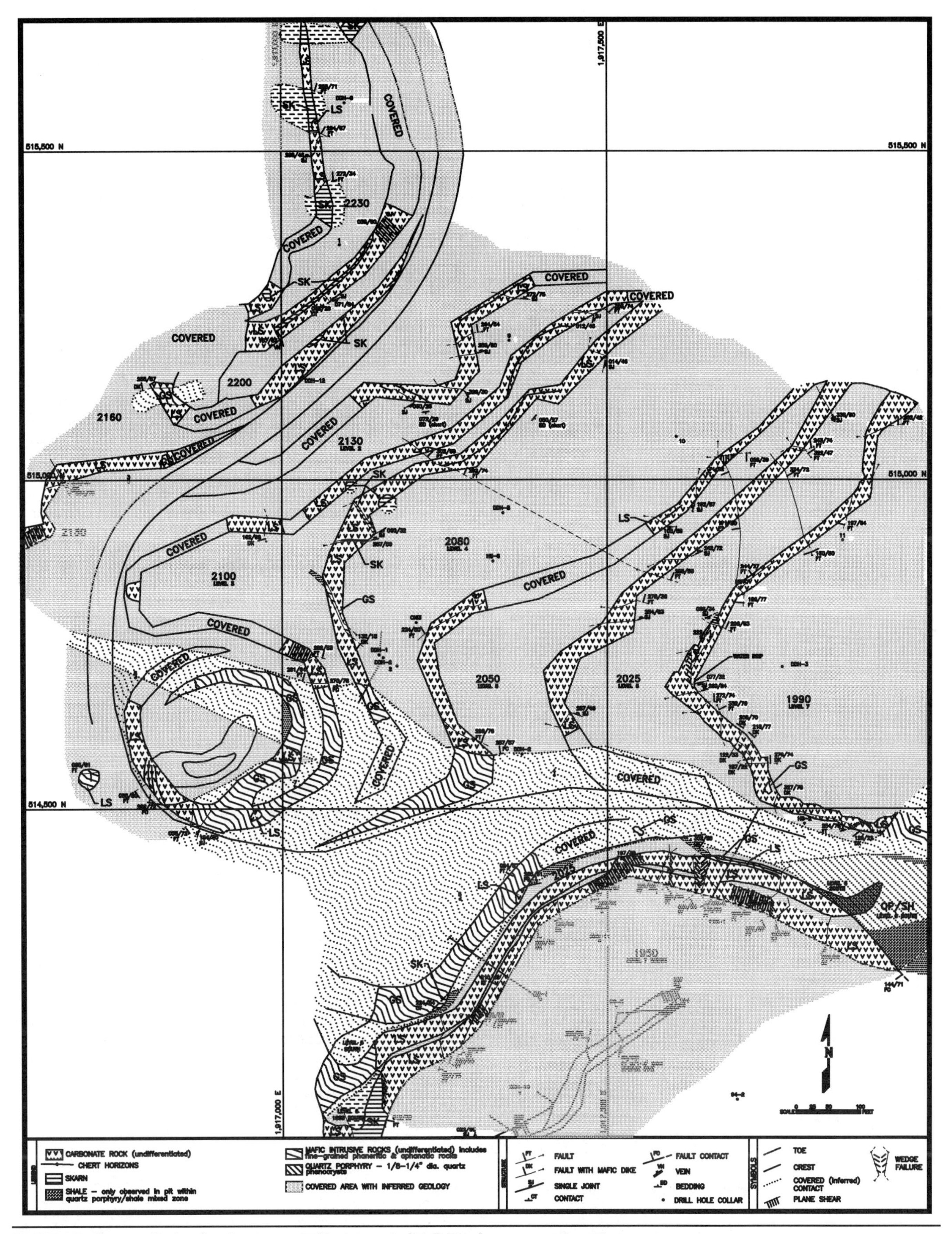

FIGURE 2.7 Face method major structure map for a current pit topography

- *Oriented Core.* Oriented core provides fracture orientation and spacing data, but length data cannot be determined with this technique. The oriented core is similar to the scan-line mapping method. Oriented coring is used when the rock types of interest are not exposed. It is also used to determine whether the geologic structural domains, which were mapped on the surface, extend back behind the pit walls.

The fracture-set mapping method is a general mapping method, and to the authors' knowledge, has not been described in detail in the literature. The scan-line method has been described in detail by numerous authors (Call, Savely, and Nicholas 1976; LaPointe and Hudson 1985; Warburton 1980). The reader is referred to these and other papers for further discussion of the scan-line mapping method. While the cell-mapping method has been summarized in previous literature (Call, Savely, and Nicholas 1976; Call 1992), it has never been described in detail. A detailed explanation for the cell-mapping method is presented in Appendix A. There are a number of oriented-core techniques, including Clay imprint (Call, Savely, and Pakalins 1982), Craelius (Rostrom 1961), and the Scribe method.

All three of the surface mapping methods will provide the basic data on orientation, spacing, and length. Cell mapping and fracture-set mapping are preferred because more data can be collected over a larger area to better define (1) the limits of structural domains and (2) the variability of the joint characteristics with the structural domain. The choice of the technique depends on type and amount of staffing available. The scan-line method requires little to no judgment in the data collection; cell mapping and fracture-set mapping require geologic judgments to be made. The scan line represents detailed information at one location equivalent to one or two cells. It would take three to seven times longer to map enough scan lines to cover the same area using the cell or fracture-set methods. The normal scan line is horizontal and has the inherent problem of mapping those joint sets that do not intersect that horizontal line, such as flat-dipping joint sets or sets that strike parallel to the wall orientation. The only way to map those sets is to map a vertical scan line or a face perpendicular to the wall. Fracture-set and cell mapping permit mapping all sets in all directions. The scan-line method can be used when confirming the distribution of the structures and also when individuals collecting the data lack geologic training.

Oriented coring is used either to collect data where surface data are limited or to determine whether the structural domains mapped at the surface extend behind the pit walls. Oriented core does not provide length data. Additionally, the oriented data is more scattered than is the surface mapping data because the oriented core represents only 7- to 15-cm^3 of the fracture plane. Consequently, it does not represent an average orientation. Also, the oriented core has a definite blind zone, which must be considered when analyzing the data.

2.3.2 Fracture-Data Processing

Fracture data are very amenable to computer processing. The process is to (1) develop Schmidt plots to define zones of similar structure orientations and structure domains (which are discussed in the next section) and then (2) define the distributions of the geologic or design sets. Design sets are those structures that define a certain potential failure mechanism, whereas a geologic set is defined because of a geologic condition. In general, a geologic set has less dispersion in its orientation. For slope analysis, we generally use the design set, except when the geologic set is defined by bedding or foliation that have length or spacing characteristics significantly different than those of the other joints mapped (Figure 2.8).

For each of the design sets, the defined distributions are the dip direction, dip, minimum dip, spacing, length, and overlap. In general, the dip direction and dip have a normal distribution, where the mean and standard deviation describe the distribution. The length and spacing generally are a Weibull function of which the negative exponential is a unique form.

Weibull Equation.

$$Y = 100 \times e^{-aXb} \qquad \textbf{EQ. 2.1}$$

Where:

Y = probability that X is greater than or equal to X
X = characteristic being defined, such as spacing or length
b = constant
a = constant; if $b = 1$, then $a = 1/$(mean of the characteristic)

Figure 2.9 shows the impact of using a negative exponential versus the Weibull for fault data from a porphyry copper deposit. The length distribution is a minimum length distribution because we cannot always measure the total length of the structure. Also, we do not always measure the extreme short lengths because of difficulty in observation; statistical techniques are applied to truncated distributions to provide an estimate of the "true" distribution. There are also statistical techniques for extrapolating measured lengths given the available window size, but the extrapolation becomes unrealistic when the measured length of the structure approaches the size of the mapping window.

Overlap cannot be measured using the cell-mapping and fracture-set methods. When using the detail-line or scan-line methods, the amount of structure above or below the line can be measured. In our experience, the overlap is uniformly distributed for many fracture sets. This distribution is not applicable in bedded or foliated deposits where many of the fractures are terminated against the bedding or foliation. It is important not to focus on which distribution is most common but rather to focus on using the best distribution that fits either the data or the observations. In the probabilistic analysis, the resulting answer is dependent on the distribution used.

Shear strengths from fractures are determined from direct-shear test of natural fractures or fault gouge material. It is important to measure the range of shear strengths so that the distribution of strengths can be defined. For each structural domain, we measure at least three samples; six samples per rock type are preferred. If one joint set is more pervasive than another joint set, the shear strength of this set should be measured as an individual group.

2.4 STRUCTURE DOMAINS

The structure model comprises any number of individual structure domains. Each structure domain contains geologic structure characteristics that are different from the geologic structure characteristics within neighboring domains. Boundaries that often define structure domains are engineering rock-type contacts and major structures, such as regional faults and fold axes.

The first level of structure-domain division is to separate the deposit into regions with different engineering rock types based on rock shear strength and fracture shear-strength properties. Rocks with similar strength values, regardless of petrogenesis, are considered to be unique engineering rock types, and the engineering rock-type boundaries act as the primary structure domain boundaries. Usually, the rock strength is related to the primary lithology or to secondary alteration; therefore, engineering rock types can be directly related to either lithology or alteration.

The second level of division for structure domains is regional structures. Fracture and intermediate structure orientations may vary significantly on either side of a regional fault. Fold axes are almost always structure domain boundaries because they define a boundary between areas where bedding and bedding-related fractures change in orientation.

Geologic structure orientation, length, and spacing may either be consistent across engineering rock-type boundaries and

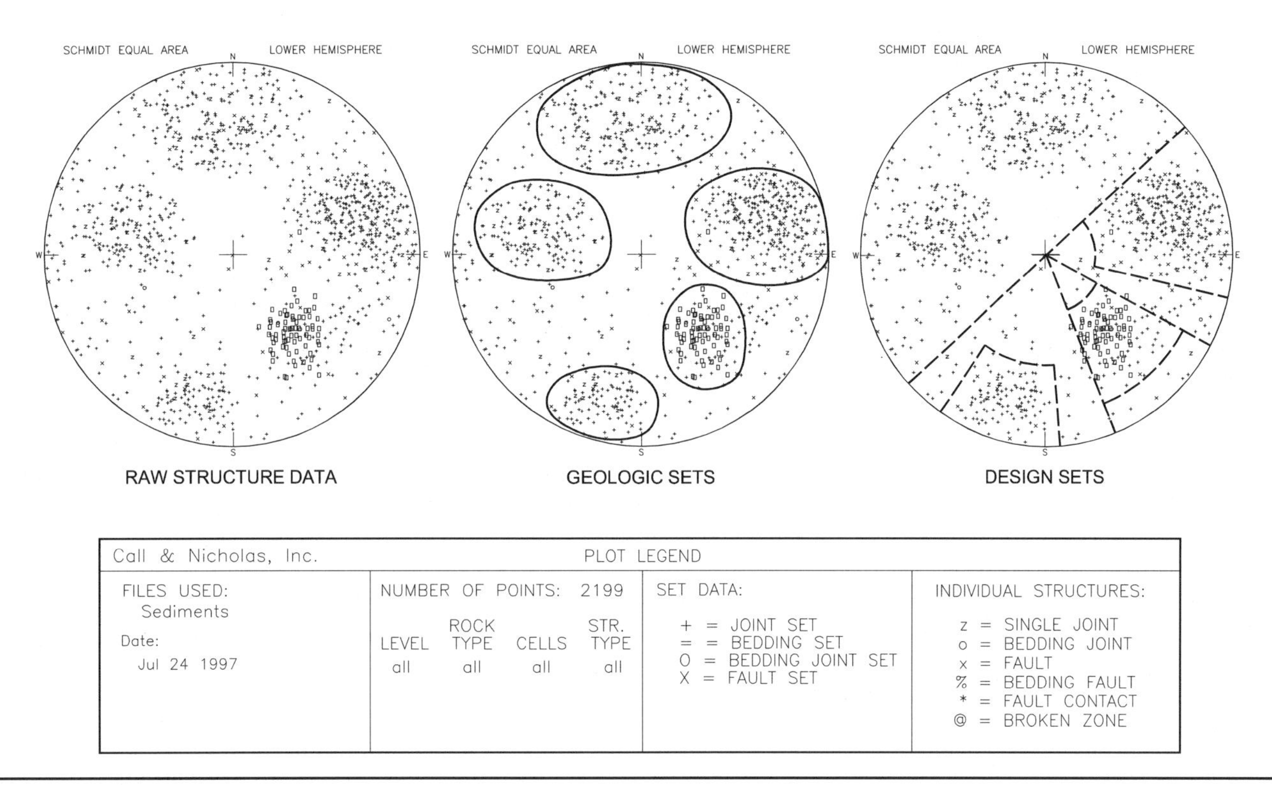

FIGURE 2.8 Geologic sets versus design sets

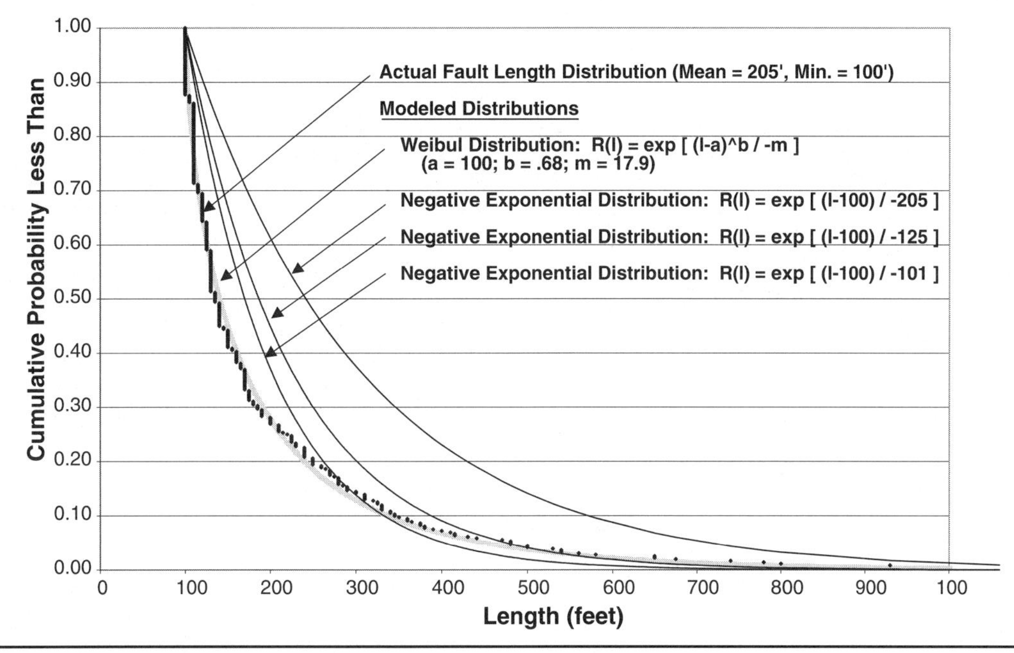

FIGURE 2.9 Structure length distributions

regional structures or they may vary between the engineering rock types and regional structures. The structure orientations are usually the best indicators of the similarities or differences among structure characteristics. Structure orientations are the easiest and most reliable structure characteristic to measure, because structure orientations can be measured at the surface from mapping and at depth from oriented core. Orientation analysis is relatively simple because variations in orientation can be easily detected on stereonets. Once the basic structure domains have been interpreted, they should be subdivided into regions both by elevation and plan. If the Schmidt plots are similar, one structure domain exists; if they are not similar, subdividing may be required. The engineer or geologist has to determine which feature is causing the need to subdivide the structure domain.

2.5 ENGINEERING ROCK TYPES

Engineering rock types are defined by intact rock and fracture shear strength; these are both critical characteristics that are used in a structural analysis. Geologic rock types with similar strength parameters are grouped into engineering rock types.

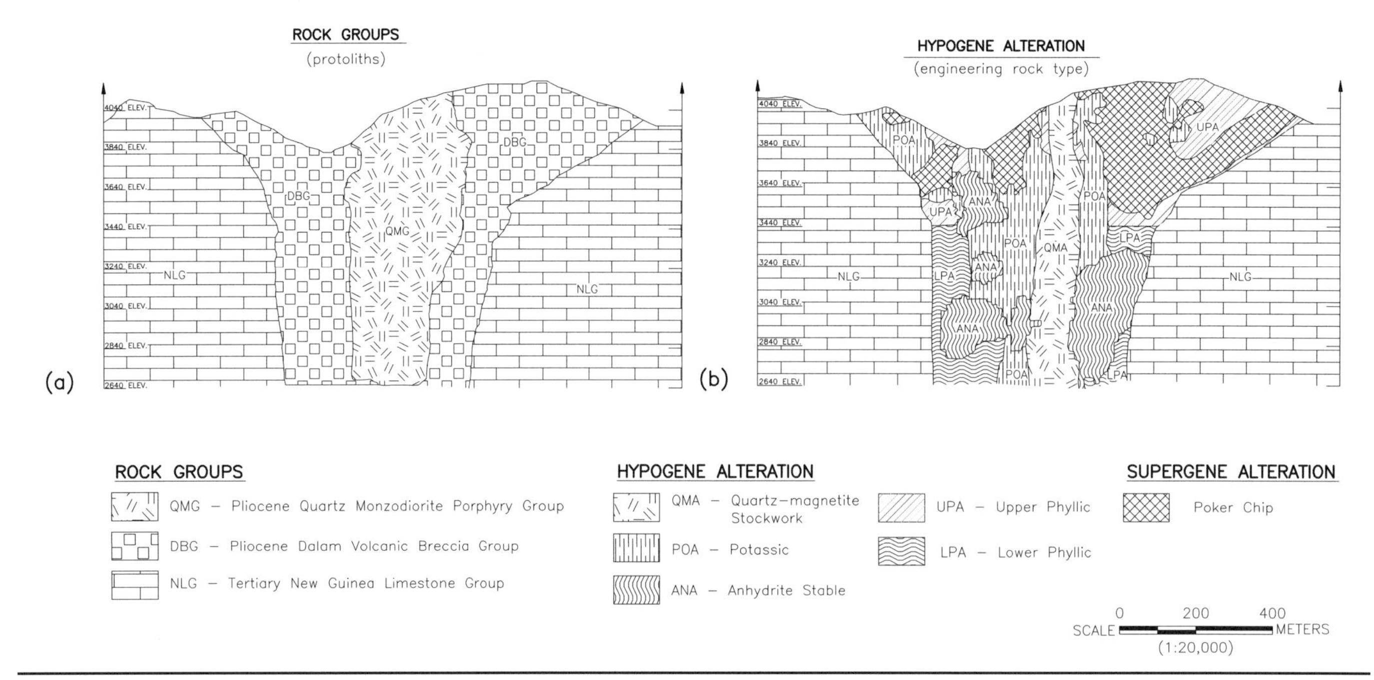

FIGURE 2.10 Grasberg deposit cross section for (a) protoliths and (b) engineering rock types

Strength test results may require dividing engineering rock types on the basis of protolith (irrespective of alteration), alteration type (irrespective of protolith), or a combination of both. Strength parameters for a given geologic rock type may also vary with depth or with geographic location within the deposit, requiring the division of one geologic rock type into two or more engineering rock types.

The engineering rock-type model for the Grasberg Igneous Complex (GIC) is presented as an example of the relationship between protolith, alteration, rock-quality designation (RQD), relative depth, and rock strength. The GIC is a diatreme, hosted within limestone. Many individual igneous phases have been identified within the GIC, and the igneous rocks can be broadly divided into the intrusive rocks and the relatively older volcanic rocks (MacDonald and Arnold 1994). Figure 2.10a demonstrates the relationship of the protoliths in cross section.

Rock-strength testing was conducted for all protoliths and alteration types, and the results were compared against all of the variables that were recorded for each test sample. The other variables included location, protolith, alteration type, and RQD. Protolith, alteration type, RQD, and relative depth ultimately differentiated the engineering rock types.

All of the igneous protoliths have been hydrothermally altered; the alteration types include quartz-magnetite stockwork, potassic, phyllic, and anhydrite stable alteration. Some of the hydrothermal alteration types overlapped so that there is some subjectivity in attributing a particular alteration type to a given rock sample.

Alteration anhydrite was leached in the near-surface environment to depths of 600 m. The leached rock is referred to as *pokerchip*. The pokerchip is characterized in core by a measured RQD of 5% or less in continuous runs. The anhydrite stable rock is best distinguished in core as rock with continuous measured RQD values of 95% or greater. The interface between pokerchip and anhydrite stable rock is often only a few meters thick.

Limestone was consistent in rock-strength properties, and all limestone was assigned as the limestone engineering rock type. Igneous rocks with an RQD of 5% or less, regardless of protolith or hypogene alteration, had similar rock-strength characteristics and were assigned to the pokerchip engineering rock type. All igneous rocks with an RQD of 95% or greater had similar rock-strength characteristics, regardless of protolith (intrusive phase or volcanic rock type in this case), and were assigned to the anhydrite stable engineering rock type. All quartz-magnetite-altered rocks, regardless of intrusive phase or volcanic rock type, had similar strength values and were assigned to the quartz-magnetite engineering rock type. All potassic-altered rocks, regardless of intrusive phase or volcanic rock type, had similar rock-strength values and were assigned to the potassic engineering rock type. Rock-strength properties for phyllic-altered rocks showed variability with depth; the phyllic-altered rocks were divided into the lower phyllic and upper phyllic engineering rock types to reflect this difference. The resulting engineering rock type model is presented in Figure 2.10b.

2.6 ROCK-MASS STRENGTH

The purpose of this chapter is not to review all classification techniques and all methods to estimate rock-mass strength; however, they must be discussed as part of the geologic structure data analysis. Chapters 3 and 4 discuss the use of these data in the catch bench, interramp, and overall slope analyses.

2.6.1 Rock-Mass Classification

All drill holes and bench faces should be mapped using one of the rock-mass classification techniques. Two of the more popular systems used today are Bieniawski's rock-mass rating (Bieniawski 1974) or Barton's rock quality (Barton, Lien, and Lunde 1974). Although these classification systems have empirical data to correlate ground behavior, such as slope angles, the empirical data should only be used if structure data are not available.

There are techniques to estimate the rock-mass strength from these rock-mass classifications (Hoek and Brown 1992; Laubscher 1977), and using a classification to estimate the rock-mass strength is both appropriate and prudent. However, the user must be aware that there is a directional condition to the rock-mass strength. For example, a rock mass with a low RQD or with a high fracture frequency will have the same classification whether the joint set is dipping back into the pit wall or dipping into the pit. Although the classification techniques allow adjustments to their values for directional considerations, they still do not provide good

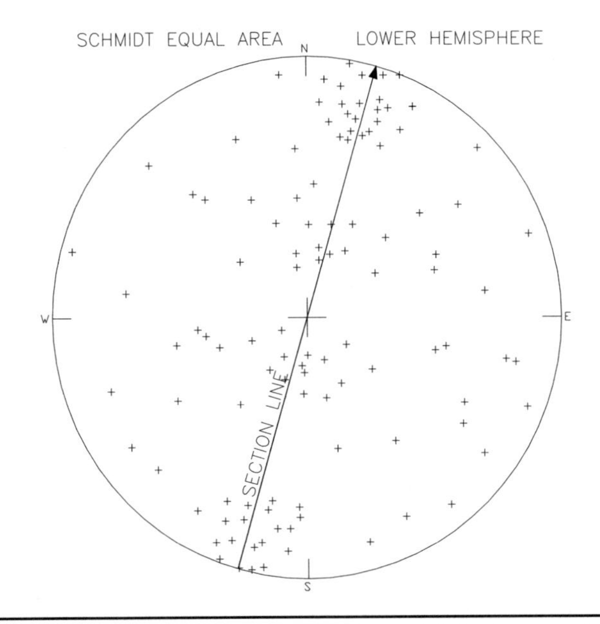

FIGURE 2.11 Schmidt plot of data used in scratch fracture model

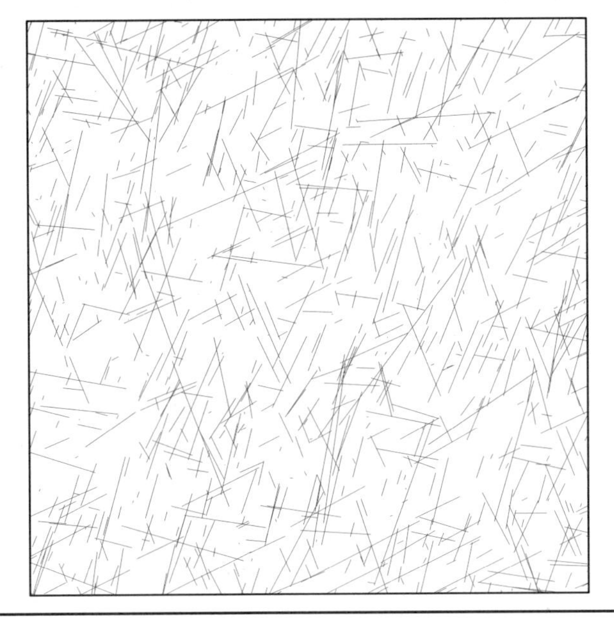

FIGURE 2.12 Scratch fracture model for three joint sets

estimates of directional strengths. If possible, the geologic structure data should be used to modify those strengths.

2.6.2 Fracture Models

Because the fabric of the rock mass is comprised of several intersecting joint systems, the strength of the rock mass will be anisotropic. This directional rock-mass strength can be estimated using fracture-modeling techniques. A step-path model (Call and Nicholas 1978) was one of the early efforts to quantify this variability. At the time, available computer memory was a serious constraint, and the analysis was limited to the evaluation of only the flattest step-path angle, which was not necessarily either the path at the critical angle or the path with the lowest percentage of intact rock. Because the failure path of interest is problem specific, Call & Nicholas, Inc. has developed a fracture simulation-modeling program that enables the engineer to visually evaluate the potential for the formation of step-path failure geometries. Initially, the fracture simulations were used to predict block sizes

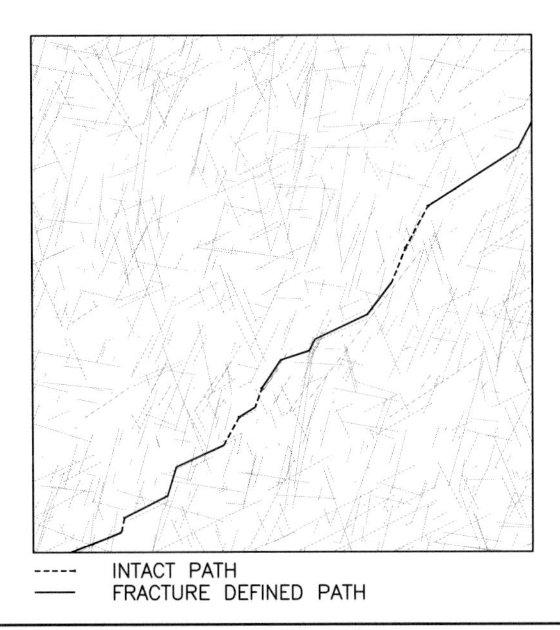

FIGURE 2.13 Path with least percent intact rock

in a cave and, subsequently, were used for the evaluation of failure geometries in underground (Nicholas and Miller 1984) and surface excavations. A similar program, FracMan, has been used by Golder Associates to model geologic structures in either two or three dimensions. The remainder of this discussion refers to the fracture-modeling program used by Call & Nicholas, Inc.

The input to the model includes the orientation of the section of interest and the statistical distributions of the dip, dip direction, spacing, length, and overlap for the fracture sets of interest. Structures sampled from these distributions that do not meet the minimum length requirements specified by the user are suppressed in the rendering of the simulation. Geologic sets selected for the analysis usually have a dip direction that is within ±40° from the bearing of the section. The orientation sampling that is conducted is based on the distribution statistics for the geologic set being modeled. However, the minimum and maximum range of sampled values is constrained by the limits of the observed population, as shown on the Schmidt plot (Figure 2.11).

Individual structures are sampled in window strips, the narrow dimension of which is a function of the mean length of the set. The sampling windows are oriented horizontally for structures with dips (45° and vertically for structures with dips <45°). Once a simulation has been generated for the section (Figure 2.12), a separate module is used to verify that the structure overlap from the window sampling routine has produced the correct apparent spacing for each set.

After the section has been generated (Figure 2.12) and verified, potential step-path failure geometries are identified and traced on the section. The critical path may be the one with the flattest possible inclination, the one with a predefined inclination, or the one with the least intact rock (Figure 2.13). Plots of percent intact rock versus step-path angle or plots of other parameters of interest can then be produced for any step-path geometries identified.

To estimate the rock-mass strength along the section analyzed, the intact and fracture shear strengths are weighted based on the percentage of the failure path that must pass through intact rock. This modeling technique has worked well in cases where the strength of the rock mass in front of a non-daylighted major structure, such as a fault, must be determined to see whether the remaining buttress is strong enough to support the rock above it (Figure 2.14).

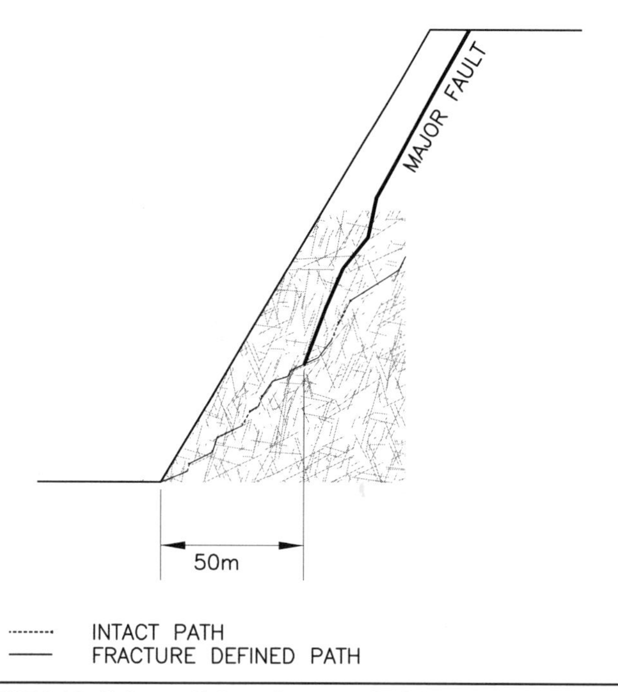

FIGURE 2.14 Failure path to evaluate non-daylighted major structure

2.7 CONCLUSIONS

The geologic structure characteristics and the orientation of the pit wall relative to the structure orientations usually determine the attainable slope angle in rock. The delineation of structural domains and the definition of the structure characteristics within each domain are essential to ensure that the structure characteristics are applied only to those areas of the current and future pits where they will actually exist.

Major-structure data are the most important and the most difficult structure data to collect and interpret. Structure length is the most difficult major-structure characteristic to measure, because the exposure in a pit requires conducting mapping on a bench-by-bench sequence, either on an existing highwall or on pushbacks of consecutive phases. Appropriate survey, mapping, and interpretation techniques are necessary to accurately measure the major-structure lengths. We strongly recommend mapping structure data for geotechnical analysis using the face-mapping method.

The major-structure data should be plotted on current pit topography maps, and the data also should be compiled on level maps and in a database. Level maps can be used to determine vertical and lateral continuity of major structures as the mine development expands. Regional structures should be extended along strike to the limits of the level maps so that they can be easily plotted on all future pit plans. The database can be used for statistical analysis for the intermediate structures.

Relatively short structures that comprise the rock fabric are too numerous to map and analyze individually, so they must be mapped with techniques that provide the appropriate input parameters for statistical analysis. There are several techniques for collecting the rock fabric data, and each has its own advantages and disadvantages. We recommend the cell-mapping method, which is described in detail in Appendix A.

Rock-mass strength is one of the most difficult parameters to quantify in slope analysis because, given the large size of the rock mass, there are no practical means to directly measure the rock-mass strength. Determination of rock-mass strength with common rock classification systems can be useful, but the classification systems usually do not account for the directional dependence of the rock-mass strength imposed by the rock fabric. A fracture model is required so that the directional aspect of the rock-mass strength can be quantitative.

2.8 ACKNOWLEDGMENTS

The *Chicken Scratch Fracture Model* was developed with the help of many people. Major contributors include David Nicholas, Dan White, John Marek, Richard Call, and Paul Pryor. We thank Chuck Brannon and George MacDonald of Freeport McMoRan Copper and Gold Company for granting permission to present and review the Grasberg rock-type data.

2.9 REFERENCES

Barton, N., R. Lien, and J. Lunde. 1974. Engineering classification of rock masses for the design of tunnel support. *Rock Mechanics*, 6:4:189–236. Originally published in 1974 as Analysis of rock mass quality and support practice in tunneling. *Norwegian Geotechnical Inst.*, Report No. 54206.

Bieniawski, Z.T. 1974. Geomechanics classification of rock masses and its application in tunneling. *Proceedings Third International Congress on Rock Mechanics*, vol. 11A, 27–32. Denver, Colorado:ISRM.

Call, R.D. 1992. Slope stability. In *SME Mining Engineering Handbook* 2nd ed. vol. 1. ed. H.L. Hartman. Littleton, Colorado:SME.

Call, R.D., J.P. Savely, and D.E. Nicholas. 1976. Estimation of joint set characteristics from surface mapping data. *Proceedings 17th US Symposium on Rock Mechanics.* Utah Engineering Experiment Station, University of Utah, Salt Lake City, pp. 2B2-1 to 2B2-9.

Call, R.D., and D.E. Nicholas. 1978. Prediction of step path failure geometry for slope stability analysis. *Proceedings 19th Rock Mechanics Symposium*.

Call, R.D., J.P. Savely, and R. Pakalins. 1982. A simple core orientation technique. *Stability in Surface Mining.* vol. 3. ed. C.O. Brawner. AIME.

Deere, D.U. 1963. Technical description of rock cores for engineering purposes. *Felmechanik und Ingenieurgeologie*. 1:16.

Hoek, E., and E.T. Brown. 1988. The Hoek–Brown failure criterion—1988 update. In *Rock engineering for underground excavations, Proceedings 15th Canadian Rock Mech. Symp.*, ed J.C. Curran. 31–38. Toronto, Ontario:Dept. Civil Engineering, University of Toronto.

La Pointe, P.R., and J.A. Hudson. 1985. Characterization and interpretation of rock mass joint patterns. *Geological Society of America Special Paper 199*. Boulder, Colorado:SA.

Laubscher, D.H. 1977. Geomechanics classification of jointed rock masses—mining applications. *Trans. Instn. Min. Metall.* 86, A1–8.

MacDonald, G.D., and L.C. Arnold. 1994. Geologic and geochemical zoning of the Grasberg Igneous Complex, Irian Jaya, Indonesia. *Journal of Geochemical Exploration*, vol. 50, p. 143–178.

Nicholas, D.E., and S.M. Miller. 1984. Geotechnical considerations. In *Northwest Mining Association Short Course Mine Feasibility—Concept to Completion*, ed. G.E. McKelvey.

Peters, P.C. 1987. *Exploration and Mining Geology*. 2nd ed. New York:John Wiley & Sons.

Rostrom, E. 1961. Craelius automatic core orientator. *Canadian Mining Journal*. pp. 60–61.

Warburton, P.M. 1980. A stereological interpretation of joint trace data. *Intl. Jour. Rock Mech. Min. Sci. & Geomech. Abstr.* 17:181.

Appendix
Cell-Mapping Procedure

2.10 INTRODUCTION

The cell-mapping procedure involves dividing the bench faces into zones of equal length, called *cells*. The area from which structure data is recorded is referred to as the *cell window*. Cell dimensions are usually square (i.e., the width of each cell is equal to the bench height). Bench faces that are not obscured by overbank or blasted material are subdivided into cells, and the geologic structure present in each cell is then mapped. Cell mapping can also be conducted along irregular outcrops or road cuts; because of the physical constraints of the rock exposures, these cells may be rectangular in shape.

In an open-pit environment where multiple benches are developed, cell mapping is generally conducted using two different approaches. First, if personnel, budget, or time are not limiting factors, entire benches are mapped with contiguous cells. This type of mapping campaign allows for a complete sampling of the geologic setting in the study area, including jointing and major faults and all other structures. A second approach may be used if resources are limited or if there are time constraints; this involves mapping strings of contiguous cells, from 2 or 3 contiguous cells to as many as 10 or more, located in different parts of the study area. With this approach, individual strings of cells are selected to spot-check for potential changes in structure characteristics across the study area. A sufficient number of cell strings have to be mapped to identify structural domain boundaries across the study area.

When mapping road cuts or isolated outcrops, an attempt should be made to map strings of at least two to three contiguous cells.

A standard data sheet is used to record the following information for each cell (Figure 2.A.1). (Note: The letter C in Column 1 of the data sheet denotes information pertaining to the entire cell; the letter S denotes data pertaining to geologic structures.)

2.11 C LINE DATA ENTRY—MAPPING CELL INFORMATION

Level. The elevation of the bench, outcrop, or road cut being mapped is indicated. If surveying of cell locations is not possible, the elevation should be estimated, for example, from a topographic map.

Cell Number. The cells within a string are numbered consecutively where mine benches are developed, including those covered by muck. A break in the cell numbering sequence should be used when starting a new string of cells that is not contiguous with the previous string.

Cell Width and Height. The recorded width and height should not include obscured or nonmappable areas of the cell. For example, if the lower 3 m of a 15-m bench are covered by muck, a cell height of only 12 m is recorded.

When mapping in outcrops or partially covered road cuts, estimates of the *average* width and height of the exposure is made.

Rock Type. The two or three dominant rock types in the cell are listed in descending order of percent occurrence.

Ideally, an individual detail cell *should not cross* major (relative to pit-scale) lithologic contacts or major faults, since fracture characteristics and orientations may differ in large bodies composed of distinct rock types or fault-bound blocks.

On the other hand, if mapping is being conducted along fairly closely spaced, alternating lithologic units, for example, in a sedimentary sequence, rock types should be listed according to descending order of percent occurrence.

Face Strike and Dip. The strike of the bench face comprising the cell (using the right-hand convention) and the bench face angle are measured. Face strike is an *average* strike along the observed cell and *should always be recorded in such a way that the strike plus 90° represents the dip direction of the bench face (right-hand rule)*. The average strike of a *muck cell* must be recorded if it is located within a string of cells where structural information has been collected. This is necessary to allow for the calculation of coordinates of the remaining cells in the string.

The face-dip or bench face angle should represent not the average dip but rather the dip of the flattest plane observed along the cell. For example, if the back of a wedge failure located along the crest represents the farthest back break in the cell, a plane should be projected from that point to the actual (or implied) toe of the bench.

The average bench face angle is also often collected, although not specifically required in the computer programs used to process the data and analyze back break. If the average bench face angle is recorded, it should be in addition to, not instead of, the minimum bench face angle.

If intact rock within muck cells is obscured because of raveling of the bench face following mine bench cleanup, the minimum bench face angle (and optionally the average bench face angle) should be recorded. However, if the material comprising the "muck cell" *is obviously* shot muck that has not yet been mined, only the strike should be recorded. In the latter case, the average dip is not valid since no attempt has been made to clean up blasted material.

When mapping an outcrop, the *face dip* may be difficult to determine because of the irregular nature of the exposure. In this case, the face dip is less relevant since it does not represent a mined face but rather a long-term erosional surface. Discretion should be used when comparing outcrop face dips to existing or predicted mine bench face angles.

Control. When strings of contiguous cells are mapped, survey coordinates should be taken every 5 to 10 cells; these control points are later used to determine the midpoint coordinates of each individual cell. If a cell is to be used for survey control, the level and cell number should be recorded in the control field on the data sheet. When the survey data becomes available, the coordinates should be listed in the remarks column of the field data form for bookkeeping purposes. Ideally, the first and last cell in every string should be surveyed.

When mapping individual cells, such as outcrops, each cell should be surveyed if possible.

Remarks. The "Remarks" column is used to describe pertinent features of the cell; the back of the field form can also be used. Additional comments might include specific potential failure geometries (i.e., plane shear or wedge) or actual failures that have occurred along the mapped faces. Information should include a sketch, direction and angle of discrete wedge-geometry plunge intersections, and the orientation of structure(s) involved. Remarks on nature of groundwater seepage (i.e., amount of water, structure control, etc.) would also be important information. If mapping underground, the support being used could be described.

2.12 S LINE DATA ENTRY—STRUCTURE (SET) INFORMATION

For each cell, the following structure information is recorded:

Type. The geologic structure (e.g., joint set, bedding set, foliation, a single joint, etc.) is recorded using a two-character code. Major structures (e.g., fault, shear zone, etc.) are identified in the same manner.

The structure types indicated at the bottom of the cell-mapping forms are not absolute. If some site-specific structure

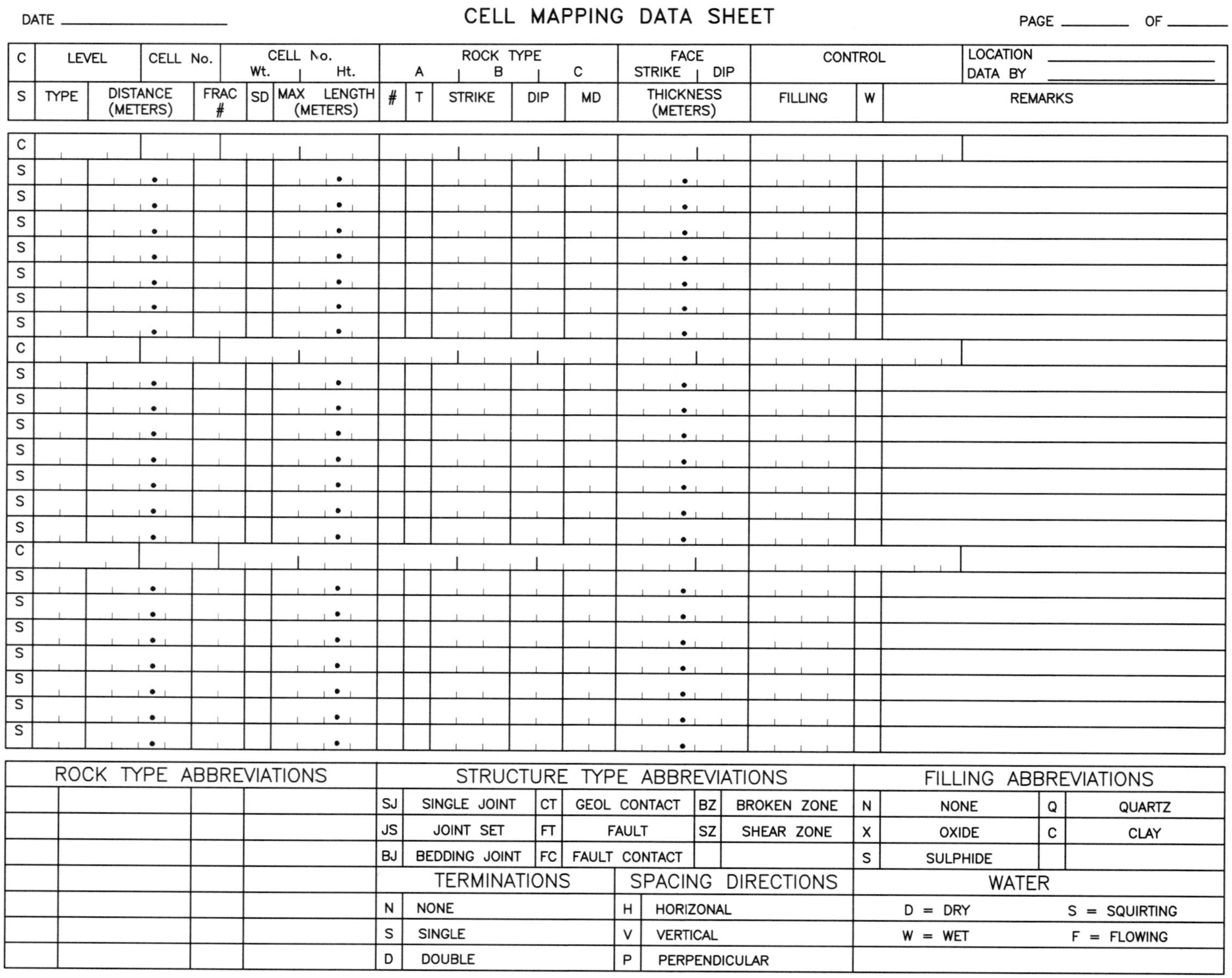

DATE ________ **CELL MAPPING DATA SHEET** PAGE ____ OF ____

C	LEVEL	CELL No.		CELL No. Wt. | Ht.		ROCK TYPE A | B | C					FACE STRIKE | DIP	CONTROL		LOCATION ____ DATA BY ____
S	TYPE	DISTANCE (METERS)	FRAC #	SD	MAX LENGTH (METERS)	#	T	STRIKE	DIP	MD	THICKNESS (METERS)	FILLING	W	REMARKS
C														
S														
S														
S														
S														
S														
S														
S														
C														
S														
S														
S														
S														
S														
S														
S														
C														
S														
S														
S														
S														
S														
S														
S														

ROCK TYPE ABBREVIATIONS			

STRUCTURE TYPE ABBREVIATIONS					
SJ	SINGLE JOINT	CT	GEOL CONTACT	BZ	BROKEN ZONE
JS	JOINT SET	FT	FAULT	SZ	SHEAR ZONE
BJ	BEDDING JOINT	FC	FAULT CONTACT		

FILLING ABBREVIATIONS			
N	NONE	Q	QUARTZ
X	OXIDE	C	CLAY
S	SULPHIDE		

TERMINATIONS	
N	NONE
S	SINGLE
D	DOUBLE

SPACING DIRECTIONS	
H	HORIZONAL
V	VERTICAL
P	PERPENDICULAR

WATER	
D = DRY	S = SQUIRTING
W = WET	F = FLOWING

FIGURE A.1 Cell mapping data sheet

type is not covered on the form, add to it. However, be consistent during mapping and provide an explicit definition of the new structure type.

The designation "MC" should be recorded in the first "S" line whenever a muck cell is laid out and traversed in a string of cells. The software program used later to process this data recognizes the cell as one having no structures mapped.

2.12.1 Individual Structures Versus Structure Sets

When *individual* structures, such as a discrete fault, a vein, or a single joint, are recorded, only the strike, dip, minimum dip, length (the maximum length equals the length in this case), thickness, and filling type(s) should be recorded.

When structural trends are visibly indicated, but open fractures are not present, only the strike and dip of these features are recorded. A good example is massive bedding in sedimentary rocks where the bedding attitude can be estimated because of the banded nature of that rock. (Note: The recorded symbol for this type of bedding attitude should be different from the symbol used for a bedding *fracture* or *fracture set*.)

Conversely, for any structure *set* (e.g., three or more individual structures with apparent similar attitudes) all parameters in the following discussion must be recorded:

Distance. This is the distance over which the number of fractures in a structure set is counted (used later to determine average spacing of the structure set within the cell). The *distance* recorded is the length of an arbitrary counting line visibly projected across the cell window being mapped. The actual placement of the counting line for each set usually depends on site-specific physical conditions, such as the development of the structure sets across the bench face being mapped or portions of the cell that are covered or displaced.

The distance over which the number of individual fractures are counted within a set should generally be the same as the cell dimensions. For example, if the count is made horizontally for a 15-m-wide cell (SD = "H"), the distance recorded would be 15 m. Similarly, for a 12-m-high cell for which the counting direction is vertical (SD = "V"), the distance would usually be 12 m.

There are several exceptions to this rule. One case, as indicated above, would be if part of the cell is obscured by muck. The effective distance recorded would be that portion of the cell where the mapper can clearly view the structures within that set.

A second exception to the rule would occur when rock is very strongly fractured (i.e., > 50 fractures/set/cell). It is acceptable, under these conditions, to take a count of at least *30* structures in a set and then to record that distance over which the 30 fractures were observed. A count of at least 30 structures within a geologic set is considered to provide adequate information to calculate the required statistical data. In this case, the counting line should be positioned to cross the longest fracture in the set, if possible.

A third exception would be when measuring structures that are parallel to the face, using either a perpendicular or a true spacing direction, in which case the counting distance is very limited.

Fractures. The number of fractures counted in the set is recorded. A fracture is a continuous open planar feature that has some minimum designated length. We commonly use either a 0.3-m or 1-m cutoff length. Thus, only those structural features (with an open fracture) having a length of the cutoff value or greater would be included in the count. The minimum cutoff used *should always be recorded* on the field sheets, either at the top of the form or under the "Remarks" column.

It is very important that only those fractures that intersect the counting line be counted.

Spacing Direction (SD). This is the direction, in a physical sense, in which the fracture count or spacing of the fractures is measured.

V = vertical spacing (counting line projected vertically, perpendicular to the strike of the bench face), usually implemented for flat-lying structures.

H = horizontal spacing (counting line projected horizontally, parallel to the strike of the bench face), usually implemented for steeper structures oriented oblique to perpendicular to bench face.

P = perpendicular spacing (counting line projected into bench face horizontally, perpendicular to the strike of the face), implemented on structure sets that strike subparallel to parallel to bench face. This count is more difficult since accurate structure count and distance parameters rely on sufficient offset in bench face (in the third dimension perpendicular to the face) to identify individual joints.

T = true spacing (counting line in the true spacing direction, perpendicular to the plane of the structure); in this case, an attempt is made to project the counting line normal to the strike of individual sets. The true spacing can be determined directly by dividing the total number of structures by the counting distance, without the need to correct for a difference in angle between the counting direction and the true spacing direction.

In summary, the SD refers to the direction, relative to the bench face (or structure set in the case of true SD), in which the structure count (to calculate true spacing) is made. The procedure involves projecting a counting line across the cell face in the most effective orientation or direction to count the total number of structures within any given geologic set. For sets parallel to the face, neither horizontal nor vertical count will work. In which case, the counting line should be oriented either perpendicular to the bench face (SD = perpendicular) or perpendicular to the structure set (SD = true). Usually, the face is irregular enough to obtain a count of structures parallel to the face.

Maximum Length. The length of the longest fracture in a joint set or the maximum length of a major structure is measured. This is the greatest observed *trace* length and does not have to be measured at any orientation or inclination along the plane. Those structures that cross more than one cell are measured only once, with the data recorded in the first cell in which the structure is encountered. To reiterate, long structures that can be traced for more than one cell should only be recorded *once*.

Number at Maximum Length (#). The number of fractures in the cell that are at the maximum length are recorded. Quite often, there may be only one structure at the maximum length, but in the case of nonterminated fractures or fractures that are terminating against another set, more than one structure equal to the maximum length can occur in the cell.

Termination (T). The type of termination of the longest fracture.

D = doubly terminated, i.e., termination of the fracture can be identified at both ends. Fractures commonly terminate against other fractures. Fractures are also observed to *die out* within the rock mass itself.

S = singly terminated, i.e., only one termination of the longest fracture can be identified, while the other end exits the bench (cell) at the crest or in the muck pile at the toe of the bench.

N = no termination, i.e., one end of the joint extends to the bench crest while the other end exits at the toe or in muck.

Strike. The strike is measured using a conventional 0° to 360° azimuth compass, with the strike or azimuth reading 90° counterclockwise from the direction of dip of the fracture or fracture set.

The following is another way of describing the "right-hand" convention: when looking in the strike direction (with the front of the compass pointing in that direction), the structure(s) will always be dipping to your right. The average strike of the structure or set should be recorded.

Dip. The dip of a major structure or the average dip of a joint set is measured.

Minimum Dip. A dip on the flattest observable portion of the fracture surface or major structure should be recorded to compare with the average dip. This deviation in dip is a quantitative measure of the roughness of the fracture surface.

Thickness. Average thickness should be recorded for both individual structures and structure sets mapped. *Joint* surfaces commonly have an average thickness of less than 0.16 cm. The software used to process the cell data allows for three decimal places for structure thickness. Therefore, the thickness of even the very narrow joint structures should be estimated.

Fracture Filling. Filling material between fractures is recorded with a single letter abbreviation. More than one filling type should be present. Filling type designations may be site specific; however, all unique filling types have to be defined at the bottom of one or more of the cell-mapping field sheets.

Water (W). This field is used to indicate the presence of water along the joint or structure surface. The following categories are typically used:

D = dry

W = wet

S = squirting

F = flowing

When water is not present, this field is often replaced with a field for other pertinent information, such as measures of joint roughness or condition. Rock hardness can also be substituted but is typically placed on the cell (C) line rather than the structure (S) line because the hardness value would typically represent the average rock hardness within the cell and would not be associated with a specific structure.

Remarks. The "Remarks" column is used to describe pertinent features of the structure or structure set. Additional comments might include specific potential failure geometries with which the structure is associated (e.g., plane shear structure or left side of wedge geometry) or actual failures that have occurred along the mapped faces. As with the remarks on the cell line, sketches are very useful. Remarks on nature of groundwater seepage associated with the structure or set would also be important information.

For cells in which the ground is altered and jointing is either lacking or too intense to measure, the cell data should be recorded and the character of the face described in the "Remarks" column.

2.13 CELL-MAPPING GUIDELINES

1. Cell mapping should be conducted by two-person crews, not only because of safety considerations but to increase productivity and accuracy. Once a line of cells is laid out, the most efficient cell-mapping procedure is to have one mapper work near the bench face on the muck pile and the second mapper on the bench level. For both flat-lying structure sets and sets that are subparallel to parallel to the bench face, it is very difficult to determine the orientation without approaching the bench face and measuring one or more individual fractures.

   For the steeper structures oriented oblique to perpendicular to the face, it is easier to record information from the bench. The number of structures in a flat-lying set is generally more easily counted from the bench level. In other words, both members may provide statistical information for the same set. The team can periodically switch mapping positions.

   The determination of an average structure set orientation to be recorded on the field sheet should involve checking the strike and dip of the set at several locations along the cell.

   Identifying the longest structure length for sets is best done from the bench floor. The key to identifying all structure sets is to be mobile and scan the cell from all possible orientations, e.g., completing a 180° arc along the bench.
2. Cell mapping should *not* be conducted along benches that appear to have been displaced or rotated. This would include zones of blasted muck that have not been cleaned up or areas that have failed.
3. As previously indicated, emphasis should be placed on identifying as many distinct structure sets as possible when cell mapping. Structure sets observed in the field should be considered *distinct* if a difference of more than 20° for the average bearing and/or dip is indicated between individual candidates. Data can be later combined in the office, if desired, but should be recorded as distinct sets in the field.
4. In addition to the recording of parameters for individual or sets of open fractures, the orientation of stratification features, such as bedding in sedimentary rocks, lamination features such as foliation in metamorphic rocks, or flow-banding in volcanic rocks, should be routinely recorded during mapping. The orientation of lineation features, such as alignment of clasts in sedimentary rocks or alignment of the long dimensions of minerals, should also be collected.
5. It may be impossible to define specific rock types while mapping. However, descriptive criteria for identifying unique rock types should be established in the field and recorded on the field sheets. This criteria would then be used in future mapping for identifying the same rock types at other locations.

CHAPTER 3

Designing Catch Benches and Interramp Slopes

Thomas M. Ryan* and Paul R. Pryor*

3.1 INTRODUCTION

Individual mine bench and interramp slope stability in rock is governed primarily by the rock's geologic structural characteristics. Because of the complex structural variability in most rocks, slope stability cannot be adequately described by a single parameter or index. Instead, rock slopes must be evaluated using probabilistic methods, whereby the full range of geologic variability can be incorporated into the stability analysis. Mine-slope-stability evaluation lends itself readily to risk-management methods, where the criteria (e.g., catch bench width or number of multiple bench failures) are developed using a reliability-based approach. This chapter provides a general overview of the approach that Call & Nicholas, Inc. (CNI) recommends for analyzing structurally controlled failures. The chapter also provides specific details regarding CNI's in-house computer-based analyses.

3.1.1 Applicability of Structure-Based Models

Experience tells us that no single model or form of analysis is applicable in every case. This is particularly true for rock slopes where the materials encountered range from weak soil-like units with very little structure to extremely competent units that are highly fractured. Therefore, a wide range of analyses is available, and each analysis has its own unique areas of application.

Limit-equilibrium methods can be used to look at discrete plane-shear or step-path failures. However, these methods are not suitable for wedge analyses because they underestimate stability based on a two-dimensional analysis of plunge of the intersection. Using limit-equilibrium methods, the number of runs required to account for variations in geologic structure (orientation, length, and spacing) and rock-strength parameters is prohibitive, except where discrete structures are modeled or where failures are controlled by overall rock-mass strength rather than by structure.

Elastic/plastic modeling methods, such as finite element, finite difference, and boundary element, also have their place in slope-stability evaluations (Cicchini and Barkley 2000). These models are very applicable for evaluating high slopes where significant stresses are generated, resulting in strain and yielding in the rock mass. However, these models cannot take the place of structural analyses in situations where potential failure modes are structurally controlled. They do not handle the variation in orientation, length, and spacing of several major structure sets and they do not deal with three-dimensional wedge-failure geometries.

Therefore, for slopes that consist of materials that have reasonable rock substance or rock-mass strengths and a moderate to high degree of fracturing or faulting, three-dimensional structural analyses of potential plane-shear and wedge-failure geometries are critical to understanding the slope behavior.

3.2 CATCH BENCH STABILITY

3.2.1 Individual Bench Configuration

Development of catch bench criteria in mine slopes is necessary in areas of rockfall because the catch benches prohibit rocks from rolling from upper portions of the pit slope to the working areas where personnel and equipment are located. Often overlooked, the bench geometry defines the steepest interramp slope that can be mined while maintaining adequate catch bench widths. The two primary factors that control bench configuration are the type of mining equipment that is used and the bench face angles that can be achieved. The type of mining equipment determines the height at which the bench can be adequately scaled. The achievable bench face angles are controlled by rock strength, geologic structure characteristics, and the mining techniques used to construct the slope (the blasting and digging practices, for example). The local stability of the benches is, largely, a construction issue; the stability condition is created contemporaneously with the mining as the equipment digs toward the final slope limits. It is the one area of the slope where the mining practices exert a large degree of control over slope stability. Conversely, slope stability analysis is needed to optimize the mining methods. It is necessary for the mine operator to know where the final blasthole rows should be located and where to place the flags that define the final digging limit for the shovel and loader operators (the *drop-flag* line).

The final, benched slope configuration is a function of the bench height, the bench face angle, and the required catch bench widths (Figure 3.1).

Bench Heights. Currently, most large mining operations drill and blast on 12- to 15-m intervals (40 to 50 ft), with 15-m intervals being the most common. The mining equipment used to drill and blast the rock determines the bench height. Catch benches can be left either at every mining level (single benching) or at every other mining level (double benching).

Bench Face Angles. In weaker rocks, face angles are often controlled by the equipment and digging technique used while mining. In hard rock, which cannot be mechanically ripped but must be blasted, the stable bench face angles are controlled primarily by the stability of the local geologic structure. In reality, in a large open-pit mine slope, the local bench face conditions are a complex mix of both hard, jointed rock and weaker, altered rock. Bench face angles are not unique; because of the variable geologic character of rock, the stable face angle takes the form of a statistical distribution or a probability density function (PDF). Since the rock structure controls the achievable face angles, bench slope stability also varies as a function of the slope orientation because the mode of sliding for structures is dependent on the wall orientation.

* Call & Nicholas, Inc., Tucson, Arizona.

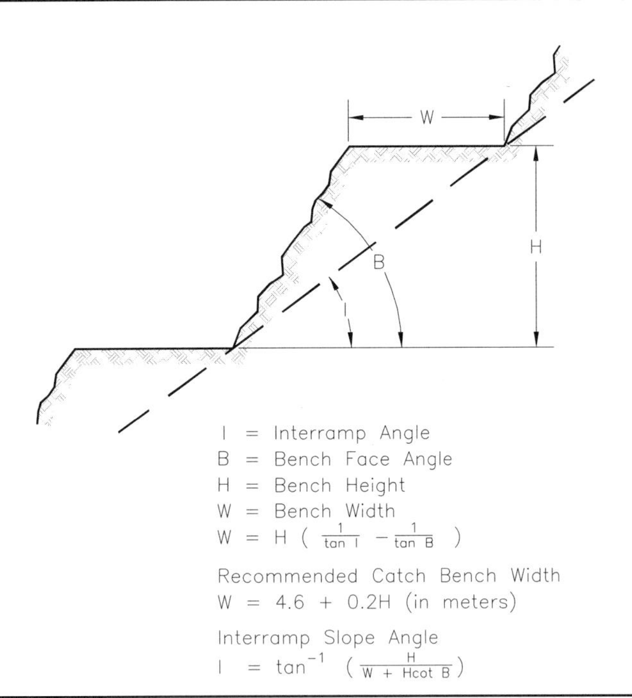

FIGURE 3.1 Relationship between interramp and bench face angles

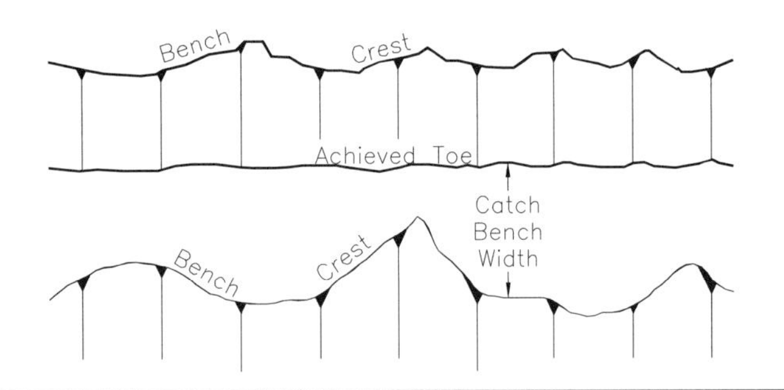

FIGURE 3.2 Catch bench width variability in plan view

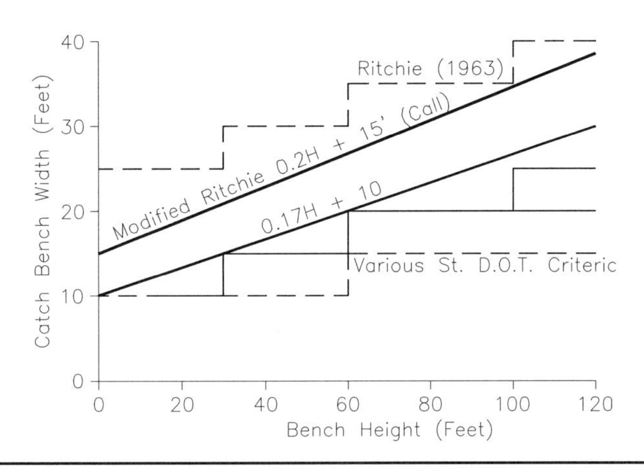

FIGURE 3.3 Catch bench width criteria for rockfall containment

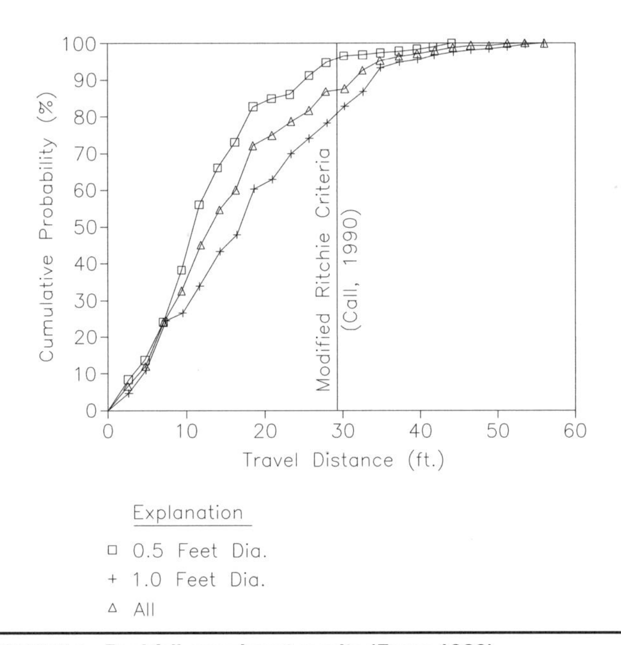

FIGURE 3.4 Rockfall experiment results (Evans 1989)

Bench Widths. The catch bench widths vary from point to point within the slope because of the variability in geologic structure, which produces varying amounts of *back break* along each mining level. The final face angles are a distribution rather than a single value therefore the achieved bench widths will also be highly variablc (Figurc 3.2).

Rockfall Mitigation. CNI employs a Modified Ritchie Criteria as a *guide* for developing catch bench width criteria. The Modified Ritchie Criteria evolved following publication of a paper by Ritchie in 1963 on the evaluation of highway shoulders for catching rockfall off excavated and natural slopes. His investigation was limited to a relatively small number of slope-angle/slope-height geometries and therefore required extrapolation for use in open-pit mining. Since slope (bench) height is one of the most important controls on the distance that a rock will travel when detached from the bench, we derived an empirical relationship between the slope height and the average, or preferred, catch bench width as follows:

$$\text{bench width (m)} = 0.2 \times \text{bench height} + 4.5\ \text{m} \qquad \textbf{EQ. 3.1}$$

This equation is published in several papers by Dr. Richard Call and in the *SME Mine Engineering Handbook* (1992). Other researchers have developed rock catchment criteria (Pierson, Davis, and Pfeiffer 1994) from field tests and from computer simulation methods (Colorado rockfall simulation), and most catch bench width criteria include some relationship between block size, bench width, and slope height (Figure 3.3).

For many years, we have tried to refine the bench width equations for pit slope stability but have found that the problem is too complex for any single criterion to be 100% effective. This position was supported effectively by the research conducted at the University of Arizona (Evans 1987; Yost 1995). In field tests we have been able to demonstrate that Call's catch bench width criteria is effective in benched mine slopes (Figure 3.4).

In those field tests, the Modified Ritchie Criteria was found to be very conservative for containing rockfall in a limestone slope benched on 40- to 50-ft heights for blocks anywhere from 30 cm to 2 m in size. Because of the complexity of the problem, we have been forced to approach the rockfall problem from a risk-management perspective, and that is why we have increasingly used the reliability-based approach for evaluating benched mine slopes. For the purpose of discussion, the method that follows uses the Modified Ritchie Criteria, as presented by Call, but other catch bench width criteria could be substituted.

3.2.2 Catch Bench Criteria Using a Reliability Approach

In evaluating benched slopes between haul roads, the primary functional requirement is to maintain adequate catch bench widths while optimizing the interramp slope angle on economic criteria. Final bench slope heights (between catch benches) can be any multiple of the drill and blast height. The catch bench widths vary from point to point within the slope because of the variability in

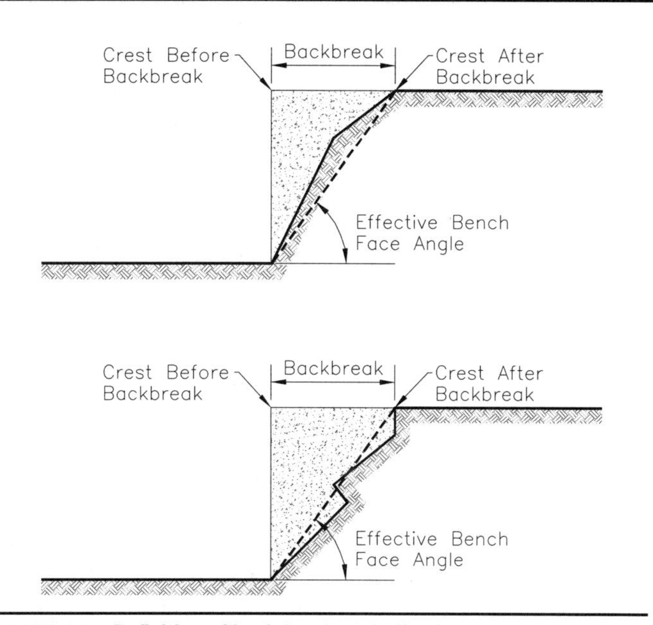

FIGURE 3.5 Definition of back break and effective bench face angle

geologic structure, which produces varying amounts of back break along the bench crest at each mining level. Back break generally is defined as the horizontal distance between the planned toe and the actual mined crest of the final bench slope (Figure 3.5).

This horizontal distance defines the unstable zone within the catch bench. Under ideal conditions (controlled blasting with vertical drill holes in unfractured rock), bench face angles would be nearly vertical. Under actual conditions, however, back break occurs throughout the bench slopes along preexisting jointing and blast-induced fractures. Uncontrolled blasting reduces rock integrity, resulting in further back break.

Because of the variability in geologic structure, namely jointing, catch bench widths vary considerably within the slope for any given interramp slope angle. In such an environment, using modal, median, or mean values for back break (or bench face angle) to evaluate and construct the slope can be misleading if there is a high degree of variability present in the slope. Similarly, it is not appropriate to assume that a *minimum* catch bench width can be incorporated everywhere into the design of the benched slope since, in a highly complex geologic environment, 100% reliability is usually not practical.

CNI has developed a reliability approach in which the analysis is structured to evaluate the percentage of the slope area that meets or exceeds a chosen catch bench width criteria. We have found that this is a more useful risk-management approach for rockfall containment and for slope management. We use a combination of structure modeling, bench face stability analysis, and the Modified Ritchie Criteria to determine the catch bench reliability, which refers to the percentage of benches having final widths equal to or greater than the Modified Ritchie Criteria. The selection of the proper reliability for maintaining a catch bench of a certain width is dependent on many factors including, but not limited to, the following:

- The potential for slope raveling
- The proximity to large slope failures
- The decision to contain overbank from a higher push-back on the benches
- The length of time the benches are expected to be functional
- The climate
- The type of blast control
- The operator's experience

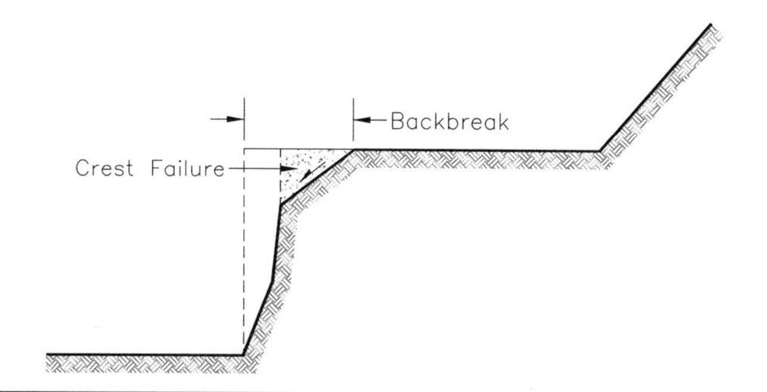

FIGURE 3.6 Crest failures often make up a majority of back break

Since the bench configuration is based on the reliability of the catch bench width, it is the amount of local back break rather than the bench face angle that is of most concern. Although we often use the term "bench face angle" to describe the bench slope, in most cases the bench slope cannot be adequately described by a bench face angle since the actual bench face slope is a complex, three-dimensional surface composed of numerous joint surfaces. Compound bench face slopes composed of several surfaces of different inclinations are not only common, they are the rule in most rock masses, whereas faces formed by one continuous joint face tend to be rare. In fact, practical experience indicates that the majority of back break occurs along short structures near the bench crest. Even where the majority of the bench face slope is quite steep, one flatter joint near the crest can result in significant back break (Figure 3.6).

Therefore, to evaluate the probability of achieving a specified catch bench width, the probability density function for the back break is required. This can be defined by a probabilistic stability analysis of the local rock structures relative to a vertical bench face.

3.2.3 Probabilistic Analysis of Back Break

The local structure data for a region in a mine can be complex. Often, there are many families of structures that show persistence in a region, and these families of structures, or joint sets, must be characterized thoroughly if a probabilistic slope stability analysis is to be conducted for the bench (Call and Savely 1990; Sims and Nicholas 2000). The structural factors that control the stability analysis include the number of joint sets in the region; their probability of occurrence; and their orientation, length, spacing, and frictional shear strength. Each attribute for each structure set must be characterized by mapping if a probabilistic analysis of the back break is to be conducted along the mine benches. From the mapping data, statistical distributions of each parameter for each structure set are developed. The statistical representations are then used in the back break stability model to evaluate bench stability. At CNI, our statistical modeling procedures for the structure can be summarized as follows:

Parameter	Statistical Model(s)	Comments
Occurrence	Deterministic, Cell Method	Probability of occurrence generally is developed using methods similar to CANMET (1977).
Orientation (strike, dip or dip direction, dip)	Normal, Fisher, Spherical	A vector summation is employed more often than simple arithmetic summation.
Length	Neg-Exponential, Weibull, Log-Normal	In cases where structure lengths exceed the outcrop that is being mapped, some extrapolation is required.

(continues)

Parameter	Statistical Model(s)	Comments
Spacing	Neg-Exponential, Weibull, Log-Normal	Spacing must account for length truncation bias (many of the short structures are not usually mapped).
Shear Strength	Deterministic-Gaussian	At low normal stress, shear strength along structures is best modeled as a power curve (Call 1985; Barton 1976).

Statistical models other than those mentioned can be used; however, it should be pointed out that the quality of the final stability model is, to a great extent, determined by the quality of the data that are input.

In the bench slope, stresses are relatively low, and experience shows that the majority of back break occurs as the sliding of blocks along daylighted geologic structure. Numerous papers (Savely 1987) have outlined the essential, sliding-block failure modes for bench slopes. As noted previously, the distribution of back break along each bench is best described by a probabilistic stability analysis. Although there are several different approaches to probabilistic stability analysis, in this case, a Monte Carlo method is used because there is a high degree of complexity to the solution. (There are six to seven key geologic parameters; all key geologic parameters are statistical distributions; several types of statistical models are being simultaneously employed in the solution; multiple failure models are evaluated.) The general procedure for a particular slope orientation is as follows:

1. Determine the potential failure modes in the slope (wedge and plane-shear). Failure modes are defined as structure sets or combinations of structure sets that have a kinematically viable sliding relationship to the slope. (There can be multiple wedge failure modes in a slope, for example.)
2. Determine the maximum back break for each cell or Monte Carlo model *run*.
   a) The following should be determined for each failure mode:
      - Evaluate the probability of occurrence for each structure set in the failure mode.
      - Calculate a theoretical back break for each failure mode occurrence. Sample the statistical model for each key parameter (dip direction, dip, length, spacing) for each structure set involved in the failure mode.
      - Construct the local geometry of the sampled failure mode.
      - Calculate the stability for the constructed geometry. If the geometry is unstable, add it to the theoretical back break model; otherwise, if it is stable, the back break is considered to be zero.
   b) The back break for the other failure modes should be evaluated.
   c) The maximum back break from all of the failure modes should be noted; it should be added to the theoretical back break model for the slope.
3. Combine all of the results once all possible plane-shear and wedge combinations have been evaluated to produce an estimated cumulative PDF for back break. Due to the complexity of the problem, the cumulative PDF for back break tends to have a complicated form that cannot be adequately represented by a statistical model.
4. Convert the back break distance distribution to an equivalent bench face angle distribution.

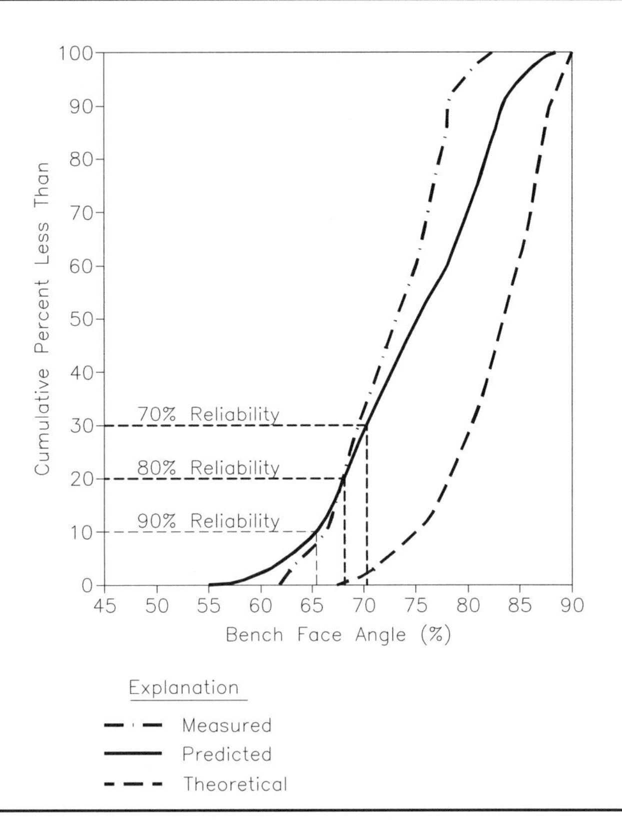

FIGURE 3.7 Effective bench face angle distributions, theoretical vs. predicted vs. measured

This procedure is well suited to computer programming since there are many variables and many model runs are required to achieve a consistent result. A computer-based numerical analysis called *Backbreak* has been in development at CNI since the early 1980s. This computer program uses rock strength and structural fabric data from pit mapping programs to predict back break and bench face angle distributions, based on the size and type of structural failures that can be expected along the benches as they are mined. The output from *Backbreak* is a probability distribution of the back break distance to be expected in a mine sector. The back break distance is then converted to an equivalent bench face angle (Figure 3.5) and is compared to measured values for calibrating the analysis. We like to convert the back break distance to an equivalent bench face angle since it provides us with a benchmark distribution, which we can compare to the face angle estimates made by the geologists who are conducting the routine bench mapping in the pit (Figure 3.7).

Where both the toe and the crest of the bench can be adequately surveyed, actual back break distances can also be measured and compared to the model results. Both the measured benches and the computer analysis are required to assess bench slope stability since the operational practices can reduce the stable face angles that are achieved. An empirical correction factor is then applied to the calculated (theoretically achievable) bench face angle distribution to account for the effects of controlled blasting and excavation. This correction factor is based on experience at mines during the past 15 years; ideally, it would be developed for each rock type in each sector of the mine where controlled blasting has been implemented.

3.2.4 Double or Triple Benching

The bench increment is the distance between mining levels; in most cases, the bench increment is chosen based on either the length of either the blasthole drill rods or the reach of the digging equipment. Most porphyry copper mines currently use a 15-m mining bench increment. In the precious metal, coal, and industrial mineral industries, bench increments vary from 6 to 15 m.

Final bench slopes can be constructed using (1) a single bench technique, where catch benches are left in the slope at every mining level, or (2) a multiple bench method, where catch benches are left in the slope every two to three mining levels. Multiple benching is usually done for the following reasons:

- To provide a steeper interramp or overall slope angle. In a typical scenario, the geologic control to the bench face angle would be strong and the variability in achieved bench face angles would be tightly bounded (10°). With strong geologic control, the bench face angles are relatively fixed and predictable. To steepen the interramp slope angle, the bench height can be doubled and the bench width can be adjusted accordingly. Because bench widths do not have to be doubled for a higher final bench, the interramp slope angle can be steeper while maintaining similar levels of rockfall containment.
- To improve working conditions for the same interramp slope angle. Leaving wider, more reliable catch benches in the slope is often useful where overbank from one pushback may interfere with another pushback lower in the slope. Wider catch benches can also contain overbank material originating from a slope failure higher in the slope.
- To improve the reliability of the catch benches in areas where the geologic structure has short lengths. The majority of bench back break occurs as small failures of the bench crest. These small failures have less effect on the reliability of a wider bench.
- To minimize the occurrence of areas where there is no catch bench left in the slope. When geologic structure has intermediate lengths (less than 60% of the bench height), it is more likely that some catch bench will be available everywhere along a working level if the bench is designed to be wider.

Our experience indicates that single benching is not always safer than double benching. Double benching can actually improve safety in certain areas depending on the local geology. Double benching is not appropriate in the following situations:

- The geologic structures are quite variable in dip. In this case, it is often difficult to get final dig faces to match from one working level to another, creating offsets between working levels, local overhangs, and other safety hazards.
- The rock mass is so fractured that it can be freely dug without the aid of blasting.
- There are many long joints or faults that daylight in the slope, which increase the potential for large failure geometries to develop when mining the lower bench level. Such failure geometries can be readily accommodated in the operation when the slope height, in failure, is 15 m because the equipment can readily scale the full bench height. However, when daylighted structures cause a slope failure of up to 30 m in height, the entire face cannot be adequately scaled, which may pose a safety hazard. For a given sliding plane, the failure volume actually increases geometrically rather than proportionally. The failure volume for a double bench can be three to four times the failure volume of the single bench for the same structures.

3.2.5 Containment Berms

In Ritchie's study, containment ditches cut into the shoulder of the road near the slope toe were found to be a very reliable method of containing rocks that were moving with horizontal momentum toward the road surface. For open-pit mining, excavation of ditches on benches is impractical, but the same effect can be achieved by constructing a berm along the edge of the expected impact zone. The criteria for berm height are based on experience and will vary from mine to mine depending on bench height and rock block size.

3.2.6 Interramp Slope Angle

Once the back break model has been completed, the recommended geometry of the benched slope can be determined by converting the back break distribution (or the equivalent bench face angle distribution) to a catch bench width distribution. The catch bench widths depend on the interramp slope angle; as the interramp slope angle becomes progressively steeper, the catch bench widths decrease proportionally. Once the back break distribution is known, the percentage of the slope that the target catch bench width has met or exceeded can be readily calculated for any interramp slope angle. An example of this is shown on Figure 3.8, where the reliability of achieving the catch bench width is plotted against the interramp slope angle for an estimated back break distribution.

As shown on Figure 3.8, the reliability of achieving a specific catch bench width (in this case a catch bench width of 7.6 m using the Modified Ritchie Criteria) throughout the slope is shown to readily decrease with an increasing interramp slope angle. The average bench width (based on the median back break distance) and a minimum bench width are also shown for each interramp slope angle. The minimum bench width is estimated from the cumulative back break PDF at the 3 to 5% level (3 to 5% of the bench face angles are flatter than this angle). All of these design factors play a role in the design of the benched slope. The average catch bench width is needed for laying out the drop-flag line to help operators assess where to stop digging. The minimum width is useful for evaluating the probability of failure for the entire catch bench for identifying the interramp slope angle at which loss of the entire catch bench will begin to occur consistently. Note that the term "catch bench reliability" refers to the reliability of 7.6-m-wide benches in this example. A catch bench reliability of 90% does not indicate that 10% of the mine benches have been totally lost; instead, it indicates that 10% of the benches are less than 7.6 m wide. All three factors (catch bench reliability, median bench width, minimum bench width) are used to choose the appropriate slope angle for the benched mine slope. There are no unique bench configuration criteria, since each slope is different. Normally, our recommendations are based on 50 to 90% catch bench reliability, depending on the site-specific operational requirements and the geologic conditions encountered. Once the catch bench analysis is complete, the next step in the benched mine slope design is to evaluate larger, structurally controlled failures that exceed the bench scale.

3.3 INTERRAMP SLOPE STABILITY

An often-neglected portion of the process of evaluating slope stability is the analysis of intermediate-size structurally controlled failures that exceed the bench scale but that cannot be modeled well by typical overall slope analyses, such as limit-equilibrium methods or elastic/plastic models (finite element, finite difference, or boundary element). As with the bench-scale analysis discussed previously, the analysis of these potential multiple bench or interramp failures lends itself to a reliability-based approach. This is due to the complexity of simultaneously handling variations in structure orientation, length, and spacing, as well as variations in rock strength. In our experience, a high percentage of failures in open-pit mines are structurally controlled, consequently, it is critical to have a tool to predict the number and size of potential failures that can be expected for various wall orientations and interramp slope angles.

Interramp Angle Based on Catch Bench Reliability

Bench Height = 12m (Single bench)

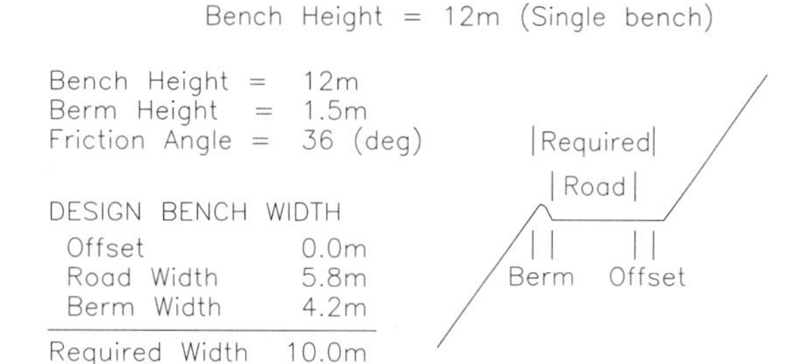

	AVERAGE	MAXIMUM	MINIMUM
Reliability	50%	5%	95%
Face (deg) =	65.0	81.0	39.0
Backbreak (m) =	5.6	1.9	14.8

INTER RAMP ANGLE (deg)	RELIA–BLILITY (%)	DESIGN AVERAGE WIDTH (m)	MAXIMUM WIDTH (m)	MINIMUM WIDTH (m)	BFA AT 10.0m WIDTH (deg)
27	93	18.0	21.7	8.7	41.4
28	90	17.0	20.7	7.8	43.6
29	89	16.1	19.8	6.8	45.7
30	86	15.2	18.9	6.0	47.9
31	83	14.4	18.1	5.2	50.1
32	80	13.6	17.3	4.4	52.4
33	76	12.9	16.6	3.7	54.6
34	72	12.2	15.9	3.0	56.9
35	66	11.5	15.2	2.3	59.1
36	60	10.9	14.6	1.7	61.3
37	54	10.3	14.0	1.1	63.5
38	48	9.8	13.5	0.5	65.7
39	41	9.2	12.9	0.0	67.9
40	35	8.7	12.4	−0.5	70.1
41	31	8.2	11.9	−1.0	72.2

Interramp Angle Based on Catch Bench Reliability

Bench Height = 24m (Double bench)

Bench Height = 24m
Berm Height = 2.1m
Friction Angle = 36 (deg)

DESIGN BENCH WIDTH
Offset 0.0m
Road Width 6.6m
Berm Width 5.8m
Required Width 12.4m

|Required|
| Road |
| | | |
Berm Offset

	AVERAGE	MAXIMUM	MINIMUM
Reliability	50%	5%	95%
Face (deg) =	71.0	81.0	49.0
Backbreak (m) =	8.3	3.8	20.9

INTER RAMP ANGLE (deg)	RELIA–BLILITY (%)	DESIGN AVERAGE WIDTH (m)	MAXIMUM WIDTH (m)	MINIMUM WIDTH (m)	BFA AT 10.0m WIDTH (deg)
27	100	38.8	43.3	26.2	34.7
28	99	36.9	41.3	24.3	36.3
29	99	35.0	39.5	22.4	37.9
30	99	33.3	37.8	20.7	39.5
31	99	31.7	36.2	19.1	41.1
32	98	30.2	34.6	17.6	42.7
33	98	28.7	33.2	16.1	44.4
34	97	27.3	31.8	14.7	46.0
35	96	26.0	30.5	13.4	47.7
36	95	24.8	29.2	12.2	49.4
37	94	23.6	28.1	11.0	51.0
38	92	22.5	26.9	9.9	52.7
39	90	21.4	25.8	8.8	54.4
40	88	20.3	24.8	7.7	56.0
41	86	19.4	23.8	6.7	57.7
42	84	18.4	22.9	5.8	59.3
43	81	17.5	21.9	4.9	61.0
44	77	16.6	21.1	4.0	62.6
45	73	15.7	20.2	3.1	64.3
46	69	14.9	19.4	2.3	65.9
47	63	14.1	18.6	1.5	67.5
48	57	13.3	17.8	0.7	69.1
49	51	12.6	17.1	0.0	70.6
50	45	11.9	16.3	−0.7	72.2
51	38	11.2	15.6	−1.4	73.7
52	32	10.5	15.0	−2.1	75.2

FIGURE 3.8 An analysis of catch bench reliability for increasing interramp slope angle

3.3.1 Data Requirements

While structural analysis of bench-scale stability is typically based on the orientation of joints and other discontinuities that make up the "rock fabric," interramp analysis is generally based on the evaluation of faults and other major structures that have lengths adequate to define failure geometries ranging from double bench to full slope height. Although regional or pit-scale structural features, such as major faults, are generally well defined and are included in the geologic model, faults of intermediate length are typically less well defined. These structures are often part of the bench mapping that is done at most properties; however, a concerted effort must be made to tie structures together from bench to bench to define the overall continuity of each structure. This step is critical to predicting the size of failures that may occur.

Also, while bench-scale analysis is typically based on sliding along rock-on-rock joint surfaces, larger interramp-scale failures are typically assumed to be sliding along fault gouge or clay infilling of the major structures. Therefore, it is necessary (1) to note the filling when mapping each structure and (2) to sample and test representative filling materials as part of the laboratory strength-testing program.

3.3.2 Analytical Approach

Probability Distributions. As indicated previously, because of the number of parameters involved and the variation within each parameter, a reliability-based approach is recommended for structural analysis of interramp slopes. The determination of parameters for the interramp analysis is very similar to the approach used for the bench-scale analysis. The major structures are grouped into sets; the sets are based on orientation relative to the slope face, with left and right wedge-forming sets and with a plane-shear set for orientations nearly parallel to the face. However, since the orientation and length of each major structure are measured individually, there is more flexibility in selecting the type of distribution to be used than there is with the rock fabric data. If enough structures are available, the discrete value of each structure orientation is used rather than a distribution function. This is generally the approach used by CNI. An example of fault

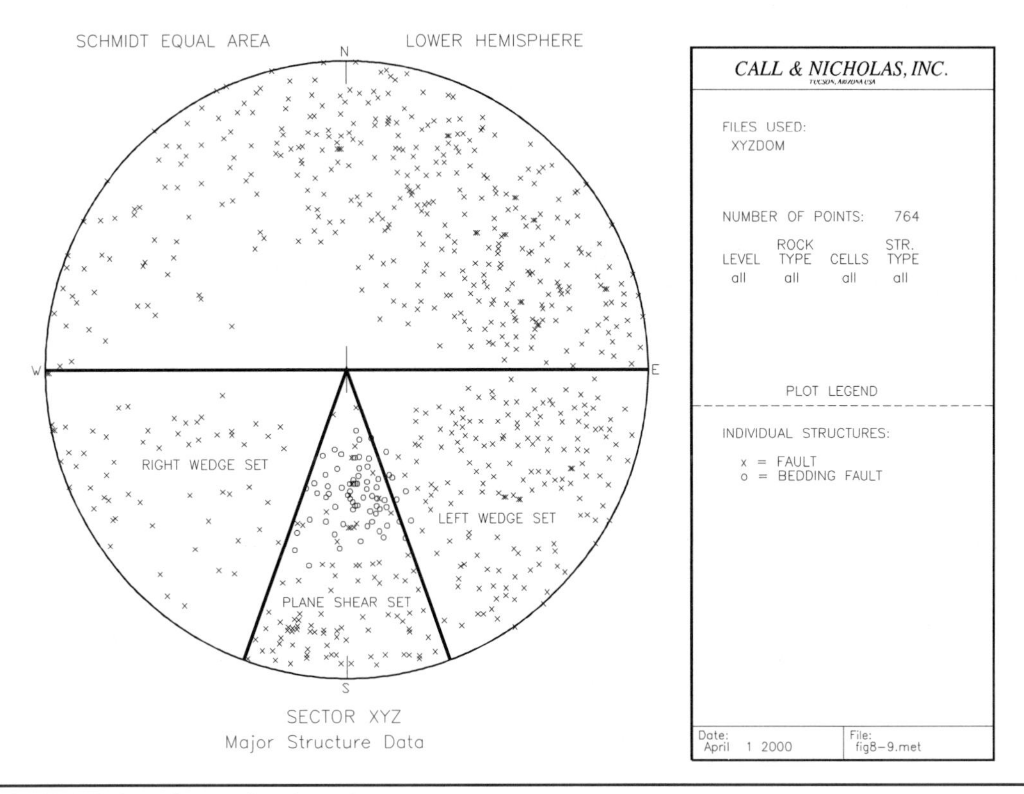

FIGURE 3.9 Schmidt plot of plane shear and wedge sets

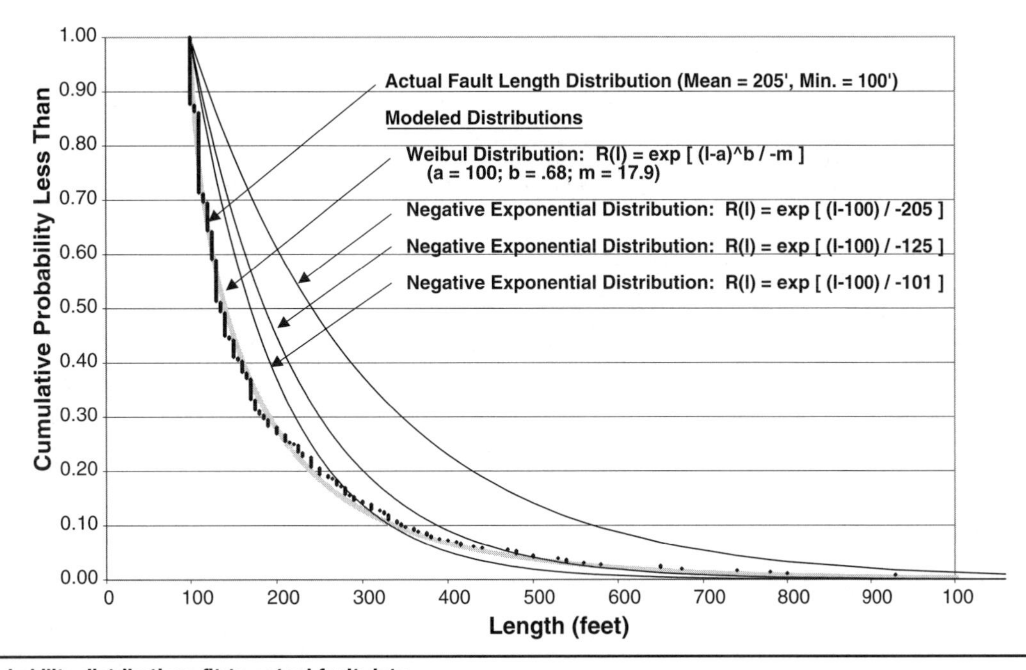

FIGURE 3.10 Probability distributions fit to actual fault data

data divided into simple plane-shear and left and right wedge sets is presented in Figure 3.9.

Lengths and spacings are typically modeled using a Weibull or negative exponential distribution, but a discrete probability distribution based on the values obtained from mapping can also be used. Our experience indicates that the Weibull distribution generally provides the best fit if a discrete probability distribution is not used. The advantage to the Weibull is that it predicts extreme lengths better than the discrete probability distribution, which has a finite upper end. An example of actual data fit with a Weibull distribution compared to several negative exponential distributions is presented in Figure 3.10.

Probability Analysis. In contrast to the bench-scale analysis, CNI uses a closed-form probability analysis rather than a Monte Carlo simulation for the interramp analysis. With a Monte Carlo analysis, the orientation and length of the structure (or structures, in the case of a wedge analysis) are sampled, and the maximum height of the failure is calculated based on the sampled values. However, the interramp analysis is typically performed for a series of heights ranging from the double bench height up to the total slope height, usually in increments equal to the double bench height. Since the various failure heights being analyzed are fixed, it is preferable to calculate the probability (P_l) that a given structure is long enough based on the distribution of lengths, rather than by sampling until the appropriate structure length is obtained. For the wedge analysis, calculate the joint probability that both the left and right structures are long enough. Given the discrete structure orientation(s), required length(s), unit weight of the material, and water conditions, the potential failure geometry is evaluated using the appropriate three-dimensional wedge (Hoek and Bray 1981) or plane shear (Marek and Savely 1978) analysis to determine the minimum strength required for a factor of safety equal to 1.0. The weight of the failure in tons (T_f) for the geometry evaluated is recorded to be used for estimating the expected failure tonnage. The probability (P_s) that the actual strength is less than the minimum value required for stability is then calculated from the distribution of fault gouge strengths. The probability of failure (P_f) is determined by combining the probabilities of the length and strength components as follows:

$$P_f = P_l \times P_s \qquad \textbf{EQ. 3.2}$$

This process is repeated for every orientation in the plane-shear set or, in the case of the wedge analysis, for every combination of left and right orientations. The resulting individual probabilities of failure and failure tonnages are summed and divided by the number (N_d) of daylighted orientations or wedge combinations evaluated to determine the expected probability of failure (P_F) and the expected failure tonnage (T_F), as follows:

$$P_F = (\Sigma P_f) / N_d \qquad \textbf{EQ. 3.3}$$

$$T_F = (\Sigma (T_f * P_f)) / \Sigma P_f \qquad \textbf{EQ. 3.4}$$

As with the bench-scale design, a cell approach is used in which the height and width of the cell are each equal to the height (H) of the failure being analyzed. The number of cells (N_C) within the sector for a given failure height is calculated by dividing the sector area (A) by the cell area (H^2). The probability of occurrence (P_O) of a plane-shear structure within the cell is based on the distribution of fault spacings for structures in the plane-shear orientation. For the wedge analysis, the probability of occurrence of a wedge intersection within the cell is based on the distributions of fault spacing for both wedge sets. The expected total number of failures (N_E) and expected failure tonnage (T_E) of a given type (plane shear or wedge) within the entire sector for a given failure height is calculated as follows:

$$N_E = P_F \times P_O \times N_C \quad \text{or} \quad N_E = P_F \times P_O \times A/(H^2) \qquad \textbf{EQ. 3.5}$$

$$T_E = T_F \times P_O \times N_C \text{ x} \qquad \textbf{EQ. 3.6}$$

To avoid double counting by evaluating small failures that are contained within areas associated with larger failures, the process is started with the largest failure height (H), the corresponding cell area (H^2), and the total sector area (A) and is worked downward. For each subsequent failure height evaluated, the total sector area is reduced by the area of expected failures for the previous height, as follows:

$$A = A - (H^2 \times N_E) \qquad \textbf{EQ. 3.7}$$

The process, which is illustrated in simplified form in Eq. 3.7, is repeated for all heights to be evaluated. For the plane-shear or wedge analysis being performed, this process results in a composited value for the expected number of failures and the expected failure tons in the sector. The expected failure height is also calculated at this time. The entire process described thus far is for the analysis of a single interramp angle. This process is repeated for each interramp angle being considered, with the typical range being from 30° to 60°. When both the plane-shear and wedge analyses have been completed, the results are combined, again discounting to avoid double counting, to obtain the totals for expected number of failures and expected failure tonnages at each interramp slope angle evaluated. The entire process may also be repeated for various groundwater conditions. Typically, if groundwater is present, a wet and a dry case are run to determine the effects of dewatering. The analysis can also be run for various seismic loading conditions.

Obviously, this is not an analysis that you would want to perform by hand. Fortunately, this highly iterative process is well suited for the computer. Sample output from an interramp analysis performed with CNI's in-house software shows the detailed results obtained for each height evaluated for a given sector and interramp slope angle (see Table 3.1).

The total expected number of failures and expected failure tonnages for these sample data for all interramp angles evaluated for both saturated and dry cases are presented in Table 3.2, along with an accompanying graph.

3.3.3 Cost–Benefit Analysis

The results from the interramp analysis do not provide a specific reliability or a factor of safety associated with a given slope angle. What the analysis does provide is an estimate of the number of failures and the total failure tonnages expected for each interramp slope angle evaluated. If there is a sharp increase in either the predicted number of failures or failure tonnage from one angle to the next, it may be possible to select the optimum interramp slope angle using these data alone. Ideally, the optimum interramp angle should be selected (1) by comparing the benefits gained (less stripping, more ore) versus the costs incurred (failure cleanup, downtime, buried ore, etc.) by steepening the slope and (2) by determining the angle that provides the economic optimum. At this point, the slope stability analysis is no longer a standalone process but is an integral part of the mine planning process. This approach is generally referred to as a cost–benefit (or benefit–cost) analysis (Call and Kim 1978), and it has been used successfully by CNI at several large porphyry copper mines to assist the on-site mine planners in maximizing economic recovery.

Two general types of cost–benefit analysis have been used by CNI over the years: a long form and a short form, with the short form being much more frequently used. The long form requires development of a fully sequenced pit, which is then modeled to predict the occurrence of failures at specific times through the course of the mine life. This very detailed process requires significant work by the mine planners prior to the stability modeling process. The short form looks only at the final wall and therefore requires much less work on the part of the mine planners. Since it is often preferable to mine the working slopes at somewhat flatter angles than those of the final walls, the short form is generally all that is required for optimizing the ultimate slopes. If the interramp slope angles in some sectors are constrained by bench-scale or overall stability problems, it may not be necessary to perform the cost–benefit analysis for all sectors of the pit. The remainder of this discussion will be limited to the short-form approach. For a detailed discussion of the long-form approach, refer to the *Pit Slope Manual, Supplement 5-3* (CANMET 1977).

Economic Input Requirements. In addition to the results of the interramp stability analysis, significant economic information is required to perform a cost–benefit analysis. Ultimate pits, including haul roads, must be developed by the mine planners for each of the interramp angles to be analyzed. The net present value of each pit is determined using standard mine planning procedures. Once these plans are available, it is necessary to determine the net present value of the entire sector being evaluated, as well as the value of the last blocks mined on each level. Also required are the failure cleanup costs per ton; the cost per

TABLE 3.1 Sample output from CNI interramp analysis

```
Sector XYZ - DRY
 RESULTS FROM PROGRAM ADD.F

     INPUT DATA:          AVERAGE SECTOR HEIGHT:  1900.
                          AVERAGE DOMAIN HEIGHT:  1900.
                           AVERAGE DOMAIN WIDTH:  2100.
                EXP. SPACING "PLANE SHEAR" SET:    76.
             EXP. SPACING OF WEDGE SETS - LEFT:    80.    RIGHT:   112.
<><><><><><><><><><><><><><><><><><><><><><><><><><><><><><><><><><><><><><><><><><><><><><><><><><><><><><><><><><><><><><>
 SLOPE ANGLE = 46.
```

HT.	COMPOSITE EXPECTED NUMBER	COMPOSITE EXPECTED TONS	STATIC EXPECTED NUMBER	STATIC EXPECTED TONS	0.060G EXPECTED NUMBER	0.060G EXPECTED TONS	0.160G EXPECTED NUMBER	0.160G EXPECTED TONS	0.400G EXPECTED NUMBER	0.400G EXPECTED TONS
TOTAL	0.3834E+02	0.2853E+06	0.3834E+02	0.2853E+06	0.0000E+00	0.0000E+00	0.0000E+00	0.0000E+00	0.0000E+00	0.0000E+00
1900.	0.1160E-13	0.5451E-06	0.1160E-13	0.5451E-06	0.0000E+00	0.0000E+00	0.0000E+00	0.0000E+00	0.0000E+00	0.0000E+00
1800.	0.6940E-13	0.2776E-05	0.6940E-13	0.2776E-05	0.0000E+00	0.0000E+00	0.0000E+00	0.0000E+00	0.0000E+00	0.0000E+00
1700.	0.4236E-12	0.1424E-04	0.4236E-12	0.1424E-04	0.0000E+00	0.0000E+00	0.0000E+00	0.0000E+00	0.0000E+00	0.0000E+00
1600.	0.2586E-11	0.7121E-04	0.2586E-11	0.7121E-04	0.0000E+00	0.0000E+00	0.0000E+00	0.0000E+00	0.0000E+00	0.0000E+00
1500.	0.1587E-10	0.3577E-03	0.1587E-10	0.3577E-03	0.0000E+00	0.0000E+00	0.0000E+00	0.0000E+00	0.0000E+00	0.0000E+00
1400.	0.9826E-10	0.1823E-02	0.9826E-10	0.1823E-02	0.0000E+00	0.0000E+00	0.0000E+00	0.0000E+00	0.0000E+00	0.0000E+00
1300.	0.6152E-09	0.8978E-02	0.6152E-09	0.8978E-02	0.0000E+00	0.0000E+00	0.0000E+00	0.0000E+00	0.0000E+00	0.0000E+00
1200.	0.3900E-08	0.4530E-01	0.3900E-08	0.4530E-01	0.0000E+00	0.0000E+00	0.0000E+00	0.0000E+00	0.0000E+00	0.0000E+00
1100.	0.2508E-07	0.2198E+00	0.2508E-07	0.2198E+00	0.0000E+00	0.0000E+00	0.0000E+00	0.0000E+00	0.0000E+00	0.0000E+00
1000.	0.1642E-06	0.1069E+01	0.1642E-06	0.1069E+01	0.0000E+00	0.0000E+00	0.0000E+00	0.0000E+00	0.0000E+00	0.0000E+00
900.	0.1098E-05	0.5208E+01	0.1098E-05	0.5208E+01	0.0000E+00	0.0000E+00	0.0000E+00	0.0000E+00	0.0000E+00	0.0000E+00
800.	0.7526E-05	0.2477E+02	0.7526E-05	0.2477E+02	0.0000E+00	0.0000E+00	0.0000E+00	0.0000E+00	0.0000E+00	0.0000E+00
700.	0.5316E-04	0.1138E+03	0.5316E-04	0.1138E+03	0.0000E+00	0.0000E+00	0.0000E+00	0.0000E+00	0.0000E+00	0.0000E+00
600.	0.3883E-03	0.5333E+03	0.3883E-03	0.5333E+03	0.0000E+00	0.0000E+00	0.0000E+00	0.0000E+00	0.0000E+00	0.0000E+00
500.	0.2948E-02	0.2246E+04	0.2948E-02	0.2246E+04	0.0000E+00	0.0000E+00	0.0000E+00	0.0000E+00	0.0000E+00	0.0000E+00
400.	0.2345E-01	0.8930E+04	0.2345E-01	0.8930E+04	0.0000E+00	0.0000E+00	0.0000E+00	0.0000E+00	0.0000E+00	0.0000E+00
300.	0.2009E+00	0.2954E+05	0.2009E+00	0.2954E+05	0.0000E+00	0.0000E+00	0.0000E+00	0.0000E+00	0.0000E+00	0.0000E+00
200.	0.2033E+01	0.8215E+05	0.2033E+01	0.8215E+05	0.0000E+00	0.0000E+00	0.0000E+00	0.0000E+00	0.0000E+00	0.0000E+00
100.	0.3608E+02	0.1618E+06	0.3608E+02	0.1618E+06	0.0000E+00	0.0000E+00	0.0000E+00	0.0000E+00	0.0000E+00	0.0000E+00

```
  EXPECTED HEIGHT =    107.
<><><><><><><><><><><><><><><><><><><><><><><><><><><><><><><><><><><><><><><><><><><><><><><><><><><><><><><><><><><><><><>
```

TABLE 3.2 Sample summary results for all angles, dry and saturated

INTERRAMP RESULTS

SECTOR XYZ

Structural Domain: ESDDOM(I) Wall DDR: 250

Average Sector Height (ft): 1900
Average Domain Height (ft): 1900
Average Domain Width (ft): 2100

	Plane Shear	L. Wedge	R. Wedge
Fault Spacing:	76	80	112
Fault Length:	152	151	152

INTERRAMP ANGLE	DRY		SATURATED	
	No. Failures	Tons	No. Failures	Tons
60	186.0	1,341,216	262.5	2,166,336
58	140.0	1,110,912	219.1	1,886,208
56	132.1	912,576	201.8	1,654,848
54	98.8	737,856	166.1	1,436,544
52	79.6	591,264	142.2	1,239,072
50	73.3	479,904	127.1	1,057,920
48	55.7	367,142	103.3	884,352
46	38.2	284,266	81.6	740,064
44	32.8	233,827	70.4	621,888
42	28.7	181,891	61.0	501,888
40	24.8	137,222	52.0	391,296
38	22.2	93,965	44.4	288,384
36	20.0	55,930	37.2	193,690
34	3.8	12,374	12.3	103,104
32	2.7	8,893	9.1	74,438
30	1.6	6,086	6.0	51,590

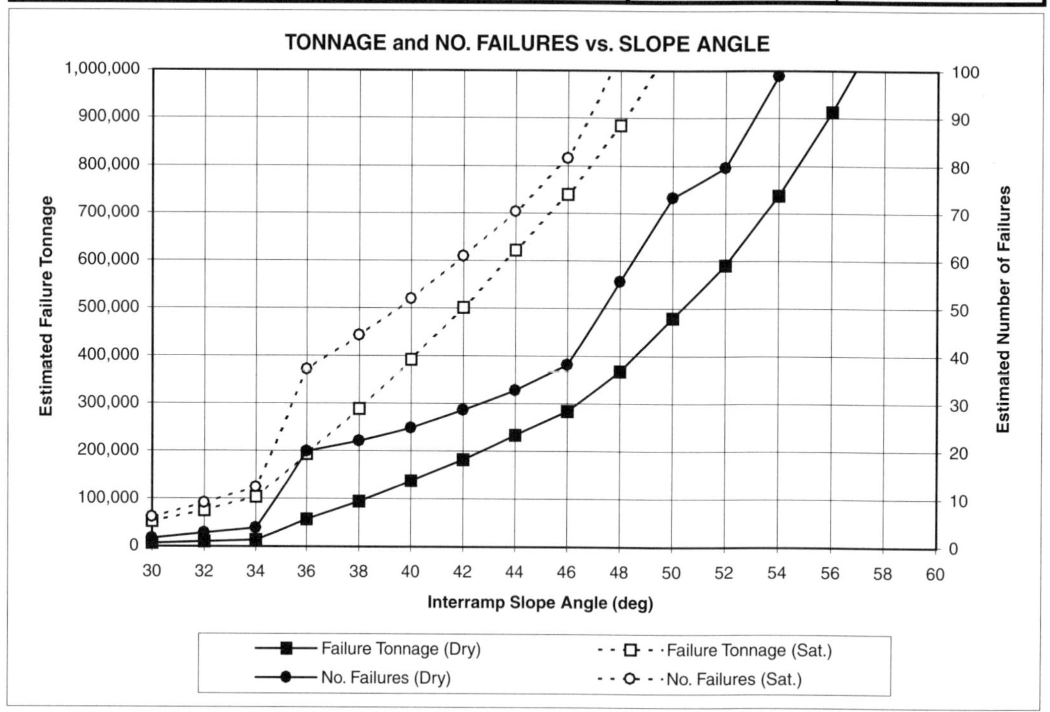

day of lost production due to downtime; the cost per ton of backfilling to restore failed haul roads; and the costs associated with repairing utility lines, dewatering collection systems, etc. Other facilities that may be impacted by failures and that should be considered might include rail lines, in-pit crushers, conveyors, access to underground workings, or any other structure that will be required for ongoing operations.

Cost Calculations. Once this information is compiled, it is combined with the interramp results in a spreadsheet; the spreadsheet is customized for each sector to account for the various haul roads and facilities that may be impacted by slope failures. The

TABLE 3.3 Interramp failure cost calculations

SECTOR XYZ					DATE: 4/1/00	HAUL ROAD												
S HEIGHT: 1900 D HEIGHT: 1900 WIDTH: 2100			CLEANUP CLEANUP COST/TON: $1.20			FAILURES LENGTH: 3900		LOST PRODUCTION FIX COST: FIX DAYS: 3 COST/DAY:				LOST ORE NET VALUE BELOW ROAD WIDTH: 1250			BELOW ROAD FAILURES TOTAL			TOTAL HAUL ROAD
								ON ROAD		BELOW ROAD								
ANGLE (deg)	NO.FAIL	FAIL HGT (ft)	TONS (x1000)	AVE TONS (x1000)	CLEANUP COST (x1000)	ANGLE (deg)	NO.FAIL	COST/FAIL (x1000)	TOTAL (x1000)	BACKFILL TONS	COST/FAIL (x1000)	THICK (ft)	WIDTH (ft)	COST/FAIL (x1000)	TOTAL PER FAIL (x1000)	NO.FAIL	TOTAL COST (x1000)	TOTAL COST (x1000)
42	40.55	115	181.9	4.5	218	42	4.56	$4,805	$21,901	15	$4,805	45	173	$6,102	$10,907	4.56	$49,714	$71,616
40	31.30	111	137.2	4.4	165	40	3.40	$4,805	$16,317	8	$4,805	35	167	$4,337	$9,142	3.40	$31,044	$47,362
38	22.25	108	94.0	4.2	113	38	2.35	$4,805	$11,286	3	$4,805	25	162	$2,843	$7,648	2.35	$17,964	$29,250
36	15.03	105	55.9	3.7	67	36	1.54	$4,805	$7,412	0	$4,805	20	158	$2,078	$6,883	1.54	$10,617	$18,029
34	13.78	103	12.4	0.9	15	34	1.39	$4,805	$6,666	0	$4,805	20	155	$1,958	$6,763	1.39	$9,382	$16,048
32	12.76	102	8.9	0.7	11	32	1.27	$4,805	$6,113	0	$4,805	20	153	$1,913	$6,718	1.27	$8,546	$14,658
30	11.65	101	6.1	0.5	7	30	1.15	$4,805	$5,526	0	$4,805	20	152	$1,868	$6,673	1.15	$7,674	$13,201

LAST BLOCK							
	LENGTH (ft)	NET VALUE (x1000) 30 Deg	34 Deg	36 Deg	38 Deg	42 Deg	WIDTH (ft)
ALL RDS	3900						1
UPPER RD	3900	$77,050	$79,200	$82,450	$87,450	$98,500	1250
CONVEYOR	500	$78,150	$79,850	$80,500	$85,600	$97,600	1100
CRUSHER: (1 OR 0 FOR WIDTH)							1

CONVEYOR										CRUSHER & TUNNEL			SECTOR XYZ					
FACTOR: 1.00 LOST PROD $/FAIL: $20,900 LENGTH: 500			LOST ORE NET VALUE WIDTH: 950			BELOW CONVEYOR			TOTAL CONVEYOR	AREA: CRUSHER 70000, TUNNEL 60000 FAIL $: CRUSHER $95,800, TUNNEL $175,000			FAILURE COST SUMMARY					
ABOVE CONVEYOR																		
ANGLE (deg)	NO.FAIL	TOTAL COST (x1000)	THICK (ft)	WIDTH (ft)	ORE COST (x1000)	PER FAIL COST (x1000)	NO. FAIL	BELOW COST (x1000)	TOTAL COST (x1000)	NO.FAIL	TOTAL COST (x1000)	ANGLE (deg)	CLEANUP (x1000)	HAUL ROAD (x1000)	CONVEYOR (x1000)	CRUSHER & TUNNEL (x1000)	SECTOR TOTAL (x1000)	INTERRAMP ANGLE (deg)
42	0.58	$12,213	45	173	$7,955	$28,855	0.58	$16,862	$29,075	0.71	$192,648	42	$219	$71,616	$29,075	$192,648	$293,558	42
40	0.44	$9,099	35	167	$5,622	$26,522	0.44	$11,547	$20,646	0.55	$148,702	40	$165	$47,362	$20,646	$148,702	$216,876	40
38	0.30	$6,294	25	162	$3,662	$24,562	0.30	$7,396	$13,690	0.39	$105,707	38	$113	$29,250	$13,690	$105,707	$148,760	38
36	0.20	$4,133	20	158	$2,669	$23,569	0.20	$4,661	$8,794	0.26	$71,406	36	$67	$18,029	$8,794	$71,406	$98,296	36
34	0.18	$3,717	20	155	$2,597	$23,497	0.18	$4,179	$7,897	0.24	$65,467	34	$15	$16,048	$7,897	$65,467	$89,427	34
32	0.16	$3,409	20	153	$2,545	$23,445	0.16	$3,824	$7,233	0.22	$60,621	32	$11	$14,658	$7,233	$60,621	$82,523	32
30	0.15	$3,082	20	152	$2,493	$23,393	0.15	$3,449	$6,531	0.20	$55,348	30	$7	$13,201	$6,531	$55,348	$75,087	30

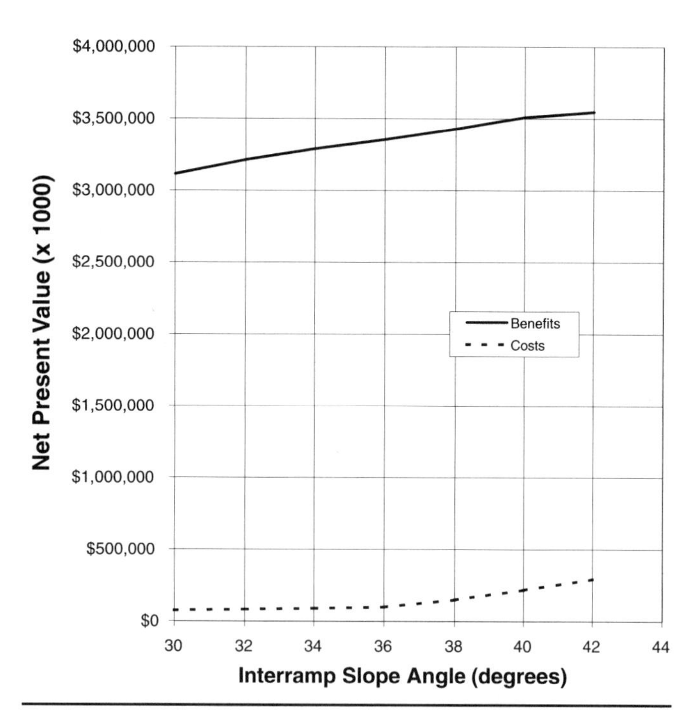

FIGURE 3.11 Benefits and costs versus slope angle

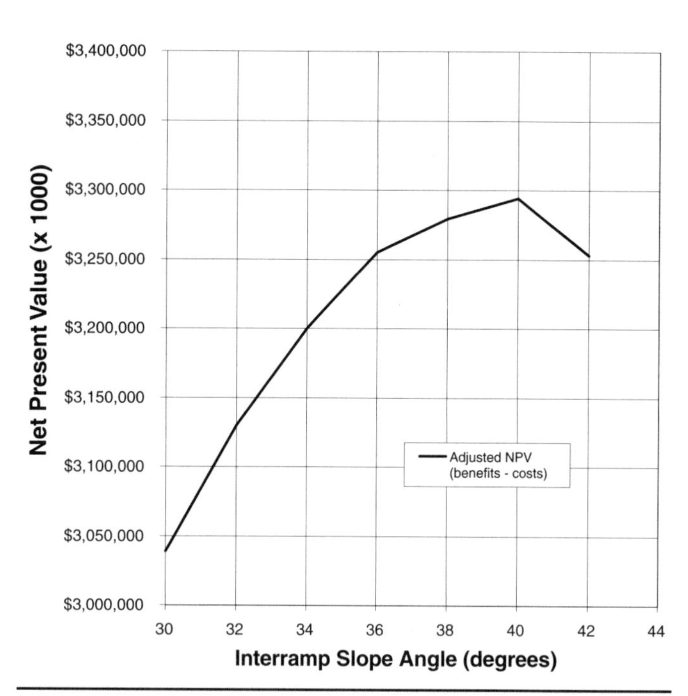

FIGURE 3.12 Adjusted net present value versus slope angle

expected number of failures, failure tonnages, and failure heights are used to estimate the total cleanup costs, as well as the number of failures on haul roads, below haul roads, and on or below other critical structures. Failures on haul roads result in blocked haulage and lost production in addition to cleanup costs. Failures below haul roads require backfilling to restore the haul roads and may require stepouts, which result in lost ore. Failures on or below other critical structures would result in similar consequences and would have specific costs associated with replacing or restoring the impacted structure. A sample spreadsheet showing the cost calculations is presented in Table 3.3.

Cost-Benefit Curves. The net present value versus slope angle is then plotted along with the total cost of failure versus slope angle (see Figure 3.11).

The difference between these two curves represents the adjusted net present value incorporating the economic consequences of slope instability into the mine planning process. A plot of the adjusted net present value versus interramp slope angle (see Figure 3.12) illustrates that this curvc pcaks at thc cconomically optimal angle.

3.4 CONCLUSIONS

Interramp analysis is a critical, if often neglected, tool in the evaluation of pitward-dipping major structures that are too small or too numerous to model individually and too large to be evaluated using bench-scale analyses. As with all other analytical methods used for slope stability evaluation, interramp analysis is not applicable in all situations because either the structures do not exist to make this a viable failure mode or the database is inadequate (due to limited exposure or economic constraints) to justify an analysis of this magnitude. However, in large open-pit mines, where the economic implications of minor changes in slope angle are significant, this analysis can be extremely valuable for determining the economically optimal interramp slope angle, particularly when combined with a cost–benefit analysis.

3.5 ACKNOWLEDGMENT

Although the Backbreak and Interramp analyses discussed herein have been modified and updated through the years, Call & Nicholas, Inc., acknowledge Dr. R.D. Call and Paul J. Visca for their work at CNI in developing the original framework for both analyses.

3.6 REFERENCES

Barton, N. 1976. The shear strength of rock and rock joints. *Intl. Jour. Rock Mech. Min. Sci. & Geom. Absts.*, 13:255–279.

Call, R.D. 1985. Evaluation of material properties. In *Design of Non-Impounding Mine Waste Dumps*, ed. M.K. McCarter, Chapter 5. New York:SME.

Call, R.D. 1992. Slope stability. In *SME Mining Engineering Handbook*, ed. H.L. Hartman, 2nd ed., Vol. 1, Chapter 10.4. Littleton, Colorado:AIME.

Call, R.D., and Y.C. Kim. 1978. Composite probability of instability for optimizing pit slope design. In *Preprint Proceedings 19th U.S. Symposium on Rock Mechanics*. 4 pp.

Call, R.D., and J.P. Savely. 1990. Open pit rock mechanics. In *Surface Mining*, ed. B.A. Kennedy, 2nd ed., Chapter 6.8. Littleton, Colorado:SME.

CANMET. 1977. *Pit Slope Manual.* Pit Slope Project of the Mining Research Laboratories, Canada Centre for Miner and Energy Technology (CANMET). Ottawa, Ontario:Department of Energy, Mines, and Resources.

Cicchini, P.E., and R.C. Barkley. 2000. Designing overall pit slopes, this volume. Littleton, Colorado:SME.

Evans, C.L. 1989. *The design of catch bench geometry in surface mines to control rockfall.* M.S. thesis. University of Arizona.

Hoek, E., and Bray. 1981. *Rock slope engineering.* 3rd ed. London:Inst. Min. Met.

Marek, J.M., and J.P. Savely. 1978. Probabilistic analysis of the plane shear failure mode. In *Preprint Proceedings 19th U.S. Symposium on Rock Mechanics,* Vol. 2. pp. 40–44.

Pierson, L.A., S.A. Davis, and T.J. Pfeiffer. 1994. *The nature of rockfall as the basis for a new fallout area design criteria for 0.25:1 slopes.* Oregon:Department of Transportation.

Savely, J.P. 1987. *Probabilistic analysis of fractured rock masses.* Ph.D. diss. University of Arizona.

Sims, D.B., and D.E. Nicholas. 2000. Geologic structure analysis for pit slope, this volume. Littleton, Colorado:SME.

Yost, R.R. 1995. *An analysis of rockfall in open-pit mines.* Technical report prepared for Cyprus-Climax Minerals Corp., Tucson, Arizona:Call & Nicholas, Inc.

CHAPTER 4

Managing and Analyzing Overall Pit Slopes

Richard D. Call,* Paul F. Cicchini,* Thomas M. Ryan,* and Ross C. Barkley*

4.1 INTRODUCTION

Many large porphyry copper mines, such as the Chuquicamata Mine and the Bingham Canyon Mine, have slopes in excess of 600 m. Several of these mines have final mine plans that include sectors with overall slope heights of more than 1,200 m. Because of their large size and the corresponding large stripping volumes, there is often an economic incentive to design working pit slopes near the optimum overall slope angle. However, experience has shown that a significant amount of slope displacement can be expected at economically optimum slope angles. Because of their size, most large open pits have enough operational flexibility to tolerate considerable slope displacement, provided the slope "failures" can be adequately modeled and predictions regarding their size and behavior can be made.

The analytical tools used to predict the expected number and volume of multibench, structurally controlled failures (wedge, step-path, plane shear, and step-wedge) are well developed. Probabilistic models have been developed to estimate the number, size, and displacement resulting from structurally controlled rock-slope failures (CANMET 1977). However, overall slope displacements related to yield in weak rock masses require more complicated numerical models to estimate the degree of instability. These models should consist of accurate characterizations of the rock units, the major-structure locations, the rock-mass strengths and elastic properties, the hydrologic conditions, and the in situ stress. If reliable and predictive rock-mechanics models of strength, stress, and displacement can be constructed for the overall pit slopes, mine plans and operational procedures capable of tolerating overall slope instability can be developed, thus (1) improving operational safety, (2) increasing production efficiency, (3) improving mine economics, and (4) extending the mine life.

The geotechnical analysis of large-scale overall slope instability involves these primary areas of study:

- Slope monitoring and kinematic displacement modeling
- Geological, geotechnical, and hydrological analysis
- Stress, displacement, and stability modeling

4.2 FAILURE CLASSIFICATION

Post-failure models are primarily based on empirical relationships derived from slope-monitoring data (Cruden and Mazoumzadeh 1987; Voight, Orkan, and Young 1989). Examination of displacement data led to the development of two general displacement models: a progressive failure model, for which slope displacement will continue to accelerate to a point of collapse (or greatly accelerated movement), and a regressive failure model, for which the slope will decelerate and stabilize. Zavodni and Broadbent (1982) defined regressive and progressive failure stages for several large-scale open-pit slope failures and related these stages to failure geometry. Savely and Call (1991) expanded this work into a useful description of failure characteristics and expected regressive or progressive slope behavior (Figure 4.1).

A progressive slope becomes less stable with time and can exhibit sudden, large movements. Ryan and Call (1992) reported a wide range of precollapse velocities for progressive slope failures. Savely and Kastner (1982) and Kennedy and Niermeyer (1970) discuss operational and monitoring procedures for minimizing the impact of progressive slope failure on mining.

Regressive slope failures often occur as a result of rock mass yielding near the toe of the slope. In the United States, case histories of regressive slope failures in open-pit mining have been documented in the literature since the early 1950s (Bisbee, Arizona, and Butte, Montana). Many rock-mass movements in high open-pit slopes decelerate with time, provided the ratio of driving to resisting forces decreases with displacement. Surface and subsurface monitoring data from several slope failures display displacements that are extremely variable in direction and in magnitude.

Slope displacements, whether progressive or regressive, must be monitored carefully to ensure operational safety and to determine the impact on slope stability of changes to the mine design and operations.

The remainder of this chapter discusses the characteristics, behavior, and analysis of regressive, overall pit-slope failures that tend to stabilize with displacement and time. Geotechnical analysis and experience with several regressive slope failures demonstrate that mining can continue in areas of large-scale displacement, provided the failure is regressive in character, the failure mechanism is well defined, monitoring procedures are established and enforced, and an effective slope-management program is in place.

4.2.1 A Displacement Model of Regressive Slope Failures

Slope-monitoring data consistently indicates that most regressive slope failures occur in response to mining activity. A typical response is shown in Figure 4.2.

As mining proceeds along a level within a zone that is in active yield, confining stresses are reduced, excess strain energy is induced, and displacement of the rock face is produced. These slope displacements can be either elastic or plastic, depending on the state of stress and the strength of the nearby field rock mass. Monitoring data often demonstrate that slope displacement initially develops in lower-strength, highly fractured rocks near the toe of an active mine slope. Toe displacement directions are typically at very low angles (5° to 20° from horizontal). Displacements higher in the slope do not develop instantaneously with displacement at the toe and can have highly variable directions. Since the slope is not acting as a rigid block, displacements will tend to develop in a time-dependent manner, resulting in differing rates of

* Call & Nicholas, Inc., Tucson, Arizona.

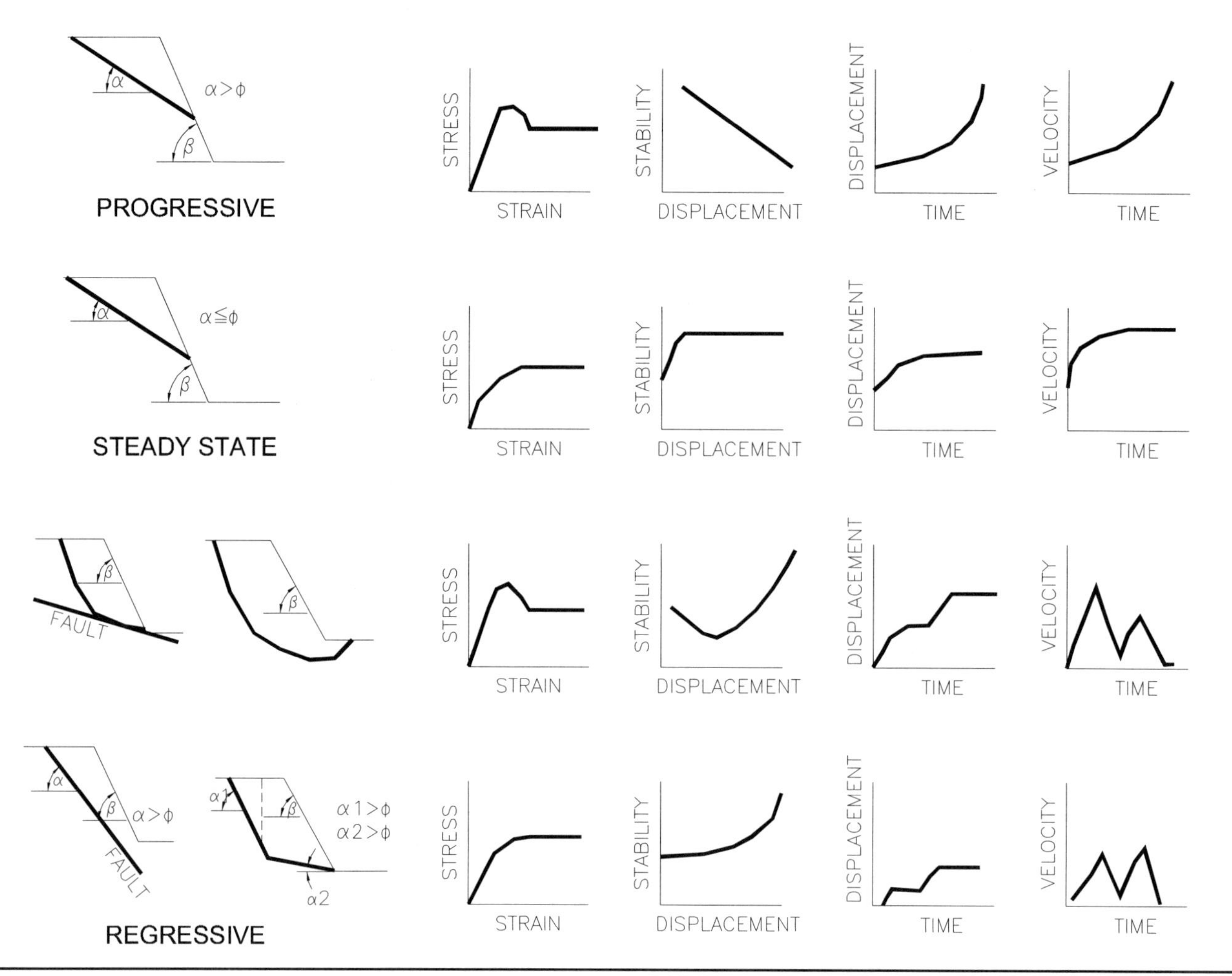

FIGURE 4.1 Classification of progressive and regressive failure

movement from point to point within the slope (Figure 4.3). This time dependency can also be demonstrated in regard to surface and subsurface displacement.

4.3 GEOLOGICAL CHARACTERISTICS OF LARGE-SCALE REGRESSIVE SLOPE MOVEMENTS

The following geologic characteristics are common to documented, large-scale regressive slope failures (Figure 4.4):

- Low rock-mass strength in the toe
- A ubiquitous joint set that dips into the pit
- High-angle faults or continuous joints that form back and side releases for the slope movement
- Saturated toe, excess hydraulic gradients, and compartmentalized groundwater conditions
- High in situ horizontal stresses

4.3.1 Low-Strength Rock Mass

Unfortunately, thick zones of low-quality rock are associated with the processes of ore emplacement. Regional-scale tectonics, contact metamorphism, and the release of volatiles from the magmatic melt all contribute to the degradation of the rock within and surrounding porphyry deposits.

In slopes that have experienced large-scale regressive displacements, the zones of low rock-mass strength generally exceed 50 m in thickness; these zones have a rock-mass rating (RMR) value of less than 40 and a rock-quality designation (RQD) of less than 30. Rock-mass characteristics range from low RQD zones with clay-filled fractures and moderate wall-rock alteration to pervasively clay-altered zones in which the original rock has been obliterated by mechanical and hydrothermal actions.

Varying degrees of clay alteration are present in zones of low rock-mass strength, as well as in the bordering transition zones. In the low rock-mass strength zone, clay alteration varies from a pervasive fracture filling with weak to moderate wall-rock alteration to complete alteration of the rock to clay by mechanical and hydrothermal processes.

Clay alteration extends beyond the low rock-mass-strength zone but is usually confined to larger structures. In some cases, a transition occurs between weak clay-altered rock and fresh rock. Transition zones are characterized by clay-filled fractures and faults; however, the wall-rock alteration is diminished and the RQD improves.

Clay adversely affects the stability of slopes for two reasons. First, clay alteration of intact rock and clay infilling of fractures reduces rock-mass strength. Second, clay impedes the natural or induced drainage of the slope, resulting in higher water pressures within the rock mass.

4.3.2 Ubiquitous Joint Set

Ubiquitous joints that dip toward the pit at 35° to 45° are a common characteristic of these regressive failures. In general, such structures do not daylight in the overall slope but provide a plane of weakness within the slope.

The principal stress aligns with the dip direction of the ubiquitous joint as mining progresses (Amadei, Savage, and Swolfs

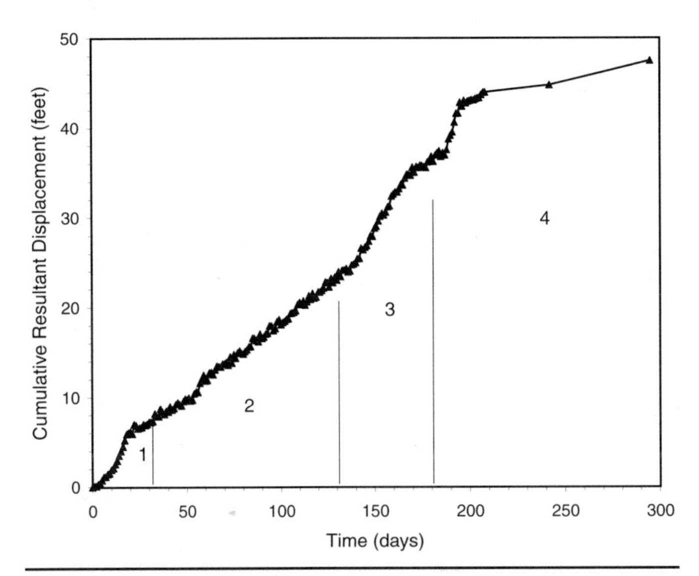

FIGURE 4.2 Four regressive slope displacement cycles displayed on a graph of cumulative resultant displacement

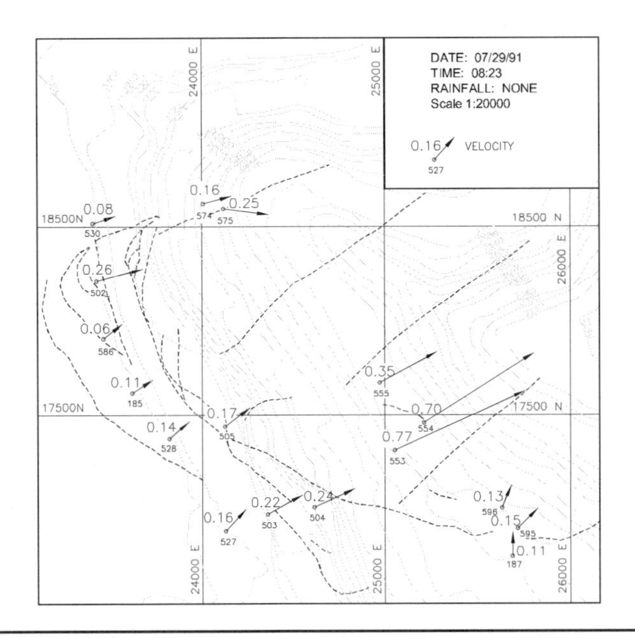

FIGURE 4.3 Measured slope velocities of a large-scale slope failure (Note the high velocity measured directly above the active mine bench at the 2,680 level.)

1987). This stress alignment enhances slope movement along the ubiquitous joints and may extend the joints.

4.3.3 High-Angle Back and Side Releases Formed by Faults

High-angle faults that strike parallel to subparallel to the trend of the pit wall form a back release for downward slope movement. These structures diminish the rigidity of the rock mass, form discrete kinematic blocks, and compartmentalize groundwater. In addition, because of the large contrast in rigidity between the fault structures and the surrounding rock, the maximum principal stress aligns parallel to these structures on the upslope blocks.

Faults or continuous joints, which strike from perpendicular to oblique to the trend of the pit wall, provide side releases for the slope movement.

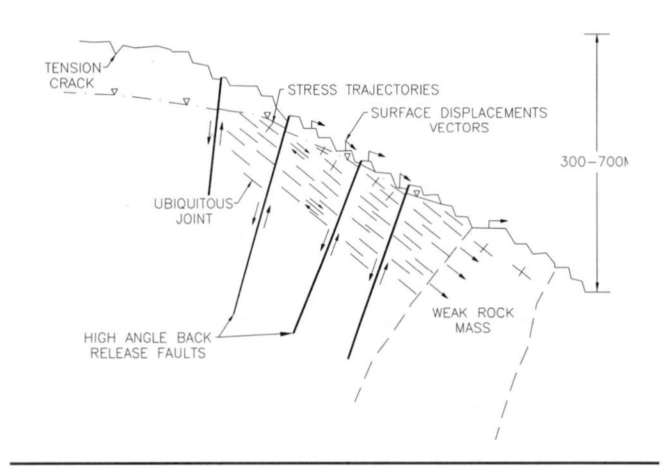

FIGURE 4.4 Geologic conditions observed in large-scale regressive slope failures

4.3.4 Groundwater Conditions

Groundwater causes a destabilizing effect in these large-scale slope failures. In a free-draining slope, the hydraulic gradients are generally low, as evidenced by a gentle drawdown cone and a dry slope. However, clay-filled faults and fractures greatly reduce the permeability of the rock mass comprising the slope. Excess pressure gradients can occur, because the clay inhibits free drainage when the stress state in the rock is changed by mining. In these cases, the drawdown cone is very steep, resulting in groundwater close to the pit face. This pressure creates a significant buoyant force on potential failure surfaces, which reduces the ability of the slope to resist shear. Driving forces increase further if the water pressure acts along a vertical back release.

Slope movements are quite sensitive to changes in water pressure. The onset of slope failure dilates the rock mass, creating additional storage. This added storage results in a lowering of the water level and a temporary increase in stability. In very dry regions, the lack of recharge may stabilize the slope or may cause diminished slope movements for several pushback cycles. However, in areas of high recharge, the increase in stability is short-lived, and mine personnel should continue to monitor the slopes for increased pore pressure.

4.3.5 Horizontal Stress

When designing overall pit slopes, high horizontal stresses related to regional tectonics should be evaluated. As mining progresses, high horizontal stresses accentuate the deviatoric stress in the rock mass near the toe of the slope. If the rock mass is not strong enough to withstand the deviatoric stress, it will yield. Higher horizontal stresses will cause more rock-mass yielding, leading to larger slope displacements.

4.4 FAILURE MODELING AND STABILITY ANALYSIS

The geologic characteristics listed in this chapter often combine to cause the large-scale regressive slope failures observed. The following is a suggested sequence of events that explains the slope displacements:

1. Initiation of movement by active mining in the weak toe rock.
2. Propagation of movement upslope due to successive release of kinematic blocks.
3. Deceleration to a more stable state.

4.4.1 Initiation of Movement

Mining through low-strength rock in the toe reduces the confining stress and creates an overstressed condition. Excess stress is relieved by movement in the toe of the slope and by stress transfer into rock that can carry additional load. The maximum principal stress is commonly horizontal at the toe, resulting in strong horizontal displacement.

Due to the pervasive clay fillings and wall-rock alteration in the weak toe rock, natural or induced drainage is limited. Pore-water pressures in this low-permeability saturated rock can be high due to stress redistribution. The high pore-water pressure reduces the ability of this weak rock mass to resist shear. In some cases, this material becomes amenable to drainage as the displacing rock mass dilates.

4.4.2 Propagation of Movement Upslope

Displacements propagate upslope as confinement is reduced in successive upslope kinematic blocks. Upslope movements occur parallel to the non-daylighted joint sets and along high-angle release faults.

A common characteristic of these failures is a differential shear along the high-angle faults that form the back releases of identifiable kinematic blocks. Along a particular fault structure, the upslope block drops relative to the downslope block (Figure 4.4). This differential movement results from a significant vertical stress gradient across the faults. Due to the low strength and modulus of the fault structure, the maximum principal stresses on the upslope block align with the fault. This, coupled with the relief of lateral confinement due to downslope displacements, can result in a stress condition analogous to active earth pressure. The additive effects of the deep shear often result in the observed differential displacement along the release faults. Field observation of this post-deformation expression along high-angle faults has misled many investigators into referring to this failure mechanism as deep-seated toppling. As such, it implies that deep-seated toppling of these upslope blocks initiates the slope failure when, in fact, the toppling of the blocks is a secondary effect due to shear deformation initiating in the toe of the slope.

The high-angle release faults that form the kinematic blocks also tend to compartmentalize groundwater. Groundwater lowers the resistance of the rock mass to shear, resulting in a deeper zone of shearing along ubiquitous joint sets. In addition, water pressure acting on the backplanes increases the driving forces.

Lowering pore-water pressure in the upslope kinematic blocks greatly reduces overall slope displacements by reducing the driving forces and increasing the shear resistance. Additionally, stabilizing the upslope blocks imparts less stress on the weak toe material. Installation of dewatering systems usually pays dividends because slightly steeper slopes can be achieved and delays in mining operations caused by a moving slope can be reduced.

4.4.3 Reduction of Slope Movement

Slope displacements decrease as a result of the following:

- Halting or lowering the production rate in the weak toe
- Lowering of the pore water pressure due to dilation of the displacing rock mass
- Displacing to a more stable geometry
- Unloading the active block of the failure

4.4.4 Stability Modeling

Simple rigid-block models are not appropriate for analyzing rock-slope stability when the primary mechanism of failure is plastic (toe) to pseudoplastic (shear along ubiquitous joints) yielding of the rock mass. Limit-equilibrium methods have limited application because they cannot predict either strain or the extent of slope deformation. For these reasons, the combination of discrete element and continuum numerical methods is finding widespread application for the overall stability analysis of these high, open-pit slopes. The sophistication and efficiency of these models has improved rapidly.

When using these numerical methods, it is critical to understand the mechanisms producing the slope failure, the strength characteristics of the rock mass, and the hydrology of the pit. If modeling is performed for a failing slope, it is refined to correspond to the historical slope displacements and the measured in situ stresses.

Parameters that influence the model response include in situ stress, rock-mass strength, elastic properties, structure, and hydrology. Because displacements are often not elastic, superposition is not valid, and a careful simulation of historical mining is required for stability analysis. A historical match between the model and the slope often can be obtained through several stress paths. Therefore, careful reasonability checks on (1) the loading and unloading of the model; (2) the geomechanical, geological, and hydrological parameters used in the model; and (3) the displacement, stress, strain, and constitutive state at each model step.

It is also important to understand the limitations and assumptions inherent to the particular numerical method selected. A continuum model may be suitable for the back-analysis of an existing failure but its usefulness in predicting the stability of future mining geometries may be severely limited. In a jointed rock mass, both the deformation characteristics of the rock mass and the ability of the rock to transfer and relieve stress are influenced by joint slip. Appropriate deformation characteristics of the rock mass can be simulated in a continuum model by reductions to the intact rock modulus of deformation. However, stress relief due to mining cannot be properly simulated since the effects of stress transfer due to joint slip cannot be accounted for in a continuum. This limitation becomes particularly acute when a large volume of material must be excavated, such as in the case of mining from a premine topography (known stress state) to an ultimate pit geometry (simulated stress state).

Another important issue to consider when setting up a numerical model is the manner in which the in situ horizontal state of stress will be initialized. The state of stress can be initialized through application and adjustments to a distributed load applied to the boundary of the model or through direct initialization of a stress state within each element. In the latter case, the model will have a zero displacement boundary, and in the former, it will not. Although identical initial stress states can be created with either approach, the induced stresses from the simulation of future mining cuts will be very different. The boundary-loaded model simulates an active, driving horizontal force, creating much greater stress concentrations around the vicinity of the excavation and more adverse stability conditions than those resulting from the zero displacement boundary conditions. Therefore, the user must have a clear understanding of the stress environment to select the appropriate boundary conditions. In the absence of strong evidence to support an active stress regime, zero displacement boundary conditions will generally be assumed.

A large geomechanical database must be developed to define the parameters used in the model. This database should include the following:

1. Rock types observed from surface and subsurface geologic mapping
2. Block sizes
   a. Cell mapping
   b. RQD logging of core
   c. Surface classification mapping of weak rock masses
3. Intact shear strength
   a. Uniaxial compression tests
   b. Brazilian disk tension tests
   c. Triaxial compression tests
   d. Point load tests

4. Fracture shear strength
   a. Small-scale direct-shear tests of core
   b. Large-scale direct-shear tests of rock blocks
5. Orientation of geologic structure
   a. Rock fabric mapping
   b. Oriented coring
   c. Geologic mapping of major structures
6. Hydrological characteristics
   a. Geologic mapping and drilling to define character of water-bearing rock
   b. Measuring water levels in exploration and geomechanical drill holes
   c. Reporting wet blastholes
   d. Reporting seeps observed on the slope
   e. Slug testing of exploration and geomechanical holes to provide a rough estimate of transmissivity
   f. Performing pump tests with observation holes for definition of transmissivity and storage in more homogeneous aquifers
7. In situ stress measurements

Rock-mass strength and elastic properties and hydrologic parameters are defined from the data mentioned in items 1 through 7. These data are used to zone the rock into discrete domains that possess similar engineering characteristics. Quantitative methods have been developed to classify rock masses according to rock type, block-size distribution, intact rock shear strength, fracture shear strength, orientation of the geologic structures, pore-water pressure, and in situ stresses. The RMR (Bieniawski 1993) and Q-system (Barton, Lien, and Lunde 1974) rock-mass classifications have been used extensively to choose appropriate underground mining methods and to estimate support requirements for underground openings, settlements in rock-mass foundations, and the rock-mass shear strength.

Of particular importance in constructing numerical models of pit slopes is the definition of the rock-mass shear strength and the deformation modulus. Several empirical classification schemes have been developed to estimate the rock-mass strength and the elastic properties. The majority of these methods rely on assessments of the RMR. Experience with numerical slope modeling has shown that the rock-mass strength and elastic properties derived from these classification schemes do not always provide model responses that match the slope performance or the measured slope displacements. An alternative method has been developed that relates the strength of the rock mass directly to the degree of fracturing present, through a combination of the intact rock strength and the natural fracture strength as a function of RQD.

To determine the modulus of deformation for the rock mass, Bieniawski (1978) proposed the following relationship for the correlation between RQD and the ratio of the rock-mass modulus to the intact rock modulus:

$$r = E_m/E_i \qquad \textbf{EQ. 4.1}$$

Where:

E_m = rock-mass deformation modulus
E_i = intact rock deformation modulus

$$r = \alpha e^{\beta(\%\text{RQD})} \qquad \textbf{EQ. 4.2}$$

For:

$\alpha = 0.225$
$\beta = 0.013$

Deere and Miller (1966) demonstrated that the elastic modulus for intact rock can be related to the intact compressive strength and defined a narrow range of observed ratios between elastic modulus and compressive strength for both brittle and soft materials. Consequently, it seems reasonable to expect that a similar relationship may exist between the rock-mass modulus and the rock-mass strength. Back-analysis of slope failures indicated that the estimation of rock-mass strength does follow Bieniawski's relationship for predicting deformation modulus; however, the strength properties were found to vary according to the square of the modulus ratio (r^2). For example, if the square of the modulus ratio (r^2) is 0.3, the estimated rock-mass strength is derived by compositing 30% of the intact rock strength with 70% of the natural fracture strength. The resulting equations for predicting the rock-mass friction angle and cohesion are

For RQD > 50% to 60%:

$$C_m = \gamma[r^2 c_i + (1 - r^2)c_j] \qquad \textbf{EQ. 4.3}$$

$$\phi_m = \tan^{-1}[r^2 \tan(\phi_i) + (1 - r^2)\tan\phi_j] \qquad \textbf{EQ. 4.4}$$

Where:

ϕ_m = rock-mass friction angle
C_m = rock-mass cohesion
ϕ_i = intact rock friction angle
c_i = intact rock cohesion
ϕ_j = joint friction angle
c_j = joint cohesion

And:

$\gamma = 0.5$ to 1.0
$\gamma = 0.5$, jointed medium to strong rock (>60 Mpa)
$\gamma = 1.0$, massive weak to very weak rock (<15 Mpa)

For simplicity, the rock-mass strength equations have been presented for a linear Mohr–Coulomb failure envelope. The rock-mass shear strength can be mapped to a power envelope by regression techniques using the calculated percentage of intact rock (r^2 * 100) and the power strength envelopes of both the intact rock and the fracture shear data.

The intact compressive strength exerts the primary control on the constant gamma (γ) in Eq. 4.3. However, the appropriate gamma (γ) value is also influenced by the degree of fracturing. In general, the gamma (γ) value increases as the intact compressive strength decreases, and as the fracture intensity becomes greater, the gamma (γ) value lessens.

Subsequent application of these equations indicated that for RQD values less than approximately 50%, Eq. 4.4 tended to overpredict the rock-mass friction angle. Consequently, the constants alpha (α) and beta (β) in Eq. 4.2 were revised to provide a better fit to back-calculated rock-mass friction angles for lower RQD rock masses.

For RQD < 40 to 50% (estimation of rock-mass friction angles only),

$\alpha = 0.475$
$\beta = 0.007$

The two relationships presented for predicting rock-mass friction angle do not follow a smooth transition between 40 and 60% RQD. Modifications to the equations in this RQD range are currently being investigated, and the authors hope to publish the results of this work in the near future.

Because the above relationships for predicting rock-mass strength are based on RQD, it is important to recognize that RQD can be an imprecise indicator of the degree of fracturing at RQD values below approximately 20% and above approximately 80%. To overcome this deficiency, a relationship based on fracture frequency is preferable. However, existing mine databases typically either lack these data or the information is very limited. Fortunately, this situation is rapidly changing as greater focus is being placed on the compilation of a comprehensive geomechanical database during exploration drilling. Once more extensive databases of fracture frequency are available, these relationships can be readily converted and extended to find wider applicability in strongly fractured, as well as in massive, rock units.

Empirical-based methods for predicting rock-mass properties have been widely used and often successfully applied in the field of rock mechanics. However, more rigorous methods for determining rock-mass strength characteristics are still needed. The use of three-dimensional numerical methods to model the mechanical response of the rock mass to various loading conditions shows some promise. To be effective, these models must accurately portray the intact rock properties, the orientation of stress fields, the orientation and spacing of discontinuities, and the strength and elastic properties of the discontinuities and/or discontinuity fillings.

Although numerical methods have proven valuable in defining the failure mechanism of these large-scale slope failures and in predicting the response of the slope to future mine plans, they are not adequate for modeling the combined responses of mechanical loading, deformation, and water flow. The large size of these slope models places a constraint on the number of sensitivity analyses for various static water conditions that can be performed in a reasonable period. Future versions of the numerical methods must efficiently represent the mechanical/hydrological interaction of large pit models. The following topics should be addressed by updated numerical methods:

- Consolidation
- Efficient establishment of hydraulic gradients in complex hydrological conditions
- Change in hydraulic gradients resulting from rock-mass dilation during failure

4.5 DISPLACEMENT RATES AND OVERALL SLOPE ANGLE

A range of overall slope angles can be excavated in regressive slope failures. Steeper slopes experience greater velocities and larger overall displacements. However, there is a maximum limit to the angle that can be excavated before a progressive accelerating slope failure will occur. Steepening the upslope geometry beyond this critical angle results in driving forces that are greater than the resisting forces in the upslope kinematic blocks. These upslope blocks can drive the toe to an accelerated failure condition. A fundamental role of stability modeling is to determine the critical slope angle and identify whether this accelerated condition is beginning to develop. Figure 4.5, a plot of model velocity histories for three slope geometries, illustrates the use of modeling in defining the critical slope angle.

4.6 OPERATIONAL PROCEDURES TO MINIMIZE THE IMPACT OF SLOPE DISPLACEMENT ON MINING

Because of the operational flexibility in most large open-pit mines, slope displacement does not necessarily constitute "failure" from the standpoint of mine management. This relationship between theoretical and operational slope failure in mining has been discussed (Munn 1985). In particular, the real hazard to mine operations is often the potential for greatly accelerated movement occurring near equipment and personnel. If this can be mitigated, a significant amount of slope displacement can often be tolerated with routine mine operations such as dozing and additional shovel shifts for cleanup. Mining in areas of large-scale slope instability generally results in unsteady production. Provided a mine slope failure is regressive in character, slope displacement can be controlled using specific operational procedures: dewatering, additional stripping, control of the excavation geometry, and control of the excavation rate. The effect of these controlling measures can be assessed and predicted with geotechnical analysis.

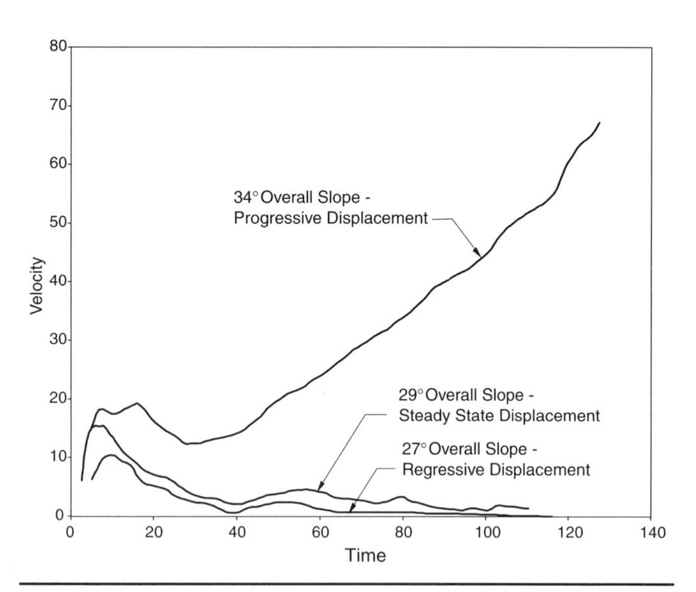

FIGURE 4.5 Velocity histories for three modeled overall slope geometries

4.6.1 Stepout Versus Unload

At Twin Buttes during the past five years, it has been demonstrated that periodic, small (less than 20 ft) stepouts into the pit could effectively decrease slope displacement in an area of historical slope failure (Ness 1992). There is often a trade-off between the cost of cleanup or additional stripping and the value of ore lost due to a stepout. There also may be an optimum location for a stepout in a pushback, and this can be defined by the geotechnical engineer through a stability analysis.

4.6.2 Controlling Excavation Rates

Controlling the mining rate is another way to maintain displacement velocity to minimize its impact on mine operations. This technique is difficult to implement and requires significant flexibility from mine operations. Once displacement velocity of a slope region reaches a limit defined by the rock mechanics staff (through analysis and experience), mining is discontinued until an acceptable relaxation limit is achieved. Savely (1993) discusses how this approach is optimized by maximizing operational efficiency in relation to the moving slope area. At several properties, short mining periods and shorter delays have been found to be more cost-effective than long mining periods and longer delays. This can also be demonstrated mathematically with displacement modeling (Figure 4.6).

Since the vast majority of accelerating and decelerating displacement curves for rock slopes approximate either an exponential or a power function, the relationship between the time spent in excavation and the time required for relaxation is not linear. The higher displacement rates associated with longer mining periods require a substantially greater proportion of time for deceleration. Stability analysis and slope-monitoring data can be used to assess optimum extraction rates.

4.6.3 Pushback Width

Optimum pushback widths can be defined from a geotechnical perspective as well as a mining perspective. Practical mining experience suggests that narrow pushbacks within failed slopes are difficult to maintain. This can also be demonstrated numerically with detailed stress and energy analyses of rock slopes. The role of the geotechnical engineer should be to determine whether there is an advantage in changing the pushback geometry because of either an existing slope instability or the location of a major geologic structure.

Stability analysis of overall slope failures using continuum models demonstrates that stress concentration occurs at the toe

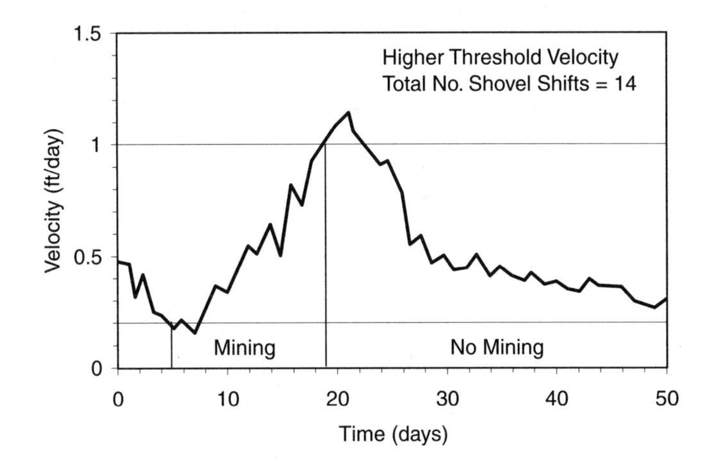

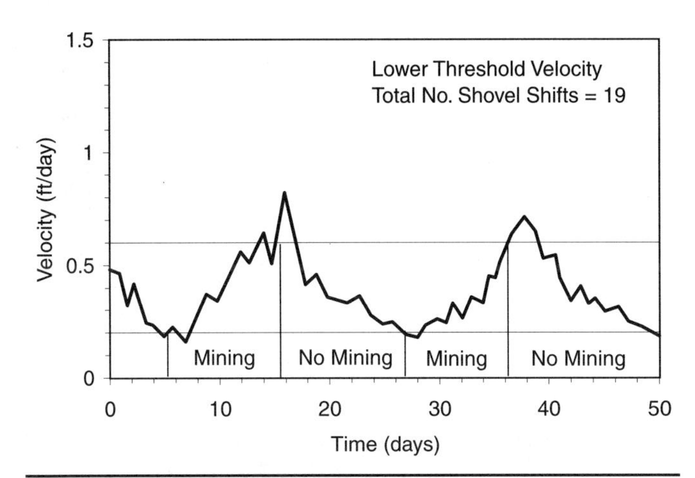

FIGURE 4.6 Optimizing production in a displacing slope (Note the lower "no mining" threshold velocity of 0.6 ft/d allows 19 shovel shifts, while the higher "no mining" threshold velocity of 1.0 ft/d allows 14 shovel shifts.)

of slopes as they are excavated; this has been validated by in situ stress measurements at several mines. In the case of the excavation occurring within a plastic zone, a change in the stress and energy state occurs, which results in slope displacement. The stress path for excavation is a combination of lateral extension and axial compression. Strain energy for underground excavations has been analyzed in detail by others (Salamon 1984; Farmer 1986), and similar methods can be applied to pit slopes that are at yield. In general, smaller excavations result in less energy changes and less displacement. However, because of the stress concentrations that develop in the toe of a slope, strain-energy potential within the slope is not uniform. This leads to a nonlinear relationship between pushback width and strain energy (or maximum shear stress) for varying sizes of incremental mining cuts. When narrow pushbacks are mined, excavation takes place within the zone of stress concentration, and the resulting displacements are typically a high percentage of the overall pushback width. This renders narrow pushbacks difficult to maintain and results in either excessive time spent in additional slide cleanup or frequent, unplanned stepouts into the pit. If stepouts must be taken too often, it may not be possible to complete the pushback, and ore production may be lost. Additionally, if strain softening of the rock mass occurs with displacement, there is considerable geotechnical incentive to *mine out* the existing failed

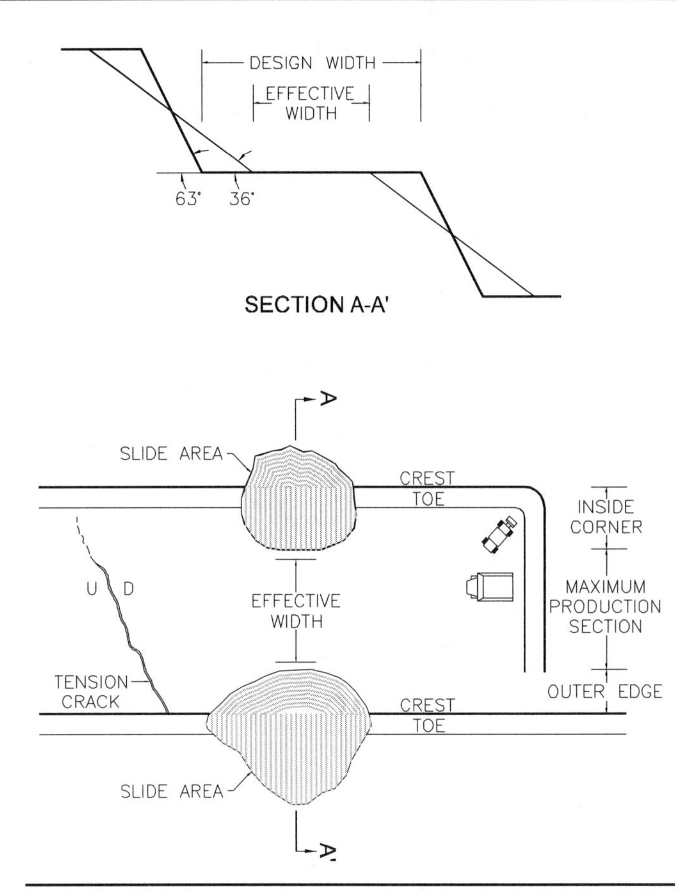

FIGURE 4.7 Operational considerations for pushback width

rock as much as possible to take advantage of the higher strength associated with a less disturbed rock mass.

For maximum production efficiency, a pushback must be wide enough for double spotting trucks and three lanes of traffic. For a normal rectangular cut with the digging face at right angles to the pit wall, the inside corner is relatively inefficient because either a wider swing or single spotting is required. The outer edge is similarly inefficient, in addition to having a less-than-complete digging face. Thus, a narrow pushback, with the majority of the digging time in the inside corner and the outer edge, will produce fewer tons per shovel shift (Figure 4.7).

Where there is overall slope displacement, the effective pushback width will be reduced by bench sloughing. For example, if a 15-m-high bench dug at 63° sloughs to a 36° angle of repose, 6.5 m of the pushback width will be lost. If sloughing of the working bench and the bench below occurs in the same section, the effective width of the pushback would be reduced by 13 m. Thus, in this case, pushbacks for a displacing slope should be designed at least 13 m wider than pushbacks for a stable slope.

In particular, the following sequence is to be avoided:

1. A production schedule is planned with a minimum width pushback and full shovel and truck efficiency.
2. Slope displacement reduces the effective width of the pushback, and the tons per shovel shift decreases.
3. At the lower production rate, ore will not be uncovered in time, so the pushback width is decreased in an attempt to uncover ore.
4. The reduction in pushback width results in one-way traffic, more single-side loading, and delays from tension crack offsets. This further reduces the tons per shovel shift.
5. Steps 3 and 4 are repeated until the pushback is not minable.

Another factor that must be considered in production scheduling is the horizontal displacement. If the slope experiences horizontal displacement on some level between one pushback and the next, additional material will have to be mined to achieve the design toe.

4.6.4 Slope Dewatering

Dewatering has been demonstrated in several cases as an effective control for displacing slopes (Argall and Brawner 1979). Cost–benefit analyses indicate that dewatering programs are one of the most cost-effective mechanisms available for improving slope stability for both stable and unstable ground. In many low-permeability rock masses, reduction of pore pressure prior to failure is difficult with conventional methods but is readily achievable after failure. Dewatering below the pit bottom is often required to achieve an acceptable pore pressure condition for the overall slope, and an analysis of aquitards and compartmentalization of groundwater must be completed by the geotechnical engineer to focus the proper level of effort on dewatering.

4.7 CONCLUSIONS

Improvements in continuum and discrete element models have enhanced the ability of engineers to design high overall pit slopes that often experience significant slope instability. These models can be used to predict the development of rock-mass movements, which previously could only be evaluated empirically. As the sophistication of the stability models improves, more data are required to extend the model; particularly high-quality data are necessary for calibrating the model. Better estimates of the state of stress and pore pressure are needed in addition to improvements in the techniques used for rock-mass strength prediction. In particular, stability models that are capable of modeling dynamic processes should continually be developed.

Provided enough data can be gathered to acquire a firm understanding of the failure mechanism, stability models that will enable engineers to provide operations personnel with reliable predictions of the impact of mine excavation on the stability of the overall pit slopes can be developed. If pit slope analyses can be developed in a spirit of cooperation between the geology, engineering, and planning staffs, geotechnical analyses can be run in concert with mine planning to create an optimum slope design.

Experience with several, extensive open-pit slope failures indicates that mining can be successful in areas of large-scale slope stability if variance in production can be accepted in the unstable pit sectors. If such fluctuating production rates can be tolerated, specific operational procedures can be used to minimize the impact of slope displacement on mine operations. Geotechnical analysis can be used to identify the procedures required to allow mine operations to work either within or near areas of large-scale slope instability. Such analyses can greatly improve the possibility of successful mining in these areas.

4.8 REFERENCES

Amadei, B., W.Z. Savage, and H.S Swolfs. 1987. Gravitational stresses in anisotropic rock masses. *Intl. Jour. Rock Mech. Min. Sci. & Geomech. Abstr.*, 24:5–14.

Argall, G.O., and C.O Brawner. 1979. Mine drainage. *Proceedings 1st International Mine Drainage.* Denver, Colorado.

Barton, N., R. Lien, and J. Lunde. 1974. Engineering classification of rock masses for the design of tunnel support. *Rock Mech.*, 6:183–236.

Bieniawski, Z.T. 1973. Engineering classification of jointed rock masses. *Trans. South African Inst. Civ. Eng.*, 15:335–344.

Bieniawski, Z.T. 1978. Determining rock mass deformability—Experience from case histories. *Intl. Jour. Rock Mech. Min. Sci.*, 15:237–247.

Call, R.D., and J.P. Savely. 1991. Open pit rock mechanics. In *SME Mining Engineering Handbook*. New York:AIME, pp. 860–882.

Cruden, D.M., and S. Mazoumzadeh. 1987. Accelerating creep of the slopes of a coal mine. *Rock Mech. & Rock Engrg.*, 20:123–135.

Deere, D.U., and R.P. Miller. 1996. *Engineering classification and index properties of intact rock.* Air Force Laboratory Technical Report No AFNL-TR-65-116. Albuquerque, New Mexico.

Farmer, I.W. 1986. Energy based rock characterization. In *Application of Rock Characterization Techniques in Mine Design.* ed. M. Karmis. New York:AIME, pp. 17–23.

Kennedy, B.A., and K.E. Niermeyer. 1970. Slope monitoring systems used in the prediction of a major slope failure at the Chuquicamata Mine, Chile. *Proceedings on Planning Open Pit Mines with Special Reference to Slope Stability.* Amsterdam:AA. Balkema, pp. 215–225.

Munn, F.J. 1985. Coping with wall failures. *Canadian Mining and Metallurgical Bulletin*, No. 78-884, 9–62.

Ness, M.E. 1993. Personal communication.

Ryan, T.M., and R.D. Call. 1992. Applications of rock mass monitoring for stability assessment of pit slope failure. *33rd U.S. Symposium on Rock Mechanics*, Santa Fe, New Mexico.

Salamon, M.D.G. 1984. Energy considerations in rock mechanics. Fundamental results. *Jour. South African Inst. Min. Metal.*, 84:8:233–246.

Savely, J.P. 1993. Slope management strategies for successful mining. *Proceedings International Congress on Mine Design*. Ontario, Canada

Savely, J.P., and V.L. Kastner. 1982. Slope instability at Inspiration's Mines. In *Proceedings 3rd International Conference on Stability in Surface Mining.* ed. C.O. Brawner. New York:AIME, pp. 609–634.

CHAPTER 5

A Slope Height Versus Slope Angle Database

Jonny Sjöberg*

5.1 INTRODUCTION

5.1.1 Previous Studies

Case studies are an invaluable source of information, because they reflect the actual conditions at a site. Theoretically, it should be possible to transfer this information into design guidelines for rock slopes in similar geological and geomechanical conditions. The problem often lies in how to describe (quantitatively) the prevailing conditions at a particular site so that the information becomes useful in forward design.

Previous work in this area includes that by Hoek and Bray (1981), which in turn was partly based on the work by Lutton (1970). No classification scheme was used in these studies, rather, slope heights were simply plotted versus slope angles. Rock-mass classification, which incorporated the MRMR system (Laubscher 1977), was used by Haines and Terbrugge (1991) to sort cases with respect to rock-mass quality. From this, design curves that could be used to obtain stable slope angle were derived. This approach was also the basis for a recent paper by Duran and Douglas (1999). More cases were added to the database of Haines and Terbrugge (1991) and new design curves were derived.

Common to most of the above studies is the fact that the design curves have similar shapes. An example, using hypothetical data, is shown in Figure 5.1. The asymptotical shape tends to indicate that there is a lower limit to the stable slope angle. This can be interpreted as the slope angle approaching the friction angle of the rock mass (Duran and Douglas 1999). However, it can also be an effect of simply having too little data for higher slopes. The design curve in Figure 5.1 illustrates where it is easy to distinguish between stable and failed cases. In reality, it is often difficult to determine a distinct divider between stable and unstable regions. Modification of previous design curves is also symptomatic for many of the above studies. As new data is compiled, previous predictions often have to be revised. Duran and Douglas (1991) concluded that this was partly true because certain factors controlling slope stability had been neglected. Examples of these are strong structural control and high water pressures. This implies that sorting of cases with respect to controlling factors is very important before attempting to derive even approximate design curves.

5.1.2 Approach, Objective, and Scope of Work

Some of the case study data presented above have not been updated to reflect current, worldwide pit-slope geometries. Other studies focused on deriving design curves and did not cover a large range of slope conditions. Consequently, there is a need to compile updated information for rock slopes in varying geological settings to provide a more complete database from which design information can be extracted.

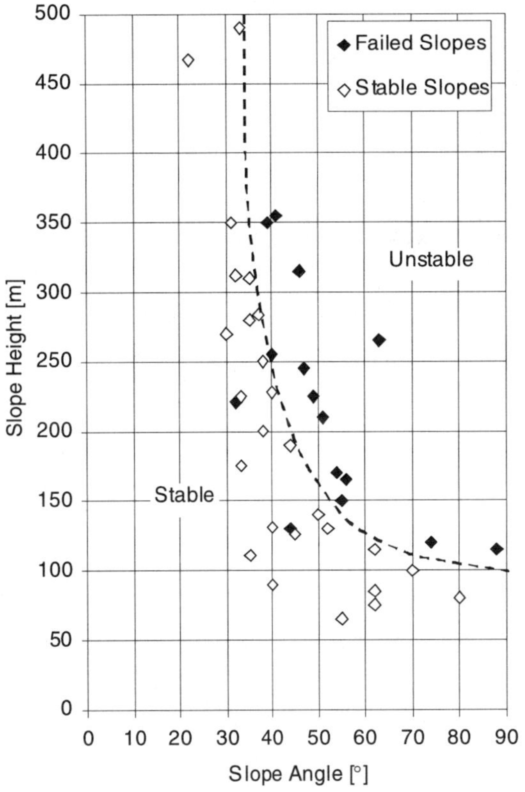

FIGURE 5.1 Typical design curve fitted to empirical slope data (hypothetical data)

This is also the objective of the work presented in this chapter. The design problem for the Aitik Mine (located in northern Sweden) was the starting point for this work. The mine is presented in more detail in Chapter 22. In summary, Aitik is a hard-rock, open-pit mine, currently nearly 300 m deep and with overall slope angles of around 45°. Final pit depth will be in excess of 450 m. Since there are no signs of incipient large-scale failure, precedent can be used for design of future slopes. One approach to solve this problem was to gather information from pits in similar rock conditions to try to establish empirical design guidelines.

Numerous case studies were compiled and analyzed. Cases were chosen to cover slopes of significant height, primarily in hard rock. However, due to the limited amount of data on hard-rock slopes, additional cases with weaker rock and a long history of stability problems were included. Furthermore, data on natural slopes were added to extend the database coverage. Thus, the

* Boliden Mineral AB, Boliden, Sweden.

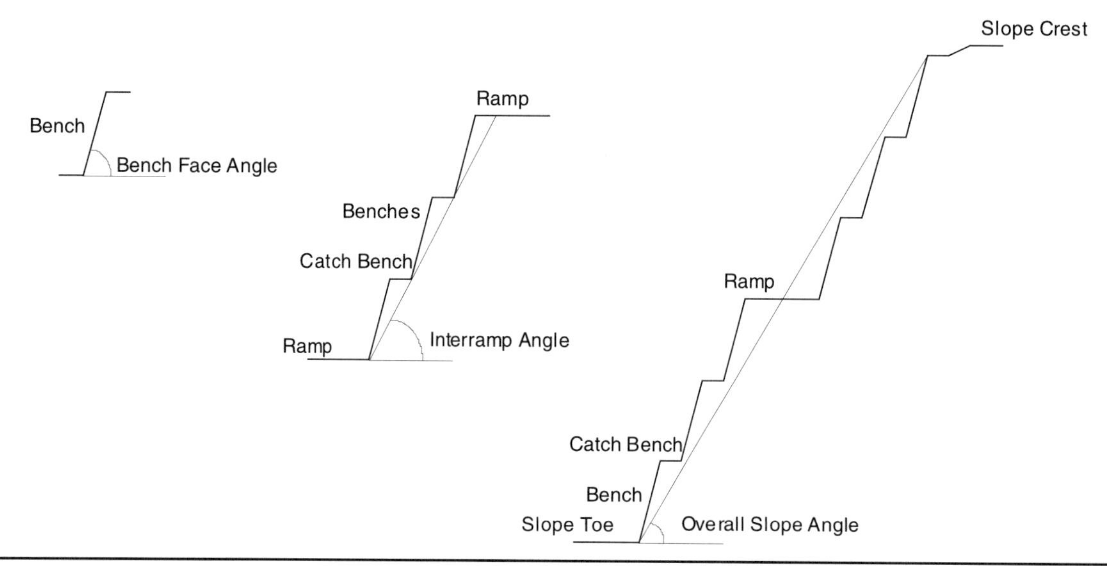

FIGURE 5.2 Definition of slope angles

final database covers a large range of geological and geomechanical conditions, different types of observed slope failures, as well as widely different slope geometries. The development of empirical design guidelines was not the primary objective of this work. However, some attempts at this were made for application to the Aitik open pit.

Data comes primarily from two sources: (1) site visits and contacts with mine staff and (2) existing literature. The data reflect relatively recent conditions; however, since most of the data were collected from 1995 to 1997, slope geometries of open pits may have changed slightly since then. The database is by no means complete. It was not possible to visit all interesting pits, and the referenced cases were selected from those readily available in the literature. There will be more cases and more detailed information available, e.g., from consultants working on the properties. It is desirable that more information be added continuously to the database to keep it updated.

5.2 CASE STUDY DATABASE

5.2.1 Collected Case Studies

The case study database includes 20 visited open pits and another 34 from the literature, 3 cases of sublevel caving mine slopes (of which 2 were visited), 117 cases of natural and engineering slopes in China, and 179 natural rock slopes in Norway. Data from the latter two categories were taken from the literature (Chen 1995a, 1995b; Broch and Nilsen 1977; Nilsen 1979; Dahlø 1976, 1996) and recompiled and reanalyzed. A full description of all cases (including all references) and the collected data is given in Sjöberg (1999).

The amount of data available for each case varied significantly. For the visited mines, good descriptions of the geomechanical environment were available, as were descriptions of slope failures. For the Chinese slopes and the natural slopes in Norway, existing data was limited and only included slope height and slope angle as well as a rough geological description. However, the natural slopes in Norway are interesting because the geomechanical environment is similar to that of the Aitik Mine.

For the open-pit slopes, slope geometry in the form of maximum stable slope height and/or minimum unstable slope height (for slopes that have failed) as well as interramp and overall slope angle was compiled. The different slope angles are defined in Figure 5.2. The overall slope angle (from slope toe to slope crest, including all ramps) was used for plotting and interpretation, unless otherwise noted. Summary plots of slope

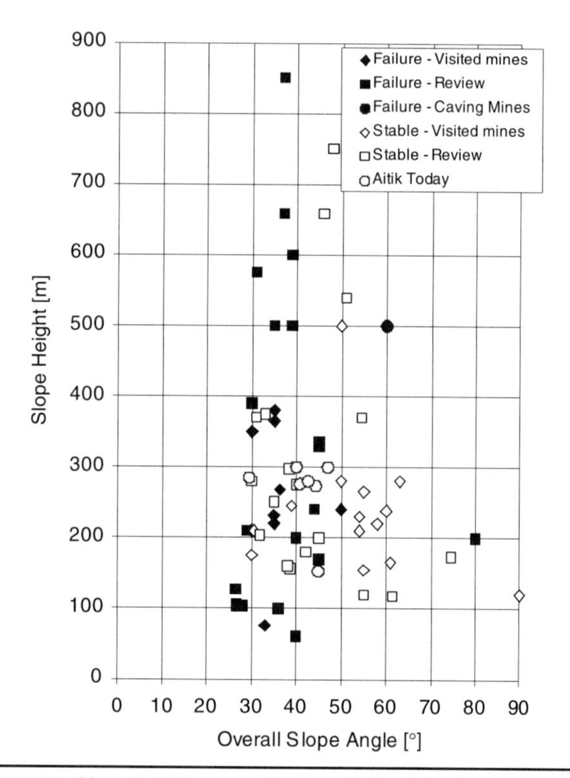

FIGURE 5.3 Slope height versus slope angle for visited and referenced open pits and caving mines

height versus slope angle for each of these categories are presented in Figures 5.3 through 5.6.

The collected data in the present study covers a wide range of slope heights and slope angles. The open pit slopes ranged from 60 to 850 m in height, with overall slope angles varying from 26° to 90°. However, very few of the open-pit slopes had interramp angles exceeding 55°. This is probably due to the difficulties associated with mining such a steep slope (requires careful blasting and safe benches; unsafe, failing benches are too dangerous in steep slopes). Slopes from China displayed slope angles of 8° to 90° and were between 40 and 1,340 m high. For the Norwegian natural slopes, slope heights were in the range of 100 to 1,540 m and slope angles were between 30° and 86°.

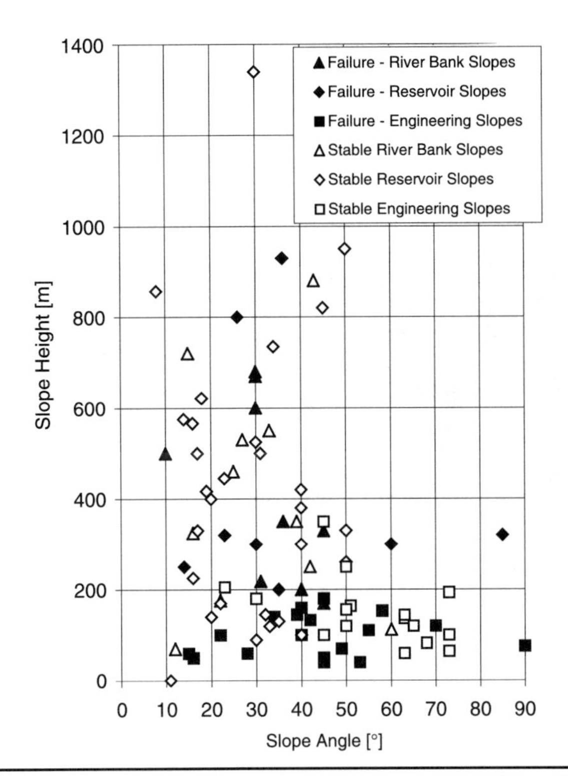

FIGURE 5.4 Slope height versus slope angle for stable and unstable natural and engineering slopes in China (data from Chen 1995a, 1995b)

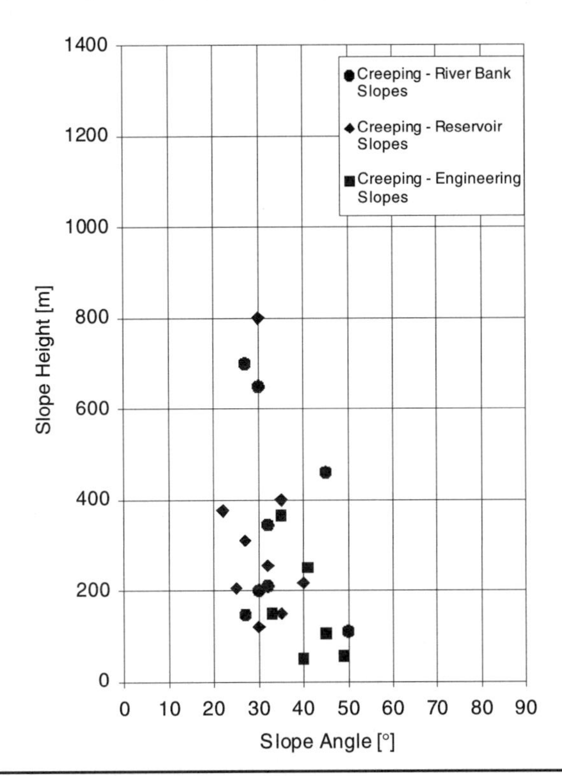

FIGURE 5.5 Slope height versus slope angle for creeping natural and engineering slopes in China (data from Chen 1995a, 1995b)

A few cases of pit slopes included detailed monitoring of slope behavior and/or back-analyzed strength values for large-scale failures. Among the best documented were two of the visited mines—the Aznalcóllar open-pit mine in Spain and the Kiirunavaara sublevel caving mine in Sweden. The Aznalcóllar Mine had

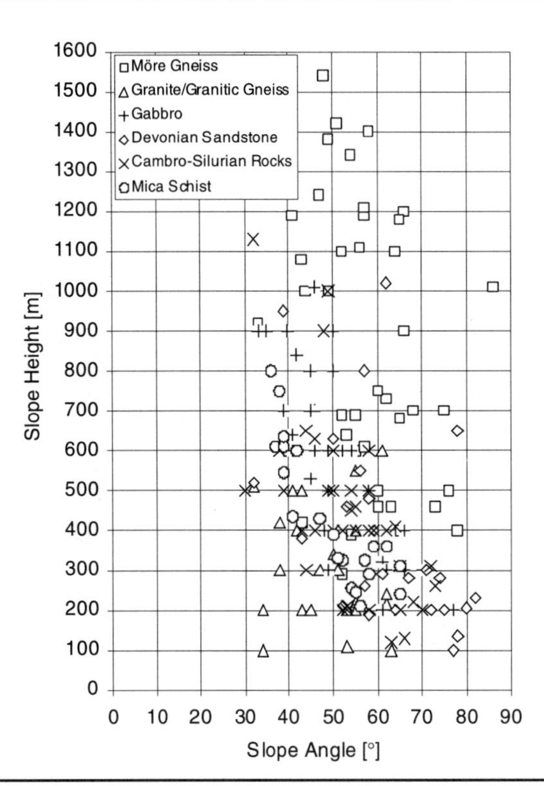

FIGURE 5.6 Slope height versus slope angle for natural slopes in Norway (data from Dahlø 1976; Broch and Nilsen 1977; Nilsen 1977)

exhibited multiple failures, although in relatively weak rock, but monitoring data from the site was of very high quality. The Kiirunavaara Mine is a hard-rock, underground, sublevel caving mine of particular interest since large-scale failures have occurred in the hanging wall and footwall (Dahnér-Lindqvist 1992, 1998). Although associated with a different mining method, these failures share many similarities with failures observed in open pits. In addition, this is one of the few cases known to the author with a nonstructurally controlled large-scale failure in hard and strong rock. One potential uncertainty regards the effects of the caved rock between the footwall and the hanging wall in this "slope." The Aznalcóllar case is presented in detail in Chapter 21. It was further used as a means of verifying rock-mass strength estimates and application of numerical modeling to open-pit slopes. The Kiirunavaara case was also used in this respect (see Sjöberg 1999).

Large-scale slope failure has occurred in approximately half of the studied open pits. The most common failure modes were circular shear failure (not structurally controlled) and large-scale toppling failure (see Figure 5.7). Circular shear failure is a collective term used to denote shear failures in the rock mass displaying a curved failure surface. A few cases of large-scale wedge and plane failures were also noted for cases with preexisting large-scale structural features.

Most of the large-scale failures have been slow and progressive; hence, it has been possible for these mines to "live with failure." Successively deeper open pits are being tested using careful monitoring and follow-up. This approach rests on the assumption that failure would not occur without sufficiently early warning signs (e.g., measurable displacements). However, two cases of rapid and uncontrollable failure were also found. Both of these occurred in relatively steep slopes and involved slip along a large-scale preexisting structure combined with failure through the rock mass at the slope toe. Measured displacements before final collapse were very small (centimeters) in both cases. Consequently, one cannot always assume that failure occurs nonviolently, in particular for steep slopes in hard and brittle rock.

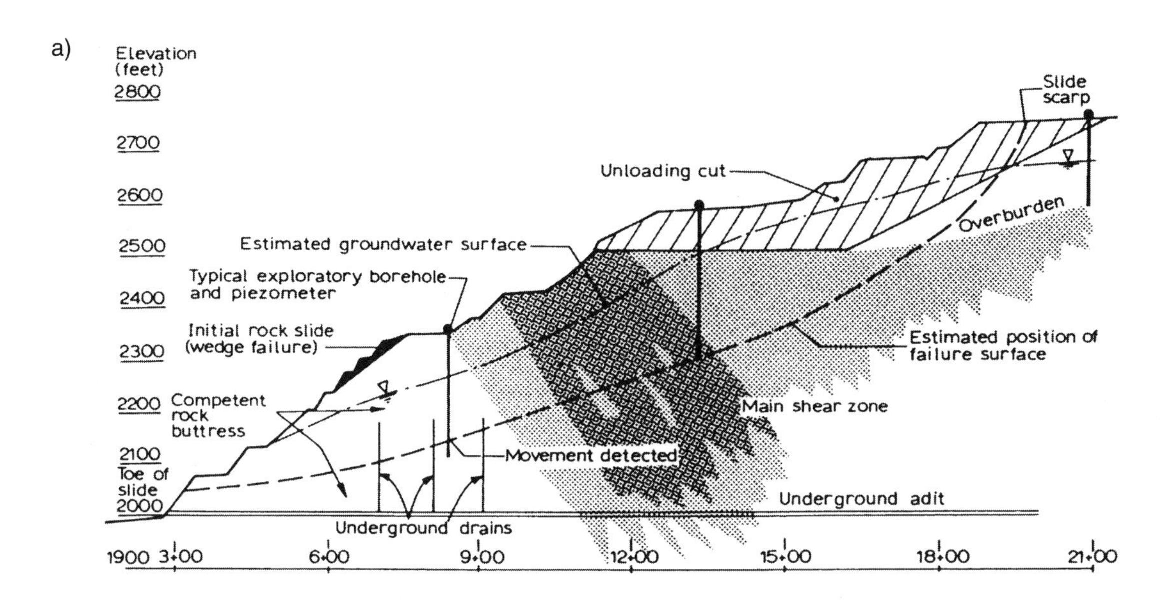

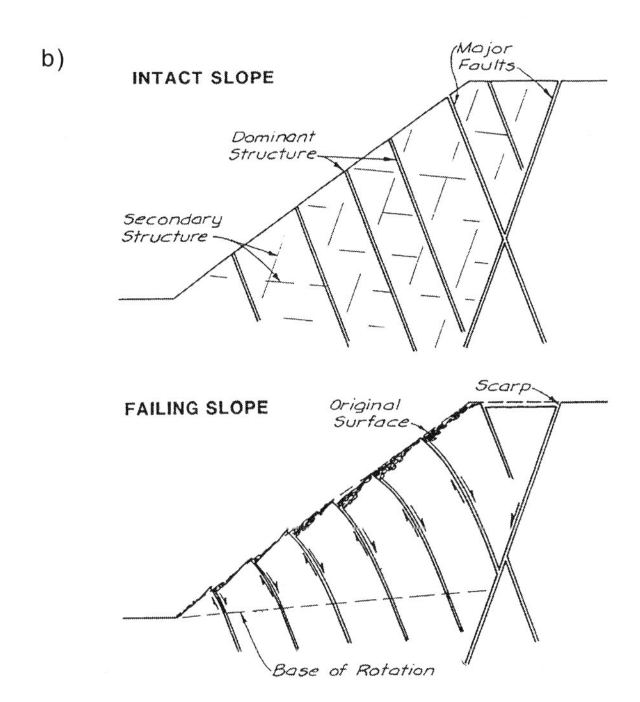

FIGURE 5.7 Example of observed large-scale slope failures in open pits: (a) Circular shear failure in the Jeffrey Mine (Pariseau and Voight 1979) and (b) Large-scale toppling failure in the Lornex Pit (Daly, Muno, and Stacey 1988)

5.2.2 Interpretation of Data

The data collected in the present study did not permit rock-mass classification of all cases. Instead, cases were sorted with respect to intact rock strength (tested and/or estimated), geological similarities, and jointing pattern (information only available for some cases). Since laboratory strength data was not available for all cases, strengths were estimated according to the following limits for ISRM index tests (Brown 1981):

1. R2: Weak rock ($\sigma_c \approx$ 5–25 MPa)
2. R3: Medium strong rock ($\sigma_c \approx$ 25–50 MPa)
3. R4: Strong rock ($\sigma_c \approx$ 50–100 MPa)
4. R5: Very strong rock: ($\sigma_c \approx$ 100–250 MPa)

It should be noted that this refers to the intact rock strength only. The slope angles used were overall slope angles (slope toe to slope crest), including all ramps. For cases that failed, the unstable slope height and the flattest slope angle were used. For stable slopes, the maximum stable height and the steepest slope angle were used. Examples of slope heights versus slope angles for these classes of strength for the open-pit slopes and one sublevel caving slope are shown in Figure 5.8. Similar plots were generated for the natural slope cases. For the natural slopes in Norway, a further distinction could be made with regard to the jointing pattern of the slopes. Data were available on the orientation of the foliation or the dominant joint set relative to the strike of the slope. Five different jointing patterns were noted

- Joints striking parallel to the slope but dipping into the slope
- Joints striking parallel to the slope but dipping out of the slope

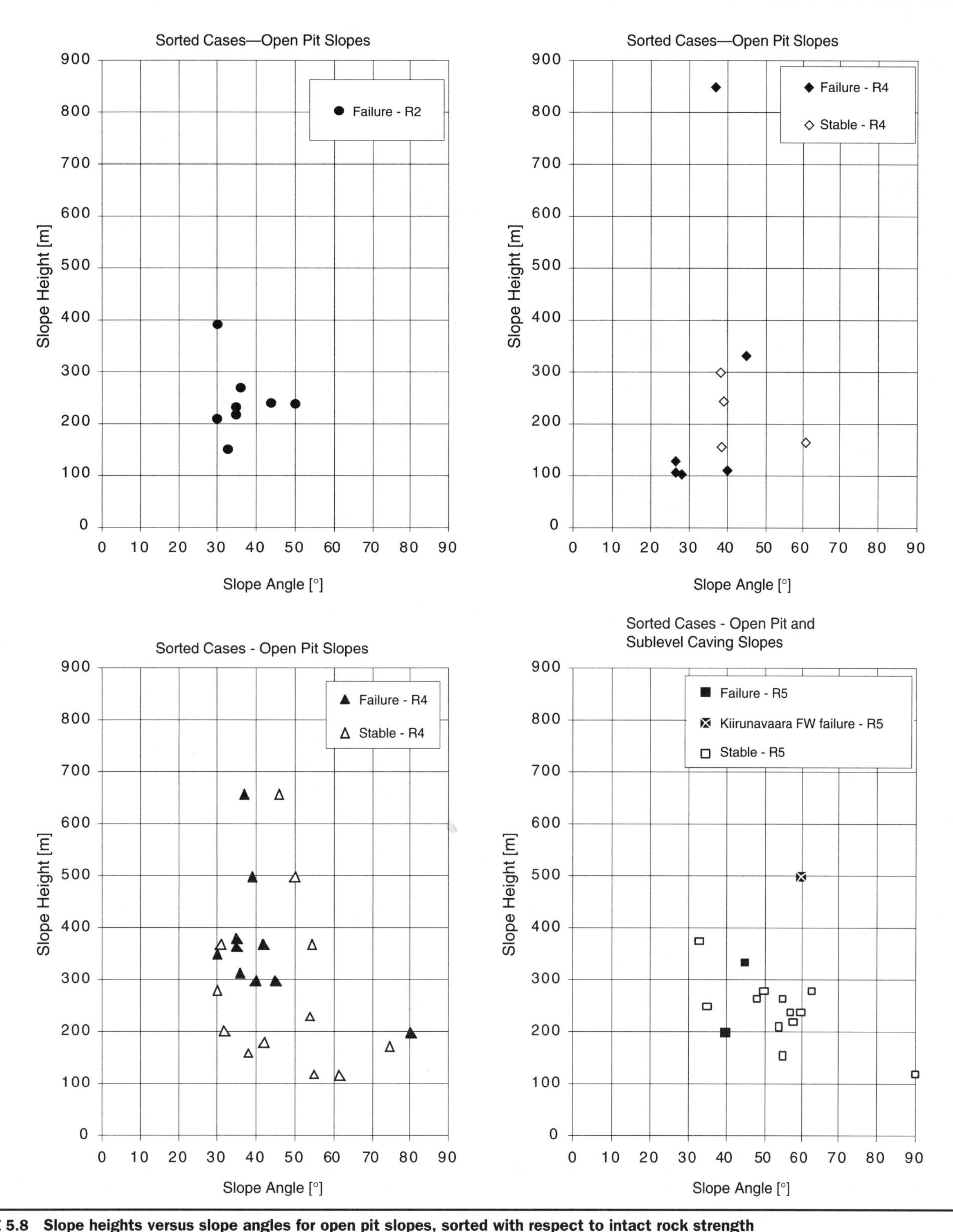

FIGURE 5.8 Slope heights versus slope angles for open pit slopes, sorted with respect to intact rock strength

- Joint striking perpendicular to the slope
- Joints striking oblique to the slope
- Joints oriented horizontally

Example plots of slope heights and slope angles for different rock types, rock strengths, and jointing patterns are shown in Figure 5.9 (for igneous and metamorphic rocks). For the mica schist, data on the dip of the dominant joint sets were also available and are shown in Figure 5.10. From this, some broad trends could be distinguished. Higher and steeper slopes were found for higher rock strengths. However, there were also several cases of failed slopes with flatter slope angles than those of stable slopes (most evident for strength classes R4 and R5 in Figure 5.8). The latter is an indication that factors other than "rock strength" govern failure (and stability) of these slopes. In addition to the obvious, such as pure structural control, this includes, block size, joint

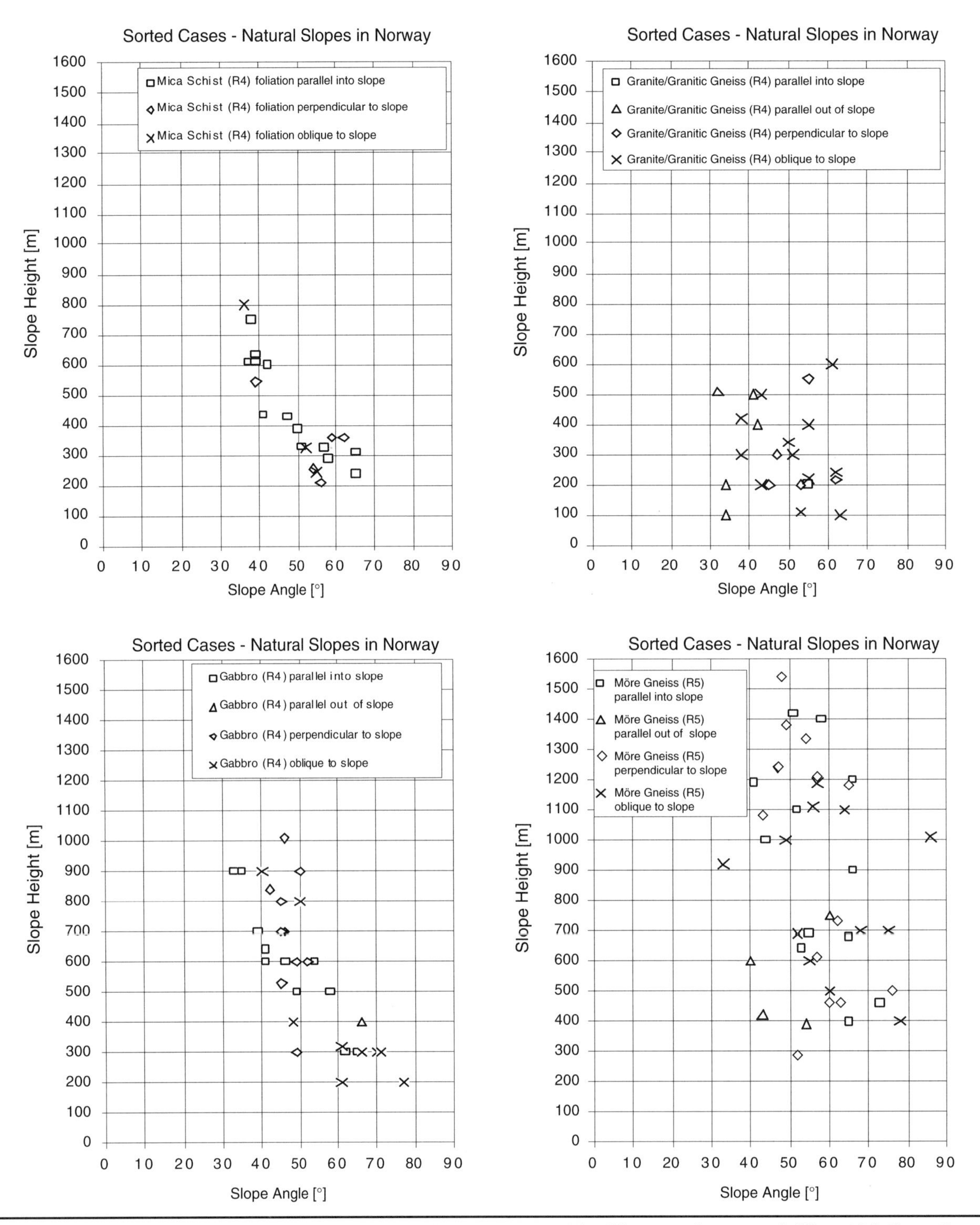

FIGURE 5.9 Slope heights versus slope angles for natural slopes in Norway, plotted for different rock types and different jointing patterns in the slopes

strength versus rock strength, and the ratio between joint lengths and intact rock bridges (in relation to the slope height). Other studies, such as those by Coates (1977, 1981), have suggested that the difference between slopes in different rocks decreases as the slope height increases. Although this appears reasonable, such conclusions cannot be drawn from the currently available material, as there is too little data on higher slopes.

In addition, slope angles were flatter for joints striking parallel to the slope face and dipping out of the slope (high potential for plane shear failure); this is compared to the steeper slopes with joints striking perpendicular to the slope (Figures 5.9 and 5.10). However, even with this sorting, the resulting scatter was large. This indicates that the broad classification used here did not provide enough information to positively identify all the various factors contributing to failure. A more refined sorting process, one involving more quantitative rock-mass classification as well as better descriptions of the failure mechanism, could conceivably give a better resolution.

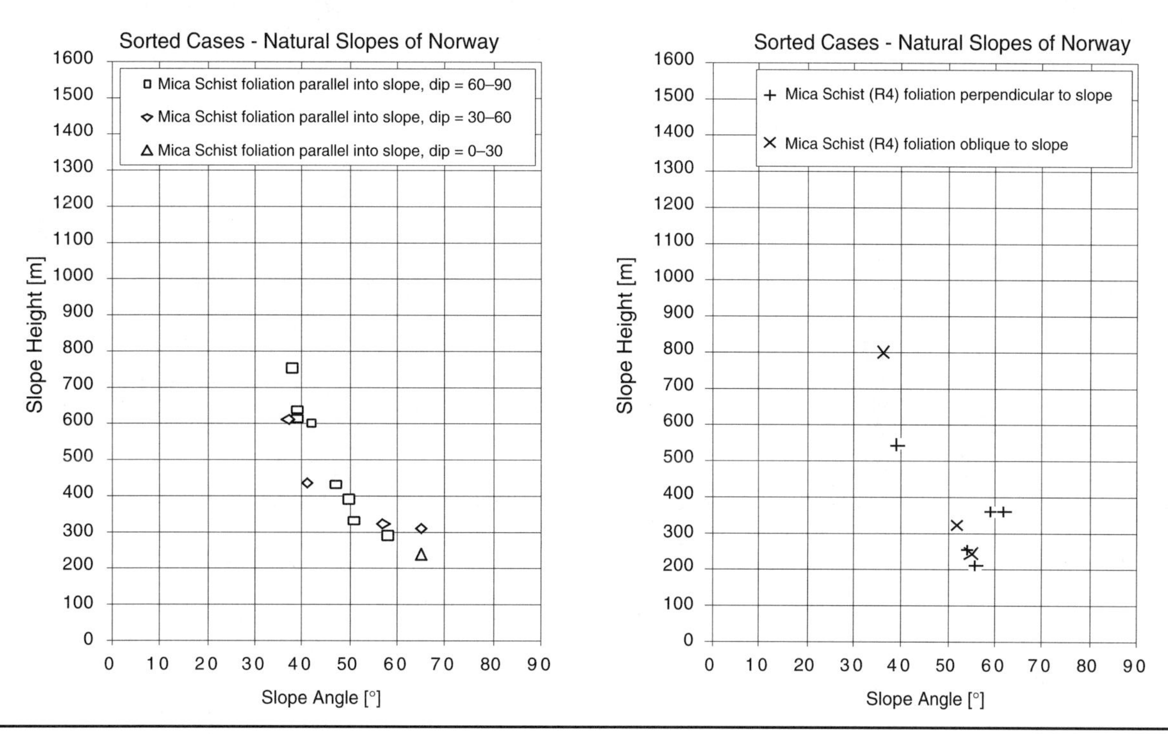

FIGURE 5.10 Slope heights versus slope angles for natural slopes in mica schist and for different orientation of the dominant joint set

5.2.3 Application to Slope Design at Aitik

The collected case studies were used to establish preliminary empirical guidelines for the Aitik pit slopes. The most relevant cases in terms of similarity to Aitik were extracted from the database. The selection of similar cases was based on rock types (igneous and/or metamorphosed, hard rocks) and moderate to high intact rock strength (strength classes R4 to R5). The resulting selection incorporated 17 open-pit slopes, 1 sublevel caving mine slope, and 75 cases of natural slopes in mica schist, gabbro, and granitic gneiss in Norway.

A further distinction could be made with respect to the orientation of the dominant foliation relative to the pit slope. Cases from pit slopes could be divided into footwall and hanging wall environments (foliation dipping out of or into the slope, respectively). Furthermore, the natural slope cases could be sorted with respect to (1) foliation striking parallel to and dipping out of the slope ("footwall"), (2) foliation striking parallel to and dipping into the slope ("hanging wall"), and (3) foliation being oriented perpendicular and/or oblique to the slope face. The last category can tentatively be compared to the corners and end sections of the Aitik pit.

Slope heights versus slope angles for the footwall and hanging wall environments are shown in Figure 5.11. A design curve delineating stable and unstable regions has been sketched into these figures (not curve-fitted). There are few cases of failure for both the footwall and hanging wall environments. Footwall failures are limited to the Kiirunavaara sublevel caving mine and the south wall of the Brenda Mine. The latter was a very rapid failure (as described previously), involving sliding along a large-scale structure and failure through the rock mass at the toe. Thus, it does not resemble the conditions at Aitik and has therefore been given less weight in determining the limiting curve for the footwall. More weight was put on the natural slopes in Norway. These slopes are all stable and can be considered a "best case," since everything affected by "dangerous" structures has already failed.

Hanging wall failures are all large-scale toppling failures in open-pit mines. These have occurred in moderately strong rock masses and in slopes that are flatter than the current hanging wall slope at Aitik. Furthermore, these slopes exhibited pronounced jointing and/or faulting with steeply dipping discontinuities (typically more than 70°), whereas the dominant foliation in the Aitik hanging wall is dipping flatter. Consequently, two design curves have been sketched into Figure 5.11b—one based on toppling failure cases and the other based on the stable natural slopes in Norway.

The derived design curves should be used with caution and only when more detailed information cannot be obtained. However, judging from these, it appears that the current slope angles at Aitik are conservative and that steeper slopes can be considered, even with the planned future increase in slope height. With a final slope height of around 450 to 500 m, it may be possible to have overall slope angles of around 50° to 55° for both the footwall and the hanging wall without jeopardizing the overall slope stability.

5.2.4 Back-Analyzed Rock-Mass-Strength Data

The Hoek–Brown failure criterion. No methods exist to directly determine the large-scale rock-mass strength. Currently, rock-mass-strength estimates rely on the empirical Hoek–Brown failure criterion used in conjunction with rock-mass classification (Hoek and Brown 1988, 1997; Bieniawski 1976). This methodology is well established but its application to slope stability is not well proven. The criterion was originally proposed by Hoek and Brown (1980) and described in detail by Hoek (1983). Since then, it has undergone several minor and major revisions, with the latest version presented in 1997 (Hoek and Brown 1997). The criterion is written

$$\sigma_1 = \sigma_3 + \sigma_c\left(m\frac{\sigma_3}{\sigma_c} + s\right)^a \qquad \textbf{EQ. 5.1}$$

Where:

σ_1 = major effective principal stress
σ_3 = minor effective principal stress
σ_c = uniaxial compressive strength of intact rock
m = constant depending on the characteristics of the rock mass
s = constant depending on the characteristics of the rock mass
a = constant depending on the characteristics of the rock mass

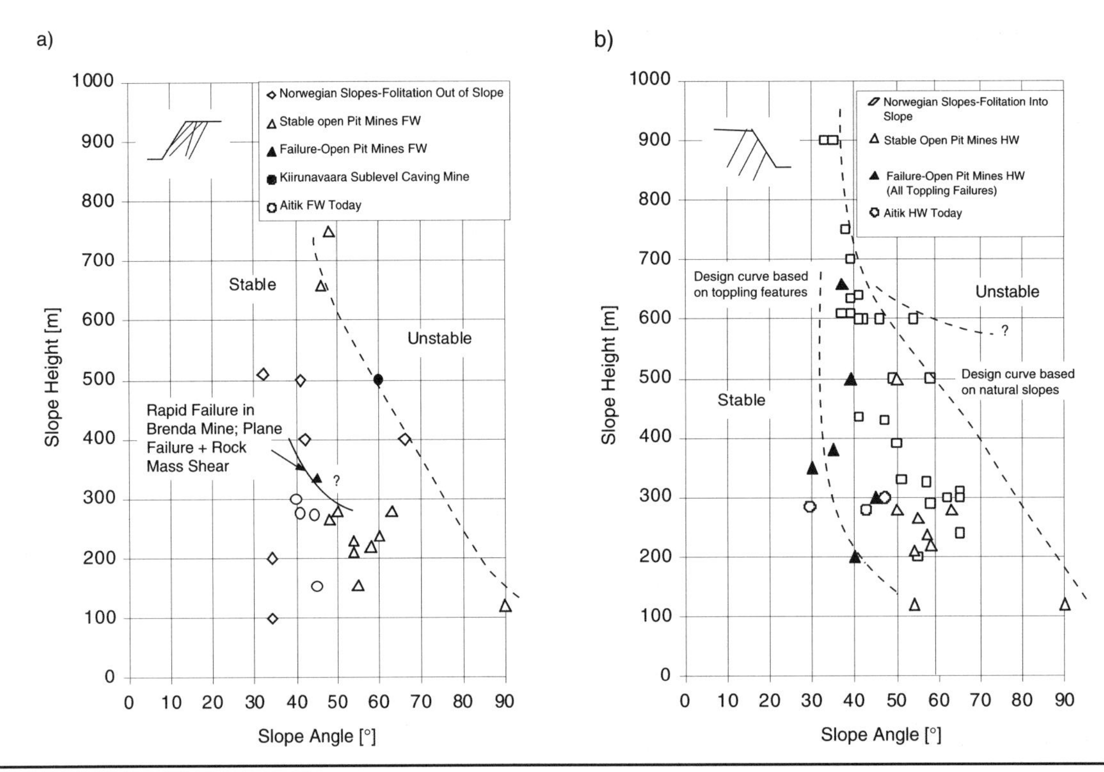

FIGURE 5.11 Slope heights versus slope angles for selected case studies similar to: (a) The Aitik footwall and (b) The Aitik hanging wall

Constants m, s, and a can be determined from rock-mass classification ratings. In early versions of the criterion, the constant a was always set to 0.5. Furthermore, Hoek and Brown (1988) distinguished between undisturbed (interlocking) rock masses and disturbed rock masses when proposing formulas for calculating m and s, as follows:

Undisturbed (or Interlocking) Rock Masses

$$m = m_i e^{\frac{RMR-100}{28}}$$ **EQ. 5.2**

$$s = e^{\frac{RMR-100}{9}}$$ **EQ. 5.3**

Disturbed Rock Masses

$$m = m e^{\frac{RMR-100}{14}}$$ **EQ. 5.4**

$$s = e^{\frac{RMR-100}{6}}$$ **EQ. 5.5**

Where:

m_i = the value of m for the intact rock
RMR = rock-mass rating (Bieniawski 1976)

In later versions, only the category *undisturbed rock masses* was used and *RMR* was replaced by GSI (geological strength index). GSI can, however, be directly assessed using the 1976-year version of *RMR* (Bieniawski 1976). Depending on what version of the criterion one uses, large differences in final strength values are obtained. This applies particularly to the choice of *undisturbed* versus *disturbed* strength values. Unfortunately, the meaning of these categories is not clearly defined in the published works by Hoek and co-workers.

5.2.5 Comparison with Back-Calculated Strengths

For the cases in which slope failures had occurred, the possibility of back-calculating strength values at failure exists, provided sufficient descriptions of the mechanism of failure exist. Back-analyzed strengths were available for a few of the more well-documented cases in the database. This included failures of footwalls and hanging walls in sublevel caving mines in Sweden as well as a few open-pit slope failures at the Aznalcóllar Mine. None of these failures were purely structurally controlled. Rather, they displayed the typical characteristics of a rock-mass shear failure, often with curved, near-circular failure surfaces. Consequently, these cases are interesting to compare with the Hoek–Brown approach to strength estimation for rock masses.

All back-calculated strengths were obtained from limit-equilibrium analyses assuming a linear Mohr–Coulomb failure criterion. This also means that it is assumed that the shear strength is fully mobilized along the entire failure surface at the time of failure (as a consequence of rigid body motion assumed in limit equilibrium analysis). This can be interpreted as the equivalent of residual strength for a fully developed slope failure. As pointed out already by Jennings and Steffen (1967), back-analyzed shear strengths for failures will be a lower limit to the actual strength. In the above cases, there are also some uncertainties regarding the location of the failure surface and the influence of the caved rock in sublevel caving mine slopes (different approaches were used). Overall the back-analyzed failure strengths are approximate and considered representative of residual failure strengths (probably lower limit strength).

The back-calculated strengths were expressed in terms of cohesion and friction angle. To facilitate comparisons, an equivalent compressive strength of the rock mass σ_{cm} was calculated for the Mohr–Coulomb failure envelope

$$\sigma_{cm} = \frac{2c \cdot \cos\phi}{1 - \sin\phi}$$ **EQ. 5.6**

Where:

σ_{cm} = rock-mass compressive strength
c = rock-mass cohesion
ϕ = rock-mass friction angle

TABLE 5.1 Summary of back-calculated strengths from observed rock-mass failures

Mine and Mining Method	Element	Intact Rock Strength, σ_c : low– high (average) (MPa)	Rock-Mass Strength, σ_{cm} : low– high (average) (MPa)	Volume (m^3)	RMR low–high (average)
Sublevel Caving Mines					
Malmberget	Hanging wall	(108)	(2.94)	2.9×10^7	(80)
Malmberget	Hanging wall	180–230 (200)	(1.54)	2.9×10^7	(80)
Grängesberg	Hanging wall	(186)	(1.41)	2.8×10^8	53–67 (60)
Grängesberg	Hanging wall	(186)	(2.08)	2.8×10^8	53–67 (60)
Grängesberg	Hanging wall	(186)	(1.04)	2.8×10^8	53–67 (60)
Grängesberg	Hanging wall	(186)	(2.27)	2.8×10^8	53–67 (60)
Grängesberg	HW - Brewery fault	(186)	(0.29)	—	15
Kiirunavaara	Hanging wall	(186)	(2.05)	1.9×10^8	60–63 (60)
Kiirunavaara	Hanging wall	170–215 (200)	2.65–5.09 (4.10)	1.9×10^8	60–63 (60)
Kiirunavaara	Hanging wall	170–215 (200)	1.39–3.46 (2.36)	1.9×10^8	60–63 (60)
Kiirunavaara	Footwall	140–300 (230)	4.13–7.53 (5.05)	5.5×10^7	(60)
Kiirunavaara	Footwall	140–300 (230)	3.98–5.20 (4.95)	5.5×10^7	(60)
Open Pit Mines					
Aznalcollar	Central footwall	35–50 (25)	0.370–0.510 (0.440)	3.3×10^7	44–58 (51)
Aznalcollar	Central footwall	35–50 (25)	0.134–1.249 (0.692)	3.3×10^7	44–58 (51)
Aznalcollar	Central footwall	35–50 (25)	0.067–1.085 (0.505)	3.3×10^7	44–58 (51)
Aznalcollar	Central footwall	35–50 (25)	0.130–0.486 (0.333)	3.3×10^7	44–58 (51)
Aznalcollar	Eastern footwall	35–50 (25)	0.519–0.943 (0.755)	3.3×10^7	44–58 (51)

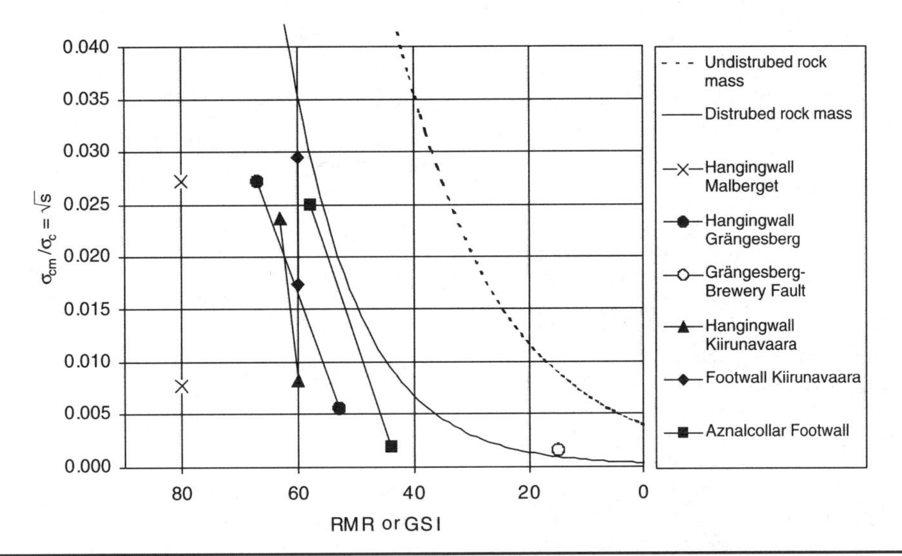

FIGURE 5.12 Comparisons of back-calculated strength values and predicted strengths using the Hoek–Brown criterion for normalized rock-mass strength (with respect to intact rock strength) versus RMR (rock-massrating; Bieniawski 1976)

A summary of the back-calculated strength values and rock-mass classification ratings for these cases is given in Table 5.1. A full description of each of the studied cases (including relevant references) is given in Sjöberg (1999). The rock-mass compressive strength can be normalized with respect to the uniaxial compressive strength of intact rock σ_c. The resulting ratio is expressed (Hoek and Brown 1980)

$$\frac{\sigma_{cm}}{\sigma_c} = \sqrt{s} \quad \textbf{EQ. 5.7}$$

Using Eq. 5.7, it is possible to calculate the value of the parameter s in the Hoek–Brown failure criterion directly from failure strengths (cf. Eq. 5.2 through 5.5). The square root of s can be interpreted as a scale factor for the rock-mass strength. (It must be noted, however, that the parameter s also depends on the jointing of the rock mass.)

The normalized strength values versus RMR rating are shown in Figure 5.12. These values are compared with strength estimates assuming *undisturbed rock mass* (Eq. 5.3) and *disturbed rock mass* (Eq. 5.5) conditions. The scatter of the back-calculated strength values is considerable, but all data are grouped near, or to the left of, the curve defining the category *disturbed rock mass*. Values for *undisturbed rock mass* are significantly higher and would have led to overly optimistic stability predictions for these slopes.

Similar findings have been reported in some of the literature. Pender and Free (1993) calculated the mobilized strength envelopes of stable natural slopes in greywacke and found that the category *undisturbed rock mass* predicted strength values that were significantly higher than the back-calculated strengths. Read and Perrin (1999) reached similar conclusions. Helgstedt (1997) compared predicted strengths with back-calculated values from a dam foundation and a large-scale natural slope, as well as from tests on rockfill, and concluded that category *undisturbed rock mass* consistently predicted shear strengths that were too high. Ulusay and Aksoy (1994) found good agreement between back-calculated and estimated strength values when assuming *disturbed* conditions for an open-pit coal mine.

Taken together, these are strong indicators that the category *disturbed rock masses* is better suited for rock slopes. One explanation is that the physical scale of the rock mass in these cases is very large; thus, it cannot be assumed that the same strength parameters should apply for large-scale rock slopes (with volumes exceeding 1 million m^3) compared to, e.g., pillars or tunnels in underground constructions (where the Hoek–Brown criterion has in most cases been applied). Furthermore, the back-calculated strengths are representative of post-failure conditions, i.e., fully mobilized shear strength and failure fully developed. As such, they represent a lower limit to the rock-mass strength, which makes the *disturbed rock mass* parameter values more reasonable. The resulting strength values are probably conservative, particularly for brittle materials, but are, nevertheless, useful for first estimates of the rock-mass strength.

These findings have been used as part of a design methodology developed for large-scale rock slopes. Rock-mass strength estimates using the Hoek–Brown approach and assuming *disturbed rock masses* were used as input to numerical modeling. The methodology was applied to the Aznalcóllar Mine (see Chapter 21) and the Aitik Mine (see Chapter 22).

5.3 SUMMARY AND CONCLUSIONS

The case study database presented here is a significant update of previously available case study information. The database primarily includes information on slope heights and slope angles for stable and failed rock slopes. From this, preliminary guidelines for the design of the Aitik hard-rock mine in northern Sweden were derived. Similar guidelines could be developed for other geological environments using the information in the database. This was outside the scope of this work; however, the database information is available from the author for others to pursue this work.

The database information was also used to compare back-analyzed rock-mass strengths to estimated strengths using the Hoek–Brown failure criterion and rock-mass classification. The results showed that for large-scale rock slopes (volumes typically exceeding 1 million m^3) the resulting rock-mass strengths are very low. Hence, it is necessary to use the *disturbed rock mass* category in the Hoek–Brown criterion to assess strengths with some certainty. These strengths are, however, representative of post-failure conditions with fully developed failure. They are probably conservative but can still be used as first estimates of input data to design.

This study showed that, even after sorting cases with respect to rock strength, geology, and jointing pattern, the resulting scatter was large. This indicates that not all controlling factors with respect to overall slope stability were adequately included. It is necessary to sort and classify the different rock slopes better; however, this also requires additional data collection. It is envisioned that large amounts of data are available at each mine site, but significant joint efforts are required to pull this information together in a useful form. More work on quantifying governing factors and failure mechanisms is also necessary. Furthermore, the number of slope failures is limited, in particular in hard-rock environments. To be able to provide better predictions, it is necessary to have more cases of observed failures. Only time will tell if such cases will arise (i.e., if failure will occur).

5.4 ACKNOWLEDGMENTS

The majority of the work presented in this chapter is the result of a four-year joint research project between Boliden Mineral AB and the Division of Rock Mechanics, Luleå University of Technology. The research project was sponsored exclusively by Boliden Mineral AB, which is gratefully acknowledged. The author is also indebted to the project supervision group for their great interest and fruitful discussions. This group consisted of Dr. Erling Nordlund, Luleå University of Technology; Mr. Norbert Krauland, formerly at Boliden Mineral AB; and Professor William Hustrulid, University of Utah (Professor Hustrulid was Head of Mining Research and Development at LKAB, Kiruna, during the main portion of the project). Additional work has been conducted recently as part of ongoing research and development within Boliden Mineral AB. The support for this through corporate research and development funding is gratefully acknowledged. Finally, mine staff at the visited properties and consultants working on those properties have been tremendously helpful in sharing information to this study (none mentioned, none forgotten).

5.5 REFERENCES

Bieniawski, Z.T. 1976. Rock mass classification in rock engineering. *Proceedings Exploration for Rock Engineering*, Johannesburg. 97–106.

Broch, E., and B. Nilsen. 1977. Undersøkelser av stabilitetsforholdene for Ørtfjell dagbrudd i Rana (Norge). *Proceedings Bergmekanikdagen 1977 – Papers presented at Rock Mechanics meeting*, Stockholm, Sweden. 19–36 (in Norwegian).

Brown, E.T., ed. 1981. *Rock characterization testing & monitoring. ISRM Suggested Methods*. Oxford: Pergamon Press.

Chen, Z. 1995a. Keynote lecture: Recent developments in slope stability analysis. *Proceedings 8th International Congress on Rock Mechanics*, Tokyo. 3:1041–1048.

Chen, Z., editor-in-chief. 1995b. *Transaction of the stability analyses and software for steep slopes in China. Volume 3:1: Rock classification, statistics and database of failed and stable slopes.* China Institute of Water Resources and Hydroelectric Power Research (in Chinese).

Coates, D.F. 1977. *Pit slope manual Chapter 5–Design*. CANMET (Canada Centre for Mineral and Energy Technology), CANMET REPORT 77-5.

Coates, D.F. 1981. *Rock mechanics principles*. CANMET, Energy, Mines and Resources Canada, Monograph 874. pp. 6-1 to 6-70; B-1 to B-12.

Dahlø, T.S. 1976. *Empirisk analyse av stabilitet i naturlige fjellskråninger*. Examensarbeid, Norges tekniske høgskole, Trondheim (in Norwegian).

Dahlø, T.S. 1996. Personal communication.

Dahnér-Lindqvist, C. 1992. Liggväggsstabiliteten i Kiirunavaara. *Proceedings Bergmekanikdagen 1992–Papers presented at Rock Mechanics meeting*, Stockholm, Sweden. 37–52 (in Swedish).

Dahnér-Lindqvist, C. 1998. Personal communication.

Daly, S., W.K. Munro, and P.F. Stacey. 1988. Slope stability studies for the Lornex Pit. *12th CIM District Six Meeting*, Fernie, Sept. 28 to Oct. 1, 1988.

Duran, A., and K. Douglas. 1999. Do slopes designed with empirical rock mass strength criteria stand up? *Proceedings 9th International Congress on Rock Mechanics*, Paris. 1:87–90.

Haines, A, and P.J. Terbrugge. 1991. Preliminary estimation of rock slope stability using rock mass classification systems. *Proceedings 7th International Congress on Rock Mechanics*, Aachen. 2:887–892.

Helgstedt, M.D. 1997. *An assessment of the in-situ shear strength of rock masses and discontinuities*. Master of Science Thesis 1997:178 CIV, Division of Rock Mechanics, Luleå University of Technology.

Hoek, E. 1983. Strength of jointed rock masses. *Géotechnique*. 33:3:187–223.

Hoek, E., and J.W. Bray. 1981. *Rock slope engineering*. London:Institution of Mining and Metallurgy.

Hoek, E., and E.T. Brown. 1980. *Underground excavations in rock*. London:Institution of Mining and Metallurgy.

Hoek, E., and E.T. Brown. 1988. The Hoek–Brown failure criterion–A 1988 update. *Proceedings 15th Canadian Rock Mechanics Symposium*, Toronto. 31–38.

Hoek, E., and E.T. Brown. 1997. Practical estimates of rock mass strength. *Intl. Jour. Rock Mech. Min. Sci.*, 34:8:1165–1186.

Jennings, J.E., and O.K.H. Steffen. 1967. The analysis of the stability of slopes in deep opencast mines. *Trans. S. Afr. Inst. Civil Eng*. 9:3:41–54.

Laubscher, D.H. 1977. Geomechanics classification of jointed rock masses–Mining applications. *Trans. Inst. Min. Metall. (Sect. A: Min. Industry)*. 86:A1–A8.

Lutton, R.J. 1970. Rock slope chart from empirical design. *Trans. Society of Mining Engineers*. 247:160–162.

Nilsen, B. 1979. *Stabilitet av höye fjellskjæringer*. The University of Trondheim, The Norwegian Institute of Technology, Reports from the Department of Geology (in Norwegian).

Pariseau, W.G., and B. Voight. 1979. Rockslides and avalanches: Basic principles and perspectives in the realm of civil and mining operations. In *Rockslides and avalanches. 2. Engineering sites*, ed. B. Voight. Amsterdam: Elsevier Publishing Company.

Pender, M.J., and M.W. Free. 1993. Stability assessment of slopes in closely jointed rock masses. *Proceedings Safety and Environmental Issues in Rock Engineering, Eurock '93*, Lisbon, Portugal. pp. 863–870.

Read, S.A.L., and N.D. Perrin. 1999. Applicability of the Hoek–Brown failure criterion to New Zealand greywacke rocks. *Proceedings 9th International Congress on Rock Mechanics*, Paris. 2:655–660.

Sjöberg, J. 1999. *Analysis of large scale rock slopes*. Doctoral thesis 1999:01, Division of Rock Mechanics, Luleå University of Technology.

Ulusay, R., and H. Aksoy. 1994. Assessment of the failure mechanism of a highwall slope under spoil pile loadings at a coal mine. *Engineering Geology*. 38:117–134.

CHAPTER 6

Rock-Mass Properties for Surface Mines

Evert Hoek* and Antonio Karzulovic†

6.1 INTRODUCTION

Reliable estimates of rock-mass strength and deformation characteristics are required for almost any analysis used to design surface excavations. Hoek and Brown (1980a, 1980b) proposed a method for obtaining estimates of the strength of jointed rock masses based on an assessment of the interlocking of rock blocks and the condition of the surfaces between these blocks. This method was modified over the years in order to meet the needs of users who applied it to problems that were not considered when the original criterion was developed (Hoek 1983, Hoek and Brown 1988). Applying the method to poor-quality rock masses required further changes (Hoek, Wood, and Shah 1992) and, eventually, led to the development of a new classification called the geological strength index (GSI) (Hoek 1994; Hoek, Kaiser, and Bawden 1995; Hoek and Brown 1997; Hoek, Marinos, and Benissi 1998). A review of the development of the criterion and of the equations proposed at various stages in this development is given in Hoek and Brown (1997).

This chapter presents the Hoek–Brown criterion in a form that is practical in the field and that appears to provide the most reliable set of results for use as input for methods of analysis currently used in rock engineering.

For surface excavations, the rock-mass properties are particularly sensitive to stress relief and blast damage and these two factors are discussed in his chapter.

6.2 GENERALIZED HOEK–BROWN CRITERION

The generalized Hoek–Brown failure criterion for jointed rock masses is defined by

$$\sigma'_1 = \sigma'_3 + \sigma_{ci}\left(m_b \frac{\sigma'_3}{\sigma_{ci}} + s\right)^a \qquad \textbf{EQ. 6.1}$$

Where:

σ'_1 and σ'_3 = the maximum and minimum effective stresses at failure

m_b = the value of the Hoek–Brown constant m for the rock mass

s and a = constants that depend on the rock-mass characteristics

σ_{ci} = the uniaxial compressive strength of the intact rock pieces

The Mohr envelope, relating normal and shear stresses, can be determined by the method proposed by Hoek and Brown (1980a). In this approach, Eq. 6.1 is used to generate a series of triaxial test values, simulating full-scale field tests, and a statistical curve fitting process is used to derive an equivalent Mohr envelope defined by the equation

$$\tau = A\sigma_{ci}\left(\frac{\sigma'_n - \sigma_{tm}}{\sigma_{ci}}\right)^B \qquad \textbf{EQ. 6.2}$$

Where:

A and B = material constants

σ'_n = the normal effective stress

σ_{tm} = the "tensile" strength of the rock mass

In order to use the Hoek–Brown criterion for estimating the strength and deformability of jointed rock masses, three "properties" of the rock mass have to be estimated. These are

- The uniaxial compressive strength σ_{ci} of the intact rock elements
- The value of the Hoek–Brown constant m_i for these intact rock elements
- The value of the GSI for the rock mass

6.3 INTACT ROCK PROPERTIES

For the intact rock pieces that make up the rock mass, Eq. 6.1 simplifies to

$$\sigma'_1 = \sigma'_3 + \sigma_{ci}\left(m_i \frac{\sigma'_3}{\sigma_{ci}} + 1\right)^{0.5} \qquad \textbf{EQ. 6.3}$$

The relationship between the principal stresses at failure for a given rock is defined by two constants: the uniaxial compressive strength σ_{ci} and a constant m_i. Wherever possible, the values of these constants should be determined by statistical analysis of the results of a set of triaxial tests on carefully prepared core samples. When laboratory tests are not possible, Table 6.1 and Table 6.2 can be used to obtain estimates of σ_{ci} and m_i.

In the case of mineralized rocks, the effects of alteration can have a significant impact on the properties of the intact rock components, and this should be taken into account when estimating the values of σ_{ci} and m_i. For example, the influence of quartz-seritic alteration of andesite and porphyry is illustrated in Figure 6.1. Similar trends have been observed for other forms of alteration and, where this type of effect is considered likely, the geotechnical engineer would be well advised to invest in a program of laboratory testing to establish the appropriate properties for the intact rock.

* Consulting Engineer, Vancouver, British Columbia, Canada.
† Department of Mining Engineering, University of Chile, Santiago, Chile.

TABLE 6.1 Field estimates of uniaxial compressive strength

Grade*	Term	Uniaxial Comp. Strength (MPa)	Point Load Index (MPa)	Field Estimate of Strength	Examples
R6	Extremely strong	>250	>10	Specimen can only be chipped with a geological hammer	Fresh basalt, chert, diabase, gneiss, granite, quartzite
R5	Very strong	100–250	4–10	Specimen requires many blows of a geological hammer to fracture it	Amphibolite, sandstone, basalt, gabbro, gneiss, granodiorite, peridotite, rhyolite, tuff
R4	Strong	50–100	2–4	Specimen requires more than one blow of a geological hammer to fracture it	Limestone, marble, sandstone, schist
R3	Medium strong	25–50	1–2	Cannot be scraped or peeled with a pocket knife, specimen can be fractured with a single blow from a geological hammer	Concete, phyllite, schist, siltstone
R2	Weak	5–25	†	Can be peeled with a pocket knife with difficulty, shallow indentation made by firm blow with point of a geological hammer	Chalk, claystone, potash, marl, siltstone, shale, rock salt
R1	Very weak	1–5	†	Crumbles under firm blows with point of a geological hammer, can be peeled by a pocket knife	Highly weathered or altered rock, shale
R0	Extremely weak	0.25–1	†	Indented by thumbnail	Stiff fault gouge

* Grade according to Brown (1981).
† Point load tests on rocks with a uniaxial compressive strength below 25 MPa are likely to yield highly ambiguous results.

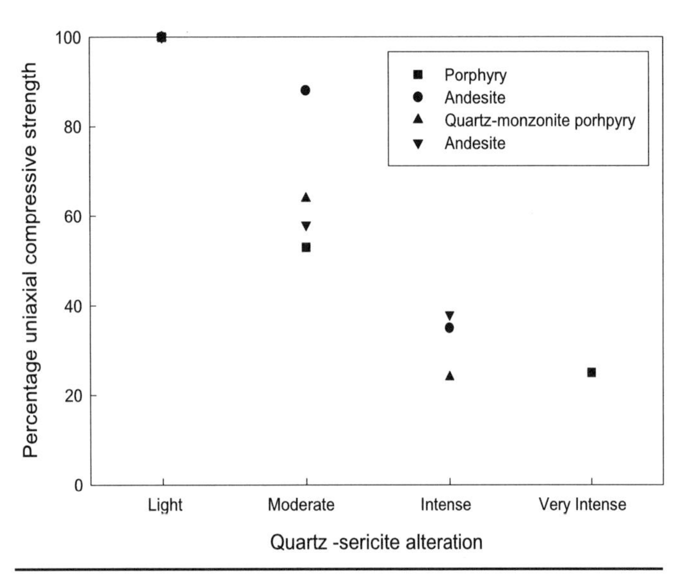

FIGURE 6.1 Influence of quartz-seritic alteration on the uniaxial compressive strength of "intact" specimens of andesite and porphyry

The Hoek–Brown failure criterion, which assumes isotropic rock and rock-mass behavior, should only be applied to those rock masses in which there are a sufficient number of closely spaced discontinuities, with similar surface characteristics, that isotropic behavior involving failure on multiple discontinuities can be assumed. When the structure being analyzed is large and the block size small in comparison, the rock mass can be treated as a Hoek–Brown material.

Where the block size is of the same order as that of the structure being analyzed or when one of the discontinuity sets is significantly weaker than the others, the Hoek–Brown criterion should not be used. In these cases, the stability of the structure should be analyzed by considering failure mechanisms involving the sliding or rotation of blocks and wedges defined by intersecting structural features. Figure 6.2 summarizes these statements in a graphical form.

6.4 GEOLOGICAL STRENGTH INDEX

The strength of a jointed rock mass depends on the properties of the intact rock pieces and on the freedom of these pieces to slide and rotate under different stress conditions. This freedom is controlled by the geometrical shape of the intact rock pieces as well as the condition of the surfaces separating the pieces. Angular rock pieces with clean, rough discontinuity surfaces will result in a much stronger rock mass than one that contains rounded particles surrounded by weathered and altered material.

The GSI, introduced by Hoek (1994) and Hoek, Kaiser, and Bawden (1995), provides a system for estimating the reduction in rock-mass strength for different geological conditions. This system is presented in Table 6.3 for blocky rock masses and Table 6.4 for schistose metamorphic rocks.

Once the GSI has been estimated, the parameters that describe the rock-mass-strength characteristics are calculated as follows:

$$m_b = m_i \exp\left(\frac{GSI - 100}{28}\right) \quad \textbf{EQ. 6.4}$$

For GSI > 25, i.e., rock masses of good to reasonable quality

$$s = \exp\left(\frac{GSI - 100}{9}\right) \quad \textbf{EQ. 6.5}$$

and

$$a = 0.5 \quad \textbf{EQ. 6.6}$$

For GSI < 25, i.e., rock masses of very poor quality

$$s = 0 \quad \textbf{EQ. 6.7}$$

and

$$a = 0.65 - \frac{GSI}{200} \quad \textbf{EQ. 6.8}$$

For better-quality rock masses (GSI > 25), the GSI value can be estimated directly from the 1976 version of Bieniawski's rock-mass rating (RMR), with the groundwater rating set to 10 (dry)

TABLE 6.2 Values of the constant m_i for intact rock, by rock group

Rock Type	Class	Group	Texture: Coarse	Texture: Medium	Texture: Fine	Texture: Very fine
SEDIMENTARY	Clastic		Conglomerates (21 ± 3) Breccias (19 ± 5)	Sandstones 17 ± 4	Siltstones 7 ± 2 Greywackes (18 ± 3)	Claystones 4 ± 2 Shales (6 ± 2) Marls (7 ± 2)
SEDIMENTARY	Non-Clastic	Carbonates	Crystalline Limestone (12 ± 3)	Sparitic Limestones (10 ± 2)	Micritic Limestones (9 ± 2)	Dolomites (9 ± 3)
SEDIMENTARY	Non-Clastic	Evaporites		Gypsum 8 ± 2	Anhydrite 12 ± 2	
SEDIMENTARY	Non-Clastic	Organic				Chalk 7 ± 2
METAMORPHIC	Nonfoliated		Marble 9 ± 3	Hornfels (19 ± 4) Metasandstone (19 ± 3)	Quartzites 20 ± 3	
METAMORPHIC	Slightly foliated		Migmatite (29 ± 3)	Amphibolites 26 ± 6		
METAMORPHIC	Foliated*		Gneiss 28 ± 5	Schists 12 ± 3	Phyllites (7 ± 3)	Slates 7 ± 4
IGNEOUS	Plutonic	Light	Granite 32 ± 3	Diorite 25 ± 5		
IGNEOUS	Plutonic	Light	Granodiorite (29 ± 3)			
IGNEOUS	Plutonic	Dark	Gabbro 27 ± 3 Norite 20 ± 5	Dolerite (16 ± 5)		
IGNEOUS	Hypabyssal		Porphyries (20 ± 5)		Diabase (15 ± 5)	Peridotite (25 ± 5)
IGNEOUS	Volcanic	Lava		Rhyolite (25 ± 5) Andesite 25 ± 5	Dacite (25 ± 3) Basalt (25 ± 5)	Obsidian (19 ± 3)
IGNEOUS	Volcanic	Pyroclastic	Agglomerate (19 ± 3)	Breccia (19 ± 5)	Tuff (13 ± 5)	

Note that values in parenthesis are estimates.
* These values are for intact rock specimens tested normal to bedding or foliation. The value of m_i will be significantly different if failure occurs along a weakness plane.

and the adjustment for joint orientation set to 0 (very favorable) (Bieniawski 1976). For very poor-quality rock masses, the RMR value is very difficult to estimate, and the balance between the ratings no longer gives a reliable basis for estimating rock-mass strength. Consequently, Bieniawski's RMR classification should not be used for estimating the GSI values for poor-quality rock masses (RMR < 25) and the GSI charts should be used directly.

If the 1989 version of Bieniawski's RMR classification (Bieniawski 1989) is used, then GSI = $RMR_{89}' - 5$ where RMR_{89}' has the groundwater rating set to 15 and the adjustment for joint orientation set to zero.

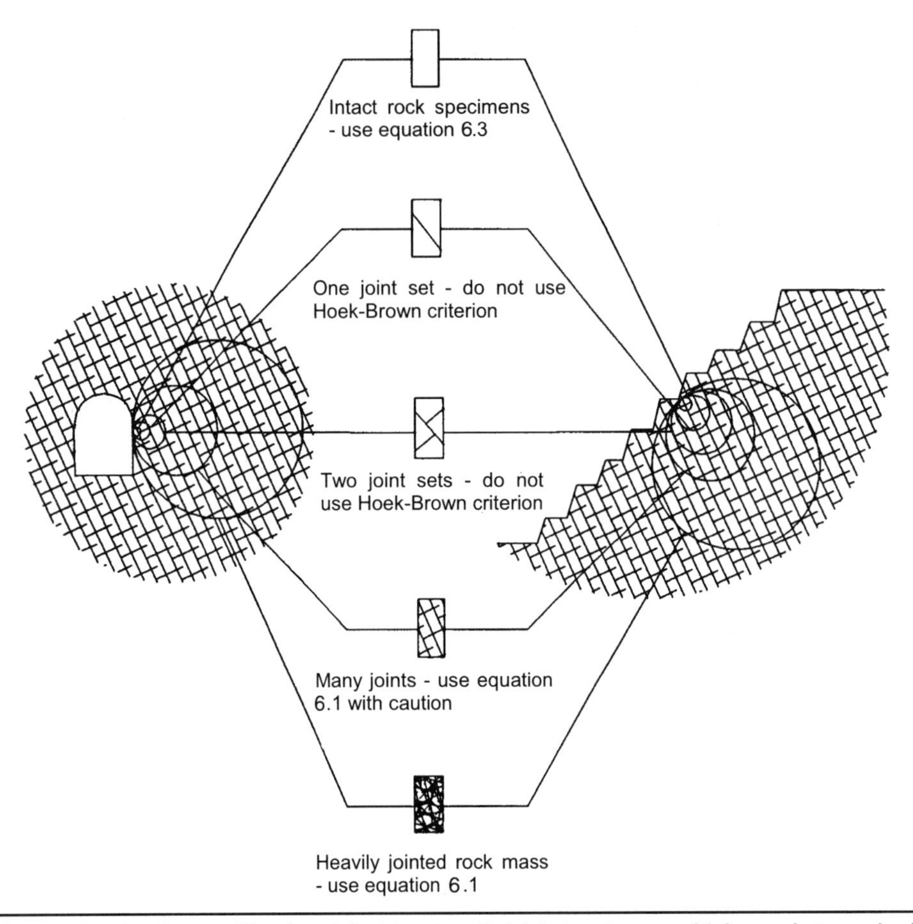

FIGURE 6.2 Idealised diagram showing the transition from intact to a heavily jointed rock mass with increasing sample size

6.5 MOHR–COULOMB PARAMETERS

Most geotechnical software is written in terms of the Mohr–Coulomb failure criterion, in which the rock-mass strength is defined by the cohesive strength c' and the angle of friction ϕ'. The linear relationship between the major and minor principal stresses, σ'_1 and σ'_3, for the Mohr–Coulomb criterion is

$$\sigma'_1 = \sigma_{cm} + k\sigma'_3 \qquad \textbf{EQ. 6.9}$$

Where:

σ_{cm} = the uniaxial compressive strength of the rock mass
k = the slope of the line relating σ'_1 and σ'_3

The values of ϕ' and c' can be calculated from

$$\sin\phi' = \frac{k-1}{k+1} \qquad \textbf{EQ. 6.10}$$

$$c' = \frac{\sigma_{cm}(1-\sin\phi')}{2\cos\phi'} \qquad \textbf{EQ. 6.11}$$

There is no direct correlation between Eq. 6.9 and the nonlinear Hoek–Brown criterion defined by Eq. 6.1. Consequently, determination of the values of c' and ϕ' for a rock mass that has been evaluated as a Hoek–Brown material is difficult.

Having considered a number of possible approaches, it has been concluded that the most practical solution is to treat the problem by analyzing a set of full-scale triaxial strength tests. The results of such tests are simulated by using the Hoek–Brown Eq. 6.1 to generate a series of triaxial test values. Equation 6.9 is then fitted to these test results by linear regression analysis, and the values of c' and ϕ' are determined from Eq. 6.11 and 6.10. A full discussion on the steps required to carry out this analysis is presented in the appendix, together with a spreadsheet for implementing this analysis.

The range of stresses used in the curve-fitting process described above is very important. For the confined conditions surrounding tunnels at depths of more than about 30 m, the most reliable estimates are given by using a confining stress range from zero to 0.25 σ_{ci}, where σ_{ci} is the uniaxial compressive strength of the intact rock elements. For this stress range, the uniaxial compressive strength of the rock mass σ_{cm}, the cohesive strength c, and the friction angle ϕ are given in Figures 6.3 and 6.4.

For slopes and shallow excavations, the user is given the choice of the stress range for this curve-fitting process. This is discussed in full in the appendix.

6.6 DEFORMATION MODULUS

Serafim and Pereira (1983) proposed a relationship between the in situ modulus of deformation and Bieniawski's RMR classification. This relationship is based on back-analysis of dam foundation deformations and it has been found to work well for better-quality rocks. However, for many poor-quality rocks it appears to predict deformation modulus values that are too high. Based on practical observations and back-analysis of excavation behavior in poor-quality rock masses, the following modification to Serafim and Pereira's equation is proposed for $\sigma_{ci} > 100$:

$$E_m = \sqrt{\frac{\sigma_{ci}}{100}}\,10^{\left(\frac{GSI-10}{40}\right)} \qquad \textbf{EQ. 6.12}$$

Note that GSI has been substituted for RMR in this equation and that the modulus E_m is reduced progressively as the value of σ_{ci}

TABLE 6.3 Characterization of blocky rock masses based on the particle interlocking and discontinuity condition

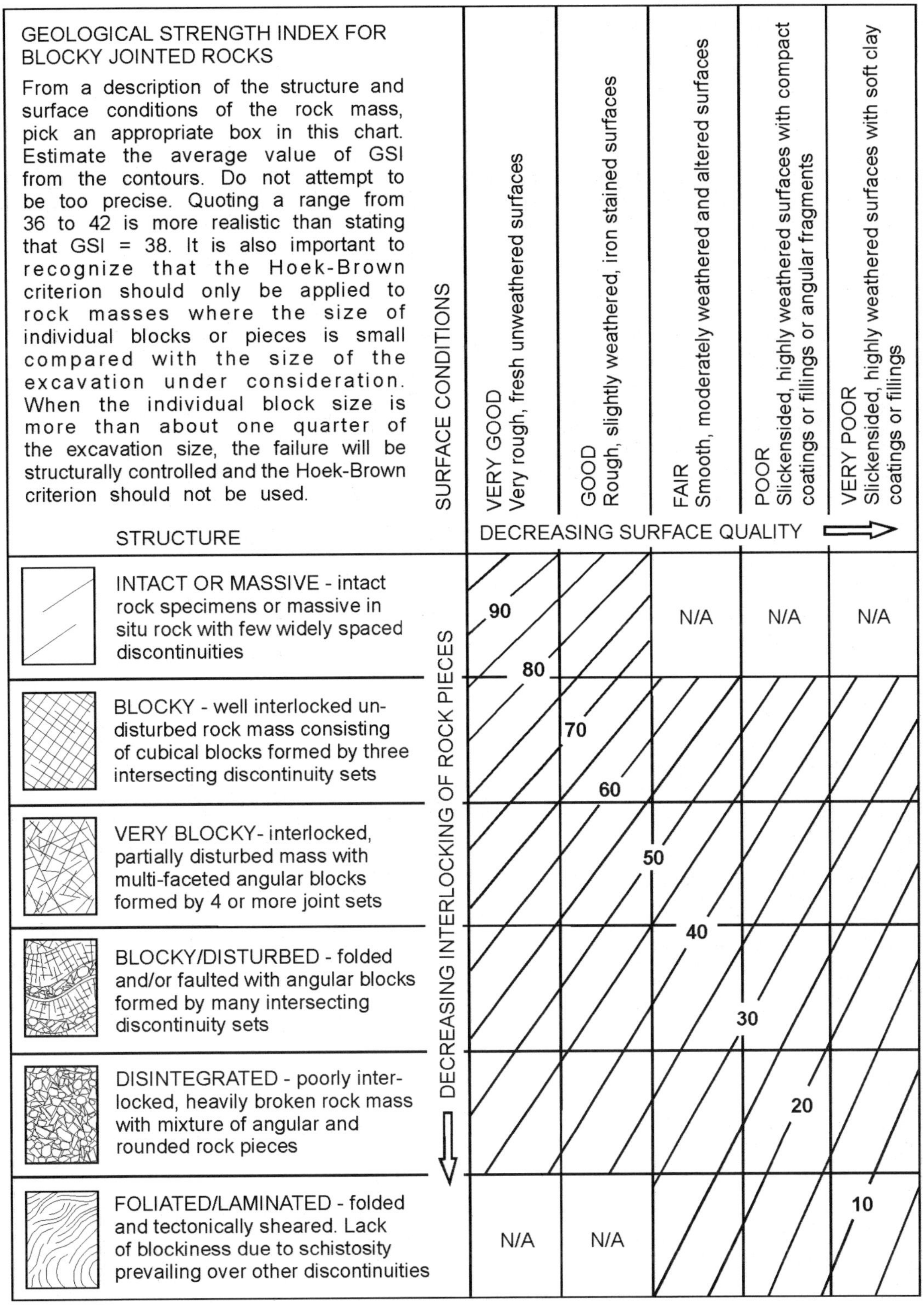

GEOLOGICAL STRENGTH INDEX FOR BLOCKY JOINTED ROCKS

From a description of the structure and surface conditions of the rock mass, pick an appropriate box in this chart. Estimate the average value of GSI from the contours. Do not attempt to be too precise. Quoting a range from 36 to 42 is more realistic than stating that GSI = 38. It is also important to recognize that the Hoek-Brown criterion should only be applied to rock masses where the size of individual blocks or pieces is small compared with the size of the excavation under consideration. When the individual block size is more than about one quarter of the excavation size, the failure will be structurally controlled and the Hoek-Brown criterion should not be used.

SURFACE CONDITIONS — DECREASING SURFACE QUALITY ⇒

STRUCTURE — DECREASING INTERLOCKING OF ROCK PIECES ⇓

STRUCTURE	VERY GOOD Very rough, fresh unweathered surfaces	GOOD Rough, slightly weathered, iron stained surfaces	FAIR Smooth, moderately weathered and altered surfaces	POOR Slickensided, highly weathered surfaces with compact coatings or fillings or angular fragments	VERY POOR Slickensided, highly weathered surfaces with soft clay coatings or fillings
INTACT OR MASSIVE - intact rock specimens or massive in situ rock with few widely spaced discontinuities	90		N/A	N/A	N/A
BLOCKY - well interlocked undisturbed rock mass consisting of cubical blocks formed by three intersecting discontinuity sets	80	70 60			
VERY BLOCKY- interlocked, partially disturbed mass with multi-faceted angular blocks formed by 4 or more joint sets			50		
BLOCKY/DISTURBED - folded and/or faulted with angular blocks formed by many intersecting discontinuity sets			40	30	
DISINTEGRATED - poorly interlocked, heavily broken rock mass with mixture of angular and rounded rock pieces				20	
FOLIATED/LAMINATED - folded and tectonically sheared. Lack of blockiness due to schistosity prevailing over other discontinuities	N/A	N/A			10

After Hoek, Marinos, and Benissi (1998).

falls below 100. This reduction is based on the reasoning that the deformation of better-quality rock masses is controlled by the discontinuities, while for poorer quality rock masses, the deformation of the intact rock pieces contributes to the overall deformation process (Figure 6.5).

Based on measured deformations, Eq. 6.12 appears to work reasonably well in those cases where it has been applied. However, as more field evidence is gathered, it may be necessary to modify this relationship.

TABLE 6.4 Characterization of schistose metamorphic rock masses based on foliation and discontinuity condition

GEOLOGICAL STRENGTH INDEX FOR SCHISTOSE METAMORPHIC ROCKS

From a description of the structure and surface conditions of the rock mass, pick an appropriate box in this chart. Estimate the average value of GSI from the contours. Do not attempt to be too precise. Quoting a range from 36 to 42 is more realistic than stating that GSI = 38. It is also important to recognize that the Hoek-Brown criterion should only be applied to rock masses where the size of individual blocks or pieces is small compared with the size of the excavation under consideration. When the individual block size is more than about one quarter of the excavation size, the failure will be structurally controlled and the Hoek-Brown criterion should not be used.

STRUCTURE (DECREASING INTERLOCKING OF ROCK PIECES ⇩) / SURFACE CONDITIONS (DECREASING SURFACE QUALITY ⇨)	VERY GOOD Very rough, fresh unweathered surfaces	GOOD Rough, slightly weathered, aperture < 1 mm hard filling	FAIR Slightly rough, moderately weathered, aperture 1 - 5 mm, hard and soft filling	POOR Smooth, highly weathered surfaces, aperture > 5 mm, predominantly soft fillings	VERY POOR Slickensided, highly weathered surfaces, aperture > 5 mm, soft fillings
INTACT OR MASSIVE - complete lack of foliation and very few widely spaced discontinuities	90	80	N/A	N/A	N/A
SPARSELY FOLIATED - partially fractured, massive intervals prevail over foliated intervals		70 60			
MODERATELY FOLIATED - fractured rock mass formed by massive and foliated intervals in similar proportions			50		
FOLIATED - folded and/or faulted rock mass with occasional massive intervals			40	30	
VERY FOLIATED - folded and/or faulted rock mass, highly fractured, formed by foliated rocks only				20	
FAULTED/SHEARED - very folded and faulted, tectonically disturbed rock mass	N/A	N/A			10

After M. Truzman (1999).

6.7 STRESS RELAXATION

When the rock mass adjacent to a tunnel wall or slope is excavated, a relaxation of the confining stresses occurs and the remaining material is allowed to expand in volume or to dilate. This has a profound influence on the strength of the rock mass since, in jointed rocks, this strength is strongly dependent on the interlocking between the intact rock particles that make up the rock mass.

As far as the authors are aware, there is very little research evidence relating the amount of dilation to the strength of a rock mass. One set of observations that gives an indication of the loss

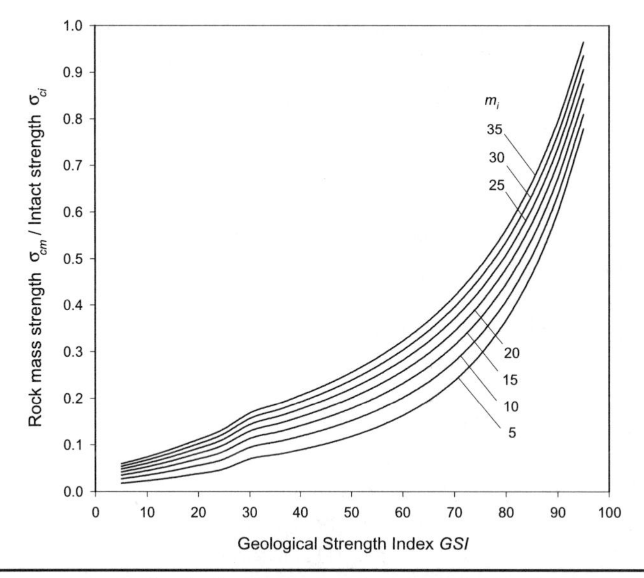

FIGURE 6.3 Ratio of uniaxial compressive strength of rock mass to intact rock versus GSI for depths of more than 30 m

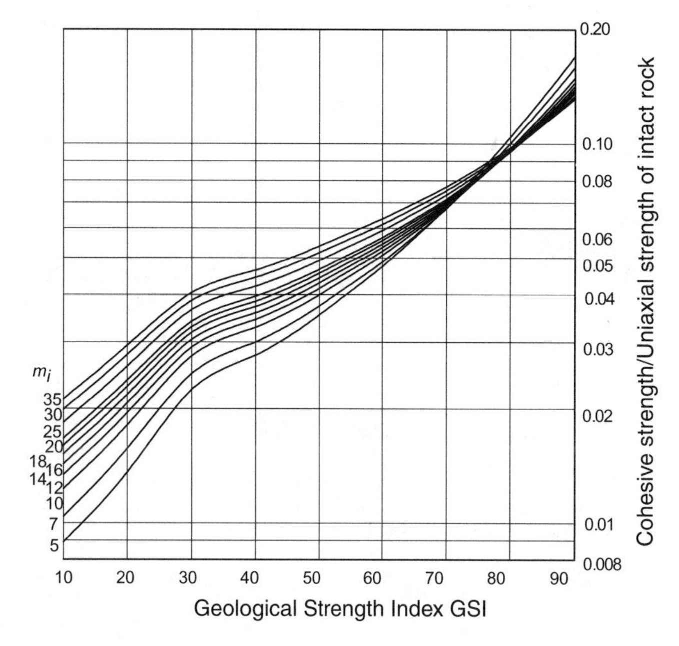

a) Plot of ratio of cohesive strength c' to uniaxial compressive strength σ_{ci} for depths of more than 30 m.

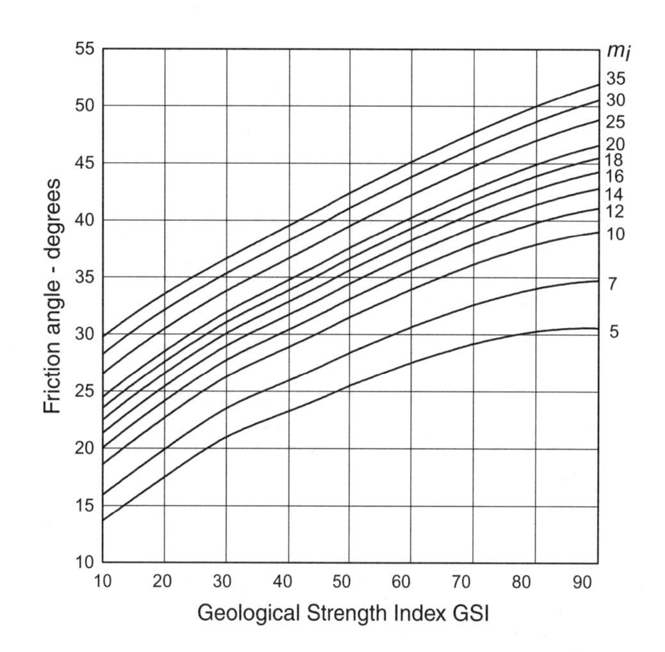

b) Plot of friction angle ϕ'

FIGURE 6.4 Cohesive strengths and friction angles for different GSI and m_i values for depths of more than 30 m

of strength associated with dilation is derived from the support required to stabilize tunnels. Sakurai (1983) suggested that tunnels in which the "strain," defined as the ratio of tunnel closure to tunnel diameter, exceeds 1% are likely to suffer significant instability unless adequately supported. This suggestion was confirmed in observations by Chern, Yu, and Shiao (1998) who recorded the behavior of a number of tunnels excavated in Taiwan. They found that all of the tunnels that exhibited strains of greater than 1 to 2% required significant support. Tunnels exhibiting strains as high as 10% were successfully stabilized but the amount of effort required to achieve this stability increased in proportion to the amount of strain.

While it is not possible to derive a direct relationship between rock-mass strength and dilation from these observations, it is possible to conclude that the strength loss is significant. An unconfined surface that has deformed more than 1 or 2% (based on Sakurai's definition of strain) has probably reached residual strength in which all of the effective "cohesive" strength of the rock mass has been lost. While there are no similar observations for rock slopes, it is reasonable to assume that a similar loss of strength occurs as a result of dilation. Hence, a 100-m-high slope that has suffered a total crest displacement of more than 1 m (i.e., more than 1% strain) may start to exhibit significant signs of instability as a result of loss of strength of the rock mass.

6.8 BLAST DAMAGE

Blast damage results in a loss of rock-mass strength due to the creation of new fractures and the wedging open of existing fractures by the penetration of explosive gasses. In the case of very large, open-pit mine blasts, this damage can extend as far as 100 m behind the final row of blastholes.

In contrast to the strength loss due to stress relaxation or dilation, discussed in the previous section, it is possible to arrive at an approximate quantification of the strength loss due to blast damage. This is because the blast is designed to achieve a specific purpose, which is generally to produce a fractured rock mass that can be excavated by means of a given piece of equipment.

Figure 6.6 presents a plot of 23 case histories of excavation by digging, ripping, and blasting published by Abdullatif and Cruden (1983). These case histories are summarized in Table 6.5. The values of GSI are estimated from the data contained in the paper by Abdullatif and Cruden, while the rock-mass-strength values were calculated by means of the spreadsheet given in the appendix, assuming an average slope height of 15 m.

These examples shows that rock masses can be dug, obviously with increasing difficulty, up to GSI values of about 40 and rock-mass-strength values of about 1 MPa. Ripping can be used up to GSI values of about 60 and rock-mass-strength values of about 10 MPa, with two exceptions where heavy equipment was used to rip strong rock masses. Blasting was used for GSI values of more than 60 and rock-mass-strength values of more than about 15 MPa.

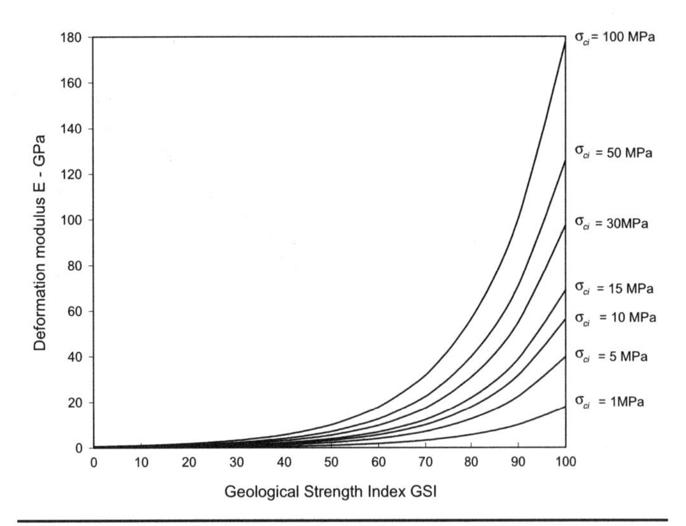

FIGURE 6.5 Deformation modulus versus GSI

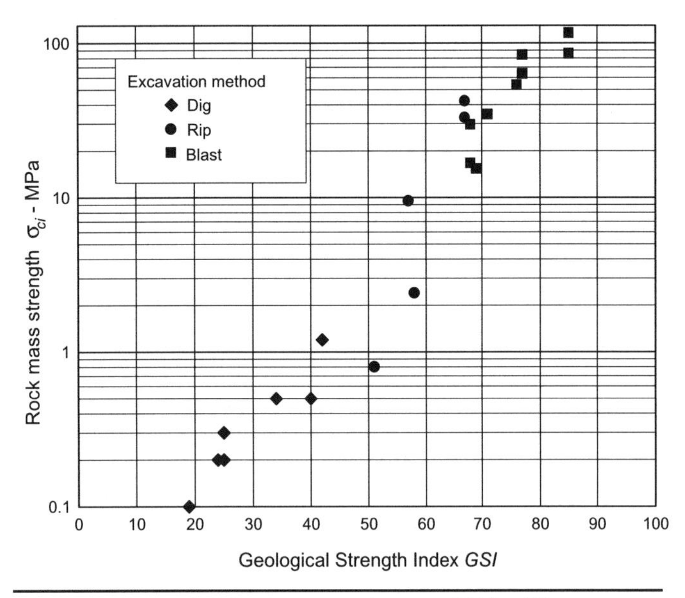

FIGURE 6.6 Plot of rock-mass strength versus GSI for different excavation methods, after Abdullatif and Cruden (1983)

Consider the case of an open-pit slope excavated in granodiorite. The uniaxial compressive strength of the intact rock is σ_{ci} = 60 MPa and GSI = 55. For granodiorite, Table 6.2 gives the value of m_i = 30. Substitution of these values into the spreadsheet given in the appendix, for a single 18-m-high bench, gives a rock-mass strength σ_{cm} = 5.7 MPa. To create conditions for easy digging, the blast is designed to reduce the GSI value to below 40 and/or the rock-mass strength to less than 1 MPa. In this case, the controlling parameter is the rock-mass strength and the spreadsheet given in the appendix shows that the GSI value has to be reduced to about 22 in order to achieve this rock-mass strength.

In another example of a 15-m-high slope in weak sandstone, the compressive strength of the intact rock is σ_{ci} = 10 MPa, m_i = 17, and GSI = 60. These values give a rock-mass strength σ_{cm} = 1.4 Mpa, and this is reduced to 0.7 by reducing the GSI to 40. Hence, in this case, both the conditions for efficient digging in this soft rock are satisfied by designing the blast to give a GSI value of 40.

Figure 6.7 summarizes the conditions for a muck pile that can be dug efficiently and the blast-damaged rock mass that lies between the digging limit and the in situ rock mass. The properties

TABLE 6.5 Summary of methods used to excavate rock masses with a range of uniaxial compressive strength values. Based on data published by Abdullatif and Cruden (1983).

GSI	Rock-Mass Strength σ_{cm}–MPa	Excavation Method
85	86	Blasting
85	117	Blasting
77	64	Blasting
77	135	Blasting
77	84	Blasting
76	54	Blasting
71	35	Blasting
69	15	Blasting
68	17	Blasting
68	30	Blasting
67	42	Ripping by D9L bulldozer
67	33	Ripping by D9L bulldozer
58	2.4	Ripping by track loader
57	9.5	Ripping by 977L track loader
51	0.8	Ripping by track loader
42	1.2	Digging by 977L track loader
40	0.5	Digging by wheel loader
34	0.5	Digging by hydraulic face shovel
25	0.3	Digging by 977L track loader
24	0.2	Digging by wheel loader
25	0.2	Digging by hydraulic backhoe
19	0.1	Digging by D9 bulldozer
19	0.1	Digging by 977L track loader

of this blast-damaged rock mass will control the stability of the slope that remains after digging of the muck pile has been completed.

The thickness *D* of the blast-damaged zone will depend on the design of the blast. Based on experience, the authors suggest that the following approximate relationships can be used as a starting point in judging the extent of the blast-damaged zone resulting from open-pit mine production blasting:

- Large production blast, confined and with little or no control — *D* = 2 to 2.5 H
- Production blast with no control but blasting to a free face — *D* = 1 to 1.5 H
- Production blast, confined but with some control, e.g., one or more buffer rows — *D* = 1 to 1.2 H
- Production blast with some control, e.g., one or more buffer rows and blasting to a free face — *D* = 0.5 to 1 H
- Carefully controlled production blast with a free face — *D* = 0.3 to 0.5 H

6.9 REFERENCES

Abdullatif, O.M., and D.M. Cruden. 1983. The relationship between rock mass quality and ease of excavation. *Bull. Intl. Assoc. Eng. Geol.*, 28:183–187.

Balmer, G. 1952. A general analytical solution for Mohr's envelope. ASTM, 52:1260–1271.

Bieniawski, Z.T. 1976. Rock mass classification in rock engineering. In *Exploration for Rock Engineering*, Proceedings of the Symposium, Ed. Z.T. Bieniawski, 1, 97–106. Cape Town, Balkema.

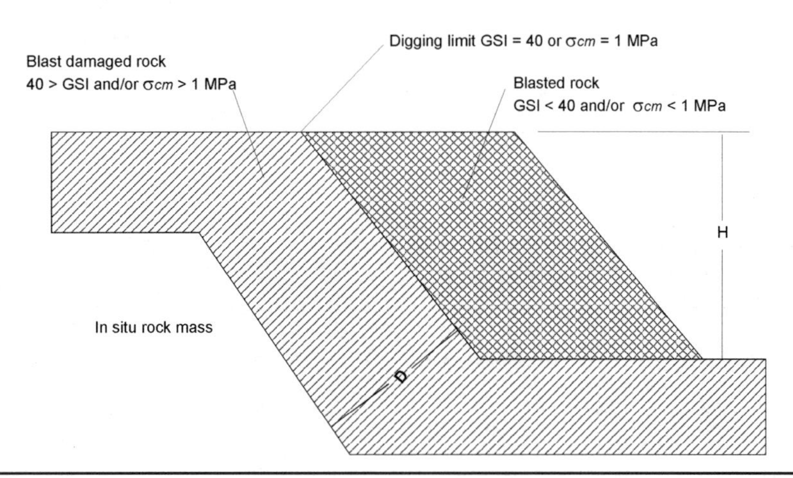

FIGURE 6.7 Diagrammatic representation of the transition between the in situ rock mass and blasted rock that is suitable for digging

Bieniawski, Z.T. 1989. *Engineering Rock Mass Classifications*. New York:John Wiley. p. 251.

Brown, E.T., ed. 1981. *Rock Characterization, Testing and Monitoring–ISRM Suggested Methods*. Oxford: Pergamon. p. 171–183.

Chern, J.C., C.W. Yu, and F.Y. Shiao. 1998. Tunnelling in squeezing ground and support estimation. *Proceedings Regional Symposium on Sedimentary Rock Engineering, Taipei*. pp. 192–202.

Hoek, E. 1983. Strength of jointed rock masses. 23rd Rankine Lecture. Géotechnique 33:3:187–223.

Hoek, E. 1994. Strength of rock and rock masses. *ISRM News Journal*, 2:2:4–16.

Hoek, E., and E.T. Brown. 1980a. *Underground excavations in rock.* London, Inst. Min. Metall. p. 527.

Hoek, E., and Brown, E.T. 1980b. Empirical strength criterion for rock masses. *Jour. Geotech. Engrg. Div., ASCE,* 106:GT 9:1013–1035.

Hoek, E., and E.T. Brown. 1988. The Hoek-Brown failure criterion–A 1988 update. In *Rock Engineering for Underground Excavations*, Proceedings 15th Canadian Rock Mechanics Symposium. Ed. J.C. Curran. Toronto, Dept. Civil Engineering, University of Toronto. pp. 31–38.

Hoek, E., and Brown, E.T. 1997. Practical estimates or rock mass strength. *Intl. Jour. Rock Mech. & Mining Sci. & Geomechanics Abstracts,* 34:8:1165–1186.

Hoek, E., P.K. Kaiser, and W.F. Bawden. 1995. *Support of Underground Excavations in Hard Rock*. Rotterdam:Balkema. p. 215.

Hoek, E., P. Marinos, and M. Benissi. 1998. Applicability of the geological strength index (GSI) classification for very weak and sheared rock masses. The case of the Athens Schist Formation. *Bull. Engrg. Geol. Env.*, 57:2:151–160.

Hoek, E., D. Wood, and S. Shah. 1992. A modified Hoek-Brown criterion for jointed rock masses. *Proceedings Rock Characterization, Symposium, Intl. Soc. Rock Mech.: Eurock '92*. Ed. J.A. Hudson. London, Brit. Geotech. Soc. pp. 209–214.

Sakurai, S. 1983. Displacement measurements associated with the design of underground openings. *Proc. Intl. Symp. Field Measurements in Geomechanics, Zurich*, 2:1163–1178.

Serafim, J.L., and J.P. Pereira. 1983. Consideration of the geomechanics classification of Bieniawski. *Proc. Intl. Symp. Engrg. Geol. and Underground Construction.* Lisbon, Portugal, Vol. 1, Part 11, pp. 33–44.

Truzman, M. 1999. Personal communication.

Appendix
Determination of Mohr–Coulomb Constants

The steps required to determine the parameters A, B, c', and ϕ' are given below. A spreadsheet for carrying out this analysis, with a listing of all the cell formulae, is given in Figure 6.A.1.

The relationship between the normal and shear stresses can be expressed in terms of the corresponding principal effective stresses as suggested by Balmer (1952)

$$\sigma'_n = \sigma'_3 + \frac{\sigma'_1 - \sigma'_3}{\partial\sigma'_1/\partial\sigma'_3 + 1} \qquad \textbf{EQ. 6.A.1}$$

$$\tau = (\sigma'_1 - \sigma'_3)\sqrt{(\partial\sigma'_1)/(\partial\sigma'_3)} \qquad \textbf{EQ. 6.A.2}$$

For the GSI > 25, when $a = 0.5$

$$\frac{\partial\sigma'_1}{\partial\sigma'_3} = 1 + \frac{m_b\sigma_{ci}}{2(\sigma'_1 - \sigma'_3)} \qquad \textbf{EQ. 6.A.3}$$

For GSI < 25, when $s = 0$

$$\frac{\partial\sigma'_1}{\partial\sigma'_3} = 1 + am_b^a\left(\frac{\sigma'_3}{\sigma_{ci}}\right)^{a-1} \qquad \textbf{EQ. 6.A.4}$$

The tensile strength of the rock mass is calculated from

$$\sigma_{tm} = \frac{\sigma_{ci}}{2}\left(m_b - \sqrt{m_b^2 + 4s}\right) \qquad \textbf{EQ. 6.A.5}$$

The equivalent Mohr envelope, defined by Eq. 6.2, may be written in the form

$$Y = \log A + BX \qquad \textbf{EQ. 6.A.6}$$

where

$$Y = \log\left(\frac{\tau}{\sigma_{ci}}\right),\ X = \log\left(\frac{\sigma'_n - \sigma_{tm}}{\sigma_{ci}}\right) \qquad \textbf{EQ. 6.A.7}$$

Using the value of σ_{tm} calculated from Eq. 6.A.5 and a range of values of τ and σ'_n calculated from Eq. 6.A.1 and 6.A.2 the values of A and B are determined by linear regression where

$$B = \frac{\sum XY - (\sum X \sum Y)/T}{\sum X^2 - (\sum X)^2/T} \qquad \textbf{EQ. 6.A.8}$$

$$A = 10\text{^}(\sum Y/T - B(\sum X/T)) \qquad \textbf{EQ. 6.A.9}$$

and T is the total number of data pairs included in the regression analysis.

The most critical step in this process is the selection of the range of σ'_3 values. As far as the authors are aware, there are no theoretically correct methods for choosing this range and a trial-and-error method, based on practical compromise, has been used for selecting the range included in the spreadsheet presented in Figure 6.A.1.

For a Mohr envelope defined by Eq. 6.2, the friction angle ϕ'_i for a specified normal stress σ'_{ni} is given by

$$\phi'_i = \arctan\left(AB\left(\frac{\sigma'_{ni} - \sigma_{tm}}{\sigma_{ci}}\right)^{B-1}\right) \qquad \textbf{EQ. 6.A.10}$$

The corresponding cohesive strength c'_i is given by

$$c'_i = \tau - \sigma'_{ni}\tan\phi'_i \qquad \textbf{EQ. 6.A.11}$$

and the corresponding uniaxial compressive strength of the rock mass is

$$\sigma_{cmi} = \frac{2c'_i\cos\phi'_i}{1 - \sin\phi'_i} \qquad \textbf{EQ. 6.A.12}$$

The values of c' and ϕ' obtained from this analysis are very sensitive to the range of values of the minor principal stress σ'_3 used to generate the simulated full-scale triaxial test results. On the basis of trial and error, it has been found that the most consistent results for deep excavations (depth > 30 m below surface) are obtained when eight equally spaced values of σ'_3 are used in the range $0 < \sigma_3' < 0.25\sigma_{ci}$. For shallow excavations and slopes, the user should input the depth below surface of the anticipated failure surface and the unit weight of the rock mass. For typical slopes, the depth of the failure surface can be assumed to be equal to the slope height.

Input:

sigci =	30	MPa	mi =	15		GSI =	55	
Depth of failure surface or tunnel below slope =				25	m	Unit wt. =	0.027	MN/n3

Output:

stress =	0.68	MPa	mb =	3.01		s =	0.0067	
a =	0.5		sigtm =	-0.0672	MPa	A =	0.7086	
B =	0.7263		k =	9.19		phi =	53.48	degrees
coh =	0.494	MPa	sigcm =	3.00	MPa	E =	7304.0	MPa

Calculation:

									Sums
sig3	1E-10	0.10	0.19	0.29	0.39	0.48	0.58	0.68	2.70
sig1	2.46	3.94	5.04	5.96	6.78	7.52	8.21	8.86	48.77
ds1ds3	19.32	12.74	10.31	8.95	8.06	7.41	6.91	6.51	80.20
sign	0.12	0.38	0.62	0.86	1.09	1.32	1.54	1.76	7.70
tau	0.53	1.00	1.38	1.70	2.00	2.28	2.54	2.78	14.21
x	-2.20	-1.83	-1.64	-1.51	-1.41	-1.34	-1.27	-1.21	-12.42
y	-1.75	-1.48	-1.34	-1.25	-1.18	-1.12	-1.07	-1.03	-10.21
xy	3.85	2.71	2.19	1.88	1.66	1.49	1.36	1.25	16.41
xsq	4.85	3.35	2.69	2.28	2.00	1.78	1.61	1.47	20.04
sig3sig1	0.00	0.38	0.97	1.72	2.61	3.63	4.75	5.98	20
sig3sq	0.00	0.01	0.04	0.08	0.15	0.23	0.33	0.46	1
taucalc	0.53	1.00	1.37	1.70	2.00	2.28	2.54	2.79	
sig1sig3fit	3.00	3.88	4.77	5.65	6.54	7.42	8.31	9.20	
signtaufit	0.66	1.00	1.33	1.65	1.97	2.28	2.58	2.88	

Cell formulae:

```
stress = if(depth>30, sigci*0.25,depth*unitwt*0.25)
mb = mi*EXP((GSI-100)/28)
s = IF(GSI>25,EXP((GSI-100)/9),0)
a = IF(GSI>25,0.5,0.65-GSI/200)
sigtm = 0.5*sigci*(mb-SQRT(mb^2+4*s))
sig3 = Start at 1E-10 (to avoid zero errors) and increment in 7 steps of stress/28 to stress/4
sig1 = sig3+sigci*(((mb*sig3)/sigci)+s)^a
ds1ds3 = IF(GSI>25,(1+(mb*sigci)/(2*(sig1-sig3))),1+(a*mb^a)*(sig3/sigci)^(a-1))
sign = sig3+(sig1-sig3)/(1+ds1ds3)
tau = (sign-sig3)*SQRT(ds1ds3)
x = LOG((sign-sigtm)/sigci)
y = LOG(tau/sigci)
xy = x*y          x sq = x^2
A =   acalc =   10^(sumy/8 - bcalc*sumx/8)
B =   bcalc =   (sumxy - (sumx*sumy)/8)/(sumxsq - (sumx^2)/8)
k = (sumsig3sig1 - (sumsig3*sumsig1)/8)/(sumsig3sq-(sumsig3^2)/8)
phi = ASIN((k-1)/(k+1))*180/PI()
coh = sigcm/(2*SQRT(k))
sigcm = sumsig1/8 - k*sumsig3/8
E = IF(sigci>100,1000*10^((GSI-10)/40),SQRT(sigci/100)*1000*10^((GSI-10)/40))
phit = (ATAN(acalc*bcalc*((signt-sigtm)/sigci)^(bcalc-1)))*180/PI()
coht = acalc*sigci*((signt-sigtm)/sigci)^bcalc-signt*TAN(phit*PI()/180)
sig3sig1= sig3*sig1        sig3sq = sig3^2
taucalc = acalc*sigci*((sign-sigtm)/sigci)^bcalc
s3sifit = sigcm+k*sig3
sntaufit =  coh+sign*TAN(phi*PI()/180)
```

FIGURE 6.A.1 Spreadsheet for calculation of Hoek–Brown and equivalent Mohr–Coulomb parameters for shallow excavations and slopes

CHAPTER 7

Failure Mechanisms for High Slopes in Hard Rock

Jonny Sjöberg*

7.1 INTRODUCTION

7.1.1 Background, Objective, and Scope of Work

Mining depths in open pits worldwide have increased steadily during the last few decades. Currently, many mines have reached, or plan to reach, mining depths of 500 m or more (a few mines are even planning to reach mining depths in excess of 1,000 m). With increased mining depths comes an increased risk of large-scale stability problems. This is further exacerbated by the desire to mine the steepest slopes possible to reduce costly waste stripping. A large-scale failure can be disastrous both to the operation and to the personnel working in the mine. Consequently, precise design of slope angles for open-pit slopes has become extremely important, particularly since very small changes in the slope angle have large economical consequences.

The design of pit slopes involves determining the stability state of a slope. Knowledge of the potential modes of failure in a slope forms the basis for this. Unfortunately, failure mechanisms in high slopes are, in general, poorly understood. This is particularly true for hard-rock slopes, since there are few cases of large-scale slope failures in high (300–500 m or more) hard-rock slopes. Even for slopes in weaker rocks, which have experienced large-scale failure, several fundamental issues are still largely unresolved. These include

- Conditions for the occurrence of different failure mechanisms
- Conditions for failure initiation
- Shape and location of the failure surface
- Development and propagation of a large-scale failure

Consequently, there is a need to study these aspects for commonly observed and assumed failure mechanisms. In this chapter, some typical (observed and assumed) failure modes in large-scale rock slopes are presented. The objective is to provide a better description of the governing mechanisms for these types of failures, including how they develop and propagate in a typical open-pit slope. A second objective is to develop guidelines for conducting stability analysis for different failure modes. Special focus is placed on the application of numerical modeling for analyzing complex failure modes in high rock slopes.

7.1.2 Approach and Analyzed Failures

Numerical modeling was used to investigate failure mechanisms. Models were used as a "laboratory" in which extensive parameter studies were conducted to identify critical factors governing various failure mechanisms. These analyses were not intended to simulate a particular case in detail but rather to explore the general characteristics of slope failures; hence, some typical failures were simulated for "generic" slope geometries. Models were verified through comparisons with reported laboratory model tests, comparisons with other analysis methods, and qualitative comparisons with generally observed failure behavior (for cases in which such descriptions existed). For the latter, data from the case study database (Chapter 7) was used.

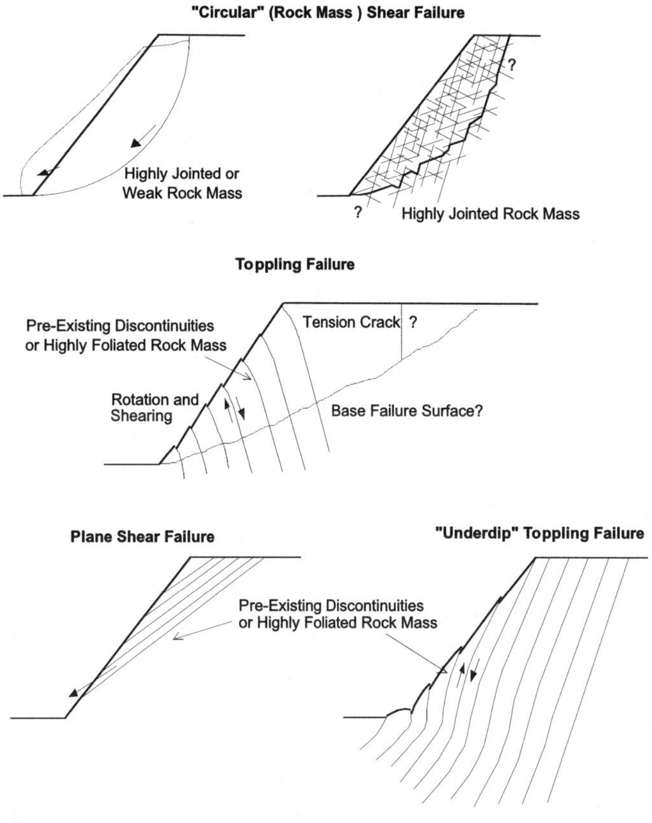

FIGURE 7.1 Analyzed failure modes

The work involved simulating the following failure modes: (1) circular (rock mass) shear failure in continuum slopes or in highly jointed or weak rock masses and (2) large-scale toppling failure in foliated, or persistently jointed, discontinuum slopes (Figure 7.1). These failure modes are commonly observed and/or assumed for large-scale slopes, but detailed descriptions of the mechanisms of these are lacking. Other structurally controlled failures in foliated rocks were also investigated. This included (1) plane shear failure, which can occur with preexisting joints striking parallel to, but dipping less than, the slope angle, and (2) "underdip" toppling failure, which can develop when joints strike parallel to, but dip steeper than, the slope face (Figure 7.1).

* Boliden Mineral AB, Boliden, Sweden.

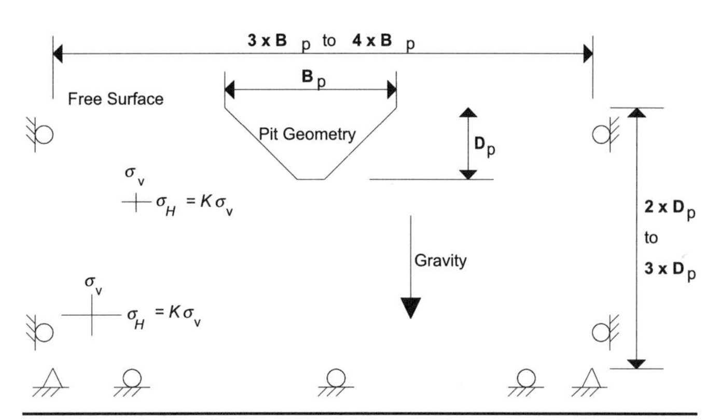

FIGURE 7.2 Model set up (size, boundary conditions, and virgin stresses) to simulate slope failures

However, these were studied in less detail and are only briefly discussed in this chapter.

Circular (or rock mass) shear failure is commonly assumed for large-scale slopes without any obvious structural control (i.e., heavily and randomly jointed rock masses). The failure surface need not be circular, but this collective term is used here to denote shear failures with curved failure surfaces. Circular shear failures have frequently been observed in weak rocks (Hoek and Bray 1981; Stewart and Reid 1986; Pariseau and Voight 1979), but cases have also been observed in hard-rock slopes (Dahnér-Lindqvist 1992, 1998). The actual failure mechanism probably involves slip along preexisting discontinuities coupled with failure through intact rock bridges (cf. Figure 7.1). This is difficult to simulate numerically, particularly for a large-scale rock slope (Sjöberg 1999). Rather, a pseudo-continuum approach with equivalent strength properties for the rock mass as a whole is often taken. However, even with this simplified approach, elementary questions, such as where failure initiates and how it progresses, are left unanswered.

Large-scale (or deep-seated) toppling failure has been observed relatively frequently in high pit slopes during the last few years (Daly, Munro, and Stacey 1988; Martin 1990; Pritchard and Savigny 1990; Board et al. 1996). This failure mode typically occurs in rock masses with steeply dipping large-scale discontinuities or in highly foliated rock masses. The mechanism leading to such failures, particularly the formation of a base failure surface (cf. Figure 7.1), is not clearly defined. This is partly due to insufficient observational data and partly due to inadequate analytical treatment of this failure mode.

7.1.3 Analysis Method and Model Set Up

Analyses were conducted using the finite difference program Fast Lagrangian Analysis of Continua (FLAC) (Cundall 1976; Itasca 1995) and the distinct element code Universal Distinct Element Code (UDEC) (Cundall 1971; Itasca 1997). FLAC is best suited for continuum problems, whereas UDEC is a discontinuum program that can handle large numbers of blocks and discontinuities. Both are two-dimensional programs, thus requiring an assumption of plane strain conditions. It is believed that many fundamental issues regarding failure mechanisms can be studied without entering the complexities of three-dimensional modeling. Both FLAC and UDEC use an explicit time-marching scheme to find the solution to a problem. The advantage of an explicit formulation is that the numerical scheme stays stable even when the physical system being modeled is unstable. Thus, it is particularly well suited for modeling nonlinear, large-strain behavior and physical instability, although calculation times are longer than for implicit codes.

A number of models with different final pit depths (100, 300, and 500 m) and with several different overall slope angles were analyzed. Most models were mined in one mining step. To minimize inertial effects, strengths were first set to high values and the model allowed to come to equilibrium after excavation. Strengths were then reduced to their correct values and the model run to a new steady state. Sequential mining (several mining steps) was simulated in selected models. In total, some 300 FLAC models and around 100 UDEC models were run.

Input data to the models (material strength, joint configuration, material stiffness, and stress and groundwater conditions) were varied to simulate a range of geomechanical environments. Values were chosen to produce failure and to study the conditions for which different failures can occur. The majority of the models were run using an elastic–perfectly plastic constitutive model (Mohr–Coulomb model) for the continuum. Several FLAC models were also run using a ubiquitous joint model (weak joints embedded in a Mohr–Coulomb solid). For the joints, both ubiquitous joints in FLAC and discrete joints in UDEC, an elastic–plastic model with a Coulomb slip criterion was used. Plastic strain-softening models were used for phenomenological studies of progressive failure. However, because the results are dependent on the finite difference-grid used, strain-softening models must be used with caution and are not recommended for quantitative analyses.

Initial model results showed that, for accurate analysis of pit slopes using plastic models, the size of the model should be roughly three times the pit width and three times the final pit depth (Figure 7.2). Roller boundaries (zero velocity) were found to be the most appropriate for the vertical sides and the bottom of the model. In this case, stresses were initialized in the model—this works best for a flat, or gently undulating, ground surface (Figure 7.2). To minimize effects from the grid in FLAC, a uniform quadratic grid with equal zone size should be used. Some caution should be taken when generating the slope geometry. Adjustments of the nodes of the grid to conform to a predescribed geometry should only be used for the slope face—not for material boundaries, since overestimated yielding can result.

7.2 CIRCULAR (ROCK MASS) SHEAR FAILURE

7.2.1 Failure Characteristics

Circular shear failure could be successfully simulated in FLAC, using a perfectly plastic constitutive model. For a homogeneous slope, shear failure occurred along a nearly circular failure surface (Figure 7.3). Tensile failure could be observed at the slope crest (Figure 7.4) and was more pronounced for steeper slopes. These results agreed well with other studies (Adhikary 1995; Dawson and Roth 1999), thus confirming the ability of FLAC to simulate circular shear failure.

Comparisons with limit-equilibrium analyses showed that FLAC predicted failure for approximately the same strength values as Bishop's routine method did (Bishop 1955). When compared to the chart method of Hoek and Bray (1981), FLAC predicted failure for lower strengths. More important, neither of the limit-equilibrium methods could predict tensile failure for steep, saturated slopes, although this is a conceivable failure mechanism for these conditions. The location of the failure surface in the FLAC models differed substantially from that in the limit-equilibrium analyses. The failure surface indicated by FLAC did not pass through the toe and extended farther back at the slope crest. (This was the case for both dry and saturated slopes.) The former is partly a grid discretization effect and partly an effect of the high confining stress at the slope toe, which "pushes" the failure surface slightly above the toe.

Furthermore, it was shown that the critical slip surface obtained from a limit-equilibrium analysis does not necessarily represent the final failure surface. A new and more critical failure surface can develop behind the first one if the failing material slides away (retrogressive failure). This phenomenon was most

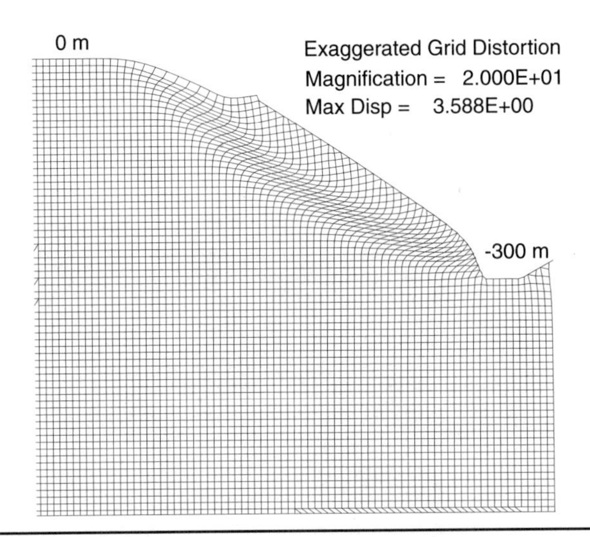

FIGURE 7.3 Calculated displacements (using FLAC) shown as exaggerated grid distortion for a 300-m-deep pit with a 40° slope angle in a perfectly plastic material and for drained conditions (entire model not shown)

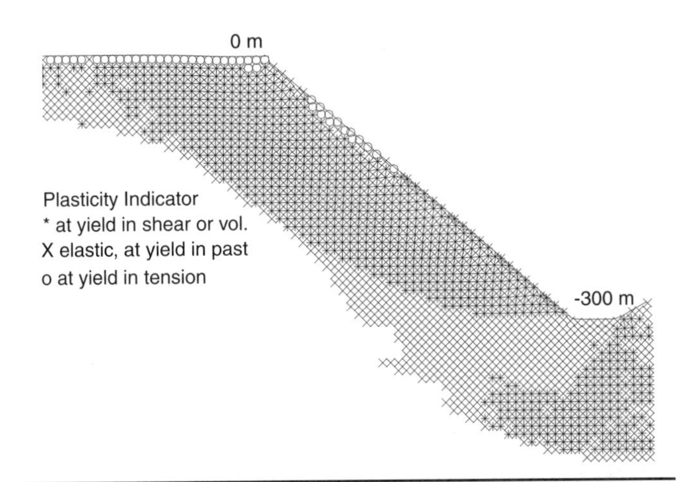

FIGURE 7.4 Plasticity indicators (from FLAC) for a 300-m-deep pit with a 40° slope angle in a perfectly plastic material and for drained conditions

pronounced for flatter slopes and for cases when all, or almost all, of the failed material can be assumed to have slid away from the slope. This can be explained by the fact that the resulting failure surface after slope failure creates a local slope that is much steeper than the original slope profile (near the crest). This new slope becomes more unstable than the original, flatter geometry.

Comparisons of the resulting failure surface from successive retrogressive failures using a limit-equilibrium analysis, with inferred failure surface in the FLAC models, showed good agreement. This is primarily due to differences in analysis method—it does not imply that FLAC automatically handles retrogressive failure behavior. FLAC allows the material to deform under failure, thus letting the failure surface develop successively in response to induced stresses and movements. Furthermore, tensile failure is accounted for in FLAC but not in limit-equilibrium methods. Both of these factors can explain some of the differences found for critical failure surfaces in FLAC and slip circle methods, respectively. A practical implication of this is not to blindly trust predicted failure surfaces from limit-equilibrium analysis without comparisons with other analysis methods.

7.2.2 Failure Initiation and Propagation

Numerical modeling allowed the failure development to be studied in detail. It was shown that failure occurred in several distinguishable phases and that significant displacements occurred before the failure surfaces had developed fully (Figure 7.5). Consequently, failures could not be characterized as rigid body movements (as assumed in all limit-equilibrium analyses). The different failure stages are summarized in Figure 7.6. Before failure, only elastic displacements result from the removal of rock. When mining to a new (and critical) slope height, yielding occurred, starting at the toe and spreading upward. Once a band of actively yielding elements has formed, shear-strain accumulation starts at the toe and progresses upward, accompanied by slightly increasing displacements. For the cases studied, the displacements at this stage were of the order of 0.2 to 0.4 m (i.e., before failure has developed fully). When the shear-strain accumulation has reached the crest, it can be said that a failure surface has formed, and larger displacements develop. This starts at the toe, followed by the middle and the crest of the slope (cf. Figure 7.5). In the final stage, the failing mass can slide away from the slope, but this could not be simulated in the models.

7.2.3 Strain-Softening and Progressive Failure

The above is valid for a perfectly plastic rock mass or a rock mass exhibiting a small difference between peak and residual strengths. For a strain-softening material, failure could be produced for significantly higher peak strengths, compared to the perfectly plastic model, provided the residual strength was significantly lower than the peak strength. The more brittle the material is (steeper slope of the stress-strain curve in the post-failure stage), the higher the peak strength that can still result in failure. A parameter study was conducted in which brittleness, peak strength, residual strength, and tensile strength was varied. It was found that failure initiated at the toe also for these cases but did not propagate directly to the slope crest. Rather, several intermediate failure surfaces developed, which eventually may reach the crest (Figure 7.7). This phenomenon only occurred for very brittle materials that also possess low tensile strength.

Similar findings were discussed in a study by Zhang, Bandopadhyay, and Liao (1989). Unfortunately, field observations are lacking; hence, it is difficult to judge whether this phenomenon also occurs in reality and how common (or uncommon) it is. The practical consequences can be considerable, since a perfectly plastic model can give unconservative results for brittle materials. This may be the case even if residual strength values are used in the modeling. The strong dependence of the above phenomenon on the softening parameter (the strain after which complete strength loss has occurred) implies that strain-softening constitutive models would be required. However, in current strain-softening models, the softening strain is also dependent on the grid size of the model; hence, calibration is necessary in each case. This limits the practical applicability of this type of models and does not allow quantification of the limits to this behavior. These shortcomings are well known within the modeling community, and ongoing research attempts to rectify this situation. Some applications to slope stability in soil and rock are those by Larsson, Runesson, and Axelsson (1994) and Tano (1997), but none of these can yet be considered practical tools for stability analysis.

7.3 LARGE-SCALE TOPPLING FAILURE

7.3.1 Failure Characteristics

Large-scale (flexural) toppling failure could be simulated using UDEC, alternatively using FLAC and a ubiquitous joint constitutive model. Comparisons with laboratory model tests by Adhikary (1995) showed good agreement. Furthermore, typical observed failure behavior in open pits (Martin 1990) was satisfactorily

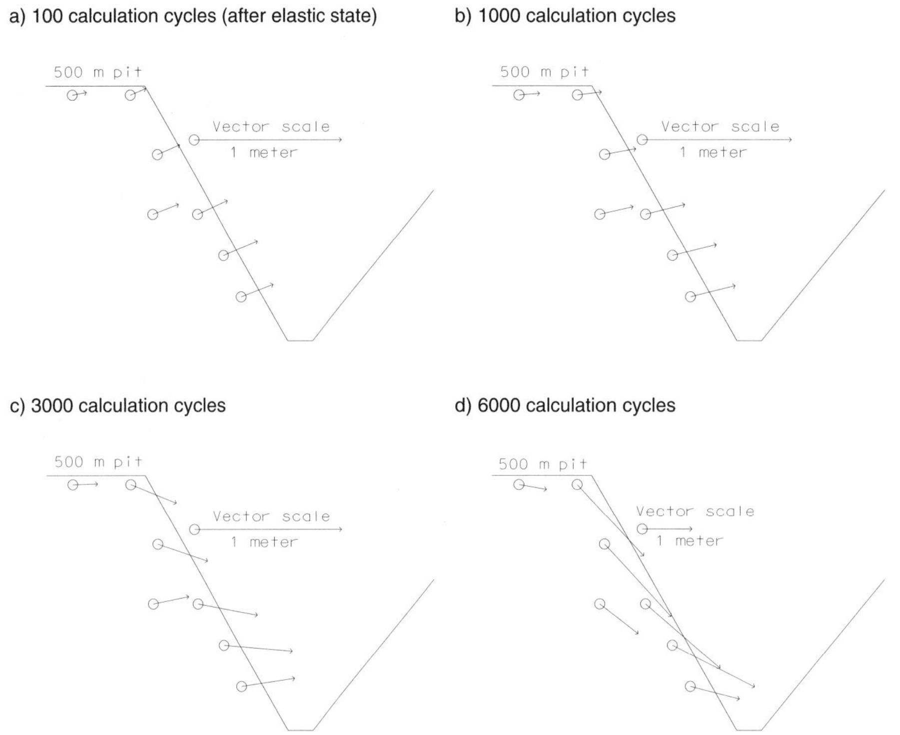

a) 100 calculation cycles (after elastic state)

b) 1000 calculation cycles

c) 3000 calculation cycles

d) 6000 calculation cycles

500 m pit

Vector scale

1 meter

FIGURE 7.5 Calculated displacement vectors using FLAC at selected points in a 300-m-high, 40° pit slope in a Mohr–Coulomb material at four different timesteps. (Note the different vector scale in figure d compared to the others.)

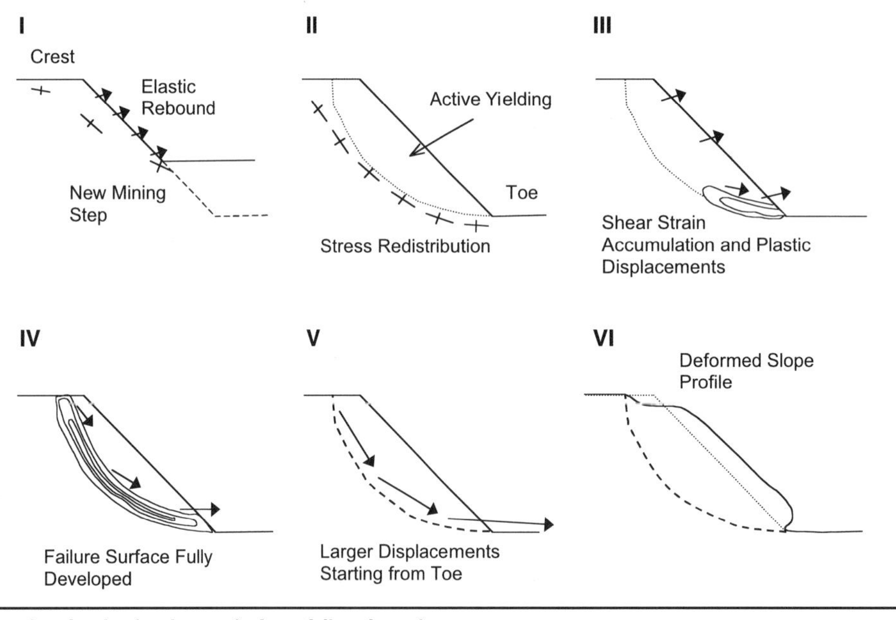

FIGURE 7.6 Failure stages for circular (rock mass) shear failure in a slope

reproduced in the models. Obviously, only the discontinuum UDEC models could simulate the kinematics of joint shearing and block movements correctly. However, the pseudo-continuum FLAC models (using a ubiquitous joint model) successfully replicated the overall response of a toppling slope.

As seen in Figure 7.8, toppling failure involves reverse shearing along the discontinuities, significant compression of the rock columns (most visible in the upper portion of the slope), and bending of the columns at the base of reverse shearing. The models showed that this resulted in tensile bending failure and the formation of a base failure surface, along which slip can occur. The model tests by Adhikary (1995) and Adhikary et al. (1997) confirmed that flexural toppling could result in the formation of a base failure surface. This is very different from the mechanism of block toppling, which involves rotation and sliding of rock columns along a preexisting basal failure plane (Goodman and Bray 1976).

7.3.2 Conditions for Large-Scale Toppling

The conditions for the occurrence of large-scale toppling failure were quantified through parameter studies. The main factors were

- Orientation and shear strength of joints
- Strength of intact rock material
- Rock-mass deformability

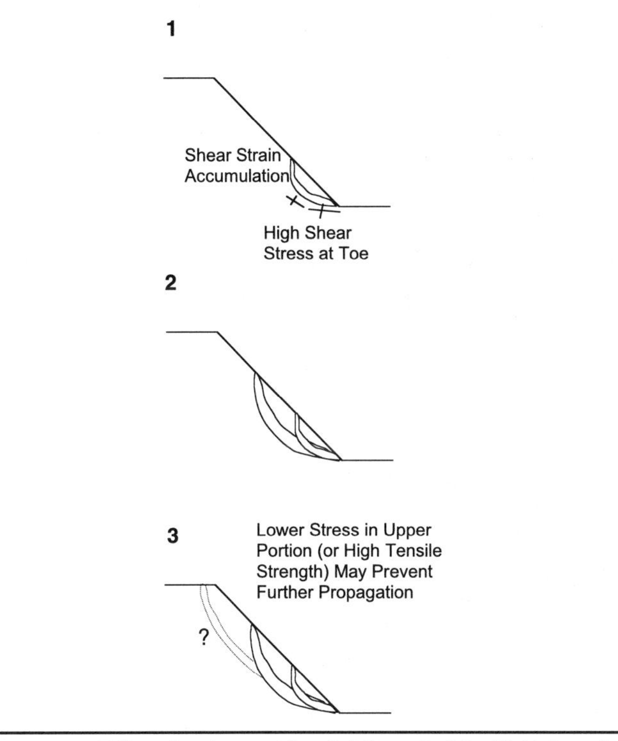

FIGURE 7.7 Failure development in a brittle material with high peak strength

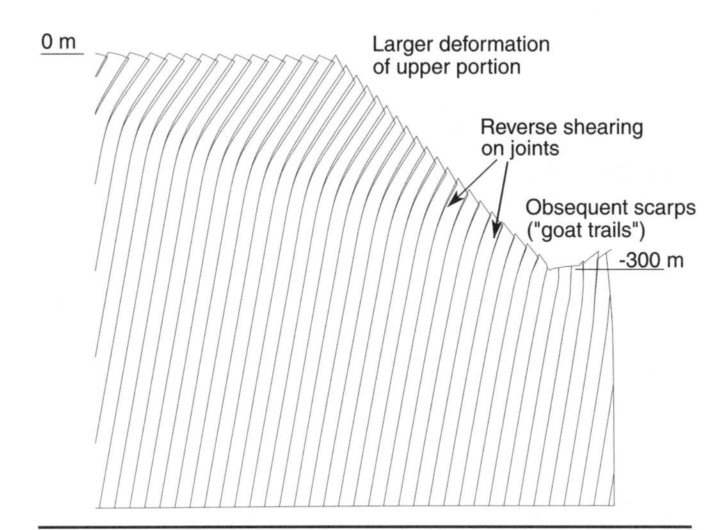

FIGURE 7.8 Calculated block movements using UDEC (magnified 100 times) for a 300-m-deep pit with a 40° slope angle and with steeply dipping joints (80°)

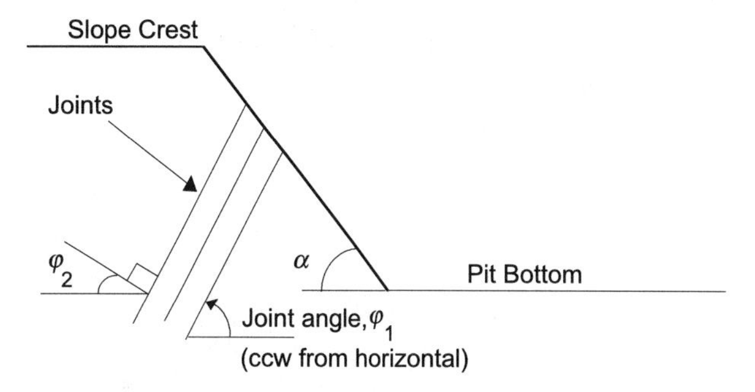

FIGURE 7.9 Definition of joint angles

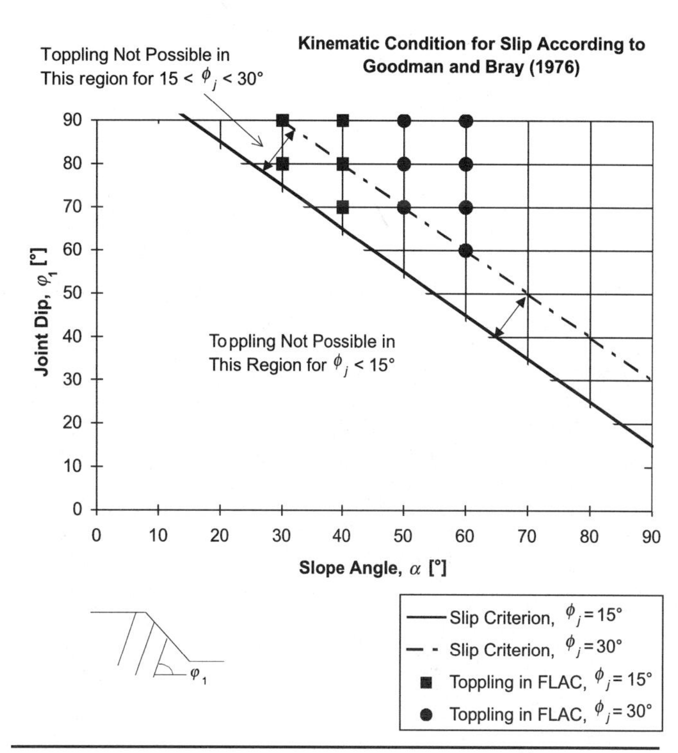

FIGURE 7.10 Limiting conditions for toppling according to Goodman and Bray (1976) and results from numerical analysis using a ubiquitous joint model in FLAC

Large-scale toppling failure developed only for certain combinations of slope angle, joint dip, and joint friction angle. Goodman and Bray (1976) proposed a criterion expressing the kinematic condition that for toppling to occur there must first be slip along the joints. This criterion is written as (see also Figure 7.9)

$$\alpha \geq \phi_j + (90 - \varphi_1) = \phi_j + \varphi_2 \quad \text{EQ. 7.1}$$

Where:

α = slope angle
ϕ_j = friction angle of the joints
φ_2 = angle between the horizontal and the normal to the joint planes
φ_1 = dip of the joint

This criterion agreed very well with the modeling results (Figure 7.10), although it appears to be slightly conservative, as was also noted by Adhikary et al. (1997). In any case, the criterion can be used for first estimates of the potential for toppling failure.

Goodman and Bray's kinematic condition only applies at the slope face, where the major principal stress is parallel to the face. As one moves away from the slope face, it becomes progressively more difficult to induce slip along the joint, as the normal stress on the joint plane increases and the driving shear stress decreases. The models showed that reverse shearing on the joint planes only extended to a certain depth into the slope. With higher joint friction angle, toppling was confined to a shallow zone close to the slope face. Furthermore, with less steeply dipping joints, only the

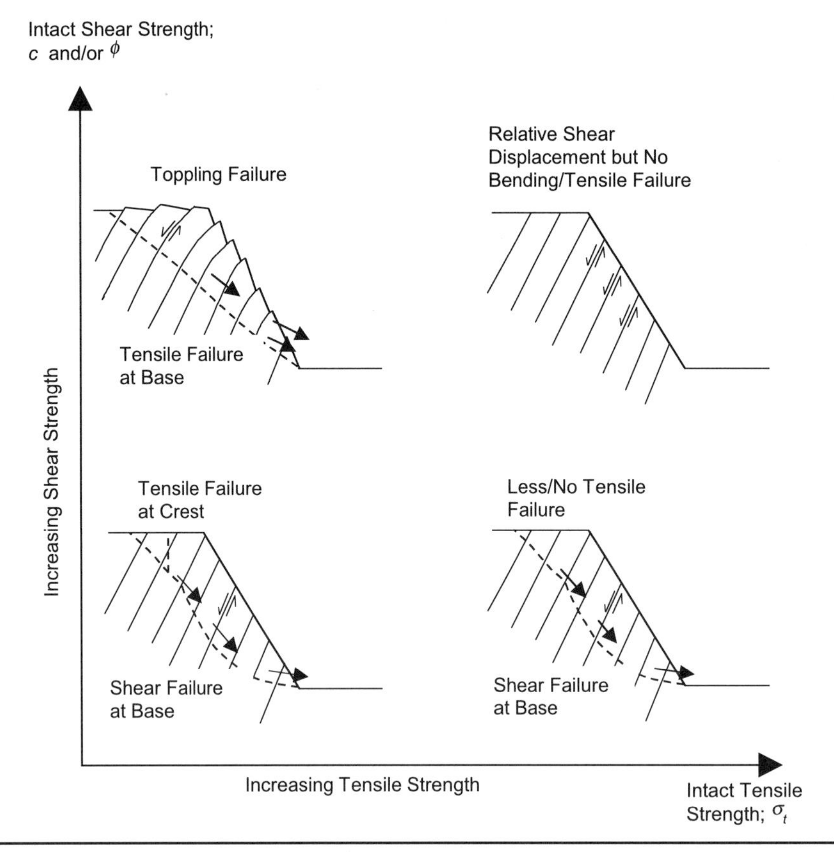

FIGURE 7.11 Failure mechanisms in slopes with steeply dipping joints as a function of intact material strength (kinematic condition for slip satisfied for the above cases)

upper portion of the slope failed. These results can all be linked to stress conditions within the slope in relation to the joint orientation and strength. Recently, Lee, Wang, and Huang (1999) derived analytical criteria describing the depth of reverse shearing, which seem to agree reasonably well with the modeling results presented here.

The second important factor for large-scale toppling failure was found to be the strength of the intact rock material (between the joints). Because the base failure surface is formed in tension, the tensile strength of the rock columns must be sufficiently low. With high tensile strength of the intact rock, slip along the joints can still occur, but no base failure surface forms. The limiting tensile strength depends on the slope height and, to some extent, the slenderness of rock columns. A precise estimate could not be derived, but for the cases analyzed here, the tensile strength limit was of the order of 1 to 2 MPa.

It was also found that toppling failures were more pronounced in rocks with high intact shear strength (much higher than the limiting strength for circular shear failure). For lower shear strength values of the intact material, the base failure surface changed shape from nearly planar to more curved. It is important to note that, for these cases, the rock failed in shear at the location of the failure surface, as opposed to tensile bending failure for the case of high intact shear strength (described earlier). Reverse shearing still occurred along the steeply dipping joints. Thus, this behavior is a form of transition between pure toppling failures and circular shear failures. The effect of intact material strength (shear versus tensile strength) on the occurrence of toppling failure is summarized in Figure 7.11. Similarly, toppling failure mode as a function of intact shear strength versus joint shear strength is shown in Figure 7.12.

The third critical factor governing large-scale toppling is the deformability of the rock mass. The rock columns do not behave as rigid bodies but deform substantially. Deformability increases with lower stiffness of the intact rock and/or smaller spacing of the rock joints. A highly deformable rock mass allows larger deformations to develop. Thus, the rock columns formed by the steeply dipping joints can start to bend more easily than in a less compressible material. Even the elastic deformations of the rock mass can be enough to create the "room" necessary for the rock columns to bend slightly, particularly for a high slope. Starting from the slope toe, the blocks are compressed slightly, thereby creating the necessary space for a slight rotation of the neighboring block up the slope. This block is also compressed, creating "room to move" for the next block, and so on. The accumulated compression over a high slope can be quite large; therefore, the blocks near the crest can bend significantly (cf. Figure 7.8).

In the above models, joints were assumed to be both persistent and continuous (an ideal situation for the occurrence of toppling failure). Although not explicitly simulated, it is believed that toppling also can develop for noncontinuous, steeply dipping, and closely spaced joints, e.g., in foliated rock masses where rock bridges often are small in relation to joint lengths.

To summarize, for toppling to occur, the following conditions must be satisfied (as verified by parameter studies):

1. The joints must dip relatively steeply into the slope and they must be able to slip relative to each other. The simple kinematic condition for slip proposed by Goodman and Bray (1976) proved satisfactory for assessing this (Eq 7.1).
2. The rock mass must be able to deform substantially for toppling to have "room" to develop.
3. The rock-mass tensile strength must be low to allow tensile bending failure at the base of the toppling columns.

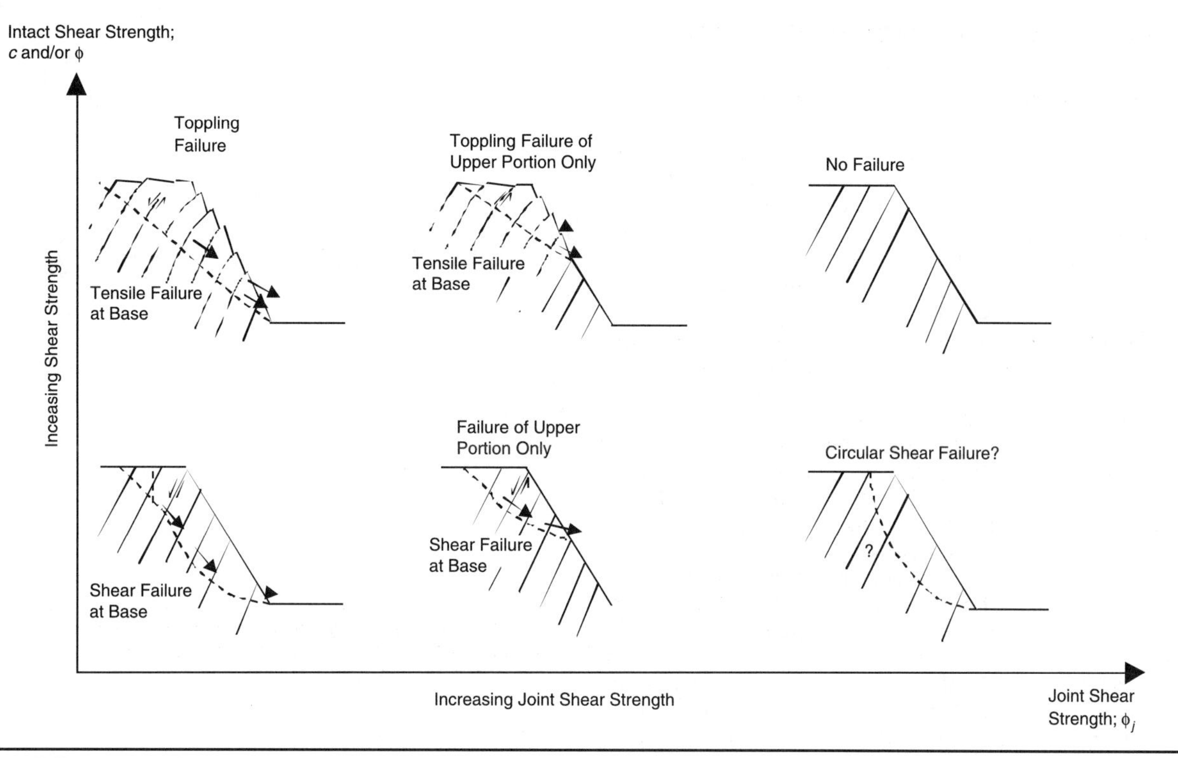

FIGURE 7.12 Failure mechanisms in slopes with steeply dipping joints as a function of intact material shear strength and joint shear strength (low tensile strength of the intact rock is assumed for the above cases)

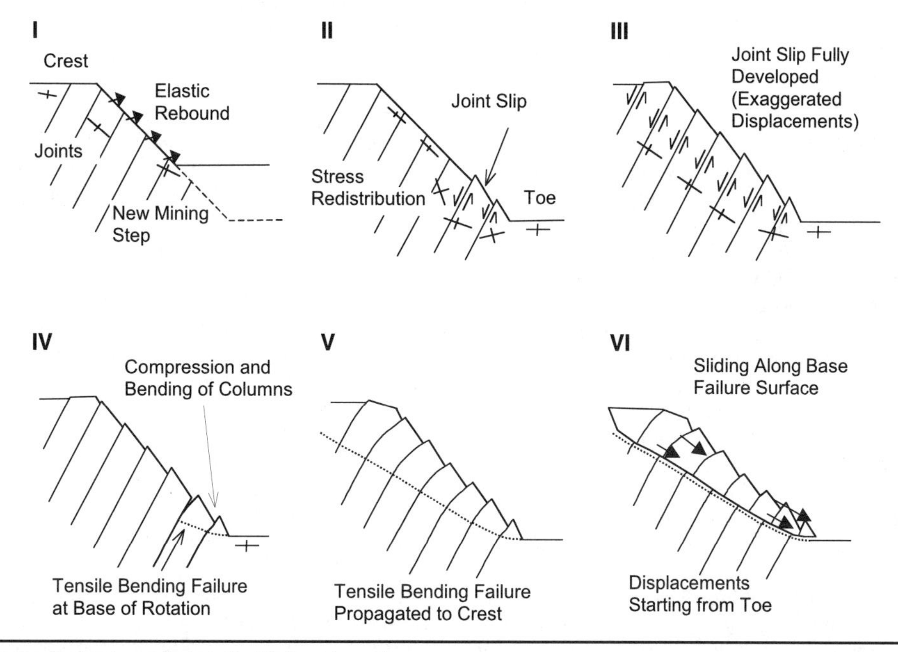

FIGURE 7.13 Failure stages for large-scale toppling failure in a slope

7.3.3 Failure Initiation and Propagation

Similar to circular shear failure, toppling failure occurred in several identifiable stages, as quantified by the modeling. These are summarized in Figure 7.13. Following the elastic rebound, failure starts in the form of slip along the steeply dipping joints in the slope. Joint slip starts at the toe and progresses toward the crest, with accompanying stress redistribution around this region. The depth to which slip along the joints develops is determined by the slope angle, the friction angle of the joints, and the stress state. Rock columns are compressed, which creates the necessary space for a slight rotation of the columns, starting at the toe. For a high slope, even the elastic deformations of the rock mass can be enough to allow a small rotation. This is followed by tensile bending failure at the base of the rotating column, which subsequently progresses toward the crest. Finally, a base failure surface has developed along which the failed material can slide (although this was not explicitly simulated in the models).

7.4 PLANE SHEAR AND "UNDERDIP" TOPPLING FAILURES

7.4.1 Plane Shear Failure in Large-Scale Slopes

Plane shear failure (Figure 7.1) can occur in slopes with continuous joints that daylight the slope face. In large-scale rock slopes, this failure mode is closely associated with the existence of large-scale structural features (of the same scale as the slope itself). While this is well known and understood, modeling of this failure mode revealed some interesting new findings.

In the simplest case of a pervasive, continuous joint plane that daylights the slope face, failure occurs if the friction angle is less than the dip of the joint plane (kinematic criterion for slip). However, this is under the assumption that the rock mass is non-deformable (rigid body movements). The numerical models (both FLAC and UDEC) showed that shear displacements are not uniformly distributed along the joint plane. Rather, the largest displacements occur near the slope face. For slope with multiple joint planes or foliated slopes, slip did not occur on all joint planes that were kinematically free to move but was concentrated to joint planes close to the slope face.

This is an effect of the rock mass being deformable, which, in turn, is most pronounced for high slopes (in reality, slope height in relation to rock-mass stiffness). The acting stress within the slope has significant influence on the ability of the joints to slip. As one moves away from the slope face, the increase in confining stress makes it more difficult to induce joint slip; hence, joint shearing is most pronounced near the slope face. In this study, only a handful of cases were analyzed. It was not possible to define the limits (in terms of joint spacing and continuity) to when failure becomes more limited, rather than developing along the most deeply seated joint plane. However, the practical consequences can be significant, as failure volumes can be less than what is often assumed; hence, more studies to confirm this and quantify the limits are warranted.

7.4.2 "Underdip" Toppling Failure

Flexural toppling cannot develop for the case of joints dipping parallel to, or steeper than, the slope. For this joint configuration, a different type of behavior was observed in the models, termed "underdip" toppling failure (see Figure 7.1). Shearing along the steeply inclined joints was followed by bending of rock columns and heaving of the slope toe. An example is shown in Figure 7.14. The term *underdip toppling* was first proposed by Cruden (1989), who also showed that the kinematic condition for joint slip was similar to that in Eq. 7.1.

However, the fact that joint slip occurs does not imply that slope failure will develop. The numerical modeling showed that small displacements (approx. 0.5 m) developed for this failure mode. More important, these displacements stabilized with increasing calculation time, indicating stable slopes. As such, this phenomenon is more of a "deformation mode" than a slope failure mode.

In a recent study by Kieffer (1998), these types of failures were referred to as "rock slumping" or, in this particular case, "kink-band slumping," in which a kink band forms at the toe of the slope. Kieffer (1998) also presented a few cases where this had been observed in the field. There is also the possibility that this kink band develops into a discontinuity shear surface. This would provide additional kinematic freedom to the overlying rock columns and result in slope failure. A modeling study by Söderhäll (1999) showed that this could indeed occur but only for undrained conditions, i.e., when effective stresses are low. Söderhäll (1999) also showed that a potential base failure surface develops in tension (similar to flexural toppling). For these cases, the models became unstable. In this study, only a limited range of conditions leading to "underdip" toppling was studied. More detailed studies were outside the scope of work. However, this failure phenomenon is important to study more, since the geomechanical environment in which it can develop is common for, e.g., footwall slopes of open pits.

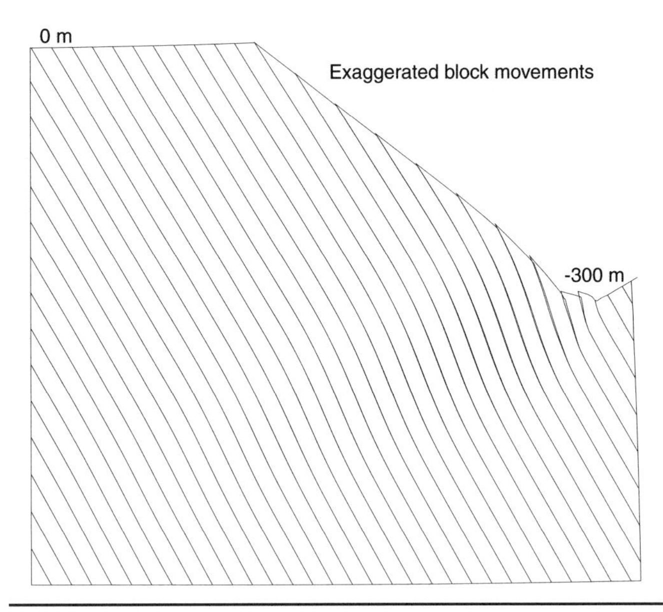

FIGURE 7.14 Calculated block movements using UDEC (magnified 100 times) for a 300-m-deep pit with a 40° slope angle and with steeply dipping joints (120°)

7.5 SUMMARY AND CONCLUSIONS

This study has shown that slope failures can be successfully simulated using numerical modeling. Thus, numerical analysis can be considered a useful tool for slope design. In fact, modeling is the preferred analysis method, since it can account for deformations. The modeling showed that deformability is a key parameter for several failure mechanisms; hence, rigid body movements cannot be assumed to apply for high rock slopes. Limit-equilibrium analysis may work satisfactorily in cases where the failure surface is predetermined. In other cases, there is a real risk that one may fail to recognize a more critical mode of failure if a certain failure surface is assumed. A drawback of current numerical models is that modeling of strain-softening and brittle material behavior is difficult and is not recommended for quantitative application until further developed.

As this study also showed, numerical modeling allows the simulation of the development of a failure and enables relevant comparisons with measured deformations in a pit slope. Modeling can also help to distinguish the types of failures that can occur in different geomechanical environments. Two-dimensional analysis is still considered sufficient for most cases, given the current limited knowledge of rock structure and failure mechanisms in three dimensions. Furthermore, two-dimensional analyses will be slightly conservative in most cases. Although only a few failure modes have been described quantitatively, this work has shown that slightly different modeling tools are required for different types of failures. Based on this, guidelines for stability analysis of large-scale open-pit slopes in different geomechanical environments were formulated and are presented in Table 7.1.

It was also shown that circular shear and large-scale toppling failures occur in several recognizable stages. The conditions for the occurrence of large-scale toppling failure, including the formation of a base failure surface, have also been quantified. In addition, this study has shown that failure initiates, in some form, at the toe of a slope. This is an example of where visual observations can sometimes be misleading, since it is often stated that the first signs of failure are observed at the crest. Also, the

TABLE 7.1 Simulated failure mechanisms, their occurrence, and recommended method of analysis

Failure Mechanism	Geomechanical Environment	Recommended Method of Analysis
Circular (rock mass) shear failure	■ Weak and/or weathered rock masses ■ Rock masses with many joint sets with different orientation and relatively closely spaced joints	*Limit-equilibrium methods*: Only for circular slip surfaces and (1) moderately steep slopes with or without groundwater or (2) steep, dry slopes. *Numerical analysis*: ■ Complex conditions (those not listed above) ■ FLAC to study failure development
Large-scale toppling failure	**1.** Rock masses with one dominant joint set, dipping steeply into the pit wall, e.g., (a) foliated rock masses and (b) rock masses with large scale structures **2.** Higly deformable rock masses (densely jointed and/or low stiffness) **3.** Rock masses with relatively low tensile strength	*Kinematic slip criterion:* ■ For estimating slip potential, (Eq 7.1) *Numerical analysis*: ■ FLAC for foliated rock masses with closely spaced joints/foliation planes ■ UDEC for rock masses with discrete, large-scale structures
"Underdip" toppling failure	**1.** Rock masses with one dominant joint orientation, dipping steeper than the pit wall, such as — Foliated rock masses — Rock masses with many large scale structures **2.** Higly deformable rock masses (densely jointed and/or low stiffness)	*Kinematic slip criterion:* ■ For estimate of slip potential *Numerical analysis:* ■ FLAC or UDEC to study failure development
Plane shear failure	Rock masses with one dominant joint orientation, dipping flatter than the slope, such as ■ Foliated rock masses ■ Rock masses with one or several large-scale structures	*Kinematic slip criterion:* ■ For estimate of slip potential *Limit-equilibrium methods:* ■ Analysis of sliding along joint planes (for one or two distinct joint planes) *Numerical analysis:* ■ FLAC or UDEC (for foliated rocks, noncontinuous joints)

observable appearance of different failure modes can be confusingly alike. Often, the observed behavior on the ground surface is not sufficient to state with certainty which failure mechanism is really occurring.

These are important findings, as they influence the choice of monitoring system to be used to detect failure at an early stage. It is clear that, e.g., mapping of tensile fractures at the crest is not sufficient and that more effort should be devoted to monitoring activities at the toe and inside the slope (despite the higher cost of such methods). This study also showed that significant displacements can occur before a failure is fully developed, which implies that slope movements before a failure are definitely measurable.

The simulated mechanisms presented here are believed to be some of the most important ones in large-scale rock slopes. There are, without a doubt, many other possible mechanisms, and there is also the possibility that these currently unknown (or poorly investigated) mechanisms are crucial for higher and steeper slopes than those presently existing. However, this study serves as the first step toward understanding potential failure mechanisms in rock slopes. By conducting more studies of this type and extending them to other potential failure modes, it is possible to improve on the current knowledge.

7.6 ACKNOWLEDGMENTS

The work presented in this chapter is the result of a four-year joint research project between Boliden Mineral AB and the Division of Rock Mechanics, Luleå University of Technology. The research project was sponsored exclusively by Boliden Mineral AB, which is gratefully acknowledged. The author is also indebted to the project supervision group for their great interest and fruitful discussions. This group consisted of Dr. Erling Nordlund, Luleå University of Technology; Mr. Norbert Krauland, formerly at Boliden Mineral AB; and Professor William Hustrulid, University of Utah, who was head of the Mining Research and Development at LKAB, Kiruna, during the main portion of the project.

7.7 REFERENCES

Adhikary, D.P. 1995. *The modelling of flexural toppling of foliated rock slopes*. Doctoral Thesis, University of Western Australia, Department of Civil Engineering.

Adhikary, D.P., A.V. Dyskin, R.J. Jewell, and D.P. Stewart. 1997. A study of the mechanism of flexural toppling failure of rock slopes. *Rock Mech. Rock Engrg*. 30:75–93.

Bishop, A.W. 1955. The use of the slip circle in the stability analysis of earth slopes. *Géotechnique*, 5:7–17.

Board, M., E. Chacón, P. Varona, and L. Lorig. 1996. Comparative analysis of toppling behavior at Chuquicamata open-pit mine, Chile. *Trans. Inst. Min. Metall. (Sect. A: Min. industry)*. 105:A11–A21.

Cruden, D.M. 1989. The limits to common toppling. *Canadian Geotechnical Jour*., 26:737–742.

Cundall, P.A. 1971. A computer model for simulating progressive large scale movements in blocky rock systems. *Proceedings Symposium Intlernational Society of Rock Mechanics*, Nancy, France. Paper II-8.

Cundall, P.A. 1976. Explicit finite difference methods in geomechanics. *Proceedings 2nd International Conference on Numerical Methods in Geomechanics*, Blacksburg, Virginia. 1:132–150.

Dahnér-Lindqvist, C. 1992. Liggväggsstabiliteten i Kiirunavaara. *Proceedings Bergmekanikdagen 1992—Papers presented at Rock Mechanics meeting*, Stockholm, Sweden. pp. 37–52 (in Swedish).

Dahnér-Lindqvist, C. 1998. Personal communication.

Daly, S., W.K. Munro, and P.F. Stacey. 1988. Slope stability studies for the Lornex Pit. *12th CIM District Six Meeting*, Fernie, Sept. 28–Oct. 1, 1988.

Dawson, E., and W.H. Roth. 1999. Slope stability analysis with *FLAC*. *Proceedings International FLAC Symposium on Numerical Modeling in Geomechanics*, Minneapolis, Minnesota. pp. 3–9.

Goodman, R.E., and J.W. Bray. 1976. Toppling of rock slopes. *Proceedings Rock Engineering for Foundations and Slopes. ASCE Specialty Conference*, Boulder. 2:201–234.

Hoek, E., and J.W. Bray. 1981. *Rock Slope Engineering*. London: Institution of Mining and Metallurgy.

Itasca. 1995. *FLAC Version 3.3. Manual*. Minneapolis: ICG.

Itasca. 1997. *UDEC Version 3.0. Manual*. Minneapolis: ICG.

Kieffer, D.S. 1998. *Rock slumping: A compound failure mode of jointed hard rock slopes*. Doctoral Thesis, Department of Civil and Environmental Engineering, University of California at Berkeley.

Larsson, R., K. Runesson, and K. Axelsson. 1994. Slope failure with plastic localization captured by discontinuous approximation. *Proceedings 8th International Conference on Computer Methods and Advances in Geomechanics*, Morgantown. 3:2465–2470.

Lee, C.F., S. Wang, and Z. Huang. 1999. Evaluation of susceptibility of laminated rock to bending-toppling deformation and its application to slope stability study for the Longtan Hydropower Project on the Red Water River, Guangxi, China. *Proceedings 9th International Congress on Rock Mechanics*, Paris. 1:119–122.

Martin, D.C. 1990. Deformation of open pit mine slopes by deep seated toppling. *Intl. Jour. Surface Mining and Reclamation*, 4:153–164.

Pariseau, W.G., and B. Voight. 1979. Rockslides and avalanches: Basic principles and perspectives in the realm of civil and mining operations. In *Rockslides and avalanches. 2. Engineering Sites*, ed. B. Voight. Amsterdam: Elsevier Publishing Company.

Pritchard, M.A., and K.W Savigny. 1990. Numerical modelling of toppling. *Canadian Geotechnical Jour*., 27:6:823–834.

Sjöberg, J. 1999. *Analysis of large scale rock slopes*. Doctoral thesis 1999:01, Division of Rock Mechanics, Luleå University of Technology.

Söderhäll, J. 1999. *Understjälpning i höga bergslänter*. Master of Science Thesis 1998:361 CIV, Division of Rock Mechanics, Luleå University of Technology (in Swedish).

Stewart, D.H., and G.J. Reid. 1986. *Afton. A geotechnical potpourri*. Unpublished report. Kamloops: Afton Operating Corporation.

Tano, R. 1997. *Localization modelling with inner softening band finite elements*. Licentiate Thesis 1997:26, Department of Civil and Mining Engineering, Division of Structural Mechanics, Luleå University of Technology.

Zhang, Y., S. Bandopadhyay, and G. Liao. 1989. An analysis of progressive slope failures in brittle rocks. *Intl. Jour. Surface Mining*, 3:221–227.

CHAPTER 8

Time-Dependent Movements of Open-Pit Slopes

Zavis M. Zavodni*

8.1 INTRODUCTION

Experience shows that all natural and human-made rock slopes deform with time in response to excavation. The challenge for the practicing slope engineer is to design slopes in the range of acceptable, safe movement and properly identify, manage, and limit the development of slopes in the unacceptable movement domain. The most commonly observed evidence of time-dependent deformation of open-pit slopes includes development of tension cracks behind the slope crest, formation of cracks on the slope, and toe heave.

A slope that exhibits no significant displacement may be designed too conservatively, whereas an unstable slope does not necessarily indicate a design that is too steep and that is compromising safety. In mining, instability can be economically advantageous if the costs incurred, including the costs of ensuring safe working conditions, are exceeded by the benefits achieved by mining at a steep slope angle.

It has been demonstrated that mining operations can proceed safely with minimum interruption if failure mechanisms are understood and slopes are properly monitored. Many examples exist where mining has been performed safely and successfully even while the slope was showing movement. A serious slope-instability condition is usually accompanied by gradual development of one or more tension cracks behind the slope crest, allowing for time-displacement monitoring. Surface-displacement measurements employing prism targets or extensometers with attendant prompt analysis of slope-movement velocities are usually adequate for predicting slope behavior. Inclinometers have also been successfully used to monitor small-scale deep-seated slope movements not evident at the ground surface.

Recent advances in survey equipment and technology have enabled increasingly precise monitoring of mined rock slopes. Greater awareness of the potential for deep-seated deformation of high slopes has resulted in a requirement for precise monitoring systems, including inclinometers, to be installed as part of the mining process. Most mines now regularly survey slope deformation for safety reasons and to assess long-term slope performance.

8.2 OPEN-PIT SLOPE DEFORMATION

There are many similarities in the documented and observed time-dependent behavior of open-pit slopes. Time-dependent response to excavation is a function of the mechanical properties of the rock mass, the structural geology, the slope geometry, the excavation rate, the ultimate failure mechanism, and the effects of external forces such as groundwater pressures, seismic activity, and in situ stresses. In general, rock slopes are expected to experience three distinct phases of time-dependent deformation behavior, as follows:

Phase I	Initial Response
Phase II	Regressive Failure Stage (short-term decelerating displacement)
Phase III	Progressive Failure Stage (strain-softening and large-scale deformation leading to collapse)

8.2.1 Initial Response

All slopes are expected to experience a period of initial response as a result of elastic rebound, relaxation, and/or dilation of the rock mass due to changes in stresses induced by excavation. This initial deformation is expected to occur without the development of a defined failure surface or failure mechanism. Slope extension can develop by spreading along joints in the rock mass or along a low-strength discontinuity.

Martin (1993) notes that movement rates range from 0.1 to 4 mm/d during initial response, although much higher instantaneous velocities may be encountered at the exact time of excavation. Following each excavation cycle, the displacement rate (velocity) of a point on the slope is expected to decrease based on a negative exponential relationship. During initial response, movement rates can be expected to result in displacements greater than 35 mm/yr (>1.5 in./yr). When pit slopes have developed over 20 to 30 years, total movements of more than 300 to 500 mm (>1 to 2 ft) might be anticipated during the initial response phase.

These magnitudes of displacement may not significantly affect rock-mass strength (i.e., strain hardening may take place) or the stability of the slope; losses of overall rock-mass strength could lead to progressive failure and collapse. Strain hardening has been documented for rock material and fractured rock masses during which both experience an increase in strength with increasing strain under an applied stress or, alternatively, require an increase in applied stress for continued strain.

Martin (1993) reports that precise slope monitoring at three open-pit mines in different geological terrains has shown considerable deformation during initial response. Total displacements ranging from 150 mm in a competent rock mass at Palabora, South Africa (Figure 8.1) (rock mass rating [RMR] 75–90) to more than 500 mm in a highly fractured and altered rock mass at the Goldstrike Mine, Nevada (RMR 30–50) were documented. Generally, movement rates and total movement during initial response appear to be related to rock-mass quality, with the higher-quality rock experiencing lower initial and ongoing movement rates and lower total movement.

Movement rates during initial response decrease from an initial value of up to several millimeters per day to no movement over a defined period. Results from the various case histories that

* Rio Tinto Technical Services, Bingham Canyon, Utah.

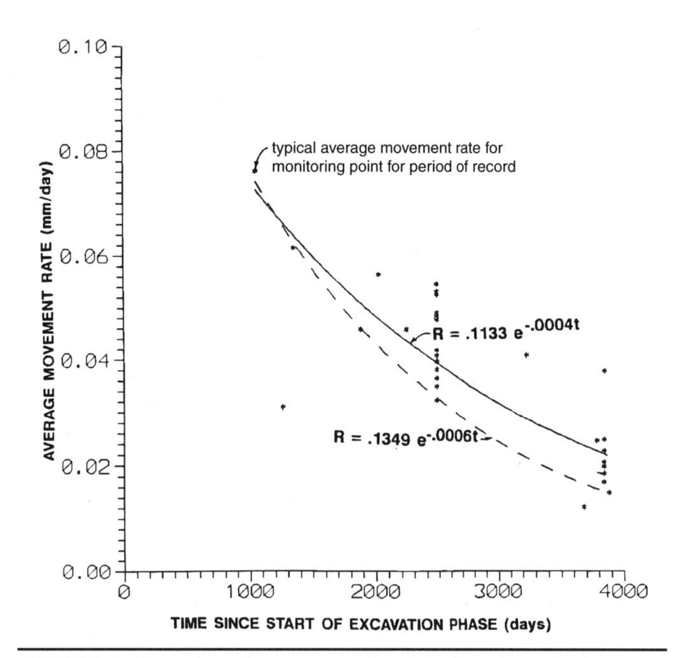

FIGURE 8.1 Summary of average movement rates of monitoring points at Palabora Mine in a competent rock mass during initial response (from Martin 1993)

Martin studied indicate that this decrease conforms to a negative exponential relationship of the form

$$R = Ae^{-bt}$$

Where:

R = the movement rate
t = time
A and b = constants that are a function of the rock-mass properties, slope geometry (height and angle), mining rate, external influences, and ultimate failure mechanism

Martin (1993) reports values for A to range from 0.113 to 2.449 and b from −0.0004 to 0.00294 during the initial response phase.

Measurement of total movement during slope excavation is difficult because of logistical problems related to establishing and maintaining permanent monitoring stations and/or instrumentation devices at or near an active working area. Monitoring stations cannot generally be established until well after the slope is exposed, thereby losing part of the initial displacement record.

In an effort to overcome this problem, Small and Morgenstern (1992) documented the response of a highwall slope at Highvale Coal Mine near Edmonton, Alberta, by installing inclinometers perpendicular to and ahead of an advancing highwall. Relaxation and/or rebound movements were recorded up to 200 m from the existing slope crest. Total horizontal movement ranged from 6 to 14 mm at a point 200 m behind the existing slope crest to 150 to 400 mm at the highwall crest. Beyond 50 m from the highwall face, the movements stabilized in a few days following excavation; near the face, the movements tended to creep at rates of 1 to 3.5 mm/d, often continuing until the next highwall was cut, 50 to 70 days later.

The translational rebound movements documented at Highvale occurred along weak and probably presheared bentonitic mudstones that dip at 0.5° into the slope. It is eye-opening to note that movements can extend over 200 m from a slope face that is only 20 to 23 m high at an angle of 40 to 50°.

Monitoring of slope dilation during mine steepening was conducted by Kennecott for the Kimbley Pit porphyry copper mine in Nevada during the 1960s. The mine had a maximum slope height of 150 m (500 ft) and slopes were steepened from 45 to 60°. During this slope-steepening program, displacements were measured in two adits. The pit wall moved upward a maximum 12 mm (0.48 in.) as the rock was removed some 30 m (100 ft) from the slope face; documented rebound was noted 120 m (400 ft) into the slope (near the pit toe) without evidence of surface cracking. Finite element modeling confirmed the deep-seated and near-surface slope dilations observed at the Kimbley Pit.

More recently, numerical slope models employing FLAC (Fast Lagrangian Analysis of Continua) and UDEC (Universal Distinct Element Code), both programs marketed by Itasca Consulting Group, Inc., have likewise predicted deep-seated dilation of high, steep, open-pit rock slopes. At Palabora, UDEC models have successfully been calibrated to monitor surface displacements of up to 200 mm; the models predict measurable wall dilations some 1,200 m from the slope toe for a wall height of 670 m.

8.2.2 Regressive and Progressive Failure Stages

Initiation of a conventional open-pit mine "slope failure" occurs after initial response and is normally associated with the creation of one or more visible tension crack(s) along the slope surface or at the slope crest. Proper monitoring and analysis of tension-crack time-displacement records is usually adequate for predicting slope behavior and evaluating the safety of mining a failing slope.

Technically speaking, a slope failure develops when the driving stress exceeds the resisting stress and yielding movement develops. Stability is decreased by increasing the driving stress, decreasing the resisting stress, or changing both driving and resisting stresses.

As noted by Call (1982), defining a slope failure is not as simple as it would first appear. From a theoretical standpoint, if the rock is considered an elastic material, any displacement beyond recoverable strain constitutes failure. This, however, is not a satisfactory failure definition for an open-pit mine where slopes that have "failed" from an elastic standpoint can still be mined successfully. Call suggests a distinction between theoretical and operational "failure" and proposes that an "open-pit mine failure" results when the rate of displacement is greater than the rate at which the slide material can be mined safely and economically or the movement produces unacceptable damage to a permanent facility.

The Concept of Regressive Versus Progressive Failure. During analyses of many rock-slope failures, it has become apparent to the author that all existing or potential failures must be either regressive or progressive, depending on the tendency of the condition to become more stable or more unstable. A regressive failure is one that shows short-term decelerating displacement cycles if disturbing events external to the rock are removed from the slope environment. A progressive failure, on the other hand, is one that will displace at an accelerating rate, usually an algebraically predictable rate, to the point of collapse unless active and effective control measures are taken (Zavodni and Broadbent 1980).

Figure 8.2 (Broadbent and Zavodni 1982) defines the conditions that must exist for each type of large-scale failure (greater than 100,000 t). In the figure, column 1 shows the relative geometry of slope and structure, where the structure line represents either a dip or plunge of a single or combination of several discontinuities; column 2 relates the geometries and the effective mechanical properties semiquantitatively; and columns 3 and 4 describe some well-documented failures that typify these conditions. For this purpose, a simple structural control (column 3) regards a condition where one or several continuous or near-continuous features form the primary mechanical system.

Complex structural control, on the other hand, regards a mechanical system comprised of numerous discontinuous surfaces of one or more structural systems. The foregoing classification covers all possible relationships of structure, slope, and

STRUCTURE ATTITUDE		PRIMARY STRUCTURE	
α = Mean Structure Dip or Plunge β = Slope Angle ϕ = Angle of Friction		SIMPLE CONTROL One or Two Surfaces	COMPLEX CONTROL Multiple Structures
Type I Regressive 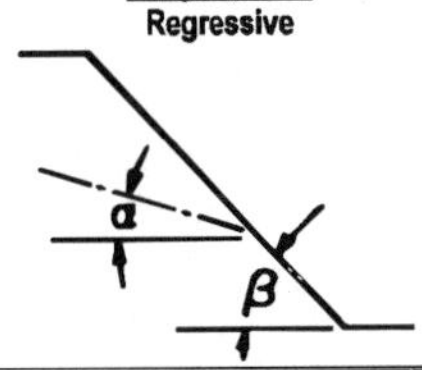 α β	α < β < ϕ	MINERAL CREEK Structural Control – Single Fault with Gouge Weight – 11 Million T Height – 50 M Mean Structure Dip Approx. 10° Overall Slope (β) Variable External Stimuli – Blasting, Water Unique Feature – Predictable & Regular Response to Blasting	Uncommon Type
Type II Progressive 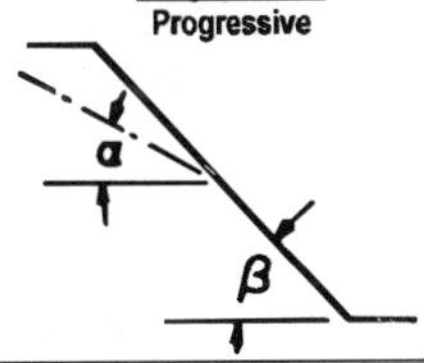 α β	α < β > ϕ	LIBERTY PIT Structural Control – 2 Intersecting Faults Weight – 7 Million T Height – 175 M Overall Slope (β) - 33° Mean Plunge (α) - 19° External Stimuli - ? Unique Feature – Regularity of Fault Surfaces & Attitudes	KIMBLEY PIT Structural Control – Single Joint System Weight – 1.1 Million T Height – 160 M Overall Slope (β) - 58° (Initial) Mean Structure Dip (α) - 65° External Stimuli - None Unique Feature – Induce Failure by Explosive Undercut, (β) 69° (Final)
Type III Regressive / Progressive 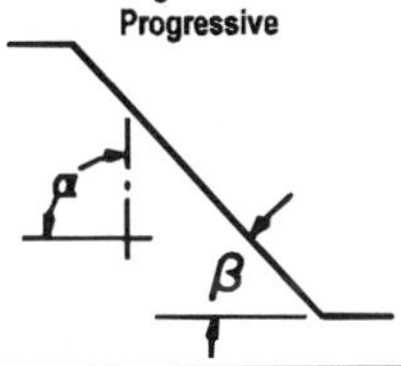 α β	α ≶ β ≶ ϕ	BINGHAM PIT C+ BEDS Structural Control – Prominent Faults Dip into Slope Weight – 20 Million T Height – 260 M Overall Slope (β) - 39° Unique Feature – Extreme Sensitivity to Groundwater Levels – Low Rock Mass Strength	CHUQUICAMATA Structural Control – Multiple Fault Joints Weight – ? (Est. 1 Million T) Height – 151 M Overall Slope (β) - 42° Mean Structure Dip (α) - Various External Stimuli – Heavy Pit Blasting, Frequent Earthquakes, Pore Pressure Unique Feature – Rubble Confined Above Pit Bottom

FIGURE 8.2 Rock failure types based on structure/slope characteristics (after Broadbent and Zavodni 1982)

strength, where strength refers to the rock discontinuities as opposed to the rock substance.

The value of differentiating between these three failure types and of properly identifying them in the field should be apparent. With this information, an optimum slope design can be achieved or an effective correction program can be designed if a failure exists.

Once a failure situation is recognized, the determination of its regressive or progressive status will enable the proper and most effective correction efforts to be initiated. A regressive condition would be corrected by removing or minimizing the external cause. A progressive failure, on the other hand, requires immediate planning for the likely collapse. In either case, monitoring systems and data interpretation must be oriented toward assessing the effectiveness of control programs or identifying a reversal of status, i.e., a regressive system becoming progressive, as often happens, or a progressive system reverting to a regressive mode, an unlikely event.

The following sections, partially taken from the paper by Broadbent and Zavodni (1982), report general displacement characteristics of several types of inherent conditions of instability.

8.2.3 Displacement Characteristics of the Structural Types

Type I Regressive Condition. A common condition of instability is represented by the Type I (regressive) structural system in the "simple control" mode. There are numerous examples of Type I instability, and their displacement characteristics are typified by curve A in Figure 8.3. On the curve, as shown, cycle initiation points 1, 2, and 3 plot as linear displacement assuming the discontinuous data. In reality, these points describe either an accelerating or decelerating displacement trend as revealed from continuous monitoring programs. The characteristic that qualifies this curve as regressive is the deceleration of each cycle between external stimuli at points 1, 2, and 3.

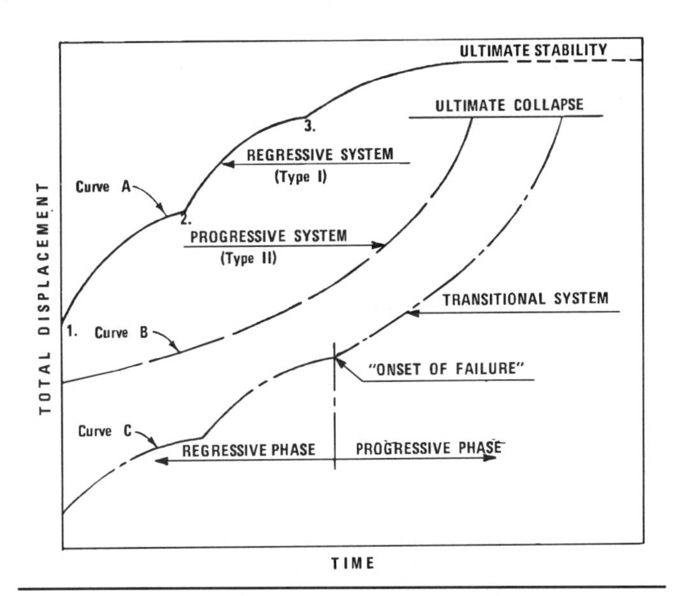

FIGURE 8.3 Typical regressive/progressive stage displacement curves (Broadbent and Zavodni 1982)

Broadbent and Ko (1971) show that each displacement cycle (e.g., 2 to 3) is predictable in a relatively uniform environment and that, generally, the points (1, 2, 3) will describe acceleration. These cycles have been observed in both the regressive and progressive failure stages and have been postulated to follow the behavior of a Kelvin or Voigt rheologic model, which employs both elastic and viscose properties.

The cycles are believed to be initiated when the driving force temporarily exceeds the resisting force, thereby causing the rock slope condition to drop slightly below a factor of safety of 1. The velocity of movement will decay if the external disturbance is eliminated. The excess force is usually related to an external event such as a mine blast, earthquake, precipitation, temperature change, groundwater pressure change or excavation of buttress rock. For example, in the case of a water-filled tension crack, displacement will cause the crack to widen and the water level to drop, reducing the driving force. As the crack fills again, the driving force will increase. Ko and McCarter (1975) have demonstrated a positive correlation between the scaled distance of pit blasts, precipitation, and cyclical response amplitude of a failing rock slope in the progressive stage. A number of case histories relate accelerating progressive stage movement with the onset of thawing temperatures, and Zavodni and McCarter (1976) have demonstrated a deceleration to dormancy of a regressive stage slide through the use of a successful slope dewatering program.

Characteristics of regressive stage failures are the

- Ratio of driving stress to resisting stress decreases with displacement
- Slope will tend to become more stable with time and show decelerating or stick-slip behavior

Stability of many slopes in the regressive failure stage can be controlled if only small displacement is allowed before mining is stopped and the slope is allowed to relax. If mining continues until the slope is at a high velocity, it will usually take a longer time for the slope to return to equilibrium. If the slope is pushed too far or if hydrologic conditions change, the slope may accelerate to the progressive stage and collapse. When mining a failure in the regressive stage, more frequent pullouts with short mining periods and short time delays may be preferable to long mining periods with fewer pullouts and extended time delays (Savely 1993). Experience has shown that acceptable failure velocities for safe mining can be as high as one meter per day to as low as a few centimeters per day, depending on the failure characteristics. Call and Nicholas (1998) suggest that for a predictable regressive failure displacement history, safe mining can be continued up to about 30 cm/d.

Type II Progressive Condition. The qualifying characteristic of a Type II condition is a structure dip greater than the effective structure strength, which manifests itself in positive exponential displacement, as shown in Figure 8.3 (curve B). Decelerating cycles may be present but would be subtle and nearly indistinguishable from the long-term trend.

The time period over which progressive displacement of a large-scale failure takes place is usually short, 4 to 45 days. It appears to be somewhat related to the mass of the failure, although efforts to quantify a relationship (Zavodni and Broadbent 1980) have not been successful.

Transitional Condition. Data are available to support the contention that many, if not most, economically significant failures began life as regressive and because of varied elements became progressive. This situation, which is described by displacement curve C of Figure 8.3, led to the formula for failure collapse estimation reported by Zavodni and Broadbent (1980). The prediction relies upon characteristics and relationships of the regressive phase and the "onset of failure" point in the time history of the failure.

The transitional condition failures are significant because Type I structural conditions are often controlled or the highly unstable Type II condition fails shortly after construction. However, transitional conditions (curve C) require extensive engineering, monitoring, control, protection, and reconstruction.

Type III Regressive/Progressive Condition. Toppling or "wedge-induced" (Calder and Blackwell 1980) failures would be characteristic of Type III structural conditions. These types of failures have highly varied characteristics. They can be regressive, particularly in the case of wedge-induced instability, or progressive, as typified by domino-type falls referred to as toppling. Type III conditions demand intensive study and analysis if sensible improvement is to be achieved. The so-called large-scale, deep-seated toppling failures (Martin 1993) are usually a slowly developing failure phenomenon; catastrophic slope failures seldom occur. Movements typically undergo a "ratchet" type of displacement involving acceleration followed by deceleration in response to mining at the toe or fluctuation in groundwater pressure. The C+ beds failure at the Bingham Canyon Mine in Utah is an example of a deep-seated yielding mechanism that was strongly influenced by transient pore-pressure conditions. Movement rates reached approximately 90 cm/d without catastrophic failure and total movement was documented at over 9 m. This failure was successfully managed by monitoring, controlling the mining rate and location, and reducing the transient pore pressure with an active pumping program.

Savely (1993) notes that a slope that continues to creep at constant velocity with a gradual increase in displacement with time is considered to be in a "steady state." A steady-state slope behavior can change into the progressive stage after considerable movement has taken place due to a reduction in shear strength with time. Steady-state behavior usually occurs when

- Well-defined failure surface(s) have a dip approximately equal to the sliding friction angle and
- Shear strength shows no peak strength and the driving force is close to the residual strength.

Steady-state behavior is also noticed with slope instabilities that have undergone large displacements over time, where failure surfaces have become well defined and the surface worn down to a residual strength.

Failures that show toe heave, such as non-daylighted failures with horizontal or upward movement at the toe, are typically regressive. Savely (1993) notes that regressive failures can become progressive when

- Mining daylights a previously non-daylighted failure surface.
- The rock mass at the toe breaks up and a shear surface or shear zone is developed.
- Water pressures increase.
- Mining is continued or increased without letting the slope regain equilibrium and the slope accelerates beyond recovery.

8.2.4 Time-Displacement Rate Monitoring

One of the most often reported structural-controlled mine slope failures was the Liberty Pit failure in Nevada. The data suggest this failure initiated as a Type I regressive failure but became a progressive (Type II) failure, probably as gouge was created and lubricated and as asperities were sheared off (i.e., strain softening). The time-displacement rate history of the Liberty Pit failure is presented in Figure 8.4. This figure, which was first published by Zavodni and Broadbent (1980), shows the basis for an empirical relation for estimating the time to total failure collapse. The following paragraphs are largely taken from that paper.

The overall displacement record in both the regressive and progressive failure stages was found to be of simple exponential form, with a definite break occurring at the onset-of-failure point (Figure 8.4). The straight-line semi-log curve fit for the progressive stage satisfied empirical data near to the immediate collapse point, where it became impractical to monitor displacement as the asymptotic nature of slide displacement took place. The rate of movement, rather than the magnitude of movement, was selected for the plot because it provided the most sensitive indication

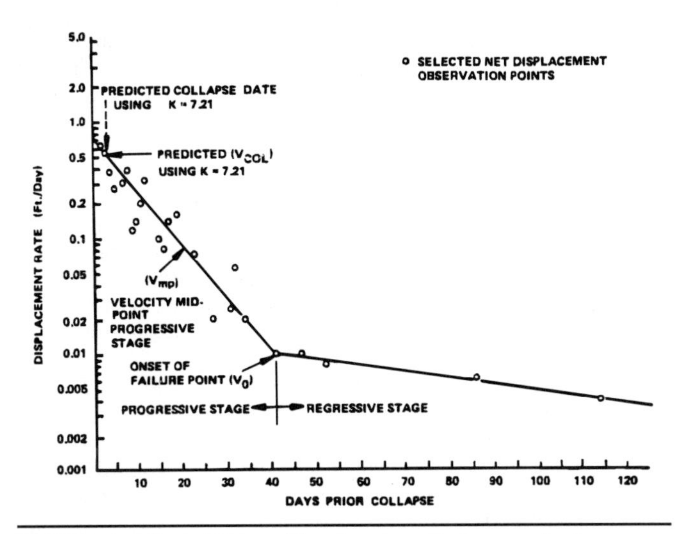

FIGURE 8.4 Liberty Pit transitional system displacement rate curve and failure collapse prediction (Zavodni and Broadbent 1980)

slope behavior and allowed for a greater range of displacements to be plotted.

A close examination of nine major transitional condition open-pit rock slides progressing to total collapse (and several more since 1980) revealed a semiquantitative empirical relationship for failure collapse estimation. It was observed that

$$\frac{V_{mp}}{V_o} = K \qquad \textbf{EQ. 8.1}$$

Where:

V_{mp} = velocity at mid-point in the progressive failure stage (Figure 8.4)
V_o = velocity at onset of failure point
K = constant (ave. −7.21, σ = 2.11, range 4.6–10.4 for 6 failures)

Knowing that the general equation for a semi-log straight-line fit has the form

$$V = Ce^{St} \qquad \textbf{EQ. 8.2}$$

Where:

V = velocity (ft/d)
S = slope of line (days = $^{-1}$)
C = constant
t = time (days)
e = base of natural logarithm

and assuming t = 0 at the collapse onset point, Eq. 8.2 takes the following form for the progressive failure stage:

$$V = V_o e^{St} \qquad \textbf{EQ. 8.3}$$

From this equation and the empirical relationship of Eq. 8.1, one can estimate the velocity at the collapse point (V_{col}) as

$$V_{col} = K^2 V_o \qquad \textbf{EQ. 8.4}$$

Equation 8.4, in conjunction with a semi-log plot such as in Figure 8.4, enables one to estimate the number of days until total collapse once the failure onset point is reached. The progressive stage failure displacement rate pattern is established from the monitoring record. This appears to be a useful empirical relationship for estimating the approximate time of failure collapse.

The above relationship can be effectively applied provided daily monitoring records are maintained. Because the data show

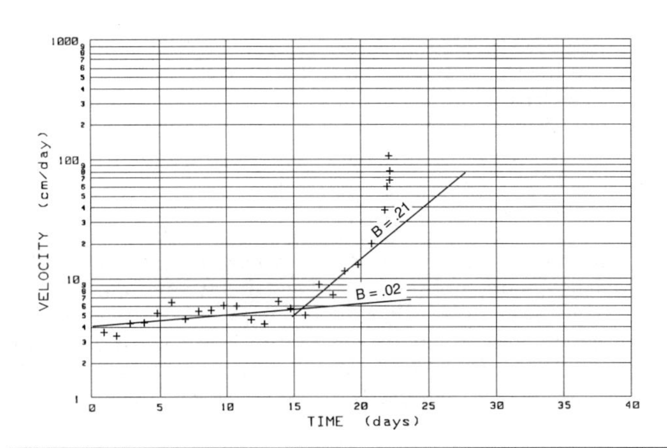

FIGURE 8.5 Extensometer displacement—regressive and progressive stages (from Call 1982)

that large-scale failures can collapse from this point in the relatively short time span of four days, it is possible to quickly recognize the failure onset point.

"Rapid" large-scale failures have been documented at the Aguas Claras Mine in Brazil and the Brenda Mine in Canada (Stacey 1996). The Aguas Claras failure occurred some 21 days after the first sign of instability was noted, and the failure at Brenda Mine occurred only five days after the first crack was observed. Both failures showed only limited measurable displacement (0.1 to 0.2 m) before collapse and developed along prominent weak discontinuities subparalleling the slope face (Stacey 1996). Sjöberg (1999) suggests that rapid and uncontrollable failures are more prominent in (1) steep rock slopes and (2) slopes in brittle and high-strength rock masses. The difference between peak and residual strengths appears to be a key factor. A large difference means that there is more energy that can be released in the event of failure. In a very brittle rock mass, the residual strength can be manifested quite suddenly.

The author is familiar with two medium-size block glide failure collapses that developed in coal highwalls within one day after drilling of blastholes for the next cut. These failures occurred along a practically horizontal, low-strength clay interface and a prominent joint backplane possibly created by air fracturing during drilling. An increase in pore pressure (precipitation) and slope strain softening that developed late in the initial response phase is also likely to have contributed to the failures.

Experience using the exponential predicting equation after its publication in 1978 (Zavodni and Broadbent 1978) suggests that the average value for the constant K should be reduced from 7.21 to 6.70 (range 3.7 to 10.4) and that there may be a continuous acceleration rather than an abrupt collapse point just prior to collapse (see Figure 8.5). Also, as the acceleration increases, extrapolation from linear regression becomes less precise. This means that there is probably more than one linear progressive stage prior to collapse. Call (1982) suggests that additional regressions should be done with a new evaluation completed after each new data point is plotted for better velocity predictions at failure collapse. Ryan and Call (1992) indicate that acceleration appears to be a more definitive slope collapse indicator in the final 48 hours than velocity magnitude.

Ryan and Call (1992) note that not all slope failures have readily identifiable displacement stages separated by a distinct onset of failure point. For the 11 failures studied, they found a mean initial velocity of 1.2 cm/d for the progressive stage and suggest that a minimum of 48 hours exist before slope collapse when velocity reaches 5 cm/d. Martin (1993) found a movement rate in excess of 1 cm/d to be indicative of progressive failure.

These progressive-stage velocity values generally support findings reported by Zavodni and Broadbent (1980).

Large-scale block glide failures (greater than 20 Mt) have been documented at the U.S. Borax Mine, California, in bedded soft rock at overall slopes of 24°. These failures extend over long distances behind the slope crest (75 m for a 145-m slope height) and can demonstrate a very long "stand up time" in the initial response and regressive failure stages (up to 10 years). These Type III failures can run out over very long distances, up to 425 m beyond the original slope toe. The failures enter the progressive stage through a reduction in lateral confinement (stress relief due to mining), an increase in transient pore pressure, and strain softening along the shallow, low-strength basal failure plane that dips into the pit. These failures have demonstrated a dual linear progressive stage displacement, as suggested in Figure 8.5. Approximate failure collapse estimates have successfully been made employing the exponential predicting equation along with a slope strain criterion outlined below. These methodologies have allowed for safe mining at U.S. Borax.

Current critical velocities used by U.S. Borax to guide slope management include

Incremental Velocity	Response
15 mm/d (0.05 ft/d)	Careful evaluation of stability/ mining continues
60 mm/d (0.20 ft/d)	Shoot prisms daily; berm-off area
150 mm/d (0.50 ft/d)	Abandon area

The U.S. Borax slope typically yields in initial response to a point of showing tension cracks and then moves in a fairly steady-state manner (incremental slope displacement velocities around 3 mm/d [0.01 ft/d or less]) until the displacements either stabilize the slope (regressive/strain hardening) or approximately 0.5% strain levels are attained. Once the 0.5% strain level is exceeded, the slope displacements typically accelerate to incremental velocities between 6 to 15 mm/d (0.02 to 0.05 ft/d), until strain levels approach about 1 to 2%. After the 1 to 2% strain level is reached, the slope accelerates again and typically results in a rapid collapse.

A slope dilation strain criterion has also been documented at the Highvale coal mine (Small and Morgenstern 1992). Much like U.S. Borax, it was found that slopes experienced maximum strains of 0.6 to 1% before moving into progressive failure.

8.2.5 Application of Time-Dependent Deformation

An understanding of the time-dependent behavior of rock slopes enables rational evaluation of slope performance and management of slopes. The processes of initial response and strain hardening may take several days to several years during which slope deformations proceed at decreasing rates. Depending on the slope geometry and rock-mass characteristics, total movements between 0.1 and 0.5 m might be anticipated.

Regressive and progressive slope-failure behavior is only expected after a period of initial response. Regressive/progressive failure of large open-pit mine slopes has been documented to range from a few days to 700 days.

Careful monitoring of slopes using state-of-the-art surveying equipment and processing technology enables time-dependent slope deformation behavior to be precisely defined and the slope to be managed. Several generalizations can be made from the existing empirical data that can assist the open-pit engineer when dealing with a large-scale (+100,000 t) slope dilation:

1. Slope movements on the order of 0.1 mm/d are indicative of initial response or ongoing failure without acceleration to collapse. In the case of poor-quality rock masses, movement rates of 1 to 4 mm/d may be documented during initial response (Martin 1993).
2. The onset-of-failure point can occur between 4 and 45 days prior to total collapse. This illustrates the importance of continuous monitoring, even of dormant slope failures, which might suddenly be reactivated and transferred into the progressive stage by a significant external impulse. The onset-of-failure point is practically impossible to predict while the slide is still in the regressive stage. It does not appear to be a function of days of movement or amount of total displacement in hard rock. It is likely a function of some "slope fatigue" factor that is related to the number and size of external slide impulses, the break up of the sliding rock mass with attendant strength reduction (strain softening), possible soil pressures developing in the crest-bounding tension cracks, and sudden increase in transient pore pressure.
3. The displacement record of instabilities does not indicate that there is a universally applicable critical velocity or total displacement criteria for predicting large, rapid movements.
4. It is unlikely that total failure collapse will take place within 24 hours if the displacement velocity is below 1.7 cm/d (0.055 ft/d) or within 48 hours if it is below 1.5 cm/d (0.050 ft/d).
5. A displacement rate above 5 cm/d (0.17 ft/d) usually indicates that a failure is in the progressive stage and that total collapse could occur within 0 to 48 days.
6. For a progressive failure geometry and a progressive velocity, the mining area should be cleared when the velocity exceeds about 10 cm/d.
7. For a regressive (Type I) displacement geometry, the mining area should be cleared when the velocity exceeds 15 cm/d. When the displacement rate stabilizes below 15 cm/d, mining can be continued. With a predictable displacement history, mining can be continued up to about 30 cm/d (Call and Nicholas 1998).
8. Regressive-stage velocities have a very large range, from 0 to 90 cm/d (0 to 3 ft/d). A broad velocity variety is revealed even for practically identical slide conditions. Somewhat deceptively, low regressive stage velocities can exhibit very short times to total collapse once the onset-of-failure point is reached.
9. Projections of the velocity curve should be made on a regular basis so that advance notice can be given to operations, predicting the time when an area may have to be cleared.
10. Two distinct ranges of slope values S (Eq. 8.2 and 8.3) are exhibited for the regressive and progressive failure stages:

    Regressive stage $S = -0.100$ to 0.013
    Progressive stage $S = 0.054$ to 1.171
    (assuming a more conventional plot with the time axis increasing to the right)

    The overall regressive failure stage velocity can decelerate to dormancy, remain constant, or slightly accelerate.
11. Monitored displacements in the progressive stage before slope collapse can be very small; as little as 8 cm (0.25 ft) was noted in the progressive stage of one large Bingham Canyon Mine slope failure. At the other extreme, total displacements of 9 m (30 ft) have been noted for slides that were still in the regressive failure stage.
12. The concept of slope strain appears to be a useful guide for predicting slope behavior in some soft-rock environments. Based on a few documented failures, slope collapse can develop after a 1 to 2% strain.

8.3 MONITORING

To ensure safety, a comprehensive monitoring program must be conducted throughout the life of an open-pit mine. The principal monitoring activity is measuring ground movement, which includes (1) regular visual inspections of early signs of instability, such as loose rock and cracks; (2) monitoring tension crack dilation; (3) prism displacement monitoring employing EDM and more recently GPS units; (4) continuous surface displacement monitoring using wire-line extensometers linked to automatic warning devices designed to turn on a flashing light in the event of excessive slope movement; (5) acoustic monitoring to document rock noise generated by excess stress buildup; and (6) borehole inclinometer and extensometer monitoring to record deep-seated dilation. A diligent ground movement monitoring program, coupled with prompt analysis of movement velocities and a sound understanding of the dilation mechanisms, has proven to be adequate for large-scale slope failure management (Zavodni 1997).

8.4 RESPONSES TO SLOPE FAILURE

When slope dilation beyond the initial response phase occurs, Savely (1993) provides the following list of possible responses to the slope instability:

- Leave the unstable area alone
- Continue mining (but at a controlled rate) without changing the mine plan if the velocity is low and predictable and the failure mechanics are well understood
- Unload the unstable area through additional stripping (only efficient under certain conditions)
- Leave a stepout in the mine design
- Do a partial cleanup
- Mine out the instability
- Support the unstable ground with artificial reinforcement
- Dewater the unstable area
- Revise the blasting procedures

The choice of option or combination of options depends on the nature of the instability and its operational impact. Each case should be evaluated individually, safety aspects carefully examined, and cost–benefit comparisons conducted.

8.5 CLOSING COMMENTS

All slopes deform in response to excavation. The mechanism of deformation and the deformation behavior of a slope varies depending on the engineering geology, structural geological conditions, slope geometry, rock-mass properties, groundwater conditions, and other transient environmental influences, such as climate or seismic events.

Rock slopes are expected to experience three distinct types of time-dependent deformation behavior. These include

- Initial response
- Regressive-failure stage
- Progressive-failure stage

Given the complexity of slope displacement, no single mathematical relationship is sufficient for predicting slope behavior. This does not mean, however, that safe working conditions cannot be maintained within the open-pit mine. The onset of a slope failure may come relatively rapidly but not "without warning" if the slope is properly monitored and the failure mechanisms are understood. Major displacements are preceded by small but measurable displacement and other indications of instability, such as tension cracks, rock noise, and changes in pore pressure. As stated by Terzaghi (1950), "...if a landslide comes as a surprise to the eyewitnesses, it would be more accurate to say that the observers failed to detect the phenomena which preceded the slide."

8.6 REFERENCES

Broadbent, C.D., and K.C. Ko. 1971. Rheologic aspects of rock slope failures. *Proceedings of 13th U.S. Symposium on Rock Mechanics*, University of Illinois, Urbana, Aug. 30–Sept. 1, 1971.

Broadbent, C.D., and Z.M. Zavodni. 1982. Influence of rock structure on stability. *Stability in Surface Mining*, Volume 3, Society of Mining Engineers, Chap. 2.

Calder, P.N., and G. Blackwell. 1980. Investigation of a complex rock slope displacement at Brenda Mines. *CIM Bulletin*, Vol. 73, No. 820.

Call, R.D. 1982. Monitoring pit slope behavior. *Stability in Surface Mining*, Volume 3, Society of Mining Engineers, Chap. 9.

Call, R.D. and Nicholas. 1998. Monitoring and slope management. Internal Memorandum, April 1998.

Kennedy, B.A., and K.E. Niermeyer. 1970. Slope monitoring systems used in the prediction of major slope failure at the Chuquicamata Mine, Chile. *Proceedings of Symposium on Planning Open Pit Mines*, Johannesburg. Amsterdan:A.A. Balkema.

Ko, K.C., and M.K. McCarter. 1975. Dynamic behaviour of pit slopes in response to blasting and precipitation. *Applications of Rock Mechanics*, ASCE, New York.

Martin, D.C. 1993. Time dependent deformation of rock slopes. Ph.D. Thesis, University of London, August 1993.

Ryan, T.M., and R.D. Call. 1992. Application of rock mass monitoring for stability assessment of pit slope failure. *Proceedings of 33rd U.S. Rock Mechanics Symposium*, pp. 221–229.

Savely, J.P. 1993. Slope management strategies for successful mining. In *Innovative Mine Design for the 21st Century*. Kingston, August 23–26, 1993, pp. 25–34.

Sjöberg, J. 1999. Analysis of large scale rock slopes. Ph.D. Thesis, Lulea University of Technology.

Small, C.A., and N.R. Morgenstern. 1992. Performance of a highwall in soft rock Highvale Mine, Alberta. *Canadian Geotech. Jour.*, 29:3, June 1992.

Stacey, P. 1996. Second Workshop on Large Scale Slope Stability. Las Vegas, September 13.

Terzaghi, K. 1950. Mechanism of landslides. *Application of Geology to Engineering Practice*, Berkey Volume, G.S.A.

Zavodni, Z.M. 1997. Elements of effective ground control in open pit operations. *28th Annual Institute on Mining Health, Safety and Research*, Virginia Polytechnic Institute and State University.

Zavodni, Z.M., and C.D. Broadbent. 1978. Slope failure kinematics. *Proceedings 19th U.S. Symposium on Rock Mechanics*, MacKay School of Mines, Reno, Nevada, May 1–3.

Zavodni, Z.M., and C.D. Broadbent. 1980. Slope failure kinematics. *CIM Bulletin*, Vol. 73, No. 816.

Zavodni, Z.M., and M.K. McCarter. 1976. Main Hill Slide Zone, Utah Copper Division. *Proceedings of 17th U.S. Symposium on Rock Mechanics*, Snowbird Resort, Alta, Utah.

HAPTER 9

he Role and Mitigation of Groundwater in Slope Stability

e C. Atkinson*

.1 INTRODUCTION

ater can have very deleterious effects on surface mining operans, with direct rainfall and associated runon being significant ctors in regions of high precipitation. This chapter, however, cuses strictly on groundwater—the water that already is in or is ing recharged to the soil and rock near a mine.

The most significant groundwater-related problem is the fect that water pressure has on the angle at which slopes can be cavated. Water pressure in the discontinuities (i.e., joints, fracres, bedding planes, etc.) in a rock mass reduces the effective esses on such discontinuities with a consequent reduction in ear strength. The problem is exacerbated where there are portunistic features such as bedding, foliation, or a wedge ucture dipping into a highwall. As described in other chapters this section, fully depressurized slopes can usually be excated more steeply by 10 or more degrees. Although flattening of 'wet" slope is sometimes an option, dewatering of the slope is ually a more desirable and economic alternative.

For the sake of completeness, other water-related problems surface mines, some of which are not necessarily less signifint, are

- Problems associated with wet ore (e.g., material bind-up in crushers, excessive moisture content in coal, disintegration of kimberlite) or overburden/waste rock (e.g., stacking problems)
- The direct costs of managing water (i.e., wells, pumps, pipelines, electric power, dewatering drifts)
- The need for waterproof explosives
- Higher equipment maintenance (e.g., excessive tire wear on wet ground)
- Periodic slowdowns/shutdowns due to pump failures or winter freeze up
- Engineering costs (the need for a specialized staff or consultants)

Mine personnel who are given the less-than-enviable task of lving water problems typically come from one of three backounds: (1) hydrogeologists who traditionally have been involved developing water supplies and might have derived most of their derstanding of groundwater flow from well hydraulics, (2) geochnical engineers who are interested primarily in the location of e phreatic surface in a highwall, and (3) mining engineers or ologists who have been to the "school of hard knocks" with spect to water at one mine but whose experience might not be plicable at another mine with different hydrogeologic conditions. It is hoped that this chapter will provide some additional sight into "mining hydrology" for all three groups.

Mine-water problems tend to be unique, especially in hard-rock deposits. The geology and, hence, the hydrogeology are usually quite complex. It is often a race against time (and the shovels!) to solve a particular dewatering problem, and one has to work around the mining operations. Consequently, the person in charge of dewatering is usually not the most popular—nor most well-understood—person around a mine site.

One concept that clearly needs to be understood as part of solving groundwater problems associated with surface mines is that we usually are not dealing with "aquifers" as originally defined in the field of water supply. In sedimentary sequences such as coal cyclothems or in carbonate-hosted orebodies, the idea of a laterally continuous water-bearing unit with a top, bottom, and essentially constant hydraulic properties might be appropriate. In almost any crystalline rock-hosted orebody or a structurally complex setting (which is the rule rather than the exception in mining), however, the applicability of the aquifer concept breaks down. We are usually dealing with a heterogeneous continuum of geologic materials with significantly different hydraulic properties.

Recognition of a potential mine-water problem and mitigating it before it becomes a problem is the "art" of mine dewatering. The following sections will define the tools of this art by focusing on dewatering of pit slopes where water poses a potential stability problem.

9.2 THEORETICAL CONSIDERATIONS

9.2.1 Basic Groundwater Hydraulics

To develop an understanding of the basic groundwater hydraulics that will be applied when designing a dewatering system, let's first examine the "anatomy" of the groundwater flow system in a pit slope, as shown in Figure 9.1. The boundary between the saturated and unsaturated soil or rock, where the pressure in the pores is zero ($P = 0$) relative to atmospheric pressure, is the phreatic surface or, as it is sometimes called, the water table. The point where the phreatic surface intersects or crops out on the pit slope, which is also a $P = 0$ surface, is the top of the seepage face. The height of this seepage face is a function of the hydraulic conductivity and the horizontal-to-vertical anisotropy (both of which will be defined below) of the material(s) comprising the slope. Material with low hydraulic conductivity and large lateral-to-vertical hydraulic conductivity ratios will have higher seepage faces than those in more permeable and isotropic rock masses. It should also be noted that the height of the seepage face is always >0 (unless we have lowered the phreatic surface below the highwall by dewatering), even if we cannot see it! If conditions are very dry or windy and evaporation is high and if the amount of seepage is small, it is sometimes possible that the only indication of a

*Hydrologic Consultants, Inc. of Colorado, An HC Itasca Company, Lakewood, Colorado.

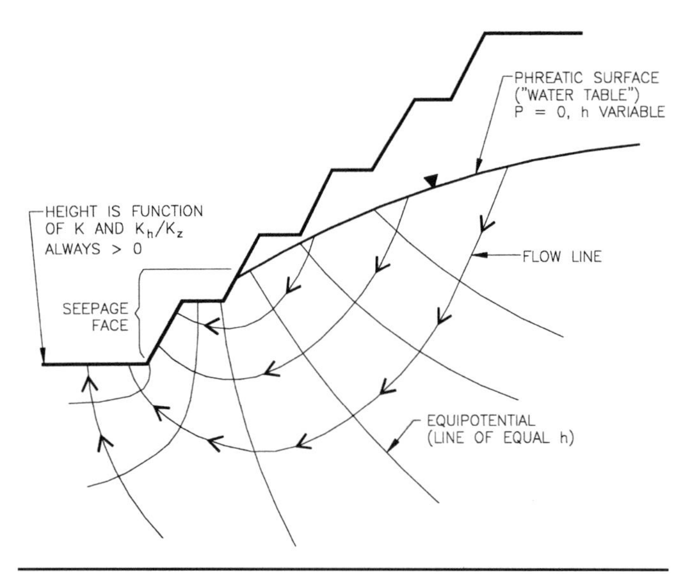

FIGURE 9.1 "Anatomy" of the groundwater flow system in a pit slope

seepage face is the buildup of evaporative residues, such as calcite or gypsum or ice if the temperature is below freezing.

Figure 9.1 also includes a flow net (something that took many hours to draw by hand by those of us from another generation!), indicating the equipotential lines and the theoretical flow lines. Equipotential lines describe the location where the hydraulic head, defined by

$$h = \frac{P}{\rho g} + z = h_p + h_z \qquad \textbf{EQ. 9.1}$$

Where:

h = hydraulic head
P = pressure
ρ = density of water
g = gravitational constant
z = elevation (usually with respect to the datum $z = 0$ at sea level)
h_p = pressure head
h_z = elevation head

is the same in a flow field. When we measure a water level (expressed in elevation) in a point piezometer with a standpipe, we are measuring the hydraulic head; and the pressure (expressed in meters of water) at the intake point is the hydraulic head less the elevation of that point. Conversely, if we measure pressure in a transducer, the hydraulic head is the elevation of the transducer plus the pressure (again expressed in meters of water). In addition to being the basis from which we can calculate pore pressures, hydraulic head is important to quantify because it is the form of potential energy that causes groundwater to flow. The change in head Δh over the distance ΔL is referred to as the hydraulic gradient i or

$$i = \frac{\Delta h}{\Delta L} \qquad \textbf{EQ. 9.2}$$

Before leaving the generic flow system depicted in Figure 9.1, some points are worth noting. Flow tends to converge to geometric singularities, such as the bottom of the highwall. In plan view, you will most often see seeps in the corners of pits or at the ends of long elliptical pits for the same reason. Finally, it should be noted that much of the flow from the base of the highwall and the floor of the pit is from vertically upward flow. Mines create "flowing artesian" conditions from nonflowing artesian groundwater flow systems.

The most significant property of a soil or rock with respect to groundwater flow is its hydraulic conductivity. For the simplest case of a material with hydraulic conductivity that is homogeneous (i.e., uniform throughout) and isotropic (i.e., the same in all directions), the hydraulic conductivity K is derived from Darcy's law

$$Q = K \cdot i \cdot A \qquad \textbf{EQ. 9.3}$$

Where:

Q = the measured volumetric flow rate [L^3/T]
A = the known cross-sectional area [L^2] of flow
i = the measured hydraulic gradient that causes the flow
K = an empirical constant of proportionality

Hydraulic conductivity is defined as the flow rate per unit area under a hydraulic gradient of 1 or

$$K = \frac{Q}{A} \qquad \textbf{EQ. 9.4}$$

A quick dimensional analysis of Eq. 9.4 indicates that hydraulic conductivity has the dimension of [L/T], the most commonly used unit being meters per day and the non-SI unit of centimeters per second.

Geologic materials have about 13 orders of magnitude range in their hydraulic conductivities, and it is not at all unusual to have materials differing by as much as 4 orders of magnitude adjacent to each other in a mine area. Figure 9.2 indicates the ranges of hydraulic conductivity for both unconsolidated and consolidated materials.

In hard-rock mines, we are usually dealing with fractured rock masses that have hydraulic conductivities that are heterogeneous (i.e., varies from rock unit to rock unit and often within a rock unit) and anisotropic (i.e., varies with direction) and almost entirely attributable to the fractures. In other words, we can usually ignore the contribution to the total hydraulic conductivity of the rock mass from the rock matrix.

The hydraulic conductivity of a single idealized fracture described by two smooth parallel plates is

$$K = \frac{\rho g a^2}{12\mu} \qquad \textbf{EQ. 9.5}$$

Where:

a = aperture, or width, of the fracture
μ = viscosity of water

Obviously, we cannot measure or even test (as we will discuss below) every fracture in a rock mass. However, we can utilize the concept of an effective or average hydraulic conductivity defined by

$$K_{eff} = \frac{\Sigma K_i}{L} \qquad \textbf{EQ. 9.6}$$

Where:

K_i = hydraulic conductivity of fracture i
I = number of fractures in "sampling" interval
L = length of sampling interval

The concept of heterogeneity in rock masses in a mine area is easy to understand because it can affect ore grade as well as hydrogeologic and geotechnical properties. From a groundwater flow and dewatering standpoint, the less-universal concept of anisotropy is equally important. Figure 9.3 shows a typical rock mass. In addition to the obvious heterogeneities associated with the fracture and shear zones, there are at least three joint sets. The latter would likely result in different hydraulic conductivities in different directions within the rock mass. If one of the joint

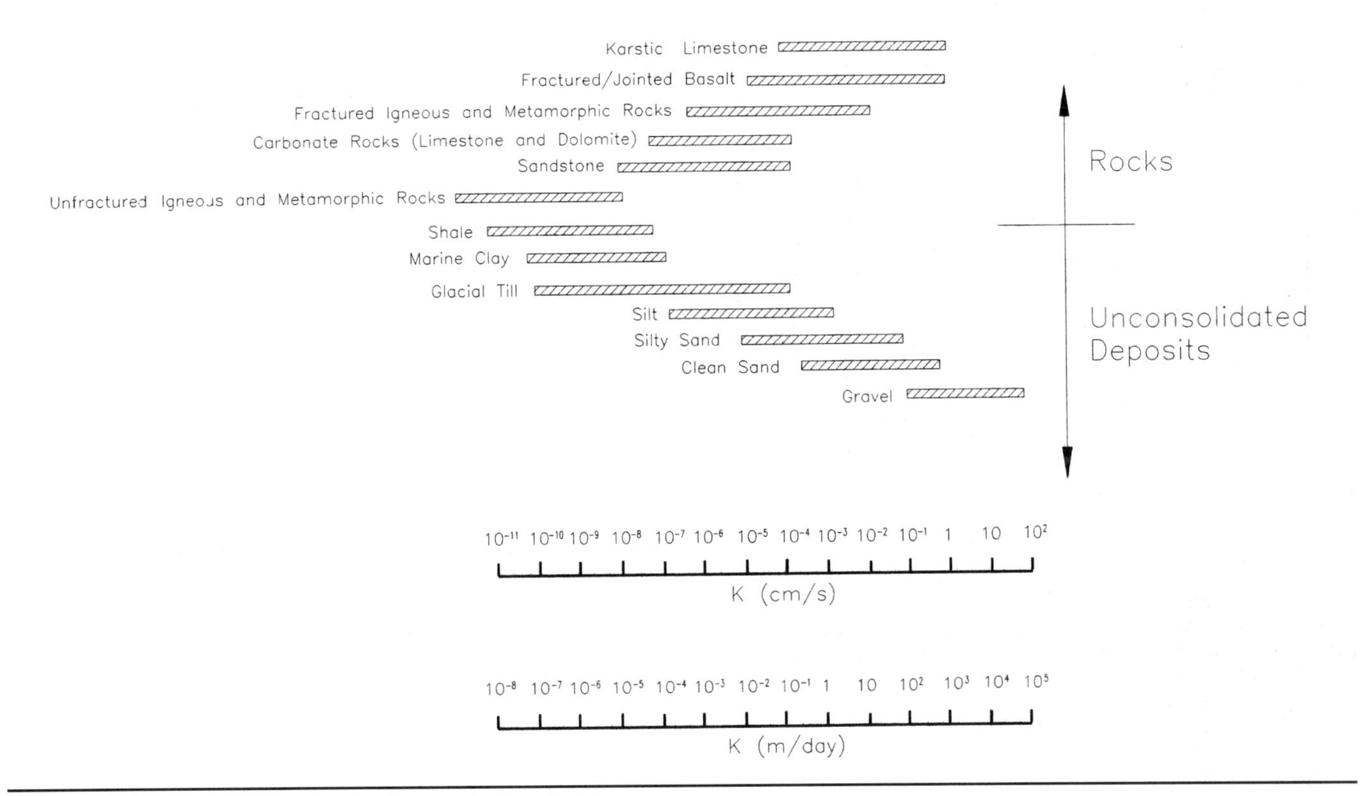

FIGURE 9.2 Hydraulic conductivity of various geologic materials

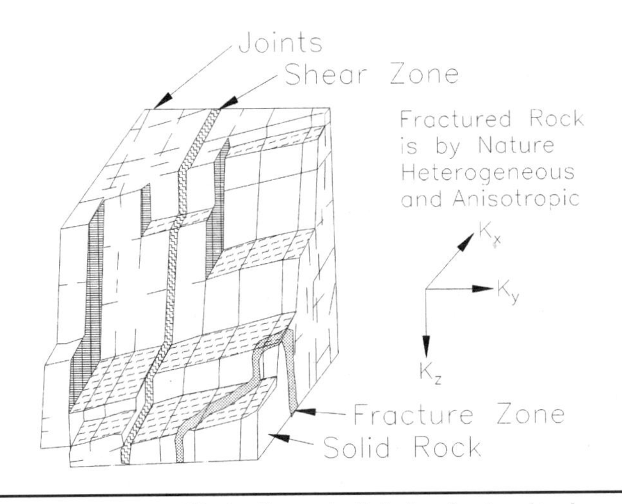

FIGURE 9.3 Nature of hydraulic conductivity in a fractured rock mass

sets is significantly more conductive than the others, we need to know this so we can exploit it with the dewatering system.

To be covered much more briefly than hydraulic conductivity, storage is the other primary hydraulic property of a geologic material. There are two types of storage–specific yield and specific storage. Specific yield is the volume of water that is released by actual drainage of pores when the phreatic surface is lowered. The specific yield S_y of geologic materials ranges from about 0.15 to 0.30 (with dimensions of L^3/L^3 or "dimensionless") in unconsolidated sedimentary materials to 0.005 to 0.02 in fractured rock. The second type of storage, specific storage, is the volume of water that is released when the saturated portion of the soil or rock mass is depressurized. The specific storage S_s is a function of both the compressibility of the rock and porosity and compressibility of the water as described by

$$S_s = \rho g(\alpha + n\beta) \qquad \textbf{EQ. 9.7}$$

Where:

α = compressibility of the material
n = porosity
β = compressibility of water in the pores

Dimensional analysis will show that S_s has the somewhat unusual SI units of m^{-1}. Using typical values for the compressibility of various soil and rock types (Freeze and Cherry 1979), typical values for S_s range from about $1 \times 10^{-4}\ m^{-1}$ for very compressible rock (highly fractured and altered) to $5 \times 10^{-7}\ m^{-1}$ for very rigid rock.

Having defined hydraulic conductivity and storage, we now have all of the factors to mathematically describe groundwater flow in a fully three-dimensional, anisotropic flow field

$$\frac{\partial}{\partial x}\left(K_x\frac{\partial h}{\partial x}\right) + \frac{\partial}{\partial x}\left(K_y\frac{\partial h}{\partial y}\right) + \frac{\partial}{\partial x}\left(K_z\frac{\partial h}{\partial z}\right) = S_s\frac{\partial h}{\partial t} + \sum Q_i \qquad \textbf{EQ. 9.8}$$

This partial differential equation is really not that complicated. It is merely an expression of the continuity equation and says that the change in flow (the first three terms on the LHS are simply the changes in flow defined by the Darcy relationship, Eq. 9.3, in all three directions) is equal to the change in storage (first term on the RHS) plus any sources or sinks such as pumping, recharge, etc., represented by the Q_i term.

9.2.2 Propagation of Drawdown During Dewatering

Most problems involving application of Eq. 9.8 to a specific groundwater situation are solved by numerical methods, and we will give an example of one such problem below. However, we may get a better understanding of how dewatering works if we first consider an analytical solution. For one-dimensional flow without any sources or sinks, Eq. 9.8 reduces to

$$\frac{\partial}{\partial x}\left(K_x\frac{\partial h}{\partial x}\right) = S_s\frac{\partial h}{\partial t} \qquad \textbf{EQ. 9.9}$$

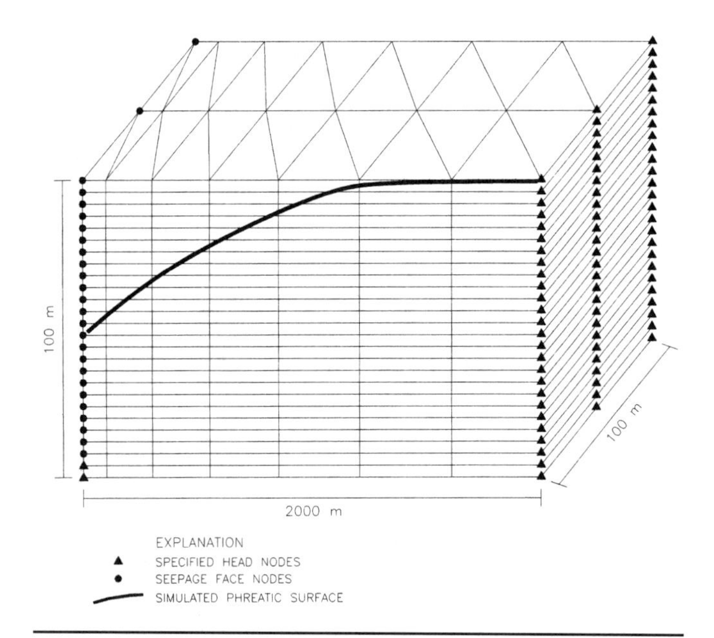

FIGURE 9.4 Model for estimating drawdown with time and distance

The analytical solution to Eq. 9.9 is

$$\Delta h(x,t) = h_o \cdot erfc\left(\frac{4K_x t}{S_s x^2}\right)^{-0.5} \qquad \textbf{EQ. 9.10}$$

Where:

$\Delta h(x,t)$ = drawdown at distance x from the source of the drawdown and time t
Δh_o = drawdown at the source (x = 0) starting at t = 0
K_x = hydraulic conductivity in the x-direction
S_s = specific storage

The argument of the *erfc* in Eq. 9.10 is one of those truly remarkable dimensionless terms that arise in the various physical sciences. It incorporates time, space, and the hydraulic properties. Because the *erfc*(ϕ) decreases when ϕ increases, it is easy to demonstrate (and it should be intuitive) that

- if x increases, then Δh decreases
- if K is larger, then Δh at x and t increases
- if t increases, then Δh at x increases
- if S_s increases, then Δh decreases

Thus, Eq. 9.10 gives us a quantitative understanding of how drawdown propagates through time from a drawdown source as a function of the hydraulic properties K and S_s of the geologic material. Because we cannot change the physical properties–although we can, by selection of our dewatering method, exploit the maximum hydraulic conductivity–we can use this relationship to tell us how far ahead in time and at what spacing we have to begin dewatering to achieve a certain drawdown at a certain point at a certain time.

Unfortunately, Eq. 9.10 cannot address complex, heterogeneous and anisotropic hydrogeology or boundary conditions such as a seepage face or release of storage from lowering of the phreatic surface. Consequently, as the next step in our heuristic journey, we will examine the results of a numerical model–albeit a very simplistic one–to understand how drawdown is propagated through a rock mass when there is a seepage face and lowering of the phreatic surface.

As shown in Figure 9.4, the model consists of a block of material 100 m wide and 100 m high with hydraulic conductivity

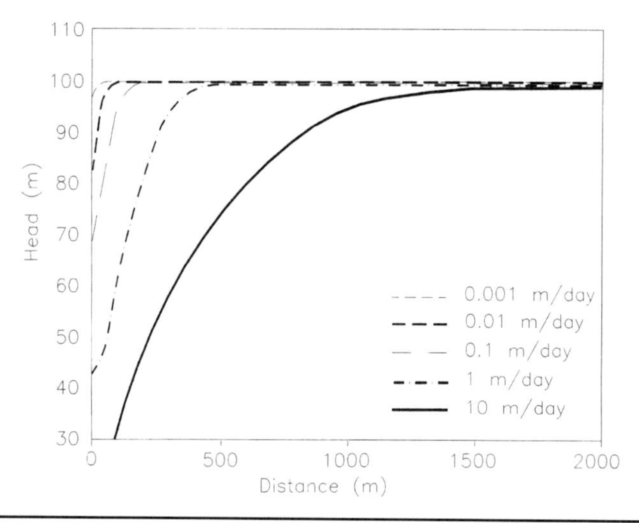

FIGURE 9.5 Drawdown versus distance at 100 days for various hydraulic conductivities

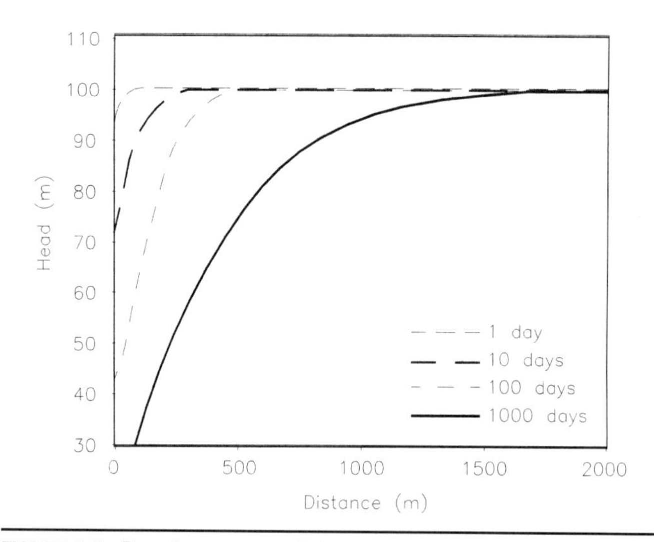

FIGURE 9.6 Drawdown versus distance at various times for K = 1 m/d

that is changed in a series of numerical simulations. In all simulations, the initial water level is at the very top (h = 100) and is maintained throughout all time at that level at the model boundary at a distance of 2,000 m. In each of the simulations, the head at the left boundary is instantaneously (at t = 0) lowered to h = 4 m (equivalent to a drawdown of 96 m) and the model calculates the propagation of the drawdown with time for several different values of hydraulic conductivity. Figure 9.5 is a plot of drawdown versus distance after 100 days for K values ranging from 0.001 to 10 m/d. Figure 9.6 is another drawdown versus distance plot at various times from 1 to 1,000 days for a single K value of 1 m/d.

Although these numerical simulations were for a very idealized situation, a few general, but important, conclusions can be drawn. First, the results depicted in Figure 9.5 indicate that a significant seepage face can develop along a surface of drainage (which could be a highwall under passive inflow conditions or the well bore of a dewatering well) when the hydraulic conductivity is less than about 1 m/d. This seepage face limits the drawdown that can be propagated from the drawdown source. Both Figure 9.5 and 9.6 indicate that the drawdown takes a significant amount of time to propagate over distance and that, at K values less than 0.01 m/d, the drawdowns that can be propagated in 100 days are very small.

9.3 PRACTICAL CONSIDERATIONS

Having reviewed some of the theoretical considerations associated with groundwater flow, particularly in fractured rocks, we now turn to the practical considerations of dewatering a rock mass. First, we need to decide what our goal is: are we trying to *depressurize* a rock mass in order to reduce the pore pressures in the discontinuities to a geotechnically acceptable level for a given rock type and slope design or do we have to actually *drain* a water-sensitive material (e.g., kimberlite or clayey overburden at a coal mine)? Together with the hydrogeologic conditions, the answer to that question will direct us as to how we dewater and when we have to get started.

9.3.1 Collecting Hydrogeologic Data and Planning for Dewatering

Collecting hydrogeologic data and designing and revising dewatering plans is a process that should start at the prefeasibility stage and continue almost to the end of mining. However, in this section, we are going to address only the types of information necessary at the early phases of planning and designing a dewatering system.

To determine whether there is the potential for a water-related problem and, if so, its nature and magnitude and then to decide how to mitigate it, the minimum data and information that need to be collected include

- Geologic maps and cross sections indicating the general geologic framework
- Hydraulic heads (i.e., water levels) within the major geologic units, hopefully showing any significant differences both laterally and vertically
- Hydraulic conductivity (including the anisotropy) of the various rock units
- The mine plan showing depth and extent of excavation by year

The collection of hydrogeologic data can and should be incorporated into the exploration program because so much of the information can be obtained from the exploration boreholes. It is unfortunate when separate hydrogeologic data-collection programs have to be conducted later when the data could have been collected during the exploration program.

Almost any exploration borehole can yield some information on water levels and relative hydraulic conductivity, unless it has been drilled with very heavy mud that obliterates the hydraulic characteristics of the in situ material. A simple way to obtain both water level data and estimates of hydraulic conductivity if the holes are drilled with air is to record the airlift production while drilling (Kauffman and VanDell 1983) and periodically, and certainly at total depth, conduct a short (no longer than 1-hr duration) airlift recovery test (Smith 1983).

Packer tests, which are most cost effective when a wireline packer system is used while coring, can provide useful estimates of variations in hydraulic conductivity along the length of the core hole providing that heavy muds are not required during the drilling. A major advantage of packer testing is that it can be conducted in coreholes that have been specially oriented to intersect a specific set of discontinuities and, hence, help define the anisotropy of the system. In addition to the packer tests, a useful indirect indicator of variations in hydraulic conductivity can be derived from the core logs, specifically the rock-quality designation (RQD), fracture orientation, fracture density (i.e., fractures per meter), and descriptions of the nature of the fractures (e.g., clay alteration, staining, etc.).

Traditional pumping tests are still very useful for determining the hydraulic conductivity of a large volume of rock and identifying the boundaries of any significant water-bearing units (again, I avoid the term "aquifer"). If the rock has a relatively low hydraulic conductivity, it might be necessary to conduct a rather elaborate test involving recycling of the discharge water with a constant-head, or Hurr, tank and installing inserts in monitoring wells to reduce the piezometric time lag (Hvorslev 1951). The reader is referred to Weeks (1977) and Kruseman and de Ridder (1991) for detailed information on the planning and analysis of pumping tests.

A select number of exploration boreholes should be completed as point piezometers with which to define and monitor changes in groundwater levels. The traditional point piezometer, implying a short (say 1 to 3 m) completion or screened interval, involves installing small-diameter pipe, sand packs around the intake (usually a short screen), and seals (either bentonite or cement) above the sand packs. Such construction can be quite problematic in deep or unstable boreholes. As an alternative, a limited number of investigators have experimented with grouting-in of transducers (McKenna 1995) and reported favorable results.*

9.3.2 Preliminary Analysis of Data

After hydrogeologic data have been obtained, the next step is to develop a conceptual hydrogeologic model. This model is simply a schematic diagram or "cartoon" that shows the

- Various geologic/hydrogeologic units that are known or presumed to have distinct hydraulic properties
- Hydraulic conductivity and, ideally, any known anisotropy either in terms of all three directions–K_x, K_y, and K_z– or at least in the horizontal K_h and vertical K_v directions
- Recharge to the system both from precipitation and any other natural (e.g., lakes) or constructed (e.g., tailings) water bodies
- Excavation of the pit over time

The next step is to quantitatively evaluate the potential problem to the extent possible and reasonable (in other words, do not overanalyze the problem with limited or uncertain data). Although analytical solutions exist for Eq. 9.8 under simplifying assumptions and can be used to address very simple hydrogeology and mine geometry (Singh and Atkins 1984), most analyses of mine-water problems today utilize a numerical model. A detailed model can be costly (in excess of $100,000 US), but those costs are usually insignificant compared to the cost of an ineffective or poorly conceived dewatering system or, worse, a major slope failure. A groundwater model allows a relatively complex hydrogeologic setting and time-variable hydraulic stresses (e.g., the excavation of the pit and implementation of various dewatering systems) to be simulated, tasks beyond the reach of analytical solutions. In the final section of this chapter, a case history utilizing a three-dimensional groundwater flow model will be summarized.

9.3.3 Dewatering Methods

Before describing the most common methods of pit dewatering, a few general "rules" might be appropriate. These are

- In materials of low hydraulic conductivity (say less than about 0.01 m/d), the drawdown from any reasonable number of dewatering points (e.g., wells, drain holes, etc.) is going to be propagated slowly, so start early!
- Always cut off any recharge, particularly recharge to materials with low hydraulic conductivity. It is usually easier to keep recharge away from such materials than to try to depressurize or drain them when they have a large source of recharge.

---

* This author has no personal experience with this potentially advantageous method and would appreciate any information from readers who have such experience.

TABLE 9.1 Comparison of common pit dewatering methods

Method	Advantages	Disadvantages	Best Application
Perimeter Wells	Usually have long life (life of mine); have large space for drilling; can be installed prior to mining; can intercept lateral inflow; logistically simple	Impacts on center of pit may be limited; usually must be deeper (hence, higher cost); might be in less permeable rock	Where flow to pit is lateral for small pit; where hydraulic conductivity is primarily horizontal
In-Pit Wells	Creates most drawdown in pit; can located in zone of potentially large hydraulic conductivity (often the ore zone); relatively shallow; can intercept vertical inflow through bottom of pit	Hard to mine around; short life; drilling logistics; need to deliver water and power to and from well; potential drilling problems; can be installed only after mining commences	"Compartmentalized" rock mass; very asymmetric pits; mine in which large permanent benches are established early
In-Pit Horizontal Drain holes	Increase pit slope stability; can drain-depressurize through targeted structures; passive dewatering; no special location needed; inexpensive	Winter freezing; drains/depressurizes only limited area; can only be installed after mining begins; water delivery	Deep pits; low permeability rock masses; highly anisotropic rock masses; to breach groundwater "dams" (e.g., gouge zones)
Drainage Galleries	Dewaters from below the mine; can intercept structures at optimum angles; can handle large quantities of water	High cost of excavation; large lead time for construction	Long mine life; good tunneling conditions; where construction can be done for dual purposes (e.g., exploration, high-grading)

- Make sure the depths of the drawdown centers (again, wells or drain holes) are sufficiently deep to be able to propagate sufficient drawdown over distance. In some hydrogeologic settings, however, the hydraulic conductivity decreases significantly with depth, so there can be a limitation to the most appropriate depth.
- Almost all fractured rocks have anisotropy. Determine what it is and exploit it! As indicated by Eq. 9.6, the most effective dewatering system will be the one that intersects the greatest number of water-bearing fractures. Unless you have other information indicating the joints are not conductive (e.g., filled with clay gouge), orient wells and drain holes as orthogonal as possible to the main fracture sets.

With few exceptions, such as the planned use of a perimeter freeze wall in a unique hydrogeologic setting (Hanna et al. 1999), the most common methods of dewatering a surface mine are to simply collect the passive inflow to the pit in sumps or to install an active dewatering system that uses wells (perimeter and/or in-pit), horizontal drain holes from pit benches, or underground drainage galleries with drain holes. It is not unusual to utilize more than one method at a mine. Simply sumping passive inflow to the pit does nothing to reduce pore pressure in a highwall and is usually done only in small pits or where the rock is unusually competent. Table 9.1 focuses on the four most common active dewatering methods and summarizes the general advantages and disadvantages of each.

Based on both theoretical and practical considerations, the reader should now be aware that the choice of dewatering system must be based on the site-specific hydrogeology and the mine plan. Obviously, economics and risk factors must also be considered. I hope that the reader will have some new insight into the mine-water problem and can be a valuable member of any decision-making team.

9.4 CASE HISTORY

This chapter closes with a brief case history. This example shows how a numerical model is being used to design the long-term dewatering system for a surface mine that will ultimately have a highwall height of approximately 700 m and a bottom nearly 650 m below the original phreatic surface.

As shown in Figure 9.7, the east highwall of the pit is comprised of three main rock units: an upper rhyolite with a hydraulic conductivity of about 0.1 m/d, an intermediate rhyolite with a hydraulic conductivity of about 0.05 m/d, and a lower andesite with a hydraulic conductivity of about 0.006 m/d. The situation is complicated by recharge from seepage from a tailings pond located about 500 m from the pit. Fortunately, the upper rhyolite has a relatively high hydraulic conductivity and appears to have a volcanoclastic layering that imparts a horizontal anisotropy that has been exploited by a line of five vertical wells in effectively intercepting the recharge from the tailings. Currently, only a limited number of seeps and a few small slope failures exist in the highwall.

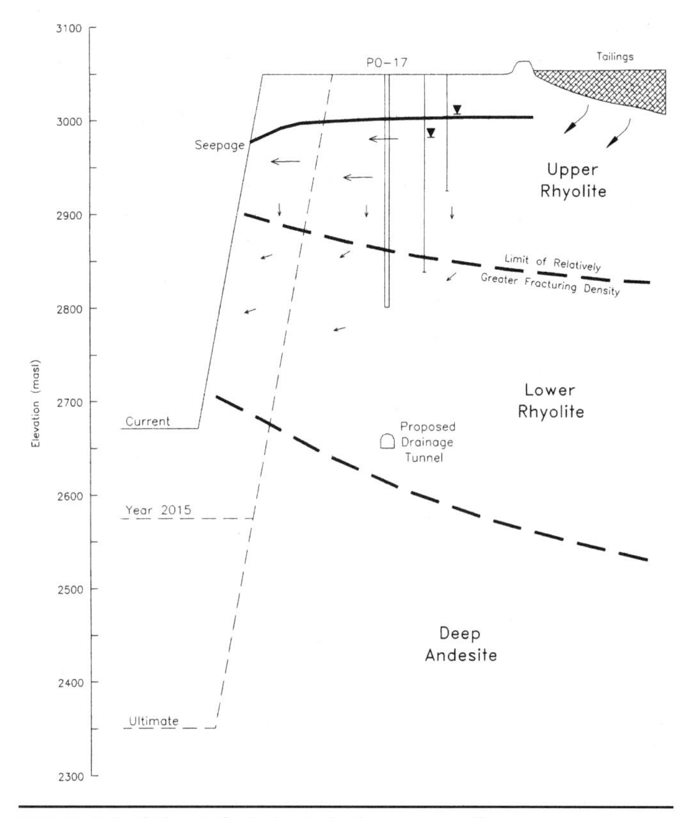

FIGURE 9.7 Schematic hydrogeologic cross section

A three-dimensional finite model using the code *MINEDW* (Azrag, Ugorets, and Atkinson 1998) has been developed to help evaluate various alternative future dewatering systems. The model grid is shown in map view in Figure 9.8 and in cross section in Figure 9.9. The model (Figure 9.10) predicts that, in the future when the pit deepens and penetrates the less-permeable andesite, a high phreatic surface and a seepage face nearly 200 m high will develop without any dewatering in the lower rhyolite and andesite highwall.

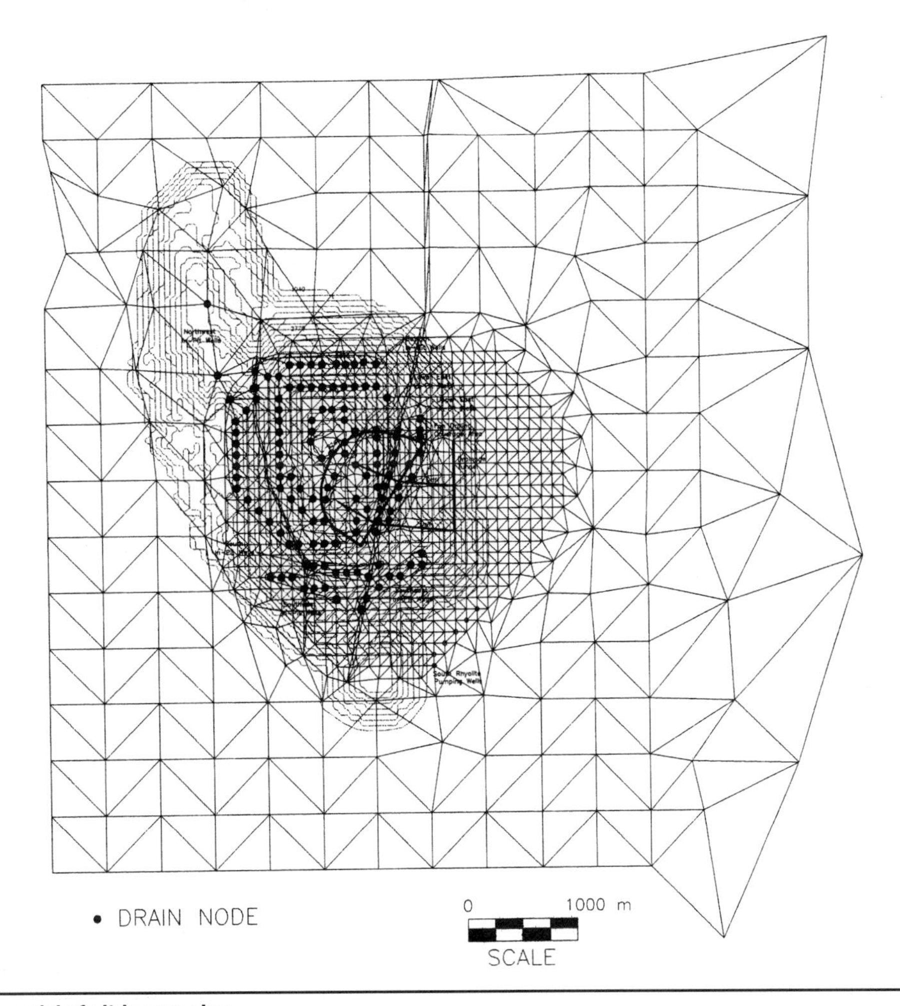

FIGURE 9.8 Numerical model of pit in map view

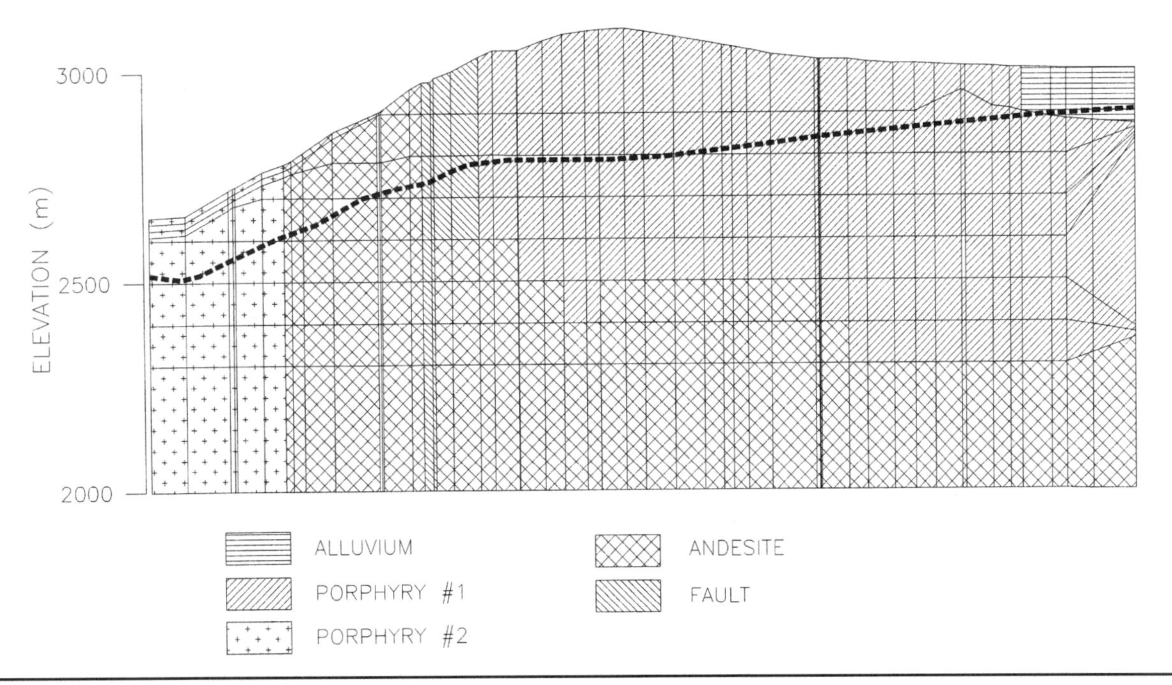

FIGURE 9.9 Cross section through numerical model

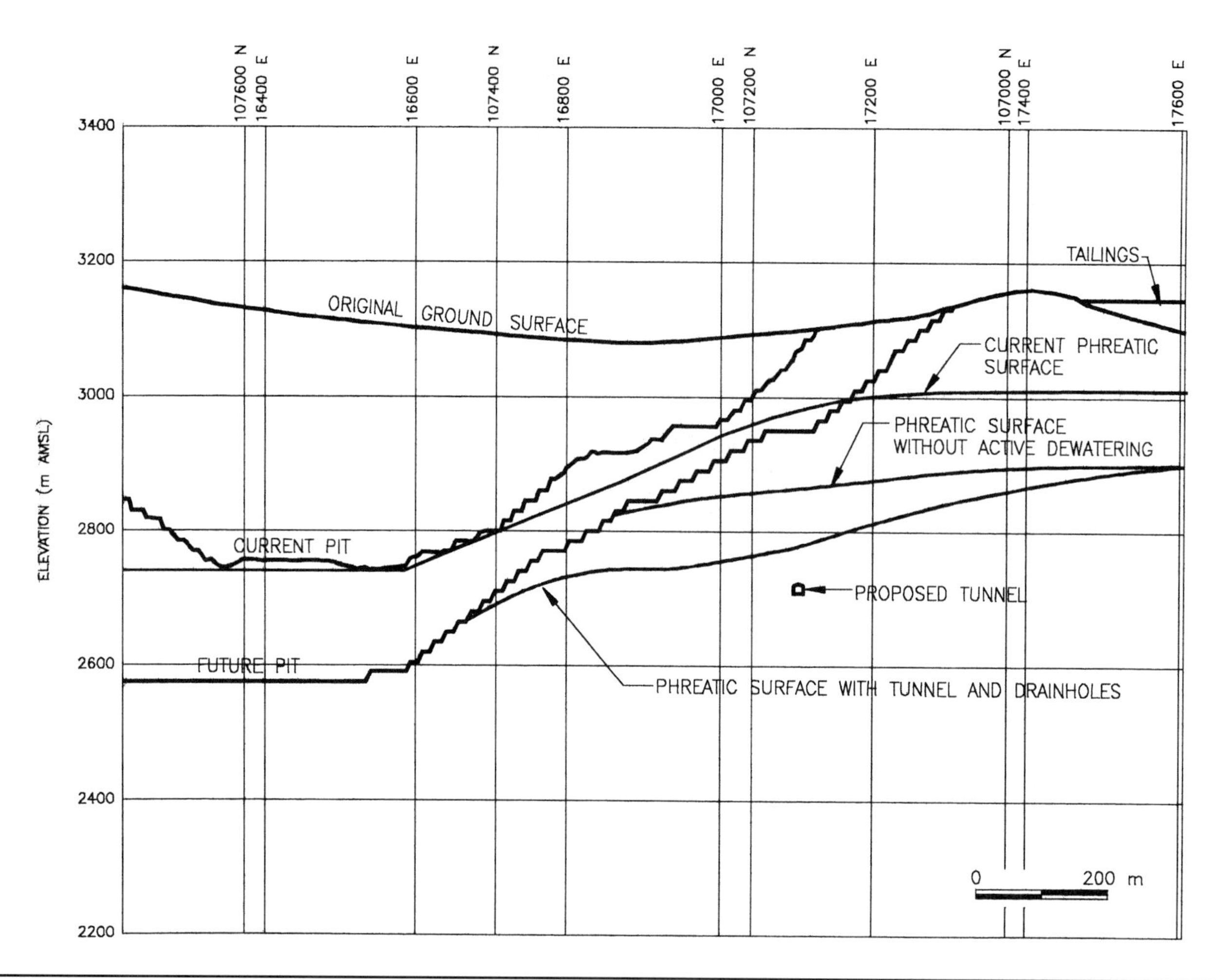

FIGURE 9.10 Predicted effect of dewatering on phreatic surface in highwall

In this ongoing work, a series of wells and a drainage gallery from which numerous drain holes would be drilled (represented by the nominal drain nodes in Figure 9.8) has been simulated. Figure 9.10 shows the predicted lowering of the phreatic surface by this proposed dewatering plan. As previously stated, this is an ongoing study. Better estimates of hydraulic conductivity are being obtained by calibration of the model to measured water levels in the field, and additional dewatering scenarios based on engineering design and economic factors will be simulated in the future.

9.5 REFERENCES

Azrag, E.A., V.I. Ugorets, and L.C. Atkinson. 1998. Use of a finite element code to model complex mine water problems. *Proceedings of Symposium on Mine Water and the Environment.* International Mine Water Association, Johannesburg, South Africa. Vol. 1, 31–41.

Freeze, R.A., and J.A. Cherry. 1979. Groundwater. Englewood Cliffs, New Jersey: Prentice-Hall, Inc.

Hanna, T.M., R.L. Howell, V.I. Ugorets, T. Ternes, and J. McCarter. 1999. Installation of a freeze wall. *Sudbury '99.* Canadian Institute of Mining and Metallurgy. Sudbury, Ontario.

Hvorslev, J. 1951. Time lag and soil permeability in groundwater observations. *U.S. Army Corps of Engineers, Waterways Experiment Station.* 36:50.

Kauffman, J.M., and T.D. VanDell. 1983. Integrating a groundwater data reconnaissance program into a mineral exploration program. *Mining Engineering.* 35(2).

Kruseman, G.P., and N.A. de Ridder. 1991. Analysis and evaluation of pumping test data. Institute for Land Reclamation and Improvement. Wageningen, The Netherlands. 377.

McKenna, G.T. 1995. Grouted-in installation of piezometers in boreholes. *Canadian Geotechnical Journal.* 32:1:355–363.

Singh, R.H., and A.S. Atkins. 1984. Application of analytical solutions to simulate some mine inflow problems in underground coal mining. *International Journal of Mine Water.* 3:4:1–27.

Smith, P.C. 1983. Determination of aquifer transmissivity from airlift and recovery tests. *The Geological Survey of South Australia.* No. 86.

Weeks, E.P. 1977. Aquifer tests. In U.S. Geological Survey, National Handbook of Recommended Methods for Water-Data Acquisition. 2-115 to 2-149.

CHAPTER 10

The Influence of Seismic Events on Slope Stability

Charles E. Glass*

10.1 INTRODUCTION

Landslides triggered by earthquakes can cause widespread damage, damage that can, in some instances, exceed damage due to earthquake ground shaking and fault displacement (Jibson, Harp, and Michael 1998). The October 1989 Loma Prieta earthquake in northern California, for example, triggered approximately 1,500 landslides, and as many as 4,000 existing landslides may have experienced additional displacement (Keefer and Manson 1998).

The Northridge, California, earthquake in 1994 triggered more than 11,000 landslides (USGS 1996), most of which were shallow slides and rockfalls (Harp and Jibson 1995, 1996), indicating that slope displacements caused by the seismic shaking were relatively small. Correlation of seismic shaking with mine slope failures is far less compelling, perhaps because small, shallow slides and rockfalls seldom disrupt mining operations. Few large, deep, coherent landslides occurred during the Loma Prieta and Northridge earthquakes, and we are aware of no large, deep, coherent mine slope failures that can be attributed to earthquake shaking.

The degree to which a seismic event influences a mine slope depends on a number of factors, not the least of which is the infrequent occurrence of earthquakes. Unlike gravity and other static forces, earthquake forces are *not* likely to influence a slope during its design lifetime. If a large earthquake should occur within the vicinity of a mine slope, however, the effects can be profound. To provide motivation for our study of the effects of seismic events on slopes and to aid in visualization, a fictitious mine, SME Mine, has been created in south central Idaho (Figure 10.1). At the northeast corner of SME Mine is a rock slope defined by a plane fracture zone daylighting at the base of the slope. We designate it Slope 24NE and show its cross section in Figure 10.2. Our charge is to evaluate the stability of Slope 24NE subjected to earthquake shaking. We shall refer to SME Mine and to Slope 24NE in the discussions of seismic analysis techniques that follow.

10.2 EARTHQUAKE HAZARD

The first step in assessing the influence of seismic forces on Slope 24NE is to assess the likelihood that an earthquake will occur in the vicinity of the slope during the life of the mine. A reasonable approach is to determine whether earthquakes have occurred there in the past. This task is divided into the following two parts: (1) evaluate historical seismicity within the mine environs and (2) evaluate evidence of prehistoric seismicity on nearby active faults. The strategy is divided into two steps, because the lack of a large earthquake during the short historical period does not preclude one from occurring in the near future. Conversely, the occurrence of a large earthquake during the historical period does not imply that one is likely to occur again soon.

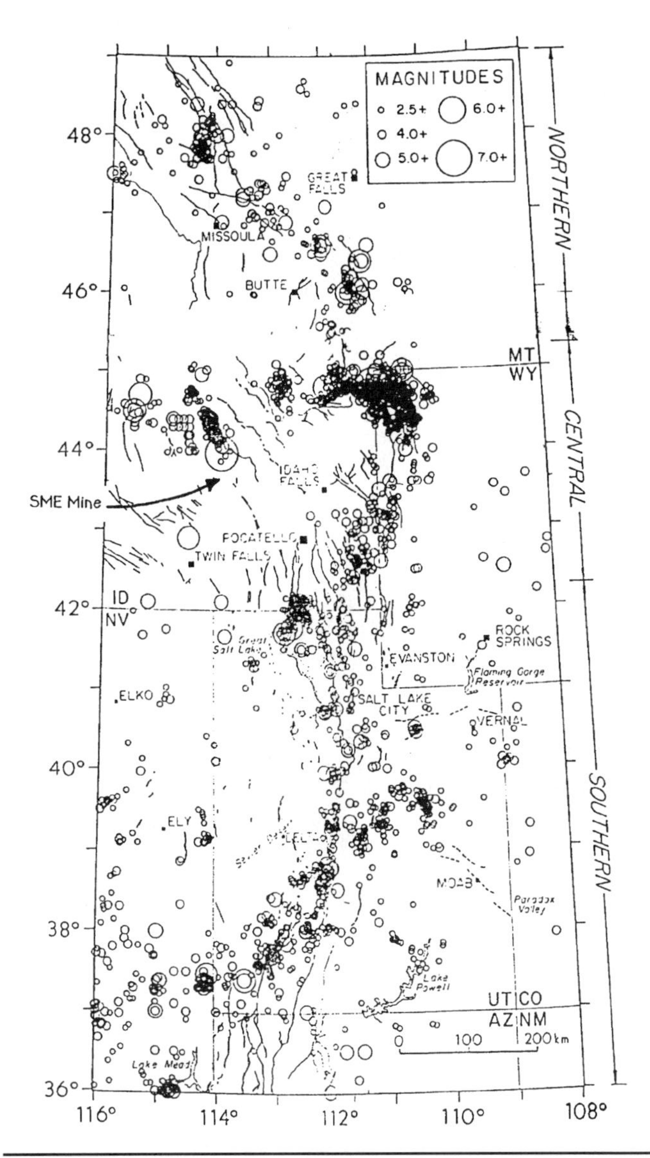

FIGURE 10.1 Location of SME Mine within the intermountain seismic belt (Smith and Arabasz 1991)

10.2.1 Historical Seismicity

The historical seismicity provides a background level of seismic activity. This background activity usually consists of relatively small earthquakes that cannot be related confidently to known

* Department of Mining and Geological Engineering, The University of Arizona, Tucson, Arizona.

TABLE 10.1 Historical earthquake catalog for the SME Mine, Idaho

Date	Magnitude	Distance (km)	Acceleration (%g)	Date	Magnitude	Distance (km)	Acceleration (%g)
1975	4	156	0.42	1984	4.4	91	0.86
1976	4.4	158	0.51	1984	4	90	0.7
1977	4.3	124	0.61	1985	4.7	106	0.87
1978	4.1	142	0.49	1985	4.5	106	0.79
1978	5	152	0.73	1985	4.1	88	0.76
1982	4.1	117	0.58	1986	4	61	1.02
1983	4	172	0.38	1986	4.1	83	0.8
1983	7.3	52	6.68	1986	4	79	0.8
1983	4.6	64	1.32	1986	4	75	0.84
1983	4	80	0.79	1986	4	76	0.83
1983	4	71	0.88	1986	4.3	76	0.97
1983	4.1	71	0.93	1986	4.6	78	1.11
1983	5.8	53	2.99	1987	4.4	77	1.01
1983	4	79	0.8	1988	4.1	98	0.69
1983	5.8	70	2.3	1988	4.2	142	0.51
1983	5.5	76	1.82	1988	4.8	141	0.71
1983	4.8	53	1.74	1988	4	167	0.4
1983	4.7	70	1.29	1988	4.9	95	1.08
1983	4	76	0.82	1991	4.5	107	0.78
1983	4.2	74	0.95	1992	4	185	0.36
1983	4.3	77	0.95	1992	4.3	117	0.65
1983	4.6	60	1.4	1992	4.3	110	0.69
1983	4	77	0.81	1992	4.7	101	0.92
1983	4.5	92	0.9	1992	4.4	97	0.81
1983	4.1	85	0.78	1992	4	85	0.74
1983	4.4	77	1.01	1993	4	115	0.56
1983	4	75	0.83	1993	4.1	121	0.56
1984	4	76	0.82	1993	4	116	0.56
1984	4.5	61	1.31	1993	4.6	114	0.78
1984	4.5	84	0.98	1994	5	99	1.1
1984	4.2	131	0.55	1994	4.3	99	0.76
1984	4	129	0.5	1994	4.5	104	0.8
1984	4.1	135	0.51	1995	4.7	117	0.8
1984	4	82	0.77	1995	4.2	174	0.42
1984	5.8	96	1.71	1995	4	166	0.4
1984	4	99	0.64	1996	4	166	0.4
1984	4.1	97	0.69	1996	4.3	102	0.74
1984	5	94	1.15	1998	4.2	100	0.71
1984	4	91	0.7	1999	5.1	163	0.72
				1999	4	163	0.4

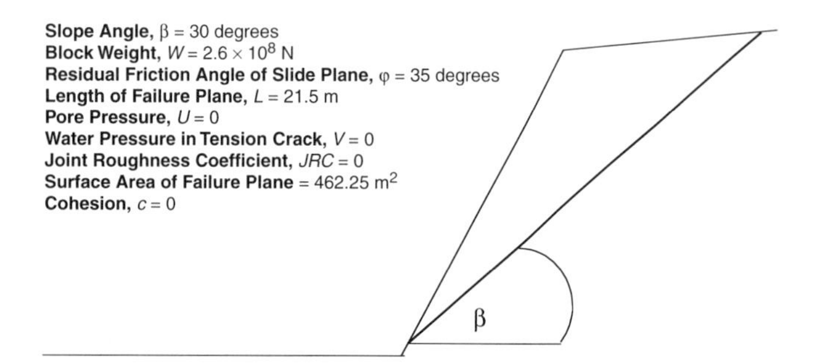

FIGURE 10.2 Cross section of slope 24Ne, SME Mine, Idaho

geologic faults. Historical seismicity within the slope vicinity can be assessed using one of many excellent earthquake databases. The U.S. Geological Survey's National Earthquake Information Center (NEIC) provides several national and worldwide earthquake databases on its web site (wwwneic.cr.usgs.gov), and many states have similar services.

SME Mine is arbitrarily located at 43.6° north latitude and 114° west longitude (see Figure 10.1), approximately half way between Boise, Idaho, and Idaho Falls, Idaho. This location was chosen because it is approximately 50 km south of the 1983 Borah Peak earthquake, which had a magnitude of 7.3. A search of the NEIC database reveals 79 earthquakes having magnitudes greater than 4.0 within 200 km of the SME Mine site since 1975 (Table 10.1).

10.2.2 Prehistoric Seismicity

The likelihood of future large earthquakes on known faults (these earthquakes have been termed *characteristic earthquakes* by Schwartz and Coppersmith [1984]) in the site vicinity is estimated using geologic information on the rupture length, fault offset, and recurrence rates for characteristic earthquakes.

The largest historical earthquake to occur within 200 km of the site was the 1983 Borah Peak earthquake, approximately 50 km to the north of SME Mine. This earthquake occurred on the Lost River Fault in central Idaho. To assess the rate of occurrence of large earthquakes on the Lost River Fault, Crone et al. (1987) mapped three sets of striations along the Thousand Springs segment of the fault. The latest set was due to rupture of the fault causing the 1983 Borah Peak earthquake. The other two occurred during previous ruptures on the same segment. The authors suggest that such small-scale striations would not be preserved long (10,000 to 100,000 years), hence the recurrence rate along this segment of the Lost River Fault probably ranges between 3,000 and 30,000 years. Morphological dating of pre-1983 fault scarps on the Lost River Fault, using diffusion modeling (Hanks and Schwartz 1987), indicates that the rupture before the 1983 earthquake occurred between 6,000 and 8,000 years ago.

These geologic studies provide an estimate of the recurrence rate on the Lost River Fault (approximately 7,000 years) and the time since the last rupture (17 years). Let us assume, for the purposes of this discussion, that careful study reveals no other seismogenic faults within 200 km of the site. All of the important information necessary for assessing the likelihood that Slope 24NE will experience a future seismic event is now present.

10.2.3 Hazard Analysis

The goal is to estimate the effects that earthquakes will have on mine slopes. Slopes respond to elastic waves generated by rupture along a fault. We will use knowledge of earthquake size (represented by the magnitude) and distance from the site to estimate the characteristics (amplitude, duration, and frequency content) of the elastic waves that may propagate across the site.

Ground motion at SME Mine is a function of earthquake magnitude, distance of an earthquake (hypocentral distance) from the mine site, and attenuation of seismic energy as it propagates from an earthquake source to the site. The first two aspects are available in the NEIC data file (Table 10.1). The latter aspect involves using an attenuation function, an empirical relationship relating earthquake size and distance to maximum peak particle acceleration (ppa) in the far field of an earthquake.

Matching the attenuation function to the site region is important, because earthquake vibrations are attenuated differently in different parts of the world. The attenuation function chosen for the SME Mine site was developed by Spudich et al. (1997) using data from extensional regime earthquakes having moment magnitudes greater than 5.0 and epicentral distances less than 105 km. Many of the earthquakes used in developing the attenuation function were selected from within the intermountain seismic belt (ISB), a belt of earthquakes extending from southeast California through Nevada, Utah, and Idaho, and into Montana (Figure 10.1). Extensional tectonic regimes usually display some or all of the following features:

- A mixture of normal dip–slip and strike–slip fault displacements (the Lost River fault displays an oblique slip mechanism combining normal and strike–slip components)
- Recent volcanism (SME Mine is on the northern edge of the Snake River Plain)
- Aligned volcanic features
- Lithosphere thinning
- High heat flow

Observations indicate that an extensional state of stress within a region affects the amplitude of earthquake ground motion occurring there (Spudich et al. 1997). Normal dip–slip fault displacements, for example, tend to produce lower ground-motion amplitudes than do strike–slip fault displacements. Because extensional regimes also display a degree of similarity throughout the world, earthquakes from extensional regimes in other countries were used to improve the attenuation function. Earthquake records used to construct the attenuation function included the 1983 Borah Peak, Idaho, earthquake and earthquakes associated with the Yellowstone hot spot. Data were divided into two categories depending on whether the recording instrument was on rock or soil. For this example, choose the attenuation function for rock, given by

$$\log_{10}Y = b_1 + b_2(M-6) + b_3(M-6)^2 + b_4R + b_5\log_{10}R + b_6\Gamma \quad \textbf{EQ. 10.1}$$

Where:

$R = \sqrt{r^2 + h^2}$
$\Gamma = 0_{Rock} = 1_{Soil}$
$Y = PPA(g)$
M = Magnitude
$b_1 = 0.156$
$b_2 = 0.229$
$b_3 = b_4 = 0.0$
$b_5 = -0.945$
$b_6 = 0.077$
$h = 5.57$
r = the epicentral distance, in kilometers

The earthquake ground motion at the mine site is estimated using Eq. 10.1 for each of the historical earthquakes occurring between April 4, 1975, and August 26, 1999. The estimates produce a catalog of earthquake ppa's at the mine site (Table 10.1).

Attenuation relationships, such as Eq. 10.1, are usually derived empirically using measurements of earthquake ground motion recorded at numerous locations. The function so derived provides a minimum error fit to a distribution of ground-motion measurements. Hence, attenuation functions present mean values. Distribution of the ground-motion measurements about the mean can be quite large. The Spudich et al. (1997) attenuation function has a listed standard deviation of $\sigma_{\ln} = 0.216$ ($\sigma_{\ln}$ is the standard deviation of Log_{10} Y). Consequently, the one-sigma variation in the ground motion at SME Mine due to the October 28, 1983, magnitude 7.3 Borah Peak earthquake ranges from 4%g to 11%g, with an expected value of 6.8%g.

We can use equations similar to Eq. 10.1 to estimate the peak velocity, predominant period, and duration of strong motion. The Borah Peak earthquake, for example, would have caused vibration at SME Mine having duration of strong motion of approximately 15 s and a predominant period of approximately 0.4 s.

The hazard analysis proceeds by computing the annual rate (λ) of earthquakes having a given ppa at the mine. Figure 10.3 presents a histogram of earthquake ground-motion rates at the site. Once the acceleration rate is established, the earthquake hazard may be computed using a zero-memory Poisson model, given by

$$P(a \geq a_i) = 1 - e^{-\gamma t} \quad \textbf{EQ. 10.2}$$

Where:

$P(a \geq a_i)$ = the probability that a site acceleration, a, will exceed a given level, a_i
t = the time from present (a time period of 100 years in increments of 10 years is chosen for this example)

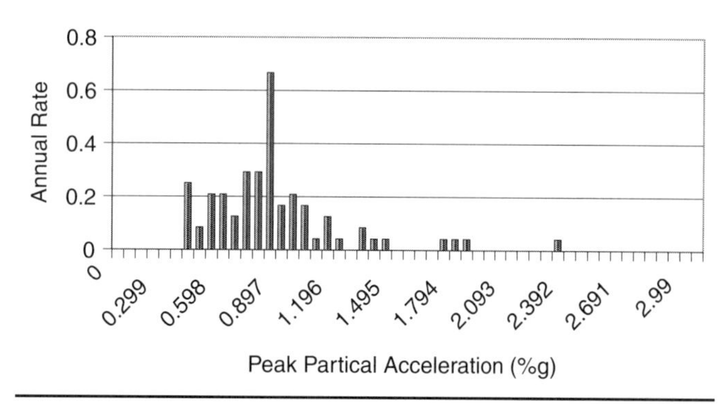

FIGURE 10.3 Annual rate of earthquake ppa's at SME Mine, Idaho

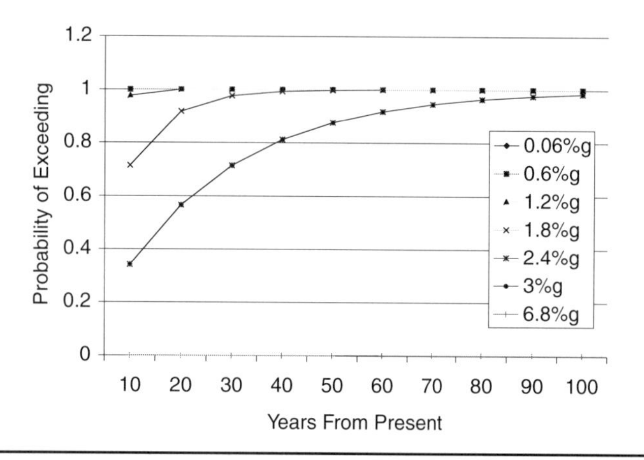

FIGURE 10.4 Earthquake hazard at SME Mine, Idaho. The earthquake hazard is the probability that a given ppa will be exceeded in the given period of years.

By using a zero-memory model, one assumes that historical earthquakes are uncorrelated in time, so that knowledge of the occurrence of an earthquake today indicates nothing about the likelihood that another will occur tomorrow. This is a reasonable assumption for the historical record, which includes earthquakes from many different sources. Later, when we deal with individual fault zones, we shall use a model that considers earthquake occurrence more likely as the time interval since the last earthquake (the *holding time*) increases.

Figure 10.4 presents the historical earthquake hazard for SME Mine computed using Eq. 10.2. The hazard computation reveals a 60% chance that SME Mine will experience earthquake vibrations having ppa's exceeding 3%g within the next 20 years. No historical earthquake produced ppa's at the mine site between 2.4%g and 3%g, so the probability curve for 2.4%g overlaps the 3%g curve. In computing Figure 10.4, the Borah Peak earthquake was deleted from the historical data. The Borah Peak earthquake is a characteristic earthquake that may not recur for another 7,000 years. If it is included with the historical data, its occurrence within the past 17 years will cause the probability of its occurrence within the next 100 years to be unrealistically high. Hence, the probability of a recurrence of this earthquake will be treated differently.

To assess the probability of experiencing ground vibrations from another Borah Peak earthquake within the next 100 years, a Markov approach is used (see Glass 1990). It seems reasonable that a fault that has just released a significant amount of stress in the form of an earthquake will require some time to accumulate sufficient stress to generate another one. The Markov approach considers the likelihood of earthquake occurrence to increase as the holding time for a given fault increases.

The Markov probability that the site will experience a ground acceleration exceeding 6.8%g within the next 100 years is 9.3×10^{-4}. This is shown in Figure 10.4, but the probability is too low to be seen above the axis. Similar analyses should be completed for other active faults within 200 km of the mine site. We assume that no other characteristic earthquake sources can be found, so we shall arbitrarily choose the Borah Peak earthquake as the design earthquake for slope 24NE, even though the probability of it occurring within the next 100 years is low.

10.3 SLOPE STABILITY

Determining the response of Slope 24NE to the design earthquake will require, in some cases, more detailed information regarding the design earthquake motion. Specifically, it will be helpful to estimate the general form of the actual earthquake accelerogram that would have been measured at SME mine had an accelerometer existed there at the time of the Borah Peak earthquake. A design accelerogram can be simulated either by scaling a recorded accelerogram from another site (a deterministic approach) or by creating an accelerogram anew using filtered and modified segments of white noise (a probabilistic approach). We shall choose the latter approach so that a large number of accelerograms can be simulated if desired. The technique, which follows that suggested by Jennings, Housner, and Tsai (1968), generates a sequence of uncorrelated random numbers having a mean of zero and a standard deviation of one. The sequence is modified by an envelope function to provide the proper shape, filtered to provide the proper frequency content, and finally scaled to provide the proper ppa. The design accelerogram for SME Mine is shown in Figure 10.5.

Given the design earthquake motion, the problem now is to determine the stability of Slope 24NE subjected to this vibration. The following three approaches are commonly used today to compute the stability of slopes subjected to seismic shaking: pseudo-static analysis, Newmark analysis, and dynamic analysis.

10.3.1 Pseudo-Static Analysis

The pseudo-static analysis is somewhat of a misnomer, because there is nothing *pseudo* about it; it is a static, limit-equilibrium analysis that treats earthquake motion as a static force similar to gravity. As we can see in Figure 10.5, earthquakes are not static, so the analysis is conservative, often predicting failure for modest slope angles. Civil construction practices tend to embrace conservatism, however; so this procedure has acquired substantial momentum in current construction practice and has even been used recently for mine slopes (Abel 1997).

The basic assumption underpinning the pseudo-static analysis (Nash 1987; Janbu 1973; Bromhead 1986; Chowdhury 1978; Morgenstern and Sangrey 1978; Hunt 1984; Duncan 1996) is that earthquakes can be modeled as a static force acting on the center of mass of a potential landslide. To perform the analysis, one adopts a ground-motion parameter, referred to as the *seismic coefficient, k*. The seismic coefficient, when multiplied by the gravitational force, g, acts in a horizontal direction to produce an inertial force out of the slope. The factor of safety for a pseudo-static analysis, assuming no dynamic pore pressure effects, is given by the following limit-equilibrium equation

$$FS = \frac{cL + [W(\cos\beta - k\sin\beta) - U - V\sin\beta]\tan\phi}{W(\sin\beta + k\cos\beta) + V\cos\beta} \qquad \textbf{EQ. 10.3}$$

Where:

FS = factor of safety
c = cohesion along the slide plane
L = length of the slide plane
W = weight of the slide block
U = pore pressure

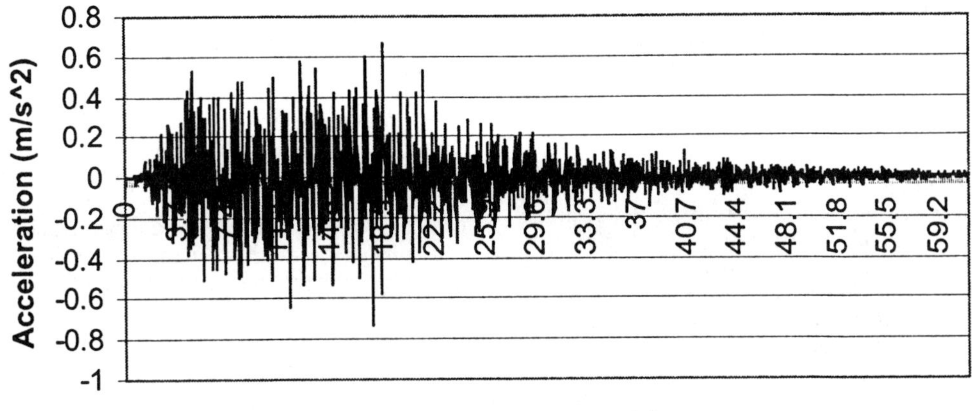

FIGURE 10.5 Design accelerogram for SME Mine, Idaho

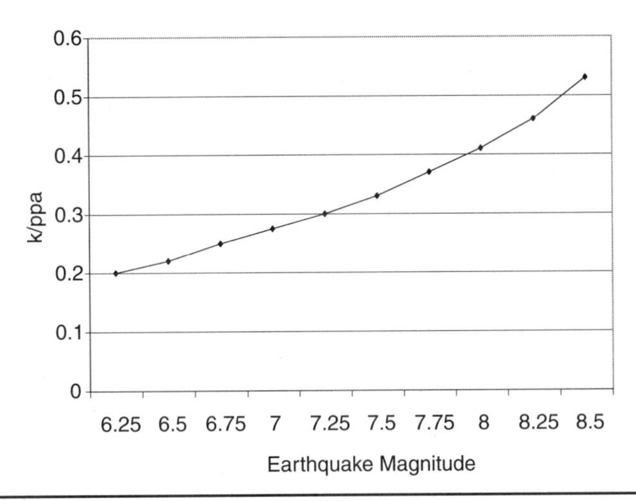

FIGURE 10.6 Recommendations for choosing seismic coefficient, *k* (Pyke 1999)

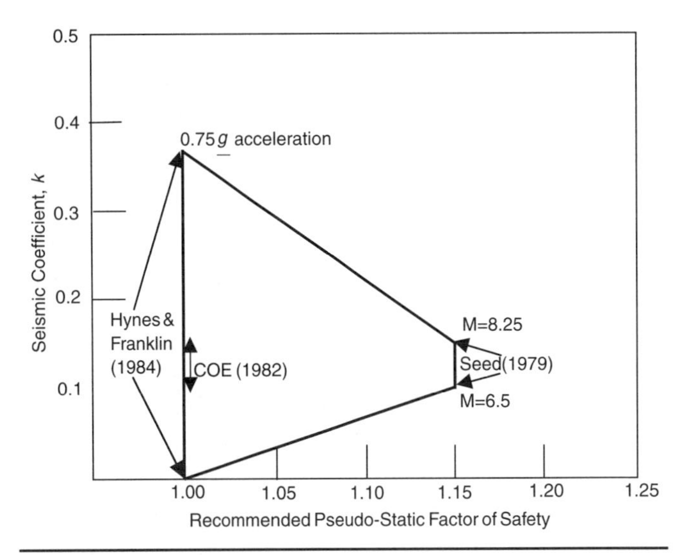

FIGURE 10.7 Summary of published criteria for choosing seismic coefficient, *k*

β = angle of the slide plane
V = pore pressure in vertical tension cracks at the head of the slide
k = seismic coefficient

Values for these parameters are given in Figure 10.2. If a "method of slices" technique is employed, a suitable seismic coefficient must be applied separately to each slice.

The seismic coefficient is fictitious–as fictitious as SME Mine. It is *not* the ppa estimated in the above analysis. In fact, it is not measurable at all, and there is no simple, universally accepted way to determine an appropriate value. Therefore, selecting the seismic coefficient has relied heavily on "engineering judgment" in the past. The U.S. Army Corps of Engineers' (COE) manual for seismic design of dams stipulated in 1982 that slopes in seismic zone 3 must have a minimum factor of safety of 1.0 using a seismic coefficient of 0.1, and slopes in seismic zone 4 must use a seismic coefficient of 0.15. Today, many California State agencies stipulate that slopes must be designed for a seismic coefficient of 0.15 and have a minimum factor of safety of from 1.0 to 1.2 (CDMG 1997).

Seed (1979) used a Newmark analysis, to be discussed below, to develop a recommended seismic coefficient of 0.1 for magnitude 6½ earthquakes and 0.15 for magnitude 8¼ earthquakes. These seismic coefficients are to accompany factors of safety of 1.15. Seed believed that these guidelines ensured that permanent slope deformations were acceptably small. The earthquake magnitude is used in his guidelines to modulate the seismic coefficient to account for longer-duration strong motion accompanying large earthquakes.

More recently, Pyke (1999) used Seed's work to develop the recommendations included in Figure 10.6. Figure 10.6 shows the seismic coefficient : ppa ratio versus earthquake magnitude. A summary of these and other approaches is provided in Figure 10.7, adopted from CDMG (1997). Based on these recommendations, the seismic coefficient at SME Mine, for the Borah Peak earthquake, ranges from k = 0.02 for Pyke's recommendation (Figure 10.6) to k = 0.15 for California State agencies. Assuming that Slope 24NE can be modeled using Eq. 10.3, with parameters given in Figure 10.2, the factor of safety under static conditions is 1.213–a stable slope. Using the seismic coefficient suggested by Pyke (1999) of 0.02, the factor of safety drops to 1.159–still a stable slope. If we use Seed's recommended seismic coefficient of 0.125, the factor of safety drops to 0.925–a slope failure. This latter seismic coefficient requires a decrease in slope angle to 27° to again achieve a factor of safety of 1.0.

Despite the drawbacks and conservatism of the pseudo-static analysis, numerous popular software packages use the technique today to perform seismic and static analyses on both soil and rock slopes. Some of the most popular software tools using pseudo-static analyses today include UTEXAS3 (Shinoak 1996), STABL (Siegel 1975; NISEE 1999), and SLOPE/W (Scientific Software 1999).

10.3.2 Newmark Analysis

The Newmark analysis (Newmark 1965) transcends the limit-equilibrium approach to provide an estimate of the *displacement* of a landslide block subjected to seismic motion. In this approach, that portion of the design accelerogram above a *critical acceleration, a_c*, is integrated twice to obtain the displacement. The critical acceleration is defined using Eq. 10.4.

$$a_c \equiv (FS - 1)g \sin\beta \qquad \textbf{EQ. 10.4}$$

Where:

a_c = critical acceleration, in meters per second squared (1.04 m/s^2 for Slope 24NE)
FS = static factor of safety (1.213 for Slope 24NE)
β = slope angle

A rigorous Newmark analysis involves subtracting the critical acceleration, a_c, from the accelerogram in Figure 10.5 and integrating the difference twice to compute the total displacement. Since the ppa at the SME Mine (ppa = 0.657 m/s^2) is less than the critical acceleration (a_c = 1.04 m/s^2), the earthquake acceleration does not overcome the initial limit equilibrium of the slope, and the rigorous Newmark analysis predicts no slope displacement.

Newmark, correctly assuming that many engineers would not wish to perform a rigorous integration of accelerograms, derived a simple approximation using a single pulse of motion. In Newmark's approximation, the total slope displacement, D_n, is a function of the peak particle velocity (ppv) and the critical acceleration, a_c, as shown in Eq. 10.5.

$$D_n \cong \frac{ppv^2}{2a_c} Max(6, \sqrt{\tau}) \qquad \textbf{EQ. 10.5}$$

Where:

ppv = peak particle velocity (0.038m/s, computed by integrating the accelerogram of Figure 10.5)
τ = a dimensionless constant equal to the duration of strong motion.

The factor on the right-hand side of Eq. 10.5 is included to account for longer-duration strong motion in large earthquakes. The total displacement of Slope 24NE, predicted by Eq. 10.5, is 0.004 m. It is unlikely that a 4-mm displacement of the 21.5-m block would result in failure of the slope, as the pseudo-static analysis predicts.

If one approximates the accelerogram with a harmonic, rather than a single, pulse, a new approximation to the rigorous Newmark technique can be derived.

$$D_n \cong \frac{(ppa - a_c) Max(6, \sqrt{\tau})}{(2\pi f)^2} \qquad \textbf{EQ. 10.6}$$

Where:

ppa = peak particle acceleration
f = the harmonic frequency

Using the predominant period to compute the frequency, f, the total displacement for Slope 24NE predicted by Eq. 10.6 is zero, because a_c exceeds ppa.

Another approximation to the Newmark method is presented by Jibson, Harp, and Michael (1998) for mapping landslide hazard in southern California. The development of this approximation was motivated by the difficulty of using a rigorous Newmark approach within the geographic information system (GIS) framework, commonly adopted for regional hazard mapping. In this approximation, 280 earthquake accelerograms were collected after the Northridge, California, earthquake. For each accelerogram, the authors computed the Arias intensity (see Eq. 10.7, below) and conducted a rigorous Newmark analysis for several values of critical acceleration. The results were fit using a regression equation relating Newmark displacement, Arias intensity, and critical acceleration. The Arias intensity has recently found use in representing earthquake shaking likely to cause landslides. It is given by

$$I_a = \frac{\pi}{2g} \int_0^{\infty} [a(t)]^2 dt \qquad \textbf{EQ. 10.7}$$

Where:

I_a = the Arias intensity
$a(t)$ = the digital accelerogram values

Observations indicate that earthquake-triggered landslides are more common in areas where the Arias intensities exceed about 0.54 m/s (ABAG 1998). The Arias intensity at SME Mine due to the Borah Peak earthquake is 0.163 m/s, well below the earthquake-triggered landslide threshold of I_a = 0.54 m/s.

The regression model proposed by Jibson, Harp, and Michael (1998) is given by

$$\log D_n \cong (1.521)\log I_a - (1.993)\log a_c - 1.546 \qquad \textbf{EQ. 10.8}$$

The displacement, D_n, calculated for the slope using Eq. 10.8 is 1.016×10^{-3} m.

A similar regression equation for estimating a Newmark displacement has been suggested recently by Ambraseys and Menu (1998), and is given as

$$\log D_n \cong 0.9 + \log\left[\left(1 - \frac{a_c}{ppa}\right)^{2.53}\left(\frac{a_c}{ppa}\right)^{-1.09}\right] \qquad \textbf{EQ. 10.9}$$

The displacement of Slope 24E, using Eq. 10.9, is zero—a stable slope. Actually, the displacement is a complex number if the critical acceleration exceeds the ppa. We assume this indicates no displacement.

Guidelines relating Newmark displacement estimates, D_n, to the overall stability of slopes (CDMG 1997) are shown in Table 10.2.

10.3.3 Dynamic Analysis

A dynamic analysis typically incorporates a finite difference approximation (the Fast Langrangian Analysis of Continua [FLAC] and FLAC3D codes are examples; Itasca 1999), a discrete element approximation (the Universal Distinct Element Code [UDEC] and 3DEC codes are examples; Itasca 1999), or a finite element approximation. These numerical approaches compute slope stresses and strains using earthquake accelerograms as input. During the analysis, the strains in each zone or element are summed to obtain a permanent deformation of the slope. Glass (1982, 1985) presented simple one-dimensional dynamic techniques for analyzing single and multiple degree-of-freedom systems to model rock and soil slopes under earthquake loads. These simple techniques compared favorably with shaking-table studies using rock blocks and demonstrated that the simple dynamic analyses performed better than the Newmark analysis over a wide range of earthquake frequencies. The technique presented by Glass (1982) uses the equation of motion for a single degree-of-freedom system, similar to Figure 10.2, subjected to a design

TABLE 10.2 CDMG (1997) guidelines for Newmark displacement estimates

Newmark Displacement, D_n (cm)	Description
0–10	Unlikely to correspond to serious landslide movement and damage
10–100	Slope deformation may be sufficient to cause serious ground cracking or enough strength loss to result in continuing failure. Determining whether displacements in this range can be accommodated safely requires good professional judgment that takes into account issues such as landslide geometry and material properties.
>100	Very likely to correspond to damaging landslide movement, and such slopes should be considered unstable.

earthquake accelerogram. The algorithm uses a linear acceleration approximation for the accelerogram.

Laboratory shaking-table experiments (Wilson 1979) demonstrate that displacement time histories of rock blocks cannot be replicated using traditional $f = tan\varphi$ friction models. Analyses of rock slopes, and some soil slopes, should instead use friction models that more accurately portray true joint shear behavior. Recent research (Kulatilake et al. 1999) demonstrates that fractal and anisotropy terms may be needed to adequately represent joint asperity influence on shear behavior. We shall use a friction model proposed by Barton and Bandis (1980), given by

$$f = \tan\left[T(JRC)\log_{10}\left(\frac{JCS}{J}\right) + \phi_r\right] \quad \textbf{EQ. 10.10}$$

Where:

T = a scaling factor to model post-peak shear displacement
JRC = a joint roughness coefficient (comprising the fractal and anisotropy terms of Kulatilake et al. 1999)
JCS = joint compressive strength
J = dynamic normal force
φ_r = the residual friction angle on the failure plane

The design accelerogram in Figure 10.5 was input into the linear acceleration dynamic approach. Using a joint roughness coefficient of zero to match the previous analysis assumptions and a residual friction angle of 35°, the slope did not displace. This result is compatible with the rigorous Newmark analysis, the harmonic approximation to the Newmark analysis, and the pseudo-static analysis using the Pyke (1999) estimate for the seismic coefficient.

There are advantages to using dynamic analyses. First, they incorporate realistic earthquake ground motions. Second, they enable the use of realistic physical properties for rock or soil slopes. Third, they compute the displacement time history of the slope, enabling an assessment to be made of the impact of the displacement on the ultimate behavior of the slope. Further, the simple one-dimensional approach of Glass (1982) enables a large number of slopes to be analyzed rapidly for optimum slope designs.

10.3.4 Post Failure Flow

Stability analyses should not stop at the base of a failed slope; rather, they should attempt to assess downstream hazard if it exists. To do this may require consideration of the slide as a debris flow. Work on debris flows (Glass and Klimmek 2000) demonstrates that an extension of the single degree-of-freedom model mentioned above successfully models the important behavior of debris flows, including downstream inundation area and travel time. We model a debris-flow center of mass using

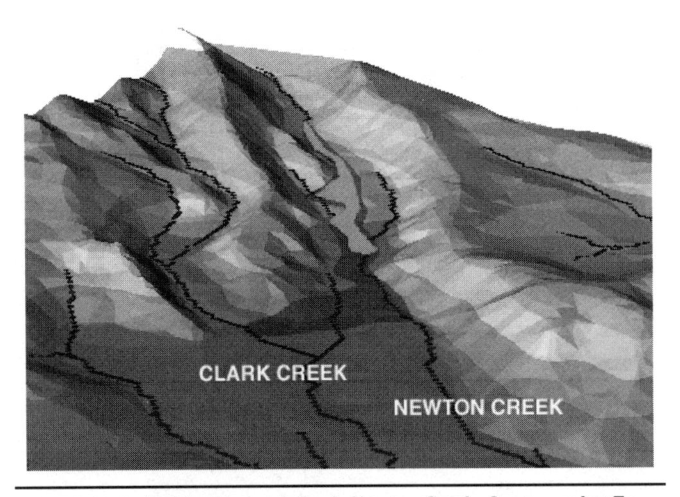

FIGURE 10.8 Debris-flow modeling in Newton Creek, Oregon, using Eq. 10.11. NOTE: Two flows beginning in the upstream reach of Newton Creek are displayed. The small debris flow has a volume of 100,000 m^3, whereas the larger debris flow has a volume of 400,000 m^3. Channel conveyance in Newton Creek allows the smaller flow to flow slightly farther downstream, but the larger flow inundates a larger portion of the stream channel.

Newton's second law of motion. The debris-flow equation of motion is

$$\frac{d(mv)}{dt} = mg\sin\theta - a - bv - cv^2 \quad \textbf{EQ. 10.11}$$

Where:

m = the mass of the flow
v = the velocity of the flow
g = the acceleration of gravity
θ = the stream gradient angle
a = a constant frictional resisting force
bv = a viscous resisting force
cv^2 = a turbulent resisting force

Equation 10.11 is a simple force-equilibrium equation in which the rate of change of momentum (left side of the equation) is equal to the gravitational force driving the flow minus forces resisting that movement.

Using Eq. 10.11, we model the behavior of debris flows as they progress down stream channels. The simplicity of the technique derives from modeling the center of mass of the flow using straightforward, one-dimensional force-equilibrium concepts. The versatility of the approach stems from its ability to (a) estimate over-bank flow, (b) scour the stream channel, (c) account for downstream changes in debris-flow viscosity and mass density, (d) account for the effects of stream channel geometry, (e) estimate debris-flow travel times, (f) provide a protocol for estimating initial debris-flow volumes, and (g) permit a probability protocol to be overlain. An example of debris-flow routing using Eq. 10.11 is shown in Figure 10.8.

10.4 CONCLUSIONS AND RECOMMENDATIONS

Table 10.3 summarizes the results of stability analyses of Slope 24NE using the techniques presented above. Rows three through seven of Table 10.3 display the pseudo-static results. All pseudo-static approaches, except that recommended by Pyke (1999), predict a slope failure. A pseudo-static analysis results in a recommendation to lower Slope 24NE to an angle ranging from 21° to 29°, depending on the choice of the seismic coefficient and the required factor of safety. Pyke's recommendation of $k = 0.02$ indicates that the slope angle can be increased to 33°. The pseudo-static analysis has outlived its usefulness. It is an approach that,

TABLE 10.3 Summary of slope stability analyses

Analysis Technique	Factor of Safety	Disp. (m)	Results
Pseudo-Static			
COE (1982) k = 0.1	0.974	NA	Slope failure. Reduce slope to 29°.
COE (1982) k = 0.15	0.879	NA	Slope failure. Reduce slope to 26°.
CDMG (1997) k = 0.15	0.879	NA	Slope failure. Reduce slope to 21–26°.
Seed (1979) k = 0.125	0.925	NA	Slope failure. Reduce slope to 22°.
Pyke (1999) k = 0.02	1.159	NA	Stable slope. Increase slope to 33°.
Newmark			
Rigorous	NA	0.0	Stable slope. Increase slope to 34° (displacement 23 cm for a 34° slope).
Newmark approximation	NA	0.004	Stable slope. Increase slope to 34° (displacement 2.1 cm for a 34° slope).
Harmonic approximation	NA	0.0	Stable slope. Increase slope to 34° (displacement 1.1 cm for a 34° slope).
Jibson (1998) regression	NA	0.002	Stable slope. Increase slope to 34° (displacement 2.5 cm for a 34° slope).
Ambraseys & Menu (1998) regression	NA	0.0	Stable slope. Increase slope to 34° (displacement 10.5 cm for a 34° slope).
Dynamic	NA	0.0	Stable slope. Increase slope to 34° (displacement 11 cm for a 34° slope).

because it has no physical basis, relies on a fictitious parameter (the seismic coefficient), which cannot be derived using logical or physical principles. If, for some reason, a project requires a pseudo-static analysis, we recommend choosing a seismic coefficient using the guidelines proposed by Pyke (1999) and shown in Figure 10.6.

Results for the Newmark analysis are displayed in rows 9 through 13 in Table 10.3. The rigorous approach indicates a stable slope at 30° with no displacement. At 34°, the slope is predicted to displace 23 cm, a marginally stable slope according to the guidelines presented in Table 10.2. The displacement time history for the first 30 s of motion (calculated displacements do not change beyond 30 s) on the 34° slope using the rigorous analysis is shown in Figure 10.9. The four approximations to the Newmark technique predict a stable slope at 30°, with from 0 to 4 mm of displacement. Furthermore, the approximations indicate a stable slope up to a slope angle of 34°. The Newmark analysis has proved to be useful and has exhibited remarkable staying power. All of the approaches described above, with the exception of the harmonic approximation (presented here for the first time), have been thoroughly tried and tested. Equation 10.8 and Eq. 10.9, however, may exhibit regional dependence.

The last row in Table 10.3 displays the results of the dynamic analysis of Glass (1982). The dynamic analysis predicts a stable slope at 30° with no displacement. The dynamic analysis further predicts a stable slope up to a slope angle of 34°, with only 11 cm of displacement at that angle, this assuming a smooth failure plane (JRC = 0 in Eq. 10.10) or significant post-peak shear (T = 0 in Eq. 10.10). The displacement time history for the first 30 s of motion on the 34° slope is shown in Figure 10.9.

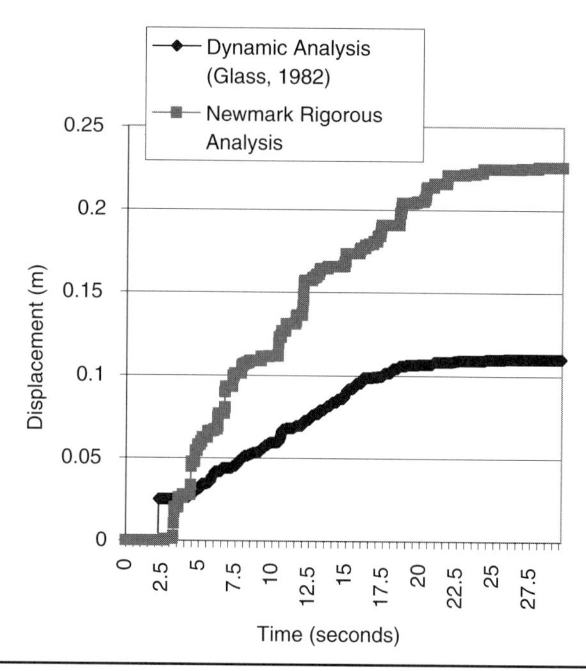

FIGURE 10.9 Displacement time history for slope 24NE

We recommend a dynamic analysis for both soil and rock slopes, because it enables more accurate modeling of slope geometry, strength, and deformation characteristics (see also Lorig, in this volume, for additional examples). In choosing a dynamic analysis approach for rock slopes, however, we recommend choosing one that incorporates a realistic joint friction model.

10.5 REFERENCES

ABAG. 1998. Part I –What does ground shaking intensity really mean? On Shaky Ground, a publication of the Association of Bay Area Governments, Earthquake Home Page, www.abag.ca.gov/bayarea/eqmaps/doc/.

Abel, J.F., Jr. 1997. Slope stability analysis supplement, Soledad Mountain project, Mojave, Kern County, California. Soledad Mountain Project Environmental Impact Report/Environmental Impact Statement, US Bureau of Land Management/Kern County.

Ambraseys, N.N. and J.M. Menu. 1998. Earthquake-induced ground displacements. Soil Dynamics and Earthquake Engineering, 16, pp. 985–1006.

Barton, N.K. and S. Bandis. 1980. Technical note: Some effects of scale on the shear strength of joints. Intl. Jour. of Rock Mech., Mining Sci., and Geomech. Abs.tr., 17, pp. 69–73.

Bromhead, E.N. 1986. The stability of slopes. New York: Chapman and Hall, 373 p.

CDMG. 1997. Guidelines for evaluating and mitigating seismic hazards in California. California Division of Mines and Geology Special Publication 117, www.consrv.ca.gov/dmg/pubs/sp/117/.

Chowdhury, R.N. 1978. Slope analysis. Amsterdam: Elsevier Scientific Publishing Company, 423 p.

Crone, A.J., M.N. Machette, M.G. Bonilla, J.J. Lienkaemper, K.L. Pierce, W.E. Scott, and R.C. Bucknam. 1987. Surface faulting accompanying the Borah Peak earthquake and segmentation of the Lost River fault, central Idaho. BSSA, 77:3:739–770.

Duncan, J.M. 1996. Soil slope stability analysis. In A.K. Turner and R.L. Schuster, eds., Landslides–Investigation and mitigation. Transportation Research Board, National Research Council, Special Report 247, pp. 337–371.

Glass, C.E. 1982. Influence of earthquakes on rock slope stability. In C.O. Brawner, ed., Third international conference on stability in surface mining. SME, New York, pp. 89–112.

Glass, C.E. 1985. Influence of earthquakes. In M.K. McCarter, ed., Design of non-impounding mine waste dumps. SME, New York, pp. 123–132.

Glass, C.E. 1990. Earthquake hazard estimation in areas of low historical seismicity: A focus on the northern Rio Grande Rift, Colorado and New Mexico. Earthquake Spectra, 6:4:657–680.

Glass, C.E. and R. Klimmek. 2000. Routing debris flows and lahars. Environmental & Engineering Geoscience. In review.

Harp, E.L. and R.W. Jibson. 1995. Inventory of landslides triggered by the 1994 Northridge, California earthquake. USGS open-file report 95-213, 17 p. 2 plates.

Harp, E.L. and R.W. Jibson. 1996. Landslides triggered by the 1994 Northridge, California earthquake. BSSA, 86:1B:S319–S332.

Hunt, R.E. 1984. Geotechnical engineering techniques and practices. New York: McGraw-Hill Book Co. 729.

Itasca. 1999. Software for mining, civil, petroleum, waste isolation projects. Itasca Consulting Group, Minneapolis, Minn.

Janbu, N. 1973. Slope stability computations. In R.C. Hirschfeld and S.J. Poulos, eds., Embankment-dam engineering. New York: John Wiley & Sons, pp. 47–86.

Jennings, P.C., G.W. Housner, and N.C. Tsai. 1968. Simulated earthquake motions. Earthquake Engineering Research Laboratory, California Institute of Technology, Pasadena, 52 p.

Jibson, R.W., E.L. Harp, and J.A. Michael. 1998. A method for producing digital probabilistic seismic landslide hazard maps: An example from the Los Angeles, California area. USGS Open File Report 98-113, 22 p.

Keefer, D.K. and M.W. Manson. 1998. Regional distribution and characteristics of landslides generated by the earthquake. In The Loma Prieta, California earthquake of October 17, 1989—landslides. USGS Professional Paper 1551-C, pp. C7–C32.

Kulatilake, P.H.S.W., J. Um, B.B. Panda, and N. Nghiem. 1999. Development of new peak shear-strength criterion for anisotropic rock joints. Journal of Engineering Mechanics, 125(9).

Morgenstern, N.R. and D.A. Sangrey. 1978. Methods of stability analysis. In R.L. Schuster and R.J. Krizek, eds., Landslides—Analysis and control. Washington, D.C.: National Academy of Sciences, pp. 155–172.

Nash, D.F.T. 1987. A comparative review of limit-equilibrium methods of stability analysis. In M.G. Anderson and K.S. Richards, eds., Slope stability—Geotechnical engineering and geomorphology. New York: John Wiley & Sons, pp. 11–76.

Newmark, N.M. 1965. Effects of earthquakes on dams and embankments. Geotechnique, 15:2:139–160.

NISEE. 1999. STABL software. National Information Service for Earthquake Engineering Software Library, University of California, Berkeley, www.eerc.berkeley.edu/software_and_data/eng_soft/software/stabl.html.

Pyke, R. 1999. Selection of seismic coefficients for use in pseudo-static slope stability analyses. TAGAsoft Limited, www.tagasoft.com/opinion/article2.html.

Schwartz, D.P. and K.J. Coppersmith. 1984. Fault behavior and characteristic earthquakes: Examples from the Wasatch and San Andreas faults. Jour. Geophys. Res., 89, pp. 5681–5698.

Scientific Software. 1999. SLOPE/W—Slope stability software. The Scientific Software Group, Washington, D.C., info@scisoftware.com.

Seed, H.B. 1979. Considerations in the earthquake-resistant design of earth dams. Geotechnique, 29:3:215–263.

Shinoak. 1996. UTEXAS3 software. Shinoak Software, www.shinoak.com, or sales@shinoak.com.

Siegel, R.A. 1975. Computer analysis of general slope stability problems. Joint Highway Research Project, Report No. JHRP-75-8, Project No. C-36-36K, File No. 6-14-11, Purdue University, West Lafayette, Indiana.

Smith, R.B. and W.J. Arabasz. 1991. Seismicity of the Intermountain Seismic Belt. In D.B. Slemmons, E.R. Engdahl, M.D. Zoback, and D.D. Blackwell, eds., Neotectonics of North America. Geological Society of America, Decade Map Volume 1, pp. 185–228.

Spudich, P., J.B. Fletcher, M. Heilweg, J. Boatwright, C. Sullivan, W.B. Joyner, T.C. Hanks, D.M. Boore, A. McGarr, L.M. Baker, and A.G. Lindh. 1997. SEA96–A new predictive relation for earthquake ground motions in extensional tectonic regimes. Seismological Research Letters, 68:1:190–198.

USGS. 1996. USGS response to an urban earthquake—Northridge '94. USGS Open-file Report 96-263, pp. 43–47.

Wilson, J.A. 1979. Physical modeling to assess the dynamic behavior of rock slopes. M.S. Thesis in Geological Engineering, The University of Arizona.

CHAPTER 11

Coupled Geomechanic–Hydrologic Approach to Slope Stability Based on Finite Elements

William G. Pariseau*

11.1 INTRODUCTION

This chapter outlines the essential features of "wet" rock slope stability whereby rock-mass deformation affects fluid flow and fluid flow affects deformation. In coupled analyses, these interactions occur concurrently. After each excavation advance, time is required for the "ground water" to adjust to the new slope geometry and for the rock mass to reach a new equilibrium state. The basic concepts of these coupled processes are easily understood in one dimension and are readily generalized to three dimensions, where a coupled computer code is essential to problem solving.

Coupled programs have been used in soil mechanics for many years, particularly for analysis of consolidation problems in association with foundation stability. Although an extensive research and development effort for coupled analyses of underground storage of high-level radioactive waste has been underway for many years (including temperature effects), applications to practical rock-mechanics problems have been slower to develop. However, if one views a saturated, jointed rock mass as an equivalent porous medium, then coupled analyses of "wet mine" design problems are certainly possible.

The subject has been under study at the University of Utah for the past decade; details of this work may be found in the references. The example problems in two dimensions illustrate the main features of coupled stability analysis of slopes in wet rock masses and the pitfalls of improvised, uncoupled analyses. The main conclusion is that a coupled analysis is essential for reliable stability analyses of wet rock slopes.

11.2 BASIC CONCEPTS OF COUPLED ANALYSIS

Water is a major detriment to slope stability, as is well known from experience. The reason is easily seen symbolically in the Mohr–Coulomb strength concept combined with the famous Terzaghi concept of *effective* stress. According to these concepts, shear strength is given by

$$\tau = (\sigma - p)\tan(\phi) + c \qquad \text{EQ. 11.1}$$

Where:

- τ = shear strength on a potential failure surface passing through a point of interest
- σ = the *total* normal stress acting perpendicular to the potential failure surface
- ϕ = the angle of internal friction
- c = the cohesion and compressive solid stress and fluid pressure are positive.

The angle of internal friction and cohesion are strength properties of the material at the considered point. The Terzaghi effective (normal) stress σ' is simply the difference between total stress and fluid pressure p. Thus, $\sigma' = \sigma - p$. Effective stress mobilizes a "frictional" component of shear strength that is transmitted through the solid skeleton; total stress is required in consideration of equilibrium. Clearly, reducing water pressure increases strength and thus slope stability. For this reason, drainage or, more accurately, depressurization is one of the main approaches to improving slope stability.

Drainage is usually a slow process of fluid flow that is described by Darcy's law. Thus,

$$v = -kh \qquad \text{EQ. 11.2}$$

Where:

- v = a nominal seepage velocity
- h = the hydraulic gradient
- k = a hydraulic conductivity

The negative sign indicates that flow is in the direction of decreasing hydraulic head. The actual seepage velocity is greater than the nominal or "Darcy" velocity. If Q is the volumetric flow rate, that is, discharge occurring through a cross section of area A, then the nominal velocity $v = Q/A$, while the actual velocity of flow $v' = Q/A_v$, where A_v is the cross-sectional area of the pore space. The total cross-sectional area $A = A_v + A_s$, where A_s is the cross-sectional area of the solids. In this regard, the ratio of pore area to total area is often assumed to be the same as porosity n, which, by definition, is the ratio of pore volume to total volume of a representative sample.

Although not intuitively evident, seepage along rock joints is mathematically similar to seepage in porous rock such as sandstone. Both are described by Eq. 11.2. This result is obtained from analysis of seepage along joints as a problem of viscous flow between parallel plates. The analysis leads to an explicit expression for joint hydraulic conductivity that indicates hydraulic conductivity of joints is proportional to the square of joint aperture b, the distance between plates, that is, $v \propto b^2$. Discharge is the product of normal velocity and cross-sectional area, i.e., $Q = vA$. So in the case of joints, Q is proportional to the cube of the aperture (per unit width) because cross-sectional area $A = (b)(1)$, so $Q \propto b^3$. This result is the basis for the "cubic law" of joint seepage.

Deformation of a porous, jointed solid is accompanied by changes in the volume of solids, the volume of pore space, and the space between opposing joint surfaces (joint volume). Here, "joints" may be microcracks in an intact laboratory test specimen, small fractures in a large block of rock, or field-scale joints observed in an outcrop of rock. The size of a representative sample increases accordingly. Regardless, joint volume means void volume associated with planar structures in contrast with pore

* Department of Mining Engineering, University of Utah, Salt Lake City, Utah.

volume that occurs between grains in an otherwise intact rock sample. In this context, joint volume and pore volume together constitute void volume. Only the connected void volume is available for fluid flow, of course.

In the elastic range of deformation, stress σ and strain $\in$ are related by Hooke's law. If a porous sample is tested dry or submerged with provision for free drainage, the fluid pressure is zero. The total and effective stresses then are the same, and

$$\sigma = E\varepsilon \quad \text{EQ. 11.3}$$

When present, fluid pressure adds to the total stress, hence

$$\sigma = E\varepsilon + p \quad \text{EQ. 11.4}$$

Equation 11.4 shows that deformation of the solid skeleton is governed by effective stress, i.e., $\sigma - p = \sigma' = E\varepsilon$. In some treatments, the concept of effective stress is modified to read $\sigma' = \sigma - \alpha p$, where α is an effective stress coefficient with a value between porosity and unity. In any case, strain of the solid skeleton and fluid pressure are related. A change in one affects the other; they are coupled. In this regard, the Terzaghi concept of effective stress applies only to normal stresses. Shear stress τ is unaffected by fluid pressure; so in the elastic range of deformation $\tau = G\gamma$, where G is the shear modulus and γ is engineering shear strain.

11.3 THREE-DIMENSIONAL MODEL

In three dimensions, the poroelastic relationship, in matrix notation for compactness of thought, is

$$\{\sigma\} = [E]\{\varepsilon\} + \{\alpha\}(p) \quad \text{EQ. 11.5}$$

Where:

$\{\sigma\}$ = a 6 × 1 column matrix of total stress with elements, $(\sigma_{xx}, \sigma_{yy}, \sigma_{zz}, \tau_{yz}, \tau_{zx}, \tau_{xy})$
$[E\,]$ = a 6 × 6 matrix of elastic properties
$\{\alpha\}$ = a column matrix of effective stress or coupling coefficients with elements $(\alpha, \alpha, \alpha, 0, 0, 0)$ in the isotropic case

If the rock mass has directional properties, then $[E]$ is a matrix of anisotropic elastic moduli. The effective stress coefficients may also vary with direction and fluid pressure may affect shearing action. However, the form of Eq. 11.5 remains the same.

Darcy's law in three dimensions in matrix notation has the form

$$\{v\} = -[k]\{h\} \quad \text{EQ. 11.6}$$

Where:

$\{v\}$ = a 3 × 1 column matrix of velocity components
$[k]$ = a 3 × 3 matrix of hydraulic conductivities
$\{h\}$ = a 3 × 1 column matrix of hydraulic gradients

If the z-direction is positive upward and pressure is positive, then the elements of $\{h\}$ may be expressed as derivatives of hydraulic (piezometric) head $H = p/\gamma_w + z$ in the x, y, and z directions. Here, γ_w is the specific weight of water. Alternatively, H may be considered a potential function, $p + \gamma z$. Differentiation then provides elements of $\{h\}$ as $(\partial p/\partial x, \partial p/\partial y, \partial p/\partial z + \gamma_w)$, while the hydraulic conductivity matrix is divided by γ_w.

A subtlety that arises with coupling between solid deformation and fluid pressure concerns the meaning of fluid velocity in Darcy's law. If the displacement of the solid skeleton is u and the nominal displacement of a fluid point is U, then the displacement of fluid relative to the deforming solid $u' = U - u$; the nominal velocity of the fluid relative to the solid is du'/dt, which is just the one-dimensional Darcy velocity, v in Eq. 11.2.

Another consideration is fluid compressibility. Fluid and solid volumetric strains e_f and e_s are related to the bulk modulus of fluid K_f and solid K by the elastic relations: $e_f = (1/K_f)p$ and $e_s = (1/K)\sigma_m$, where σ_m is the mean normal stress in the solid. The volumetric strain of fluid relative to the solid $e = (1/K_f)p - (1/K)\sigma_m$ in the isotropic case, whereas the rate of strain $de/dt = du'/dt = v$ is linked to Darcy's law in one dimension. The three-dimensional anisotropic form for volumetric fluid strain relative to the solid is $\xi = cp - \{\alpha\}^T\{\in\}$, where c is the reciprocal of the fluid bulk modulus and super-T means transpose. When a modified effective stress concept is used, the α's are combinations of the fluid and solid elastic moduli, although the most common assumption in application is that the α's are ones. In soil mechanics, water is usually considered incompressible because of a high value of bulk modulus compared with the bulk modulus of the soil skeleton. A reasonable value for the bulk modulus of water is about 1/100th of mild steel, or about 1.7 GPa (250,000 psi). Corresponding moduli for soil and rock are easily an order of magnitude less and more, respectively. The assumption of fluid incompressibility in rock mechanics is therefore not justified.

11.4 FINITE ELEMENT MODEL

A numerical model and reliable computer code are essential for solving practical problems involving coupled rock-mass deformation and fluid-flow phenomena. The most direct path to a finite element numerical model is by applying the principle of virtual work, which begins with the divergence theorem in the form

$$\int_A \{u\}^T\{T\}dA + \int_V \{u\}^T\{\gamma\}dV = \int_V \{\varepsilon\}^T\{\sigma\}dV \quad \text{EQ. 11.7}$$

Where:

$\{\ \}$ = a 3 × 1 column matrix (vector)
T, u, γ, $\in$, and σ are surface traction, displacement, specific weight, strain, and total stress, respectively
A = the surface area of the sample volume V

Equilibrium and symmetry of stress are implied; the strains are obtained from the displacements by differentiation in the usual way and no discontinuities are present.

The finite element approximation begins by specifying the shape of the sample in V. In two dimensions, the shape of this element is often a triangle or quadrilateral; tetrahedrons and bricks are common element shapes in three dimensions. An approximation to the displacement field over the element is then made. A simple approximation is a linear one. One has, over a triangle,

$$\begin{bmatrix} u \\ v \end{bmatrix} = \begin{bmatrix} a_0 & a_1 & a_2 \\ b_0 & b_1 & b_2 \end{bmatrix} \begin{bmatrix} 1 \\ x \\ y \end{bmatrix} \quad \text{EQ. 11.8}$$

where u and v are the displacement components in the x- and y-directions, respectively.

When Eq. 11.8 is applied to the corners of the triangle (element nodes), the coefficients can be found in terms of the node displacements and then back substituted into Eq. 11.8 to obtain

$$\begin{bmatrix} u \\ v \end{bmatrix} = [N]\{\delta\} \quad \text{EQ. 11.9}$$

Where:

$[N]$ = a 2 × 6 matrix that depends on (x, y)
$\{\delta\}$ = a 6 × 1 column matrix of node displacements (three nodes with two displacements each)

Application of the definitions of strain in terms of displacement to Eq. 11.9 gives the element strains in terms of the node displacements. Thus,

$$\{\varepsilon\} = [B]\{\delta\} \qquad \text{EQ. 11.10}$$

In the linear displacement approximation, entries of $[B]$ are constants.

Substitution of Eq. 11.9 and Eq. 11.10 into Eq. 11.7 leads to an integral expression with $\{\delta\}$ as a common non-zero factor that implies

$$\int_A [N]^T\{T\}dA + \int_V [N]^T\{\gamma\}dV = \int_V [B]^T\{\sigma\}dV \qquad \text{EQ. 11.11}$$

The terms on the left of Eq. 11.11 are simply the numerical expression of the external mechanical forces $\{f_m\}$, which is a 9×1 column matrix, applied to the triangular example element. Substitution of the stress–strain law Eq. 11.7 into Eq. 11.11 leads to

$$\{f_m\} = [k_{11}]\{\delta\} + [k_{12}]\{\pi\} \qquad \text{EQ. 11.12}$$

Where:

$[k_{11}]=\int [B]^T[E][B]dV$

$[k_{12}]=\int [B]^T\{\alpha\}[N']dV$

and where the pressure over the element is interpolated from the pressures at the node points by $p = [N']\{\pi\}$.

Inspection of Eq. 11.12 shows that the unknowns are the node displacements and pressures. Other approaches are possible, but this is the most common one.

There are more unknowns than equations in Eq. 11.12, so there is a deficit of information at this juncture. The complete system is obtained by considering the motion of the fluid relative to the solid in somewhat the same manner, i.e., by applying the principle of virtual power to the divergence theorem in the form

$$\int_A \{T_p\}^T\{\dot{w}\}dA = \int_V \left[\frac{\delta(p\dot{w}_x)}{\delta x} + \frac{\delta(p\dot{w}_y)}{\delta y} + \frac{\delta(p\dot{w}_z)}{\delta z}\right]dV \qquad \text{EQ. 11.13}$$

Where:

$\{T_p\}$ = a vector of applied water forces

$\{dw/dt\}$ = a vector of relative fluid velocities with components in the coordinate directions indicated by a corresponding subscript

The integral on the left of Eq. 11.13 is the work rate of fluid pressure on relative fluid velocity and is work (per unit time) in addition to the work of external forces associated with the left side of Eq. 11.7. Pressure over an element is interpolated from node pressures as before, while the pressure derivatives with respect to the coordinate axes are interpolated by $\{p'\} = [B']\{\pi\}$. The fluid velocities on the right side of Eq. 11.13 are eliminated using Darcy's law and the expression for relative fluid volume strain rate. After consideration of fluid equilibrium and mass conservation, the result has the form

$$\{f_w\} = [k_{21}]\{\dot{\delta}\} + [k_{22}]\{\dot{\pi}\} + [k'_{22}]\{\pi\} \qquad \text{EQ. 11.14}$$

Where the 3×1 vector on the left is a fluid "force," $-\int[N']^T(v_n)dA + \int[B']^T[k]\{\gamma\}dV$, $[k_{21}]$ is just the transpose of $[k_{12}]$, $[k_{22}] = -\int [N']^T(c)][N']dV$, $[k_{22}'] = -\int [B']^T[k][B']dV$. Here (v_n) is the velocity normal to the surface A.

The element system of equations, Eq. 11.12 and Eq. 11.14, is extended over a practical region of interest by subdividing the region of interest into a number of elements and then adding the forces contributed to each node from elements sharing a node. The process is one of "assembly"; the collection of elements and nodes is a "mesh." In three dimensions, there are four unknowns at each node (three displacements, one pressure), so the total system unknown is four times the number of nodes used in constructing the mesh.

Time integration of the assembled system of equations then allows for determination of the changes in solid displacement $\{\Delta\delta\}$ and fluid pressure $\{\Delta\pi\}$ at all nodes in the region, subject to conditions prescribed on the region boundary and at the start of the time period. Boundary conditions include surface forces or displacements and fluid velocities normal to a boundary or fluid pressures.

Mine excavations always occur in initially stressed ground, so provision for initial stress is necessary. Usually, no provision is made for initial displacement (and strains) because interest is focused on displacements induced by excavation. In wet ground, provision for initial fluid pressures and velocities is also necessary. Stresses, fluid pressures, and velocities after mining are simply obtained by adding the changes induced by mining to the initial values. When an excavation is mined in a sequence of steps, the data at the end of one step are the starting values for the next step. Because the coupled system of equations, Eq. 11.12 and Eq. 11.14, involves rates, the time between steps is important to an analysis.

At the end of an interval of time, which may be composed of a number of smaller time steps, the finite element output includes displacements and pressure at the nodes and stresses and velocities in the elements of the mesh. These data are relevant to questions of slope stability, for example, "Does the state of stress exceed strength at some point?", "Is the safety factor at some point of interest less than one?" In this regard, a useful method for evaluating the stability of a slope is to examine contours of the local (point) factor of safety and identifying zones that may be stressed to the elastic limit and are yielding. Another simple technique is to plot the deformed mesh to see where displacements may be potentially excessive. Questions concerning fluid pressures and velocities can also be addressed. Discharge changes may be of interest and can be estimated from the fluid velocity field. All change in time, so stability is an evolving condition. In this regard, the role of analysis is to provide quantitative data to assist in engineering evaluation of stability.

The governing equations were programmed into several of the writer's two- and three-dimensional finite element codes. The one used in this example is U98C. This code and several modifications are the writer's intellectual property. They are not in the public domain.

11.5 PRELIMINARIES

The types and values of material properties needed in the following examples are given in Table 11.1. Comparison of Young's modulus with the unconfined compressive and tensile strength in the table shows that uniaxial strains to failure are 6% and 0.6%, respectively. These limits seem reasonable for rock masses. A ratio of 10 for compressive strength:tensile strength is also reasonable for rock. The strengths in Table 11.1 are reduced an order of magnitude (1/10th) from laboratory values. The shear strength in Table 11.1 is computed from compressive and tensile strengths. The use of a quadratic yield criterion that has the shape of a parabola—open to the compressive side of the normal stress axis when plotted in the conventional normal stress shear stress plane, where Mohr circles representing stress states at the elastic limit—may also be plotted. The specific gravity of rock corresponds to 22.8 kN/m^3 (144 lbf/ft^3), whereas that of water is 9.8 kN/m^3 (62.4 lbf/ft^3). Hydraulic conductivity at 0.003 m/day is considered moderate to low. The Terzaghi effective stress concept is implied by the value of the effective stress coefficient.

11.5.1 Gravity Loading

Application of the force of gravity to a finite element mesh results in an instantaneous elastic displacement thåat is followed by a time-dependent "consolidation" phase. If the pore fluid is assumed to be incompressible, then the instantaneous elastic

TABLE 11.1 Material properties for example problems

Property	Metric	Engineering
Young's modulus	16.6 GPa	2.4 Mpsi
Shear modulus	6.9 GPa	1.0 Mpsi
Poisson's ratio	0.2	0.2
Compressive strength	10.0 Mpa	1450 psi
Tensile strength	1.0 Mpa	145 psi
Shear strength (quadratic)	1.9 Mpa	265 psi
Specific gravity, rock	2.31	2.31
Specific gravity, water	1.0	1.0
Bulk modulus, water	1.66 GPa	0.24 Mpsi
Hydraulic conductivity	0.003 m/d	0.01 ft/d
Effective stress coefficient	1.0	1.0

displacement at the surface is zero. Additional surface "settlement" of the mesh then occurs during the consolidation phase. When the pore fluid is relatively compressible, the surface "rebounds" upward during the consolidation phase. Whether additional settlement or rebound occurs depends on the ratio of solid to fluid elastic moduli. An analytical solution for this problem of poroelastic consolidation due to self-weight is available for comparison with finite element solutions of the same problem (Pariseau 1998).

Figure 11.1 shows surface settlement as a function of time using the properties in Table 11.1, with the product of fluid compressibility and solid elastic modulus (ratio of solid elastic Young's modulus to fluid modulus) as a parameter. The normalized displacement in Figure 11.1 is simply the ratio of current displacement to the long-time displacement; the dimensionless time $t^* = c_v t/H^2$, where c_v is the consolidation coefficient, t is real time, and H is depth of the mesh. As the fluid compressibility increases from zero (modulus high relative to the solid, cE = 0, i.e., INC case) to very compressible (modulus low relative to the solid, cE = 1000), the initial elastic settlement increases from zero to almost 1.8 times the long-time settlement. When the fluid and solid moduli are nearly equal, the initial elastic settlement (cE = 1) is nearly equal to the long-time settlement. Also shown in Figure 11.1 are finite element results (FEM case, cE = 10) using Table 11.1 input data. The agreement with the exact analytical solution (ANA, cE = 10) is excellent and indicates proper functioning of the finite element code U98C for the problem.

11.5.2 Excavation Unloading

Removal of portion of an initially stressed and saturated column of material is of interest because the problems encountered in this process are similar to those encountered when excavating porous, jointed rock masses where "unloading" and "rebound" occur. One may suppose a column is initially stressed by the force of gravity, as in the previous discussion. Excavation to some depth h is then done instantaneously, as if blasting. The initial response will be an instantaneous change in fluid pressure followed by a time-dependent phase of fluid pressure adjustment and displacement. If the initial depth of the mesh is $H + h$ and excavation is to a depth h, then depth after excavation is simply H. Before excavation and long after excavation, the fluid pressure p at depth z, in this example, is simply the product of specific weight of fluid times depth below surface, i.e., $p = \gamma_f z$. Although adjustments in fluid pressure, effective stress, strain, and displacement in time and space are not intuitively obvious, a closed form solution for the problem (pressure change associated with excavation of an initially stressed fluid-saturated column under gravity load) is possible (Pariseau and Schmelter 2000). A portion of the results is presented here to illustrate the nature of fluid pressure change accompanying excavation.

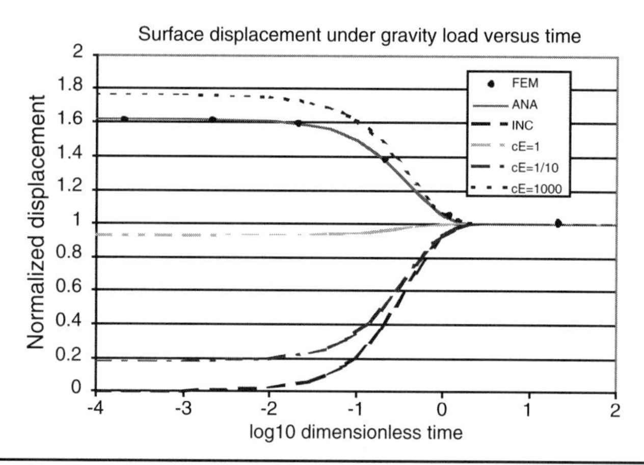

FIGURE 11.1 Gravity force consolidation at various fluid compressibilities and finite element results

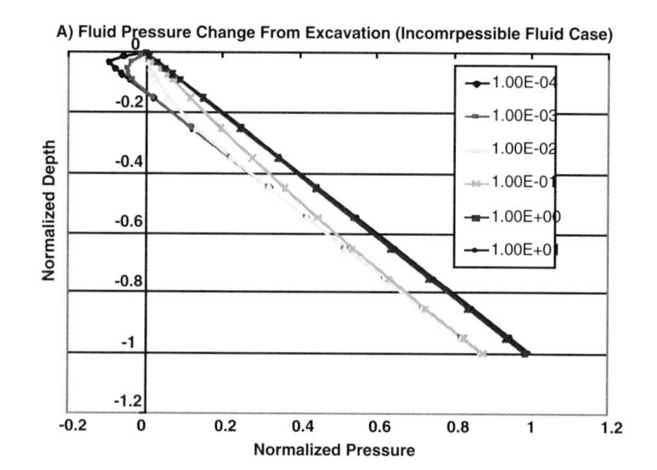

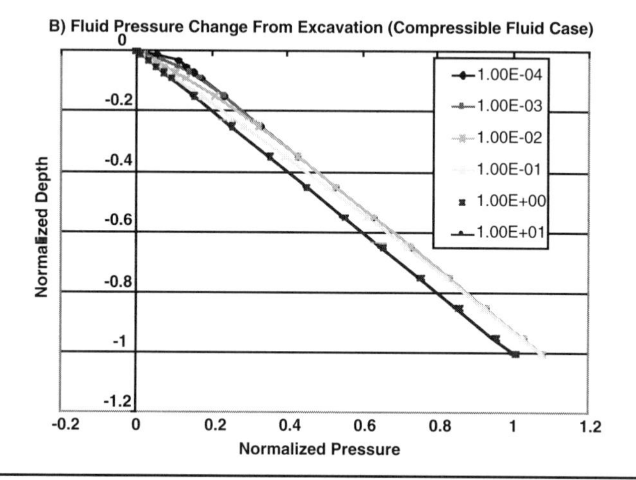

FIGURE 11.2 A. Fluid pressure versus column depth after excavation (incompressible fluid case). B. Fluid pressure versus column depth after excavation (compressible fluid case).

Figures 11.2A and B show fluid pressure versus depth at various times after excavation. The pressure is normalized by dividing the actual pressure by the static pressure p_o at the bottom of the excavated column ($\gamma_f H$). Depth z is normalized into dimensionless form by dividing the actual depth by the depth of the column H. The parameter in Figure 11.2 is the previous dimensionless time, $t^* = c_v t/H^2$. Figure 11.2A shows results in the case of an incompressible fluid, and Figure 11.2B uses data from Table 11.1 (compressible fluid case). Compression is positive in Figure 11.2 and depth of excavation h is one-tenth the final column depth H, about 30 m and 300 m, respectively, in this example.

The effective stress normal to the excavation surface, total normal stress, and fluid pressure are zero at the excavation surface at the instant of excavation and remain zero as time passes. In the case of incompressible fluid, there is an instantaneous change in fluid pressure just below the excavation surface that is equal in magnitude but opposite in sense to the total stress acting just prior to the cut. When this pressure change is added to preexcavation pressure (which is zero) at the excavation surface, a tensile fluid stress is generated that is transmitted down the entire column. The curve at the smallest dimensionless time (1.E-04) in Figure 11.2A shows this phenomenon quite clearly. The calculated fluid tension persists to a dimensionless depth (z/H) of 0.15 but decays with time. Indeed, the pressure change diminishes with time and stops changing for practical purposes near the dimensionless time of 1.E+0. Compressibility of the fluid reduces the magnitude and extent of the induced fluid tension, as seen in Figure 11.2B, which shows a momentary rise in pressure with depth that subsequently decays in time to the expected hydrostatic pressure distribution. The long-time distribution of fluid pressure, effective stress, and total stress are the same in any case. The induced fluid tension is $\gamma h/(1+cE')$, where E' is an elastic parameter that depends on Young's modulus and Poisson's ratio of the solid, γ is total specific weight, and c is fluid compressibility.

In soil mechanics, the enhancement of soil strength through surface tension of soil water in the capillary zone is well understood and quite consistent with the concept of effective stress. Whether the same phenomenon is present in fluid-filled rock joints is open to question. In this regard, the volumetric strain of the fluid $\epsilon^{f}_{v} = cp$, where c and p are fluid compressibility and pressure, respectively, and compression is positive. Clearly, if the "pressure" is tensile, then the fluid expands. In the limit of fluid incompressibility, the fluid volume does not change and the fluid pressure must be determined from the requirement for equilibrium alone. Certainly there is some physical limit to the tension a fluid (water) is capable of supporting. In rock mechanics, capillary tension may be considered relatively small and thus negligible. If no limit is imposed in calculations for the incompressible fluid case, then an overly optimistic estimate of strength may occur in consequence of the generation of unrealistically high fluid tensions.

The instantaneous fluid pressure change caused by excavation in this example is a *tensile* $\gamma h(1 + cE')$. In the rock-mechanics case using the data from Table 11.1, this expression evaluates to (9/109) γh. In a soil-mechanics case using a reasonable Young's modulus of 1/100th of the rock-mass modulus, the result is (9/10)γh. However, if incompressibility is assumed, then in either case the result is the same, γh. Thus, the assumption of incompressibility in soil mechanics makes little difference in the calculations, while in rock mechanics the difference is quite large. Hence, the assumption of fluid incompressibility should be avoided in rock mechanics. In rock mechanics, use of a realistic fluid compressibility greatly reduces the generation of unrealistically high tensile-fluid "pressures" and overly optimistic effective stresses even without imposing an explicit limit to fluid tension. However, in soil mechanics or in problems where the solid mass has a Young's modulus much lower than the fluid, a limit to fluid tension should be imposed, especially when analyzing excavation problems where unloading and rebound are likely to occur.

An example would be excavation of a saturated waste rock dump or old tailings heap. Another example where a limit to fluid tension would be important to calculations is stripping of saturated, unconsolidated overburden.

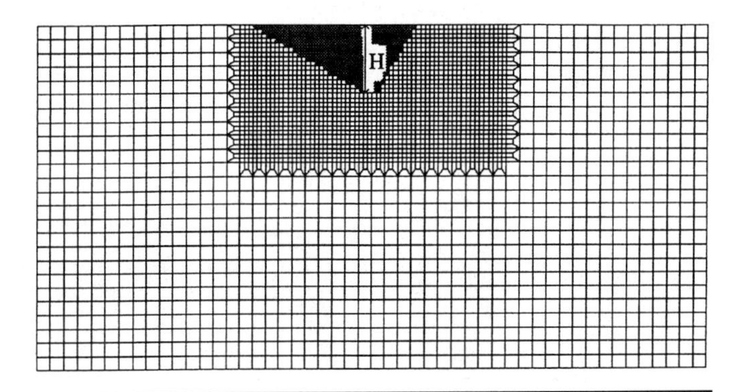

FIGURE 11.3 Finite element used in a hypothetical open-pit mine excavation. Left wall slope is 30 degrees; right wall slope is 60 degrees. Bench height is 24 m; depth H is 360 m.

11.6 HYPOTHETICAL COUPLED EXAMPLE

An example of a hypothetical open-pit mine excavated to a depth H of 360 m (1,200 ft) at slope angles of 30° and 60° illustrates some of the important features of coupled rock-mass slope-stability analysis. Bench height h is about 24 m (80 ft). Mesh width is $10H$; mesh depth is $5H$, as shown in Figure 11.3, where the larger elements are 72 m × 72 m. Dimensions of the central refined mesh are $4H \times 2H$. These dimensions provide bench-scale detail in the region near the excavation walls and some economy of computation. The outer boundaries are sufficiently remote from the excavation to have little influence on stress concentration near the excavation. In this example, the initial state of stress in the saturated rock mass is due to gravity alone (weight of rock and water combined) and is obtained by applying the force of gravity to a finite element mesh that contains the future pit. Mechanical and hydraulic properties are given in Table 11.1. Left wall slope is 30°; right wall slope is 60°. Bench height is 24 m; depth H is 360 m.

Figure 11.4A shows safety factor contours in the vicinity of the pit walls immediately after excavation. The factor of safety here is a ratio of strength to stress. Because strength is stress dependent, this ratio varies from element to element. A safety factor of 1 indicates the elastic limit is reached and yielding is possible. The black elements in Figure 11.4A are yielding; the grey elements are excavated elements. In this example, the entire pit is mined in one cut to illustrate the effect of water pressure on slope safety and stability. This effect is clearly seen in the left pit wall where excessive fluid pressure causes uplift and tensile failure of the wall over a large region. The same phenomenon occurs to a lesser extent in the steeper right pit wall. Dissipation of fluid pressure in time allows most of the yielding region to return to the elastic domain. In the long term (about 11 days), as seen in Figure 11.4B, very little of the pit toe remains at yield.

Although excavation immediately elevates fluid stress relative to strength (effective stress) at the excavation surface, stability returns in time as depressurization naturally ensues. Depressurization done in advance of excavation may therefore preclude the transitory slope instability observed in Figure 11.4A. Figure 11.5A shows the safety factor contours that develop in the case where the fluid pressure is zero at all times throughout the rock mass ("depressurized case"). The same weight of material is present in both cases, so the comparison is based on the same total stress. Overall stability is somewhat greater with complete depressurization than that achieved in the long term through natural depressurization.

If water were removed so the material were not only depressurized but also desaturated, there would be less weight. How much less would depend on the volume of fluid removed and thus on the connected void volume. In a porous rock mass, the

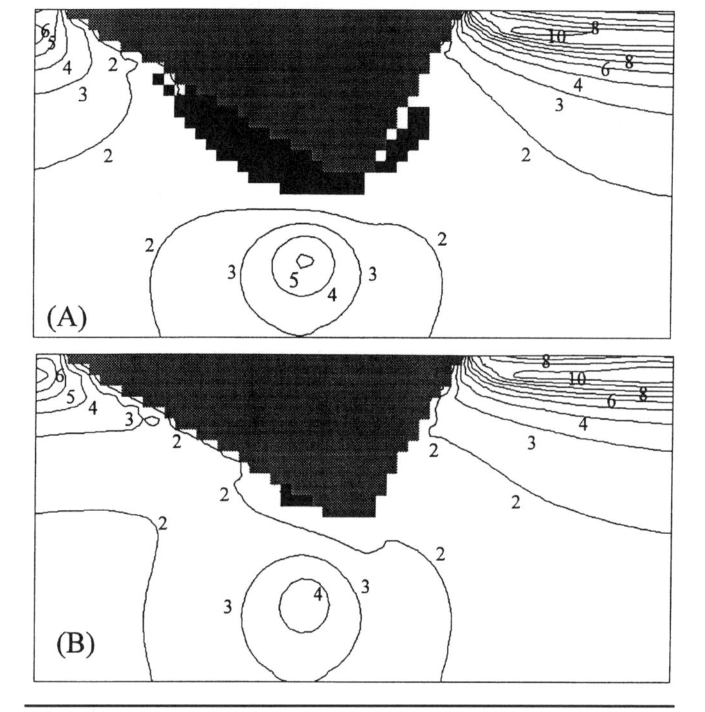

FIGURE 11.4 (A) Safety factor contours immediately after excavation. (B) Safety factor contours in the long term after excess fluid pressure has dissipated. Contoured area is the zone of small elements in the mesh shown in Figure 11.3. Grey elements are excavated. Black elements are yielding. Left wall slope is 30°, right wall slope is 60°, pit depth is 360 m (1,200 ft), bench height is 24 m (80 ft). Material properties are from Table 11.1.

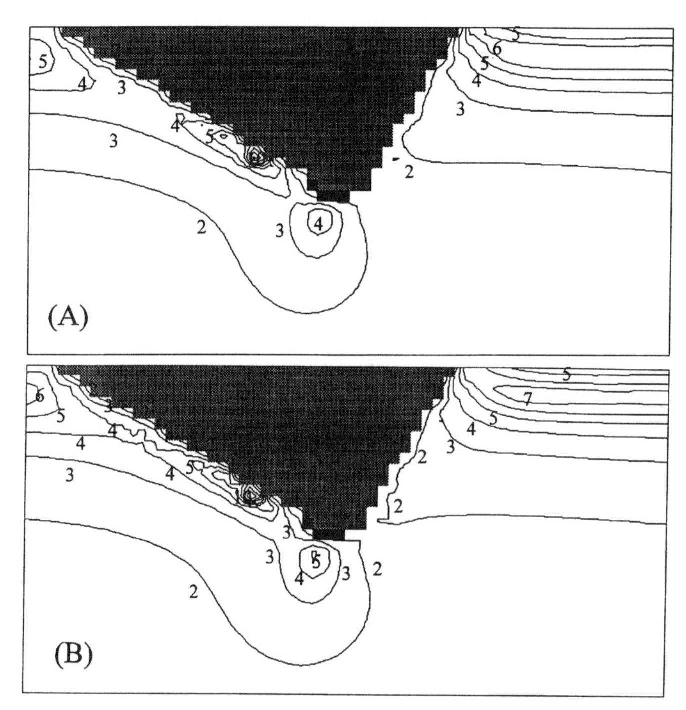

FIGURE 11.5 (A) Safety factor contours after excavation in the depressurized case. (B) Safety factor contours after excavation in the desaturated case involving 10% reduction in weight by water removal in advance of mining. No yielding occurs in the vicinity of the pit walls in either case.

fluid volume could be substantial, whereas in a fractured rock mass the volume of water removed may be inconsequential relative to total weight. Figure 11.5B shows the result of a 10% weight reduction associated with desaturating a rock mass having a generous porosity of 23.1%, i.e., 23.1% of the total volume is fluid-filled voids. The weight percent of water in the saturated rock mass is the ratio $n/s.g.$, where n is porosity and $s.g.$ is specific gravity of the saturated rock mass. Although the results in Figure 11.5 are for this example only, the data suggest that the difference between the two cases, depressurization and desaturation, is not large in rock slope-stability analysis. This observation has important practical consequences because depressurization is usually possible, whereas desaturation may require an extraordinary drainage effort, such as vacuum pumping.

Depressurization may be visualized with the aid of pressure contours, as shown in Figure 11.6. In Figure 11.6A, a high-pressure gradient is present near the pit walls where pressure contours tend to be closely spaced immediately after excavation. In time (about 11 days), elevated pressures are reduced and the pressure contours move away from the pit walls, as shown in Figure 11.6B. Generally, the pressure contours follow the excavation topography, as expected. In this example, depressurization occurs naturally.

In practice, avoidance of slope failure caused by excessive water pressure would necessitate pumping. In this regard, one may suppose that, if prior to excavation, the 100-pressure contour was moved to the long-time equilibrium position shown in Figure 11.6B, the transitory yielding shown in Figure 11.4A would be avoided. The U98C coupled finite element code allows one to investigate such a possibility and to quantitatively assess the outcome of a proposed pumping or drainage plan for depressurization.

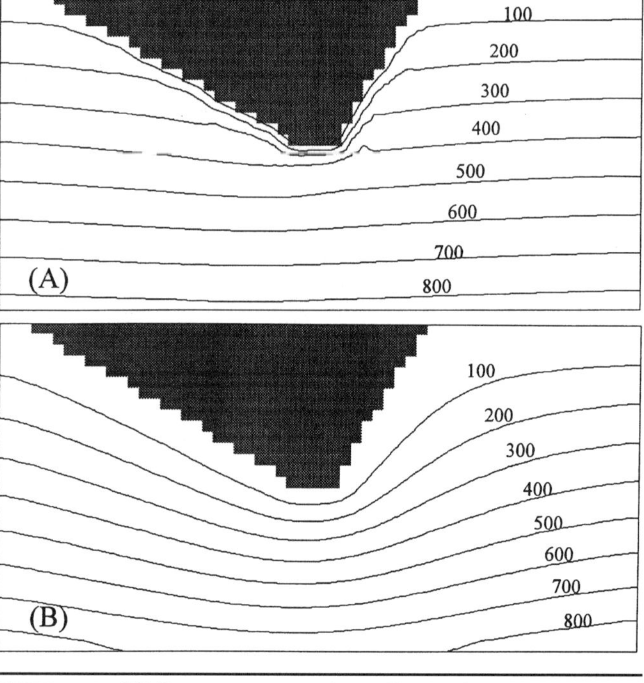

FIGURE 11.6 (A) Pressure contours immediately after excavation. (B) Pressure contours a long time after excavation. Units are psi; 145 pis = 1 Mpa.

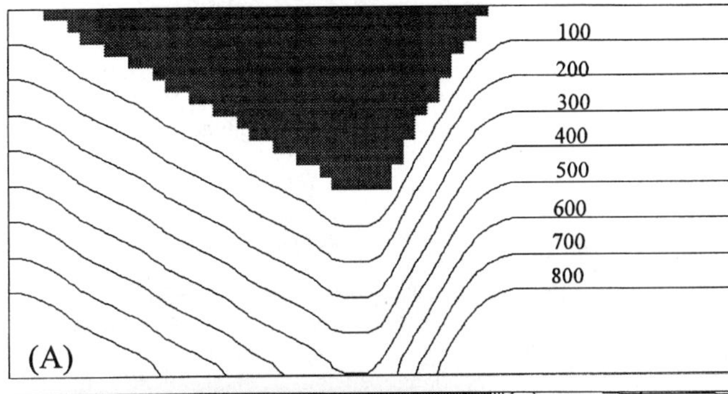

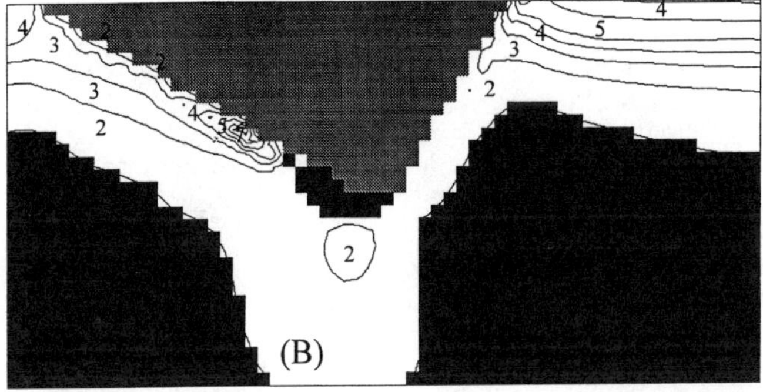

FIGURE 11.7 (A) Water pressure distribution fixed at unit weight times depth before and after excavation using U98C. (B) Long-term safety factor contours from the uncoupled approach using UTAH2PC and the water pressure distribution from Figure 11.7A.

11.7 HYPOTHETICAL UNCOUPLED EXAMPLE

A common practice in slope-stability analysis of "wet" rock slopes is to use a "dry" computer program. In a dry computer program, the total and effective stresses are the same. Water pressure is then taken into account by simply subtracting the water pressure from the previously computed "dry" normal stresses to obtain effective stresses and, subsequently, a distribution of safety factor throughout the proposed slope. The distribution of water pressure may be obtained from a separate analysis using a seepage program or simply by assuming a groundwater pressure distribution in the cut slope. The analyses of stress and fluid pressure thus do not interact and are "uncoupled." Neither is allowed to affect the other. Such practice is not physically defensible but is nevertheless done in the absence of a coupled code. As a consequence, the transitory and potentially dangerous instability that occurs immediately after excavation when water pressure is high is completely overlooked. Sequential excavation and water pressure updating does not change the problem. Although there may be an expectation that the net results are reasonable in the long term, there seems to be no logical justification for such practice other than expediency or lack of a proper coupled program.

In this regard, a common approximation to water pressure distribution in the vicinity of a cut face is the assumption that water pressure is proportional to depth below the face. This assumption is illustrated in Figure 11.7A, which shows pressure contours generated as unit weight of water times depth. The future pit profile becomes a "contour" of zero pressure. The region to be excavated is depressurized in the process of moving from a gravity-loaded equilibrium state to the newly prescribed pressure distribution. In a coupled program analysis, a relatively small settlement occurs as a consequence of the imposed change in water pressure from the hydrostatic preexcavation state. The distribution in Figure 11.7A was obtained using the poroelastic-plastic U98C finite element program.

Figure 11.7B shows safety factor contours in the vicinity of the pit walls obtained through the uncoupled procedure. The pit

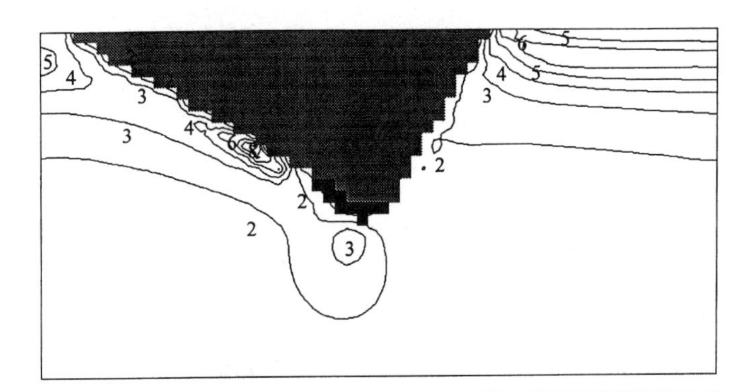

FIGURE 11.8 Long-term safety factor contours obtained using a modified uncoupled approach.

was first excavated from a "dry" premining state using a "dry" elastic-plastic finite element program, UTAH2PC, and properties from Table 11.1. Water pressure from the assumed gravity distribution in Figure 11.7A was then subtracted from the dry normal stresses to obtain effective stresses and the corresponding safety factor distribution in Figure 11.7B. The huge region of potentially yielding elements (in black) below the pit indicates that the results are erroneous. The results near the pit walls may appear "reasonable," but the evidence is clear that they are highly unreliable. In this regard, the yielding in Figure 11.7B extends below and to the sides of the whole mesh and is not confined to the section contoured. Moreover, the extensive yielding in Figure 11.7B is associated with high horizontal tensions.

There is a simple explanation for the erroneous and unreliable results obtained using the uncoupled approach. In the dry case, the vertical stress S_v beyond the zone of influence of the excavation is γz where γ is specific weight of rock and z is depth (compression is positive). The horizontal stress S_h is $K_o \gamma z$, where K_o is the ratio of horizontal to vertical stress that is often estimated as $\nu/(1 - \nu)$ where ν is Poisson's ratio. These are total stresses (dry case). Water pressure p is simply $\gamma_f z$, where γ_f is specific weight of fluid (water). The effective stresses are then $S'_v = S_v - p = (\gamma - \gamma_f)z$ and $S'_h = S_h - p = (K_o\gamma - \gamma_f)z$, which is tensile. In the wet case, the vertical stresses, total and effective, are the same as before, $S'_v = S_v - p = (\gamma - \gamma_f)z$. However, the constant K_o now relates the effective stresses so that $S'_h = K_o S'_v = K_o(\gamma - \gamma_f)z$, which is compressive. Detailed comparisons between the uncoupled improvisation and the rigorous coupled approach show these analytical results to be the case.

Although the uncoupled approach is fundamentally flawed, especially when the water pressure distribution after excavation is assumed rather than calculated, there is an improvement of sorts suggested by the previous explanation. Instead of subtracting the full fluid pressure value from the dry normal stresses, one should subtract only $K_o p = K_o \gamma_f z$. The horizontal effective stresses away from the zone of excavation influence are then the correct values, given the assumed water pressure distribution after excavation. Figure 11.8 shows the safety factor distribution using this modified uncoupled procedure. This result is somewhat similar to the coupled, long-term solution shown in Figure 11.4B. Although more reliable than the usual uncoupled approach to long-term stability evaluation, the modified result still misses the potentially dangerous transitory condition immediately after excavation, e.g., Figure 11.4A.

11.8 CONCLUSION

The detrimental effect of water on the stability of soil and rock slopes is well known. This effect is quantified in coupled stability analysis where rock-mass deformation and water flow interact instantaneously with a mining cut and subsequently in time as a new equilibrium position is approached. A coupled finite element computer program allows for site-specific analyses that include consideration of preexcavation stresses and fluid pressures, given rock-mass elastic moduli, strengths, fluid compressibility, hydraulic conductivities, and coupling constants. Different rock types, anisotropic rock, and rock joints may be taken into account. Generally, fluid compressibility needs to be included; the assumption of incompressibility is not usually justified in rock mechanics, unlike the situation in soil mechanics. Computed data for a specified mining step are changes induced by the advance and include changes in displacement, strain, stress, pressure, fluid velocity, and fluid content. These data provide guidance to stability evaluation, especially in the form of local safety factor contours in the vicinity of the slope walls. A "local" safety factor is simply the ratio of strength to stress at a considered point in the region of analysis.

In a coupled analysis, the most dangerous time is immediately after a blast when the fluid pressure is momentarily elevated and effective stresses are reduced. In time, depressurization occurs and greater stability is achieved. There seems to be little difference in rock slope stability based on depressurization in comparison with desaturation, which would require a much greater drainage effort. As a matter of computational economy, two program runs may suffice, one using a time step of zero and a second using a very large time step.

Uncoupled analysis that uses a conventional "dry" computer program for stress analysis of a mining cut and then imports a fluid pressure distribution attributed to steady-state flow through the cut slope geometry was shown to be generally erroneous and an approximation of unknown reliability. An analysis of this all-too-common approach demonstrated the nature of the problem and led to a proposed modification that alleviates the error to some extent. Of course, the technically sound coupled approach is preferable.

11.9 REFERENCES

Pariseau, W.G. 1991. Estimation of permeability in well-jointed rock masses. *Proceedings 8th International Conference on Computer Methods and Advances in Geomechanics.* Balkema. 1567–1572.

Pariseau, W.G. 1993. Equivalent properties of a jointed biot material. *Int. Jour. Rock Mech. Min. Sci. & Geomech. Abstr.,* 30:7:1151–1157.

Pariseau, W.G. 1994. Design considerations for stopes in wet mines. *Proceedings 12th Annual GMTC Workshop–Mine Systems Design and Ground Control.* Virginia Polytechnic Institute and State University, Blacksburg. 37–48.

Pariseau, W.G. 1995. Coupled 3D FE modeling of mining in wet ground. *Proceedings 3rd Canadian Conference on Computer Applications in the Mineral Industry.* McGill University, Montreal. 283–292.

Pariseau, W.G. 1996. Finite element analysis of water pressure and flow on shaft and tunnel stability. *Trans. SME,* 298:1839–1846.

Pariseau, W.G., and C. Forster. 1993. Progress and problems in wet mine design. Proceedings. 11th Annual GMTC Workshop–Mine Systems Design and Ground Control. Virginia Polytechnic Institute and State University, Blacksburg. 85–96.

Pariseau, W.G., and S.C. Schmelter. 1995. Progress in wet mine measurements for stability. *Proceedings 13th Annual GMTC–Workshop Mine Safety and Environmental Engineering.* Virginia Polytechnic Institute and State University, Blacksburg. 71–81.

Pariseau, W.G., S.C. Schmelter, and A.K. Sheik. 1997. Mine slope stability analysis by coupled finite element modeling. *Intl. Jour. Rock Mech. & Min. Sci.,* 34:3-4, paper no. 242.

Pariseau, W.G., and S.C. Schmelter. 1997. Coupled finite element modeling of slope stability. *Trans. SME,* 304:8–15.

Schmelter, S.C., and W.G. Pariseau. 1996. Case study of safety and wet mine slope stability. *Proceedings 14th Annual GMTC Workshop–Mine Safety and Environmental Engineering.* Virginia Polytechnic Institute and State University and Pittsburgh Research Center, Pittsburgh, Oct. 28–29.

Sheik, A.K., and W.G. Pariseau. 2000. Role of water in stability of a shallow underground mine. *Trans. SME* (in press).

CHAPTER 12

Practical Slope-Stability Analysis Using Finite-Difference Codes

Loren Lorig* and Pedro Varona†

12.1 INTRODUCTION

Numerical models are computer programs that attempt to represent the mechanical response of a rock mass subjected to a set of initial conditions (e.g., in situ stresses, water levels), boundary conditions, and induced changes (e.g., slope excavation). The result of a numerical model simulation is typically either equilibrium or collapse. If an equilibrium result is obtained, the resultant stresses and displacements at any point in the rock mass can be compared with measured values. If a collapse result is obtained, the predicted mode of failure is demonstrated. In either case, the factor of safety can be calculated.

Numerical models divide the rock mass into elements. Each element is assigned a material model and properties. The material models are idealized stress/strain relations that describe how the material behaves. The simplest model is a linear elastic model, which uses the elastic properties (Young's modulus and Poisson's ratio) of the material. Elastic–plastic models use strength parameters to limit the shear stress that an element may sustain.

The elements may be connected together (a continuum model) or separated by discontinuities (a discontinuum model). Discontinuum models allow slip and separation at explicitly located surfaces within the model.

Numerical models tend to be general purpose in nature, i.e., they are capable of solving a variety of problems. While it is often desirable to have a general-purpose tool available, each problem must be constructed individually. The elements must be arranged by the user to fit the limits of the geomechanical units and/or the slope geometry. Hence, numerical models often require more time to set up and run than special-purpose tools (such as limit-equilibrium methods.)

Numerical models are used for slope-stability studies for a variety of reasons, including

- Empirical methods cannot confidently be extrapolated outside their databases.
- Other methods (e.g., analytic, physical, limit equilibrium) are not available or tend to oversimplify the problem, possibly leading to overly conservative solutions.
- Key geologic features, groundwater, etc. can be incorporated into numerical models, providing more realistic approximations of real slope behavior.
- Observed physical behavior can be explained.
- Multiple possibilities (e.g., hypotheses, design options) can be evaluated.

12.2 EXPLICIT FINITE-DIFFERENCE COMPUTER PROGRAMS

The specific numerical models discussed in this chapter are called explicit finite-difference codes, or computer programs. Finite-element programs are probably more familiar, but the finite-difference method is perhaps the oldest numerical technique used to solve sets of differential equations. Both finite-element and finite-difference methods produce a set of algebraic equations to solve. While the methods used to derive the equations are different, the resulting equations are the same. The programs referred to in this chapter use "explicit" time-marching schemes to solve the equations, whereas finite-element methods usually solve systems of equations in matrix form.

Although we tend to focus on a static solution to a problem, the dynamic equations of motion are included in the formulation of finite-difference programs. One reason for doing this is to ensure that the numerical scheme is stable when the physical system being modeled is unstable. With nonlinear materials, there is always the possibility of physical instability—for example, the failure of a slope. In real life, some of the strain energy is converted to kinetic energy. Explicit finite-difference programs model this process directly, because inertial terms are included. In contrast, programs that do not include inertial terms must use some numerical procedure to treat physical instabilities. Even if the procedure successfully prevents numerical instability, the *path* taken may not be realistic. The consequence of including the full law of motion in finite-difference programs is that the user must have some physical feel for what is going on. Explicit finite-difference programs are not black boxes that will "give the solution." The behavior of the numerical system must be interpreted.

FLAC (Fast Lagrangian Analysis of Continua) and UDEC (Universal Distinct Element Code) are two-dimensional finite-difference programs developed by Itasca Consulting Group (1998, 2000) specifically for geomechanical analysis. All codes discussed here can simulate varying loading and water conditions and have several predefined material models. They are unique in their ability to handle highly nonlinear and unstable problems. FLAC is formulated to study continuum problems, although a limited number of discontinuities (in the form of interfaces) can be included. UDEC is formulated to study discontinuum problems involving large numbers (hundreds) of explicit discontinuities that divide the rock mass into blocks. The three-dimensional equivalents of these codes are FLAC3D and 3DEC (Itasca 1997, 1999).

12.3 MODELING METHODOLOGY

The primary objective of numerical modeling at any mine site is to understand the factors controlling present slope behavior and to make predictions about future slope behavior. There is usually concern about the ability to provide stable slopes at reasonable angles for increased mining depths. Numerical modeling

* Itasca S.A., Santiago, Chile.
† Itasca Consultores S.L., Asturias, Spain.

attempts to address such concerns through simulation of the mining and mechanical response of the slopes. However, some uncertainty in numerical modeling arises from the generally poor knowledge of the material behavior of rock masses. Rock is a complex material that behaves in a nonlinear and anisotropic manner. For problems such as the design of steel structures, where the material behavior is well known, it is usually assumed that a model can predict the behavior of the system accurately, without calibration.

Owing to the complexity of a rock mass and the inherent uncertainty of its behavior, it is very important that any model used for design first be evaluated so that model predictions are compared with the observed and measured response of the slopes. Observed response includes appearance of tension cracks, toppling behavior, etc. Measured response includes instrumentation results from prisms, tape extensometers, inclinometers, etc. This process is known as calibration, or validation, of the modeling approach and is necessary before the model can be used with confidence in predictive studies. One problem associated with the calibration process is nonuniqueness, i.e., more than one set of material parameters, for example, will satisfy the calibration conditions.

A detailed discussion of modeling methodology is beyond the scope of this chapter. The interested reader is referred to a paper by Starfield and Cundall (1988), which includes three case studies that were examined using finite-difference codes.

12.4 CREATING MODELS

Modeling requires that the real problem be idealized (i.e., simplified) in order to fit the constraints imposed by available material models, computer capacity, etc. Analysis of rock-mass response involves different scales. It is impossible–and undesirable–to include all features and details of rock-mass response mechanisms into one model. In addition, many of the details of rock-mass behavior are unknown and unknowable; the approach to modeling is not as straightforward as it is, say, in other branches of mechanics.

12.4.1 Two-Dimensional Analysis Versus Three-Dimensional Analysis

The first step in creating a model is to decide whether to perform two-dimensional or three-dimensional analyses. Until recently, three-dimensional analyses were relatively uncommon, but advances in personal computers have permitted three-dimensional analyses to be performed routinely. Strictly speaking, three-dimensional analyses are recommended/required if the

- Direction of principal geologic structures does not strike within 20° to 30° of the strike of the slope.
- Axis of material anisotropy does not strike within 20° to 30° of the slope.
- Directions of principal stresses are thought to be not parallel or not perpendicular to the slope.
- Distribution of geomechanical units varies along the strike of the slope.
- Slope geometry in plan cannot be represented by two-dimensional (i.e., axisymmetric or plain strain) analysis.

12.4.2 Continuum Versus Discontinuum Model

The next step is to decide whether to use a continuum code or a discontinuum code. This decision is seldom straightforward. There appear to be no ready-made rules for determining which type of analysis to perform. All slope-stability problems involve discontinuities at one scale or another. However, useful analyses, particularly of global stability, have been made by assuming the rock mass can be represented as an equivalent continuum. Therefore, many analyses begin with continuum models. If the slope

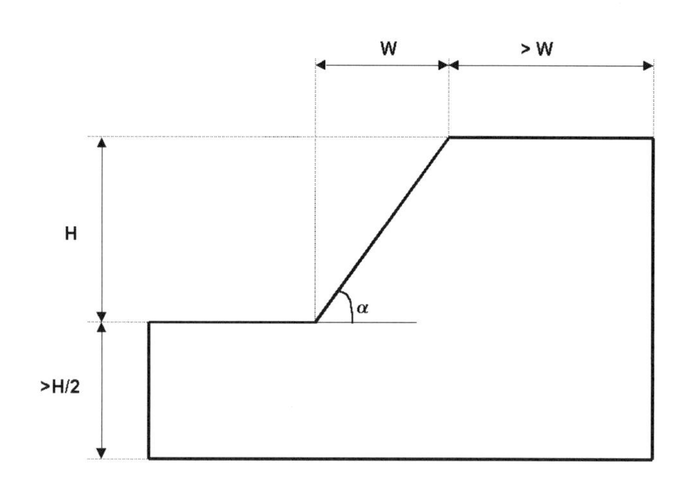

FIGURE 12.1 Minimum dimensions for slope analysis model

under consideration is unstable without structure, there is no point in continuing to discontinuum models. If, on the other hand, a continuum model appears to be reasonably stable, explicit incorporation of principal structures should give a more accurate estimation of slope behavior.

Selection of joint geometry for input to a model is a crucial step in discontinuum analyses. Typically, only a very small percentage of joints can actually be included in a model in order to create models of reasonable size for practical analysis. Thus, the joint geometry data must be filtered to select only those joints that are most critical to the mechanical response. This is done by identifying those that are most susceptible to slip and/or separation for the prescribed loading condition. This may involve determining whether sufficient kinematic freedom is provided (especially in the case of toppling) and comparing observed behavior to model response (i.e., calibration).

12.4.3 Selecting Appropriate Element Size

The next step in the process is to select an appropriate element size. The finite-difference zones in the programs listed previously assume that the stresses and strains within each element do not differ with position within the element, in other words, the elements are the lowest-order elements possible. In order to accurately capture stress and strain gradients within the slope, it is necessary to use relatively fine discretizations. By experience, the authors have found that at least 20 (and preferably 30) elements are required over the slope height of interest.

12.4.4 Initial Conditions

Initial conditions are those conditions that existed prior to mining. The initial conditions of importance at mine sites are the in situ stress field and the groundwater conditions. Since there are usually no measurements available regarding the initial or current stress field, different stress fields can be tried to determine their effect on model results. The effect of initial stresses is discussed separately in the last section of this chapter.

12.4.5 Boundary Conditions

Boundaries are either real or artificial. Real boundaries in slope-stability problems correspond to the natural or excavated ground surface and are usually stress free. Artificial boundaries do not exist in reality. All problems in geomechanics (including slope-stability problems) require that the infinite extent of a real problem domain be artificially truncated to include only the immediate area of interest. Figure 12.1 shows typical recommendations for locations of the artificial far-field boundaries in slope-stability problems. Artificial boundaries can be of two types: prescribed displacement or prescribed stress. Prescribed-displacement boundaries inhibit displacement in either the vertical direction or horizontal direction, or both. Prescribed-displacement boundaries are used to represent the condition at the base of the model and toe of the slope. Displacement at the base of the model is always fixed in both the vertical and horizontal directions to inhibit rotation of the model. Two assumptions can be made regarding the displacement boundaries near the toe of any slope. One assumption is that the displacements near the toe are inhibited only in the horizontal direction. This is the mechanically correct condition for a problem that is perfectly symmetric with respect to the plane or axis representing the toe boundary. Strictly speaking, this condition only occurs in slopes of infinite length, which are modeled in two dimensions assuming plane strain, or in slopes, which are axially symmetric (i.e., the pit is a perfect cone). In reality, these conditions are rarely satisfied. Therefore, some models are extended laterally to avoid the need to specify any boundary condition at the toe of the slope. It is important to note that the difficulties with the boundary condition near the slope toe are usually a result of the two-dimensional assumption. In three-dimensional models, this difficulty usually does not exist.

The far-field boundary location and condition must also be specified in any numerical model for slope-stability analyses. The general notion is to select the far-field location so that it does not significantly influence the results. If this criterion is met, it is not important whether the stress is prescribed displacement or prescribed stress. In most slope-stability studies, a prescribed-displacement boundary is used. The authors have used a prescribed-stress boundary in a few cases and found no significant differences with respect to the results from a prescribed-displacement boundary. The magnitude of the horizontal stress for the prescribed-stress boundary must match the assumptions regarding the initial stress state in order for the model to be in equilibrium. However, following any change in the model (i.e., an excavation increment), the prescribed-stress boundary causes the far-field boundary to displace toward the excavation while maintaining its original stress value. For this reason, a prescribed-stress boundary is also referred to as a "following" stress, or constant stress, boundary, because the stress does not change and follows the displacement of the boundary. However, following stresses are most likely where slopes are cut into areas where the topography rises behind the slope. Even where slopes are excavated into an inclined topography, the stresses would flow around the excavation to some extent, depending on the effective width of the excavation perpendicular to the downhill topographic direction.

In summary,

- A fixed boundary causes both stresses and displacements to be underestimated, whereas a stress boundary does the opposite.
- The two types of boundary condition "bracket" the true solution, so that it is possible to do tests with smaller models to get a reasonable estimate of the true solution by averaging the two results.

A final point to be kept in mind is that all open-pit slope-stability problems are three dimensional in reality. This means that the stresses acting in and around the pit are free to flow both beneath and around the sides of the pit. It is likely, therefore, that, unless there are very low strength faults parallel to the analysis plane, a constant stress or following stress boundary will overpredict the stresses acting horizontally.

12.4.6 Rock-Mass Material Models

As noted previously, it is impossible to model all discontinuities in a large slope. This may be possible for a limited number of benches, but not large slopes. Therefore, much of the rock mass must be represented by an equivalent continuum in which the effect of the discontinuities is to reduce the intact rock elastic properties and strength to those of the rock mass. This is true whether or not a discontinuum model is used. Thus, the discontinuities are not explicitly modeled; rather, they are assumed to be smeared throughout the rock mass. The process for initially estimating the rock-mass properties is usually based on empirical relations as described, for example, by Hoek and Brown (1997). These initial properties are then modified, as necessary, through the calibration process.

The rock-mass material model used in most finite-difference analyses is a linear elastic–perfectly plastic model. The shear strength is limited by the Mohr–Coulomb criterion. The tensile strength is limited by the specified tensile strength (taken to be 10% of the rock-mass cohesion in most analyses). Using this model, the rock mass behaves in an isotropic manner. Strength anisotropy can be introduced through a ubiquitous joint model, which limits the shear strength according to a Mohr–Coulomb criterion in a specified direction. The direction often corresponds to a predominant jointing orientation. A more complete equivalent-continuum model that includes the effects of joint orientation and spacing is a micropolar (Cosserat) plasticity model. This model, as implemented in FLAC, is described in the context of slope stability by Dawson and Cundall (1996). The approach has the advantage of using a continuum model while still preserving the ability to explicitly consider realistic joint spacing. Unfortunately, the model has not yet been incorporated into any publicly available code.

Real rock masses often appear to exhibit progressive failure, i.e., the failure appears to progress over time. Progressive failure is a complex process that is poorly understood and difficult to model. It may involve one or more of the following component mechanisms:

- Gradual accumulation of strain or principal structures and/or within the rock mass
- Increases in pore pressure with time
- Time-dependent deformation of material under constant load (i.e., creep)

Each of these components is discussed briefly below in the context of slope behavior.

Gradual accumulation of strain or principal structures within the rock mass usually results from excavation, and "time" is really related to the excavation sequence. In order to study the progressive failure effects due to excavation, one must either introduce characteristics of the post-peak or post-failure behavior of the rock mass into a strain-softening model or introduce similar characteristics into the explicit discontinuities. In practice, there are at least two difficulties associated with strain-softening rock-mass models. The first is estimating the post-peak strength and the strain over which the strength reduces. There appear to be no empirical guidelines for estimating the required parameters. This means that the properties must be estimated through calibration. The second difficulty is that, for a simulation in which the response depends on shear localization and in which material softening is used, the results will depend on the element sizes. However, it is quite straightforward to compensate for this form of mesh-dependence. In order to do this, consider a displacement applied to the boundary of a body. If the strain localizes inside the body, the applied displacement appears as a jump across the localized band. The thickness of the band contracts until it is equal to the minimum allowed by the grid, i.e., a fixed number of element widths. Thus, the strain in the band is

$$\varepsilon = u/n\Delta z$$

Where:

n = a fixed number
u = the displacement jump
Δz = the element width

If the softening slope is linear (i.e., the change in a property value Δp is proportional to strain), the change in property value with displacement is

$$\Delta p/\Delta u = s/n\Delta z$$

where s is the input softening slope.

In order to obtain mesh-independent results, we can input a scaled softening slope, such that

$$s = s'\Delta z$$

where s' is constant. In this case, $\Delta p/\Delta u$ is independent of Δz. If we define the softening slope by the critical strain, ε^s_{crit}, then

$$\varepsilon^s_{crit} \propto 1/\Delta z$$

For example, if we double the zone size, we must halve the critical strain for comparable results.

Strain-softening models for discontinuities are much more common than similar relations for rock masses. Strain-softening relations for discontinuities are built into UDEC and 3DEC and can be incorporated into interfaces in FLAC and FLAC3D via a built-in programming language (i.e., FISH functions). Strain-softening models require special attention when computing safety factors. A technique to account for strain-softening behavior in factor of safety calculations is discussed in the section on safety factors.

Increases in pore pressure with time are not common in rock slopes for mines. More commonly, the pore pressures reduce due to deepening of the pit and/or drainage. However, there are cases in which the pore pressures do increase with time. In such cases, the slope may appear to progressively fail.

Time-dependent deformation of material under constant load (i.e., creep) is not commonly considered in the context of slope stability. It is much more common in underground excavations. Several material models are available to study creep behavior in rock slopes. These include classical viscoelastic models, power-law models, and the Burger-creep viscoplastic model. Work is under way to apply creep models to the study of slope behavior at Chuquicamata Mine in Chile (Calderon 2000).

Of the models listed above, the Mohr–Coulomb plasticity model is used in approximately 90% of all slope analyses; the ubiquitous joint model is used approximately 10% of the time. Strain-softening models have been used, but rarely.

12.4.7 Joint Material Models in UDEC and 3DEC

The material model used most commonly to explicitly represent joints in UDEC and 3DEC is a linear elastic–plastic model. A peak and residual shear strength can be specified for the joints. The residual strength is used after the joint has failed in shear at the peak strength. The elastic behavior of the joints is specified by joint normal and shear stiffnesses.

12.5 EFFECT OF FLOW ANALYSIS VERSUS WATER TABLE ON PREDICTED SLOPE STABILITY

The effect of water pressure in reducing effective stresses and, hence, slope stability is well understood. However, the effect of various assumptions regarding specification of pore-pressure distributions in slopes is not as well understood. Two methods are commonly used to specify pore-pressure distributions within slopes. The most rigorous method is to perform a complete flow analysis and use the resultant pore pressures in the stability analyses. A less rigorous, but more common, method is to simply specify a water table. The resulting pore pressures are then assumed to be given by the product of the vertical depth below the water table, the water density, and gravity. In this sense, the water table approach is equivalent to specifying a piezometric surface. Both methods use similar phreatic surfaces, but the second method underpredicts pore-pressure concentrations that actually occur near the toe of a slope and slightly overpredicts the pore pressure behind the toe by ignoring the inclination of equipotential lines. Finally, seepage must be considered. The difference in water pressure that exists between two points at the same elevation (i.e., hydraulic gradient) results from seepage forces (or drag) as water moves through a porous medium. Flow analysis automatically accounts for seepage forces.

To evaluate the error resulting from specifying a water table (without doing a flow analysis), two identical problems were run. In one case, a flow analysis was performed to determine the pore pressures. In the second case, the pressures were determined using only a piezometric surface that was assumed to be the phreatic surface taken from the flow analysis. The material properties and geometry for both cases are shown in Figure 12.2. The right-hand boundary was extended to allow the far-field phreatic surface to coincide with the ground surface at a horizontal distance of 2 km behind the toe. Permeability within the model was assumed to be homogeneous and isotropic. The error caused by specifying the water table can be seen in Figure 12.3. The largest errors (up to 45%) are found just below the toe. Errors in pore-pressure values behind the slope are generally less than 5%. The errors near the phreatic surface are not significant, as they result from the relatively small pore pressures just below the phreatic surface (i.e., small errors in small values result in large relative errors).

For a phreatic surface at the ground surface at a distance of 2 km, a factor of safety of 1.1 is predicted using circular failure chart number 3 (Hoek and Bray 1981). The factor of safety determined by FLAC is approximately 1.15 for both cases. The FLAC analyses give similar safety factors because the distribution of pore pressures in the area where failure occurs (i.e., behind the slope) is very similar for the two cases. The conclusion drawn here is that there is no significant difference in predicted stability between a complete flow analysis and simply specifying a piezometric surface. However, it is not clear if this conclusion can be extrapolated to other cases involving, for example, anisotropic flow.

12.6 EXCAVATION SEQUENCE AND INTERPRETATION OF RESULTS

12.6.1 Excavation Sequence

Simulating excavations in finite-difference models poses no conceptual difficulties. However, the amount of effort required to construct a model depends directly on the number of excavation stages simulated. Therefore, most practical analyses seek to reduce the number of excavation stages. The most accurate solution is obtained using the largest number of excavation steps, since the real load path for any element in the slope will be followed closely. In theory, it is impossible to prove that the final solution is independent of the load path followed. However, for many slopes, the stability seems to depend mostly on the condition of the slope (i.e., geometry, pore pressure distribution) at the time of analysis and very little on the load path taken to get there.

A reasonable approach regarding the number of excavation stages has evolved over the years. Using this approach, only a few (one, two, or three) excavation stages are modeled. For each stage, two calculation steps are taken. In the first step, the model is run elastically to remove any inertial effects caused by sudden removal of a large amount of material. Next, the model is run allowing plastic behavior to develop. Following this approach, reasonable solutions to a large number of slope-stability problems have been obtained.

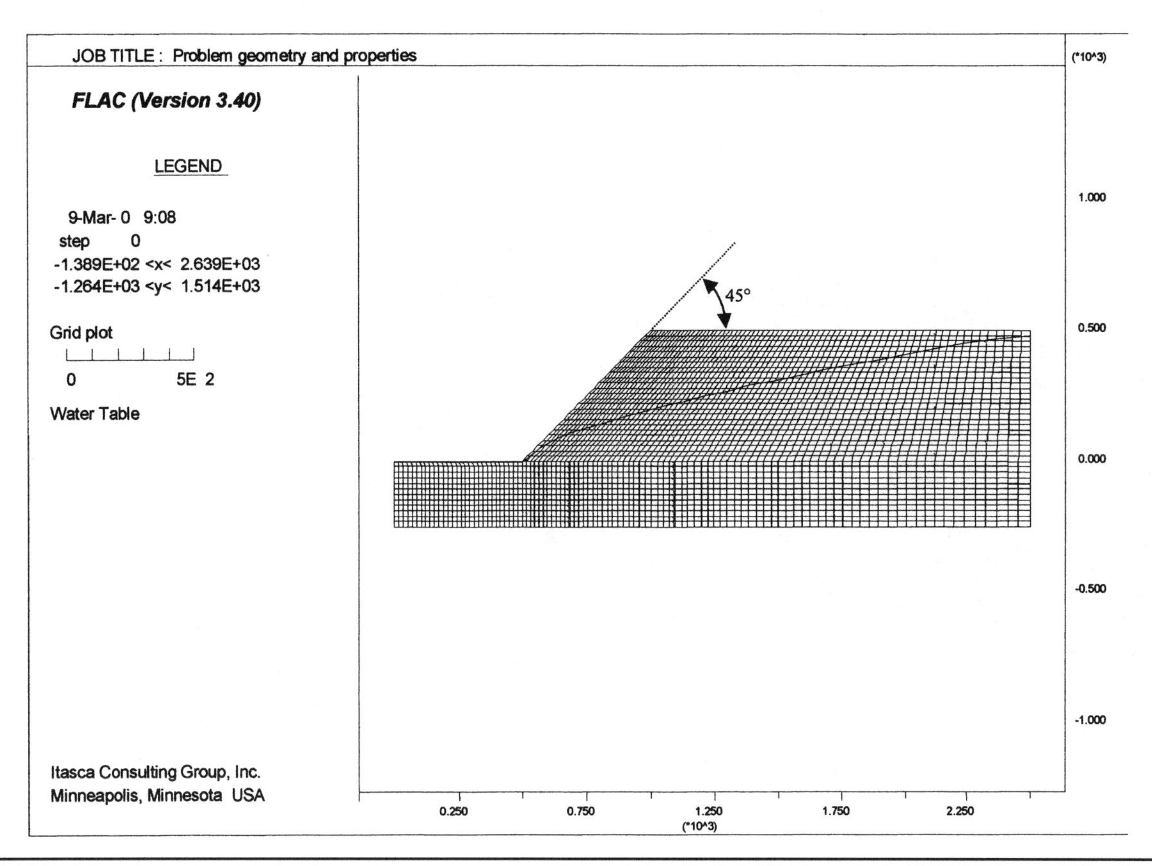

FIGURE 12.2 Problem geometry and rock-mass properties used to evaluate effect of pore pressure assumptions on slope stability

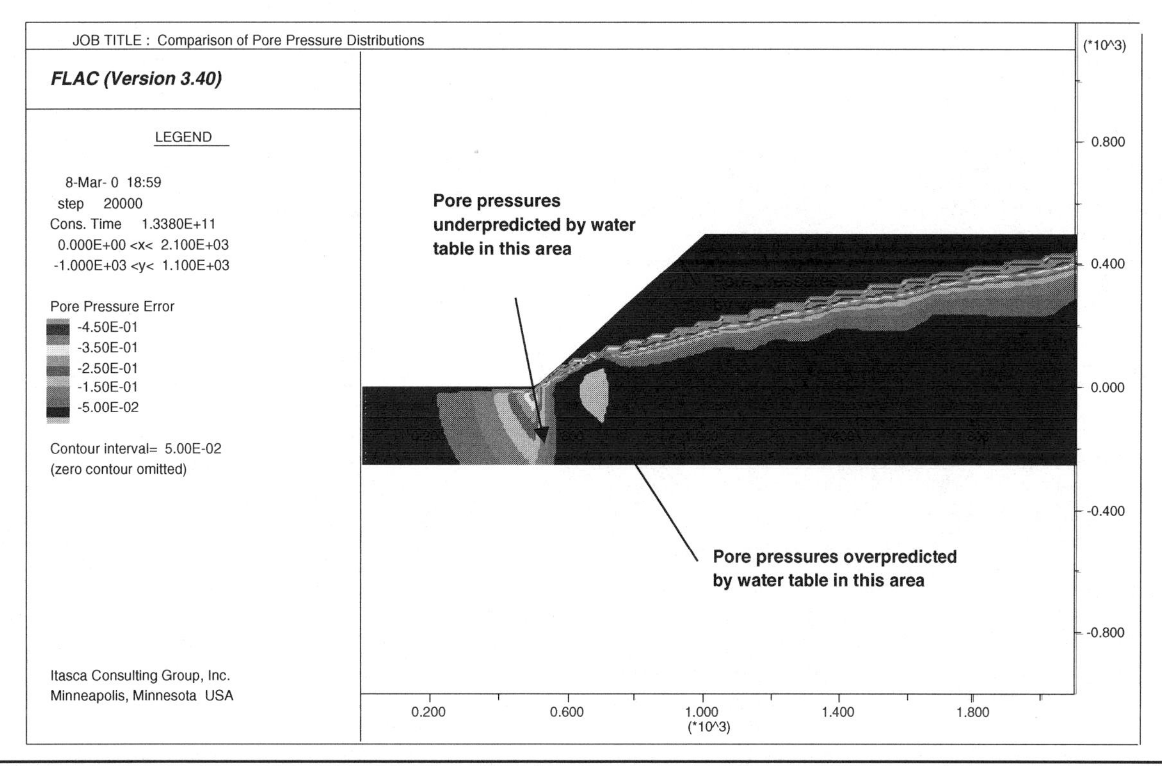

FIGURE 12.3 Error caused by specifying water table instead of performing a flow analysis

12.6.2 Interpretation of Results

As noted in the introduction, finite-difference programs are not black boxes that "give the solution." The behavior of the numerical system must be interpreted, and results from finite-difference models are interpreted in much the same way as prism data are interpreted. Finite-difference programs record displacements and velocities at nominated points within the rock mass. During the analysis, the recorded values can be examined to see if they are increasing, remaining steady, or decreasing. Increasing displacements and velocities indicate an unstable situation; steady displacements and decreasing velocities indicate a stable situation. In addition, velocity and displacement vectors for every point in the model can be plotted. Fields of constant velocity and displacement indicate failure.

The authors have found that velocities below 1e-6 indicate stability in FLAC and FLAC3D; conversely, velocities above 1e-5 indicate instability. Note that no units are given for velocities. This is because the velocities are not real, due to the damping and mass scaling used to achieve static solutions. While the velocities are not real, the displacements are, although we cannot say at what "time" the displacement occurs.

The failure (plasticity) state (i.e., elastic, failed in tension, failed in shear) of points within the model can also be examined. Care must be used in examining the failure state indicators. For example, local overstressing (e.g., at the base of toppling columns) can appear to form a deep-seated slip surface when, in reality, it is just compressive failure of toppling columns. Therefore, the failure (plasticity) indicators must be reviewed in the context of overall behavior before any definitive conclusions can be drawn.

12.6.3 Calculating Factor of Safety

Evaluation of Safety Factor Using the Shear-Strength Reduction Technique. For slopes, the factor of safety is often defined as the ratio of the actual shear strength to the minimum shear strength required to prevent failure. A logical way to compute the factor of safety with a finite-element or finite-difference program is to reduce the shear strength until collapse occurs. The factor of safety is the ratio of the soil or rock's actual strength to the reduced shear strength at failure. This shear-strength reduction technique was first used with finite elements by Zienkiewicz, Humpheson, and Lewis (1975) to compute the safety factor of a slope composed of multiple materials.

To perform slope-stability analysis with the shear-strength reduction technique, simulations are run for a series of increasing trial factors of safety, f. Actual shear strength properties—cohesion, c, and friction, ϕ, are reduced for each trial according to the following equations:

$$c^{trial} = (1/f)c \quad \textbf{EQ. 12.1}$$

$$\phi^{trial} = \arctan\{(1/f)\tan\phi\} \quad \textbf{EQ. 12.2}$$

If multiple materials and/or joints are present, the reduction is made simultaneously for all materials. The trial factor of safety is gradually increased until the slope fails. At failure, the safety factor equals the trial safety factor (i.e., f = F). Dawson, Roth, and Drescher (1999) show that the shear-strength reduction factors of safety are generally within a few percent of limit-equilibrium solutions when an associated flow rule is used.

If a strain-softening constitutive model is used, the softening logic should be turned off during the shear-strength reduction process or the factor of safety will be underestimated. When the slope is excavated, some zones will have exceeded their peak strength and some amount of softening will have taken place. During the strength reduction process, these zones should be considered as a new material with lower strength, but no further softening should be allowed due to the plastic strains associated to the gradual reduction of strength.

The shear-strength reduction technique has two advantages over slope-stability analyses with limit equilibrium. First, the critical failure surface is found automatically and it is not necessary to specify the shape of the failure surface (e.g., circular, log spiral, piecewise linear, etc.) in advance. In general, the failure mode for slopes is more complex than simple circles or segmented surfaces. Second, numerical methods automatically satisfy translational and rotational equilibrium; not all limit equilibrium methods do. Consequently, the shear-strength reduction technique will usually determine a lower safety factor compared to other methods. For example, Zienkiewicz, Humpheson, and Lewis (1975) showed comparisons between factors of safety calculated by limit-equilibrium and finite-element methods for a slope composed of three materials. They reported a factor of safety of 1.33 using Bishop's method and 1.165 using the shear-strength reduction technique. Donald and Giam (1988) reported results of factor of safety analyses for a homogeneous slope with a weak layer in the foundation. They reported a factor of safety of 1.5 using Bishop's method with a simplex optimization and 1.34 using the shear-strength reduction technique.

12.7 SOME USEFUL RESULTS

12.7.1 Effect of Radius of Curvature on Slope Stability

Most design analysis for slopes assumes a two-dimensional problem geometry (i.e., a unit slice through an infinitely long slope, under plane-strain conditions). In other words, the radii of both the toe and crest are assumed to be infinite. This is not the condition encountered in practice—particularly in open-pit mining, where the radii of curvature can have an important effect on safe slope angles. Concave slopes are inherently more stable than convex slopes due to the lateral restraint provided by material on either side of a potential failure in a concave slope.

Despite its potential importance in slope stability, very little has been done to quantify this effect. Jenike and Yen (1961) presented the results of limit theory analysis of axisymmetric slopes in a rigid-perfectly plastic material and determined S-shaped critical profiles that describe the theoretical failure shape for different radii of curvature. They showed that, as the radius of the slope increases, the profile of the stable slope in axial symmetry approaches the profile of the slope in plane strain. However, their method does not permit any evaluation of safety factors. In addition, their assumptions resulted in slopes that, in plane strain, took no account of cohesion. Hoek and Brown (1981) concluded that the analysis assumptions were not applicable to rock-slope design.

Piteau and Jennings (1970) studied the influence of plan curvature on the stability of slopes in four diamond mines in South Africa. As a result of caving from below the surface, slopes were all at incipient failure (i.e., a safety factor of 1). The average slope height was 100 m. Piteau and Jennings found that the average slope angle for slopes with radii of curvature of 60 m was 39.5° as compared to 27.3° for slopes with radii of curvature of 300 m.

Hoek and Bray (1981) summarized their experience with the stabilizing effects of slope curvature as follows. When the radius of curvature of a concave slope is less than the height of the slope, the slope angle can be 10° steeper than the angle suggested by conventional stability analysis. As the radius of curvature increases to a value greater than the slope height, the correction should be decreased. For radii of curvature in excess of twice the slope height, the slope angle given by a conventional stability analysis should be used.

To better quantify the effects of slope curvature, a series of generic analyses were performed. All analyses assumed a 500-m-high dry slope consisting of an isotropic homogeneous material

TABLE 12.1 Effect of radius of curvature on factor of safety for 45° slope

	R = 100 m	R = 250 m	R = 500 m	R = □
□ = 45°	F = 1.75	F = 1.65	F = 1.55	F = 1.3

TABLE 12.2 Effect of radius of curvature on slope angle for safety factor = 1.3

	R = 100 m	R = 250 m	R = 500 m	R = □
F = 1.3	□ = 75°	□ = 60°	□ = 55°	□ = 45°

with 35° friction, 0.66 MPa cohesion, and 2,600-kg/m^3 density (i.e., similar to Figure 12.2). Initial in situ stresses are assumed to be lithostatic, and the excavation is made in 40-m decrements beginning from the ground surface. For these conditions, a factor of safety of 1.3 is predicted using circular failure chart number 1 (Hoek and Bray 1981). A factor of safety of 1.3 is a value that is frequently used in the design of slopes for open-pit mines. Two series of analyses were performed using FLAC. In the first series, the safety factor was calculated for axisymmetric conditions with various radii of curvature. The results are shown in Table 12.1. In the second series, the slope angle was increased until a safety factor of 1.3 was achieved. The results of this series of analyses are shown in Table 12.2.

The results shown for R = 500 m in Table 12.2 confirm the notion that, when the radius of curvature of a concave slope is less than the height of the slope, the slope angle can be 10° steeper than the angle suggested by conventional (i.e., two-dimensional) stability analysis.

One reason designers are reluctant to take advantage of the beneficial effects of slope curvature is that the presence of discontinuities can often negate the effects. However, for massive rock slopes, or slopes with relatively short joint trace lengths, the beneficial effects of slope curvature should not be ignored—particularly in open-pit mines, where the economic benefits of steepening slopes can be significant.

TABLE 12.3 Rock-mass properties used in Boinas East Pit study

Material	K (GPa)	G (GPa)	ϕ (°)	Cohesion (kPa)	Density (t/m^3)
Strong dolomite	7.92	5.09	43	1120	2.70
Weak sandstone	1.31	0.75	21	270	2.30
Weathered granite	1.17	0.66	20	250	2.28
Marble	2.45	1.41	33	500	2.43
Ore	1.11	0.62	19	250	2.21

12.8 CASE STUDY: BOINAS EAST PIT

Boinas East is one of the three open pits currently operated by Narcea Gold Mines in northwestern Spain. Clifford (2000) gives a general description of the mining operation. The east slope of the pit consists of a strong dolomite overlying weak sandstone, weathered granite, and the mineralized skarn, as shown in Figure 12.4. The slope height is 280 m and its radius of curvature at the base is 70 m. The slope angle is 65° in the dolomites and 45° in the sandstone and granite. The pore pressure is negligible in the pit. The properties used for the analysis appear in Table 12.3.

A plane-strain FLAC analysis predicts a factor of safety of 1.05, with shear failure in the sandstone and weathered granite and tensile failure in the dolomite (Figure 12.5). In an axisymmetric analysis, the low radius of curvature relative to the height inhibits the tensile failure in the limestone, and the factor of safety increases to 1.70 (Figure 12.6).

However, the results of the axisymmetric analysis are too optimistic. While the geometry of the pit can be assumed to be conical, the sedimentary units dip east and do not follow an axisymmetric pattern. A more elaborate FLAC3D model (Figure 12.7) shows a similar failure pattern (Figure 12.8) in the east slope but with a factor of safety of 1.35.

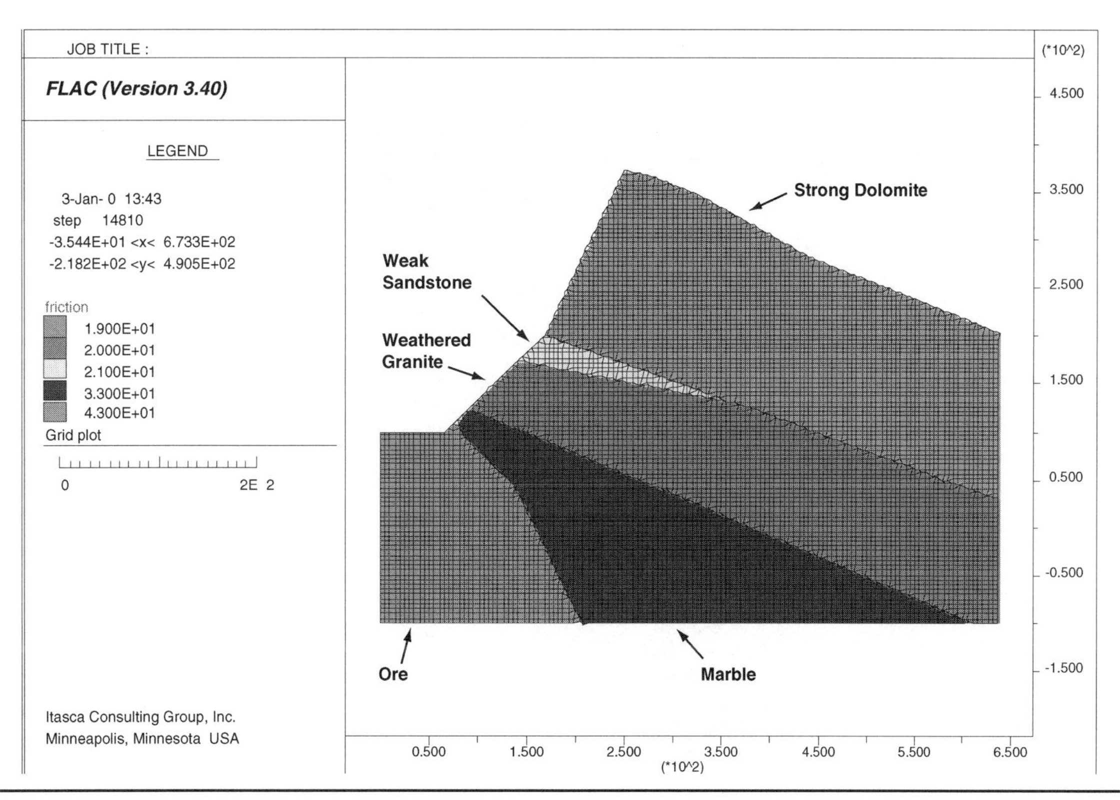

FIGURE 12.4 Geometry of Boinas East Pit with axial symmetry

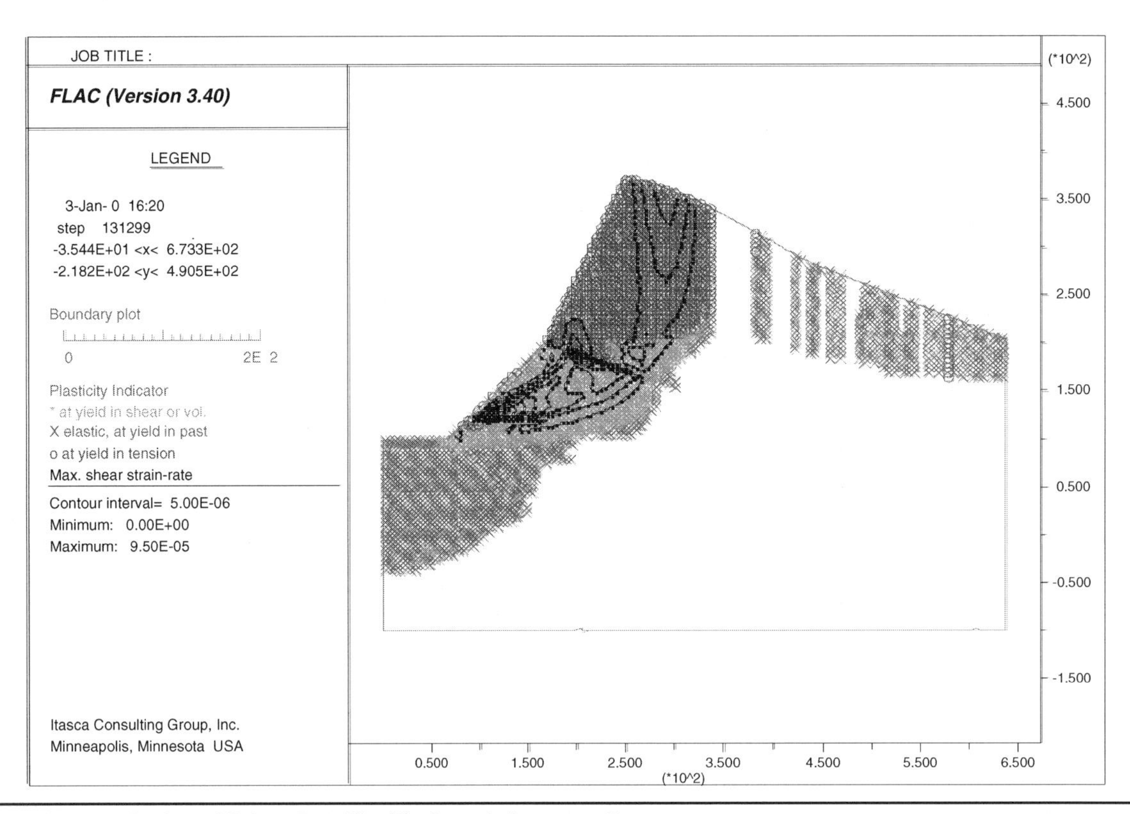

FIGURE 12.5 Failure mechanism of Boinas East Pit with plane-strain assumption

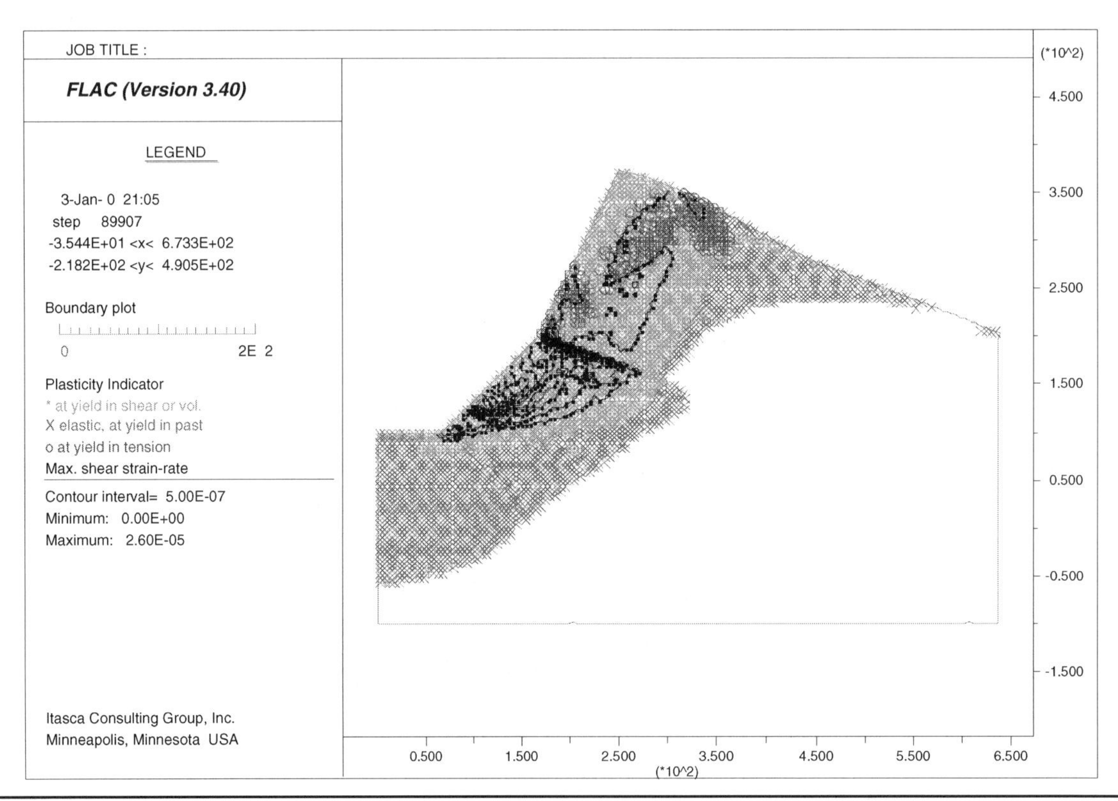

FIGURE 12.6 Failure mechanism of Boinas East Pit with axial symmetry

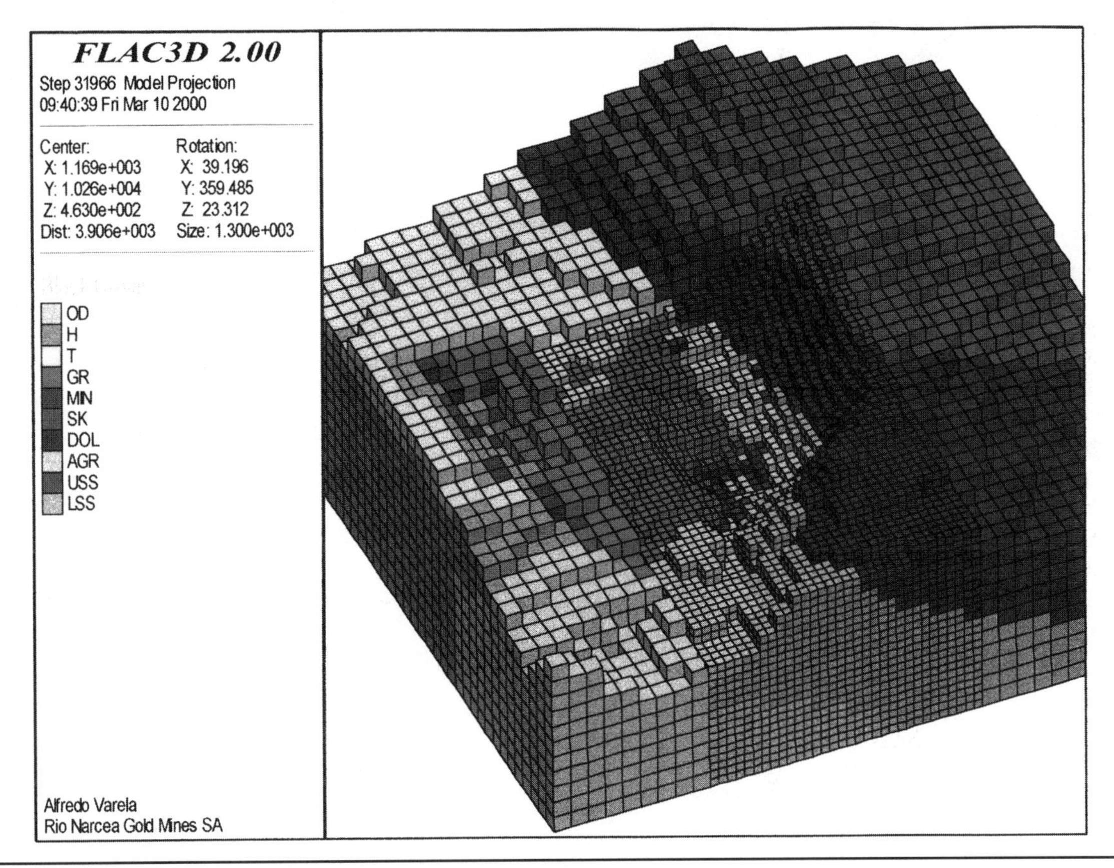

FIGURE 12.7 Boinas East Pit FLAC3D model

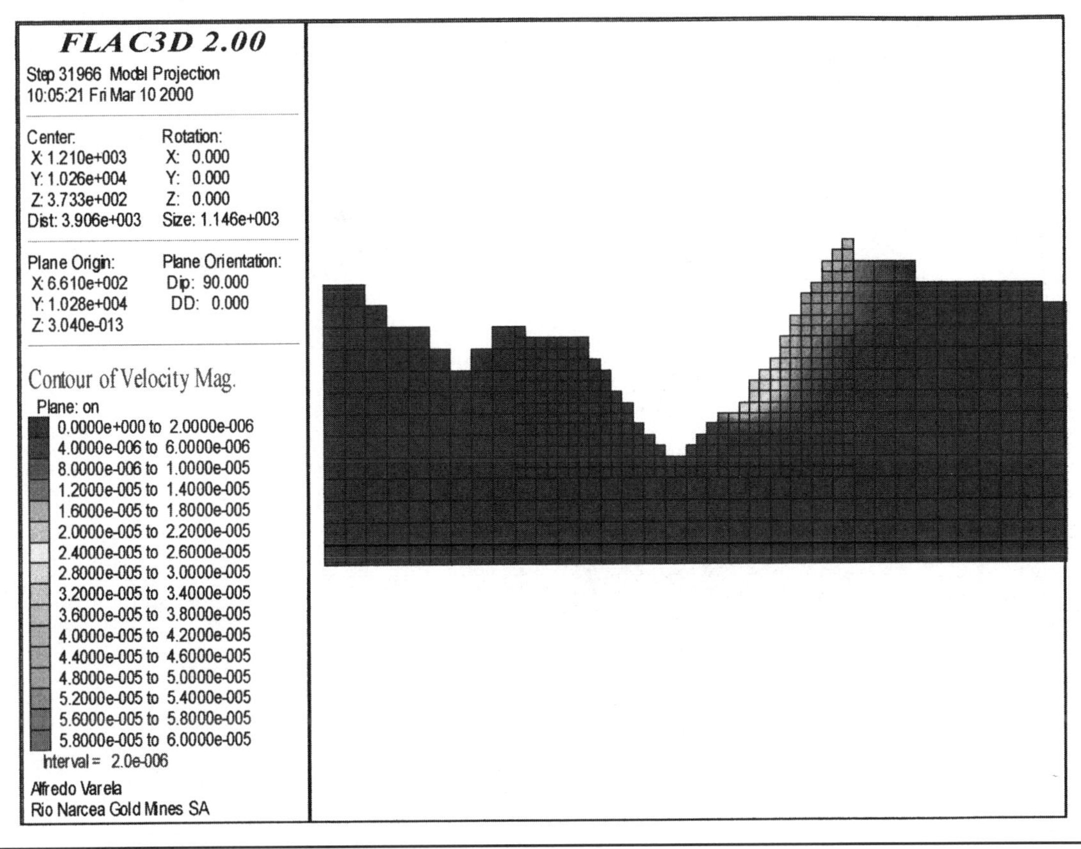

FIGURE 12.8 Cross section through Boinas East Pit FLAC3D model showing failure mode

12.9 IMPACT OF INITIAL STRESS STATE

The role of stresses has been traditionally ignored in slope analyses. There are several possible reasons for this, including

- Limit-equilibrium analyses, which are widely used for stability analyses, cannot include the effect of stresses in their analyses. Nevertheless, limit-equilibrium analyses are thought to provide reasonable estimates of stability in many cases, particularly where structure is absent (e.g., slopes in soil).
- Most stability analyses have traditionally been performed for soils, where the range of possible in situ stresses is more limited than for rocks. Furthermore, many soil analyses have been performed for constructed embankments (e.g., dams), where in situ stresses do not exist.
- Most slope failures are gravity driven, and the effects of in situ stress are thought to be minimal.
- In situ stresses in rock masses are not routinely measured for slopes and their effects are largely unknown.

One particular advantage of stress analysis programs (i.e., numerical models) is their ability to include premining initial stress states in stability analyses and to evaluate their importance.

Five cases were run using a model similar to the model shown in Figure 12.2 in order to Tao-evaluate the effects of in situ stress state on stability. For each of the five cases, the slope angle was 60°, the slope height was 400 m, the material density was 2,450 kg/m^3, the friction angle was 32°, and the cohesion was 0.92 MPa. The results of FLAC analyses are shown in Table 12.4.

Although, in general, it is impossible to say what effect the initial stress state will have on any particular problem, as behavior depends on several factors (i.e., orientation of major structures, rock-mass strength, water conditions), the following observations can be made:

The larger the initial horizontal stresses are, the larger the horizontal elastic displacements will be. This is not much help, as displacements that are of elastic magnitude are not particularly important in slope studies.

- *Initial horizontal stresses in the plane of analysis that are less than the vertical stresses tend to slightly decrease stability and reduce the depth of significant shearing with respect to a hydrostatic stress state.* This observation may seem counter-intuitive; we would expect smaller horizontal stresses to increase stability. The explanation lies in the fact that the lower horizontal stresses actually provide slightly decreased normal stress on potential shearing surfaces and/or joints within the slope. This observation was confirmed in UDEC analysis of a slope in Peru where in situ horizontal stresses lower than the vertical stress led to deeper levels of joint shearing in toppling structures compared to cases involving horizontal stresses that were equal to or greater than the vertical stress.

TABLE 12.4 Effect of in situ stress on slope stability

In-Plane Horizontal Stress	Out-of-Plane Horizontal Stress	Factor of Safety
$\sigma xx = \sigma yy$	$\sigma zz = \sigma yy$	1.30
$\sigma xx = 2.0 * \sigma yy$	$\sigma zz = 2.0 * \sigma yy$	1.30
$\sigma xx = 0.5 * \sigma yy$	$\sigma zz = 0.5 * \sigma yy$	1.28
$\sigma xx = 2.0 *.\sigma yy$	$\sigma zz = 0.5 * \sigma yy$	1.30
$\sigma xx = 0.5 * \sigma yy$	$\sigma zz = 2.0 * \sigma yy$	1.28

- It is important to note that the regional topography may limit the possible stress states, particularly at elevations above regional valley floors. Three-dimensional models have been very useful in the past in addressing some regional stress issues.

12.10 REFERENCES

Calderon, A. 2000. Codelco Chile, Division Chuquicamata. Personal communication.

Clifford, D. 2000. Rio Narceas's ongoing growth. *Mining Magazine*. 820–86.

Dawson, E.M., and P.A. Cundall. 1996. Slope stability using micropolar plasticity. *Rock Mechanics Tools and Techniques. Proceedings of the 2nd North American Rock Mechanics Symposium, Montréal, June 1996.* 1:551–558. ed. M. Aubertin et al., Rotterdam: A.A. Balkema.

Dawson, M.D., W.H. Roth, and A. Drescher. 1999. Slope stability analysis by strength reduction. Submitted for publication to *Géotechnique.*

Donald, I.B., and S.K. Giam. 1988. Application of the nodal displacement method to slope stability analysis. *Preprint Proceedings of the Fifth Australia-New Zealand Conference on Geomechanics–Prediction versus Performance (Sydney, 1988).* pp. 456–460.

Hoek, E., and J.W. Bray. 1981. *Rock slope engineering,* 3rd Ed. London: The Institute of Mining and Metallurgy.

Hoek, E., and E.T. Brown. 1997. Practical estimates of rock mass strength. *Intl. Jour. Rock Mech. Min. Sci. & Geomech. Abstr.* 34:8:1165–1186.

Itasca Consulting Group, Inc. 1997. FLAC3D (Fast Lagrangian Analysis of Continua in 3 Dimensions), Version 2.0. Minneapolis: ICG.

Itasca Consulting Group, Inc. 1998. FLAC (Fast Lagrangian Analysis of Continua), Version 3.4. Minneapolis: ICG.

Itasca Consulting Group, Inc. 1999. 3DEC (Three-Dimensional Distinct Element Code), Version 2.0. Minneapolis: ICG.

Itasca Consulting Group, Inc. 2000. UDEC (Universal Distinct Element Code), Version 3.1. Minneapolis: ICG.

Jenike, A.W., and B.C. Yen. 1961. Slope stability in axial symmetry. *Proceedings of the 5th Rock Mechanics Symposium, University of Minnesota,* pp. 689–711. London: Pergamon.

Piteau, D.R., and J.E. Jennings. 1970. The effects of plan geometry on the stability of natural slopes in rock in the Kimberley area of South Africa. *Proceedings of the 2nd Congress of the International Society of Rock Mechanics (Belgrade),* 3:Paper 7-4.

Starfield, A.M., and P A. Cundall. 1988. Towards a methodology for rock mechanics modelling. *Intl. Jour. Rock Mech. Min. Sci. & Geomech. Abstr.* 25:3:99–106.

Zienkiewicz, O.C., C. Humpheson, and R.W. Lewis. 1975. Associated and non-associated viscoplasticity and plasticity in soil mechanics. *Géotechnique.* 25:4:671–689.

CHAPTER 13

Blast Designs to Protect Pit Walls

Timothy N. Hagan* and Brad Bulow†

13.1 INTRODUCTION

Blasting has less of an influence on pit-wall stability than geology and groundwater. Of course, the geology of a pit cannot be changed, yet a mine operator can usually exert more control over blasting than over groundwater. Although blasting is usually the third most influential factor, it is the factor that can be controlled most.

When one considers the immense power of explosive charges in normal production blasts, it is not surprising that the rock mass adjacent to the blast block is weakened. Rock-mass strength can be almost eliminated where a rock mass is very close to charges. New fractures and planes of weakness are created in the adjoining rock, and fissures (i.e., natural cracks) are opened and sometimes extended. New, dilated and extended cracks are manifested as overbreak.

Overbreak has a deleterious effect on the inclination of final bench faces and a lesser effect on the overall inclination of pit walls. If a blast is not well designed, overbreak can contribute to pit-wall instability. The extra expertise, effort, and care required to produce a more stable and, consequently, safer pit wall usually cost less than the corrective measures that must be taken once a hazardous pit wall has been formed. Therefore, it is important to optimize (not minimize) overbreak, especially as blasts approach the designed wall of the pit. The successful application of overbreak-control blasting techniques reduces not only the quantity of rock to be removed, it lessens the hazard and cost of rockfalls and may reduce the need for pit-wall support. The predicted cost savings resulting from less damage help to determine the expenditure that can be allocated to improving blasting techniques.

Where ground vibration fails to produce new fractures or to extend fissures or blast-induced fractures significantly, its contribution to instability is usually less than that of overbreak.

Because the design, implementation, and effects of blasts can be controlled within wide limits, mine operators are able to promote stability by controlling blast-induced damage. To fully appreciate the relationship between damage and blasting, it is necessary to understand how the design of blasts affects overbreak and ground vibrations.

13.2 PIT-WALL BLASTING RATIONALE

As production blasts approach a designed pit wall, there is usually a distance within which mine operators should be concerned with protection rather than production. Accordingly, provided there is sufficient working area, blasts on a given bench should consist of production blasts that are relatively remote from the designed pit wall and pit-wall blasts that abut or are close to the designed pit wall. If overbreak by production blasts is minimized, the volume of pit-wall blasts can be reduced.

In pit-wall blasts, there is a strong economic incentive to achieve the required levels of soundness and smoothness at the lowest possible cost. The cardinal way to lower cost is to implement economies of scale into the design of pit-wall blasts. Economies of scale can be best achieved by utilizing blasting knowledge and expertise to control the adverse effects of using large-diameter production blastholes charged with bulk explosives as close as possible to the designed pit wall.

13.3 SELECTING PIT-WALL BLASTING TECHNIQUES

As a normal production blast is progressively modified to reduce overbreak and ground vibrations, it first becomes a modified production blast, then a cushion blast, then a postsplit blast, and finally a pit-wall blast that features both cushion blastholes and postsplit blastholes. Presplit blasts are often fired in association with cushion blasts as an alternative to postsplitting. Presplit blasts are different in that

- Presplit blastholes are fired *before* and immediately in front of the blastholes, sometimes in a separate blast.
- Where the height of a final face is to be two or three times the normal bench height, the depth of presplit blastholes is sometimes two or three times that of the blastholes in front of the presplit.

In this chapter, pit-wall blasts are divided into three categories

- Modified production blasts
- Smoothwall blasts
- Blasts that incorporate presplitting (see Table 13.1)

Smoothwall techniques and presplitting are often used to produce pit walls with great integrity and/or smoothness. The satisfactory (sometimes spectacular) results that are obtained tend to explain the popularity of such methods. However, the subject of damage control is greater than that of smoothwall blasting and presplitting.

The blasting technique selected to control damage must first be safe and then, very importantly, cost efficient. The aesthetic aspect of a presplit face should not bias judgment on the cost efficiency of this technique. If sufficient success can be achieved with modified production blasts, the need for smoothwall blasting or presplitting can be reduced or even eliminated.

13.4 MODIFIED PRODUCTION BLASTS

The overbreak caused by production blasts is reduced by lowering the concentration of explosive energy in the adjacent ground. Contrary to popular belief, the redesign of a production blast to reduce wall damage usually improves the fragmentation, displacement, and looseness of the muck pile.

A modified production blast is a blast that has been modified uniformly over its entire area, not just the back-row blastholes. A modified production blast should shoot to a free face and have only one charge detonating per delay.

* Tim Hagan Blasting Pty. Ltd., The Gap, Queensland, Australia.
† Orica Explosives, Mt. Pleasant, Western Australia, Australia.

TABLE 13.1 Pit-wall blasts options

Blast Type	Features	Relative Cost
1. All blasts	▪ Reduced number of rows of blastholes ▪ Shoot to a free face ▪ Longer inter-row delays ▪ One-hole-per-delay firings	
2. Modified production	▪ Optimize "stand-off" distance to designed pit wall	Low

3. Smoothwall	▪ Smaller burden and spacing and correspondingly higher powder factor
	▪ Effective subdrilling over final berms is zero or has negative value
	▪ Geometry of back-row blastholes matched to designed inclination of final face by using
	  ▪ stab blastholes or
	  ▪ angled blastholes on or parallel to designed final face
		Medium to high

3a. Cushion	▪ Most basic type of smoothwall blast
	▪ For back-row blastholes:
	  ▪ smaller burden and spacing
	  ▪ lighter charges, achieved by conventional decking or, preferably, air decking
	  ▪ normal powder factor
	  ▪ possibly smaller diameter
		Medium

3b. Postsplit	▪ Back-row blastholes that
	  ▪ are drilled on or parallel to the designed final face
	  ▪ usually have a smaller diameter
	  ▪ are more closely spaced
	  ▪ have a spacing to burden ratio of about 0.8:1
	  ▪ ideally contain decoupled charges
	  ▪ detonate after the blasthole in front of them
		High

4. Presplitting	▪ Back-row blastholes that
	  ▪ are drilled on the designed final face
	  ▪ usually have a smaller diameter
	  ▪ are more closely spaced
	  ▪ are loaded even more lightly than postsplit blastholes, ideally with a decoupled charge
	  ▪ are unstemmed unless mine is airblast sensitive
	  ▪ detonate before the blastholes in front of them, sometimes in a separate blast
	  ▪ can be drilled and fired over a double or even triple bench
	▪ Optimizing the "stand-off" distance between the presplit blastholes and the blastholes immediately in front of them is critical to success
		Usually highest

Compared with a normal production blast, a modified production blast should have fewer rows of blastholes and longer interrow delays. The blasthole pattern, charge weight, and powder factor should be the same as for normal production blasts. A stand-off distance should be allowed between the back-row blastholes and the designed final face (Figure 13.1). Digging equipment is required to dig back through the overbreak zone to form the final face. While the material in the overbreak zone may be considered bonus production, the digging rate normally decreases as the designed final face is approached.

The greatest challenge lies in optimizing the stand-off distance so that there is neither an excessive amount of damage nor excessively difficult digging back to the designed final face.

The optimum stand-off distance is larger for choke blasting (compared to free-face blasting) and increases with increases in blasthole diameter, effective burden distance, number of rows of blastholes, effective subdrilling, and bench height; it increases with decreases in rock-mass strength, powder factor, and the delay interval between dependent charges.

13.5 SMOOTHWALL BLASTING

The blasts that are located directly in front of postsplit or cushion blastholes should shoot to a free face rather than to a choked face and have

- The minimum practicable number of rows of blastholes
- *High* powder factors and correspondingly small burden distances and blasthole spacings
- Interhole and interrow delays that ensure single-hole firing
- Interrow delays that result in good progressive relief of burden
- Zero or, preferably, negative subdrilling for the blastholes that are above a designed final berm or located just in front of a designed final crest

Usually the optimum powder factor for these blasts is expected to be low. However, low powder factors normally result in little forward displacement and overbreak from explosion

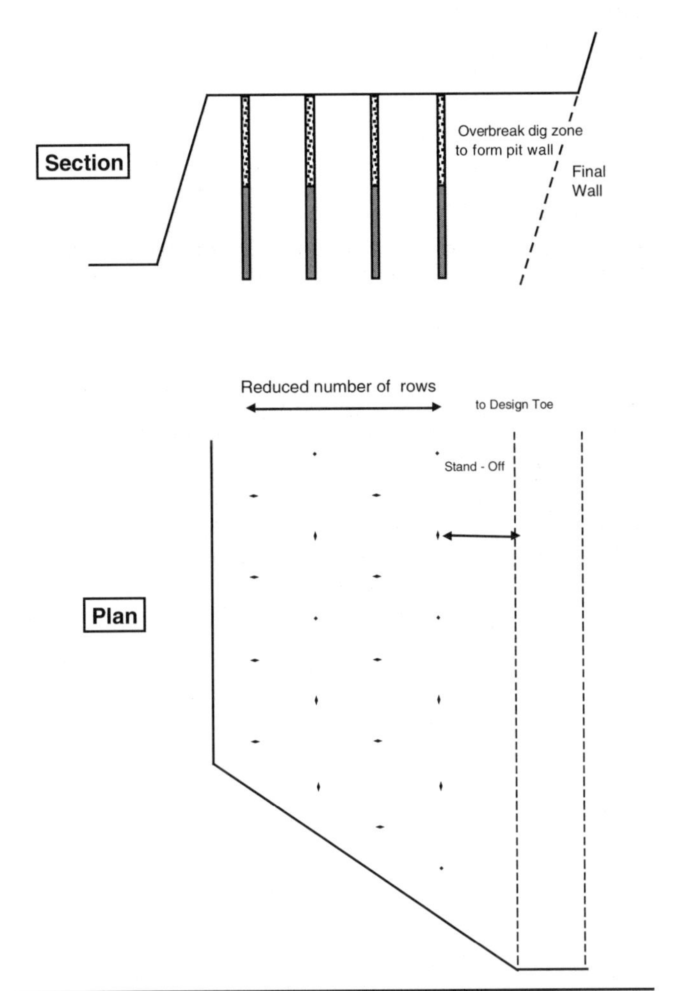

FIGURE 13.1 Modified production blast

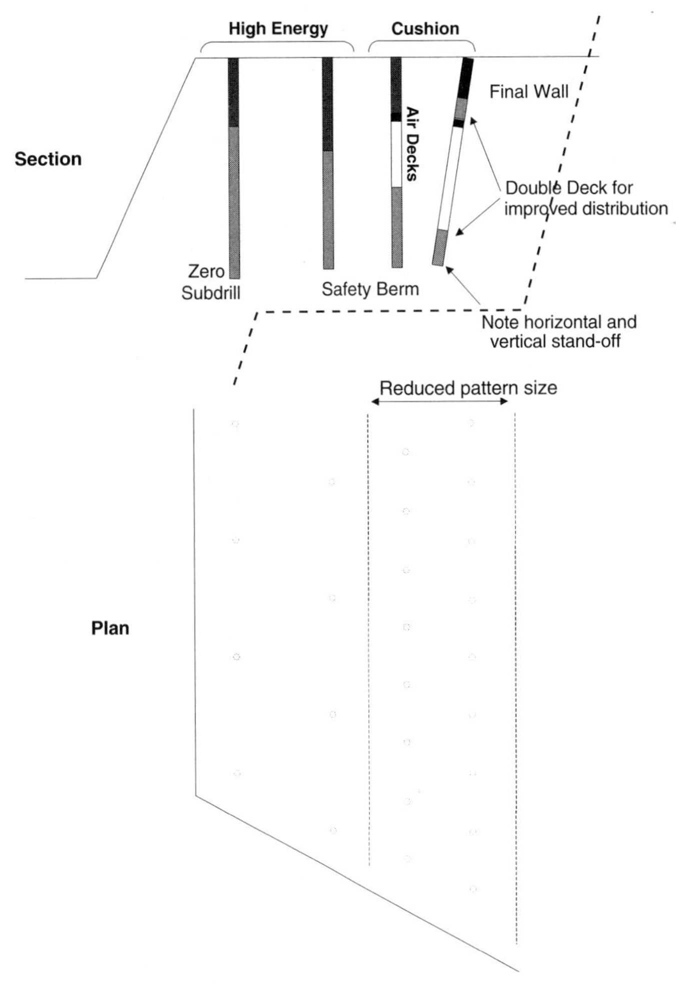

FIGURE 13.2 Smoothwall blast with vertical and angled cushion blastholes

gases jetting backward into fissures and blast-generated cracks. Overbreak is limited by using *high* powder factors, provided relatively small blasthole patterns are employed, not heavy charges.

Where the designed final face is steeply dipping (say >70°), a vertical back-row blasthole is usually satisfactory to form the final face. Where the designed inclination of a final face is flatter than 70°, the best solution is to angle the back-row blastholes, if possible, or to add vertical stab blastholes to help fragment the increasingly large volume of rock behind the full-depth blastholes.

13.5.1 Cushion Blasts

Cushion blasting is the simplest and least expensive smoothwall-blasting technique. A cushion blast is a pit-wall blast in which back-row blastholes contain lighter charges and are drilled in a correspondingly smaller pattern (Figure 13.2). The diameter of cushion blastholes is usually the same as that of production blastholes in front of them. The charge weight for the cushion holes is commonly reduced by about 45% and both burden distance and blasthole spacing by about 25%. Therefore, the powder factor is similar throughout the pit-wall blast. The lighter charges are better distributed within the blasthole by air decking and conventional decking, by using a lower-density explosive, or by decoupling. Because of the smaller burden and spacing for back-row blastholes, cushion blasting increases the costs of drilling, priming, initiation, and blast crew labor.

Cushion blastholes should detonate in a delayed sequence after the more heavily charged blastholes in front of them. Every cushion blasthole should be well relieved by the proper performance of adjacent earlier-firing blastholes. Due to the reduced spacing of cushion blastholes, it is often necessary to fire two adjacent blastholes with a short delay between them. Otherwise, the back-row production blastholes will detonate far enough ahead of the cushion blastholes to have sufficient time to dislocate their charge(s).

Cushion blasting is used without pre- or postsplitting where the rock is strong or only minor reductions in damage are required or for forming pit walls with relatively short lives. The use of smaller-diameter cushion blastholes reduces and may even eliminate the need for decked charges and correspondingly small pattern sizes. As the diameter of cushion blastholes decreases, cushion blasting merges into postsplitting.

13.5.2 Postsplit Blasts

Postsplitting consists of

1. Drilling a row of parallel, closely spaced blastholes with a suitable burden to spacing ratio (usually about 1.25:1) along the designed final face.
2. Loading every blasthole with a light, well-distributed charge.
3. Initiating these charges after the charges in front of them have detonated.

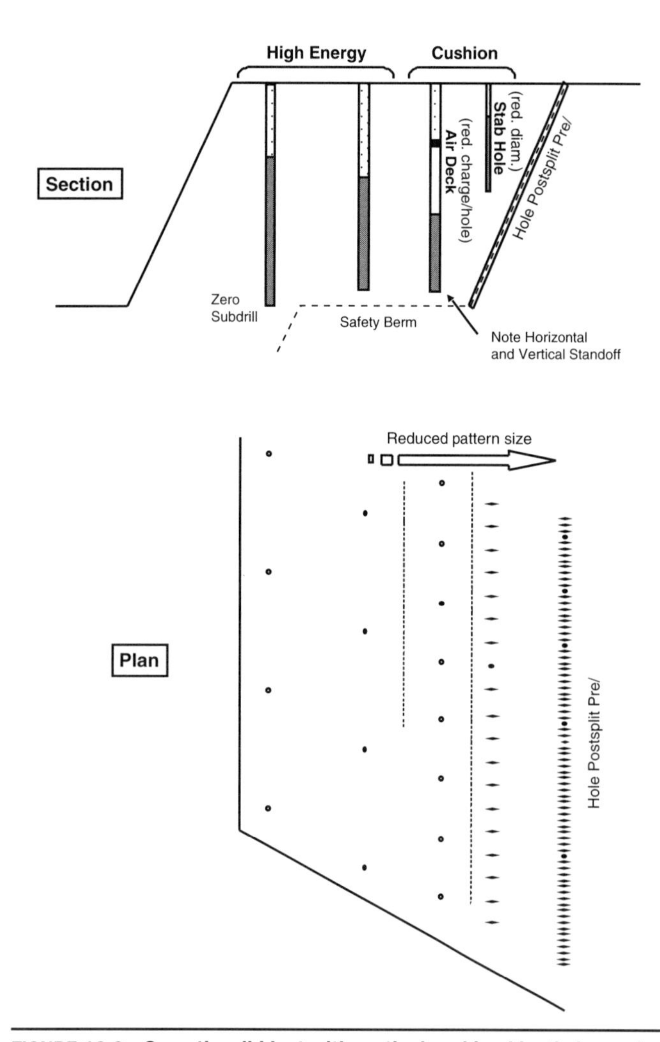

FIGURE 13.3 Smoothwall blast with vertical cushion blastholes and angled presplit or postsplit blastholes

Postsplit blastholes are drilled, charged, and fired to split the rock between them. This produces a sound, smooth face with minimal damage. In the interest of overall cost efficiency, the diameter of postsplit blastholes may need to be equal to or only slightly smaller than that of the blastholes in the front of them.

To achieve the optimum charge distribution, continuous columns of highly decoupled explosive are ideal. However, such explosives are expensive and usually available only for blastholes with diameters up to about 140 mm. For larger-diameter blastholes, a low-density explosive (e.g., ANFO or, better still, a lower-energy, ANFO-based explosive) should be used in conjunction with one or more air decks. The optimum length of top stemming increases for weaker rock types.

Cushion blasting is frequently used in conjunction with postsplitting. If two rows of cushion blastholes are used, the later-firing row usually has a smaller charge than the earlier-firing row (Figure 13.3).

In massive rocks, postsplitting gives a considerable reduction in damage, but the final face is rarely as sound as that produced by presplitting. In closely fissured rocks, however, the final face formed by postsplitting is more sound than that produced by presplitting. Because the optimum spacing of postsplit blastholes is larger than that for presplit blastholes, the cost of postsplitting is usually lower.

13.6 PRESPLITTING

13.6.1 Background

Presplitting involves

1. Drilling a row of parallel, closely spaced blastholes along the designed final face.
2. Charging these blastholes very lightly.
3. Detonating these blastholes *before* the blastholes in front of them.

Firing of the presplit charges splits the rock along the designed final face, producing an internal surface to which the later-firing blastholes in front of them can break. The presplit plane acts as a pressure release vent for the explosion gases generated by charges in the back one or few rows of blastholes in front of the split. It also partially reflects the blast-generated strain waves and so reduces the strains that are experienced beyond the split. The result is a relatively undisturbed face with less shattering, rock movement, and damage.

13.6.2 Effect of Rock-Mass Properties on Presplitting

Presplitting rarely gives impressive results in closely fissured rock. When overcharged, presplitting can damage appreciable volumes of rock as explosion gases vent through the fissures. Under favorable geological conditions (massive rock of moderate to high strength), presplitting can provide improved results over postsplitting but is generally more costly. However, if the presplit blastholes are too close together or overcharged, they themselves will produce damage.

13.6.3 Limitations of Presplitting

The few limitations of presplitting are primarily restricted to the ground and air vibrations produced by firing the presplit blastholes. The levels of ground vibration per kilogram of explosive for these blasts are usually much higher than those for production blasts because the rock that is remote from the bench face is relatively undilated.

In theory, there is no limit to the length of presplit that can be fired in a single blast. In practice, however, the possibility of rock properties changing and the charge either failing to produce the split or causing excessive damage is good reason not to fire presplits far in advance of the pit-wall blasts. By keeping the presplit about three regular spacings (see Figure 13.4) in advance, any knowledge of changes in rock properties gained from the results of pit-wall blasts can be promptly applied to subsequent presplit blasts. The risk of presplitting the entire length of the designed pit wall in a standard manner is high when rock properties vary frequently over short distances.

13.6.4 Diameter of Presplit Blastholes

There is no limit to the diameter of presplit blastholes; diameters of up to 406 mm have been used in large open pits. To achieve economies of scale, there is an incentive to use large-diameter rather than small-diameter presplit blastholes. Unfortunately, the diameter of *angled* presplit blastholes is limited by the rigs that can drill angled blastholes beneath the body of the drill and/or at 90° to the drill's tracks. Most large rigs can only drill angled blastholes where there is sufficient berm width to provide tail room for the drill. Therefore, maximum economies of scale can often be achieved only where vertical presplit blastholes are geotechnically acceptable.

13.6.5 Length of Presplit Blastholes

Theoretically, there is no limit to the depth that can be presplit in a single blast. In common with other pit-wall blasting techniques, however, presplit results depend on good blasthole alignment. Blasthole deviation usually improves with increased hole diameter.

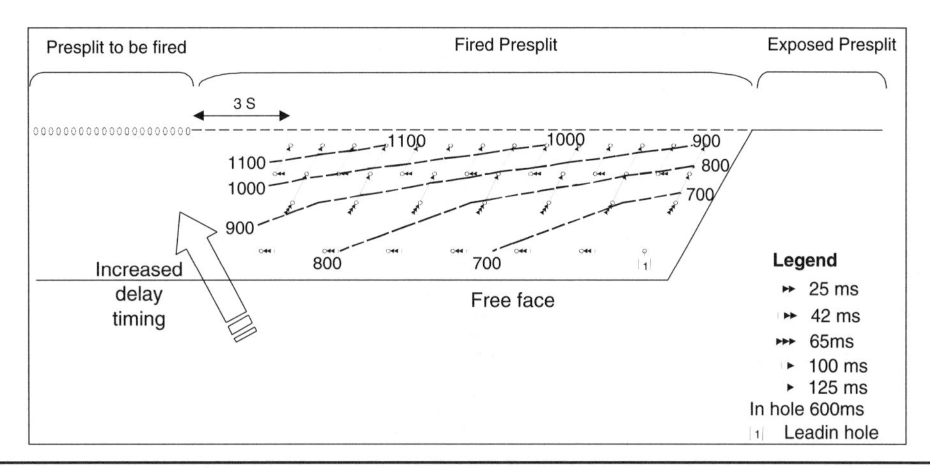

FIGURE 13.4 Typical delay timing for pit-wall blast in front of a previously fired presplit

TABLE 13.2 Presplit and postsplit blast designs for rock of average strength

	Presplitting		Postsplitting		
Blasthole Diameter (mm)	**Average Charge (kg/m)**	**Blasthole Spacing (m)**	**Average Charge (kg/m)**	**Burden Distance (m)**	**Blasthole Spacing (m)**
89	0.5	1.2	0.7	1.8	1.35
102	0.65	1.4	1.0	2.1	1.6
127	0.95	1.7	1.45	2.5	1.9
165	1.55	2.2	2.6	3.3	2.5
200	2.3	2.7	3.9	3.9	3.0
251	3.25	3.4	4.9	4.9	3.8
270	3.6	3.6	5.3	5.3	4.1
311	4.85	4.2	6.1	6.1	4.7

13.6.6 Blasthole Spacing and Charge Concentration

The spacing and charge load for presplit blastholes normally increase with the blasthole diameter, as shown in Table 13.2. But because rock properties have a dominant effect on blasthole spacing and charge load, these data, based largely on experience in average rocks, should be considered only as recommended starting values. The optimum spacing and charge load for a particular rock should be determined by field trials. The spacings shown in Table 13.2 are conservative, and satisfactory results could be achieved with spacings larger than those listed, especially where pronounced fissures are parallel or nearly parallel to the designed final face.

13.6.7 Charging Presplit Blastholes

Over the last several years, the labor-intensive task of attaching cartridges of explosive to a detonating cord downline has been changed by the development of continuous columns of small-diameter explosive for presplitting. Continuous charges are optimum from the energy distribution viewpoint. However, because their energy yield per meter of charge length is constant, their use does not allow small changes in powder factor, which are achievable with spaced cartridges on a downline. With continuous charges, the effective concentration of energy can be varied by

- Taping two or more continuous charges (of equal or different diameter) alongside each other
- Doubling over different lengths of the charge at the base of the blasthole
- Changing the stemming length or (uncharged) collar length of the blasthole

Changing the blasthole diameter is a fourth but much less practical way to vary the effective energy concentration.

With large-diameter blastholes, presplitting can sometimes be carried out using a low-density bulk explosive. The energy concentrations of these explosives are too high for long, continuous, fully coupled presplit charges but are usually acceptable for fully coupled charges that have a long column of air above them. With these charges, the best presplit results are achieved by

- Lowering the density and energy concentration of the explosive
- Top priming
- Avoiding the use of stemming

If stemming is necessary to control air vibrations, the air deck above the charge should be long enough to provide a sufficiently low mean energy concentration over the unstemmed part of the blasthole.

13.6.8 Stemming Presplit Blastholes

If there is no need to control air vibrations, presplit blastholes should not be stemmed, allowing the explosion gases to jet into the atmosphere very rapidly. Because they are not "bottled" up at high pressure below a stemming column, the explosion gases are less likely to jet into cracks that intersect the upper wall of the blasthole. This will reduce damage to the crests of final berms.

Where long, highly decoupled presplit charges are used in massive rock, every blasthole should generally be charged to within about 10 blasthole diameters of its collar. In closely fissured rock (rock in which the mechanical efficiency of presplitting is lower), the uncharged collar should be as long as about 15 diameters.

13.6.9 Firing Presplits

Firing Presplits as a Separate Blast. Presplit blasts should be fired separately and ahead of the adjacent pit-wall blast; the pit-wall blast is located where the total burden distance is so large that the presplit charges could not possibly push the entire block of rock to be broken by the adjacent pit-wall blast forward. In this situation, the presplit defines the designed final face before blastholes in the adjacent blast are drilled. Once satisfactory blasthole spacing and energy concentration have been determined in one or more trial blasts, presplitting can be carried out at any convenient time. Presplit planes should not be subjected to heavy rainfall or severe groundwater conditions for prolonged periods. The accumulation of water and silt in presplit fractures significantly reduces their ability to act as vents for backward-jetting explosion gases, as reflectors of strain waves, and, hence, as damage barriers.

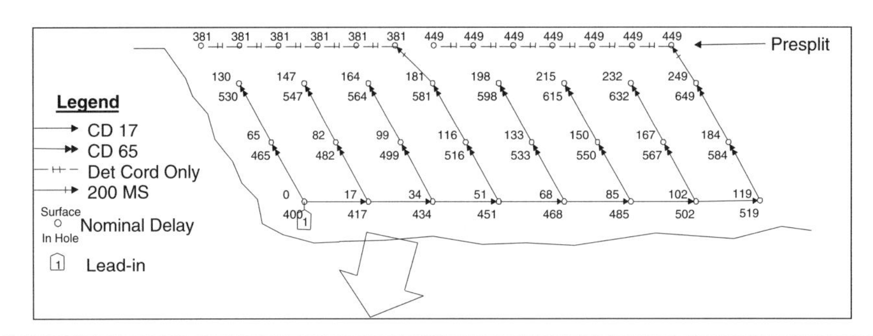

FIGURE 13.5 Tie-in combining a presplit with a pit-wall blast to minimize air/ground vibration

If optimum presplit results are to be achieved, charges should be initiated simultaneously. This is successfully achieved by joining all downlines from presplit blastholes to a detonating cord trunkline. Where ground vibrations are likely to cause damage or disturb residents, however, delays should be inserted into the trunkline at intervals to obtain the consecutive firing of groups of blastholes. The number of blastholes in each group should be sufficient to achieve a satisfactory splitting action while not exceeding the maximum charge weight that can be fired per delay. The major advantage of firing presplits separately is that it simplifies drilling and blasting operations. Where a pit-wall blast is already quite complex, the addition of presplit blastholes further increases the complexity. There is a greater probability of errors and, hence, suboptimum results with blasts that are more complex.

Firing Presplit Blastholes with the Adjacent Pit-Wall Blast. Presplit blastholes should not be fired separately where the total burden distance is smaller than about 150 times the diameter of the presplit blastholes, since ground movement could cause drilling, charging, or cut-off problems. Where the total burden distance is inadequate, the presplit blastholes should be fired *with* the adjacent blast. Such ground movement could make the subsequent drilling of production (or cushion) blastholes difficult or impossible. If blastholes for the adjacent pit-wall blast have been drilled before firing the presplit, the blastholes just in front of the presplit could be damaged to the extent that they cannot be charged properly. Where these blastholes have been drilled *and* charged before the presplit, the downline could be severed before it has detonated.

Combining the presplit blast and adjacent pit-wall blast into a single firing is more complex but it eliminates a step in the drilling-and-blasting cycle. Composite blasts can be drilled, charged, and fired in a single cycle, thereby reducing labor requirements and the (unproductive) movement of personnel and equipment out of and back into the blast area.

As shown in Figure 13.5, a key requirement in firing a presplit with a pit-wall blast is to time the presplit to fire *after* the adjacent surface delays have fired to minimize the risk of a surface cut-off but *before* the in-hole delays fire to enable the presplit to develop completely. Allow at least 50 ms between the presplit and the earliest-firing production blastholes in their vicinity for the latter. In the above example, there is more than a 200-ms gap between all blastholes, allowing plenty of time for the presplit to form.

As many presplit charges as possible should detonate simultaneously as blast-generated vibrations or as the initiation timing will allow. The example in Figure 13.5 shows six to seven blastholes being fired together to reduce that risk.

If the time interval between the presplit detonation and the blastholes immediately in front of them is excessive, ground movement could cause cut-offs in the production or cushion blastholes. This is especially likely where

- Adverse fissures exist
- Presplit blastholes are overcharged
- An inadequate stand-off distance between the presplit blastholes and the row of blastholes immediately in front of them is used

The optimum delay interval between the presplit detonation and the blastholes immediately in front of them should be at least 50 ms and typically in the 100-ms to 200-ms range. For this reason, as illustrated in Figure 13.5, 200-ms delays were used to initiate the presplits to force a sub 200-ms gap between them and the adjacent production holes.

13.6.10 Protecting Presplit Faces

Presplit faces will be damaged or even destroyed if production or cushion blastholes are drilled too close to them. If the stand-off distance is excessive, a wedge of unbroken rock will be left at the base of the presplit face, possibly requiring re-blasting.

The back-row charges of the pit-wall blast, not the presplit charges, break the rock between these two rows of blastholes. Heavier toe charges in presplit blastholes do not assist in preventing large wedges or monuments of rock but, in most cases, damage the presplit face at its base. Although the stand-off distance is optimized by trials, it is typically 40 to 70% of the burden distance for the back-row production or cushion blastholes, depending on whether the rock mass is strong and massive or weak and closely fissured.

It is most important that back-row production or cushion blastholes be well relieved. The burden distance, energy factor, and delay allocation should be such that each back-row charge "sees" an effective free face that is reasonably near and as extensive as possible. Where such relief is achieved, back-row charges find it reasonably easy to break and heave their burdens laterally and, therefore, do not subject the presplit plane to prolonged back pressure.

13.7 CONCLUSION

Although blasting does not influence pit-wall stability as much as geological factors or groundwater, it is more easily controlled. The aim of pit-wall blasting is to achieve the optimum balance between damage control and cost effectiveness. The most effective and expensive technique is to use presplit holes in conjunction with cushion blastholes. The simplest technique is to modify production blast design adjacent to the pit wall. Whatever design is selected, it will require optimization using a series of engineered field trials.

CHAPTER 14

Use of Blast Timing to Improve Slope Stability

Claude Cunningham*

14.1 INTRODUCTION

The link between slope stability and blasting practice is complex—it relates to geological conditions, the needs of the excavation, the scale of operations, and the economics of the situation. However, intelligently applied blasting procedures will always reduce the extent of back break and leave the final rock mass in a more stable condition. The ideal outcome is to leave a solid wall of rock delineated by the barrels of the blastholes, from the top of the bench to its toe, with neither serious cracking of the face nor unbroken rock stuck against it after the main blast. If this outcome is not achieved, it is necessary to determine the reason for the failure and whether the need is sufficiently great for additional effort and expense to be invested for improving the next blast.

The most common approach to final limit blasting is to specify presplitting techniques for the final wall with trim blasting of the rock burden in front, so as to protect it from damage by heavily charged production holes. Presplitting provides a preferential fracture plane behind the blast to terminate cracks growing from the blastholes, while trim blasting reduces the rate of energy release against the final wall.

However, even when these well-known techniques are applied, slope failures and back damage can persist. The key parameters within the control of the blasting engineer are type and amount of energy in the hole, drilling pattern, hole depth, hole diameter, hole angle, bench geometry, and blast timing. This last parameter, which is the most readily addressed, is the focus of this paper.

14.2 EFFECT OF BLAST TIMING ON ROCK BEHIND THE BURDEN

Blast damage is created by the transfer of explosive energy into the rock mass. Two distinct mechanisms operate, creating quite different kinds of damage:

1. Strain waves and expanding gases of detonation lead to crushing and crack development, shown in Figure 14.1. These phenomena are over within a few milliseconds.
2. Inertial mechanisms lead to ground shift, as shown in Figure 14.2. Since ground mass reaction is relatively slow (face velocities are typically in the range of 8 to 30 m/s), the time frame for these mechanisms lasts for tens and even hundreds of milliseconds, in effect for the duration of the blast event.

Remedial steps must counter both mechanisms. Unfortunately, methods that benefit the mechanisms of item 1 above sometimes exacerbate those of item 2 above. For example, presplitting is the most common way to achieving a smooth final wall. This harnesses a well understood and widely publicized mechanism of favorable stress alignment that arises when all blastholes in a line are fired simultaneously. Tensile crack development

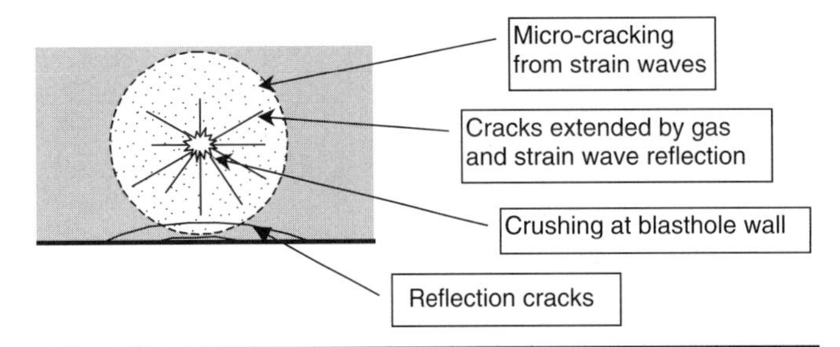

FIGURE 14.1 Damage resulting from strain wave intensity and gas penetration

between the holes is favored, while tensile crack development perpendicular to the row of holes is suppressed (Figure 14.3).

However, as indicated in Figure 14.3, when holes are fired simultaneously, even if the charge in the holes is greatly reduced compared to normal blasting, a massive, low-frequency, lateral impulse is delivered to the surrounding block of ground. The body of rock is thrust back by the sustained pressure of explosive gases expanding into the forming split. As this pressure is released, the whole body in which the joints have been loosened and the bedding uplifted slumps back. A split has been formed, but the quality of the rock is no longer what it was. It has been shaken and dislocated by high vibration levels, the upheaval of the surface, and, sometimes, significant forward movement of the whole block of ground between the presplit holes and the nearest free face. Holes that may have been drilled in this block will tend to be closed by shearing (Figure 14.4).

The lateral impulse of splitting also has particular potential to upset joint integrity in the block ahead of the split. Previously intact ground may now be able to shift during the production blast and even lead to sympathetic detonation if explosives gases stream through the loosened joint planes. Moreover, the creation of the split may not achieve its aim of preventing damage when the buffer blast is taken, since the split itself presents a free face to which the back holes can break. If the split is tightly closed, the compressive and shear stress waves (and to some extent even the tensile waves) from these holes will cross the split with little attenuation. If the split has created a gap, the solid wall is impacted by the reactionary thrust of fragmenting rock from the back holes.

Trim blasting is undertaken to reduce the energy released close to this split plane and therefore minimizes the impact across the plane. However, lack of appreciation of the mechanisms at work can still result in poor outcomes.

To summarize, it is not unusual to find that presplitting and trim blasting, which are the normal methods adopted for limiting back damage and promoting slope stability, do not, on their own,

* AECI Explosives Limited, Modderfontein, South Africa.

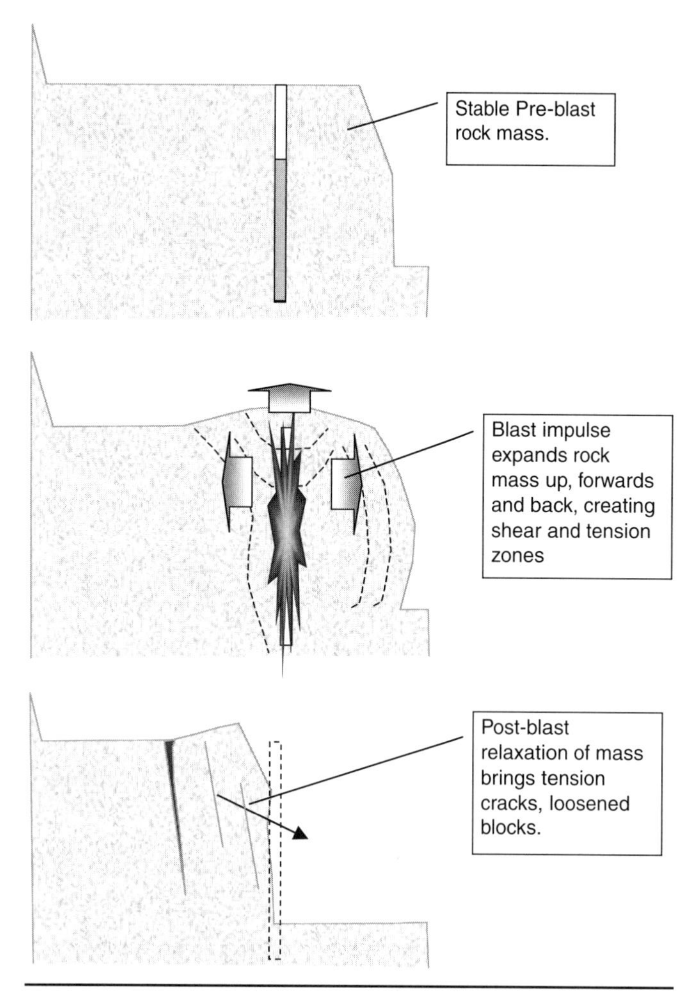

FIGURE 14.2 Creation of relaxation cracks by post-blast inertial mechanisms at free face

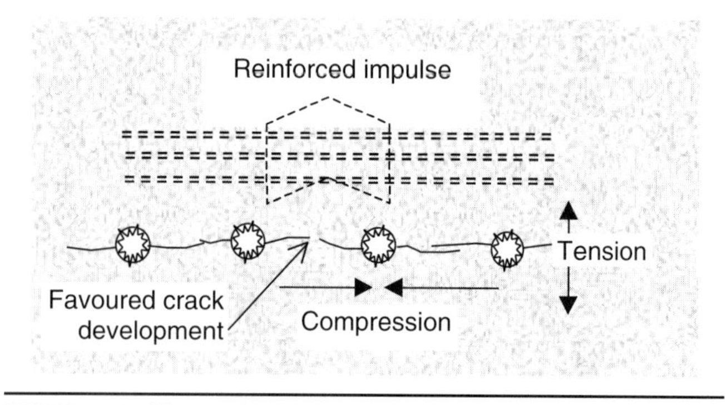

FIGURE 14.3 Theory of crack development between simultaneously initiated blastholes in presplit theory

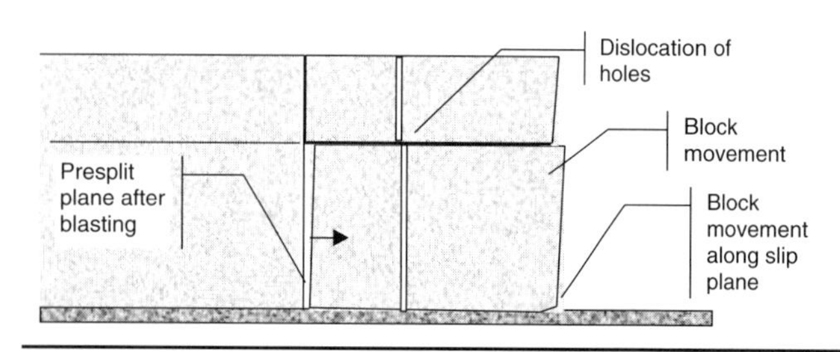

FIGURE 14.4 Ground dislocation resulting from high presplit energy

deliver what is required. Barrels of split blastholes that are left after the blasting may sometimes collapse some time after exposure, and what was thought to be solid rock begins to slide off the face. Therefore, the blasting engineer must

- Create a preferential split at the back wall so as to provide a clean break between damaged and undamaged rock
- Reduce energy flow into the back wall

While much can be done to address this problem in terms of drilling and loading of holes, blast timing plays a key role in controlling the release of energy and enabling blast layouts to deliver sound back walls. The following discussion assumes that everything has been done to limit damage in terms of hole spacing and depth and loading of holes.

14.3 OPTIONS FOR REDUCING DAMAGE

Clearly, damage is maximized when the maximum amount of explosive is detonated at the same instant. Timing is the key to controlling both the rate at which the available energy is released and the direction of thrust of the blast. Thus, the need is to monitor

- Rock conditions to prevent slope failure
- The amount of explosive detonating in any given time window
- The direction in which the rock mass is loaded

The worst conditions occur when the rock mass is generally very incompetent and liable to collapse into the open pit under its own weight and when the rock mass is strong but intersected by pronounced off-vertical slips and joints running parallel to and out of the face. Under both of these conditions, normal, controlled blasting measures tend to give disappointing results.

The issues that need to be faced are the technical requirements of blasting and the capability of the available explosives initiation systems to deliver these requirements.

14.3.1 Technical Requirements

In optimizing back wall conditions, the key is to record as closely as possible what is done and the results achieved. This information is crucial in applying principles of blasting to improve results. In terms of blasting results, the simplest tools are photography and vibration monitoring. High vibration levels are a direct symptom of blast energy channelled into the wall, and even simple capture systems set up consistently behind the wall will provide useful information on success in protecting the wall. If at all possible, proper slope-stability/ground-movement monitors should be deployed and examined after each blast.

Presplitting. Splitting is highly desirable but not always possible, especially where the desired split plane is within 20° of a major joint or slip plane. When conditions are difficult, the engineer needs to improve the distribution of energy along the split plane. This requires adoption of increasingly light and more decoupled charges and closer spacing of blastholes. However, to reduce the vibration level in the rock mass, timing may need to be introduced to reduce the explosive mass firing instantaneously.

The simplest approach is to break up the split blast into groups of holes that fire simultaneously, each group separated by a short delay that is only long enough for the strain waves from the previous group to disperse. This can be determined by monitoring, but typically 17 to 25 ms is more than adequate.

Decks occurring within the blastholes need to be fired simultaneously.

"Simultaneously," means "within 1 ms," since cooperation of the strain waves is crucial and it has been widely demonstrated that the quality of split is greatly reduced when the timing between holes is in excess of 1 ms. This is, to some extent, scale dependent, but it is a good rule to follow.

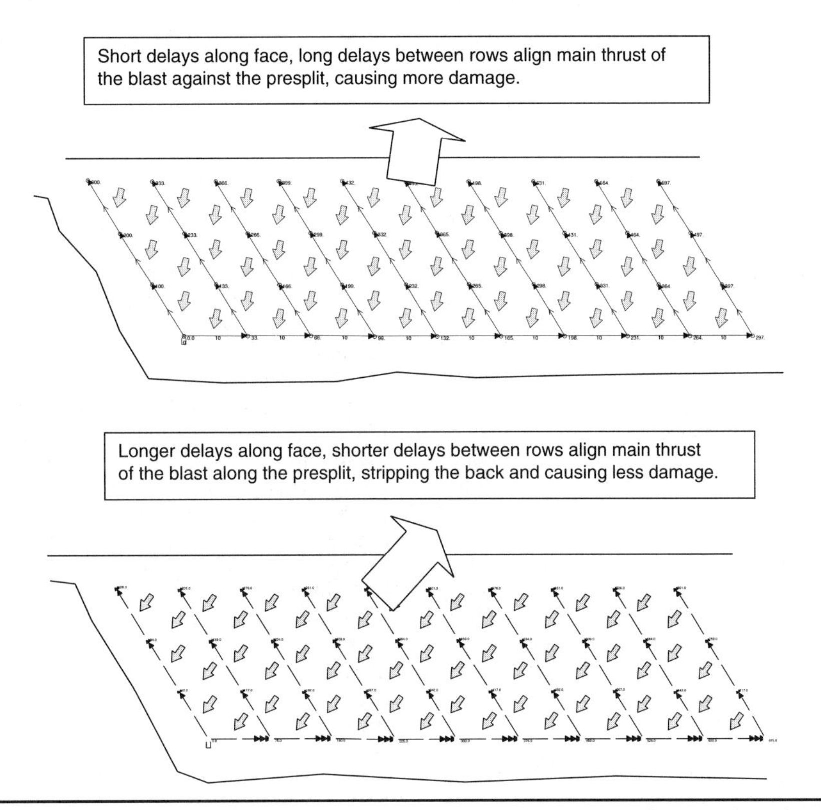

FIGURE 14.5 Line blasting towards presplit creating excessive thrust and resulting in rock accretions stuck against the split. It is better for the trim blast to be timed to thrust at an angle to the split.

Trim Blast Timing. In keeping with the need to reduce the energy transfer across the split plane, trim holes usually have reduced drilling patterns and charge masses and are timed to achieve single hole firing. Where decking is introduced, it is conceivable to introduce timing between the decks, but this practice is fraught with problems in terms of preventing some decks from dislodging or desensitizing those around them. The rough rule is that at least 8 ms are needed between shots so that the shots do not reinforce each other in creating vibration. However, if the holes are physically well separated from each other (by 100 hole diameters or more), the travel time and buffering of any intervening broken rock tends to negate this requirement.

Since the impact of the developing inertial thrust of a trim (or production) blast is substantial, it is important to arrange the blast so that it is parallel, rather than perpendicular, to the final split surface. This is achieved by developing the firing sequence with shorter delays perpendicular to the split and longer delays parallel to the split. The length of the delays is dependent on the prevailing conditions and the available delays and systems. However, a good starting point is to have, say, 42-ms delays running back towards the split, and 75- or 100-ms delays running parallel to the face (see Figure 14.5). This deflects the thrust of the blast along, rather than into, the split, which reduces the impact. A real benefit of this procedure is that any rock sticking to the split becomes sheared off by the movement of the trim blast, leaving a clean face.

The time per meter of face traversed by the blast is controlled by the delay across the front of the blast and is a useful measure of the thrust on the back wall. For example, if the hole spacing is 5 m and 75-ms delays are used between the holes, the rate is (75/5) or 15 ms/m of face. If 25-ms delays are used, the rate is (25/5) or 5 ms/m of face, and if the final row is shot with detonating cord with a velocity of 6,000 m/s, the rate is (5/6/5) or 0.14 ms/m of face. These represent quantum leaps in the thrust delivered and the extent of potential damage. To get an

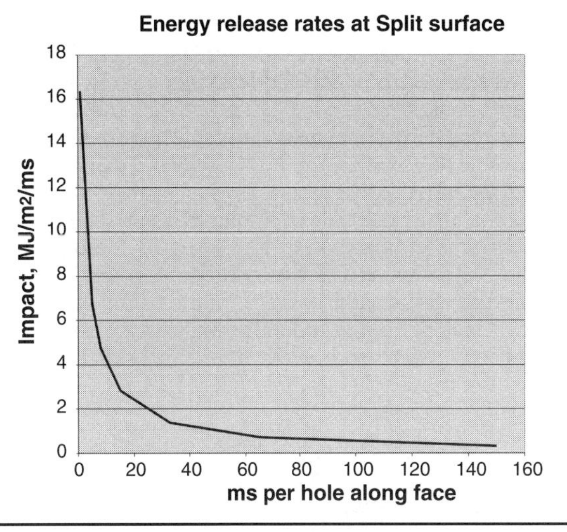

FIGURE 14.6 Example of effect of choice of delay between holes on energy release rate impacting on split surface

idea of the actual power being delivered, divide the mass of explosive per hole by the delay per square meter of face and multiply by 3. The results report in MJ per millisecond per square meter. This material, which is open for further comment, illustrates the principle. Figure 14.6 portrays how, for a particular blast layout, increasing the delay across the face reduces the rate at which energy release takes place opposite the split. Clearly, short delays result in very high release rates, which are likely to be detrimental to the condition of the finished wall.

The delay of holes from the free face toward the split needs to be sufficient to allow relief of burden, i.e., to permit the tightest holes against the split to move their rock burden outward from the split. Although shorter timing between these holes is effective in directing the thrust beneficially, tight confinement of the innermost holes can lead to excessive thrust. It is permissible to delay the back holes by an extra period of time in order to achieve better relief, but this has to be considered in relation to the overall design and constraints.

14.3.2 Capability of Explosives Initiation Systems

By far, most initiating systems currently available are based on shock tube systems with pyrotechnic timing. Pyrotechnic delays are exceptionally robust and simple to use but suffer from intrinsically restricted precision. This has an important effect on blasting results, since it removes the ability to predict the achieved interval between shots. Basically, the average nominal delay will be achieved, but scatter will be distributed within a range of about ±5 to 9% of the in-hole nominal delay.

Typically, in-hole detonators for opencast mining conditions have delays of 500 to 800 ms, with surface timing units providing the interhole and interrow delays. These will result in a normally distributed system scatter of ±30 ms (which will vary from batch to batch). This means that, if the surface delay is e.g., 42 ms, a few shots can fire out of sequence and some can fire as much as 70 ms apart. Although this phenomenon is well understood, there was little that could be done until recently. As a consequence, there has been a tendency toward inconsistent blasting results, which is half expected owing to the vagaries of geology. However, with the introduction of electronic digital-delay detonators, a new capability has become available.

The author has noted that, with regard to splitting, conference papers have been presented demonstrating that even quite small deviations from instantaneous firing result in split surfaces, which are rougher, with increased scalloping of the wall. Typically, these papers call for holes to fire within 1 ms of each other. Normally, the use of detonating cord, which fires at velocities in excess of 6,000 m/s, or 0.17 ms/m, is the best way to achieve this. Holes are frequently less than 2 m apart, which means that they fire within 0.3 ms of each other, while in-hole delay detonators usually have scatter over a range of 30 ms or more. The greatest hole spacing encountered for splitting is about 5 m, and even with this, detonating cord will deliver a delay of only 0.7 ms. The significant scatter achieved with in-hole delay detonators is clearly unacceptable, consequently, the only options are for detonating cord initiation or for zero-delay detonators in the hole with detonating cord on the surface.

However, with digital delay detonators, the split holes can be fired without the noise and performance loss associated with detonating cord and, if desired, these can be fired as the last holes after the trim shot.

What I have found with these accurate systems is that much of the prevailing wisdom around delays can be rewritten. Present guidelines have developed from studies in which significant delay scatter was intrinsic and wide variation in results was factored into recommendations. With tight control of timing, I have found that much shorter delays are not only possible but desirable for reducing back break, enhancing fragmentation, and controlling movement. It is gratifying to note that precise timing brings consistent and reasonably predictable blasting results, with much enhanced control of back break.

14.4 CONCLUSION

The timing needed to deliver a stable slope is highly dependent on the conditions and needs at the site, but the general principles discussed here are the correct starting point.

First, the geometry of blasting needs to favor energy release away from the critical final wall. Timing influences this by directing the ground reaction of both trim and production blasting away from the back wall. Line blasting parallel to the wall is entirely inappropriate and is likely to cause not only loosening of previously tight joints but freezing of significant volumes of cracked rock against the split. It has been proven time and again that where blast progression is toward a site, vibration levels can be double the levels on the opposite side, with propagation away from the site.

Second, the packets of explosive energy released need to be appropriate for the conditions. Not only should the loading in each hole be addressed but serious attention must be given to achieving a delay long enough for the strain waves from neighboring holes to disperse individually.

Third, the precision of the timing system is crucial to achieving consistent results. With pyrotechnic timing systems, the lack of precision demands that fairly long interhole delays be employed in order to avoid crowding or out-of-sequence shots. Long delays can result in ground movement interfering with the functioning of holes around each shot and in excessive fracture and movement between shots. Digital electronic timing is indispensable for achieving strong, consistent final wall conditions without compromising production blasting.

Fourth, close monitoring of blasting results is crucial to achieving sound designs for each domain of final limit blasting. Photographs, which are the simplest and most convincing proof of what has been achieved, need to be filed for easy reference and must show the extent of deterioration with time. The blasting parameters, geological analysis, and any meaningful records of vibration levels behind the wall should also be filed for future reference.

It is not always appreciated that changing any one parameter in a blast can have serious implications for other parameters and blast results. The task must be approached holistically, so that broad input and learning can be achieved.

CHAPTER 15

Large-Diameter and Deep-Hole Presplitting Techniques for Safe Wall Stability

Rodney Burke*

15.1 INTRODUCTION

Bridger Coal Company is a subsidiary of Pacificorp and Idaho Power Company. The Jim Bridger Coal Mine is 35 miles northeast of Rock Springs, Wyoming. The mine, which began coal production in 1974, provides coal exclusively to the Jim Bridger Power Plant. The Bridger Mine is a strip mine that is presently 13 miles long and excavates approximately 45 to 50 million bcy's (Bank cubic yards) of overburden and produces 6.5 to 7.5 million tons of sub-bituminous, low-sulfur coal annually. Average coal composition is 9,400 btu, 0.6% sulfur, 19.0% moisture, 10.0% ash, and 2.75% sodium.

The mining process at Bridger Coal uses four draglines: two Marion 8200s, one Page 757, and a Page 732 that excavate the majority of the overburden and innerburden. The remaining overburden, innerburden, and coal is removed with two 195B Bucyrus Erie shovels, a Caterpillar 992 loader, a Caterpillar Dart 600C loader, a Caterpillar 994 loader, and a fleet of Caterpillar scrapers. Drilling of the presplit and production patterns is accomplished using two Driltech D90Ks with 10⅝-in. bit diameter and an Ingersol DML with a 7⅞-in. bit diameter. The coal drilling is done with a Schroeder twin-auger drill system using a 5⅛-in. claw bit. The annual amount of bulk explosives to shoot presplit, overburden, innerburden, and coal is approximately 43 to 48 million pounds.

The mine is in the Fort Union Formation, which consists of sandstone, siltstone, claystone, and mudstones, with some carbonaceous content in the siltstone and thin filaments in the bedding planes of the sandstone. The highest carbonaceous content is in the claystone. The compressive strength of the rock ranges from 1,000 to 8,000 psi. Major faults are present throughout the mine, having a vertical displacement of more than 20 ft. The mined coal seams lay in an area known as the Deadman Coal Zone, which lies near the bottom of the Fort Union Formation.

Cast blasting is a critical sequence/function of the Bridger Coal drilling and blasting program. Since highwalls are 200 ft high, it was necessary to develop presplitting to ensure employee safety and highwall stability.

15.1.1 Previous Loading Procedures

Our presplit program began in 1984 and evolved in 1988 with the aid of Calder and Workman, Inc. The vertically drilled presplit initially had disastrous results due to wall failures. We then looked at 20° angle drilling of the presplit, which has produced our current highwalls. We loaded the presplit using a toe load of heavy ANFO or ANFO in dry conditions. The explosives load amounts were based on a chart provided by Lyall Workman. The chart was broken down into 10-ft increments to a depth of 200 ft, hole sizes range from 6 in. to 12¼ in. diameter, at 14-ft spacings.

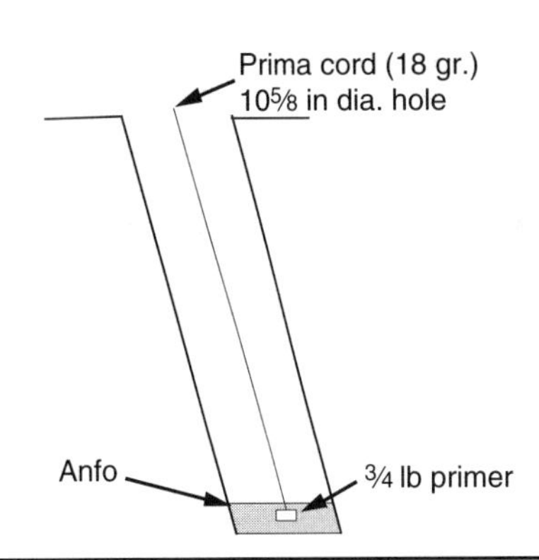

FIGURE 15.1 Presplit loading of a dry hole 75 ft deep or less

At depths of 125 to 175 ft, we began to see cratering of the toe and no visible crack at the surface. This cratering caused an overhang of rock once we cast-blasted the pit. The exposed wall was too high for a dozer to clean the overhanging material, meaning the dragline would have to clean the wall as it passed. The cast-blast program worked extremely well, but draglines had to replace material in order to get high enough to reach the top of the wall. Using the dragline to clean the wall created maintenance problems because of the unnecessary stress on the boom and the very tight swinging conditions against the wall that developed. This motivated us to experiment and perfect new techniques to solve our presplit problem.

15.2 CURRENT LOADING PROCEDURES

The first step in solving our presplitting problem was to determine if the hole was dry or if there was water in the hole. This was done by marking the water surface and total depth of the hole on a lathe. Parameters on how to load each hole based on how much water, if any, was in the hole were tested and developed at Bridger Coal. We are currently using a powder factor of 0.11 lb/sq ft on 12-ft spacing, 10⅝-in. hole diameter.

The steps and criteria for loading presplit holes are listed below.

1. **Dry Hole, 75 ft or Less**
   - A. If the drilled hole is in competent rock, use a toe charge of ANFO and use 18–25 gr. cord and a cord-sensitive ¾-lb primer (Figure 15.1).
   - B. If the hole has been drilled where there is a large amount of alluvium or unconsolidated material, use a foam plug and split the charge in half as close to the transition plane of solid or unconsolidated material

* Bridger Coal Company, Rock Springs, Wyoming.

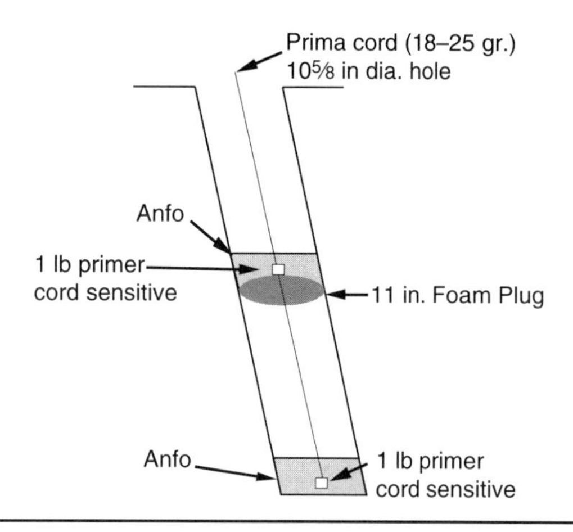

FIGURE 15.2 Presplit loading of a dry hole 75 to 150 ft deep

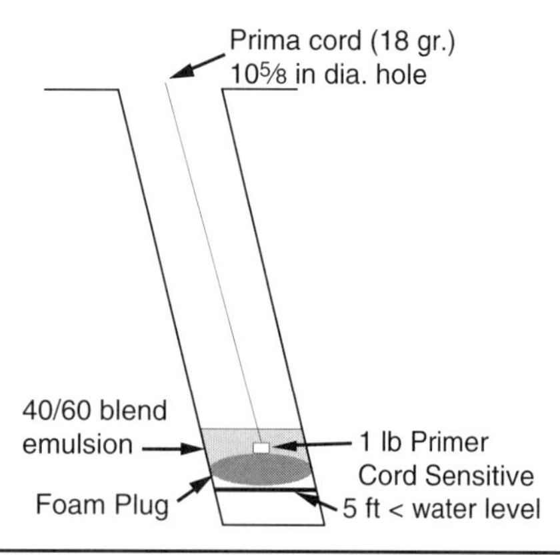

FIGURE 15.3 Presplit loading of a hole that is 75 ft deep or less with water less than 5 ft deep

as possible. This will cause a crack to form bottom-to-top of the borehole. Use 18–25 gr. cord and a cord-sensitive ¾-lb primer in both decks (Figure 15.2).

2. **Wet Hole, 75 ft or Less**
   A. If there is water in the hole less than 5 ft deep, place a foam plug on top of the water. If dewatering is not possible or a liner cannot be used, calculate the total pounds of a water-resistant emulsion blend. Use 18–25 gr. cord and a cord-sensitive 1-lb primer (Figure 15.3).
   B. If the water is deeper than 5 ft, dewater the hole, if possible, and load with a water-resistant product. If dewatering is not possible, use a liner (7 in. diameter, 8–12 ft in length) and load a minimum of 75 lb at the bottom of the hole. This toe load is only to eliminate the water with minimal presplitting expected of the charge. Then, place a foam plug on top of the water with a water-resistant emulsion blend to achieve desired results for the remaining length of hole. Use 18–25 gr. cord for the downline and cord-sensitive primers–2-lb primer with the toe load and 1-lb primer in the top deck (Figure 15.4).

3. **Dry Hole, 75 ft to 150 ft**
   A. If the hole is dry, calculate the total pounds of ANFO needed. The total load is split in half and two smaller loads are used with a foam plug at a point half the depth of the total depth of the hole. Use 18–25 gr.

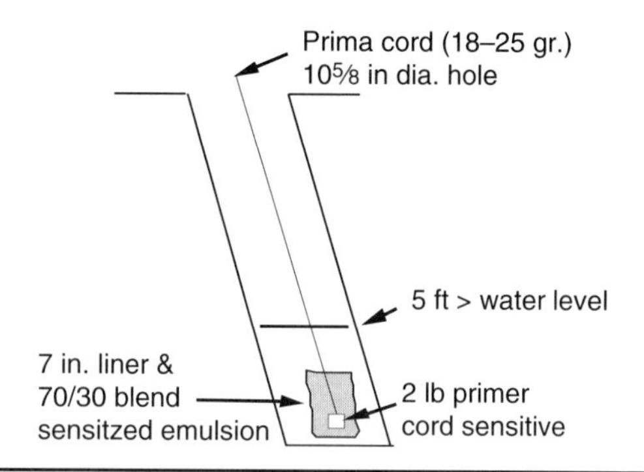

FIGURE 15.4 Presplit loading of a hole that is 75 ft deep or less with water 5 ft deep or more

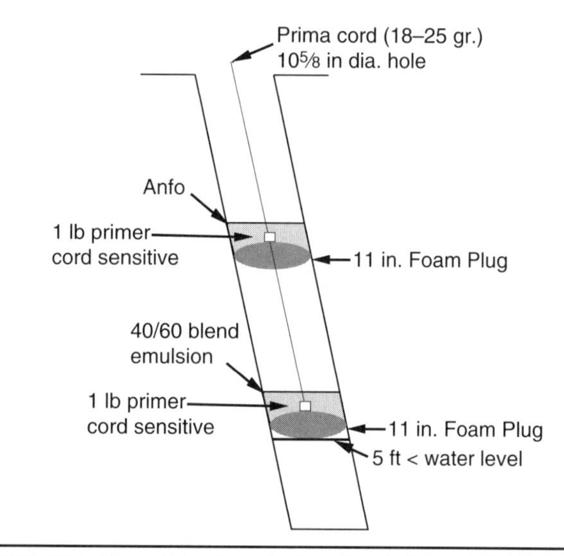

FIGURE 15.5 Presplit loading of a hole that is 75 to 150 ft deep with water less than 5 ft deep

cord for the downline and cord-sensitive primers–¾-lb primer with the toe load and ¾-lb primer in the top deck (Figure 15.4).

4. **Wet Hole, 75 ft to 150 ft**
   A. If there is water in the hole less than 5 ft deep, place a foam plug on top of the water. (If dewatering is not possible or a liner cannot be used.) This toe load is intended to eliminate the water with minimal presplitting expected of the charge. Then place a foam plug on top of the water with a water-resistant emulsion blend to achieve desired results for the remaining length of hole. A third deck may need to be placed in the hole when the hole depth is close to 150 ft. Consider the total length of the hole when calculating the total pounds. Then split the total pounds in half and load the deck above the water with a water-resistant emulsion. Place a foam plug half the distance of the hole depth measured above the water; use ANFO for this deck load. Use 18–25 gr. cord for the downline and cord-sensitive primers–2-lb primer with the toe load, 1-lb primer in the second deck, and ¾-lb primer in the third deck (Figure 15.5).
   B. If the water is deeper than 5 ft, dewater the hole, if possible, and load with a water-resistant product. If dewatering is not possible, use a liner (7 in. diameter,

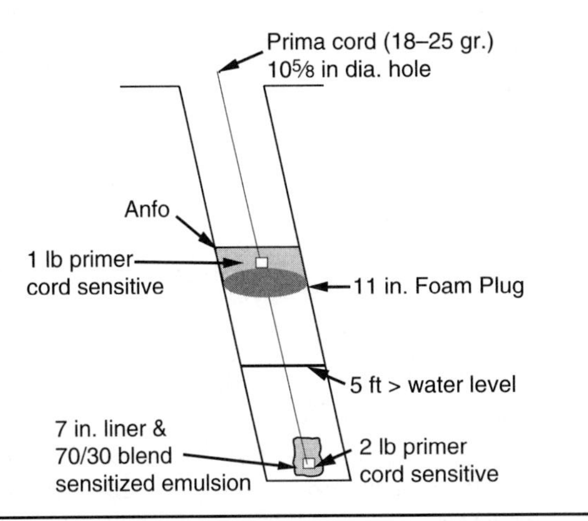

FIGURE 15.6 Presplit loading of a hole that is 75 to 150 ft deep with water deeper than 5 ft

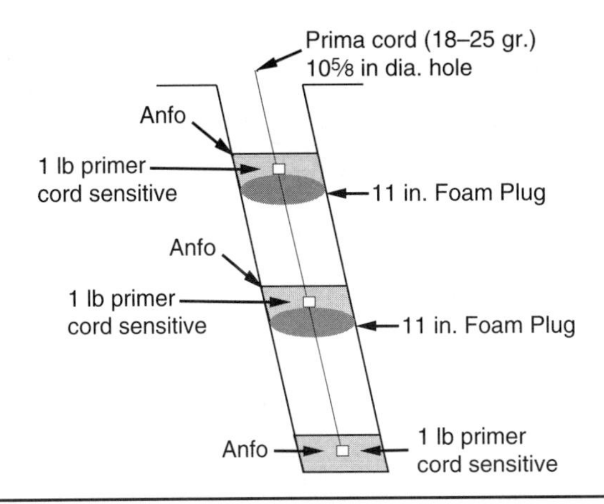

FIGURE 15.7 Presplit loading of a dry hole 150 to 220 ft deep

8–12 ft long) and load a minimum of 75 lb at the bottom of the hole. This toe load is intended to eliminate the water with minimal presplitting expected of the charge. Then place a foam plug on top of the water with a water-resistant emulsion blend to achieve desired results for the remaining length of hole.

A third deck may be needed in holes when the hole depth is close to 150 ft. Consider the total length of the hole when calculating the total pounds. Then split the total pounds in half, loading the deck above the water with a water-resistant emulsion and placing a foam plug half the distance of the hole depth measured above the water using ANFO for this deck load. Use 18–25 gr. cord for the downline and cord-sensitive primers–2-lb primer with the toe load, 1-lb primer in the second deck, and ¾-lb primer in the third deck (Figure 15.6).

5. **Dry Hole, 150 ft to 220 ft**
   A. If the hole is dry, calculate the total pounds of ANFO needed. Split the total load into three decks utilizing ANFO loads, placing a foam plug at a point two thirds of the depth of the total depth of the hole. Then place the second plug at one third of the depth remaining for the third charge. Use 18–25 gr. cord for the downline and cord-sensitive ¾-lb primers in all decks (Figure 15.7).

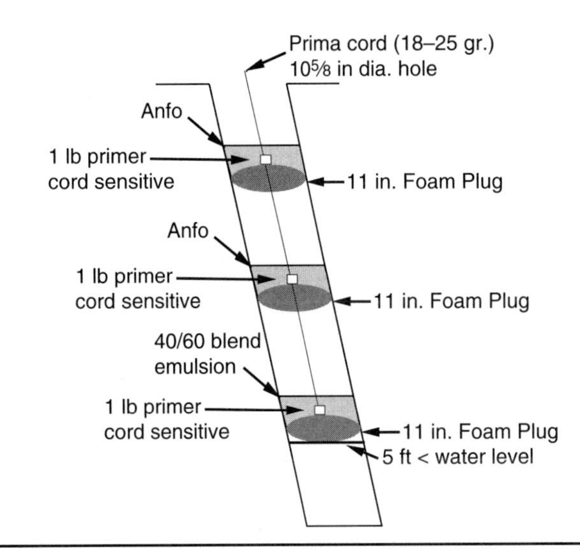

FIGURE 15.8 Presplit loading of a hole that is 150 to 220 ft deep with water less than 5 ft deep

6. **Wet Hole, 150 ft to 220 ft**
   A. If there is water in the hole, the water is less than 5 ft deep, and dewatering or a liner cannot be used, place a foam plug on top of the water with a water-resistant emulsion blend. Evaluate the possibility of the water flowing in from above the water level. The total explosive pounds needed for this load is calculated using the entire length of the borehole and splitting the total pounds in half. A third deck may be necessary in the hole. Take the total length of the hole above the water and divide the length in half for the plug location. Load the other one third of the total explosive pounds with ANFO at the third deck. Use 18–25 gr. cord for the downline and cord-sensitive primers–2-lb primer with the toe load, 1-lb primer in the second deck, and ¾-lb primer in the third deck (Figure 15.8).
   B. If the water is deeper than 5 ft, dewater the hole, if possible, and load with a water-resistant product. If dewatering is not possible, use a liner (7 in. diameter, 8–12 ft long) and load a minimum of 75 lb at the bottom of the hole. This toe load is intended to eliminate the water with minimal presplitting expected of the charge. Place a foam plug on top of the water with a water-resistant emulsion blend, be aware that water may flow in from above the water level. The total explosive pounds needed for this load is calculated using the entire length of the borehole and splitting the total pounds in half. A third deck may be necessary in the hole. Take the total length of the hole above the water and divide the length in half for the plug location. Load the other one third of the total explosive pounds with ANFO at the third deck. Use 18–25 gr. cord for the downline and cord-sensitive primers–2-lb primer with the toe load, 1-lb primer in the second deck, and ¾-lb primer in the third deck (Figure 15.9).

15.3 CONCLUSIONS

Our presplit program, which began more than 10 years ago, is a continuing process. We have tested numerous presplitting methods, which include vertical drilling, angle drilling, and varying hole diameters (7⅞ in., 10⅝ in., 12¼ in.). Other methods

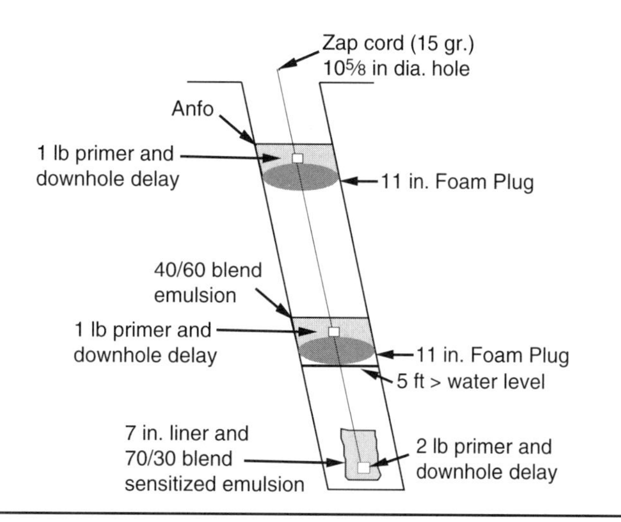

FIGURE 15.9 Presplit loading of a hole that is 150 to 220 ft deep with water deeper than 5 ft

include increased hole spacings (8–16 ft), toe loading with an open hole, air decking with a toe load, hanging or suspending ANFO bags, and loading with liners or decking using foam or gas plugs.

Our best results occurred when we used foam or gas plugs in decking with a combination of ANFO and emulsion blends. The air deck technique, which allows the energy of the explosive charge to be contained for a longer period and lower the total amount of explosive charge used, has proven to be beneficial in our presplit techniques. We are using a modified version of this technique that does not use stemming to contain any of the explosive energy. We are leaving an open borehole and using the foam bags as our containment and taking advantage of the added ability to separate the charges throughout the borehole with excellent results. This current practice will allow us to experiment and possibly increase our hole spacings.

There is no single method for loading and shooting presplit over the length of any mine. By continuing to develop and evaluate our program, we are determined to establish the appropriate presplit procedure that will ensure a safe, stable, and productive presplit wall at the lowest possible cost.

15.4 ACKNOWLEDGMENTS

The author thanks the drilling and blasting crews at Bridger Coal Company for their continued effort in making our highwalls some of the safest and most stable in the country and for being patient in the development of the presplit loading techniques. We also thank Lyall Workman who provided the technology to begin our presplit program.

15.5 REFERENCES

Teller, A.E. 1993. Blasting techniques III. *Controlled Blasting*, July.

Workman, J.L., and P.N. Calder. 1995. Considerations in pre-split blasting for mines and quarries. Calder and Workman, Inc., Washburn, North Dakota, March 16, 1995.

Workman, J.L., and P.N. Calder. 1995. A method for calculating the weight of charge to use in large hole presplitting for cast blasting operations. Calder and Workman, Inc., Washburn, North Dakota, March 16, 1995.

SECTION 2

Case Studies in Rock Slope Stability

CHAPTER 16

The Role of the Geotechnical Group in an Open Pit: Chuquicamata Mine, Chile

Germán Flores* and Antonio Karzulovic†

16.1 INTRODUCTION

As shown in Figure 16.1, the Chuquicamata Mine is in the second region of northern Chile, in the province of El Loa, about 16 km from the town of Calama. Currently, Chuquicamata is one of the biggest open-pit mines in the world (Figure 16.2), with a length of 4.5 km in the north–south direction, a width of 2.7 km in the east–west direction, and a depth from 720 to 780 m. This mine began to be worked in 1913, and, currently, it removes 60 Mton of mineral and 140 Mton of waste annually. The Current Base Plan for year 2000 anticipates the mine to reach a depth of 1,100 m by year 2020, which is a major challenge from a geotechnical point of view.

Management of a mine this size as well as the future challenges have encouraged the Chuquicamata Division to establish a relatively important Geotechnical Group for the geological, geotechnical, and hydrogeological characterization of the rock mass, the analysis and design of slopes, the control of slope excavation to develop the proposed designs, and the monitoring and control of slope behavior. This is in addition to the necessary interaction with the planning and mine operation groups. Today, the Geotechnical Group consists of six engineers, two geologists, four technicians, and one secretary; this organization is illustrated in Figure 16.3. This chapter describes the work developed by this group in the Chuquicamata Mine. It is important to point out that this group is also in charge of the geotechnical aspects of South Mine, the leach piles, waste dumps, tailings deposits, and the mining–metallurgical projects of the Chuquicamata Division.

16.2 GEOLOGICAL, GEOTECHNICAL, AND HYDROGEOLOGICAL CHARACTERIZATION

A basic task of the Geotechnical Group is to characterize the Chuquicamata Mine from geological, geotechnical, and hydrogeology viewpoints. The "geotechnical-geometry" below the surface and the properties of the rock mass and the structures is required information for developing a slope-stability analysis.

At the Chuquicamata Mine, the rocks consist, mainly, of granodiorites and porphyries. The contact is defined by the west fault with a north–south orientation. It defines a shear zone that is 150 to 200 m wide. Porphyries present quartz-sericitic, potassic, and chloritic alteration. Metasedimentary rocks also appear in some sectors of the mine but with small areal extent. Rocks outcrop in the northern part of the pit, but the southern part is covered by about 50 m of paleogravels. For engineering geology purposes, the Chuquicamata Mine can be defined by the 12 geological units detailed in Table 16.1 and whose areal distribution is shown in Figure 16.4.

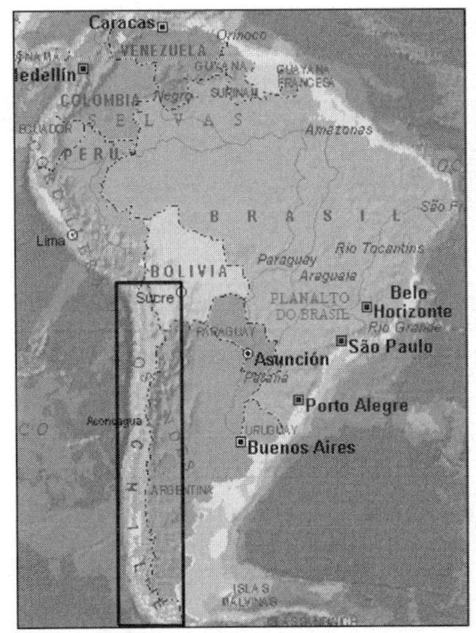

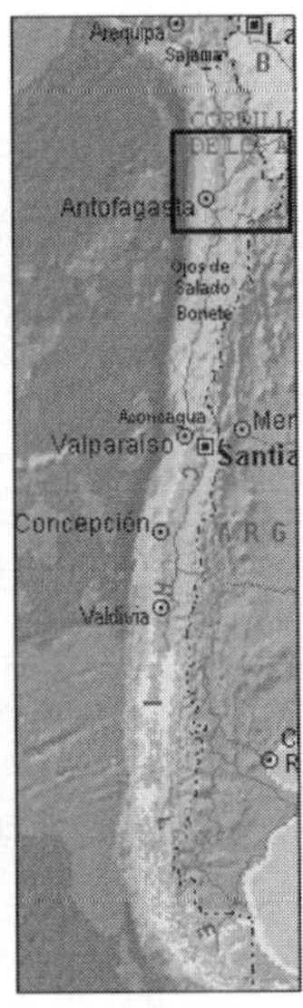

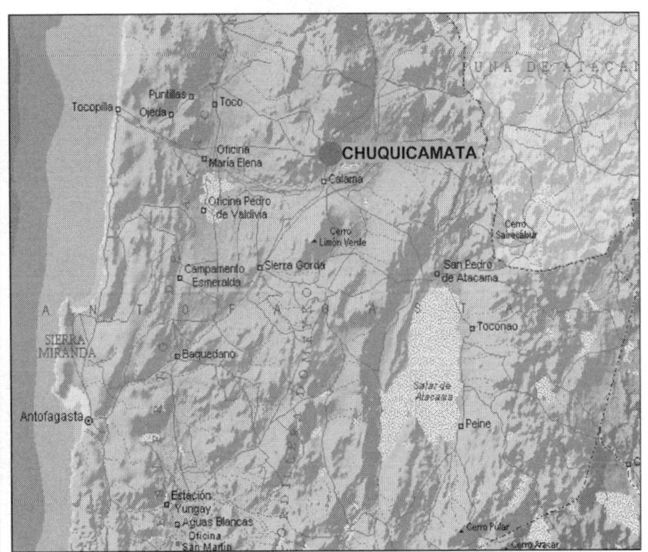

FIGURE 16.1 Map showing the location of the Chuquicamata Mine

* Geotechnical Group, Chuquicamata Division, Codelco, Chile.
† A. Karzulovic & Associates, Ltd., Santiago, Chile.

TABLE 16.1 Geological units present in Chuquicamata Mine

Unit	Description
Paleogravels	Gravels of Tertiary age (end of), with subangular shapes and a large variation in sizes, with a silty-sand/sandy-silt matrix and, sometimes, with sulfates as cement.
Fortuna Granodiorite	Intrusive mass with an important areal extent in Chuquicamata, located to the west of the west fault. It does not have mineralization or hydrothermal alteration. Age: about 36 to 38 My. Can be described as a blocky rock mass.
Fortuna Granodiorite, Moderately Sheared	Fortuna Granodiorite affected by the west fault, which shows a kind of tectonic alteration and is moderately sheared. Can be described as a very blocky rock mass.
Fortuna Granodiorite, Highly Sheared	Fortuna Granodiorite very affected by the west fault, which shows a kind of tectonic alteration and is highly sheared. Can be described as a very blocky and disturbed rock mass.
Quartz-Sericitic Rock	East porphyry with a quartz-sericitic alteration so intense that it has completely obliterated the original rock texture and most of the joints. Can be described as a massive rock mass.
East Porphyry with Sericitic Alteration	East porphyry (mozonitic-granodioritic porphyry) with a coarse texture, primary sulfide mineralization, and quartz-sericitic alteration. Can be described as a blocky rock mass.
East Porphyry with Potassic Alteration	East porphyry (mozonitic-granodioritic porphyry) with a coarse texture, primary sulfide mineralization, and potassic alteration. Can be described as a blocky rock mass.
East Porphyry with Chloritic Alteration	East porphyry (mozonitic-granodioritic porphyry) with a coarse texture, primary sulfide mineralization, and chloritic alteration. Can be described as a blocky to very blocky rock mass. Age: about 36 to 38 My.
East Granodiorite	Intrusive mass that outcrops in the east side of Chuquicamata, with an important areal extent. Can be described as a blocky rock mass. Age: about 230 to 240 My.
Elena Granodiorite	Intrusive mass that outcrops in the east side of Chuquicamata, with a north-south trend. It is intruded by the east porphyry. Can be described as a blocky rock mass. Age: about 120 to 160 My.
West Porphyry	Monzo-granitic porphyry with a fine texture that outcrops in the northeast sector of Chuquicamata, with a minor areal extent. Primary sulfide mineralization and quartz-sericitic alteration. Can be described as a blocky rock mass. Age: about 31 My.
Metasediments	Sedimentary rocks (shales, sandstones, and limestones of the Quejita Formation) affected by contact metamorphism. The sandstones are green, the shales are dark gray to black, and the limestones are dark brown and black. The shales have a high fissility. Can be described as a very blocky to very blocky and disturbed rock mass.

FIGURE 16.2 Panoramic view of Chuquicamata Mine at the end of 1999

The main structural feature of Chuquicamata Mine is the west fault, shown in Figure 16.5. It is a regional fault with a thickness of 4 to 6 m and an associated shear zone 150 to 200 m wide. The predominant structures are subvertical, but in the west wall, some joints dip 40° to 50° into the pit. Structures define the typical morphology of the bench faces in the west and east walls of the mine. So, as shown in Figure 16.6, in the west wall the benches show step-path failures, defined by joints dipping 40° to 45° into the pit and structures dipping 70° to 75° away from the pit. In the east wall, the benches show planar and wedge failures, as shown in Figures 16.7 and 16.8. Structural data are obtained from bench mapping and also from oriented boreholes. Based upon the analysis and interpretation of the structural data, the Chuquicamata Mine has been defined by the eight structural domains described in Table 16.2; locations are shown in Figure 16.9.

TABLE 16.2 Characteristics of the structural domains at Chuquicamata Mine

			Structural Domain							
System	Dip	Dip Dir.	Fortuna Sur	Fortuna Norte	Zaragosa	Estanques Blancos	Mesabi	Balmaceda	NorOeste	Americana
S1	80° ± 10°	115° ± 20°			≤		≤		≤	≤
		295° ± 20°	≤	≤	≤		≤		≤	≤
S2	75° ± 5°	180° ± 10°						≤		
		360° ± 10°	≤	≤			≤	≤		≤
S3	75° ± 5°	045° ± 20°			≤				≤	
		225° ± 20°	≤		≤	≤	≤		≤	≤
S4	75° ± 5°	150° ± 10°				≤				
		330° ± 10°				≤			≤	
S5	70° ± 10°	275° ± 20°	≤					≤		
S6	40° ± 5°	185° ± 10°						≤		
S5	35° ± 5°	275° ± 10°		≤						

The geotechnical characterization of the rock mass is based on the geological strength index (GSI) (Hoek et al. 1985), which depends on the rock-mass fabric and the condition of the structures. The characterization does not consider a unique GSI value

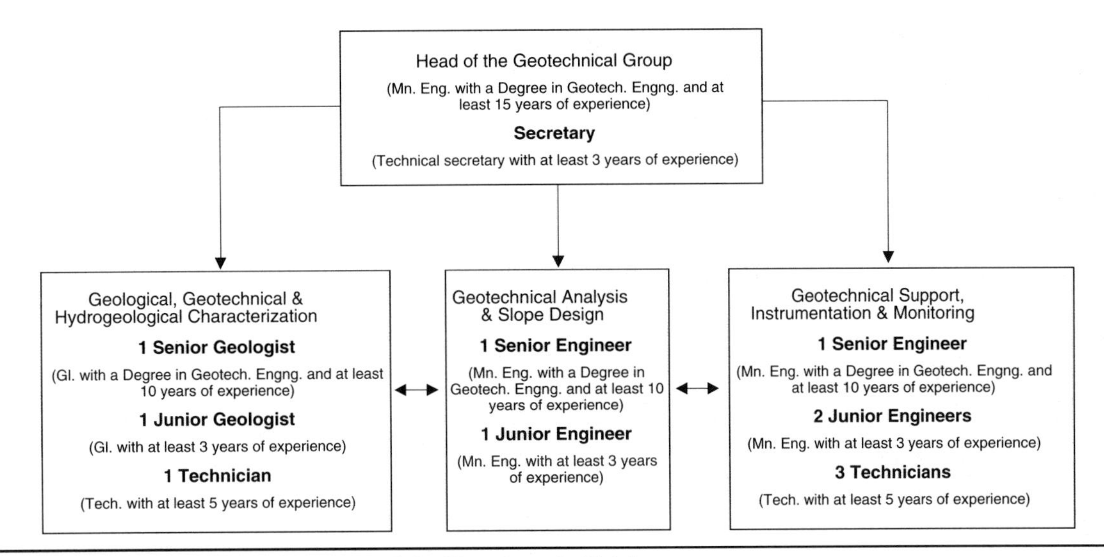

FIGURE 16.3 Organization of the geotechnical group of the Chuquicamata Division

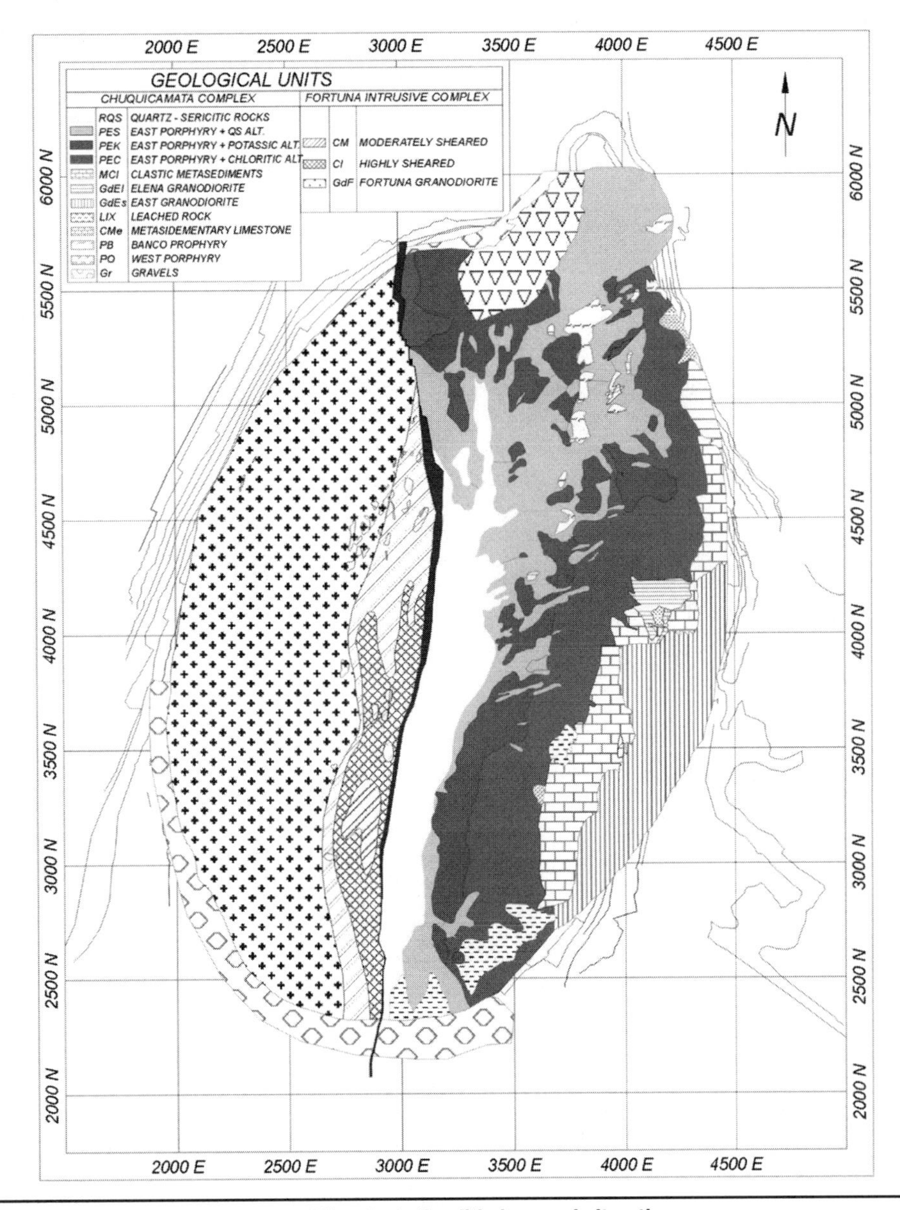

FIGURE 16.4 Geological units present in Chuquicamata Mine, including lithology and alteration

FIGURE 16.5 West fault trace on the west wall of Chuquicamata Mine

FIGURE 16.6 Step-path of failure in a bench of the west wall of Chuquicamata Mine, defined by joints dipping 40° to 45° toward the pit and larger structures dipping 70° to 75° against the pit, in the Fortuna Sur structural domain

but a "typical" range for each sector of the mine. According to this, the geotechnical zonation of the Chuquicamata Mine was developed as shown in Figure 16.10. Taking into account this GSI zonation and the results of laboratory tests, rock-mass properties were estimated using the Hoek–Brown methodology (Hoek and Brown 1997; Hoek 1998). These properties were "adjusted" using engineering judgment and the observed in situ behavior. The "adjusted" properties are summarized in Table 16.3. The properties of joints and other structures, summarized in Table 16.4, have been estimated taking into account the back-analysis of structurally controlled bench failures and, also, the results of direct shear tests.

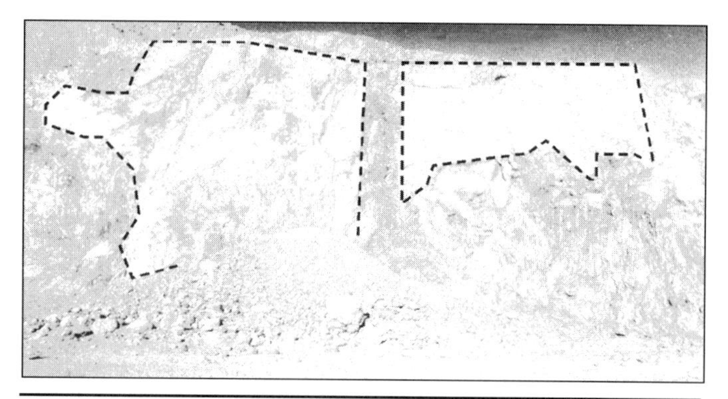

FIGURE 16.7 Planar slide in a bench of the upper part of the east wall of Chuquicamata Mine, defined by subvertical structures of the Mesabi structural domain

FIGURE 16.8 Wedge slide in a bench of the middle part of the east wall of Chuquicamata Mine, defined by subvertical structures of the Nor Oeste structural domain

The hydrogeological characterization of the Chuquicamata Mine is supported by observation wells and 32 piezometers (vibrating wire) and has allowed the development of a hydrogeological model of the mine. This model predicts the phreatic surface shown in Figure 16.11. These results have been used to design and construct the current system for slope depressurization and mine drainage, which includes a 1,200-m drainage tunnel and two

TABLE 16.3 Properties of the rock masses and paleogravels (typical values)

Geotechnical Unit	γ (ton/m^3)	m_i	σ_{ci} (Mpa)	E_i (Gpa)	GSI	c (kPa)	ϕ (degrees)	E (Gpa)	ν	B (Gpa)	G (Gpa)
Paleogravels	2.10					100	42	0.60	0.25	0.40	0.24
Fortuna Granodiorite	2.66	31.6	82	33	40–60	675	43	9.0	0.26	6.3	3.6
Fortuna Gd. Moderately Sheared	2.51	24.0	30	30	25–40	210	33	2.3	0.30	1.9	0.9
Fortuna Gd. Highly Sheared	2.30	20.3	15	7	15–25	125	25	0.7	0.33	0.7	0.3
Quartz-Sericitic Rock	2.49	17.9	15	22	70–100	825	34	4.5	0.25	3.0	1.8
East P. with Quartz-Sericitic Alt.	2.52	19.7	31	18	40–60	365	39	4.9	0.27	3.6	1.9
East P. with Potassic Alt.	2.58	31.3	85	22	55–65	770	45	16.6	0.24	10.6	6.7
East P. with Chloritic Alt.	2.62	17.2	84	52	45–60	560	45	10.5	0.26	7.3	4.2
East Granodiorite	2.62	26.1	62	34	50–60	565	47	10.5	0.25	7.0	4.2
Elena Granodiorite	2.62	26.5	77	40	45–60	575	48	10.1	0.26	7.0	4.0
West Porphyry	2.52	19.1	59	29	45–60	480	43	8.9	0.26	6.2	3.5
Metasediments	2.67	24.5	45	30	25–40	245	35	5.8	0.30	4.8	2.2

NOTES:
γ unit weight.
m_i Parameter m of the Hoek–Brown criterion for "intact" rock.
σ_{ci} Uniaxial compressive strength of the "intact" rock.
E_i Deformability modulus of the "intact" rock.
GSI Geological strength index of the rock mass.
c Cohesion of the rock mass or the paleogravels (for confining stresses in the range from 0 to 4 MPa).
ϕ Angle of friction of the rock mass or the paleogravels (for confining stresses in the range from 0 to 4 MPa).
E Deformability modulus of the rock mass or the paleogravels.
ν Poisson's ratio of the rock mass or the paleogravels.
B Bulk modulus of the rock mass or the paleogravels.
G Shear modulus of the rock mass or the paleogravels.

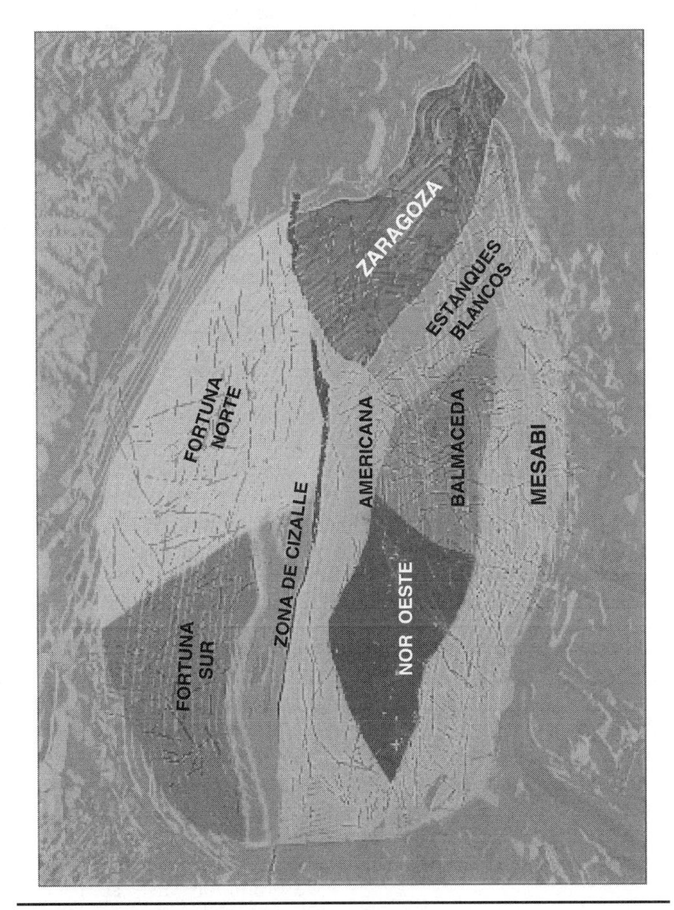

FIGURE 16.9 Structural domains of Chuquicamata Mine, west fault and shear zones

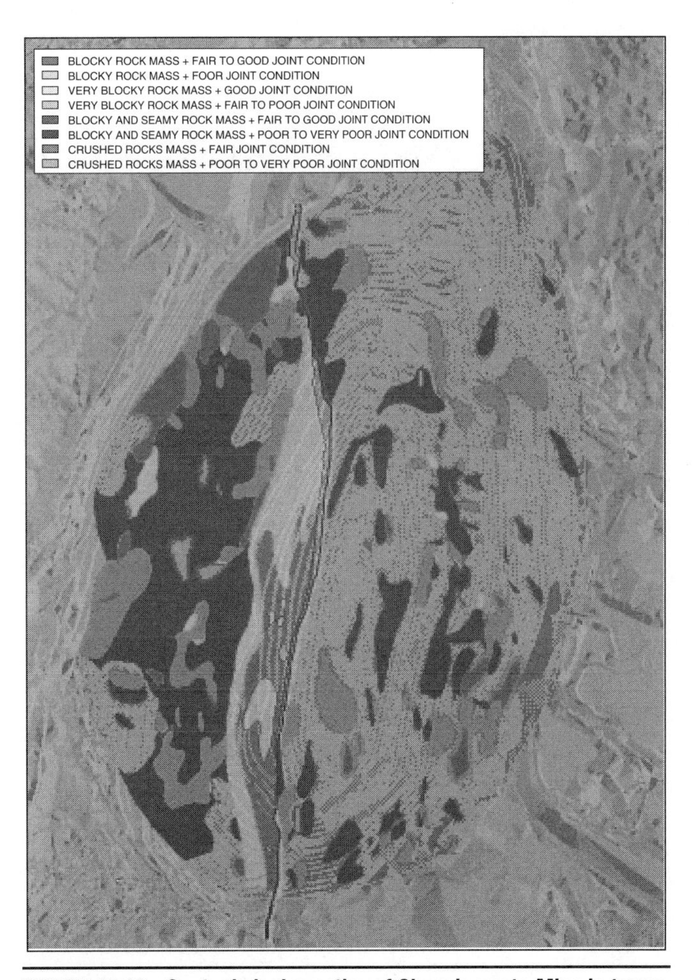

FIGURE 16.10 Geotechnical zonation of Chuquicamata Mine in terms of the GSI

TABLE 16.4 Properties of the structures (typical values)

Type of Structure	Cohesion (kPa)	Angle of Friction (degrees)
Major faults with clayey gouge	0–50	18
Continuous major structures	25–75	32–38
Continuous minor structures (bench scale)	25–100	30–40
Minor structures with rock bridges (interramp or overall scale)	100–250	30–45

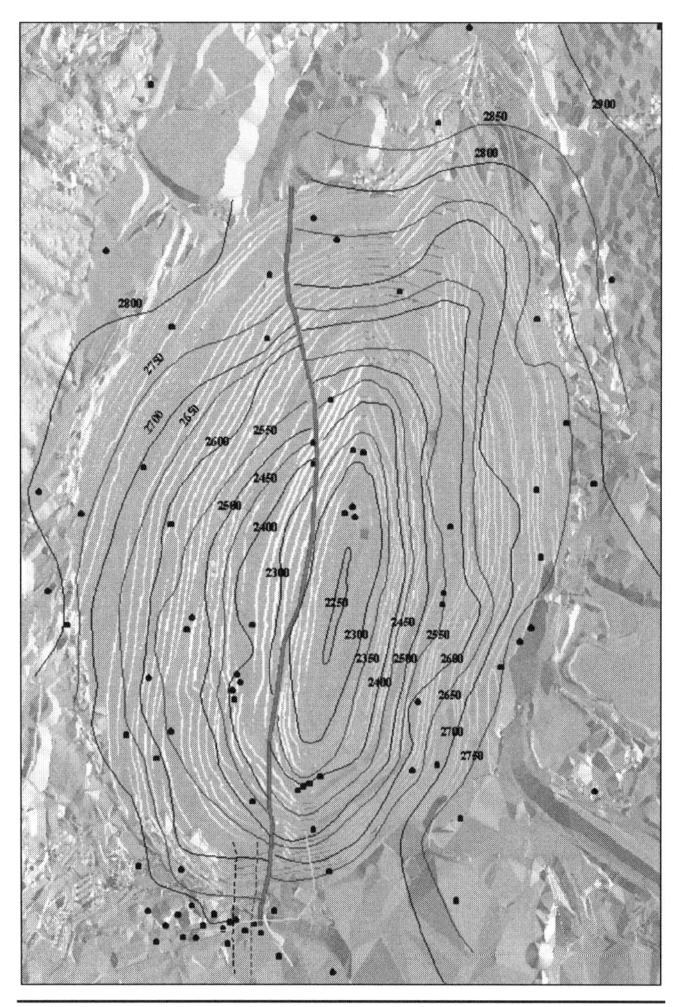

FIGURE 16.11 Phreatic levels at Chuquicamata Mine

pumping stations (each one working with three pumping wells). The location of the drainage system, piezometers, and some observation wells is shown in Figure 16.12.

Taking into account all of the above data and the orientation of the slopes, it is possible to define eight design sectors in the Chuquicamata Mine in such a way that each design sector has typical geotechnical, structural, and hydrogeological characteristics. In each sector, vertical sections are defined and used for slope stability analysis and slope design.

16.3 GEOTECHNICAL SLOPE ANALYSIS AND DESIGN

To perform slope-stability analyses, the Geotechnical Group uses both limit-equilibrium and numerical methods. For limit-equilibrium analyses, the computer codes SWEDGE and PWEDGE are most often used for examining wedge stability at the bench scale and XSTABL is used for interramp and overall slope-stability analyses. It is important to indicate that when XSTABL is used, the effect of structures is taken into account by defining an anisotropic rock-mass strength. The search for the critical sliding surface is done using the Janbu method and, once the critical sliding surface has been defined, the factor of safety is calculated using the Spencer method. The probability of failure is also evaluated using the method proposed by Rosenblueth (Harr 1987). Limit-equilibrium methods are used for the analysis of short-term or contingent problems and, also, when a large number of cases are to be analyzed for mid- and long-term conditions.

Analyses by numerical methods are performed using the Fast Langrangian Analysis of Continua (FLAC), FLAC^{3D}, Universal Distinct Element Code (UDEC), and 3DEC computer codes. From 1994 to 1997, FLAC was the most commonly used software for slope-stability analysis. In order to achieve a better representation of the real conditions, it was necessary to include explicitly in the model numerous major structures with several intersections. As the number of these explicit structures and their intersections increased, it was more and more difficult to construct the model. Due to this and the need to include explicitly all major structures, in 1998 the numerical analyses began to be done using UDEC, which allows an easier "handling" of the structures.

In certain special cases, three-dimensional numerical models are used. Due to the larger engineering resources required by these three-dimensional models, their use is less frequent than the two-dimensional models. In 1998, 3DEC was used to develop a three-dimensional model of the southern sector of the Chuquicamata Mine. This was used, together with two-dimensional models and in situ observations, to predict the evolution of the subsidence that will affect the sector from 1999 to 2008.

Numerical models constitute the main tool for slope-stability analysis of the Chuquicamata Mine. These models allow the user to calculate the factor of safety (using the shear strength reduction technique described by Dawson and Roth [1999]) and also give information about possible failure mechanisms, the slope-displacement pattern, eventual zones of stress concentration, etc. Figure 16.13 shows an example of the results obtained using UDEC that illustrate how the presence of the shear zone adjacent to the west fault affects the slope-displacement pattern.

On the other hand, there are situations when it is required to analyze a large number of cases (e.g., to study how sensitive the design is to changes in one or more variables), when these analyses are performed using limit-equilibrium methods. These limit-equilibrium models have been previously calibrated with the results of numerical models. These calibrations indicate that limit-equilibrium analyses, with the methodology used in Chuquicamata, overestimate the factor of safety by 4 to 9%, as compared to numerical models.

For each design sector, slopes are designed according to the following procedure: the bench-berm system is designed, defining initial values for the interramp angle; the resulting interramp and overall slopes are analyzed and, if they do not fulfill the acceptability criteria, the berm widths are increased, thereby decreasing the interramp angle until the acceptability criteria are fulfilled by the interramp slopes. Overall slopes are treated in the same way, but by increasing the ramp widths if necessary.

The bench height is defined by considering the efficiency of the shovels and also the volumes of eventual unstable wedges (from an operational point of view the steeper the bench face the better). Until 1998, the Chuquicamata Mine used 26-m-high benches, but since 1999 all new developments use 18-m-high benches both to optimize shovel efficiency and to limit the eventual unstable small wedges to relatively small volumes. The inclination of the bench faces depends on the structures and the quality of blasting; therefore, it varies from 60° to 70°, depending

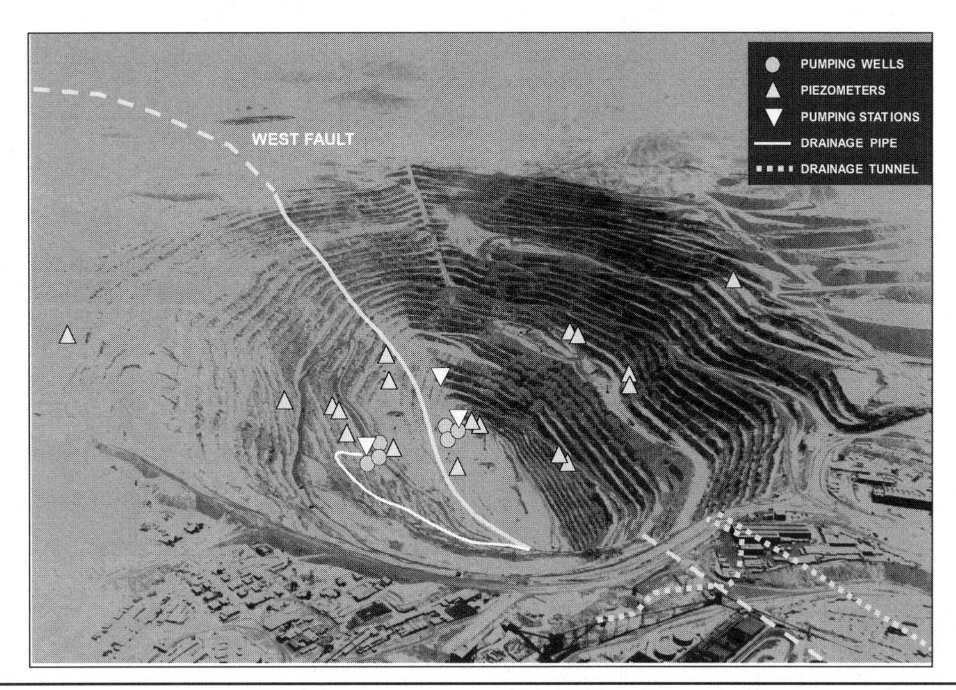

FIGURE 16.12 Piezometers, observation wells, pumping stations and drainage tunnel in Mina Chuquicamata, at the end of 1999

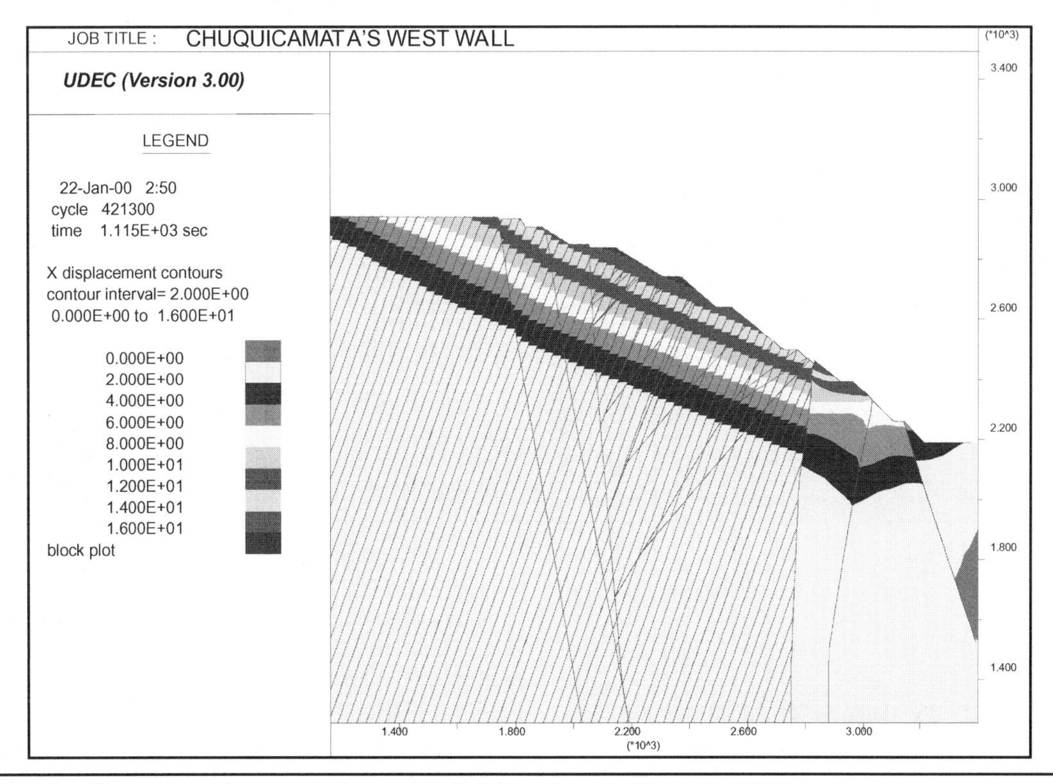

FIGURE 16.13 Results of the analysis of a slope in the west wall of Chuquicamata Mine using UDEC

on the design sector. The minimum berm width is determined to be the width required to contain the debris associated with an unstable wedge with a size that has a 15% probability or more of occurring.

Currently, the design of interramp and overall slopes at the Chuquicamata Mine considers the following acceptability criteria:

- The factor of safety in an operational condition (with no earthquake, with the expected or typical phreatic level, and with the use of controlled blasting) must be equal to or larger than 1.30.
- The factor of safety in an extreme condition (with earthquake, with a higher phreatic level, and with poor-quality blasting) must be equal to or larger than 1.10.
- For interramp slopes, the probability of failure in an operational condition must be equal to or smaller than 10% and for overall slopes it must be equal to or smaller than 5%.

TABLE 16.5 Current slope design at Chuquicamata Mine

<table>
<tr><th rowspan="2">Design Sector</th><th colspan="3">Bench-Berm</th><th colspan="4">Interramp</th><th colspan="2">Overall</th><th rowspan="2">Comments</th></tr>
<tr><th>h_b (m)</th><th>α_b (degrees)</th><th>Q (m)</th><th>b (m)</th><th>α_r (degrees)</th><th>h_r (m)</th><th>r (m)</th><th>α_o (degrees)</th><th>h_o (m)</th></tr>
<tr><td>1</td><td rowspan="8">18</td><td rowspan="3">63</td><td rowspan="3">9.0–9.5</td><td>15.5</td><td>36</td><td rowspan="8">162</td><td rowspan="5">40</td><td>31</td><td>700</td><td rowspan="5">Slope behavior is highly affected by the presence of the west fault and its zone of moderately and highly sheared rock. These slopes have a typical displacement rate of about 0.5 to 1.0 cm/d.</td></tr>
<tr><td>2</td><td rowspan="2">13.0</td><td rowspan="2">39</td><td rowspan="3">32</td><td rowspan="4">750</td></tr>
<tr><td>3</td></tr>
<tr><td>4</td><td>61</td><td>10.0–10.5</td><td>15.0</td><td rowspan="2">36</td></tr>
<tr><td>5</td><td>63</td><td rowspan="4">7.0–7.5</td><td>18.0</td><td>31</td></tr>
<tr><td>6</td><td rowspan="3">69</td><td>9.0</td><td rowspan="2">48</td><td rowspan="3">30</td><td rowspan="2">42</td><td rowspan="2">780</td><td rowspan="3">These slopes do not show displacements, but in 1969 suffered a large instability (12 Mton), defined at least partially by major structures.</td></tr>
<tr><td>7</td><td>9.0</td></tr>
<tr><td>8</td><td>10.5</td><td>46</td><td>41</td><td>730</td></tr>
</table>

h_b Bench height α_b Bench face inclination Q Backbreak
h_r Interramp height α_r Interramp angle b Berm width
h_o Overall height α_o Overall angle r Ramp width

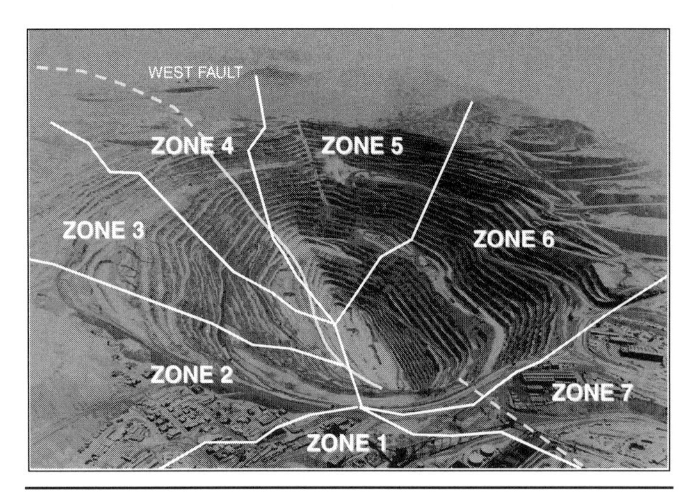

FIGURE 16.14 Design sectors of Chuquicamata Mine, defined according to the geological, structural, and geotechnical characteristics and the orientation of the slopes

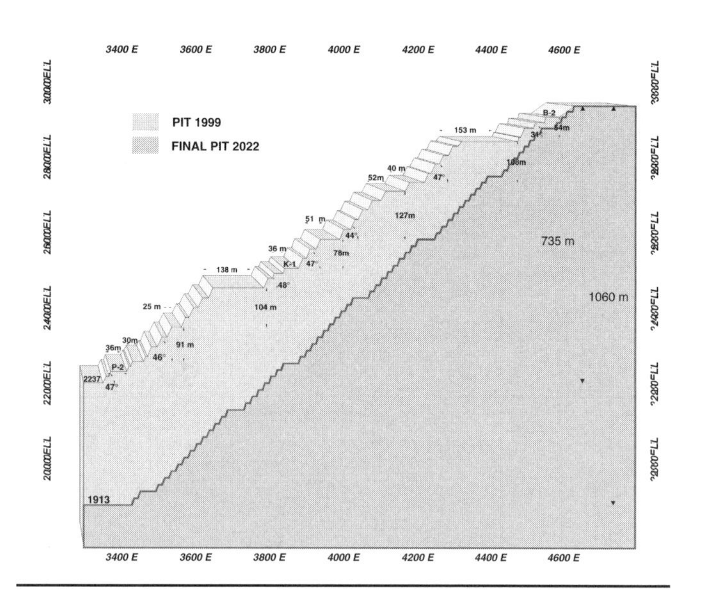

FIGURE 16.15 Vertical section that shows the slope design for the east wall of Chuquicamata Mine at the end of 1999

- The displacement vectors obtained from numerical models must not have an important downward vertical component.

The slopes for each one of the design zones shown in Figure 16.14 have been designed according to this procedure. The current design of Chuquicamata Mine slopes is summarized in Table 16.5 and illustrated in Figures 16.15 and 16.16 for the east and west walls, respectively.

16.4 INTERACTION WITH MINE PLANNING

A good mining plan can only be achieved if there is a proper interaction between the geotechnical and mine planning groups. At the Chuquicamata Mine, this interaction is developed. First, because of geotechnical analysis, a slope geometrical design is given to the mine planning group for each sector of the mine. Then recommendations are made with regard to requirements of controlled blasting and/or drainage (to take into account eventual investments in equipment), pushback sequences, in plan view and in vertical sections, and extraction rates. This makes it possible to develop adjacent pushbacks that do not produce eventual conditions of potential instability (e.g., this could occur if a shovel with larger capacity is working in a lower pushback and a shovel of smaller capacity is working in a higher pushback).

Based on this, the planning group develops short-, mid-, and long-term mining plans. These plans are sent to the geotechnical group, which then checks that the plans fulfill the geotechnical restrictions and, eventually, proposes options to optimize the plans (e.g., to optimize the position of some ramps, to "soften" the closure of some pushbacks, etc.).

Once checked and eventually improved, the plans return to the mine planning group, which then develops the definitive short-, mid-, and long-term mining plans for the Chuquicamata Mine. Beginning in 1994, the geotechnical aspects of the mining plans have been checked every two years.

16.5 INTERACTION WITH MINE OPERATION

An open-pit slope design can only be accomplished if there is an effective interaction between the geotechnical and mine operation groups. Otherwise, there will be no control over the actual fulfillment of requirements, such as the location of the program line, bench face inclination, berm width, quality of blasting, etc. In order to achieve this interaction, the following procedure is used at the Chuquicamata Mine:

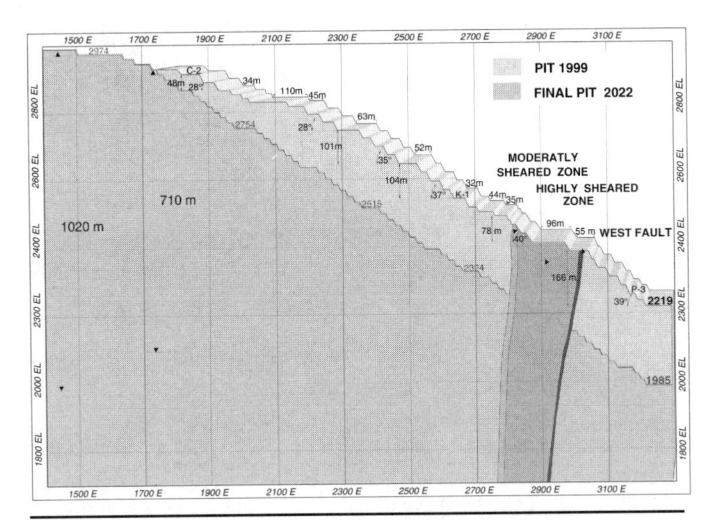

FIGURE 16.16 Vertical section showing the slope design for the west wall of Chuquicamata Mine at the end of 1999

- At the beginning of each year, there is a meeting at which the mine operation group participates. At the meeting, the geotechnical group presents the basis of the proposed slope design, the geotechnical restrictions for each sector of the mine, and the eventual geotechnical hazards that could affect the production program. The purpose of this meeting is to arrive at a common understanding, to make compromises, and to define basic contingency plans. Also presented at this meeting is the strategy and program for the implementation of controlled blasting to minimize the blast-induced damaged in the rock mass (at the Chuquicamata Mine the geotechnical group is in charge of implementing the use of controlled blasting).
- At the end of each month, there is a meeting at which the staffs of the geotechnical group, short-term planning, and the mine operation group participate. At this meeting, the aspects of the monthly production program that could not be achieved are discussed and solutions to problems are proposed. Moreover, the program for the next month is discussed in order to prevent the eventual occurrence of geotechnical hazards (e.g., a wedge instability that could affect one access ramp) and, if possible, to optimize the program (e.g., if a pushback is leaving wide platforms, it might be possible to reduce some berm widths to gain access to more mineral). Also the designs for controlled blasting to be performed during the next month are discussed at this meeting.
- At the end of each week, there is a meeting at which the staffs of the geotechnical group, short-term planning, and the mine operation group participate. At this meeting, the aspects of the weekly program that could not be achieved are discussed and solutions to problems are proposed. The program for the coming week is discussed in order to prevent the eventual occurrence of geotechnical hazards and, if possible, to optimize the program. Also the results of controlled blasting are discussed.
- The geotechnical group performs daily inspections of the Chuquicamata Mine to detect any signs of an eventual instability or unforeseen slope behavior, to verify the performance of the slope design, and to provide in situ support to the mine operation group. This daily work includes the following activities:
  - Visual inspection of benches and berms to detect cracking or other signs of instability or unexpected slope behavior
  - Revision of those factors where installed instrumentation indicates an increase in the displacement rates and/or other signs of slope activity or potential instability
  - Detection of eventual new seeps and/or changes in the already known seeps
  - Local checking, with tape, of the program line locations and the berm widths; in cases of doubt, surveying support is provided
  - Inspection of those sectors where it has been noted that a condition of potential instability could exist and, if necessary, definition of monitoring and/or stabilizing measures
  - Support to the mine operation staff to define the best digging sequence for the shovel in each pushback
  - Revision of the production blastings to make sure that the permissible maximum tonnage is not exceeded (200,000 tons in the vicinity of the west fault and 600,000 tons in the rest of the mine)
  - Revision of the controlled-blasting design and evaluation of the blast-induced damage on the rock mass

16.6 INSTRUMENTATION AND MONITORING

At any large open-pit mine, if it is necessary to stabilize a wall by changing the slope geometry, it is advisable to take advantage of the pushbacks considered in the mining plan to achieve this purpose. Arriving with a pushback in the sector to stabilize and extract enough material to control the problem can take several months; therefore, it is important to know well in advance if the stability of the sector is evolving in a unfavorable way. One way to deal with this problem is to develop pushbacks and remove material as soon as any crack or other sign of instability is detected. However, this probably will waste resources and negatively impact the mining business, because the presence of cracks does not always mean a short-, or even mid-, term stability problem. Therefore, a control system that makes it possible to define the evolution in time of the slope's stability condition by recording one or more variables is necessary. The system must also be able to resort to pushbacks and material removal when it is necessary, and not before. The following cases are examples:

- In 1995, in the southwest sector of the Chuquicamata Mine, the displacement rate increased from 0.5 to 1.2 cm/d and the infrastructure located along the periphery of the pit began to suffer deformations and damage, especially to the foundations and parts of the steel structures of the stockpile facility. To stabilize the sector, part of Pushback 38S was advanced, removing material from the upper part of the slope and defining a 100-m-wide platform in the upper third of the slope. In order to initiate a decrease in the displacement rate, it was necessary to remove 6 Mton of material, which required seven months. To complete the stabilization program, 14 Mton of material were removed, which required 16 months (a 17-cu yd shovel was used). Once the sector was stabilized, the displacement rate was reduced to 0.2 cm/d; since then it has not changed.

- In 1997, the northwest sector of the Chuquicamata Mine, which had presented stability problems in 1994, was reactivated, reaching displacement rates of about 2 cm/d. To stabilize this sector, Pushback 34N was accelerated in order to remove the two upper benches of the unstable sector (52 m). This required the removal of 4 Mton of material over four months (a 34-cu yd shovel was used). Once the sector was stabilized, the displacement rate was reduced to 0.5 cm/d; since then it has not changed.

Given the size of the Chuquicamata Mine and the fact that the west wall displaces at a rate of 0.5 to 1.0 cm/d, it is necessary to have a control system that allows one to detect, with enough warning, an eventual increase of this sliding rate. With such a system, the necessary measures to avoid a large stability problem can be taken at the correct time. Although the east wall does not present displacements of the same magnitude, it has been affected by major stability problems (Voight and Kennedy 1979). The east wall could eventually suffer large stability problems, making it convenient to have a control system in this wall of the Chuquicamata Mine as well.

In addition, slope design at the Chuquicamata Mine is based on analyses that assume a certain geology, structural geology, geotechnical condition, phreatic level, material properties, and boundary conditions. These assumptions have a degree of uncertainty. Consequently, the only way to assure the proposed design is feasible in practice and, at the same time, improve the design is to validate the predictions of these analyses with the observed slope behavior. This constitutes what is currently known as the observational method, which was proposed in 1945 by Terzaghi (Peck 1969). Obviously, if a validation that is not purely qualitative is desired, it becomes necessary to make measurements with a monitoring system. Such a system must be capable of recording the magnitude of one or more variables that define the slope behavior; these magnitudes are predicted by the geotechnical analysis (e.g., displacements, pore pressures, etc.).

Therefore, the geotechnical group has installed a complete instrumentation system in the Chuquicamata Mine. The system allows monitoring of the slopes and the area adjacent to the pit periphery and enables the geotechnical group to check the effectiveness of eventual stabilization measures. This instrumentation system includes:

- 2 Leica APS-Win robotic systems to measure the inclined distance and coordinates of prisms located in the mine walls. One system measures the west wall and the other the east wall. In each wall there are about 100 prisms. This system is schematically shown in Figure 16.17.
- 14 inclinometers installed in the southern sector of the mine, with depths of about 200 m.
- 60 observation wells, most of them located inside the mine, with depths from 100 to 200 m.
- 32 vibrating-wire piezometers; 18 are installed inside the pit at depths of about 150 m and 14 are installed in the periphery of the pit at depths from 130 to 180 m.
- 20 GPS control points placed in the periphery of the southern sector of the mine.
- 60 topographical control points placed in the periphery of the southern sector of the mine.
- 7 wire line extensometers, like the one shown in Figure 16.18, that are used to monitor local stability problems, usually with structural control and affecting no more than six benches.

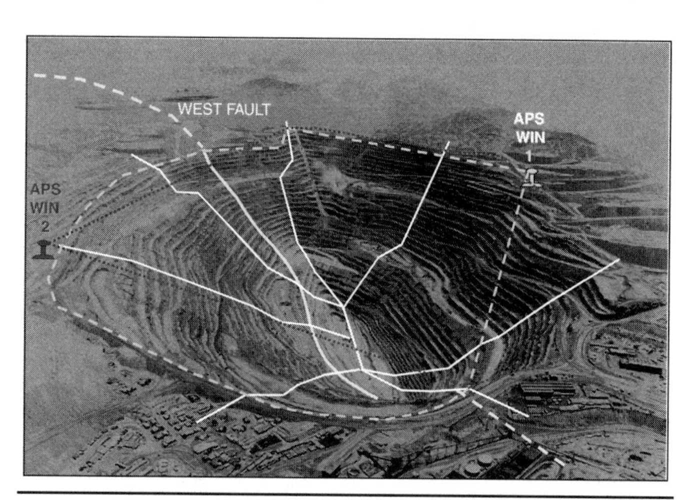

FIGURE 16.17 Scheme showing the location of the Leica APS-Win robotic system (two stations), used to measure the displacements of prisms placed on the east wall (100 prisms) and west wall (100 prisms) of Chuquicamata Mine

FIGURE 16.18 Wire line extensometer used by the geotechnical group of Chuquicamata Mine to monitor local instabilities, usually with structural control and affecting no more than six benches

- 5 surface extensometers, like the one shown in Figure 16.19, that are used to monitor local stability problems, usually with structural control and affecting no more than six benches.

It is important to point out that since robotic systems allow a near-continuous record of displacements, they produce such a quantity of information that it makes the use of a computerized system an absolute necessity. This system stores the information and, even more importantly, allows a daily data interpretation, in situations of potential instability, and a weekly data interpretation, under normal conditions. This data presentation capability is important because any time period without data interpretation will produce a volume of data that could eventually become unmanageable and, therefore, useless.

FIGURE 16.19 Surface extensometer used by the geotechnical group of Chuquicamata Mine to monitor local instabilities, usually with structural control and affecting no more than six benches

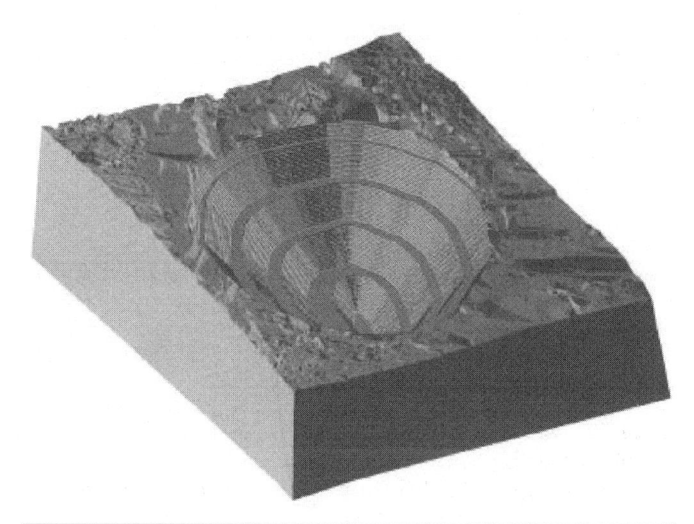

FIGURE 16.20 Three-dimensional view of Chuquicamata Mine with a decoupled slope design

Accumulated experience to date, five years of operating the robotic systems has allowed development of the following empirical criteria for the west wall of the Chuquicamata Mine:

- If the displacement rate of the sector under study does not exceed 1 cm/d, the condition is considered normal and there is no risk of a potential instability.
- When the displacement rate exceeds 1 cm/d, it is necessary to begin a more detailed monitoring of the sector under study. Usually, displacement rates of 1 to 3 cm/d are associated with the appearance of notorious cracks, which dictates that additional instrumentation be installed in the sector (e.g., wire line extensometers and/or surface extensometers).
- If the displacement rates are from 3 to 5 cm/d for longer than two weeks, the sector is considered to have a condition of potential instability. If the problem does not affect more than two benches, mine equipment and/or secondary blasting are used to remove the unstable material. If the problem affects more than two benches, removal of material is initiated using, if possible, a pushback currently being developed.
- If the displacement rate exceeds 5 cm/d, no operation is allowed at the slope toe and all equipment and personnel must leave the affected sector. At the same time, rock removal activities are accelerated in the upper part of the slope.

16.7 APPLIED RESEARCH AND TECHNOLOGICAL INNOVATION

Geotechnical engineering is one branch of engineering in which continuous updating and improvement is necessary. This is due to inevitable geotechnical uncertainty and the limitations of current models for rock-mass behavior. However, due to the size of the Chuquicamata Mine and the fact that the current mine plan calls for the pit to reach a depth of 1,100 m, it is necessary to go beyond current practice. It becomes necessary to implement an applied research program that extends current concepts of slope analysis and design to very deep slopes in rock masses of poor to fair geotechnical quality. In addition, increasing competition and the current low prices of copper are forcing technological innovations that yield more and more improvements for the mining business.

16.7.1 Continuous Updating and Improvement

One geologist and three engineers from the geotechnical group of the Chuquicamata Division are currently studying to obtain their master's degrees. One employee is studying full time in the United States and the other three are studying part time in Chile.

16.7.2 Applied Research

The geotechnical group developed the concept of decoupled slopes in 1998. This concept involves segmenting a large slope by using wide platforms that subdivide it into smaller slopes. In this way, when an eventual instability occurs, it will affect only one of the segments or smaller slopes. For the west wall of the Chuquicamata Mine, this concept presents the additional advantage that the lowest platform can be used to decouple the upper part of the wall from the effect of the shear zone adjacent to the west fault. Currently, a conceptual mining plan based on a design with decoupled slopes is being developed, as are the geotechnical studies required for a basic engineering stage. This concept, illustrated in Figure 16.20, shows a preliminary design of the Chuquicamata Mine with decoupled slopes.

16.7.3 Technological Innovation

The geotechnical group is working on two tasks. The first is the development of a risk model to rationalize the definition of an acceptability criteria and to allow economical comparison of different options for slope design. The second is the development of a procedure for applying the observational method in mine slope design in order to optimize this design by incorporating factors that usually are not considered in slope-stability analysis.

The development of a risk model applied to Chuquicamata Mine slopes makes it possible to quantitatively evaluate the risk level in each slope of the mine, considering its sensitivity to eventual changes in the different geotechnical and mining parameters. This will yield a more reliable slope design that includes geotechnical considerations as well as contributions from the other groups at the Chuquicamata Mine. This way of working will optimize the use of resources, since the model will show explicitly the costs associated with different slope design options, their impact on other components of the mine business, and the expected effect of eventual stabilization measures. The model is expected to be implemented for practical use at the end of the year 2000.

Given the typical uncertainties of geology and geotechnical engineering, it is not possible, in practice, to define precisely the geotechnical geometry (i.e., lithological contacts, location and orientation of major structures, phreatic levels, etc.), nor is it possible to determine accurate properties for the structures and the rock mass. Therefore, geotechnical engineers in charge of slope design can only evaluate an "expected" or "more probable" condition, and they know a priori that geotechnical parameters have a variability that can be important. Because of this, it is not uncommon for changes to be found in the geological–geotechnical conditions that severely affect the slope design when a proposed slope design is being implemented in practice. Therefore, a "tool" based on the observational method will make it possible to deal with eventual changes in the geological–geotechnical conditions according to a preestablished plan that has accounted for slope design stage, with the consequent improvement of the mining business.

A procedure to apply the observational method at the Chuquicamata Mine is being developed by the geotechnical group. It has the following objectives:

- To explain the fundamental concepts of the observational method and develop a methodology for its practical application to slope design at the Chuquicamata Mine.
- To design slopes for the "expected" condition, taking into account (a) what could be the "worst credible condition," (b) its possible effects on the proposed slope design, and (c) what can be done if this "worst credible condition" is encountered (i.e., contingency plans).
- To improve the current knowledge regarding the observed slope behavior at the Chuquicamata Mine.
- To define beforehand what conditions could trigger instabilities and when it would be necessary to implement the contingency plans.

This work will be completely developed and implemented by the end of the year 2000.

16.8 CONCLUSIONS

A large open-pit mine requires a vast quantity of geotechnical work. At the Chuquicamata Mine, this work is performed by the geotechnical group. This work is primarily related to geological, geotechnical, and hydrogeological characterization; slope analysis and design; control of the practical implementation of slope designs; controlled blasting and blast-induced damage; and slope instrumentation and monitoring.

This work is oriented toward optimization of slope design, with a consequent improvement of the mining business. It also includes development of applied research and technological innovation, such as the development of decoupled slopes and a methodology to apply the observational method to slope design at the Chuquicamata Mine. In addition, development of an effective interaction between the geotechnical, mine planning, and mine operation groups has turned out to be necessary for the Chuquicamata Mine.

Geotechnical engineering is considered a practical tool of fundamental importance for mine slope design, mine planning, and mine operation at the Chuquicamata Mine. The mine has improved the design of mine slopes and optimized the relationship between the economical aspects of design and the capacity of a rock mass to maintain its stability while minimizing the likelihood of major instabilities that eventually could affect staff, equipment, and the productive process.

In this way, the Chuquicamata Mine is using geotechnical engineering to face important challenges such as developing slopes with depths of approximately 1,100 m, where small variations in the overall angle can have a major impact on the mining business.

The geotechnical group at Chuquicamata Mine will continue to optimize mine slope design, with emphasis on the economical and safety issues, in order to achieve a productive process and fulfill the short-, mid-, and long-term mining plans.

16.9 ACKNOWLEDGMENT

The authors thank Codelco-Chile, Chuquicamata Division, for authorizing the presentation of this work and, especially, Juan H. Rojas for his support and contributions in the revision of this text. The authors also thank the staff of the geotechnical engineering group of the Chuquicamata Division. Without their spirited, dedicated, and continuing efforts, papers like this would not be possible.

16.10 REFERENCES

Dawson, E.M., and W.H. Roth. 1999. Slope stability analysis with FLAC. In *FLAC and Numerical Modeling in Geomechanics*, ed. C. Detournay and R. Hart, Rotterdam: A.A. Balkema.

Esterhuizen, G.S. 1999. *PWEDGE. Probabilistic Analysis of Wedge Failure in Rock Slopes*, Littleton, Colorado, USA.

Harr, M.E. 1987. *Reliability-Based Design in Civil Engineering*. New York: McGraw-Hill.

Hoek, E. 1998. Reliability of Hoek-Brown estimates of rock mass strength properties and their impact on design, *Intl. Jour. Rock Mech. Min. Sci.*, 35:1:63–68.

Hoek, E., and E.T. Brown. 1997. Practical estimates of rock mass strength, *Intl. Jour. Rock Mech. Min. Sci.*, 34:8:1165–1186.

Hoek, E., P.K. Kaiser, and W.F. Bawden. 1995. *Support of Underground Excavations in Hard Rock*. Rotterdam: A.A. Balkema.

Itasca. 1993. *UDEC. Universal Distinct Element Code*, Version 2.0, Minneapolis, Itasca Consulting Group, Inc.

Itasca. 1994. *3DEC. 3-Dimensional Distinct Element Code*, Version 1.5, Minneapolis, Itasca Consulting Group, Inc.

Itasca. 1997. *FLAC^{3D}. Fast Lagrangian Analysis of Continua in 3 Dimensions*, Version 2.0, Minneapolis, Itasca Consulting Group, Inc.

Itasca. 1998. *FLAC. Fast Lagrangian Analysis of Continua*, Version 3.4, Minneapolis, Itasca Consulting Group, Inc.

Peck, R.B. 1969. Advantages and limitations of the observational method in applied soil mechanics, Ninth Rankine Lecture, *Geotechnique*, 19:2:171–187.

Rocscience. 1999. *SWEDGE. Probabilistic Analysis of the Geometry and Stability of Surface Wedges*, Toronto, Rocscience, Inc.

Sharma, S. 1997. *XSTABL. An Integrated Slope Stability Analysis Program for Personal Computers*, Version 5, User Manual, Moscow, Idaho, Interactive Software Designs, Inc.

Voight, B., and B.A. Kennedy. 1979. Slope failure of 1967–1969, Chuquicamata Mine, Chile. In *Rockslides and Avalanches. 2. Engineering Sites*, ed. B. Voight. Amsterdam: Elsevier.

CHAPTER 17

Slope Stability at Escondida Mine

Cristián Valdivia* and Loren Lorig†

17.1 INTRODUCTION

The Escondida Mine was discovered on March 14, 1981, as a result of an exploration program initiated by Utah International and Getty Mining. The current owners of the Escondida Mine are The Broken Hill Proprietary Inc. (BHP), owning 57.5% of the property; the British company Rio Tinto Zinc (RTZ), with 30% ownership; a Japanese consortium lead by Mitsubishi, owning 10%; and the International Financial Corporation (IFC) of the World Bank, which owns the remaining 2.5%.

The mine is in the Second Region of Chile, 160 km southeast of the city of Antofagasta, at an elevation of 3,100 m above sea level. The south latitude is 24°, 15', 30"; the west longitude is 69°, 4', 15" (Figure 17.1).

At this site, the company owns a porphyry copper deposit mined as an open pit, a concentrating plant with capacity for processing some 127,000 tons per day, an ES-OX plant for processing oxide minerals up to 125,000 tons per year of cathodes, and two camp sites for lodging personnel from Escondida and contractors.

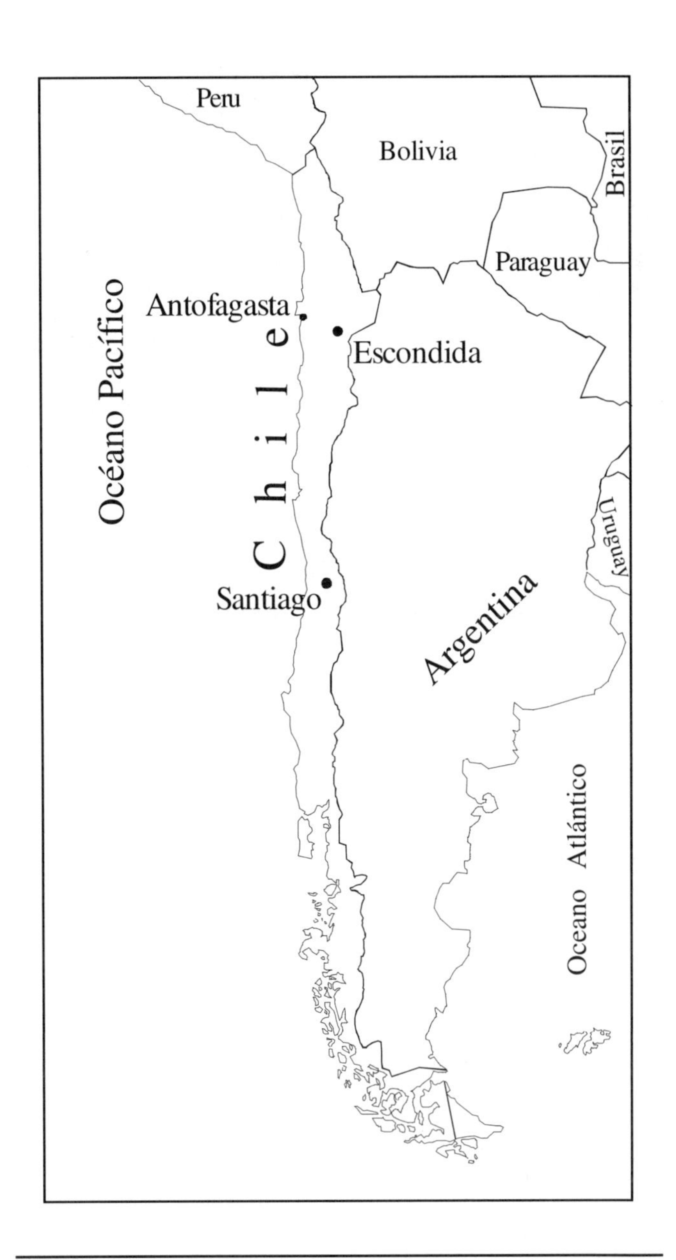

FIGURE 17.1 Escondida Mine location

17.2 SLOPE DESIGNS AT ESCONDIDA

17.2.1 Background

From the mine's start, it was recognized that the rock masses at Escondida had a rather medium-to-low competence, that the structural condition of the orebody was quite complex, and that there was a natural water level approximately 70 m below the ground surface. The mine was designed with a single 40° inter-ramp angle for the entire pit, with a mining scheme based on 15-m-high single benches.

Beginning in 1992, wedge failures were identified on the north wall. A report of the failures that occurred in the North-2 Pushback in December 1992 showed a vertical displacement of at least 2 m. The movement was sudden and initially very fast. This instability affected a stack of 11 benches (165 m) in andesite.

At that time, the failure mechanism was considered to be a function of (1) the medium-to-low competence of andesite and (2) the structural control of wedges and blocks.

17.2.2 Failure Mechanism at Escondida

The typical failure mechanism at Escondida involves strong structural control with formation of non-daylighting wedges. Within the structural limits defining the wedges, it is possible to find rock masses of medium-to-low competence that, together with the presence of water in the slopes, are capable of generating instabilities. Since these wedges usually do not daylight in the slope, important settlements and/or vertical displacements can be observed in the upper portion of the wedges, together with breakage of the rock mass at the toe of the slopes. A typical wedge formation can be observed in Figures 17.2a and 17.2b.

* Minera Escondida Limitada, Antofagasta, Chile.
† Itasca S.A., Santiago, Chile.

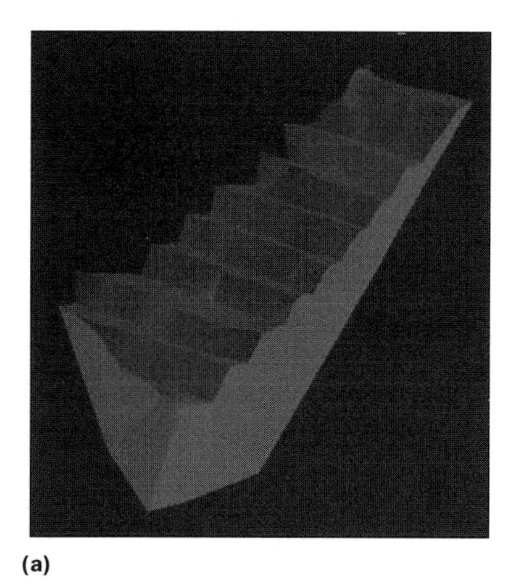

(a)

(b)

FIGURE 17.2 Wedge failure in the north wall: (a) as interpreted by triangulation of structures in the VULCAN program; (b) showing vertical displacement of 2 to 3 m in the upper part

Usually, the north wall and northeast corner of the pit show repeated instabilities from one pushback to the next (Figure 17.3a), very often resulting in the formation of "multi-wedges" (i.e., wedges inside wedges). Important slope angle changes were made to the pit during 1995, when the interramp angle was reduced from 40° to 32° in the northeast corner. The current interramp angle designs can be observed in Figure 17.3b.

17.2.3 Water Conditions in the Slopes

Using the data from wells drilled during the exploration stage, it has been possible to determine that before the mining operation a rather flat groundwater level was located approximately 70 to 80 m below the surface.

From the start of treatment plant operations in 1991, the tailings have been deposited at the Salar de Hamburgo, located on the east side of the pit. Observation wells located between the mine and the tailings have shown an increase in the phreatic levels on the order of 26 m after nine years of operation at this tailings deposit area.

Infiltration of water toward the pit through northwest-striking fault systems has caused an increase in the presence of water in the slopes, as well as an increase in the pore pressures. The latter situation has contributed negatively to the problems of slope stability in the east and north walls, as well as in the northeast corner of the pit. Figures 17.4a (plan view) and 17.4b (cross section) show a conceptual hydrogeological model of the pit.

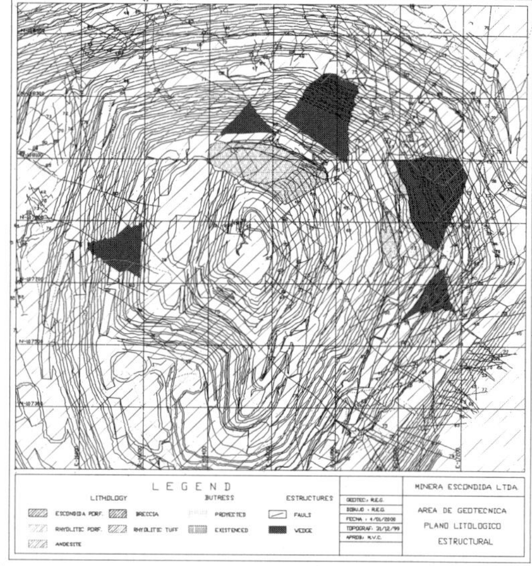

(a)

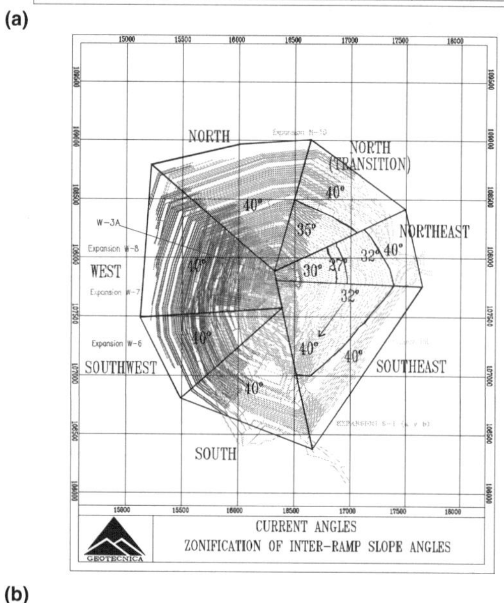

(b)

FIGURE 17.3 Location of wedge failures within the last two years: (a) current interramp design angles in various mine sectors

17.2.4 Handling Instability Problems

In order to solve the instability problems, a series of remedial actions have been implemented, including

- Changes in the interramp angles
- Inclusion of "stepouts" in the slope designs
- Use of buttresses to stabilize slopes (In July 1998, Dr. Evert Hoek visited Escondida and recommended using backfill as a stabilization method at the toe of the wedge in the northeast corner.)
- Improvements in the operational handling of water (construction of ponds and channeling of water away from the pushback advance)
- Horizontal drains as a means to efficiently depressurize the most critical slopes (this turns out to be the most effective short-term action)
- Vertical drains to intercept the recharge of water infiltrating from the tailings at the Salar de Hamburgo

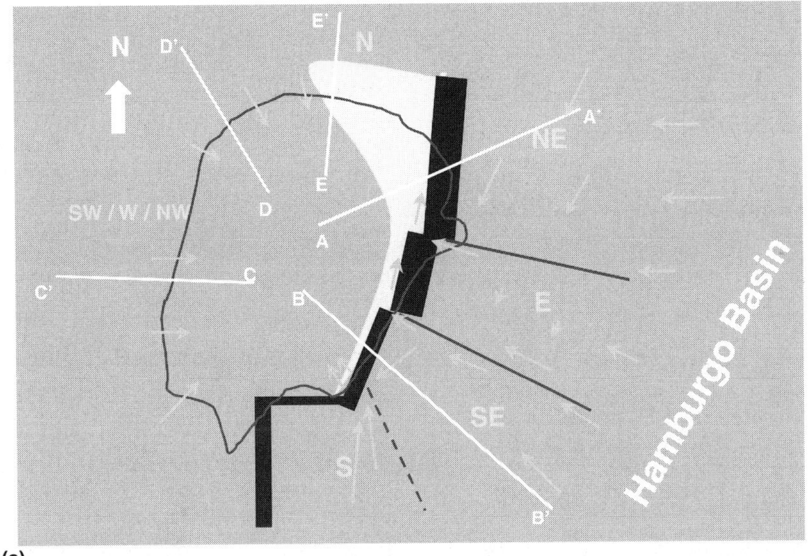

(a)

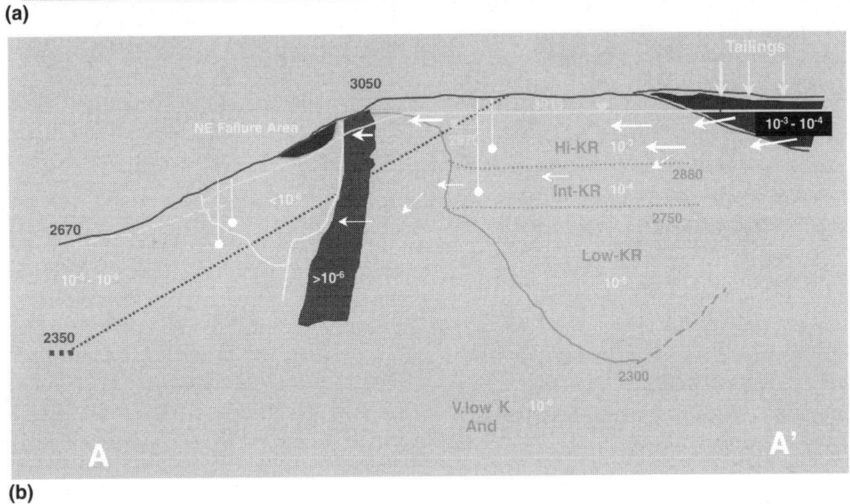

(b)

FIGURE 17.4 Conceptual hydrological model: (a) plan view; (b) cross-sectional view (permeabilities = cm/sec)

- Recommendations to construct a drainage gallery
- Changes in blasting design to minimize the damage to slopes due to vibration and gas pressures (use of pre-split with lower diameters [140 mm] than those used in production blasting [270 mm], use of trim and production blasting)
- Regrading of the slopes

17.2.5 Rock-Mass Characterization at Escondida

The geotechnical characterization of the rock masses at Escondida has been carried out using Bieniawski's (1989) rock-mass rating (RMR_{89}). In general, the RMR interpretations have been derived both from surface mapping and geotechnical drill holes. Important geotechnical drilling campaigns have been performed in the last four years to obtain a geotechnical characterization at depth.

Samples from the geotechnical drill holes have been tested in rock-mechanics laboratories to characterize the rock matrix. The main laboratory tests for intact rock characterization have been the unconfined compression strength (UCS) and triaxial tests to determine the m_i parameter, required when using the Hoek–Brown failure criterion.

From the drill hole data, it has been possible to define the strength parameters for each sector of the pit of Escondida Mine. Table 17.1 summarizes typical values for the main lithological units in the north sector of the pit.

17.3 METHODS OF SLOPE-STABILITY ANALYSIS

Three types of slope-stability analysis are routinely performed at Escondida. Preliminary analysis is usually made using limit-equilibrium methods with the XSTABL program (Sharma 1999). More detailed analyses are often made using the continuum finite-difference program, FLAC (Fast Langrangian Analysis of Continua) (Itasca 1998). Three-dimensional analyses that include explicit representation of faults are performed using the distinct element program, 3DEC (Itasca 1999). Attempts are made to learn as much as possible from simpler models. Models that are more complex are used to explore the neglected aspects of geology (e.g., faults) that are most likely to affect the behavior of simple models.

17.3.1 Limit-Equilibrium Analysis

Two-dimensional analyses using limit-equilibrium methods through the XSTABL program are performed at Escondida to provide rapid preliminary slope designs. Together with review consultant Peter Stacey (Golder Associates), attempts have been made to improve the accuracy of the limit-equilibrium analyses to the point where the results are as accurate as possible. Because the slopes at Escondida are relatively inhomogeneous, analyses

TABLE 17.1 Typical values for the main lithological units in the north sector of the pit

Sector	Lithologic Unit	Type of Alteration*	RMR_{89}	m_i	UCS (MPa)
North	Escondida Porphyry	A3/A4	48	12.2	37.1
	Andesite	A4	50	14.7	42.0
	Riolitic Porphyry	A4	50	16.0	55.0

*Type of alteration given by ISRM.

usually start with the Janbu noncircular method. The objective of the Janbu analysis is to provide candidate critical-failure surfaces for further analysis using the more rigorous Spencer method. Adjustments (e.g., addition of tension cracks) for indicated negative stress slices in the Spencer analysis are made as required, and the initial Janbu analysis is rerun. The refined Janbu analysis is again checked using the Spencer method. Each of the 10 indicated critical surfaces reported for the Janbu method are analyzed using the Spencer method, and the surface with the lowest factor of safety is reported.

Where structure dipping is toward the pit but the dipping does not undercut the slope, anisotropic strengths are applied with a range of ±5° around the mean dip of the fracture set. Although more than one fracture set may be included in the analyses, this is typically not done.

The following guidelines have evolved for Janbu analyses of global slope stability:

- Initiation points should be restricted to the toe of the slope and any lithological/alteration contacts in the lower 50% of the slope.
- There should be a minimum of 5,000 surfaces for the Janbu noncircular analyses from each initiation point.
- The minimum elevation of the failure surfaces should be no more than 50 m below the toe of the slope.
- The default segment length (i.e., one-third the slope height) in the program should be used.
- Initial upper and lower angles for the toe failure surfaces should be +20°/25° and −30°, respectively, although the upper angle is reduced if potential failure surfaces exit the slope before the termination point range.

Figure 17.5a shows the results of a typical XSTABL analysis. The particular analysis indicates instability (factor of safety less than 1) for a section through the northeast corner of the pit for the case in which no engineered drainage is provided. The results suggest the need to provide engineered drainage.

17.3.2 Two-Dimensional Stress Analysis

FLAC is probably the most widely used, general-purpose geomechanical numerical model in the world; it has been used routinely at Escondida since 1996. FLAC simulates the behavior of structures (such as slopes) built of soil, rock, or other materials that undergo plastic flow when their yield limits are reached. Each element in the two-dimensional grid behaves according to prescribed nonlinear stress–strain law in response to applied forces or boundary restraints. The material can yield and flow, and the grid can deform and move with the material it is representing. The explicit, Lagrangian calculation scheme and mixed-discretization zoning ensure that plastic flow and collapse are modeled very accurately. In addition to modeling rock slopes at Escondida, FLAC has also been used to study the static and dynamic stability of waste dumps, tailings-retention dikes, and leach piles.

One important aspect of slope-stability analyses is the determination of the safety factor. Limit-equilibrium methods can make thousands of safety factor calculations almost instantaneously. Numerical methods require longer times to make just one safety factor calculation. However, recent advances in the computational speed of personal computers now permit safety factor calculations with numerical models to be made routinely. Numerical models have an advantage over limit-equilibrium methods in that they do not presuppose a failure surface; rather, the failure surface corresponding to the minimum factor of safety develops naturally as a function of the combination of material properties and geometry. At Escondida, the shear-strength reduction technique has been used with FLAC to calculate safety factors for the last two years. The technique is described below.

For slopes, the factor of safety, F, is often defined as the ratio of the actual shear strength to the minimum shear strength required to prevent failure. Therefore, a logical way to compute the factor of safety with a finite element or finite difference program is to reduce the shear strength until collapse occurs. The factor of safety is the ratio of the soil's or rock's actual strength to the reduced shear strength at failure. This shear-strength reduction technique was first used with finite elements by Zienkiewicz, Humpheson, and Lewis (1975) to compute the safety factor of a slope composed of multiple materials.

To perform slope-stability analysis with the shear-strength reduction technique, simulations are run for a series of increasing trial factors of safety, f. Actual shear-strength properties (cohesion, c, and friction, ϕ) are reduced for each trial according to the equations

$$c^{trial} = (1/f)c \qquad \text{EQ. 17.1}$$

$$\phi^{trial} = \arctan\{(1/f)\tan\phi\} \qquad \text{EQ. 17.2}$$

If multiple materials and/or joints are present, the reduction is made simultaneously for all materials. The trial factor of safety is gradually increased until the slope fails. At failure, the safety factor equals the trial safety factor (i.e., f = F). Dawson, Roth, and Drescher (1999) show that the shear-strength reduction factors of safety are generally within a few percent of limit-equilibrium solutions when an associated flow rule is used for problems involving homogeneous materials.

The shear-strength reduction technique has two advantages over slope-stability analyses with limit equilibrium. First, the critical failure surface is found automatically, and it is not necessary to specify the shape of the failure surface in advance. In general, the failure mode for slopes is more complex than simple circles or segmented surfaces. Second, numerical methods automatically satisfy translational and rotational equilibrium. Not all limit-equilibrium methods satisfy translational and rotational equilibrium. Consequently, the shear-strength reduction technique will usually determine a lower safety factor compared to other methods, particularly if the slopes are composed of material that is neither homogeneous nor isotropic. For example, Zienkiewicz, Humpheson, and Lewis (1975) show comparisons between factors of safety calculated by limit-equilibrium and finite element methods for a slope composed of three materials. They report a factor of safety of 1.33 using Bishop's method and 1.17 using the shear-strength reduction technique. Donald and Giam (1988) report results of factor-of-safety analyses for a homogeneous slope with a weak layer in the foundation. They report a factor of safety of 1.5 using Bishop's method with a simplex optimization and 1.34 using the shear-strength reduction technique. For Escondida, we find the following approximate relation between XSTABL (Spencer) results and FLAC results for nonhomogeneous slopes:

$$FS_{FLAC} = 0.87FS_{XSTABL} + 0.08 \qquad \text{EQ. 17.3}$$

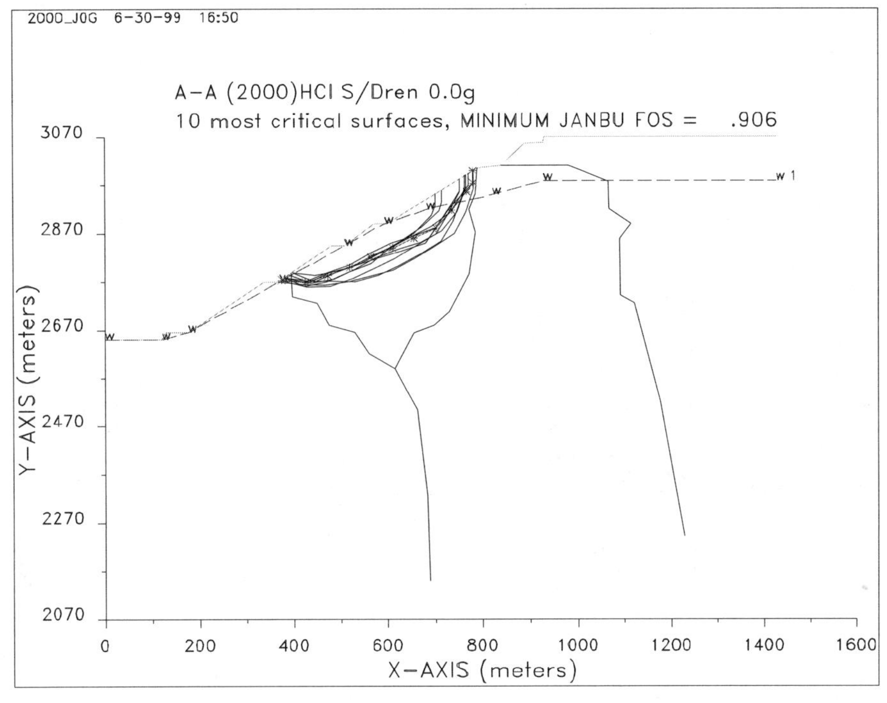

(a)

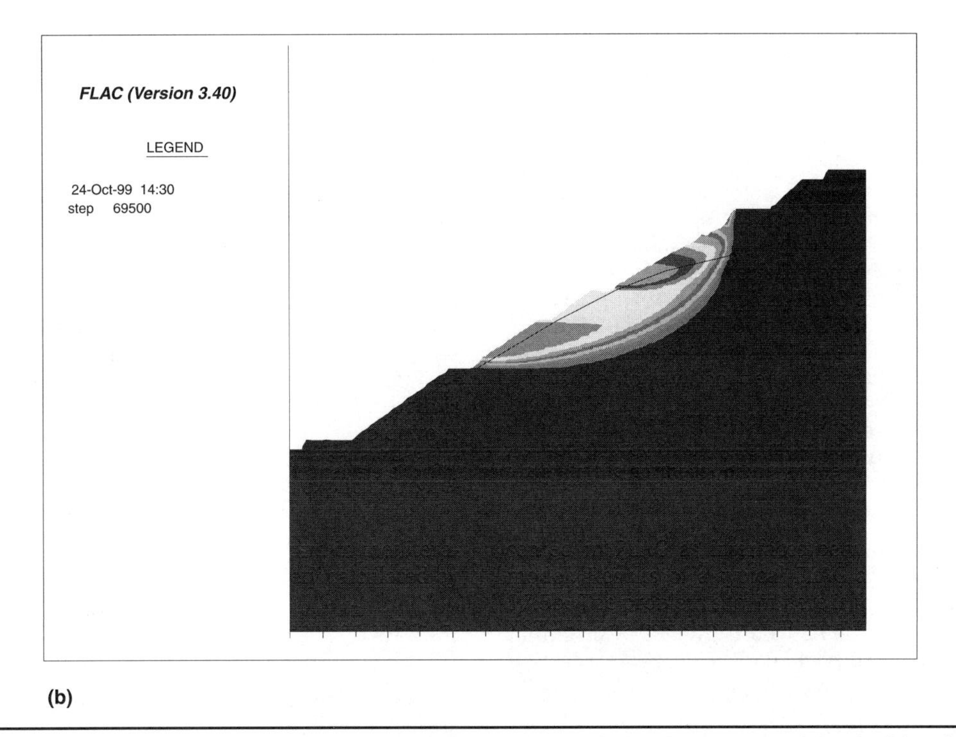

(b)

FIGURE 17.5 Results of analysis: (a) typical XSTABL analysis showing instability in the northeast corner of the pit; (b) typical FLAC analysis showing instability in the northeast corner of the pit

This relation agrees well with the values found in the literature. For homogeneous slopes, the factors of safety calculated by the two methods are not as different.

Two elastic-plastic material models have been used at Escondida. The most commonly used model assumes isotropic behavior with a Mohr–Coulomb yield criterion. Anisotropic strength is represented using a ubiquitous joint model. In the ubiquitous joint model, a joint direction and shear strength are specified, but the joint spacing and location are not. Faults are explicitly represented in FLAC as interfaces between different regions of the model. Results of a typical FLAC analysis are shown in Figure 17.5b. The problem shown here is for the same conditions as the XSTABL analysis shown in Figure 17.5a. The FLAC analysis also shows the slope without engineered drainage to be unstable.

17.3.3 Three-Dimensional Stress Analysis

As noted previously, the most common failure mode for larger instabilities at Escondida involves non-daylighting wedges (i.e.,

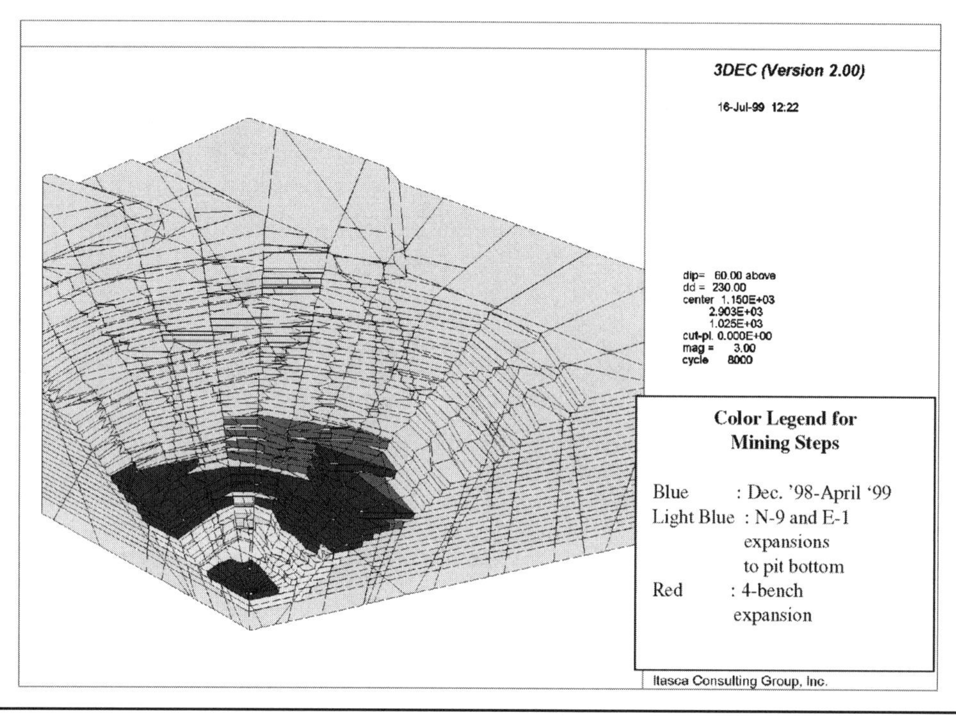

FIGURE 17.6 3DEC model of northeast corner of Escondida pit showing mining steps

wedges whose line of intersection dips more steeply than the interramp slope angle). In order for such wedges to fail, there must be sliding on major discontinuities forming the back of the wedge, as well as failure fail through the rock mass near the level of the toe of the failure. In reality, the toe failure mechanism may involve some form of step-path failure (i.e., failure along discontinuous joint surfaces and through intact rock). However, on the scale of problems involved, these failures are treated as rock-mass failures. The stress analysis program 3DEC (3-Dimensional Distinct Element Code) is used to compute the stability of such wedges. 3DEC treats the rock mass in much the same way as FLAC does, using an elastic-plastic stress-strain relation to describe the continuum behavior. In addition, faults are explicitly represented as planar failure surfaces that allow slip, rotation, and/or separation of adjacent deformable blocks of material. Pore pressures are specified for both the rock mass and faults, so that effective normal stresses are used to determine whether shear failure occurs. An example is given in the following section to illustrate the use of 3DEC in typical stability analyses at Escondida.

17.3.4 Illustrative Three-Dimensional Stability Analysis

The following sections describe the use of 3DEC to study stability problems at Escondida. As noted previously, stability problems at Escondida have three factors in common

- Strong structural control resulting in non-daylighting wedges
- A weak/altered rock mass
- Relatively high water pressures

3DEC is capable of considering all of these factors in one analysis.

Introduction. Stability problems occurred in the northeast corner of the mine during 1998 and 1999. The major stability problems were large unstable areas involving non-daylighting wedges in the northeast and north walls. The largest displacements within these unstable areas were of the order of 10 m. In December 1998, a recommendation was made to not mine below level 2845 in the area below the northeast wedge. However, this effectively restricted extraction of high-grade ore in the area. By mid 1999, various remedial measures (stepouts, buttresses, regrading, i.e., smoothing, vertical and horizontal drainage wells) had slowed prism movement in the northeast wedge above level 2845, and a mining plan involving excavation of four benches with an interramp angle of 27° was proposed below the northeast wedge between level 2845 and level 2785. There was concern, however, that instabilities in the north and northeast wedges could be reinitiated as a result of excavation near their toes. Expansion of the failure areas could adversely impact the overlying ramps. A three-dimensional (3DEC) model was created to evaluate future stability conditions, including the four-bench excavation. Before analysis of future conditions could be made, it was first necessary to calibrate the model. This section of the chapter describes the calibration studies and analysis of the proposed four-bench excavation.

Model Description. The model was 1,300 m in the east–west direction and nearly 2,000 m in the north–south direction. The total height of the model was nearly 500 m. Model construction was initiated by dividing the model region into radial sectors. Coordinates where each bench toe and crest are intersected by the radial lines were used to develop the three-dimensional pit geometries. The model shown in Figure 17.6 included geometries for December 1998, April 1999, completion of expansions E-1 and N-9 to the bottom of the pit, and the four-bench excavation between level 2845 and level 2785.

Lithology. The lithology included the following four lithologic units:

- Porfido Escondida (Escondida Porphyry)
- Porfido Escondida Silicificado (Escondida Porphyry With Siliceous Alteration)
- Porfido Riolitico (Rhyolite Porphyry)
- Andesita (Andesite)

The Porfido Escondida was further subdivided into two geomechanical units based primarily on alteration and appearance in the pit. One unit was termed Porfido Escondida Alterado and was the most highly altered and weakest material. It was assumed to be located mainly within the northeast wedge and

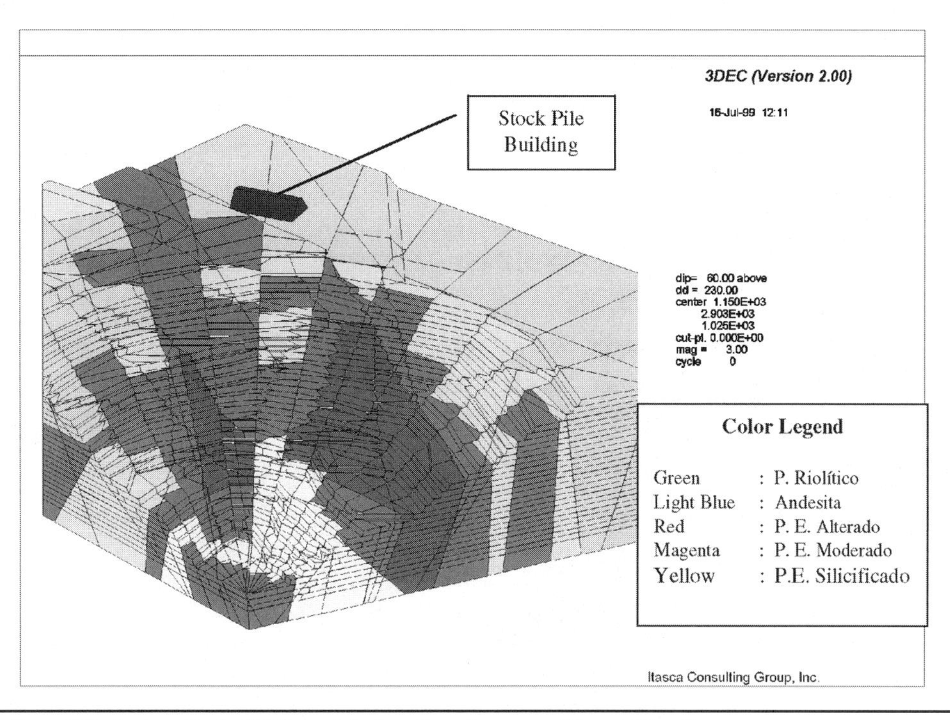

FIGURE 17.7 Distribution of lithologic units in 3DEC model of northeast corner of Escondida pit

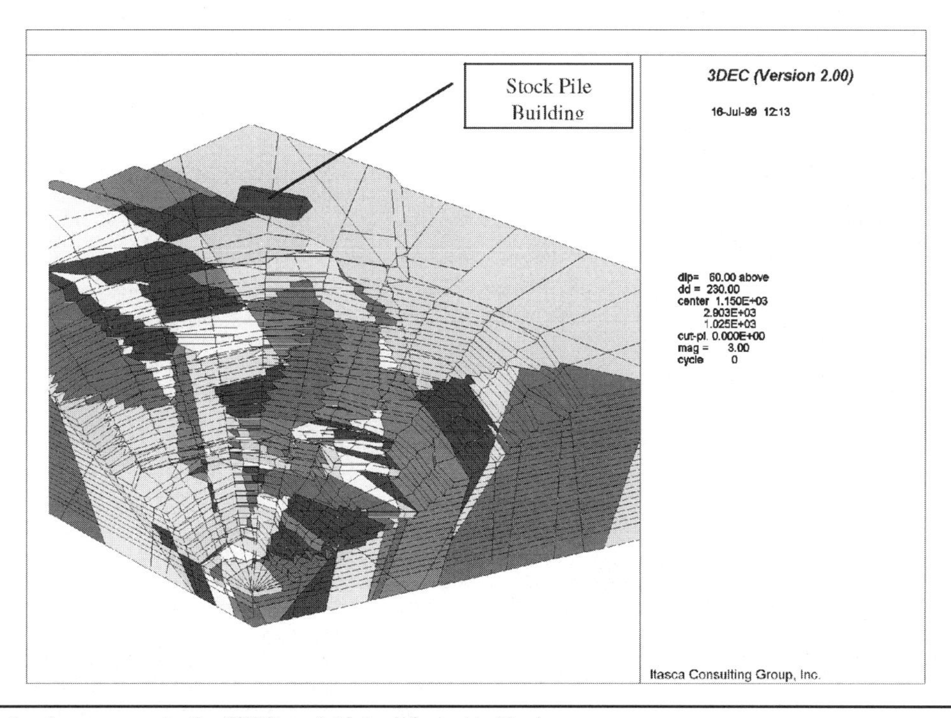

FIGURE 17.8 Major structures separate the 3DEC model into deformable blocks

the north wedge. The remaining Porfido Escondida that was neither Porfido Escondida Alterado nor Porfido Escondida Silicificado was termed Porfido Escondida Moderado. Figure 17.7 shows the distribution of materials in the 3DEC model. Of particular interest is the location of the Porfido Escondida Silicificado. This is the strongest of the Porfido Escondida units. The block model shows that the Porfido Escondida Silicificado reaches its highest elevation in the pit between the two wedge failures. This suggests that the Porfido Escondida Silicificado between the two wedges may have inhibited movement between the two wedges. Uncertainty regarding the location of the Porfido Escondida Silicificado at the base of the northeast wedge was addressed by performing additional studies.

Major Structures. The main structural features included in the model were obtained from surface mapping. Figure 17.8 shows how the structures divide the model into blocks. It should be noted that, due to the limited number of colors available in 3DEC (six), some adjacent blocks have the same color, although they are structurally separated in the model. All structures are assumed to be planar. Most structures extend from the initial pit

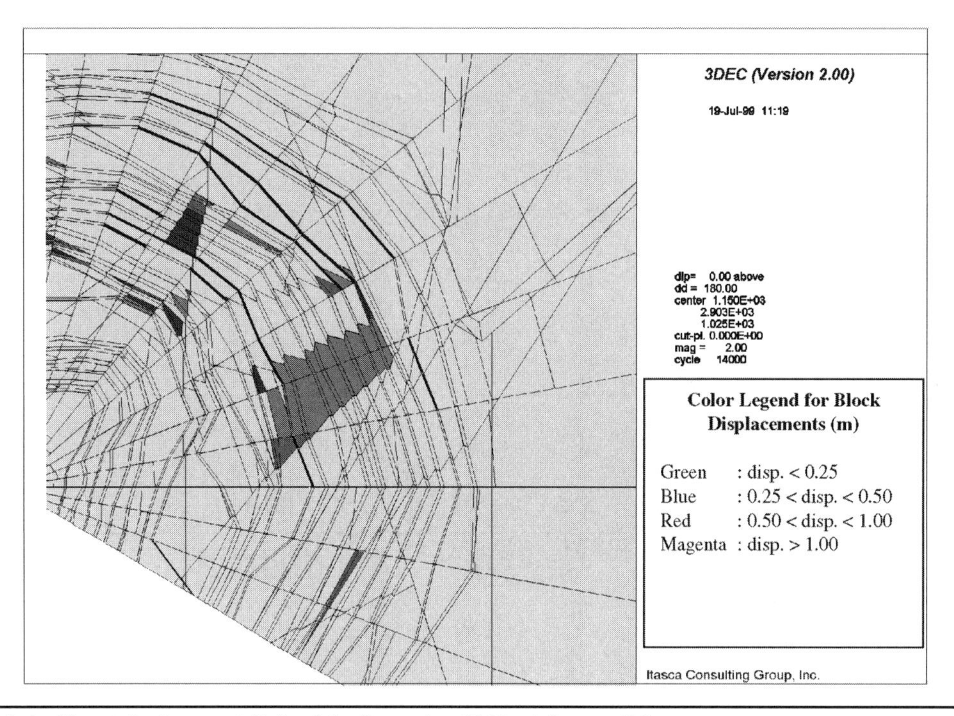

FIGURE 17.9 Areas of significant displacement (failure) for December 1998 mining conditions

surface (i.e., the December 1998 pit) to the bottom of the model—unless terminated by another structure. All structures were assumed to be cohesionless with an 18° friction angle. This friction value represents the residual strength for fault gouge taken from the E-1 east wedge below the main ramp, as measured by laboratory testing.

Assumed Phreatic Levels. Water pressures are calculated in 3DEC by computing the vertical distance between any point and the phreatic surface. Points above the phreatic surface have zero water pressure. Where piezometric data were not available (including future excavation stages), the phreatic surface was estimated based on nearby data, preserving the general trends in other profiles. However, considerable uncertainty remained with regard to both current and future phreatic levels. Where uncertainties existed, aggressive levels (i.e., essentially connecting bench toes) were assumed.

Modeling Sequence and Calibration Criteria. The calibration modeling was performed in two stages. In the first stage, the model was brought to elastic equilibrium using the December 1998 pit geometry. In the second stage, the rock-mass and discontinuity properties are reduced to their actual (i.e., realistic) values and displacements are recorded. This modeling sequence did not reproduce any of the actual mining sequence prior to December 1998.

The model allowed unstable areas to be identified, as the unstable areas exhibit significant displacement relative to stable areas. Unstable areas are also characterized by their velocity distributions and failure (i.e., plasticity) indicators.

Two main criteria were used to calibrate the model. First, the model needed to reproduce the wedge failures in both the northeast and north that existed for the December 1998 condition. The failure in the northeast was characterized by large (5-m to 10-m) displacements that continued until a buttress was placed between level 2830 and level 2860 in March 1999. Second, the model should reproduce the general prism movements in the area, including lack of prism movement below about level 2815 to level 2830 near the northeast wedge.

TABLE 17.2 Adopted rock-mass properties for lithologic units in 3DEC model

	Bulk (GPa)	Shear (GPa)	Friction (°)	Cohesion (kPa)
Porfido Riolitico	5.0	4.0	37	840
Andesita	3.0	2.5	31	150
Porfido Escondida Alterado	1.5	1.0	25	40
Porfido Escondida Moderado	1.5	1.0	28	150
Porfido Escondida Silicificado	1.5	1.0	36	780

Initial Calibration Results. One of the main uncertainties in the model setup was the rock-mass strength properties. Accordingly, these properties were adjusted by trial to satisfy the calibration criteria. It should also be noted that other combinations of rock-mass and discontinuity strength properties could satisfy the calibration criteria. Therefore, the calibrated properties are not unique. The rock-mass properties adopted from the calibration study are shown in Table 17.2. Figure 17.9 shows the areas of significant displacement for the December 1998 condition when the properties in Table 17.2 are used. Figures 17.10a and 17.10b show the displacements in cross-section plots through the north and northeast wedges, respectively.

Analysis of Proposed Mining of Four Benches on the Northeast Wall. Analyses were performed to evaluate the effect of excavating the proposed benches between level 2845 and level 2785. Previous analysis demonstrated that excavation of the N-9 and E-1 expansions as originally planned would result in significant expansion of the north and northeast wedges. Therefore, the design was modified by jumping from level 2845 to level 2785 in the E-1 expansion. This jump was made possible because of a wide access available at the 2785 level. The N-9 and E-1 expansions were continued together to the bottom of the pit. Excavation of four benches between level 2845 and level 2785 was proposed to occur after completion of the N-9 and E-1 expansions. Figure 17.6 shows the location of the four benches to

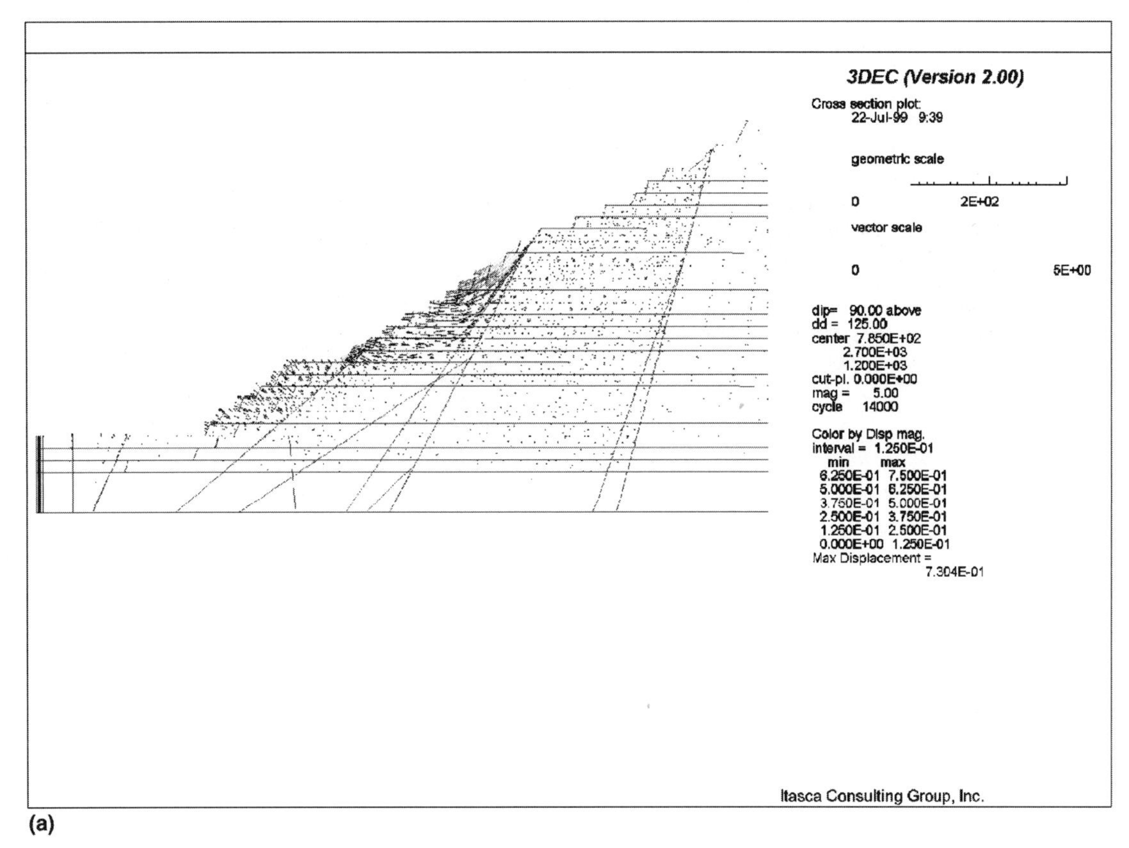

(a)

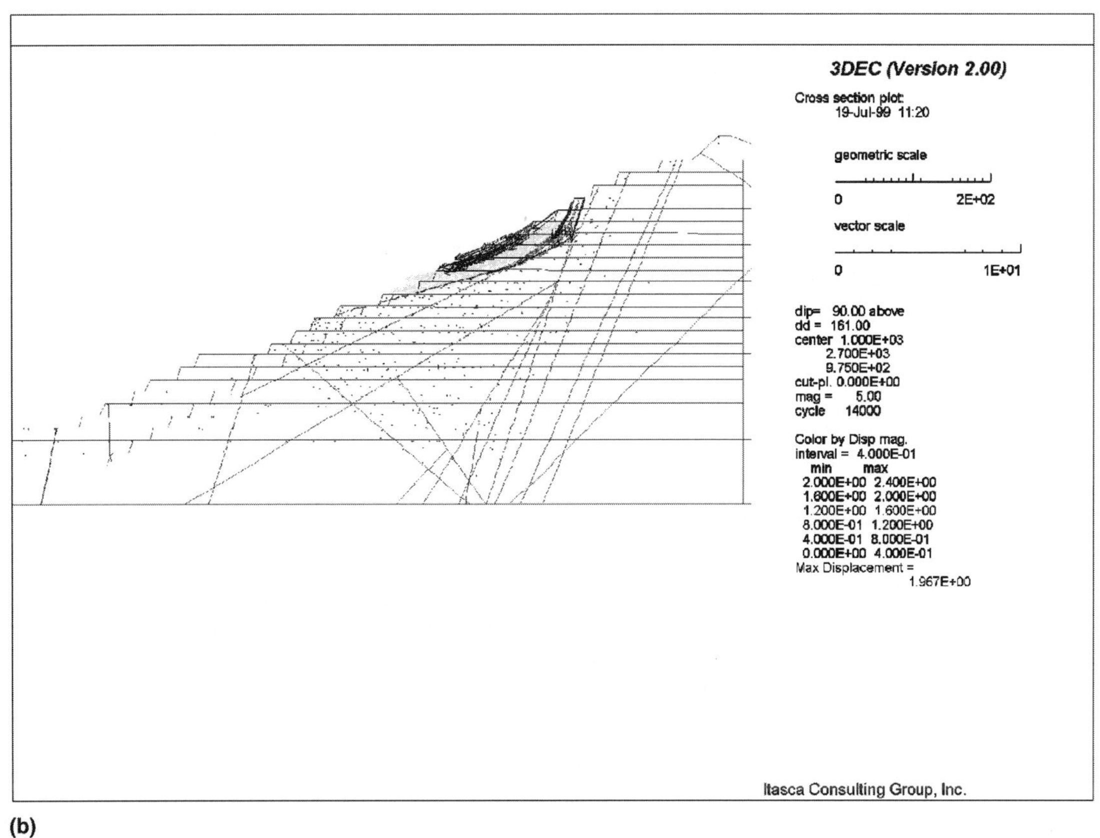

(b)

FIGURE 17.10 Vertical cross-section plots of 3DEC results: (a) near north wall failure; (b) near northeast failure

be excavated. The slope design consisted of benches at a 27° interramp angle with a wide bench at the 2785-m level.

17.4 CONCLUSIONS

The primary objective of the analyses presented here was to evaluate whether the four benches between level 2845 and level 2875 could be mined below the northeast wedge without significantly expanding the previous unstable area. Of particular concern was potential for further excavation in the area to undercut the ramp that crosses the upper section of the wall and that will eventually carry the conveyor from the crusher on the north wall. The analyses suggested that the benches could be mined without significantly expanding the previous unstable area. The four benches were successfully mined in 1999 using mobile equipment without any significant stability problems.

17.5 REFERENCES

Bieniawski, Z.T. 1989. *Engineering Rock Mass Classifications.* New York:Wiley.

Dawson, M.D., W.H. Roth, and A. Drescher. 1999. Slope stability analysis by strength reduction. Submitted for publication to *Géotechnique*.

Donald, I.B., and S.K. Giam. 1988. Application of the nodal displacement method to slope stability analysis. *Preprint Proceedings of the Fifth Australia-New Zealand Conference on Geomechanics–Prediction versus Performance (Sydney, 1988).* 456–460.

Itasca Consulting Group, Inc. 1998. *FLAC (Fast Lagrangian Analysis of Continua),* Version 3.4. Minneapolis: ICG.

Itasca Consulting Group, Inc. 1999. *3DEC (Three-Dimensional Distinct Element Code),* Version 2.0. Minneapolis: ICG.

Sharma, S. 1999. *XSTABL, An Integrated Slope Stability Analysis Program for Personal Computers,* Version 5.2. Moscow, Idaho: Interactive Software Designs, Inc.

Zienkiewicz, O.C., C. Humpheson, and R.W. Lewis. 1975. Associated and non-associated visco-plasticity and plasticity in soil mechanics. *Géotechnique,* 25:4:671–689.

CHAPTER 18

Slope Stability at Collahuasi

Graham Swan* and Ricardo S. Sepulveda†

18.1 INTRODUCTION

The Collahuasi district is located in the First Region of Chile, 180 km southeast of Iquique and 15 km west of the Bolivian border. The region, known as the Chilean Altiplano, is a high plain characterized by broad valleys bounded on the east by a chain of volcanoes, forming the crest of the Andean Cordillera. Exploration and exploitation of the Collahuasi copper deposits began with the Incas; commercial exploitation occurred by the 1880s and continued until 1930. Exploration of the current reserves began in 1977. While there are several mineral deposits in the area, the Ujina Pit was developed first, producing 60,000 tons per day of sulfide ore beginning in 1999 after stripping 166 million tonnes of waste. This chapter considers the evolution of slope design principles as developed for the west wall of the Ujina Pit.

18.1.1 History

The initial design work for the Ujina Pit was contracted to SRK in 1992; they used the MRMR empirical classification method of Laubscher (1990) modified for pit slopes by Haines and Terbrugge (1991). At the time, only unoriented drill core was available; therefore, the results from this method were seen as very preliminary. Subsequently, following several independent reviews and audits, Compania Minera Dona Ines de Collahuasi S.A. (CMDIC) proceeded with the formation of a Collahuasi Geotechnical Group to be responsible for the definitive short- and long-term design basis for the Ujina Pit. Local and international consultants, as required, would assist this group. Finally, a Geotechnical Review Board was established to meet twice a year to both audit and advise on slope design and stability issues.

18.2 GEOLOGY AND HYDROGEOLOGY

18.2.1 Mineralization

Ujina is a classic porphyry copper deposit, overlain by a 70- to 80-m-thick volcanic flow referred to locally as ignimbrite (Figure 18.1). Leaching and supergene enrichment processes have led to the development of a significant blanket of high-grade copper sulfide mineralization. Chalcocite, the predominant copper mineral in the blanket, has an average thickness of 60 m. Oxide ores occur within and at the base of the leached capping, with an average thickness of 5 to 10 m, locally up to 50 m. Chalcopyrite is the principal copper mineral contained in the primary ore, which underlies the secondary enrichment blanket. The cross section shown in Figure 18.1 is a typical mineral cross section at Ujina. The total minable ore reserves‡ at Ujina are 1.27 billion tonnes grading at 0.78% copper.

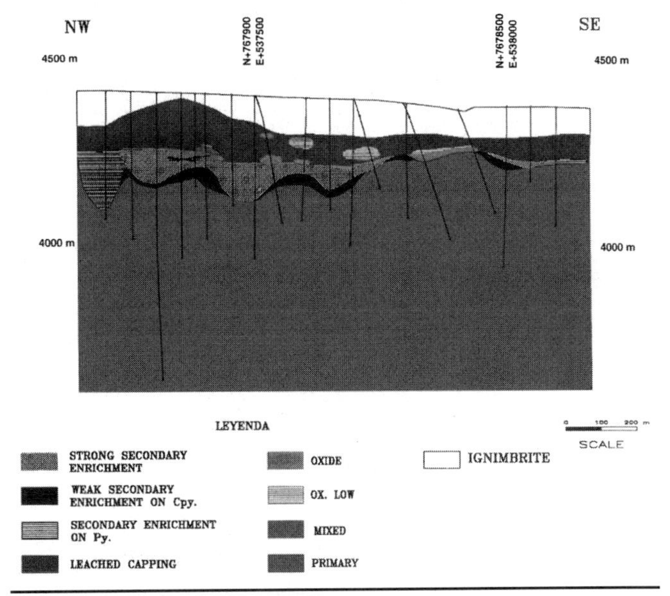

FIGURE 18.1 Cross section showing the mineralization of the Ujina deposit

18.2.2 Geological Structures

Characterization of structures at Ujina has proceeded using a variety of observational resources, including satellite images, an underground exploration shaft, oriented borehole camera images, and detailed face mapping of exposures in the starter pit. The ignimbrite cover, being of recent volcanic origin, features thermal cooling fractures, which conform to tectonically derived structures only coincidentally, if at all. For this reason, most of the underlying structure in the pit has been directly observed only after first stripping the ignimbrite. A summary of the faults and joints observed in the various west wall lithologies and identified by genetic parentage is given in Figure 18.2.

18.2.3 Hydrogeology

Work to relate the response of groundwater flow to mining activity began by characterizing the porosity and permeability of the rock sequence, beginning with the overlying alluvial material, ignimbrite, paleogravels, and weathered rocks (SRK 1994). Because of the naturally high permeability of the rocks and/or intensity of discontinuities within this sequence, it was considered that the surface rock mass would likely drain naturally on exposure. However, the question of how deeply the phreatic surface would be moved into the pit wall and bottom by this process was the subject of a theoretical study using the MODFLOW code (Montgomery 1997). This model attempted to include the effects of pervasive fault zones, such as the SF1 set observed in the west wall of Ujina (Figure 18.2). Depending on the assumed hydraulic

* Falconbridge Limited, Canada.
† Collahuasi Mining Company, Chile.
‡ Defined as the proven and probable ore reserves that are contained within the ultimate pit limits at a copper price of $0.90 per pound.

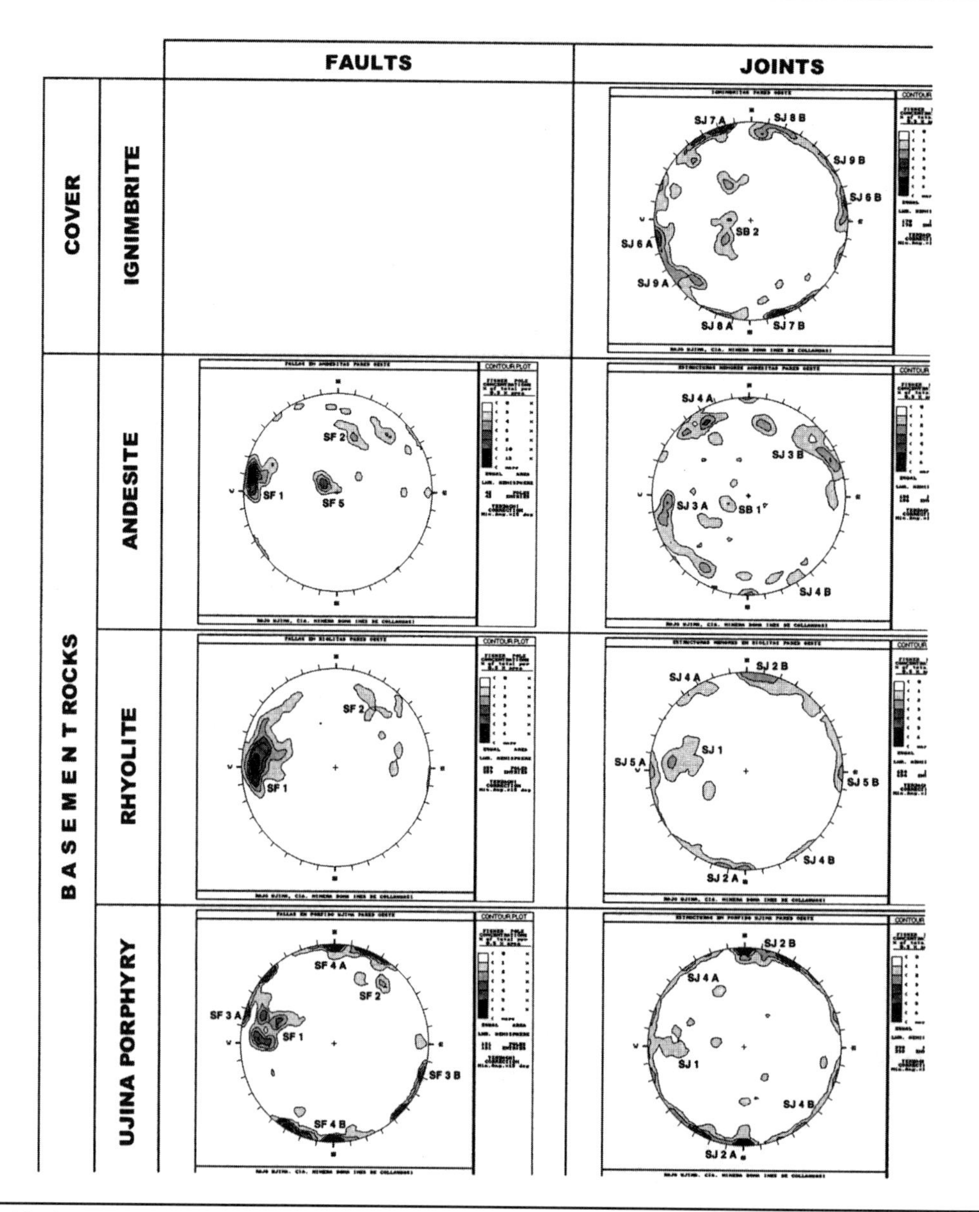

FIGURE 18.2 Mapped faults and joint sets in the west wall of the Ujina Pit

conductivity of the faults (0.2–2 m/d, compared to 0.024 m/d for the rock mass), this work concluded, in summary, that

- Flow rates into the pit decrease from 63 to 172 lps, year 1 to 61 to 122 lps at end of year 8.
- Affected areas will be mainly in the pit bottom and, to a lesser extent, in the lower slopes.
- In the interest of drainage efficiency, horizontal holes drilled into the pit wall are preferred over vertical wells with pumps drilled outside the pit perimeter.
- The design of a detailed and effective drainage program will require a better description of the hydraulic characteristics of the jointed rock mass (Figure 18.3) and the fault zones.

FIGURE 18.3 Appearance of SJ1 joints in the west wall starter pit

TABLE 18.1 Rock-mass properties of the principal rocks, Ujina Pit

Geotechnical Name	Intact Rock Properties		GSI	Rock-Mass Properties				
	m_i	σ_{ci} (Mpa)		Property	Mean	Std. Dev.	Max.	Min.
Primary Inca Porphyry	12	80–110	52–62	m_b	2.60	0.30	3.39	1.97
				s	0.008854	0.002812	0.01466	0.004829
				a	0.500	–	0.500	0.500
				φ (deg.)	35	1	37	33
				c (kPa)	3674	398	4774	2766
				E (GPa)	14.7	2.5	19.9	10.1
Primary Ujina Porphyry	12	70–90	48–58	m_b	2.25	0.26	2.92	1.69
				s	0.005677	0.001803	0.009401	0.003096
				a	0.500	–	0.500	0.500
				φ (deg.)	34	1	36	32
				c (kPa)	2823	261	3555	2209
				E (GPa)	10.8	1.8	14.9	7.5
Primary Andesite	18	60–80	42–52	m_b	2.73	0.32	3.52	2.08
				s	0.002915	0.000926	0.004827	0.001590
				a	0.500	–	0.500	0.500
				φ (deg.)	36	1	34	38
				c (kPa)	2406	222	3046	1880
				E (GPa)	7.1	1.2	10.0	5.0
Ignimbrite	16	75–100	42–52	m_b	2.42	0.29	3.16	1.83
				s	0.002915	0.000926	0.004826	0.001589
				a	0.500	–	0.500	0.500
				φ (deg.)	35	1	37	33
				c (kPa)	2927	276	3691	2289
				E (GPa)	8.0	1.4	11.1	5.5
Argillized Rhyolite	14	25–40	20–30	m_b	.97	0.11	1.26	0.73
				s	0.000161	0.000165	0.000419	0.000
				a	0.519	0.019	0.550	0.500
				φ (deg.)	27	1	30	25
				c (kPa)	661	131	982	399
				E (GPa)	1.4	0.2	2.0	0.9
Fault Zones	16	1–15	22–32	m_b	0.54	0.06	0.70	0.41
				s	0.000	0.000	0.000	0.000
				a	0.585	0.014	0.610	0.560
				φ (deg.)	21	1	24	18
				c (kPa)	116	34	210	46
				E (GPa)	0.4	0.1	0.6	0.2

Subsequent efforts to validate this hydrological model have resulted in limited monitoring of piezometric heads and in drilling horizontal drain holes as part of a forced dewatering test program. Seven holes with a total length of 1,630 m were drilled into the west wall starter pit and were observed to cumulatively produce approximately 5.5 lps over eight days of testing. Although more work is needed, it was agreed that the MODFLOW water-level predictions would form the basis for pore pressure levels in the slope-stability analyses.

18.3 ROCK-MASS AND STRUCTURAL PROPERTIES

A detailed and comprehensive approach was taken to arrive at the rock-mass and structural properties for the Ujina Pit, involving triaxial and shear testing of core, MRMR empiricism, field observations and assessments, literature case studies, and expert opinions. The work began by dividing the west wall into eight discrete geotechnical sectors and mapping four, independent geologically based factors: lithology, mineralization, alteration, and MRMR. The geological classification of rocks was reduced to a geotechnical classification, resulting in 11 discrete units under three groupings:

- Group 1, Lithology–Ignimbrite, andesite, rhyolite, Ujina porphyry, Inca porphyry, fault zones
- Group 2, Mineralization–Primary, secondary, tertiary rocks
- Group 3, Alteration–Argillized, non-argillized rocks

For the above geotechnical classifications, the properties of faults and joints were assigned. The cohesion was assumed to be zero; the friction angle for the faults ranged between 18° and 25° and those for the joints ranged between 40° and 65°. Of all 11 units, joints in argillized rocks were considered to be only marginally stronger than fault zones. Finally, the rock-mass strength of the various units was assigned using the most recent

TABLE 18.2 Summary results from back-analysis of two instabiliites in the south sector of Ujina Pit

Case	Slope	Material	Results from Anaylsis	
			ϕ	c (kPa)
Sector SE Failure in the South Ramp, in South Zone of Ujina Pit		Ignimbrite	35 (Δ = 0°)	2900 (Δ = 0 kPa)
		Secondary Ujina Porphyry	30 (Δ = 0°)	700 (Δ = 0 kPa)
		Argillized Ujina Porphyry (No Aparece en la Pared Oeste)	21	200
Sector SW Failure in the South Ramp, in South Zone of Ujina Pit		Argillized Adnesite	28 (Δ = 0°)	600 (Δ = 0 kPa)
		Argillized Rhyolite	27 (Δ = 0°)	600 (Δ = –60 kPa)
		Fault Zones	21 (Δ = 0°)	91 (Δ = –9 kPa)

Values in brackets refer to differences between the result from back-analysis and the assigned property, Table 18.1.

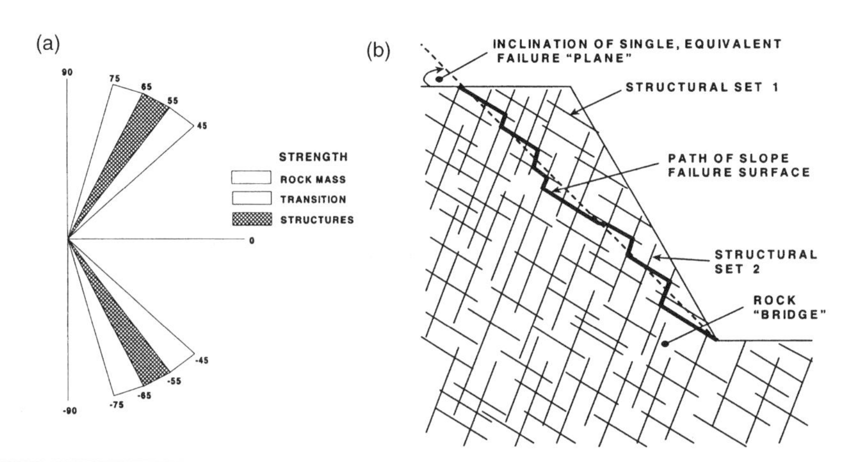

FIGURE 18.4 (a) A rock mass with anisotropy due to structure; (b) definition of failure surface in a tightly jointed rock mass

formulation of Hoek and Brown (1998) and recommendations from Hoek (1998). Table 18.1 provides a summary of rock-mass strength values for the most important properties and geotechnical units. For each property, the distribution of typical values has been characterized through estimates of the mean, standard deviation, minimum, and maximum values.

18.3.1 Failure Back-Analysis

In an attempt to verify some of the above rock-mass strength variables, a back-analysis was made of two slope instabilities that were observed in the Ujina starter pit local to the temporary south ramp. Spencer's method was used to calculate the failure condition. Table 18.2 shows the results, including differences between assumed and actual values for rock-mass cohesion and friction angle.

18.3.2 Major Structures and Joints

Where fault zones are known to occur in the rock mass, the method of assigning strength properties was modified using an anisotropic description. This is particularly appropriate for slope designs on the interramp and global scale. Using the code XSTABL, an anisotropic form of rock-mass strength is defined, as shown in Figure 18.4a. Here, a fault with a dip angle of 60° is considered to have an angular variation of ±5°, such that its properties are assigned over 10°. Next, a transitional strength is assigned over an angular sector of 10° before the continuum rock-mass strength is assigned (see Figure 18.4a). Alternatively, in a tightly jointed rock mass, a plane failure surface may be defined with properties due to some combination of rock bridges and rock joints derived using a code called STEPSIM.

18.4 SLOPE DESIGN

18.4.1 Acceptable Criteria

Prior to proceeding with the slope design, it was necessary to define those criteria that, from a management point of view, served to define and manage acceptable risk with respect to slope failure or instability. Such criteria are commonly expressed in terms of factor of safety, probability of failure (P_p), accumulated slope displacement (D), and slope displacement rate (V). A

TABLE 18.3 Acceptable failure criteria for the Ujina Pit slopes

		Characteristics of Instability		Acceptability Criterion		
Slope Type	Case	Loss of Ramp Berm (%)	Material Affected (ktons/m)	Factor of Safety	Prob. of Failure (%)	Comments
Bench	Expansion, not adjacent to a ramp	<25	<0.5/<1.0			Berms should have a nominal width to contain unraveling wedges whose probability of occurrence is >30%; controlled blasting will be used to minimize induced damage + presplitting for final wall slopes
		25–50	<1.0/<2.0		<45	
		>50	>1.0/>2.0		<35	
	Expansion, adjacent to a ramp	<25	<0.5/<1.0			
		25-50	<1.0/<2.0		<40	
		>50	>1.0/>2.0		<30	
	Final Wall, not adjacent to a ramp	<25	<0.5/<1.0			
		25–50	<1.0/<2.0		<35	
		>50	>1.0/>2.0		<25	
	Final Wall, adjacent to a ramp	<25	<0.5/<1.0			
		25–50	<1.0/<2.0		<30	
		>50	>1.0/>2.0		<20	
Interramp	Expansion	<25	<5	>1.20	<30	Stability analysis must include explicit effect of rock mass structures; two independent access ramps will be made to pit bottom; measures will be implemented for slope drainage
			>5	>1.25	<25	
		25–50	<5	>1.25	<25	
			5–10	>1.30	<22	
			>10	>1.35	<20	
		>50	<10	>1.30	<22	
			10–20	>1.35	<20	
			>20	>1.45	<18	
	Final Wall	<25	<5	>1.20	<25	
			>5	>1.25	<20	
		25–50	<5	>1.30	<22	
			5–10	>1.35	<20	
			>10	>1.40	<18	
		>50	<10	>1.35	<20	
			10–20	>1.40	<18	
			>20	>1.50	<15	
Global	Expansion		<25	>1.30	<15	Stability analysis must include rock mass structures; all infrastructure lie outside pit perimeter limits
			25–50	>1.40	<12	
			>50	>1.50	<10	
	Final Wall		<25	>1.30	<12	
			25–50	>1.45	<10	
			>50	>1.60	<8	

review of the various criteria published in the technical literature for mining applications led to the following conclusions:

1. The minimum acceptable factor of safety range is from 1.2 to 1.5, with a typical value of 1.3.
2. The maximum P_p ranges from 5 to 30%.
3. In general, the acceptable permissible value chosen depends on the consequences of an eventual instability, the size of equipment, and the collective confidence in any interpretation of the slope-monitoring systems.

Consistent with the above, CMDIC has defined the acceptable criteria for design according to Table 18.3, noting the following:

- The occurrence of bench-scale failures is inevitable and permissible provided the acceptable contained volumes of material on berms are unlikely to be exceeded. In general, the larger the volume, the smaller the acceptable probability of failure. Also, benches located immediately above ramps and those in the final wall must have lower tolerance for failure.
- In the case of interramp instabilities, acceptance depends on the amount of ramp loss and the overall volume affected (Figure 18.5). The minimum permissible values can be defined in terms of factor of safety and a maximum limit to P_p. Final wall interramp slopes must have an operational life in excess of those for the purpose of an expansion.
- Finally, global instability must consider the possibility of loss of ramps in the affected sector(s), given the likelihood that the volumes will be substantially greater than those affecting interramp failures. Acceptance is defined in terms of a minimum permissible value for the factor of safety. In addition, because of the great uncertainty associated with the geotechnical parameters, a maximum limit is also defined for the probability of a permissible failure (Table 18.3). Other considerations are (a) global slopes only reach their maximum condition at the

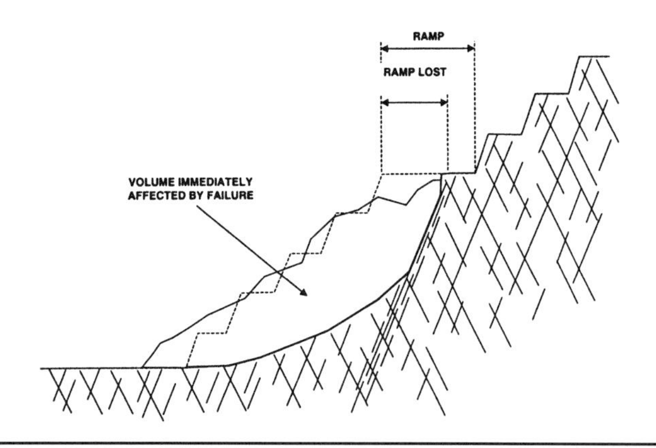

FIGURE 18.5 Interramp instability condition/geometry

completion of the final wall pushback; (b) for Ujina, two access ramps will be carried to the pit bottom; and (c) there will be no important infrastructure either within the pit or on the surface close to the perimeter.

Based on operational considerations and the availability of equipment, CMDIC management has determined that the removal of up to 800,000 tonnes could be handled without major implications to operations or planning. If a major instability occurs on an interramp or global scale, causing an extension of the wall by the order of 100 to 150 m, it would be possible to handle up to 5,300 t/m. Considering Ujina's loading equipment fleet of 57-yd^3 and 28-yd^3 shovels and 240-t trucks, this tonnage is still manageable without serious losses in production. CMDIC management has also agreed to the rigorous implementation of slope-drainage measures, controlled blasting to minimize damage to wall rock, and adherence to design and operating standards in each bench.

18.4.2 Failure Analysis

Slope failure analysis began by considering permissible bench/berm geometry, followed by interramp slopes, and finally global slopes. In each case, a systematic probabilistic approach was used in the analysis of structure/geometry for each of the five sectors in the west wall: north, northeast, central, southeast, and south (Figure 18.6). In principle, each sector was analyzed for the possible occurrence of three types of structurally controlled failure: plane, wedge, and toppling. In each case, the procedures used involved calculating a probability of failure P_p as illustrated below for the bench/berm scale.

For the case of plane failure

$$P_p = P_1 \times P_2 \times P_3 \times P_4$$

Where:

- P_1 is the probability that the necessary structure(s) exists
- P_2 is the probability that the structure forms an angle <20° with wall bearing
- P_3 is the probability that the structure is present in the bench face
- P_4 is the probability that the dip>friction angle

After considering the various probabilities/rock types/sectors, the results may be summarized as follows:

- P_p = 20% in andesite or argillized andesite for a 15-m bench and 75° face angle, 15% for 30-m bench and 70° face angle.
- P_p = 30 to 40% in rhyolite or argillized rhyolite, sector northwest, 60 to 80% central sector. It appears that the set SJ1 causes most problems with its dip of 50° to 70°. The most probable slip volume of 35 m^3/m is calculated for 15-m benches, 95 m^3/m for 30-m benches. By considering the maximum possible volumes of 45 m^3/m and 125 m^3/m, respectively, a safe containing berm width of 7 m and 10 m, respectively, is calculated.

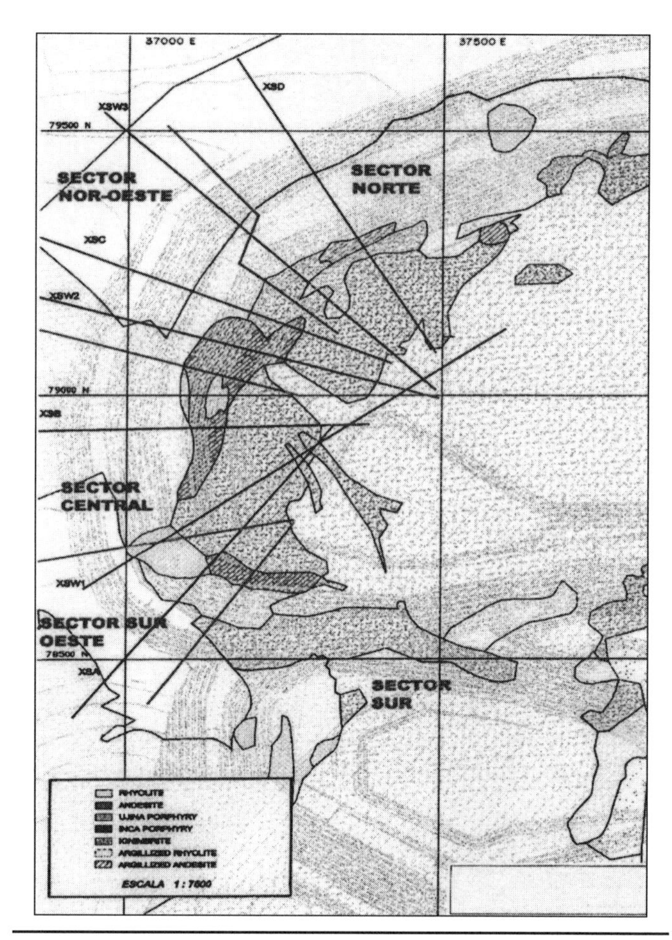

FIGURE 18.6 Ujina Pit west wall sectors

- P_p = 30 to 35% in Ujina porphyry, northwest sector, 60 to 70% central sector, with SJ1 responsible for the high failure probability.
- P_p = 5 to 10% in ignimbrite, with the worst case applying to 30-m benches.

Consequently there appears to be a relatively low probability of plane-type failure on a bench/berm scale in the west wall of Ujina, with only small volumes anticipated.

Next, the case of wedge failure was considered using the necessary conditions that (a) planes of weakness intersect to form a wedge, (b) the line of intersection outcrops in the wall, and (c) the dip of the structures exceeds the friction angle. By making the necessary assumptions (no tension cracks, horizontal berms, 4° variation in bench face angles, no pore pressure, frictional resistance only, triangular bench height variation, standard deviation of dip, dip direction, and friction angle given by 0.5 [max-min], etc.), wedge stability was evaluated using SWEDGE. The general procedure was to vary bench height, angle, and berm and calculate, for all unstable combinations, the P_p, maximum affected tonnes, and berm loss. For all cases of instability, two acceptance conditions then had to be met: (1) P_p<10% or that there was a 30% probability of exceeding the wedge weight and (2) there was a 50% probability of the assumed berm width retaining the wedge. If these conditions were not met, then the assumed berm width was increased in

TABLE 18.4 Recommended slope design options for the west wall of the Ujina Pit

Slope Parameter	Option 1	Option 2
Bench height, m	15	30
Bench face angle	75–80	70–75
Berm width	7.0–8.5	10.0–11.5
Interramp angle	53–57	53–59
Interramp height, m	150	150
Overall slope angle	45–51	45–51
Overall slope height	315	315

0.5-m increments and the calculation was repeated. In summary the results of this exercise showed that

- There would be no wedge problems in ignimbrite.
- A universal bench/berm design was not possible for all sectors.
- In general, bench face angles should vary from 70° to 80°, with berm widths from 7 to 11.5 m, depending on sector and bench height (Table 18.4).

Finally, for the case of toppling failure, the necessary conditions are that (a) the structure must be present, (b) its strike angle must be ≤30° to that of the wall, (c) it must dip into the wall at an appropriate angle (Goodman 1989), and (d) additional structures with shallow dips of 30° to 50° to define base of toppling are required. After reviewing the various probabilities of failure for the various geometries and geotechnical units, it was evident that toppling failure is very unlikely on a bench scale with double benches.

18.4.3 West Wall Recommendations

After completing probabilistic analyses similar to that described above for the interramp and global slope scales and discounting slope designs that failed to meet agreed acceptance criteria, it was concluded that two design options were preferred, each with a maximum interramp height of 150 m. The details of these two options are given in Table 18.4.

18.5 OPERATING CONTROLS

18.5.1 Blasting

The recommended slope designs require the use of controlled blasting to minimize the damage induced in the rock mass and, in particular, to achieve the required bench face angles. Based on results obtained from a series of presplit blasting tests in the Ujina Pit, it was concluded that

- It has been possible to achieve very good results with presplit blasting in single or double benches in ignimbrite.
- Additional trials are required in order to achieve acceptable results in rhyolite.
- From a purely economical point of view, the use of 6-1/2 in.-diameter presplitting holes and double benches for the final wall of Ujina's west sector is preferred.

18.5.2 Mine Planning and Line Control

A number of geotechnical-based recommendations have been made and agreed to by the Collahuasi mine planning and operations personnel. In summary these are

- The best bench height is 30 m and therefore a double-bench scheme should be executed wherever possible.
- The maximum interramp height should be 150 m (i.e., 10 single benches or 5 double benches). If a greater height is required, then an additional ramp or a cut bench between 30 to 35 m wide will have to be considered.
- If double benches are used, a 55° interramp angle should be considered; for single benches, a 54° interramp angle.
- The ramp width in the central sector of the west wall should be locally increased to 10 m in order to catch eventual small-scale failures in the rhyolite.
- A mine development plan should be considered such that the toes of major wedges do not daylight until most of the upper mass has been removed. Also, the blasting of benches located close to the base of critical wedges should be executed with care.
- Designing slopes that are convex in plan (i.e., noses) must be avoided.
- Operations' supervision must ensure that program lines defining the position of benches are maintained by controlling the cleaning of final pit benches and the overbreak.
- At all times in the pit bottom or in new benches, a box cut should be opened as far as possible from the wall to minimize blasting damage.

18.5.3 Monitoring

Although the slope design has considered the existence of certain structural and hydrological factors, ongoing near and far field monitoring is required for the Ujina Pit. Where gradual settlement may be expected, particularly in the argillized rock masses, remote monitoring of displacement rate has been devised and installed. Borehole extensometers will be utilized on an ad hoc basis to monitor potential structurally related failures where collapse can be sudden, particularly on an interramp scale. Finally, groundwater pore pressure will be monitored using piezometers in pit sectors where failure conditions are particularly sensitive to lack of drainage, i.e., in argillized rock masses in the toes of slopes.

18.6 CONCLUSIONS

A systematic approach to slope design has been achieved for the west wall of Collahuasi's Ujina Pit by the mine's geology and planning departments, together with several consultants specializing in hydrology, geomechanics, and blasting design. Based on the calculated likelihood of different structurally controlled slope failure mechanisms, two feasible bench and berm geometries were defined for various design sectors and geotechnical units. The proposed overall slope designs require the use of presplit blasting techniques in final walls, dewatering of the pit slopes and bottom, control of the program line, and a global instrumentation and monitoring program.

18.7 ACKNOWLEDGMENTS

The authors thank the management of the Collahuasi Mining Company (CMDIC) for supporting the detailed geotechnical design work for the Ujina west wall; this work forms the basis of this chapter. A special thanks goes to the following people, whose expertise contributed to this work: Antonio Karzulovic of A. Karzulovic & Ass. Ltd.; Manuel A. Duran; Pedro T. Sanhueza and Guillermo S. Albornez of CMDIC; and John Read and Alan Guest, members of Collahuasi's Geotechnical Review Board.

18.8 REFERENCES

Goodman, R.E. 1989. *Introduction to Rock Mechanics,* 2nd ed. John Wiley & Sons, New York.

Haines, A., and P. Terbrugge. 1991. Preliminary estimation of rock slope stability using rock mass classification schemes. *Proceedings 7th International Congress on Rock Mechanics,* Vol. 2. pp. 887–892, Aachen, Germany.

Hoek, E., and E.T. Brown. 1998. Practical estimates of rock mass strength. *Intl. Jour. Rock Mech. & Mining Sci.,* Vol. 34, No. 8.

Hoek, E. 1998. Reliability of Hoek-Brown estimates of rock mass properties and their impact on design, Technical Note. *Intl. Jour. Rock Mech. & Mining Sci.,* Vol. 35, No. 1.

Laubscher, D. 1990. A geomechanics classification system for the rating of rock mass in mine design. *Jour. South African Inst. Min. Metall.,* 90:10:257–273.

Montgomery, E. 1997. Preliminary projections of hydrologic conditions during dewatering of the proposed Ujina open pit copper mine, Collahuasi Project, Chile. Technical Report, Errol L. Montgomery & Associates Inc., November.

SRK 1994. Dewatering Investigation for Ujina Open Pit, Chile, Part 1, Summary Report. Technical Report No. 190437/3, SRK, Cape Town, South Africa.

CHAPTER 19

The Sur Sur Mine of Codelco's Andina Division

Reinaldo Apablaza,* Emilio Farías,* Ricardo Morales,* Jaime Díaz,† and Antonio Karzulovic†

19.1 INTRODUCTION

The Andina Division belongs to the Chilean Copper Corporation, Codelco. Its mines are in the Andes Mountains at elevations from 3,500 to 4,200 m, about 40 km southeast of Los Andes City, Region V, as shown in Figure 19.1. Currently, the Andina Division is operating two mines: Rio Blanco, a panel caving operation, and Sur Sur, an open pit.

The annual operation of Sur Sur Mine involves 7 million tonnes of ore and 15 million tonnes of waste rock. In the upper part of the east and west walls of this pit there are rock glaciers that move toward the pit, in the east, and parallel to the pit wall, in the west.

Due to climatic constraints, the pit can operate only 320 days each year, which necessitates stockpiling the ore in an old pit that has two ore-pass shafts that feed an underground plant in its bottom.

19.2 GEOLOGY

The Sur Sur orebody corresponds to a hydrothermal breccias complex; it contains important reserves of copper and molybdenum and is covered by rock glaciers. Several breccia types can be defined according to differences in characteristics such as clast size and composition, matrix and/or cement type, clast:matrix ratio, mineralization, and alteration.

From an economical point of view, the most important unit is the tourmaline breccia (BXT), which includes 90% of the mineralization and defines the main part of the orebody, with a north–south trend. The other rock types, all of them located to the west of the tourmaline breccia, have low grades or are waste rock: Cascada granodiorite (GDCC), Monolito breccia (BXMN), Monolito breccia with tourmaline (BTBXMN), rock flour breccia (BXTO), and rock flour breccia with tourmaline (BXTTO). These rock types are summarized in Table 19.1, and their location is shown in Figure 19.2.

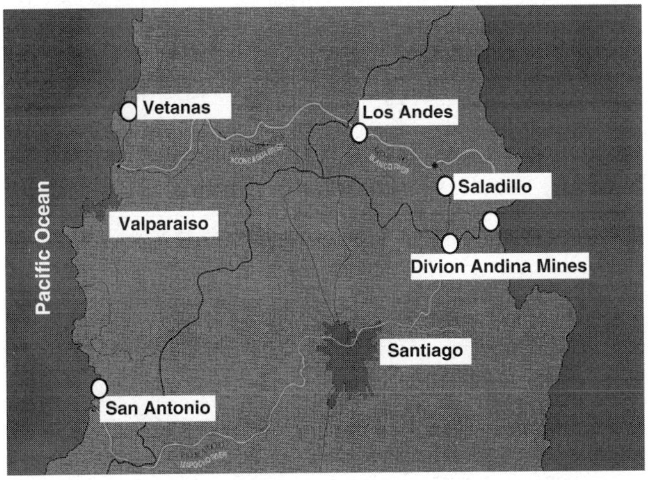

FIGURE 19.1 Map showing the location of Andina Division's mines

19.2.1 Structural Geology

The structural mapping includes all the structural features whose persistence is such that they affect at least one bench and/or

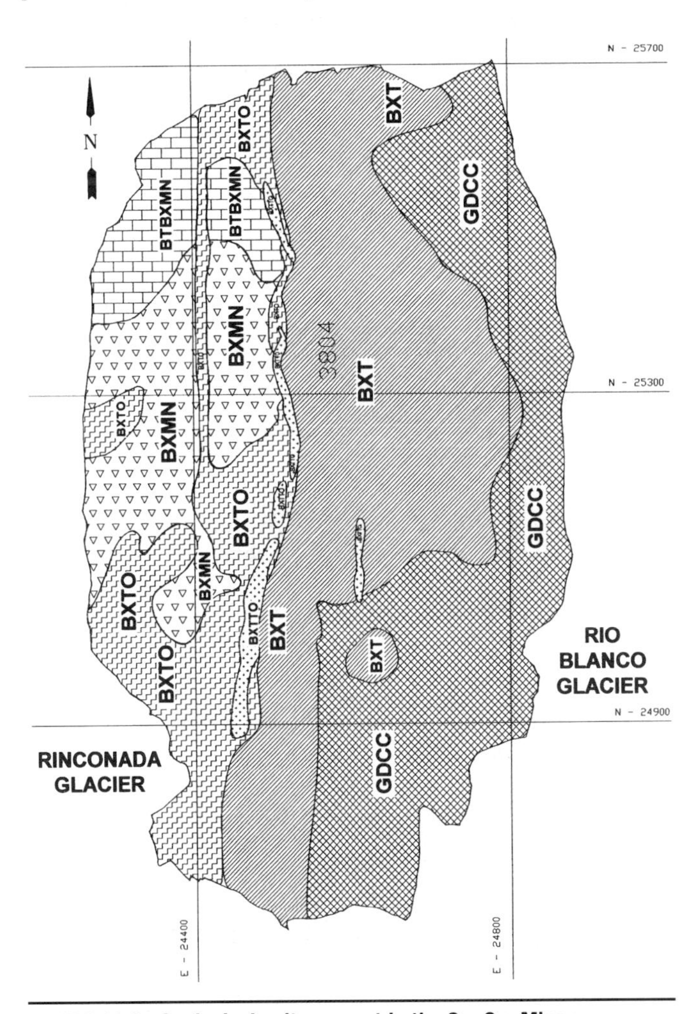

FIGURE 19.2 Geological units present in the Sur Sur Mine

* Geotechnical Group, Andina Division, Codelco, Chile.
† A. Karzulovic & Associates Ltd., Santiago, Chile.

TABLE 19.1 Geological units present in the Sur Sur Mine

Rock Type	Designation	Location in the Pit	Ore Grade	Alteration
Tourmaline breccia	BXT	Central part	Good	Intense to moderate
Cascada granodiorite	GDCC	East wall	Poor to fair	Moderate to weak
Monolito breccia	BXMN	West wall	Poor	Weak
Monolito breccia with tourmaline	BTBXMN	Northwest sector	Poor	Weak
Rock flour breccia	BXTO	Southwest sector	Poor to fair	Intense
Rock flour breccia with tourmaline	BXTTO	Central part	Poor	Intense

TABLE 19.2 Characteristics of the structural domains at the Sur Sur Mine

	Structural Sets (Dip/Dip Direction)				
Domain	S-1	S-2	S-3	S-4	S-5
I	77° ± 7° / 139° ± 10°	45° ± 6° / 340° ± 7°	80° ± 10° / 311° ± 5°	80° ± 3° / 034° ± 5°	Subhorizontal
II	85° ± 5° / 130° ± 15°	82° ± 8° / 132° ± 10°	Subhorizontal	87° ± 3° / 223° ± 12°	
III	85° ± 5° / 150° ± 15°	78° ± 8° / 041° ± 8°	86° ± 4° / 320° ± 7°	Subhorizontal	
IV	80° ± 10° / 338° ± 9°	49° ± 5° / 344° ± 9°	75° ± 7° / 070° ± 8°	Subhorizontal	
V	77° ± 13° / 140° ± 10°	85° ± 5° / 044° ± 8°	37° ± 8° / 353° ± 13°	48° ± 3° / 310° ± 6°	30° ± 5° / 089° ± 6°
VI	76° ± 11° / 140° ± 16°	84° ± 6° / 052° ± 5°	78° ± 9° / 315° ± 7°	70° ± 5° / 281° ± 4°	
VII	80° ± 10° / 142° ± 12°	80° ± 10° / 243° ± 10°	80° ± 10° / 311° ± 6°	30° ± 5° / 048° ± 6°	

TABLE 19.3 Properties of the intact rock (typical values)

Rock Type	γ (ton/m^3)	UCS (MPa)	TS (MPa)	E (GPa)	ν
Tourmaline breccia	2.65	114	9	30	0.20
Cascada granodiorite	2.59	115	9	41	0.24
Monolito breccia	2.55	127	10	33	0.18
Monolito breccia with tourmaline	2.60	120	9	31	0.20
Rock flour breccia	2.47	70	8	27	0.13
Rock flour breccia with tourmaline	2.56	65	8	28	0.15

NOTES:
γ unit weight
UCS unconfined compressive strength
TS tensile strength
E deformability modulus
ν Poisson's ratio

those structures associated with bench-scale instabilities. The interpretation of these data made it possible to define the structural domains summarized in Table 19.2. These structural domains are shown in Figure 19.3.

19.3 GEOTECHNICAL CHARACTERIZATION

The rock mass is classified according to Codelco's Standard INB-CMR-CL-01-97 using the geological strength index (GSI), which takes into account the rock-mass fabric and the quality of the structures. The typical range for the GSI is 50 to 60, but there are also sectors with lower and higher values. Figure 19.4 shows the zonation of the Sur Sur Mine in terms of the GSI. The mechanical properties of the different intact rock types are summarized in Table 19.3, and the typical strength of the structures is presented in Table 19.4.

19.3.1 Geotechnical Slope Analysis and Design

The geological, structural, and geotechnical data are used to define the design of sectors in the pit, each one of them having specific characteristics that should be considered in the geotechnical analysis and design, that will define the slope geometries. The geotechnical analysis includes an evaluation of the possible types of slope failure, including the effect of major structures.

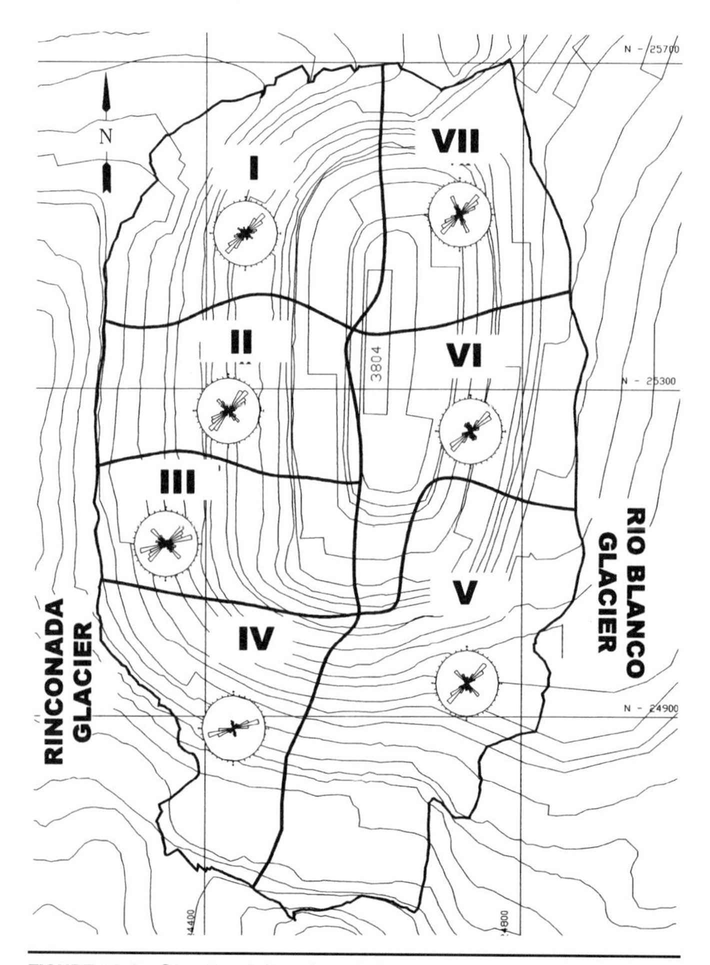

FIGURE 19.3 Structural domains of the Sur Sur Mine

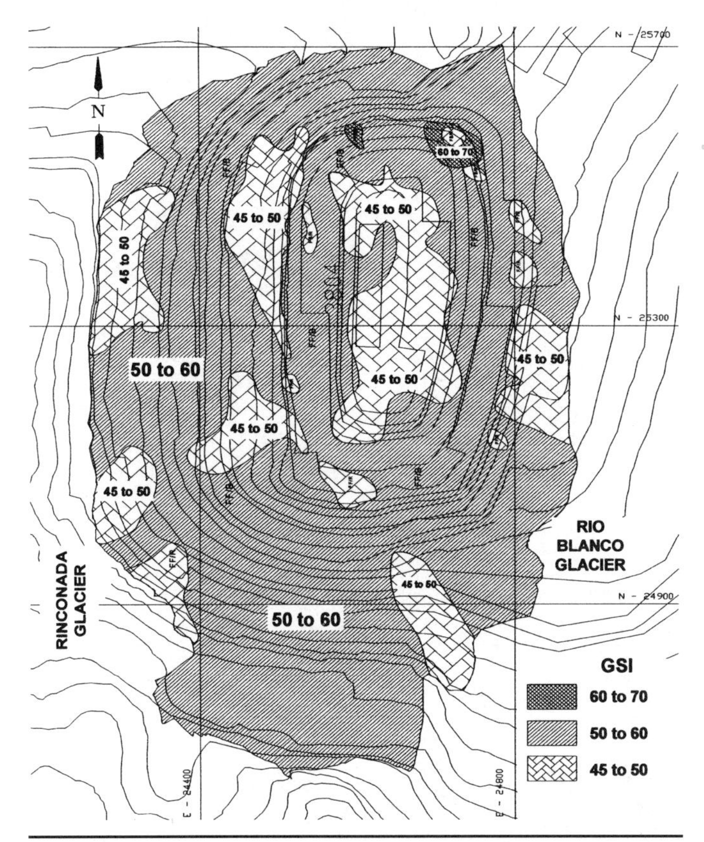

FIGURE 19.4 Geotechnical zonation of the Sur Sur Mine in terms of the GSI

TABLE 19.4 Properties of the structures (typical values)

Type of Structure	Cohesion (kPa)	Angle of Friction (degrees)
Major faults with clayey gouge	0 to 25	18 to 25
Continuous minor structures (bench scale)	0 to 100	30 to 45
Minor structures with rock bridges (interramp or overall scale)	100 to 500	30 to 35

The bench height is defined by considering the efficiency of shovels and the volumes of eventual unstable wedges. From an operational point of view, the steeper the bench face, the better. The berm width is first defined using the criteria of Ritchie modified by Call & Nicholas and then checked by a probabilistic analysis of the structurally controlled instabilities that could affect the benches.

The stability of interramp and overall slopes is analyzed by two-dimensional limit-equilibrium methods, using Spencer's method for noncircular failure surfaces.

Currently, the Sur Sur pit is in a transition stage, because the current design of 12-m-high single benches is being changed to 32-m-high double benches, with a 70° bench face inclination, to allow a more efficient use of today's large mining equipment. This new design allows considerable savings and an important decrease in the stripping ratio, because it increases the interramp and overall slope angles.

The current design for the Sur Sur Mine is summarized in Table 19.5 and Figure 19.5, while the same data for the final pit condition are presented in Table 19.6 and Figure 19.6.

The acceptability criterion for interramp and overall slope design is that the factor of safety under operational conditions (i.e., dry slopes, no earthquake, and good-quality blasting) must be equal or larger than 1.3, while for a seismic condition the factor of safety must be equal or larger than 1.1.

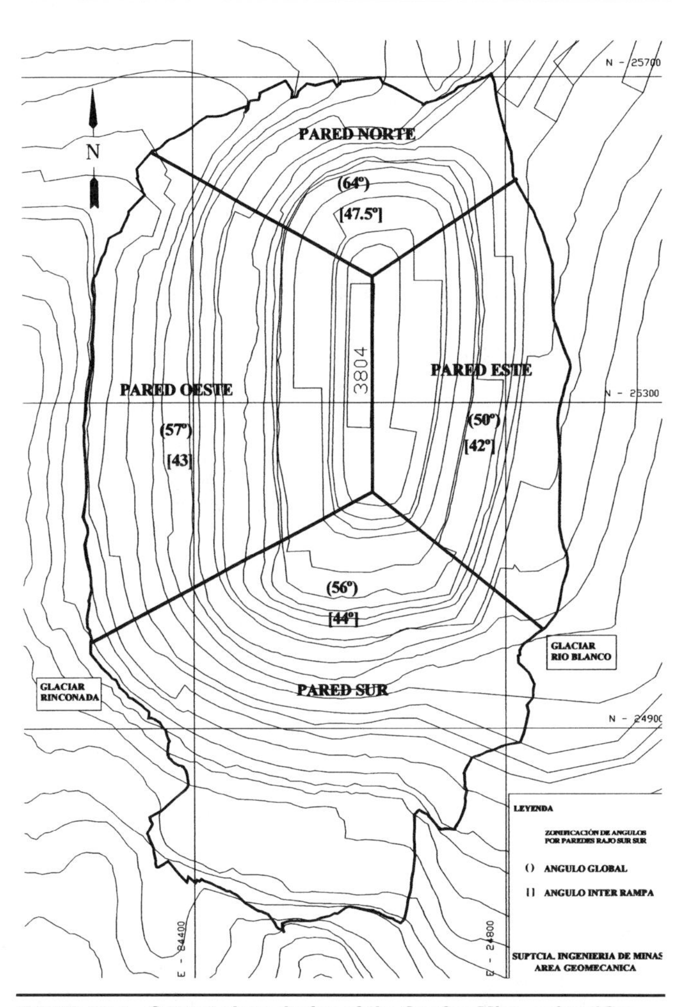

FIGURE 19.5 Current slope design of the Sur Sur Mine, using 16-m-high single benches (α_i is the interramp slope angle and α_o is the overall slope angle)

TABLE 19.5 Current slope design at the Sur Sur Mine

	Bench-Berm			Interramp			Overall	
Pit Sector	h_b (m)	α_b (degrees)	B (m)	α_r (degrees)	h_r (m)	r (m)	α_o (degrees)	h_o (m)
North	12/16	70	12.5	54	48	35	48	144
East	12/16	70	12.6	50	64	35	42	176
South	12/16	70	11.6	56	64	35	44	304
West	12/16	70	11.4	57	80	35	43	320
Moraine	12/16	70	12.0	29	80	40	—	80

NOTES:
h_b bench height (single benches)
h_r interramp height
h_o overall height
α_b bench-face inclination
α_r interramp angle
α_o overall angle
b berm width
r ramp width

19.4 INSTRUMENTATION AND MONITORING

Perhaps the most distinctive characteristic of the Sur Sur Mine is the fact that the upper part of the pit is surrounded by active rock

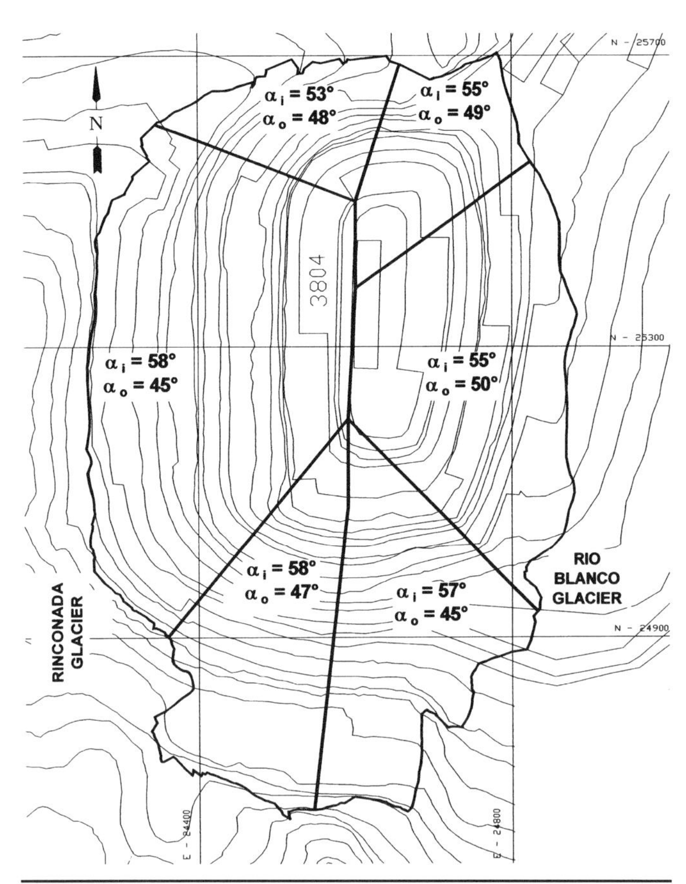

FIGURE 19.6 Final pit design of the Sur Sur Mine, using 32-m-high double benches (α_i is the interramp slope angle and α_o is the overall slope angle)

TABLE 19.6 Final pit slope design at the Sur Sur Mine

	Bench-Berm			Interramp			Overall	
Pit Sector	h_b (m)	α_b (degrees)	B (m)	α_r (degrees)	h_r (m)	r (m)	α_o (degrees)	h_o (m)
North	32	70	12.5	53	160	35	48	300
Northeast	32	70	10.8	55	160	35	49	330
East	32	73	12.6	55	160	35	50	520
Southeast	32	70	11.6	57	160	35	45	590
Southwest	32	75	11.4	58	160	35	47	590
West	32	75	11.4	58	160	35	51	590
Moraine	16 (SB)	74	25.0	29	80	35	—	80

NOTES:
h_b bench height (double benches)
h_r interramp height (maximum)
h_o overall height
α_b bench-face inclination
α_r interramp angle
α_o overall angle
b berm width
r ramp width
SB single benches

glaciers, as shown in Figure 19.7. These are the Rio Blanco Glacier, on the southern and eastern sectors, and Rinconada, on the southern and western sectors of the pit. Figure 19.8 shows a typical bench on these glacier materials, in the east wall of the pit.

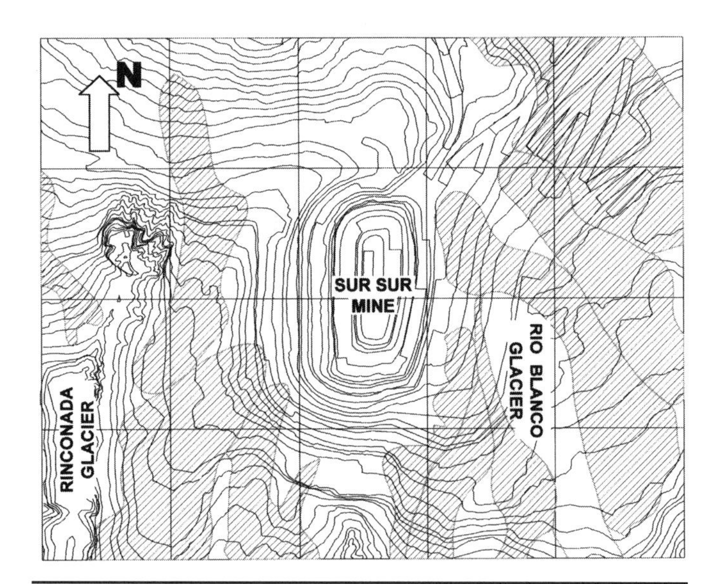

FIGURE 19.7 Active rock glaciers in the Sur Sur area

FIGURE 19.8 Benches in rock and rock glacier on the east wall of the Sur Sur Mine

During the initial stages of the pit and until 1998, waste, low-grade ore, and material removed from the rock glaciers were dumped on the surface of the glaciers. These waste dumps, with heights up to 30 m, overloaded and increased the rate of deformation of the glaciers, which reached maximum values of 30 to 35 m/yr with displacements in the 30° N to 50° W direction, causing instabilities in the upper benches of the pit and affecting its operation.

This problem made necessary the monitoring of the slopes to detect potentially unstable sectors and deal with them before the triggering of a slope failure. To do this, a robotic monitoring system was installed, which includes the following:

- 1 ATS-PM geodimeter
- 30 prisms located on the upper part of the pit walls

Figure 19.9 shows the station where this robotic equipment is installed. This system is currently operative and is being used mainly for daily control of the benches excavated in the rock glacier, which shows a displacement rate typically of the order of 5 to 6 cm/d. This monitoring allows a safer mine operation and the development of remedial measurements to stabilize those sectors that show signs of potential instability. Additionally, two

FIGURE 19.9 Station where the robotic slope-monitoring system is installed

FIGURE 19.10 Structurally controlled instability in a rock bench on the east wall of the Sur Sur Mine (January 1999)

piezometers have been installed in the pit to control the groundwater level, at depths of 208 and 430 m.

19.5 SLOPE MANAGEMENT

In spite of the fact that some bench-scale instabilities like the one shown in Figure 19.10 have occurred in the Sur Sur Mine, the slope management at the mine deals basically with those benches excavated in the rock glaciers because they show the largest rates of displacement. Therefore, the geotechnical group of Andina

FIGURE 19.11 Benches deformed by the cumulative displacement of rock glaciers in the upper part of the east wall of the Sur Sur Mine

Division developed a detailed study in 1998 to define the following characteristics of these glaciers:

- Areal extent and surface morphology
- Thickness and stratigraphy
- Rates of displacement at surface
- Rates of displacement at depth

As a result of this study, a two-dimensional numerical model was developed to predict the rate of displacement of these glaciers, taking into account the ice thickness, the slope of the rock–glacier contact, the surcharge due to the waste dumps, and the temperature. This model, called MDGR, predicted that the expansion of the mine to the final pit condition will trigger an increment in the rates of displacement of the glaciers, which reached up to 35 m/yr in 1998. Therefore, the same model was used to study different options to reduce these displacement rates to values not larger than 20 m/yr, which is a condition that mine operation can handle. As a result of this analysis, a sequence for the removal of material from the glaciers was developed and included in the long-term mine plan.

Every year the benches excavated in the rock glaciers need maintenance to remove the material accumulated due to the displacement of the glaciers and recover the program lines (currently the accumulated yearly displacement is about 20 m). Figure 19.11 shows some benches excavated in the rock glacier, deformed by the displacement of the glaciers.

19.6 CONCLUSIONS

A large open-pit mine requires a significant quantity of geotechnical work, not only to develop the geotechnical analysis and slope design but also to manage the slopes. This is especially true in the case of Andina Division's Sur Sur Mine, because the upper part of the pit is affected by two rock glaciers that move up to 35 m/yr, and the use of robotic slope monitoring is required in order to anticipate the possible occurrence of stability problems, allowing a safe operation.

19.7 ACKNOWLEDGMENT

The authors thank Codelco-Chile, Andina Division, for authorizing the presentation of this work. We also thank the engineers and geologists of Andina Division, since, without their permanent work, chapters like this would not be possible.

19.8 REFERENCES

Call, R.D., A.A. Hernandez, W.K. Walker, and E.E. Bates. 1997. Slope Design for the Andina Sur Sur and Don Luis Pits Plans. *Technical report prepared for Andina Division of Codelco-Chile.*

Hoek, E. 1998. Reliability of Hoek-Brown estimates of rock mass strength properties and their impact on design, *Intl. Jour. Rock Mech. Min. Sci.,* 35(1):63–68.

Hoek, E., and J.W. Bray. 1981. *Rock Slope Engineering.* London: IMM.

Hoek, E. and E.T. Brown. 1997. Practical estimates of rock mass strength. *Intl. Jour. Rock Mech. Min. Sci.,* 34(8):1165–1186.

Hoek, E., P.K. Kaiser, and W.F. Bawden. 1995. *Support of Underground Excavations in Hard Rock.* A.A. Balkema.

Itasca. 1999. *UDEC. Universal Distinct Element Code.* Version 3.0. Minneapolis: Itasca Consulting Group, Inc.

Itasca. 1999. *FLAC. Fast Lagrangian Analysis of Continua.* Version 3.4. Minneapolis: Itasca Consulting Group, Inc.

Karzulovic, A. 1991. Rajo Sur Sur. Efectos de las tronaduras Fase I: Revisión, análisis y evaluación de la información actualmente existente. *Technical report prepared for Andina Division of Codelco-Chile.*

Karzulovic, A. 1992. Rajo Sur Sur. Efectos de las tronaduras Fase II: Evaluación de daños e inestabilidades inducidas por las faenas de tronadura. *Technical report prepared for Andina Division of Codelco-Chile.*

Marangunic, C. 1998. Estudio de glaciares de roca del area del Rajo Sur Sur. *Technical report prepared for Andina Division of Codelco-Chile.*

Marangunic, C. 1999. Remociones de glaciares en las fases del Rajo Sur Sur. *Technical report prepared for Andina Division of Codelco-Chile.*

Seegmiller, B.L. 1985. Phase I Stability Study: Initial assessment of Sur Sur Pit–Rio Blanco Mine. *Technical report prepared for Andina Division of Codelco-Chile.*

CHAPTER 20

Design, Implementation, and Assessment of Open-Pit Slopes at Palabora over the Last 20 Years

Alan Stewart,* Fanie Wessels,† and Susan Bird*

20.1 INTRODUCTION

The Palabora Copper open pit is located in the Northern Province, near the town of Phalaborwa, South Africa. Mining at Palabora began in 1964, and production was initiated in 1966. The current design, which was implemented in 1980, involves the excavation of more than 2 billion tonnes of material, with open pit mining continuing to the end of the year 2002. Present (early 2000) slopes have a height of up to approximately 665 m (Figure 20.1), with the proposed ultimate pit having an overall slope height of about 830 m. The final overall slope angle (toe to crest) will range from approximately 45° to 50°, making it one of the steepest and deepest pits in the world. The upper half of the pit is accessed via a dual-ramp system, which reduces to a single ramp in the lower portion of the pit.

20.1.1 Engineering Geology

The Palabora Igneous Complex is approximately 7 km long and 3 km wide, consisting of three alkalic carbonatite pipes that have intruded the surrounding Archean granite. The central Loolekop Pipe hosts the Palabora Copper open pit. The main rock types, progressing concentrically inward from the micaceous pyroxenite (MPY) host, include a shell of foskorite and a central core of banded carbonatite, which was later intruded by transgressive carbonatite. Fenite occurs along the upper benches of the south wall and there is a localized intrusion of glimmerite in the southwest corner of the pit. Several dolerite dykes up to 50 m wide as well as numerous narrower dykes cross the pit. A simplified plan of the final pit geology is illustrated in Figure 20.2.

Several steeply dipping to subvertical major faults can be traced through the pit. Geologic contacts are subvertical and primarily gradational. The dolerite dykes generally strike east–northeast and dip steeply to the west. Eight major joint sets have been recognized in the main rock types (Piteau Associates 1980). The jointing demonstrates an increase in both dip and intensity of the peak orientations towards the center of the pit. Four distinct joint sets occur in the dolerite dykes and appear to be related to the later-stage intrusion and cooling of the dykes.

Laboratory testing of the various rock types indicates that all unweathered rocks, with the exception of glimmerite, are very hard with average unconfined compressive strengths (UCS) being greater than 80 MPa. Upper limits of the UCS are 150 MPa for the major rock types and 350 MPa for the dolerite dykes. Glimmerite has a moderate UCS of 37 MPa; however, its rock-mass strength is generally much lower due to the abundance of mica-coated joints and micaceous parting planes. The weathered MPY is also relatively weak, with a UCS ranging from 0.2 to 71 MPa. Shear strength of the discontinuities in all rock types indicates little to no cohesion, with friction angles ranging from 15° to 35°, depending on rock type and infilling.

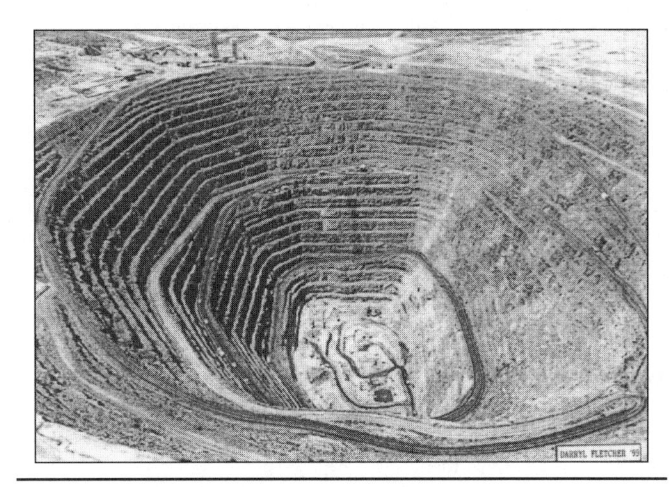

FIGURE 20.1 Aerial photo of open pit looking east

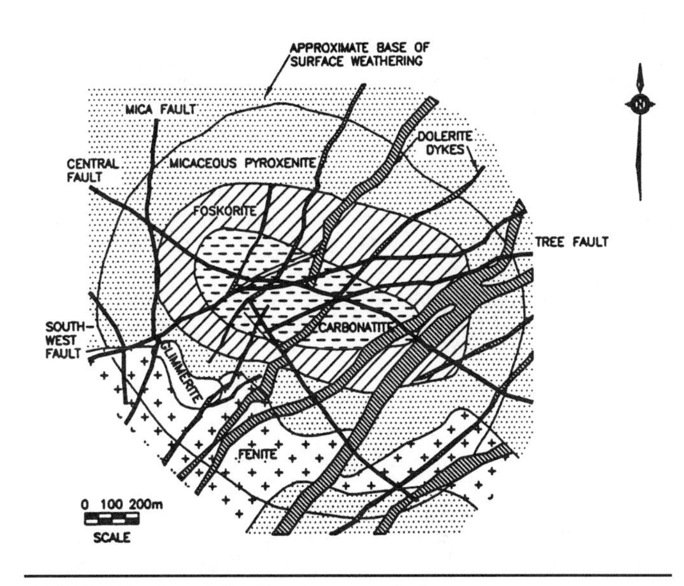

FIGURE 20.2 Generalized final pit plan geology

Hydrogeology assessments of the pit area in 1980 (Piteau Associates 1980) indicated that the rock mass has a low hydraulic conductivity (10^{-12} to 10^{-9} m/s) that is essentially independent of rock type. At that time, total steady-state flow into the pit was about 49 lps, one third of which originated from the north wall. The water table was at a depth of between 10 and 40 m, with pit

* Piteau Associates Engineering Ltd., North Vancouver, British Columbia, Canada.
† Palabora Mining Company, Phalaborwa, South Africa.

deepening not significantly reducing this level. The dolerite dykes, which appeared to be relatively permeable zones, seemed to act as diversions, influencing the direction of flow into the pit.

20.2 ORIGINAL (1980) SLOPE DESIGN

Piteau Associates Engineering Ltd. (Piteau) originally undertook the design of the open pit slopes in 1980. The recommended slope angles, which incorporated 30.5-m-high double benches, resulted in a lower stripping ratio with a net financial gain. The wider safety berms that were recommended with the double benching also provided greater access to the benches than was available with single benches.

The initial step in the slope-design process was to divide the open pit area into 16 structural domains, based on rock type and location with respect to the major faults. Each domain was then subdivided into design sectors, based on the orientation of the slopes, giving rise to 51 design sectors for the final pit slopes.

Due to the steep orientation of all major structures, wedges formed from major faults or fault sets were determined to be an unlikely mode of failure. It was also determined that only the southwest corner of the pit was subject to wedge-type failures that could involve several benches. However, the potential for involvement of an entire interramp slope was considered low. Furthermore, because of the generally high strength of the majority of rocks comprising the final pit wall, failure through intact rock was considered unlikely. Therefore, analysis and design of the pit slopes in the various design sectors involving the stronger rock types (fenite, MPY, foskorite, and carbonatite) was based on bench-design kinematics. In the weaker rock (glimmerite and weathered MPY), slope design was based on rotational rock-mass failure.

20.2.1 Statistical Kinematic Analysis

Kinematic analyses concentrated on bench-scale wedge and plane failures, which could develop through combinations of joints, faults, or shears, and daylight at the pit face. Design of the slope geometry in the stronger rock areas was devised to provide catch benches of sufficient width to contain slope failures of one or two benches and to allow access for removal of debris, if necessary, while maximizing the interramp angle (IRA).

To analyze the numerous joint set combinations for each design sector, a unique statistical technique developed by Piteau (Martin and Piteau 1978; Martin, Steenkamp, and Lill 1986; Piteau et al. 1985) was used. The technique assumes that all possible failures are wedges (i.e., wedges sliding on one or more planes) in order to calculate the apparent plunge of lines of intersection that could daylight at the bench face (assuming vertical bench face angles [BFAs]). For a particular design sector (i.e., slope orientation), all kinematically possible wedges and planes were identified and their factors of safety calculated. Those wedges having a factor of safety of less than 2.0 were considered in the statistical analysis. The relative frequency of a particular combination of structures was based on the number of actual discontinuities for the average or peak orientation of the discontinuity sets involved compared to the total number of possible failures for that design sector. Cumulative frequency plots of the apparent plunge of wedge intersections were then plotted to determine the "probability" of wedge failures for a given BFA. Based on engineering experience, it was determined that a design BFA equivalent to the apparent plunge at a cumulative frequency of 20% of all wedge failures would result in an acceptable amount of back break. For dry or depressurized slopes, this meant that 80% of all wedges must have a factor of safety greater than 1.2, while for undrained slopes the BFA was reduced so that 80% of the possible wedges have a factor of safety greater than 2.0. The width of the catch benches (safety berms) were calculated based on the width required to catch this amount of material, lying at the angle of repose, for the given bench height.

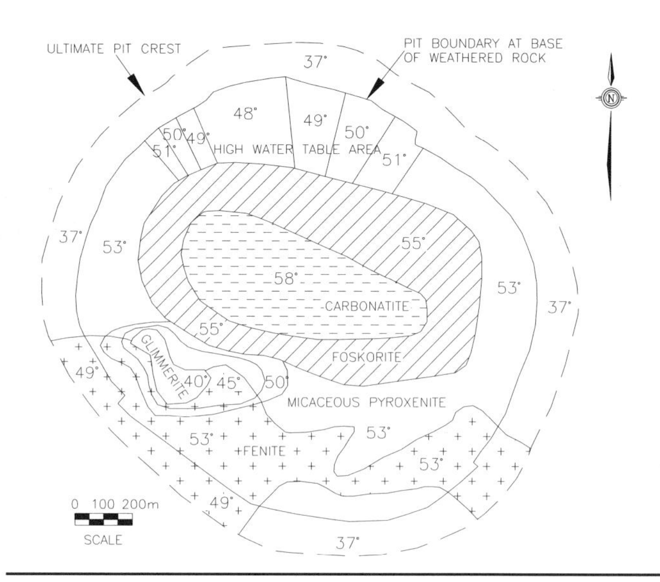

FIGURE 20.3 Interramp slope angles for final pit design

The BFAs determined from the statistical design method were checked against existing slopes, where it was found that the average actual BFA in unweathered rock was about 80° for double (30.5-m-high) benches. This BFA incorporated the effect of a 3-m "mini" bench at mid height, which was due to the inability of the blasthole drill to work sufficiently close to the slope. It was recommended that this 3-m mini bench be reduced as much as possible, allowing for more effective catchment and a steeper BFA of about 84°.

20.2.2 Deep-Seated Failure Analysis

Deep-seated instability involving large sections of the interramp slopes was considered to be the controlling factor in the design of the pit face within the weaker glimmerite and weathered MPY rock types. A back analysis of a November 1979 failure involving three benches in the glimmerite (about 10,000 tonnes of material) was the basis for an interramp design in this rock unit. Instability due to fault interaction that resulted in sliding or overturning of wedges and instability due to crushing of the rock mass resulting in rotational instability were each analyzed. Within the weathered MPY and extending one bench below the weathered/unweathered contact, rotational failures and raveling appeared to control the design of the slope. In each of these cases, it was found that the presence of water was the controlling factor.

20.2.3 Implementation of Design

The final interramp slopes (37° to 58° IRAs) adopted at Palabora in 1980 are illustrated schematically in Figure 20.3. The final slopes for the design sectors were smoothed and blended to take into account various operational considerations. Double (30.5-m-high) benches were recommended in all areas, except in the glimmerite and weathered MPY rock units where single (15.2-m- or 12.2-m-high) benches and interramp angles of 37° to 40° were implemented.

To maintain the integrity of the catch benches, control blasting was recommended. Blasting techniques developed at Palabora (Fauquier 1983) included trim blasts, which were decoupled by using small-diameter tubes placed within the larger-diameter blastholes. For crest protection, subgrade drilling was eliminated on the bench immediately above the final pit walls. Furthermore, geomechanical information and joint and fault orientations were used to aid in blast design.

20.3 ONGOING GEOTECHNICAL ASSESSMENTS

Updates of geotechnical assessments have been carried out once or twice a year since 1988 and were conducted periodically from 1980 to 1988. For each update, geologic structural and kinematic analyses were carried out and monitoring data (slope movement and piezometric data) were reviewed. Numerical modeling was first utilized as an assessment tool in 1989.

20.3.1 Kinematic Assessments

Ongoing kinematic assessments have typically been based on updated mapping data obtained from the newly exposed benches. A similar procedure to that utilized in the original slope-design study has been used in the update assessments. That is, the average or peak orientations of the various discontinuity populations identified in each structural domain have been used to identify the kinematically possible mechanisms of instability that could affect the various design sectors in the open pit. Kinematic and statistical analyses, as described above, were then conducted, with the results of these analyses compared to the original slope design and the documented conditions for each area of the pit.

20.3.2 Slope Displacement Monitoring

A survey monitoring system was implemented in 1984 to detect slope displacements in response to mining. Initially, the surveying was conducted with precise leveling and electronic distance measurements. However, this was a cumbersome exercise, with monitoring points only being surveyed 8 to 10 times a year. A global-positioning system (GPS) was implemented in August 1993. Sixty monitoring points are now surveyed within a two-week period, generally to an accuracy of 5 mm. However, the accuracy of monitoring points deeper in the pit is less because of the limited satellite window available. In this regard, GPS surveys can be conducted down to approximately 560 m below the pit perimeter, with the vertical component of movement below this depth becoming too inaccurate for detailed analysis. The horizontal and vertical movements are plotted for each monitoring point, with horizontal movement vectors also plotted on a pit plan. Movement trends are then identified and movement rates tabulated and color-coded for easy identification of any changes. The results of the slope-monitoring program are important not only for determining general trends in pit response but also for calibrating numerical models.

Surface extensometers installed at Palabora are of the wireline tripod type, in which a weight is suspended between two trigger mechanisms. If excessive movement takes place and the weight moves past the triggers, an alarm is activated to warn mining personnel of possible instability.

To monitor deep-seated movement, time domain reflectometry (TDR) and inclinometers have been used. For TDR monitoring, coaxial cables have been grouted into vertical holes in three critical areas in the pit. To date, monitoring results from the readings have not indicated any movement. Three 200-m-deep inclinometers have recently been installed. While the depth of installation and the magnetic environment have resulted in several complications at the time of writing, the system is expected to be fully operational in the near future.

20.3.3 Piezometric Monitoring

Open standpipe piezometers were installed in selected areas on the pit perimeter as part of the design study in 1980. Additional piezometers have been installed along ramps as mining progressed, with monitoring conducted on a monthly basis.

Generally, the piezometric levels have gradually declined in response to mining, although the piezometers on the perimeter of the pit have remained static. Some piezometric levels have dropped dramatically in response to the installation of drain holes and to the mining of the in-pit crusher conveyor tunnel. Depressurization, which plays an important role at Palabora, is achieved with subhorizontal drain holes between 200 m and 300 m in length. Drain holes are drilled with a down-the-hole percussion drill rig from the final faces and are drilled to intersect specific targets, such as the dolerite dyke contacts, which are known water conductors. Where possible, seepage from drain holes is collected in pipes to prevent it from re-entering the slope.

20.3.4 Numerical Modeling

Numerical modeling of the Palabora pit slopes has been carried out on a number of occasions to develop a more comprehensive understanding of the slope-deformation behavior and to assess the potential for deep-seated slope-deformation mechanisms to adversely affect the current mine design. Initial numerical modeling was conducted in 1989 by Palabora staff utilizing the Fast LaGrangian Analysis continua (FLAC) program (du Plessis and Martin 1991). FLAC was chosen due to its ability to undergo plastic flow when the yield criterion is exceeded, allowing displacements to be modeled as excavation proceeds. The west wall, which had accumulated the largest horizontal displacements at the time (up to 200 mm since 1984), was modeled to determine if steepening of the slope was feasible. The rock mass was modeled as a Mohr–Coulomb elastic, perfectly plastic continuum. Through a calibration of the model with documented slope movements since 1984, a range of rock-mass strengths was developed for which the modeled slope response matched the documented slope-monitoring data. It was also determined that a horizontal side stress equal to the vertical stress (i.e., a stress ratio of 1.0) was required to match both the horizontal and vertical recorded displacements. However, due to the model's inability to account for discrete discontinuities and plastic behavior such as strain softening or hardening, the model was not considered as a predictive tool.

To update the modeling of the west wall of the pit, Universal Distinct Element (UDEC) modeling of this wall began in 1991 (Piteau Associates 1991). The UDEC code is similar to the FLAC code, in that it provides a numerically stable solution even when the modeled condition is statically unstable. However, distinct modeling of discontinuities is also possible. Toward this end, the effect of a continuous, steeply westward-dipping fault set was considered, with other discontinuities accounted for in the rock-mass strength and deformation parameters. Results from this modeling indicated the importance of dewatering to slope stability and indicated that the horizontal to vertical stress ratio is more likely between 1.5 and 2.0.

UDEC modeling of the west wall was updated in 1997, with initial UDEC models also constructed for the southwest wall and the north wall. The rock-mass strength parameters were adjusted using the updated Hoek–Brown 1997 (Hoek and Brown 1997) criterion. West wall modeling now included two major fault sets—the continuous west-dipping fault set previously modeled and a discontinuous subhorizontal set as well as the two major faults in the west wall—the Glimmerite fault and the Mica fault (Piteau Associates 1998a). Modeling results indicated a complex deformation mechanism involving a combination of elastic rebound and differential slip along the discontinuities, representing a toppling-dominated mechanism with overall pitward rotation. The complex mechanism also included block sliding on the flatter discontinuities and down dropping of blocks along the major faults. Again, the importance of dewatering was evident from the predictive modeling results, indicating horizontal drains of at least 200-m length would greatly reduce movements.

A preliminary model of the southwest corner of the pit (Piteau Associates 1997a) was built using UDEC to determine the rock-mass strengths in this area, which were thought to be lower than in other portions of the pit due to the micaceous zones in the foskorite. This modeling, along with the results of pit slope documentation, helped to determine the requirement for remedial measures between two major faults that intersected this corner of the pit.

Initial north wall UDEC modeling (Piteau Associates 1998b) included two continuous fault sets and the Southwest, Central, and Tree faults. The rock-mass strength parameters were the same as those used in the west wall model. Sensitivity with respect to the spacing of the continuous fault sets and on the stress ratio were conducted. Modeling results indicated that final north wall slopes (to bench 58) did not have a potential for deep-seated failure, with predicted displacements significantly reduced by dewatering to 200 m behind the face.

20.4 ONGOING REMEDIAL MEASURES AND BLASTING TECHNIQUES

In general, performance of the 1980 slope design has been relatively good, largely due to the favorable structural conditions and the competence of most rock at Palabora. Slope remediation to maintain the integrity of the pit wall slopes has been restricted primarily to local measures, typically in the vicinity of major faults and in other less competent areas of the rock mass.

Remedial measures such as dewatering (Piteau Associates 1996a), mini-benches/stepouts, rail and cable dowels (Piteau Associates 1984), mechanical anchors, steel fiber, wire-mesh–reinforced shotcrete (Piteau Associates 1995), and even the use of a tire wall (Piteau Associates 1996b) have been locally implemented. The philosophy behind the stepouts is to provide some form of buttressing in less competent areas. A stepout is normally implemented on the wall trim (bottom of a double bench) and is typically 5 to 7 m wide, with the length dependent on the specific structure. Steel rail and shovel cable have been grouted as dowels into inclined and vertical drill holes up to 15 m deep. Dowels have primarily been used for stabilization of bench crests and along the edge of the ramp, with a number of dowels being regularly implemented in recent years in the southwest corner of the pit. Installation of the dowels is typically dependent on the localized structure and resulting failure mechanism. A schematic of a typical dowel installation combined with a stepout is illustrated in Figure 20.4.

Changes to the design on a larger scale have been implemented on two occasions–on the north wall in 1985 and on the west wall. While mining the final cut on the north wall in 1985, it became evident that the influence of the Mica fault would require modification of the slope design. The Mica fault strikes obliquely to the final wall and dips toward the pit at approximately 70°. Single benches were required in this area in order to flatten the slopes to 43° from the original 53° design. Mesh-reinforced shotcrete, dowels, and drain holes also formed part of the stabilization measures. When the west wall was mined in the foskorite, the Southwest fault required extensive shotcrete and doweling to maintain stability. However, this was found to be insufficient and single benches were again required (Piteau Associates 1997b) in this rock unit. In recent years, steel-fiber–reinforced shotcrete has replaced mesh-reinforced shotcrete. No further design changes have been required.

Blasting practices have evolved and improved over the years. In 1994, a decision was made to presplit all final faces. Initially, the presplits were drilled at 80°, both on the crest and wall trims. However, this configuration resulted in narrower catchment berms, with the back break not improving significantly. Because inclined presplitting was also more difficult in terms of the accuracy of the holes, presplits are now drilled vertically, with very encouraging results.

In 1999, initiation of explosives in trim and production blasts was changed from detonation cord and booster to a shock-tube system (Palabora Mining 1999a). The latter system, which is simpler, more reliable, and flexible with respect to timing, still makes use of bottom-primed boosters and is initiated by detonators. A major advantage of shock-tube or nonel blasting is that it

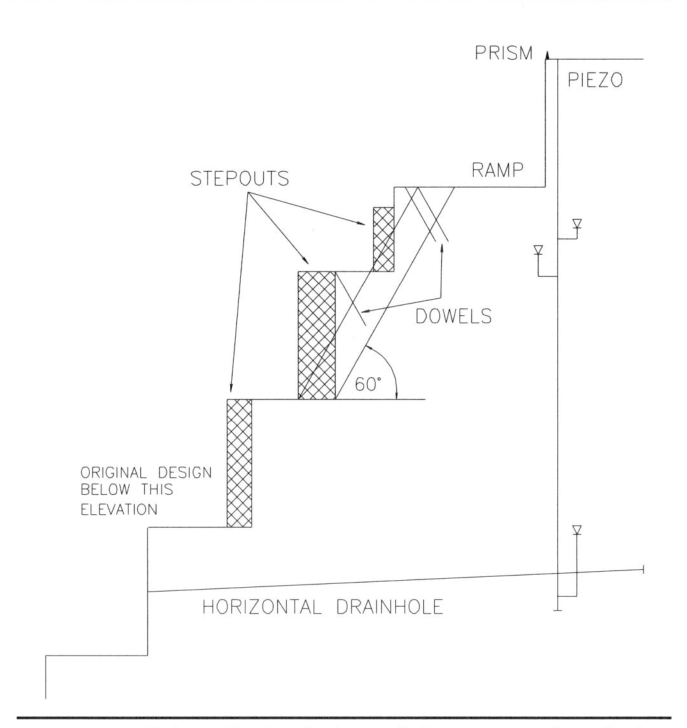

FIGURE 20.4 Schematic of typical remedial measures and monitoring methods

is immune to impacts found in a normal mining environment and to initiation from stray currents, static electricity, and radiofrequency energy. An improvement in the fragmentation, muck pile movement, and looseness has been noted since the introduction of the shock-tube system. Control of the throw of a blast in the relatively small pit bottom that presently exists at Palabora is important, since safe blast positions for equipment are limited. The shock-tube system has also been found to be a less expensive blasting option.

20.5 FUTURE MINING—RAMP NARROWING AND PIT DEEPENING

A preliminary assessment was carried out in early 1999 to identify extra ore that could be mined in addition to that contained within the present ultimate pit shell (Palabora Mining 1999b). Results of the study indicated that by narrowing the ramp and slightly increasing the IRA, 10 million tonnes of extra ore would be available from the open pit operation. Based on these results, two ramp-narrowing pit design options were proposed for further assessment through both kinematic and numerical modeling methods. The first design option utilized a 26.5-m-wide ramp below bench 51 and assumed mining to bench 64 (six benches below the originally planned pit bottom), while the second option incorporated a 31-m-wide ramp below bench 51 and assumed mining to bench 63.

20.5.1 Numerical Modeling of Pit Deepening

Using the 1998 model as a basis for assessing the ramp narrowing options, the UDEC program was again used to model the north wall of the pit (Piteau 2000). Sensitivity analyses that evaluated the length and spacing of the fault sets, groundwater conditions, and the effects of strain softening were undertaken. Undisturbed rock-mass strengths remained the same as during previous UDEC modeling, with residual strengths based on an equivalent Hoek–Brown 1997 "disturbed" criterion.

The model was calibrated by comparing the monitored slope displacements since 1984 with the model response. A good correlation between the two was indicated for both magnitude and direction of the horizontal and vertical displacements, with the various UDEC models (nine in total for one ramp-narrowing option) bracketing the monitoring data. A slightly reduced applied horizontal stress ratio of approximately 1.2:1 (H:V) matched the most recent underground in situ stress measurements (Gash 1999) and also yielded a better correlation to movement data than previous models.

Predictive or forward modeling of the proposed mining indicates that the degree of faulting (i.e., fault set spacing and length) required for instability is much more intense than observed on the exposed north wall. The inclusion of strain softening does not greatly increase failure in the rock mass. Furthermore, the stress levels in the lower portions of the north wall do not appear to be of sufficient magnitude to cause crushing of the carbonatite. Therefore, deepening of the pit through ramp narrowing is not expected to adversely affect the north wall from an overall slope-stability standpoint.

20.5.2 Kinematics of Pit Deepening

Based on kinematic analyses of the most recent geologic structural mapping data from the lower portions of the pit slopes, it would appear that bench-scale, structurally controlled instabilities will be the controlling factor in the pit-deepening design (Piteau Associates 1999a). Due to the decreased ramp width that will be required, it will be even more important to minimize damage to the ultimate walls by optimizing all final wall blasting and excavation procedures. Mapping of exposed bench faces and assessment of possible requirements for remedial measures to control slabs, wedges, and/or raveling will also be imperative as each bench is excavated below the ramp.

To consolidate and stabilize the ramp crest, a pinned wall will be constructed along the outside edge of the ramp. As well as taking up less space than the windrow of waste rock that is normally placed alongside the ramp for safety, the wall will be an integral part of a crest-reinforcement system (Piteau Associates 1999b). The wall will be anchored to solid bedrock by way of grouted I-beams or bars and will be designed to withstand impact from mine traffic and to prevent vehicles from riding over the wall. To maximize their effectiveness, the dowels (to be installed vertically) will be installed prior to taking the trim blast alongside the ramp and may be tied back to dowels on the other side of the ramp. Depending on conditions exposed on the bench face immediately below the ramp, additional dowels or other remedial measures could also be utilized.

Notwithstanding the above measures, should breakback into the ramp occur near the bottom of the pit or only involve a few meters of the ramp, it may be acceptable to operate the affected portion of the ramp as a single lane without significantly impeding normal mining operations. However, if the instability causes too great an inconvenience to mining operations or if it does not allow the safe operation of even single-lane traffic, more sophisticated remedial measures may be required, such as the construction of a tied-back, reinforced-concrete wall.

20.6 CONCLUSIONS

Ultimate pit slopes up to about 665-m high at interramp and with overall slope angles of up to about 58° and 50° have been developed over the past 20 years at the Palabora Mine. It is one of the steepest and deepest open pits in the world, with the proposed ultimate slope height now being approximately 830 m. Although new analytical tools have been used in recent years to conduct slope-stability assessments for the design of these slopes, basic geologic structural and kinematic analyses still form the basis of any geotechnical assessments of the pit slopes. While no major alterations of the open-pit slope designs have been made since the initial slope-design study in 1980, a number of minor design modifications and remedial measures have been implemented in local areas to allow the original slope design to be maintained.

20.7 DEDICATION

This chapter is presented in memory of Dr. Dennis Martin, past president of Piteau Associates Engineering Ltd. In addition to being a recognized leader in his chosen field of rock mechanics and engineering geology, Dr. Martin directed the pit slope-design assessments at the Palabora Mine from 1980 until he passed away in August 1998.

20.8 REFERENCES

du Plessis, L., and D.C. Martin. 1991. Numerical modeling studies for design of high rock slopes at Palabora Copper Mine. *Seventh Congress of the International Society for Rock Mechanics.* 799–804.

Fauquier, G.P. 1983. Trim blasting and double benching for steeper slopes and competent walls at PMC. *Engineering Mining Journal,* April.

Gash, P.J.S. 1999. *Analysis and interpretation of the in situ stress measurements carried out in RAW #1 on the production level.* Report prepared for Palabora Mining Company.

Hoek, E. and E.T. Brown. 1997. Practical estimates of rock mass strength. Submitted to the *International Journal of Rock Mechanics and Mining Sciences.*

Martin, D.C. and D.R. Piteau. 1978. Select berm width to control local failure. *Engineering and Mining Journal,* June.

Martin, D.C., N.S.L. Steenkamp, and J.W. Lill. 1986. Application of a statistical analysis technique for design of high rock slopes at Palabora mine, South Africa. *Mining Latin America.* 241–255.

Palabora Mining Company. 1999a. Shock tube trial at Palabora Mining Company. Internal report prepared by J.G. Hermann.

Palabora Mining Company. 1999b. Additional open pit tonnage—preliminary investigation. Internal report.

Piteau, D.R., A.F. Stewart, D.C. Martin, and B.S. Trenholme. 1985. A combined limit equilibrium and statistical analysis of wedges for design of high rock slopes. ASCE Specialty Conference, Denver, Colorado.

Piteau Associates Engineering Ltd. 1980. Slope stability analysis and design of the open pit slopes. Report prepared for Palabora Mining Company.

Piteau Associates Engineering Ltd. 1984. Review of geotechnical and hydrogeological conditions in the copper open pit. Report prepared for Palabora Mining Company.

Piteau Associates Engineering Ltd. 1991. Geotechnical assessments and numerical modelling for the west wall. Report prepared for Palabora Mining Company.

Piteau Associates Engineering Ltd. 1995. 1995 geotechnical review. Report prepared for Palabora Mining.

Piteau Associates Engineering Ltd. 1996a. June 1996 geotechnical review. Report prepared for Palabora Mining Company.

Piteau Associates Engineering Ltd. 1996b. November 1996 geotechnical review. Report prepared for Palabora Mining Company.

Piteau Associates Engineering Ltd. 1997a. July 1997 geotechnical review. Report prepared for Palabora Mining Company.

Piteau Associates Engineering Ltd. 1997b. November 1997 geotechnical review. Report prepared for Palabora Mining Company.

Piteau Associates Engineering Ltd. 1998a. Numerical modelling of the west wall with the UDEC Program. Report prepared for Palabora Mining Company.

Piteau Associates Engineering Ltd. 1998b. Numerical modelling of the north wall with the UDEC Program. Report prepared for Palabora Mining Company.

Piteau Associates Engineering Ltd. 1999a. May 1999 geotechnical review. Report prepared for Palabora Mining Company.

Piteau Associates Engineering Ltd. 1999b. October 1999 geotechnical review. Report prepared for Palabora Mining Company.

Piteau Associates Engineering Ltd. 2000. Updated numerical modelling of the north wall with the UDEC Program. Report prepared for Palabora Mining Company.

CHAPTER 21

Slope Stability at Aznalcóllar

Jonny Sjöberg*, John C. Sharp,† and David J. Malorey†

21.1 INTRODUCTION

The Aznalcóllar open-pit mine is in southern Spain, 40 km northwest of Seville. Open-pit mining started in the early 1970s, but portions of the orebody were mined underground as early as Roman times. The mine was acquired by Boliden Mineral AB in 1987 and was operated by Boliden Apirsa S.L. (a subsidiary of Boliden Mineral AB) until its closure in September 1996. As the Aznalcóllar pit was closed, the nearby Los Frailes pit was put into production.

Despite the relatively moderate slope heights (270 m at the end of mining) and the shallow slope angles (30°–38°), the footwall slope of the mine has suffered several large-scale failures. This chapter describes the stability conditions of the pit and how they were handled so as to continue mining without serious production disturbances. This included extensive instrumentation and monitoring of pit slopes during mining. Some attempts to quantify the failure mechanisms through analysis and numerical modeling are also described.

21.2 MINE DESCRIPTION

21.2.1 Geological Setting

Geology. The geology surrounding the Aznalcóllar Mine comprises a series of metamorphosed sediments, reworked sediments of volcanic origin, slates and shales, and phyllites (schists) from the Paleozoic Era. The general setting of the Iberian Pyrite Belt is shown in Figure 21.1. The major rock-mass structure, the cleavage/foliation, strikes east–west and dips to the north between 40° and 60° with local shallower and occasional steeper variations. The Paleozoic rocks are overlain by Miocene deposits of partially welded cap rock (1–2 m thick) and stiff brown and red clays with inclusions of slate (approximately 10 m thick).

In the footwall, slate is the predominant rock type along with the more schistose phyllite. The slate is dark gray or black and has strong cleavage, which strikes essentially parallel to the footwall. The phyllite is grey with a light-green tint. The cleavage is less pronounced than in the slate and is less planar, although the overall northward-dipping structure still dominates. The distribution of the various rock types (including pyroclastics) throughout the footwall is highly variable, with transition zones being evident. The interpretation is thus subject to a significant degree of uncertainty.

The rock mass is generally slightly weathered, i.e., the only evidence of staining is on the surfaces of the joint structures. The degree of weathering increases toward the contact with the orebody, adjacent to the contact with the Miocene and also to the west of the open pit, where a fault cuts through the sequence.

Discontinuities–Cleavage. The dominant structure throughout the footwall is associated with the cleavage where master joints have developed along this structure. The cleavage has a dip direction essentially toward the north. The dip of the cleavage varies throughout the extent of the footwall and ranges typically from 45° to 70°. The observed continuity of the cleavage is more than 80 m along strike, although in reality it is likely to be greater.

Spacing of the cleavage planes, developed as discontinuities, was recorded to range from 0.5 to 2 m. The cleavage itself would be expected to be on a much closer spacing (perhaps largely incipient). The surface roughness of the slate cleavage is limited. The phyllite cleavage surfaces are mineralized and smooth, probably as a result of the presence of mica on the cleavage planes.

Discontinuities–Joints. The discontinuities are typically suborthogonal to the cleavage, with a continuity of the order of 5 to 20 m. The roughness of all these joint sets is similar, with local asperities on the surface of the joints and slight undulation on a larger scale. The typical spacing of all joints is from 0.5 to 2.0 m. The degree to which the cleavage has controlled or dictated the development of the master joint sets is likely to have been significant.

The range of effective friction angles estimated from Barton roughness criteria (JRC) for the discontinuities at Aznalcóllar, assuming a stress across the joint equivalent to 20 m of rock overburden, is as follows:

Geological Unit	Effective Joint Friction Angle
Slate/phyllite (cleavage)	32°
Other joints	42°–51°

The above friction angles assume a basic friction angle of 25°.

21.2.2 In Situ Stress State

The magnitude and orientation of the in situ stresses within a geological unit depend on the geotectonic history (i.e., the stress history) and the rock type, which may influence the ability of the rock mass to retain a "locked in" stress. In areas of the Earth's crust that have undergone tectonic convergence, such as in southern Spain, the principal stress is typically horizontal and often significantly greater than the vertical stress (which remains primarily a function of the overburden depth).

While several different theories on the precise geotectonic setting of the South Portuguese Zone have been proposed, they all have the same inference in terms of an induced north–south maximum horizontal stress for the Palaeozoic sedimentary cover, in which the footwall slopes are excavated. The deformation history comprised folding and thrusting of the sediments into a series of imbricated thrust-bounded units. The orientation of the major thrusts is clearly shown in Figure 21.2 as a predominately

* Boliden Mineral AB, Sweden.
† Geo-Engineering, Goin Varin, St. Peter, Channel Islands, United Kingdom.

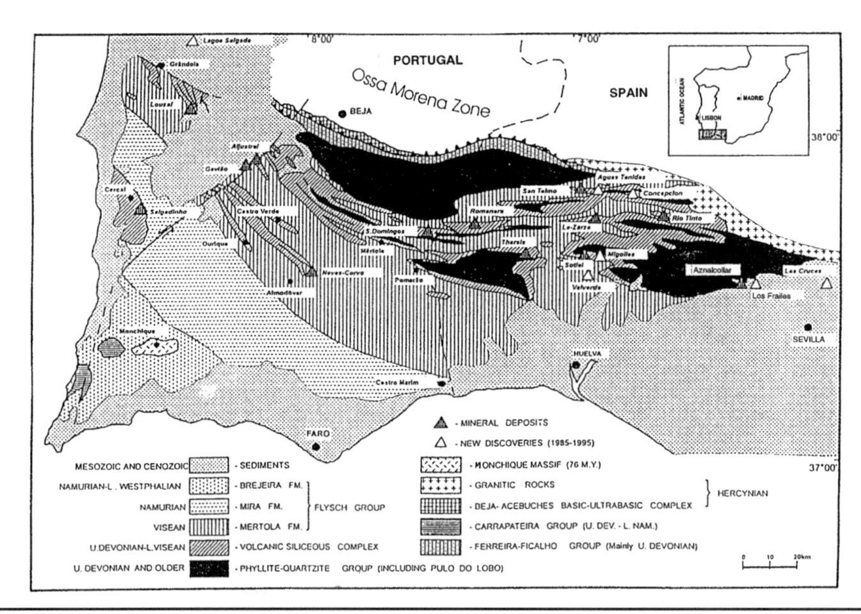

FIGURE 21.1 Iberian Pyrite Belt—location and general geology (adapted from Cavalho et al. 1997)

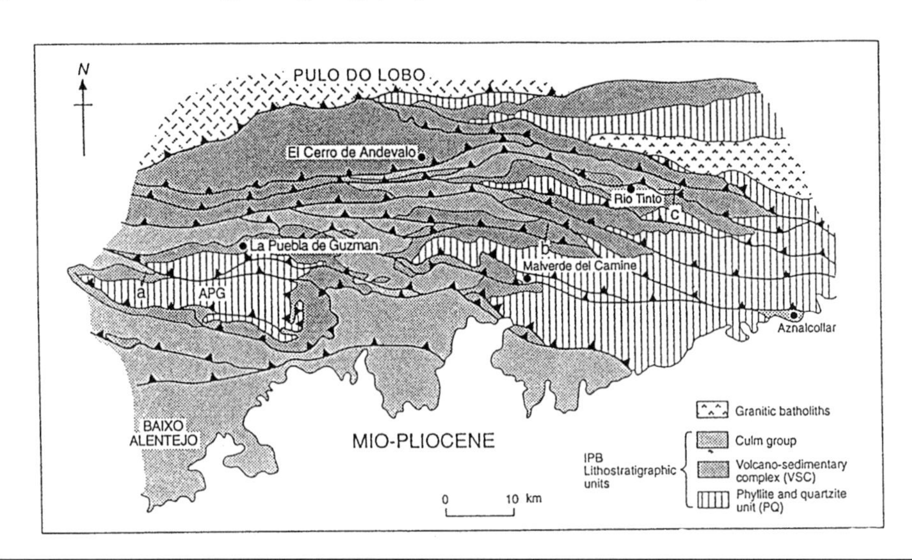

FIGURE 21.2 Iberian Pyrite Belt Spanish sector—lithological contacts (after Instituto Geologico y Minero de España 1982)

east–west trend, which implies a north–south principal compressive stress field. There is no evidence to suggest that this stress state would have been subject to any more recent variation, based on the published geotectonic data for the region.

21.2.3 Summary Footwall-Stability Model

In general, the footwall comprises a layered sequence with weak rock types. The rock type of the parent rock mass is a layered sequence of both stronger, less-foliated volcanic rocks and weaker, foliated and cleaved metasedimentary materials. The sequence has undergone several phases of deformation, and the strain of the deformation is more evident in the weaker rock types, hence the more pervasive foliation and cleavage. The cleavage and foliation within the weaker footwall rocks is pervasive and strongly developed. This structure has a variable dip with a recorded variation of between 45° and 70°. The primary structure is largely developed along the cleavage and foliation. In a fully confined nondeviatoric stress state, the cleavage and foliation would remain an incipient weakness. However, upon excavation of the footwall slope, the fundamental change in in situ stress would assist in opening and developing these features. Secondary jointing is also present but can be considered as subordinate to the cleavage.

21.2.4 Mine Geometry and Production

The east-west-striking orebody of the Aznalcóllar mine was mined for lead and zinc. The orebody dips to the north at an angle of 45° to 60° (becoming flatter with depth); hence, the upper portion of the footwall is located in waste rock, whereas some ore was left in the lower portion. Production from the mine during the last few years of operation amounted to approximately 2.3 Mton of ore, with a stripping ratio of 8.7:1 (waste:ore).

At the end of mining, the pit measured approximately 1,300 m by 700 m. A simplified map of the mine is shown in Figure 21.3. The western and central portion of the mine (approximately to section E213 000) was mined until 1994. The mine was then extended toward the east (section E213 300 to E213 410). Final mining of this portion of the pit was conducted during 1994 to the end of September 1996 and involved deepening of the pit (final depth of 270 m) as well as steepening of the overall slope angle from approximately 34° to 38°.

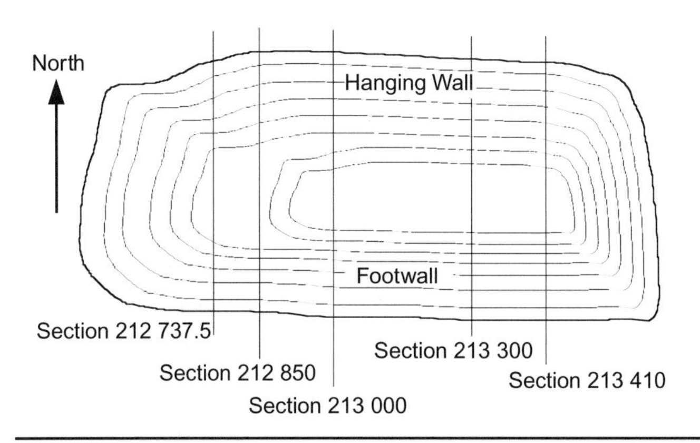

FIGURE 21.3 Schematic map of the Aznalcóllar open pit (figure not to scale)

Both the slope geometry and the behavior of the footwall were somewhat different in these two portions and are therefore treated separately here. During mining of the central footwall, several large-scale failures occurred in the footwall slope. These were (with one important exception) slow-moving failures, and mining could continue with only slight interruptions in the production. Final mining of the eastern footwall was accompanied by extensive slope monitoring, but unlike the central footwall, no overall slope failure developed during active mining. However, subsequent to the finalization of mining, a large-scale failure developed in this portion of the footwall slope. The mining sequence and stability conditions are reviewed in more detail for the central and eastern footwall, respectively.

21.3 CENTRAL FOOTWALL STABILITY RESPONSE, 1987 TO 1994

21.3.1 Introduction

A summary of the stability response of the Aznalcóllar Central Wall Zone over the period 1983 to 1992, when slope failures were experienced, is as follows.

January 1983	Central footwall: EL +90 m to +5 m; 335-m crest length
December 1987	Central footwall: EL +95 m to –80 m; 700-m crest length
November 1988–April 1989	Central footwall: EL +90 m to –90 m; 760-m crest length
June 1992	Central footwall: EL +95 m to –105 m; 760-m crest length

Key comparative data on the failures are shown in Figure 21.4. The principal failures of 1987 to 1992 are outlined below.

21.3.2 The 1987 Central Footwall Failure

In mid December 1987, the greater part of the footwall slope failed during and after a period of continuous and heavy rainfall. A series of crest-line cracks approximately 700 m long was created as well as cracking of the slope face down to the –50-m level. The surface geometry of the failed slope was reliably defined with the aid of an airborne photogrammetric survey. Records of precipitation indicated that a period of heavy rainfall occurred between December 1 and December 15, 1987, immediately prior to the failure. In total, 313 mm of rain fell during this period.

The maximum observed movement vector was 4,000 mm in the upper third of the slope, compared with an average overall movement of approximately 1,000 mm. An interpretation of the slope failure is shown in Figure 21.5. As recorded during the 1983 failure, there was no record or inference of any obvious geological structure coincident with the inferred failure plane profile. However, the heavy and continuous rainfall experienced prior to this failure event would have resulted in a raised groundwater level and been a contributing factor to the movements. Geometric changes in the slope profile resulting from the mine deepening and precipitation events were considered the major contributory factors in relation to the failure event.

21.3.3 The 1988 Central Footwall Failure

In November 1988, an area similar in extent to that which failed in December 1987 moved after a heavy rainstorm of 200 mm that lasted for three days. The surface geometry of the failed slope was defined following an airborne photogrammetric survey in late November 1988. Daily records of precipitation indicated that, in addition to the three-day rainstorm immediately prior to the failure (200 mm), rain persisted for the first 11 days of November 1988 and precipitation totaling 220 mm was recorded. A limited assessment of the groundwater profile was undertaken in the form of cored boreholes drilled into the mid-section of the footwall slope to depths of between 30 and 40 m normal to the slope profile. No groundwater was reported in these boreholes.

At the crest of the slope, the tension cracks opened to such an extent that large (1–2-m deep) troughs (grabens) developed. The tension cracks were near vertical to 70° in dip. Cleavage planes in the slate were observed as the controlling geometrical feature. In the main body of the footwall, a series of cracks became apparent. These indicated dilation of the superficial rock, local shear movement along the structure trending normal to the slope, and crack opening along the major cleavage structures. There was limited evidence of crack development in the lower levels of the open pit below the –25-m level, although at the –45-m level, a 90-m-long crack exhibiting the effects of a rock buttress being pushed upward and outward by the rock mass higher up the slope was observed. A more resistant section of the footwall in the lower portion of the slope was thus inferred, based on the general lack of significant surface distress.

An interpretation of the probable failure profile through Section E213 000 is shown in Figure 21.6. The position of the failure profile has been defined from the mapped crest and slope crack locations and from the inclinometer results obtained after the failure. The groundwater profile is based on a transient response (the form of which is based on later monitoring undertaken in 1989), superimposed on the inferred background profile.

As with prior failures, there was no obvious geological structure to which the overall failure geometry could be attributed. The failure was interpreted as a development of the 1987 failure; consequently, significant prior movements were likely to have occurred along the same basal failure surfaces. The increased level of groundwater and the resulting geometrical changes to the slope caused by mine deepening were considered the major contributing instability factors.

21.3.4 The 1989 Central Footwall Evaluation

Following the 1988 footwall failure and the subsequent evaluation, subsurface monitoring instruments, i.e., inclinometers and piezometers were installed into the slope to monitor the failure and record ongoing movements. The subsurface monitoring program comprised installation of five, 50-m-deep inclinometers on two cross sections through the slope and the installation of three 50-m-deep piezometers along the lower part of the slope (Figure 21.7).

From the inclinometer monitoring, the depth to the major movement zones and the movement magnitude with time were inferred. Typical daily movement rates of 3 mm/d were recorded during the evaluation period, although this increased to 9 mm/d when the groundwater level rose above 7 m following rainfall in early April. Summary results from two of the inclinometers are

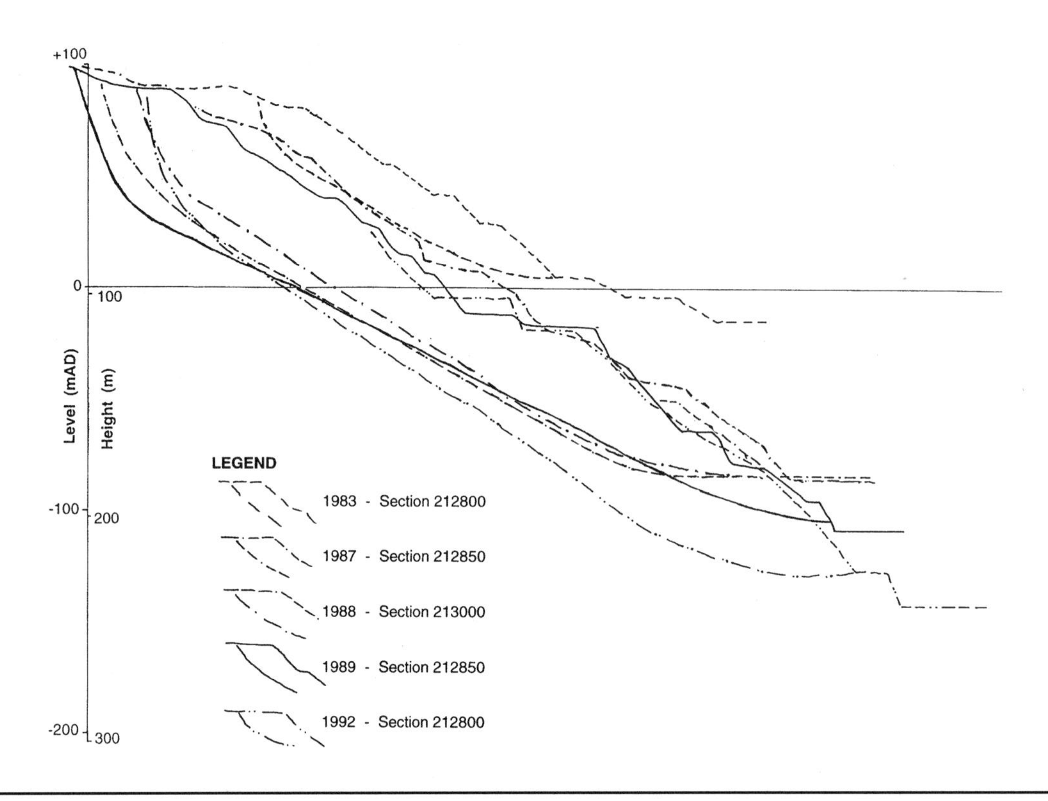

FIGURE 21.4 Central footwall failures—comparative failures

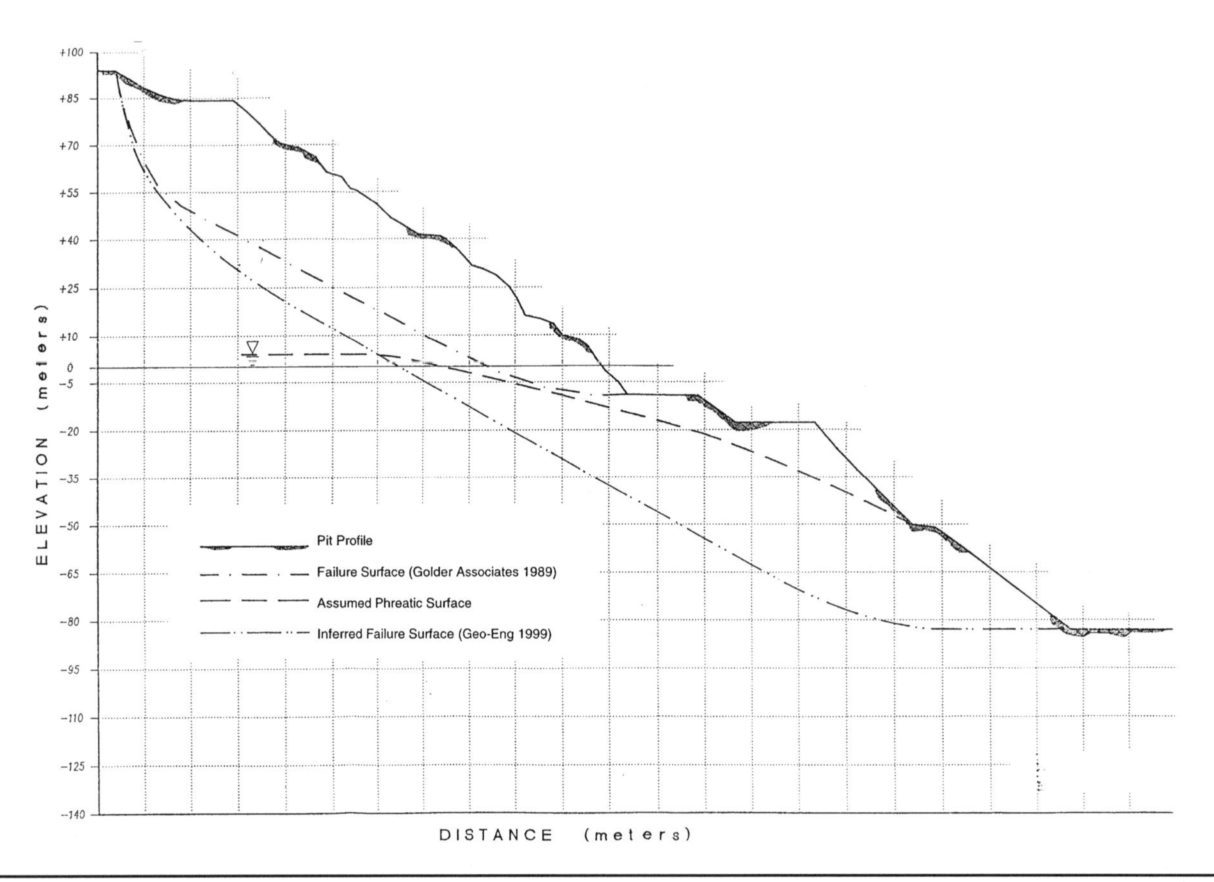

FIGURE 21.5 Central footwall failure 1987—interpreted failure geometry

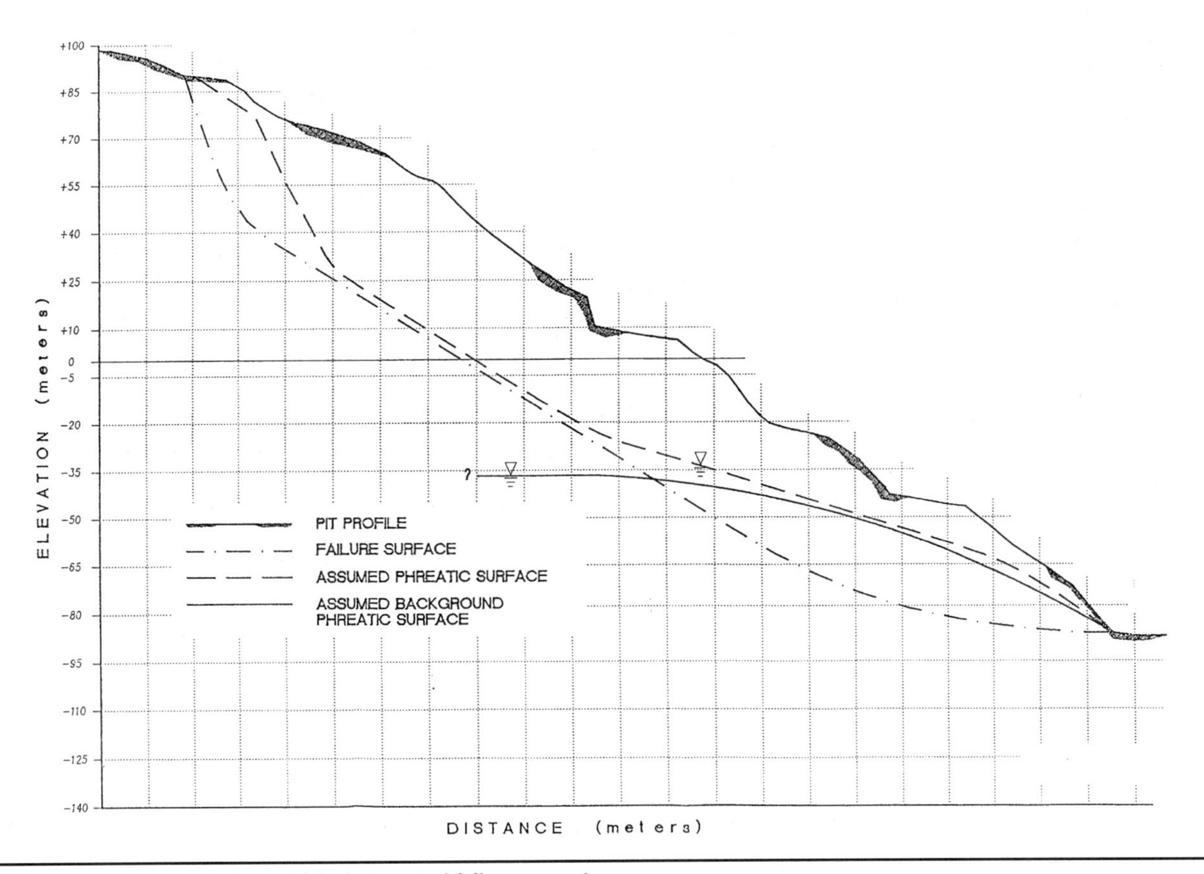

FIGURE 21.6 Central footwall failure 1988—interpreted failure geometry

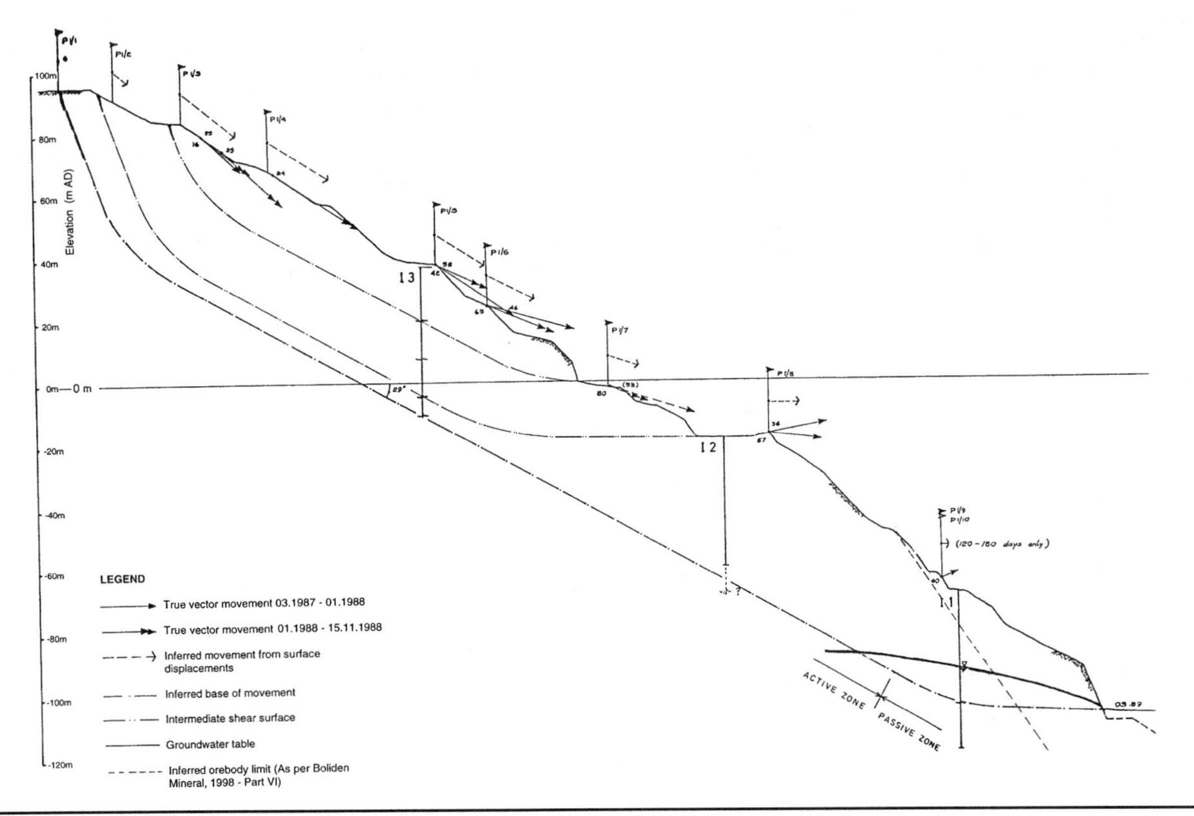

FIGURE 21.7 Central footwall failure 1989—interpreted section 212.737

given in Figure 21.8. From these, the following significant findings were deduced working downslope for Section E212 737:

- Inclinometer I-3 (upper slope) indicated a series of pronounced and discrete movement zones (0–17 m, 17–29 m, 29–41 m, 41–47 m) with distinct shear displacement evident, particularly at 17 m, 28.5 m, and 41.5 m. Within each zone there was no displacement gradient with depth, i.e., simple lateral translation occurred.
- Inclinometer I-2 (mid slope) indicated simple lateral translation of the slope section down to a depth of 40 m with no displacement gradient.
- Inclinometer I-1 (lower slope) showed a discrete shear movement at a depth of approximately 36 m below surface. A progressive displacement gradient with depth is evident between the surface and a depth of about 30 m, indicating possible rotation (toppling?) of the rock strata above this level. In addition, a complex zone of enhanced differential shearing termed the "boot" effect is evident over a thickness of some 5 m above the basal shear zone.

Three piezometers were installed in March 1989. These piezometers indicated a moderate phreatic surface profile in the toe zone of the slope that increased eastward. The interpreted profile is shown in Figure 21.7. Water-level measurements and drilling observations from inclinometer holes further up the slope indicated a general phreatic surface that was 50 m or more below the slope profile and hence generally below the zone of movement.

Sections through the failed slope were produced at the locations of primary information, such as installed surface and subsurface instrumentation. Figure 21.7, illustrating through the failed slope at E212 737, shows the key features of the slope-movement response. The inclinometer data from the upper slopes are particularly noteworthy in that they indicate a number of discrete failure surfaces in this zone, each interval being defined by a zone of consistent translational movement. In the toe zone of the failure (below the −50-m mining level), there is evidence of outward rotation of the rock mass, which is considered to represent differential shear along cleavage planes, accommodating overall movement downslope to give a "passive" overturning movement zone. Upward components of movement can also be attributed to dilation of the toe zone under ongoing movement.

With the positions of the failure surfaces identified and the behavior of the groundwater profile in relation to rainfall established, a model of active and passive movement zones could be inferred. The active zone, consisting of the majority of the slope, was recorded as moving parallel to the failure plane, while the passive zone was defined as the toe region with lesser movements being inferred coupled with the development of complex shear/dilational processes, as reflected by the surface-movement profile.

21.3.5 The 1992 Central Footwall Failure

The background slope movement rates recorded for the month of June 1992 were approximately 2 to 4 mm/d, which increased up until the week ending July 16 to a rate of 3 to 5 mm/d. However, between July 16 to July 22, the rate increased generally to 10 to 15 mm/d over the majority of the slope. Between July 22 and 0800 hours on July 24, a completely unexpected movement trend developed with a very significant movement rate of 1.21 m/d (50 mm/hr). Readings reached a maximum recorded movement rate by 1800 hours on July 24 of 1.68 m/d (66 mm/hr). The rate of movement then gradually decreased to about 0.48 m/d (20 mm/hr) by 1800 hours on July 25, then, over the next few days, it reduced further to 0.27 m/d (11 mm/hr) by July 27 and to 0.23 m/d (9 mm/hr) by July 31. Thereafter, the rate continued to decrease to about 25 mm/d by the end of August 1992.

At the time of the 1992 failure, the toe of the main slope was at the −125-m level. (This compares with a toe level of −105 m at the time of the 1988/1989 movements.) Daily records of precipitation showed that for the three years prior to this movement event there was limited rainfall, in particular no storm events recorded for the previous wet season. There is no record of any rainfall between June 16, 1992, and August 29, 1992.

The main movement observations showed that the maximum recorded movement of 7 to 8 m occurred on the western part of the central footwall, while the average range of movement for the overall slope profile was 4 to 5 m. The dip of the movement (in section) was generally the same as that for previous movements (November 1989 to January 1990). The available information collected after the 1992 failure enabled an interpretation of the possible failure profile to be made. The profile through Section E212 800 is shown in Figure 21.9.

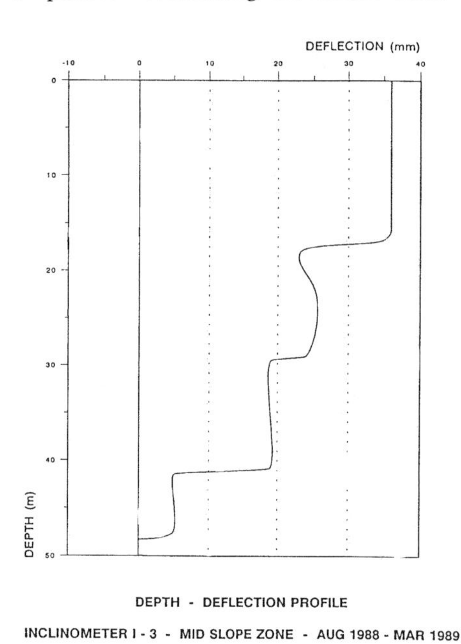

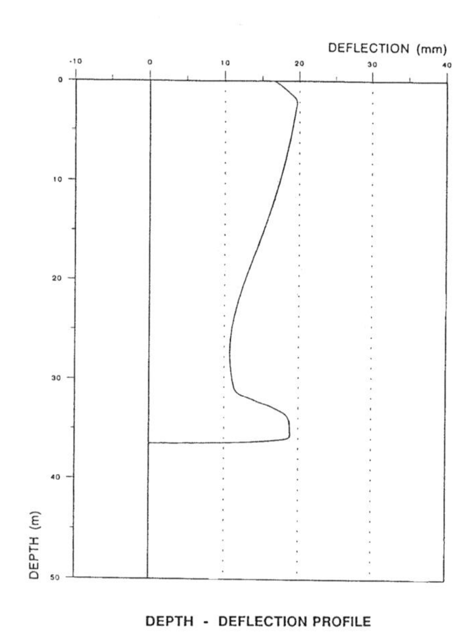

FIGURE 21.8 Central footwall failure 1989—summary inclinometer data

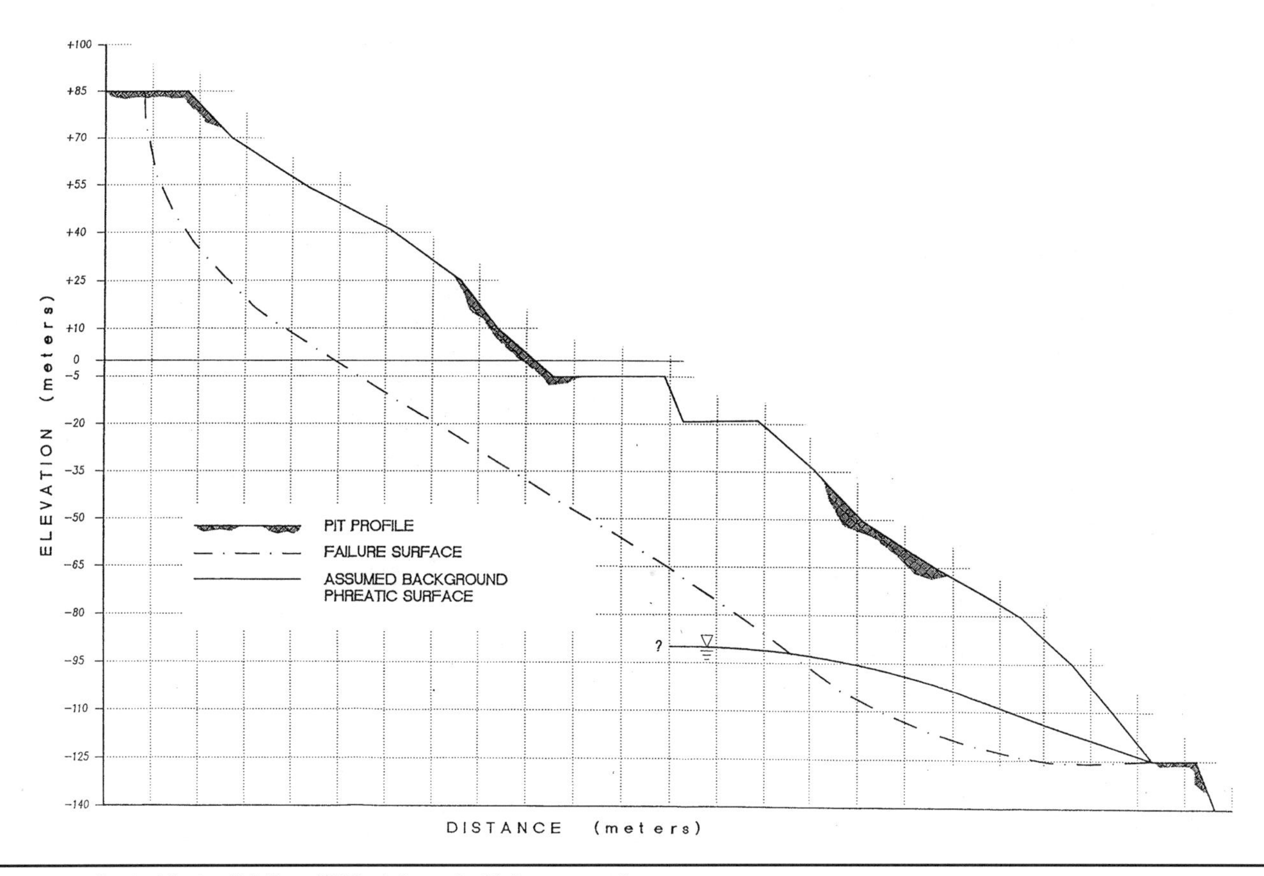

FIGURE 21.9 Central footwall failure 1992—interpreted failure geometry

TABLE 21.1 Predicted stability conditions—P95 slopes—calculated safety factors and associated slope heights

Profile	Section Location	
	E213 110	E213 310
August 1994 [DR] <NORM>	1.00 [230]	1.00 [240]
PROJECT 95 [DR] <NORM>	0.89 [260]	0.85 [285]
PROJECT 95 [UDR] <NORM>	0.77 [260]	0.75 [285]

NOTES:
DR drained (favorable stability state)
UDR undrained (most adverse, plausible following rainfall)
NORM normalized factor of safety referred to August 1994 datum
[ ] slope height in meters

The 1992 failure occurred when the groundwater profile could be regarded as extremely low, following the reduced levels of rainfall prior to the failure. The slope behavior was characterized as similar to the previous failures, with an active and passive zone model. The failure cause in relation to the high rates of movement observed remains unclear. Minor changes in slope geometry would have led to a progressive reduction in stability levels, but this does not explain the basis for the observed sudden, but finite, movement patterns.

21.4 EASTERN FOOTWALL STABILITY RESPONSE, 1994 TO 1996

21.4.1 Mining Plans and Sequence

The eastern footwall development at Aznalcóllar (also termed Project P-95) was an extension of the prior central footwall phase, as described above. The key stages of development in terms of slope stability took place between 1994 and 1997. The project was planned to extract the eastern section of the Aznalcóllar orebody and involved relatively steep slopes compared with the prior central footwall zone.

21.4.2 Preliminary Stability Assessment

A comprehensive analysis of slope-stability conditions for the footwall slopes at Aznalcóllar was undertaken from which the stability conditions of the proposed slopes for P-95 were evaluated. The findings showed adverse stability states compared with past experience based on the central footwall behavior. The predictions (from limit-equilibrium analysis) are summarized in Table 21.1 and outlined below.

- The planned P-95 slope profiles would not, on the basis of both precedent study and current monitoring trends, remain sufficiently stable to permit safe and efficient development of the planned slope profiles with the proposed mining sequencing.
- The risks of a significant slope failure were considered very severe, in particular during and after heavy rainfall.
- The combination of planned slope height and steeper profile could give rise to a sudden, catastrophic event that would be difficult to predict, even with continuous monitoring.
- Stabilization using slope drainage and improved blasting coupled with continuous monitoring surveillance could be used to extend the excavation from the current profile, subject to the implementation of careful and rigorous control measures.

A revised mining and stability-control approach was identified. This sequence proposal, in effect, comprised the development of an overall slope profile. This profile allowed progressive deepening and ore development (sequence I) but deferred creation of the final slope profile to as late as possible. This was done by leaving a buttress zone (sequence II) adjacent to the lower footwall profile. In addition, from a program standpoint, it was planned that all mining be completed prior to the winter of 1996/1997 to avoid adverse influence from precipitation.

21.4.3 Control Monitoring Systems

To develop the mining objectives in a reliable, controlled manner, the following monitoring systems were identified:

1. Active, continuous, alarmed systems
   - Direct extensometer monitoring of footwall from the −125-m level
   - Direct settlement monitoring of footwall from the −125-m level
2. Active global slope-monitoring system
   - Surface hub network survey using automated precision survey system with selection of critical hubs for input to monitoring control system
3. Passive information-monitoring system
   - Inclinometers to detect depth of movement
   - Piezometers to measure groundwater pressure

Systems 1 and 2 were intended to form the primary control system for safety. The intent was that system 1, via time-regulated alarms, would give an immediate warning to the working area most likely to be influenced by potential instability. System 2 would provide a second level of assessment to confirm the output of system 1 on a more extended basis.

A direct extensometer (dilation) monitoring (DDM) system was required to measure the potential outward deflection of the slope relative to a fixed point within the slope at three locations along the footwall slope (E-1, E-2, E-3), as shown in Figure 21.10. The system was comprised of high movement tolerance extensometers capable of operational validation at all times. The instruments detect critical outward movement of the lower slope (toe) zone that could indicate a change in state of the overall slope. It was recognized that the retention of a competent toe zone would be critical to overall stability.

A differential settlement monitoring (DSM) system was required to measure the vertical deflection of the slope (lower part) relative to a fixed point on the pit endwall at different locations along the slope (−125-m-level bench). The system was comprised of a differential level system capable of continuous monitoring and alarm sensing. The system identifies the consistency of movements in the toe zone over the strike length of the slope. It would then be possible to identify progressive failure extent and assess the consequence of this in terms of overall stability trends.

The continuous surface monitoring system was a standard high-precision Leica survey system (total station) with high accuracy and direct distance capability. In addition, it could operate on a programmed automated survey sequence.

The inclinometers were installed to detect the depth of movement resulting from the ongoing slope displacements as well as the movement profile with depth. The units were primarily intended to provide information on the overall slope deformation from which ongoing stability assessments could be derived.

The piezometers were installed to check water-pressure conditions in the lower slope zone. The water-pressure data would be used as part of the stability-assessment process and identify adverse changes following periods of significant precipitation.

These multiple methods of monitoring also provided some redundancy. Furthermore, measurement results from one monitoring method could be cross-checked against results using another measurement method. This enabled some validation of monitoring results and increased the reliability of the entire system.

21.4.4 Observed Movement Response

Mining activities were completed in October 1996, and the majority of the instrumentation remained operational until mid November 1996. The primary data sources, which were used for the comprehensive stability assessment, were as follows. Note that the purpose rating value for stability–movement evaluation is given in parentheses.

- Global surface monitoring network (highly significant)
  - Surface-monitoring points on the slope for which long-term displacement trends are available.
- Subsurface inclinometers (highly significant)
  - Three inclinometers (I-1, I-2, and I-3) on two sections that recorded the relative shear displacement with depth. Inclinometer I-4 was introduced to the system in August 1996 and proved valuable in assessing the lower slope behavior.
- Extensometer data (significant)
  - Three extensometers (E-1, E-2, and E-3) on three sections that recorded overall movement along the axis of the extensometer with slope development. Capable of high degree of time resolution. E-1 was lost during December 1995 and was considered to have been too shallow to effectively show any movements. E-4 was introduced in August 1996 and proved to be very reliable in terms of recording slope movements.
- Hydraulic leveling system (significant)
  - Precise continuous leveling system along the −125-m-level bench giving trends through time. This system ceased to operate after May 1996 but proved to be very valuable during the previous winter.
- Precipitation (very significant)
  - Input parameter in terms of movement potential and groundwater pressure generation. Duration and magnitude (in combination) important.
- Piezometric levels (very significant)
  - Three piezometers on two sections that provided information on water levels (piezometric heads) through time and hence inferred effective stress state within rock mass. Potentially significant in terms of stability/movement influence.
- Excavation geometry (very significant)
  - A fundamental input to the stability criterion. Flooding of the pit floor prevented mining during much of the 1995/1996 winter period, but from July 1996, geometric changes were the primary cause of movements.
- Blast-event monitoring and PPV values (significant)
  - Initially only a limited number of events were recorded but during the final months, monitoring was performed on a regular basis and produced correlatable data. Considered particularly significant in terms of the localized stability of the critical lower slope (toe) zone.

It was also found possible to define movement periods having similar geometrical and rate characteristics, which can be broadly summarized as follows:

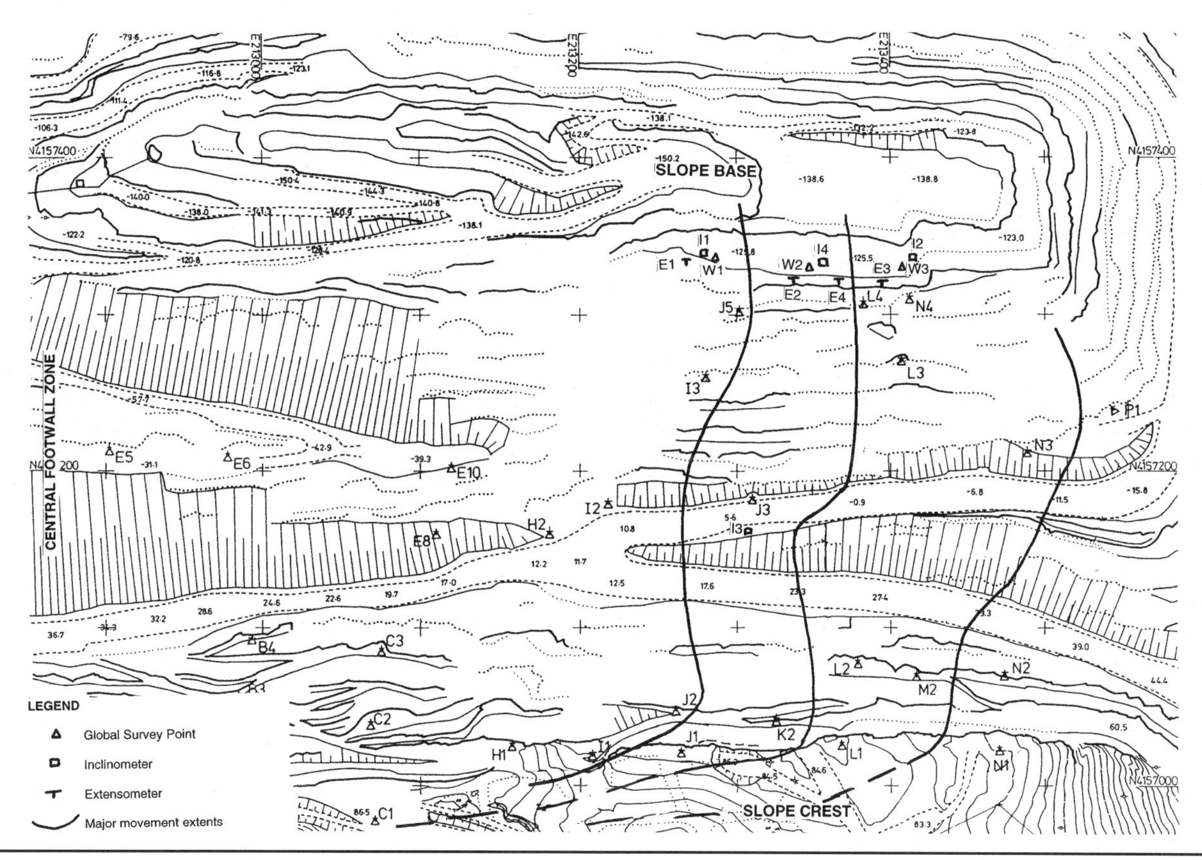

FIGURE 21.10 Eastern footwall failure—plan of slope, monitoring locations, and failure zones

Period	Date Interval
1	November 1994–end April 1995
2	May 1995–end March 1996
3	April 1996–end September 1996
4	October 1996–December 1996

Period 1 commencement is based on available data but generally takes into account the winter enhanced movement period and mining. Period 2 covers summer 1995 and winter 1995/1996 and incorporates periods of limited slope excavation (April–October 1995) and no slope development (between October 1995 and May 1996). Period 3 covers April to June 1996, when very limited slope development occurred followed by intense development (−155 to −183 m) from late June to October 1996. Period 4 covers the postmining up to cessation of detailed monitoring, which includes commencement of backfilling operations from the eastern end of the mine. Some monitoring was conducted even after this date, but the accuracy of these measurements was much lower.

From the overall movement data, the following types of movement responses were recognized and defined in terms of the particular slope response for the eastern footwall. The central footwall precedent behavior was also incorporated.

- **Consistent mass movement with depth (CTM).** Essentially translational movement of the slope that can be identified as a uniform rate of displacement with depth. Each zone is therefore bounded by a discrete shear displacement zone. The response is typified by the two-zone response of inclinometer IE-1.
- **Variable mass movement with depth (VTM).** Essentially translational movement of the slope that can be identified with a movement gradient with depth. May or may not be terminated by a discrete shear displacement zone. The response is typified by the lower zone response of inclinometer IE-3.
- **Complex shear and dilational behavior (CDB).** Movement other than represented by simple translational phenomena. Generally evident in the toe zone by outward and upward vector trends. Potentially more variable and irregular movements possibly involving block dislocations. May be accompanied by basal shear zone response. May be influenced by the orebody zone/geometry. Usually associated with significant water pressure in toe zone.

21.4.5 Movement Compliance With Geological Structure

Extensive studies from 1987 to 1997 focused on the potential compliance between movements and geologic structure. At Aznalcóllar, over both the central and eastern zones, no geologic structures that could be considered associated with significant weakness zones of limited shear strength were identified.

The main geologic structure was recognized as moderate to steeply dipping foliation within the phyllite rock mass, typically with a dip direction subparallel to the slope but at a dip angle typically in the range of 50° to 65°. Subordinate subnormal jointing (accommodation in nature) was also evident. In general, as with the central wall failures, movement compliance with existing geologic structure was limited to the upper movement zones in terms of the rear scarp geometry consistent with parting along foliation.

The ability of the foliation surfaces to accommodate internal shear is evident but not apparent from the mid-slope movement data that commonly indicate mass movement over the major proportion of the slope. The flexural deformation of the foliated rock mass is, however, a potential and important mode of response for the toe zone.

21.4.6 Intact Pyrite Orebody in Eastern Footwall Toe Zone

While the available information from drilling was not detailed, it was sufficient to allow an interpretation of the extent of pyrite ore that remained in the slope toe. The pyrite rocks have a unit weight that is significantly greater than that of the footwall rocks and foliation is not present to the same extent. It was postulated that the stronger pyrite orebody rocks could add to the strength of the footwall toe, assuming that there is a sufficient cross-sectional area to provide some resistance to the overall scale of slope movements.

The remnant orebody thickness was inferred to be of the order of 10 to 20 m, and it was clear that the contact surface varied across the footwall. It was also noted that within the pyrite ore discrete but limited occurrences of shale (phyllite) occurred that would reduce the overall flexural integrity of the ore zone as a unit.

21.4.7 Basal Shear Zones (Movement Induced)

From the inclinometer data, it was evident that, although discrete zones of significant movement change do occur, interpretable as shear movements, their extent is of note in terms of potential thickness. Detailed observations from unit IE-3 and the progressive deflection behavior with time leads to the tentative conclusion that the shear zones over which differential movements are concentrated are of the order of 5 to 8 m in thickness. The results are also supported from prior observations by the central footwall movements that indicated zone thicknesses of the order of 4 m. Such zones could potentially occur as kink bands over this width. Although this mechanism is somewhat speculative, it is supported by tentative proposals put forward by Ladanyi and Arcahmbault (1970, 1972, 1980). The zones are portrayed in terms of sectional development as basal shear zones (BSZ). Within a given slope, more than one of these zones may occur in practice at the base of individual movement zones.

21.4.8 Observed Groundwater Response

During period 2, the groundwater level was low throughout the traditionally dry summer (1995). The precipitation experienced from November 1995 to January 1996 resulted in a recorded increase in groundwater level of some 21.5 m (piezometer 3). Piezometers 1 and 2 rose concurrently with piezometer 3 up to the −125-m bench level. Piezometers 1 and 2 were both making water between mid December and mid February, after a water level rise of 10 m from the summer level. The groundwater level dropped during March and April 1996 by some 8 m at the end of the period, and the water in the mine had been pumped out.

During period 3 (April–September 1996), groundwater levels continued to fall, except for a small transient rise at the end of April/start of May associated with a heavy rainfall. By October, groundwater levels had dropped 8 m below the previous summer minimum, reflecting significant geometrical changes and drainage measures.

During the early part of period 4, rainfall remained low and the groundwater level fell an additional 5 to 6 m as the pit bottom was reduced to −183 m. No further monitoring was undertaken after November 9, 1996; however, it is known that rainfall remained low until the middle of December 1996.

21.4.9 Sectional Interpretation of Slope Movements

The sectional interpretation of slope movements based on the monitoring data is key to understanding movement response and inferred stability. From such assessments, it may be possible to derive models of behavior for use in future slope design. Such models will always be subject to a level of uncertainty, particularly where the movement conditions and associated failure mechanisms are complex. In the context of the eastern footwall, the primary intent was to interpret the observed movement characteristics and hence infer possible overall slope deformation mechanisms.

The principal basis of the movement interpretation was the discrete movement surfaces recorded by the inclinometers and the extent of crest cracking observed on site. A typical sectional interpretation of the more significant slope movements associated with period 2 is shown in Figure 21.11. Such interpretations indicate the complex nature of the slope movements in section. A progressive movement response is evident, although a fundamental appreciation of the nature and mechanisms of failure could not be derived from the available database.

21.4.10 Eastern Footwall Slopes—Tentative Failure Model

An interpretation of the failure state was made in an attempt to understand the progressive movement response based on two time responses corresponding to the end of March 1996 and the end of December 1996. The movement response and status for each time are summarized in Figure 21.12. The cross sections, which are partly diagrammatic and interpretative, indicate the following:

- Foliation characteristics and ore zone occurrence
- The primary movement zones
- The discrete surfaces of shear movement as inferred from inclinometer data
- Principal displacement vectors (measured and inferred) in relative rates
- Characterization of the slope zones in terms of movement response as follows:
  - consistent mass movement with depth (CTM)
  - complex shear and dilational behavior (CDB)
- Displacement magnitudes for key points within the slope
- Average periodic phreatic surface

By March 1996, the slope failure had developed in the form of an upper, faster-moving zone "U" and a slower-moving, more deep-seated zone "A." Both the upper and overall zones are bounded at the base by well-developed shear zones over the main central slide zone. An approximately planar failure surface is inferred dipping at 30° to 32°. As noted previously, there is no evidence of any preexisting structures of this form; however, the mechanism controlling the development of this extremely weak surface remains obscure. Above this surface, the slope movements are essentially translational, except for the upper scarp zone. The overall zone U1, U2, U3 could be considered the active zone. The upper scarps of both zones are interpreted as being structurally controlled by foliation.

From the available data, all movements are essentially above the groundwater table and hence groundwater pressures appear to have been of limited influence in the overall upper movement zone. Some influences in the toe zone can be inferred, which could have been significant in association with the changing stress field caused by excavation.

The toe zone of zone U, bounded by points U2, U3, and U4, is inferred to be a zone of complex shear and dilational behavior that cannot be easily rationalized. As of March 1996, the toe zone was undergoing (1) upward and outward movements and (2) rotation of the toe zone (as inferred from inclinometer movements). The key and primary weakness along which movements can occur is the foliation, as shown on the section. Particularly in this zone, the effective normal stress is reduced both by the influence of excavation and by elevated groundwater pressures during winter.

The outermost part of the slope is formed by the ore zone, which is considered relatively strong and blocky, in comparison

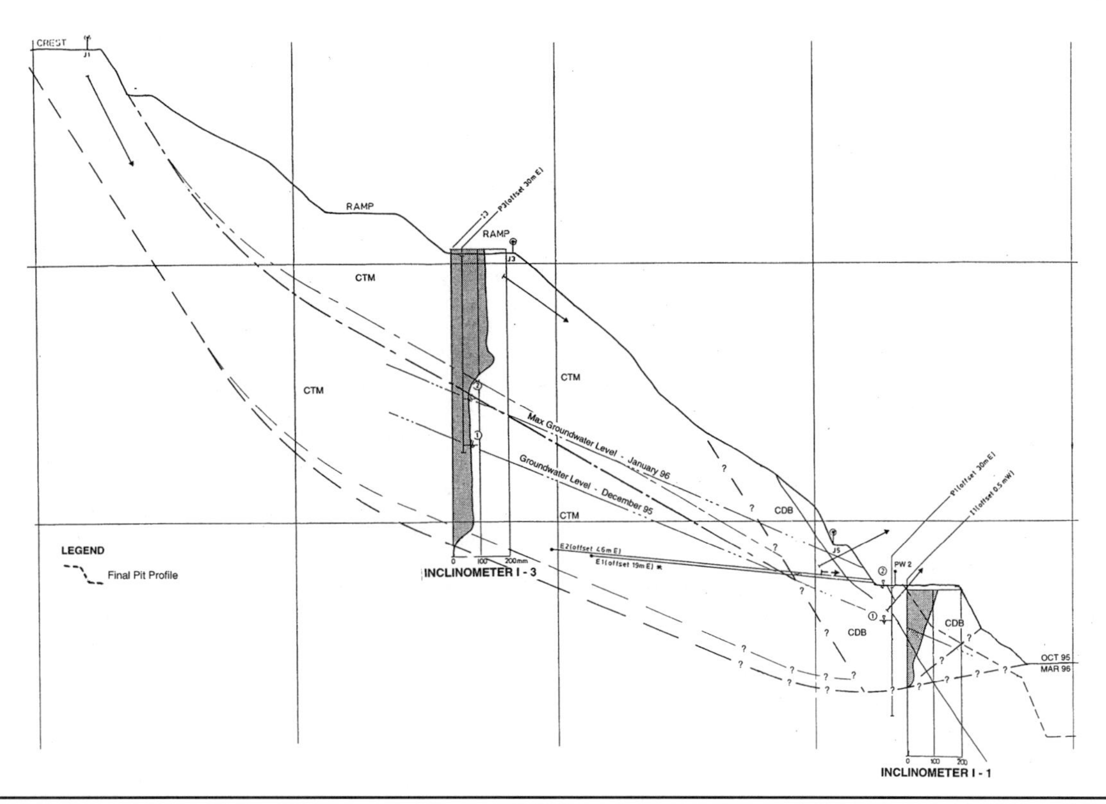

FIGURE 21.11 Eastern footwall failure—typical cross section, March 1995 to March 1996

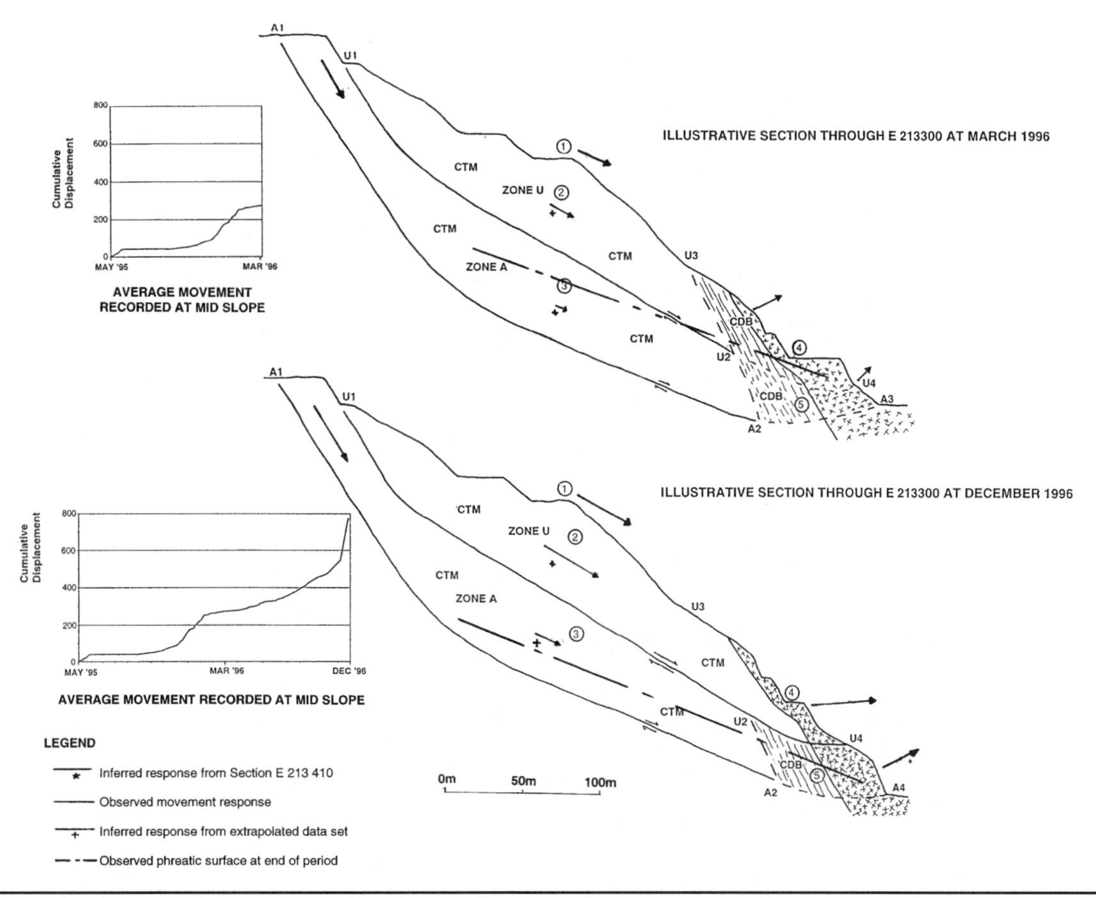

FIGURE 21.12 Aznalcóllar eastern footwall—observed failure mechanisms

to the metasedimentary footwall sequence, but also of fairly narrow thickness relative to the overall slope scale. From the data available, it is concluded that the toe zone at this time was being subjected to an overturning movement, based on the stress imparted from the upper translational block. This was manifested by uplifting of the toe zone and orebody slice, with multiple shear accommodation along the foliation surfaces. With time, a shear failure zone along the base of the overturning zone could be expected, as the individual strata ruptured under the induced flexural strains.

Zone A movements comprising the overall failure are in essence similar to the zone U response described above. The cumulative deformations are, however, significantly less—95 mm compared with 195 mm for points 2 and 3 in the central movement zones. The basal shear zone is inferred to be significantly shallower in angle than the upper zone (23° compared with 31°), and this can be attributed, in part, to the more adverse effective stress state at this depth due to the significant groundwater pressure levels (equivalent head of 50–60 m).

The toe zone of zone A, bounded by points U2, A2, and A3, is again inferred to be a zone of complex shear and dilational behavior, similar in nature to the upper movement zone. The degree of deformation is, however, markedly less and hence it is in a significantly earlier failure state. The zone is also subject to significant variation in terms of changes in geometrical influence because it is at the toe of a slope subject to active mining.

By December 1996, i.e., two months after termination of mining, the failure state evident in March had developed significantly. Available data indicated that the failure development since March had followed a similar pattern in terms of extent and depths of movement. The toe zone was, however, influenced to a significant degree by the later stages of mining, as might be expected. By December, the basal movement zones of both the upper and overall failure had become clearly established through the toe zone as a continuous shear surface.

Of particular note was the response of the toe zone, which appeared to develop from a zone of complex shear/dilational behavior to a translational shear response. This is indicated both by the inclinometer data (I-1 : CTM response) and the movements in the toe zone (principally the −125-m-level bench), which showed a near-horizontal vector response.

The movement change from March 1996 can be considered in terms of a ratio of displacement rates for the period January to March 1996 compared to those for the period October to December 1996. The monitoring data showed a fairly consistently increase in movement rates by approximately 50% throughout the movement zone. A manual survey of a few still accessible survey points was conducted in April 1997, i.e., nearly seven months after the completion of mining. This showed that movement rates had continued to increase in the postmining period, reaching values that were at least 5 to 10 times larger than those recorded in period 4 (October to December) (see also Figure 21.18). This was despite the fact that backfilling of waste rock had commenced, which is thought to add support to the lower portion of the slope.

Overall, the zone U movements could be considered governed by a post-failure condition in terms of shear strengths along the induced failure surfaces. It is considered likely that shear strengths had developed to a state close to residual in the toe zone, with cumulative discrete shear movements of the order of 200 to 250 mm along the basal movement plane.

The significant geometrical changes due to the later stages of mining had a marked effect on the zone A lower toe zone, which in turn influenced movements of the overall slope. It can be reasonably inferred that the base of the zone A movements in the central and upper parts of the movement zone did not change between points A1 and A2. It is also important to appreciate that, when first recognized, the inferred movement zone occurred to a depth significantly below the slope toe.

Until the cessation of mining by the end of September 1996, it was apparent that the toe zone of the overall slope was still responding with a complex shear/dilational response. It is inferred that a basal discrete shear surface had not fully developed by this stage and that ongoing deformations involving significant and complex strain/displacement mechanisms (as indicated by inclinometer I-4) were occurring. From the data available, it can be inferred that as of early December 1996 a controlled overall movement state still existed as a result of both the passive resistance (support) provided by the toe zone and favorable groundwater conditions. It was, however, evident that this stability reserve would be lost or severely attenuated as a result of more adverse groundwater conditions during the winter together with a decreased level of strength with ongoing movements and the generation of a discrete basal shear plane through the slope toe. The decrease in stability as a result of both the groundwater and strength changes would have led to an unacceptable stability state for ongoing mine operations.

21.5 ANALYSIS

21.5.1 Approach

Previous stability analysis of the central footwall included simple limit-equilibrium models with various groundwater conditions assumed (Golder Associates 1989, 1995; Krauland 1995), as well as numerical modeling (Proughten 1991; McCullough 1993; Hencher, Liao, and Monaghan 1996). The limit-equilibrium analyses were useful in that they provided some data on failure strengths (back-calculated). The previous numerical modeling studies were, however, of limited use, since they were not calibrated against available measurement data. Furthermore, the numerical models were very small and boundary effects could not be ruled out. None of these studies addressed the failure development of the eastern footwall.

In this section, the eastern footwall is analyzed, with a focus on its behavior during final mining of the pit. Additional analyses were conducted for the central footwall. These were performed to provide additional insight into the complex mechanisms of the footwall failures. Numerical modeling was chosen as the analysis tool. Because failures did not appear to be directly structurally controlled, a pseudo-continuum modeling approach was chosen. The two-dimensional finite difference program Fast Lagrangian Analysis of Continua (FLAC) (Cundall 1976; Itasca 1995) was used for this. The ability of FLAC to simulate slope failures has been demonstrated earlier, including necessary adaptations of the modeling technique to pit slope problems (Chapter 4 in Section 1 of this publication, and Sjöberg 1999).

21.5.2 Input Data

The majority of the models used were run with a perfectly plastic constitutive model. To estimate the rock-mass strength and stiffness, the approach suggested by Hoek and Brown (1997) was used, but with a few important adjustments. Most importantly, parameters m and s in the Hoek–Brown criterion were calculated assuming disturbed rock-mass conditions, since recent work has shown this category to be more appropriate for large-scale rock slopes (Sjöberg 1999). It is important to realize that this approach only yields approximate values of rock-mass strength to be used as first estimates. However, in this particular case, the extensive monitoring database provided additional opportunity to calibrate the models against actual observations.

A set of classification ratings (rock-mass rating [*RMR*] values) and estimated values of the uniaxial compressive strength of intact rock was used to estimate a plausible range of strength parameters. From these values, the corresponding Hoek–Brown failure envelope was calculated assuming *disturbed rock-mass*

TABLE 21.2 Estimated rock-mass strength for the footwall schist, assuming disturbed rock mass, for a stress range of σ_3 = 0 – 6 MPa

m_i	σ_c (MPa)	RMR	ϕ (°)	c (MPa)	σ_{cm} (MPa)	σ_{tm} (MPa)
8	25	58	23.8	0.25	0.75	0.06
8	50	58	28.5	0.45	1.51	0.11
8	35	44	19.7	0.12	0.33	0.02

TABLE 21.3 Estimated rock-mass strength for the pyrite ore, assuming disturbed rock mass, for a stress range of σ_3 = 0 – 15 MPa

m_i	σ_c (MPa)	RMR	ϕ (°)	c (MPa)	σ_{cm} (MPa)	σ_{tm} (MPa)
16	100	72	40.3	2.25	9.7	0.43

conditions. The curved Hoek–Brown failure envelope was then "translated" to a linear Mohr–Coulomb envelope, for use as input into the numerical models. Cohesion and friction angle for the Mohr–Coulomb model were determined using linear regression over a representative stress range of the Hoek–Brown envelope. The regression stress range was determined from elastic stress analysis. The resulting strength values (c and ϕ) are summarized in Tables 21.2 and 21.3, along with calculated compressive (σ_{cm}) and tensile (σ_{tm}) strength of the rock mass.

Young's modulus of the rock mass, E_m, was calculated from rock-mass classification ratings using the empirical relation of Serafim and Pereira (1983). For pyrite, Young's modulus was estimated rather than calculated. Previous experience from underground pyrite mines has shown that the value of Young's modulus for intact pyrite ore can be as high as 100 GPa (Stephansson 1981). Thus, values from Table 21.4 were used for the analysis. Since no stress measurements have been carried out at Aznalcóllar, the virgin stress state can only be estimated. For most models, a horizontal-to-vertical stress ratio (K) of 1.5 was used. A more detailed study on the influence of the virgin stress was outside the scope of this work.

21.5.3 Model Setup

For the eastern footwall, modeling concentrated on section E213 300 (Figure 21.3) for which plane strain conditions can be assumed. A fairly detailed model was used, with quadratic elements and an element size (distance between grid points) of 5 m. To reduce calculation times, it was decided to model only every second mining increment of the final mining sequence. Because no large-scale failures were observed until the final mining step, it was further decided to carry out analyses with a fixed water table only, representative of the final mining step. The modeled water table and pyrite ore geometry are shown in Figure 21.13. Note that the slope profile has been somewhat simplified compared to the actual profile. The position of monitoring points in the model (for comparison with observations) is shown in Figure 21.14. Also, the extent of the pyrite ore in this model is considered an upper bound in terms of thickness.

A similar model was used for the central footwall but obviously with the representative mining geometry of that portion. Analyses were conducted with varying groundwater conditions and strength parameter values, starting with the values in Tables 21.2 and 21.3. In the following, the modeling of the eastern footwall is presented first, although, in reality, mining of this portion was second to mining of the central footwall.

21.5.4 Modeling Results

Eastern Footwall. Model calibration was conducted mainly using a perfectly plastic (Mohr–Coulomb) constitutive model. Strain-softening models were tested but abandoned because of the strong grid-dependency of the softening parameters, making it impossible to "export" data from one model to

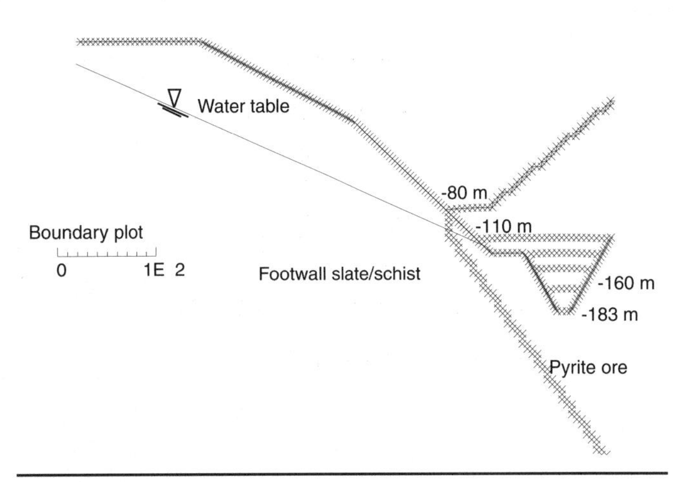

FIGURE 21.13 Close-up of model geometry showing water table, pyrite ore boundary, and mining increments

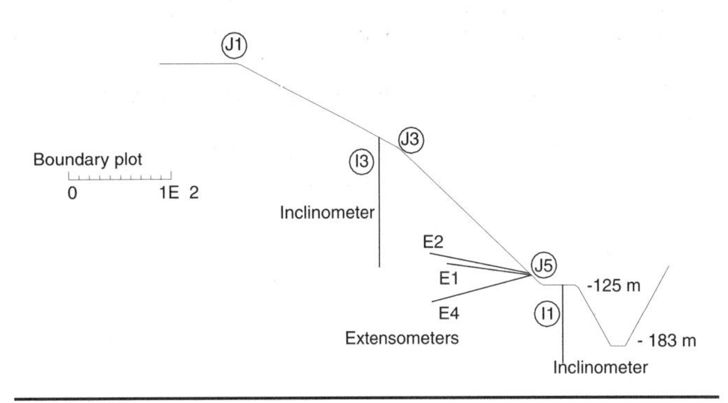

FIGURE 21.14 Position of monitoring points in the model of section 213 300

TABLE 21.4 Estimated elastic constants and density of the rock mass

Rock Type	E_m (GPa)	ν	ρ (kg/m^3)
Footwall schist	15.8	0.25	2700
Pyrite ore	100	0.30	3000

another, which is impractical and limits the applicability of the modeling technique (see also Chapter 5).

Calibration focused on achieving a good correlation with the final failure observed after termination of mining. Displacement monitoring started in 1994, and during mining, displacements were relatively small until the final mining cut (−183-m level). Therefore, it was decided to first mine the model down to −80-m depth, not permitting any failure, and then to mine the model with "correct" strength parameters down to the −183-m level. This involved mining of the −110-, −125-, −140-, −160-, and −183-m levels. Evaluation and interpretation of the eastern footwall models were done for four periods (see Table 21.5).

Model calibration started with strength values according to Table 21.2. Strengths were then adjusted until a satisfactory agreement with observed failure behavior was achieved. The best fit was obtained when the footwall strength values were set to c = 0.15 MPa, ϕ = 25°, and σ_{tm} = 0 (all other parameter values were according to Tables 21.3 and 21.4). For this case, a shear band first formed for the −160-m mining step, but this did not extend through the pyrite zone at the toe, nor did it result in increasing displacements of the slope. Rather, overall failure did not occur until a failure surface had propagated also through the stronger

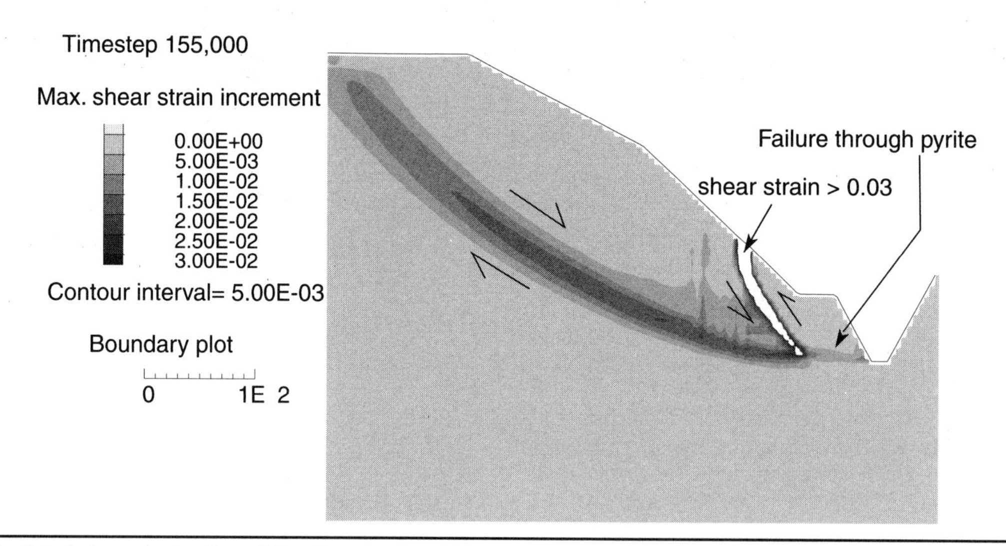

FIGURE 21.15 Calculated shear strain after mining down to the –183-m level (completed mining)

TABLE 21.5 Evaluation periods and corresponding mining steps in the FLAC model

Period	Start and Stop Date	Mining Increments	Mining Step in FLAC
1	November 1994 to and of April 1995	–110 to –140 m	–110 to –140
2	May 1995 to end of March 1996	–140 to –153 m	–140 to –183
3	April 1996 to end of September 1996	–153 to –183 m	–140 to –183
4	October 1996 to December 1996	–183 m (mining completed)	–183 m (mining completed)
4 extended	December 1996 to January 1997	–183 m (mining completed)	–183 m (mining completed)

pyrite zone, and this did not take place until after the final mining step and with continued time-stepping of the FLAC model. The failure surface, as indicated by the accumulation of shear strain, is shown in Figure 21.15. It should be noted that the sense of shearing varies, with downdip shearing inside the slope and updip shearing along the ore boundary. The failure surface in the slope is in active shear failure, whereas tensile failure occurs in the ore zone and along the ore boundary, as well as at the slope crest (Figure 21.16).

A time history of displacements versus mining step (and calculation times) is shown in Figure 21.17, indicating that the slope becomes unstable after the final mining step. This correlates qualitatively with the measured increase in displacement rates after the completion of mining (Figure 21.18). Calculated displacements at the location of the survey points also agreed quantitatively with measured displacements, as shown in Figure 21.19.

The failure surface inferred from the FLAC model was more deep seated compared to previous interpretations based on inclinometer measurements. By plotting calculated displacements along the position of I-3, one finds that the failure surface in FLAC is located at least 20 m below the bottom of the inclinometer hole (Figure 21.20). Thus, it is possible that the inclinometer could have been too short. In this particular case, I-3 ceased to function before the failure surface had developed fully. Thus, it cannot be used to deduce the location of the final failure surface.

Taken together, the FLAC model provided a fair agreement with observations and measurements in the eastern footwall (cf. Figure 21.12). The model provided important information not attainable from monitoring data alone, although it did not replicate all aspects of footwall behavior. The modeling results indicated a plausible mechanism for the observed failure in the eastern footwall. Four stages could be identified, as shown in Figure 21.21. In stage I, active yielding develops along a band going from the ore boundary to the crest of the slope for the –160-m mining step. (Displacements are only slightly larger than elastic displacements.) An upper failure surface could also have developed previously, daylighting above the ore boundary, since there is nothing restraining this portion of the slope.

In stage II, mining continues to the –183-m level. Shear strain starts to accumulate in the yielded zone, propagating from the ore boundary toward the crest. Displacements have increased slightly but are still small (less than 0.3 m). Shear strain also accumulates along the ore boundary as the ore starts to move slightly upward relative to the footwall schist. In the third stage, a failure surface has formed from the ore boundary to the crest of the slope. The upper portion of the footwall is moving downward, forcing the toe zone (pyrite ore) to rotate out toward the pit. Tensile bending failure is induced in the ore. Finally, in stage IV, the failing footwall continues to translate, thus forcing the ore zone to move out toward the open pit. Toe displacement vectors change orientation, becoming essentially horizontal. Shear failure occurs in the ore zone as it is being pushed out. Thus, a failure surface going from the toe to the crest has formed.

The restraining effect of the ore zone is a key factor in this case. However, once failure has broken through the ore buttress, there is nothing restraining the failing mass. This is also in agreement with the measured increase in displacement rate, although the backfilled waste rock prevented accelerated displacements in the toe zone. An uncontrollable and rapid failure would have been predicted had the pit not been backfilled.

Central Footwall. The failure development of the central footwall proved to be more challenging to replicate in a numerical model. It was not possible to simulate both the early (1979–1983) and later (1989–1992) failures in the same model. Furthermore, observations indicated that no new failures developed between approximately the –15 and the –80-m mining levels (see Figures 21.5 and 21.7). This could not be reproduced in a model with only slate (weak, foliated rock) making up the footwall slope. This suggested that one or several important geological features were possibly missing from the model. During drainage drilling of the central footwall, some pyrite ore was found within the footwall, starting at the –55-m level. The exact extent of the pyrite is, unfortunately, not known. Consequently, the effect of pyrite being left in the footwall was tested by modeling four different cases, ranging

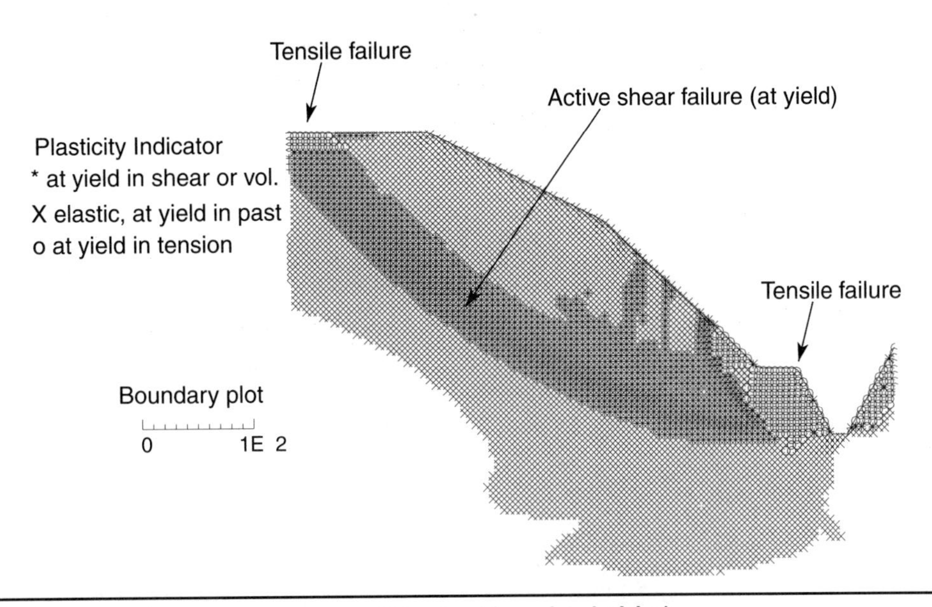

FIGURE 21.16 Plasticity indicators after mining down to the –183-m level (completed mining)

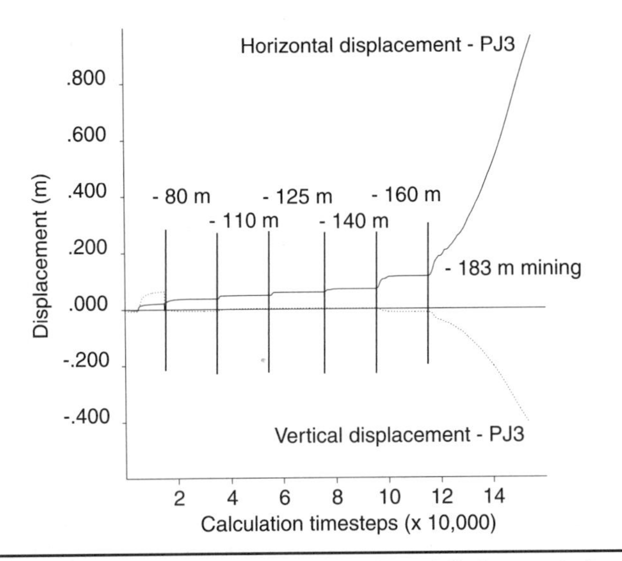

FIGURE 21.17 Calculated vertical and horizontal displacements for survey point J3 (PJ3)

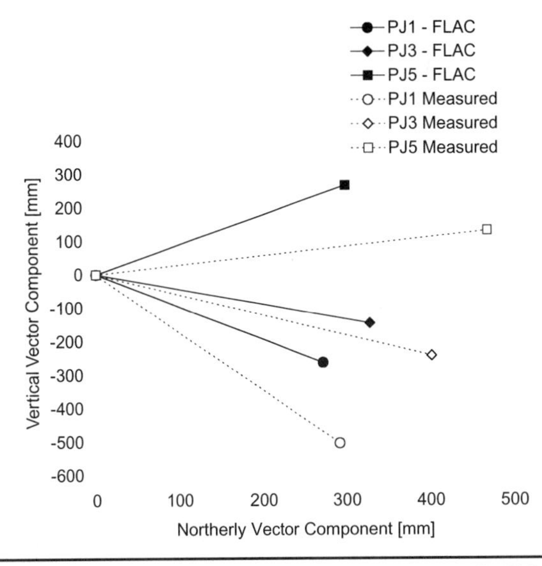

FIGURE 21.19 Measured and calculated displacements (in FLAC) for periods 2 and 3

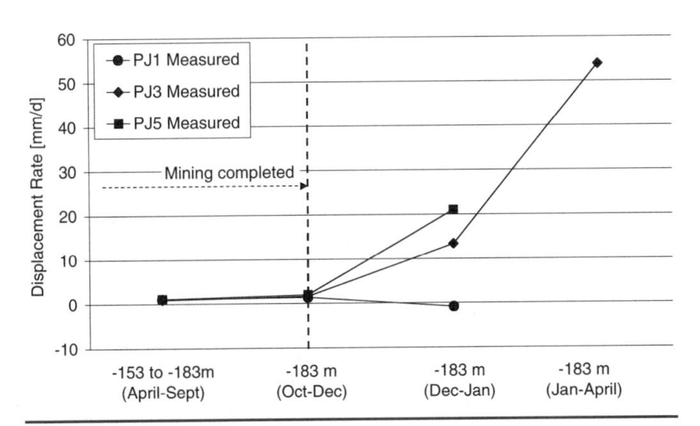

FIGURE 21.18 Measured average displacement rates in periods 3 and 4 for survey points in section 213 300

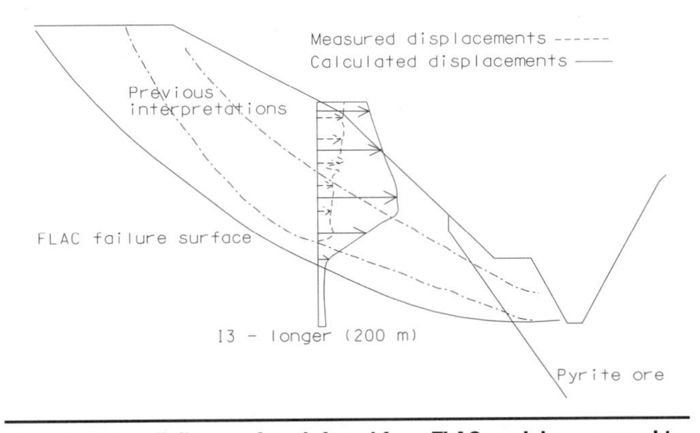

FIGURE 21.20 Failure surface inferred from FLAC models compared to previous interpretations and measured and calculated horizontal displacements (not to scale) along inclinometer I-3

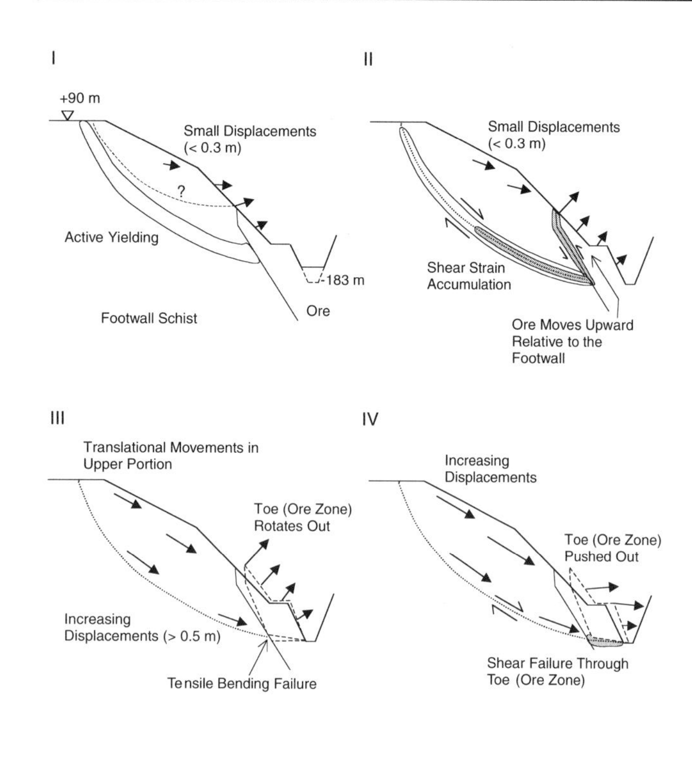

FIGURE 21.21 Summary of possible failure development in the eastern footwall

from no pyrite to slices of various thicknesses. Material properties were chosen according to Tables 21.2 and 21.3.

The modeling showed that only a thin slice of ore was enough to prevent failure from occurring. However, at some point (with increasing pit depth), the acting force on the ore buttress will force failure to occur also through the strong pyrite (Figure 21.22 [cf. Figure 21.7]). The mechanism is thus similar to that observed in the eastern footwall. The existence of a pyrite ore buttress could also explain the more rapid failure of 1992. This can be interpreted as a sudden failure of the ore zone, itself triggered by a small change in slope geometry as a result of mining activity.

The modeling conducted here could not, however, fully explain all aspects of this failure. In particular, the inclinometer measurements from the toe area of the central footwall are still to be fully explained by an analytical or numerical model. Also, although the overall behavior of the footwall is fairly well represented, the failure mechanism on the smaller (micro) scale is not resolved. The foliated rock fails in complex manners, probably involving kink band failure and similar phenomena, which are not possible to reproduce in current numerical models.

21.6 DISCUSSION AND CONCLUSIONS

21.6.1 Instability in Foliated Rock Masses

The experience gained from Aznalcóllar has been reviewed in the light of international practice related to footwall slope developments in foliated rocks. This was done in order to derive appropriate design guidelines for slopes in similar environments.

Rock-Mass Characteristics. Foliated rock masses, such as the metasediments and metavolcanics forming the footwall slopes at Aznalcóllar, Los Frailes, and other notable mines such as Jeffrey in Canada, can be characterized by three key component parameters: (1) a moderately strong to strong rock material, (2) an incipient weak fabric (foliation) resulting from mineral alignment (sometimes two sets of fabric may be present), and (3) a primary fabric dip in the same sense as the footwall with a dip of 60° to 65° or thereabouts.

The rock mass intrinsically has a significant degree of strength anisotropy controlled by the foliation, such that the strength is a primary function of the stress directions relative to the fabric orientation. The behavior of foliated rock masses in slope configurations is considered linked to this fundamental strength anisotropy, which originates and follows from the tectonic shear phenomena that originally was responsible for the foliation development. A number of phenomena can be considered particular to these rock types (or unusual in others):

- Large-scale creep of benches and multiple bench slopes (Rio Tinto mines)
- Flexural toppling and overturning effects at bench scale (Aznalcóllar footwall)
- Dilation and upward movement of toe zones forming part of larger-scale failures with associated flexural deformations of the foliated rock mass
- The contribution of foliation planes to the geometry of overall slope failure surfaces in the upper failure zones

In Situ Stress State. It can be established from published information that the principal stress direction for the Iberian Pyrite Belt region is north–south. Furthermore, the ratio of the maximum principal horizontal stress to the vertical stress can be high. From observations of slope failures in the Iberian Pyrite Belt, notably at Rio Tinto and Aznalcóllar, it is inferred that, as a result of the inferred stress state, the rock mass potentially undergoes a process that involves significant strength deterioration during slope development. Prior to mining and under a given stress state, the rock mass is in a state of equilibrium. The development of slopes leads in turn to a state of overstress and further weakening of the slope material (rock mass). These deterioration processes, based on observation, appear to be deep seated rather than being simply a surface phenomenon. The apparent extent of deterioration does not necessarily relate to the scale of the slope height, based on the limited data available.

It is also important to understand that, during the process of slope development, rotation of the principal stresses will occur that may lead to development of the foliation as a fabric into a discontinuity system, which is a result of shear overstress phenomena if the respective stress field rotations pass through a critical shear stress state. For the typical conditions at Aznalcóllar and other similar properties with a characteristic foliation dip of 50° to 65°, this may indeed be the case. The depth into the slope and the extent in the vicinity of the toe zone to which such "disturbance" effects could occur may be significant as a result of a cumulative progressive effect during the ongoing slope excavation stages.

The conditions at Aznalcóllar can also be compared with the Jeffrey Mine north wall in Canada (Sharp, Bergeron, and Ethier 1987). The footwall rock mass at Jeffrey is very similar to that at Aznalcóllar, but the encountered slopes are higher and significantly steeper. Coupled with the adverse influence of groundwater at Jeffrey, the slopes are undoubtedly much more stable than those in Spain. In comparing the Jeffrey and Iberian sites, a noteworthy difference is attributed to the occurrence of a zone of intensely sheared, extremely weak material, many meters thick, that would not have supported a high transverse (subhorizontal) stress field. Consequently, a key difference that could have a potentially negative effect on stability is seen to be the particular stress state for the Iberian mines, with their inferred high horizontal (transverse) component relative to the footwall slope sections.

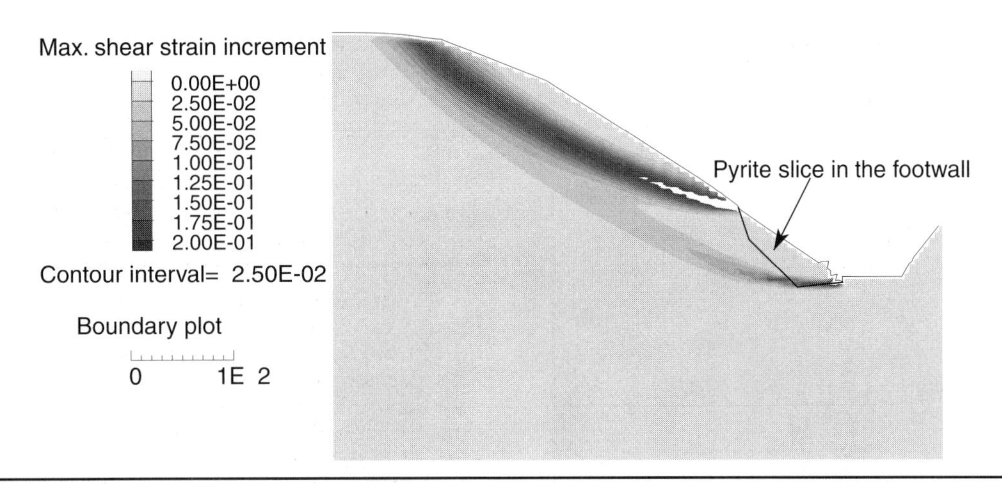

FIGURE 21.22 Calculated shear strain for mining to the –125-m level of the central footwall, for the case of a slice of pyrite ore left in the footwall

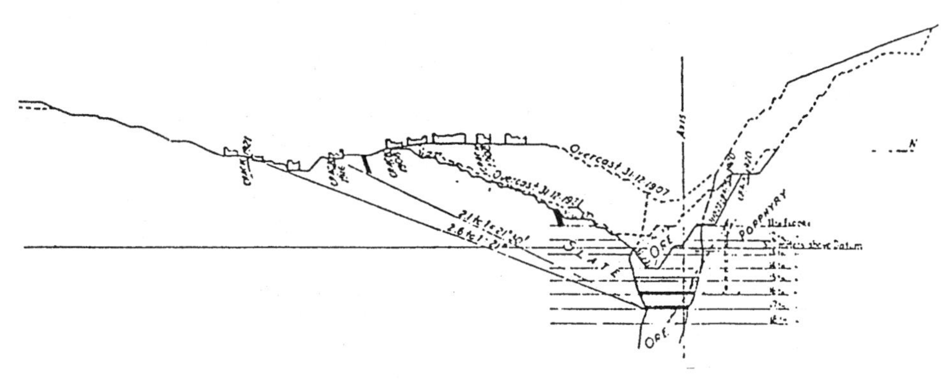

FIGURE 21.23 Iberian footwall slopes—slope height–slope angle

Precedent Experience—Rio Tinto and South Lode Footwall Slopes. The observations from Rio Tinto from 1907 to 1928, as reported by Palmer (1930), in relation to the slate slopes for the South Lode (Filon Sud) are of particular note in terms of the detail of the observations given. They are, however, only one of a number of such overall slope deformation phenomena, the others being principally associated with the Atalaya open cast.

The slopes in question are of limited height (typically 75–150 m; see Figure 21.23) and underwent progressive deformation over a period of at least 20 years. When viewed in the late 1960s, prior to backfilling, the entire slope comprised a succession of distorted, partially overturned benches that gave the appearance of long-term "mass creep" behavior. The deformations of the slope over time indicated overall translation of the slope and associated overturning (flexural distortion) of the slope in the lower toe zone. Initially, movements were believed to have occurred as a result of the open cast development alone, although later movements were associated with the combined open cast and underground orebody development. Of note, in terms of stability mechanisms, is the base angle of apparent movement derived by joining the position of the furthermost crack to the lowest excavation stage, either in the open cut or the underground development. This typically was of the order of 21°. The generation of such shallow angle failure surfaces due to a gravitationally induced failure of the intact foliated rock mass typical of the Rio Tinto slates is not explicable based on either rock strength considerations or other closely correlatable experience from slopes in similar rock types but different tectonic environments.

It is also interesting to examine whether such "failures" (or more correctly, slope deformations), once initiated, remained essentially static. Reference to Palmer's observations implied that they appeared to progress, but perhaps this was more as a result of ongoing mining rather than ongoing movement of the same slope with time. What is clear is that the depth of the failure surfaces progressed with mining depth and these appeared to be new and discrete events, as indicated by observed surface crack progressions away from the slope crest. The generation of multiple failure surfaces within the slope dipping at approximately 20° from successive slope toe elevations can thus be inferred.

The observations would appear to indicate a stress-induced discrete response of the slope that was progressive with time. Stresses were not necessarily dissipated or redistributed but appeared to remain active within the slope, leading to discrete extensions of the depth of movement with associated overall rock-mass disturbance.

Summary—Rock-Mass Response Model. From the current state of the art, it is inferred that, for rocks such as the foliated metasediments forming the footwall rocks at many of the Iberian mines, the in situ stress state and its potential resultant state of metastability, i.e., state liable to generate ongoing shear process, are important factors in terms of rock-mass strength behavior and resulting deformation processes that can occur.

A failure mechanism can be predicted for the in situ rock mass as a result of slope formation and excessive shear stresses that develop over zones where the stresses are rotated to a critical state in terms of the particular anisotropic strength characteristics of the foliated rocks. This mechanism is potentially enhanced in degree by the very significant potential reduction in the normal stresses acting across the foliation planes. It is clear that this mechanism does not lead directly to slope failure in itself. It may, however, serve to explain why the footwall rock masses in the Iberian mines

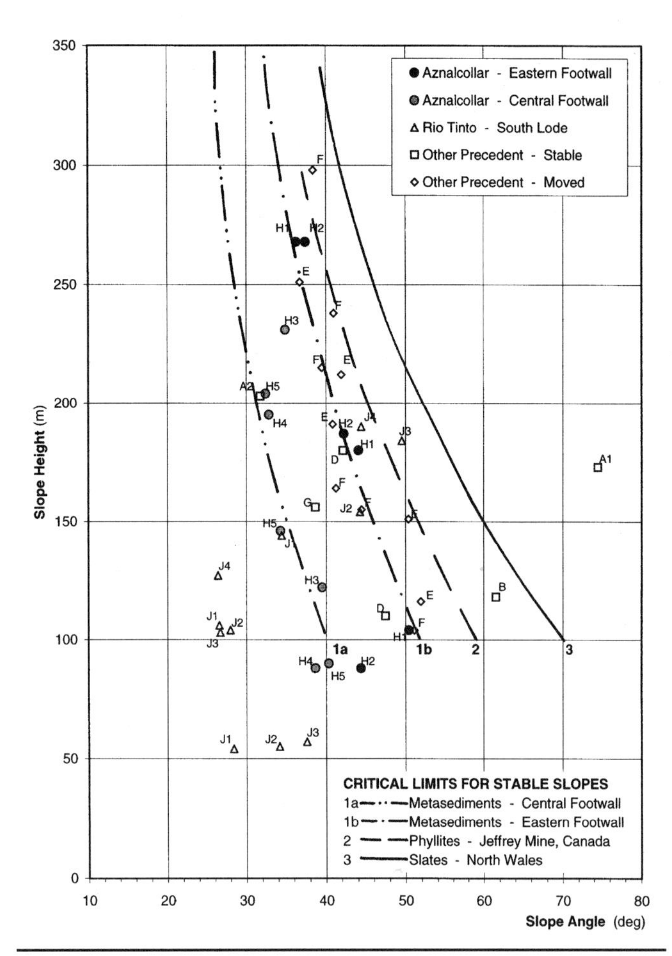

FIGURE 21.24 Slope deformation phenomena, Rio Tinto, Filon Sud

appear to be significantly weaker at depth than their overall characteristics would suggest. This inferred state can then be used to explain why the rocks apparently behave so poorly as structural elements of a slope in the high stress Iberian context compared with other global precedent.

In terms of observed response, a rational explanation is still not available for the development of well-defined failure surfaces dipping at 20° to 25°. While a process of foliation development (weakening) can be postulated from the slope excavation–stress interaction effect, this does not explain the subsequent development of a major failure surface through the slope. Studies of kink band development have not identified a plausible working mechanism, but further development of this approach may prove worthwhile.

Of note is the relationship between excavation geometry and slope response. Although it is clear that steeper and higher slopes will be less stable than flatter, shallower slopes, experience from Aznalcóllar and Rio Tinto points to an additional causative effect that is not closely related to either height or angle. This places the slopes in question beyond available precedent in terms of slope stability theory and indicates the potential difficulties associated with predictive slope design.

21.6.2 Maximum Slope Angles in Deforming Slopes

For slopes subject to ongoing deformations, predicting "safe" angles is extremely difficult. Observed data for such slopes are given in Figure 21.24. No guidelines exist and it is therefore necessary to address each case in relation to potentially critical site-specific factors. For the Iberian pyrite belt experience, it is important to note that no overall slope "failure" has occurred that has led to a catastrophic conclusion, either in terms of mine closure or, as far as can be ascertained, fatalities. For the three study examples for which data are available, the following typical slopes were operated, albeit with inevitable disruptions from time to time, notably for Aznalcóllar, eastern zone:

Slope	Max. Slope Angles	Slope Height Range
South Lode, Rio Tinto	32°–34°	70–140 m
Aznalcóllar, Central Zone	32°–34°	150–220 m
Aznalcóllar, Eastern Zone	35°–36°	240–270 m

In general slope engineering terms, catastrophic failure is inferred to be a major risk for slopes of significant height (>200 m) with angles greater than about 40°. This implies that there is no major underlying weakness such as a dipping shear or fault zone that could allow sudden massive failure under, say, adverse transient groundwater conditions or sudden strength loss with shear displacement.

For the foliated footwall slopes under consideration at Iberian sites, a maximum slope angle of 40° to 45° could be considered an initial target based on similar slope developments that have proven to be stable overall with more adverse groundwater conditions. A more appropriate range of overall angles to allow for acceptable interramp slopes would be in the range of 30° to 35°. This range would hopefully allow for an adequate degree of stability to be maintained through the design life of the slope such that ramp access could be achieved on a more or less uninterrupted basis. It should be noted that such a range of angles is some 10° steeper than the values given for inferred "stable" slopes but of course assumes that deformations will be experienced.

21.6.3 Envisaged Slope Movement Characteristics and Causes of Instability

Key characteristics of potential slope movements in the foliated footwall slopes that need to be considered in the design include

- Movements are likely to be deep seated, i.e., to a significant depth below the slope profile and possibly below the slope toe.
- Movements close to the surface are likely to be more accentuated than those at depth due to dilation of the deforming rock mass.
- Movements are likely to be progressive with time, although the rate of movement will vary significantly.
- Movements or increased movement rates are likely to be triggered by adverse groundwater states (even though these may still not be very significant in global terms) for a given mining state. (The influence of groundwater on movement rates is clearly indicated by the long-term precipitation–movement response plots given in Figures 21.25 and 21.26.)
- In the absence of movement events triggered by groundwater, progressive strain buildup is likely to occur that in turn may lead to periods of enhanced movement of the overall slope to attain a new state of equilibrium.
- While the movements will be noncatastrophic (as a result of the selected slope form with relatively shallow slopes [30°–35°]), they will be difficult to accurately predict in advance, and special contingencies for a range of plausible movements will need to be adopted in the slope design.

From a comprehensive assessment of the Aznalcóllar footwall stability state with time, the primary causes of slope instability are (1) the generated slope profile, (2) the slope geologic conditions, (3) in situ stress and induced stress conditions, and (4) groundwater and effective stress–strength considerations. In

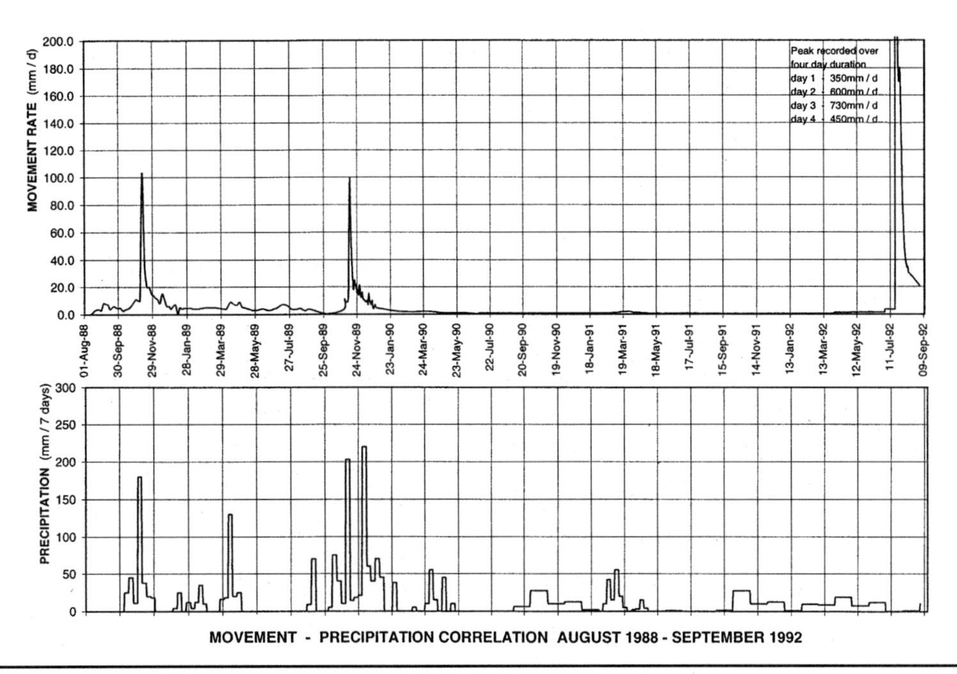

FIGURE 21.25 Aznalcóllar central footwall—slope response–precipitation data

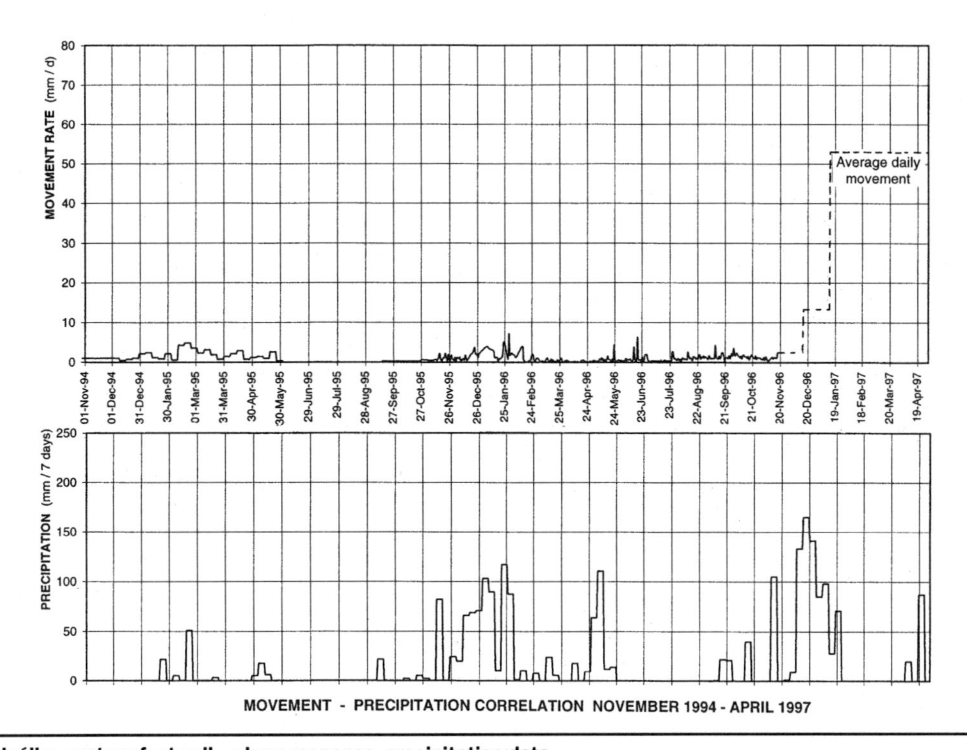

FIGURE 21.26 Aznalcóllar eastern footwall—slope response–precipitation data

summary, the slope profiles were not excessively steep when movements were initiated and the geological conditions were not unduly unfavourable. In general terms, the slope geometry in combination with the weak rock mass (as induced) resulted in an unstable configuration. The principal uncertainty in the slope response is the effective weakening of the rock mass that led to instability development. Key causative factors would appear to be related to in situ stress and water (probably interacting) which in turn responded to reactions caused by intrinsic, induced strains resulting from excavation and possibly accentuated by blasting events.

21.7 CONCLUSIONS AND LESSONS LEARNED

The excavations at Aznalcóllar were carried out with limited knowledge of the controlling geological factors; consequently, a heavy reliance was placed on monitoring, particularly for the last phases of excavation. Slope movements were inevitable and occurred for most slopes higher than approximately 75 m.

In the central footwall, multiple failures occurred. Large movements occurred along the failure surfaces and were, most frequently, in post-failure state. Consequently, movement rates were very sensitive to groundwater conditions. In the eastern footwall, total displacements up until the termination of mining were much smaller. After mine closure, displacement rates increased significantly, in spite of partial refilling of the eastern pit (with waste rock). The steeper slope angle of this portion makes it reasonable to expect higher displacement rates in the post-failure stage compared to those in the central footwall. Sudden catastrophic failure could have occurred had the pit not been backfilled.

The instrumentation program at the mine proved essential for completion of mining and for understanding the behavior and stability state of the slopes. Different types of measurements made it possible for the measured displacements to be correlated and validated. The surface survey monitoring network was the most important and reliable method for monitoring overall slope behavior and surface displacement rates. This was complemented by valuable subsurface measurements (inclinometers and extensometers) to detect the depth and nature of slope movements.

From a practical mining perspective, small-scale failures and disintegration (at bench scale) often proved a more severe hazard than the movements associated with the large-scale failure of the footwall. Controlled mining, scaling, and clean-up of the benches was necessary to provide safe working conditions.

None of the large-scale failures could be attributed to single, large-scale, preexisting discontinuities in the footwall. Rather, failures were categorized as progressive rock-mass failures–the mechanisms of which are still only partially quantified. The above findings indicate several key uncertainties that need further clarification prior to establishing a working stability model for the study of future slope designs in similar rock conditions. These uncertainties include

- The mechanism of generation and the nature of the induced shallow-dipping shear surfaces controlling stability of the upper (active) slope zone and their potential relationship to shear overstress processes
- The sequence and mechanism of movement initiation in the overall slope section (active–passive model rationalization)
- The complex nature of the lower slope movements involving shear/dilation/overturning phenomena and associated deformation processes
- The potential influence of in situ stress and induced stresses in the failure process

The case histories from southern Spain and, in particular, the recent studies at Aznalcóllar have demonstrated that conventional design methods involving a safety factor or other limiting stability concepts are not appropriate for such slope designs. For Aznalcóllar, numerical modeling helped to identify the mechanism of slope failure. Modeling is considered useful for planning, provided that it is used judiciously. The limitations of numerical models for design are related to input information (geology, rock-mass properties, etc.) and rock-mass constitutive behavior rather than to software.

The need to accept progressive deformations albeit with a clear understanding of the rock-mass model is apparent for these footwall slopes. Such measures should, however, only be adopted in conjunction with comprehensive monitoring controls to regulate safety and drainage measures to control groundwater influence.

21.8 ACKNOWLEDGMENTS

The assistance and support by the mine staff of Boliden Apirsa from 1987 to 1996 is gratefully acknowledged. Norbert Krauland (formerly of Boliden Mineral AB) provided continuous rock-mechanics support to the mine during this period and, in particular, during the final mining of the eastern footwall. The staff of Golder Associates, Dr. Laurie Richards, and Brendan Monaghan were instrumental in assessing the central footwall failures, and Arthur Crease of Geo-Engineering was involved in monitoring and evaluating of the eastern footwall failures. Their work and that of other colleagues formed an important basis in the production of this chapter.

21.9 REFERENCES

Cundall, P.A. 1976. Explicit finite difference methods in geomechanics. *Proceedings 2nd International Conference on Numerical Methods in Geomechanics,* Blacksburg, Virginia, June, 1976, 1:132–150.

Golder Associates (UK) Ltd. 1989. *Footwall slope stability Aznalcollar mine, Seville.* Report to Boliden Mineral AB, report no. 8851056, September 1989.

Golder Associates (UK) Ltd. 1995. *Slope stability review east footwall, Aznalcollar mine, Seville, Spain.* Draft report to Boliden Apirsa, January 1995.

Hencher, S.R., Q-H. Liao, and B.G. Monaghan. 1996. Modelling slope behaviour for open-pits. *Trans. Instn. Min. Metall.* (Sect. A: Min. Industry), 105:A37–47.

Hoek, E. and E.T. Brown. 1997. Practical estimates of rock mass strength. *Intl. Jour. Rock Mech. Min. Sci.,* 34:8:1165–1186.

Itasca. 1995. *FLAC Version 3.3. Manual.* Minneapolis: ICG.

Krauland, N. 1995. Letter to Golder Associates (UK) Ltd. 1995-02-17.

Ladanyi, B. and G. Archambault. 1970. Simulation of shear behavior of a jointed rock mass. *Rock Mechanics – Theory and Practice, Proceedings 11th US Symposium on Rock Mechanics.* Berkeley, 1969, pp. 105–125. New York: A.I.M.E.

Ladanyi, B. and G. Archambault. 1972. Evaluation de la résistance au cisaillement d'un massif rocheux fragmenté. *Proceedings 24th International Geological Congress.* Montreal, 1972. Sec. 13D:249–260.

Ladanyi, B. and G. Archambault. 1980. Direct and indirect determination of shear strength of rock mass. *Preprint No. 80-25, A.I.M.E. Annual Meeting,* Las Vegas, Nevada, February 24–28, 1980. Littleton, Colorado: Society of Mining Engineers of A.I.M.E.

McCullough, M.F. 1993. *Numerical modelling with UDEC of the footwall slope at Aznalcollar mine, southern Spain.* Master of science thesis, University of Leeds.

Palmer, R.E. 1930. Observation on ground movement and subsidences at the Rio Tinto Mines, Spain. *Trans. Amer. Inst. of Mining Engrg.,* 91:168–185.

Proughten, A.J. 1991. *Numerical modelling of large rock slopes with particular reference to the Aznalcollar mine, Seville, southern Spain.* Master of science thesis, Royal School of Mines, Imperial College of Science and Technology, London.

Serafim, J.L. and J.P. Pereira. 1983. Considerations of the geomechanical classification of Bieniawski. *Proceedings International Symposium on Engineering Geology and Underground Construction,* Lisbon. 1:II: 33–42. Lisbon: SPG, LNEC.

Sharp, J.C., M. Bergeron, and R. Ethier. 1987. Excavation, reinforcement and monitoring of a 300 m high rock face in slate. *Proceedings 6th International Congress on Rock Mechanics,* Montreal. 1:533–540. Rotterdam: A.A. Balkema.

Sjöberg, J. 1999. *Analysis of large scale rock slopes.* Doctoral thesis 1999:01, Division of Rock Mechanics, Luleå University of Technology.

Stephansson, O. 1981. The Näsliden project–rock mass investigations. *Applications of Rock Mechanics to Cut and Fill Mining.* Luleå, June 1–3, 1980. Pp. 145–161. London: Institution of Mining and Metallurgy.

CHAPTER 22

Slope Stability at Aitik

Jonny Sjöberg* and Urban Norström*

22.1 MINE DESCRIPTION

22.1.1 General

The Aitik Mine is approximately 60 km north of the Arctic Circle in northern Sweden, 1,200 km north of Stockholm (Norrbotten County), at latitude 67°07'N and longitude 21°E (Figure 22.1). The mine is owned and operated by the Boliden Mineral AB mining company. The low-grade copper mineralization was discovered in the 1930s, but mining did not start until 1968. Annual production during the first years of mining was fairly low, about 2 Mton. Production volumes have increased steadily ever since; currently (year 2000), 18 Mton of ore is planned to be mined, with an additional 24 Mton of waste (stripping ratio 1.33:1).

22.1.2 Geology

The Aitik Mine is located along the Kiruna-Ladoga shear zone, a major structure that extends from Lake Ladoga in Russia to Kiruna in Sweden. This structure also marks the boundary between the Karelian plate and the Svecofennian plate. The area around the Aitik Mine consists of metamorphosed plutonic, volcanic, and sedimentary rocks, all of Precambrian age. Relatively large gabbroic bodies occur farther to the west (most notably the Dundret gabbro mountain), whereas, farther to the east, granitic rocks dominate (the Lina granite series). The mine surroundings are relatively flat, with some undulating hills. The entire region is covered by a fairly uniform moraine layer, often overlain by peat. The total thickness of the soil layers is generally less than 10 to 20 m (Monro 1988; Drake 1992; SGU 1996a, 1996b).

The Aitik mineralization is enclosed by shear zones, which also divide the deposit into northern and southern sections (Figure 22.2). The low-grade copper mineralization occurs as disseminated thin veinlets of chalcopyrite, along with minor contents of silver and gold. Ore-grade mineralization is hosted in the main ore zone gneiss, with minor copper present in the footwall gneiss. The genesis of the ore has been subject to some discussion. Monro (1988) argued that the Aitik mineralization is different from classic porphyry copper deposits and should form its own group of mineralizations, while more recent investigations (SGU 1996a, 1996b) indicate the opposite.

The orebody strikes approximately N20°W, dips approximately 45° to the west, and is approximately 2,000 m long and some 300 m wide. The ore zone can be divided into two portions. In the northern section, ore grades are relatively constant with depth, whereas they decrease with depth in the southern section. The ore reserves in the southern section extend to a depth of around 300 m below the ground surface. In the northern section, the orebody is known to be more than 800 m deep (see Figure 22.2). The total size of the mineralization is approximately 700 Mton, of which around 200 Mton have been mined from 1968 until today. Average ore grades are approximately 0.40% Cu, 3.5 ppm Ag, and 0.2 ppm Au.

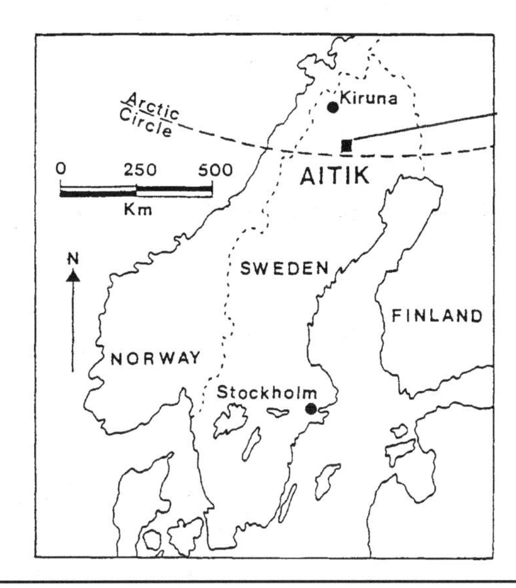

FIGURE 22.1 Map of Scandinavia showing the location of the Aitik Mine

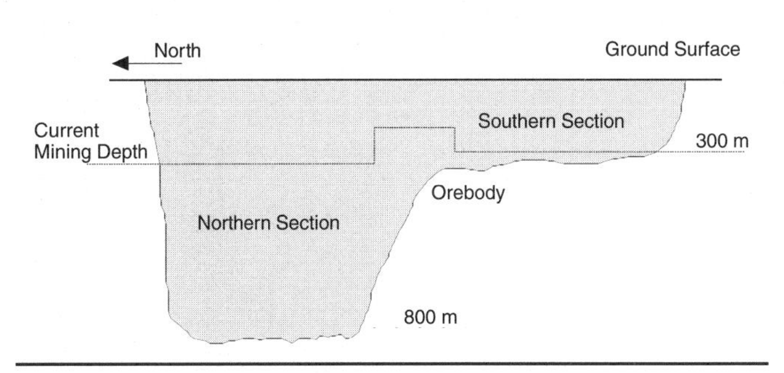

FIGURE 22.2 Longitudinal cross section showing schematic ore zone geometry and approximate current mining level at Aitik

The hanging wall at Aitik is dominated by amphibole gneiss (Figure 22.3). The hanging wall contact toward the ore zone is an old thrust contact. Consequently, the economic ore boundary toward the hanging wall is very distinct. The thrust contact has been subjected to intense shearing, thus reducing its strength significantly. Clayey material is occasionally found in this zone.

The main ore zone can be separated into two broad lithological units: (1) muscovite schists in the upper part toward the hanging wall and (2) fine-grained biotite gneiss and schist in the lower part nearer the footwall (see Figure 22.3). Lenses of muscovite schists can sometimes be found closer to the footwall. This rock unit exhibits significantly lower strength compared to the biotite schist and gneiss. There is a gradual transition between the biotite

* Boliden Mineral AB, Sweden.

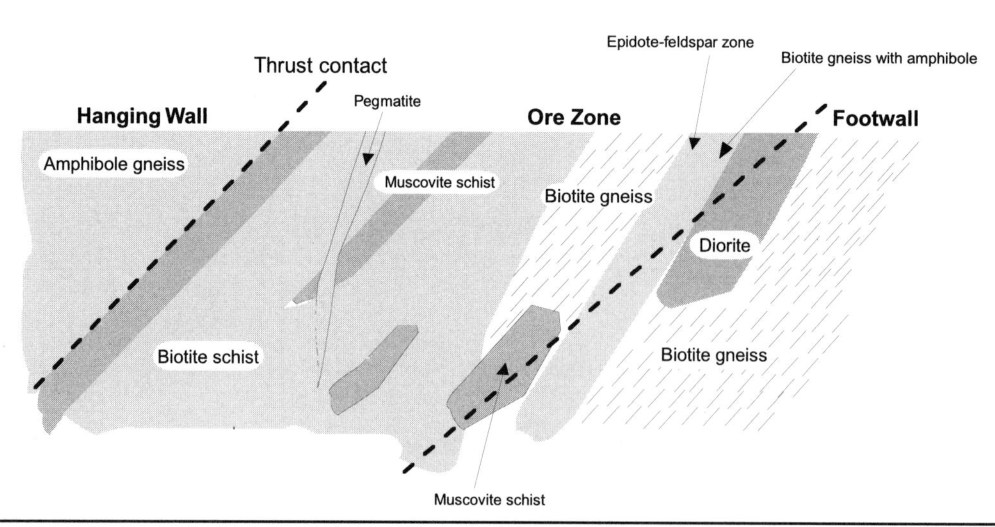

FIGURE 22.3 Simplified typical geological cross section (not to scale)

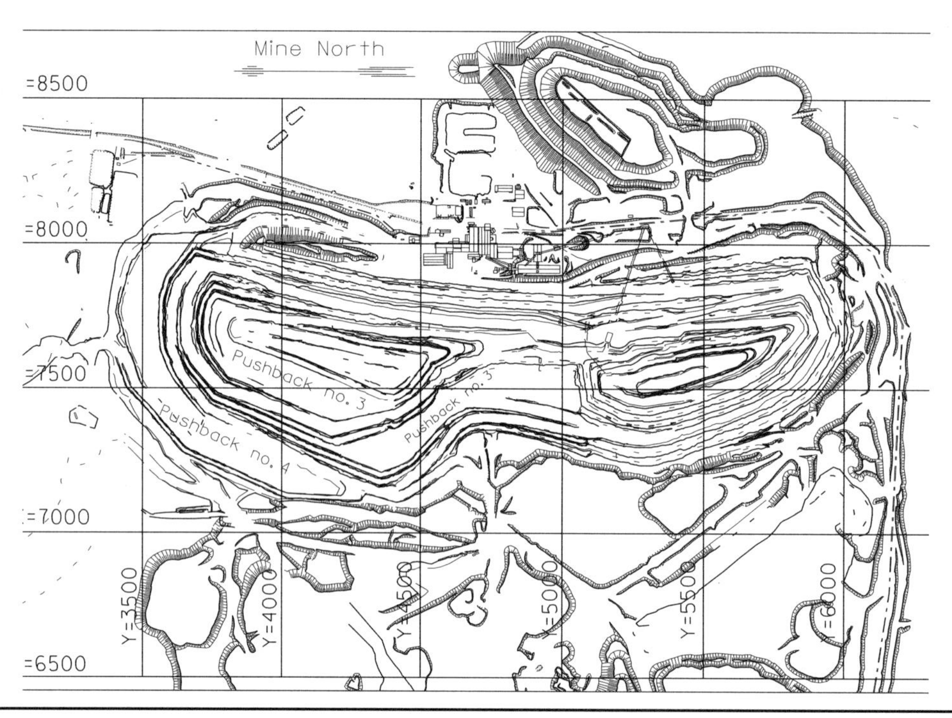

FIGURE 22.4 Horizontal map showing Aitik Mine geometry as of 2000

schist and the biotite gneiss, sometimes making it difficult to separate the two. Pegmatite dykes occur within the ore zone, more frequently nearer the hanging wall. They typically crosscut the stratigraphy at high angles and strike subparallel to the orebody.

The footwall rocks are dominated by diorite (particularly in the southern section) and biotite gneiss (Figure 22.3). The footwall biotite gneiss and the hanging wall amphibole gneiss can be very similar in appearance and also have similar geomechanical properties. A zone of biotite gneiss with more amphibole occurs close to the biotite gneiss of the ore zone. The contact between the footwall rocks and the mineralization is not distinct; it is based on cutoff grade. The coarse-grained amphibole-epidote-feldspar gneiss zone can be viewed as an approximate ore boundary. However, the economic mineralization contact generally dips steeper on the footwall side compared to the hanging wall and cuts through the stratigraphy. Consequently, the upper portion of the footwall slope is located mainly in the stronger diorite. The majority of the rocks exhibits well-defined foliation (preferred alignment of mica minerals). However, the foliation is weakly developed in the diorite and in microcline- and feldspar-rich gneisses (Monro 1988).

22.1.3 Mine Operations

Mining at Aitik is done using traditional open-pit mining with pushbacks. So far, three pushbacks have been mined and a fourth one has recently been started (Figure 22.4). According to the current mining plans, mining will continue at least until the year of 2012, with final pit depth at the −460-m level. Ground surface is at zero elevation in the southern end of the pit and slopes toward the northeast. (Actual elevation above sea level is around 350 m.) For the northern footwall, the ground surface is approximately at the −40-m mining level; for the hanging wall, it is at the −25-m level. The pit currently measures approximately 2,500 m long and 750 m wide.

The southern section (south of approximately coordinate Y5100 in Figure 22.4) was mined to its final depth at the −300-m

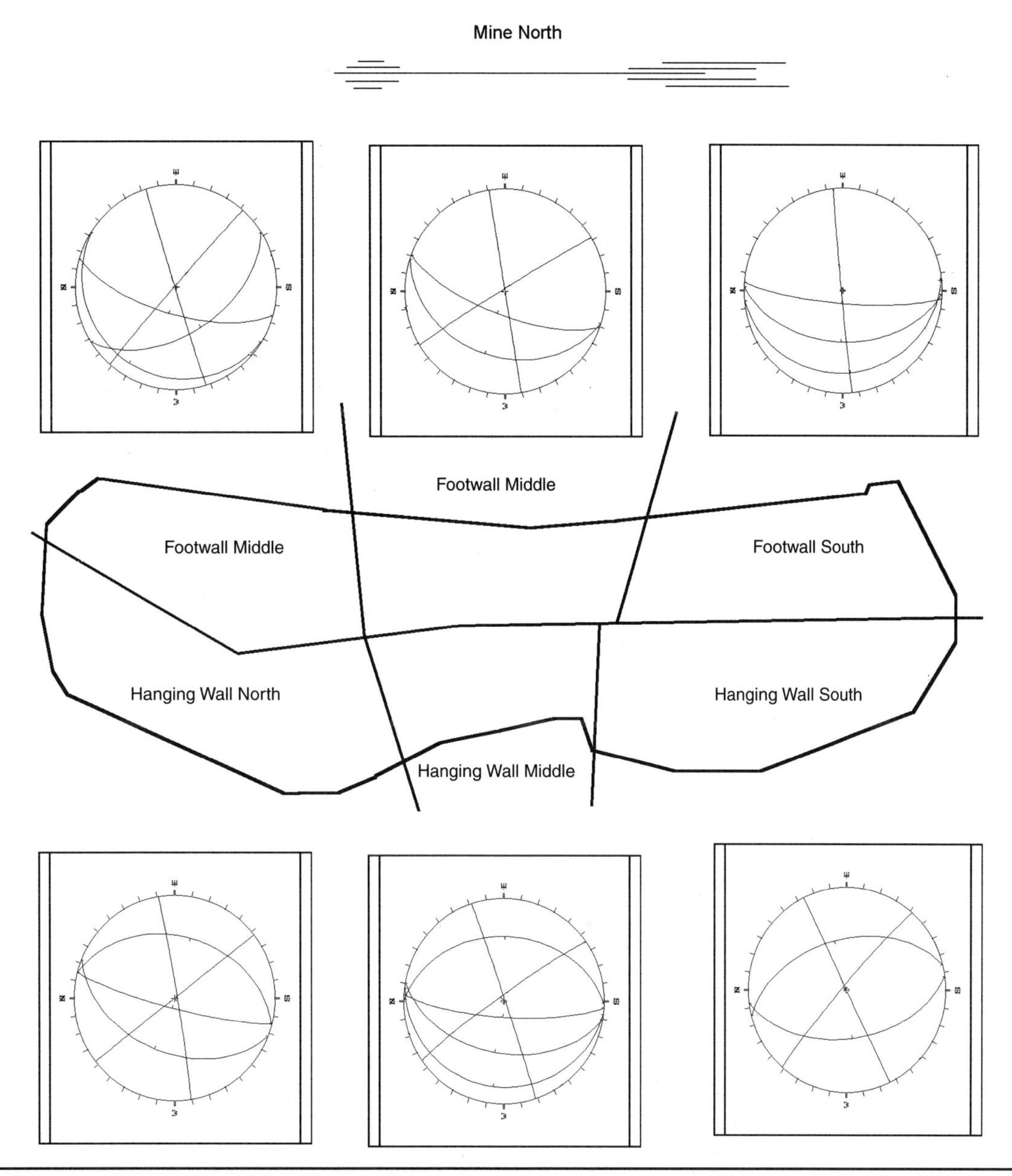

FIGURE 22.5 Design sectors and corresponding joint set orientations in the Aitik Mine (after West et al. 1985)

TABLE 22.1 Recommended interramp slope angles

Sector	FN	FM	FS	HN	HM	HS
Angle (°)	47	46	49	51	56	53

NOTES:
FN footwall north
FM footwall middle
FS footwall south
HN hanging wall north
HM hanging wall middle
HS hanging wall south

Source: West et al., 1985

mining level in 1997. Current mining level in the northern section (north of coordinate Y4600) is −315 m. The central section between the northern and southern portions is currently at the −165-m level, but it will be deepened to approximately the -300-m level in the future. The current pushback (no. 4) on the hanging wall has only come down one double bench. At least one additional pushback (no. 5) is planned, involving both the hanging wall and the footwall.

The main ramp is currently on the footwall, with additional ramps on the hanging wall. Ore is transported (by trucks) to an in-pit crusher at the −165-m mining level. The crushed rock is transported to the plant by conveyor belt through an underground drift in the southern footwall. Waste rock is trucked along the hanging wall ramp to dumps located to the west and north of the pit. There is also an old shaft in the footwall, going down to the −110-m mining level, but this is not being used today. A more detailed description of the mining operations, equipment, and milling can be found in Chadwick (1996).

22.2 SLOPE DESIGN

22.2.1 Current Slope Design

Current mine design is based on the probability of small-scale failures and was proposed in the studies by Call et al. (1976, 1977) and West et al. (1985). Recommended interramp angles, using a double-bench configuration, are shown in Table 22.1. The mine was divided into six design sectors–zones in which geologic structures, rock strengths, and pit wall orientations are similar (Figure 22.5).

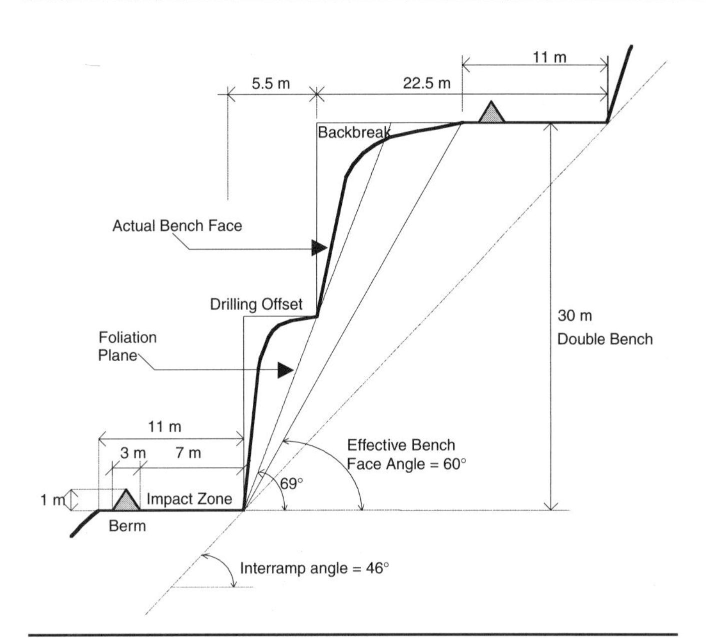

FIGURE 22.6 Double-bench and catch-bench design (after West et al. 1985) for the northern footwall sector of the Aitik Mine

The double-bench geometry used at Aitik involves two 15-m-high benches between each catch bench (see Figure 22.6). The current design criterion for the catch benches states that these should be at least 11 m wide in 90% of all cases (West et al. 1985). A small drilling offset is left between each single bench. For the footwall, this offset is chosen to 5.5 m, so that approximately the same foliation plane would cut through the toe of both single benches. This was based on a foliation dip angle of 69° in the north and middle footwalls. The final, effective bench face angle is flatter than this angle, due to backbreak from blasting at the crest. Taking the northern footwall as an example, an effective bench face angle of at least 60° must be achieved in order to have an interramp slope angle of 47°, while still satisfying the catch bench width criterion.

The actual slope geometry has been controlled through aerial photography for a number of years. In the early studies (Petersson 1988, 1990), it was found that actual interramp slope angles were, in general, slightly flatter than the design criteria in Table 22.1. In most cases, this could be attributed to the fact that required bench face angles have not been achieved due to excessive blast damage. As a consequence, catch benches were too narrow and the design criterion for these was not fulfilled.

The above studies showed that steeper bench face angles were necessary to satisfy the design criteria, which can be achieved through smooth blasting. An extensive project on this was conducted from 1992 to 1996 (Ouchterlony et al. 1997). The outcome of the project, which was a joint research effort between Boliden Mineral AB and SveBeFo (Swedish Rock Engineering Research), was a criterion for determining blast damage in terms of calculated peak particle velocity (PPV) levels. Based on this, a set of drill-and-blast plans were developed and tested in the mine. One commonality was that one to three rows of small-diameter holes were required to achieve acceptable results.

Applying of these results in the production blasting has resulted in steeper bench face angles and improved face conditions. Currently, two slimhole rows are used in the standard drill-and-blast configuration. The results from the blasting project indicated that decoupled charges in the contour row would give improved results. Furthermore, for weaker rocks, an additional slimhole row is required (a total of three). In any case, a tool for designing blasting to achieve a certain bench geometry now exists, including adjustment of smooth blasting techniques to local rock conditions.

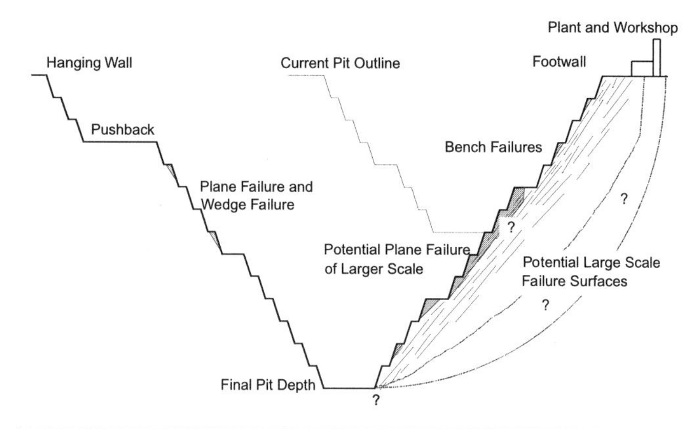

FIGURE 22.7 Potential stability problems at the Aitik Mine (figure not to scale)

More recent studies of aerial photos taken in 1994 and 1998 showed that bench face angles have, indeed, increased during the last few years. The design criterion for the bench face angles is currently satisfied for most of the design sectors of the mine. Thus, steeper interramp slopes may be achieved if the knowledge on smooth blasting techniques is implemented fully throughout the mine.

22.2.2 Stability Conditions and Future Design

Slope-stability conditions at Aitik are generally good. No signs of large-scale instabilities have been observed and only relatively few bench-scale failures have been observed. All of these have occurred along well-defined discontinuities, rarely exceeding one bench in height. Mining interruptions associated with these failures have been fairly limited. Slope displacement monitoring has been fairly regular since the mid 1980s. A total station has been used to measure the displacement of prisms installed at various locations in the pit. The system was recently updated with new hardware and software. These measurements have not revealed any significant slope movements.

The favorable stability conditions for current mining geometry do not imply that similar conditions will prevail for continued mining. Future mining of the Aitik deposit involves significant deepening of the pit. Current mining plans call for a final slope height of 450 m; however, depending on metal prices and production costs, even higher slopes are possible. In the strategic (long-term) plan for the mine, mining down to the −585-m level (until year 2018) is a viable option. The limit as to how steep the overall slopes can be made is not known, and the lack of previous failures precludes a purely empirical design for final slopes. Furthermore, since large-scale structures have not been proven to exist, the potential mode of failure is not obvious. The problem is illustrated in Figure 22.7. These issues were not adequately addressed in previous design studies.

The Aitik Mine shares this problem with many hard-rock mines worldwide. Fundamental issues pertaining to large-scale rock slopes remain unresolved, and reliable design methods are lacking for cases with no precedent in terms of large-scale failure. Correspondence with scientists, consultants, and mining staff at other operations shows that little development has taken place in this field during the last 25 years. This is despite the drastically increased requirements needed for ever-increasing slope heights. To remedy this situation, a research project was initiated to focus on advancing the knowledge in this area and, at the same time, develop a sound design methodology for the Aitik Mine.

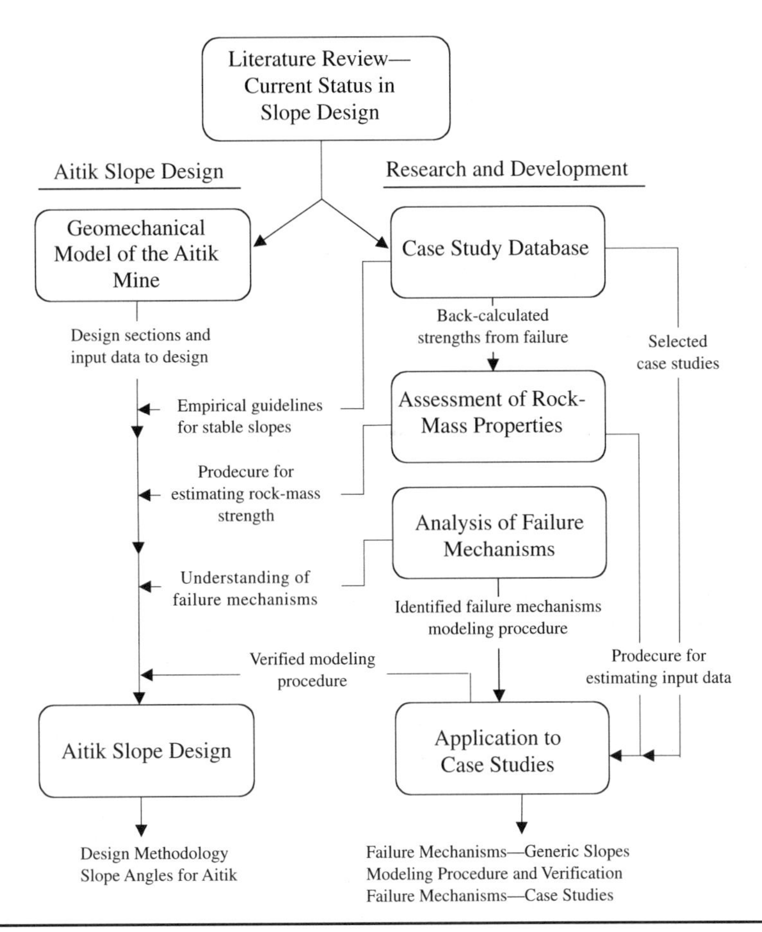

FIGURE 22.8 Outline of research project

22.2.3 Research Project

The objective of the research project was to develop an improved design methodology for hard-rock pit slopes and to apply this method to the Aitik Mine. Large-scale slope stability was treated rather broadly, as opposed to concentrating on a small topic. A combined empirical-analytical approach was used. An important assumption was that a potential large-scale failure in a hard-rock mine with fairly steep slopes (such as Aitik) will be rapid and uncontrollable with no time for remedial measures. Consequently, a conservative design approach (no failure allowed) was warranted.

The different tasks undertaken in this project are outlined in Figure 22.8. The work fell into two distinct categories–the first explicitly concerned the Aitik slope design and the second was less site specific and focused on increasing the general knowledge of rock slopes. This approach allowed production of practical results to be applied to the mine concurrent with more scientific work. It is important to note that the research and development work was required for later application to the Aitik Mine, as indicated in Figure 22.8. The work began with development of a geomechanical model for the Aitik Mine. Empirical data in the form of case studies were then collected into a database of stable and unstable slopes. Analytical work, mostly in the form of numerical modeling, was conducted to study different potential failure mechanism. Numerical models were verified by comparing them to a few selected and well-documented case studies, including validation of a methodology for estimating large-scale rock-mass strength. The project and its results are presented in full in Sjöberg (1999). In this chapter, only the results pertaining to the Aitik Mine, including the application of the design methodology, are presented.

22.3 LARGE-SCALE SLOPE STABILITY

22.3.1 Approach

The approach employs numerical modeling to assess overall pit slope stability. For the current work, the finite difference program Fast Lagrangian Analysis of Continua (FLAC) (Cundall 1976; Itasca 1995) was used. This is a two-dimensional continuum program, specifically suited to geomechanical problems. A three-dimensional analysis is not justified, considering the uncertainties regarding failure mechanisms and knowledge of structural characteristics in three dimensions. This means, however, that only the footwall and hanging wall are modeled in this work–the north area of the pit and the associated design was not addressed.

Numerical modeling is superior to many other design tools in that both deformations and stresses can be represented correctly. Modeling of different failure mechanisms of high pit slopes showed that, in particular, the deformability of the rock mass was a key factor for correctly simulating many failure mechanisms. The successful modeling of different mechanisms also gave us confidence in the modeling technique, as such. Further proof was obtained when the case studies were applied (Sjöberg 1999).

Input data to the stability analysis was obtained from rock-mass classification ratings used in conjunction with the Hoek–Brown failure criterion. This methodology is well established and was slightly modified to better suit the problem of large-scale rock-mass strength. In this procedure, the category *disturbed rock masses* was used to estimate rock-mass strength parameters, as this was found to best fit observed large-scale failures. From this, equivalent cohesion and friction angles were determined for a representative stress range. The full procedure is described in Sjöberg

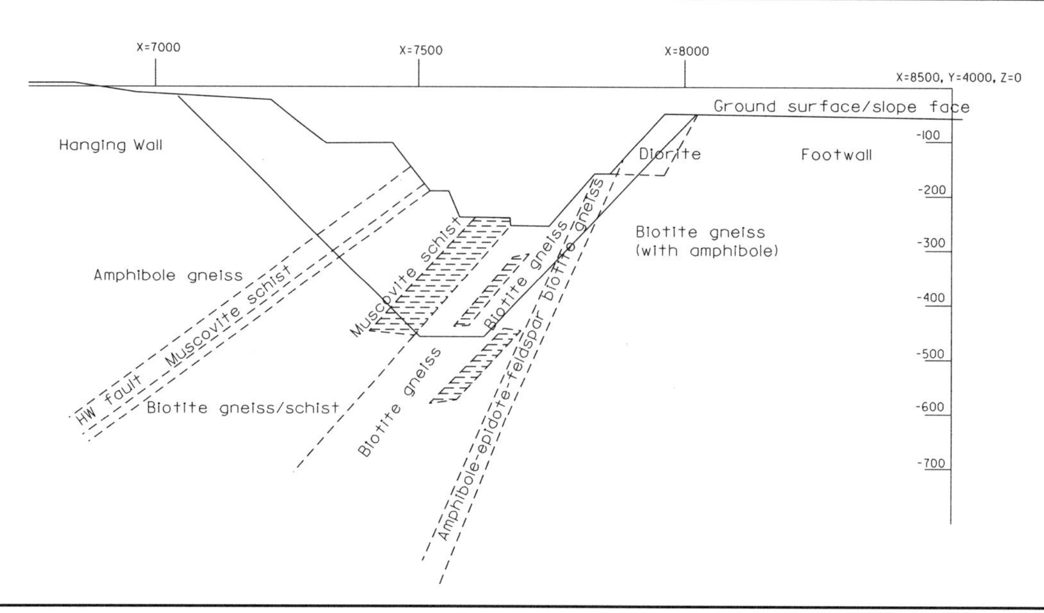

FIGURE 22.9 Representative cross section at Y4000 of the Aitik Mine (looking north), showing current and future mining geometries and major rock units

(1999). These strength values are, however, representative of the rock mass as a whole and for fully developed failure (residual strength). Thus, they are slightly conservative but nevertheless useful for design analysis and were used in the current application.

22.3.2 Geomechanical Model

A geomechanical model was developed based on previous investigations, supplemented by field mapping and testing, measurements, and observations. The model incorporated the following factors:

- *Geological units*–location and extent of the main rock types at Aitik
- *Large-scale structures*–faults, shear zones, and discontinuities of the same scale as the pit slopes
- *Joint sets*–discontinuities of smaller scale (bench scale)
- *Mechanical properties*–strength and stiffnesses for intact rock, preexisting joints, and large-scale structures
- *Groundwater conditions*–geohydrological conditions in and around the pit
- *Rock stress conditions*–premining state of stress in the Aitik area

The dominant rock types at Aitik are biotite gneiss, muscovite schist, and amphibole gneiss, along with diorite intrusions in the footwall. The rock strength is fairly high, with the exception of muscovite schist zones. The rock mass is relatively jointed, with a pronounced foliation dipping between 50° and 70° to the west and striking subparallel to the orebody. The majority of preexisting discontinuities are of limited length, typically less than 10 m. No large-scale structures have been positively identified, with the exception of a wide contact zone defining the orebody limit on the hanging wall side (the thrust contact). This zone largely coincides with the muscovite schist zone on the hanging wall side. Based on these findings, a representative design section was compiled and is shown in Figure 22.9.

Mechanical properties for intact rock were determined from laboratory tests, point load testing, Schmidt hammer tests, and strength index testing. Joint shear strength parameters were assessed primarily through field tilt tests and roughness index estimates. Based on these, representative design strength values were determined for each of the rock units and dominant joint sets. Based on these small-scale strength values and rock-mass classification ratings, the rock-mass strength was estimated using the Hoek–Brown approach (assuming *disturbed rock mass* conditions) described previously. Three sets of parameter values were determined: *typical (average)*, *high*, and *low* values for each of the geological units in Figure 22.9. The *typical* values are reproduced in Table 22.2.

Hydrological investigations and drainage drilling tests showed that the hanging wall is more or less undrained, whereas the footwall can be assumed to be drained to a horizontal distance of approximately 100 m from the current slope face (see Figure 22.10). The premining stress state was estimated from regional trends, which pointed at larger horizontal than vertical virgin stress (typical of Fennoscandia).

22.3.3 Stability Analysis—Current Slope Angles

The planned future mining geometry of the Aitik pit (down to the –460-m mining level) was subjected to a stability analysis using FLAC. The design sections of the geomechanical model were used as input to the analysis. Overall final slope angles, including all ramps, are approximately 46° for the footwall and 44° for the hanging wall. The *typical* strength parameters in Table 22.2 were used as base case parameters. Models with the *high* and *low* values were analyzed to address possible uncertainties in the input data as well as to quantify the sensitivity of the numerical model. Both perfectly plastic and ubiquitous joint (to simulate foliation joints) constitutive models were used. Two sets of groundwater conditions were analyzed: (1) no further drainage (water table coinciding with the slope face) and (2) drained slope to a distance of 100 m from the face (Figure 22.10). In total, 14 mining steps were simulated to arrive at the final pit geometry. As this was rather time consuming, additional analyses were conducted in which the pit was mined out in one step. Since plastic models are path dependent, mining in one step can give different results compared to multiple-step analyses. Nevertheless, these models could be used to indicate what might happen and interesting cases then selected for more detailed analysis with all mining steps. The different analyses are summarized in Table 22.3.

The parameter studies must be interpreted cautiously because the model has not been calibrated against failures and/or measurements. The models did show, however, that the geohydrological conditions had a drastic impact on the results. Current slope angles are probably stable provided that the slopes (both hanging wall and footwall) are drained (see Figure 22.11). For undrained conditions, a large-scale hanging wall failure

TABLE 22.2 Estimated rock-mass strength for rock types at Aitik, assuming disturbed rock mass for a stress range of σ_3 = 0–7.5 MPa

Rock Type	m_i	σ_c (MPa)	RMR	ϕ (°)	c (MPa)	σ_{cm} (MPa)	σ_{tm} (MPa)
Amphibole gneiss	33	150	69	53	1.88	11.3	0.24
Hanging wall thrust contact—muscovite schist	17	50	52	31	0.26	0.92	0.03
Muscovite schist	17	50	60	35	0.46	1.78	0.07
Biotite schist	25	65	64	43	0.71	3.24	0.08
Biotite gneiss	33	100	69	51	1.35	7.55	0.16
Amphibole-epidote-feldspar gneiss	28	70	64	44	0.73	3.49	0.08
Diorite	28	120	69	50	1.63	9.06	0.22
Biotite gneiss (with amphibole)	33	120	69	52	1.56	9.06	0.19

TABLE 22.3 Initial parameter studies conducted for the Aitik case

Analyzed Case	Model	Strength Values	Groundwater
Mining to final pit depth (–460 m) in one step			
A	Perfectly plastic	Typical	High water table
B	Perfectly plastic	Typical	Drained
C	Ubiquitous joint	Typical	Drained
D	Ubiquitous joint	Low	Drained
E	Ubiquitous joint	Typical; higher joint cohesion	High water table
F	Ubiquitous joint	Typical; higher joint cohesion	Drained
G	Perfectly plastic	Typical for hanging wall rocks Low for footwall rocks	High water table
H	Ubiquitous joint	Typical Lower joint friction angle (–2°)	High water table
All 14 mining steps			
1	Perfectly plastic	Typical	High water table
2	Perfectly plastic	Typical; higher tensile strength	High water table
3	Perfectly plastic	Low	High water table
4	Ubiquitous joint	Typical	High water table
5	Ubiquitous joint	Typical	Drained

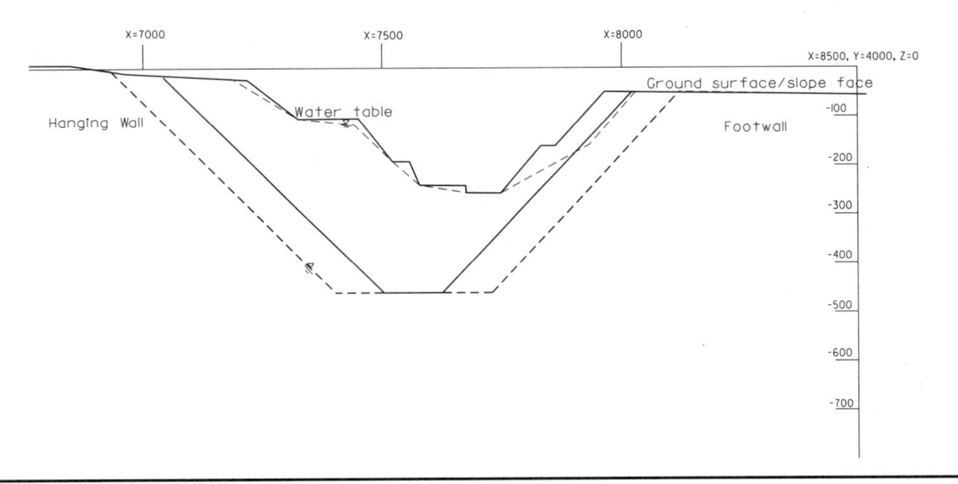

FIGURE 22.10 Representative cross section at Y4000 of the Aitik Mine (looking north), showing current and future mining geometries and probable location of the water table with continued drainage (best case)

occurred when mining to the −400- and −460-m mining levels. Failure occurred primarily as circular shear failure but with a large region of tensile failure in the upper portion (Figure 22.11). A controlling factor appeared to be the weak hanging wall contact, which yielded first.

The footwall also exhibited failure for undrained conditions and *typical* strength values, but only for the ubiquitous joint models, i.e., with foliation joints included. The resulting failure mode involved slip along the ubiquitous joints accompanied by large heaving of the slope toe (Figure 22.12). This type of failure has been termed "underdip toppling" because it occurs in slopes where the joints dip subparallel to, but steeper than, the slope face. Kieffer (1998) used the term "rock slumping" to describe the phenomenon. This mechanism is not yet fully understood; however, a failure surface only seems to develop for undrained conditions (Söderhäll 1998), which in turn stresses the importance of drainage for continued mining at Aitik. A change of the rock-mass strength from *typical* to *low* values gave somewhat different results. Footwall failure occurred for *low* strength values; however, these values are not applicable to the hanging wall because they produced failure already for the current mining depth—which has not been observed in reality.

TABLE 22.4 Additional parameter studies with lower rock-mass strength for the footwall (mining in one step to final depth for all these cases)

Analyzed Case	Model	Strength Values—Footwall Rocks	Groundwater
P1	Perfectly plastic	Typical	Drained
P2	Perfectly plastic	Typical but 50% lower cohesion	Drained
P3	Perfectly plastic	Typical but rock-mass rating 5 units lower	Drained
P4	Perfectly plastic	Typical for muscovite schist; low values for other rocks	Drained
P5	Perfectly plastic	Low for muscovite schist; typical for other rocks	Drained
P6	Perfectly plastic	Low; rock-mass rating 5 units lower	Drained
P7	Perfectly plastic	Extra low; rock-mass rating 10 units lower; typical for muscovite schist	Drained
P8	Perfectly plastic	Extra low; rock-mass rating 10 units lower	Drained
P9	Perfectly plastic	Extra low; rock-mass rating 15 units lower	Drained
U1	Ubiquitous joint	Typical	Drained
U4	Ubiquitous joint	Typical for muscovite schist; low values for other rocks	Drained
U5	Ubiquitous joint	Low for muscovite schist; typical for other rocks	Drained
U6	Ubiquitous joint	Low; rock-mass rating 5 units lower	Drained
U7	Ubiquitous joint	Extra low; rock-mass rating 10 units lower; typical for muscovite schist	Drained
U8	Ubiquitous joint	Extra low; rock-mass rating 10 units lower	Drained
U9	Ubiquitous joint	Extra low; rock-mass rating 15 units lower	Drained

TABLE 22.5 Additional parameter studies with different foliation dip (mining in one step to final depth for all these cases)

Analyzed Case	Model	Strength Values and Foliation Dip	Groundwater
U1	Ubiquitous joint	Typical strength values; foliation dip: 68° on footwall, 46° on hanging wall	Drained
U3:1	Ubiquitous joint	Typical strength values; foliation dip: 68° for all rocks	Drained
U3:2	Ubiquitous joint	Typical strength values; foliation dip: 46° for all rocks	Drained
U3:4	Ubiquitous joint	Typical for muscovite schist; low values rock-mass rating 5 units lower for all other rocks; foliation dip: 46° for all rocks	Drained

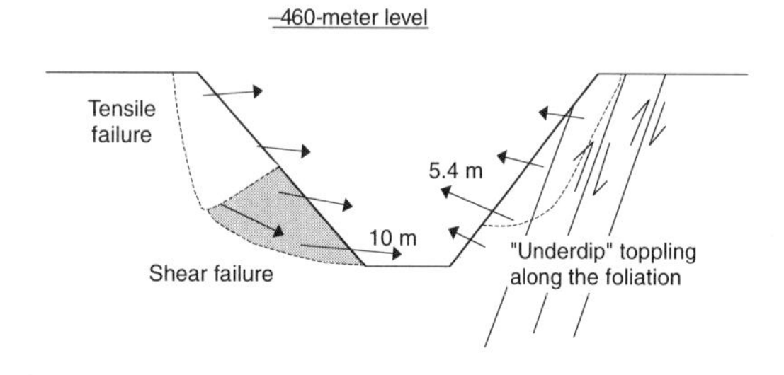

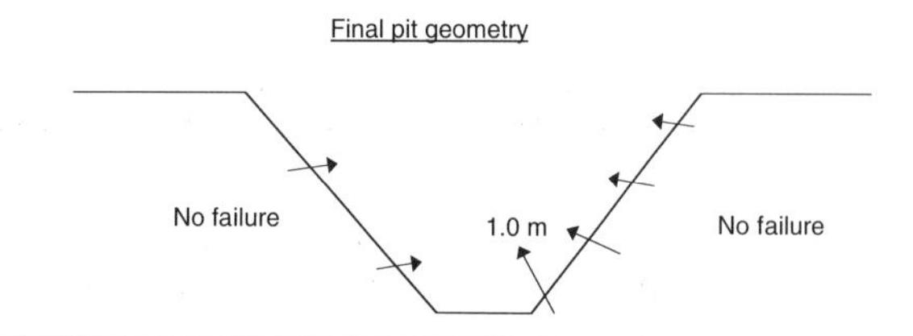

FIGURE 22.11 Example of interpreted stability conditions (from numerical modeling) for final mining depth of the Aitik pit and typical strength values for (a) high water table (no further drainage) and (b) drained conditions

To further quantify whether failure can also occur under drained slope conditions, additional parameter studies were conducted. These focused on the footwall slope, which is more critical to the mine. A set of analyses with even lower rock-mass strength values was run (Table 22.4), as well as a set with varying dip for the foliation joints (Table 22.5). These parameter studies showed that large-scale failure of the footwall only occurred for cases P9 and U9, i.e., for much lower strength values than estimated. Lower strength of the muscovite schist zones within the footwall did not result in widespread instabilities. Rather, the strength had to be low for the biotite gneiss unit for failure to develop. Even when large-scale failure did not develop fully, fairly large displacements (up to 1 m) were obtained in the toe region.

By varying the dip of the foliation joints, larger displacements resulted in the footwall for flatter foliation dip. However, the model did not become unstable for this case. With steeper foliation dip in the hanging wall rocks, large-scale toppling failure developed in the upper portion of the hanging wall. This is an important finding since it indicates that hanging wall failure can occur even under drained conditions if the foliation dip is locally steeper in the hanging wall.

22.3.4 Stability Analysis—Steeper Slope Angles

Steeper overall slope angles were analyzed for a set of parameter values. Initially, slightly steeper overall slope angles (50° rather than 46°, including all ramps) were analyzed for both the hanging wall and footwall and for a total of eight parameter combinations (see Table 22.6).

TABLE 22.6 Conducted analyses of steeper mining geometry for the Aitik pit (mining in one step to final depth for all these cases)

Analyzed Case	Model	Strength Values	Groundwater
Steeper overall slope angles (50°)			
SP1	Perfectly plastic	Typical	Drained
SU1	Ubiquitous joint	Typical	Drained
SP2	Perfectly plastic	Typical	High water table
SU2	Ubiquitous joint	Typical	High water table
SP4	Perfectly plastic	Low for footwall rocks; typical for muscovite schist; typical for hanging wall rocks	Drained
SU4	Ubiquitous joint	Low for footwall rocks; typical for muscovite schist; typical for hanging wall rocks	Drained
SP6	Perfectly plastic	Low for all footwall rocks; typical for hanging wall rocks	High water table
SU6	Ubiquitous joint	Low for all footwall rocks; typical for hanging wall rocks	High water table
Steeper footwall slope including two stepouts (Figure 22.12)			
P1 and U1	Same as in Table 22.4		
P4 and U4	Same as in Table 22.4		
P6 and U6	Same as in Table 22.4		
P7 and U7	Same as in Table 22.4		
P8 and U8	Same as in Table 22.4		
P9 and U9	Same as in Table 22.4		

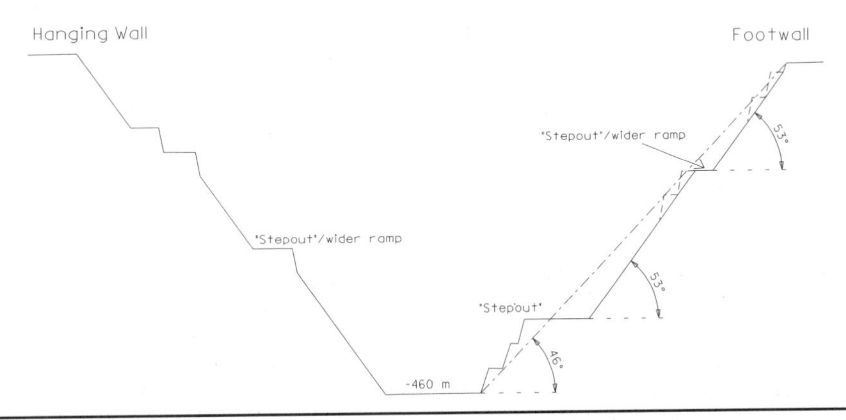

FIGURE 22.12 Proposed new slope design for the Aitik Mine, involving 5° steeper interramp angles complemented by stepouts at regular intervals

These showed that failure only occurred for undrained conditions. Under the assumption that the slopes can be drained at least to a distance of 100 m from the slope face, no large-scale failure resulted, even for *low* strength values. Following these promising results, discussions with the mine planning staff at Aitik were undertaken, and a new design was proposed. This involved 5° steeper interramp angles on both the hanging wall and footwall, combined with slightly wider ramps (Figure 22.12). The wider ramps (or "stepouts") provide catchment for overspill when mining additional pushbacks in the future. This design results in drastically decreased waste stripping for future mining at Aitik and at the same time reduced production disturbances during pushback mining. By including stepouts, the overall slope angle is slightly reduced, which reduces the risk of failure that comes with steeper interramp angles. For the footwall, there is also the possibility of recovering additional ore in the final mining sequence by steepening the slope from the lower stepout and to the pit bottom. Additional parameter studies were conducted for this mining geometry, as listed in Table 22.6.

Large-scale failure only occurred for cases P9 and U9. Signs of incipient failure could be observed for case U8. This is in fact very similar to the model results for the original slope design. It appears that the proposed new design does not result in drastically increased risks of large-scale failure. The largest uncertainty lies in the validity of the rock-mass strength values. The *typical (average)* values are judged to be the most representative, considering the current knowledge of the rock mass at Aitik. It is, however, likely that the local rock-mass strength is much lower than these values. On the other hand, it is unlikely that large portions of the rock mass are substantially weaker than the estimated strength values. Since the analyses indicated that the rock-mass strength had to be relatively low in large portions of the footwall for large-scale failure to occur, it seems that the proposed design can be implemented with some confidence.

Further indications of this can be found from the case study database. This data permitted development of preliminary empirical guidelines for the footwall and hanging wall at Aitik. Judging from these, steeper slopes (up to say, 50°–55° overall slope angles) appear viable for both for the footwall and hanging wall.

22.4 IMPLEMENTATION

The mine is currently implementing steeper interramp angles (+5°) for a new pushback on the hanging wall side. This will not be a final wall since (at least) one additional pushback will be mined on the hanging wall side. Consequently, the overall slope angles are still low and stability conditions will not be critical for some time. The footwall pushback has been postponed until later but it will also be mined using the new and steeper interramp angles if mining plans remain intact. The steeper interramp angles also imply some changes in the drill-and-blast pattern.

Work is currently ongoing to address this. A drainage program is also being planned as the footwall slope and the final pushback on the hanging wall are being mined.

An automated slope-displacement monitoring system has been installed. A surface survey displacement system was chosen to provide good areal coverage of the slopes in question. Elastic analysis was conducted to estimate the required maximum accuracy of the system. Calculated elastic displacements for mining of a single bench were around 10 mm of horizontal movement but only a few millimeters of vertical movement. Thus, the monitoring system needs to be able to resolve displacements of this order of magnitude. The chosen system consists of a high-precision total station (Leica TCA 2003A with computer software APSWin) and a number of survey prisms. The operation of the total station is controlled through APSWin, which is installed on an external standard desktop personal computer. Currently, 22 prisms are uniformly distributed on the footwall, from the crest and all the way to the pit bottom. The individual distances between the prisms are about 150 m both horizontally and vertically. Distances and angles (horizontal and vertical) to each prism are measured every half-hour, which gives enough data to eliminate any measurement errors. Four reference prisms are used for corrections and base station checks.

The achieved accuracy is approximately ±2 mm for the distance measurements and ±0.3 mgon in terms of angular accuracy (both horizontally and vertically). The distances from the base station to the prisms are, for the majority of the prisms, in the range of 500 to 900 m. The measurement error is around 4 to 5 mm for the angle measurements, which is within the required accuracy for this application. Evaluation of measurement data has shown that the footwall slope is stable, with no significant deformations detected, as predicted.

Currently, only the footwall slope is being monitored from the base station located on the hanging wall crest. However, provision is made to also monitor the hanging wall slope in the future. The early installation of this monitoring system will provide additional safety, as well as increased knowledge of how the slope behaves. The latter can be used to fine-tune and enhance the numerical models.

22.5 SUMMARY AND CONCLUSIONS

The Aitik Mine is an interesting case, in fact, it can be considered as having "too strong rock." At first glance, the lack of previous failures seems to make the mine an ideal candidate for increasing the slope angles. However, this situation is at least as difficult to handle as slope design for mines with ongoing failures, due to the high risk of a rapid and uncontrollable failure in hard and brittle rock. This case is a good example of the need for extensive research to reduce uncertainties and provide a basis for decisions. Currently, steeper interramp angles are being implemented in the mine, which will result in significant cost savings that far outweigh both the costs of drainage and slope monitoring, as well as the costs of the research project itself.

It is important to note that large uncertainties still exist, and it is essential that the choice of slope angle not be based solely on the modeling results presented here. The models must be calibrated, initially against displacement measurements. Hence, it is important that slope monitoring be conducted continuously at the site. The geomechanical model must also be continuously updated as mining progresses deeper. This includes additional investigation of the footwall, identification of weaker zones of muscovite schist, and better quantification of the properties and extent of the hanging wall contact. Additional studies on the effects of stepouts on the overall slope stability should also be conducted.

An additional uncertainty concerns the potential deepening of the pit to the −585-m mining level (pushback no. 6). This geometry has not been analyzed in this study. However, an important result of this work is that there now exists a design tool that can be used with some confidence in mine planning. Hence, it is possible to analyze and evaluate different mining geometries from a rock mechanics perspective. It should be possible to use this design methodology in other open-pit mines worldwide.

22.6 ACKNOWLEDGMENTS

A majority of the work presented in this chapter is the result of a four-year joint research project between Boliden Mineral AB and the Division of Rock Mechanics, Luleå University of Technology. The research project was sponsored exclusively by Boliden Mineral AB, which is gratefully acknowledged. The author is also indebted to the project supervision group for their great interest and fruitful discussions. This group consisted of Dr. Erling Nordlund, Luleå University of Technology; Norbert Krauland, formerly at Boliden Mineral AB; and Professor William Hustrulid, University of Utah (Professor Hustrulid was Head of Mining Research and Development at LKAB, Kiruna, during the main portion of the project). In addition, the support of the mine staff at Aitik is gratefully acknowledged, in particular the contributions by former mine geologists Stig Abrahamsson and Tua Welin.

22.7 REFERENCES

Call, R.D., D.E. Nicholas, and J.P. Savely. 1976. *Aitik slope stability study.* Pincock, Allen & Holt, Inc., Report to Boliden Aktiebolag, Gällivare, Sweden.

Call, R.D., J.P. Savely, D.E. Nicholas, and J.M. Marek. 1977. Probabilistic approach to slope design for the Aitik Mine, Sweden. *Proceedings Bergmekanikdagen 1977, Papers presented at Rock Mechanics Meeting,* Stockholm, Sweden. pp. 37–62.

Chadwick, J. 1996. Aitik plans for the 21st Century. *Mining Magazine,* July 1996, pp. 8–10.

Cundall, P.A. 1976. Explicit finite difference methods in geomechanics. *Proceedings 2nd International Conference on Numerical Methods in Geomechanics,* Blacksburg, Virginia, June 1976, 1:132–150.

Drake, B. 1992. *Aitik copper mine.* Internal Report, Boliden Mineral. ISSN:0046-5720.

Itasca. 1995. *FLAC Version 3.3. Manual.* Minneapolis: ICG.

Kieffer, D.S. 1998. *Rock slumping: A compound failure mode of jointed hard rock slopes.* Doctoral thesis, Department of Civil and Environmental Engineering, University of California at Berkeley.

Monro, D. 1988. *The geology and genesis of the Aitik Cu-Au deposit, Arctic Sweden.* Doctoral thesis, Department of Geology, University College, Cardiff, Wales.

Ouctherlony, F., S. Nie, U. Nyberg, and J. Deng. 1997. Monitoring of large open cut rounds by VOD, PPV and gas pressure measurements. *FRAGBLAST–International Journal of Blasting and Fragmentation,* 1:1–26.

Petersson, A. 1988. *Släntprofilstudier längs Aitiks ligg- och hängvägg.* Bergteknisk rapport, Bergmekanikrapport 11/88, Boliden Engineering (in Swedish).

Petersson, A. 1990. *Släntprofilstudier längs Aitikgruvans ligg- och hängvägg; 1988 års flygfotografering.* Bergmekanikrapport B5/90, Boliden Mineral (in Swedish).

SGU. 1996a. *Bedrock Map 28K Gällivare SV.* Sveriges Geologiska Undersökning: SGU serie Ai nr 100, Berggrundskartan.

SGU. 1996b. *Bedrock Map 28K Gällivare SO.* Sveriges Geologiska Undersökning: SGU serie Ai nr 100, Berggrundskartan.

Sjöberg, J. 1999. *Analysis of large scale rock slopes.* Doctoral thesis 1999:01, Division of Rock Mechanics, Luleå University of Technology.

Söderhäll, J. 1998. *Understjälpning i höga bergslänter.* Master of science thesis 1998:361 CIV, Division of Rock Mechanics, Luleå University of Technology (in Swedish).

West, R.J., N.B. Larson, P.J. Visca, D.E. Nicholas, and R.D. Call. 1985. *Aitik slope stability study.* Call & Nicholas, Inc., Report to Boliden Mineral AB, Aitik Mine.

CHAPTER 23

Practical Rock-Slope Engineering Designs at Barrick Goldstrike

Nick D. Rose* and Robert P. Sharon†

23.1 INTRODUCTION

The Betze-Post open pit, 95 km northwest of Elko, Nevada, is owned by Barrick Gold Corporation and Newmont Gold Company and operated by Barrick Goldstrike Mines Inc. (Barrick). This open-pit gold mine operates at a capacity of about 380,000 tonnes per day (ore and waste), with an average stripping ratio of 16:1 (tonnes waste:ore). Surface topography lies at an average elevation of about 1,645 m above sea level (a.s.l.), with a projected ultimate pit bottom elevation of about 1,255 m a.s.l. Ultimate slope heights are expected to range between about 390 and 460 m, with completion of open-pit mining projected to 2010.

An expansion of the southeast and east walls, called the Second East (2E) Layback, was initiated in 1995. With a design inter-ramp slope angle (IRA) of 38°, instability of the southeast wall occurred almost from the inception of mining along structural discontinuities called "Laughing-type" faults, which strike parallel to the wall and dip northward (towards the pit) at 47° to 52°. These structures, in combination with a series of "Wicked-type" faults that dip south (into the slope) at 25° to 35°, developed a heaving failure mechanism that eventually led to six individual failure zones over a period of two years. These failures ranged in size from about 0.2 M to 2.5 M tonnes. This particular heaving failure mechanism was not experienced on the previous mining phases and, therefore, was not recognized as a potential failure mechanism in the 2E design.

In mid-March 1997, exposure of the 25° south-dipping Fraud fault on the 1,415-m (4,640-ft) level led to two major instabilities known as Zone SE-96-A and Zone S-97-B (Figure 23.1). Zone SE-96-A involved the culmination of the six previous instabilities in the southeast corner of the pit, defining a failed mass of 18 M tonnes over a slope height of 280 m. Zone S-97-B occurred to the west of zone SE-96-A in an area previously absent of large-scale instability. This failure (approximately 5 M tonne) occurred over a slope height of 170 m.

Instability in zones SE-96-A and S-97-B led to a significant modification of the 2E slope design in March 1997. Design studies were undertaken between March and October 1997 to develop a remedial slope design for a new mining phase named the Second Southeast (2SE) Layback. This seven-month design study was conducted using two-dimensional and three-dimensional limit-equilibrium stability analyses and two-dimensional numerical modeling. This chapter focuses on the process of first defining and understanding the mechanism of instability in Zone S-97-B, followed by development of an empirical and analytical approach for the final slope design in this area of the open pit. A similar but more detailed design approach to that for Zone S-97-B was utilized for Zone SE-96-A and other areas of the 2SE Layback. Due to the level of detail required to adequately explain these assessments, they have not been included in this chapter.

23.2 ASSESSMENT OF SOUTHEAST WALL SLOPE CONDITIONS

23.2.1 Engineering Geology

Intrusive bedrock on the south and southeast walls of the Betze-Post open pit is interpreted to have been influenced by extensive faulting, hydrothermal alteration, and surface weathering. The pertinent rock types and geotechnical units that relate to this study include fresh granodiorite (FGD), sheared and altered granodiorite (SAG), and extremely altered and weathered granodiorite (BAG). Small amounts of fresh and altered limestone of the Popovich Formation (FL and AL, respectively) and mudstones and siltstones of the Vinini Formation (Ovicb and Ovb) also occur at the boundaries of the study area.

23.2.2 Structural Geology

Detailed stability assessments were conducted to investigate the effects of possible variations in the structural populations of Laughing (set C)- and Wicked (set D)-type faults mapped on the May 1997 2E pit slopes. Figure 23.2 is a lower-hemisphere equal-area projection of fault populations in the southeast-wall intrusive rocks. This figure shows the distribution of specific populations of Laughing- and Wicked-type faults mapped on the 2E Layback. Laughing-type faults were indicated to have peak dip angles of 47° and 52°, a mean dip of 49°, and a standard deviation of 9°. Wicked-type faults were indicated to have peak dip angles of 25° and 32°, a mean dip of 34°, and a standard deviation of 10°. Both of these structural sets strike parallel (east–west) to the main pit slope orientation.

23.2.3 Fault and Rock-Mass Strength Properties

Back-analyses conducted for Zone S-97-B indicated peak fault shear strength conditions on moderately north-dipping Laughing-type faults with shear strengths comprised of a friction angle (ϕ) of 21° and cohesion (c) of 50 kPa. For the shallowly south-dipping Wicked-type faults, back analyzed shear strength parameters of ϕ of 18° and c of 50 kPa were derived (section 23.2.2). Residual fault strengths were indicated to range in friction from 14° to 21°, with a cohesion of zero.

Rock-mass-strength characterization for the southeast wall intrusive rock masses and specifically Zone S-97-B was derived from geomechanical core logging data and from rock-mass classification of open-pit benches. Table 23.1 provides a summary of Hoek–Brown (1988) rock-mass-strength parameters for the geotechnical units at the southeast wall. These rock-mass-strength

* Associate Consultant, Piteau Associates Engineering Ltd., North Vancouver, British Columbia, Canada.
† Barrick Goldstrike Mines Inc., Elko, Nevada.

FIGURE 23.1 Southeast wall of the Betze-Post pit showing areas of instability in March 1997

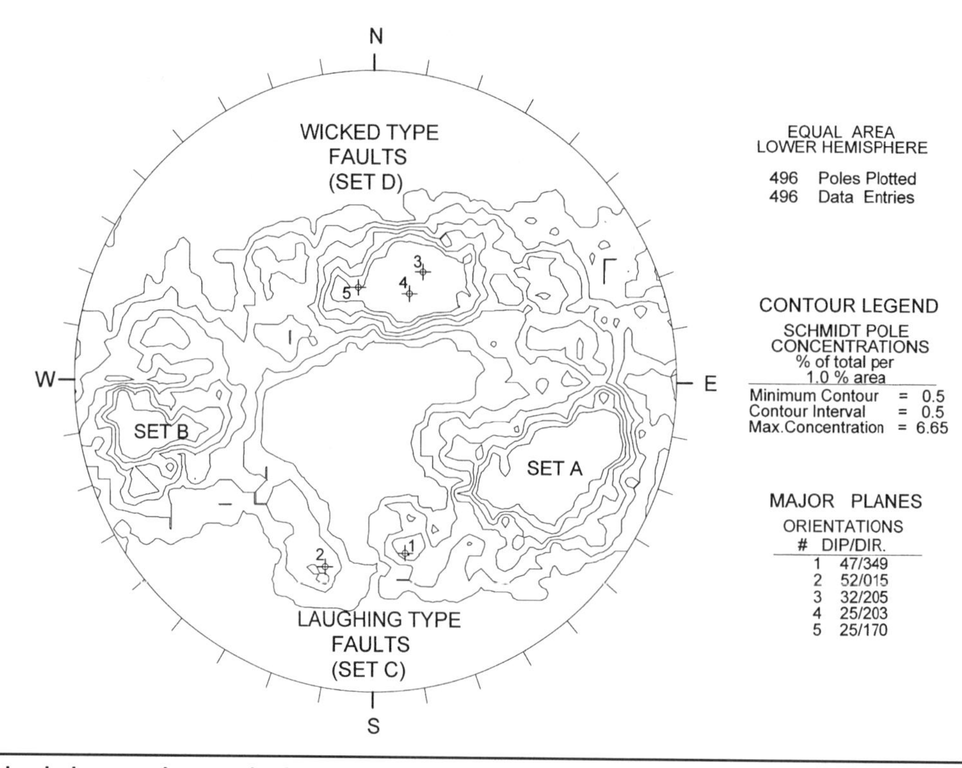

FIGURE 23.2 Lower hemisphere equal area projection of faults mapped on southeast wall

TABLE 23.1 Rock mass strength parameters for southeast wall intrusives

Geotechnical Unit	RMR	UCS (MPa)	m_i
SAG	29	14	25
Zone S-97-B, lower bound	30	24	25
Zone S-97-B, mean	36	33	25
Zone S-97-B, upper bound	42	42	25
FGD	48	77	25

parameters consist of rock-mass rating (RMR) according to Bieniawski (1976), unconfined compressive strength (UCS), and the Hoek-Brown material constant m_i.

The Zone S-97-B lower-bound and upper-bound rock-mass strengths reported in Table 23.1 represent one standard deviation in strength values below and above the mean, respectively.

23.2.4 Documentation of Previous Slope Instabilities

Documented instability on the southeast wall was interpreted to have developed by block and/or wedge sliding along northward-dipping

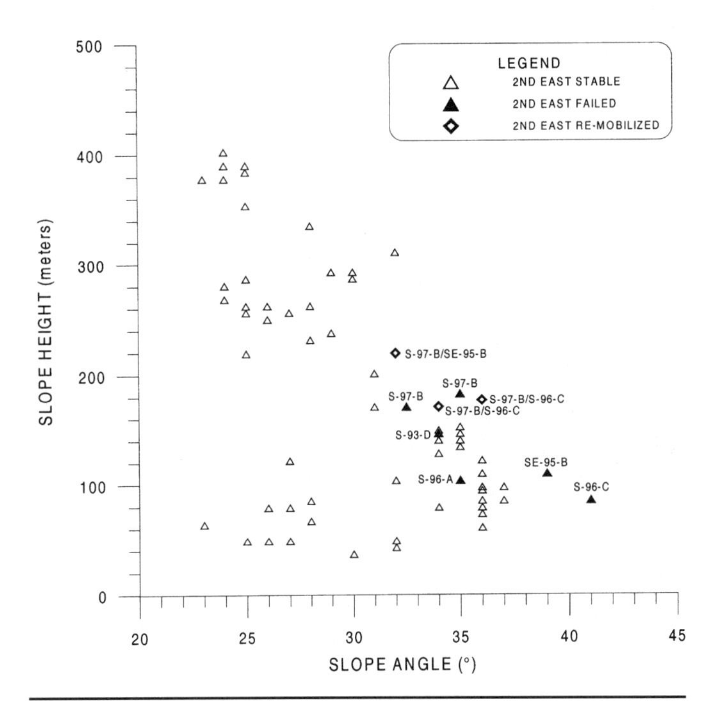

FIGURE 23.3 Documented slope height versus slope angle relationships for stable and unstable slopes on the 2E Layback

Laughing-type faults, with associated heave along shallowly south-dipping Wicked-type faults. Figure 23.3 illustrates the slope height–angle relationships for mined slopes in the area of Zone S-97-B on the 2E Layback. These documented slope geometries represent crest-to-toe measurements on 60-m-spaced sections oriented normal to the pit slopes.

Measured slope geometries shown in Figure 23.3 are identified as "stable," "failed" (unstable), or "remobilized" slopes. Stable slope geometries represent slope heights and angles that did not result in slope failure. Failed slope geometries represent the highest slope height (and angle) that was achieved prior to the point where instability developed and slope remediation or a change in the slope design was required. Remobilized slopes represent slope geometries that experienced further instability after remedial measures were implemented, such as crest unweighting or dozing.

In general, 35° and 38° interramp slopes in the area of Zone S-97-B experienced instability at documented heights of 85 to 185 m. Remobilized slopes are characterized by slope geometries that were 170 to 220 m high at angles of 36° to 32°. For the other areas of the southeast wall, 90- to 280-m-high slopes defined unstable slope angles of 34° to 27°. In the post-failure state, progressive breakback of the slope crest and accumulation of slope debris near the toe of the failure resulted in an overall slope angle as low as 23° in the area to the east in Zone SE-96-A (Figure 23.1).

23.2.5 Hydrogeology

The southeast wall intrusive rock mass is structurally complex. This complexity along with associated clay alteration results in a very compartmentalized rock mass with low hydraulic conductivity. As a result, removal of perched groundwater from the rock mass requires the implementation of intensive slope depressurization measures. Enhanced slope depressurization has been achieved by installing horizontal drain holes, vertical drains, and pumping wells, although groundwater yields from these installations tend to be low.

The implementation of groundwater depressurization measures and monitoring systems (piezometers) on the 2E Layback

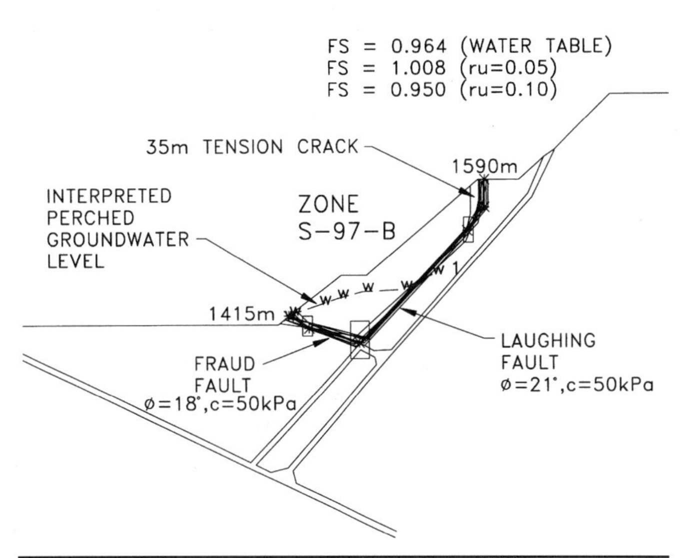

FIGURE 23.4 Back analysis of Zone S-97-B using XSTABL

was complicated due to safety concerns in unstable areas. This led to the inability to adequately depressurize the rock mass and also led to uncertainty regarding the distribution of groundwater in the more critical areas of the slope. As a result, simplifying assumptions had to be made with respect to the distribution of groundwater in slope-stability assessments. For this purpose, a pore pressure ratio (r_u), which relates the pore pressure to overburden stress, was utilized in limit-equilibrium stability analysis models. Based on back analyses in areas where adequate piezometric information existed, r_u values of between 0.05 and 0.10, representing 13 to 26% saturation of the rock mass, were considered representative of slope conditions.

23.3 SLOPE-STABILITY ANALYSES

23.3.1 Assessment of Kinematically Possible Failure Modes Involving Discontinuity Sets

The stability of kinematically possible wedge, plane, and toppling failures involving discontinuities that could daylight on the slope was assessed based on peak orientations of the various discontinuity populations identified from structural geological mapping on the 2E Layback. Lower hemisphere equal area projections of structural mapping data were utilized to identify the peak (average) orientations of the various discontinuity sets in individual structural domains on the southeast wall.

The results of kinematics assessments indicated that IRAs ranging from 28° to 38° could be achieved under fully depressurized slope conditions.

23.3.2 Back-Analysis of Zone S-97-B Using XSTABL Limit-Equilibrium Software

A two-dimensional limit-equilibrium analysis approach was considered appropriate for back-analysis of Zone S-97-B due to the lateral extent of the instability in this area of the pit and the parallel orientation of major faults to the pit slope. Figure 23.4 is a back-analysis model conducted in XSTABL$_{TM}$ (version 5.1) showing the geometry of the yielded portion of slope between the 1,415-m (4,640-ft) and 1,590-m (5,220-ft) levels. Based on field observations, a 35-m tension crack was placed in the analysis model at the crest of the failure and groundwater levels were estimated using available and interpreted hydrogeological information.

Back-analyzed material properties derived from this model consisted of the peak fault strengths, discussed in section 23.2.3 and shown on Figure 23.4, and the Zone S-97-B mean rock-mass strength, reported in Table 23.1. As seen in Figure 23.4, an r_u of

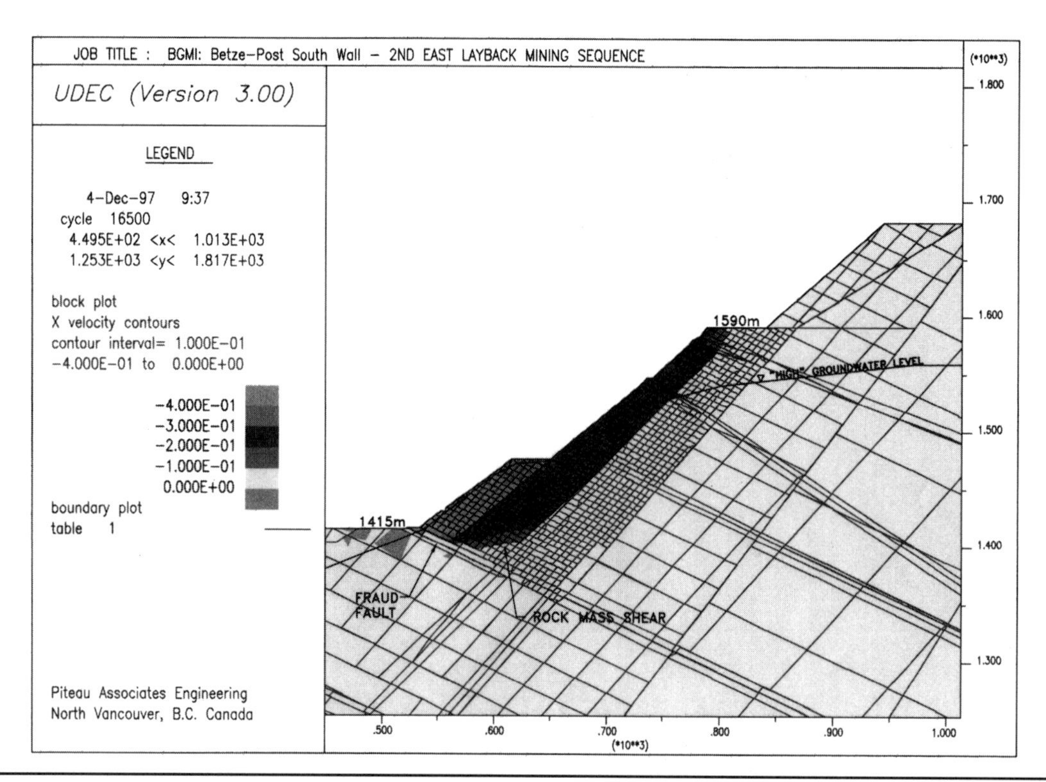

FIGURE 23.5 UDEC model of Zone S-97-B showing horizontal velocity contours

between 0.05 and 0.10 correlates reasonably well with the interpreted water levels within this zone at the time of failure.

One observation from the back-analysis model was that the minimum factor of safety resulted from shearing along the faults rather than a combination of structural and rock-mass failure. This was due to the contrast in shear strength between the weak fault zones and the Zone S-97-B mean rock-mass strength. The likelihood of the acute shear path geometry resulting from this analysis is investigated further in the following sections.

23.3.3 Numerical Modeling of Zone S-97-B Using the Universal Distinct Element Code

Numerical modeling of Zone S-97-B was conducted using the Universal Distinct Element Code (UDEC) (version 3.0) to provide an additional assessment of the possible mechanisms of instability that were identified using structural kinematics and limit-equilibrium methods of analysis. This numerical code was used to investigate the heaving failure mechanism that was interpreted based on documented field behavior and two-dimensional and three-dimensional limit-equilibrium analysis results. This numerical modeling approach was also used to assess the possible depth of deformation that may have occurred in the slope.

Although the UDEC model required a higher groundwater level (Figure 23.5) than interpreted based on available hydrogeological information and a narrow discontinuity spacing of 5 m to calibrate field performance, the model of Zone S-97-B provided a comprehensive understanding of the mechanism of failure in Zone S-97-B. The following points summarize the model behavior:

- Deformation of the slope may not have extended as deep as the interpreted "keel" of the Laughing and Fraud faults, as indicated in XSTABL (Figure 23.5). The interpreted heaving mechanism was supported by the UDEC modeling results, with indicated shearing through the rock mass at depth and inclined displacement (heave) at the toe of the failure above the Fraud fault.
- In order to calibrate this model, the Zone S-97-B rock-mass strengths had to be lowered to "lower bound" values (Table 23.1) and a groundwater level higher than what was used in the limit-equilibrium models was required.

23.3.4 Generalized Limit-Equilibrium Stability Analyses

Two limit-equilibrium analysis approaches were used to assess the sensitivity of proposed slope designs for the 2SE Layback to variations in major structural orientations, groundwater, and rock-mass conditions based on experience in Zone S-97-B. These approaches included:

- Generalized analyses of potential block shear instability associated with variations in structural orientations of Laughing- and Wicked-type faults (Figure 23.6a).
- Generalized analyses involving strength anisotropy related to major structural orientations and rock-mass conditions (Figure 23.6b).

Generalized Analyses of Block Shear Instability on Laughing- and Wicked-Type Faults. Two-dimensional limit-equilibrium analyses of the potential for heaving-type instability involving Laughing- and Wicked-type faults were conducted in XSTABL (Figure 23.6a). Sensitivity analyses were conducted for varying IRAs and slope heights using strengths, geometries, and groundwater conditions determined from the S-97-B back-analysis. Average dips of 48° and 25° were assigned to the Laughing- and Wicked-type faults, respectively, with a 35-m tension crack placed at the ramp crest, based on observed field conditions in Zone S-97-B. Peak fault strengths consisted of those reported in section 23.2.3.

The sensitivity analyses results for possible block shear instability indicate that an IRA of 31° would provide a factor of safety of 1.2 for an interramp slope height of 120 m with an assumed groundwater condition defined by an r_u of 0.05. With the incorporation of 37-m-wide ramps or stepouts at 120-m vertical increments in the design, slope heights of 120 m to 480 m are indicated

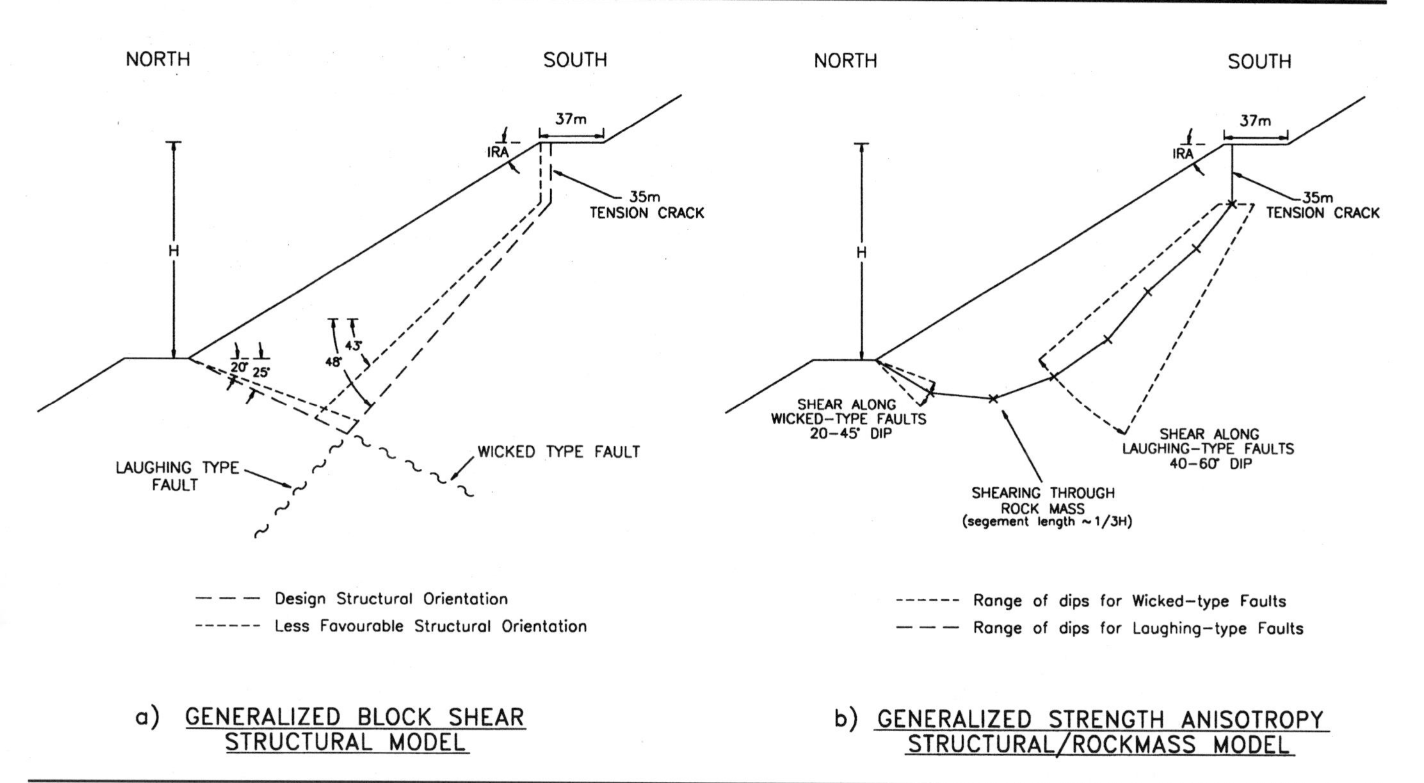

FIGURE 23.6 Generalized stability analysis models for southeast wall intrusives

to have a range in a factor of safety of 1.15 to 1.2. With the incorporation of ramp switchbacks in the slope designs, these factors of safety increased to the design factor of safety of 1.2.

Sensitivity analyses to assess the effects of groundwater in the analysis models indicated a 2° reduction in allowable IRA (for comparable factor of safety) for an increase in r_u from 0.05 to 0.10. The analyses results also showed a difference of about 4° in allowable IRA between a factor of safety of 1.0 and 1.2. For a reduction in the average dip of 5° on the Laughing- and Wicked-type faults in the stability analyses, a reduction in allowable IRA of 4° was noticed ("Design" versus "Less Favorable" structural cases on Figure 23.6a).

Generalized Analyses of Strength Anisotropy for Combined Fault and Rock-Mass Instability. Generalized stability analyses were conducted for Zone S-97-B to assess the effects of strength anisotropy related to major structural orientations and the strength of the rock mass. As shown on Figure 23.6b, the generalized strength anisotropy stability-analysis model incorporated a range in dip of 20° to 45° for Wicked-type faults and 40° to 60° for Laughing-type faults. As in the previous models, a 35-m tension crack was placed near the ramp crest. Fault strengths for these structural orientations consisted of those defined in the S-97-B back-analysis model (Figure 23.4) and reported in section 23.2.3. The analyses of strength anisotropy involved noncircular shear assuming a design r_u of 0.05 and a design interramp slope height of 120 m. Sensitivity analyses were run to assess ranges in IRA and rock-mass strength. The Hoek–Brown (1988) failure criterion for "disturbed" rock masses was used to estimate linear instantaneous Mohr–Coulomb strength parameters for the geotechnical units reported in Table 23.1, at a normal stress of 1 MPa.

The rock-mass strengths used in the analyses ranged from SAG (ϕ = 19.2°, c = 200 kPa) to FGD (ϕ = 46.9°, c = 450 kPa). Analyses results for the Zone S-97-B lower-bound strengths (ϕ = 24.2°, c = 235 kPa) indicated a factor of safety of 1.2 for an IRA of 31°. These results were consistent with the block shear analyses discussed previously. Results for the S-97-B mean, S-97-B upper-bound, and FGD strengths indicate stable slopes (i.e., factor of safety ≥ 1.0) for IRAs greater than 38°. Based on experience from the 2E Layback, where instabilities were experienced (i.e., factor of safety < 1.0) on slopes with IRAs of 35° to 38° and similar heights, the results for S-97-B lower-bound strengths were considered to be consistent with previous slope performance. These lower-bound parameters are consistent with the rock-mass strengths used to calibrate the UDEC model discussed in section 23.3.3.

23.4 DEVELOPMENT OF DESIGN CRITERIA BASED ON SLOPE HEIGHT–ANGLE RELATIONSHIPS AND STABILITY-ANALYSES RESULTS

The generalized block shear and strength anisotropy analyses (section 23.3.4) indicate an allowable IRA of 31° at a factor of safety of 1.2, an assumed r_u of 0.05, and 37-m-wide ramps or stepouts defining a maximum 120-m interramp slope height.

Figure 23.7 provides a comparison of the documented stable and unstable slope geometries on the 2E Layback, with slope geometry curves representing slope designs defined by 120-m interramp heights and 37-m-wide stepouts. The "Design Case" consists of the slope geometry curve with a 31° design slope angle at a slope height of 120 m, a factor of safety of 1.2, and an r_u of 0.05. Based on observations made from stability analyses results discussed in section 23.3.4, where a 2° reduction in the allowable IRA resulted from an increase in r_u from 0.05 to 0.10 and a difference of 4° in allowable IRA occurred between a factor of safety of 1.0 and 1.2, slope geometry curves representing a factor of safety of 1.0 and 1.2 were generated for r_u's of 0.05 and 0.10, as shown on Figure 23.7.

As seen on Figure 23.7, the "factor of safety = 1.0" slope geometry curves for an r_u of 0.05 and 0.10 provide good correlation with the documented failures in the area of Zone S-97-B. The "factor of safety = 1.0 – r_u = 0.05" curve provides a reasonable average approximation to the documented unstable slope geometries. This curve defines an IRA of 35° at an interramp height of 120 m, which corresponds to the previous design IRAs of 35° to 38° that resulted in instability (factor of safety < 1.0) on

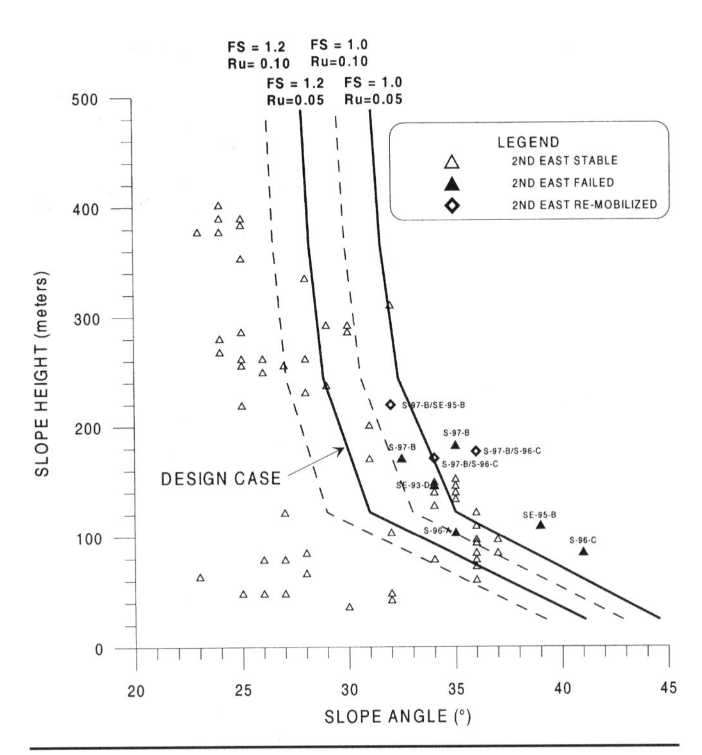

FIGURE 23.7 Comparison of slope geometry design curves to documented stable and unstable slopes on the 2E Layback

the 2E Layback. Therefore, assuming that this curve provides calibration for the unstable condition, the "factor of safety = 1.2 – r_u = 0.05" curve was considered appropriate to be applied as the design case for the 2SE Layback. This geometry curve indicates allowable overall slope angles of about 29° at a height of 480 m and an IRA of 31° at a maximum slope height of 120 m.

23.5 SUMMARY AND CONCLUSIONS

This chapter provides a relatively brief summary of the process that was utilized to identify and produce a mitigation design for a complex, structurally controlled failure mechanism that resulted in a 5-M tonne failure in Zone S-97-B. This mitigative design was developed through the combined use of traditional rock-slope engineering design tools, including structural kinematics, limit-equilibrium analysis, and empirical slope height–angle relationships. A numerical modeling approach was used to confirm the interpreted failure mechanism and complement the more traditional design assessments. This overall combination was used to develop a practical slope-design methodology for proposed mining of the 2SE Layback.

An iterative design process was undertaken with the Barrick mine planning department to develop the 2SE Layback design for the southeast portion of the open pit. This design process involved:

1. Processing and analysis of 2,500 individual geological structural measurements mapped on 2E Layback benches on the southeast wall.
2. Extensive two-dimensional limit-equilibrium analyses on specific geological cross sections as well as generalized analyses of structural and rock-mass conditions.
3. Twelve three-dimensional CLARA limit-equilibrium analysis models of multi-planar complex wedge geometries on the south and southeast walls.
4. UDEC modeling of Zone S-97-B.
5. Development of approximately 20 different mine plan options to optimize the 2SE Layback pit slope design with respect to geotechnical constraints.

It took approximately seven months to complete the design process. It is postulated that this design study was conducted in a fraction of the time and cost that an alternative full-scale numerical modeling design study would have required.

In the area of Zone S-97-B, 2SE design interramp slope angles ranged from 31° to 33°, as compared to the 35° to 38° IRAs that resulted in instability on the previous 2E Layback. Assumed groundwater conditions in the 2SE design were approximated by an r_u of 0.05, which is equivalent to 13% saturation of the rock mass. In order to achieve this assumed condition, intensive groundwater depressurization measures consisting of horizontal drain holes and vertical pumping wells were considered necessary, due to the compartmentalized nature and low hydraulic conductivity of the southeast wall intrusive rock mass.

At the time this paper was written, mining of the 2SE Layback had occurred to approximately the 1,390-m (4,560-ft) level, or below the level of previous instabilities on the 2E Layback. No instability has been experienced in the area of Zone S-97-B, other than at the bench scale. From the initiation of mining the 2SE Layback, investigations were carried out to better define the groundwater conditions through the installation of shallow angle and horizontal (slightly inclined) piezometers to target critical areas of the slope beneath the previous failure zones. Ongoing assessments have been conducted to assess the implications of groundwater on slope stability and to confirm the design assumptions. Ongoing stability assessments to investigate the occurrence and orientation of Laughing- and Wicked-type fault sets mapped on the southeast wall have also been conducted. These assessments along with positive slope performance to date have confirmed that the 2SE Layback slope designs were successful.

23.6 ACKNOWLEDGMENTS

The permission from Barrick Goldstrike Mines Inc. for the geotechnical information used in this paper is greatly appreciated. This paper is dedicated to the late Dr. Dennis Martin of Piteau Associates Engineering Ltd., who provided many years of geotechnical advice and insight into slope stability problems at the Betze-Post open pit. His presence and contribution in the field of surface rock mechanics is greatly missed.

23.7 REFERENCES

Bieniawski, Z.T. 1976. Rock mass classification in rock engineering. *Proceedings of the Symposium on Exploration for Rock Engineering,* ed. Z.T. Bieniawski, Rotterdam: Balkema, pp. 97–106.

Hoek, E. and E.T. Brown. 1988. The Hoek–Brown failure criterion–A 1988 update. Rock Engineering for Underground Excavations, *Proceedings 15th Canadian Rock Mech. Symposium,* Toronto, pp. 31–38.

CHAPTER 24

Slope Stability and Operational Control at Barrick Goldstrike

Robert Sharon*

24.1 INTRODUCTION

The Barrick Goldstrike Mines Inc. Betze-Post open pit is in the Carlin Trend in northern Nevada, as shown in Figure 24.1. Betze-Post open-pit operations date from 1987 with the development of the Post Oxide deposit. Approximately 136 M tonnes (150 M tons) of ore and waste rock are mined annually from the open pit. About 35.7 tonnes (1.15 M oz) of gold were produced from this ore body in 1999. Open-pit dimensions are approximately 2,450 m by 1,500 m (8,000 ft by 5,000 ft) and slope heights vary from 335 to 460 m (1,100 to 1,500 ft). Overall slope angles vary from 24° to 36° and interramp slope angles vary from 25° to 44°.

Slope instability has been attributed to the presence of adverse structural geology, generally weak rock-mass conditions, and perched groundwater. Several years of experience at this operation have confirmed that a comprehensive knowledge of the structural geology is essential to the development of practical slope designs. Reduction of groundwater pressurization, as a result of regional dewatering of the more permeable sedimentary rocks in the west pit area, has resulted in significant improvements to slope stability. Enhanced drainage of less-permeable south and east wall rock masses, including small-diameter pumping wells and horizontal and vertical drains, has been required to reduce groundwater levels to design tolerances. Controlled blasting adjacent to the highwalls is routinely used to minimize the effects of blast vibration on slope stability. An integrated slope-monitoring system, featuring real-time, accurate measurement of movement using two motorized theodolites, is used to ensure safe mining conditions adjacent to instabilities and below mine highwalls.

This chapter presents a review of the open-pit geology, surface-water control, groundwater and slope depressurization, slope monitoring, and highwall blast control. A discussion of local slope stability and the development of successful slope designs are then presented in two case histories, followed by a review of safety aspects.

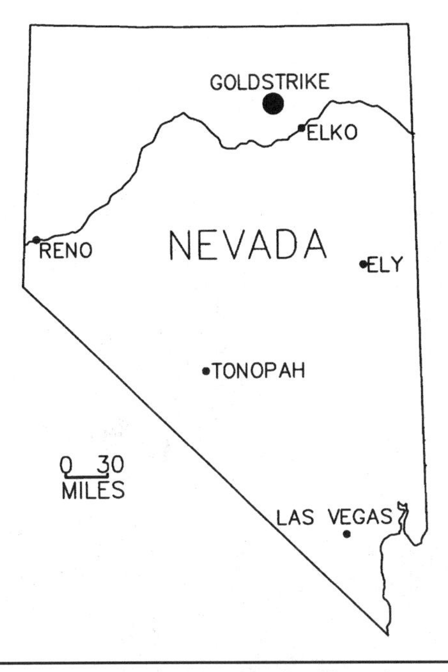

FIGURE 24.1 Location of Barrick Goldstrike Mines, Inc.

24.2 STRUCTURAL GEOLOGY

24.2.1 General Geology

The Betze-Post orebody occurs at and near the contact of a Jurassic-age dioritic intrusive sill and sedimentary rocks of Devonian and Ordovician age. The rock masses are highly fractured and altered due to faulting, contact metamorphism, hydrothermal activity, and surface weathering. Poorly consolidated silty sands of Tertiary age, called the Carlin Formation, cover these rocks in significant thickness in the north and east sections of the open pit.

24.2.2 Geology Mapping

Knowledge of the structural and clay alteration geology is essential to the understanding of slope-stability controls and the development of appropriate slope designs at Goldstrike. All highwall benches are mapped according to rigorously defined procedures.

To ensure accuracy, discontinuities such as faults, joints, and contacts are mapped on each bench face from crest to toe onto level plans in the field. The geologist then enters the field mapping data into a PC using an AutoCAD-based computer program called PITMAP, which was developed in-house. Structural data is connected between adjacent levels across the individual benches to create a comprehensive geology map of the pit. The PITMAP program requires entry of the structure type (i.e., fault, joint, bedding, folding, dike, or vein) and descriptive information for each structure (i.e., contact rock, hardness, alteration, infilling type, thickness, oxidation, brecciation and bedding or joint continuity, spacing, and roughness). Output from PITMAP defines 20 geological units plus overlay patterns for alteration types, including calc-silicate and hornfels. Faults are represented by different colors and placed into four categories, based on intensity and continuity.

24.2.3 Geological Model

Field mapping data entered into PITMAP is periodically compiled to produce a geological plan for the open pit. This plan is then

* Barrick Goldstrike Mines Inc., Elko, Nevada.

TABLE 24.1 Summary of strength and material parameters

Rock-Mass Unit (Symbol)	Intact Rock Strength (MPa)	Nonlinear		Linear		Unit Weight (kN/m^3)
		RMR	mi	c (kPa)	φ (deg)	
FGD/GD-CS	77	48	25	450	47	26.7
SAG/MSC	14	29	25	200	19	22.8
FRC	51	45	15	340	36	23.6
ARC	13	34	7	145	12	22.8
FMS	30	43	10	260	26	23.6
AMS	7	31	10	135	11	22.8
FL	55	53	12	410	40	25.1
AL	16	35	7	160	14	22.0
ICB	34	35	10	230	22	22.0
OVB	15	N/A	N/A	60	21	22.8
Gravel	2.1	N/A	N/A	170	35	18.9
Silt	2.8	N/A	N/A	105	30	17.3
Sand (ash)	1.4	N/A	N/A	105	35	14.1
Waxy-silt	N/A	N/A	N/A	35	9	17.3
Fault	N/A	N/A	N/A	0–60	14–21	22.8

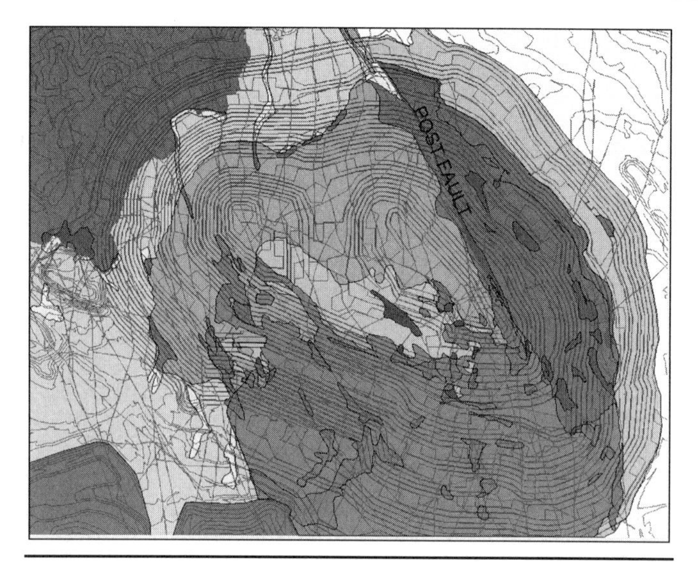

FIGURE 24.2 Geological model of the Betze-Post open pit

incorporated into the three-dimensional geological model using VULCAN. A recently updated version of the geological model projected to the open-pit shell is illustrated in Figure 24.2. The PITMAP structural data is loaded into VULCAN and sorted by structural domain for stereo-net and kinematic analysis. The highest emphasis is placed on PITMAP data in the development and refinement of the geological model. Core data from exploration and rock-mechanics drilling programs supplement PITMAP information. Most rock-mechanics core drilling is oriented using scribe technology, due to the importance placed on understanding the nature and orientation of geological structure. The geological model includes 12 "geological units" and contains more than 240 faults. These units and their clay-altered equivalents are correlated with a total of 10 rock-mass or geotechnical units that are used for slope-stability analysis. Strength and material properties for these geotechnical units, Carlin Formation subunits, and faults are summarized in Table 24.1.

24.3 SURFACE AND GROUNDWATER

24.3.1 Surface Water Control

Water Management. Surface water is controlled to induce flow to planned locations so as to minimize infiltration into the haulage ramp crests and the slopes. Surface water from precipitation events that flows toward the pit from the south and east is directed to a network of lined sumps. In-pit and perimeter haul roads are typically graded to allow surface runoff to flow inboard, away from the pit crest, into ditches and to the lined sumps. Highwall benches and other flat areas where water is observed to pond are ditched to allow drainage downhill and ultimately to the lined sumps. Water collected in sumps and transfer tanks is typically pumped out of the pit area through 200-mm (8-in.) collector pipes and 300-mm (12-in.) main lines using a series of 60- to 225-kW (80- to 300-hp) transfer pumps. Goldstrike uses a three-stage booster-pump system to lift water from the 4,280 to 5,520 levels for a maximum elevation gain of 380 m (1,240 ft).

Recharge. Surface deformation monitoring indicates that stability can be affected by local groundwater recharge, as a consequence of precipitation and surface runoff. Surface water management is of primary importance to ensure that infiltration into the slopes is minimized. The horizontal drainage system is planned, in part, to control seasonal pore pressure buildup.

24.3.2 Dewatering and Slope Depressurization

Groundwater Distribution Overview. Significant mine dewatering operations at Goldstrike commenced in 1989. Much of the ore deposit extends below the original water table elevation in high-permeability rock. The south wall intrusive rock mass and sedimentary rocks exposed in the east wall are of generally low permeability. The Post fault, a major ore boundary exposed on the lower east wall of the pit, forms the boundary between high-permeability rocks to the west and low-permeability rocks to the east.

Pit Dewatering System. As many as 25 deep, large-diameter perimeter wells have been used to lower the water table below the pit floor in advance of mining. Large-scale high-head pumping of groundwater in the temperature range of 52° to 60°C (125° to 140°F) has been achieved using electric submersible and conventional line-shaft pump systems. Over a 10-year period since 1989, the water level in the mine area has been lowered more than 490 m (1,600 ft) by dewatering at a rate of up to 4,400 L/s (70,000 gpm). Slope stability in sedimentary rocks on the north and west slopes has improved dramatically as a result of dewatering in advance of mining, where it has been possible to steepen interramp slope designs by up to 7°.

Enhanced Drainage. Areas of lower-permeability intrusive rocks exposed on the south wall and sedimentary rocks east of the Post fault (Figure 24.2) have not drained freely as a result

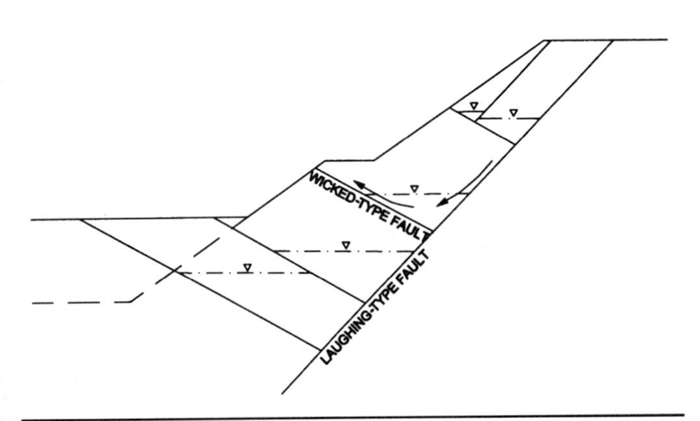

FIGURE 24.3 Simplified view of the south wall structural control and perched groundwater distribution

of lowering the water table, as described above. Enhanced drainage methods, including small-diameter low-yield pumping wells and vertical and horizontal drains, have been necessary to minimize the risk of instability and associated costs of mining through the lower-permeability rocks.

Horizontal Drains. Hundreds of horizontal drains have been drilled into the less-permeable south and east walls over the last 10 years to achieve groundwater depressurization objectives. Holes 150 mm (6 in.) in diameter are drilled up to 360 m (1,200 ft) in length for the purpose of depressurizing the developing highwalls and to drain future laybacks and ultimate highwalls. Each drain that yields water is completed with 50-mm (2-in.) slotted polyvinyl chloride (PVC) pipe over the entire length. A 3-m- (10-ft-) length of 150-mm- (6-in.-) diameter steel surface casing is grouted at the collar to control flow. Drain-hole spacing varies depending on depressurization targets and objectives. All significant production is contained and diverted to 11,350- to 22,700-L (3,000- to 6,000-gal) transfer tanks through vacuum hoses attached at the drain collars or through a 200-mm (8-in.) high-density polyethylene (HDPE) trunk-line into which a number of horizontal drains may be plumbed. Water collected in transfer tanks is pumped out of the pit through 300-mm (12-in.) diameter steel or HDPE pipes using a series of transfer pumps.

Intrusive Rocks. The south wall intrusive is a nonhomogeneous, highly fractured rock mass. Drainage across certain faults and associated argillization is very poor. As a consequence of slope instability experienced on the southeast wall in 1996 and 1997, it became necessary to investigate the groundwater distribution and develop a depressurization strategy for remediation purposes. Interpretation of the perched groundwater distribution and the geology that controls the distribution has been possible by superimposing hydrological information onto the geological model to develop an interpreted "hydrogeological model." A simplified view of the primary structural control and major perched water distribution is shown in Figure 24.3. The hydrogeological model indicated that the most important locations of perched groundwater occurred above the intersection of certain Wicked- and Laughing-type faults, which were also identified as being important slope-stability controls, as described in section 24.7.5. From the resulting interpretation, it was possible to confirm the slope-design groundwater tolerances and develop and justify a specific depressurization plan, including placement of additional small-diameter dewatering wells and horizontal drains.

East Wall Sedimentary Rocks. Effective depressurization of lower-permeability east wall sedimentary rocks of Tertiary age (Carlin Formation) and of Paleozoic age (Vinini Formation, or bedrock) has been achieved by using both conventional regional dewatering and localized drainage-enhancement methods. Most of the depressurization effort for the east wall was implemented prior to and during development of the Second East Layback, which started in 1995. Mining of the ultimate highwall through this low-permeability ground, which started in January 2000, has benefited from up to four years of advance slope depressurization. The dewatering strategy for the east wall sedimentary rock is shown in Figure 24.4. Water levels in bedrock have been successfully lowered to or beyond targets defined by the ultimate highwall designs using a number of deep pumping wells. Typical wells are 150 to 300 mm (6 to 12 in.) in diameter to depths of 300 m (1,000 ft) and have historically pumped at rates of 3 to 30 L/s (50 to 500 gpm). Most of these wells are screened over most of their length, allowing them to double as vertical drains where they pierce the Carlin Formation. Dewatering of the overlying, less-permeable Carlin Formation using vertical drains became viable as a consequence of lowering the water level in the bedrock, thus creating an underdrain.

Carlin Formation Groundwater Distribution. Depressurization of the Carlin Formation has been impeded by the occurrence of subhorizontal, very- low-permeability, clayey-silt beds. Interbedded fine sandy ash layers of the Carlin Formation that are of relatively higher permeability (averaging 10^{-4} cm/s) are conducive to long-term depressurization and have been targeted by horizontal drilling to produce underdrains for the less-permeable layers. Interpretations of the Carlin Formation groundwater distribution on five cross sections were developed to analyze slope stability, design highwalls, and develop a dewatering strategy. This was achieved by modeling stratigraphy and structure and superimposing hydrological information from piezometric records.

24.3.3 Dewatering Performance Monitoring

Confirmation of Groundwater Distribution and Dewatering Rates. Monitoring of groundwater levels and dewatering rates is necessary to confirm the adequacy of the highwall designs and evaluate the effectiveness of the slope-depressurization program. If it is determined through performance monitoring that dewatering targets assumed by the designs may not be achieved, implementation of additional dewatering measures may be justified or other options may be considered, including delaying mining or proceeding on design with the acceptance of increased risk. Once the dewatering targets are achieved, piezometric data can be used to justify reducing the dewatering rate by shutting off certain pumping wells to minimize cost.

Groundwater Monitoring. Groundwater-level monitoring is achieved using both standpipes and vibrating wire transducers. In most rotary holes drilled for piezometer construction, two 50-mm (2-in.) diameter standpipes are installed. Screened intervals are usually 12 m (40 ft) long. Standpipes are commonly installed in the intrusive rocks and Vinini Formation. Recognizing the development of multiple vertically stacked perched water tables in the Carlin Formation, rotary holes of the same standard size (nominal 15 cm, or 6 in. diameter) have been successfully completed with up to seven vibrating wire transducers, each isolated by a gravel pack and bentonite seal. The resulting installations are capable of monitoring multiple intervals, versus the limitation of only two standpipes per standard 150-mm- (6-in.-) diameter drill hole.

Piezometer Installations. Piezometers are typically installed in vertical holes, which are usually the least expensive and easy to install. However, due to specific structural targets and limited access and/or safety considerations, some standpipes and vibrating wire transducers have been installed in shallow angle holes and also into horizontally drilled holes. To determine the groundwater distribution beneath the southeast wall instabilities se-96-a and s-97-b in 1998 (see also section 24.7.5), it was necessary to install piezometers into horizontally drilled holes to target locations up to 425 m (1,400 ft) from the nearest available safe drilling location. The design for these piezometers, illustrated on Figure 24.5, includes a double packer installation, a 30-m (100-ft) grout seal, and a pressure gauge at the collar.

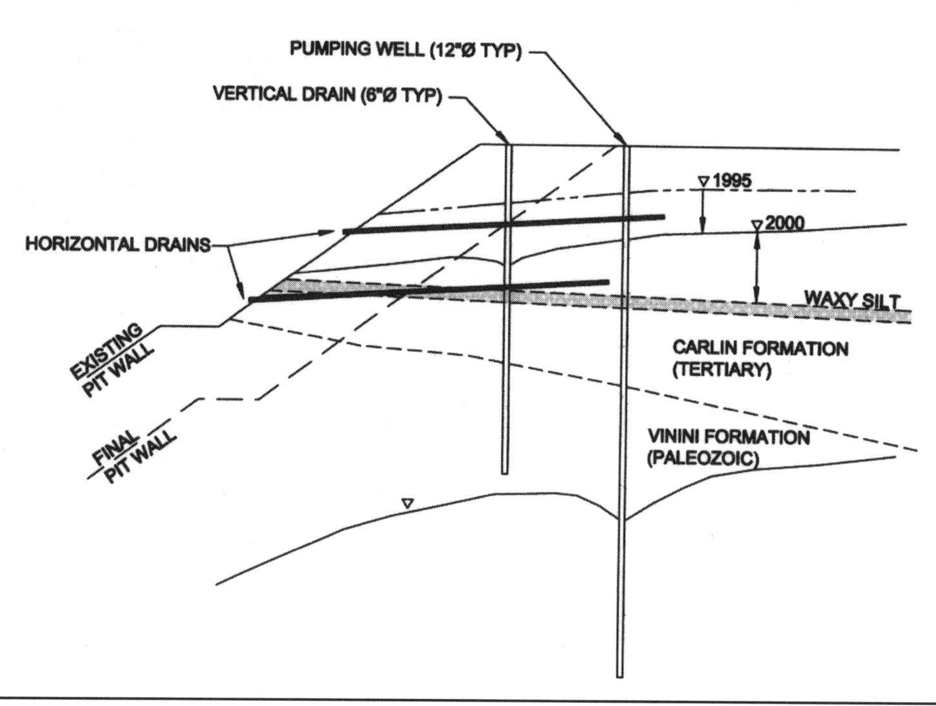

FIGURE 24.4 Dewatering strategy for the east wall

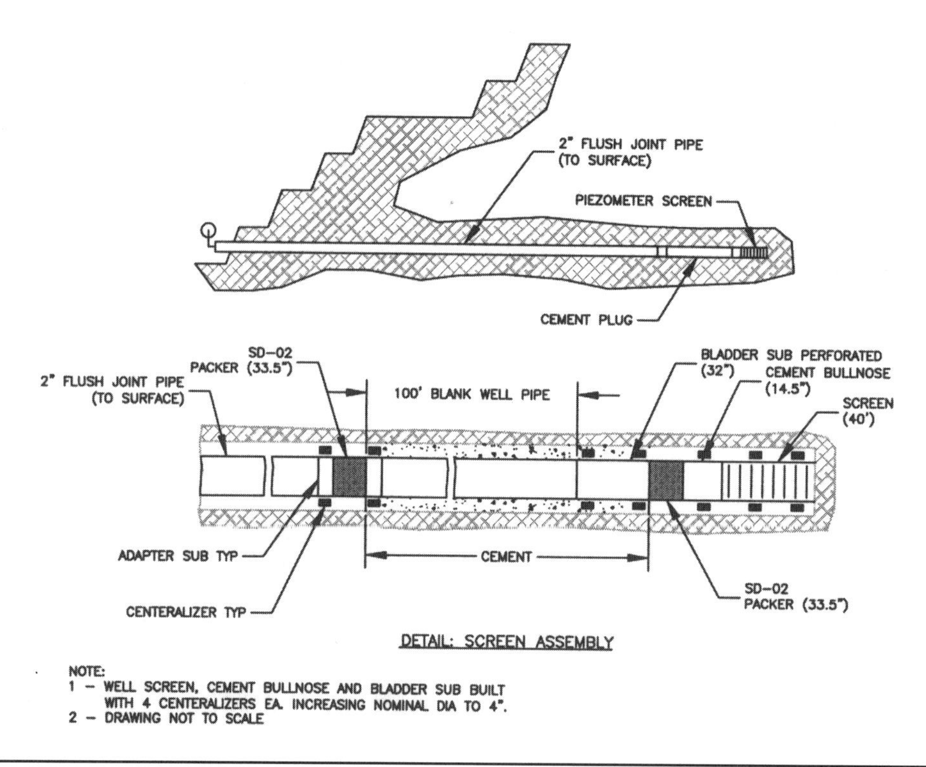

FIGURE 24.5 Horizontal piezometer design

24.4 SLOPE MONITORING

Slope movements are monitored primarily with an automated system. Two motorized theodolites are used around-the-clock to track movement of more than 100 prisms over the entire pit area in real-time. The density of prisms is greater in active movement areas, particularly in the vicinity of active mining. Components include a Leica TM3000D theodolite and a DI2002 EDMI, which are housed in a climate-controlled building. Communications between both survey stations and the base station, located in the mine engineering offices, are achieved using a RF115 Blackbox Industrial Modem. Survey accuracy is on the order of 4.8 mm/km (0.30 in./mi.) for distance and 19 mm/km (1.2 in./mi.) for horizontal and vertical angles.

The remote system does most of the slope-monitoring work but is only part of an effective monitoring system. SMS Modular Mining wireline extensometers supplement the system primarily

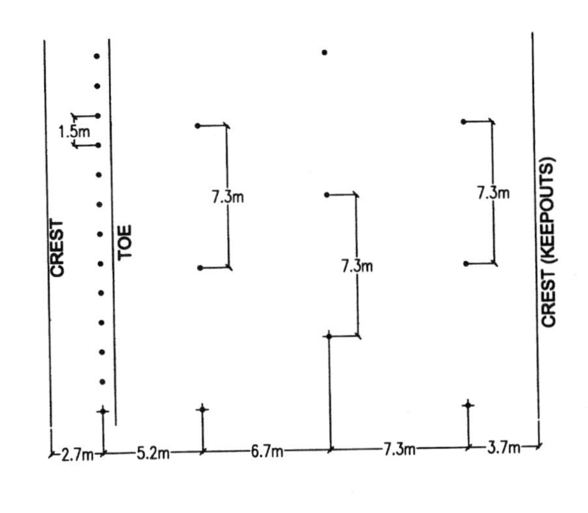

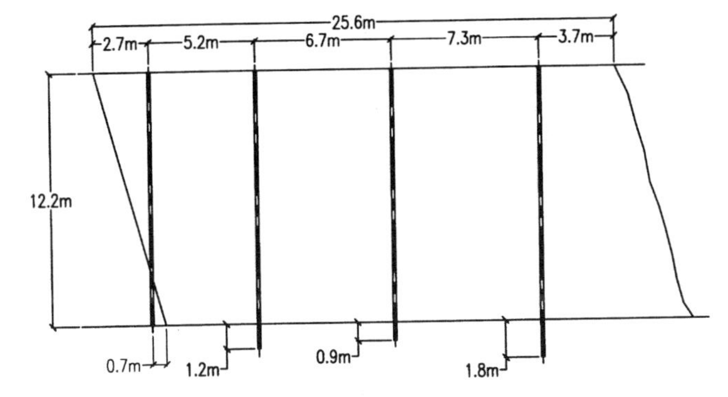

FIGURE 24.6 Standard four-row trim blast pattern

by monitoring movement in critical areas where tension cracks have developed, such as on active haulage ramps. An alarm will alert mine dispatch and mine operations personnel in the event that programmed movement thresholds are exceeded. Inclinometers are used to assess deep-seated deformation in specific locations where surface monitoring alone is insufficient. Time domain reflectometry (TDR) technology, in which a two-conductor cable is installed in selected holes drilled for piezometer installation, is used to monitor for subsurface shear displacement. Goldstrike application uses a cable tester to locate a break in a coaxial cable, which would indicate shear displacement.

24.5 BLAST CONTROL

24.5.1 Trim-Wall Blasting

Trim-wall blasting procedures are used at Goldstrike to produce adequate fragmentation to achieve efficient excavation and minimize the effects of blast vibrations and the potential loss of shear strength along structure in the highwalls. ANFO is typically used, except in areas producing water where slurry is substituted. Blast effects from production shots are isolated from the highwall using three- or four-row blast trim patterns. When the mine sequence requires the trim area for temporary haulage, a four-row pattern is used. When there is no haulage requirement, a three-row pattern is used, which results in a relative cost savings. Modifications to trim blast-hole spacing and loading have recently been made to increase the width of the trim-pattern designs; this accommodates the conversion of much of the Goldstrike 170 tonnes (190 ton) Haul Pack haulage fleet to the wider 300 tonnes (330 ton) Komatsu 930E. Blast trim patterns are used whenever possible throughout the pit. This procedure has been used effectively in pit operations with no significant impacts on overall mine productivity.

TABLE 24.2 Typical blast data—intermediate hardness ground

Toe row	171-mm hole; 22.7 kg ANFO
Buffer row	222-mm hole; 152 kg ANFO; 9.8-m stem
Outer buffer row	222-mm hole; 243 kg ANFO; 7.3-m stem
Modified production row	222-mm hole; 272 kg ANFO; 7.6-m stem
Powder factor	0.26
Energy factor	98 kcal/tonne
Primers	One 0.34-kg (¾-lb) cast primer in bottom of each hole

24.5.2 Standardized Designs

Trim-pattern designs have been standardized to accommodate almost all highwall design configurations for the purpose of simplifying the planning process and field implementation. Most slope designs at Goldstrike produce interramp slopes between 35° and 44°. For a limited range of slope designs, the design standard burden between rows could damage the crest of the next lower design bench. Under those conditions, modifications may be made to the bench design or an adjustment may be made to the standard trim-pattern design as necessary to minimize bench-crest blast damage.

24.5.3 Trim-Blast Design Criteria

The standard design for a single benched 12-m- (40-ft-) high four-row trim pattern is shown in Figure 24.6. All blast holes are drilled vertically. The design features a toe row that is shot as a presplit. These holes are 170 mm (6¾ in.) in diameter and are drilled on 1.5-m (5-ft) centers. Alternatively, toe row holes may be 220 mm (8¾ in.) in diameter and drilled on 2.4-m (8-ft) centers. Toe row holes are drilled 0.7 m (2.2 ft) inside the design toe of the target bench with no subgrade. Holes on the outside (modified production row) are loaded heavier to effectively move rock at the toe of the free face and provide sufficient relief for the rest of the shot. The typical staggered pattern is shot en echelon. Surface delays are used to prevent premature ignition due to hot-hole conditions resulting from oxidizing sulfides. The effective delay between holes in a row is 42 ms and between the rows is 59 ms. A 25-ms delay is placed between every fifth hole along the toe (presplit) row. Blast-hole loading criteria is variable, depending on rock hardness, local geology, and experience from mining the previous lift. Typical trim blast data for ground with intermediate hardness is summarized in Table 24.2.

24.6 DISCUSSION ON SLOPE STABILITY

24.6.1 Highwall Designs

Successful highwall designs at Goldstrike depend largely on the accuracy of the interpreted critical fault structure and clay alteration distribution represented in the geological model. Additionally, for highwalls that have been dewatered in advance of mining, slope performance has been remarkably good compared to experience in similar rocks that had not been completely dewatered.

24.6.2 Slope-Failure Mechanisms

The most common slope-failure mechanism experienced at Goldstrike involved instability along faults that formed complex wedge geometries usually modified by an associated argillized rock mass. Other slope-failure mechanisms experienced include translational movement along bedding planes, complex toppling modes, or combinations of these modes.

24.6.3 Slope-Design Criteria

All highwalls are designed to generally satisfy a nominal factor of safety of 1.2. Detailed slope-stability analyses may identify localized areas where the predicted stability could be marginal. Exceptions to the standard design criteria may be acceptable if the risks of potential slope instability can be clearly identified and

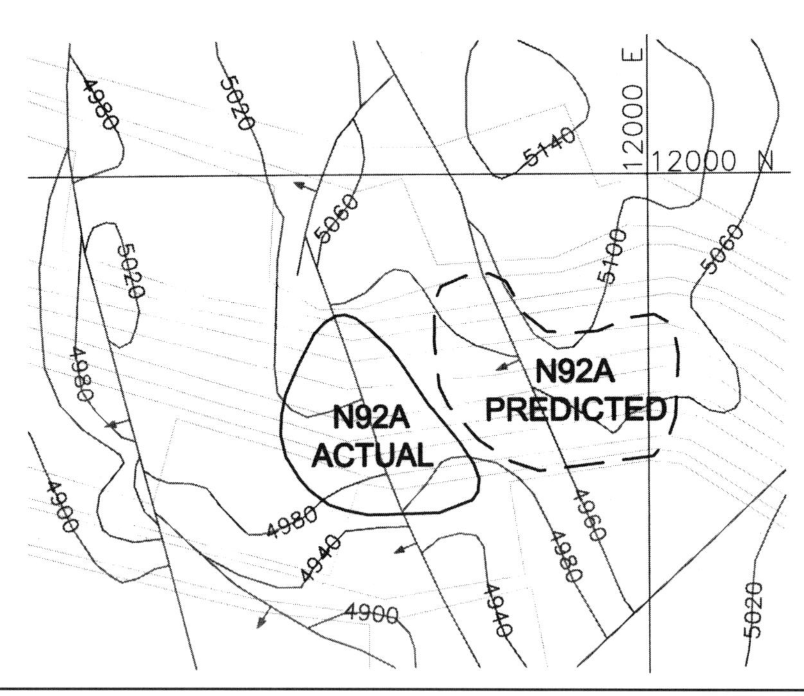

FIGURE 24.7 N-92-A instability on the CBC layer

if contingencies can be incorporated in the long-range mining plan that will satisfactorily mitigate the predicted condition. For example, if an instability was to occur, does a secondary access exist that would allow mining to continue and are the potential effects on ore production satisfactorily addressed? More conservative slope designs may be required where the risks are determined to be unacceptable. Two Goldstrike experiences are presented below.

24.6.4 Case History 1: N-92-A Instability

Structural Geology and Characterization. In advance of north wall mining in mid-1992, structural geology interpretations identified the presence of very weak bedding planes that dipped toward the pit. The bedding structure, called the CBC layer, was composed of graphitic slickensided bedded clays associated with the contact between the Rodeo Creek and Popovich Formations. Core and rotary drilling results were used to interpret the CBC distribution. A top structure contour plan of the CBC was developed to identify the limits of potential impact to the design slope. The contour plan indicated a southward dip of 12° to 15°. Laboratory direct shear testing produced strength parameters, including an internal friction angle of 7° and a possible cohesion range of 0 to 50 KPa (0 to 1,000 psf).

Mining Options and Results. Two options for mining the north wall, determined from limit-equilibrium analyses using the program XSTABL, included mining out the entire CBC or finding a less onerous alternative to reduce waste rock stripping. It was determined that a mining cut that predicted manageable slope failure in waste rock could be designed. The plan featured a ramp design at the nominal elevation where the CBC would be daylighted and would provide at least temporary extended access for slide cleanup. The ramp would also isolate the predicted slide area from the lower slope. This option was implemented, and a section of the highwall between the 5,000 and 5,200 levels failed (called the N-92-A instability), generally as predicted in September 1992. Figure 24.7 shows the mine plan superimposed on the interpreted CBC top structure plan and the predicted location of the potential instability. The FS determined for the design highwall through the critical region was 0.89 to 1.26 for the range in cohesion of the CBC layer derived from the laboratory tests. The actual instability, which proved to be about the same size as that predicted, occurred in an adjacent location to the west; this was attributed to differences between the interpreted and actual geology. The consequences of the instability, however, were as predicted, and further layback development proceeded safely and on design.

24.6.5 Case History 2: Second Southeast Layback

SE-96-A and S-97-B Instabilities. Significant instability occurred in the southeast wall during development of the Second East Layback between March and May 1997. Two coalescing slide masses, called SE-96-A and S-97-B, were back-analyzed to confirm the plan for safe advancement of the layback below the instabilities and develop a remedial design and safe operating plan for the final highwall. The height of the wall in failure was up to 300 m (1,000 ft) and the slide mass was estimated at 19 M tonnes (21 M tons). Slope failure was almost entirely confined to the intrusive rocks. Observation, geological assessments, back analysis results, and slope-monitoring records were analyzed to identify the complex structural control for the southeast wall, illustrated conceptually in Figure 24.8, and determine shear strength parameters for fault surfaces and the rock mass. The original slope designs for this highwall failed to recognize the contribution of structural control, particularly the combined effects of the Laughing- and Wicked-type faults, as shown on Figure 24.3.

Final Wall Designs. A final design for the southeast wall was developed using a combination of slope-stability analysis programs, including XSTABL, CLARA, and UDEC (Piteau 1998). Analysis profiles were developed from sections cut from the geological model. The final highwall design will produce slopes with an overall height of 425 to 500 m (1,400 to 1,600 ft), interramp slopes that vary from 25° to 33°, and overall slope angles of 24° to 28°. The failed wall was originally designed at an average overall 35°.

Highwall Design Criteria. The design FS for the final wall is shown in plan on Figure 24.8. Although the overall highwall satisfies a nominal factor of safety of 1.2, the design includes some local exceptions. Analyses indicated that a small triangular area produced a factor of safety < 1.0, defined by the intersection of two wedge-forming faults (Dormant and Laughing), and instability

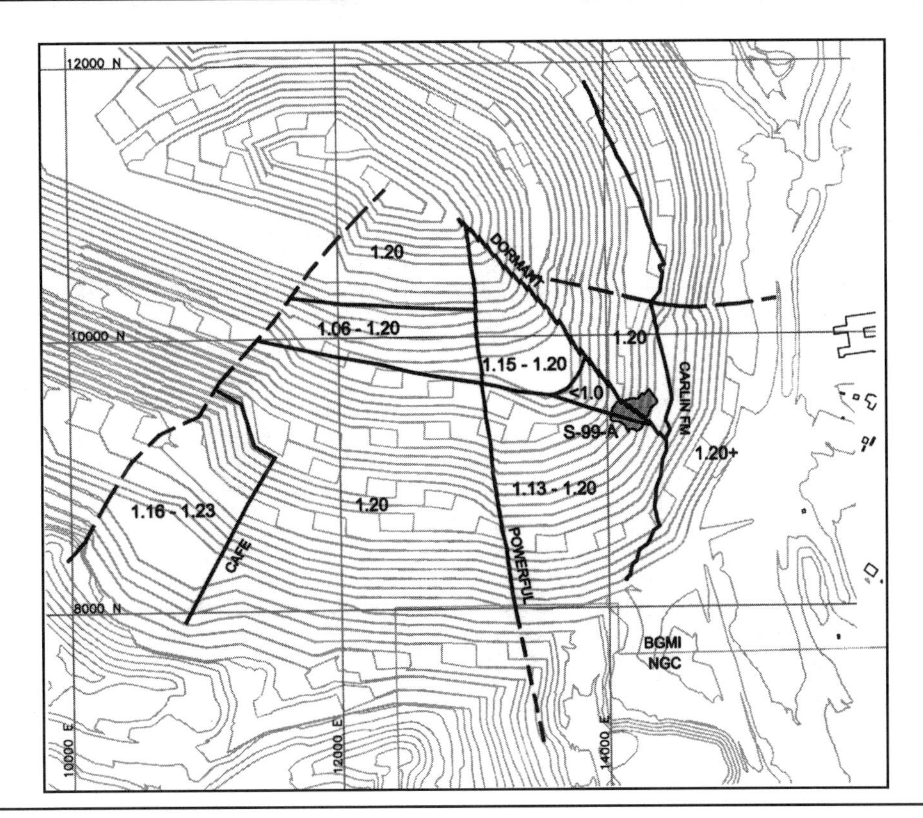

FIGURE 24.8 FOS plan for the southeast wall ultimate

was therefore predicted. Consequences of a comparatively small instability were considered to be insignificant down slope, since the design included a significant stepout that would capture any failed debris. However, instability could jeopardize a proposed ramp originally designed above it. An alternative mitigating design would have added considerably to waste rock stripping. Ultimately, the design ramp width was reduced to permit one-way haulage and its importance was reduced to optional. Additionally, a remnant of the failure mass in a portion of the final design wall was predicted. The design factor of safety varied from 1.06 to 1.20 for this region, which included a 25° to 30° interramp slope. Contingency planning includes dozing and excavating to design in the event that failure was to occur.

Results. Slope instability, called SE-99-A, occurred along the Dormant fault and in highly altered hanging wall rocks in the location predicted by the analysis. The Laughing fault actually projected further down slope than the model predicted. The instability was contained in the area predicted and did not affect the mining plan down slope. Development of the planned half-ramp above the slide was considered to be too much of a safety risk and was abandoned. No significant instability has yet developed in the yield zone lower down the slope. Mining through this region was in progress at the time this paper was written. A view of the Second Southeast Layback, as it appeared in August 1999, is shown on Figure 24.9.

24.7 SAFETY ASPECTS

Achievement of safe and successful operating conditions for mine personnel and equipment adjacent to slope instabilities is attributed to a high level of interaction between operations, maintenance, engineering services, and mine engineering personnel. Operations meetings are held daily and mine planning meetings are held weekly, at which times stability issues are addressed. Operations personnel are to be credited for their continuing demonstrated awareness of slope-stability conditions, including recognizing indications of changing conditions and reporting of

FIGURE 24.9 Second Southeast Layback in August 1999

their findings immediately to mine engineering personnel and site management.

The movement-rate threshold criteria for safe operation adjacent to unstable ground are typically 50 to 75 mm/day, based on local experience. Higher threshold rates were acceptable for mining in most areas above and below the SE-96-A and S-97-B instabilities in 1997 and 1998, which were justified by continuous observation coupled with continuous real-time surface-monitoring records that generally indicated a long period deceleration trend. During the course of mine development, sections of the pit have been closed when movement thresholds were exceeded, as measured by wireline extensometer or by the motorized theodolites.

24.8 CONCLUSIONS

Highwall stability and slope designs at Goldstrike are controlled primarily by the structural geology, distribution of clay-altered rocks, and groundwater. Experience has demonstrated that a thorough knowledge of these factors is necessary in order to design stable and economic slopes. Utilization of a blast trim pattern, as developed through local experience, yields an effective balance between minimizing the effects of blasting near the highwalls and achieving of acceptable mine productivity and associated costs. Detailed geological mapping and model development, implementation of an extensive surface and groundwater management system, and a sophisticated ground-monitoring system are necessary to produce optimum highwall designs, operate safely and efficiently, and are well justified at Goldstrike.

24.9 ACKNOWLEDGMENTS

The author thanks the management of Barrick Goldstrike Mines Inc. for their support and permission to prepare and publish this chapter. The support and assistance of mine engineering, mine operations, water management, and environmental personnel are appreciated.

24.10 REFERENCES

Bassier, T. January 2000. Engineering services–Water management. Personal communications. Barrick Goldstrike Mines Inc.

Hoek, E. and E.T. Brown. 1988. The Hoek–Brown failure criterion–A 1988 Update. *Rock Engineering for Underground Excavations. Proceedings 15th Canadian Rock Mechanics Symposium.* Toronto, pp. 31–38.

Penick, M. January 2000. Mine engineering–Geology. Personal communications. Barrick Goldstrike Mines, Inc.

Piteau Associates Engineering Ltd. March 1988. Slope stability assessments for the second *Southeast Layback of the Betze-Post pit.* Internal Engineering Report.

CHAPTER 25

Pit Slope Design at Orapa Mine

J. Jakubec,* P.J. Terbrugge,† A.R. Guest,‡ and F. Ramsden**

25.1 INTRODUCTION

The Orapa diamond mine is in central Botswana, approximately 220 km west of Francistown (Figure 25.1). The climate is semi-arid and temperatures range from above 45°C in summer to below 0° at night in winter. The average annual rainfall is less than 450 mm. Orapa is situated on the southern edge of the Makgadikgadi basin. The morphology of the surrounding area is flat, apart from a few low hills and ridges. The semi-arid climate supports low savannah-type vegetation.

The A/K1 kimberlite pipe, with a surface area of 118 ha, is the largest of a group of more than 50 known kimberlites in the area. Included among these are Orapa A/K2 and Letlhakane D/K1 and D/K2 pipes, which, like A/K1, are mined for diamonds.

Orapa A/K1 kimberlite was discovered in 1967, and from 1971 to date more than 121 million carats have been extracted. The bottom of the open pit is currently 130 m below the surface, at 830 m above mean sea level (a.m.s.l.). To date, all mining faces of cut 1 in Orapa Mine are confined within the kimberlite pipe boundary. It is currently estimated that the Orapa kimberlite pipe will be exploited by the open-cast mining method to the year 2030 at an increased rate of production resulting from the commissioning of the number 2 plant. The ultimate pit depth will reach more than 500 m below surface.

It is obvious that for such a large pit the slope angle design will have enormous economical consequences. It is therefore necessary for every aspect of the open-pit design to be examined to minimize the stripping ratio and meet safety requirements. This chapter discusses the approach to slope design for the Orapa pit, with emphasis on the gathering of geotechnical data, input parameters, numerical analysis, and slope geometry.

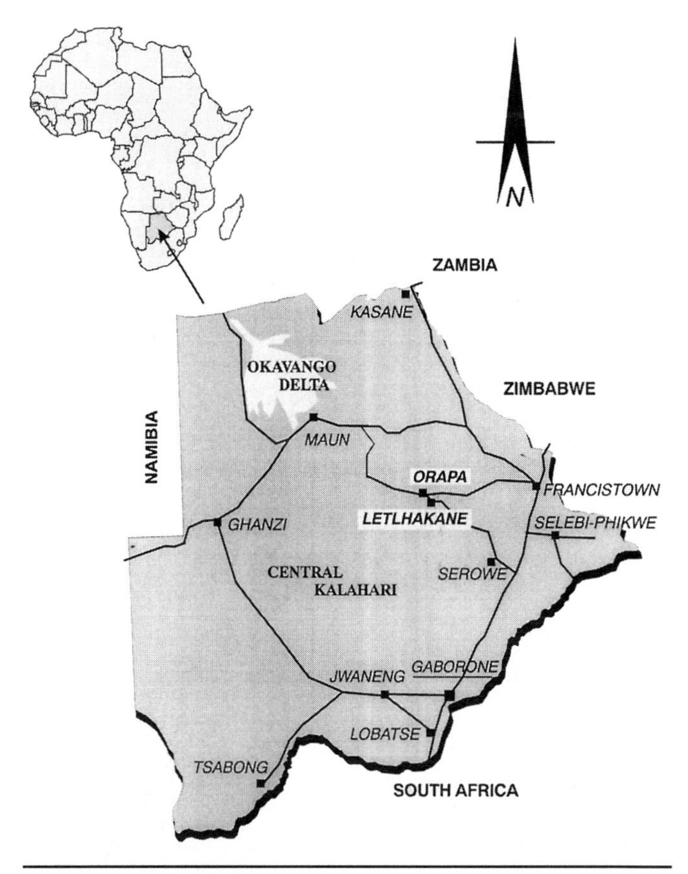

FIGURE 25.1 Geographical location of the Orapa Mine

25.2 GEOLOGY

25.2.1 Orapa A/K1 Geology

Exploration has indicated that the Orapa A/K1 kimberlite pipe comprises two diatremes (northern and southern lobes) that, above 700 m a.m.s.l., form a single body (Figure 25.2). The surface intersection of the pipe has an ovoid shape 1,500 m long and 1,000 m wide. The results of in-pit mapping and borehole logging indicate that the northern lobe is older than the southern lobe, which was emplaced at a later stage. The slopes of the wall rock in the northern lobe are considerably steeper (approximately 75°) than those of the southern lobe (approximately 50°).

The latest dating (U-Pb method on zircons) indicates that the age of emplacement of the kimberlite pipe is approximately 93 million years (late Cretaceous).

All the kimberlite exposed by the mining operation at Orapa is classified as crater-facies kimberlite within which epiclastic and pyroclastic varieties have been distinguished. Epiclastic kimberlite types include talus deposits, debris flow breccias, basal and wall rock breccias, boulder beds, shales, grits, and numerous transitional units. The individual rock types are variably weathered with a large range of strength properties. At a depth of approximately 250 to 300 m below the surface, the diatreme-facies kimberlite has been intersected in both the northern and southern lobes. While the transition between pyroclastic crater-facies kimberlite and massive tuffisitic kimberlite breccia (TKB) below is gradual in the northern lobe, the contact is sharp between the two units in the case of the southern lobe. The lithological units within Orapa kimberlite pipe were recently described (Field et al. 1995) and their spacial extent is illustrated in Figures 25.3 and 25.4.

* Steffen, Robertson and Kirsten, Vancouver, British Columbia, Canada.
† Steffen, Robertson and Kirsten, South Africa.
‡ Debeers Corporate Headquarters, South Africa.
** Debswana Diamond Company (Pty.) Ltd., Orapa and Letlhakane Mines, Botswana.

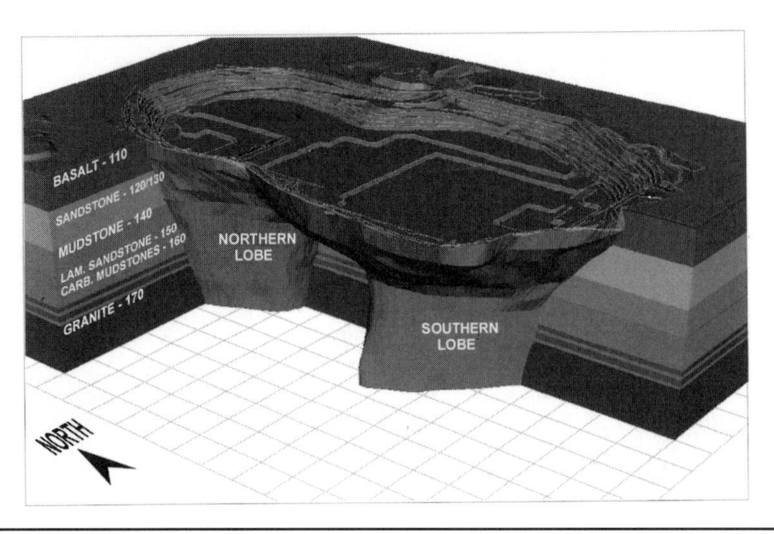

FIGURE 25.2 Isometric view of the A/K1 kimberlite pipe

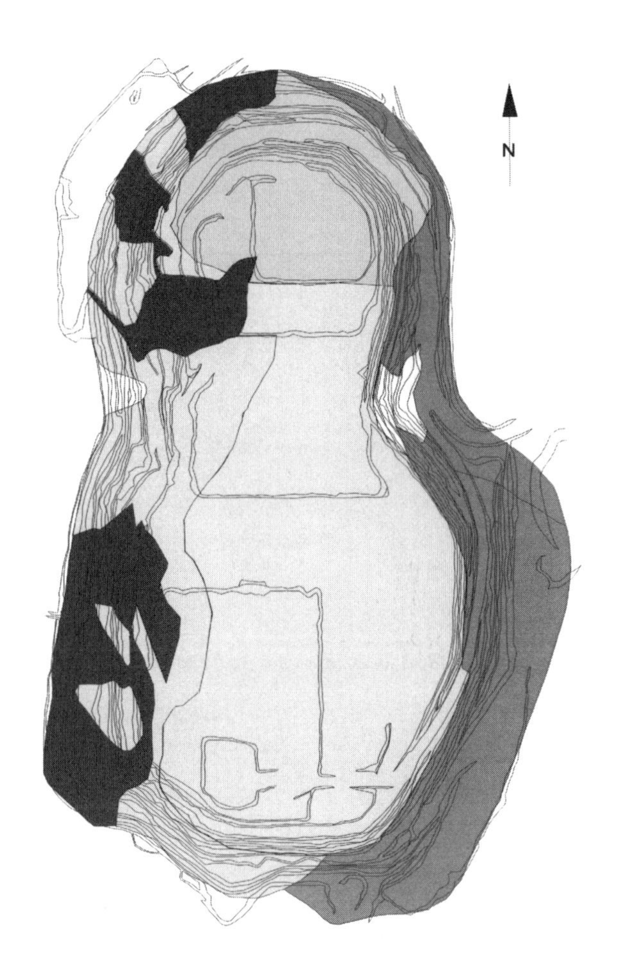

FIGURE 25.3 Exposed kimberlite rock types in the Orapa Mine

25.2.2 Country Rock Geology

The kimberlite orebody has intruded subhorizontally bedded volcanic and sedimentary rocks of the Karoo Supergroup, which lie unconformably on a basement of Archaean granite. The top of the sequence consists of 70 to 110 m of Stormberg-aged basalt lavas, which unconformably overlie the Ntane sandstone. The basalt sequence consists of a series of basalt flows of various thicknesses. An amygdaloidal zone and sometimes a weathered horizon can distinguish the top of each flow. The contact zone with underlying sandstones is usually weathered and pervaded with calcite-filled joints.

The Ntane formation consists of fine- to medium-grained sandstones with variable thickness of 15 m to more than 60 m. The variability of the thickness can be explained by the presence of sand dunes in the paleorelief.

The underlying Mosolotsane unit consists of 70 to 100 m of fine- to medium-grained sandstones intercalated with red mudstones, the proportion of mudstones increases toward the base of the unit.

The Mosolotsane formation lies unconformably on the Tlhabala formation and is comprised of grey, massive mudstones of 85 to 95 m thickness, including grey-green mudstones with occasional siltstone and fine sandstone horizons. Red mudstones of variable thickness occur at the top of the Tlhabala formation.

The underlying rocks of the Tlapana formation consist of a sequence of fine- to medium-grained sandstones and siltstones, with coarse arkoses usually formed at the top of carbonaceous shale-coal horizons. The unit is approximately 150 m thick and includes three distinct carbonaceous mudstone-coal horizons.

Although the drill holes did not intersect any major structures in the country rocks, there are indications of northwest-southeast running subvertical faults in areas surrounding the A/K1 kimberlite pipe (Gibson 1987; Shaw 1991; Bush, Hoffmann, and Van Rensburg 1995).

In the design of the pit slope, the geological profile in the country rock was divided into seven main units, as shown in Figure 25.2.

25.3 HYDROGEOLOGY

Various drilling programs defined a limited secondary aquifer within the basalt group, with the phreatic surface at approximately 50 m below ground level.

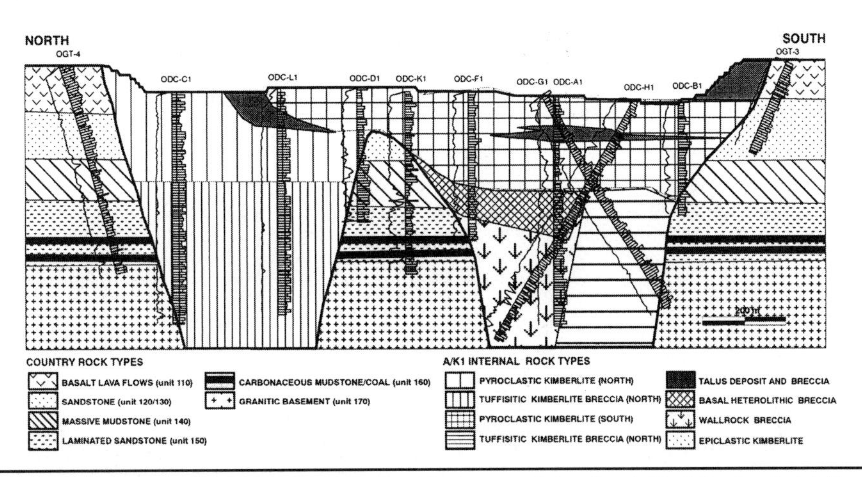

FIGURE 25.4 North-south geological section through A/K1 kimberlite. The bar graph (on the right of the borehole) illustrates the mining rockmass rating, and the line graph (on the left of the borehole) illustrates the rock-quality designation.

The Ntane sandstone forms the most important aquifer of the upper section of Karoo sediments (Bush, Hofmann, and Van Rensburg 1995). The low-yield primary and high-yield secondary aquifers have the most significant zone occurring at the contact with the basalt flows.

The Mosolotsane formation, especially its coarse fluviatile arkoses, is also believed to form important aquifer horizons.

The massive mudstones of the Tlhabala unit form a highly impermeable zone and therefore slope-stability considerations are insignificantly influenced by water pressure.

Very little is known about groundwater conditions below the Thlabala unit. From core logging, two possible aquifers were defined in coarse arkose units, one at the top of the unit with a 10-m head and a second with a 30-m head in the middle of the unit.

25.4 FIELD WORK

Field work consisted of gathering data from the six inclined geotechnical diamond drill holes located on the circumference of the A/K1 orebody and from eight geological diamond drill holes collared within the open pit but penetrating the country rock formations as well as kimberlitic rocks.

The following activities were carried out during geotechnical and geological drilling:

- Geotechnical logging of the drill core
- Downhole core orientation and structural measurements
- Point load testing on core samples
- Density measurements on core samples
- Analysis of downhole camera surveys
- Core sample collection for laboratory testing
- Weathering observations
- Photographic documentation

25.4.1 Geotechnical Logging

Detailed geotechnical logging was carried out on drill core utilizing Laubscher's rock mass classification system (Brown 1981). The following geotechnical parameters were collected:

- Total core recovery (TCR)
- Solid core recovery (SCR)
- Rock-quality designation (RQD)
- Intact rock strength (IRS)
- Fracture frequency per meter (FF/m)
- Joint frequency per meter (JF/m)
- Joint condition (JC) with respect to waviness, roughness, alteration, and filling

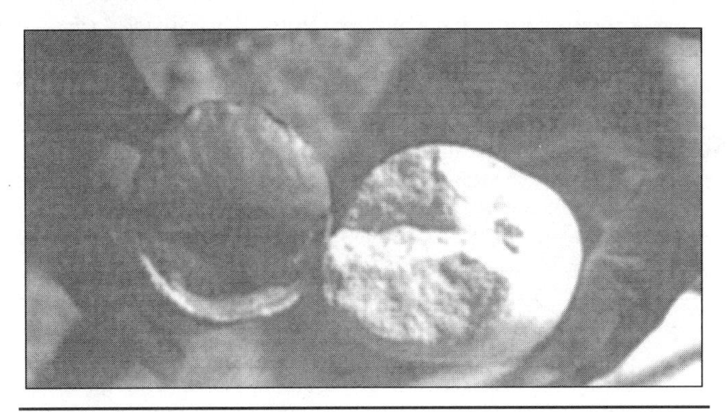

FIGURE 25.5 Plasticine imprint on the core (bottom) from the use of the clay imprint orientator, and the spear mark on the core (top) from the use of the point orientator

25.4.2 Core Orientation

A clay imprint orientator was used to determine core orientation. The disadvantage of this method is that, when joints or fractures are more or less perpendicular to the core axis or when the surface of the joint is covered with mud or alteration products, accurate orientations cannot be measured. For this reason, the core orientator was modified with a sharp steel point to mark the core (Figure 25.5). This modification also enabled the drilling crew to mark the core in the absence of geotechnical personnel, and the orientor could therefore be lowered after every run.

The orientated cores were measured using a core goniometer. The dip angle (alpha) was measured in all the joints and the angle dip direction (beta) was measured for those joints that had been oriented. In total, 1,447 joints were measured, of which 398 were fully orientated.

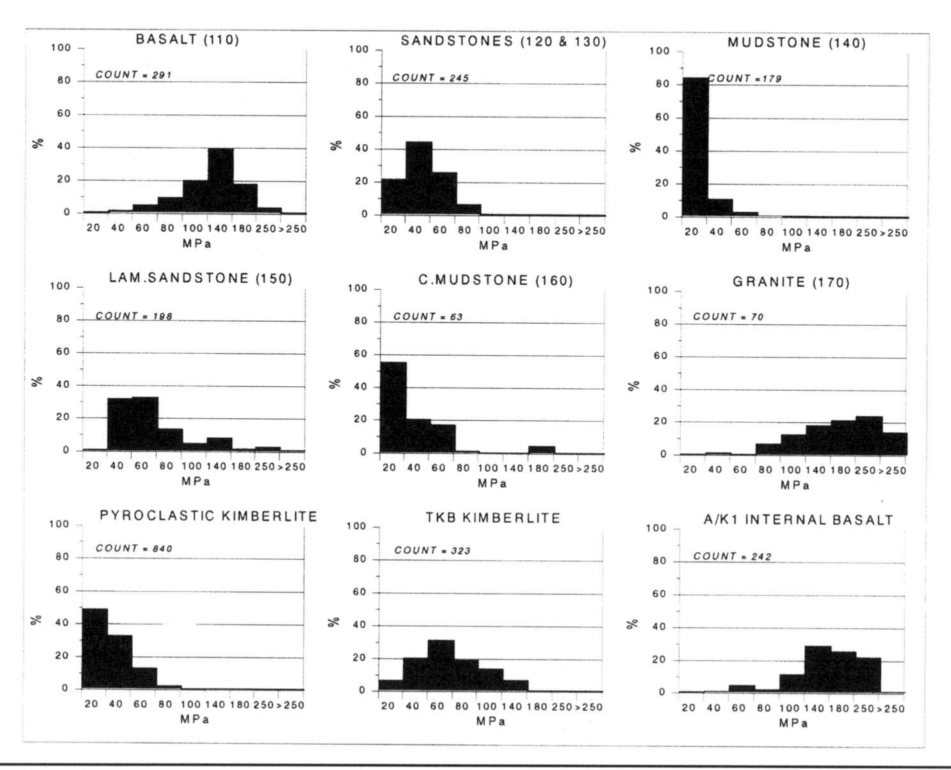

FIGURE 25.6 Frequency distribution in the uniaxial compressive strengths of different types of rock

A spreadsheet was developed for the calculation of the true dip and dip direction of the discontinuities, based on information from downhole survey. The orientations of the discontinuities were analyzed from stereographic projections (by use of Dips software), and statistical methods were used to assess joint spacing and roughness. These data were used for kinematic analyses of the potential failures.

25.4.3 Point Load Testing

The standard procedure for point load testing of drill core (Brown 1981) was followed so that the strengths of the different rock types could be described. The results of 5 to 10 tests per sample were corrected for size and core diameter, and the corrected figures were then equated to the uniaxial compressive strength (UCS), according to the formula

$$UCS = Is(50)*24$$

Frequency distributions of the results of the point load tests are shown on Figure 25.6. The average values from the laboratory and in situ point load tests were compared, and good correlation was found, except for the mudstone (Figures 25.6, 25.7, 25.8). This is due to the sensitivity of mudstones to weathering and rapid disintegration of the rock. A bias in the selection of the rock samples for the laboratory tests has also influenced the final results.

25.4.4 Density Measurements

Density measurements were carried out on the different types of rock encountered in the drill core. The dimensions and weights of the core samples were taken every 10 m (or from representative samples), and the approximate relative densities were then calculated.

25.4.5 Down-Hole Camera Survey

To obtain information about the deflection of drill holes, the Sperry Sun downhole survey system was utilized to show the true orientation of the discontinuities.

25.4.6 Weathering Observations

The reaction of different rock types to wetting and drying was observed, and the results were used in the weathering adjustment of the rock-mass rating. Selected samples were subjected to long-term weathering observations. The following scale was set up to describe the rate of weathering.

1. No weathering
2. Slightly weathered
3. Weathered
4. Highly weathered
5. Disintegrated

25.4.7 Photo Documentation

Drill cores, as well as laboratory test samples, were photographed. Representative samples selected for ongoing weathering were also photographed on a regular basis.

25.5 LABORATORY TESTS

Representative samples of the various rock types encountered during the drilling were sent to the CSIR (Drescher and Vogler 1994a, 1994b) geotechnical laboratories for testing. The following tests were carried out for each lithological unit:

- Uniaxial compressive strength test with deformation measurements
- Uniaxial compressive strength test without deformation measurements
- Triaxial compression test
- Uniaxial tensile strength test by the indirect Brazilian method
- Sonic wave velocity
- Shear test on natural joint
- Shear test on artificial joint (saw cut)

A summary of the results is shown in Table 25.1.

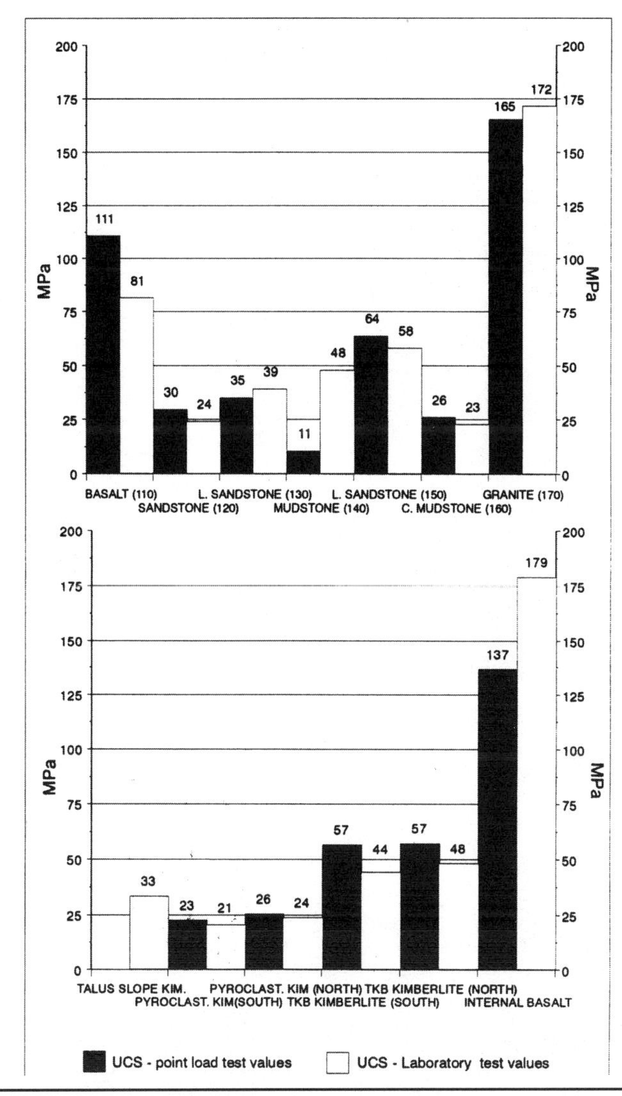

FIGURE 25.7 Correlation of in situ point-load results with those obtained in the laboratory for different types of rock

25.6 GEOTECHNICAL ASSESSMENT

The geotechnical assessment of the different rock types was based on rock-mass classification, rock strengths obtained from the laboratory tests, and weathering observations made on the core and in the A/K1 pit. The input parameters for numerical models were obtained from geotechnical assessment of the rock properties.

25.6.1 Rock-Mass Classification

From the parameters collected during the core logging and from those obtained in the laboratory and point load tests, rock-mass ratings (RMR) and rock-mass strengths (RMS) were calculated. Adjustments in terms of the influence of weathering, joint orientations, blasting, and stress environment were applied to give the mining rock-mass rating (MRMR) and the design rock-mass strength (DRMS). Adjustments to the RMR were based on drill core properties as well as on mining experience from both Orapa and Letlhakane Mines.

The average values of the geotechnical parameters are summarized in Table 25.2.

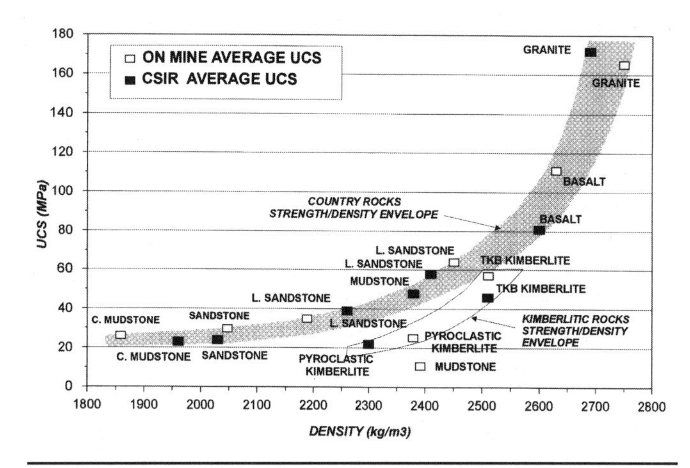

FIGURE 25.8 Relationship between the average uniaxial compressive strength and density

25.6.2 Strength Assessment

The material properties used in the analyses were derived from laboratory tests, field tests, and the rock-mass characterization studies carried out on the borehole cores.

- Intact rock strengths were derived from the laboratory tests and from the mine point load tests. Values for the cohesion and friction angles were determined by triaxial testing in the laboratories.
- Joint strengths were determined from laboratory shear tests. Shear tests on natural joint surfaces, as well as on artificial saw cut surfaces, were evaluated.
- The RMR was used with the Hoek and Brown failure criteria to determine the instantaneous friction angle and cohesive strength for given normal stress values.

Table 25.3 shows the values of intact rock and natural joint parameters used in the stability analyses, and Figure 25.8 shows the relationship between the uniaxial compressive strength and the density.

25.6.3 Characterization of the Rock Mass

Based on the rock-mass classification, strength assessment, and weathering assessment, the following rock-mass characterization was established:

1. **Basalt Lavas (Unit 110).** Most of the basalt encountered is strong to very strong and has good rock-mass quality, similar to the basalt at Letlhakane Mine. Except for some small sections of weathered basalt close to the sandstone contact and on the top of individual lava flows, the basalt is not susceptible to weathering. The joints are mostly closed and filled with calcite. There are occasional sandstone lenses on top of the lava flows.
2. **Massive Sandstone (Unit 120).** The massive sandstone unit underlying the basalt is medium strong and has a fair-quality rock mass. The sandstone has variable thickness and is not susceptible to weathering.
3. **Laminated Sandstones (Unit 130).** Layers of massive sandstone, which is a medium strong rock, are intercalated with weak to very weak mudstones. The amount of mudstones increases toward the base of the unit. In general, the unit has a fair to good rock-mass quality. While the sandstones are not susceptible to weathering, the mudstone units will disintegrate very quickly. As a result of subhorizontal bedding, the strength of the rock in a vertical direction is greater than in a horizontal direction.

TABLE 25.1 Average values of laboratory test results for different rock types

Rock Type/Test	Density (kg/t)	UCS (MPa)	TS (MPa)	PWV (m/s)	SWV (M/s)	Young's Modulus (Gpa)	Poisson's Ratio
Basalt	2,600	81	9	4,660	2,630	42	0.20
Sandstone	2,030	24	2	3,330	1,910	17	0.32
Mudstone	2,380	48	4	2,820	1,540	6	0.14
L. sandstone	2,410	58	—	3,330	2,250	16	0.24
C. mudstone	1,960	23	7	2,600	1,580	7	0.22
Granite	2,690	172	11	5,480	3,260	66	0.23
P. kimberlite	2,300	22	3	2,470	1,320	66	0.23
TKB	2,510	46	5	3,410	1,860	12	0.23

NOTES:
UCS uniaxial compressive strength
TS tensile strength
PWV P wave velocity
SWV S wave velocity
L laminated
C carbonaceous
P pyroclastic
TKB massive tuffisitic kimberlite breccia

TABLE 25.2 Average values of geotechnical parameters for different rock types

Rock Type/Parameter	RQD	JC	IRS	FF/m	RMR	MRMR	DRMS MPa
Basalt	14	25	12	25	62	57	58
Sandstone	14	20	4	26	50	46	16
Mudstone	14	14	3	22	39	28	4
L. sandstone	14	16	4	26	46	37	14
C. mudstone	10	17	2	10	41	33	4
Granite	15	26	8	26	60	55	54
P. kimberlite	14	22	4	15	56	45	10
TKB	14	26	5	17	66	56	18

NOTES:
RQD rock-quality designation
JC joint condition
IRS intact rock strength
FF/m fracture frequency per meter
RMR rock-mass rating
MRMR mining rock-mass rating
DRMS design rock-mass strength

TABLE 25.3 Intact rock and natural joint parameters used in stability analyses

				Intact Rock		Joints	
Rock Type	Density (kg/t)	Young's Modulus (Gpa)	Poisson's Ratio	Friction (degree)	Cohesion (MPa)	Friction Rating	Cohesion (MPa)
Basalt	2,600	42	0.2	33.1	23.9	34.6	0
Sandstone	2,100	17	0.3	43.2	10.9	35.7	0
Mudstone	2,350	6	0.2	25.7	15.2	30.5	0
L. sandstone	2,400	16	0.3	43.2	51.5	33.6	0
Granite	2,650	66	0.2	37.5	44.4	31.0	0

4. **Massive Mudstones (Unit 140).** The mudstones are competent medium to strong rocks that have a fair rock-mass quality if not exposed to weathering. The joints are mostly closed and filled with calcite. Except for occasional siltstone-sandstone horizons, which are also generally stronger, the whole unit is extremely susceptible to weathering and will disintegrate rapidly if exposed (Figure 25.9). The smectites and illites are the main phyllosilicates in the mudstone horizon and they pose a risk for slope stability (Buhmann and Atanasova 1997).
5. **Laminated Sandstones (Unit 150).** Medium strong to strong sandstones and siltstones are intercalated with three weak carbonaceous mudstone horizons of unit 160. Sandstones-siltstones can be described as fair to good rock-mass quality and are not susceptible to weathering. Strength in the vertical direction is again greater than in the horizontal direction because of subhorizontal bedding and lamination.
6. **Carbonaceous Mudstones-Coal (Unit 160).** There are three major carbonaceous horizons. Lower, weak horizons are richer in coal, while the upper medium-strong horizon is siltier, with variable amounts of sulfates and sulfides, which make the unit more susceptible to weathering (Buhmann and Atanasova 1997). In general, units are highly jointed and are described as poor rock-mass quality.
7. **Granitic Basement (Unit 170).** Tonalite granite is strong to very strong and has good rock-mass quality, with some weaker amphibolitic zones. Very little is known about joint directions, and the rocks of this unit are not susceptible to weathering.
8. **Pyroclastic Kimberlite.** In general, the unit consists of weak to very weak rocks of the crater facies. These kimberlitic rocks are higher than 600 m a.m.s.l. and are described as poor to fair rock-mass quality. These rocks

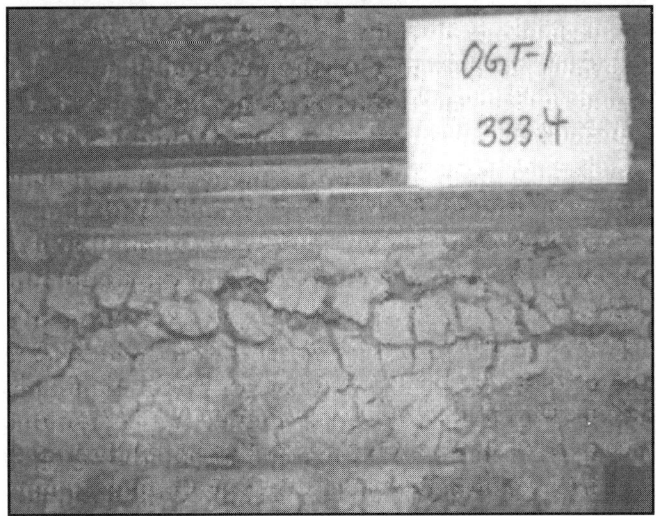

FIGURE 25.9 The laminated red mudstone and massive grey mudstone of unit 140 are highly susceptible to weathering. The photographs illustrate drill core before (upper) and after (lower) a short exposure to water.

are highly susceptible to weathering. From in-pit measurements, the weathering rate of intact kimberlite is 50 to 200 mm/y.

9. **TKB Kimberlite (Unit 220).** The kimberlitic diatreme facies below 600 m a.m.s.l. are medium strong to strong rocks with fair to good rock-mass quality. Like the crater-facies kimberlite, the diatreme-facies types of kimberlite rock are very susceptible to weathering.

25.7 SLOPE ANALYSIS

In the assessment, the MRMR criteria and Haines–Terbrugge slope design chart were applied, along with comparisons of similar rock formations at the Letlhakane Mine (Haines and Terbrugge 1991).

A series of relationships between composite slope angles versus slope height were derived for each of the critical lithological units, with the empirical curves forming the base case. Structural data from the pit and the drilling program were processed, and kinematic analyses of the slopes were made so that the potential for structurally controlled instability could be evaluated. Finally, a rigorous analysis was undertaken utilizing the Fast Lagrangian Analysis of Continua (FLAC) computer software, for which a further series of slope heights versus slope angle were defined for the various lithological units. Composite slope angles were derived from an empirical analysis.

TABLE 25.4 Overall pit slope angles based on RMR system

Rock Types	Slope Height (m)	Slope Angle (degrees)
Basalt	120	63
Sandstone	120	56
Mudstone	100	33
L. sandstone	150	39
Granite	60	63
Pyroclastic kimberlite	60	50
TKB	60	55

25.7.1 Empirical Analysis

The results from the rock-mass classification, together with slope heights, were used for empirical slope analysis. The Terbrugge–Haines chart was used to define slope angles for a factor of safety of 1.2 for particular lithological units. The overburden stress for each unit was taken into account. Overall slope angles for the different rock units incorporating the ramp system are shown in Table 25.4.

25.7.2 Kinematic Analysis

The results from the measurements of core orientation were analyzed using stereographic projections (Figure 25.10) so that the critical joint sets within each of the lithological units could be determined. Although the quantity of the structural data obtained from the core is limited, there is a suggestion that the main joint set (J1) dips away from the pipe in all directions. This phenomenon can be explained as a result of pipe emplacement in the host rocks. The data were used for kinematic analyses.

Although all the joint combinations were considered for kinematic analyses, together with the relevant strength parameters, the set of subvertical joints expected in the sediments is not particularly evident from the data. This is due to the steep angle of the drill holes that resulted in sampling bias on the subvertical features.

The results from discontinuity analyses indicate that kinematic failures are possible but that, in all cases, the inherent strength of the discontinuities generally precludes such failure. Furthermore, experience of sedimentary sequences suggests that the continuity and spacing of joint planes preclude the possibility of large-scale wedge failures, although localized single or double bench failure, 15 m and 30 m high, respectively, can occur.

25.7.3 Rigorous Analyses

FLAC, a two-dimensional explicit finite difference computer software package, was used for rigorous slope analysis. The software is capable of simulating the nonelastic behavior of a material that may undergo plastic flow when its yield limit is exceeded.

The following constitutive material models were used in the slope-stability analyses:

- **Elastic, isotropic model.** Hook's law in plane strain is used to express the relationship between stress and strain. This model was used for equilibration of initial structures under gravity (profile of the west wall was chosen as the representative profile).
- **Mohr-Coulomb plasticity model.** The Mohr–Coulomb yield condition and nonassociated flow rule was used to represent the elastic properties in plane strain. This model was used for failure simulations using constant rock-mass parameters (cohesion, friction, and dilation) based on the RMR system.

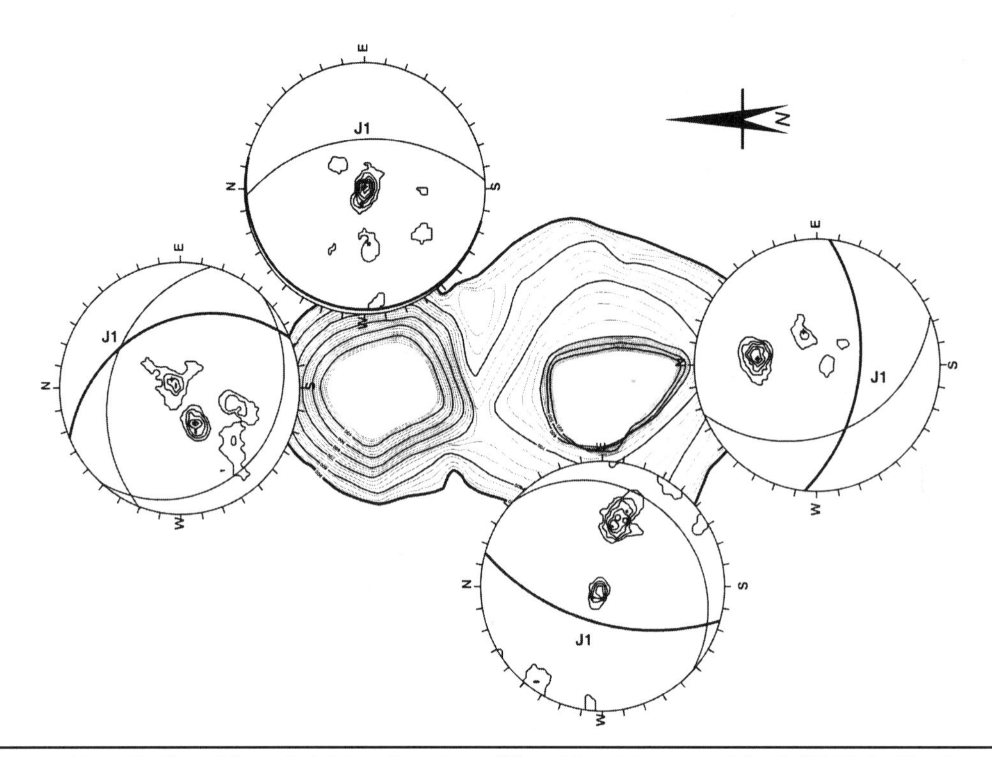

FIGURE 25.10 Stereographic projection of the main joint systems from different boreholes around the A/K1 kimberlite pipe

- **Ubiquitous joint model.** This is an anisotropic plasticity model that exhibits strength anisotropy on a series of assumed weakness planes. The model was used for failure simulations using intact and natural joint parameters, as illustrated in Table 25.3.
- **Strain-softening model.** This model is based on Mohr–Coulomb elasto-plasticity, with the nonassociated flow rule. Cohesion, friction, and dilation need not be constant but can evolve with increasing plastic strain. The model was used for failure simulations using rock-mass parameters based on the RMR system with "softening" of friction and cohesion due to blasting and weathering (Figures 25.11 to 25.14).

Three general-behavior models based on horizontal slope displacement versus computer-simulated time steps were also analyzed:

- **Progressive behavior model.** The slope displacement continues to accelerate to the final collapse.
- **Creeping behavior model.** The rate of slope displacement remains constant with time.
- **Regressive behavior model.** The slope displacement decelerates and finally stabilizes.

A maximum horizontal slope displacement of 100 mm after 4,000 time steps was accepted as an additional criterion of regressive behavior. The validity of this criterion for individual rock units will be investigated further.

25.7.4 Graphs of Slope Height Versus Slope Angle

From empirical and rigorous analyses, empirical and theoretical curves reflecting the relationship between slope height and slope angle were drawn. By applying field observations and practical experience, the practical curve for each lithological unit was developed (Figures 25.11 to 25.14).

The practical curve is believed to represent the steepest slope angles currently achievable. However, through improved blasting techniques and the utilization of different equipment, etc., the slope angles can be steepened, but the theoretical curve should not be exceeded.

The curves are based on current knowledge and will be periodically reviewed as mining progresses and new information becomes available.

Empirical Slope Angles. The empirical curve is based on RMR data and was developed using the Terbrugge–Haines slope-design chart. The overburden stress for each unit was taken into account.

Theoretical Slope Angles. So that the range of slope angles for different lithological units could be determined, a large number of stability simulations were conducted with variable slope heights and angles. Based on the criterion of regressive behavior, graphs of theoretical slope height versus slope angle were developed for each individual lithological unit.

The theoretical curves are based on the following conditions:

- The overall geometry is based on south-west face lithology.
- No aquifers were included; slope dewatering will be required. The rock-mass parameters were based on the RMR system.
- A theoretical reduction of strength due to blasting was applied in all lithological units, and a further theoretical reduction of shear-strength parameters due to weathering was applied to the mudstone unit (Figures 25.11 to 25.14).
- For regressive behavior, the criterion of 100 mm at 4,000 time steps was applied.

Practical Slope Angles. Based on the development of the empirical and theoretical slope angle versus slope height relationship for the lithological units, a practical relationship was derived taking the following into account:

- Mining constraints on the design profile, such as access, equipment, etc.
- Configuration of the pit slope

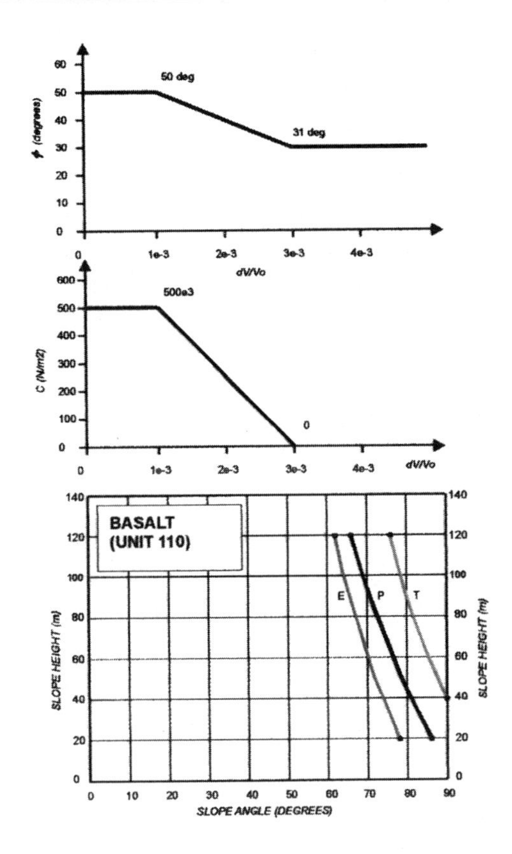

FIGURE 25.11 Slope angles, friction angle, and cohesion for basalt lavas

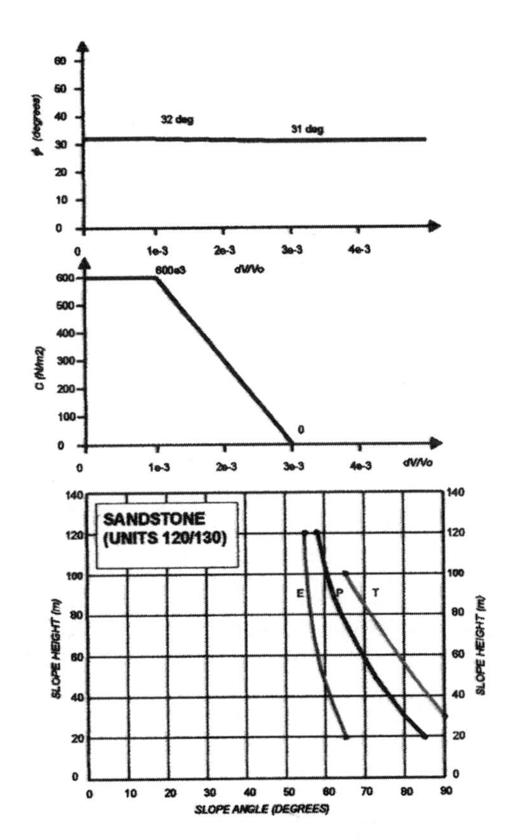

FIGURE 25.12 Slope angles, friction angle, and cohesion for sandstones

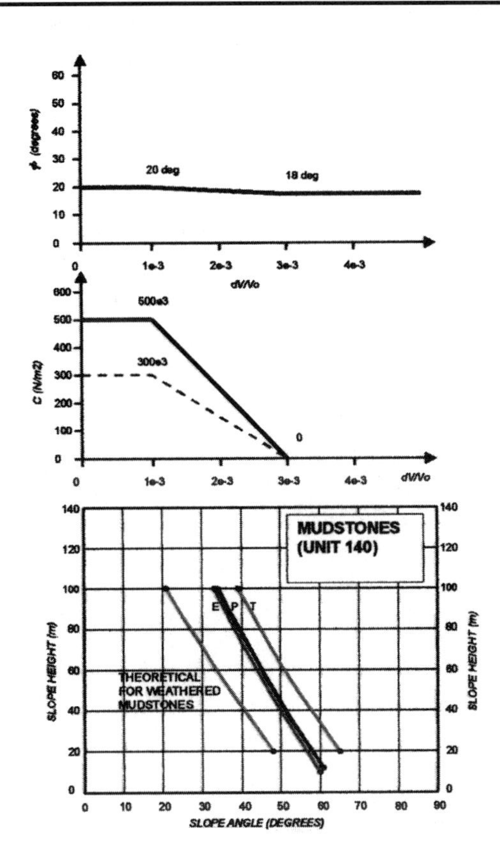

FIGURE 25.13 Slope angles, friction angle, and cohesion for massive mudstone

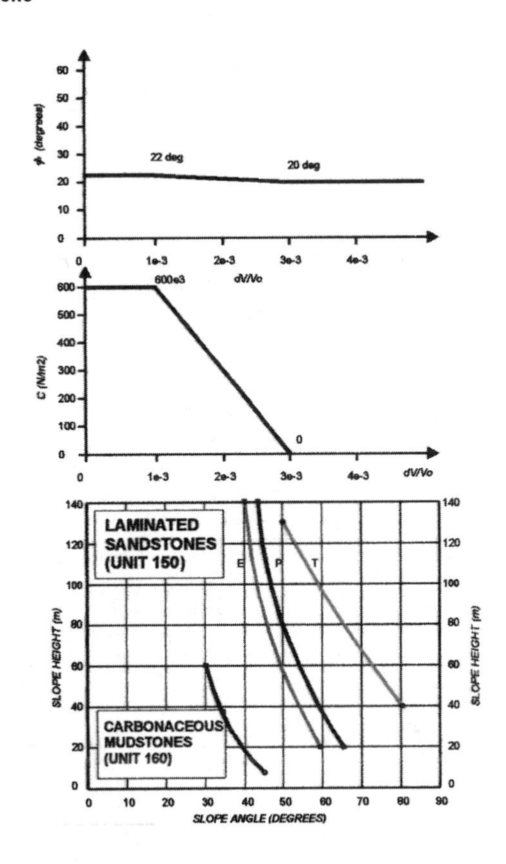

FIGURE 25.14 Slope angles, friction angle, and cohesion for laminated sandstones and carb. mudstones

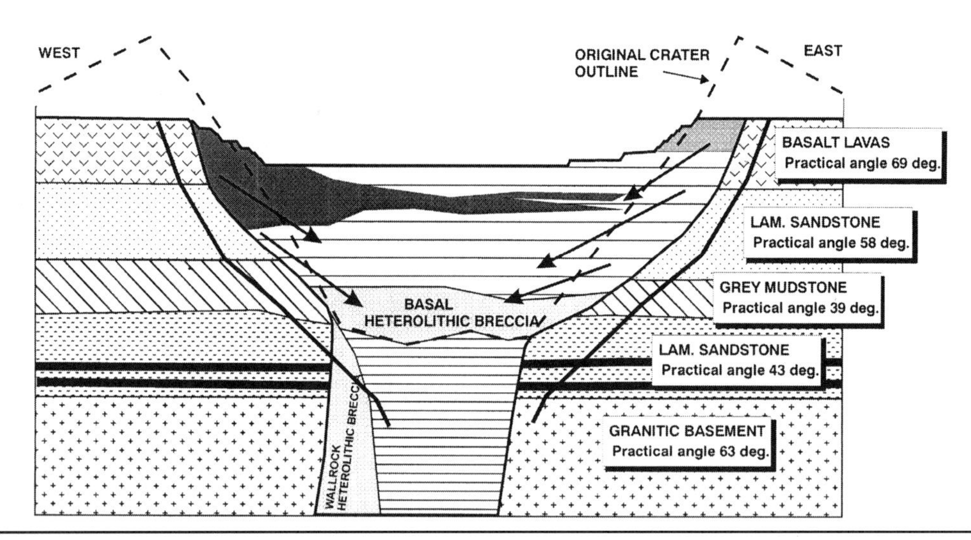

FIGURE 25.15 Correlation of the slope angles in contact between the A/K1 kimberlite (southern lobe) and the country rock, and the slope angles derived from geotechnical studies (indicated by thick lines)

- Quality of perimeter blasting
- Experience from the Letlhakane Mine
- Experience with sedimentary units in strip mines
- Behavior of carbonaceous shale at the Koffiefontein Mine
- General behavior of shale and mudstone slopes in southern Africa

Since only limited exposure of granitic and kimberlitic rocks may be anticipated in the perimeter walls (depending on the cut design geometry), the graphs for these rock types were not constructed. Recommended toe-to-crest slope angles based on the empirical Terbrugge–Haines slope design chart were as follow:

- Granite (approximately below 500 m a.m.s.l.) -63°
- Crater-facies kimberlite (above approximately 600 m a.m.s.l.) -50°
- Diatreme-facies kimberlite (below approximately 600 m a.m.s.l.) -55°

It was noticed that when the slope angles for different lithological units derived from numerical modeling are superimposed on the geological section there is a similarity with the slope angles of the pipe sidewalls (Figure 25.15).

Unique heterolithic breccias, which mark the base of the former crater in the southern lobe, have been intersected approximately 250 m below the surface by several boreholes. The thicknesses of breccia encountered vary between 3 and 85 m. The breccia consists of Karoo rock types concentrated into unique layers. It is believed that the breccia was formed as a result of post-eruption rockfalls of the sidewalls of the crater.

Since the side of the former crater might have been exposed for a long time, spalling of the sidewalls progressed until stability was reached. During magmatic events in later stages, the crater was eventually filled with volcaniclastic kimberlite and crater lake sediments. The sidewalls of the southern lobe, as we can interpret them from the drill holes, above elevation 650 m a.m.s.l., may be the inner slopes of the "maar" type of crater that reached stability (Lorenz 1973).

25.8 SLOPE DESIGN

The practical curves illustrated in Figures 25.11 through 25.14 were used for detailed slope design in each lithological unit. From the graphs, the slope for variable heights in individual lithological units can be maximized. Using these, the mine designer has greater flexibility compared with the old method in which the angle is fixed for a particular unit height and has to be recalculated for each different height.

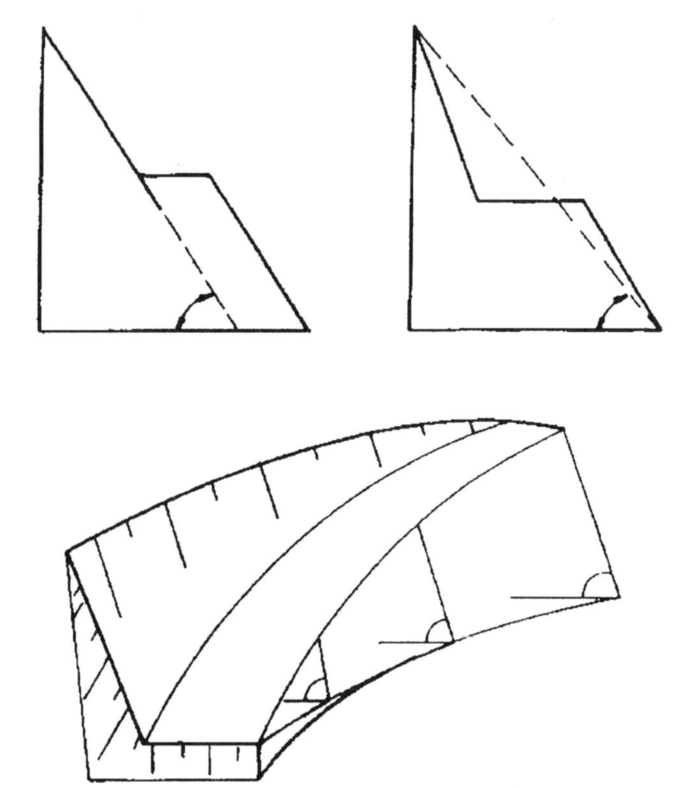

FIGURE 25.16 Difference between "push ramp" design (top left) and optimized slope-angle design (top right). The lower diagram illustrates the variable slope angle above and below the access ramp.

The access ramp that intersects a lithological unit divides the total height of the unit into segments of variable heights, depending on the position of the ramp (see Figure 25.16). By using fixed angles in a "push-ramp" design, the ramp width increases with slope depth. This results in a shallower overall slope angle for that particular unit (Figure 25.17). Using curves of slope height

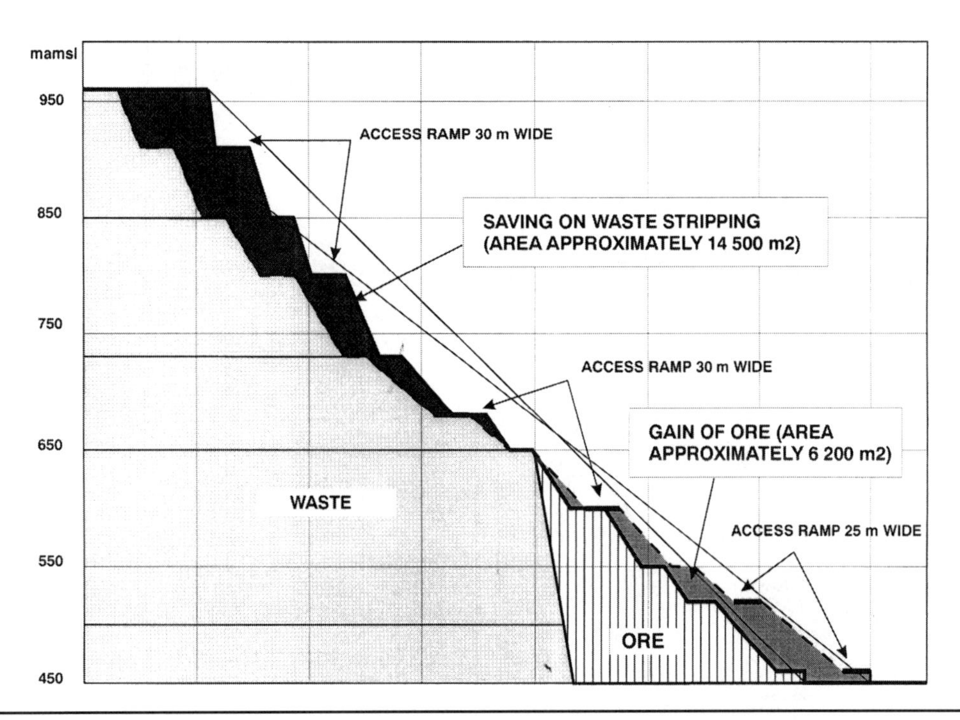

FIGURE 25.17 Pit wall section illustrating the difference between the "push ramp" design (thick line). The dark shaded area illustrates savings achieved on waste stripping, and the light shaded area illustrates the ore gain due to steeper angles.

versus slope angle, one can steepen the angle of each segment, providing the maximum slope angle for the unit is not exceeded. This will, however, mean that the bench width will vary according to the angle of the slope segment.

In basalt, sandstone, laminated sandstone, and granite, double benching (30 m) on the pit limits should be considered, half benching (15 m operating height) is recommended in the mudstones and carbonaceous shales. The angle of the bench face will also be calculated using the practical curves.

25.9 CONCLUSIONS AND RECOMMENDATIONS

The following conclusions can be drawn, based on the geotechnical program carried out to date:

- Based on the analyses of the composite slope designs, it is evident that overall slope angles exhibit regressive behavior, indicating stable conditions.
- Results from the discontinuity analyses indicate that kinematic failures are possible, but that, in all cases, the inherent strength of the discontinuities generally precludes such failures.

Based on the above conclusions, the following recommendations can be made:

- Under the existing conditions and based on experience, the practical curves for the slope height and slope angle relationship should be used for long-term planning purposes.
- An effort should be made to avoid leaving isolated kimberlite "slivers" in the perimeter walls since they might become highly unstable.
- Where the contact traverses the haulroad at an oblique angle, a 20-m-wide catch berm should be designed to cater for precontact failure.
- In basalt, sandstone, laminated sandstone, and granite, double benching (30 m) on the pit limits should be considered, while half benching (15-m operating height) is recommended in the mudstones and carbonaceous shales.
- An effective surface-perimeter dewatering program is essential, and, as the mine progresses to deeper levels, dewatering with subhorizontal toe drains to a depth 50 to 100 m will be required.
- The development of a good drainage system throughout the pit is essential to minimize weathering of the laminated sandstones and mudstones.
- Pit sumps should not be located next to the perimeter walls because of the high susceptibility to weathering of laminated sandstones and mudstones.
- It is essential to develop cautious blasting techniques for each lithological unit. Inclined double-bench presplit blasting should be considered in order to reduce crest damage and reduce the possibility of leaving "hard toes." Poor blasting practices will increase the weathering of the rock mass by creating cracks and voids into which water will seep. As a result the desired steep slope angles may not be achieved.
- The prevention of weathering will have a marked economic impact, since the slope angles that can be achieved in a weathered rock mass differ significantly from those achieved in a rock mass not affected by weathering.
- Protection of the mudstone units against weathering has to be economically evaluated.

- Protection similar to that for the mudstones may be required in carbonaceous horizons to prevent self-combustion of the coal and subsequent undercutting, with the failure of overlying strata.

25.10 ACKNOWLEDGMENTS

The authors acknowledge the management of Debswana Diamond Company (Pty.) Ltd., and the management of Orapa and Letlhakane Mines for their permission to publish this chapter. Appreciation is expressed to P. Opelokgale for his assistance in field data collection.

25.11 REFERENCES

Brown, E.T. 1981. *Rock Characterization, Testing and Monitoring.* International Society for Rock Mechanics.

Buhmann, D., and M. Atanasova. 1997. Karoo core samples from Orapa and Letlhakane, Botswana: A mineralogical report. *Council of Geoscience report no.*:1997–0347.

Bush, R.A., R.H. Hofmann, and H.J. Van Rensburg. 1995. Dewatering plan for Orapa Mine A/K1 pit. Unpublished AAC Civil Engineering Report.

Drescher, K., and U.W. Vogler. 1994a. Laboratory determination of rock properties - A/K1 - in pit boreholes. CSIR Report no.: EMAP-C-94007.

Drescher, K., and U.W. Vogler. 1994b. Laboratory determination of rock properties - Geotechnical borehole - I. CSIR Report no.: EMAP-C-94064.

Drescher, K., and U.W. Vogler. 1994c. Laboratory determination of rock properties - Geotechnical borehole - II. CSIR Report no.: EMAP-C-94006.

Field, M., J.G. Gibson, J. Gababotse, and P. Khutjwe. 1995. The geology of the Orapa A/K1 kimberlite, Botswana: Further insight into the emplacement of kimberlite pipes. *Extended abstracts from Proceedings of the Sixth International Kimberlite Conference.* pp. 155–157.

Gibson, J.G. 1987. Internal memorandum. Geological structures within the Orapa plant area. Geology Department, Orapa and Letlhakane Mines. Debswana Diamond Company.

Haines, A., and P.J. Terbrugge. 1991. Preliminary estimation of rock slope stability using rock mass classification systems. *7th International Congress on Rock Mechanics,* Aachen.

Laubscher, D.H. 1977. Geomechanics classification of jointed rock masses–mining applications. *Trans. Instn. Min. Metall. (Sect. A: Min. Industry),* vol. 86, pp. A1–A8.

Lorenz, V. 1973. On the formation of maars. *Bull Volcanol.,* 37: 183–204.

Shaw, A.L. 1991. Internal Memorandum. A geological model for the Orapa kimberlite 2125A/K1 for use in mine planning. Geology Department, Orapa and Letlhakane Mines. Debswana Diamond Company.

CHAPTER 26

Slope Stability Assessment at the Main Cresson Mine

M. Pierce,* T. Brandshaug,* and M. Ward†

26.1 INTRODUCTION

The Cresson Project is an open-pit gold mine in the Cripple Creek Mining District in Teller County, Colorado. Indications are that the current reserve is about 168 Mt at an average stripping ratio of 1.7:1. The low grade of the ore (1-g/tonne average grade) makes the economics very sensitive to operating costs and strip ratio. The company wishes to maintain the pit-slope angles to be as steep as possible to maximize the recovery of the resource and, at the same time, operate safely and effectively.

As of January 2000, the east wall of the Main Cresson Mine was 230 m high and had an overall slope angle of 60°; the west wall was 145 m high and had an overall slope angle of 35°. The proposed final design for the pit is shown in Figure 26.1. The expansion would result in a final depth of 440 m. The highest and steepest proposed slopes would be in the east and west walls. The east wall would be 440 m high, with slope angles of 60° in the upper 230 m and 50° in the lower 210 m. The west wall would be 380 m high, with slope angles of 60° in the upper 230 m and 50° in the lower 150 m. Although slopes of this combined height and angle are outside the range of experience at other open-pit mines, the favorable structural geology, adequate rock-mass strength, and lack of water make the proposed slopes worth considering.

Two-dimensional numerical models were used to analyze stability of the proposed east and west walls of the Main Cresson Mine. Because no slope failures have occurred at the Main Cresson Mine to date, it was not possible to back-analyze slope failure to guide the choice of material models and model parameters through calibration between response predictions and physical observations (qualitative and quantitative). Consequently, a forward-analysis that relied on rock-mass strength and stiffness parameters determined solely from rock-mass-quality estimates and empirical relations to rock-mass shear strength was carried out. The program FLAC (Fast Lagrangian Analysis of Continua [Itasca 1999]) was used to determine the factors of safety against circular-type failure in the east and west walls, and the program UDEC (Universal Distinct Element Code [Itasca 2000]) was used to determine the potential for slope-scale structurally controlled failure in the east wall.

26.2 GEOTECHNICAL PROPERTIES OF THE MAIN CRESSON MINE ROCK MASS

The geotechnical properties of the rock mass at the Main Cresson Mine were determined from a combination of pit-wall mapping and logging and testing of oriented core from boreholes drilled near the proposed east and west walls. Data from lithologic and video logging of several exploration holes were also used to help determine the extent and location of strongly oxidized and poor-quality zones. The locations of all boreholes used to obtain information for stability analysis are shown in Figure 26.1.

26.2.1 Geology

Mineralization in the Main Cresson Mine is hosted in breccias and intrusive rocks that erupted from the Cripple Creek Diatreme 28 to 32 million years ago. This diatreme is located at the intersection of Precambrian terrains and covers about 18 km^2. The Main Cresson Mine is in the southwestern portion of the diatreme. Breccias are the dominant ore host and consist of angular to subrounded clasts of Precambrian rocks, phonolites, and volcanic sedimentary rocks in a matrix composed of quartz, microcline, and very fine-grained rock fragments (Thompson, Trippel, and Dewelly 1985). These breccias were intruded by irregular bodies of alkaline magmas of phonolite-to-phonotephrite composition and were, in turn, intruded by lamprophyre dikes and pipe-like bodies. The gold mineralization is thought to have occurred shortly after intrusion of the lamprophyres (Kelly et al. 1998). Although lamprophyre dikes and pipes make up a small portion of the rock volume in the mine area, they are weaker than the surrounding rock mass and are often associated with areas of poor rock quality. The dikes are generally thin (1–2 m in width), nearly vertical, and generally intersect the mine walls at high angles. In contrast, the Cresson Pipe is a large (100 m × 200 m) lamprophyre breccia pipe, the upper extents of which are contained within the mine and intersect the toe of the western wall. Associated with the pipe is a zone of argillic alteration, which adversely affects the rock strength.

Mineralization occurred in two styles: high-grade veins and disseminations within the diatremal breccias and other volcanic rocks. Historic production has been in excess of 21 million troy ounces, primarily from high-grade gold and gold-silver-telluride veins, and extracted from underground drifts and stopes. Current mining is focused on low-grade disseminated ore in near-surface deposits. The Main Cresson Mine is the largest surface mine in the district and has been in production since 1995.

26.2.2 Structure

East Wall. Stereonet analysis of joint data from wall mapping in the east wall of the pit carried out to date indicates three major joint sets: two vertical sets striking N22E and N12W and one horizontal set (Figure 26.2). In addition, several widely spaced persistent features have been observed dipping into the pit at approximately 45° in the northern section of the east wall. Although they have not resulted in major movement of slope material to date, the possibility of large-scale slope failure resulting from combination of slip and intact rock failure along these structures in the future wall was examined as part of the numerical analysis.

Oriented core from hole GT98-1 indicates primarily vertical and horizontal jointing near the proposed east wall. However, stereonet analysis of joint data from vertical boreholes GT99-4 (245 m north of GT98-1) and CR-2057 (just east of GT98-1) indicates a

* Itasca Consulting Group, Inc., Minneapolis, Minnesota.
† Cripple Creek & Victor Gold Mining Company, Cripple Creek, Colorado.

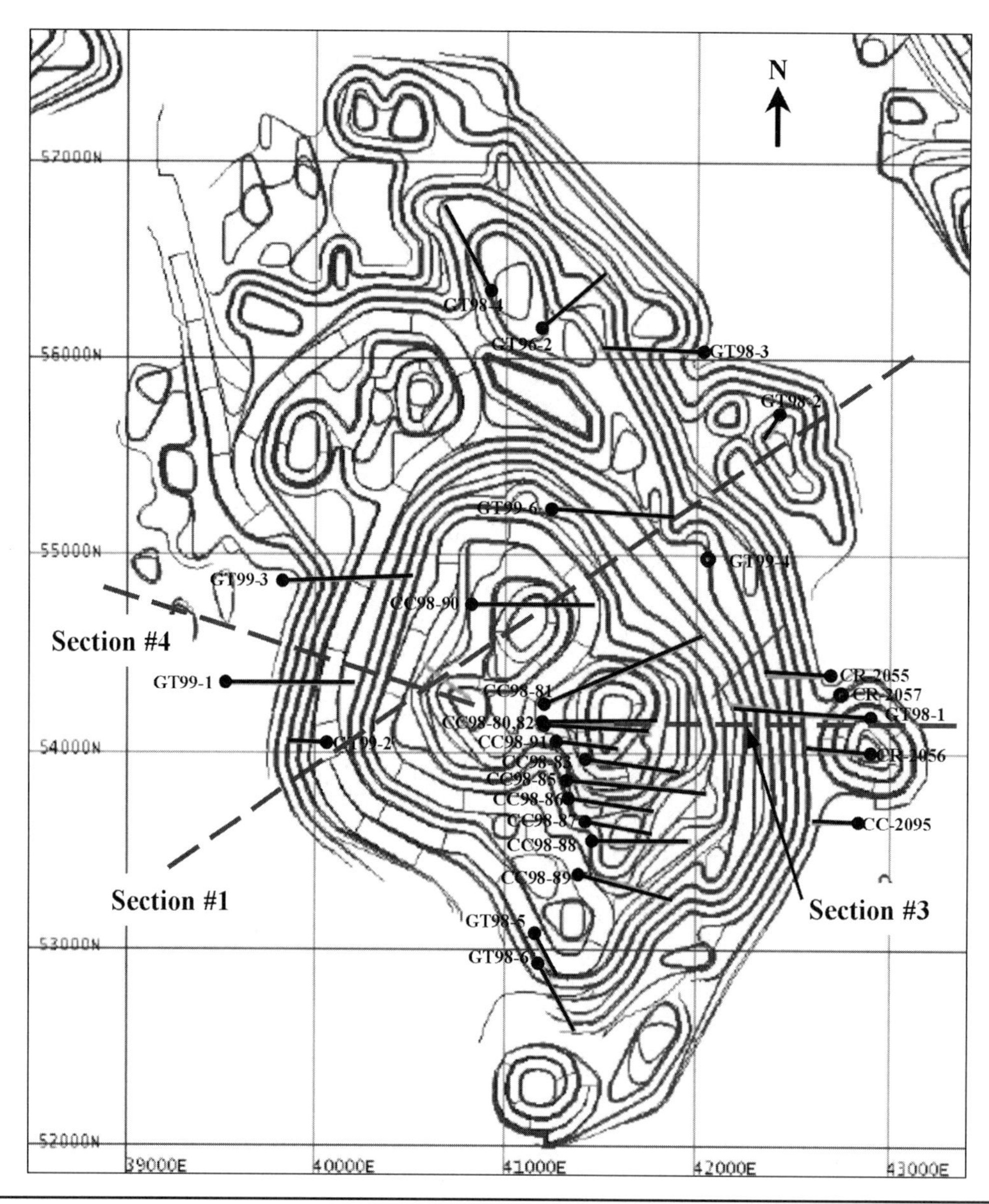

FIGURE 26.1 Plan view of the proposed final design for the Main Cresson Mine

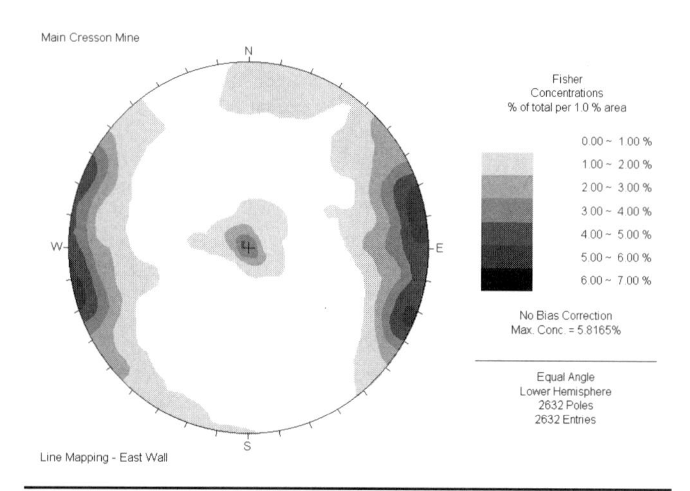

FIGURE 26.2 Contoured stereonet plot of poles to structures mapped in the east wall of the Main Cresson Mine

joint set dipping 50° to 60° into the wall. The potential for slope-scale toppling failure on persistent structures of this orientation (a worst-case scenario) in the future east wall was examined as part of the numerical analysis.

West Wall. Stereonet analysis of joint data from wall mapping of the west wall indicates two major joint sets: one vertical joint set striking N27W and one horizontal set (Figure 26.3). Overall, the joint orientations are favorable with respect to stability, which is in agreement with observations of the current west wall. Oriented core from holes GT99-1 and GT99-3 (in the central and northern sections of the future west wall, respectively) indicates primarily vertical and horizontal jointing. This represents stable orientations and is consistent with structures mapped in the current west wall of the mine. Structural data from GT99-2 (in the southern section of the future west wall) indicate several minor faults with thin clay fillings (<1 mm) at a depth of 30 to 55 m below ground surface. The orientation of these faults suggests that there may be some local structurally controlled instability in the top of the wall in the southern section when it is pushed back. In addition to the faults, a joint set was identified in GT99-2 that dips 50° to 60° into the proposed west

TABLE 26.1 Estimates of average RMR in the east wall

	Phonolite		Phonolite (Broken Zone)		Cripple Creek Breccia		Cripple Creek Breccia (Toe of Slope)	
	Avg. Value	Avg. Rating	Avg. Value	Avg. Rating	Avg. Value	Avg. Rating	Avg. Value	Avg. Rating
UCS	60 MPa	7	45 MPa	5	90 MPa	8	65 MPa	7
RQD	70%	13	25%	6	82%	17	50%	9
Joint frequency	6.5/m	15	13/m	10	1.5/m	23	10/m	10
Joint condition	Slightly rough, ~1-mm clay/ oxide filling, <1-mm separation	12	Slightly rough, ~1 mm clay/ oxide filling, 1-mm to 5-mm separation	10	Rough, <1-mm oxide/clay filling, <1-mm separation	15	Rough, <1-mm oxide filling, <1-mm separation	17
Groundwater	Dry	10	Dry	10	Dry	10	Dry	10
		57		41		73		53

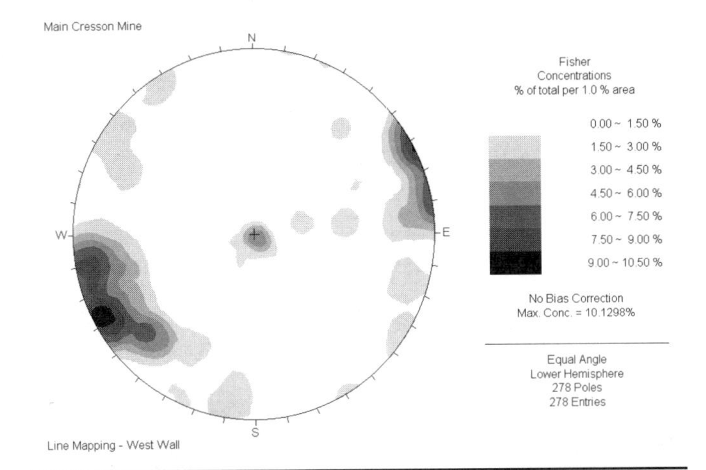

FIGURE 26.3 Contoured stereonet plot of poles to structures mapped in the west wall of the Main Cresson Mine

wall. With a joint set of this orientation, the potential exists for toppling failure. The fact that this joint set is not identified in GT99-1, 90 m to the north, suggests that it does not have sufficient persistence to result in large-scale toppling failure of the entire west wall. Geotechnical mapping of the wall as it is pushed back will allow the persistence and character of the faults and joints in the south end of the wall to be studied more closely.

26.2.3 Rock-Mass Quality

The rock-mass quality near the proposed walls was estimated from core using the rock-mass rating (RMR) system of Bieniawski (1976). Unconfined compressive strength (UCS) measurements obtained from point-load and unconfined compression testing were averaged for 15-m intervals, and measurements of rock-quality designation (RQD), fracture frequency, joint filling, and joint roughness obtained from the geotechnical core logs were averaged for 1.5-m intervals. Joint persistence and separation were assumed to be similar to what is observed in the current pit walls.

East Wall. Most of the rock to be exposed in the proposed east wall is good-quality breccia. Deeper in the wall, there is a breccia/phonolite contact that strikes roughly N25W and dips 85° to the east (Figure 26.4). A zone of poor-to-fair-quality rock with a thickness of approximately 30 m is associated with this contact. This zone was oriented from an intersection logged in oriented core from borehole GT98-1 and from intersections logged in borehole videos from boreholes CR-2055, CR-2056, and CR-2057. The phonolite to the east of this contact zone is of good quality. A zone of fair-quality breccia (referred to as the West Breccia) is found in the bottom of the east wall. The transition between it and the good-quality breccia above (referred to as the Center Breccia) has been roughly approximated as a contact dipping 60° to the east, based on differences in rock quality from holes GT98-1 and CC98-82. The average RMR values calculated for each of these units are summarized in Table 26.1.

West Wall. In the west wall (Figure 26.5), a fair-to-good-quality phonolite sill will be exposed near the top of the proposed wall, with breccia above and below. The Upper Breccia is of similar quality to the phonolite, but the Lower Breccia is of poor-to-fair quality, as evidenced by the core from GT99-1, GT99-2, and CC-1954. The near-cylindrical-shaped Cresson Pipe extends vertically upward to the east of the proposed west wall, intersecting the toe of the slope. It is of poor quality. The average RMR values calculated for each of these units are summarized in Table 26.2.

26.2.4 Hoek–Brown Parameter m_i

Triaxial tests were conducted on core obtained from several core holes to obtain estimates of the Hoek–Brown parameter m_i (a measure of how the intact rock strength changes with confinement). The m_i estimates are combined with RMR estimates of each unit to calculate m_b, a constant in the Hoek–Brown equation describing rock-mass strength. Tests were conducted on samples of breccia and phonolite up to confining pressures equal to half the unconfined compressive strength, as specified by Hoek and Brown (1997). A regression was performed on all triaxial test data for each rock type to obtain m_i values of 7.6 and 10.8 for the phonolite and breccia, respectively.

26.2.5 Joint Shear Strength

In order to define the peak and residual shear strengths of the persistent joints dipping approximately 45° into the pit in the northern end of the east wall, direct shear tests were conducted on joint surfaces of this orientation obtained from borehole GT99-4. Iron oxide coatings approximately 5 mm thick were observed on these persistent joints in the field, so samples with similar coatings were chosen for laboratory testing. In addition to laboratory testing, field estimates of peak joint shear strength were made using the empirical relation developed by Barton (1973). Based on laboratory and field analyses, the following estimates of peak and residual shear strengths were made:

	Peak	Residual
Joint cohesion (MPa)	0.17	0.1
Joint friction angle (degrees)	38	34

26.3 STABILITY ANALYSIS OF EAST AND WEST WALLS

The objective of the numerical analyses was to assess the stability of the future east and west highwalls, given current plans for the final pit geometry and current geologic and geotechnical information. Highwall stability was analyzed for two vertical sections

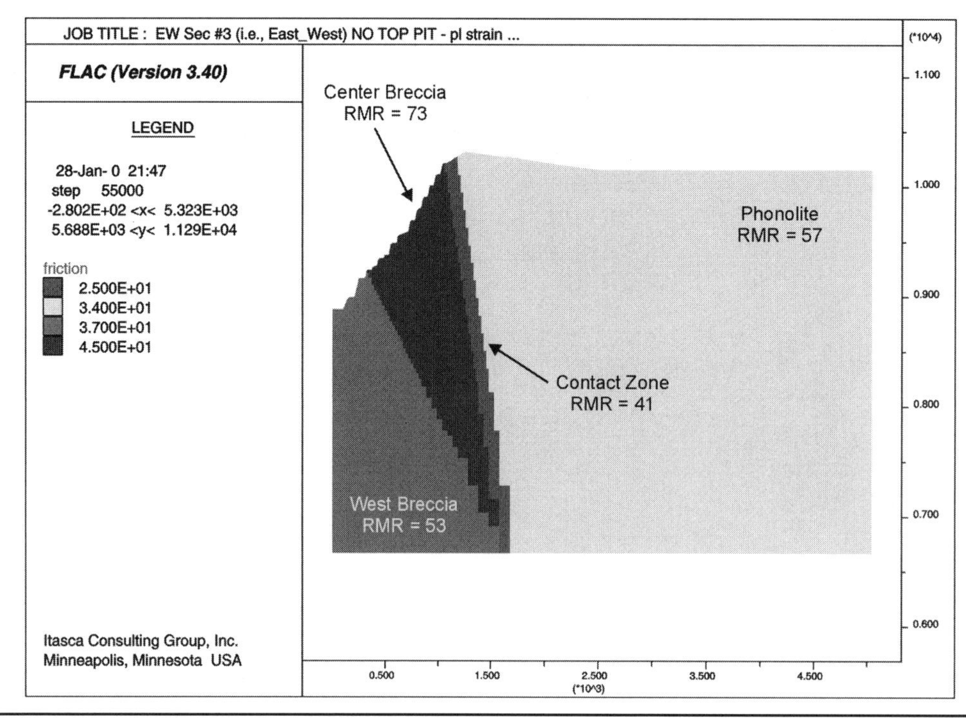

FIGURE 26.4 Rock mass in the proposed final east wall of the Main Cresson Mine

TABLE 26.2 Estimates of average RMR in the west wall

	Phonolite		Cripple Creek Breccia (Above Phonolite)		Cripple Creek Breccia (Below Phonolite)		Cresson Pipe	
	Avg. Value	Avg. Rating	Avg. Value	Avg. Rating	Avg. Value	Avg. Rating	Avg. Value	Avg. Rating
UCS	55 MPa	6	55 MPa	6	55 MPa	6	65 MPa	7
RQD	67%	14	74%	14	25%	6	20%	3
Joint frequency	2.5/m	16	3.5/m	16	13/m	10	16.5/m	10
Joint condition	Rough, ~1-mm oxide/clay filling, <1-mm separation	14	Rough, ~1-mm oxide/clay filling, <1-mm separation	14	Rough, 1-mm to 5-mm clay/oxide filling, <1-mm separation	12	Slightly rough, 1-mm to 5-mm clay/oxide filling, <1-mm separation	10
Groundwater	Dry	10	Dry	10	Dry	10	Dry	10
		60		60		45		40

TABLE 26.3 Slope angles of proposed walls at sections analyzed

East Wall Section 1	Upper 1/3	60°
	Lower 2/3	50°
East Wall Section 3	Upper 1/2	60°
	Lower 1/2	50°
West Wall Section 1	Upper 1/3	60°
	Middle 1/3	50°
	Lower 1/3	40°
West Wall Section 4	Upper 1/3	60°
	Lower 2/3	50°

in the east wall and two sections in the west wall. The locations of these sections are shown in the mine plan view in Figure 26.1. The slope angles of the proposed walls at these sections are given in Table 26.3.

FLAC models emphasize rock-mass response, in a continuum sense, by incorporating the mechanical effects of the discontinuities of the rock fabric, as well as major joint sets, into an average response in terms of rock-mass strength and stiffness. Because this is a continuum model, with the rock-mass response characterized by an isotropic elasto-plastic constitutive material model (Mohr–Coulomb), only a "circular-" type of failure mechanism will occur in the highwall. FLAC was used to model sections in the east and west walls to determine the factors of safety against this type of failure. UDEC models, on the other hand, emphasize the structural aspects of the rock mass (i.e., joint sets) explicitly in the context of estimating stability. It is more suited to investigating stability where the mode of failure in the slope is expected to be dominated largely by slip along structure or by a combination of slip along structure and shearing through intact rock.

26.3.1 FLAC Modeling

FLAC was used in this study to determine the factor of safety against circular-type failure along section 3 in the east wall and sections 1 and 4 in the west wall. Section 3 in the east wall and section 4 in the west wall correspond to the highest and straightest portions of the proposed walls. These were analyzed assuming plane-strain conditions in the model. "Plane strain" implies

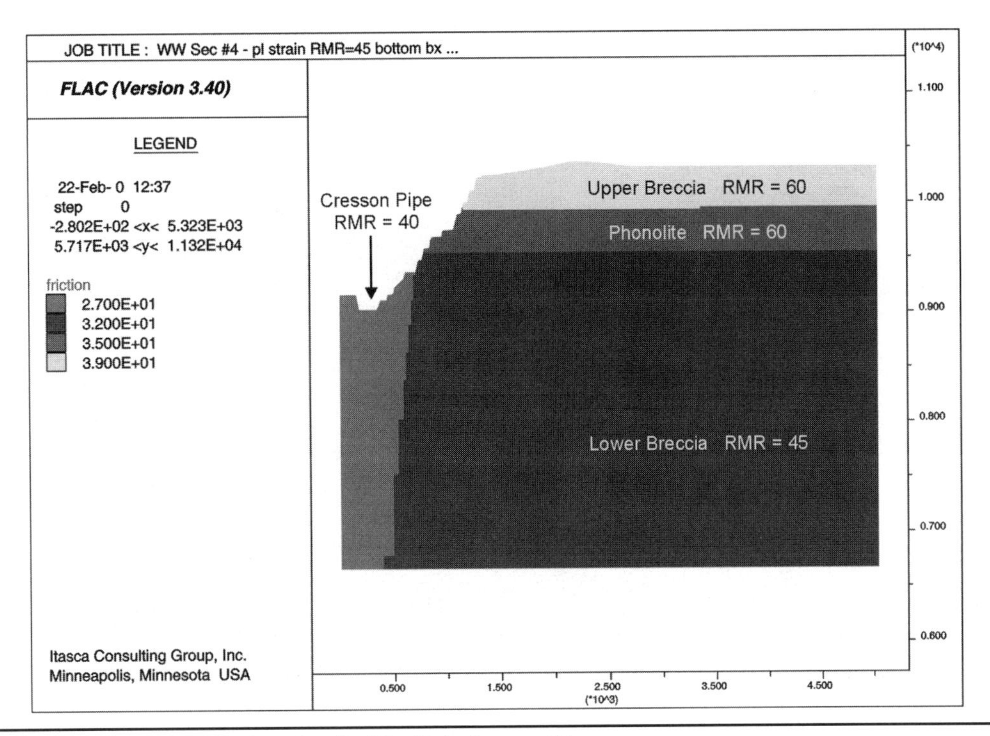

FIGURE 26.5 Rock mass in the proposed final west wall of the Main Cresson Mine

that the highwall is infinitely long and straight in the out-of-plane direction and represents a conservative assumption. The southern end of the west wall has significant horizontal curvature, which, because of confinement, can contribute substantially to highwall stability (Lorig 1999). Therefore, west wall section 1 was analyzed with an axisymmetric model. Axisymmetry implies that the highwall is curved in the out-of-plane direction, representing the future mine design as a circular "cone," which, for west wall section 1, is much closer to reality than is the assumption of plane strain.

Initial conditions are specified in the model by assigning a vertical stress equal to the weight of the overburden rock mass and assuming the horizontal stress to be a factor of 1.2 higher (i.e., k_o = 1.2). The assumption of k_o of 1.2 was made based on previous stress measurements in the region (Hoek and Brown 1980). The model extents and degree of discretization were adjusted to obtain reasonable run times while minimizing artificial effects due to boundary placement and zone size. Figure 26.6 shows the FLAC model for section 3 in the east wall. The size of the finite-difference zoning used in the vicinity of the slope is approximately 7.5 m × 7.5 m.

For a stable highwall, the FLAC model is used to provide a measure of "safety" with regard to stability. This is done by gradually reducing the rock-mass strength (cohesion and friction angle) by a constant factor until slope failure occurs. The final factor provides a relative measure of strength reduction (strength reduction factor, SRF) that produces slope failure in the models. This is illustrated in Figure 26.7. Assuming that the estimated shear strengths of the rock mass are correct, the SRF is similar to the more conventional factor of safety.

FLAC Model Input Parameters. In the FLAC model, the rock mass was prescribed an elastic–perfectly-plastic response. The shear strength was limited by a simple Mohr–Coulomb criterion. Rock-mass cohesion and friction angle (i.e., internal angle of friction) define this criterion. Empirical relations between RMR and the Hoek–Brown strength criterion were used to estimate a rock-mass cohesion and friction angle. The Hoek–Brown criterion is given in Eq. 26.1; the Mohr–Coulomb criterion is given in Eq. 26.2. The empirical relations between the Hoek–Brown parameters, m_b and s, and RMR distinguish a disturbed rock mass from an undisturbed rock mass (Priest and Brown 1983). Based on analyses using a similar approach, Sjöberg (1999) found that the relations for a disturbed rock mass provided a better match between the predicted and observed behavior of large rock slopes. These relations were used for these analyses and are given in Eq. 26.3 and 26.4. Figure 26.8 illustrates the Hoek–Brown and Mohr–Coulomb criteria in a principal stress ($\sigma_1 - \sigma_3$) coordinate system. The Mohr–Coulomb criterion in this figure represents a best-fit in a least-squares sense of the Hoek–Brown criterion, within the constraints of $0 < \sigma_3 < 2.5$ MPa, and $\sigma_1 = \sigma_c(s)^{1/2}$ for $\sigma_3 = 0$ (where σ_c is the uniaxial compressive strength of the intact rock).

$$\sigma_1 = \sigma_3 + (m_b{*}\sigma_3{*}\sigma_c + s{*}\sigma_c^2)^{1/2} \quad \text{EQ. 26.1}$$

$$\sigma_1 = k{*}\sigma_3 + \sigma_c{*}(s)^{1/2} \quad \text{EQ. 26.2}$$

$$s = \exp(RMR - 100)/6 \quad \text{EQ. 26.3}$$

$$m_b = m_i{*}\exp(RMR - 100)/14 \quad \text{EQ. 26.4}$$

Several empirical relations exist that attempt to correlate the stiffness of the rock mass with the rock-mass quality, as expressed by RMR (Bieniawski 1978; Serafim and Pereira 1983; Hoek and Brown 1997). A common factor in these relations is that the presence of joints and their conditions have a dominant affect on the rock-mass stiffness. For a competent rock mass, these relations suggest a rock-mass stiffness of about 25% of the intact rock stiffness.

The FLAC model parameters used in the analyses of the east and west walls are listed in Tables 26.4 and 26.5, respectively. These parameter values reflect average conditions except for those listed in parenthesis for section 3, which represent reasonable lower-bound values (approximately one standard deviation below the average).

Results of Analysis—East Wall. The future east wall was analyzed along section 3 in a direction due east of the approximate

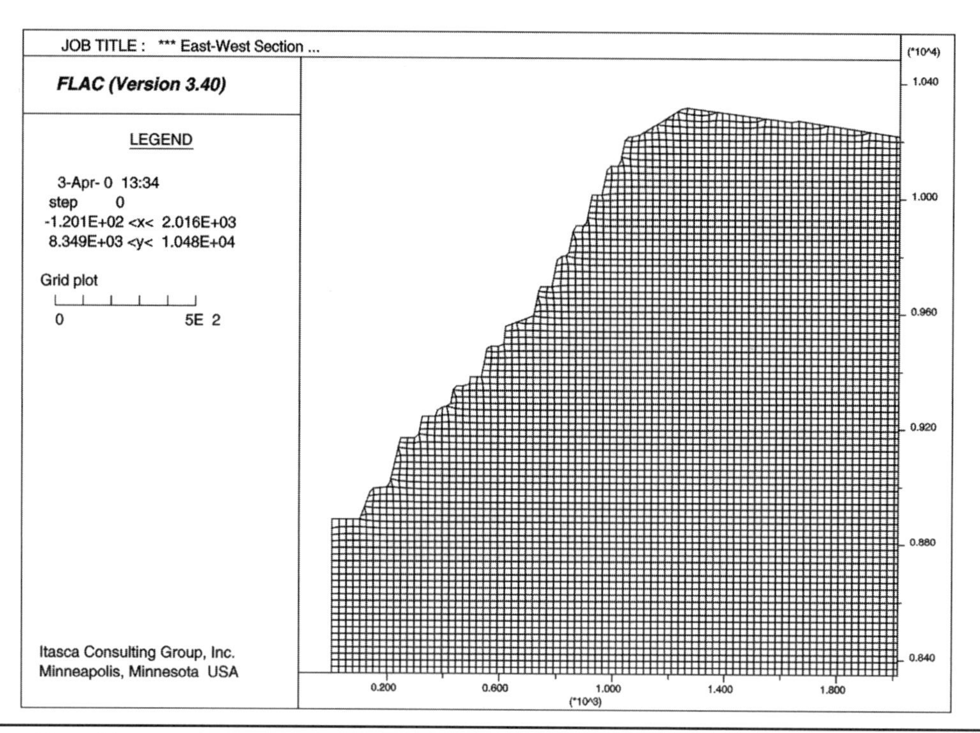

FIGURE 26.6 FLAC model used for analysis of east wall stability

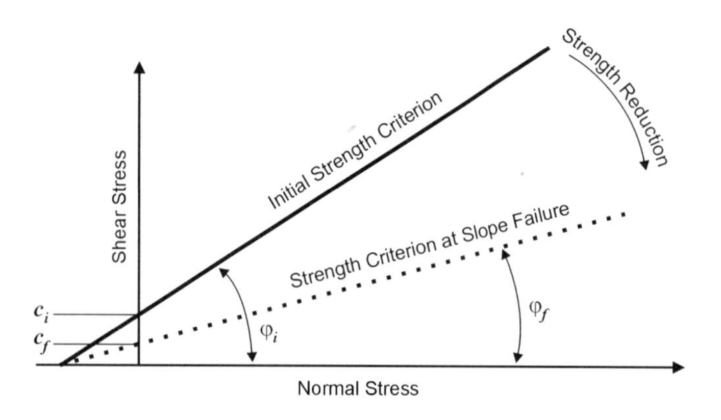

Strength-Reduction-Factor (SRF)

$$\text{SRF} = \frac{c_i}{c_f} = \frac{\tan(\varphi_i)}{\tan(\varphi_f)}$$

FIGURE 26.7 Definition of strength reduction factor used to estimate factor of safety against circular-type failure in a FLAC model

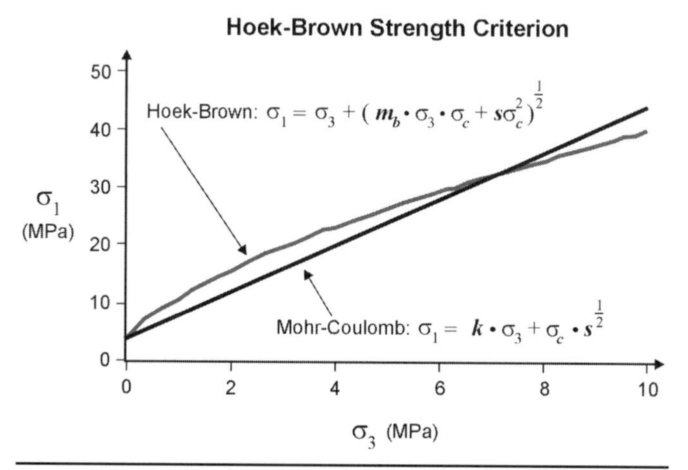

FIGURE 26.8 Illustration of the Hoek–Brown and Mohr–Coulomb rock-mass strength criteria

center of the mine (see Figure 26.1). The factor of safety was obtained through successive FLAC model evaluations in which the rock-mass strength was decreased incrementally until overall slope failure occurred. The ratio of the estimated strength to the failure strength defines the factor of safety in this context. Using average strengths (see Table 26.4), the east wall was predicted to have a factor of safety of 1.5. When the slope angle in the entire high wall was increased to 60°, a factor of safety of 1.4 was predicted. A factor of safety of 1.3 or higher is considered acceptable for future slopes in the Main Cresson Mine. The highwall as designed along section 3 was also evaluated for lower-bound strength (approximately one standard deviation below average), where the factor of safety was predicted to be 1.1, i.e., the highwall was stable. Although this factor of safety is less than 1.3, it reflects lower-bound strength and a conservative analysis approach (i.e., plane-strain model). Based on the FLAC model evaluations, it is reasonable to expect the future east wall to remain stable. Figure 26.9 illustrates the predicted stable future east wall (section 3), with regions of inelastic deformations indicated. That such regions occur is normal, as the wall adjusts to changes in stress during mining. Figure 26.10 shows the same wall for conditions of reduced strength, to the point at which circular-type failure occurs. The region marked with dark symbols is actively deforming into the pit. Tensile failure is indicated along the top of the ground surface and extends a distance of about 225 m behind the highwall crest. This is consistent with the development of vertical surface cracks in these failure types. The black contours indicate concentration of shear strain in the rock mass and are located near the circular failure surface. To achieve the results of Figure 26.10, the estimated rock-mass strength was reduced by a factor of 1.5.

Results of Analysis—West Wall. For average strength properties, the axisymmetric FLAC model predicts a factor of safety of 1.35 for section 1, which is located in a curved section in

TABLE 26.4 East wall FLAC model parameters

	East Wall Section 1			East Wall Section 3			
Rock-Mass Parameters	**Phonolite**	**Upper Breccia**	**Lower Breccia**	**Phonolite**	**Contact Zone**	**Center Breccia**	**West Breccia**
Density (kg/m^3)	2,395	2,395	2,395	2,395	2,395	2,395	2,395
Bulk modulus (GPa)*	17.5	17.5	17.5	17.5	17.5	17.5	17.5
Shear modulus (GPa)†	6.25	6.25	6.25	6.25	6.25	6.25	6.25
Cohesion‡ (MPa)	1.44	1.94	0.43	0.44	0.1	1.94	0.33
				(0.26)	(0.1)	(0.72)	(0.18)
Friction‡ (degrees)	40	45	38	34	25	45	37
				(31)	(25)	(42)	(33)
Tensile stress (MPa)	0	0	0	0	0	0	0
RMR‡	65	73	58	57	41	73	53
				(50)	(41)	(60)	(45)
m_i	7.6	10.8	10.8	7.6	7.6	10.8	10.8

*Intact bulk modulus taken as 70 GPa.
†Intact shear modulus taken as 25 GPa.
‡Average values, lower bound in parenthesis.

TABLE 26.5 West wall FLAC model parameters

	West Wall Section 1			West Wall Section 4			
Rock-Mass Parameters	**Phonolite**	**Upper Breccia**	**Lower Breccia**	**Phonolite**	**Upper Breccia**	**Lower Breccia**	**Cresson Pipe***
Density (kg/m^3)	2,395	2,395	2,395	2,395	2,395	2,395	2,395
Bulk modulus (GPa)†	17.5	17.5	17.5	17.5	17.5	17.5	17.5
Shear Modulus (GPa)‡	6.25	6.25	6.25	6.25	6.25	6.25	6.25
Cohesion (MPa)	0.49	0.46	0.11	0.49	0.46	0.15	0.14
Friction (degrees)	35	39	29	35	39	32	27
Tensile stress (MPa)	0	0	0	0	0	0	0
RMR§	60	60	40	60	60	45	40
m_i	7.6	10.8	10.8	7.6	10.8	10.8	7.4

*Cresson Pipe parameters used in both west wall sections.
†Intact bulk modulus taken as 70 GPa.
‡Intact shear modulus taken as 25 GPa.
§Average values.

the southern end of the west wall. Figure 26.11 illustrates the stable future west wall along section 1, showing regions (marked in symbols) along the slope where inelastic deformations have been predicted. Section 4 through the west wall is outside the curved section and was conservatively analyzed using a plane-strain FLAC model. For average strength properties, the slope is predicted to be unstable, as shown in Figure 26.12. This prediction is identified by increasing horizontal and vertical displacements at the highwall crest (monitored during the analyses) and by an actively yielding rock mass along the highwall, as indicated by the dark symbols. The circular sliding plane is located near the black contours shown in the figure. To achieve a stable slope, the slope angles must be reduced. By reducing the slope angle by 15° in both the upper one third of the slope (designed at 60°) and the lower two thirds of the slope (designed at 50°), a factor of safety of 1.3 was achieved. The new slope incorporating these changes is shown in Figure 26.13, with regions of inelastic deformations indicated. It should be noted that these are somewhat conservative results because the assumption of "plane strain" in the model is not entirely consistent with the true geometry of the mine wall. The west wall south of section 4 begins to be affected by the horizontal curvature of the mine wall, which contributes stability. In addition, because the extent of low-quality rock in the lower half of the slope north of section 4 is not well known, it was assumed in the model that it exists through the length of the slope. If further site investigation indicates that the region of low-quality rock is limited to a certain portion of the slope, more stable conditions can be expected.

Cripple Creek & Victor Gold Mining Company (CC&V) staff predict that a 15° slope angle reduction in the area of the west wall surrounding section 4 would result in a waste tonnage increase of less than 2% and a decrease in ore tonnage of 1.2% in the current life of mine reserve. Given the conservative nature of the two-dimensional analysis, a smaller reduction in slope angle may be possible that would limit these increases in waste tonnage while maintaining a stable slope. The authors believe that there is potential to obtain a more accurate estimate of the required layback with additional characterization and modeling. Additional drilling in the west wall would help to further delineate and characterize the region of low-quality rock. A three-dimensional analysis of the west wall could then be carried out that would more accurately account for the extent of fair-quality rock along the slope, the cylindrical shape of the Cresson Pipe, and the increase in wall curvature to the south.

26.3.2 UDEC Modeling

The program UDEC was used to investigate the potential for slope-scale structurally controlled instability in the east wall. Potential sliding features were modeled along section 1 in the north end of the wall (Figure 26.14), and potential toppling features were

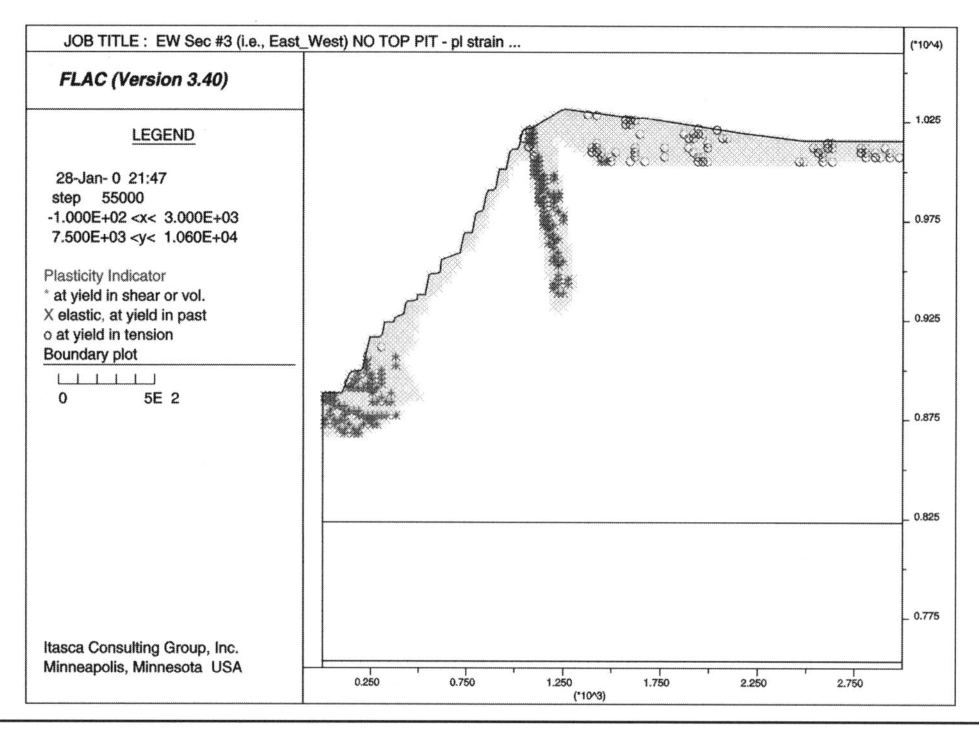

FIGURE 26.9 FLAC model of the stable proposed east wall along section 3 showing areas of inelastic deformations in the rock mass

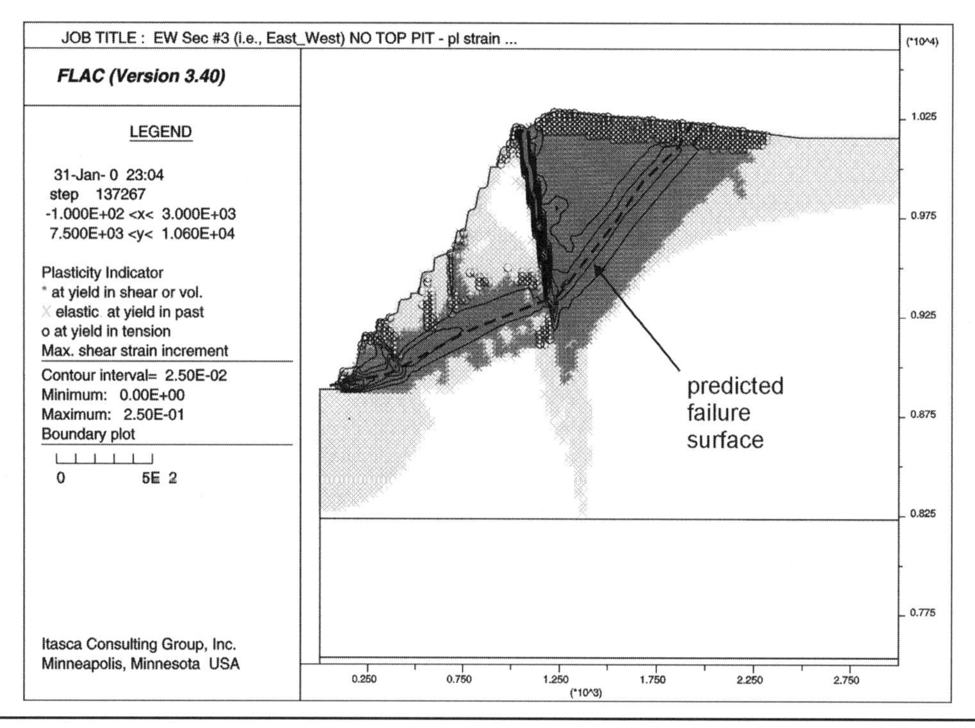

FIGURE 26.10 FLAC model of the proposed east wall along section 3 in the case where the estimated average rock-mass strength is reduced by a factor of 1.5 (Dark symbols indicate area of slope actively deforming; black contours show the vicinity of sliding surface.)

modeled along section 3 in the middle of the wall (Figure 26.15). The potential sliding features were input in the model with a dip of 45° to the southwest and a random persistence within the limits of a maximum length of 90 m and a minimum length of 30 m. The potential toppling features were input in the model as continuous features with a dip of 60° to the east. In both cases, the horizontal and vertical joint sets were included in the model as continuous structures spaced at 15 m. While the in situ joints have a spacing much closer than that used in the model, the larger spacing in the model is still considered sufficient to investigate the overall high-wall stability as it is affected by the structure.

UDEC Model Input Parameters. The UDEC model evaluates the overall high-wall stability in the context of the dominant joint sets of the rock mass. Therefore, in the UDEC model, the rock-mass strength is directly associated with the joint strength. The rock between the dominant joint sets (i.e., blocks) is assumed to behave elastically, while a Coulomb slip criterion is used to describe the joint strength. The joint strength properties

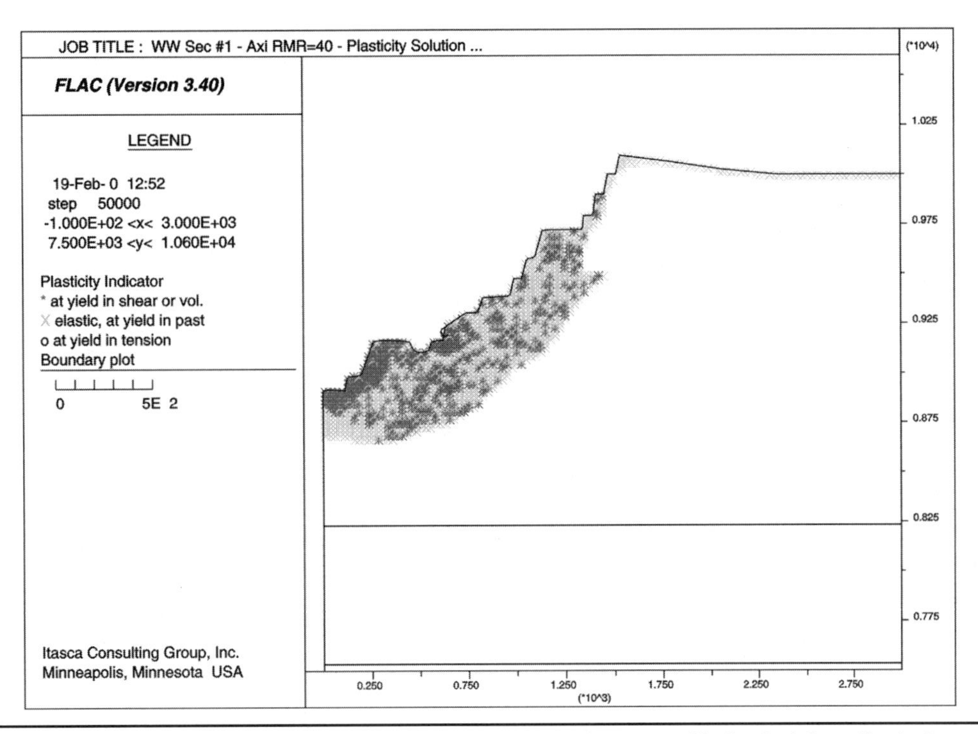

FIGURE 26.11 FLAC model of the stable proposed west wall along section 1 showing areas of inelastic deformation in the rock mass

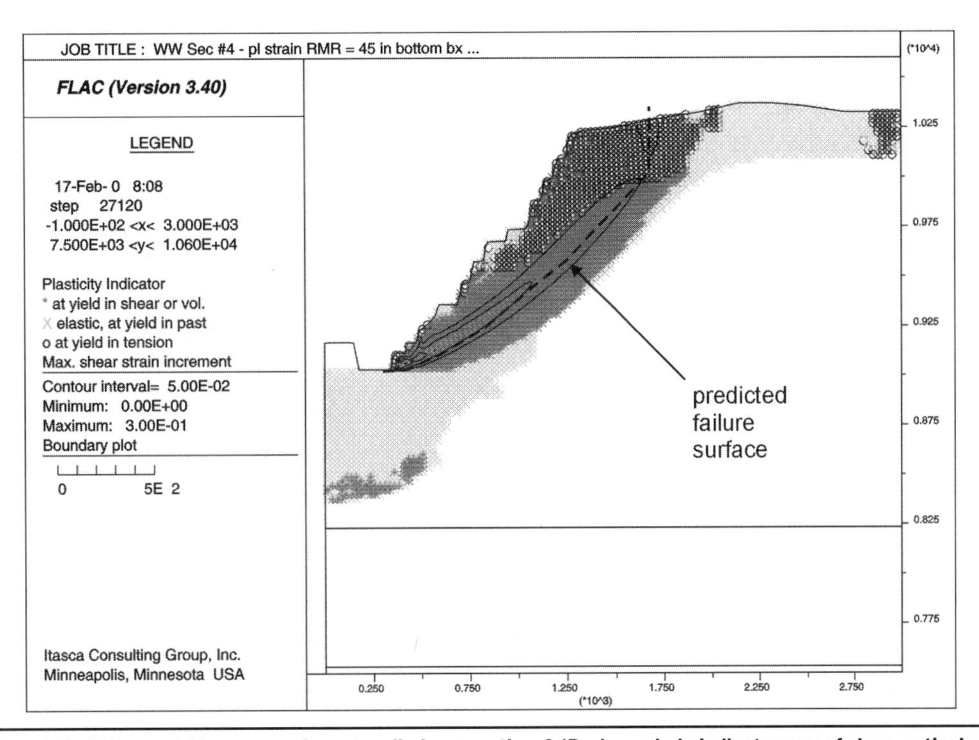

FIGURE 26.12 FLAC model of the unstable proposed west wall along section 4 (Dark symbols indicate area of slope actively deforming; black contours show the vicinity of sliding surface.)

used in the UDEC models are listed in Table 26.6. Note that the joint shear strength is expressed in terms of peak and residual (i.e., post peak) values, as estimated from the results of direct shear testing and empirical estimation of joint shear strength. If the peak strength is reached, the strength will subsequently be reduced to that defined as the residual strength and will remain at this level.

Results of Analysis. In the UDEC models of the east wall, joint shear displacements occur throughout the highwall, even to the point that the joints develop residual strength, but there is no evidence of unstable blocks. Figure 26.16 illustrates the joint shear displacements predicted in the case of the 45° in-dipping joints. Each line thickness in the plot represents a shear displacement of 3 cm (0.1 ft). A maximum joint shear displacement of about 30 cm (1 ft) occurs at the toe and in the crest region of the slope. Figure 26.17 shows the same for the case of the 60° out-dipping joints. A maximum joint shear displacement of about 21 cm (0.7 ft) is also found to occur at the toe and in the crest region of the slope in this model. Relative sliding on the 60°

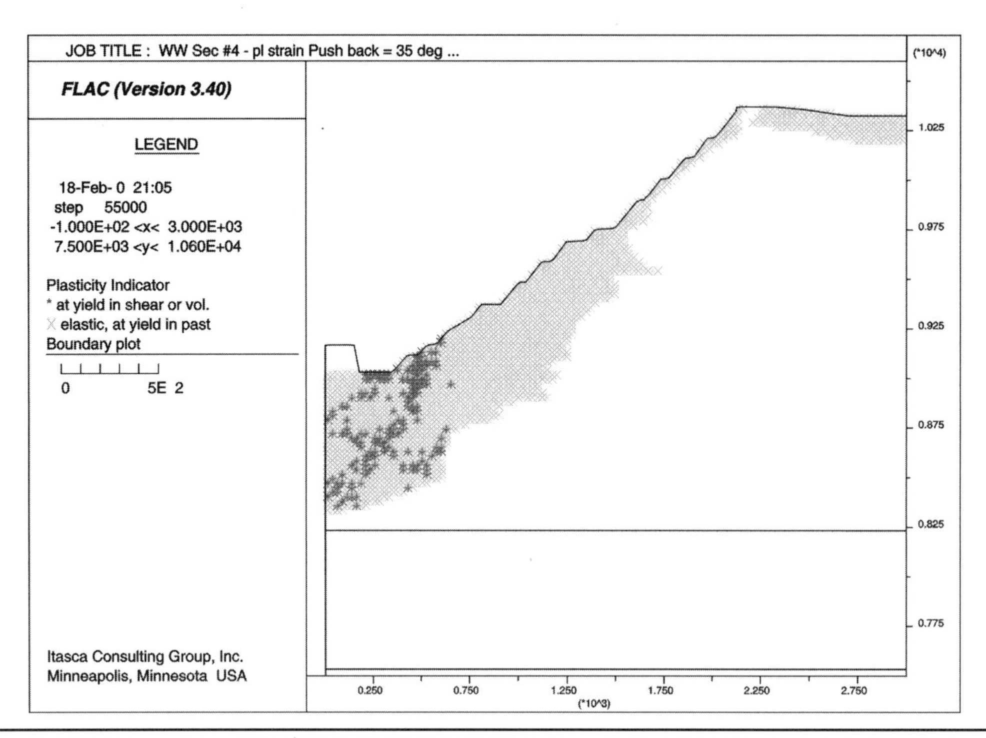

FIGURE 26.13 FLAC model of the stable proposed west wall through section 4 after slope angle is reduced to produce a factor of safety of 1.3

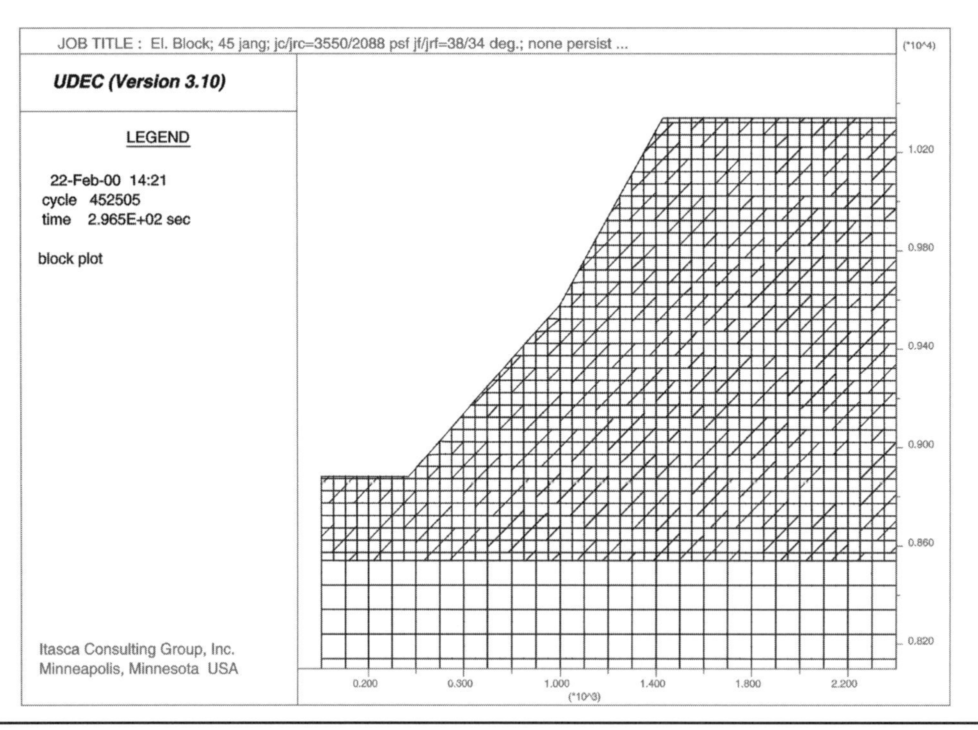

FIGURE 26.14 UDEC model of east wall at section 1 used to investigate potential for slope-scale instability due to slip on joints dipping at 45° into the pit

joints is evident, but the joint system has sufficient strength to resist a toppling-type failure. Because these analyses focused on the potential for slope-scale instability, the UDEC models do not incorporate the amount of structural detail that exists on the bench scale. As a result, the prediction of stability cannot be extended to smaller blocks and wedges that may form in the pit walls. If the 45° in-dipping structures are found to persist at lower elevations as the mine expands, it would be prudent to expect local failure, with blocks sliding across one to three benches. Bench-scale UDEC models could contribute to an evaluation of local stability.

26.4 CONCLUSIONS AND RECOMMENDATIONS

The FLAC model evaluated slope stability in terms of the potential for an overall circular-type highwall failure, while the UDEC model evaluated stability in terms of unstable blocks sliding on joint surfaces. For the rock-mass conditions as currently understood, the model results show that the future east wall will

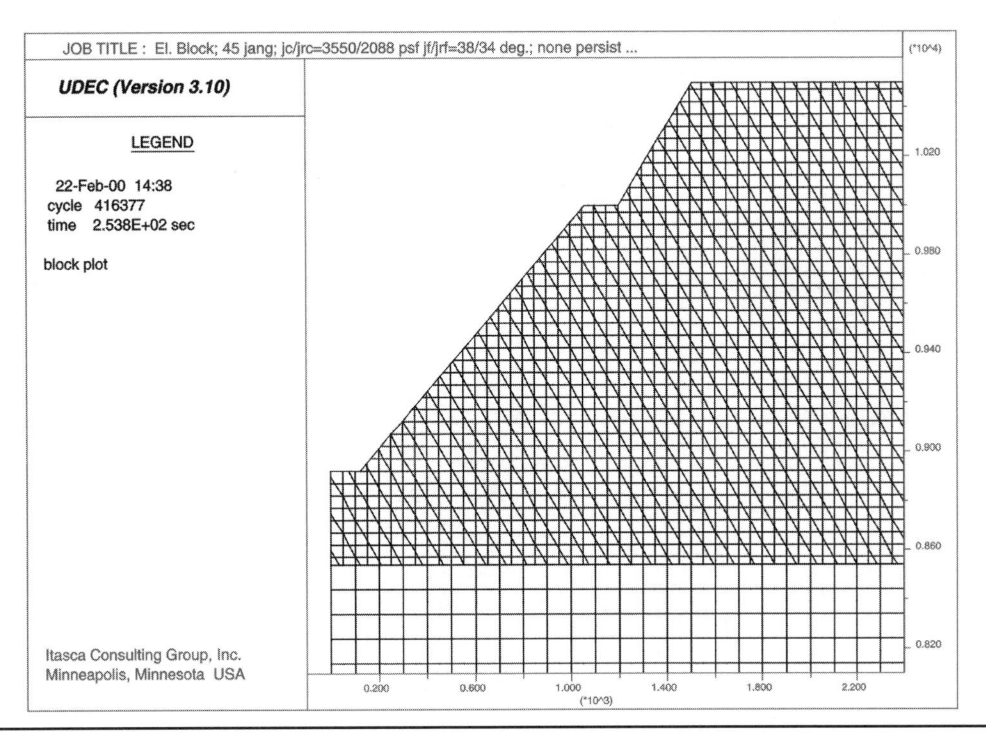

FIGURE 26.15 UDEC model of east wall at section 3 used to investigate potential for slope-scale instability due to slip on joints dipping at 60° into the wall

TABLE 26.6 UDEC model parameters

Parameter	Peak	Residual
Bulk modulus (GPa)	17.5	17.5
Shear modulus (GPa)	6.25	6.25
Joint cohesion (MPa)	0.17	0.1
Joint friction angle (degrees)	38	34
Joint tensile strength (MPa)	0	0
Joint normal stiffness (GPa/m)	5	5
Joint shear stiffness (GPa/m)	5	5

remain stable as designed, with a factor of safety of at least 1.3 against a circular-type failure. Although unstable blocks are not predicted in the UDEC models, the models do not incorporate the amount of structural detail that exists on a local scale in the mine and therefore have some limitations in predicting local block sliding. If the 45° in-dipping structures are found to persist at lower elevations as the mine expands, it is possible that small local failures could occur, with blocks sliding across one to three benches. Toppling associated with slip along the 60° joint set (dipping east out of the mine wall) is not expected to present a slope-scale stability problem.

For local instability issues, vigilant monitoring of the highwalls is appropriate for continued safe operation of the Main Cresson Mine. The mine's geologic staff places survey prisms on the mine wall as it evolves, with special attention to potential problem areas. These prisms currently are measured twice weekly using an automated total station EDM instrument. Daily visual inspections and biweekly crest inspections are also part of the highwall-monitoring program.

The FLAC models show that the current design of the future west wall along section 1 (i.e., in the southwest corner) will produce a stable highwall with a factor of safety of 1.35 against a circular-type failure. However, further north along section 4, the current west wall design results in a slope that is possibly too steep to remain stable. In order to produce a factor of safety of 1.3 against circular failure in this region of the west wall, the slope of the upper one third of the wall was reduced from the current 60° to 45° and the lower two-thirds was reduced from the current 50° to 35°. Because the assumption of plane strain used in this model is not entirely consistent with the wall geometry and rock-mass conditions, these results are somewhat conservative.

In light of these findings, further analysis of the west wall will be completed by CC&V. The potential instability issues in the northern section of the wall are linked to a low rock-mass quality in the lower portion of the wall and the configuration of the highwall in that sector. If further data acquisition and increased definition better resolve the extent of this zone, stability of the west wall will be reexamined. Reengineering of this sector of the western highwall may involve realignment to slope angles, as recommended, and/or a new configuration of haulage ways to reduce interramp heights. Catch-bench configuration or other appropriate engineering changes may also be used to help obtain a stable slope profile and an acceptable factor of safety.

26.5 REFERENCES

Barton, N.R. 1973. Review of a new shear strength criterion for rock joints. *Eng. Geol.* 7:287–332.

Bieniawski, Z.T. 1976. Rock mass classification in rock engineering. In *Exploration for Rock Engineering (Proceedings of the Symposium on Exploration for Rock Engineering, Johannesburg, November 1976)*, pp. 97–106. Rotterdam: A.A. Balkema.

Bieniawski, Z.T. 1978. Determining rock mass deformability: Experience from case histories. *Intl. Jour. Rock Mech. Min. Sci. & Geomech. Abstr.*, 15: 237–247.

Hoek, E., and E.T. Brown. 1980. *Underground Excavations in Rock.* London: The Institution of Mining and Metallurgy.

Hoek, E., and E.T. Brown. 1997. Practical estimates of rock mass strength. *Intl. Jour. Rock Mech. Min. Sci.*, 34(8):1165–1186.

Itasca Consulting Group, Inc. 1999. *FLAC (Fast Lagrangian Analysis of Continua)*, Version 3.40. Minneapolis: ICG.

Itasca Consulting Group, Inc. 2000. *UDEC (Universal Distinct Element Code)*, Version 3.10. Minneapolis: ICG.

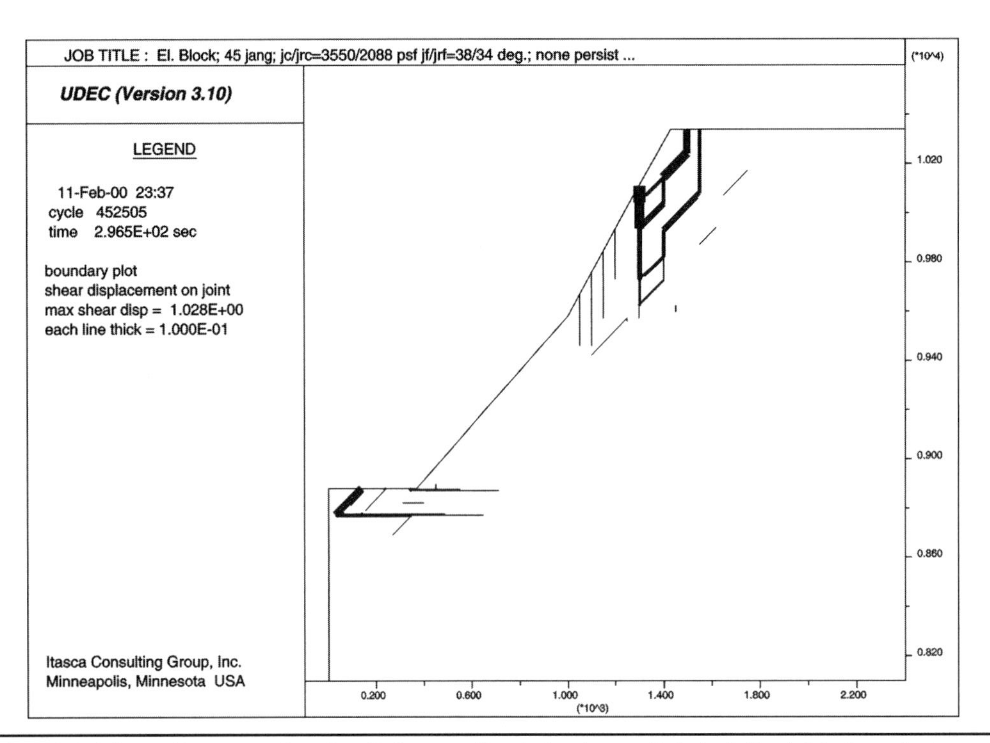

FIGURE 26.16 UDEC model of the stable proposed east wall along section 1 showing predicted relative joint shear displacements (units in feet)

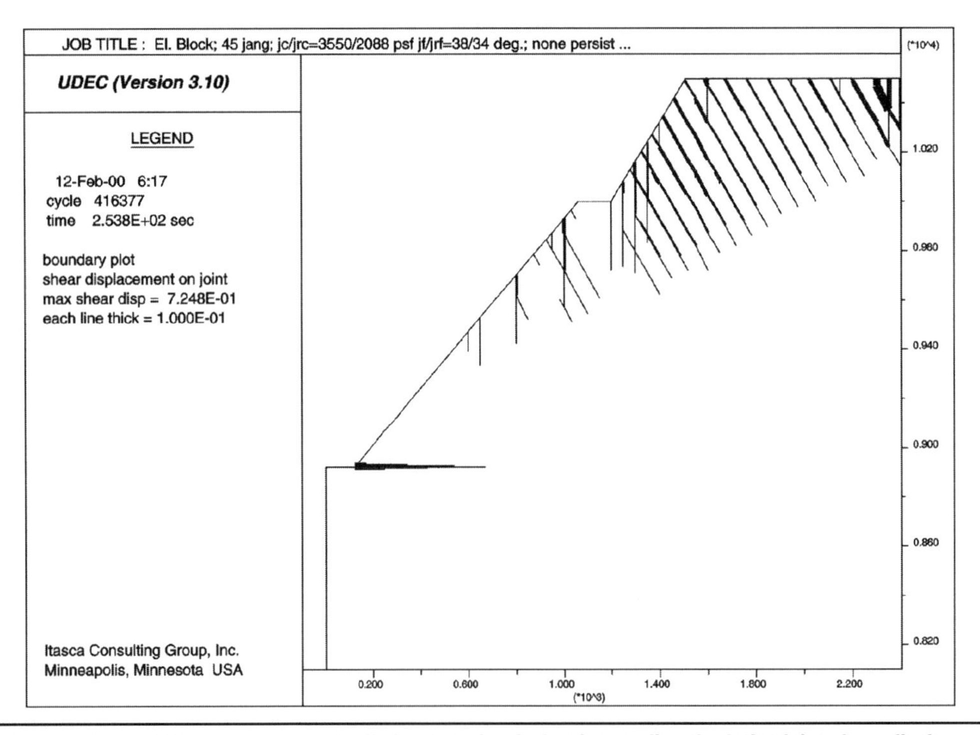

FIGURE 26.17 UDEC model of the stable proposed east wall along section 3 showing predicted relative joint shear displacements (units in feet)

Kelley, K.D., S.B. Romberger, D.W. Beatty, J.A. Pontius, L.W. Snee, H.J. Stein, and T.B. Thompson. 1998. Geochemical and geochronological constraints on the genesis of Au-Te deposits at Cripple Creek, Colorado. *Economic Geology,* 93:981–1012.

Lorig, L. 1999. Lessons learned from slope stability studies. In *FLAC and Numerical Modeling in Geomechanics (Proceedings of the Conference, Minneapolis, September 1999),* pp. 17–21. C. Detournay and R. Hart, eds. Rotterdam: A.A. Balkema.

Priest, S.D., and E.T. Brown. 1983. Probabilistic stability analysis of variable rock slopes. *Trans. Instn. Min. Metall., Section A,* 92: A1–A12.

Serafim, J. Laginha, and J. Paulino Pereira. 1983. Considerations on the geomechanical classification of Bieniawski. In *Proceedings of the International Symposium on Engineering Geology and Underground Construction (Lisbon, 1983),* Vol. 1, II.33–II.42. Lisbon: SPG/ LNEC.

Sjöberg, J. 1999. Analysis of failure mechanisms in high rock slopes. In *Proceedings of the 9th ISRM Congress on Rock Mechanics (Paris, 1999),* Vol. 1, pp. 127–130. Rotterdam: A.A. Balkema.

Thompson, T.B., A.D. Trippel, and P.C. Dewelly. 1985. Mineralized veins and breccias of the Cripple Creek district, Colorado. *Economic Geology,* 80:1669–1688.

CHAPTER 27

Analysis of Stability Loss in Open-Pit Slopes and Assessment Principles for Hard, Tectonically Stressed Rock Masses

Anatoly A. Kozyrev,* Sergei P. Reshetnyak,* Victor A. Maltsev,* and Vadim V. Rybin*

27.1 INTRODUCTION

Stability loss in hard-rock mass slopes occurs in many open pits throughout Russia and the CIS. For example, slope deformation was observed in the iron ore open pits of the Kursk magnetic anomaly (the Olkon Mining Company, Russia) and in some open pits of the Krivorozhsky deposit (Ukraine). A number of open pits mining nonferrous minerals and chemical raw materials have faced the problem of stability loss in hard- and medium-hard rocks. Slides involving 200,000 m^3 have been observed in the Koashva, Uchalinsk, and Sibaisk open pits of Russia and the Zyryanovsk (Kazakhstan), Kurgashinsk (Uzbekistan), and Mandeulsk (Georgia) open pits.

In ferrous mineral open pits, about 25% of all cases connected with slope instability occur in fractured/weathered hard and medium-hard rocks. Landslides involving sedimentary, inundated masses composed of loose rock and talus resulting from rock weathering are observed everywhere. Most stability losses on slopes in hard-rock masses are due to the intersection of weakness planes (fractures, faults, bedding surfaces, zones of fractured rocks, and others) by the slopes.

Currently, during the transition from mining the upper parts of a deposit located in sedimentary cover to mining parts of a deposit located in medium-hard or hard rocks, it is possible to improve both the technical and economic parameters of mining by determining the optimal angles for the bench and pit wall slopes.

27.2 THE PRINCIPAL OPEN PITS IN THE KOLA PENINSULA

The Murmansk region is in northwest Russia in the Kola Peninsula. It is one region of Russia with a highly developed mining industry. The area is rich in mineral and raw material resources. Over 200 deposits contain 40 different types of minerals; of these, 29 types of useful minerals are mined in more than 70 deposits. Geological surveys identify more than 200 ore occurrences as predicted reserves. Significant hydrocarbon reserves have been discovered in the Barents and Kara sea shelves.

The mineral resources of the region include profitable sources of chemical and iron ore raw materials, nonferrous and rare metals, vermiculite, phlogopite, muscovite, feldspar, and other nonmetallic minerals. The Kola Peninsula reserves are significant to the total Russian reserve base. The mineral resources of the Kola Peninsula are characterized by the complex nature of their composition.

Mineral operations have generated an effective scientific technical and industrial potential within the Kola Peninsula. This region supplies about 80% of the phosphate raw materials; half of the nickel, cobalt, and copper; a significant share of iron ore concentrates and aluminum; and 100% of the rare-metal concentrates for Russia.

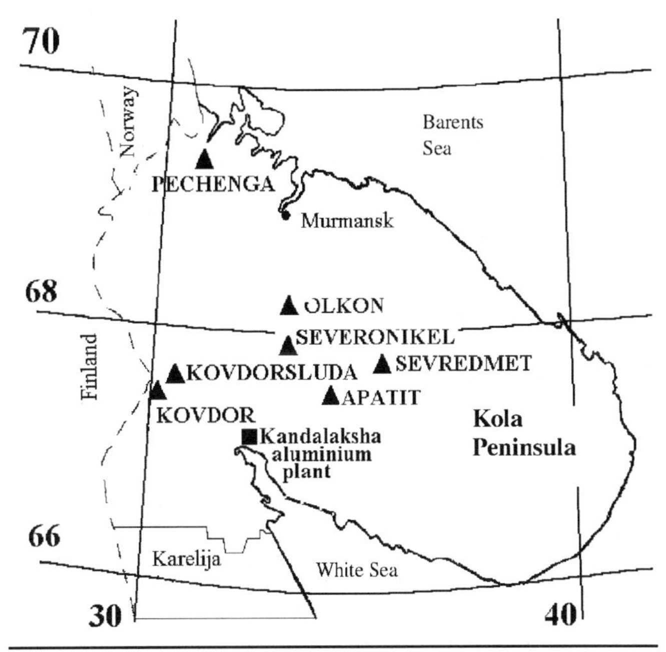

FIGURE 27.1 The mining companies in the Kola Peninsula

Presently, there are large, integrated mining and processing companies producing the iron-ore concentrates (Olkon and Kovdorsky GOK); apatite-nepheline concentrates (Apatit); loparite (rare metal) concentrates (Sevredmet); and vermiculite, muscovite, phlogopite, and olivinite (Kovdorslyuda). In addition, there are smelting enterprises (Severonikel, GMK Pechenganickel) and the Kandalaksha aluminum plant producing metallic nickel, cobalt, aluminum, and precious metal concentrates (Figure 27.1).

Six apatite-nepheline deposits in the Khibiny massif are mined by the Apatit Company. There are two underground mines, Kirovsky and Rasvumchorr, and two open pits, Tsentralny and Vostochny. The company also has two processing plants. According to standard reserve classifications, the proven reserves of the deposits currently being mined total 2.4 billion tons. Of these, more than 400 million tons can be mined by open-pit methods. The potential reserves of the explored deposits total 1.1 billion tons.

* Mining Institute, Kola Science, Russian Academy of Sciences, Apatity, Russia.

By the late 1980s, the annual ore output was 60 million tons, with the capacity of the Tsentralny Mine's open pit being as high as 28.2 million tons and that of the Koashva open pit of the Vostochny Mine being 6.7 million tons. In 1999, ore output was 18.4 million tons mined from open pits (9.4 million tons from the Tsentralny open pit and 9 million tons from the Vostochny open pit) and 10.5 million tons from underground mines. Open-pit mining accounts for 62% of the total output.

The design depth of the Tsentralny open-pit mine is more than 600 m. Presently, it is about 400 m deep. The design depth of the Koachva open pit is 350 m, although there are some prerequisites for it being as high as 570 m. The stripping-to-ore ratio is 3 to 4 tons/ton.

Four iron ore deposits, with total reserves of 147.5 million tons, as of January 1, 1997, are being mined as open pits by the Olkon Company. In the future, the company proposes to mine eight new deposits with 340.2 million tons of reserves and to mine the remaining 970.6 million tons using underground methods. From the beginning of their operation, the four operating open pits have extracted more than 75% of their reserves. The design ore output, amounting to 13 million tons/yr, has not yet been reached. At present, it is equal to 9.8 million tons.

One of the largest-capacity open pits is the Olenegorsky with a production of 14 million tons/yr. Currently, its output has been significantly reduced. The design depth of the open pit is 450 m; today, it is more than 250 m deep. The stripping-to-ore ratio is 1.5 to 2.0 tons/ton.

A complex baddeleyite-apatite-magnetite ore deposit is mined by the Kovdorsky GOK Company using open-pit methods. The deposit's reserves are 421.9 million tons, as of January 1, 1997. In addition, there are low-grade ferruginous apatite ore reserves of 90.1 million tons and a reserve apatite-staffelite ore deposit of 41.6 million tons. Since the start of mining, 45% of the reserves have been mined.

The design ore output of the Zhelezny open pit is 19 million tons/yr. Of this, 3 million tons have not been processed but rather stockpiled for use in years to come. Like other open pits, the output has been cut in about half at present. The design depth of the open pit is about 700 m. The present pit is more than 250 m deep. The overburden stripping-to-ore ratio is about 2.5 tons/ton.

The raw-materials base for the Kovdorslyuda Company are phlogopite and vermiculite deposits. Their reserves have a favorable geological setting for mining. The phlogopite and vermiculite reserves are large enough to ensure the operation of the open pit for more than 50 and 100 years, respectively.

From 1997 to 1999, the production capacity for vermiculite was 60,000 to 70,000 m^3 and for phlogopite it was 15,000 to 20,000 tons. From 2000 to 2005, they are supposed to increase the production capacity for vermiculite and phlogopite to 100,000 to 120,000 m^3 and 20,000 to 30,000 tons, respectively.

The GMK Pechenganikel Mining Company was established for nonferrous and precious metal production. It is a daughter company of the Russian company JC RAO Norilsky nickel. The raw materials are its own copper–nickel sulfide ore and, in part, high-grade ore from the Russian JC Norilsky nickel. The main product produced by the company is converter matte, which is processed at the Severonikel JC to marketable metals and sulfuric acid. The Murmansk region is second in Russia in its copper–nickel sulfide ore reserves (19.7% and 17.8%, respectively). Ten deposits with remaining reserves are registered. Of these, the Zhdanovskoye deposit (the Tsentralny open pit) is mined as an open pit and the Zapolyarny, Kotselvaara, and Semiletka deposits are mined as underground mines. Within the boundaries of the deposits, about 80% of reserves are concentrated, ensuring the operation of the GMK Rechenganikel for the next 25 years.

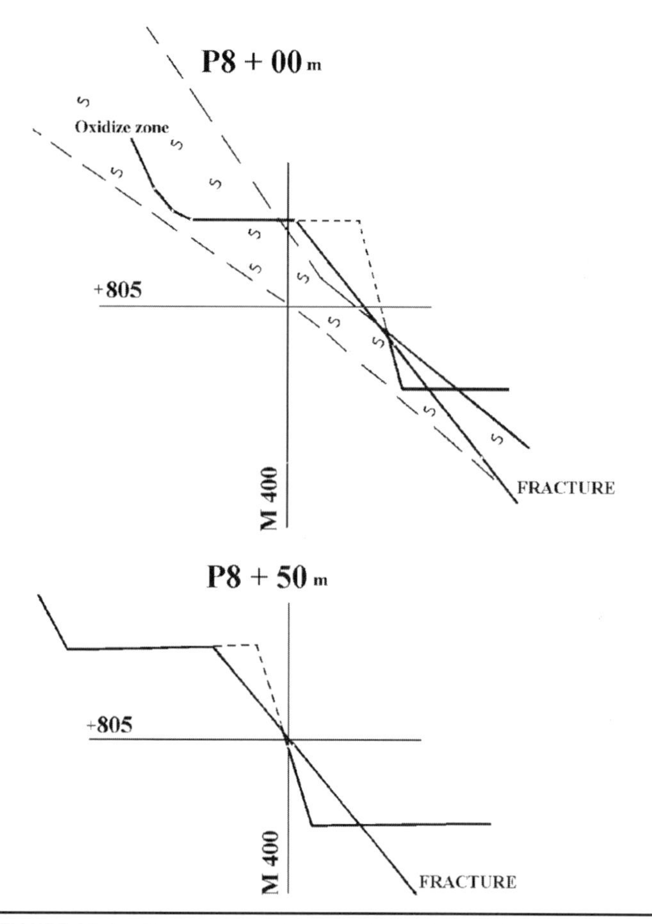

FIGURE 27.2 Cross section of the southern pit wall of the Tsentralny open pit, the Apatit Mining Company, after failure

27.3 OPEN-PIT SLOPE FAILURES

27.3.1 Tsentrany Open Pit—Apatit

The benches of the southern wall at this open pit, which is constructed along the lower contact of the orebody with the underlying footwall rocks, are inclined to have local failures. The wall is composed of hard rock–urtite. Along the contact of the orebody with the underlying rocks, however, an oxidized zone with an east-west strike dipping practically parallel to the wall at an angle of 45° to 50° occurs. The zone is 20 m thick in the upper near-surface levels and 1 to 3 m thick at the lower levels. Near the oxidized zone and parallel to it, intensive feathering/fracturing that significantly affects the bench stability is observed. The fractures have smooth surfaces and the openings are sometimes filled with hydrothermal minerals. The fractures are spaced at 0.3 to 1.0 m. This system of fractures, in combination with the other systems of steeply dipping fractures, occurs along the full length of the southern wall. Over a number of years, local failures have been observed in the benches (benches: 1000/970; 1045/1030; 820/790). All the failures are of a similar nature.

One of the most representative and recent of the slope zones was bench 820/790. The bench face angle was designed at 75°. The failure occurred simultaneously with the bulk blast, along a fracture dipping at an angle of 40° toward the open pit (Figure 27.2). The length along the front was 250 m and the thickness was on average 6 m, with the maximum being 10 m. The failure volume was 10,400 m^3. This failure was not dangerous for mining operations, causing only additional rock clearing.

It was characteristic that bench failures were developed only to a depth of 200 m, i.e., within the near-surface zone of the rock mass.

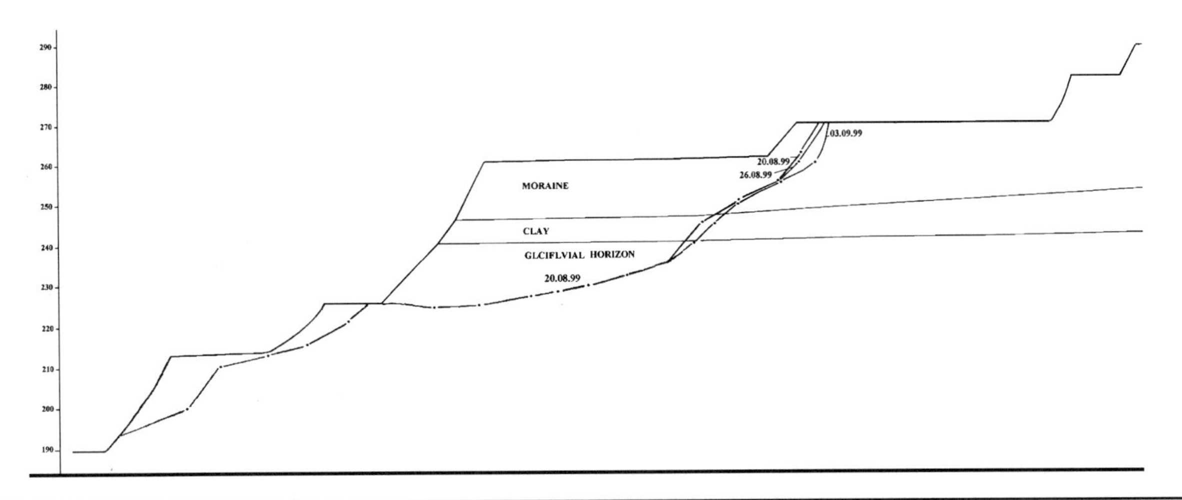

FIGURE 27.3 The moraine ground landslide in the eastern wall of the Koashva open pit of the Vostochny mine

No failures have been observed in walls or benches composed of hard rock and not undercut by unfavorably oriented fractures.

27.3.2 The Koashva Open Pit of the Vostochny Mine—Apatit

In August 1999, a failure (Figures 27.3 and 27.4) in moraine ground occurred as a landslide, 75,000 m^3 in total volume, in the eastern wall of the Koashva open pit. Prior to the failure, the actual parameters were as follows: bench between levels 260/230 m was 30 m high, slope angle was 49°, and berm was 70 m. The upper bench between levels 270/260 m was 10 m high and the slope angle was 49°. The design parameters for the benches between levels 230/270 m were different. There were four benches, each 10 m high, the slope angle was 35° and the berms were each 15 m. The service life of the benches, before the failure, was about two years.

In early August 1999, as a result of torrential rains, the water intake and flooding of the ground surface was significantly higher than normal. At the location of the future landslide, there was a groundwater stream discharge at the bottom of bench 260/230 m, which, due to rains, had been significantly increased. As a result of undermining, a failure and evacuation of about 100 m^3 of rock occurred at the bench bottom on August 5, 1999. Four days later, the main stage of the landslide began, continuing for about a month. The main part of failure lasted for about the first 10 days.

The overall dimensions of the landslide after the failure were as follows: the length along the front, 85 m; average width of development, 45 m; maximum width of development, 85 m; average value of displacement, 300 m, with maximum value of displacement, 750 m. The landslide stopped operation in an area of the open pit where stripping operations were being conducted. The distance to which the rock mass was transported has increased by 2 km. To remedy the situation, an embankment of 8,000 m^3 of rock was formed, followed by an additional 3,000 m^3 of rock to load the moraine pit wall. In addition, some water drainage boreholes were put into operation. The hard-rock pit wall was beyond the failure area.

27.3.3 Olenegorsky Open Pit—Olkon

In the Olenegorsky open pit of the Olkon Mining Company, the benches located at the footwall are subject to failure. The footwall of the orebody is a schist-gneiss complex composed of various kinds of gneisses and crystalline schists. The rocks of the metamorphic complex are interrupted by dikes and veins of different composition. Of most abundance are the granite-pegmatite, diabase, and gabbro dikes.

The dip angle of the dikes varies from 10° to 90°, and the thickness varies from a few tens of centimeters to a few tens of meters. Along some diabase dikes, as a result of secondary tectonic movements, zones of melonitization occur that significantly weaken the contact zone between the dikes and the host rocks.

FIGURE 27.4 The moraine ground landslide in the Koashva open pit of the Vostochny mine

Stability loss occurs in the form of rock blocks sliding along the underlying contacts and planes of weakness in the rock mass, resulting in unsafe mining operations. The cause of bench failure is an unfavorable orientation of their slopes (formed at an angle of 60°) relative to the contacts and planes of weakness in the rock mass (in particular the diabase dikes), most of which occur at an angle of 40° to 60°. To solve the problem of bench instability, special measures have been suggested for relieving the dynamic effect of bulk blasts on the bench. These measures are a shield (presplit) slot and a limitation on the explosive in the perimeter boreholes. Rock bolting to provide local support to the bench faces was also performed.

27.3.4 Zhelezny Open Pit—Kovdorsky GOK

The upper benches located within the morainic deposits in the northeastern pit wall at this open pit are more susceptible to increased deformation and local failure. The northeastern pit

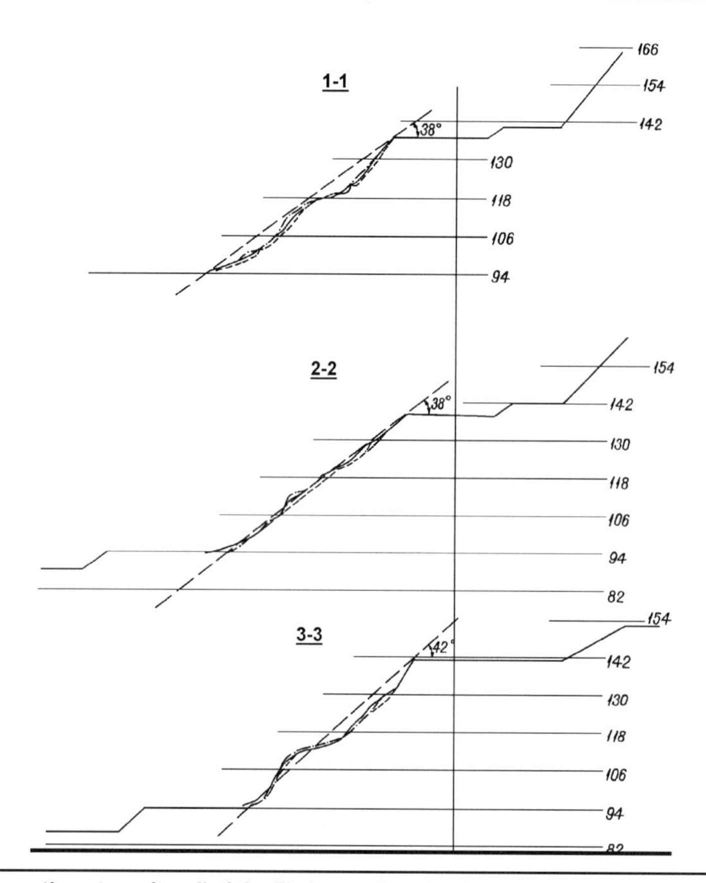

FIGURE 27.5 Cross-section of the southwestern pit wall of the Zhelezny mine, the Kovdorsky GOK, after stability loss in a set of benches

wall is characterized by a high water content resulting from a stream running too close to the open pit. Prior to mining, this stream ran through the deposit. Bench stability is also greatly affected by dynamic effects from blasting.

To prevent landslides and ensure bench stability, a decision was made to reduce the overall slope angle of the upper pit wall to 30°. This is lower than the angle of repose of the moraine. The benches located in the southwestern part of the open pit where loose apatite-staffelite ores occur also exhibit poor stability. Within this area, significant bench deformations have occurred (Figure 27.5).

The volume of the deformed mass amounts to thousands of cubic meters. It should be noted that apatite-staffelite ores are characterized by low strength (the uniaxial compressive strength is as low as 3 to 5 MPa) and are unstable, as a whole. To ensure their stability, a decision was made to give the benches a slope angle not greater than 38° to 40° and to reduce their height to 4 to 6 m.

In the eastern pit wall, just below the inpit crushing and inclined skip hoisting system, a dangerous event occurred. If the bench deformation had developed, the hoist would have been destroyed. The zone of deformation was studied and it was noted that three steeply dipping radial faults occur near this zone. These are spaced closely enough together to create a high degree of rock failure. In addition, the stability of the weakened bench was adversely affected by the blasting. At present, the zone of deformation is localized, and measures to prevent its spreading have been implemented. It is now necessary to monitor the stabilized benches and develop special techniques for constructing the underlying and adjacent benches.

In 1998, the Mining Institute, KSC RAS, initiated a program to evaluate the possibility of steepening the waste walls of the Zhalenzy open pit operated by the Kovdorsky GOK Mining Company. This would be accomplished by creating benches with vertical slopes instead of the design ones of 70°. In 1999, benches with vertical slopes extend to a length of about 300 m. In the year 2000, they plan to substantially increase the volume of similar operations. In the future, they plan to test the possibility of increasing the final wall slope angles by several degrees while preserving the safety of mining operations.

27.4 ANALYSIS OF STABILITY LOSS IN OPEN-PIT SLOPES

The situations related to stability loss in open-pit slopes are summarized in Table 27.1.

Despite the different mining, geologic, and geomechanical settings different open pits and in the different zones of instability, one can describe the most representative forms of stability loss. Of the five cases connected with stability loss, two were landslides and three were failures of bench sets. The main reasons the landslides occurred within the moraine deposits were: groundwater, undermining, and the formation of benches with slope angles greater than that of the angle of repose of loose rocks. The main reasons for collapse of the bench sets in hard and medium-hard rocks were: bench slopes being intersected by weakness planes, bench locations within weakened zones, and the dynamic affects of blasting.

27.5 ASSESSMENT PRINCIPLES FOR OPEN-PIT SLOPE STABILITY

It follows from the examples given above that the mechanism of stability loss in slopes composed of loose sedimentary rocks differs from that in slopes composed of strong, medium-hard and hard rocks. Stability loss in slopes within a sedimentary mass occurs, as a rule, as landslides, i.e., as a slow movement of highly saturated rock masses. Slope stability can be calculated in this case based on the theory of limiting equilibrium using the Mohr–Coulomb failure criterion. The design characteristics for rocks in

TABLE 27.1 The loss of stability situations in open-pit slopes

Place	Brief Characterization of the Open-Pit Zone	Types of Stability Loss	Most Probable Reason	Measures Taken
Tsentralny open pit, Apatit Mining Company; southern wall	Upper bench 30-m-high slope angle at 75°	Failure of a set of benches	Bench undercutting by unfavorably oriented planes of weakness in a rock mass	Only cleaning of failed rock
Koashva open pit, Apatit Mining Company; eastern wall	Upper bench 30-m-high slope angle at 50°, within loose morainic deposits	Landslide	Moraine ground with water undermining; bench face angle larger than design	Loading the pit wall with spoil-bank rocks; additional measures on water-level lowering
Olenegorsk open pit, Olkon Mining Company	The zone of a pit wall in the final pushback from the foot wall of the deposit within the schist-gneiss complex rocks; contacts between rocks of 40° to 60° in dip; bench face angles of 60°	Failure of a set of benches	Bench undercutting by unfavorably oriented planes of weakness in a rock mass	Local stabilization of benches; reduce the dynamic effect of blasting on the rock mass
Zhelezny open pit, Kovdorsky GOK Mining Company; northeastern pit wall	Upper benches with a slope angle of 40° within the loose morainic deposits	Landslide	Moraine groundwater undermining; bench slope angle larger than the design; adverse, dynamic effect produced by blasting in the rock mass	Formation of the pit wall with an slope angle not greater than 30°
Zhelezny open pit, Kovdorsky GOK Mining Company; southeastern pit wall	The pit-wall zone within apatite-staffelite rocks	Failure of a set of benches	Location of the benches within the weakened zone	Formation of the benches with a slope angle not greater than 38° to 40°; decreasing the height of benches to 4 to 6 m
Zhelezny open pit, Kovdorsky GOK Mining Company; eastern pit wall	The zone of the pit wall is within carbonatite rocks crossed by several extended structural dislocations	Failure of a set of benches	Location of the benches within the weakened zone of failed rocks	Decrease the dynamic affect of blasting on the rock mass

this case are cohesion (C) and the angle of internal friction (φ). These parameters are determined using Mohr's circle representation of stress. Mining practice in such rock masses shows that, under favorable conditions, an overhung slope (slope angle greater than 90°) can be formed even at its rather significant height. For instance, as a result of underground mining in the zone of failure at the Rasvumchorr Mine (the Apatit Mining Company), a 200-m-high rock wall with an overhung slope has formed. It has remained stable for about 10 years. This example confirms the fact that the main reason for stability loss in rock slopes is the presence of unfavorably oriented planes of weakness in the rock mass.

In assessing the stability of open-pit slopes, it is necessary to distinguish between "stability loss in a pit wall" and "stability loss in a bench or a set of benches." It is also necessary to distinguish the fractured near-surface zone of a rock mass whose rock properties influence the stability of a bench and the deep hard-rock mass whose strength properties and stress state influence the stability of the pit wall as a whole.

Stability loss in a pit wall is the failure in the pit wall structure due to rock blocks sliding relative to each other, both along existing rupture fractures and those that are newly formed in the inner zones of the rock mass. This results in the formation of a new, much gentler slope angle of the pit wall.

Stability loss in a bench (a set of benches) is the failure in the bench structure (a set of benches) manifested in rock block fall outs along planes of weakness, taluses, and other events characteristic of the surface zones. The slope angle of the pit wall itself is unchanged.

The *failed near-surface zone* is the zone within which a rock mass is in a discrete (divided into blocks) state. Its fracturing exerts the main influence on stability of some rock-mass zones within this area. The weathering properties, dynamic effect on the rock mass exerted by blasting, and the level of the actual stresses in the rock mass determine the thickness of the failed zone. The most dangerous final stage of stability loss in the open-pit slope (pit wall as a whole or a separate bench) is rock block sliding. For rock block sliding to occur, a plane of weakness in the rock mass that is a potential sliding surface must be present. For rock blocks to slide, it is enough to have an excess of sliding forces over the retaining ones on the plane of weakness in the rock mass.

The plane of weakness in a rock mass can be of a natural or synthetic origin. In the first case, its parameters depend on the geologic structure of a rock mass and the basic design characteristics being the coefficient of sliding friction and the coefficient of static friction expressed through the cohesion and the angle of internal friction for the plane of weakness. In the second case, the parameters for the plane of weakness in the rock mass depend on the stress state near the open-pit excavation. This can be determined by applying the strain criterion of failure, according to which the failure occurs by rupture when one of the components of the principal strains (ε_1, ε_2, ε_3) reaches the limiting tensile strain of the rocks (ε_t): $\varepsilon_i > \varepsilon_t$.

For the three-dimensional case, the principal strains in a rock mass are determined from the formulae:

$$\varepsilon_1 = (\sigma_1 - \nu(\sigma_2 + \sigma_3))/E$$

$$\varepsilon_2 = (\sigma_2 - \nu(\sigma_1 + \sigma_3))/E$$

$$\varepsilon_3 = (\sigma_3 - \nu(\sigma_1 + \sigma 2))/E$$

where σ_1, σ_2, and σ_3 are the components of principal stresses and E and ν are the Young's modulus and Poisson's ratio, respectively.

The synthetic plane of weakness in the rock mass is oriented in the plane that is parallel to two components of the principal stresses (σ_2, σ_3) and, correspondingly, perpendicular to the largest component (σ_1).

The formation of a synthetic plane of weakness in a rock mass is possible only under significant absolute values of actual principal stress components and corresponding strains. This is possible at a shallow depth during open-pit mining in tectonically stressed rock masses.

Slope stability insurance is gained at the expense of maximum preservation of the rock mass beyond the pit limits, minimization of the thickness of the failed zone, and their local consolidation, if necessary to minimize the thickness of the failed zone. This is only possible by decreasing the dynamic effect exerted beyond the perimeter. This is possible by applying special smooth blasting schemes when the pit wall is at its final limit.

27.6 CONCLUSIONS

In the open pits of the Kola Peninsula, the most significant cases of stability loss in open-pit slopes known at present are either landslides occurring within moraine deposits, or failures of a bench (or a set of benches) induced by blocks sliding along the unfavorably oriented extended planes of weakness in the rock mass.

Stability loss in pit walls formed in unmined rock masses was not observed. In deep open pits in hard rock, fracture zones form in the walls beyond the limits of near-surface zones. Stability must be assessed on the basis that rock block sliding along unfavorably oriented weakness planes of either a natural or synthetic origin must be avoided. The main design characteristics of the planes of weakness are the coefficient of sliding friction and the coefficient of static friction. One way for the synthetic planes of weakness to be assessed is by using the strain failure criterion. Further investigations into the degree of open-pit slope stability in rock masses should focus on identifying the extended, unfavorably oriented planes of weakness both in operating open pits and in those under construction and to analyzing the stress–strain state in the vicinity of the existing open-pit excavations.

27.7 ACKNOWLEDGMENTS

The authors are very grateful to the engineers and technical personnel of the mining companies for the materials provided.

CHAPTER 28

Coal Mine Highwall Stability

Ben L. Seegmiller*

28.1 INTRODUCTION

The stability of coal mine highwalls is addressed in this chapter. The intent of the chapter is not to present a primer on how to collect stability data or how to perform stability analyses. Methodologies for developing stability data and producing stability criteria (i.e., safety factors and failure probability) are addressed in other rock slope design chapters. The intent of this chapter is to illustrate what makes coal mine highwall stability unique and how it contrasts with hard-rock mine slopes.

In a word, many coal mine highwalls are simply not as complex as their hard-rock counterparts. The geology is generally much more straightforward. The discontinuities are more easily defined and may not play as significant a role as they do in a metal mine. The rock/rock mass strength is somewhat similar from mine to mine because the same rock types are usually found. In most cases, the rock/rock mass strength is much less than that encountered in hard-rock deposits composed of massive carbonates, quartzites, igneous lithologies, or metamorphic rocks. Groundwater pressurization parameters are not as variable due to the more simplified geologies encountered above the coal seam. An audit of surface coal operations in the western United States reveals that, with the exception of rock falls, more than 80 percent of the mines have few, if any, significant highwall stability problems.

We will first examine stability elements and describe their relationship and importance to highwall instability. Stability evaluations are then reviewed and the most significant failure modes discussed. Methods of improving the stability of a coal mine highwall are then outlined. The chapter concludes by summarizing a variety of actual problem examples and related solutions. These examples are presented for flat or gently dipping seams, moderately dipping seams, and steeply dipping and highly variable seams.

28.2 STABILITY ELEMENTS

28.2.1 Overview

Existing surface coal mines throughout the world show that at a particular mine only a few stability elements may be significant. This is in contrast to metal or hard-rock surface mines where many stability elements may become critical simultaneously. The geotechnical nature of surface coal deposits is, in many cases, one of simplicity. The geology is generally very straightforward and may only consist of a sequence of relatively flat layers of rock and coal. Exceptions exist to this generalization because some coal deposits are very steeply dipping, have many fault discontinuities, and are dramatically folded. These exceptions, however, constitute only a small percentage of the actively mined surface coal deposits in the world. Stability elements, which commonly affect coal mine highwalls, include the mine plan geometry, geologic discontinuities/structure, rock/rock-mass shear strength, groundwater, and seismic acceleration.

28.2.2 Mine Plan Geometry

The highwall geometry of a surface coal mine may be dictated by a number of factors including seam depth, seam dip, mining method, and mine boundary limits. Flat or gently dipping seams with cover of less than about 15 m (50 ft) usually do not have highwalls with stability concerns. Even when several seams are mined in sequence, a significant highwall may not be developed. This is particularly true when the mining method is by truck and shovel. For seams with thick cover, on the order of 30 m (100 ft) or more, a significant highwall is usually developed. Where draglines are used and cover thicknesses are approaching the maximum economic limits, very significant highwalls may be developed. Where coal deposits have moderately steep dips, in the range of 10° to 25°, it is usually necessary to mine down-dip to avoid undercutting potentially unstable bedding planes. In these cases, a very significant highwall may be developed and its stability can be of great concern. In very steeply dipping seams, with the removal of the coal, it may be difficult to avoid undercutting the bedding planes or structure developed during seam tilting and folding. Such cases may require mining methods and procedures similar to those used in hard-rock metal mines. Surface coal mining with bucket wheel excavators may eliminate the need for major highwalls in many cases, except near mine boundary areas. In these cases, as in any case where a mine boundary is approached, a significant highwall may be developed, and related stability concerns should be addressed. In summary, the mine plan geometry is dictated by a number of variables relating to the seam characteristics, mining method, and property boundaries. The results may be the development of highwalls that are up to 61 m (200 ft) or more in vertical height. The acceptable stable slope angle and slope profile at which these highwalls may be mined becomes a most-important stability element.

28.2.3 Geologic Discontinuities/Structure

This stability element includes all naturally occurring two-dimensional geologic features such as faults, bedding, joints, cleats, and rock contacts. In addition, dikes/sills and folds are considered either discontinuities or structure. These elements are important because they control or usually strongly influence the potential highwall failure mode. Based on the frequency of coal mine highwall instabilities, the most important geologic discontinuities may be the bedding planes and their dip magnitude. This is particularly true for moderate to steeply dipping or highly variable seams. The occurrence of faults in a flat seam deposit may not present any stability problems, as shown in Figure 28.1. In moderately dipping seams, steeply dipping seams, or seams that are highly variable and folded, faults may completely dictate the stability. Joints, cleats, and rock contacts are generally of much

* Principal Consultant, Seegmiller International, Salt Lake City, Utah.

FIGURE 28.1 Steeply dipping fault displacing a 0.3-m (1-ft) coal seam approximately 3 m (10 ft) downward to the right in a mine highwall

less importance and may have only minor influence on stability. Dikes/sills do not often occur in coal deposits, but when they do, they may influence stability adjacent to their highwall location.

28.2.4 Rock/Rock-Mass Shear Strength

Both the rock and rock-mass shear strengths are much lower in coal deposits than those found in hard-rock mines. The rock types are generally limited to sandstones, siltstones, claystones, and the coal seam. In some deposits, limestone or even rock of igneous origin may be present. However, in the majority of cases, sands, silt, and clays form the highwall and any interburden rock units. In some Tertiary age coals, the overburden or cover materials may be more accurately described as stiff soils rather than rocks. Bedding shear strength for moderate to steep or highly variable seam dips may be the most important strength magnitude for surface coal mines. If a coal deposit has a dip in excess of about 10°, shear strength in the bedding direction may completely dictate the ability to mine in an updip direction. Strikeline mining in an updip direction, with spoiling on the downside, is economically very beneficial in most cases. An additional advantage to updip mining includes good drainage from the mining area at the highwall toe. Unless the seam has a very steep dip or is highly variable due to folding, the shear strength along faults and joints may be of only minor importance.

28.2.5 Groundwater

For flat or gently dipping seams, located in wet environments, groundwater may be the only natural stability variable impacting highwall stability. Groundwater seepage from the base of a major sandstone channel is shown in Figure 28.2. For moderately to steeply dipping seams or highly variable seams, groundwater effects will probably aggravate already-existing stability problems. On the other hand, for many surface mines such as those found in the western United States in semiarid environments,

FIGURE 28.2 Groundwater seepage from the base of a wide sandstone channel drains down the highwall face

groundwater may be completely absent or may affect only limited highwall sectors. The removal of groundwater for operational considerations may have positive benefits relative to highwall stability. That is, mines may dewater so that equipment operation and efficiency can be improved. Not only is groundwater removed from the working areas, but also piezometric levels are lowered in the pit walls, effectively improving stability. Depressurization to improve stability is usually found to be a more viable economic solution than reducing the highwall slope angle to maintain needed stability. Consequently, most surface coal mines make a major effort to remove groundwater to improve both equipment operation and highwall stability.

28.2.6 Seismic Considerations

The affects of earthquakes on highwall stability need only be considered where there is significant seismic potential. In most surface coal mines located in North Dakota or Texas, the seismic potential is negligible (UBC 1991). In other US states, including Colorado, Arizona, northwest New Mexico, Wyoming, Montana, much of the Midwest, and Appalachia, the seismic potential is low. For coal mines located in the Pacific Northwest and portions of southern Illinois, there is enough significant earthquake activity such that a seismic coefficient of at least 0.2 may be necessitated for highwall design. Coal mine operations in other countries need to consider the seismic coefficient applicable to that region when designing the highwall profile.

28.3 STABILITY EVALUATION

28.3.1 Trial and Error

Owing to the fact that many surface coal mines, particularly truck-and-shovel operations, may have limited major highwall exposures, the design of highwall slope angles and profiles may be done by trial and error. That is, these mines do not follow a geotechnical program to determine their slope profiles. They simply pick a profile convenient for mining purposes and proceed with it. Mines, which advance on the strikeline and place the spoil behind them, may create significant endwalls. Again, as in the case of highwalls, the endwall slope angles may be selected based on experience and convenience and not determined by

FIGURE 28.3 Planar mode failure of soft, weak materials in a mine footwall dipping at 20°

rigorous geotechnical evaluation. Although it is usually desirable to have steep endwalls to maximize resource recovery, trial and error may be the only design method employed. If an operation has repeated stability problems or if the highwall or endwall is near a permanent structure that needs protection, a detailed geotechnical analysis may be used for slope angle design. Otherwise, many surface coal mines use trial and error as their basic design methodology.

28.3.2 Planar

This failure mode most commonly occurs in moderate to steep-dipping seams where the bedding planes are undercut. It also may occur in folded or highly variable seams, and it almost always involves the undercutting of geologic discontinuities. In some flat or gently dipping deposits, a geologic anomaly may unexpectedly cause the seam dip to increase from a few degrees to more than 20°. In such cases, planar mode failure may abruptly occur as mining removes the in-place material buttressing the anomaly toe. Mines that have moderately low seam dips, in the range of about 8° to 15°, may or may not have planar mode failure depending on the sliding friction angle magnitude in the bedding plane direction. A seam dipping steeper than about 20° will likely have planar displacement when undercut. A seam with a dip of about 6° or less will likely be stable. Consequently, a geotechnical program should be undertaken if the seam dip is in the range of 7° to 20° and seam undercutting is planned as part of the mining procedure. Planar mode failure of soft, weak materials in a mine footwall dipping at 20° is demonstrated in Figure 28.3.

28.3.3 Rotational Shear

The rotational shear failure mode is the most common failure mode to occur in flat or gently dipping seams in wet environments. Unless multiple seams exist, the economic surface mining depth, in many cases, may be up to about 61 m (200 ft). The overburden cover usually has enough shear strength to prevent slope failure, even for highwall slope angles of 65° to 70°, unless groundwater is present. Depending on the shear strength, the groundwater pressurization, and the success of groundwater removal, a highwall or endwall slope angle may have to be significantly reduced to maintain acceptable stability. A highwall or endwall located adjacent to a permanent structure, such as a highway, should have a detailed stability evaluation performed if groundwater is present. The slumping nature of a rotational shear failure impacted by groundwater is shown in Figure 28.4.

FIGURE 28.4 Slumping nature of a rotational shear failure containing groundwater

28.3.4 Other Failure Modes

A number of other highwall failure modes exist, including wedge, toppling, bilinear, ploughing, and buckling. For flat or gently dipping seams, these failure modes would not be very common, although they can occur in isolated or unusual circumstances. Moderate, steep-dipping, or highly variable seams would have a greater frequency of these failure modes. This is due to the greater number of associated discontinuities occurring in tilted or folded seams.

FIGURE 28.5 Smooth highwalls are obtained using proper presplit blasting

FIGURE 28.6 Rockfalls can be a safety hazard as well as a dilution problem

28.4 STABILITY IMPROVEMENT METHODS

28.4.1 Mining Sequence/Method Change

When a highwall instability problem develops, one method that may remedy the situation is to modify the planned mining sequence or mining method. Such changes could include a slope angle change or a highwall slope strike change. Another example of a mine plan change would be to mine downdip if updip mining creates unstable highwall sectors in a moderately dipping seam. A radical mine method change to improve highwall stability could be a change from dragline mining to truck-and-shovel mining if the conditions were warranted.

28.4.2 Dewatering/Depressurization

The removal of groundwater from a highwall may be the quickest and most effective way to prevent further slope displacements. Highwall stabilization can generally be achieved by depressurizing of the slope materials. Vertical wells are the most common method of removing groundwater from surface coal mines. Depending on the overburden rock types and their permeabilities, vertical wells may be very effective, particularly where sandstones and siltstones are encountered. In other cases, the use of horizontal drains may be very beneficial. They can be placed to intercept the zone just above an existing or potential failure plane.

28.4.3 Blast-Damage Reduction

The improvement of near-surface rock-mass integrity may be done by lessening the damage caused by blasting. Methods to do this include reducing the charge weight per delay on blasts adjacent to the highwall. This reduction is commonly done by using more delays in the blast pattern. In certain cases, each single hole may be individually detonated. Usual maximum particle velocities are 635 mm/sec (25 in./sec) but may need to be as low as 380 mm/sec (15 in./sec) in very weak, blocky materials. Another blast-damage-reduction technique is to use presplit blasting. In some mines, this technique is used to prevent damage from blast casting detonations. Holes are drilled down the highwall and detonated simultaneously to produce a smooth rock face. The method works quite well in highwalls having a predominance of sandstone/siltstone. An example of excellent presplit blasting in a Wyoming mine is shown in Figure 28.5.

28.4.4 Buttressing

A natural in-place rock buttress or a buttress composed of overburden materials can be effective in preventing or minimizing a slope instability. In moderately dipping seams there may be planar mode failure developed in the bedding plane direction. An undisturbed block of overburden and coal may be left to prevent down-dip creep of weaker materials. Similarly, a buttress could be used to prevent down dip movement where there are folded seams. A highwall failing in rotational shear may be stabilized by placing a rock buttress at the highwall toe.

28.4.5 Artificial Support

The use of artificial support is always an intriguing proposition at a surface coal mine. However, only in isolated cases can the economics justify the use of expensive support devices. Nevertheless, artificial support may be effective and economically viable under certain circumstances. Among these circumstances may be those where a not-to-be-disturbed permanent structure exists near the mining area. A rock bolt and wire mesh scheme placed on a portion of the highwall may be justified to prevent any possible slope displacement. In most circumstances, the economic loss of the coal resource would be much less than the artificial support cost. In the downdip mining of moderately steep seams, the placement of low-cost, used, wire-rope dowels to improve footwall stability may be justified. In very steep downdip mining, the use of rock anchors, rock bolts, and mesh may be economically viable.

28.4.6 Rockfall Reduction/Elimination

Rockfalls may be a concern in any surface coal mine but are most likely to be a problem in pits with moderately strong to strong overburden rocks and slope heights greater than about 30 m (100 ft). Rockfall debris from very weak near-surface materials can be a safety hazard and a dilution problem, as shown in Figure 28.6. Rockfall problems may be lessened using a variety of methods. The most common methods are slope geometry modification, blast damage control, use of catch benches, and scaling.

Slope geometry modifications may include a change in the slope strike direction as well as a change in the dip magnitude. Such changes may help to eliminate undercutting joints striking approximately parallel to the highwall face. Blast-damage control through the use of reduced charge weight per delay and/or modified presplit/buffer blasting may be very effective. The addition of catch benches in very high slopes serves to protect personnel and equipment working in the lower pit levels. Use of scaling procedures can literally remove loose potential rockfalls. In many surface mines, one or more of these remedial methods are commonly used where problems exist or may potentially exist.

28.4.7 Displacement Monitoring

Monitoring of potential or actual slope displacements helps to improve stability conditions by allowing mine personnel to know if a highwall is displacing, the surface limits of displacement, the direction of movement, the rate of movement, the mode of failure, whether or not catastrophic failure may occur, and when such potential catastrophic failure may occur. For these reasons, all significant planar and rotational shear failures should be monitored. The most effective monitoring method is the use of precision surveying equipment to monitor a set of fixed survey points. Data collection and handling techniques, as well as methods to interpret the meaning of the data relative to stability, are available in the literature (Seegmiller 1973).

28.5 STABILITY PROBLEMS AND SOLUTIONS

28.5.1 Overview

When a highwall stability problem develops, the mine operator needs an immediate solution. What can be done and what success have other mining operations had when they have faced similar problems? To provide some help in answering these questions, a series of actual mine problems and solutions are presented in the following paragraphs. Where possible, the actual mine names and locations are given. When the mine cannot be identified because permission could not be obtained to do so, a general geographical location is used. The described problems and solutions are presented in three groups. The first is for flat or gently dipping seams (dips up to 8°). The second is for moderately dipping seams (dips between 8° and 30°). The third grouping is for steeply dipping and highly variable seams (dips in excess of 30°).

28.5.2 Flat or Gently Dipping Seams

Classical Rotational Shear Failure

United States—Wyoming, Powder River Basin. In excess of 15 million metric tons of bituminous coal are produced annually at this mine using truck-and-shovel methods. The coal occurs in the Fort Union Formation of Paleocene age. Its thickness is approximately 30 m (100 ft) and it dips to the southwest at about 2°. Overburden is composed of poorly indurated undifferentiated sandstones, siltstones, and claystones. A main highway route passes within approximately 61 m (200 ft) of the pit highwall, which is composed, in part, of a saturated sand channel. The basic problem centered around the fact that the mining company could not afford, from a public relations standpoint, to allow any displacements to occur in the highway as a result of potential highwall instability. Consequently, a geotechnical investigation of stability variables, including highwall shear strength, groundwater pressurization, and blast vibration, was undertaken. The study concluded that slopes above the saturated sand channel could safely be mined at 30°, while slopes below the water table should be excavated at approximately 18°. Dewatering wells and survey monitoring points were placed between the highway and the slope crest. The slope angle design and remedial depressurization efforts have proven effective in that no displacements have been detected in the highway or slope crest area.

Germany—Rhenish Brown Coal District. This district contains the largest brown coal deposit in Europe. Mining is done using bucket-wheel excavators to remove lower strength overburden materials consisting of gravels, sands, silts, and clays of Tertiary and Quaternary ages. Generally, two brown coal seams are mined with mining depths ranging from 320 to 500 m (1,050 to 1,640 ft). Strata dips are flat to gently dipping in most cases. Many major angled faults dipping about 50° to 60° cut through the deposit and have offsets of 10 to 20 m (33 to 66 ft). Groundwater is generally below the coal, except where trapped near faults or in synclines. According to investigators (Pierschke 1979), the most significant stability problem has been rotational shear failure in long-term slopes. Some slopes were cut too steep and failed and others were cut too flat and the stripping ratio was not optimal. The solution to the problem was to undertake stability investigations, particularly to determine the effective shear strengths. Detailed stability analyses were then conducted using standard limit-equilibrium methods, including Janbu and Bishop. A backfilling program was implemented to decrease the time a given slope needed to stand. The results were very good, indicating considerable improvement in the stripping ratio and many less slope failures.

United States—Northern Great Plains and Gulf Coast Lignite. Lignite mines typically have Tertiary-age weak overburden composed of unconsolidated sands, silts, and clays varying in thickness from 12 to 45 m (40 to 148 ft). Substantial groundwater is usually present, and surface runoff is high, with typical annual precipitation varying from 1,016 to 2,032 mm (40 to 80 in.). Mining is usually done by dragline. Rotational shear failures in the highwalls result in substantial coal revenue losses. Solutions to these problems have been developed (Calder and Workman 1982; Sutphin 1982; Bhattacharyya, Vakili and Chi 1982; Jaworski and Zook 1978) including slope angle guidelines for different overburden material types. Standard limit-equilibrium methods were applied using material properties developed by vane shear tests in boreholes, field scale testing, and standard laboratory tests. Adverse groundwater pressurization is lessened by vertical well dewatering and surface runoff diversion. Improved stability has resulted in most operations where dewatering has been undertaken and the design guidelines have been used. The design slope angles have varied up to 65°, but could be as low as 45° depending on slope height and material type.

Block and Structural Failure

Canada—Alberta, Plains Region. At a dragline operation west of Edmonton, six approximately flat coal seams have been mined (Brawner 1989). Bentonitic layers, having a low friction angle of only 10° to 12°, are found in the sand, silt, and clay overburden. Tension cracks formed behind the highwall crest due to mining stress relief. When groundwater from infiltration or surface runoff entered the cracks, block glide failure occurred into the pit. Remedial measures have included drilling to locate the bentonitic layers, controlling surface water flow, and dewatering with vertical wells. In addition, it has been found beneficial to raise the dragline bench elevation and thereby increase the vertical load on the bentonitic layers. The greater load increases the shear strength, as per the Mohr–Coulomb failure criterion. The results have been fewer block glide failures.

United States—New Mexico, Gallup Coal Field. Some 7 million metric tons of coal are annually mined by both dragline and truck-and-shovel at this mine. Sandstone and shale of Late Cretaceous age form the overburden and interburden rock units. As many as 11 different seams are mined. Highwalls average about 40 m (130 ft) in height but may reach 61 m (200 ft). Many faults cross the property, particularly a major set striking northwest-southeast. These faults dip about 60° to 70° in most cases but could dip as low as 20° to 30°. The layout of the highwalls had originally been in the northwest-southeast direction and many slope failures occurred as a result of faults being approximately parallel to the highwalls. The solution to these structural failures was to completely reorient the highwalls to the northeast-southwest direction so that the faults would intersect the pit slopes at high angles. The results have been that the high frequency of major slope failures has been dramatically reduced, providing improved safety and a reduction in coal loss.

Australia—Morwell Brown Coal Project. More than 100 billion metric tons of brown coal reserves are known to exist in the Latrobe Valley, Victoria. These Tertiary deposits have overburden consisting of sands, silts, and clays. Coal thicknesses vary from 107 to 137 m (350 to 450 ft) and mining is done by bucket-wheel and ladder excavators. A low-strength clay layer is typically found beneath the coal seam. The main instability problem has been the block glide of coal into the pit as a result of excessive groundwater pressures behind the exposed coal face (Hutchings 1979). Cleats in the coal and stress relief cracks provide access for the groundwater to percolate into the slope walls. With water being the most significant factor in slope stability, remedial action has principally involved diverting of surface runoff and reducing pressurization through drainage. Horizontal drains placed in the slope toe area have been extensively used. These drains have been between 200 and 300 m (656 and 985 ft) long and spaced 50 to 70 m (164 to 230 ft) apart. The drain holes are lined with 100-mm (4-in.) slotted PVC pipe. The success of the remedial drainage measures has been manifested by improved safety, and a reliable and orderly mine operation has been possible.

Slumping and Rock Falls

United States—Wyoming, Powder River Basin. Several mines in the central south portion of this coal field have highwalls on the order of 70 to 75 m (230 to 246 ft) high and slope angles of 63° to 70°. Coal forms the lower 20 m (66 ft) of the highwall and is overlain, in part, by a weak and permeable sandstone. The sandstone transmits groundwater from beyond the crest to the highwall face, where it slowly discharges into the pit. As the water exits the highwall sandstone face, it slowly erodes the weak sand materials from the highwall, creating an overhang in the dry strata above. Owing to weakness planes created by semivertical joints, the overlying strata fail into the pit, usually in a dribbling fashion. The problems created by this failure mode are safety and coal dilution. The solution has been to dewater behind the highwall. Haul roads passing the undercut failures are coned off, in part, to increase safety. Complete effectiveness of the remedial action is not always achieved due to difficulties in groundwater removal. If significant problems still occur, even after dewatering, the highwall toe area must be permanently bermed off to prevent injuries and equipment damage.

United States—Colorado, Yampa Coal Field. Some eight seams are being mined at this operation where Upper Cretaceous Williams Fork overburden is removed by dragline. An annual tonnage of approximately 5.5 million metric tons of high-quality coal is mined. Highwalls range from about 107 to 137 m (350 to 450 ft) in height and slope angles are cut at 70°. Typical overburden rock types are sandstones, siltstones, and claystones. Mining is above the groundwater table and only an occasional perched water table is found. Few, if any, faults exist, but the strata have minor rolls and some sandstone channels. Slope stability is excellent at this operation and the only concern has been rockfalls from the high slopes. To minimize rockfalls, all highwall faces are presplit and one or two catch benches are placed on the highwalls. Dozers are also used to scale loose rocks from the slopes during stripping.

Rockfalls had not caused any injuries or equipment damage, but concern by regulatory authorities that the catch bench widths were not great enough prompted the mining company to further study the situation. A study using the Colorado Rockfall Simulation Program (CRSP) was undertaken to evaluate acceptable minimum catch bench widths. This program (Pfeiffer et al. 1993) allows a detailed simulation of rockfalls to be made for varying catch bench widths and inclinations (i.e., a flat bench or one with a slight angled outslope). For this mine, bench widths of 6.1 m (20 ft), 7.6 m (25 ft), and 9.2 (30 ft) were evaluated. The analyses showed that two 9.2-m (30-ft) wide benches placed 31 m (100 ft) and 56 m (185 ft) from the base of a 81-m (265-ft) slope would catch all rocks on either the higher bench or the lower bench. No rocks would reach the highwall toe. Implementation of the 9.2-m (30-ft) wide benches has been done and the results are that the regulatory agency is satisfied and safety is improved.

United States—New Mexico, San Juan Basin. Mines in this coal field are located in the Cretaceous age Fruitland Formation. Vertical heights of highwalls vary from about 31 m (100 ft) to as high as 62 m (200 ft) and slope angles are about 65° to 70°. One catch bench is usually placed on the higher slopes. Major faults are few but they do occur and can adversely impact highwall stability. Overburden is composed of sands, silts, and clays. Sandstone channels are not uncommon. The sands and silts are stronger than the clays, which may be quite weak. Groundwater is essentially absent and surface water is kept out of the pits using berms. Slope stability is generally good, except in the claystones where differential compaction and related "slicks" are common near the margins of sandstone channels. Such zones are unstable due to the slicks intersecting other slicks or other discontinuities. Failure can occur in planar or wedge modes. The result is that slope undercut zones up to 3 m (10 ft) or more in height and 10 m (33 ft) or more in length can develop. The larger of these undercut failures are locally referred to as "Mesa Verde" failures due to their resemblance to the shape of the national park in Colorado by the same name. Presplit blasting and dozer scaling helps to preserve the rock-mass integrity and lessen the number of Mesa Verde failures. When such failures appear to be creating adverse safety conditions, the mines berm off the highwall toe areas. An example of a Mesa Verde failure is presented in Figure 28.7.

Moderately Dipping Seams

United States—Wyoming, Powder River Basin. Approximately 13.5 million metric tons of coal are annually mined with truck and shovel from a 33.5-m (110-ft) seam containing a 1.5-m (5-ft) parting. The dip of the Paleocene sediments in the mine area is approximately 6° to the north. The overburden sediments consist of sandstone, siltstones, and claystones with some sandstone channels being encountered. Some groundwater is found in most of the sand channels. Highwalls are up to about 55 m (180 ft) high. As mining proceeded in a north-northwest direction, an anomaly was encountered. The anomaly was in the form of an uplift that warped the sediments, including the coal, upward as much as 30 m (100 ft) or more. The uplift appeared to radiate out from a central point, giving the strata a semicircular shape, which measured as much as 230 m (750 ft) across. The coal dip changed from a 4° north dip to a south, southeast and east dip, which steepened up to as much as 27°. As mining proceeded to the north-northwest, the overburden toe buttress was removed and planar mode failure occurred along the top of the coal seam. The slippage had not been expected by mine operators and was thus a surprise development. As much as 10 m (33 ft) of displacement occurred in a roadway crossing the front of the anomaly. Remedial measures included monitoring slide displacement to determine failure extent, direction, and velocity. In addition, the upper portion of the anomaly was stripped to remove sliding debris as well as in-place overburden that could potentially slide. The result was that no safety problems developed from the slide and displacements were brought under control. However, some coal dilution was experienced as a result of the overburden displacements.

United States—Colorado, Yampa Coal Field. Draglines are used to strip at this mine, where 1.8 million metric tons of coal

FIGURE 28.7 Mesa Verde-type failure developing on the underside of a sandstone channel

are produced each year. Upper Cretaceous age coal from the Williams Fork Formation is mined from two seams that have a hard, strong 6-m (20-ft) parting between them. Highwalls are cut at approximately 45° over heights up to 55 m (180 ft). The overburden is composed primarily of weak sandstone, but no significant faults are known to occur. Groundwater is minimal. At present, dipline mining is used to extract the coal, which dips at about 9°. No significant slope stability problems have been encountered in the past. A new pit, currently in the planning stages, is being engineered. Owing to site characteristics and required mining procedures, the new pit would be much more economical if it were mined along the strikeline. Mining could begin with a large box cut and then proceed updip with overburden spoiling on the downdip side. Such strikeline mining is very desirable except for one major unknown–if the shear strength along the bedding direction weakness (contacts, clay layers, etc.) is great enough to resist sliding on a 9° plane. The mine is currently examining this question and plans to have outside consultants evaluate the problem and make suitable recommendations as to how they should proceed. The problem with just trying strikeline mining and then switching to dipline mining is that a major capital expenditure is required to create the initial box cut. If strikeline mining does not work, the initial monetary expenditure would be a significant economic loss. The results of the planned geotechnical study are not yet known but should provide a basis for a sound decision by the mining company.

Colombia—El Cerrejon Coal Mine. This South American strip mine produces about 12.3 million metric tons of steam coal per year. Mining of multiple seams is done downdip on straight dip slopes varying from 14° to 20°. The highwalls may reach 270 m (886 ft) in height and are composed of sandstones, siltstones, and claystones. Three major fault sets with dips ranging from 55° to 75° are known. Groundwater also occurs in the footwall slopes. The major stability problem (Jaramillo 1996) encountered at this mine is footwall instability. To address the problem, a major stability investigation was undertaken. The investigation produced various strategies to control slope failure. These strategies include placing cable bolt dowels along the footwall slope and flattening the slope at the crest. Another strategy has been to bench the footwall toward the pit to the next coal seam. A unique cost-benefit model was developed to provide the proper bolting design. Overall, some 11,800 bolts are planned for installation during mine life. In addition to bolting, slope crest flattening, and footwall stepouts, a major dewatering program was put in place and includes the use of horizontal drains, pipe wicks, and piezometers. Slope displacement monitoring was used to measure stabilization success. The result is that only minimal slope displacements have been detected. Furthermore, no failure has occurred in the bolted footwall slopes.

United States—Wyoming, Powder River Basin. This truck-and-shovel operation produces approximately 15 million metric tons of coal per year. The overburden consists of Paleocene Fort Union Formation sandstones, siltstones, and claystones. Two coal seams occur at the mine and are separated by 1-m (3-ft) parting. The upper seam is 10.7 m (35 ft) thick, and the lower is 30.5 m (100 ft) thick. A weak fireclay occurs at the base of the lower seam. Mining had been proceeding in a south to north direction updip at about 1½°. A limiting mining factor, as mining proceeded north, was a mainline railway running in an east-west direction. As mining continued, a sand channel was detected under the lower seam, warping the sediments upward. The upward warping increased the seam dip to about 10° to 15° to the south. Thereafter, whenever the entire lower seam was removed down to the fireclay, downdip planar mode failure occurred in the fireclay. If mining were to continue northward as it had been, the displacement of the mainline railway was a distinct possibility. Consequently, a different mining plan had to be devised or the remaining coal resource had to be abandoned. One key factor noted was that a time lapse of about two weeks took place before the undercut fireclay actually failed. Using this knowledge, a mining plan was developed whereby strikeline strips some 150 m (500 ft) long were excavated. The overburden, upper coal, parting, and 27.4 m (90 ft) of the lower seam were stripped and mined. A 3-m (10-ft) block of coal was left along the entire 150-m (500-ft) excavation zone. As long as the 3-m (10-ft) thick coal block remained in place, the fireclay was not undercut and the updip slope remained stable. The 3-m (10-ft) thick coal block was then rapidly mined and the slope toe backfilled with a spoil material buttress, all within two weeks. Section after section, 150 m (500 ft) long, was removed in this manner and backfilled. Displacement monitoring along the upslope and along the rail line showed that no significant movement was occurring in the downdip direction. The result was that all the desired coal was successfully mined and the rail line suffered no adverse displacement.

United States—Washington, Centralia Coal Mine. Historically, this mine has used a variety of stripping equipment, including truck/shovel, dragline, scrapers, dozers, and even a bucket-wheel excavator to produce about 4 million metric tons of coal annually. Highwall heights are up to about 122 m (400 ft) and highwall slope angles vary from 45° to 55°. The mine is geologically complex (Paul 1979) with a variety of features, including folds, reverse faults, upthrusts, and overthrusts. Three seams with dips ranging from 11° to about 20° are mined. At times, elevated pore pressures may be found in highwalls, but high surface runoff is common. Many partings are found in the coal seams and represent layers of volcanic ash. These ash layers have altered to soft, weak, expansive clays, primarily montmorillonite. When a highwall failure occurs, it is in the planar mode and the slip plane is usually along a parting. Typically, the partings are extruded and can be found in large sheets at the slide toe. Solutions to these stability problems have included a reorientation of the mining direction to be at an angle to the formational strike. This reduces the effective slope on any undercut planes by causing potential sliding to occur at a lesser angle than the formational dip angle. In addition, surface water diversion has reduced water inflows and related stability problems. Scaling of highwalls has reduced rock fall potential.

Steeply Dipping and Highly Variable Seams

Canada—Western Coal Fields. A feasibility study revealed that a mining excavation 488 m (1,600 ft) deep would be economical in dipping sediments (Brawner, Pentz, and Campbell 1975). The deposit was approximately 4.8 km (3 miles) long and the coal seams dipped at 39°. The major potential problem was the stability of the footwall back slope. An uneconomical coal seam, 0.3 m (1 ft) thick and 9 m (30 ft) below and parallel to the final design slope, was revealed by geologic investigation. There was concern that the 9-m (30-ft) slab, which was competent sandstone, may fail on or in the coal layer and slide into the pit as it was deepened. The safest procedure would be to remove the sandstone, but that would require more than 30.6 million cubic meters (40 million cubic yards) of excess stripping. A detailed geotechnical investigation was undertaken to determine the shear strength characteristics of the sandstone and coal and the surface irregularities along potential sliding surfaces. Stability analyses were then used to develop design curves for various shear strengths, discontinuity geometries, and groundwater conditions. Final recommendations included dewatering and mining along the base of the lowest seam with no final benches on the footwall back slope.

United Kingdom—English, Scottish, and Welsh Coal Fields. The geology in these coal fields is very complex with numerous folds and faults. The faults have a significant influence on stability and are often complex and unpredictable (Cobb, Scoble, and Stead 1982). Normal faults with dips of 60° or more comprise 85% of the fault population. However, low-angle reverse faults, thrusts, and overthrusts are common in the tectonically disturbed Welsh and Scottish coal fields. Offsets are typically less than 10 m (32 ft) and low shear strength gouge infilling is generally less than 1 m (3.3 ft) thick. Rock-mass quality is reduced in badly faulted areas. The impermeable nature of the fault gouge often creates barriers and, consequently, allows elevated groundwater pressures to exist. Mining depths average 37 m (121 ft) but have a maximum depth of 220 m (722 ft) in the Westfield area of the Scottish Region (Bish 1980). With many radically folded seams, mining is performed with shovels on the anticlines and hydraulic backhoes on the synclines. Draglines are used to strip overburden above the folds. Slope instability problems have necessitated various remedial procedures. These include keeping the mining plan flexible so that any particular fault can be handled according to its individual characteristics. Pit wall faces are oriented perpendicular to faults if possible. Faults are not allowed to undercut any portion of the working face or other slopes. Convex slope zones, or "noses," are not allowed to occur. Drilling and geophysical methods are used to predict fault occurrence in advance of mining. Dewatering is undertaken behind faults. Owing to the fact that 81% of the instabilities occur when faults are within 40 m (131 ft) of slope crests, proper stand-off distances from slope crest to fault are used. Finally, the placement of spoil materials is kept back from pit slopes in order to reduce surcharge loading when faults are found beyond the crest line. Slope failures have been significantly reduced using these remedial measures in these steeply dipping and/or irregular seam orientations.

Canada—Alberta, Smoky River Coal Mines. These coal mines are located in the Rocky Mountains near Grande Cache, Alberta. Low-volatile metallurgical coal is mined by truck-and-shovel methods, with a backhoe being used in narrow pit bottoms. The most desirable surface mining areas are where the coal seam is overthickened in syncline and anticline hinge zones and in dip slope situations. The no. 4 seam, which is about 6 m (20 ft) thick, is the main seam exploited. In order to mine the footwall of the no. 4 seam, which dips at 65° in the Upper East Limb Pit, a choice of benching or using artificial support in the form of anchoring was evaluated (Fawcett, Sheehan, and Martin 1988). A risk evaluation of the benching/anchoring alternatives revealed that a combination of both benching and anchoring was superior to either benching or anchoring. Detailed geotechnical studies showed that the most likely footwall failure modes would be buckling, bilinear failure, ploughing, and planar slab sliding. The artificial support system was designed using 6.1-m- (20-ft-) long, 35-mm- (1-in.-) diameter, grade 150 Dywidag rock anchors, chain link fencing, 1.8-m (6-ft) rock bolts, groundwater drain holes, and displacement monitoring devices. Actual implementation of the support project started in September 1985 and completed was in February 1987. The pit was successfully finished in April 1987. A total of 1,764 rock anchors and 600 mechanical rock bolts was placed in the lower footwall. Economic computations revealed that a savings of $500,000 (CN) for the anchoring alternative versus using only slope benching.

28.6 REFERENCES

Bhattacharyya, K.K., J. Vakili, and S.Y. Chi. 1982. Geotechnical considerations of the design of highwall and spoil slopes in lignite mines. *Lignite Mining and Stability, SME Mine Symposium,* No. 82 Coal/M&E-03, pp. 1–7.

Bish, M. 1980. Surface mining in difficult conditions. *World Coal,* July, pp. 48–51.

Brawner, C.O. 1989. Examples of instability at surface mining projects. *SME Preprint,* No. 89–27.

Brawner, C.O., D.L. Pentz, and D.B. Campbell. 1975. Ground stability in surface coal mines. *Mining Congress Journal,* June, pp. 42–53.

Calder, P.N., and J.L. Workman. 1982. Practical aspects of stable pit slope design in U.S. lignite mining. Lignite Mining and Stability, *SME Mine Symposium,* No. 82 Coal/M&E-03, pp. 19–34.

Cobb, Q., M. Scoble, and D. Stead. 1982. Faulting and surface coal mine design. *CIM 4th Open Pit Operators Conference,* Preprint No. 23, 15 pp.

Fawcett, D.A., D.L. Sheehan, and D.C. Martin. 1988. Footwall anchoring at Smoky River Coal Limited. *CIM Bulletin,* Vol. 81, No. 915, July, pp. 78–85.

Hutchings, R. 1979. Slope stability in Australian brown coal open cuts, In *Stability in Coal Mining.* Miller Freeman, Chapter 5, pp. 65–74.

Jaramillo, M. 1996. Slope stabilization at El Cerrjon Coal Mine, Colombia, South America. *Mining Engineering,* December 1996, pp. 49–54.

Jaworski, W.E., and R.L. Zook. 1978. Considerations in the stability analyses of highwalls in tertiary rocks, In *Stability in Coal Mining.* Miller Freeman, Chapter 4, pp. 47–63.

Paul, R.A. 1979. The effects of geologic structures on slope stability at the Centralia Coal Mine, *Stability in Coal Mining.* Miller Freeman, Chapter 2, pp. 29–34.

Pfeiffer, T.J., J.D. Higgins, R.D. Andrew, R.K. Barrett, and R.B. Beck. 1993. Colorado rockfall simulation program, users manual for Version 3.0. Colorado Transport Institute, Report No. CDOT-DTD-ED3-CSM-89-2B, 66 pp.

Pierschke, K.J. 1979. Assessment of the stability of open pit mine slopes in the Rhenish Brown Coal District, In *Stability in Coal Mining.* Miller Freeman, Chapter 3, pp. 35–46.

Seegmiller, B.L. 1973. Time dependent output from in-situ measurements: Its meaning with respect to stability. *Pre-Symposium Seminar, Ninth Symposium on Rock Mechanics.* Ecole Polytechnique, Montreal, Quebec, December 13.

Sutphin, B.L. 1982. Geotechnical investigations performed during planning for lignite mines in North Dakota and Texas. *Lignite Mining and Stability, SME Mine Symposium,* No. 82 Coal/M&E-03, pp. 9–17.

Uniform Building Code (UBC). 1991. pp. 23-2.

SECTION 3

Stability of Waste Rock Embankments

CHAPTER 29

Site Selection, Characterization, and Assessment

P. Mark Hawley*

29.1 INTRODUCTION

Rational waste rock embankment site selection and design constitute an iterative process. Prospective sites are first identified and preliminary site characterization studies are undertaken. Possible alternative design concepts are developed, evaluated, and compared, and the optimum site and design concept are then selected. Additional site characterization and design studies are conducted to confirm feasibility and to prepare detailed design specifications.

29.2 BASIC SITE-SELECTION CONSIDERATIONS

Basic factors that should be considered when selecting a waste rock disposal site can be divided into five general categories, as described in the following: mine planning considerations, physical constraints, environmental impacts, stability, and social and political considerations.

29.2.1 Mine Planning Considerations

Mine planning considerations encompass aspects of the waste rock disposal plan that relate to materials handling and mine scheduling. Haulage costs (whether by truck, rail, or conveyor) often comprise the largest proportion of the waste rock disposal costs; hence, it usually is desirable to locate the disposal site as close as possible to the open pit. Level or downhill hauls are generally preferred; uphill hauls can be prohibitively expensive.

Scheduling flexibility can be an important factor, too, particularly for large mines where several alternative sites may be required or are desirable. Equipment requirements also may vary, depending on the type and location of the embankment. And costs for access development also have to be considered. All of these mine planning considerations often drive initial site-selection criteria.

29.2.2 Physical Constraints

The capacity of a given site may be limited by topographic features such as streams, steep foundation slopes, or other mine infrastructure or land uses. Depending on the quantity of waste rock that must be accommodated, one site may have an advantage over another. Alternatively, more than one site may have to be developed. Site configuration also may dictate the optimum construction technique.

Feasibility studies for most new mines are typically evaluated on the basis of known (proven and probable) reserves. As the mine develops and exploration advances, additional reserves may be added to the reserve base. An increase in reserves usually means more waste rock. Changing commodity prices also may impact mineable reserves and stripping ratios. Depending on the nature of the orebody and potential for expansion of reserves, flexibility of the disposal site to accommodate additional volumes of waste rock can be an important consideration in site selection.

29.2.3 Environmental Impacts

The importance of early assessment of potential environmental impacts and mitigation strategies for waste rock embankments cannot be overemphasized. The potential for development of acid rock drainage (ARD) from the embankment has to be considered in the site-selection process. Other chemical or physical processes that could impact surface water and groundwater quality, such as degradation and erosion of fine-grained rock materials or foundation soils, also have to be considered. Some sites are more amenable to mitigation, containment, and treatment of ARD and suspended sediment in surface runoff. Potential impacts on fish and wildlife habitat and sensitive plant ecosystems also have to be addressed. Finally, reclamation and closure requirements may vary from site to site and have to be considered during site selection.

29.2.4 Stability

The physical stability of a waste rock embankment depends on a variety of site-specific factors, including geometry of the site, foundation conditions, nature of the waste rock materials, regional seismicity, climatic conditions, and hydrology. Site geometry also might dictate construction methodology, as it can affect stability. If the risk of instability is too great, proximity to critical infrastructure may preclude the use of a given site.

Site selection also should consider the potential consequences of failure. Would site geometry exacerbate failure runout or favor containment? Operational safety also has to be considered. For example, embankments constructed on steep foundations typically present a higher risk to equipment and personnel than embankments constructed on flat or moderate foundations, all other things being equal.

29.2.5 Social and Political Considerations

Mining projects worldwide are being subjected to increasingly stringent permitting and regulatory requirements. Public perception and acceptance of a specific site also can be critical considerations. Issues that have to be addressed in the site selection process include environmental protection, resource conservation, competing land uses, aboriginal land claims, displacement of indigenous people, preservation of archaeological sites, and visual landscape aesthetics.

29.3 SITE CHARACTERIZATION

Rational site selection first requires an understanding of the basic biophysical characteristics of the site. Characteristics that are considered most important are listed in Table 29.1. This list is not exhaustive. Unusual site conditions or characteristics may override other concerns. Not all of these characteristics will be important or even relevant at all sites, and relative impact can vary widely. The list in Table 29.1 is intended as a preliminary guide or checklist for experienced project planners and reviewers.

* Piteau Associates Engineering Ltd., North Vancouver, British Columbia, Canada.

TABLE 29.1 Key biophysical characteristics of the site

Characteristic	Important Attributes
Physiography	Location, topography, size, shape
Geomorphology	Landforms, geological hazards, glacial history
Surficial geology	Soil types, distribution, stratigraphy, depth to bedrock
Bedrock geology	Rock types, alteration, geologic structure, mineralization
Tectonics	Seismicity
Climate	Precipitation, temperature, prevailing winds
Hydrology	Runoff characteristics, catchments, streams
Flora and fauna	Fish/wildlife habitat, plant ecosystems, endangered species
Air and water quality	Surface water and groundwater chemistry, dust
Land ownership	Surface rights; forestry/agricultural sources
Archaeology	Culturally significant sites
Visual impact	Viewscape, tourism resources
Waste rock characteristics	Competency, durability, chemical stability

Evaluation of the relative importance of each of the various factors in terms of site selection and design requires experience and application of good engineering judgment.

29.3.1 Physiography

The physiography of a site refers to its location, shape, size, and topography. As discussed above, site location affects transportation (haulage) costs. Other mining activities, such as open pit or access development and location of critical infrastructure, likewise influence site selection. Size and shape determine the ultimate capacity and also influence selection of the type of embankment and construction techniques. Topographic constraints, such as steep slopes, major drainages, and divides, place additional physical limitations on the site and also can affect embankment type and construction.

Basic topographic mapping and cross-sectioning at a scale appropriate to the stage of project evaluation is required. For purposes of preliminary site selection at the prefeasibility level, 1:5000 scale mapping with 10- to 25-m contour intervals may be sufficient. As the project advances, more detailed mapping will be required. Development of three-dimensional topographic models using established computer software such as Vulcan, Surfer, or Mine Sight is becoming more common. These models enable rapid estimates of site geometry and capacity, as well as facilitate classification of foundation areas based on steepness. They are also helpful in visualizing possible alternative embankment configurations and in communicating site characteristics to the rest of the project team, regulators, and the public.

29.3.2 Geomorphology

Geomorphology of the site refers to the geological origin of various landforms and active geologic processes. Understanding the geomorphology provides insight into the nature of the surficial soils. The occurrence of landslides, debris flows, avalanches, or other geologic hazards may be important indicators of potential foundation stability problems that could require special stabilization or mitigative measures. Some landforms, such as terraces, gullies, and irregular topography, may have positive influences on embankment stability.

Basic terrain mapping should be carried out on all prospective sites early in the selection/characterization process. Preliminary terrain maps may be interpreted from good-quality aerial photographs and topographic maps with limited site reconnaissance. Stereo aerial photographic coverage of the mine site and all prospective waste disposal sites is desirable at an early stage in project evaluation. Depending on the nature of the site and stage of project evaluation, additional detailed terrain studies, including comprehensive ground reconnaissance and possibly subsurface investigations (e.g., test pits, drilling), may be required to adequately define specific geomorphic features that could impact site selection and development.

29.3.3 Surficial Geology

An understanding of the surficial geology of the site is essential to be able to evaluate foundation conditions and overburden material characteristics for stability analyses, and to determine foundation preparation requirements. Special emphasis must be placed on determining the characteristics and extent of soft, loose or incompetent soils that may affect foundation or embankment stability.

A preliminary indication of the nature of site soils may be gleaned from terrain maps and good-quality aerial photographs, supplemented by limited ground reconnaissance. At least some test pitting, sampling, and rudimentary soils index testing and classification are generally required at an early stage of site evaluation. More comprehensive investigations, such as intensive test pitting, trenching, drilling, or sophisticated in situ or laboratory testing, may be required for advance feasibility and design, to define the stratigraphy, extent and character of weak or unusual soils, or the depth to bedrock or competent soil strata.

29.3.4 Bedrock Geology

An understanding of the bedrock geology and competency/durability of the rock mass underlying the site is necessary. Adversely oriented geologic structures (e.g., faults, bedding, or foliation discontinuities) or deep weathering or alteration zones may affect foundation stability or limit bearing capacity. Geology also influences surface drainage patterns and groundwater flow systems. Knowledge of the site geology also may be useful when assessing the potential for economic mineral deposits occurring beneath the site (i.e., condemnation investigations).

Regional geologic mapping with limited ground reconnaissance is usually sufficient for preliminary site selection studies. This scale of mapping should identify basic lithologic types and major alteration facies, as well as regional structures that could influence overall embankment stability, groundwater flow, and the like. More detailed geologic studies including trenching and drilling may be required for more advanced feasibility and design phases. Condemnation drilling usually is carried out in concert with advanced design studies. In cases where the rocks that comprise the foundation are weak, highly altered, or degradable, specific rock mechanics investigations (e.g., geomechanical drilling and core logging, intact and shear strength testing, durability testing) may also be required to help characterize the competency of the bedrock foundation.

29.3.5 Seismicity

Regional seismicity sometimes plays an important role in selecting the optimum site and approach to development. Depending on the nature of the waste rock materials, foundation characteristics, and other factors, some sites are not suitable for waste rock disposal in seismically active areas. Other sites require special design considerations, or capacity might be severely affected. Regional probabilistic or deterministic seismic risk studies usually are carried out for the overall project during feasibility studies. Early in the site-selection process, project planners need to be aware of the general level of seismic activity. During the detailed design phase, specific seismic design criteria usually are required for active sites.

29.3.6 Climate

Prevailing wind patterns, precipitation, frequency and severity of storm events, snowpack, and temperature influence surface runoff and groundwater levels in both the embankment and foundation. Areas with high or intense precipitation may require special construction methods to control runoff and minimize infiltration and to control erosion. Heavy snow accumulations and extended periods of sub-zero temperatures may lead to seasonally adverse conditions within the embankment or foundation, or both. On the other hand, prevailing winds may prevent significant snow accumulation. Basic climatic data also are necessary to prepare a site water balance.

Baseline climate studies have become an integral part of most mining feasibility studies and typically are required by regulatory authorities to support project Environmental Impact Assessments (EIAs). Climate stations now are routinely established at the site early in the exploration phase. In terms of waste rock embankments, regional climatic data available from property-wide baseline studies are usually sufficient for site selection and preliminary feasibility studies. However, microclimate effects related to site location, aspect, elevation, and the like may result in significant deviations from the regional norm and may have to be considered during advanced feasibility and design studies. In these cases, site-specific monitoring or calibration of regional data, or both, may be required.

29.3.7 Hydrology

The size and shape of the drainage basin, steepness of the topography, and nature of the surficial soils and vegetation influence the surface runoff and drainage characteristics of a site. A basic understanding of site hydrology is necessary to determine the need for diversions or flow-through rock drains and to enable rational sizing of downstream sedimentation facilities, if these are required. The impact of waste rock embankment construction on the runoff characteristics of the site also must be taken into consideration.

Basic topographic mapping, aerial photography, and regional climatic data such as described above are probably sufficient for preliminary hydrologic studies required for site selection and prefeasibility. Site-specific studies, including reconnaissance and mapping to confirm watershed boundaries, surface drainage and erosion patterns, and runoff characteristics, normally would be required for advanced feasibility and design studies. In cases where major diversions or flow-through rock drains are contemplated, more detailed topographic mapping and climatic data, as well as focused stream-flow monitoring, may be required. When sedimentation facilities must be designed, laboratory soil gradation and sedimentation studies are normally required.

29.3.8 Hydrogeology

Foundation conditions, and hence stability, may be directly influenced by the hydrogeology of the site. In addition, waste rock embankment construction can directly impact groundwater levels and quality. To be able to evaluate potential impacts, a basic understanding of the groundwater flow system and hydrogeologic characteristics of the site are essential.

Preliminary groundwater studies may be limited to general reconnaissance and mapping of discharge zones. If preliminary analyses indicate that stability is sensitive to piezometric pressures within the foundation and the groundwater table appears high, additional investigations could be warranted. These investigations might include installation of piezometers, packer testing, and possibly aquifer testing.

29.3.9 Flora and Fauna

Careful documentation and evaluation of the flora and fauna that occupy a site are required early in the site-selection and evaluation process to be able to assess potential impacts and the viability of possible mitigative measures. Baseline fisheries and wildlife habitat studies, with particular reference to endangered species, have to be conducted. Native plant species also should be inventoried to assess their uniqueness and to serve as a guide for future reclamation objectives.

A preliminary site reconnaissance by a qualified biologist/botanist/ecologist familiar with the regional and local ecology may be sufficient for purposes of site-selection studies and conceptual/preliminary design. For ecologically sensitive areas, more detailed plant and habitat inventories may be necessary for advance feasibility/detailed design studies, to develop mitigative strategies and realistic reclamation objectives. These studies normally are conducted for the mining property as a whole and include prospective waste rock disposal sites.

29.3.10 Air and Water Quality

The impact of waste rock disposal on air quality and surface water and groundwater quality also have to be considered. Waste rock embankment construction can result in considerable generation of dust. Depending on prevailing wind directions, dust clouds could pose environmental and health hazards alike. Siting of disposal areas upwind of facilities that are sensitive to dust, such as camps, settlements, offices, and sensitive agriculture, among others, should be avoided.

Potential impacts to surface water and groundwater quality also have to be identified and quantified. Runoff and groundwater recharge from waste rock embankment sites can contain significant levels of suspended sediments and heavy metals or other minerals (e.g., salts, nitrates), which could adversely affect downstream aquatic life, as well as irrigation and industrial and domestic uses of surface water and groundwater. Documenting baseline air and water quality in and around the site is an important first step in evaluating potential impacts and for long-term monitoring.

Qualitative assessments of potential impacts to air and water quality probably are sufficient for preliminary site-selection purposes. A basic understanding of prevailing wind patterns is helpful. Prospective disposal sites have to be evaluated in the context of the overall project layout. Where is the site in relation to possible locations of other facilities/settlements? Can runoff be controlled? Is the site amenable to collection and treatment of contaminated runoff? More detailed, quantitative studies may be required to enable design of mitigative measures, such as diversions and sedimentation/treatment ponds, if preliminary screening studies indicate the potential for significant impacts.

29.3.11 Land Use and Ownership

Past and present land uses have to be identified, and the value of existing and potential settlements and forestry, agricultural, and fisheries resources evaluated, to facilitate comparison with other viable waste rock disposal sites and to help assess opportunity costs. Land ownership issues, including surface rights such as timber harvesting, grazing, fishing, hunting, and water use, also have to be identified. In some jurisdictions, unresolved land claims can cause significant project delays and may render some sites untenable even if the underlying legal issues have been technically addressed. Project planners need to be aware of potential land use/ownership issues at an early stage of site evaluation.

29.3.12 Archaeology

Any archaeological sites within a proposed disposal site have to be documented and the cultural significance evaluated. If important sites are identified, the prospective disposal site may have to be discarded or modified to protect the archaeological find. For less important sites, excavation, documentation, and removal of artifacts may be necessary before site development can proceed.

Early in the evaluation process, prospective sites should be screened for archaeologically/culturally important sites. Screening should be conducted by a qualified archaeologist or anthropologist who is familiar with the region. If important sites are

TABLE 29.2 Basic types of waste rock embankment

Embankment Type	Schematic Illustration	Basic Characteristics
Valley fill		Partially or completely fills valley Graded to prevent impoundment of water at the head of the valley
Cross-valley fill		Extends across valley Fill slopes both upstream and downstream Requires diversions, culverts, or rock drains to convey water around or through embankment
Sidehill fill		Constructed on sloping terrain Does not block any major drainages Embankment slopes generally in same direction as foundation
Ridge-crest fill		Fill slopes formed on both sides of ridge
Heaped fill		Flat or gently inclined foundation slopes Slopes on all sides of embankment

Modified after Wahler 1979.

identified, advanced feasibility and design studies may have to include detailed site assessments and mitigation plans.

29.3.13 Visual Impact

Is the site highly visible to local settlements? Does it negatively impact the scenery along a recognized scenic route or from a viewpoint? The impact of site development on the viewscape and existing or potential tourism values also may have to be assessed.

29.3.14 Waste Rock Characteristics

Though not technically site characteristics, the physical and chemical characteristics of the waste rock often have an important, and sometimes controlling, influence on both site selection and construction methodology; hence they are included in this discussion and the list in Table 29.1. Sites that may be suitable for containment of coarse, angular waste rock with a high degree of chemical and physical stability may be inappropriate for weak, degradable, or chemically reactive waste rock. The shear strength characteristics of the waste rock may restrict the way in which the site is developed and could limit ultimate capacity.

Basic geologic descriptions of the waste rock, including lithology alteration/weathering and mineralogy, have to be prepared during the exploration stage in advance of disposal site-selection studies. In particular, the presence/absence of sulfides and clay mineralization has to be documented. The basic geomechanical characteristics of the waste rocks (i.e., intact strength, degree of alteration, degree of fracturing) also should be documented. This type of geotechnical data can easily be obtained from the drill core. For this reason, even early exploration drilling programs should include provisions for basic geomechanical core logging.

Detailed design studies may require intensive characterization of waste rock mineralogy, including whole rock geochemistry, acid-base accounting, humidity cells, and other sophisticated laboratory testing, to further characterize chemical stability and the potential for generation of ARD. Blast fragmentation modeling may be required to help assess the likely gradation of the waste rock. Specific geotechnical testing to assess waste rock durability and shear strength also may be required for advanced feasibility and detailed design studies.

29.4 CONCEPTUAL DESIGN

Once a preliminary waste rock disposal site has been selected and characterized, conceptual design alternatives should be considered. More than one approach to development of a given site may be feasible; therefore, comparative evaluations of possible alternatives may have to be carried out.

The first step in developing a design concept is to determine which types of waste rock embankments are compatible with the site. Most existing classification schemes (OSM 1989; MESA 1987; USBM 1982) classify waste rock embankments into a few typical types based on foundation geometry and overall embankment configuration. Basic types of waste rock embankment and some of their important characteristics are summarized in Table 29.2.

Once the basic type of embankment type has been chosen, a decision has to be made on how the embankment will be constructed. As illustrated in Figure 29.1, waste rock embankment construction can take two basic approaches: (a) ascending construction, whereby the embankment is constructed from the bottom up in a series of lifts, and (b) descending construction, whereby the embankment is constructed from the top down using

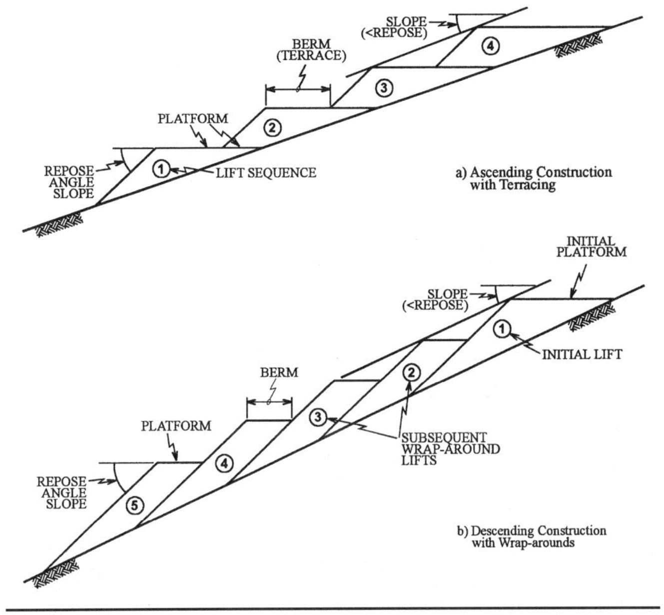

FIGURE 29.1 Ascending versus descending embankment construction

descending platforms and wrap-arounds. Hybrid design concepts that incorporate both of these approaches are also possible.

The optimum approach to construction depends on several factors, including the shape of the site, its proximity and relative elevation with respect to the open pit, and the nature of the waste rock and foundation. In mountainous terrain, where the disposal site is situated at a lower elevation than the pit, descending construction is usually the most economical approach; however, waste rock quality and foundation characteristics may preclude or restrict this type of construction. In subdued terrain or where the waste rock or foundations are poor, ascending construction may be the only option. Ascending construction generally allows more control over placement of the material and, hence, is usually superior to descending construction in terms of stability.

29.5 CLASSIFICATION AND COMPARATIVE EVALUATIONS

Classification schemes provide a convenient tool for comparing alternative disposal sites and embankment design concepts, as well as evaluating relative risk. One classification system that was developed specifically for this type of application is the dump stability rating (DSR) scheme summarized in Table 29.3 (BCMDC 1991). The DSR system recognizes 11 key factors that affect overall waste rock embankment stability. Each factor is assigned

TABLE 29.3 Waste rock embankment stability rating scheme

Key Factors Affecting Stability		Range of Conditions	Point Rating
1 Embankment Height	Low	<50	0
	Moderate	50–100 m	50
	High	100–200 m	100
	Very high	>200 m	200
2 Embankment Volume	Small	<1 million BCMs	0
	Medium	1–50 million BCMs	50
	Large	>50 million BCMs	100
3 Embankment Slope	Flat	<26°	0
	Moderate	26°–35°	50
	Steep	>35°	100
4 Foundation Slope	Flat	<10°	0
	Moderate	10°–25°	50
	Steep	25°–32°	100
	Extreme	>32°	200
5 Degree of Confinement	Confined	Concave slope in plan or section	0
		Valley or cross-valley fill with buttressed toe	
		Incised gullies or ridges	
	Moderately Confined	Natural benches or terraces	50
		Even slopes, limited topographic diversity	
		Heaped, sidehill, or broad valley or cross-valley fills	
	Unconfined	Convex slope in plan or section	100
		Sidehill or ridge-crest fill with no toe confinement	
		No gullies, ridges, or benches	
6 Foundation Type	Competent	Foundation materials as strong or stronger than waste rock	0
		Not subject to adverse pore pressures	
		No adverse geologic structure	
	Intermediate	Intermediate between competent and weak	100
		Foundation soils gain strength with consolidation	
		Adverse pore pressures dissipate if loading rate controlled	
	Weak	Limited bearing capacity, soft soils	200
		Subject to adverse pore pressures, groundwater	
		Shear strength sensitive to strain, liquefaction potential	

(continues)

TABLE 29.3 Waste rock embankment stability rating scheme (Continued)

Key Factors Affecting Stability		Range of Conditions	Point Rating
7 Quality of Waste Rock	High	Strong, durable, <10% fines	0
	Moderate	Moderate strength, variable durability, 10%–25% fines	100
	Poor	Weak, subject to slaking or degradation, >25% fines	200
8 Method of Construction	Favorable	Thin lifts (<25–50 m thick), wide platforms Advancement along contours Ascending construction	0
	Mixed	Moderately thick lifts (25–100 m) Mixed construction methods Descending construction with wide wrap-arounds	100
	Unfavorable	Thick lifts (>100 m thick), narrow platforms, sliver fills Advancement down fall line of slope Descending construction with narrow wrap-arounds	200
9 Piezometric and Climatic Conditions	Favorable	Free-draining embankment, no phreatic surface or perching Limited precipitation, negligible snow/ice inclusion Minimal infilration	0
	Intermediate	Limited development of phreatic surface or perched zones Moderate precipitation, discontinuous snow/ice lenses Moderate infiltration	100
	Unfavorable	High phreatic surface or extensive perched zones High precipitation, continuous lenses of snow or ice High infiltration	200
10 Advance Rate	Slow	<25 BCMs per lineal meter of crest per day Crest advancement rate <0.1 m per day	0
	Moderate	25–200 BCMs per lineal meter of crest per day Crest advancement rate 0.1–1.0 m per day	100
	High	>200 BCMs per lieneal meter per day Crest advancements rate > 1.0 per day	200
11 Seismicity	Low	Seismic risk zones 0 and 1 (Canada) <10% probability of ground acceleration exceeding 0.08 g in 50 yrs	0
	Moderate	Seismic risk zones 2 and 3 (Canada) <10% probability of ground acceleration exceeding 0.16 g in 50 yrs	50
	High	Seismic risk zone 4 or higher (Canada) >10% probability of ground acceleration exceeding 0.16 g in 50 yrs	100

Source: Full BCMDC

a point rating based on qualitative and/or quantitative descriptions that encompass a wide range of possible characteristics. The overall DSR is calculated as the sum of the individual ratings for each of the various parameters.

As summarized in Table 29.4, possible DSRs range from 0 to 1800 and are subdivided into four broad categories called dump stability classes (DSCs). Each DSC represents a different level of instability hazard.

In applying the DRS system and interpreting the results, the behavior of a waste rock embankment and its potential for instability depend on a wide range of diverse and interrelated factors. Not all of these factors lend themselves to easy quantitative measurement or assessment; consequently, some judgment is required in selecting the most appropriate point rating. The weighting of the key stability factors, as presented in Table 29.3, is intended to represent their relative importance or contribution to overall stability. The ranges and descriptions are intended to strike a reasonable balance between range of applicability and ease of use.

Although the DSR is intended to cover a wide range of possible waste rock embankment configurations and site conditions, it may not be applicable or appropriate in all cases. A new concept, the DSR is subject to testing, verification, and calibration. Additional details regarding the DSR classification system and example applications are given in BCMDC (1991).

29.6 RISK ASSESSMENT

Although classification schemes such as the DSR are helpful in defining and comparing potential hazards, they do not provide a complete measure of risk. A comprehensive assessment of risk requires consideration of not only the potential hazard or likelihood of an adverse event but also the consequences or impact if the event occurs. For example, a given site and embankment configuration may be subject to a relatively high failure hazard but the consequence of instability may be negligible. In this case, the overall “risk” of failure may be low. Alternatively, the failure hazard may be low but the consequences of instability may be severe, resulting in an overall high “risk” of failure.

TABLE 29.4 Waste rock embankment stability classe

Dump Stability Class	Relative Failure Hazard	DSR Range	Guidelines for Preliminary Site Characterization
I	Negligible	<300	▪ Basic site reconnaissance and baseline documentation ▪ Minimal laboratory testing of foundation soils and waste rock materials
II	Low	300–600	▪ Basic site reconnaissance and baseline documentation, followed by: — thorough site investigation, including terrain mapping — test pitting and sampling may be required to supplement mapping — limited laboratory index testing
III	Moderate	600–1,200	▪ Basic site reconnaissance and baseline documentation, followed by: — detailed phased site investigation — initial investigation phase including test pitting — advanced investigation phases possible including drilling, undisturbed sampling — some instrumentation (e.g., piezometers) likely required — comprehensive laboratory testing program
IV	High	>1,200	▪ Basic site reconnaissance and baseline documentation, followed by: — detailed phased site investigation — initial investigation phase including intensive test pitting/trenching — advanced investigation phases likely requiring drilling, undisturbed sampling — comprehensive in situ testing and instrumentation program — comprehensive, phased laboratory testing program

Analytical or numerical assessments of risk usually require detailed sensitivity or probabilistic event analysis. These analyses generally require detailed statistical knowledge of a wide variety of parameters, such as shear strength of the waste rock materials and foundation soils, geochemistry and reactivity of the rock, return period precipitation or seismic events, and so on. During preliminary site selection and characterization studies, available data typically are insufficient to support detailed risk assessment. Qualitative risk assessments, however, often can be used effectively to compare and assess the relative risk of alternative sites and to help guide advanced site assessment studies.

A generic approach to assessing relative risk is illustrated in Table 29.5. For the purposes of this approach, "risk" is defined as the product of hazard and consequence. Potentially adverse events are first identified and a rating is assigned based on the perceived level of hazard. Possible consequences then are considered and also assigned a rating. The overall "risk" rating then is determined from the hazard/consequence matrix given in Table 29.5. Selection and application of appropriate rating scales for hazards and consequences alike requires experience and application of good engineering judgment.

TABLE 29.5 Risk rating matrix

Hazard Rating		Consequence Rating		Risk Rating
High	×	High	=	Very high
High	×	Moderate	=	High
Moderate	×	High	=	High
High	×	Low	=	Moderate
Moderate	×	Moderate	=	Moderate
Low	×	High	=	Moderate
Moderate	×	Low	=	Low
Low	×	Low, Moderate	=	Low
Negligible	×	High	=	Low
Negligible	×	Low, Moderate	=	Very Low

The key to successful application of this approach is in considering and defining possible adverse events and their potential consequences. For waste rock embankments, there are three primary sources of hazard: physical instability (e.g., failure of the embankment), erosion and sedimentation (e.g., degradation and erosion of waste rock materials), and chemical instability (e.g., development of ARD). In terms of physical instability, the DSR system provides a convenient semiquantitative measure of the potential for overall, large-scale embankment instability.

Comparable indices for erosion and sedimentation hazards or chemical instability hazards, though not presented here, also could be developed. Erosion and sedimentation hazard ratings could be based on the mineralogy or overall fines content of the waste rock, or the results of slake durability testing. Chemical instability hazard ratings could be based on geological or mineralogical descriptions, estimated or assayed pyrite content, acid-base accounting, or the results of kinetic testing, if these are available.

Consequence ratings must be defined and applied based on site-specific conditions and should take into account potential mitigative measures. In assessing the consequences of large-scale embankment instability, particular consideration should be given to potential runout distances and possible impacts to critical infrastructure, settlements, habitat, and present and future land uses. Can runout be reduced or controlled by construction of impact or deflection berms? Consequence assessments for erosion and downstream sedimentation and ARD hazards have to consider the potential impacts to potable water supplies and fish habitat. Can sedimentation control and/or water treatment facilities be constructed to mitigate these hazards, and, if so, at what cost?

29.7 SUMMARY AND CONCLUSIONS

Site selection and characterization studies have to consider a wide range of factors, including overall mine planning and scheduling requirements, physical site constraints, potential environmental impacts, embankment stability, and the social and political implications of development. Important biophysical characteristics that have to be documented include site physiography, geomorphology, surficial and bedrock geology, seismicity, climate, hydrology, hydrogeology, flora and fauna, air and water quality, land use and ownership, archaeology, and visual quality. In addition, the characteristics of the waste rock materials have to be defined.

Once the basic site characteristics have been defined, alternative design concepts and construction methods have to be considered and compared. Classification systems such as the DSR system provide a convenient tool for comparing alternative disposal sites and design concepts. Qualitative risk assessments that consider both hazard and consequence also provide a useful tool for comparative evaluations. Reliable application of classification systems and risk assessments requires experience and good engineering judgment.

29.8 REFERENCES

British Columbia Mine Dump Committee (BCMDC). 1991, May. Investigation and Design of Mine Dumps–Interim Guidelines. *Report prepared for B.C. Ministry of Energy, Mines and Petroleum Resources by Piteau Associates Engineering Ltd.*

Mining Enforcement and Safety Administration (MESA). 1975. Engineering and Design Manual–Coal Refuse Disposal Facilities. *Report prepared for U.S. Dept. Int. by D'Appolonia Consulting Engineers.*

Office of Surface Mining Reclamation and Enforcement (OSM). 1989, Nov. Engineering Design Manual for Disposal of Excess Spoil. *Report prepared for U.S. Dept. Int. (Contract No. J5110084) by CTL/Thompson.*

U.S. Bureau of Mines (USBM). 1982, June. Development of Systematic Waste Disposal Plans for Metal and Nonmetal Mines. *Report prepared for U.S. Dept. Int. (Contract No. J0208033) by Goodson and Associates.*

Wahler, W.A. 1979. A perspective—Mine waste disposal structures–Mine dumps, and mill and plant impoundments. *Proceedings 6th Pan-American Conf. Soil Mech. Fndn. Eng., Vol. III, Lima, Peru.*

CHAPTER 30

Assessment of Embankment Parameters

David J. Williams*

30.1 INTRODUCTION

Embankments constructed using mine waste rock typically involve side slopes formed at the angle of repose of the material, in the range of 35° to 40°. As a result, the side slopes cannot be compacted. Even where slopes sufficiently flat to allow compaction are employed, compaction of the side slopes is unusual. Mine site embankments typically are raised progressively in a series of lifts as the development of the mine and mineral processing dictate. This may result in the use of differing construction materials depending on the waste rock materials available at the time of the raising. It can also lead to seepage and stability problems at the interface between raises.

The density of the waste rock used for embankment construction ranges from about 1.6 to 2.2 t/m^3, depending on whether the material is placed loose or compacted to some degree. Over this range, density will have little impact on embankment stability. The natural gravimetric moisture content of waste rock typically ranges from 3% to 7%, and its optimum moisture content for compaction is likely to be on the order of 10% to 15%.

Location of the phreatic surface within a waste rock embankment depends on the use to which the embankment is put and any zoning of construction materials. Where the water level against the upstream face of the embankment is low or is maintained at a low level by the inclusion of drainage within the embankment, the phreatic surface will be low and will have little impact on the stability of the embankment. On the other hand, where the water level against the upstream face of the embankment is high, the phreatic surface will be high and will have a substantial impact on the stability of the embankment.

The foundation beneath a waste rock embankment should be selected to provide adequate strength and low permeability. Proper selection should eliminate foundation failure and excessive foundation seepage.

In the following section, the possible long-term behavior of a waste rock embankment is considered in the light of natural slope angles and hill slope formation processes. Subsequent sections discuss the other waste rock parameters.

30.2 NATURAL SLOPE ANGLES AND HILLSLOPE FORMATION PROCESSES

Distribution of natural slope angles has received some attention in the past and is of relevance in assessing the stability of waste rock embankments in the long-term. Studies during the 1950s, 1960s, and early 1970s in the United Kingdom (Carson and Petley 1970) suggested that natural hillslope angles are grouped about different "threshold values." These are the angle of repose (35° to 40°), half the angle of repose (about 18°), and "slopewash" in the range from 18° down (Figure 30.1). Figure 30.1 highlights the variability of natural hillslope angles. The threshold slope angle depends as much on the diameter of the material on a slope as on the energy of the hillslope formation processes. As the diameter of the material diminishes from weathering over geological time, it becomes possible for the threshold slope angle to diminish (Holtz 1960).

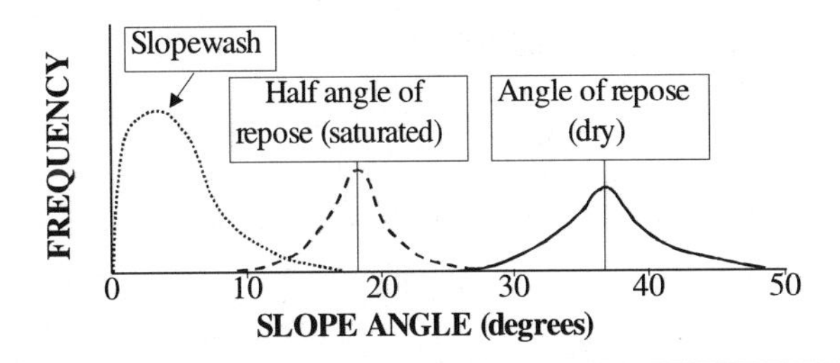

FIGURE 30.1 Schematic frequency distribution for natural hill slope angles

Principles of soil mechanics can be used to explain the observed grouping of slope angles. Loose-dumped, dry, or free-draining granular materials form at their angle of repose. Such slopes subjected to seepage parallel to the angle of repose surface can sustain a slope angle of only about half the angle of repose of the dry material, because of buoyancy effects. Further reduction in the slope angle occurs with the production of further fines. The flow of erosion products on a floodplain results in deposition slopes in the range from 10% (coarse material) to 1% (fine material).

In a similar way, the distribution of slope length, vegetation types and coverage, gully frequency and depth, and so on, may be determined for a given landform. These distributions could be determined both prior to and after mining. It then may be possible for spoil rehabilitation to be designed to achieve distributions similar to premining conditions, based on minimizing the earthworks involved.

The processes involved in hillslope formation include the following:

- *Weathering of the surface materials.* This takes place to a depth and at a rate dependent on the materials and the climate. In the rocks of the United Kingdom coalfields, weathering takes place to a depth of 0.3 to 1.0 m in about 100 years. The fines produced by weathering penetrate the near surface voids, increasing the density of the material, and also flow downslope, resulting in a slight flattening of the slope angle. The buildup of fines also allows the formation of perched water tables, and the possibility of minor surface slumping, exposing underlying material to weathering.

* Department of Civil Engineering, University of Queensland, Australia.

- *Development of sinkholes as a result of rainfall infiltration through dispersive materials.* This can take place rapidly in susceptible materials but is localized. Apart from the obvious surface expressions, sinkholes result in the localized piping of fines with depth, aiding in the buildup of water.
- *Weathering at depth because of water table fluctuations.* This results in the buildup of fines and water, and densification, with accompanying surface settlement.
- *Increasing saturation of the hillslope material.* As the saturation level rises, the hillslope angle cannot be sustained at the angle of repose (dry) and will revert to about half this value.
- *Accumulation of erosion products out from the base of the hillslope.* The erosion products will be deposited on the floodplain beyond the base of the hillslope.

These processes are likely to occur relatively rapidly (perhaps on a 100-year timescale) in steep, erodable terrain, in a high-rainfall climate; and gradually (perhaps on a 10,000- to 100,000-year timescale) in undulating, stable terrain, in a temperate or dry climate.

30.3 STRENGTH PARAMETERS

There is ongoing confusion about the relationship between the angle of repose of coarse-grained mine waste rock and other granular materials and their strength parameters. The angle of repose is often adopted as the friction angle of the material (Williams 1996).

30.3.1 Angle of Repose

On end-dumping from a truck, waste rock ravels (fails) downslope at the angle of repose of the material. The angle of repose of granular material is affected by the following (Rowe 1962):

- Particle size, shape, and surface roughness (increasing with increases in size, angularity, and roughness)
- Specific gravity of the particles (increasing with increasing specific gravity)
- Height of fall (decreasing with increasing height of fall)
- Amount of water present (increasing with the addition of a little water, but decreasing with further saturation)
- Curvature of the slope in plane (concave slopes being more stable than planar and convex slopes because of arching effects)
- Base conditions
- Whether the slope is natural or artificial

Simons and Albertson (1960) presented angle of repose data showing an increase in the angle of repose of granular materials with increasing mean particle size. The more angular the particles, the higher the angle. The data show, for instance, that the effect of scalping to allow laboratory testing may reduce the angle of repose of the scalped material by 6° compared with the field value for the full-scale material. The field value typically is in the range of 37° to 40° for durable, angular mine waste rock.

Rapid dumping leads initially to over-steepening of the upper part of the slope. With continued rapid dumping, the over-steep section of the slope will lengthen, possibly leading to a slip. Over-steepening results from the cohesion of fines hanging up near the crest of the slope, while the coarse particles ravel to the base of the slope.

The angle of repose also is affected by the weathering of particles over time. The production of fines on weathering is accompanied by an increase in the density of the material and a slight flattening of the slope, making it more stable with time. Provided that the slope does not fully saturate, it will remain at this angle in the long-term; although some erosion likely will occur. Saturation gradually reduces the slope angle to perhaps half the initial angle of repose (Stratham 1974).

At large strains (typically greater than 10%) and in a loose state, such as occur on the raveling of coarse-grained mine waste rock down an angle-of-repose slope, the friction angle of the material reverts to its ultimate value. The associated void ratio is termed the *critical state void ratio* (Roscoe, Schofield, and Wroth 1958) and the friction angle is designated the critical state friction angle ϕ_{cv}, numerically equivalent to the angle of repose at which the material ravels.

30.3.2 Friction Angle

The strength of waste rock is characterized by a friction angle and zero cohesion. The value of the friction angle a function of the following:

- Particle size distribution (reducing with decreasing particle size)
- Particle shape and surface roughness (increasing with increasing angularity and surface roughness)
- Strength and specific gravity of individual particles
- State of packing (increasing with increasing density)
- Applied stress level (decreasing with increasing stress, resulting in a curved strength envelope passing through the origin)

It has long been recognized (Holtz and Gibbs 1956; Holtz 1960) that an increase in the proportion of coarse material in an otherwise fine-grained granular soil can result in an increase in friction angle. Alternatively, when the voids within a coarse granular material are filled with fines, its friction angle is increased by as much as 10° (Stratham 1974). The amount of fines required to have a significant effect on the friction angle is relatively small.

Limiting the maximum particle size of a coarse-grained spoil material that can be strength-tested readily in the laboratory will result in a low estimate of the friction angle of the coarser whole material. Over time, the friction angle will increase further as a result of infilling by fine-grained weathering products.

Leps (1970) presented friction-angle data based on triaxial strength testing of large size (up to 200 mm) rockfill particles. These data suggest that the friction angle of durable spoil could be as high as 55° at low stress and is likely to be at least 50° at moderate stress levels. Waste rock would be expected to have a friction angle in the range of 40° to 50°, the lower end of the range corresponding to weathered or crushed, fine-grained material, and the upper end of the range corresponding to fresh, coarse-grained material.

For medium-dense sandy gravel materials, which approximate the waste rock materials used in embankment construction, the value of ϕ_{cv} would be expected to be 4° to 6° less than the peak friction angle (Lambe and Whitman 1979). Together with Leps data, this suggests that mine waste rock would be expected to have a peak friction angle of about 45°.

30.4 OTHER STRENGTH-RELATED FACTORS

30.4.1 Effect of Segregation

Waste rock tends to segregate according to particle size during placement by end-dumping from a truck, because of a combination of Bagnold's grain dispersive pressure and particle kinematics (Middleton 1970). Since the friction angle of granular material tends to increase with particle size, segregation tends to produce alternating weaker and stronger angle of repose layers. As the permeability of the material decreases with mean particle size, alternating layers of lower and higher permeability result. The continuity of the individual layers, however, may be insufficient to produce clearly defined weak layers or preferred seepage flow paths.

30.4.2 Effect of Slope Profile

The stability of a granular slope for a given type of failure depends not only on the slope angle but also on the plan geometry of the slope (Azzouz, Baligh and Ladd 1983). Longitudinally concave slopes (steepest toward the crest and flattest toward the toe) tend to be intrinsically more stable than conventional straight, planar slopes (Schor and Gray 1995). Concave slopes are stable at an angle of about 3° steeper than convex slopes, with planar slopes in between (Rassam and Williams 1999).

30.4.3 Effect of High Water Tables and Rainfall

High water tables reduce the factor of safety for deep-seated slope failures, possibly by a factor of 2. Perched water tables may give rise to surficial failures, which sometimes arise as a result of weathering of the near surface spoil, which produces fines, preventing the infiltration of rainfall. The effect of water near the surface of a spoil slope depends on the direction of any flow. The greatest reduction in the factor of safety (by a factor of up to 2) occurs for flows parallel to the surface of the slope (Gray 1995), which is promoted by the angle of repose layering of waste rock end-dumped by truck.

Slope stability is influenced significantly by intense or prolonged rainfall (Chowdhury and Nguyen 1987). If mine waste rock remains stable on placement, subsequent failures usually are rainfall-related. However, direct correlations between rainfall intensity and the frequency and magnitude of waste rock slope instability have not been established. In dispersive waste rock (Emerson and Seedsman 1982), sinkholes can develop as a consequence of rainfall infiltration (Figure 30.2).

FIGURE 30.2 Surface expression of sinkholes

30.4.4 Effect of Tree Roots

The factor of safety for surficial failures may be increased significantly by the roots of any trees that become established on waste rock slopes (Gray 1995). According to Hubble and Hull (1996), the increase in strength because of the presence of the roots is more than sufficient to offset the extra loading that results from wind acting on the trees, although contrary views have been expressed (Brown and Sheu 1975; Barker 1995). Surficial sliding or planar failures limited to a depth of 250 mm can be stabilized by normal engineering methods, plus vegetation (Lawrance 1995). Removing inappropriate shrub and tree vegetation may result in significant benefits.

Integrated hydrology and slope stability models are required to properly account for the effects of vegetation on slope stability (Collinson, Anderson and Lloyd 1995; Anderson and Lloyd 1991). The factor of safety of bare slopes with high permeability ($> 10^{-5}$ m/s) may be increased by 20% because of the effects of tree cover. Although the factor of safety of a slope with medium permeability (10^{-6} m/s) may be increased by 20% in the long-term as a result of trees, it may be reduced by 20% in the short-term as the tree cover is becoming established. The effects of the tree cover on the slope hydrology also must be considered (Collinson and Anderson 1996).

30.4.5 Effect of Weathering

The generation of clay size fines by physical weathering may reduce the friction angle by 2° or 3° (Seedsman and Emerson 1985). This reduction does not occur gradually, as the clay fraction increases but, instead, relatively suddenly at a clay content of about 10%. At this clay content, the larger particles in the spoil are no longer in direct contact with each other but instead tend to be supported in a matrix of clay-sized particles. The weathering may occur at the surface of the spoil piles to a relatively shallow depth, or deep within the spoil piles because of a fluctuating water table. Chemical weathering reduces the friction angle by 6° to 12°. The physical breakdown of coal mine waste rock, as opposed to chemical processes, is a short-term process (Taylor and Spears 1970; Taylor 1984). Susceptibility to erosion also must be considered.

30.4.6 Progressive Failure

In waste rock slopes, shear strength may decrease significantly as a result of moisture softening. Moreover, waste rock materials exhibit characteristic brittleness overstressing also can lead to a marked decrease in shear strength. Progressive phenomena are important in such circumstances (Chowdhury, Nguyen, and Nemcik 1986).

Because moisture softening is important in promoting progressive failure, the extent of moisture infiltration, the wetting sequence, and the development of pore water pressures are of considerable importance. Other important factors are the development of tensile failure in the material and the potential for slip at the interface between old and recently placed material.

Delayed waste rock slope failure can be a source of disruption in rehabilitation works. It may be necessary to monitor waste rock slopes over a number of years and use the observational data to update stability analyses. The primary cause of delayed instability is infiltration from rainfall.

30.5 APPROPRIATE FACTOR OF SAFETY

The appropriate factor of safety or probability of failure for the stability of a waste rock embankment depends on the coefficients of variation of the strength parameters (cohesion showing considerably more variation than friction angle) and the waste rock density (which shows relatively little variation) and on the acceptable failure rate. An acceptable annual failure rate may be 1/10,000th of the area of the waste rock slopes, corresponding to "dams" defined by Whitman (1984). However, the acceptable failure rate and the appropriate factor of safety also depend on the consequences of failure, including potential loss of life, damage to infrastructure, and loss of function, both on and off lease. Consequences of the possible failure of waste rock embankments are difficult to assess and have to be site-specific.

Generally, waste rock slopes designed with a factor of safety of 1.10 to 1.15 have only a minor risk of failure (Khandelwal and Mozumdar 1992; Miller et al. 1979). Waste rock slopes designed for a factor of safety of less than 1.10 are subject to a greater risk, even if the input data used are accurate, because of variability in the height of the slope or of the strength of the waste rock material or foundation. These conditions may result in local fluctuations of about 10% in the factor of safety, leaving little safety margin.

The initial failure of a slope composed of loose waste rock is governed by the normally consolidated friction angle of the material. However, the waste rock layers exposed by a failure are overconsolidated. Consequently, following the initial failure, the dilation angle of the material becomes an important influence on the factor of safety of the exposed failure scarp. Normally, this will increase after each failure, as the degree of overconsolidation and dilation angle increase, even if the failure debris–the presence of which tends to decrease the overall slope angle–is removed. This means that stability is enhanced post-failure. It explains the apparently over-steep failure scarps that remain stable post-failure.

The magnitude of displacement of the waste rock during a slope failure can be a significant safety consideration. For example, large displacements might threaten roads, railways, or other structures. If the debris of waste rock slope failures is removed–by a stream at the toe of the slope, for example–the slope will retreat at the critical angle (the angle of repose for the material, in the absence of groundwater). Otherwise, the slope angle will gradually reduce because of the loss of material from the crest, and the accumulation of debris and erosion products at the toe.

30.6 SAFE OPERATION OF PLANT

Waste rock slope angles of 14° or flatter readily allow machinery to work along the contours, constructing contour banks, seeding, and fertilizing. A slope angle of 18° may be the upper limit for this purpose (Peterson 1987). A D9 dozer can work up and down slope angles of up to 27° (Walker 1987), but operators are more comfortable working on slope angles of less than 22°. Conventional dozers have lubrication problems on steep slopes (Williams 1997).

30.7 SETTLEMENT OF WASTE ROCK

30.7.1 Mechanisms and Components

Although the mechanisms of waste rock settlement are not well understood, they include the following:

- Particle reorientation
- Water-induced weakening of interparticle bonding
- Weathering (swell/slake) of high clay-content spoil materials
- Dispersion and transport of fine particles through the spoil or backfill

Components of the settlement of waste rock include primary settlement, creep settlement, and collapse settlement (Geoffrey Walton Practice 1991). Both primary and creep settlements are time-dependent, reducing with time. Collapse settlement is attributable to a reduction in the strength of the spoil materials when they become saturated, and probably results from partial breakdown of the material and/or crushing of weakened point contacts between rock particles. It may occur where water levels rise within the waste rock and happens immediately on wetting. It also can result from mining subsidence or surcharge loading and can be particularly severe when waste rock has been placed with no compaction.

Weathering of waste rock, an important factor in the development of settlement, is highly dependent on the amount and type of clay minerals present (Thomson et al. 1986; Naderian 1997). Rapid weathering occurs near the surface of the waste rock but also can occur at depth in the presence of groundwater.

Self-weight settlement of waste rock is rapid (Naderian 1997). Most of it occurs during placement of the material, and it continues for only a short time thereafter. Groundwater is the most significant agent causing subsequent settlement of the material and any volume change (Charles, Burford, and Hughes 1993; Naderian 1997).

30.7.2 Field Measurements

A review of the literature carried out by Naderian and Williams (1996) indicated a range of measured settlements under dry conditions of between 0.3% and 7% of the waste rock height, with further 1% to 4% settlement on groundwater recovery. Creep or consolidation settlement was shown to continue beyond 10 years after construction. Typical settlement (normalized relative to the initial fill thickness) versus time (after the end of backfilling) plots for waste rock backfilling of open pits in the Midlands coalfields of England are shown in Figure 30.3.

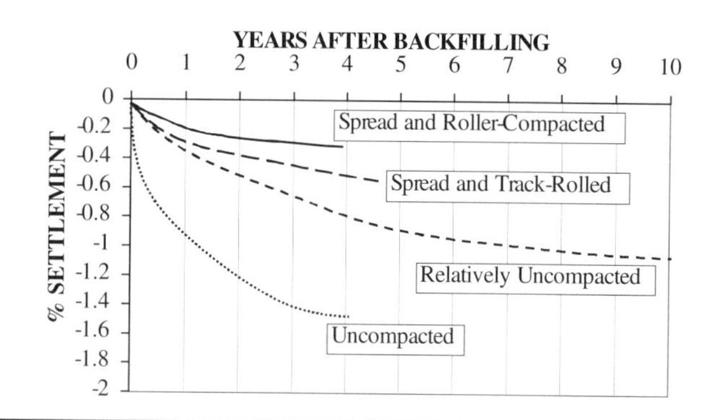

FIGURE 30.3 Backfill settlement versus time (Source: Knipe, personal communication, 1979)

FIGURE 30.4 Loose-dumping over the edge of a 260-m-deep pit

FIGURE 30.5 Differential settlement scarps in loose-dumped waste rock 260 m high

At Kidston Gold Mines in northeastern Queensland, Australia, durable waste rock loose-dumped over the edge of a 260-m-deep pit (Figure 30.4) sustained post-backfilling settlements of about 4 m (1.5% of the waste rock height) and differential settlement scarps 1 to 2 m high, which are a reflection of benches on the pit face (Figure 30.5).

30.7.3 Laboratory Measurements

Naderian and Williams (1996) carried out laboratory compression testing of initially loose sandstone and claystone waste rock

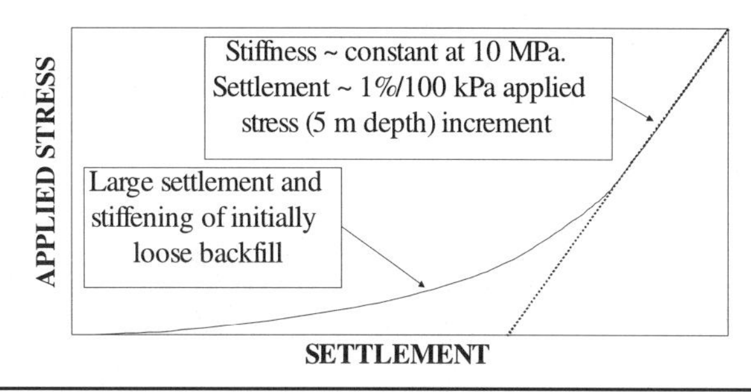

FIGURE 30.6 Applied stress versus settlement (source: Naderian and Williams 1996)

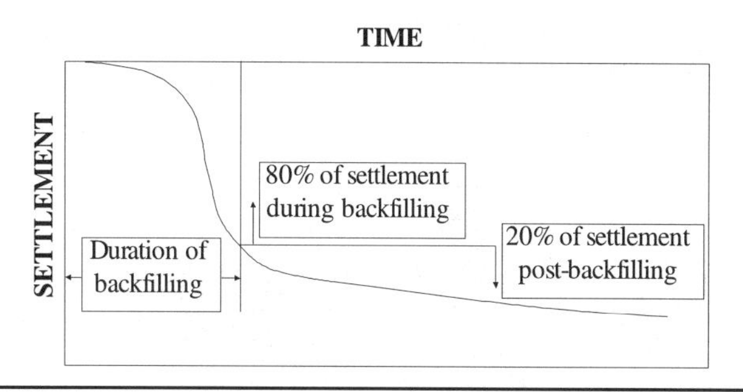

FIGURE 30.7 Settlement versus time (source: Naderian and Williams 1996)

materials from Jeebropilly Colliery, in southeastern Queensland, Australia. They showed that the material initially had little stiffness but became stiffer with increasing applied stress until a constant high stiffness was achieved above about 200-kPa applied stress (equivalent to about a 10-m depth of backfill; shown schematically on Figure 30.6). From 0- to 200-kPa applied stress, the sandstone specimens settled a total of about 4% and 5.5% for dry and saturated specimens, respectively, and the claystone specimens settled a total of about 11% and 20% for dry and saturated specimens, respectively.

The difference between the dry and saturated settlements for each material indicates the magnitude of the potential for settlement on groundwater recovery. Inundation settlement or collapse occurs rapidly. For the sandstone, inundation collapse was about 40% of the dry settlement as a result of application of the stress, whereas the corresponding figure for the claystone was about 80%.

Above 200-kPa applied stress, the sandstone and claystone specimens, whether tested dry or saturated, attained a constant stiffness of about 10 MPa, implying a settlement of 1% for each additional 100-kPa (5-m depth of backfill) increment of applied stress (Figure 30.6).

Naderian and Williams (1996) modeled numerically (using Fast Lagrangian Analysis of Continua [FLAC]) the backfilling of a typical 60-m-deep pit at Jeebropilly Colliery with randomly distributed, equal volumes of sandstone and claystone. The backfilling method employed at Jeebropilly was end-dumping from trucks without compaction. The results indicated typical total (dry) settlements of about 2.5% of the backfill depth and differential settlements on the order of 1% of the backfill depth. Inundation settlements were calculated to be on the order of 0.4% of the depth of saturated backfill but would occur gradually over the time it took for groundwater recovery.

The analysis indicated that about 80% of the total (dry) settlement would occur during the 6 months it would take for the backfill to be placed (as shown schematically on Figure 30.7); that is, total and differential (dry) post-backfilling settlements of only 0.5% and 0.2% of the backfill depth, respectively. Inundation settlements of the order of 0.4% of the saturated backfill depth would occur post-backfilling, giving a total potential post-backfilling backfill settlement of up to about 1% of the backfill depth. The settlement rate would decrease after the end of backfilling, reaching a constant rate of about 0.02% per year one year after the end of backfilling. In the long-term, settlement would continue at roughly a constant rate per log cycle of time.

Settlement measurements carried out at Jeebropilly after the completion of backfilling and surface recontouring averaged about 6.4 cm over a 2-year period (0.05%/y). This rate agrees with the results of the numerical analysis.

30.8 BEARING CAPACITY OF WASTE ROCK

Plate bearing tests conducted by Naderian and Williams (1997) showed that the bearing capacity of mine waste rock is mainly a function of the compaction it has undergone, not its age. The compaction of backfill afforded by mine vehicle traffic seems to be sufficient to provide a reasonable bearing capacity and the ability to support ordinary lightweight structures. The densification the material undergoes upon loading results in an increase in its bearing capacity. Limited testing has shown that waste rock is capable of supporting stresses of up to 200 kPa, while sustaining settlements of the order of 5 mm and 17 mm for traffic-compacted and freshly placed waste rock, respectively.

30.9 EROSION PARAMETERS

Erosion rates are difficult to measure and even more difficult to predict with any accuracy. The accuracy of estimated erosion rates is often little better than 100% (So 1999). Data on erosion rates available in the literature are presented and discussed in the following sections.

30.9.1 "Natural" Erosion Rates

The history of ancient upthrust zones of the earth's crust, such as the MacDonnell Ranges in central Australia, gives an indication of "natural" erosion rates. The MacDonnell Ranges were formed about 310 million years ago to an estimated height of about 9 km. They now are an average 1 km high. Allowing for isostasy (the raising of the crust's surface as the weight of overburden is reduced by erosion), this is equivalent to an average erosion rate of 0.05 mm depth/y (about 0.74 t/ha/y, assuming a dry density of 1.5 t/m^3). This figure fits within the range of soil formation rates typically quoted from 0.01 to 0.10 mm/y.

30.9.2 Erosion Rates from Developed Land

The average river sediment yield for Australia has been variously estimated at between 0.32 t/ha/y and 0.45 t/ha/y (0.022 mm/y and 0.030 mm/y; Rieger and Ogilvie 1988). Raudkivi (1990) indicated erosion rates from mixed pasture and woodland generally in the range of 1.5 to 3.0 t/ha/y (0.10 to 0.20 mm/y), with a maximum rate of about 4.0 t/ha/y (0.27 mm/y). The erosion rate increases for arable land, but seldom exceeds 7.5 t/ha/y (0.51 mm/y) (Raudkivi 1990). For agricultural land use in the Darling Downs of South East Queensland, Australia, the "allowable" erosion rate is 15 t/ha/y (1.0 mm/y; Bell 1994).

Erosion rates from construction sites in Maryland and Virginia, range from 120 to 500 t/ha/y (8.1 to 34 mm/y) (Chen 1974). The erosion rate of bare soil in the humid tropics on a 4° slope is 100 to 170 t/ha/y (6.8 to 11.5 mm/y; Kirkby 1980). Erosion rates for unvegetated mine waste rock could be in the same range as for construction sites, perhaps even higher for highly erodable spoil. However, few observations have been made of erosion rates from angle-of-repose waste rock slopes.

The Queensland Department of Mines and Energy (QDME) target erosion rate for rehabilitated mine sites is 12 to 40 t/ha/y (0.81 to 2.7 mm/y; QDME 1993). This range is much higher (by typically 30-fold) than natural erosion rates and river sediment yields, higher than erosion rates from agricultural land by typically

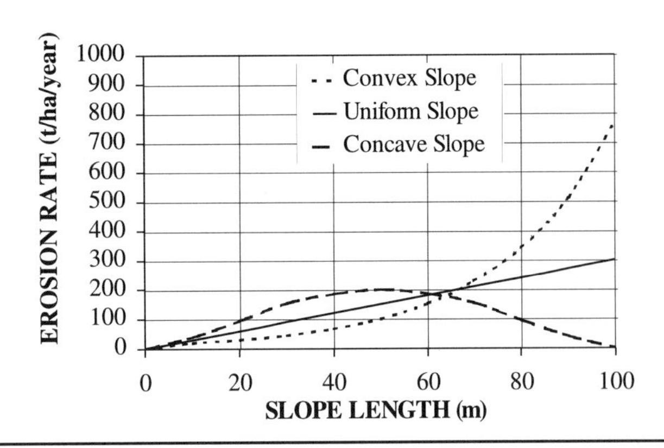

FIGURE 30.8 Form of variation of erosion rate with slope profile (source: Meyer and Kramer 1969, in Lal 1990)

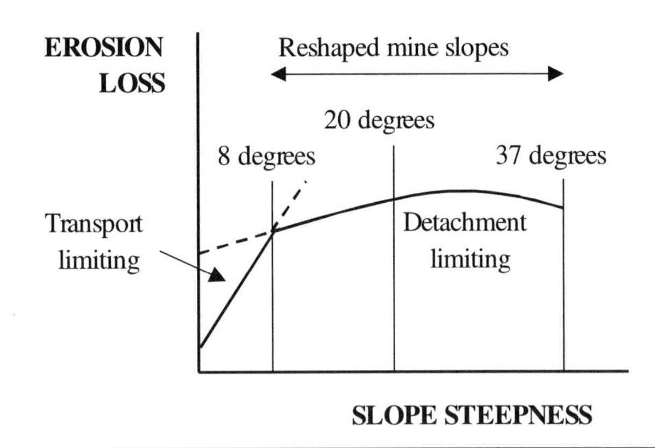

FIGURE 30.9 Erosion loss versus slope steepness up to angle of repose (adapted from Haan et al. 1982, in Lal 1990)

4-fold, but lower than erosion rates from construction sites by about 10-fold.

The erosion rate is strongly influenced by the reconstructed slope profile (Meyer and Kramer 1969 in Lal 1990). Concave slopes deliver water to the toe of the slope with less erosion than uniform slopes of the same average slope (Loch 1999). Convex slopes display an exponentially increasing erosion rate with increasing slope length. The erosion rate on uniform slopes tends to be more proportional to slope length, and the erosion rate on concave slopes peaks at intermediate slope lengths (on the order of 50 m) before decreasing for longer slopes. The effect of slope profile is shown diagrammatically on Figure 30.8.

Haan et al. (1982) in Lal (1990) suggested that the relationship between erosion loss and slope steepness over a large range of slope angles up to angle of repose, as shown schematically in Figure 30.9. In reality, there is a family of curves for different materials, stress histories, and climatic histories. Figure 30.9 identifies two regimes. For slope angles up to about 8° (14%), erosion loss is “transport-limited”; the supply of erodable materials is plentiful and erosion is limited by the carrying capacity of the runoff.

Under this regime, which applies to agricultural land use on relatively flat land, the erosion loss is directly proportional to the slope steepness. For slope angles steeper than about 8°, erosion loss is limited by the ability of the runoff to detach particles from the slope. This regime is described by a flat parabola, indicating that erosion loss is not strongly dependent on the slope steepness for slope angles greater than about 8°, which are typical of reshaped mine slopes.

Although supporting evidence for the form of the relationship given in Figure 30.9 is not readily available, the Chamber of Mines South Africa (1995) stated that erosion loss from coal mine waste rock reaches a maximum for slope angles of 25° to 35°. Relatively little erosion occurs on slopes flatter than 20° or steeper than 40°. Clearly, Figure 30.9 has implications for erosion loss from the relatively steep, final, mine waste rock slope angles.

30.9.3 Effect of % Surface Cover

The % surface cover (which could include grasses, trees, litter, and rock) on soil erosional loss is dramatic, particularly where the surface cover comprises large elements (So et al. 1998). For vegetative cover alone, 50% or more cover is required to dramatically lower erosion loss.

30.9.4 Erosion Prediction Based on USLE

The universal soil loss equation (USLE) (Hudson 1981) remains the most commonly used means of estimating soil erosion rates (Gray 1995). The equation, however, is based on US data from agricultural slope angles in the range from 0° to 5° (Barker 1995). Mine waste rock slopes typically are graded to 8° or steeper, well outside this range.

The USLE is used to predict the long-term average soil loss in runoff resulting from the combined effects of sheet and rill erosion (Wischmeier and Smith 1948). It does not take account of gully erosion and does not attempt to predict sediment deposition or transport within a catchment. The USLE is given by Evans, Aspinall, and Bell (1991) as

$$A = RKLSCP \qquad \text{EQ. 30.1}$$

Where:

A = computed annual loss per unit area (t/ha/y)
R = rainfall factor as an erosion index EI
K = soil erodability factor measured using a standard plot 22.13 m long sloping at 9%
= function of wet density, particle size distribution or texture and organic matter content
L = slope length factor
S = slope steepness factor generally lumped with L as the topographic factor LS
C = cover and management factor
P = conservation practice factor

The US Department of Agriculture Soil Conservation Service Curve Number Method (Boughton 1989; Schroeder 1994) has been traditionally used in Australia for estimating runoff from rainfall on small rural catchments and has been applied to mine waste rock by Schroeder (1994).

The factors in Eq. 30.1 may be estimated for the Bowen Basin coalfields in central Queensland, Australia, using data from Hudson (1981) and Evans, Aspinall, and Bell (1991). Average annual rainfall for the region is 600 to 750 mm, and R (or EI) may be taken as 335. Few measured data are available for K. Typical K values are 0.34 for topsoil and in the range of 0.05 for both sandy and heavy clayey spoil to 0.25 for silty clayey spoil.

The erosion loss increases with both increasing slope length and increasing slope steepness, but not linearly. Several effects are involved, and it is usual to combine the factors L and S into a single topographic factor LS. Estimates of LS derived from Evans, Aspinall, and Bell (1991) are given in Table 30.1 for slope angles of 8° and 15°, for both spoil peaks and ramps. Two cases were considered for each case. In the first, a spoil slope initially 20 m high was regraded to a maximum height of 10 m. In the second, a spoil area was regraded to a maximum height of 30 m without intermediate benches or contour banks. Johnstone (1992) reported that for the Curragh Mine, the length between benches is 70 m for a slope angle of 8°, corresponding to an LS value of 7.

Bare soil has a C value of 1, reducing to 0.005 or lower for good pasture. In the Bowen Basin, 40 to 50% vegetation cover is

TABLE 30.1 Estimates of LS applicable to the Bowen Basin spoil

Location	Slope Angle (deg)	Slope Length (m)	LS
Spoil peak to 10 m	8	7	7
Spoil peak to 30 m	8	212	12
Ramp to 10 m high	15	41	3.5
Ramp to 30 m high	15	124	6.3

TABLE 30.2 Estimates of erosion loss A for Bowen Basin spoil areas

		A (t/ha/yr)					
		Topsoil		Sandy or Heavy Clayey Spoil		Silty Clayey Spoil	
Slope Angle (%)	Slope Length (m)	Bare	50% Vegetated	Bare	50% Vegetated	Bare	50% Vegetated
14	71	399	199	59	29	293	147
14	212	683	342	101	50	503	251
27	41	199	100	29	15	147	73
27	124	359	179	53	26	264	132

reasonably stable against erosion. Erosion also is reduced by stony ground. For spoil rehabilitation in the Bowen Basin coalfields, C might range from 1 (bare ground initially) to 0.5 (vegetated). A standard 22.13-m-long, 9%, bare soil slope, cultivated up and down, has a P value of 1, reducing to 0.5 for contour cultivation and possibly even further for rehabilitated mine waste rock.

Inserting the best-guess estimates for the various factors into Eq. 30.1 yields the estimates of erosion loss A for Bowen Basin spoil areas given in Table 30.2, for two slope angles and lengths, various surface materials, and for initially bare and at least 50% vegetated conditions.

Based on the A values in Table 30.2, only the combinations highlighted would meet the QDME target erosion rate of 12 to 40 t/ha/y. Sandy or heavy clayey spoil cover comes close to achieving this in most of the cases considered. The vegetated silty clayey spoil comes close to achieving the target erosion rate for the shortest slope length, and the topsoil, whether bare or vegetated, is unlikely to achieve an acceptable erosion loss.

Table 30.2 shows that, for slope angles of 14% and 27%, increasing the slope length three-fold increases the erosion loss (for all materials) 1.7 and 1.8-fold respectively confirming that slope angle alone has little impact on erosion. The steeper the slope, the shorter it will be for a given height, decreasing the catchment size. Table 30.2 suggests that steepening–and hence shortening–the slope 1.8-fold halves the erosion loss (for all materials) for a given slope height (that is, a 3.6-fold effect overall). Based on this simplistic assessment, topsoil is the most erodable, silty clayey spoil is 73% as erodable as topsoil, and sandy and heavy clayey spoil are only 15% as erodable. Establishment of a reasonable vegetation cover reduces the erosion loss by about 50%. Although topsoil may facilitate vegetation, it is highly erodable prior to the establishment of vegetation.

30.9.5 MINErosion Program

In tropical environments, most of the erosion is the result of a limited number of high-intensity storms (Fairbairn and Wocker 1986). The spreadsheet program MINErosion is a single-rainfall, event-based model, giving an estimate of the erosion rate for a rainfall event of a particular recurrence interval (So et al. 1998). MINErosion applies to unvegetated Bowen Basin soil or spoil. The program was based on the results of rainfall simulator tests carried out in a tilting flume on 16 soil types and 17 spoil types from 16 Bowen Basin mine sites. The flume was 3 m long and tests were carried out at slope angles of 5%, 10%, 15%, 20%, and 30% (3°, 6°, 8°, 11°, and 17°). Up to three replicate tests were carried out on each material, with typically 25% agreement being achieved between replicates. Both inter-rill erosion from simulated rainfall and rill erosion from simulated runoff were measured.

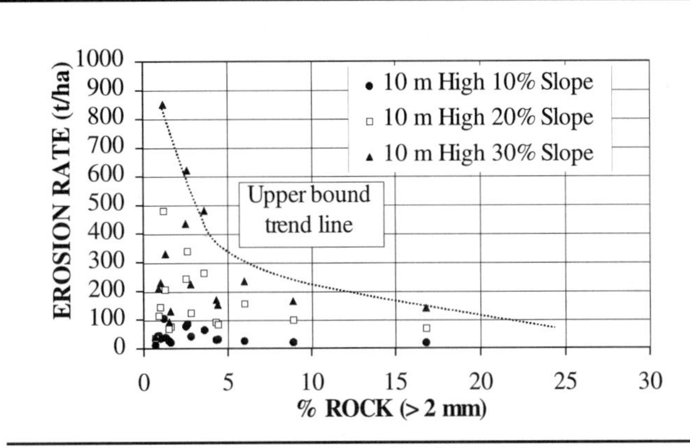

FIGURE 30.10 Erosion rate versus % rock for Bowen Basin soil materials

The measured data were extrapolated to cover slope angles from 0% to 50% (0° to 27°) and slope lengths from 0 m to 100 m, using available erosion prediction equations. The extrapolated results are presented in the program as sets of curves of erosion loss (in t/ha) versus slope length (to 100 m) for slope angles of 10%, 20%, and 30%, for a particular soil or spoil type, rainfall intensity and duration, rill spacing, and infiltration rate. The erosion rates given in the program have not been fully calibrated against large-scale field measurements, and their accuracy is unknown. However, erosion rates are difficult to predict and the accuracy of predictions is often little better than 100% (So 1999).

To gauge the effects on erosion rate of a number of key parameters, calculated data points have been taken from the sets of curves presented in the MINErosion program. Initially, the rainfall intensity was set at 100 mm/h, the rainfall duration at 30 min, and the rill spacing at 1 m. Data points then were read off the curves for all 16 soil materials and all 17 spoil materials, for 10-m-high slopes at slope angles of 10%, 20%, and 30%.

Figure 30.10 shows the predicted erosion rate versus percent rock (> 2 mm) for the soil materials. Although there is considerable scatter, the erosion rate clearly decreases with increasing percent rock, with apparently little erosion for >30% rock.

The calculated data points for a 10-m-high slope in a range of Bowen Basin spoil materials are plotted against slope angle on Figure 30.11, together with possible extrapolations to angles up to the angle of repose of the material (75% or 37°). The extrapolations include the trend lines for the calculated data points and an alternative extrapolation based on Figure 30.9. Clearly, there is a strong need to collect erosion data for slope angles in the range of 30 to 75%.

Limited erosion data from waste rock slopes at an iron ore mine in the Pilbara region of Western Australia also are included on Figure 30.11. These lend support to the flat parabola suggested by Haan and co-workers for steeper slopes; that is, a roughly constant erosion rate for a range of slope angles. These erosion data are converted to an erosion rate per unit slope width and plotted against slope height on Figure 30.12. Figure 30.12 demonstrates that, for a given slope height, steeper slopes will erode less than flatter slopes, as the catchment length decreases with increasing slope steepness. The Pilbara waste rock exhibits a threshold slope height below which there is negligible erosion. This threshold slope height is strongly dependent on the age of the material. For recently placed waste rock, the threshold slope height for no erosion is about 13 m, whereas for old spoil that has undergone self-armoring, the threshold slope is about 43 m.

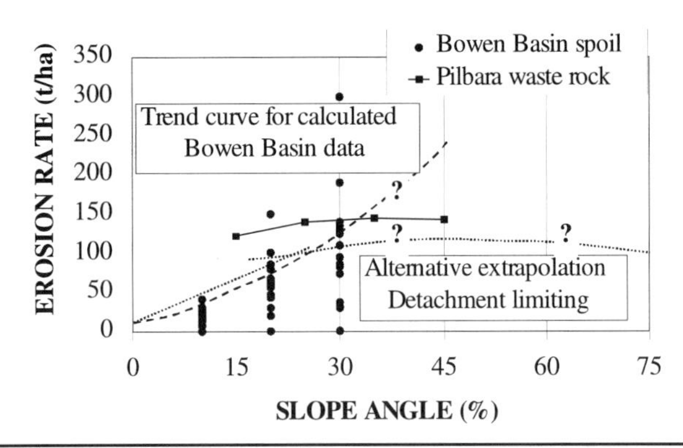

FIGURE 30.11 Erosion rate versus slope angle for Bowen Basin spoil materials

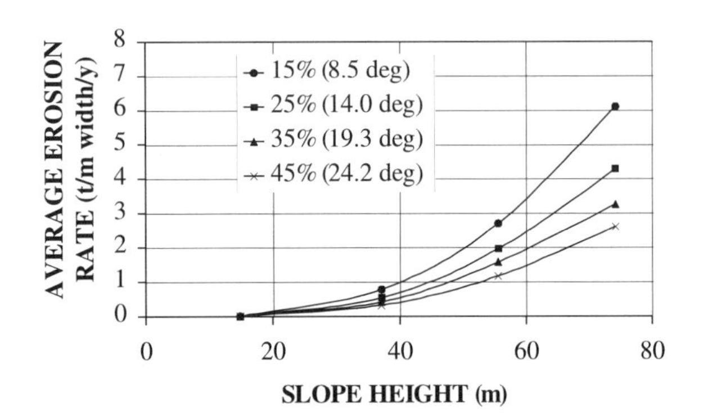

FIGURE 30.12 Erosion rate per unit slope width versus slope height for Pilbara waste rock slopes at different angles

The effects on erosion rate of some of the other parameters are highlighted in Table 30.3. The relative effect on soil and spoil materials is about the same in all cases. Overall, increasing the rainfall intensity has the most dramatic effect on erosion rate, followed by rainfall duration, rill spacing, and infiltration rate.

At Kidston Gold Mines, limited erosion rates have been monitored since 1996/1997 on various durable waste rock slopes either at angle of repose (37° to 40° or 75% to 84%) or regraded to 20° (36%), and with various surface covers. Annual rainfall totals were 524 mm, 623 mm, and 475 mm for the 1996/1997, 1997/1998, and 1998/1999 years, respectively. The results are summarized in Table 30.4, which highlights the negligible erosion from durable waste rock either at angle-of-repose or half-angle-of-repose slopes, the high erosion from erodable oxide waste rock covers without vegetation or with limited vegetation, and the greatly reduced erosion from well vegetated slopes.

30.10 CONCLUSIONS

This chapter has dealt with the assessment of material parameters relevant to the use of mine waste rock or spoil in the construction of embankments. The parameters considered include the geometry of the embankment, and the strength, density, moisture content, settlement, bearing capacity and erosional characteristics of the waste rock and cover materials. Location of the phreatic surface and the foundation conditions also have been briefly considered.

30.11 ACKNOWLEDGMENTS

I gratefully acknowledge the assistance of my colleague at the University of Queensland, Dr. Peter Morris, for his assistance in searching the literature discussed in this chapter. The Australian Coal Association Research Program (ACARP) is acknowledged for providing financial support that enabled a literature review of the subject matter.

TABLE 30.3 Effect of other parameters on erosion rate

Parameter	Effect on Erosion Rate: 10% Slope Angle	30% Slope Angle
Rainfall intensity:		
2-fold increase from 50 to 100 mm/hr	4-fold increase	3-fold increase
Rainfall duration:	Linearly related	Linearly related
Rill spacing:		
2-fold increase from 1 to 2 m	1.2- to 1.4-fold increase	1.35- to 1.55-fold increase
Infiltration rate:		
2-fold increase from 15 to 30 mm/hr	25% decrease	35% decrease

TABLE 30.4 Measured erosion rates from Kidston waste rock embankment slope

	1996/1997		1997/1998		1998/1999	
Description	Vegetative Cover (%)	Erosion (t/ha/yr)	Vegetative Cover (%)	Erosion (t/ha/yr)	Vegetative Cover (%)	Erosion (t/ha/yr)
20°, no oxide cover or vegetation	0	1	0	0	0	0
20°, oxide cover, no vegetation	0	386	0	186	0	425
20°, oxide cover, vegetation	33	359	41	112	48	68
37°, no oxide cover or vegetation	0	0	0	0	0	0
37°, oxide cover, vegetation	10	68	15	1.8	15	19
17°, oxide cover, 1991 vegetation	> 50	23	> 50	0.4	> 50	5.6
37°, oxide cover, 1991 vegetation	> 50	31	> 50	0.3	> 50	2.2

30.12 REFERENCES

Anderson, M.G., and D.M. Lloyd. 1991. Using a combined slope hydrology-stability model to develop cut slope design charts. *Proceedings Institution of Civil Engineers*, 91: 705–718.

Azzouz, A.S., M.M. Baligh, and C.C. Ladd. 1983. Corrected field vane strength for embankment design. *ASCE Journal of Geotechnical Engineering*. 109: 730–734.

Barker, D.H. 1995. The way ahead—continuing and future developments in vegetative slope engineering or ecoengineering. *Proceedings International Conference on Vegetation and Slopes—Stabilisation, Protection and Ecology,* September 1994, Oxford. 238–257. London: Institution of Civil Engineers.

Bell, L.C. 1994. Private communication.

Boughton, W.C. 1989. Evaluating partial areas of watershed runoff. *ASCE Journal of Irrigation and Drainage Engineering*. 113: 356–366.

Brown, C.B., and M.S. Sheu. 1975. Effects of deforestation on slopes. *ASCE Journal of Geotechnical Engineering*. 101, GT2: 695–699.

Carson, M.A., and D.J. Petley. 1970. The existence of threshold hillslopes in the denudation of the landscape. *Transactions of the Institute of British Geographers*. 14: 71–95.

Chamber of Mines South Africa. 1995. *Guidelines for Environmental Protection, Volume 1, The Engineering Design, Operation and Closure of Metalliferous, Diamond and Coal Residue Deposits*.

Charles, J.A., D. Burford, and D.B. Hughes. 1993. Settlement of opencast coal mining backfill at Horsley 1973–1992. In *Engineered Fills*. London: Thomas Telford. 429441.

Chen, C.N. 1974. Evaluation and control of erosion in urbanising watersheds. *Proceedings National Symposium on Urban Rainfall and Runoff and Sediment Control, University of Kentucky, Lexington*.

Chowdhury, R.N., and V.U. Nguyen. 1987. Risk updating for rainfall-triggered spoil failures. *Proceedings International Symposium on Prediction and Performance in Geotechnical Engineering*. June 1987, Calgary. 439-444.

Chowdhury, R.N., V.U. Nguyen, and J.A. Nemcik. 1986. Spoil stability considering progressive failure. *Mining Science and Technology*. 3: 127–139.

Collinson, A.J.C., M.G. Anderson, and D.M. Lloyd. 1995. Impact of vegetation on slope stability in a humid tropical environment: a modelling approach. *Proceedings Institution Civil Engineers, Water, Maritime and Energy*. 168–175.

Collinson, A.J.C., and M.G. Anderson. 1996. Using a combined slope hydrology/stability model to identify suitable conditions for landslide prevention by vegetation in the humid tropics. *Earth Surface Processes and Landforms*. 21: 737–747.

Emerson, W.W., and R.W. Seedsman. 1982. A field test to predict the behaviour of overburden materials during mining: Modified Emerson crumb test. CSIRO Australia, Division of Geomechanics Australia, Technical Memorandum 15.

Evans, K.G., T.O. Aspinall, and L.C. Bell. 1991. Erosion prediction models and factors affecting the application of the universal soil loss equation to post-mining landscapes in central Queensland. *Proceedings Queensland Coal Symposium, Brisbane*. 123–132.

Fairbairn, D.M., and G.H. Wocker. 1986. A study of soil erosion on vertisols of the eastern Darling Downs, Queensland. 1. Effect of surface conditions on soil movement within contour bays. *Australian Journal of Soil Research*. 24: 135–158.

Geoffrey Walton Practice. 1991. *Handbook on the Design of Tips and Related Structures*. London: HMSO. 100, GT4:695–699.

Gray, D.H. 1995. Influence of vegetation on the stability of slopes. *Proceedings International Conference on Vegetation and Slopes—Stabilisation, Protection and Ecology,* Sept. 1994, Oxford. London: Institution of Civil Engineers. 2–25.

Holtz, W.G. 1960. The effect of gravel particles on friction angle. *Proceedings ASCE Research Conference on Shear Strength*. 1000–1001.

Holtz, W.G., and H.G. Gibbs. 1956. Triaxial shear tests on pervious gravelly soils. *Proceedings ASCE, Journal of Soil Mechanics and Foundation Engineering Division*. 82, 867: 1–22.

Hubble, T., and T. Hull. 1996. A model for bank collapse on the Nepean River, Camden Valley, NSW. *Australian Geomechanics*. 29: 80–98.

Hudson, N. 1981. *Soil Conservation*. London: B.T. Batsford Ltd.

Johnstone, P. 1992. Rehabilitation strategies and planning at Curragh Mine. *Proceedings 3rd Large Open Pit Mining Conference,* Mackay, Queensland. 299–311.

Khandelwal, N.K., and B.K. Mozumdar. 1992. Preliminary design of spoil dumps. *Proceedings 2nd International Conference on Environmental Issues and Management of Waste in Energy and Mineral Production,* September 1992, Calgary, Canada. 2: 1031–1035.

Kirkby, M.J. 1980. *Soil Erosion*. Chichester, England: Wiley.

Knipe, C. 1979. Private communication.

Lal, R. 1990. *Soil Erosion in the Tropics, Principles and Management*. Sydney: McGraw-Hill.

Lambe, T.W., and R.V. Whitman. 1979. *Soil Mechanics, SI Version*. New York: Wiley.

Lawrance, C.J. 1995. Low cost engineering measures and vegetative measures for stabilising roadside slopes in Nepal. *Proceedings International Conference on Vegetation and Slopes—Stabilisation, Protection and Ecology,* 29–30 September 1994, Oxford. London: Institution of Civil Engineers. 142–151.

Leps, T.M. 1970. Review of shearing strength of rockfill. *Proceedings ASCE, Journal of Soil Mechanics and Foundation Engineering Division*. 96, SM4: 1159–1170.

Loch, R. 1999. Private communication.

Middleton, G.V. 1970. Experimental studies related to problems of flyash sedimentation. In *Flyash Sedimentation in North America,* ed. J. Lajoie. Geological Society of Canada. 253–272.

Miller, R.P., P.M. Douglass, R.A. Robinson, D.A. Roberts, and D.A. Laprade. 1979. *Surface Mine Spoil Stability Evaluation*. USBM-OFR 78 (1) & (2) 80.

Naderian, A.R. 1997. *Characterisation of behaviour of open-cut coal mine back-fill*. PhD Thesis, University of Queensland.

Naderian, A.R., and D.J. Williams. 1996. Simulation of open-cut coal mine back-fill behaviour. *Proceedings National Symposium on the Use of Recycled Materials in Engineering Construction,* May 1996, Sydney, Australia. Canberra: IEAust National Conference Publication No 96/06. 17–22.

Naderian, A.R., and D.J. Williams. 1997. Bearing capacity of open-cut coal-mine backfill materials. *Transactions IMM, A: Mining Industry*, 106: A30–A33.

Peterson, T. 1987. Waste management and stabilisation. *Workshop on Environmental Management in the Goldfields,* 19–20 May, Chamber of Mines of Western Australia. 15–32.

Queensland Department of Minerals and Energy. 1990. Environmental *Management for Mining in Queensland*. Brisbane: Queensland Department of Minerals and Energy.

Rassam, D.W., and D.J. Williams. 1999. Three-dimensional effects on slope stability of high waste rock dumps. *International Journal of Surface Mining, Reclamation and Environment.* 13: 19–24.

Raudkivi, A.J. 1990. *Loose Boundary Hydraulics*, 3d ed. Oxford: Pergamon.

Rieger, W.A., and L.J. Ogilvie. 1988. Channel Sediment Loads: Comparisons and Estimation. In *Fluvial Geomorphology in Australia* ed. R.F. Warner Sydney: Academic Press. 69–85.

Roscoe, K.H., A.N. Schofield, and C.P. Wroth. 1958. On the yielding of soils. *Geotechnique*. 8, 1: 2253.

Rowe, P.W. 1962. The stress-dilatancy relation for static equilibrium of an assemblage of particles in contact. *Proceedings Royal Society*, A269. 500-527.

Schroeder, S.A. 1994. Reliability of SCS curve number method on semi-arid reclaimed minelands. *International Journal of Surface Mining, Reclamation and Environment*. 8: 41–45.

Seedsman, R.W., and W.W. Emerson. 1985. The role of clay-rich rocks in spoil pile failures at Goonyella Mine, Queensland, Australia. *International Journal of Rock Mechanics and Mining Science*, 22, 113–118.

Schor, H.J. and D.H. Gray. 1995. Landform grading and slope evolution. *Proceedings of ASCE, Journal of Geotechnical Engineering Division*, 121, 729–734.

Simons, D.B. and M.L. Albertson. 1960. Uniform water conveyance channels in alluvial material. *Proceedings of ASCE, Journal of Hydrology Division*, 86, HY5, p. 33.

So, H.B. 1999. Private communication.

So, H.B., G.J. Sheridan, R.J. Loch, C. Carroll, G. Willgoose, M. Short, and A. Grabski. 1998. Post mining landscape parameters for erosion and water quality control. *Final Report on ACARP Projects C1629 and C4011*.

Stratham, I. 1974. The relationship of porosity and angle of repose to mixture proportions in assemblages of different sized materials. *Sedimentology*. 21: 149–162.

Taylor, R.K. 1984. *Composition and Engineering Properties of British Colliery Discards*. London: National Coal Board.

Taylor R.K., and D.A. Spears. 1970. The breakdown of British coal measure rocks. *International Journal of Rock Mechanics and Mining Science*. 7: 481–501.

Thomson, S., J.D. Scott, D.C. Sego, and T.M. Schulz. 1986. Testing of model footings on reclaimed land, Wabamun, Alberta. *Canadian Geotechnical Journal*. 23: 541–547.

Walker, K.J. 1987. Moonscaping, a method of slope stabilisation developed and used on Mt Whaleback, by Mt Newman Mining Company, Western Australia. *Workshop on Environmental Management in the Goldfields,* 19–20 May 1987, Chamber of Mines of Western Australia. 33–41.

Whitman, R.V. 1984. Evaluating calculated risk in geotechnical engineering. *ASCE Journal of Geotechnical Engineering*. 110, 2: 145–188.

Williams, D.J. 1996. Role of geomechanics in minesite rehabilitation. *Proceedings 7th Australia-New Zealand Conference on Geomechanics,* July 1996, Adelaide, 850856. Canberra: Institution of Engineers, Australia. NCP 96/07.

Williams, D.J. 1997. Natural versus constructed rehabilitation of open pit mine sites worldwide. *Australian Minerals and Energy Environment Foundation Occasional Paper*. 83 p.

Wischmeier, W.H., and D.D. Smith. 1948. Predicting rainfall erosion losses: A guide to conservation planning. US Department of Agriculture, Handbook No. 537. National Technical Information Service, Springfield, Virginia.

CHAPTER 31

The Mechanism Controlling Angle-of-Repose Stability in Waste Rock Embankments

David B. Campbell*

31.1 INTRODUCTION

The mechanism that controls the stability of waste rock embankments is different from the failure mechanisms that are commonly assumed and on which many of the methods of stability analyses are based. If the potential failure model employed in a stability analysis differs significantly from the true mechanism that controls stability, assessments of stability as might be inferred from the results of conventional methods of analyses are likely to be unreliable. The mechanism that controls the stability of a waste rock embankment also has a bearing on the initiation of the ensuing debris flow that may develop in the event of an embankment failure.

The failure mechanism presented here is based on my field observations during more than three decades of consulting practice relating to the stability and the performance of waste rock embankments located in the mountainous regions of North and South America, Europe, and Asia.

31.2 PHYSICAL CHARACTERISTICS OF THE WASTE ROCK EMBANKMENT

The waste rock embankments discussed in this chapter are associated with open-pit, truck-shovel mining operations. The structures consist of waste rock that is removed in the course of accessing the ore. The waste rock is deposited by end-dumping at the crest. As a result, throughout the intervals that these facilities remain active, the advancing faces remain at the angle of repose for the waste rock.

Embankment heights, which represent the differences in the elevations of the crests and the toes of the slope, may be up to 300 m or more. The topographic slopes over which the piles have been developed are as steep as 2 horizontal to 1 vertical, about 26°. The foundations over which the embankment is constructed consist of dense glacial till soils or, alternatively, of colluvium over in situ rock.

Field inspection of waste rock embankments reveals deformation characteristics that indicate that the mechanism that controls stability is not in conformance with the assumptions inherent in the conventional methods of stability analyses. In the conventional methods of stability analyses, the potential failure mass is divided into a series of segments, and the forces that act across the boundaries between adjacent segments are adjusted so the factor of safety of all of the segments comprising the potential failure mass is at the same value.

FIGURE 31.1 Crest region of a stable waste rock embankment. Note: The photograph shows the crest region of a stable embankment where the displacements that have developed on discrete shear surfaces total more than 3 meters. Displacements at the crest are measured using the "clothesline" apparatus, which consists of the stands and the wire that appear in the photo. The near end of the wire is attached to a pin on the sloping surface of the embankment. Interpretation of the rates at which the crest movements develop provides a warning of approaching instability.

31.3 MOVEMENTS

In the region of embankment crests, shear displacements totaling several meters are common. At the same time, the structures remain stable. Figure 31.1 shows an example of the movements that have developed at the crest of a disposal site that remains

* Specialist Consultant, Golder Associates Ltd., Burnaby, British Columbia, Canada.

stable. Clearly, the significant displacements that have developed on discrete shear surfaces demonstrate that, along those surfaces, the shear strength of the waste rock has been fully mobilized.

Full mobilization of the shear strength parameters on a surface corresponds to a factor of safety of 1.0 on that surface. It follows that at the perimeter of a waste rock embankment, the factor of safety for all segments of the potential failure mass cannot be the same value. This calls into question the validity of the conventional methods of stability analyses in evaluating the stability at the embankment perimeter.

When waste rock is end-dumped at the crest, it rolls and slides down the sloping face. As a result, the waste rock on the face is initially at a density that corresponds to the steady state at low normal stress. The embankment advances through the process of gradual accretion of material on the face, and the mass of waste rock below the face undergoes compression in response to the progressively increasing normal stresses imposed by the weight of the overlying waste rock. In addition, with the sloping face at the angle of repose, the waste rock extending to at least modest depth below the face is subjected to shearing stresses at levels that are only moderately lower than the maximum available shearing resistance.

For these stress conditions, the waste rock also undergoes shearing strains. Thus, compression as well as shearing strains contribute to the movements that are manifest in the region of the crests. For a high waste embankment, movements in the region of the crest may develop at rates greater than 1 m/d. Nevertheless, the structure remains stable. Movements in the region of the crests are so large, and they develop so rapidly, that the installation of instrumentation–for example, slope indicator casings, to measure the internal strains that develop in the region behind the face–is usually impracticable. Because of the high kinetic energy attained by individual boulders that roll down the faces, installation of instrumentation on the faces and in the region of the toes of active waste rock embankments is also impracticable.

31.4 THE MODEL WASTE ROCK EMBANKMENT

A model waste embankment was constructed in the laboratory to obtain some insight into the internal strains and the pattern of the deformations that develop within the perimeter regions of waste rock embankments. This model, consisting of sand and fine gravel, was confined between two sheets of Plexiglas. Horizontal bands of Ottawa sand were introduced within the section of the model to serve as markers that would show the internal displacements that developed within the model, in response to displacements of the toe. The base of the model consisted of an inclined layer of small plastic spheres to simulate a sloping foundation surface having an angle of shearing resistance lower than the angle of internal friction for the waste rock comprising the body of the structure.

The toe of the model embankment was supported by a plate oriented perpendicular to the base of the model. In turn, the plate was supported by a screw mechanism that permitted controlled yielding of the toe. Figure 31.2 (top) shows the model before any deflection of the toe had taken place. The horizontal Ottawa sand markers are clearly visible in the photograph. The size of the model is indicated by the 10-cm spacing of the square reference lines marked on the Plexiglas.

After less than 5 mm of toe displacement, it was evident that two inclined shear surfaces had developed within the model. These shear surfaces were planar, and they extended from the surface to the base of the model embankment. Figure 31.2 (bottom) shows the model after the supporting toe plate had moved downslope a distance of 40 mm. The significant items to note in the model after the 40 mm of toe displacement are the following:

FIGURE 31.2 (top) Model embankment prior to movement at the toe Note the grid spacing is 10 cm. (bottom) Model embankment after toe had moved downslope through 4 mm. Note that the planar shear surfaces that divide the region below the dump face into two wedge-shaped segments. Also note the bench where the mutual boundary between the two wedges intersects the face.

- Internal shear surfaces developed beneath the embankment face. These shear surfaces are planar, extend to the base of the model, and divide the region below the face into two wedge-shaped segments.
- Significant differential shear displacements developed on the shear surfaces that form the wedge boundaries.
- A small bench developed on the face of the model, at the location where the mutual boundary between the two wedge-shaped segments intersect the face.
- A modest reduction in inclination occurred over the face of the upper wedge, i.e., over the segment of the face extending between the crest and the bench.

Measurement of the spacing between the horizontal bands of Ottawa sand showed that the movement of the toe had been accompanied by vertical contraction of the material comprising the upper wedge. This indicates that the material comprising the upper wedge also had expanded horizontally. This infers that, as a result of the movements, the materials comprising the upper wedge had been reduced to the active Rankine state of stress. The upper wedge, therefore, is referred to as the *active wedge*. Because the materials comprising the lower wedge were not in the passive Rankine state, the lower wedge is referred to simply as the toe wedge.

FIGURE 31.3 A bench on the face of a waste rock embankment. Note the location is where the mutual boundary of the active wedge and the toe wedge intersects the dump face.

31.5 COMPARISON WITH FEATURES OBSERVED IN THE FIELD

31.5.1 Nonlinearity of Embankment Faces

When standing at the crest of a high waste rock embankment, a person will commonly find that the face is slightly convex, with the result that the toe is obscured from view. Prior to failure, the upper region of the sloping surface typically becomes flatter, while the lower region of the face becomes modestly steeper.

31.5.2 Benches on Embankment Faces

In the field, if the downslope displacement of the toe wedge is large enough to result in development of a bench on the face, the movements commonly proceed to general failure. As a result, only rarely can benches on embankment faces be observed in the field. Occasionally, however, displacement of the toe wedge stops without progressing to general overall failure. In such cases, benches on the faces are in evidence.

Figure 31.3 shows an example of a bench on the face of a waste rock pile at the Faro open-pit lead–zinc mine in the Yukon, Canada. Disturbance of the foundation soils at the toe attested to the fact that forward movement of the toe wedge had taken place as a result of shearing along its base. I have observed similar examples of benches on waste rock embankment faces at open-pit coal mines in British Columbia, Canada, and at copper mines in Peru. These examples are noted in the chart presented in Figure 31.4.

31.5.3 Planar Wedge Boundaries

In the field, it is seldom possible to trace the wedge boundaries that extend through the interior of the structure or to confirm that they intersect the foundation in a manner similar to the wedge boundaries exhibited by the model. Figure 31.5 illustrates one field example at an open-pit coal mine where the wedge boundaries were clearly evident. The facility shown in Figure 31.5 was in its initial stages of development when a failure occurred. In this early stage of development, the geometry of the forward part of the embankment resembled a segment of a truncated right vertical cone. The failure resulted in removal of the central and highest portion of the cone, leaving in place the flanks on either side of the central failure.

The curved shear surfaces behind the failure scarp are inclined steeper than the prefailure face of the slope, which was inclined at the angle of repose for the waste rock. As such, the lines of intersections of these shear surfaces with the inclined surface of the conical-shaped surface of the waste rock embankment represent segments of hyperbolae. When viewed in a direction perpendicular to the azimuth of a vertical plane through the centerline of the failure, these shear surfaces were planar in form, and they were inclined at 55° to the horizontal.

As is evident in the photo, these planar shear surfaces extended to the foundation. Traces of the mutual boundaries between the active wedge and the toe wedge are evident on the remnant of the original conical surface of the pile. The arrow on the photo marks the most prominent member of the set of mutual wedge boundaries. These mutual wedge boundaries were also planar, inclined at 50° to the horizontal, and extended to the foundation.

Figure 31.6 shows a section through the centerline of the prefailure structure. The section shows measured inclinations of the wedge boundaries. The attitudes of the wedge boundaries shown on Figure 31.6 are approximately consistent with the attitudes of the wedge boundaries that developed in the model in response to downslope deflection of the toe. In the prototype embankment, the locations of the wedge boundaries move forward as the position of the face advances. This is one of the reasons for the series of wedge boundaries as they appear in Figure 31.5.

31.6 DOWNSLOPE DISPLACEMENT OF TOE WEDGE

The stability of an embankment depends on the stability of the toe wedge. Provided that the toe wedge remains stable, it continues to provide support to the active wedge, with the result that the whole face of the structure remains stable. Stability of the toe wedge is dependent on the resistance to downslope sliding on a surface located at or near its base.

When waste rock is end-dumped at the crest, the largest of the rock fragments tend to separate from the general dumped mass and roll down the face of the embankment. Because of the kinetic energy these large rock fragments attain as they roll down the face, they come to rest within a zone extending a modest distance beyond the locus of the line of intersection of the angle-of-repose embankment face and the foundation. Consequently, the largest and most durable fraction of the waste rock accumulates within a zone that extends a modest distance ahead of the toe.

As the face advances, this zone of coarse segregated rock (CSR) becomes covered, to form a very coarse drainage layer that extends over the base of the structure. An example CSR in the toe region of a waste rock embankment is illustrated in Figure 31.7.

When failure occurs at a waste rock pile that is built on a steeply sloping foundation, the failure debris commonly becomes mobile and can travel surprisingly long distances. Examination of the distal portion of the dry avalanche debris originating from waste rock embankment failures shows that the materials consist of CSR that is virtually devoid of fine particles, except for the materials that may have been incorporated as the debris advanced along the runout path.

An example of the segregated rock material at the limit of a debris run from an embankment failure is illustrated in Figure 31.8. This material is the CSR that was extant in the region of the toe of the prefailure structure.

From examination of the gradation of the rock illustrated in Figure 31.7, strain-generated pore water pressures clearly could not develop, nor could pore pressures be sustained within this coarse material. It follows that shearing strains would have little effect on the shearing resistance within this coarse zone. It is therefore evident that the shear surface must be located either above or below this zone. That is, the position of the potential failure surface on which downslope sliding of the toe wedge occurs, and on which stability of the toe wedge depends, must be located either within waste rock or, alternatively, within the foundation at shallow depth below the base of the waste rock.

If the surface of sliding on which the stability of the toe wedge depends were located above the base zone of CSR (that is within the body of the embankment), the failure mass could be

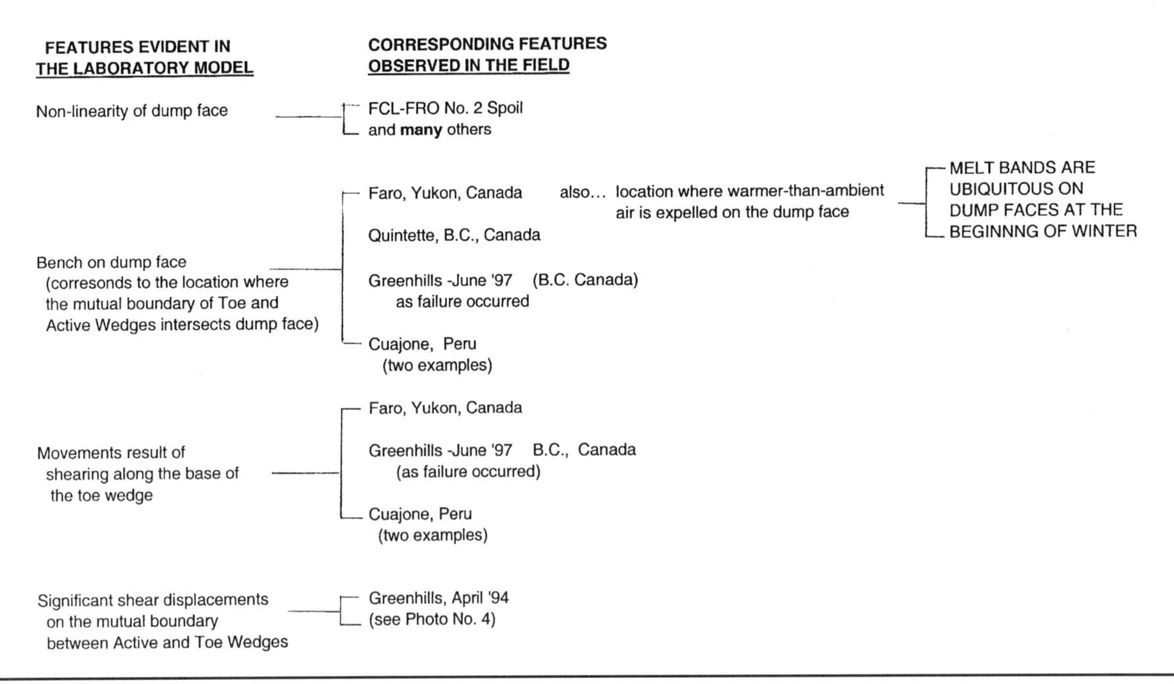

FIGURE 31.4 Correlation of features. The chart summarizes the correlation of pertinent features that develop in response to toe movements at the laboratory model embankment with features observed in the field.

FIGURE 31.5 Failed waste rock embankment. The back boundaries of the active wedge are planar, and they extend to the foundation. The arrow on the photo marks a trace of the mutual active wedgepassive wedge boundary that intersects the original conical surface of the embankment face remaining on the near side of the failure.

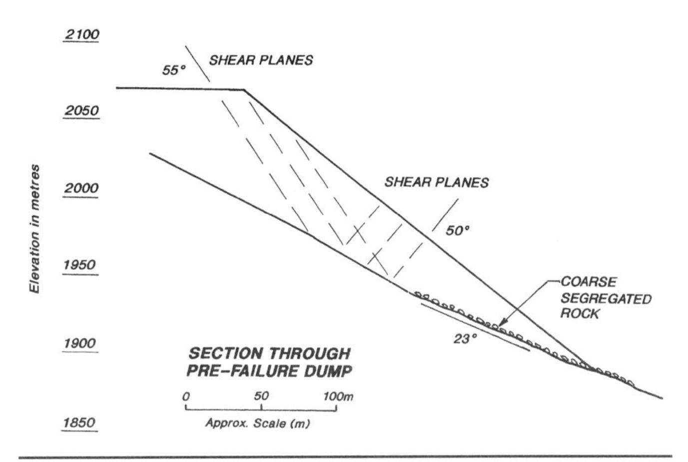

FIGURE 31.6 Section through prefailure embankment shown in Figure 31.5

expected to pass over the CSR in the region of the toe. In this case, the CSR in the region of the embankment toe would be unlikely to appear within the distal portion of the debris run. For the failure surface located at shallow depth below the base of the toe wedge, however, the CSR extant in the toe region of the pre-failure angle-of-repose face would be transported at the leading edge of the failure debris.

The fact that CSR invariably forms the distal portion of the debris originating from failed structures indicates that the toe wedge moves downslope as a result of sliding on a failure surface that is located within the foundation, at shallow depth below the base of the toe wedge. Referring to the chart on Figure 31.4, shearing within the foundation at shallow depth below the base of the toe wedge was evident in all of the field examples where benches had developed on the embankment faces.

The field observations show that the stability of the toe wedge and, in turn, the stability of the embankment face is dependent on the resistance to downslope sliding of the toe wedge on its base. The toe wedge is analogous to the retaining wall.*

31.7 VIDEO OF A WASTE ROCK EMBANKMENT FAILURE

In 1997, failure of a waste rock pile at one of the coal mines in eastern British Columbia was recorded on video. The video shows a line extending along the toe, where foundation materials were being pushed up as the toe crept forward. During the early stages of the overt movements, a bench was seen to develop on the face.

The failure mechanism took the form of the double wedge. The toe wedge failed first, thereby removing lateral support for the active wedge. After traveling a moderate distance from its prefailure position, the toe wedge slowed and appeared to have

* In the case of the retaining wall, movement through only 0.1% of the height of the wall is sufficient to put the granular backfill retained by the wall into the active Rankine state.

FIGURE 31.7 Coarse, segregated, rock fragments that have accumulated in the toe region of a waste rock pile. The white circle included as a scale is 0.5 m in diameter.

FIGURE 31.8 Coarse, segregated, rock at the terminus of a debris run that originated from failure of a waste rock embankment

nearly stopped. Within a few tens of seconds after the toe wedge had failed, the active wedge failed, moved downslope, and impacted the mass comprising the toe wedge. Transfer of energy by the impact remobilized the toe wedge, and the whole of the failure mass traveled downslope. Over more than ½ km of the debris run, the upper wedge remained largely intact, retaining its shape. The total volume of material involved in this failure was approximately 3 million m^3. The velocity attained by the failure debris approached 30 m/sec.

The delay in the failure of the active wedge following failure of the toe wedge was somewhat surprising. The most likely explanation for this delay is related to subatmospheric pressures within the capillary-held water at the particle-to-particle contacts within the waste rock. When the toe wedge failed and moved downslope, removal of lateral support would have been responsible for a reduction in the lateral stresses within materials comprising the active wedge. This reduction in lateral stress would have been accompanied by elastic expansion, which, in turn, would have been accompanied by a further reduction in the subatmospheric pressures within the capillary-held water.

Under static conditions, water that is retained by capillary forces within a granular mass is at the same value of subatmospheric pressure at all locations within the mass. When the toe wedge failed, the reduction in lateral stress that occurred at the inter-wedge boundary would have been accompanied by elastic expansion. This expansion, in turn, would have resulted in a further reduction in the negative capillary pressures, with a resulting increase in particle-to-particle effective stresses, and a temporary increase in shearing resistance. Flow would have been required for the capillary pressures within the active wedge to equilibrate.

The delay in failure of the active wedge was the result of the time required for flow to occur in order to effect the subatmospheric pressure equilibration in the capillary-held water within the active wedge. Equilibration in the subatmospheric water pressures was accompanied by a reduction in effective stress, which, in turn, resulted in a reduction in the available shearing resistance. The result was a time delay in the failure of the active wedge.

Impact loading of saturated or nearly saturated foundation soils by the fall of the active wedge is the mechanism most likely responsible for the high velocities attained by failure debris. It is also likely responsible for the long debris travel distances.

31.8 MELT-BANDS AND UNIVERSALITY OF DOUBLE WEDGE

The melt-band is a feature that appears on the faces of virtually all waste rock embankments located in areas of snowfall. The melt-band—a feature associated with the double wedge—indicates that stability at the perimeter of virtually all angle-of-repose waste rock slopes is controlled by the double-wedge mechanism.

The face of a developing disposal site advances over the layer of CSR that accumulates at the toe. This CSR serves as a pervious layer extending over the base of the structure. In addition to segregation of the coarse rock fragments that accumulate at the toe, segregation of particle sizes also occurs on the embankment face. This segregation results in development of coarser and finer layers oriented parallel to the angle-of-repose face. As the face advances, the differential shearing that takes place on the mutual boundary between the active wedge and the toe wedge is sufficiently large that the continuity of the layering parallel to the sloping face is disrupted at this boundary. This inclined layering is retained within the toe wedge. Behind the active wedge/toe wedge mutual boundary, the layering parallel to the face is completely destroyed.

Waste rock embankments undergo air exchange on a seasonal basis. As temperatures rise in the spring and early summer, colder-than-ambient air that occupies the void spaces between constituent rock particles drains from the structure by gravity. Conversely, with the onset of the winter season and the drop in ambient temperatures, convection results in upward migration and expulsion of warmer-than-ambient air from the void spaces within the waste rock. Convection currents would preferentially follow the paths of least resistance, which consist of the coarse segregated layer along the base of the toe wedge, together with layers of coarser segregated rock inclined parallel to the sloping face.

The differential shearing at the mutual boundary between the active and the passive wedges is sufficiently large that the continuity of the inclined coarse layers is disrupted at that location. As a consequence, when the warmer-than-ambient air arrives at this boundary, the increased resistance to flow causes the air to emerge at the location where the active wedge/toe wedge boundary intersects the sloping face. The bench illustrated in Figure 31.3 was inspected on a cold November day, when the air temperature was about −20° C. At that time, the slopes above and below the bench were covered by ice and snow, whereas the surface of the bench was bare, the snow having been melted by the relatively warm air emerging on the face of the embankment at the location of the bench.

The omnipresent melt-bands on embankment faces during early winter attest to the ubiquitous nature of the double wedge

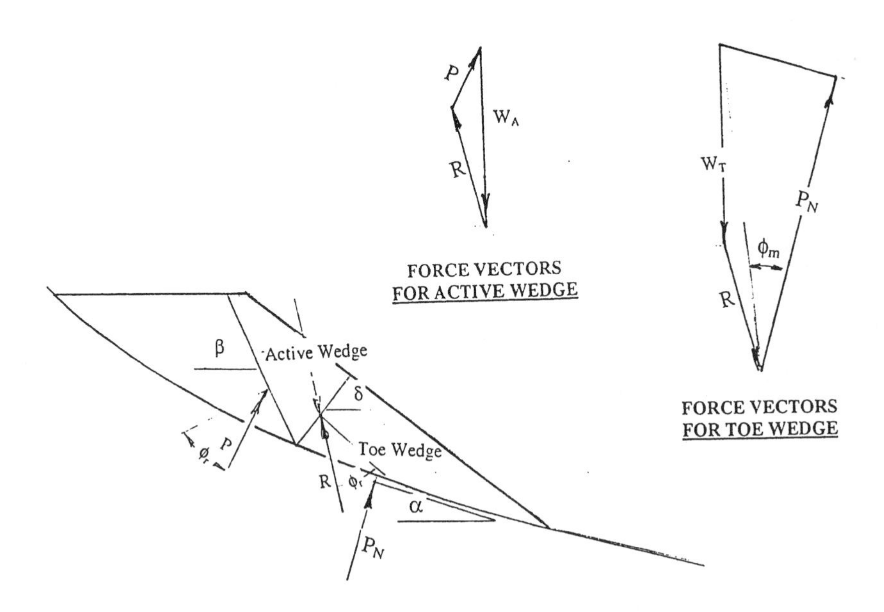

NOTATION

W_A = Weight of Active Wedge

W_T = Weight of Toe Wedge

P, R = Resultant forces that provide support to the Toe Wedge

P_N = Normal force on base of Toe Wedge

α = Inclination, base of Toe Wedge

β = Inclination, back boundary of Active Wedge

δ = Inclination of Mutual boundary

ϕ_r = Friction angle for waste rock

ϕ_m = Mobilized angle on base of Toe Wedge for limiting equilibrium

The stability of the Active Wedge and the Toe Wedge are considered separately.

The Active Wedge is in a state of limiting equilibrium. That is, its factor of safety is 1.0, so that the shear strength parameters for the waste rock are fully mobilized along the wedge boundaries.

For a trial set of wedge boundaries, the reaction forces 'P' and 'R' that support the Toe Wedge are calculated.

The calculated force 'R' is applied to the Toe Wedge, and the value of ϕ_m corresponding to limiting equilibrium is calculated. The procedure is repeated until the maximum value of ϕ_m corresponding to limiting equilibrium of the Toe Wedge has been found.

FIGURE 31.10 Method of stability analyses for the double wedge

FIGURE 31.9 Melt-bands on the face of a waste rock embankment at the location where the mutual boundaries between the active and toe wedges intersect the sloping face. The universality of melt-bands attests to the ubiquity of the double wedge as the mechanism that controls waste rock embankment stability.

and to the fact that the double-wedge mechanism is what controls the stability of angle-of-repose slopes. Examples of melt-bands on the faces of angle-of-repose waste rock embankments are illustrated in Figure 31.9.

31.1 METHOD OF STABILITY ANALYSES FOR DOUBLE WEDGE

For the results of stability analyses to have credence, the model analyzed should be in conformance with reality. Conventional methods of stability analyses that assume the following are not appropriate for assessment of the stability of angle-of-repose waste rock embankment faces:

- Potential shearing on some curved surface passes through the body of the structure
- The factor of safety on all segments of the potential failure surface are at the same value

Stability is controlled by the double-wedge mechanism, and the double-wedge model is what is appropriate for assessment of stability. The significant shear displacements evident in the crest region of a stable embankment indicate that the waste rock within the upper part of the structure is in the active Rankine state of stress. For rock in the active Rankine state, the shear strength of the rock comprising the active wedge, including at the wedge boundaries, is fully mobilized, and the factor of safety of the active wedge remains at unity. Stability of the perimeter of the structure is dependent on the ability of the toe wedge to provide support to the active wedge. Because the factor of safety of the active wedge remains at unity, the factor of safety for the toe wedge must remain greater than unity for the spoil pile to remain stable.

A method for assessing the stability of a waste rock embankment is illustrated in Figure 31.10. The active wedge and the toe wedge stabilities are considered separately. At the outset, trial positions of the wedge boundaries are selected. With the factor of safety of the active wedge at unity, the angle of obliquity of the resultant forces that act on the wedge boundaries is numerically equal to the friction angle for the waste rock. Hence, for any trial inclination of the wedge boundaries, the direction of the force vectors that support the active wedge can be calculated. This calculation provides the direction and magnitude of the force that acts across the mutual boundary between the active wedge and the toe wedge.

Applying this force to the toe wedge, and assuming limiting equilibrium, the mobilized angle of shearing resistance along the base of the toe wedge can be calculated. The ratio of the tangents of the available angle of shearing resistance and the mobilized angle of shearing resistance corresponding to limiting equilibrium of the toe wedge is a measure of the factor of safety of the toe wedge, provided that pore water pressures along the base are negligible. Alternatively, results of the stability analyses permit calculation of the pore water pressures, in terms of r_u (defined as the ratio of pore water pressure to normal stress) that would be required to reduce the toe wedge to a state of limiting equilibrium.

31.2 INFLUENCE OF FOUNDATION PORE WATER PRESSURES

In some instances, failures have developed at disposal sites that had remained dormant for a month or more. Because no additional loads were applied during the interval of dormancy, it follows that failure must have been a result of a reduction in shearing resistance that reduced the stability of the toe wedge. This reduction in stability could not be attributable to an increase in the forces imposed by the active wedge or to a reduction in the shear strength parameters of the waste rock. Such failures must be the result of increased pore water pressures resulting in reduced shearing resistance within the foundation soils at shallow depth below the base of the toe wedge. This increase in pore water pressure takes place in response to unidirectional shearing strains in the foundation soils located at shallow depth below the base of the toe wedge.

Pore water pressures in the foundation at shallow depth below the base of the toe wedge are influenced by the unidirectional strains as well as by consolidation. Unidirectional strains are responsible for increases in pore water pressure, whereas consolidation serves to decrease pore water pressures. Methods are available for estimating the rate of pore pressure reduction that results from consolidation. However, the rate of pore pressure increase attributable to unidirectional strain is dependent in part on the rate of shearing strain. Since rate of strain cannot be predicted with any degree of reliability, it follows that pore water pressures at shallow depth below the base of the toe wedge cannot be predicted reliably.

As a consequence, the designer must be aware that embankment stability assessments based on the results of stability analyses may not be reliable. For this reason, engineering assessments for waste rock piles should always include assessments of the maximum distance that failure debris might travel, in the event that instability were to develop. The results of such debris run assessments may:

- Govern the appropriate location for a proposed waste rock facility, or alternatively
- Govern the necessity for, and the height of, debris containment barriers that may be required to provide protection to infrastructure and to the environment

31.3 SUMMARY

Fortunately, waste rock embankments do not fail without warning. Failure is invariably preceded by intervals that range from several hours to a few days, during which time displacements in the crest region develop at progressively faster rates. Proper interpretation of the measured rates of displacements in this region provides a clear warning of approaching failure conditions so that workers and machines can be evacuated from the dumping platforms before failure occurs. At the same time, access to all areas that could be impacted by failure debris must be excluded.

The clothesline type of monitor that appears in Figure 31.1, together with proper interpretation of the crest movement data, provide ample warning of approaching instability so areas of danger can be evacuated before failure occurs. Monitoring and interpreting the crest movements is essential to the safe operation of angle-of-repose waste rock embankments.

CHAPTER 32

Embankment Hydrology — Surface Water Controls

Gary R.E. Beckstead,* John Slater,* Neil van der Gugten,* and Andrzej Slawinski*

32.1 INTRODUCTION

Embankment hydrology is needed to understand runoff rates and seepage volumes resulting from snowmelt and rainfall runoff and seepage water. Embankment drainage design considers the collection and treatment of stormwater from rainfall on containment areas. Usually this water is called *contact water* and therefore must be treated prior to release to the environment. Runoff from any other area that has not been in contact with the pit or the process operation and thus is not contaminated (i.e., noncontact water) can be released to the environment. Considering treatment costs, minimizing the amount of contact water is beneficial.

Other aspects of embankment design with respect to hydrology are related to the risk of overtopping water storage and tailings containment structures. Loss of water control in a mine will lead to environmental disasters that could financially ruin the mine and cause extensive damage to the downstream ecosystem.

Erosion and sediment control also is needed to improve slope stability and to limit the amount of sediment transported into the drainage systems and, eventually, receiving streams. Other factors may involve investigating the amount and effect of seepage from tailings ponds upon the groundwater regime and estimating the availability of process water. Embankment hydrology also addresses regulatory issues regarding water diversion and the discharge of water into the environment.

32.2 HYDROLOGICAL PROCESSES

Hydrological processes involve hydrological estimates of runoff or seepage and of runoff and infiltration.

32.2.1 Hydrological Estimates of Runoff or Seepage

Hydrological estimates of runoff or seepage are made in accordance with criteria established to meet acceptable risk (risk equals probability times consequences of failure) that the design event will be exceeded. Probability is often specified by a return period event, such that

$$p = \frac{1}{T} \qquad \text{EQ. 32.1}$$

Where:

p = probability that the event will be equaled or exceeded in any year (%)
T = return period of the event (years)

Design return periods in some cases are specified by regulatory agencies or in combination with the mine planners. When the design return period is not specified, the best practice is to recommend events similar to other jurisdictions in the area. These recommendations must be tempered with the level of environmental or operational risk and consequences of failure that the designer and the mine are willing to accept. For instance, the consequence of flooding in the undeveloped upland area of a mine is significantly less than the consequence of flooding a tailings impoundment resulting in release of contaminated runoff to the environment. Figure 32.1 is a graphical representation of the probability of damage related to the life span of the structure and design return period.

The hydrology of the mine site usually is based on best-guess estimates, as significant stream flow or rainfall data are generally lacking. Soil type and placement history, vegetation reclamation progress, and the mine plan itself also affect runoff. As data seldom exist at a mine site and runoff is seldom measured from embankments, the best approach is hydrologic modeling and, if possible, comparison with other sites using regional analysis.

Standard models should be used so others, including regulators, can reproduce results if desired. A universally accepted model is the US Army Corps' HEC-1 rainfall runoff model. This model uses the well-documented US Soil Conservation Service Curve Number method of estimating runoff. This is a single-event type of model that estimates higher peak flows than expected. Thus, these values will be sufficient for design of culverts, or spillways, in which a peak flow estimate is the normal design parameter. If the mine is using runoff storage for peak flow attenuation, the HEC-1 model also has flow routing routines.

Structures that involve containment of contaminated water should use a continuous simulation method of analysis that will estimate the total volume of water over an annual period or the planned operational period of the structure. Continuous simulation is a more difficult analytical process requiring much more data than single-event calculations. If the continuous simulation model cannot be calibrated accurately, more conservative factors of safety for freeboard and total storage volumes should be used.

Considering safety factors and the sometimes small consequences of flooding on mine sites, standard methods normally produce sufficiently good hydrologic estimates. Complex modeling requires more time and, considering the probable lack of significant rainfall or snowmelt data, the additional complexity will still produce only a good "hydrologic guess" at best. So the question "Does the extra effort justify the end result?" must be answered prior to embarking on a complex hydrologic study that would achieve a level of confidence that could be produced only by standard methods. Time may be better spent developing designs that minimize damage to the environment and the mine's operations, for the least cost in terms of capital expenditures and maintenance.

Design of drainage systems for the mine should be considered along with the overall site configuration and natural features. The following concepts should be considered for:

* Water Resources and Civil Projects Division, AGRA Earth and Environmental Limited.

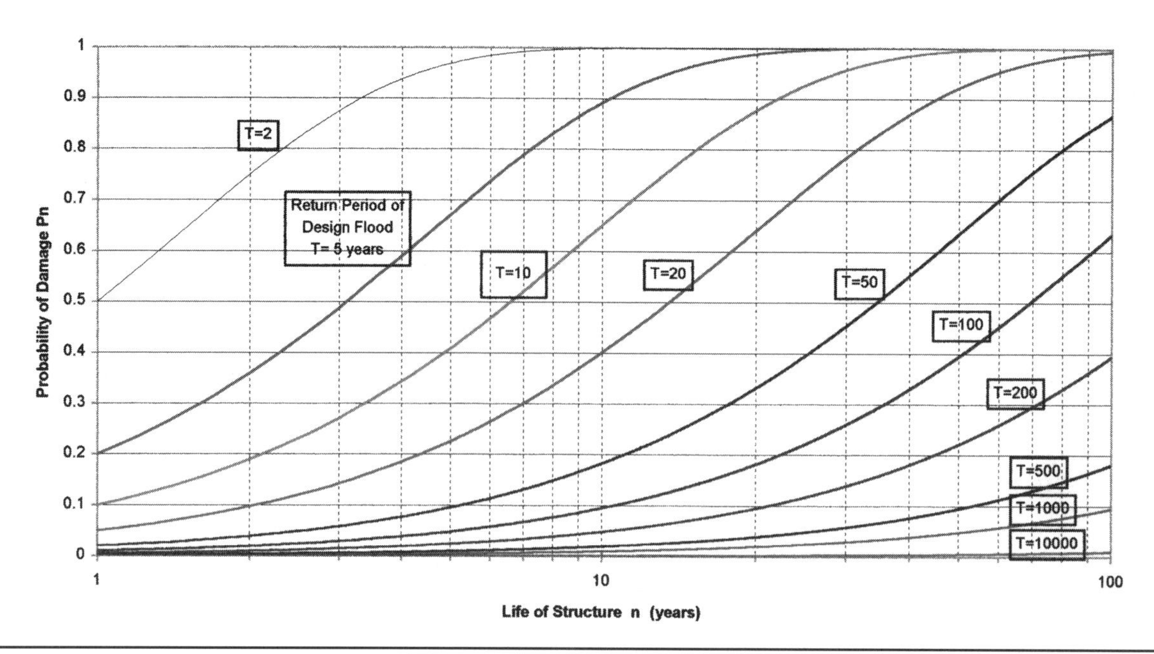

FIGURE 32.1 Probability of structural damage in terms of design flood

- Drainage systems:
  - Types of reclamation soils and schedule of placement should be planned to reduce treatment requirements for process-affected or construction water management issues
  - Location of drainage facilities relative to existing drainage features, allowing effective diversion and protection of the receiving streams
  - Design of seepage control and drainage systems appropriate to limit sediment transport and seepage into the groundwater regime
- Consideration of discharge points:
  - A small, clean creek may have a higher water quality requirement than a large stream
  - Risk and cost may warrant conveying runoff around important environmental areas

32.2.2 Runoff and Infiltration

Two streams of water come from tailings facilities: stormwater runoff and infiltrated or seepage water from the pond. Typically, this runoff and seepage water is collected in interception ditches constructed at the toe of the embankment dykes and waste rock embankments. Because seepage water is in contact with the mined material, it usually is considered to be contaminated or contact water. Storm runoff water is also considered contact water if construction material used for the embankment is built from mine tails. Because both streams of water usually are collected in one interception ditch, this water also is considered to be contaminated, regardless of dilution by storm runoff water.

As treatment and the cost of transporting runoff water can be excessive, it is reasonable to reduce this amount of water as much as possible. A current approach is to pump the runoff water back into the tailings facility. This eliminates the need to treat the water but creates a potential water-management problem in the facility, especially if the volume of the facility is limited.

Another approach is to return this water to the mine process system. This is an excellent water management approach if there are facilities to handle excess flow from unusual runoff events.

Still another approach is to prevent storm runoff from becoming contaminated and separating this runoff from the contaminated seepage water. If separated runoff water can be safely returned to the environment, treatment and pumping costs will be substantially lower because the volume of seepage water is much less than the volume of stormwater. This approach relies on the theory of unsaturated flow and the concept that reclaimed embankment slopes will not allow runoff water to become contaminated.

The unsaturated flow theory assumes that the material above the phreatic surface has a negative pore pressure, thereby creating a much higher coefficient of conductivity in the vertical direction. Because of this, infiltrated runoff water will tend to flow parallel to the surface in the contact area between the reclaimed soil and embankment construction material.

The basic concept of this approach is to reclaim the embankment slope as soon as possible, thereby preventing contamination of surface runoff. Then, with the use of subsurface drainage, draw the phreatic line of seepage water low enough to limit contact with the surface drainage system near the embankment toe. Seepage water can be transported to the treatment system or pumped back into the tailings facility. Because seepage volumes are small, the extra storage requirements in the tailings pond could be significantly reduced. A surface drainage system could be constructed to prevent mixing of the two streams and then harmlessly discharge them into the environment.

32.3 HYDROLOGICAL ASPECTS OF LANDSCAPES

32.3.1 General Landscape Concepts

Before undertaking the hydrological design of drainage features for rock embankments, it is useful to consider the embankment as a landform. The final landform must be stable, safe, and nonerodible (over the short term). Over geological time, the latter cannot be assured; rather, the landform must be sustainable over the long term.

Another useful guide is that the landform should be compatible with the land uses of the surrounding region. For instance, the elevation should be in the same range as that for the surrounding lands. If the design is undertaken successfully, the embankment will be hydrologically compatible with the surrounding land.

Landscape involves the components of topography: elevation, slope (gradient), and drainage density. Slope affects erosion potential and the ability of the final topographic feature to blend into the surrounding landscape. Aspect is another function, although little may be done to modify this parameter. Nevertheless, it is an important consideration in the design of drainage features. For instance, in northern latitudes, south-facing slopes often are drier than north-facing slopes because of the additional exposure to the sun.

Formation of an integrated drainage system is one of the most challenging aspects of the design of embankment landscapes. It requires consideration of the drainage aspects for the entire mine (and its evolving landscape). Of particular importance is the maintenance of major drainage paths to connect to receiving streams offsite.

Topographic information, usually in the form of contour plans, should be reviewed. Features such as drainage paths should be identified, including permanent and ephemeral watercourses and wetlands. Catchment boundaries should be delineated. Once this baseline information is in place, the drainage design can commence.

Small, isolated out-of pit embankment may be designed on an individual basis. Larger disposal structures, which evolve over a number of years and may be developed in several phases, or which might become amalgamated with other embankment, should be designed with the overall drainage plan for the reclaimed mine site in mind. Otherwise "dead-end" drainage paths could result. Or overly steep drainage paths may be necessary to reach "target" streams from another landform component.

Drainage design initially should be based on the location of receiving streams beyond the disposal site. Original topographic plans should be consulted to determine the drainage areas of the streams prior to mining. The drainage area following embankment construction and reclamation should not vary significantly (±30%) from the original drainage area. Otherwise erosion caused by excess water or sedimentation resulting from less water could occur.

In some instances, haul roads can be used as drainage paths. Although road gradients usually are less than overland gradients, the slopes along road ditches may be excessive for large flow volumes; erosion control along such steep channels can be challenging. If these roads are selected, tributary streams will have to be diverted to other locations or runoff stored while operations and reclamation are in progress. "Best practice requires that where a change becomes necessary to mining operations; this must be reflected in corresponding changes to landform [and drainage] design" (Environment Australia 1998).

Fluvial erosion processes are the major factor affecting the geomorphic stability of mine waste rock embankments. To minimize the fluvial erosion processes on waste rock embankments, two key actions should be followed:

- Fluvial forces can be minimized by limiting catchment runoff and maximizing available slope infiltration.
- If infiltration is exceeded by precipitation and runoff occurs, the runoff should be directed to predesignated channels capable of conveying the runoff without excessive erosion.

32.3.2 Drainage Density and Stream Order

Drainage density is a fundamental parameter used in the design of drainage systems. It depends on the climate and the physical characteristics of the drainage basin (Morisawa 1968). Drainage density is defined as the total length of drainage courses divided by the catchment drainage area (units of km^{-1}). Several examples of drainage density are illustrated in Figure 32.2. Drainage density provides a means of ensuring that receiving streams are not overwhelmed by a large upstream drainage system.

Drainage density for the premine landscape may be considered as a model or guideline for the design of postmine drainage. Changes to the landscape (elevation, aspect, slopes, runoff coefficients, and erosion potential), however, may justify adjustments to the drainage density. Land use also may be a consideration in determining a suitable drainage density for the reclaimed landscape. In no case should the postmining drainage density be less than the premining density. It is advisable to have the density greater for a couple of reasons:

- The greater density reduces the unit runoff carried by each stream, thereby reducing the erosion potential in each channel.
- The swelling of backfilled earth structures can elevate the catchment feeding drainage channels, which often produces higher gradients. Therefore, it is advisable to reduce stream velocities by reducing the runoff in the channels.

Another parameter that can be useful in landscape design is stream order, especially for large catchments. Figure 32.2 illustrates the concept of stream order, moving upstream along the drainage system and increasing the number associated with streams upstream of each confluence, moving up toward the headwaters, based on the method proposed by Strahler (1957). Stream order is an important tool as it can assist in determining the catchment area (and hence flow volume) associated with streams of each order. As illustrated in Figure 3.2, it is possible to have high drainage density with either low or high stream order; however, low drainage density is generally associated with only low stream order.

There is a relationship between drainage density and slope length. If a large number of channels are draining a given area, the channels will be close together and slope lengths draining to each channel will be short. For example, for a drainage density of 2 km^{-1}, each meter length of channel was draining 500 m^2 of area, or approximately a space of 250 m each side of the channel. From this, the spacing between watercourses can be selected.

32.3.3 Slope Design

Slope angle is an important factor in the drainage design for waste rock embankments. Not only does it directly affect the overland flow velocity and erosion potential but it also affects the drainage area contributing water to the slope. For instance, the drainage area of a 26 slope (as defined by the slope length between a given top and bottom elevation) is approximately 1.5 times the drainage area of a 40 slope.

Long slopes that otherwise would have excessive lengths (and associated drainage areas) can be broken into subcatchments by constructing back-sloped benches on the contour. These benches can direct drainage to small watercourses to convey runoff off the slope. These features are useful where steep slopes are involved, such as where external features limit the spreading distance for the embankment. In areas of high or intense rainfall, these slope breaks may be required to control water runoff volumes, flow velocities, and erosion potential. In these cases, the spacing between the benches may be relatively short. Slope lengths of 50 m may be considered as a first start in establishing drainage design. Figure 32.3 illustrates the concept of benches for slope breaks.

The slope length ultimately selected will depend on the nature of the waste rock material. If the embankment is constructed from coarse, competent (nonerosive) rock, steep slope angles may be tolerated, especially in arid climates. The embankment slopes will have to be flatter for erosive materials, perhaps in the order of 20 (36% slope).

Embankment slopes rarely have uniform gradients. Combinations of convex and concave slopes are generally used to provide

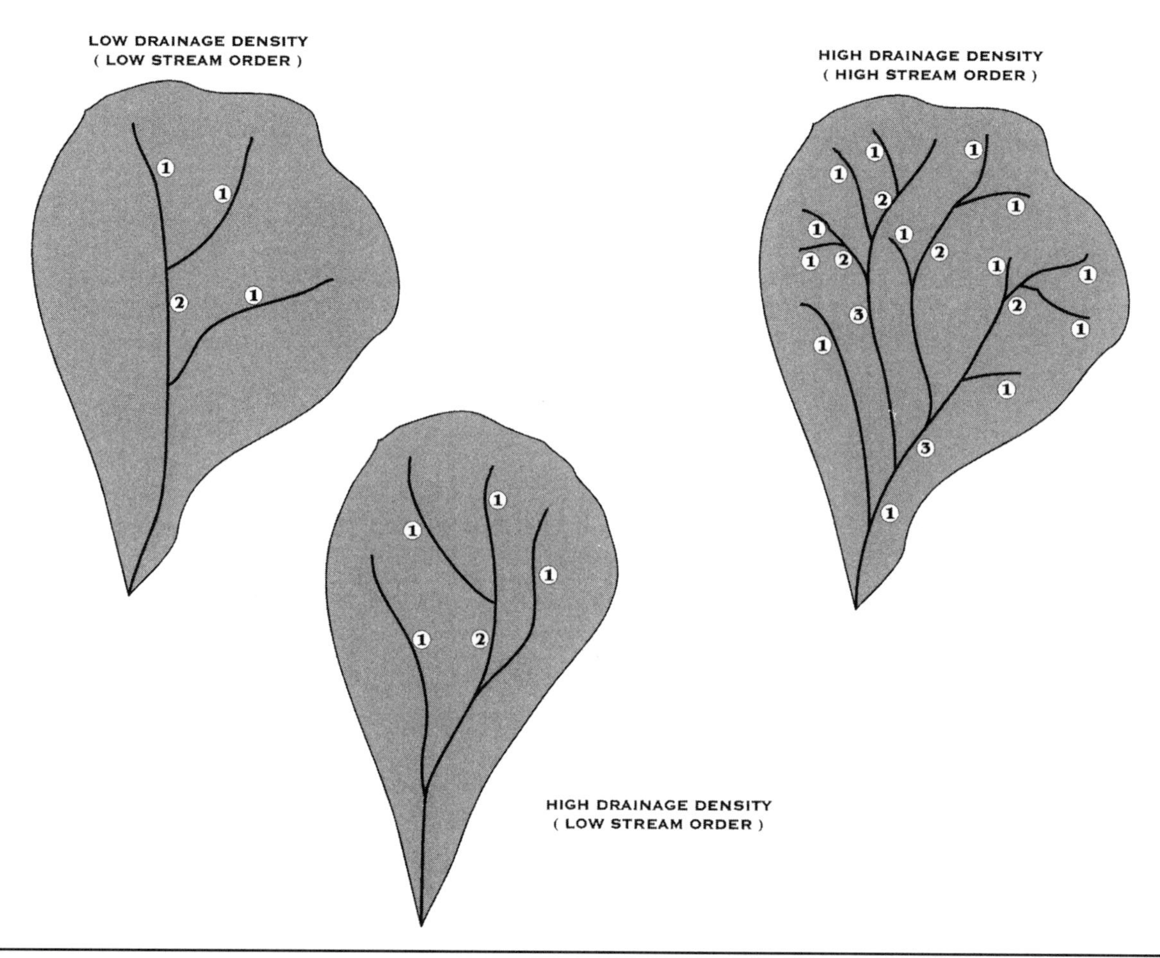

FIGURE 32.2 Examples of drainage density and stream order

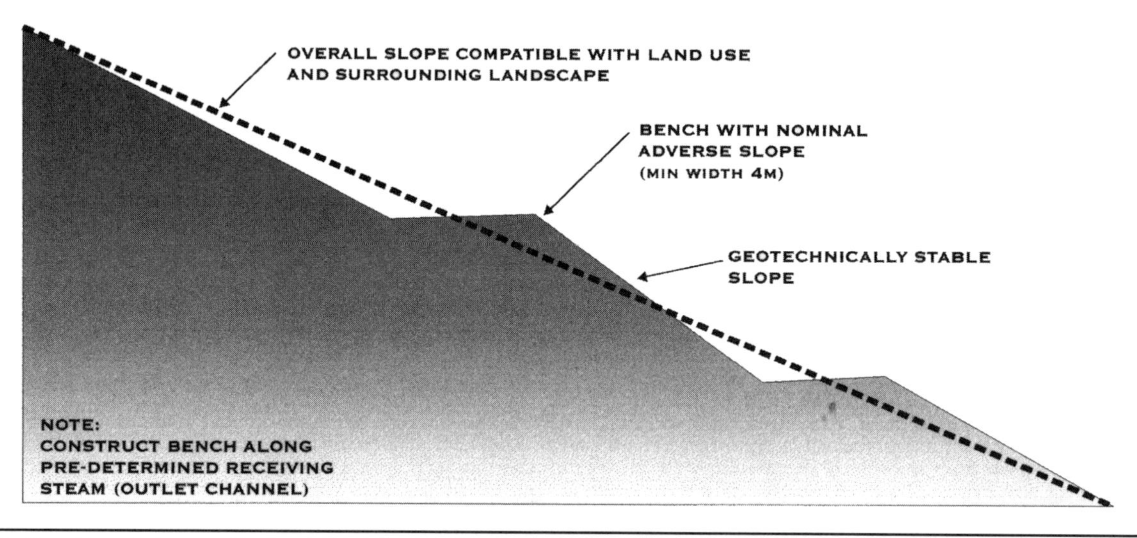

FIGURE 32.3 Benching of slope profiles

the overall final grade of the slope. The preferred profile from a drainage perspective is to start with a (short) convex slope at the top and (long) concave slope at the bottom, as shown in Figure 32.4. The reasons for this are simple: The materials can tolerate the steeper slopes provided near the top because the drainage area near the "summit" is small and as the drainage area increases (moving down the slope), the gradient flattens, lessening the erosion potential.

32.3.4 Drainage Control

A decision has to be made on how best to interface the drainage with the surrounding land. For waste rock embankments located outside of pits, receiving streams have to be identified. For embankments located in pits, the drainage can be directed either to the pit or away from the pit. If the drainage is to be toward the pit, water management to keep the pit free of excess water during

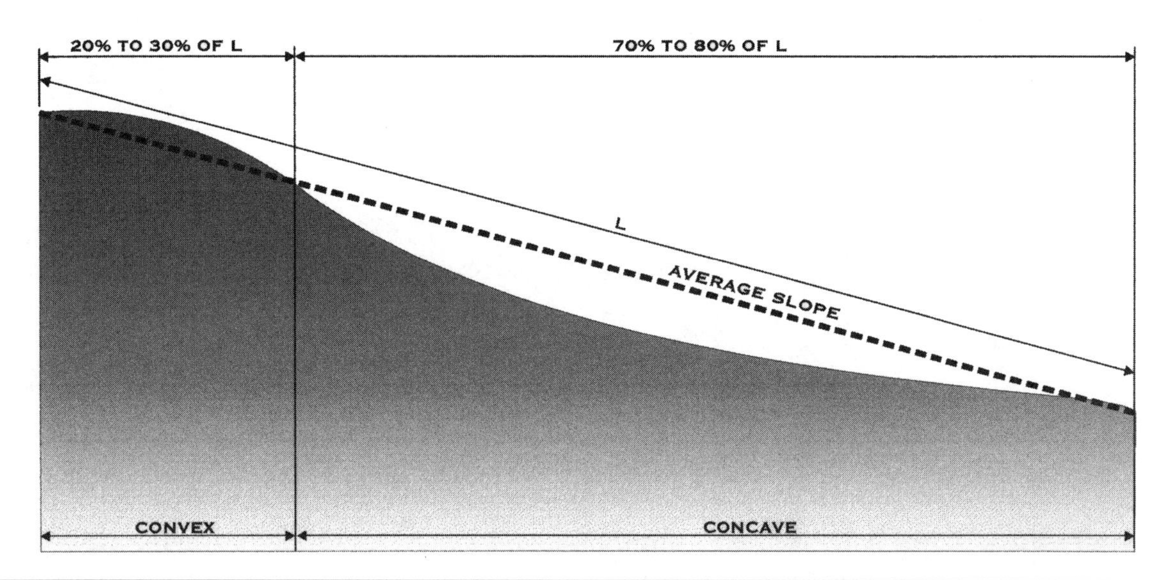

FIGURE 32.4 Preferred longitudinal slope profile

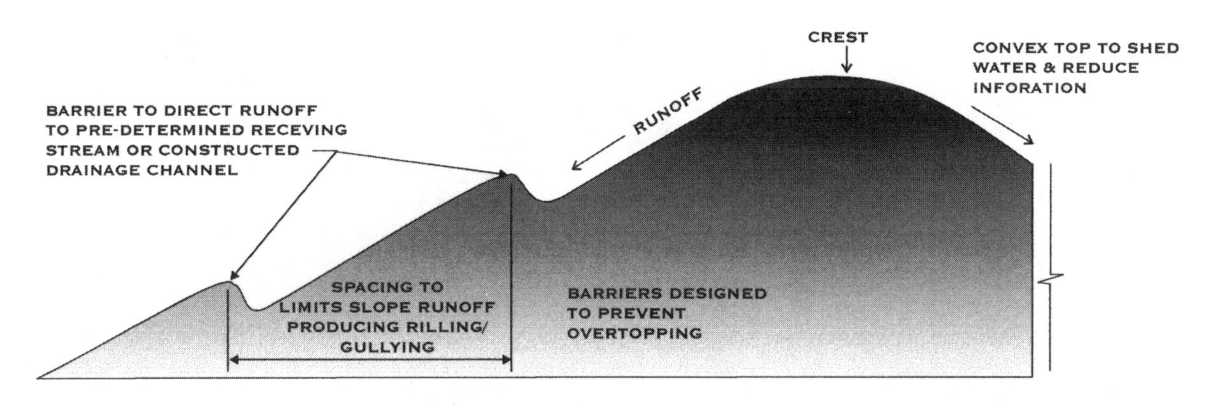

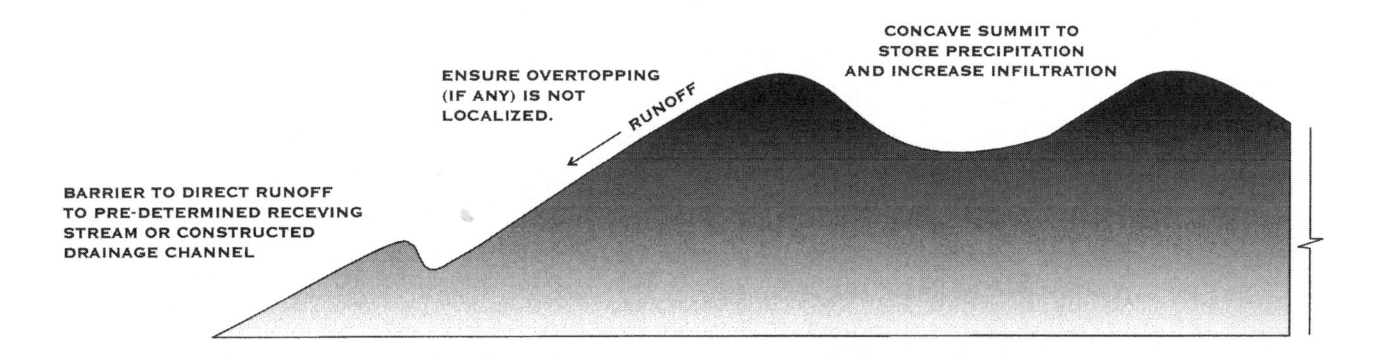

FIGURE 32.5 Waste rock embankment shaping alternatives

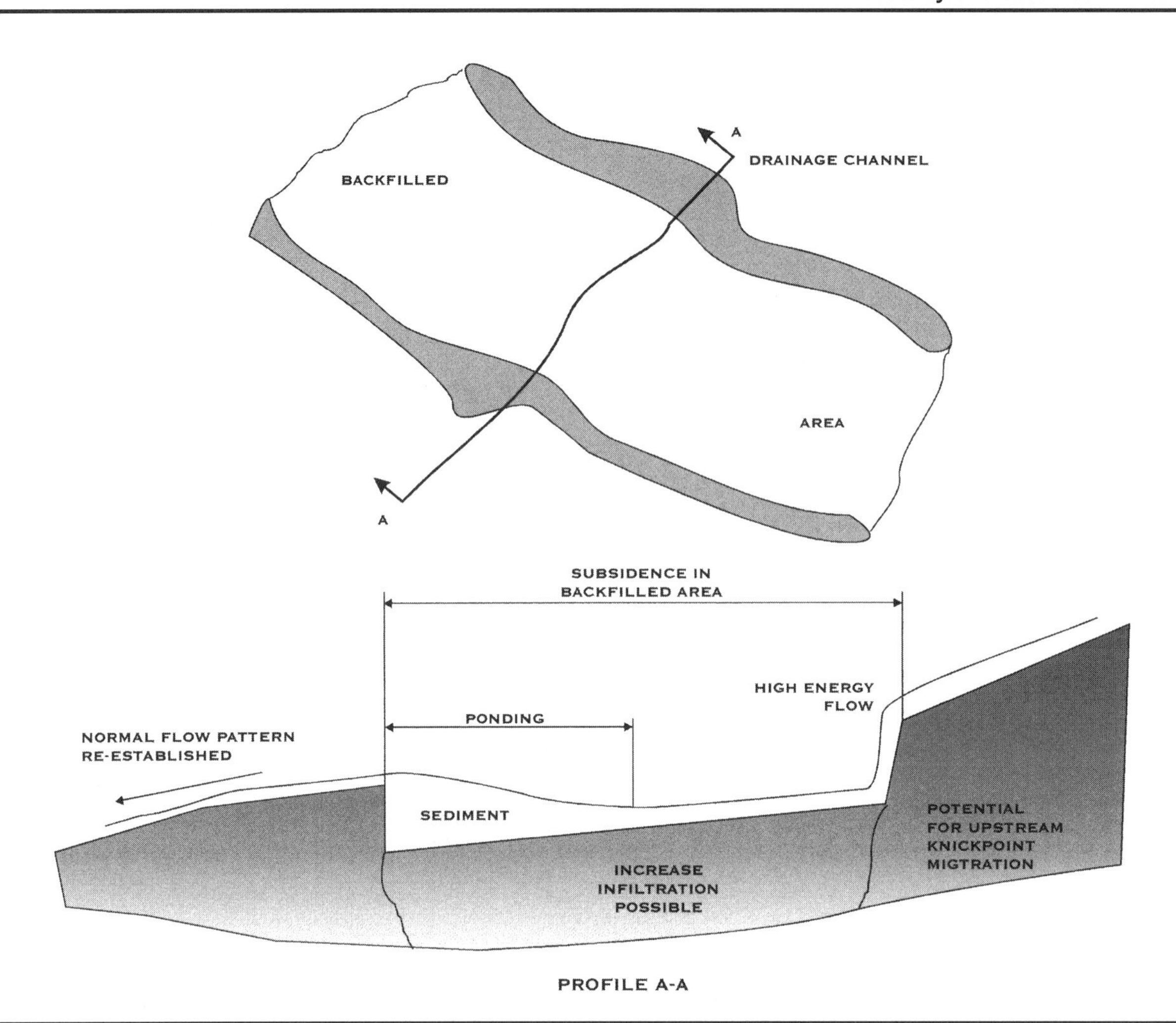

FIGURE 32.6 Effect of subsidence on drainage

mining may be a problem. To direct runoff away from pits, gradients for drainage courses have to be carefully selected.

The following aspects should be carefully considered:

- Locations where drainage may be directed to existing water courses may be selected, provided that catchment area is close to that of the original stream.
  - Several alternatives exist for controlling infiltration and runoff. At the crest of the embankment, slopes may be either concave or convex, as illustrated in Figure 32.5.
  - Concave slopes should be used only if the waste rock is chemically stable, as the "bowl" that is created will tend to increase water storage between precipitation events. The resulting increased infiltration and evaporation will reduce runoff and lower the erosion potential for flow down the sideslopes.
  - Convex slopes are advisable where it is desirable to have the embankment shed as much water as possible. An example is where the embankment contains chemically unstable materials, such as acid-generating rock. Shedding water also may also be important where it is advantageous to reduce infiltration to enhance the geotechnical slope stability of the structure.
- The effect of differential settlement of backfilled areas on drainage design is an important consideration. The interface between backfilled and undisturbed land may produce either high stream gradients or ponding, as illustrated on Figure 32.6. The anticipated amount of settlement should be identified, if possible, and integrated into the design of the drainage channels. Alternatively, reshaping of drainage channels may be required following settlement. In some cases, it may be possible to construct dams to create ponding at the interface. The potential for increased infiltration and the interim loss of surface flow in the stream should be considered.
- Low channel gradients should be provided at the downstream reaches of drainage channels to reduce the potential impacts of the receiving stream.

32.4 ROCK DRAINS

32.4.1 General Information

Mining operations have deposited large waste rock embankments in numerous mountain valleys. These are among the largest humanmade structures and may contain from tens of millions to more than a billion cubic meters of rock material. The rock embankments interfere with the premining, natural drainage systems. Stream diversions around these structures are often expensive and under some circumstances may be totally impractical. Flow-through rock drains provide technically feasible and economically practical alternatives to stream diversions.

To satisfy operational, environmental, and regulatory requirements, the flow-through drain design has to consider a large number of aspects, including

- Natural land topography
- Local hydrologic conditions, including documented and/or predicted flood analyses
- Stability of natural and humanmade slopes
- Availability of waste rock of suitable quality
- Method of waste rock selection and placement
- Assessment of sediment load to streams
- Evaluation of acid rock drainage potential
- Operation during mine life and following mine abandonment and reclamation
- Other site-specific aspects

In the second half of the 20th century, a significant amount of experience in the construction and operation of the waste rock piles has been developed in Western Canada, Australia, and the United States.

32.4.2 Design and Construction

The mining industry uses two main techniques for rock drain construction: (1) constructing the drain within the natural drainage course prior to waste rock dumping, or (2) constructing the drain through natural segregation by end-dumping suitable rock materials.

The first option, referred to as an *underdrain,* is usually accomplished by placing suitable-quality coarse rock along a preselected pathway prior to beginning waste rock dumping. The rock material must be durable, resistant to weathering, and not prone to acid generation. The advantage of this method is the high level of control over material selection, placement, and final configuration (size) of the drain. The main disadvantage of this method is the high cost of the direct, controlled placement of the drain material, as well as timing constraints.

The second option, widely used in Western Canada, involves end-dumping waste rock while relying on natural segregation of the rock material. It has been well documented that the coarsest material tends to concentrate at the toe of the slope (Figure 32.7). This option is of significant interest to the industry and regulators and, consequently, important progress has been made in understanding rock drain construction and operation. The main aspects to be considered during drain design and construction are as follows:

- Waste rock has to be of a suitable quality and readily available; in some cases, special procedures for blasting and rock selection have to be incorporated into the mining operation to meet the required rock drain quality.
- A minimum height for the dumping platform must be maintained to ensure sufficient material segregation. It has been demonstrated that direct end-dumping is considerably more efficient than pushing the rocks over the embankment crest.

Some published sources indicate that for good-quality rock material, such as hard and resistant sandstones or igneous rocks, the total embankment height should be no less than 20 m, plus the required depth of the flow-through drain. Several studies have indicated that properly installed drains constructed by end-dumping present a cost-effective and high-performing alternative to preconstructed under drains.

32.4.3 Conveyance Capacity

The coarse character of any properly installed rock drain (preconstructed or end-dumped), with the average rock size on the order of several tenths of a meter (decimeter), and the angular character of the rock material results in non-Darcy, turbulent water flow. The literature has presented a number of different formulas to assess quantitatively the flow through a rock drain, assuming a nonlinear relation between the flow velocity and the hydraulic gradient. One of the most popular equations used in the industry is Wilkin's empirical formula:

FIGURE 32.7 Slope produced by end-dumping showing natural segregation of waste rock material (surface coal mine, Rocky Mountains, Alberta, Canada, 1998)

$$V = 5.24m^{0.5}I^{0.54} \qquad \text{EQ. 32.2}$$

Where:

V = void velocity, m/s
I = hydraulic gradient
m = the hydraulic mean radius

Researchers using direct flow-velocity measurement methods, such as tracers, have demonstrated that this formula, and others, give only approximate results. Typically, the void flow velocities are reported in the range of a few to substantially more than 10 cm/s. The saturated section of the rock drain is usually limited to a few meters in depth. Increases in the water table within rock drains have been observed during flood conditions.

When taking into account a relatively high level of uncertainty in the calculations of expected performance of a given rock drain, it is recommended that a generous factor of safety be incorporated into the design based on hydrographs produced for a given drainage basin. Because of their size and character, waste rock piles are considered to be permanent structures. Therefore, in calculating the rock drain conveyance capacity, the effect of the probable maximum flood (PMF) should be considered.

FIGURE 32.8 Inlet pond at upstream end of a waste rock embankment

FIGURE 32.9 Discharge from downstream end of flow-through rock drain (surface coal mine, Rocky Mountains, Alberta, Canada, 1998)

32.4.4 Flood Attenuation

Placement of large structures such as these within mountain valleys impacts the surface stream flows conveyed via the flow-through rock drains. Of particular interest is the conveyance of flood events by the rock drains. Short-duration, high-flow rates are often characteristic of peak flows within the river or stream channels. As the flow velocity within a rock drain is reduced with respect to the flow velocities in open channels, it is logical that the travel time of a peak flow should be delayed and the instantaneous peak flow should be reduced (i.e., the hydrograph peak would be flattened) because of storage upstream of the drain.

The total flow volume passing through the rock drain as a result of a flood event, however, would not decrease significantly because the storage capacity of a coarse rock infill is believed to be relatively limited. Reducing the peak flow and distributing it over a longer time may have considerable positive impact on construction of the outlet water control facilities.

32.4.5 Sediment Load

Suspended solid load may originate from the upstream reaches of the local watercourse or it may be introduced by subsurface erosion within the rock drain. If the drain is constructed with coarse, resistant rock material, the remobilization of fines should not be extensive and, in fact, observations indicate it to be a relatively short-lived phenomenon. Conversely, turbulent flow dominating within the rock drain promotes little sedimentation and the sediment loading at the discharge face is often not much different from that measured at the inlet end of the structure (Figures 32.8 and 32.9).

32.4.6 Water Chemistry

A primary concern for the water quality impact of a rock drain is the potential for acid generation. This has to be carefully investigated, addressing the geochemical characteristics of waste rock and the buffering potential of surface and groundwaters. Each site that is considered for construction of a waste rock embankment or rock drain should be evaluated for acid rock drainage within the site-specific context.

32.5 CHANNEL DIVERSIONS

Mining operations often have to extend across creek or river channels, which then must be diverted around (or, occasionally, through) the active mining area. These diversions require consideration of the following design aspects:

- Layout and routing
- Channel capacity
- Hydraulic design
- Channel lining
- Grade control structures
- Culverts

32.5.1 Layout and Routing

The routing of a diversion channel is often dictated largely by the existing topography. The upstream starting point may have to be located some distance above the actual mine to achieve a gravity route around the mine area. The main consideration is the longitudinal slope, which should be as close as possible to that of the natural channel.

Typically, the upstream reach of the diversion will have flatter slopes, whereas the downstream reach has steeper slopes. Steeper slopes generate higher flow velocities, which may require erosion-resistant linings. If slopes become too steep, grade control structures may be needed.

32.5.2 Channel Capacity

A diversion channel should be designed to convey all discharges up to a maximum value (termed the *design discharge*) corresponding to a certain probability of occurrence, or return period deemed suitable for the project. For example, a discharge that has a probability of occurring once in 10 years (i.e., 10%) is said to have a return period of 10 years, or to be the 10-year flood.

Typically, natural streams experience one or more peak discharges throughout the year in response to snowmelt and/or rainfall events on the watershed. Calculating discharge probabilities usually is based on the data series consisting of the largest peak for each year. If discharge data are not available, discharges usually are derived from precipitation data using the watershed area and estimated runoff characteristics (see Hydrological Estimates, 32.3.1).

Once the probability analysis has been done, the main decision is the level of probability, or return period, to select. Clearly, the longer the return period, the larger will be the corresponding design discharge and the larger and more costly will be the diversion channel and the smaller will be the probability of failure (exceeding the design capacity). The risks and consequences of failure should be balanced against the capacity and cost of the

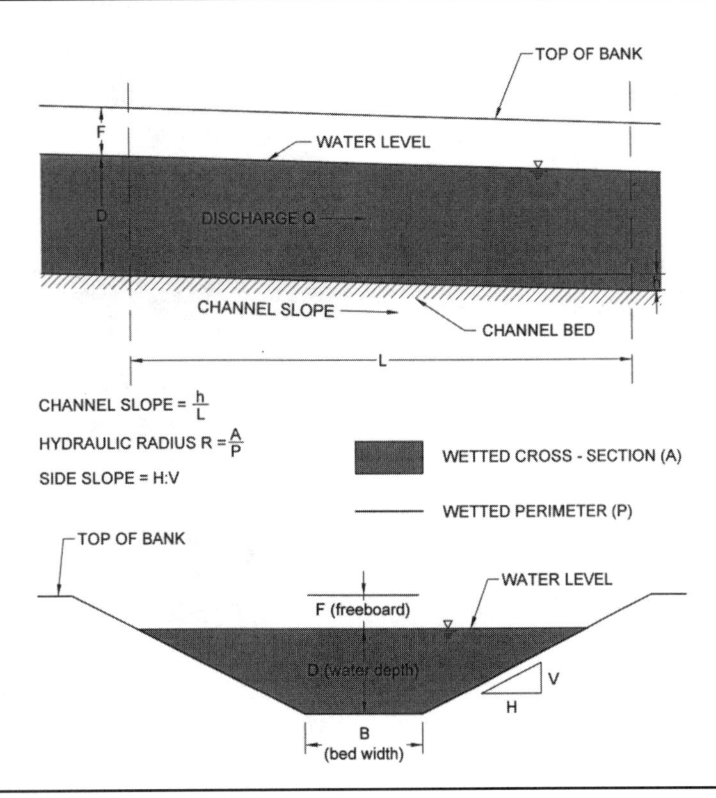

FIGURE 32.10 Diversion channel design parameters

channel. Consideration of the consequences of failure should specifically include the environmental consequences (Szymanski 1999). For situations in which no significant interruptions to mine operations or significant environmental consequences would result from failure, a design flood return period in the range of 10 years for smaller channels to 100 or 200 years for larger channels seems reasonable.

32.5.3 Hydraulic Design

Hydraulic design involves selecting the channel longitudinal slope and cross section required to convey the design discharge without significant erosion or sedimentation. The parameters to be selected are shown in Figure 32.10.

The most commonly used equation for design is the Manning equation

$$Q = AV = A\frac{1}{n}R^{\frac{2}{3}}S^{\frac{1}{2}} \qquad \textbf{EQ. 32.3}$$

Where:

Q = the design discharge, m^3/s
V = average velocity in the cross section, m/s
A = cross-sectional area of the wetted portion of the channel, m^2
R = hydraulic radius of the wetted portion = A/P, m
P = wetted perimeter of the cross section, m
S = longitudinal slope of the channel (dimensionless)
n = Manning's roughness parameter

The longitudinal slope is generally selected first, as it is strongly constrained by the existing topography, with limited room for choice. Because flow velocities are strongly influenced by the longitudinal slope, the selected slope value must fall between a minimum, at which velocities are so low that sediment deposition becomes a problem, and a maximum, at which velocities start to erode the channel bed and/or banks. The most cost-effective slope is usually the one that will produce the maximum allowable velocity, as it will correspond to the smallest required cross-sectional flow area. Minimum slopes of less than 0.0005 are rarely used, except for very large channels. If the topography is so steep that erosive velocities would result, consideration must be given to providing an erosion-resistant channel lining or incorporating drop structures.

The magnitude at which flow velocities become erosive depends on the characteristics of the native soil. Velocities as low as 0.2 m/s can be erosive in fine sands and silts, whereas velocities of up to 1.0 m/s have been found to be acceptable in some clay soils. Note that these are average velocities and apply to straight reaches in which the local velocities immediately adjacent to bed and banks are typically less than the cross-section average. At locations where local velocities adjacent to banks or bed are higher that in a straight channel reach, such as in a bend or around obstructions, erosion protection may have to be provided.

Channels through fill areas should be avoided because of the appreciably higher risk of erosion and seepage, leading to possible loss of bank integrity and avulsion (or breaking out) of the channel. If fill channel sections cannot be avoided, a lined channel should be considered, as well as increased freeboard if the consequences of failure are significantly higher than for normal cut channel sections.

The most practical section shape is almost always trapezoidal, as shown in Figure 32.10. The cross-section geometry is defined by the side slopes, the flow area, and the width-to-depth ratio B:D. Side slopes generally should be selected based on soil stability considerations, accounting for rapid drawdown and possible areas of seepage inflow, both of which reduce stability. A layer of rock (underlain by an adequate filter) can significantly improve sideslope stability where seepage inflow is a problem. Sideslopes for channels where vegetative growth may need to be controlled are sometimes governed by the requirements of maintenance equipment.

The cross-section flow area is selected to achieve the design discharge, once the channel longitudinal slope and flow velocity are approximately established. Width-to-depth ratios in the range of 4 to 8 are commonly used, although this can be somewhat less

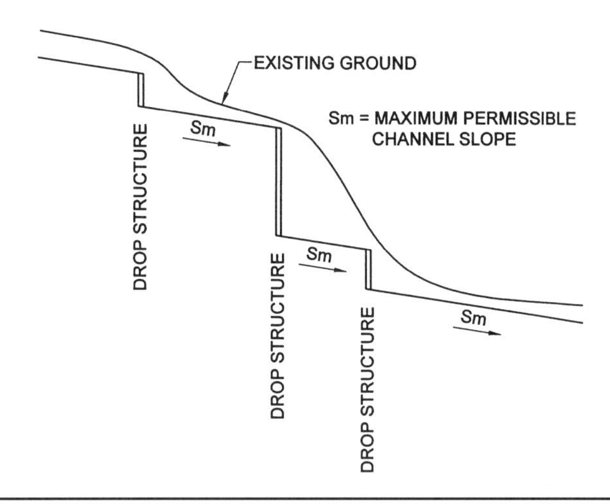

FIGURE 32.11 Schematic of stepped channel profile using drop structures

FIGURE 32.12 Rock chute under construction, slope at 4.4%

FIGURE 32.13 Gabion chute at a slope of 17% with a 5 m drop (Highvale Coal Mine, Alberta, Canada)

for small channels. The width-to-depth ratio may be governed by the capabilities of the available excavation equipment. An access road along one or both sides may be required for maintenance.

The value of the roughness parameter *n* must be selected using engineering judgment and experience. Typical values range from 0.025 for uniform smooth channels to 0.065 for non-uniform and rough channels. Various texts provide pictorial guides to assist in selecting an appropriate value (Barnes 1967; Chow 1959). The channel should be designed to function both under initial conditions, when the channel will be relatively smooth and unvegetated (and somewhat more susceptible to erosion), and later on, when vegetation, small local slumps, and animal activity will have increased the roughness, resulting in somewhat lower flow velocities but correspondingly higher water levels for the same discharge.

Freeboard (*F*), the vertical distance from the design water level to the top of the bank (see Figure 32.10), provides a safety margin against overtopping of the channel banks. Freeboard is selected on the basis of engineering judgment, taking into consideration the uncertainty in estimating design discharge and channel roughness, accuracy of construction, consequences of overtopping the banks, and cost of providing freeboard. Typical values range from 150 mm to 1,000 mm. The hydraulic design of open channels is covered in various standard texts (Chow 1959; Henderson 1966; Simons and Senturk 1992).

32.5.4 Channel Lining

A channel lining may be required to control erosion or seepage out of the channel. Seepage flow rates are usually small relative to diversion channel flows; control of seepage thus is rarely considered except where seepage out of a diversion channel flows into the mine area or otherwise requires handling by the mine water management system. An impermeable liner would be required to control such seepage. Liners have been constructed of cast-in-place concrete, shotcrete, soil cement, plastic, rubber, and asphalt. The literature on irrigation canals provides further information on seepage control with liners (e.g., USBR 1963).

Channel lining for erosion control aims to provide a surface able to resist the hydraulic forces associated with higher velocities. These liners therefore typically consist of rock or boulders (riprap) of sufficient size. Concrete or shotcrete liners also would achieve this purpose but tend to be significantly more costly than riprap. The erosion resistance of a riprap liner can be improved considerably by enclosing the rock in wire baskets, or gabions; individual baskets are wired together to form a continuous, flexible, and permeable liner.

Design of a riprap lining for erosion control involves selecting the median rock size required to resist movement at the design velocities, the gradation of the rock, the thickness of the riprap layer, and selection of one (or possibly two) filter layers to satisfy soil mechanics criteria for control of piping of the underlying native soils through the riprap. Heavier riprap may need to be placed along channel bends or at other locations where local velocities are expected to be higher than in a straight reach. Note that rounded fieldstone is less stable than blasted or broken quarry rock of the same nominal size. Design guides for riprap include Simons and Senturk (1992) and Maynord (1993).

Grade control structures are required where the channel slope becomes too steep to permit erosion control using channel linings. A grade control structure provides for a large change in elevation over a relatively short distance. Individual vertical drops of up to 2.5 m can be achieved using concrete, sheet piling, or gabion structures. Such structures can be located along the channel to create a stepped profile (Figure 32.11). Alternatively, a rock or gabion chute can be used to traverse a length of moderately steep terrain.

Rock chutes (Figure 32.12) can be used for slopes of up to about 12%, although lab tests have shown that rock chutes can be stable at slopes approaching 20% (Smith and Kells 1995). The granular filter underlying the riprap is a key component and must be carefully designed and installed. Experience suggests that two filter layers should normally be used.

A gabion chute, consisting of a continuously stepped drop (Figure 32.13), can be used for slopes of up to 17% (i.e., 1V:6H), using a standardized design developed in Alberta (Lindner 1983). Rock and gabion chutes both require a weir at the chute entrance to avoid drawdown and flow acceleration in the upstream channel. For steep drops requiring slopes of 20% or greater and heights in excess of 2.5 m, a concrete chute, with or without baffles, would be required. All drop structures require energy dissipation by means of a stilling basin at the downstream end (Peterka 1974). Gabion chutes typically incorporate a counterweir on the final step, which functions similar to a flipbucket to spread the flow laterally and upward to facilitate energy dissipation in the stilling basin (Figure 32.13).

32.5.5 Culverts

Culverts are required wherever access is needed across the diversion channel. Culverts at mine sites usually consist of corrugated steel pipe (CSP) because their light weight minimizes shipping costs and facilitates handling. CSP in the larger sizes is usually shipped in curved segments known as *structural plate* and is assembled by bolting together on site. Concrete pipe or box culverts sometimes are preferred, particularly where excessive pressures resulting from high fills or a combination of heavy equipment and shallow fills might occur, or because of corrosive conditions from either the water or the surrounding soils.

Culvert design involves both hydraulic design and structural design. CSP culverts are flexible structures, and their load-bearing capacity depends on careful compaction of the surrounding soil. A minimum depth of cover is required, depending on the applied traffic loads. The wall thickness is selected based on the structural requirements (which sometimes can be governed by handling or construction loadings) and an allowance for corrosion. CSP culvert inlets and outlets for larger sizes should be reinforced with concrete to provide extra stiffness, especially if the ends are beveled. Use of concrete at the inlet also can add sufficient weight to prevent the inlet from buckling upward because of buoyancy if the inlet becomes submerged.

Culvert hydraulics is complicated, with flow regimes being controlled by either inlet or outlet conditions, or a combination thereof, depending primarily on the slope of the culvert and the outlet water levels. Most culverts are designed to operate under inlet control, and the culvert capacity then depends solely on the inlet diameter and characteristics. Culverts on steep channels can be expected to handle flows with high sediment bed loads, and protection of the culvert invert by a concrete lining may be advisable.

Culverts should be sized to limit the design discharge headwater level at the inlet to no higher than about 0.8 of the diameter (D) (i.e., with an inlet freeboard of about 0.2 D). If considerable floating debris is expected during floods, consideration should be given to providing a greater freeboard or a debris-control device some distance upstream of the culvert. In these cases, culvert inlets should be monitored and debris removed regularly. The hydraulic design of culverts, especially of larger sizes, is best left to experienced hydrotechnical engineers.

Erosion protection of channel bed and banks is typically required at culvert outlets, where velocities are high. Protection also may be required at the inlet because of acceleration of the flow as it enters the culvert. Clay seals around the culvert barrel at the inlet, and possibly also at the outlet, are recommended to provide cutoffs for seepage along the outside of the culvert barrel, which could cause piping and lead to structural failure.

As a result of higher flow velocities through culverts, they can constitute a barrier to fish passing through. These conditions can be mitigated by using oversize culverts, selecting an arch or elliptical barrel shape, or using flow baffles in the barrel.

32.6 REFERENCES

Barnes, H.H. 1967. Roughness characteristics of natural channels. *USGS Water Supply Paper 1849.*

Chapell, B. 1986. Rock fill dams in handling mine waste material. *Proceedings International Symposium on Flow-Through Rock Drains.* Cranbrook, British Columbia.

Chow, V.T. 1959. *Open Channel Hydraulics*. New York: McGraw-Hill.

Claridge, Frederic B., Nichols, Robert S., and Diddens, Willem N. 1986. Rock drain behaviour at Byron Creek Collieries, sedimentation pond. *Proceedings International Symposium on Flow-Through Rock Drains for Mine Waste Dumps.* Cranbrook, British Columbia.

Environment Australia. 1998. Landform design for rehabilitation. *Best Practice Environmental Management in Mining*.

Henderson, F.M. 1966. *Open Channel Flow.* New York: Macmillan.

Lane, D., Berdusco R., and Jones, R. 1986. Five years experience with the Swift Creek rock drain at Fording Coal Limited. *Proceedings International Symposium on Flow-Through Rock Drains.* Cranbrook, British Columbia.

Lawson, J. 1986. Protection of rockfill dams and cofferdams against overflow and through-flow - the Australian experience. *Proceedings International Symposium on Flow-Through Rock Drains.* Cranbrook, British Columbia.

Leps, T.M. 1973. *Flow Through Rockfill*. Embankment Dam Engineering. Casagrande Volume. New York: John Wiley & Sons. 87–107.

Lighthall, Peter C., Sellars, C. David, Burton, W.D. 1986. Design of waste dumps with flow-through rock drains, *Proceedings of the International Symposium on Flow-Through Rock Drains.* Cranbrook, British Columbia.

Lindner, D.H. 1983. Application and design practice of gabion structures in Alberta, *CSCE Proceedings, 6th Canadian Hydrotechnical Conference.* Ottawa. 55–76.

Maynord, S.T. 1993. Corps riprap design guidance for channel protection. *International Riprap Workshop.* Fort Collins, Colorado.

Morisawa, M. 1968. *Streams, Their Dynamics and Morphology*. New York: McGraw Hill.

Peterka, A.J. 1974. *Hydraulic Design of Stilling Basins and Energy Dissipators.* Washington, DC: USBR.

Piteau Engineering Ltd. 1999. *Mine Rock and Overburden Piles, Rock Drain Research Program*. Final Report. -ISBN 0-7726-3831-4.

Simons, D.B., and Senturk, F. 1992. *Sediment Transport Technology*. Littleton, CO: Water Resources Publications.

Smith, C.D., and Kells, J.A. 1995. Stability of riprap channel linings on steep gradients. *Proceedings, CSCE Annual Conference.* Ottawa. 317–326.

Strahler, A.N. 1957. Quantitative analysis of watershed geomorphology. *Trans. Am. Geophys. Union*. 38: 913–920.

Szymanski, M.B. 1999. *Evaluation of Safety of Tailings Dams.* BiTech Publishers Ltd. Vancouver.

USBR. 1963. *Linings for Irrigation Canals*.

Ventura, J.D. 1986. Design and construction practices in rock drains. *Proceedings of the International Symposium on Flow-Through Rock Drains.* Cranbrook, British Columbia.

Wilkins, J.K. 1956. Flow of water through rockfill and its application to the design of dams, *Proceedings 2nd Australia, New Zealand Soils Conference*.

CHAPTER 33

Embankment Hydrology and Unsaturated Flow in Waste Rock

G. Ward Wilson*

33.1 INTRODUCTION

It is essential to characterize water transport in waste rock embankments. Fluid flow in waste rock controls the distribution of pore-water pressures within the embankments, which may in turn control stability. Furthermore, water flow and transport processes influence oxidation, weathering, and particle size evolution within waste rock systems. The development of a conceptual model to describe the flow of water through waste rock requires an understanding of the internal structures within waste rock embankments and the hydraulic properties of the materials. Wilson (1995) describes the field investigation for a large waste rock excavation program at Golden Sunlight Mine in Montana. Detailed observations, in situ measurements, sampling, and laboratory testing were carried out during the research program. The results for this investigation and subsequent seepage analyses are described here.

33.2 WASTE ROCK EXCAVATION AND FIELD OBSERVATION

Approximately 15 million tonnes of waste rock were excavated at the Golden Sunlight Mine in 1994. The waste rock material was excavated in order to stabilize a large landslide within the foundation soils of the embankment. The exposure of a 100-m vertical section of waste rock provided an excellent opportunity to investigate the physical structure, material properties, and hydraulic behavior within the profile. Wilson (1995), Herasymuik et al. (1995), and Herasymuik (1996) describe a field investigation that was initiated in October 1994. A total of 30 test excavations were examined. Samples taken from individual layers as well as *in situ* measurements for water content, temperature, matric suction, and relative humidity were obtained. Laboratory testing was carried out to determine the grain-size distribution, hydraulic conductivity, and soil-water characteristic curve for selected samples.

Figure 33.1 is a photograph of the exposed profile of the waste rock embankment. Three benches, each approximately 20 m in height, can be seen. The excavation shows all bench profiles to be highly structured. Beds dipping at the angle of repose (i.e., typically at 38°) that developed during end-dumping are apparent. Figure 33.2 provides a close-up view of the bedding planes that can be seen to vary in both texture and tone.

FIGURE 33.1 Photograph showing the excavated face for three 20-m benches of oxidized waste rock at Golden Sunlight Mine

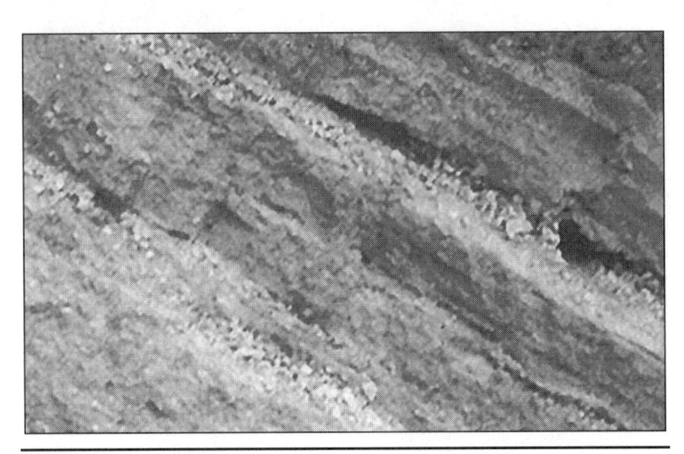

FIGURE 33.2 Photograph showing the structure of a single bench

The waste rock consisted of shale and intrusive rock with disseminated pyrite. In general, massive weathering associated with oxidation in the shales was observed. Figure 33.3 shows the condition of the fresh shale waste rock at the time of placement. This material has a relatively uniform particle size distribution at the time of placement. However, the particle size of the shale rapidly evolves to fine sand, silt, and clay with oxidation. Figure 33.4 illustrates the same waste rock that has been allowed to weather for a period of several months. The coarse layers observed in the waste rock were found to be latite that was more resistant to weathering.

All waste rock materials encountered during excavation were found to be unsaturated. In all cases, the coarse layers were fully drained. The fine layers in the upper 20 m (i.e., uppermost waste rock bench) were found to be moist, with volumetric water contents ranging between 9% and 23%; the degree of saturation ranged between 30% and 75%. The waste rock encountered at greater depth was found to be much drier, with volumetric water contents ranging between 0.4% and 5%. Temperatures up to

* Department of Mining and Mineral Processing, University of British Columbia, Vancouver, British Columbia, Canada.

FIGURE 33.3 Photograph showing the particle size distribution for fresh waste rock

FIGURE 33.4 Photograph showing highly weathered waste rock

65°C were measured in the upper 20-m waste rock bench where water contents were found to be high. Numerous surface vents corresponding to the exposure of coarse layers of waste rock were also observed in this region. Cooler temperatures, ranging between 15°C and 25°C, were observed in the deeper benches that were found to have lower in situ water content. The dry condition of the waste rock embankment at several depths has not been fully explained. One explanation is that excess water may have been evacuated through the surface vents. Furthermore, water is consumed in pyrite oxidation. These active processes, together with low recharge, may have resulted in desiccation within the deeper profile of the waste rock embankment.

33.3 CONCEPTUAL MODEL FOR UNSATURATED FLOW IN WASTE ROCK

Herasymuik (1996) proposed a general framework and conceptual flow model for the waste rock embankments at Golden Sunlight Mine. Figures 33.5 and 33.6 illustrate a typical structure for waste rock embankments constructed on a side hill. The figures show coarse and fine layers of dipping beds of waste rock as previously described. The waste rock system is unsaturated, and negative pore-water pressures dominate throughout the embankment. The coarse-grained layers are drained and form pathways for the flow of water vapor, oxygen, and other atmospheric gases through diffusion and advection. The fine-grain layers retain water under negative pore-water pressures due to the small pore sizes of the silt and clay material. Herasymuik (1996) suggested that the fine-grained layers form the principal pathway for liquid water flow, since they maintain a relatively high water content. The coarse-grained layers may form fast pathways during high infiltration events; however, flow is generally confined to the fine layers.

The coarse rubble zone that forms at the base of the embankment due to segregation serves as a drain at the toe of the embankment. Furthermore, the rubble zone provides access for the entry of oxygen through advective and diffusive airflow along the foundation of the waste rock embankment.

In general, relatively few studies have been carried out that accurately describe the hydrology of unsaturated waste rock embankments. Nichol, Smith, and Beckie (2000) are currently studying the hydraulic behavior of a constructed waste rock embankment with a contiguous lysimeter system at the base of the structure. Profile instrumentation has been installed for the measurement of volumetric water content, matric suction, temperature, and pore gases. This study is nearing completion and is expected to significantly advance our understanding of unsaturated flow systems in waste rock embankments.

33.4 UNSATURATED FLOW MODELLING IN WASTE ROCK EMBANKMENTS

There are a number of numerical models for predicting saturated and unsaturated flow in waste rock embankments. Lopez, Smith, and Beckie (1997) use a kinematic wave model to describe the hydrograph for discharge from waste rock embankments after periods of rainfall. The kinematic wave model has been used for predicting discharge from waste rock embankments. Unfortunately, the model does not rigorously describe the mechanics of saturated/unsaturated flow for fast and slow pathways. Furthermore, the kinematic wave model does not provide a description of the pore-water pressure distribution within the waste rock embankment for various infiltration events. The need to predict pore-water pressures requires an alternative approach for seepage analysis in waste rock.

Pore-water pressure distribution and seepage pathways in waste rock embankments can be described with a two-dimensional equation for saturated/unsaturated transient seepage in porous media, as follows:

$$\frac{\partial}{\partial y}\left(k_y \frac{\partial h_w}{\partial y}\right) + \frac{\partial}{\partial x}\left(k_x \frac{\partial h_w}{\partial x}\right) = -m_2^w \gamma_w \frac{\partial h_w}{\partial t} \qquad \text{EQ. 33.1}$$

Where:

h_w = the hydraulic head
k_y and k_x = the hydraulic conductivity in the x and y direction as a function of matric suction
m_2^w = the slope of the soil-water characteristic curve
γ_w = the unit weight of water

The application of the equation for flow outlined above requires Darcian flow conditions to prevail within the waste rock embankment. In general, this requires that the waste rock behave more as a soil rather than as a rock-like material. The physical observations for the waste rock excavated at the Golden Sunlight Mine support this approach, particularly in cases where the waste rock embankments have been allowed to oxidize and weather for an extended period prior to closure and reclamation.

The hypothesis that the waste rock at Golden Sunlight Mine behaves like a soil material is supported by the soil-water characteristic curves for the waste rock measured by Herasymuik (1996). Figure 33.7 shows typical soil-water characteristic curves measured by Herasymuik (1996) for the fine-grain waste rock, with more than 50% passing the 4.75-mm size, and coarse-grain waste rock, with less than 30% passing the 4.75-mm sieve. It can be seen that the fine material has an air entry value approaching 5 kPa. In other words, the waste rock will not begin to desaturate until the negative pore-water pressures exceed 0.5 m of water.

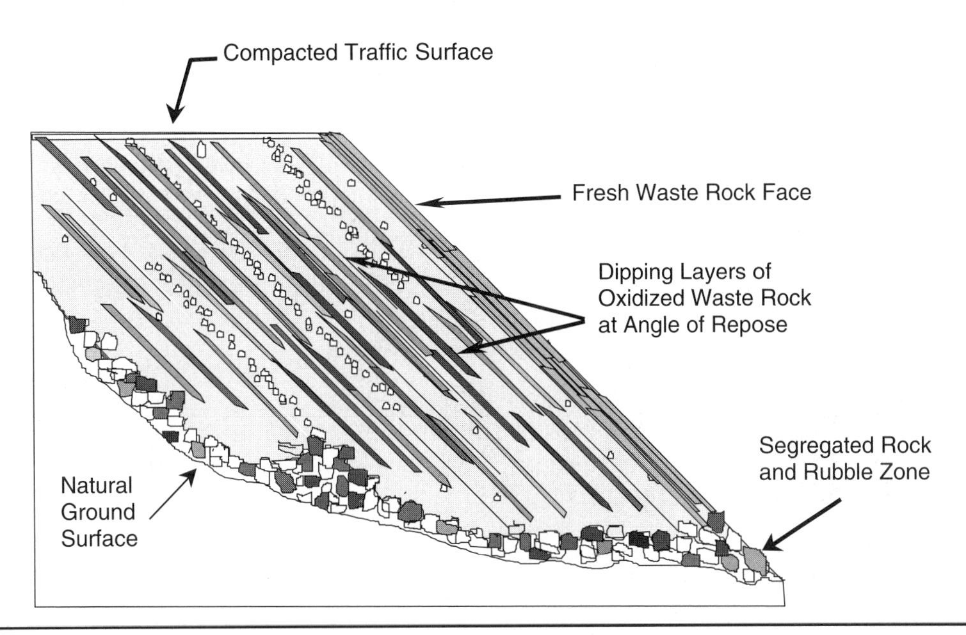

FIGURE 33.5 Typical structure of a waste rock embankment

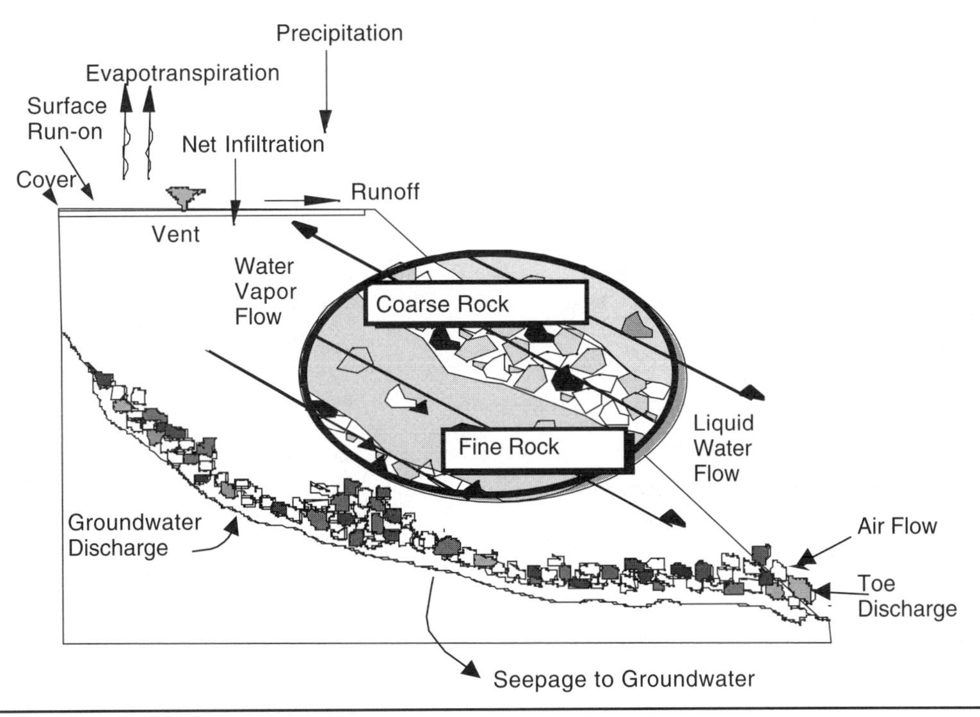

FIGURE 33.6 Typical flow regime within a waste rock embankment

Alternatively, it can be seen that the coarse-grained waste drains as soon as the value of matric suction exceeds zero. Furthermore, the coarse material is nearly fully drained at a value of matric suction equal to 5 kPa.

Figure 33.8 illustrates the hydraulic conductivity functions for the coarse and fine waste rock shown in Figure 33.7. The hydraulic conductivity versus matric suction relationships were calculated using the method described by Fredlund, Xing, and Huang (1994). It can be seen that the hydraulic conductivity for the coarse material declines rapidly as the value of matric suction approaches a small value of matric suction less than 1 kPa. The fine waste rock maintains a value of saturated hydraulic conductivity until the value of matric suction exceeds 5 kPa. It is interesting to note that at full saturation, the value of hydraulic conductivity of the coarse waste rock is 20 times greater than that for the fine material (i.e., 1×10^{-3} m/sec compared 5×10^{-5} m/sec). However, the value of hydraulic conductivity of the fine material is 5,000 times greater than the coarse waste rock for a value of matric suction equal to 1 kPa.

Newman et al. (1998) and Newman (1999) simulated and measured saturated and unsaturated flow in vertical profiles of coarse and fine waste rock layers. Figure 33.9 illustrates the results of the seepage analysis for adjacent coarse and fine vertical profiles of waste rock. It can be seen that the seepage vectors for flow cross over between the coarse and fine materials depending on position above the base of the profile. The position of the water table for the simulation corresponds to the base of

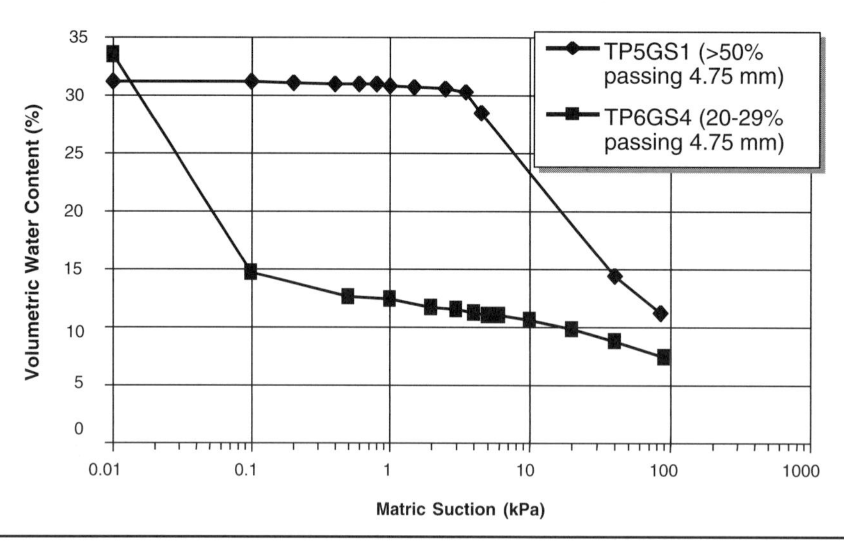

FIGURE 33.7 Typical soil water characteristic curves for weathered waste rock

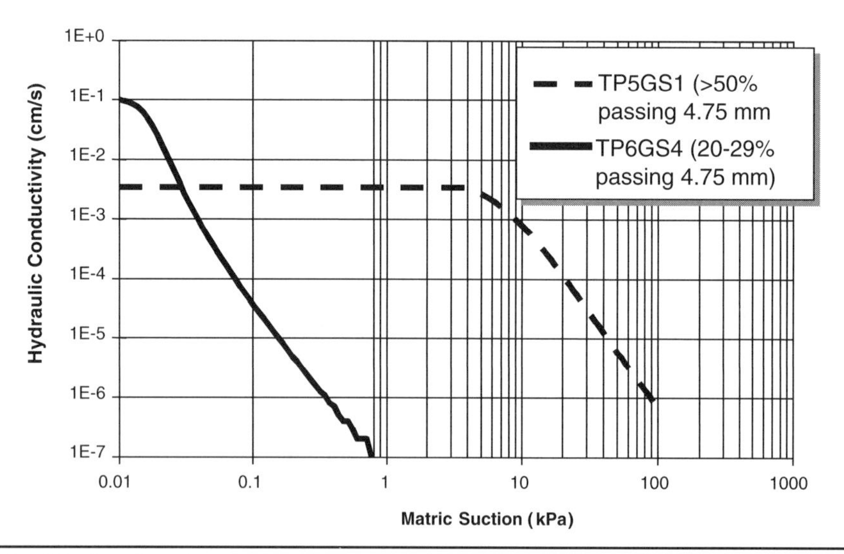

FIGURE 33.8 Hydraulic conductivity functions for weathered waste rock

the profile. A surface flux of 5×10^{-7} m/sec was applied to the 1.5-m height profile of waste rock.

It can be seen that all the flow takes place in the coarse layer on the right side of the profile immediately above the water table. Flow occurs in the coarse-grained waste rock because the hydraulic conductivity of the coarse material exceeds that of the fine material at values of matric suction less than approximately 5 kPa. Alternatively, the flow shifts to the fine profile on the left side as the elevation increases above the water table. This occurs because the hydraulic conductivity of the coarse material decreases rapidly once the value of matric suction exceeds the air entry value of the coarse material. For example, the coefficient of hydraulic conductivity for the coarse material at a value of matric suction equal to 15 kPa (i.e., at the top of the profile) is approximately 1×10^{-10} m/sec. Alternatively, the hydraulic conductivity of the fine material at the same value of matric suction is approximately 1×10^{-6} m/sec; hence, water finds the flow pathway corresponding to the least resistance in the fine layer.

The seepage results presented for the two-layer coarse and fine profile described above were confirmed with laboratory measurements described by Newman (1999). The findings are significant since they clearly show the potential development of preferential pathways in the waste rock embankment. These pathways are a function of pressure head and position above the water table. In saturated waste rock (i.e., below the water table), flow is confined to the zones with the highest saturated hydraulic conductivity, namely, the coarser waste rock. Alternately, the preferential flow path shifts to the fine layers as the value of matric suction increases and unsaturated conditions prevail. The value of suction for which this occurs depends on the saturated hydraulic conductivity and air entry value for the coarse and fine waste rock materials. The applied surface flux boundary condition will also change the flow pathway. For example, all flow will shift to the coarse material if the applied surface flux exceeds the saturated hydraulic conductivity of the fine waste rock.

Wilson, Wilson, and Fredlund (2000) investigated flow in multilayered coarse and fine rock profiles. Figure 33.10 illustrates how the flow converges to the fine layers in a waste rock system with two fine layers and two coarse layers. This work extends the study completed by Newman (1999). Wilson, Wilson, and Fredlund (2000) also studied the influence of slope angle and embankment height. Figure 33.11 illustrates the results of a seepage analysis with a vertical four-layer system and a four-layer system inclined at 45°. Significant contrast in the two flow regimes can be seen. The flow regime for the 4-m-high vertical profile is similar to that previously shown for the two-layer system. Preferential flow is divided

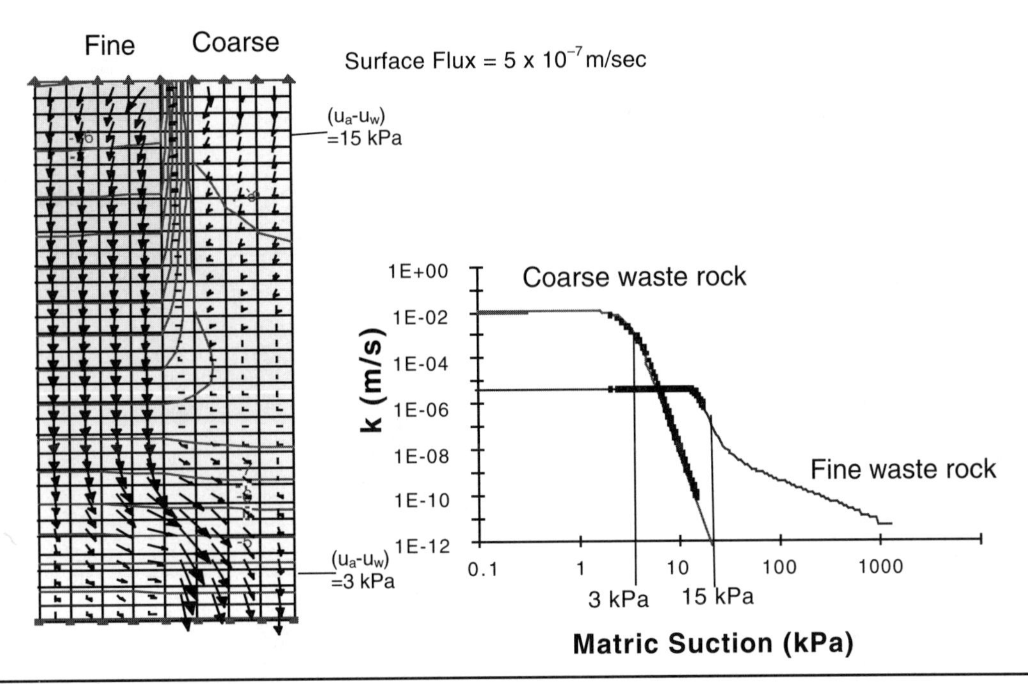

FIGURE 33.9 Example simulation for unsaturated flow in a two-layer coarse and fine waste rock system

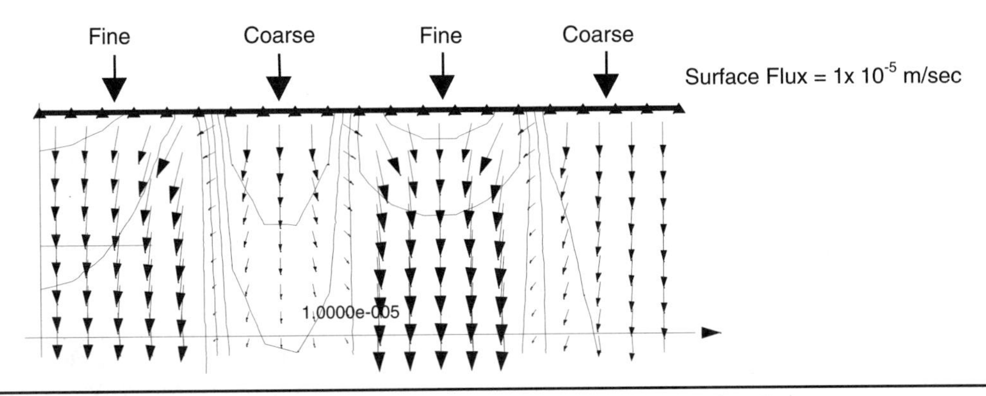

FIGURE 33.10 Example simulation for unsaturated flow in a four-layer coarse and fine waste rock system

between the two fine layers of waste rock. Flow in the inclined profiles also tends to be dominated by the fine layers; however, the distribution of flow between the two coarse layers and the two fine layers is less uniform. In general, it can be seen that the greatest flow occurs in the bottom fine layer (i.e., 6.3×10^{-6} m^3-m/sec). Alternately, seepage fluxes within the interior coarse zone are greater than those of the upper coarse zone (i.e., 1.0×10^{-6} m^3-m/sec versus 4.5×10^{-10} m^3-m/sec). It should also be noted that the distribution of pressure head and pore-water pressure vary significantly across the four layers of waste rock. For example, the negative pore-water pressures in the fine layers are smaller (less negative) than the adjacent coarse layers. Wilson, Wilson, and Fredlund (2000) repeated the analysis presented here for a variety of slope angles and slope heights. The results of the sensitivity studies showed that as much as 65% of the applied surface flux could be concentrated within a single fine layer. This trend may have implications with respect to shear strength, collapse, and physical stability.

33.5 SUMMARY AND CONCLUSIONS

A large waste rock excavation program was completed at the Golden Sunlight Mine in Montana. Observations for the waste rock materials removed showed that the waste rock embankments consist of an interbedded sequence of coarse and fine materials that have developed as a result of end dumping construction practices, segregation, oxidation, and weathering.

The waste rock profile was observed to be unsaturated at all times. Field observation suggests that the fine layers form the principal or preferred pathway for seepage through the dump due to surface infiltration. Saturated and unsaturated seepage modeling confirms this hypothesis. The distribution of flow between coarse and fine layers depends on the soil-water characteristic curve and hydraulic conductivity function for each material. Furthermore, the inclination of the dipping beds and embankment height also influence the distribution of flow. These observations and conclusions should be considered preliminary. However, the results of the numerical simulation suggest the development of complex flow regimes and the need for care when investigating unsaturated flow in waste rock embankments.

33.6 ACKNOWLEDGMENTS

The financial support provided by Placer Dome North America and the Golden Sunlight Mine to complete this research program is greatly appreciated. The field, laboratory, and analytical work completed by graduate students Greg Herasymuik, Darren Swanson, Lori Newman, and James Wilson is also gratefully acknowledged.

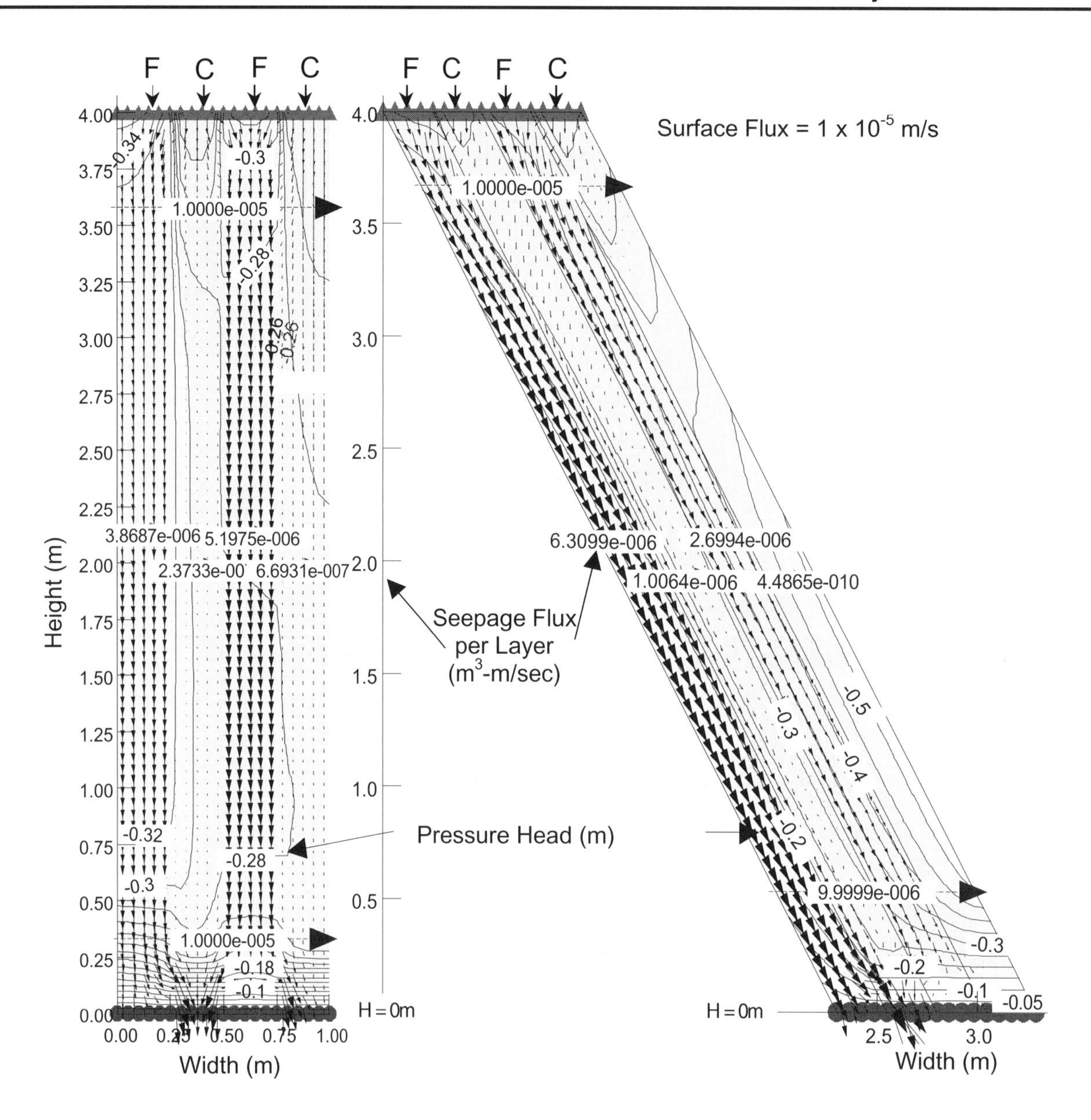

FIGURE 33.11 Example seepage simulation for unsaturated flow in dipping interbedded waste rock

33.7 REFERENCES

Fredlund, D.G., A. Xing, and S. Huang. 1994. Predicting the permeability function for unsaturated soils using the soil-water characteristic curve. *Canadian Geotechnical Journal,* 31:533–546.

Geo-slope International Ltd. 1995. *SEEP/W User's Manual*. Calgary, Alberta, Canada:Geo-Slope International Ltd.

Herasymuik, G.M. 1996. Hydrogeology of a sulphide waste rock dump. M.S. Thesis. University of Saskatchewan, Saskatoon, Saskatchewan, Canada.

Herasymuik, G.M., G.W. Wilson, S.L. Barbour, and T. Smith. 1995. The characterization of hydrologic properties and moisture migration pathways of a waste rock dump. *Proceedings of the Nineteenth Annual British Columbia Mine Reclamation Symposium*.

Lopez, D.L., L. Smith, and R. Beckie. 1997. Modeling water flow in waste rock piles using kinematic wave theory. *Proceedings Vol. II, Fourth International Conference on Acid Rock Drainage*. 497–514.

Newman, L.L. 1999. Preferential flow in vertically oriented, unsaturated soil layers. M.S. Thesis. University of Saskatchewan, Saskatoon, Saskatchewan, Canada.

Newman, L.L., S.L. Barbour, D.G. Fredlund, and G.W. Wilson. 1998. Preferential flow in vertically layered unsaturated system. *Vol. 1, Proceedings of the 2nd International Conference on Unsaturated Soils*. 586–589.

Nichol, C., L. Smith, and R. Beckie. 2000. Hydrogeologic behaviour of unsaturated waste rock: An experimental study. *ICARD 2000*. 215–224.

Wilson, G.W. 1995. Assessment of protective covers and dump behavior. 1995. *Proceedings of the 2nd Australian Acid Mine Drainage Workshop*. In progress.

Wilson, G.W., D.G. Fredlund, and S.L. Barbour. 1994. Coupled soil atmosphere modeling for soil evaporation. *Canadian Geotechnical Journal*. 31:151–161.

Wilson, J.A., G.W. Wilson, and D.G. Fredlund. 2000. Numerical modeling of vertical and inclined waste rock layers. *ICARD 2000*. In progress.

. . . .
CHAPTER 34

Operation and Monitoring Considerations from a British Columbia Mountain Terrain Perspective

Tim Eaton*

34.1 INTRODUCTION

This chapter is written from a British Columbia, Canada, perspective. Some of the province's site specifics are described to help put this chapter into context.

Monitoring of spoils at most mines in the province of British Columbia is conducted at a level not seen in many other jurisdictions around the globe. There are important reasons for this high level of management and operational control. Spoil piles can and do fail, sometimes frequently. The likelihood of failure can be significant in certain settings within the province. Failure of spoil piles in some instances is clearly not an option. Lives and infrastructure would be at risk, and in other instances fish and wildlife habitat would be compromised. To complement an appropriate level of engineering site investigation and design, monitoring has demonstrated itself to be a valuable tool for predicting failure and where conditions permit preventing a significant failure from occurring. This chapter describes the knowledge gained from experience in British Columbia.

The importance of the performance of tailings facilities and waste rock embankments in creating public impressions of mining cannot be overstated. Direct and indirect costs of a "failure" can run in the hundreds of millions of dollars and legal cases can create uncertainty for many years. Recent examples in the 1990s are the Omai and Los Frailes tailings dam failures, which created instant share value losses of 25 to 35% (Vick 1999).

Failure of waste rock piles can have similar impacts, although no recent event has been widely publicized. Coal waste embankment failures at Aberfan (1966) and Buffalo Creek (1972) resulted in 144 and 125 lives lost, respectively. In addition to loss of human life, the potential magnitude of detrimental environmental impacts could be on the same scale as those from a tailings dam failure. Waste-rock production from opencut mining operations is often many times larger than the amount of tailings produced.

Some waste embankments contain rock with potential for acid generation and metal or element leaching. Blasting residues are generally present. Some waste-rock embankments are engineered to mitigate potential detrimental impacts to varying degrees and levels of confidence. The engineered design may apply to the operating (construction) and or closure phase. Operating practices are critical to achieving the design objectives (including stability) of a disposal facility. Operating realities should be considered more carefully during design.

This chapter outlines operating and monitoring practices that can be employed to help achieve short- and long-term design objectives and postmining land-use goals.

34.2 OVERVIEW OF MINING HISTORY IN BRITISH COLUMBIA

The province of British Columbia has a long history of mining. The first "European" underground coal mine opened on Vancouver Island in the 1860s followed by underground gold, silver, and base metal mines by 1900. Surface mines began to appear in the 1950s and 1960s with advances in earth moving equipment. By the 1970s British Columbia was the largest copper-producing region in the world with production from a number of large porphyry copper deposits (medium to small scale by today's standards). The province was also becoming a significant producer of high-quality metallurgical coal from new surface mines. All of these older mines have closed, and the production of metal and coal today is dominated mainly by a small number of surface copper and coal mines. While there are underground mines in British Columbia, waste rock production from this source is very small and it is not a focus of further discussion here. To date, the historical value of all mineral production from the province in 1999 Canadian dollars is $156 billion (data on record with Ministry of Energy and Mines, British Columbia 1999).

Concurrent with the change in character of mining in the 1960s, 1970s, and 1980s was a change in the attitude of the people living and visiting the province. Increasingly the public began to recognize the stunning natural beauty of British Columbia's landscape and to desire to preserve that beauty and the wildlife for future generations to enjoy. Annually, millions of people vacation and recreate in the province's natural settings such that there is little if any "true wilderness" left. However, that does not diminish enjoyable memories imprinted on the minds of those travelers. Complementary to the recognition of the scarcity of wild temperate areas and habitats was the recognition of the importance of biodiversity. British Columbia was politically motivated and thus initiated in 1990 a long-term detailed land-use plan for the province.

Many mining regions around the globe may not be similar climatically and topographically to British Columbia, but the sociopolitical landscape defined in British Columbia is reaching into all corners of the globe. It may take several decades for it to be felt everywhere, but it is coming nevertheless. To date the land-use planning process has resulted in doubling of provincial park spaces in the past 10 years and comprehensive land-use plans for a large portion of the province. Land-use planning in British Columbia is scheduled to take another 10 years to complete. While in British Columbia important factors such as profitability and market determine the location of a mine (for a proponent), access to land for exploration, certainty of mineral tenure, and environmental assessment "approval in principle" are linked to the land-use plan or perceived land use.

* Ministry of Energy and Mines, Victoria, British Columbia, Canada.

34.3 THE BRITISH COLUMBIA EXPERIENCE

It is important to recognize the wide range of physical climate and landscape conditions that exist around the globe. Conditions in British Columbia may be harsher from a spoil stability perspective than in most mining regions. The influence of climate and landscape (term used collectively here for relief, foundation materials, and drainage) on spoil design and performance cannot be overstated. While British Columbia may have a combination of physical climate, landscape, and sociopolitical setting that is unique, some of these factors exist in other mining jurisdictions or are of increasing influence.

British Columbia's physical landscape is characterized by a series of mountain ranges with associated interior plateaus and highlands that run from southeast to northwest the entire 1,200-km length of the province. It was formed by the collision of Pacific and continental crustal plates, subduction zone volcanic activity, and erosion processes. Vertical relief of mountain ranges is typically between 1,000 and 3,000 m. The province was largely covered by alpine and continental ice sheets 15,000 years ago. Today, glaciers are found in most of the mountain ranges but only the Coast Mountains have glaciers of any significant size and continuity because of the high annual snowfall. Much of the province is covered by a variety of glacial, periglacial, fluvial, and colluvial deposits, sometimes deposited in complex sequences. The climate varies from a temperate wet coastal climate to the drier continental climate east of the Coast Mountains, where winters can be cold and summers hot. Annual precipitation averages between 1,000 and 3,000 mm along the coast and 400 to 1,000 mm in the interior. Microclimates abound because of orographic effects. A large percentage of precipitation falls in the form of snow except on low-elevation coastal areas where it is predominantly in the form of rain.

Thus, in British Columbia the following hold true:

- Climate and landscape conditions range from moderate to severe.
- Large volumes of waste rock are produced from open-cut surface mines.
- Land use is planned, specified, and highly valued by multiple stakeholders, putting resource extraction at odds with preservation, recreation, and other uses.

34.4 OPERATION

34.4.1 Introduction

Surface coal mining operations are located in steep mountain topography. By comparison, surface metal mining operations are mainly west of the Rockies and east of the Coast Mountains, typically in the interior mountain and plateau ranges where topography is more subdued. Experience has been obtained over 30 years from operation of metal and coal surface mines in British Columbia in a wide range of settings. Most of the waste rock embankment issues have arisen at the large surface coal mines. The case study presented by Zeitz in this volume illustrates "state-of-the-art" monitoring as it can be applied where the risks are significant.

Coal measures are exposed in formations in the Rocky Mountains. There are seven large surface coal mines with annual coal production ranging from 2 to 9 million tonnes of clean coal and annual waste-rock production ranging from 10 to 70 million bank cubic meters (bcm). Dumping rates can be in the higher range because of the large production volumes and confined waste-rock dumping sites offered by mountainous terrain. The coal mine overburden piles can be massive structures, among the largest manmade structures in the world–up to 400 m high, containing (when completed) in excess of 1 billion bcm of rock and form cross valley fills with rock drains at the base. The rocks are predominantly sandstones, siltstones, and mudstones (or shale) and can include smaller volumes of overburden soils, coal partings, and thinner coal seams. Sandstones are generally strong and durable; siltstones are weaker with variable durability; mudstone is weakest and friable.

A preferred waste rock embankment site and plan provide economic advantages over other site options. The primary objective is to minimize production costs. This primary objective must be achieved within the constraints of ensuring worker safety, protection of infrastructure, protection of the environment, and perceived performance of the disposal facility by the public or regulator. Construction or operation of waste rock embankments is usually conducted by the mine operations department with technical support from mine engineering and consultants.

Spoiling operations and plans constantly change to meet production and market requirements. British Columbia coal mines supply a variety of special blend coals. Meeting market requirements requires great flexibility in accessing different coal seams and the concurrent ability to spoil waste rock with significant short-range changes in spoiling rates and locations. The planning exercise is very dynamic.

A spoiling operation, to be cost effective and safe, must do the following:

- Have a monitoring program for worker safety and other operating requirements
- Have alternative spoiling locations
- Allow safe traffic access and movement for fleet to meet production supply
- Have documented and current operational guidelines and procedures
- Operate with minimal interruption
- Be amenable to monitoring instruments and visual observations
- Allow spoil platform maintenance and grading

34.4.2 Planning and Operations Flexibility

Aspects of early or design phase spoil planning including site selection criteria, risk assessment, and level of investigation and design effort have been described by Hawley in this book. Piteau (1991) also provides a general overview of the planning and design steps. This section provides an overview of spoil performance, availability, and flexibility.

High availability requirements for dumping platforms can kill what was considered a good spoil-pile plan. Every mine develops its own production efficiencies for its mine fleet based on site-specific experience and knowledge of conditions. A supplier's performance handbook or data from other sites would be used only for feasibility. If the spoil pile has low availability due to high settlement rates or frequent failure and rehabilitation losses, both can kill the plan and increase production costs significantly. Availability of dumping platforms must be carefully considered in the context of risk assessment and making practical management decisions. How much confidence does the mine have in knowledge of foundations conditions, waste rock quality, stability assessment, and assumptions? Will the production volumes increase above those envisaged? Will the actual production equipment be larger than assumed in feasibility? Will operating criteria ensure safety and meet production requirements? Are there alternative spoiling plan options?

A waste rock disposal plan developed by an engineer may be based solely on embankment volume available in the proposed spoil, haul distance, haul road grade, and the number of trucks required, without giving due consideration to geotechnical factors that will affect availability. If the spoil continually cracks and settles such that there are frequent closures, then the trucks cannot dump their loads, and ore and coal accessibility may be

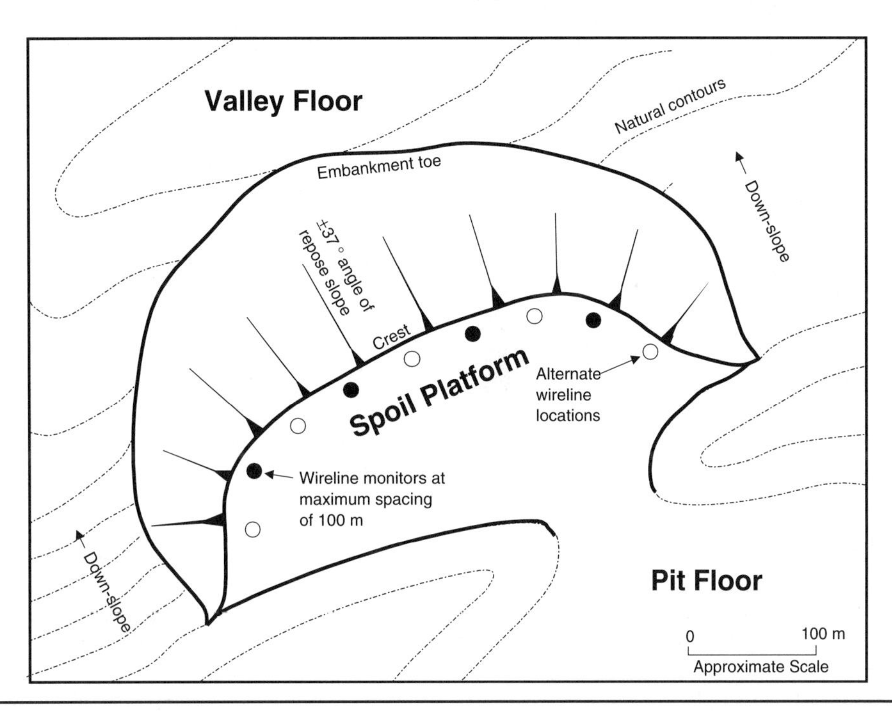

FIGURE 34.1 Plan schematic of a typical waste-rock embankment, single platform, in British Columbia's mountainous terrain

restricted. This can occur to the point of significant loss of revenue and or increased operation and remedial costs. This is poor planning.

On the other hand, a spoil pile may often exhibit high settlement rates that close the facility temporarily but that do not accelerate to failure. In this instance the settlement criteria and controlling factors should be reviewed to determine whether dumping platform safety and availability can be achieved with higher allowable settlement rates.

A rationale for a risk-based classification of mine waste dumps was provided by Mark Hawley earlier in this section. This procedure originated from a draft proposal for a risk-based classification scheme prepared for the British Columbia Ministry of Energy and Mines in November 1992. In British Columbia a waste-rock embankment can be designed with a high likelihood of failure and receive regulatory approval. The basis for issuing such an approval relies on the likelihood and consequences of failure. It is similar to the risk-assessment approach developed for dam safety in Canada. Safety of the operators on the dumping platform is paramount and must be assured in every instance. The need for the classification system and a better understanding of spoil behavior arose from a high incidence of waste-rock embankment failures spanning the previous 10 years that appear to be continuing in spite of industry claims it was doing a better job.

The risk classification is a guide in draft form, never formally adopted by the industry or regulator in British Columbia. However, it was distributed to British Columbia practitioners for comment and subsequently used (unofficially) in developing new waste dump designs and improving management tools. A series of interim guidelines and reports were produced from studies sponsored by the regulator between 1991 and 1995, and a final report on rock drains was made public in 1999 (see References for publications produced for the Ministry of Energy and Mines). The interim guidelines and reports were widely distributed in British Columbia and to a lesser degree outside the province. (Three of the earliest reports are now available on the Edumine web site.) The Mining Association of Canada produced *A Guide to the Management of Tailings Facilities* in September 1998. This is a well-organized and flexible management framework that could be readily adapted to waste-rock embankments and is a recommended reference in this regard.

Figures 34.1 and 34.2 show plan and cross-section schematics of typical waste rock embankment configurations in mountainous terrain. Some general consideration regarding performance and stability include the following:

- High embankments will settle more than low embankments.
- High waste rock embankments will place larger loads on the foundation materials and can generate pore pressures, although the low rate of crest advance may allow pore pressure dissipation.
- Low embankments will advance more quickly than high embankments and thus place higher loading rates on foundations materials and will generate pore pressures that may not dissipate fast enough.
- The steeper the foundation slope, especially beneath the embankment toe area, the greater the likelihood of instability.
- For a given spoil the higher the dumping rate and meters of crest length, the higher the settlement rate recorded by the monitors.
- Finer-grained materials (including rehandled material, poor-quality waste rock, and overburden soils) result in higher rates over longer periods and higher total displacements.
- Fine-grained and wet materials can produce oversteepened slopes at the crest that eventually slough off in sliver failures.
- Springtime snowmelt or a rainy season results in higher rates and longer periods for rates to settle to safe (acceptable) levels as there is a higher phreatic surface in the foundation (or embankment), which generates higher pore pressure.

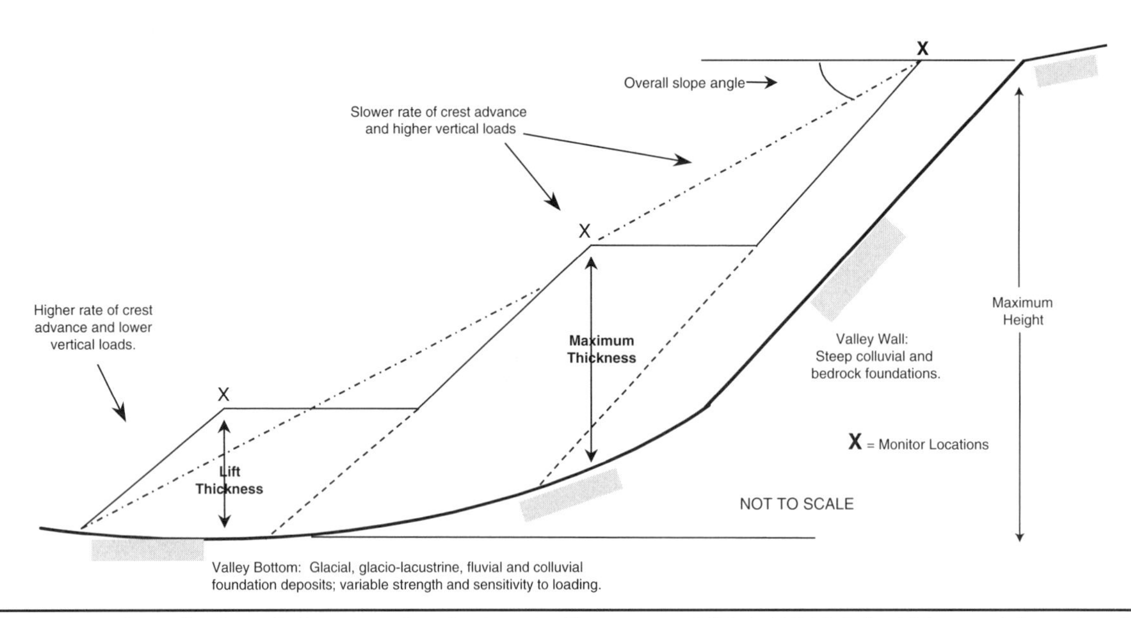

FIGURE 34.2 Section schematic of a typical waste-rock embankment with wraparound lifts in British Columbia's mountainous terrain

- Surface-water flows can erode a dumping platform and face if directed over the crest and can create deeper-seated stability concerns if flows disappear behind the spoil crest.

Disposal facility criteria should be established for the following:

- Stand-alone visual indicators of instability
- Allowable safe monitor settlement rates in millimeters per hour or day for direct dumping over crest
- Allowable rate of crest advance
- Direction of crest advance
- Quality of waste rock consigned to different parts of a spoil for safety and stability reasons
- Allowable loading rate in terms of tonnes or loads per meter of crest per shift
- Dump short and push by dozer
- Closure of the spoil and reopening of the spoil
- Evacuation of downstream areas and access to downstream areas
- Closure during hazardous conditions such as poor visibility (fog or smoke) or adverse climatic conditions (intense rainfall or snowstorm)

34.4.3 Safety

With a good spoil plan in place, successful implementation is highly dependent on resources that include committed and prepared employees. Appropriate training and education should be provided to operating personnel responsible for waste-rock disposal. Table 34.1 outlines the needs and benefits for each level of responsibility. The mine manager is part of this group. Materials can be prepared and given internally, although input from the geotechnical consultant and regulator should be sought.

A successful spoiling operation has the spoil platform at its core, and this area can be a busy place. Under optimal conditions, there is a steady and unimpeded flow of rock or earth haulers to the spoil. Earth haulers today can move a lot of material in 24 hours. Operations need to be almost continuous, rain or shine, day and night. Equipment operators need to know with confidence where they can and should go on the platform and feel completely safe. Operators need to know what to look for in terms of warning signs or unusual behavior. They also need to know that their supervisors and engineering support are active in performing their monitoring, assessments, and quality control responsibilities.

A dumping operation must be safe from small-scale local failures (which can put operators at risk) and larger deeper-seated failures (operators, infrastructure, the public, and the environment may be at risk) that can contain several million bank cubic meters of waste rock. After small or large failures occur, safe procedures are required to rehabilitate the embankment crest in order to resume normal dumping operations.

Monitoring instruments, when properly located and monitored, will provide indicators of changes in rate of settlement and crest settlement acceleration to failure. Combined with regular visual observations, features of undesirable configuration and large-scale deformations leading to instability can be identified. With experience and good knowledge of dumping platform behavior, operating criteria can be refined to maximize use and availability of the spoil while ensuring safety for the operators working on the dumping platform or below (downstream of) the embankment toe. When one starts an operation with little or no experience, conservative criteria should be used for operating and closing a spoil until enough experience is obtained to optimize operation and efficiency.

Safety criteria need to be established that consider the following:

- Scale of the operation and the physical setting
- Consequences of unplanned shutdowns
- Consequences of failure
- Direct dumping at the spoil crest when it's safe, and when to "dump short and push" or to shut down
- Size of crest berms
- When it's safe to reopen after shutdown
- Monitoring information required to put a spoil into operation that has been on standby
- When to evacuate areas downstream of an active spoil toe, within a conservative potential runout zone
- When to evacuate areas downstream of inactive spoil toes, within a conservative potential runout zone
- Dozing off settlement scarps to reopen inactive spoils

TABLE 34.1 Personnel training and education recommendations

Group	Need	Benefit
Mine Manager	Thorough understanding of risks (safety, economic, environment).	Ability to commit resources to maximize performance objectives. Ability to inform Board of Directors of risks and contingencies.
Engineering	Thorough understanding of all technical aspects of design, construction and operation. Knowledge of geotechnical properties of all relevant waste and foundation materials and effects of groundwater.	Ability to detect and recognize abnormal performance. Ability to react to changes in properties or embankment performance resulting in need for modification of design or procedures. Ability to suggest required changes for review by designer.
	Knowledge of types of monitoring, when to use, and ability to analyze monitoring results. Familiarity with guidelines pertaining to monitoring, closure, reopening, etc. Knowledge of potential magnitude, speed and effects of embankment failure.	Ability to react appropriately to movements or unusual occurrences. Appropriate monitoring techniques applied to specific situations. Ability to review operation with management, operators and regulator or public.
Operations (Managers and Supervisors)	Thorough knowledge of design specifications. Basic understanding of possible implications of variance from technical design specs.	Ability to quickly recognize variance from design specs. More rational approach to construction decisions.
	Ability to recognize signs of instability. Knowledge of potential magnitude, speed and effects of embankment failure. Knowledge of appropriate reactions to various signs.	Shorter and more timely response to potential problems. Recognition of potential injury or death, interruptions to operations, environmental damage, industry/regulator relations.
Operations (Operators)	Ability to recognize signs of instability. Thorough working knowledge of all monitoring equipment in use. Knowledge of good dumping procedures.	Faster reactions to problems due to immediate reporting by personnel working on the dumping platform. Safer working environment due to better understanding.

Adopted from Klonn (1991).

- Reestablishing a dumping platform crest after a failure by advancing from an established stable crest
- Use of a spotter for dump trucks and traffic signals
- Responsibilities of the operations superintendent, shift general foreman, the engineer, crew supervisor, truck operators, dozer operators, and overburden disposal personnel
- Frequency of monitoring and observations
- Establishing settlement rates and deformation patterns
- Interpreting settlement data, changes in rates, and establishing trends
- Nighttime operation
- Inclement weather operation

Because of the range of personnel involved, monitoring programs must be kept simple to be effective and generate high confidence levels. Judgment is required in setting up monitors. Different people will set monitors up differently (different locations, different locations of stands relative to displacement), and they will move monitors at different times onto crest noses from crest gullies. This will result in different movement rates being measured at different times given different personnel, and if a situation leads to an acceleration preceding a failure, different amounts of warning. Rates must be set conservatively enough so that there will be adequate warning in all situations. If necessary, availability can be increased with warning necessarily reduced. This should be attempted only if the skill level of personnel involved warrants such action. Eliminating judgment necessarily creates guidelines that are too complicated to cover all situations and will ultimately be useless to personnel in the field. A mine's staff must be comfortable with and cognizant of the definite tradeoffs: skill level, effort needed to train, and effort to audit procedures and responsibilities.

34.5 MANAGEMENT

34.5.1 Introduction

The Mining Association of Canada guideline mentioned earlier in this chapter could be used to develop the management framework and philosophy for a critical or high-profile spoiling operation. Most open-cut mines in British Columbia requires spoiling procedures, whether metal or coal. There is a wide range in management effort and involvement with procedures that reflects the importance of particular spoiling operations to the viability of the mine or the awareness of management. The British Columbia regulator ensures that procedures do not collect too much dust by requesting current copies and conducting informal audits during inspections. The more elaborate management procedures include the following topics:

- Introduction and definitions
- Designation of spoils (i.e., active, inactive, standby, dormant, reclaimed)
- Operating guidelines for the different designations
- Loading equipment working on spoil platforms
- Provisions for disposal of weaker materials (i.e., coal rejects, overburden soils, mud)
- Monitor placement
- Installing and moving monitors
- Automatic monitors (possible to use after careful site and trial assessment)
- Inspections
- Reporting
- Failure alerts
- Access and working below waste rock embankments
- Dumping platform maintenance and drainage control

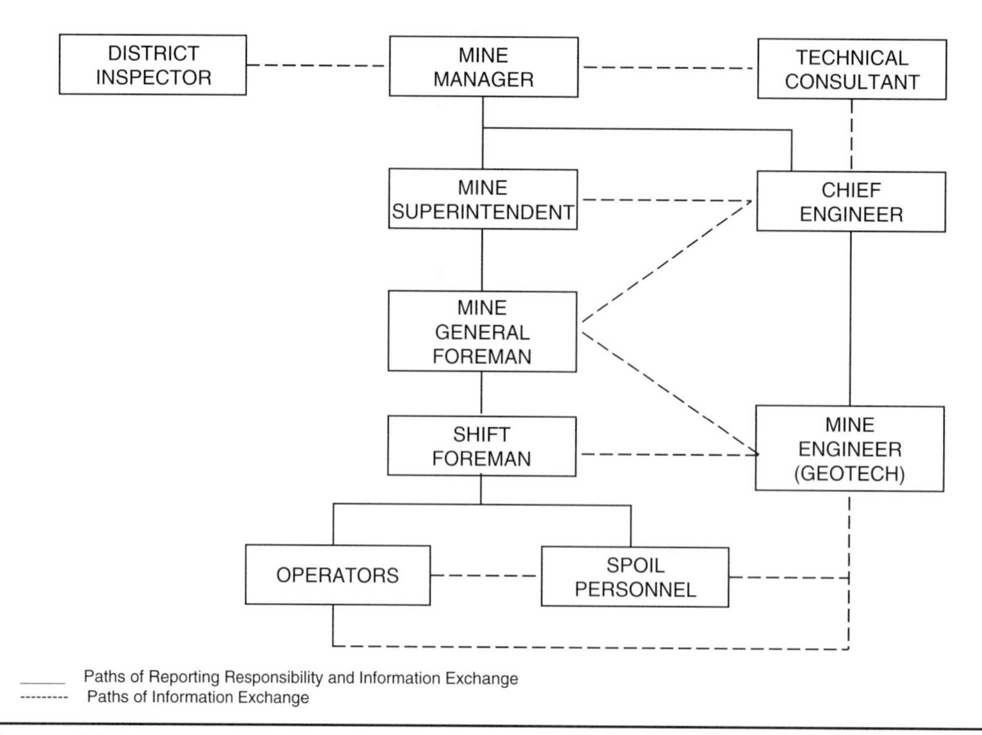

FIGURE 34.3 Typical responsibility diagram relating to waste-rock embankment operations

- Training of monitor personnel
- Responsibilities (by job title)

Management responsibilities reside with personnel at all levels (Figure 34.3), from the mine manager to the spotter on the dumping platform. Risks must be reviewed and managed for the life of the mining operation and in some cases for many years after closure. Klohn (1991) provides a general review of operating and monitoring requirements, recommends use of the observational method, and identifies areas of responsibility for different mine personnel in this regard. HBT AGRA (Cavers, Peer, and Rae 1992) also recommends the use of the observational method and gives general guidance on how it should be applied to waste rock embankments.

Conditions that can affect the stability of a waste rock embankment or conditions that can affect the consequences of spoil failure are rarely, if ever, static. The observational method has been used almost exclusively to ensure worker safety at spoil embankments in British Columbia. Extensive experience has been obtained from 30 years of operations at several surface coal mines in the Rockies. During the past 10 years, improvements in management controls have enhanced the performance of critical waste rock embankments.

34.5.2 Observational Method

The application of the observational approach is shown schematically in Figure 34.4, which also includes the author's opinions. This section is largely reproduced from HBT AGRA (Cavers, Peer, and Rae 1992).

Waste-rock embankments are frequently constructed using the observational method. In many cases, the method is abused, not understood, or is not fully implemented. It is therefore useful to review the "observational method" as originally defined by Terzaghi and discussed by Peck (1969). The observational method is well suited to mining projects where informed risk taking is an understood and necessary element of the economics of development. It is recommended that the observational method should be explicitly used in the construction of waste-rock embankments.

The basic steps in the observational method for design and construction of waste-rock embankments consist of the following:

- Undertake field and office investigations sufficient to establish conditions and design for facility. Assess likely and worst case conditions and the response of the embankment to these conditions (design stage).
- Base design on the most probable response considering the rating system in the British Columbia Waste Dump Design Guidelines. The more state-of-the-art the issue, the more useful the observational method is as a way of handling uncertainties (design stage).
- Define and select quantities, factors, and measurements to be taken as construction proceeds. Calculate or predict the anticipated values of the monitored parameters under the most likely prevailing conditions and under unfavorable conditions. These monitoring limits form the basis for caution, for stopping dumping, or for changing the design when construction is proceeding (design stage).
- Select a mitigative method or response for every foreseeable significant deviation of the observational findings from those predicted on the basis of the working hypothesis of behavior. These responses can be adjustments to dumping procedures, berms, drainage, and so forth. These responses must be defined beforehand. They cannot be left to the operational phase, although it is clear that actual field conditions will always suggest modifications to the initially conceived responses (design and operational stages).
- Devise a monitoring system that allows clear and timely observation of the selected parameters (design and operational stages).

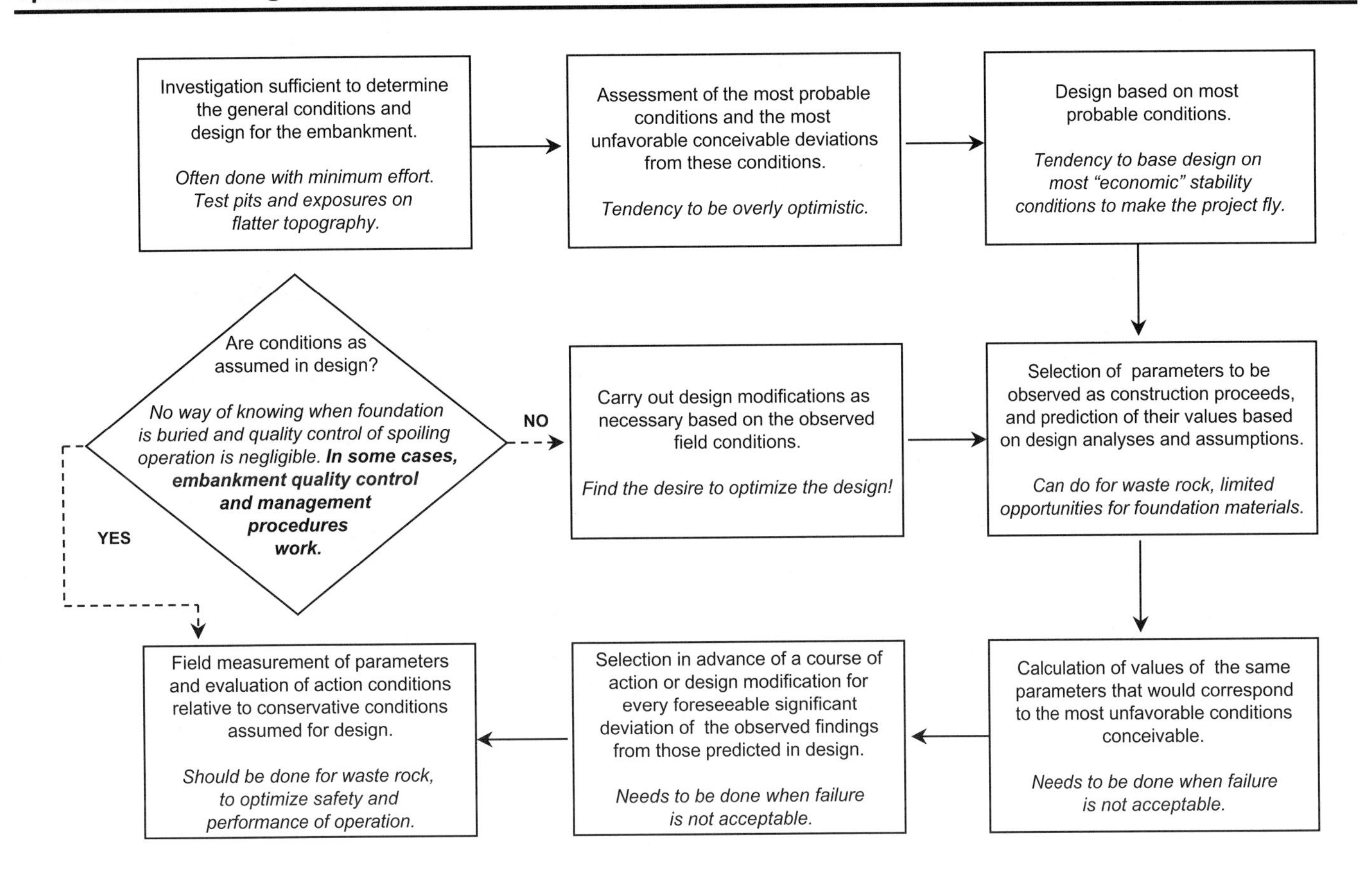

FIGURE 34.4 Suggested elements of the observational approach for waste-rock embankments

- Proceed with construction. Ensure that an appropriate system is in place not only to collect data but also to act on it. This requires close cooperation between geotechnical engineering, mine engineering, and operations staff. A management commitment to the method is mandatory to ensure that the process is in place and is respected by all those involved. Regulatory overview to ensure that this process is occurring is also important.
- As monitoring data are collected and construction proceeds, use the monitoring data to control the rate of dumping and to evaluate the overall design. Modify dumping procedures and design as appropriate and reevaluate the monitoring procedures.

It should be noted that the following aspects of the observational method are vital to its correct application:

- A feedback loop from design to monitoring and evaluating and back to design
- Mitigative procedures that are formulated at the time of design for every conceivable
- Unfavorable circumstance; these procedures may be later modified as construction proceeds
- Monitoring that furnishes the required data to evaluate conditions in the field and that is used to control modifications to dumping or to the design

The selection and application of monitoring instrumentation forms one part of the application of the observational method to the design and construction of waste-rock embankments. An example of application of this method is provided in the case study by Walker and Johnson later in this volume.

34.5.3 Mitigation Strategies

There are two areas that are considered here: one is maintaining maximum availability of the dumping platform for spoiling operations and second is ensuring the safest possible operation.

For maximum availability, a long embankment crest is desirable in order to distribute waste-rock deposition and thus reduce loading. However, operating personnel, if not directed otherwise, will tend to dump all of the waste rock in one area to minimize travel and work for the dozer, road grader, and monitor person. Consequently, one area will receive high loadings and could become more susceptible to failures. The failure and crest rehabilitation periods could become significant. An alternate plan could include having alternate disposal sites available at all times.

Safety can be maximized in several ways. Do not allow any situations to arise that could allow personnel to travel into potential embankment failure hazard zones. Eliminate access or securely block and sign access routes. Make it mandatory for persons who might travel into the area to contact a senior person in operations or engineering and obtain permission to enter. All access should be recorded in a logbook for sign-in and sign-out. Do not allow equipment maintenance work, shift changes, or equipment operators to take lunch or coffee breaks on dumping platform areas with a potential for instability. Ensure all equipment operators and other personnel who have to work on or below an embankment are fully aware of potential hazards and of their obligations to notify appropriate supervisors of unsafe practices or unusual field conditions. This must be part of orientation for new staff and contractors. Procedures should be reviewed annually and reflect current operating conditions and needs.

34.5.4 Inspection and Record Keeping

Typically, inspections can or will be performed by spoil personnel, a shift boss, a general foreman, and a spoil engineer with a decreasing level of frequency and increasing level of responsibility, respectively. Audit inspections can be performed by the design consultant geotechnical engineer. For day-to-day operating and safety, the importance of good record keeping by operations personnel cannot be overstated. Typically, every 12 hours there is a shift change, and a new set of operators is on the platform dumping, dozing, and monitoring. They need to have a good picture of what has happened on the spoil during the previous few hours or even several shifts. The data and written records need to be readily available and presented in a consistent format that is easily understood by everyone involved.

By necessity, there are different frequencies of visual observations—by personnel working at the dumping platform (the spotter or dozer operator) almost continuously, by truck drivers every few minutes, by supervisors once every 2 to 4 hours, by field engineers once per day, by geotechnical personnel (most large surface mines employ or designate one or more permanent geotechnical staff) once per week. This built-in redundancy can ensure safety and efficiency as different sets of eyes are making observations at regular intervals. The team of observers brings differing concerns, priorities, knowledge, and experience to the spoiling operation. With good operating procedures these differences should complement each other with standardized documentation, reporting loops, and communication and exchange of information.

The embankment configuration needs to be kept current and available in plan and sections, showing topography, pertinent features, facilities, weekly or monthly crest lines, monitor moves, and relocation progression. Current database management and ability to produce hourly, daily, and monthly plots of individual monitor settlement rates is considered a minimum requirement, while plots of rates versus waste-rock production volumes, precipitation, rock quality, foundation topography, or other factors can be useful. Some typical plots are shown in Figures 34.5 and 34.6

34.6 MONITORING

34.6.1 Purpose of Monitoring

Monitoring must provide the following:

- Sufficient warning to clear personnel and equipment from the crest and toe areas if necessary
- Quick and clear changes in the rate of settlement of the embankment crest so that loading rates and operating status of the spoil can be changed accordingly
- A well-documented performance history of crest settlements so that modifications to the dumping plan can be made to improve operating efficiency, stability, and safety where warranted or possible

The case study by Zeitz included in this volume illustrates how monitoring programs were developed and implemented and how data are used to make important decisions.

34.6.2 Methods of Monitoring

Monitoring can never be done as a stand-alone exercise and must never be relied upon independently of visual observations. Visual observations are at the core of any monitoring program. There is essentially one main monitoring instrument in widespread use at British Columbia surface coal mines. The manual wireline extensometer (wireline) is shown in Figure 34.7. This simple device, which must be robust, measures the settlement of the embankment crest relative to the base stand of the instrument. The base stand location is located as far back from the crest as practically possible (usually 20–30 m). An automatic version of the wireline was developed in recent years and is in use at one or more mines. An automatic wireline monitor should be used only after extensive experience is obtained with manual monitors and visual observations. Automatic and manual monitors can be used in parallel on the same spoil crest to complement each other and gain experience. Automatic monitors might be used on critical active and dormant spoils as a way of providing continuous data without the need to take manual readings every 2 hours from multiple monitors.

Other instruments are used for special needs, and these instruments include total or real surveys (global positioning systems or prisms), stakes, piezometers, and inclinometers. One must not hold a notion that instruments can replace the need for visual inspections, for they can only complement and support what is observed with the eye. However, the use of "delicate" instrument applications must be carefully planned in advance. The environment on and surrounding a waste-rock embankment is extremely punishing on buried instruments such as piezometers. The only location where they can be installed beforehand is in the foundation materials. The installations have to be durable enough to withstand the impacts of large blocks of rock (often over 1 m in diameter) bouncing and rolling to the foundation toe and withstand the vertical load, shearing, and consolidation settlement imposed by up to 400 m of rock, in some instances. The chapter by Zeitz dealing with the construction and operation of mined-rock disposal facilities at Elkview provides a detailed example of the range of monitoring instruments available.

34.7 PLACEMENT OF WIRELINE MONITORS

Wirelines are the most practical way to monitor settlements and safety of the dumping platform. HBT AGRA (Cavers, Peer, and Rea 1992) gives general guidance in this regard. However, wirelines can provide a false sense of confidence if personnel are not properly informed about their installation and operation. They must be properly positioned. They rarely, if ever, indicate the true magnitude of the movement. For large volume spoiling, several monitors (see Figure 34.1) may be required and may need to be relocated at frequent or regular intervals. On a small dumping platform, it may not be possible to position a wireline to monitor the active face. In this instance, more reliance must be place on visual observations. Wirelines should never be placed more than 100 m apart on a long embankment crest, and closer spacing may be warranted, especially if the spoiling platform is of low height and being rapidly advanced across sloping or weak foundation soils.

34.7.1 Monitoring and Indicators of Instability

Klohn (1991) describes a variety of failure modes, the chapter by Campbell in this volume describes the double wedge failure mechanism, and Dawson and Morgenstern (1995) describe brittle behaviour leading to flow failure of coal mine waste-rock embankments in mountainous terrain. Based on observation by this author, it appears that many postfailure configurations are similar to a double-wedge geometry, that debris runouts have flowed, and that other failure modes occur. With the exception of a deep-seated failure surface in the foundation or far back in an embankment, the other failure modes generally express deformations at or near the crest. An effective wireline monitoring program should record, at the embankment crest, the continual settlement that occurs as a result of consolidation and shearing that occurs within the waste rock but not necessarily the foundation. Visual inspections, however, are the frontline

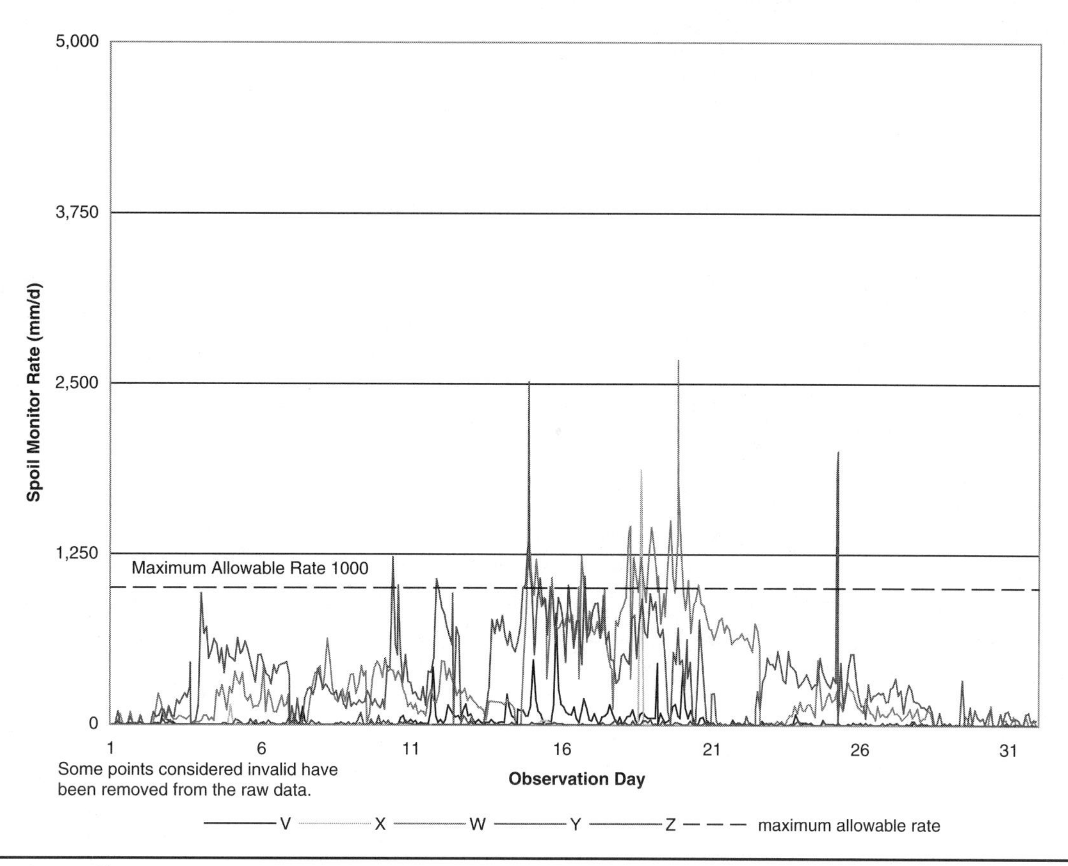

FIGURE 34.5 Two-hour rolling-average crest monitor rates for a waste-rock embankment in British Columbia

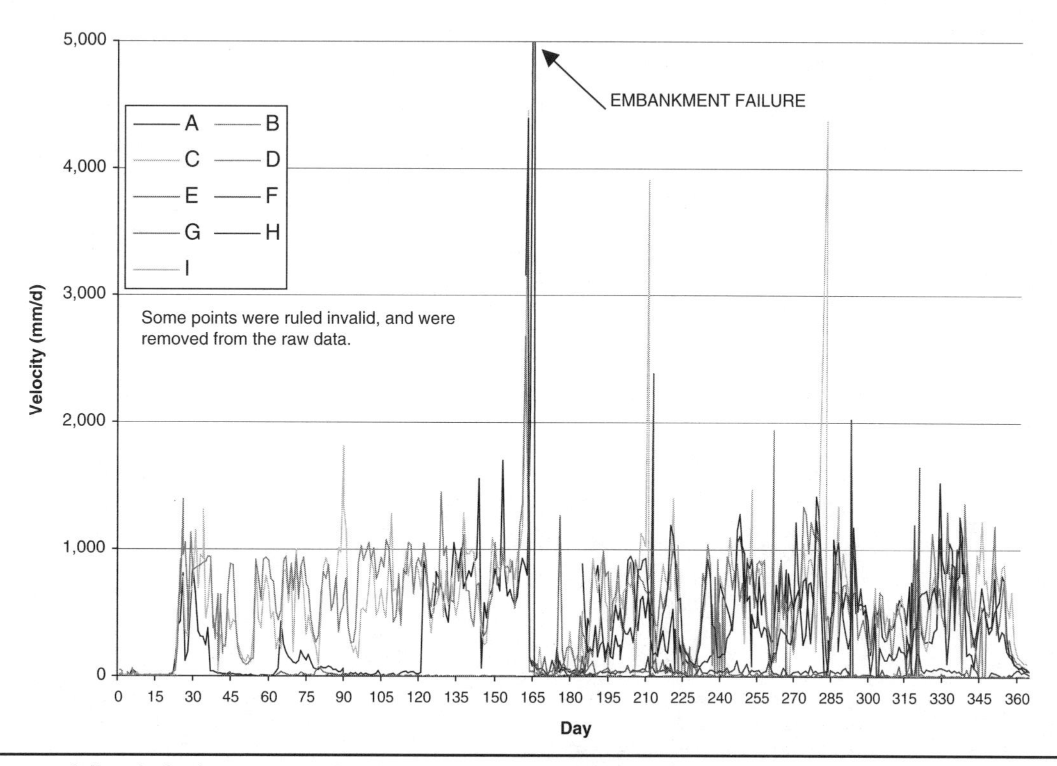

FIGURE 34.6 Average daily velocity for a waste-rock embankment in British Columbia

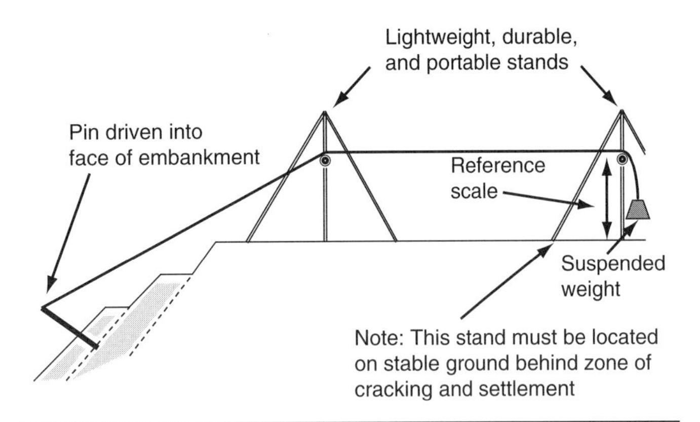

FIGURE 34.7 Wireline extensometer monitor

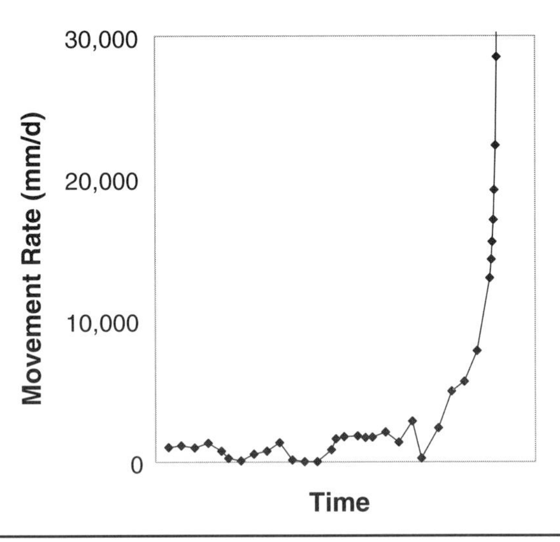

FIGURE 34.8 Spoil monitor rate plot prior to failure

tool for identifying foundation deformations. On occasion, when conditions lead to failure of a portion of a spoil pile, the monitoring program, wireline data (see Figure 34.8), and visual observations will provide early indications of the impending failure.

More than 160 waste-rock embankment failures have been recorded with the British Columbia regulator in the past 20 years, the majority of which have occurred at coal mines. This is coincident with the start-up or expansion of several mines to larger production opencut operations during that period. A much smaller number has occurred at metal mines during that same period and or older surface coal mine operations between the late 1960s and 1980. The coal mine failures have all occurred in mountainous terrain, and pore pressure generation in the foundation (or waste rock) is believed to be a trigger mechanism.

Experience in British Columbia from the past 20 years indicates that waste-rock embankments can be operated safely by relying on wireline monitoring and visual observations. Success is very dependent on good management and committed and informed employees. Wirelines when properly positioned have provided accurate (for the purposes intended) settlement data histories, indications of changes in settlement rates, and ample warning of impending failure. In general (the author has not verified the many failure records), increasing settlement rates leading to failure have provided 12 or more hours of warning and sometimes several days. Twelve hours is ample time to safely close and evacuate hazardous areas on the spoil platform or in the valley below the embankment toe. However, operating procedures need to clearly define the parties responsible for decision making and the criteria for closing and reopening the dumping platform, and for closure and reentry or access to sites downslope from the toe.

There are several instances where wireline monitors did not detect crest settlements in advance of a failure. In these instances, wireline monitoring was inadequate because of the following:

- The failure was small and located between monitors.
- The dumping platform was small and did not allow proper placement of monitors.
- There were no wireline monitors in place because the dumping platform was inactive or dormant at the time of failure.

A monitoring program should include wireline monitors (and other instruments if required) and structured visual observations. Visual observations are a critical component and cannot be replaced or omitted. It is the key tool for safe and effective operation. Personnel require training in making and recording visual observations. They require a framework within which to work in order to bring some consistency in reporting observations. Many people will be involved in this task at various responsibility levels for a mine. The framework should be established jointly by engineering and operations and include technical input from the geotechnical staff and or consultant.

Some indicators that should raise concern and that could lead to failures, small or large, include the following:

- Ravelling of the embankment face independent of dumping and dozing material over the crest.
- Cracking on the dumping platform parallel to the embankment crest. Cracks can extend laterally short distances or the length of the embankment crest and can be in series extending 20 to 30 m behind the crest. Bulldozers and graders can mask settlement cracks when pushing waste rock over the crest or during regrading of a dumping platform. Often, the location of the wireline is the only area where the surface remains undisturbed and cracks can be observed.
- Cracks on the dumping platform at or near the original natural contour representing deeper-seated shear in the embankment or foundation. This is common for structures started on steep topography and may continue for many months until the embankment toe advances onto flatter topography. It is important to recognize the possibility that instability can develop behind the location of the wireline back stand. It is impractical in some instances to increase the distance between the crest and back stand because of the need to maintain vehicular traffic access and reasonable wireline length. In such cases, visual observations must prevail.
- Oversteepening of the upper embankment face to the crest due to high fines content, moisture, and apparent cohesion. Oversteepened crests are unstable.
- Bulging near or below the middle region of the embankment face. The active wedge may be moving relative to the toe wedge in Campbell's model.
- Shoving and heaving of foundation materials in front of the embankment toe or on the lateral margins. Shearing is occurring in the foundation; features are close to the embankment toe if shearing is shallow and further away if it is deeper.

In addition, contributing factors always need to be considered:

- Preceeding recent significant rainfall and/or snowfall
- Snowmelt during spring
- Significant volumes of rehandled waste rock or overburden soils
- Knowledge of weak or pore-pressure-sensitive foundation conditions
- High embankment
- High loading rates
- Steep foundation topography
- Lack of lateral confinement or crest noses that are advanced ahead of the remaining crest line
- Surface drainage, natural or diverted

In British Columbia, mines are required annually to classify their waste-rock embankments in terms of hazard and risk and to review the monitoring program in place for every waste-rock facility, regardless of status (active, inactive, dormant, or reclaimed), on each mine property. Mines are not required to report their internal audit to the regulator and are to adjust their monitoring programs accordingly. In general, dormant and reclaimed spoils are monitored by visual observation methods only. There are four cases where waste-rock embankments failed in time frames from 1 to 10 years after active dumping was stopped. In all four instances, the facilities had not been reclaimed, and faces were at angle of repose (about 37), were constructed over moderate to steep foundations, or contained poor-quality foundation or rock materials.

There are several instances where very serious consequences occurred as a result of unfortunate timing, the type of or lack of a formal monitoring program, and lack of access restriction or evacuation plans for areas beyond the toe of the waste-rock embankment.

Visual inspections are important. Don't make the job difficult. On inactive, dormant, or reclaimed embankment, keep crests and adjacent platforms clear of free-dumped piles when a spoil is not in use. Free-dumped piles can mask the appearance of surface cracking. On active embankments, set up so that visual observations of the dumping platform are possible as well as faces and toe areas. Ensure there are locations where cracks can be observed on the platform, if they develop.

34.8 REFERENCES

Broughton, S.E. 1992. Mined rock and overburden piles: Review and evaluation of failures, interim report. *Prepared for Mine Waste Rock Pile Research Committee and Ministry of Energy and Mines, Victoria, British Columbia.*

Cavers, D.S., L.B.B. Peer, and C.E. Rea. 1992. Mined rock and overburden piles: methods of monitoring, interim report. *Prepared by HBT AGRA Earth and Environmental Limited for CANMET and Ministry of Energy and Mines, Victoria, British Columbia.*

Dawson, R.F., and N.R. Morgenstern. 1995. Liquefaction flowslides in western Canadian coal mine waste dumps–Summary report phase III, volumes I and II. *Prepared by CANMET, Edmonton, Alberta.*

Dawson, R.F., R.L. Martin, and D.S. Cavers. 1995. Review of long term geotechnical stability of mine spoil piles. *Prepared by AGRA Earth and Environmental Ltd. for Ministry of Energy and Mines, Victoria, British Columbia.*

Eldridge, T.L. 2000. Bodie 5 dump stability. *Prepared by Golder Associates Ltd. for Elkview Coal Corporation, Sparwood, British Columbia, and submitted to Ministry of Energy and Mines, Victoria, British Columbia.*

Endicott, D.R., and G.D. Warnock. 1999. Design and construction of the Bodie waste rock dump. *Proceedings May 1999, Canadian Institute of Mining Annual General Meeting, Calgary.*

Fording Coal Limited. 1996. Henretta Ridge Pit Section 10 permit application. *Submitted to Ministry of Energy and Mines, Victoria, British Columbia.*

Gold, R.D. 1998. Justification of the automated spoil monitor system at Fording River Operations. *Prepared by Fording River Operations, Elkford, British Columbia, for Ministry of Energy and Mines, Victoria, British Columbia.*

Gold, R.D. 1998. Proposed revision to Henretta Ridge East spoil toe berm design. *Prepared by Fording Coal Ltd., Elkford, British Columbia for Ministry of Energy and Mines, Victoria, British Columbia.*

Golder Associates Ltd. 1997. Report on geotechnical field investigation of proposed Henretta Ridge East spoil carried out in January 1997. *Prepared for Fording Coal Ltd. for Ministry of Energy and Mines, Victoria, British Columbia.*

Golder Associates Ltd. 1997. Geotechnical assessment for the 2000 bench, Henretta Ridge East spoil, at the Fording Coal Ltd., Fording River Operations. *Prepared for Fording Coal Ltd., Elkford, British Columbia, and submitted to Ministry of Energy and Mines, Victoria, British Columbia.*

Golder Associates Ltd. 1997. Report on a stability review of the proposed 1850-metre bench for the Henretta Ridge East spoil. *Prepared for Fording Coal Ltd., Elkford, British Columbia, and submitted to Ministry of Energy and Mines, Victoria, British Columbia.*

Golder Associates Ltd. 1999. Project construction report and July 20, 1999 external audit of Bodie Dump, Volume I and II. *Prepared for Elkview Coal Corporation and submitted to Ministry of Energy and Mines, Victoria, British Columbia.*

Hungr, O., and A. Kent. 1995. Mined rock and overburden piles: Runout characteristics of debris from dump failures in mountainous terrain stage 2: Analysis, modeling and prediction, interim report. *Report prepared for British Columbia Mine Waste Rock Pile Research Committee and CANMET, by Golder Associates Ltd., in association with O. Hungr Geotechnical Research Ltd., Vancouver, British Columbia.*

Klohn Leonoff Ltd. 1991. Mined rock and overburden piles: Operation and monitoring of mine dumps, interim guidelines. *Prepared for British Columbia Mine Dump Committee and Ministry of Energy and Mines, Victoria, British Columbia.*

Mining Association of Canada. 1998. *A Guide to the Management of Tailings Facilities.*

Peck, R.B. 1969. Advantages and limitations of the observational method in applied soil mechanics. *Géotechnique*. 19:2:171.

Piteau Associates Engineering Ltd. 1991. Mined rock and overburden piles: Investigation and design manual, interim guidelines. *Prepared for Ministry of Energy and Mines, Victoria, British Columbia.*

Piteau Engineering Ltd. 1999. Mined rock and overburden piles: Rock drain research program, final report. *Prepared with input from Mine Waste Rock Pile Research Committee for Ministry of Energy and Mines, Manalta Coal Ltd., and CANMET.*

Vick, S.G. 1999. Tailings dam safety–Implications for the dam safety community. *Proceedings of the Canadian Dam Association Annual Conference.*

CHAPTER 35

Reclamation and Surface Stabilization

René A. Renteria*

35.1 INTRODUCTION

Reclamation and surface stabilization of an overburden embankment go hand in hand. What is reclamation? A common definition is *to put into a desired condition*. What is stabilization? Once again, a common definition is *resistant to chemical or physical change*. So, before an overburden embankment can be put into a desired condition that is resistant to chemical or physical change, the desired future condition must be defined. This task will require knowledge of those chemical and physical processes that can interact with the individual site characteristics such as geochemistry, geology, climate, geometry, and natural vegetation. These processes and characteristics must be looked at using science-based data collection as well as methods that define accepted levels of uncertainty.

The US Department of Agriculture (USDA), Forest Service, Intermountain Region has adopted a regulatory process for the review and acceptance of reclamation and surface stabilization. The objectives are defined within the regional supplemental to Forest Service Handbook 7109.21:

1. Control of erosion and landslides (soil mass movement)
2. Control of water runoff
3. Isolation, removal, or control of toxic materials
4. Reshaping and revegetation of disturbed areas, where reasonably practicable

The methods presented here have been developed in cooperation of the USDA Forest Service and mining companies operating within the Intermountain Region (boundary defined as the states of Nevada, Utah, Idaho [southern], and Wyoming [western]). Since the goal of reclamation is long-term stabilization, site monitoring and improvements to these techniques are essential elements for any current or future reclamation project.

35.2 CONTROL OF EROSION AND SOIL MASS MOVEMENT

Both external and internal processes influence the long-term stability of an overburden embankment. In general, the actual processes cannot be halted, but control is influenced through design. Erosion, which is influenced by external processes, is a major concern for the design of a stable embankment. Just as important, soil mass movement is influenced by both external and internal processes. There are multiple processes that cause erosion. These processes include water, ice, wind, and gravity. In special circumstances ice and wind should be included in the design analysis. The processes that are addressed in this chapter are water and gravity. Erosional processes influenced by water that could impact an embankment include sheet, rill, and gully erosion. Examples of erosion (soil mass movement) influenced by gravity as well as soil material properties and strength, include creep, earth flows, and slides (Gray and Leiser 1992).

35.2.1 Erosion Processes

The New World Dictionary defines erosion as *to wear away*. The wearing away of soil occurs at different rates. The rates vary from very shallow (sheet erosion) to very deep and massive (gully erosion). Sheet erosion occurs when dry coarse material loses its apparent cohesion. By definition, sheet erosion is thin in depth and uniform over the slope. While sometimes designed as an acceptable erosion type, sheet erosion can be detrimental in the reclamation of overburden embankments. When accumulation of water along with weak and fine soil particles converge, small channels of water develop. Channels of water up to 300 mm (12 in.) in depth are called rills. Rills are often consistently spaced, interconnect with other rills, and collect water to create larger features known as gullies. Gullies move a larger size and quantity of material at depths exceeding 300 mm (12 in.). Gullies have a life process of their own and are not easily mitigated (Gray and Leiser 1992).

Factors to consider when designing an overburden embankment to resist erosion include velocity of runoff, particle shape and gradation, thickness of the surface material, and type of underlying material.

35.2.2 Landslide Processes

Shallow flowslides are the most common surface landslide process to occur on waste rock embankments. Flowslides may be initiated as a result of precipitation and snowmelt. According to Vandre (1980):

> The factors which can affect stability are infiltration duration and rate, soil density, soil moisture content, slope ratio, and seismic disturbances ... Shallow flow slides disrupt reclamation activities and can impact locations at considerable distances below the waste embankment.

A study conducted in the nonarid climate of western Canada indicates granular sand-sized material within a waste embankment slope can be at risk for liquefaction flowslides, triggered by mechanisms such as consolidation, creep, shearing in the toe region, and weathering (Dawson, Morgenstern, and Gu 1994).

35.2.3 Angle-of-Repose Slopes

An angle-of-repose (or natural) slope is often selected for overburden embankments due to the favorable economics of haul-truck end-dumping. Not all embankments will remain at angle of repose due to stability concerns. The acceptance of an angle-of-repose slope for final closure should consider the embankment's resistance to erosion as well as the overall mass stability.

Erosional Stability. The erosional stability of an angle-of-repose overburden embankment is essential for long-term stability as well as visual and resource protection concerns. Water

* U.S. Department of Agriculture–Forest Service, Humboldt-Toiyabe National Forest, Elko, Nevada.

flows onto the slope as a result of direct precipitation and runon from the top of the embankment. The accumulation of water flow produces erosional damage as discussed in the subsection entitled Erosion Processes. According to Vandre (1993), the necessary particle size for stability increases as the flow rate and slope gradient increase, but a decreasing size is necessary with increasing particle packing and specific gravity. Therefore, at the toe of the embankment there would be a much greater particle size requirement than near the top surface. Although gravity sorting from end-dumping generally provides for larger material at the toe, the proper-sized material may not be available as a result of the excavation process or rock material characteristics.

A stable design against erosion is created either by controlling the quantity of water flow onto the slope or by specifying a stable particle size that will not alter physically over time. Some methods of controlling water flow onto the slope include the following (USDA 1991):

- If site conditions will not create impoundment hazards, berm the crests of abandoned angle-of-repose slopes to accommodate future shear strain displacement beneath the crests and to prevent drainage from the bench areas from running onto the slopes below.
- Limit surface water on slopes to direct snowmelt or precipitation.

Other methods are discussed in the section titled Control of Water Runoff. Methods of design that consider particle size and physical durability are discussed in the following subsection.

Coarse and Durable. The term *coarse* is used to describe the particle size needed to prevent erosion and runoff. The term *durability* is used to describe the resistance of the rock material to physical alteration. According to Vandre (1993),

> Infiltration appears to be the best strategy for controlling the erosion potential of angle-of-repose slopes ... Particle coarseness will control the occurrences of erosion and shallow flow slides. Increasing particle size will: (1) increase infiltration thereby decreasing runoff, (2) provide more resistance to runoff detachment forces and (3) promote vertical hydraulic gradients thereby reducing the chance for shallow flow slides ... Durability extends the life of armor protection by maintaining coarseness for the long term. Rock durability is not determined by passing a single specification test. It is a classification based upon the correlation of results from several tests, experience and engineering judgment.

An example of a coarse and durable slope is shown in Figure 35.1.

FIGURE 35.1 Coarse and durable angle of repose slope (Renteria)

35.2.4 Designed Slopes

Improved slope stability and erosion resistance is commonly gained through the use of a designed slope. Common practice is to choose a 1 (vertical) to 3 (horizontal) slope gradient (USDA 1991). The use of the 1:3 ratio provides for equipment access, reduced gradient for particle detachment (erosion), the ability to utilize topsoil for revegetation, and improved mass stability. Oftentimes a designer may choose to represent a benched slope with an overall average slope ratio from the crest to the toe for the purposes of stability analysis. In this situation, the individual slopes between the benches should be analyzed as to their ability to resist erosion and support revegetation.

In comparing a designed slope to an angle-of-repose slope, the designed slope will have the potential for increased site impacts, especially in arid climates. A designed slope will have a greater slope length exposed to water runoff, but at a lesser gradient. In an arid climate, the lack of precipitation to support vegetation may result in significant soil loss during extreme storm events. Another site disturbance is the increased embankment footprint resulting from the flatter slope gradient. Consideration should be given to the impacts on other natural resources when considering a designed or angle-of-repose slope.

35.3 CONTROL OF WATER RUNOFF

35.3.1 Design Event

The basis for selection of a design storm event is the direct result of the regulatory objectives: provide for a long-term stable and maintenance-free reclaimed embankment. In most instances, the available precipitation and storm event data collected is from a period of less than 100 years; the determination of an extreme event such as 500- or 1,000-year return interval is highly questionable. On this basis, the use of the probable maximum precipitation (PMP) or probable maximum flood (PMF) is selected as the best available design event. The use of PMP will provide the designer the hydrologic basis for flood flow design, drainage basin analysis, and surface and potential subsurface water conditions. Methods for the determination of PMP are not discussed here, but they are available in references such as the *Hydrometeorological Reports, Probable Maximum Precipitation* produced by the National Oceanic and Atmospheric Administration (NOAA) for various locations throughout the United States.

35.3.2 Hydrologic Contributions

The most efficient method to control water runoff on the disturbed areas is to control the natural external influences to the best extent possible. A typical hydrologic cycle is shown in Figure 35.2. Since direct precipitation is not readily controlled, attention is directed to surface flow run-on and subsurface flows to the disturbed area. Methods of water control from natural contributions include the following (USDA 1991):

- Minimize chances that surface drainage will deviate from the planned locations after abandonment
- Prevent runon from adjacent areas onto the embankment and face
- Minimize potential of damage to control structures
- Provide long-term access to control structures for future maintenance
- Minimize and simplify future maintenance needs for control structures
- Limit surface water on slopes to direct snowmelt or precipitation

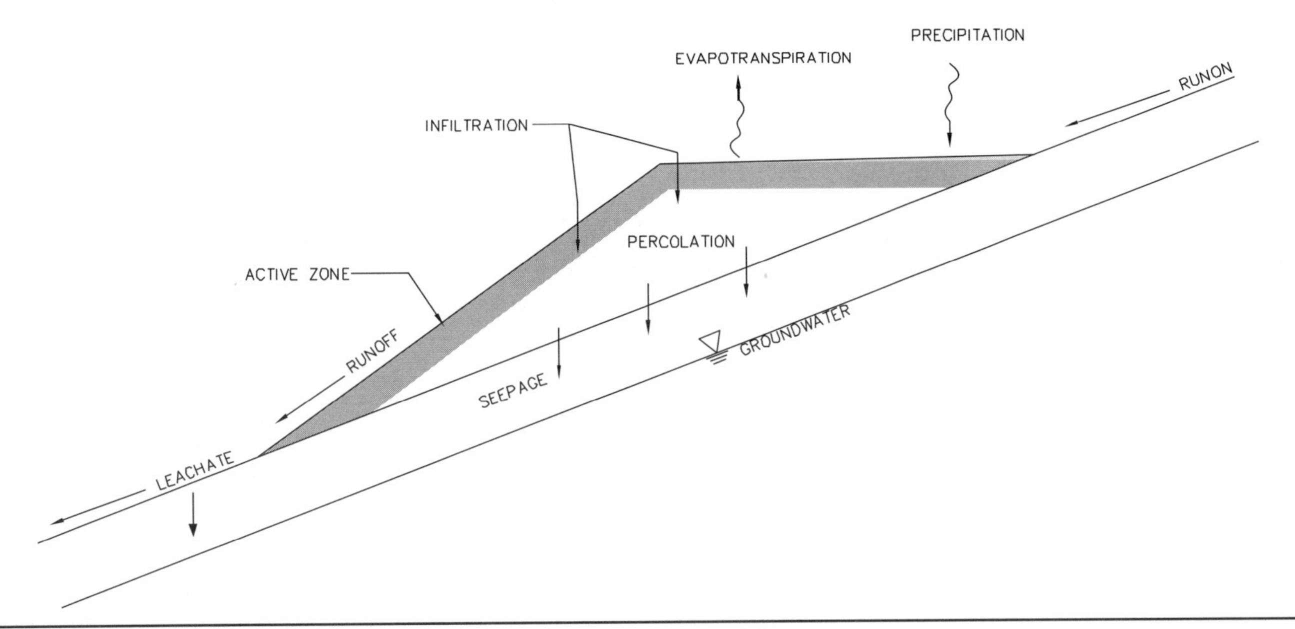

FIGURE 35.2 Typical hydrologic cycle for a mine overburden embankment (Renteria)

- Minimize chances of water concentrating on slopes
- Avoid slope conditions such as depressions and irregularities that will concentrate surface runoff
- Construct the bottom of reclamation drainage ditches equal to, or greater than, 3.6 m (12 f) to permit a dozer to be used for future maintenance
- Crown the top of dump benches along the centerline of valley fills to provide for surface runoff and to offset the disruption of the drainage directions by future settlement
- Grade the top of the reclaimed dumps so that future settlement will not disrupt the planned drainage
- Use haul road locations for long-term diversions when possible

Diversion. An efficient and effective method of surface and subsurface water control is designed diversion facilities to covey water around the disturbed area. The benefits of diversion include reduced impacts to background water quality, erosion potential, instability, and transport of chemical constituents. The management concerns of diversion facilities include long-term integrity, visual impacts, resource compatibility, and uncertainty. An example diversion location is shown in Figure 35.3.

Underdrain. An economical and efficient method to convey water through or collect water within the overburden embankment is with the use of an underdrain. An underdrain can be constructed through the normal process of placing dumped material–through gravity sorting the coarse material travels to the toe of the embankment while the finer materials remain near the top of the embankment.

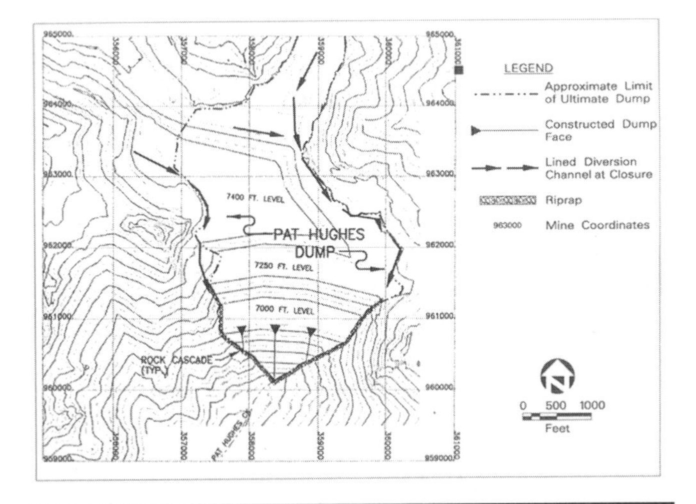

FIGURE 35.3 Plan view of example diversion location (USDA Forest Service 1999)

35.4 CONTROL OF TOXIC MATERIALS

35.4.1 Removal

Active control of toxic materials can be accomplished through removal, which if not done properly may result in the release of toxics to the environment. Removal includes physical relocation either within the mine site or off site. In assessing the use of removal as an alternative, economics and safety are major considerations. The removal of material from one location requires the placement in another location, and that does not necessarily relieve all concern for the release of toxic materials. In some situations, the identification of geochemical concerns does not occur until well into the project. In general, the site conditions are fixed, and the issue of removal should consider the ability to improve the disposal conditions. Issues to be considered include the following (Hutchison and Ellison 1992):

- Available water for transport
- Chemical buffering capacity
- Economics of relocation
- Constructability
- Increased disturbance and potential for toxic release

35.4.2 Encapsulation and Isolation

Encapsulation of materials provides control from toxic release by isolating the production or movement of toxics. For an overburden embankment, encapsulation is commonly designed by the use of cells where acid rock drainage potential overburden can be separated from air and water influences (Figure 35.4). A clay material is often chosen to provide the liner for the cell. The design of an encapsulation cell must consider the effects of settlement, slope stability, desiccation of clay material, and erosion of

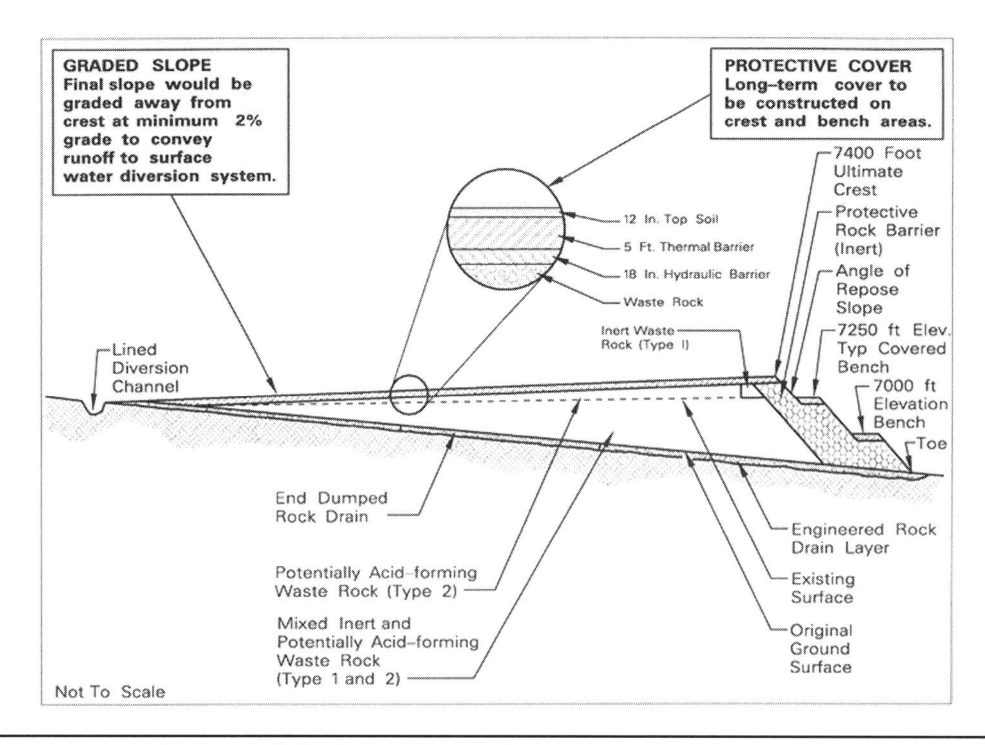

FIGURE 35.4 Example of encapsulation design (USDA Forest Service 1999)

the liner material. The objective for regulatory approval is a design that will provide the stated objective of encapsulation, without future maintenance, erosion, or instability.

35.4.3 Treatment

Control of toxic substances may be accomplished by altering the chemical composition of the stored material. Treatment includes both physical and biological agents. Issues of concern include long-term effectiveness, viability of biological agents, economics, and potential adverse effects. The effectiveness of treatment methods is improved the closer the treatment is to the source of the toxic material. The addition of limestone material to an acid potential overburden material is preferred compared to the alternative of treating the effluent for acid-rock drainage (Hutchison and Ellison 1992).

35.5 RESHAPING AND REVEGETATION OF DISTURBED AREAS

35.5.1 Angle-of-Repose Slopes

A proposed use of either reshaping, revegetation, or both, on an angle-of-repose slope produces a unique challenge. By the nature of the definition of angle of repose, the slope is at stability equilibrium. Reshaping of the slope produces a designed slope, which is discussed further in the following subsection. For successful revegetation to occur, water must be available to the roots of the plant, through a productive soil medium. The soil will require some fraction of fines (silt or clay or both) to increase the soil moisture retention. Water held in the soil can produce de-stabilizing conditions, which may result in soil landslide mass movement.

The use of revegetation on angle-of-repose slopes will require further research and study. As discussed previously in this chapter, surface erosion is greatly influenced by soil particle size, and mass stability is influenced by the presence of water in the soil material. A test study in the arid climate of Nevada (Leavitt 1998) concluded that:

> ... although the coarse textured soils on AOR [angle-of-repose] slopes provide for slope stability, they also limit revegetation by inhibiting seed germination and seedling establishment. In contrast, highly erosive fine textured soils may enhance plant establishment and survival during the short term, but unstable slopes eventually lead to plant mortality. A method, which stabilizes fine textured soils on steep slopes, would provide the highest revegetation success.

35.5.2 Designed Slopes

Reshaping and revegetation can have both aesthetic and functional purposes. Planning and implementation of reshaping and revegetation can be greatly enhanced with the assistance of a landscape architect and plant specialist.

Reshaping. Reshaping of the disturbed areas can enhance the restoration of disturbed areas with improved stability as well as visual benefits. Slopes can be undulated to break up the linearity of a designed slope, which creates a more natural appearance. Upon final spreading of soil material, mechanical equipment should proceed along contour to create minor terraces that interrupt surface water along the direct flowline (Figure 35.5). Earthmoving equipment can be used to create slope depressions that allow for additional soil moisture retention and potential support of an increased variety of vegetation. Soil berms may be designed at the base of a slope to create a foreground view, which softens the view to the background of the reclaimed slopes.

Revegetation. The selection of plant species for revegetation of disturbed areas must be made on a site-specific basis. Climate, slope, aspect, successional stages, animal grazing, and resource management will influence the long-term establishment of a plant community on the disturbed areas. For these reasons, determination of successful revegetation for reclamation bond release is one of the more difficult judgments made by a regulatory agency. For the purposes of stability, successful designed revegetation occurs when an adequate root network has been established that can support annual regrowth and recovery from minor disturbances such as initial soil erosion, invader (weed) species, grazing, and below-average precipitation, and when adequate cover as per design is demonstrated (Hutchison and Ellison 1992).

FIGURE 35.5 Contour versus vertical machine tracks (Renteria)

35.6 SUMMARY AND CONCLUSIONS

In conclusion, reclamation and surface stabilization are key to the long-term stability and longevity of mine overburden embankments. The natural processes (climate, gravity, and geochemistry) along with site characteristics (erosion, mass movement, slope geometry, and vegetation) are examined through data collected using generally accepted scientific methods. Using this site information aids in developing reclamation and surface stabilization alternatives early in the planning of final mine closure.

Some design alternatives to aid in the control of erosion and mass movement on slopes include angle of repose, coarse and durable, and designed. While angle of repose is widely accepted, more research on this method is needed to improve revegetation and surface stabilization quality. Diversions and underdrains should be considered in the aid of controlling water runoff. Finally, removal, encapsulation, isolation, and treatment are alternatives for design of the control for toxic materials.

It is important to remember that all design elements chosen for a site must work in conjunction with each other. No single design element will be able to meet all the needs to place a mine overburden embankment into a desired condition that is resistant to chemical or physical change, as well as overall mass stability.

35.7 ACKNOWLEDGMENTS

The regulatory framework currently in use by the USDA Forest Service, Intermountain Region, is the result of research and policy formulation advanced by Bruce Vandre, Regional Geotechnical Engineer (retired, 1994). The author gratefully recognizes Susan Ortiz, Geotechnical Engineer, USDA Forest Service, for her contributions and assistance in completing this document.

35.8 DISCLAIMER

The opinions expressed in this paper represent the thoughts of the author and not necessarily the policy of the USDA Forest Service.

35.9 REFERENCES

Chatwin, S.C., D.E. Howes, J.W. Schwab, and D.N. Swanston. 1991. *A Guide for Management of Landslide-Prone Terrain in the Pacific Northwest. Land Management Handbook Number 18*. Research Branch, Ministry of Forestry. Victoria, B.C.: Crown Publications, Inc.

Dawson, R.F., N.R. Morgenstern, and W.H. Gu. March. 1994. *Liquefaction Flowslides in Western Canadian Coal Mine Waste Dumps, Phase II: Case Histories*. Edmonton, Alberta: University of Alberta Department of Civil Engineering.

Gray, D.H., and A.T. Leiser. 1992. *Biotechnical Slope Protection and Erosion Control*. New York: Van Nostrand Reinhold.

Hall, D.E., M.T. Long, and M.D. Remboldt. 1994. Slope Stability Reference Guide for National Forests in the United States, Vols. 1–3. EM-7170-13. USDA Forest Service Engineering Staff. Washington, D.C.

Hutchison, I.P.G., and R. Ellison. 1992. *Mine Waste Management*. California Mining Association. Chelsea, Michigan, Lewis Publishers, Inc.

Leavitt, K.J. 1998. Plant establishment on angle of repose (AOR) dump slopes. MS thesis, University of Nevada, Reno.

USDA Forest Service. 1991. Intermountain Region 4 FSH 7109.21—Geotechnical and Materials Engineering Handbook, Effective July 24, 1991. Ogden, Utah.

USDA Forest Service. 1999. Thompson Creek Mine, Final Supplemental Environmental Impact Statement. Volume I. Salmon and Challis National Forests, Intermountain Region, January 1999. Ogden, Utah.

Turner, K.A., and R.L. Schuster. 1996. *Landslides, Investigation and Mitigation. Special Report 247, Transportation Research Board*. Washington, D.C.: National Academy Press.

Vandre, B.C. 1980. Stability of non water impounding mine waste embankments. USDA Forest Service, Intermountain Region. Ogden, Utah.

Vandre, B.C. 1993. What is coarse and durable mine waste rock? In *Proceedings 29th Symposium Engineering Geology and Geotechnical Engineering*.

CHAPTER 36

Observational Engineering for Open-Pit Geotechnics: A Case Study of Predictions Versus Performance for the Stability of a High Overburden Embankment over a Soft/Deep Soil Foundation at PT Freeport Indonesia's Grasberg Open-Pit Mine

W. Kevin Walker* and Mark J. Johnson†

36.1 INTRODUCTION

Development of the Grasberg Pit, located in the west central highlands of Irian Jaya, began in the late 1980s. The initial plans for overburden stockpiling included an embankment in the Carstenzweide Valley (Figure 36.1), which is a deep basin filled with natural soil deposits. Infrastructure in the lower reaches of the valley required careful planning for the ultimate embankment.

Initial geoengineering concerns centered on the ability of the foundation to support a large embankment under both static and dynamic (earthquake) conditions. The results of a feasibility-level investigation and stability analysis were the basis for initial planning recommendations, including embankment placement from the bottom up in controlled lift heights and laying-back the overall slope below the angle of repose.

A Carstenzweide field trial embankment program, with angle-of-repose placement for slope heights in excess of 140 m, was undertaken in the valley for operational reasons and for confirmation of aspects of the initial geotechnical predictions. Trial embankments were constructed using a geotechnical observational approach to monitor and measure trial embankment performance. This monitoring, tempered by site experience, provided early warning of any embankment instability, safeguarding personnel and equipment working in the area. This chapter describes the initial geotechnical stability evaluation and compares geotechnical predictions with actual behavior from operational embankment placement experience to date.

36.2 FIELD INVESTIGATION AND GEOLOGICAL CONSIDERATIONS

The relatively flat ground of the Carstenzweide Valley lies at an average elevation of 3,600 m above sea level in the Sudirman mountain range of Irian Jaya. The canyon walls forming the valley (Grasberg mountain to the west and very steep Tertiary limestone canyon walls to the north and east) are generally free of vegetation and soil cover. The mine site is wet; annual precipitation is approximately 3 m. The valley foundation soils were investigated in the early 1990s using test pits and geotechnical drilling (Figure 36.1). Geotechnical soil logging of all stratigraphic soil layers encountered in test pits and drill holes was performed using lithologic descriptions, including visual/manual USCS soil classifications (*Unified Soil Classification System, ASTM D-2488-90* 1988). USCS soil code and symbols are presented in Figure 36.2.

36.2.1 Test Pits and Geotechnical Core Drilling

Seven test pits (excavator dug to maximum depths <4 m) revealed near-surface conditions. Conditions at depth were investigated using a wire-line (Longyear LY44 HQ-3) diamond core drill (61.1-mm [2.406-in.] core diameter) modified for soil coring and capable of collecting standard penetration testing (SPT) blow count data. (The "blow count" is the number of blows required from a standard SPT safety hammer to advance a split-spoon 30 cm.) SPT work generally followed procedures described in *ASTM D-1586-84* (1992). Undisturbed Shelby-tube samples were occasionally taken in fine-grained soils using AW geotechnical drilling tools through the HQ core string. The drilling of five geotechnical drill holes (holes CNI-1 to CNI-5 to soil depths ranging from 19 to 117 m) supplemented existing data. The objectives of feasibility drilling included collecting stratigraphic data, sampling, SPT, and depth-to-bedrock testing where applicable.

The results of diamond coring exceeded expectations in most soil types, with recoveries averaging around 50%. Slowing drill rotation and reducing down-pressure and pump pressure increased the level of success. Core loss was higher in loose, granular sands and in gravels low in cementation; recoveries generally improved with depth.

SPT blow count data were not collected continuously in the drill holes; however, in holes CNI-3 and CNI-4, relatively more blow count data were specified. Equations (Hynes and Franklin 1989) were used to correct the raw N(60) blow counts for overburden effects and hammer energy effects to obtain "corrected N-1(60)" blow count data.

36.2.2 Sampling

Representative test pit and core samples were collected to evaluate the strength of materials for the development of the stability analysis. Soil samples were carefully prepared in the field to retain moisture. The objective was to collect representative soil samples of each soil type encountered. For the test pits, representative samples were taken for each distinct stratigraphic layer observed. Shelby tube samples were also obtained in the soft, fine-grained units in the test pits, while bag samples were generally collected in granular soil types. Core drilling provided samples of cored soils, split-spoon samples from the SPT penetration testing, and occasional Shelby-tube samples.

* Call & Nicholas, Inc., Tucson, Arizona.
† Freeport McMoRan Copper & Gold Inc., New Orleans, Louisiana.

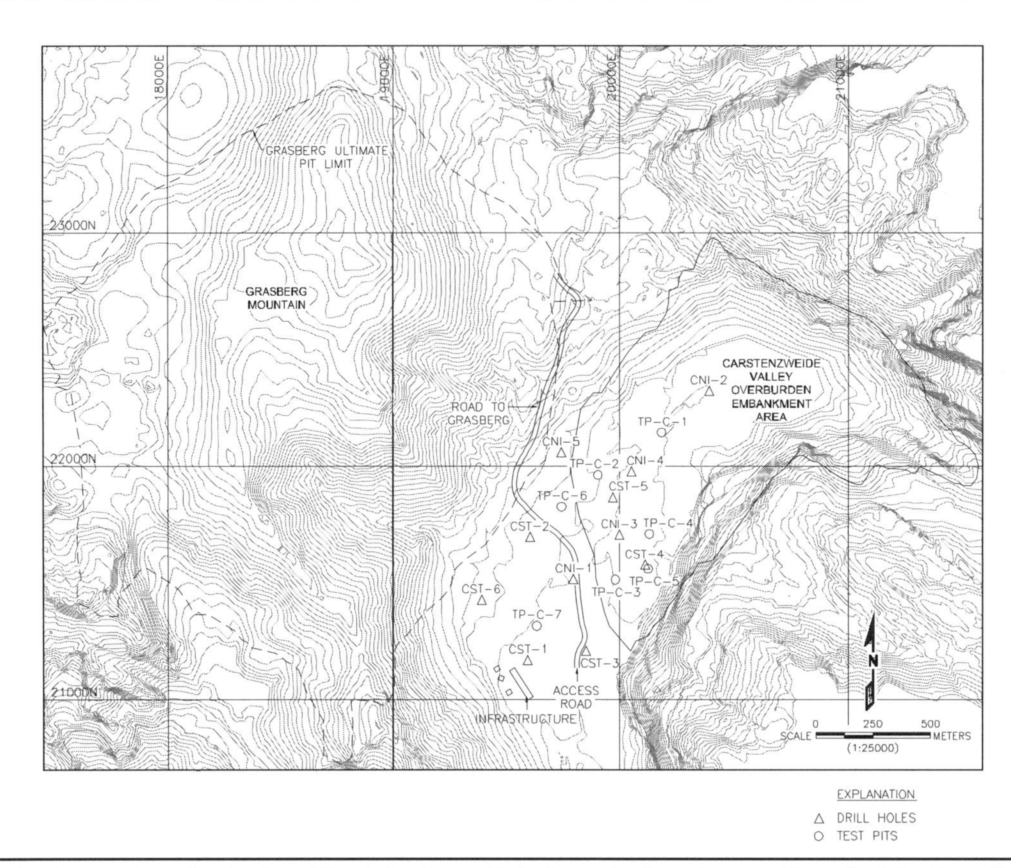

FIGURE 36.1 Premine topography of the Grasberg Pit and Carstenzweide Valley embankment areas with geotechnical investigation details

36.2.3 Geologic Model

The foundation soils in the Carstenzweide embankment area are interpreted to be deposited in a deep glacial cirque basin.

The near-surface Quaternary sediments are considered to be fluvial in origin. The east-central portion of the valley's surface is dominated by peat and muskeg horizons, interbedded with organic sands to silts. Organic silts often display a clay-like behavior. At depth and on the valley's west side, the presence of organic silts and fine sands is reduced, while interbedded sands and gravels become more abundant. Alluvial deposition of detritus, eroded mainly from the Grasberg dioritic complex, is the probable source for much of the near-surface deposits. The stratigraphy near the front of Grasberg Mountain appeared typical of a meandering streambed sequence. Gravels are interbedded with finer alluvial soils, including sands to organic silts and peats.

Sediments at depth appear glacially derived and mainly consist of detrital sands, gravels, and organic silts. Organic clays and consolidated peat units were also occasionally observed. Figure 36.3 shows interpreted contours to competent bedrock circa 1991 data. In the area of the proposed ultimate embankment toe, the alluvial sediments are approximately 125 m thick and to the south they increase in depth to over 200 m.

A geologic interpretation of the soil deposition in Carstenzweide Valley is offered:

1. Glaciation occurred, scouring a glacial cirque bowl.
2. Glaciers retreated with deposition of interfingering fine- and coarse-grained glacial deposits in the cirque valley, possibly including a somewhat continuous varved glacial lake deposit (organic silt unit).
3. Further retreat of the glacier ensued and coalescent fans formed, depositing out-wash sands and gravels from streams emanating from the melting glacier front.
4. A small glacier may have advanced back over the valley, which densified or consolidated an interpreted deep and dense soil unit.
5. The small glacier also retreated. This could have been followed by additional fluvioglacial fan activity that deposited loose granular materials from high-energy fluvial flow events and soft, fine-grained materials during low-energy times.
6. Recent alluvial processes deposited loose granular materials on basin margins that grade to soft, fine-grained organic soils in the basin center.

36.3 GEOTECHNICAL LABORATORY TESTING

Laboratory testing was performed on representative samples of both foundation and embankment materials from the investigation (Table 36.1).

Tests were performed for effective shear strength (Mohr–Coulomb strength parameters of cohesion and friction angle) and pore pressure response (or both). Tests performed included (1) direct-shear; (2) staged, consolidated, undrained triaxial compression (CU) tests with pore-pressure measurements (for fine-grained foundation soil samples); (3) lab vane shear; and (4) one-dimensional consolidation testing. Pore-pressure response data for foundation soils (consolidation data) usually were obtained from calculations with the CU triaxial consolidation stage data; however, some conventional one-dimensional consolidation tests were also performed.

Geotechnical index testing (sieves, Atterberg limits, moisture content, density) was performed to classify and correlate foundation soil types and helped to evaluate engineering properties. Special tests for classification of organic soils, such as peat and muskeg, included organic content, fiber content, and lab vane shear tests. The results from testing foundation soils were grouped by simplified Carstenzweide "soil group" codes (Figure

USCS CODES					CARSTENZWEIDE SOIL GROUPS	
MAJOR DIVISIONS				TYPICAL NAMES	CODES	DESCRIPTION
COARSE GRAINED SOILS MORE THAN HALF IS LARGER THAN #200 SIEVE	GRAVELS MORE THAN HALF COARSE FRACTION IS LARGER THAN #4 SIEVE SIZE	CLEAN GRAVELS WITH LITTLE OR NO FINES	GW	WELL GRADED GRAVELS, GRAVEL-SAND MIXTURES.	G	GRAVELS CLEAN TO LITTLE FINES
			GP	POORLY GRADED GRAVELS, GRAVEL-SAND MIXTURES.		
		GRAVELS WITH OVER 12% FINES	GM	SILTY GRAVELS, POORLY GRADED GRAVEL-SAND-CLAY MIXTURES.	GF	GRAVELS WITH FINES
			GC	CLAYEY GRAVELS, POORLY GRADED GRAVEL-SAND-CLAY MIXTURES		
	SANDS MORE THAN HALF COARSE FRACTION IS SMALLER THAN #4 SIEVE SIZE	CLEAN SANDS WITH LITTLE OR NO FINES	SW	WELL GRADED SANDS, GRAVELLY SANDS.	S	SANDS – CLEAN TO LITTLE FINES
			SP	POORLY GRADED SANDS, GRAVELLY SANDS.		
		SANDS WITH OVER 12% FINES	SM	SILTY SANDS, POORLY GRADED SAND-SILTY MIXTURES.	SF	SANDS WITH FINES
			SC	CLAYEY SANDS, POORLY GRADED SAND-CLAY MIXTURES.		
FINE GRAINED SOILS MORE THAN HALF IS SMALLER THAN #200 SIEVE	SILTS AND CLAYS LIQUID LIMIT LESS THAN 50		ML	INORGANIC SILTS AND VERY FINE SANDS, ROCK FLOUR, SILTY OR CLAYEY FINE SANDS, OR CLAYEY SILTS WITH SLIGHT PLASTICITY.	M	SILT INCLUDING ORGANIC SILT
			CL	INORGANIC CLAYS OF LOW TO MEDIUM PLASTICITY, GRAVELLY CLAYS, SANDY CLAYS, SILTY CLAYS, LEAN CLAYS.	C	CLAYS INCLUDING ORGANIC CLAYS
			OL	ORGANIC CLAYS AND ORGANIC SILTY CLAYS OF LOW PLASTICITY.	O	ORGANIC RICH PEAT/ MUSKEG
	SILTS AND CLAYS LIQUID LIMIT GREATER THAN 50		MH	INORGANIC SILTS, MICACEOUS OR DIATOMACIOUS FINE SANDY OR SILTY SOILS, ELASTIC SILTS.	M	AS ABOVE
			CH	INORGANIC CLAYS OF HIGH PLASTICITY, FAT CLAYS.	C	AS ABOVE
			OH	ORGANIC CLAYS OF MEDIUM TO HIGH PLASTICITY, ORGANIC SILTS.	O	AS ABOVE
HIGHLY ORGANIC SOILS			PT	PEAT AND OTHER HIGHLY ORGANIC SOILS.		
				BEDROCK		

NOTE : DUAL SYMBOLS ARE USED TO INDICATE BORDERLINE SOIL CLASSIFICATION

FIGURE 36.2 Unified soil classification system (USCS) and Carstenzweide "soil group" codes

36.2), which are referenced to the USCS symbols applied during the field investigation. The USCS soil symbols were used to group test results because only a small percentage of samples could be actually tested in the lab. By using a classification system referenced to the field USCS symbols, tested soil properties could be associated with soil units that were not necessarily tested.

36.4 FEASIBILITY GEOTECHNICAL STABILITY MODEL

Initially, a series of geologic cross sections through the valley were constructed to determine the likelihood of through-going correlatable soil layers. Distinct soil units generally could not be interpolated between drill holes due to heterogeneous or discontinuous conditions between holes. However, very similar stiff, laminated organic silts were observed at depth in holes CNI-3 and CNI-4, with underlying densely mixed soil types. The stiff, varved organic silt with some clay was assumed to be laterally continuous. Therefore, the geotechnical model developed for the stability analysis considered upper and lower mixed soil packages (*loose surficial unit* and *dense deep unit*) separated by the assumed continuous varved organic silt. Figure 36.4 displays a cross section through the Carstenzweide Valley with the resulting geotechnical foundation model.

36.4.1 Evaluation of Material Properties for Stability Analysis

Table 36.2 summarizes the selected shear-strength and pore-pressure response parameters associated with the geotechnical model. These parameters, in addition to the shear strength of the overburden stockpile embankment, were used in the embankment stability analysis.

36.4.2 Shear Strength

The shear strengths used in the embankment stability analysis are summarized graphically on Figure 36.5.

The foundation shear strengths for the *loose surficial* and the *dense deep* mixed soil units were calculated based on a weighted average using soil strengths assigned to the Carstenzweide soil groups (Table 36.1 and Figure 36.2) and the average soil group thicknesses by depth interval. The average soil group thicknesses were evaluated through a statistical analysis based on the results of geotechnical field logs. The weighted-average approach was considered reasonable because of the interpreted heterogeneous and relatively noncontinuous nature of these surficial and deep mixed soils. The shear strength of the assumed, continuous, organic fine silt is based on results from laboratory strength testing on samples retrieved from drill holes CNI-3 and CNI-4.

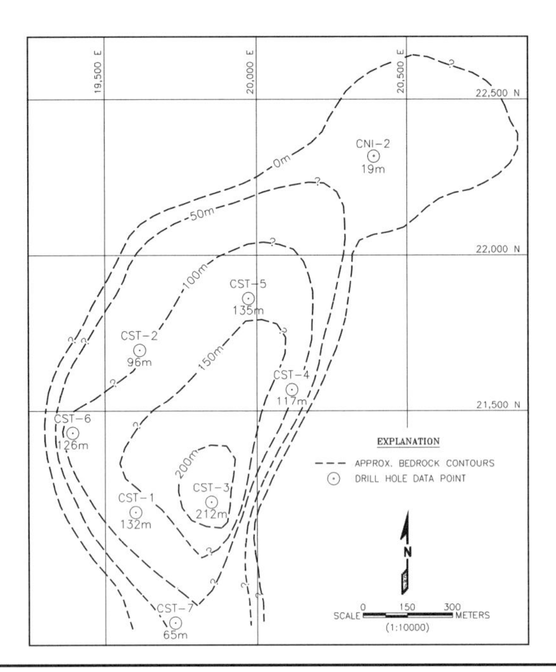

FIGURE 36.3 Carstenzweide Valley interpreted depth-to-bedrock contour map

The strength of the overburden stockpile embankment material was estimated from large-scale direct-shear and triaxial compression data from run-of-mine material (Table 36.1). A power fit was modeled for typical diorite overburden based on laboratory data for normal loads up to approximately 1.4 Mpa (200 psi) and on diorite rockfill data (Leps 1970).

36.4.3 Pore-Pressure Response in Foundation Soils from Embankment Loading

Soil permeability and pore-pressure and loading response data are required to estimate foundation water pressures induced through embankment loading. The stability analysis modeled independent pore-pressure development for the three modeled foundation soil units (Figure 36.4) using classical soil consolidation theory (Holtz and Kovaks 1981). Lab testing produced estimates of Skempton's pore-pressure parameters (used to predict a rise in pore pressure in the foundation due to embankment loading). Soil variables governing pore-pressure dissipation over time are the consolidation parameters *Cv* and Hdr: the coefficient of consolidation and the height of the drainage path, respectively. *Cv* is a measure of soil permeability, and Hdr defines the length of the path that excess pore pressures must travel to be relieved. Estimates of *Cv* were made from lab tests. The height of the drainage path is controlled by geology and drill-hole soil thickness data.

Pore-pressure response in the *loose surficial unit* and in the *dense deep unit* was modeled by estimating the *average* pore-pressure variables (A, *Cv*, and Hdr) for the more fine-grained soils in each unit, as characterized by Carstenzweide soil group symbols M, C, and O (Figure 36.2). This estimate was considered conservative since the surficial loose and deep dense units are interpreted to contain mainly granular and more free-draining materials. The pore-pressure variables for the assumed continuous organic silt were derived from specific samples tested in staged CU triaxial compression tests (Head 1986).

36.5 GEOTECHNICAL ANALYSIS AND PREDICTIONS

Various stockpile embankment configurations, as well as static and a dynamic stability analysis, were considered in the initial embankment geotechnical evaluations:

- static and dynamic (pseudostatic) limiting-equilibrium stability analysis of a 270-m-high ultimate embankment built by end-dumping at the angle of repose
- static and pseudostatic stability analysis of lifts up to a 270-m-high embankment built by staged ascending construction with a 27° overall slope angle
- evaluation of embankment displacement from foundation liquefaction, performed on the basis of potential seismic earthquake hazards

36.5.1 Limiting-Equilibrium Stability Model

Limiting-equilibrium stability was evaluated using Spencer's method of slices for both static and dynamic (pseudostatic) cases. Automatic shear surface searches were performed in an effective stress analysis with the computer program UTEXAS2 (Wright 1986) to evaluate the critical factors of safety (FOS) for both circular and noncircular shear surfaces (Figure 36.6).

Excess pore-pressure development in the foundation, related to loading the foundation with the embankment, was considered. In the pseudostatic embankment analysis, it was assumed that the strengths would not degrade as a result of cyclic loading; however, the liquefaction analysis assumed earthquake-induced reduction in foundation strength.

36.5.2 Plate Tectonics and Seismicity

The island of Irian Jaya, which is seismically active today, is the product of collision between the Australian and Pacific lithospheric plates. A historical seismic hazard analysis (probability of exceeding a given site acceleration in a specified period of years) was used in the dynamic embankment stability analysis. Based on historical records, peak site accelerations were calculated by attenuating the energy released at the epicenter of each event over its distance to the site. The attenuation relationship (Patwardhan et al. 1978) is valid for soft and hard sites adjacent to subduction zones. A Gumble extreme-value statistical analysis was then used to evaluate the resulting distribution of peak site accelerations (Glass 1981). History suggests that over the next 25 years, there is a 39% probability of occurrence of an earthquake event producing a site ground acceleration of 0.16 g (16% of the earth's gravity).

36.5.3 Comments Regarding Pseudostatic Embankment Stability Analysis

Pseudostatic dynamic slope stability is evaluated by Spencer's method of slices, whereby an additional horizontal force directed outward from the slope is applied to the center of mass of each slice. The additional force is calculated as a function of the pseudostatic seismic coefficient input parameter. For embankments under seismic loading, the crest of the embankment slope generally will experience a peak acceleration that is two to four times the peak base acceleration. This translates into Carstenzweide embankment peak crest accelerations of 0.30 to 0.80 g. A good indication of a suitable pseudostatic seismic coefficient is 35% of this crest acceleration, or 0.11 to 0.28 g. For feasibility purposes, a pseudostatic seismic coefficient of 0.16 g (157 cm/sec^2) was selected.

The pseudostatic method, although a valuable screening tool for slopes in seismically active areas, is usually considered conservative since the same horizontal acceleration is applied to all slices. An actual embankment behaves in a more complex manner under dynamic loading as follows:

- Forces may act in several different directions along the slope at the same time.
- Foundation soils may amplify or attenuate base accelerations based on their dynamic properties and on the actual seismic event.

TABLE 36.1 Summary of geotechnical laboratory test results from individual representative samples of foundation and embankment materials

SOURCE	DEPTH (m)	USCS Field	USCS Lab	INDEX DATA: SIEVE DATA D60 (mm)	D50 (mm)	D10 (mm)	- #4 (%)	- #200 (%)	PI (LL-PL) (%w)	MOISTURE CONTENT (%)	WET DENSITY (pcf)	ORGANIC CONTENT (%)	VANE STRENGTH (psf)	FIBER CONTENT (%)	STRENGTH DATA: TRIAX COMPRESSION Consolidated, Undrained Effective Strength: Cohesion (psf)	φ (deg)	DIRECT SHEAR: Cohesion (psf)	φ (deg)	CONSOLIDATION: Stress Level (psi)	Cv *** (in^2/in)
CLEAN GRAVELS (SOIL GROUP SYMBOL = G)																				
TP-C-3	0.40	GW-GM	SW	5.50	3.80	0.180	57.0	7.0	NP	-	-	-	-	-	-	-	-	-	-	-
TP-C-6	0.69	GW	-	-	-	-	-	-	-	-	116.8	-	-	-	-	-	-	-	-	-
TP-C-3	1.40	GW	SM	2.10	1.30	0.048	82.5	14.0	NP	92.1	117.6	-	-	-	-	-	1634.0	29.6	-	-
TP-C-2	1.65	GW	GP	10.00	6.30	0.210	42.0	6.0	NP	-	-	-	-	-	-	-	-	-	-	-
CNI-4	39.37	GW-GM	SM	0.44	0.18	~0.009	87.0	38.0	NP	17.9	135.0	-	-	-	-	-	-	-	-	-
CNI-3	67.97	GP	SW	4.00	2.00	0.130	63.8	7.8	NP	14.2	-	-	-	-	-	-	-	-	-	-
			AVG.	4.41	2.72	~0.115	66.5	14.6	-	41.4	123.1									
GRAVELS WITH FINES (SOIL GROUP SYMBOL = GF)																				
CNI-4	6.83	GC	GM	5.30	2.00	~0.007	56.5	21.0	NP	15.9	108.3	-	-	-	1152.0	30.0 **	-	-	-	-
CNI-4	10.02	GC-CL	SW	5.50	3.50	0.083	56.0	10.0	NP	17.0	107.0	-	-	-	-	-	-	-	-	-
CNI-4	27.91	GM-GC	SM	0.90	0.42	~0.021	90.5	25.0	NP	17.6	-	-	-	-	-	-	-	-	-	-
			AVG.	3.90	1.97	~0.037	67.7	18.7	-	16.8	107.7									
CLEAN SANDS (SOIL GROUP SYMBOL = S)																				
CNI-4	3.36	SW-SM	GP	6.80	3.50	0.130	47.0	8.5	NP	-	-	-	-	-	-	-	-	-	-	-
CNI-3	31.41	SW	SM	1.90	1.30	0.020	70.0	20.0	NP	14.8	126.6	-	-	-	-	-	-	-	-	-
			AVG.	4.35	2.40	0.075	71.0	14.3	-	14.8	126.6									
SANDS WITH FINES (SOIL GROUP SYMBOL = SF)																				
CNI-4	6.44	SC	GM	10.30	10.00	0.025	34.0	14.5	NP	-	-	-	-	-	-	-	-	-	-	-
TP-C-2	2.00	SM	SM	0.25	0.17	0.030	84.0	33.0	NP	33.9	-	-	-	-	-	-	-	-	-	-
CNI-3	20.26	SM	SM-SC	0.30	0.16	~0.020	91.7	38.0	NP	36.1	-	-	-	-	-	-	-	-	-	-
TP-C-4	0.88	SM	ML	0.09	0.07	~0.027	98.5	53.0	NP	-	-	-	-	-	-	-	748.8	27.2	-	-
TP-C-3	3.00	SM	ML	~0.06	~0.04	~0.013	99.0	64.0	NP	-	-	-	-	-	-	-	-	-	-	-
			AVG.	~2.20	~2.09	~0.022	81.4	40.5	-	35.0										
SILTS (SOIL GROUP SYMBOL = M)																				
CNI-4	15.97	ML	PT	0.20	0.09	0.014	91.0	49.0	NP	2.0	-	35.8	-	-	-	-	-	-	-	-
CNI-4	20.51	ML	PT	2.00	1.80	0.020	86.0	22.0	NP	-	-	-	-	-	-	-	-	-	-	-
CNI-4	21.85	ML	SM	1.60	0.30	0.013	71.0	42.0	NP	47.4	68.1	-	-	-	-	-	-	-	-	-
CNI-4	35.15	MH-OH	ML	0.08	~0.05	~0.010	100.0	60.0	NP	48.8	100.4	13.4	-	-	-	-	-	-	-	-
CNI-4	47.50	MH-ML	SM *	0.32	0.17	~0.002	100.0	42.5	-	119.4	87.5	24.5	-	-	5173.0	11.2	-	-	65.0	0.22200
																			95.0	0.02100
																			175.0	0.00800
																			210.0	0.00297
																			236.0	0.00407
CNI-4	53.98	ML	ML	0.02	0.01	0.001	100.0	90.0	4	38.2	116.4	-	-	-	1266.0	31.1	-	-	14.0	0.21280
																			60.0	0.09050
																			90.0	0.03440
																			124.0	0.03780
																			204.0	0.04310
			AVG.	0.70	~0.40	~0.010	91.3	50.9	4	51.2	93.1	24.6								
CLAYS (SOIL GROUP SYMBOL = C)																				
CNI-2	7.07	CH	OL	0.02	0.01	~0.001	100.0	94.5	NP	95.8	75.8	11.2	-	-	-	-	-	-	-	-
CNI-2	7.62	CH	OL	0.03	0.02	~0.001	97.5	70.0	NP	77.0	105.0	-	-	-	1602.0	20.6	-	-	8.0	0.00780
																			30.0	0.05700
																			55.0	0.05700
																			145.0	0.02780
CNI-2	10.91	CL	CL-OL	0.02	0.01	~0.002	97.0	72.0	NP	63.4	87.8	9.8	-	-	-	-	-	-	-	-
CNI-3	101.94	CL	OL	0.03	0.01	~0.001	100.0	74.0	NP	51.4	111.0	35.6	-	-	2583.0	29.2	-	-	25.0	0.02214
																			47.0	0.00370
																			85.0	0.00073
																			103.0	0.00024
																			118.0	0.00033
																			236.0	0.00685
																			118.0	0.00526
																			236.0	0.00018
CNI-3	103.22	CL	ML-CL	0.01	0.01	-	100.0	70.5	NP	65.2	98.1	20.3	-	-	-	-	-	-	-	-
CNI-3	113.13	CH	ML-CL	0.02	0.01	-	98.0	93.7	NP	18.9	130.1	-	-	-	-	-	-	-	-	-
			AVG.	0.02	0.01	~0.001	98.8	79.6	-	62.0	101.3	19.2								
PEAT/ MUSKEG (SOIL GROUP SYMBOL = O)																				
TP-C-4	0.31	OL	-	-	-	-	-	-	-	26.4	-	49.3	-	-	-	-	-	-	-	-
TP-C-5	0.90	OH	-	-	-	-	-	-	-	419.0	-	33.0	-	-	-	-	-	-	-	-
TP-C-3	1.10	OH	ML/CL	-	-	-	100.0	71.0	NP	-	-	-	-	-	-	-	0.0	33.0	-	-
TP-C-3	1.12	OH	-	-	-	-	-	-	-	439.0	96.4	44.3	490.0	48.0	115.0	35.0	-	-	10.0	0.00078
TP-C-1	1.31	OH	-	-	-	-	-	-	-	154.0	-	21.0	320.0	2.0	-	-	-	-	-	-
TP-C-4	1.40	OL	-	-	-	-	-	-	-	68-155	-	-	-	2.0	-	-	-	-	-	-
TP-C-4	1.51	OL	-	-	-	-	-	-	-	68.0	-	4-11	-	2.0	-	-	-	-	-	-
TP-C-3	2.32	OL	-	-	-	-	-	-	-	94-147	-	8-15	170.0	7.0	-	-	-	-	-	-
CNI-3	17.25	OH	-	-	-	-	-	-	-	37.0	-	2.0	-	-	-	-	-	-	-	-
TP-C-6	0.25	OL	-	-	-	-	-	-	-	-	-	20.0	-	-	-	-	-	-	-	-
CNI-5	39.13	PT	PT	-	-	-	-	-	-	124.8	94.0	26.5	-	-	-	-	-	-	-	-
			AVG.	-	-	-	100.0	71.0	-	166.7	95.2	23.9	326.7	12.2						
OVERBURDEN																				
PHYLLIC	-	-	GM	10.4	7.0	~0.020	44.0	13.0	NP	6.0	-	-	-	-	-	-	724.0	42.2	-	-
POTASSIC	-	-	GP-GM	9.5	7.8	0.060	43.0	11.0	NP	5.5	-	-	-	-	-	-	1339.0	37.6	-	-
S 16	-	-	GM	9.5	6.0	~<.001	~45.0	~19.0	1.0	11.0	122.0	-	-	-	209.0	39.9	-	-	-	-
S 17	-	-	GM	7.5	3.5	~<.001	43.0	21.0	NP	12.5	115.0	-	-	-	0.0	40.4	-	-	-	-
S 18	-	-	GM	8.0	4.0	~<.001	54.0	20.0	1.0	12.0	122.0	-	-	-	230.0	39.6	-	-	-	-
			AVG	9.0	5.7	0.017	45.8	16.8	1.0	9.4	119.7									

NOTES: (1) For sieve data, "~" indicates that data was extrapolated to determine the value.
* The sample testsed was thinly laminated with fine laminae of organic silt, silty organic fine sand with stiff calcareous stringers.
** The tiraxial membrane ruptured after initial confinement. Due to this, the shear strength was estimated from one Mohr circle.
*** Cv is the coefficient of consolidation

TABLE 36.2 Carstenzweide Valley expected-strength parameters for embankment stability analysis

Carstenzweide Foundation Soil Unit			Loose and Soft Surficial Unit	Organic Fine Silt Unit	Dense and Firm Deep Unit	Overburden
	Density	(pcf)	111	114	122	115
		(mt/m^3)	1.779	1.827	1.955	1.843
EFFECTIVE SHEAR STRENGTH	Cohesion	(psf)	1150	1250	1150	Power Curve
		(mt/m^2)	5.618	6.107	5.618	
	ϕ	(degrees)	25	29	29	
Skempton's Pore Pressure Coefficient A at Failure		(-)	0.2	0.71	0.4	-
PORE PRESSURE PARAMETERS	Coefficient of Consolidation-Cv	(in^2/min)	0.02	0.0038	0.004	-
		(cm^2/min)	0.129	0.025	0.026	-
	Height of Drainage Path-Hdr	(in.)	35	104	37	-
		(m)	0.89	2.65	0.94	-

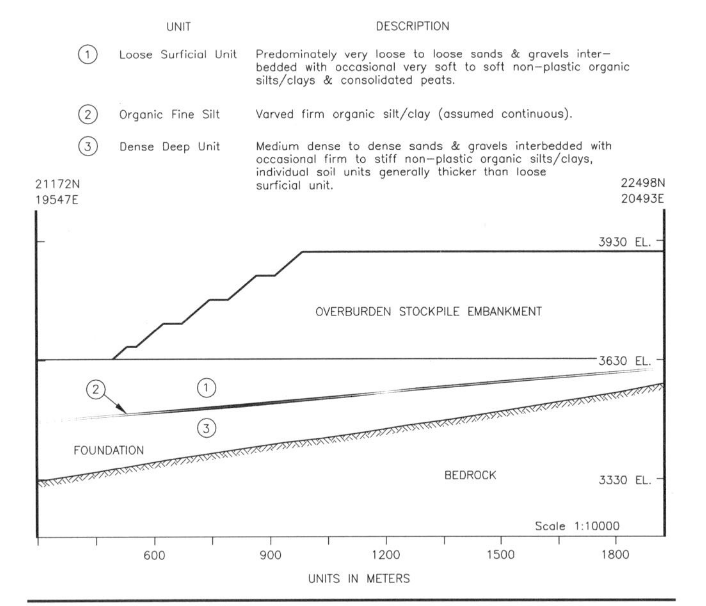

FIGURE 36.4 Cross section of Carstenzweide geotechnical foundation model with embankment

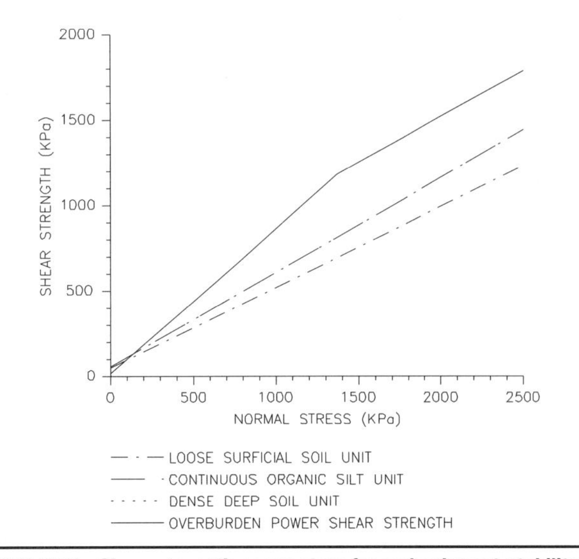

FIGURE 36.5 Shear-strength parameters for embankment stability analysis

- The fundamental period of the foundation-embankment system changes over time with embankment construction.
- The amplifying effects of soil deposition on ground accelerations are greatest when the fundamental period of the system is close to the predominant period of the base motion seismicity.

36.5.4 Modeling Foundation Excess Pore-Pressure Development with Embankment Loading

Typically, it was assumed that embankment pore pressures would not develop in the relatively free-draining rockfill overburden. However, the Carstenzweide subsurface soil types indicated there could be significant foundation pore-pressure development due to embankment loading. Independent pore-pressure development was modeled by the static and pseudostatic stability analysis for the three foundation soil units (Figure 36.4).

Pore-pressure development is controlled by the rate of foundation loading, natural water conditions, and foundation soil characteristics (Table 36.2). Foundation loading rates are a function of overburden placement rate, the affected foundation area, and embankment height (or lift height in the case of staged construction). For example, a thick impermeable clay-like soil that is loaded quickly will develop large pore pressures that take a long time to dissipate (due to low permeabilities). A thin gravelly soil will also develop excess pore pressures, but they will dissipate quickly because they are thin and more permeable.

For the embankment construction schemes considered in the stability analysis, the pore pressures for the three foundation units were calculated as a function of overburden placement rate, loading geometry (embankment height, foundation surface area, and overall embankment slope angle) and the foundation soil properties (Table 36.2). Excess pore pressures were estimated along the center line of each foundation soil unit according to Skempton's pore pressure equation using estimates of the change in stresses due to embankment loading (Perloff 1975).

The dissipation of pore pressures was modeled using one-dimensional consolidation theory. The average coefficient of consolidation (*Cv*) was evaluated from the consolidation of triaxial specimens in staged CU triaxial compression tests (Head 1986). The placement rate and resulting embankment crest and toe advance rate defined the time available to dissipate these pore pressures as the embankment is placed.

For the Carstenzweide analysis, a maximum average daily overburden placement rate of 200,000 metric tonnes per day (MTPD) was used. The placement rate defines the advance rate of the embankment over the foundation interface and is used to calculate excess pore pressures in the foundation materials. Pore pressures were calculated and tracked as a function of embankment placement sequences considered (angle of repose versus

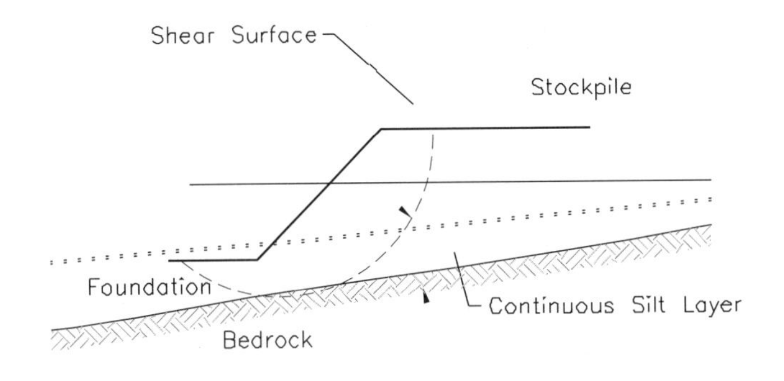

Circular Shear Surface

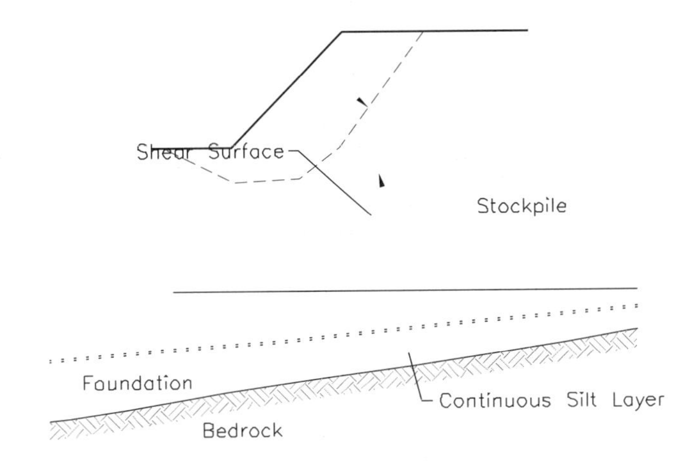

Noncircular Shear Surface

FIGURE 36.6 Possible Carstenzweide embankment stability mechanisms considered in limiting-equilibrium analysis

staged construction in controlled lifts). In the case of the staged construction, residual pore pressures were tracked from the effects of previous lifts as the embankment ascended.

36.5.5 Limiting-Equilibrium Stability Analysis of the Ultimate Embankment at Angle of Repose

Static stability and pseudostatic stability were analyzed using UTEXAS2 for a single end-dumped lift built at the ultimate 3,900-m elevation (270-m-high angle-of-repose embankment). The results indicated that this method of construction would produce large, deep-seated foundation shear surfaces with pseudostatic FOS values well below 1.0 and static FOS values at unity (Table 36.3).

The noncircular failure modes (Figure 36.6) yielded the most critical FOS for both pseudostatic and static cases, since the shear-surface searches found the weak organic silt unit. The organic silt is predicted, on average, to be slightly stronger than the loose surficial unit, but it develops significantly higher excess pore pressures.

The model predicts higher levels of stability for a fully dewatered foundation. Full dewatering of the Carstenzweide foundation was considered to be unrealistic at the feasibility stage due to the high precipitation and complex geology of this area. Nevertheless, the fully dewatered case predicts that a single lift angle-of-repose embankment would be fairly stable.

Sensitivity to pore pressure and stability levels as a function of placement rate was performed using a placement rate of 50,000 MTPD; the pseudostatic FOS was still less than 1.0. The embankment FOS for the pseudostatic case should be at least at or above unity. The aggressive overall slope angle (angle of repose) and the associated high stress concentrations under the slope of the embankment generate adverse pore-pressure in the foundation; therefore, low levels of stability were predicted for a single lift angle-of-repose embankment configuration.

36.5.6 Multiple-Lift Staged Construction

Given marginal stability of a single angle-of-repose embankment, the stability and pore-pressure interaction for the ascending, multiple-lift construction method was evaluated for the 200,000-MTPD placement rate. Pore-pressure generation versus dissipation was tracked; pore pressures were calculated at the end of each lift based on the residual pore pressure remaining from prior lifts and on the new pore pressures created by each successive lift. Stress distributions in the foundation soil units were recalculated for each lift applied (Perloff 1975). The embankment configuration for this model consisted of flattening the overall slope to 27° and building up in lifts (a 30-m-high lift followed by four 60-m lifts).

Results indicated that ascending construction produces deep-seated critical shear surfaces with pseudostatic FOS values for all lifts that were approximately $1.0; static FOS values were >1.30 (Table 36.3). All critical FOS values were obtained for the circular shear surface mode, except for the case of the ultimate embankment (post Lift-5) where the FOS was approximately the same. For a fully dewatered foundation, the predicted FOS also improves significantly over the saturated case.

36.5.7 Carstenzweide Foundation Liquefaction Analysis

The loose surficial foundation soil unit contains saturated sandy-silty soils that are potentially liquefiable. Loose gravels and soft organic silts and clays with low SPT blow counts were also observed and also may be prone to a loss of strength due to strong ground shaking (Hynes and Franklin 1989). Cyclic loading of saturated cohesionless soils can result in significant build-up of excess pore pressures. The effective shear strength may then decrease appreciably, especially with limited or no drainage. If the strengths drop low enough, large shear strains can develop. When accompanied by continued cyclic stresses, these large shear strains can lead to further reductions in the effective shear strength of the soil. After several cycles of shear straining, the soil may experience a nearly total loss of strength, at which point the soil has liquefied. The build-up of pore pressure has been shown to be a function of the cyclic shear strain and the number of significant cycles (Ladd et al. 1989).

The potential for foundation liquefaction under the proposed Carstenzweide embankment was estimated from the results of SPT penetration testing (Seed and Idriss 1982; Seed et al. 1984) and was later modified (Seed and Harder 1990; Marcuson, Hynes, and Franklin 1990). The method is based on an empirical approach from liquefaction case histories and in situ SPT N-1(60) blow count data. The loose surficial foundation soil unit (considered susceptible to liquefaction) was divided into individual soil layers based on logged drill-hole statistics. The liquefaction FOS was determined for each layer based on the site's probable range of seismic excitation levels and on average SPT N-1(60) blow count data. For modeled individual soil layers that were predicted to liquefy (liquefaction FOS less than 1.0), a residual liquefied strength (liquefied total strength without friction) was estimated based on the work of Seed and Harder (1990).

The static shear strength of units in the foundation that were not predicted to liquefy (liquefaction FOS $1.0) was reduced to account for soil deformation characteristics and earthquake-induced pore-pressure generation (after Marcuson, Hynes, and Franklin 1990). A residual strength (cohesionless material with an adjusted friction angle) was calculated for each

TABLE 36.3 Carstenzweide Valley overburden stockpile static and pseudostatic stability summary of critical shear surfaces for expected strengths

CASE	Foundation Water Condition	Overall Slope Height (m)	Overall Slope Angle (deg)	Critical Factor of Safety (FS) Static	Critical Factor of Safety (FS) Pseudo Static ($\bar{a}$=16%g)	Remarks	Graphic
Single Lift–end Dumped at Ultimate Slope Height	S	270	40	1.08	0.84	Angle of Repose Circular Shear	
	S	270	40	1.01	0.76	Angle of Repose Non–Circular Shear	
	D	270	40	1.57	1.20	Angle of Repose Circular Shear	
STAGED CONSTRUCTION:							
Lift – 2	S	90	26	1.45	1.06	60m Lift Increment Circular Shear	
Lift – 3	S	150	26	1.39	1.05	60m Lift Increment Circular Shear	
Lift – 4	S	210	26	1.38	1.02	60m Lift Increment Circular Shear	
Lift – 5	S	270	26	1.35	1.00	60m Lift Increment Circular Shear	
Lift – 5	S	270	26	1.37	0.99	60m Lift Increment Non–Circular Shear	
Lift – 5	D	270	26	1.74	1.30	60m Lift Increment Circular Shear	

S = Saturated Foundation D = Fully Dewatered Foundation --- Excess Pore Pressure Surfaces

soil layer from the static friction angle, based on the residual pore-pressure ratio due to seismic shaking (after Sykora et al. 1992). The resulting foundation shear-strength profile for the postearthquake case is shown on Figure 36.7 for the two design earthquake events considered.

Static slope stability analysis (UTEXAS2) was performed using these postearthquake liquefied strengths to evaluate the stability of the embankment following the design earthquake events. The analysis was run statically since the postearthquake residual shear strengths have already been adjusted for the effects of ground shaking, immediately following the earthquake events considered. Only circular shear-surface searches were specified since the surficial loose soil unit is interpreted to consist of a complex laterally discontinuous series of mixed soils. Noncircular geometries were not considered; they would unrealistically model the weaker liquefied residual soils as laterally continuous. Only the staged lifts were considered since static analysis for the angle-of-repose ultimate embankment produced unacceptable FOS values.

Table 36.4 presents the detailed steps to the foundation liquefaction analysis.

Pore-Pressure Conditions. The stability model for postearthquake foundation liquefaction is based on the initial static limiting-equilibrium model with the postearthquake strength profile for the loose surficial unit (Figure 36.7). For the saturated foundation case, the piezometric surfaces used for the static analysis were retained in the postearthquake analysis. A dewatered model was obtained by removing all pore-pressure surfaces while retaining the residual strengths.

Results of Postearthquake Static Liquefaction Stability Analysis. The primary case used to assess liquefaction potential was the magnitude 7.5 earthquake event with associated 0.20 g base acceleration. The analysis indicated potential for seismically induced liquefaction in foundation soils 30 to 60 m deep; therefore, there is potential for embankment displacements. The key question for cases of predicted liquefaction is the probable response of the embankment. For postearthquake slope stability FOS values of approximately 0.9 to 1.0, some limited deformations due to liquefaction can be expected. The embankment should, however, attain a postfailure geometry without large catastrophic displacements. For lower FOS values, some slope-wide instability may occur.

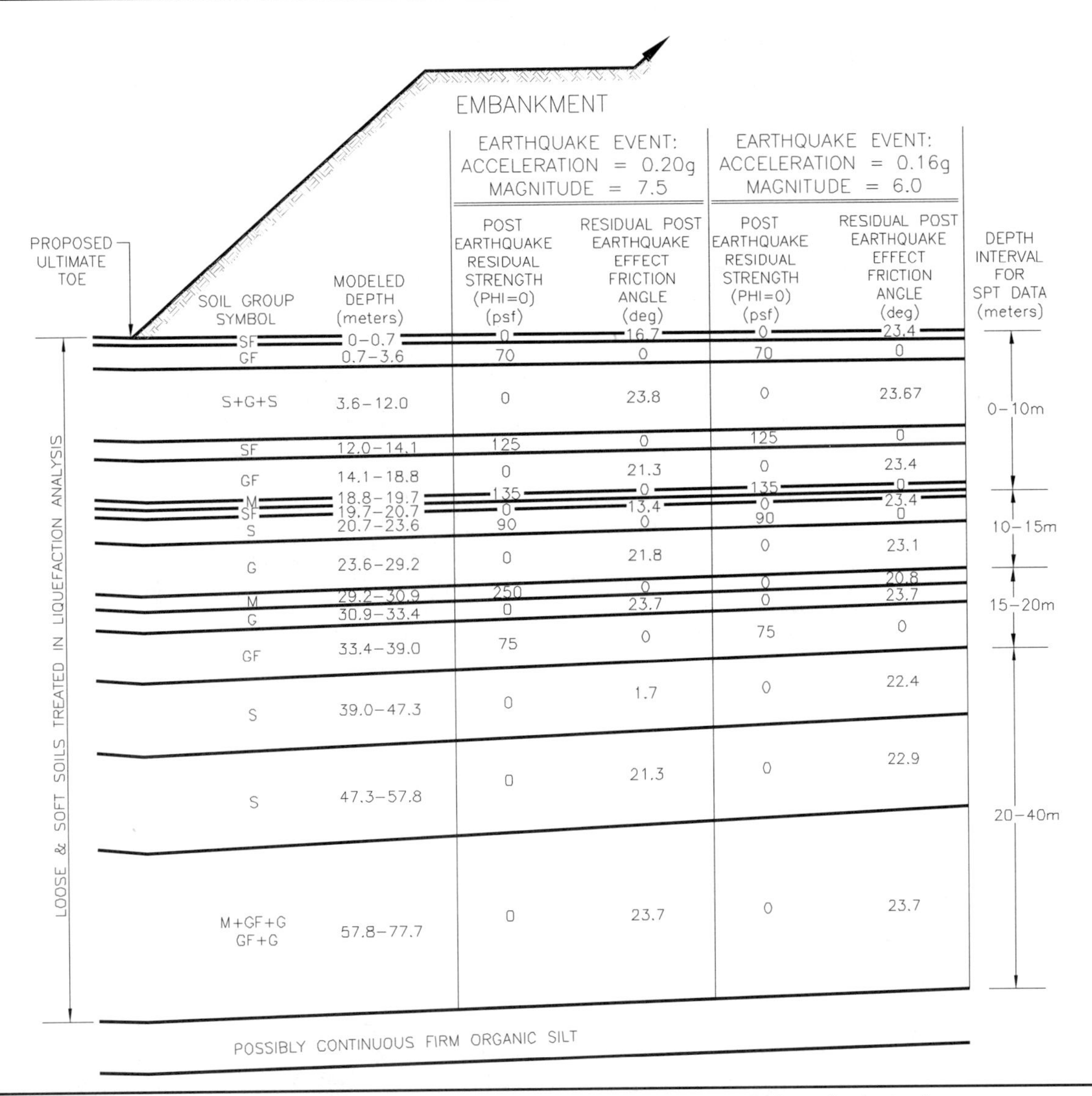

FIGURE 36.7 Postearthquake residual soil strengths for the loose surficial soil unit under two design seismic events

The UTEXAS2 postearthquake stability runs indicate a high likelihood of liquefaction in the toe area (Table 36.5).

The critical shear surface is a toe circle with a FOS of 0.7. For approximate limiting-equilibrium conditions (Run 2 FOS of 1.0), the shear surface is approximately 65 m deep and exits the foundation 160 m from the embankment toe. The third case shows the postearthquake FOS for the same shear surface presented in the post Lift-5 pseudostatic effective stress stability analysis. This indicates a FOS of 1.26 with a shear surface depth of 80 m and a foundation exit point 255 m from the embankment toe. The fourth case shows the critical FOS for fully dewatered foundation conditions. The fifth case shows the postearthquake FOS for the same shear surface as that presented for the third case, for full foundation dewatering.

Sensitivity of postearthquake stability to the magnitude 6.0 earthquake event (with 0.16 g base acceleration) indicates that there is no significant difference between the expected potential for liquefaction for the two earthquake events. However, the expected behavior of the ultimate embankment is somewhat different since the shear surface depth and foundation exit point are smaller for the magnitude 6.0 earthquake event. Seed, Makdisi and de Alba (1978) indicate that liquefaction failure has never occurred in a hydraulic fill dam or a dam on an alluvial foundation, for a seismic event with magnitude less than 6.5 and an associated base acceleration of less than 0.20 g. Historically, the site has experienced a maximum estimated base acceleration of 0.166 g, which was associated with a magnitude 6.0 event occurring 45 km from the site. Also, an event of magnitude 7.6 occurred 156 km from the site and imported accelerations of about 0.13 g. Based on the analysis, the most likely site response of the Carstenzweide foundation has been bounded by a somewhat marginal condition with respect to foundation liquefaction.

Sensitivity of the postearthquake liquefaction stability model to embankment slope angle was also evaluated by finding the critical FOS for the magnitude 7.5 event and a 30° overall slope angle. The critical FOS for this case (FOS of 0.64) with saturated foundation conditions was found to be about 7% lower than that of the associated 27° slope case for the same earthquake event (Table 36.5). The shear surface for the 30° slope possesses similar depth and exit point as the 27° case.

Given the interpreted heterogeneous and noncontinuous nature of the loose surficial soils, it is possible that some zones of the foundation may liquefy and the embankment may experience deformations. The slope will readjust at some flatter angle, which is expected to be approximately 15°. Some settling at the foundation interface may occur with embankment slumping;

TABLE 36.4 Detailed steps in the evaluation of foundation liquefaction potential

1. A critical cross section was selected for evaluating liquefaction potential and postearthquake stability. For the Carstenzweide overburden embankment, the loose surficial foundation soil unit was the only foundation unit considered susceptible to liquefaction.
2. The maximum acceleration and magnitude for earthquake-induced base excitation of the embankment and foundation was estimated based on the site-specific seismic hazard assessment. For evaluation of foundation liquefaction potential, seismic base accelerations of both 0.16 g and 0.20 g events were considered. Although not historically substantiated, the evaluation of liquefaction provided for a 0.20 g event (associated with a Richter magnitude of about 7.5), to account for the possibility of a large event occurring close to the site.
3. For soils within the foundation, the resistance to liquefaction (cyclic shear strength, Rf) was evaluated using the SPT N-1(60) blowcount data corrected for soil fines content based on Seed's work (1984). The corrected "equivalent clean sand" blowcounts, N-1(60)-ecs, were based on average fines content by the Carstenzweide soil groups that were determined as part of the lab testing program. The fines correction was made to standardize the N-1(60) values to the same equivalent fines content at the onset of the liquefaction analysis. The corrected blowcounts from individual geotechnical drill holes were combined and treated statistically by averaging the blowcounts by depth interval and Carstenzweide soil group symbol (Figure 36.2). Corrected N-1(60)-ecs blowcounts varied mostly between 10 to 30. Cyclic shear strength ratio (Rf) for a magnitude 7.5 earthquake was calculated based on the N-1(60)-ecs values (Seed 1984). For other magnitude events, the cyclic stress ratio was corrected for earthquake magnitude, overburden pressure, and initial static shear stress effects (Seed and Harder 1990; Marcuson, Hynes, and Franklin 1990). Clay soil types (Carstenzweide soil group symbol "C") were not considered susceptible to liquefaction based on geotechnical index test results and Wang (1979).
4. The cyclic stresses (Ri) which developed in the foundation from the design seismic event were evaluated based on the peak surface acceleration and estimated in situ stress conditions using equations by Seed and Idriss (1982). Average stresses in the foundation (due to embankment and foundation material loads) were estimated using elastic theory (Perloff 1975); these stress estimates were considered reasonable for feasibility level analysis. Additional investigation, lab testing and numerical stress/strain modeling would be required to obtain a more accurate estimate of the likely stresses within the foundation.
5. The Factor of Safety (FOS) against liquefaction for each soil unit was calculated by comparing the cyclic shear strength (Rf) to the cyclic shear stress (Ri) as follows: Factor of Safety (FOS) = Rf/Ri. The FOS against liquefaction was determined for site base accelerations of 0.16 g and 0.20 g for each lift in the embankment construction sequence. Sensitivity to levels of foundation saturation were also conducted. The results indicate that the potential for liquefaction increases up to the 90-m-high embankment (Lift-2) and that the potential for liquefaction is approximately the same for the 90-m- high embankment and the ultimate 270-m-high embankment.
6. For individual modeled soil layers in the loose surficial foundation unit that were predicted to liquefy (FOS against liquefaction less than 1.0), a residual liquefied strength (liquefied total strength without friction) was estimated based on the work of Seed and Harder (1990). The static shear strength of units in the foundation that were not predicted to liquefy (FOS against liquefaction greater than or equal to 1.0) were reduced to account for soil deformation characteristics and earthquake-induced pore pressure generation (after Marcuson, Hynes, and Franklin, 1990). A residual strength (represented as a cohesionless material with an adjusted friction angle) was calculated for each soil layer from the static friction angle, based on the residual pore pressure ratio due to seismic shaking (Sykora et al. 1992) (Figure 36.7).
7. Static slope stability analysis (limiting-equilibrium using UTEXAS2 with Spencer's method of slices) was then performed using these postearthquake liquefied strengths to evaluate the stability of the embankment following the design earthquake events. The analysis was run as a static analysis, as the postearthquake residual shear strengths have already been adjusted for the effects of ground shaking immediately following the earthquake events considered.

the embankment materials may sink into liquefied foundation zones.

The stockpile embankment is not expected to experience large catastrophic translations of the overall embankment as long as the embankment materials do not become saturated. Major seismic activity may result in extensive cracking of the embankment in the area of the crest. These dynamic deformations, not attributed to liquefaction, occur during an earthquake event and are expected to be on the order of tens of meters. These deformations may lead to some lateral embankment spreading, particularly in the toe area.

Potential for liquefaction can be remedied through two basic approaches. To accommodate deformation near the toe, a wide safety step-out from the toe of the slope should be planned. Conversely, measures can be taken to densify and dewater the foundation materials from under the ultimate crest to some distance out in front of the planned ultimate toe. For the loose, surficial Carstenzweide soils, the most practical measure would be to densify the foundation through blasting and to dewater aggressively. Blasting would require field tests involving iterative SPT blow count measurements to evaluate the optimum spacing, charge, and firing sequence (usually from the bottom of holes up) to arrive at the most dense conditions possible. Other more costly remedial measures include vibrocompaction, admixtures, dynamic compaction (heavy tamping), and excavation and replacement (the last of which is probably too expensive).

Dewatering the foundation soils will decrease the possible extent of liquefaction and will reduce potential deformations from seismic shaking. Effective dewatering must be achieved, however, since partially saturated loose deposits may still experience liquefaction (e.g., Lower San Fernando Dam, California).

36.5.8 Carstenzweide Geotechnical Recommendations from Initial Feasibility Study

Based on the initial embankment stability study, the following points summarize the recommended design configuration for the ultimate Carstenzweide overburden stockpile (Figure 36.8):

1. A maximum slope height (difference in elevation from crest to toe) of 270 m
2. An overall slope angle of two horizontal to one vertical (2H:1V) (27°)
3. An initial 30-m-high angle-of-repose lift, followed by four 60-m-high angle-of-repose lifts with appropriate offsets between lifts to obtain the overall slope angle
4. Embankment placement rates of around 200,000 MTPD or one 60-m lift over any one year. From a stability standpoint, uniform placement over time is preferred over short periods of rapid placement.
5. Permanent infrastructure restricted outside a 275-m-wide safety step-out from the ultimate embankment toe. This would protect against the potential for earthquake-induced embankment displacements. A 160-m-wide

TABLE 36.5 Carstenzweide Valley overburden stockpile postearthquake liquefaction stability summary

EARTHQUAKE EVENT	FOUNDATION WATER CONDITION	FACTOR OF SAFETY	REMARKS	GRAPHIC
STAGED CONSTRUCTION; 26 DEG. OVERALL SLOPE ANGLE 270m MAXIMUM SLOPE HEIGHT; 60m MAXIMUM LIFT THICKNESS				
Magnitude = 7.5 (Richter Scale) Acceleration = 0.20g	S	0.69	Critical Shear Surface.	
	S	1.07	Shear Surface for Limiting Equilibrium (Approx.).	
	S	1.26	Factor of Safety for Shear Surface Exiting Foundation 250m From Proposed Stockpile Toe.	
	D	0.89	Critical Shear Surface for Fully Dewatered Foundation.	
	D	1.65	Factor of Safety for Shear Surface Exiting Foundation 250m From Proposed Stockpile Toe for Fully Dewatered Foundation.	
Magnitude = 6.0 (Richter Scale) Acceleration = 0.16g	S	0.70	Critical Shear Surface.	
	D	0.92	Critical Shear Surface for Fully Dewatered Foundation.	
STAGED CONSTRUCTION; 30 DEG. OVERALL SLOPE ANGLE 270m MAXIMUM SLOPE HEIGHT; 60m MAXIMUM LIFT THICKNESS				
Magnitude = 7.5 (Richter Scale) Acceleration = 0.20g	S	0.64	Critical Shear Surface.	

S = Saturated Foundation D = Fully Dewatered Foundation --- Excess Pore Pressure Surfaces

safety step-out from the ultimate toe is recommended for mobile equipment and roads.

6. A concave crest line to promote three-dimensional stability
7. Surface drainage of the embankment and diversion of runoff from areas above the embankment. This should be considered because of the high annual precipitation levels and concerns for surface and subsurface water conditions.

Figure 36.9 describes various modes for overburden embankment instability (MEMPHR 1991).

The more likely Carstenzweide embankment failure modes involve deep-seated shear surfaces, which probably would be induced by dynamic shaking. Events of this nature will probably impact the crest area and areas out from the toe. Sliver failures may affect local areas at the crest of angle-of-repose lifts. The soft, near-surface foundation materials may cause localized toe bulging and possibly associated crest sloughing.

36.6 OBSERVATIONAL ENGINEERING AND PERFORMANCE

36.6.1 Terzaghi's Observational Method of Soil Mechanics and Mine Geotechnics

A common engineering approach to open-pit overburden operations consists of an initial investigation to obtain enough site-specific information to make a preliminary evaluation of stability and embankment volumes. The preliminary analysis indicates areas that should be investigated further, and the embankment

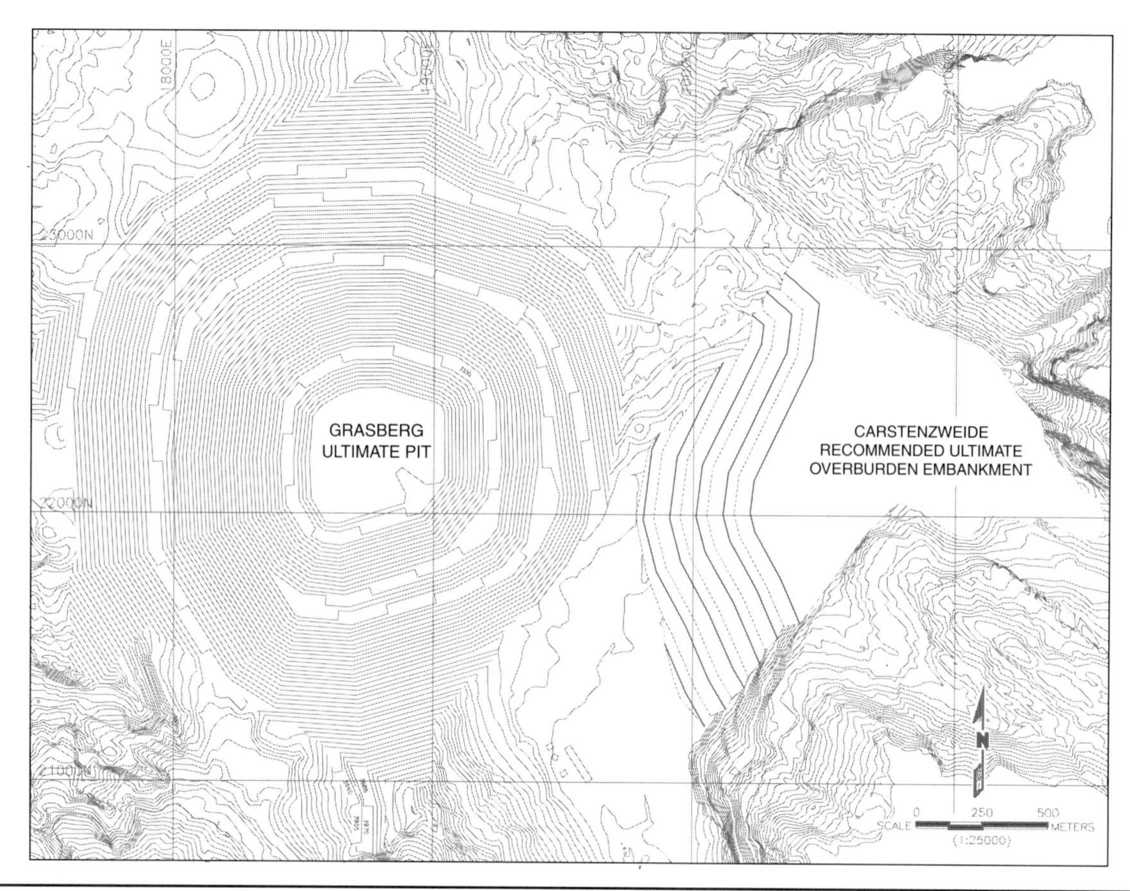

FIGURE 36.8 Ultimate embankment configuration for the recommended Carstenzweide overburden stockpile with the Grasberg ultimate pit limits (circa 1992)

plan is modified as required. Also, any key design assumptions should be verified to ensure that relatively safe operational conditions prevail.

Terzaghi (Peck 1969) described a practical approach for verifying geotechnical design assumptions and for investigating the effects of the critical parameters identified during feasibility studies. Terzaghi's observational method, applied to mine geotechnics, provides some guidance for initial embankment placement operations (field trials). Field trials are one of the few ways to gain new geotechnical experience with unfamiliar conditions for which no precedent can be found. In the early years of Carstenzweide embankment development, trial embankments were used to verify geotechnical predictions and to provide site-specific operational field experience. An observational engineering approach to embankment construction with geotechnical monitoring was implemented with the trial embankments. The monitoring provides for safe operational conditions and measures geotechnical performance.

36.6.2 Monitoring

It is essential to use monitoring data to analyze geotechnical performance over time to compare actual versus predicted behavior and to modify plans, if required. An overburden embankment-monitoring program has been in place in support of Grasberg open-pit mining since the early 1990s. Embankment instability is almost always preceded by measurable changes in embankment behavior that occurs within several hours or days prior to significant displacements. To provide operational safety to personnel and equipment, a reliable and redundant embankment monitoring system is required to detect the early warnings. There are many techniques available for monitoring overburden stockpile embankments and foundations. The following are the more applicable and practical techniques used for monitoring the Carstenzweide embankment:

- Visual monitoring to make frequent and thorough visual inspections. Visual monitoring by qualified personnel is perhaps one of the most effective embankment-monitoring practices available and is a key to early detection of overburden embankment instability. However, it should be used in conjunction with other monitoring techniques.
- Wire-line extensometers to measure embankment crest displacements in areas of active instability (Figure 36.10).
- EDM survey prism points to monitor ground response in embankment crests and foundation areas in front of the embankment toe. Prism survey data provide information on the failure mode if instability occurs. Embankment failures initiated by foundation distress are often observed first in foundation prism monitoring data.
- Piezometers to measure any pore pressures within embankments and foundations.

36.6.3 Operational Guidelines for Embankment Crest Advance Rate

Based on experience in British Columbia, Canada, with high angle-of-repose waste rock embankments (for coal and base metal mines), a relationship between embankment slope height versus maximum recommended crest advance rate was established (MEMPHR 1991). The British Columbia relationship was used during the early stages of overburden embankment development to help guide operational and geotechnical decisions. Over time,

FAILURE TYPE	DIAGRAM	USUAL CAUSE	LIKELY EFFECTS	WARNING SIGNS
Sliver Failure	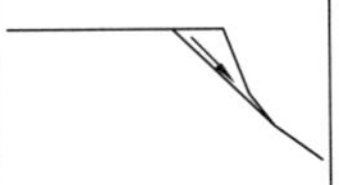	OVERSTEEPENED CREST DUE TO HIGH FINES CONTENT, RAPID PLACEMENT RATE, WET MATERIAL.	SMALL SCALE CREST FAILURE, SUBSIDENCE AT CREST OF STOCKPILE.	CREST CRACKING, SUBSIDENCE NEAR CREST, STEEP SLOPE BELOW CREST, INCREASING CREST DISPLACEMENT RATES.
Foundation Failure	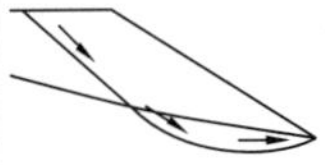	WEAK MATERIAL IN FOUNDATION, RAPID LOADING RATE, HIGH PORE PRESSURES IN FOUNDATION BLASTING OR EARTHQUAKE EFFECTS.	CAN CAUSE LARGE FAILURE INVOLVING SIGNIFICANT PART OF STOCKPILE.	SEEPAGE AT THE TOE, BULGING OR SPREADING OF STOCKPILE TOE, CRACKS WELL BEHIND STOCKPILE CREST.
Overall Failure		WEAK MATERIAL ALONG BASE OF STOCKPILE (eg LAYER OF SOIL), POOR DRAINAGE ALONG BASE OF STOCKPILE, STEEP FOUNDATION, RAPID LOADING RATE.	ENTIRE STOCKPILE FAILS ALONG BASE.	CRACKING OF STOCKPILE SURFACE BACK AS FAR AS CONTACT WITH GROUND, SETTLEMENT OF ENTIRE PLATFORM.
Rotational Failures		WEAK MATERIAL IN STOCK-PILE OR FOUNDATION, HIGH PORE PRESSURED, RAPID LOADING RATES, MAY INVOLVE STOCKPILE MATERIAL ONLY OR ALSO INCLUDE FOUNDATION, MAY BE CIRCULAR OR NON-CIRCULAR IN CONFIGURATION.	CAN CAUSE FAILURES INVOLVING MAJOR PART OF STOCKPILE, MAY INVOLVE STOCKPILE ONLY OR STOCKPILE AND FOUNDATION (TWO TYPICAL FAILURE SURFACES ILLUSTRATED).	BULGING AT TOE, CRACKING AND SETTLEMENT WELL BEHIND CREST, ROCK NOISE, POSSIBLE SCARPS WELL BEHIND CREST.
Toe Failure	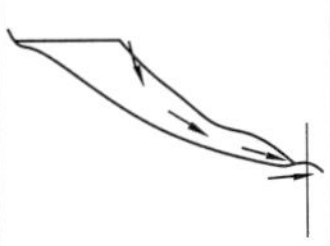	WEAK FOUNDATION MATERIALS AT TOE. HIGH PORE PRESSURES AT TOE, STEEP SLOPES AT TOE. RAPID LOADING RATE.	LOSS OF SUPPORT OF TOE, MAY LEAD TO TO PROPAGATION OF FAILURE UP-SLOPE.	SPREADING OF TOE, YIELDING AND BULGING OF FOUNDATION SOILS AND/OR STOCKPILE MATERIAL AT TOE.
Foundation Liquefaction	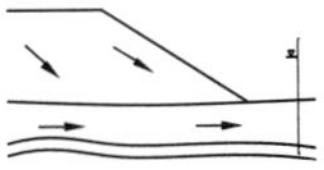	SILTY TO SANDY MATERIAL IN FOUNDATION, POSSIBLY CONFINED BY AQUITARDS, PORE PRESSURE BUILD-UP DUE TO RAPID LOADING, SEISMIC FORCES MAY BE IMPORTANT	POSSIBLE MAJOR FAILURE OF SIGNIFICANT PORTION OF THE STOCKPILE WITH LARGE RUNOUT DISTANCE, MAY OCCUR ON FLAT FOUNDATION SLOPE.	HIGH PIEZOMETRIC PRESSURES IN FOUNDATIONS IN SOME CASES. SAND BOILS MAY BE PRESENT PRIOR TO COMPLETE FAILURE.
Planar Failure		WEAK PLANE IN STOCKPILE MATERIAL APPROXIMATELY PARALLEL TO STOCKPILE FACE, DUE TO POOR MATERIAL, SNOW LENS OR RAPID LOADING.	MAY INVOLVE LARGE AMOUNT OF MATERIAL WITH LARGE RUNOUT DISTANCE.	SLUMPING OF STOCKPILE CREST, BULGING OF TOE OR FACE, CRACKS ON PLATFORM WELL BEHIND CREST.

FIGURE 36.9 Common modes of overburden stockpile embankment instability

site-specific monitoring and operational experience suggested that that guideline was perhaps slightly conservative for embankment slope heights over 100 m, and site-specific criteria were also required for slope heights in excess of 200 m. To optimize day-to-day operations and provide mine planning guidance for life-of-mine embankment plans, historical records from Grasberg mine operations and crest displacement monitoring were analyzed to develop a site-specific relationship for embankment slope height versus crest advance rate.

A database was created from mine records (including embankment slope height, average crest advance rate, foundation conditions, crest displacement monitoring from wire-line extensometers, whether or not rapid crest displacements occurred, and foundation slope angle under crest). An analysis was performed to evaluate maximum recommended crest advance rates (how fast to build) compared with allowable crest displacement rates (how fast can the embankment crest displace without creating progressive crest instability) as a function of embankment slope height. Based on the site experience available at the time, a critical crest-displacement control rate of 0.5 m per day was used to help define the "site-specific maximum crest displacement rates" versus "slope height" relationship. The data suggested that if velocity exceeded approximately 0.5 m per day, a progressive crest displacement episode was more likely to occur.

The results of the analysis concluded that there were two general relationships for the site, which depend primarily on the foundation conditions. For placement over a poor foundation (steep natural foundations or soil deposits in a valley bottoms), a lower crest advance rate was obtained compared with placement over a buttressed toe (buttressed against a natural bedrock side slope, preexisting overburden embankment slope or placement platform). The site-specific experience compared with the relationship developed in British Columbia is shown on Figure 36.11. It is important to note that application of this relationship in different geotechnical conditions could be inappropriate.

36.6.4 Operational Considerations and Trial Embankment Experience

The mine area, in terms of geography and topography, presents one of the world's most challenging environments for open-pit mining. In fact, early mine planning studies suggested that the maximum rate at which the Grasberg pit could be mined would be geotechnically and operationally limited by the ability to place the required overburden volumes. The major

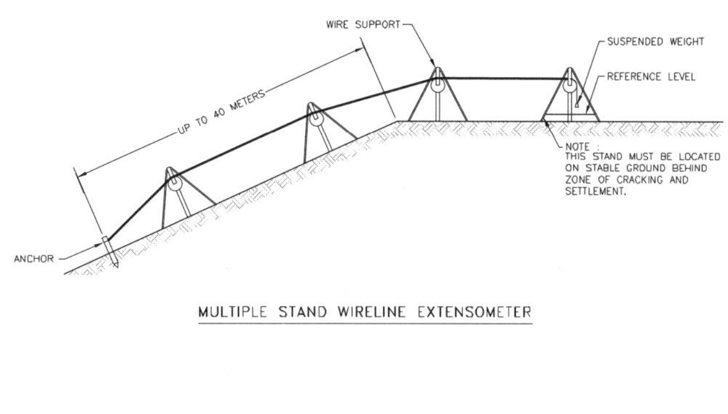

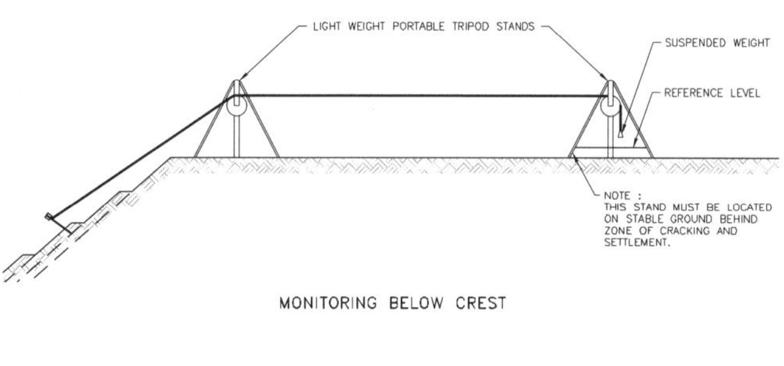

FIGURE 36.10 Types of wire-line extensometers for crest displacement monitoring

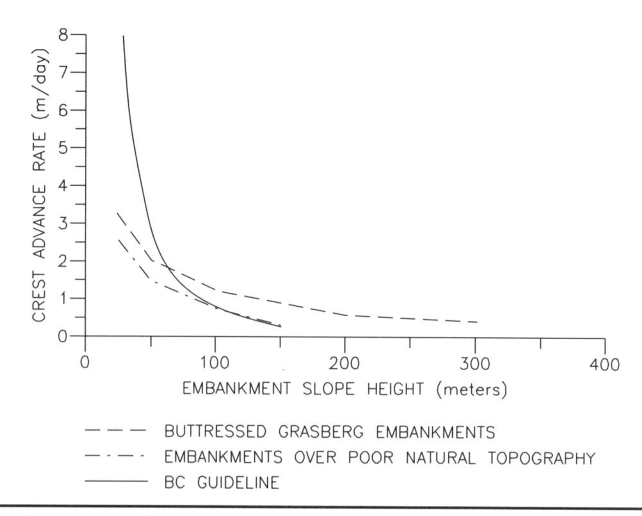

FIGURE 36.11 Site-specific embankment crest advance rate relationships compared to the British Columbia experience

operational consideration for the initial Carstenzweide embankment development included the fact that overburden was originating from pit elevations over 300 m higher than the valley floor. This created operational access restrictions for placing the initial lift across the valley floor via large haul trucks. Therefore, an operational objective of early overburden placement was to establish an embankment for production truck haulage that descended along the far north side of the Carstenzweide Valley. This operational decision set the stage for a trial embankment program guided geotechnically by the application of Terzaghi's observational method. In the early to middle 1990s, angle-of-repose embankments approximately 150 m in vertical slope height would be constructed over steep rock foundations, with the embankments toeing out over the soft soils of the Carstenzweide Valley. Geotechnically, the high trial embankments were considered susceptible to instability. To manage the potential risk to downstream infrastructure, runout distance relationships (Zavodni et al. 1984) were used, which provided a margin of comfort that any embankment instability would be far enough away. Geotechnical monitoring systems were installed to manage the risk to embankment placement operations. As another safety precaution, trucks usually dumped short of the crest, and dozers with spotters were used to push material over the crest.

A number of embankment crest instabilities were documented for these high embankments, leading up to a large instability that occurred in January 1997. The Carstenzweide embankment monitoring systems worked very well at detecting instability prior to more rapid crest displacements; relatively safe operational conditions were attained for personnel and equipment by limiting or relocating overburden placement based on the geotechnical monitoring. The observational record for Carstenzweide embankment operations and the associated instability history from two well-documented crest displacement events are detailed in the following sections, with geotechnical interpretations of cause and effect.

Carstenzweide Crest Displacement of 23 July 1996. A crest instability involving approximately 8,000 tonnes occurred in mid-1996 on a high-angle-of-repose embankment (vertical slope height of 140 m) in north Carstenzweide. Due to placement platform closures in other areas and a short haul distance to the north Carstenzweide area, the operational limits of maximum crest advance rate were tested for a poor soil foundation (soil thickness estimated to be <15 m). Crest advance rates of approximately 2 to 3 m/d resulted prior to instability over a steep natural side slope (approximately 25°) with toe advance over the Carstenzweide soil foundation. Figure 36.11 would suggest that the average maximum crest advance rate over the poor natural foundation should be less than approximately 0.7 m/d.

The instability formed in an area where the crest advance was concentrated in such a way that an unconfined nose was developed. The crest broke back approximately 5 m, defining a spoon-shaped sliver failure that developed into a flow slide; runout from the prefailure toe to the postfailure toe was estimated at approximately 90 m.

Due to the rapid crest advance in this area, direct crest displacement monitoring was not possible since there was no room to work around monitoring equipment. Instead, geotechnical staff personnel were assigned as spotters to visually monitor placement. The nearest wire-line extensometers (approximately 200 m to the east) indicated that crest displacement rates were less than 0.12 m/d for the day prior to the crest displacement. Some creep in a foundation prism station was also detected. The instability was reported to develop fairly quickly, suggesting a brittle response in the foundation or the embankment.

The interpreted causes for the instability include rapid crest advance over a steep foundation with a weak toe condition and little crest confinement. The flow-slide behavior also suggested a possible static collapse mechanism of the embankment materials, possibly triggered by the rapid crest advance or foundation yielding. Aspects of the embankment flow-slide behavior were investigated with geotechnical laboratory index tests (sieve analysis and moisture content) of embankment materials sampled in the field along the post failure profile. The embankment material involved was a relatively fine-grained weathered oxidized diorite overburden, which forms a nonsegregating habit down the angle-of-repose slope. The lab testing indicated that the overburden classified in the USCS system as clayey sand (SC-SM) to clayey gravel (GP-GC), with sand contents (−2-mm particle size) between 20% to 40%. Nonsegregating embankment materials with these types of gradations have been recognized in British Columbia experience to be associated with static embankment collapse mechanisms (Dawson and Morgenstern 1995).

Carstenzweide Crest Displacement of 16 January 1997. A relatively large crest instability involving approximately 2.5 million tonnes occurred in early 1997 (Julhendra 1997). The instability appears to have been initiated due to placement on a high-angle-of-repose embankment (vertical slope height of 110 m) in the northeast corner of Carstenzweide Valley. Operational placement was most active here, and two adjacent platform areas to the west, which shared a common soil foundation, were also affected. The placement area just to the west was closed, while the placement area further west was open for limited placement rates of less than 10,000 tonnes/d.

Operationally, the limits of maximum crest advance rate were tested again with crest advance rates over approximately 1.5 to 3 m/d for three days prior to rapid crest instability. Similar foundation conditions exist here (25° side slope with toe advance over soil foundation). Figure 36.11 would suggest that the average maximum crest advance rate over the poor natural foundation should be less than approximately 0.8 m/d.

The crest broke back approximately 20 to 55 m along the three placement platforms, defining a general overall instability that developed into a flow-slide event; runout from the prefailure toe to the postfailure toe was estimated at over approximately 110 m. Large translational embankment instability was reported to develop fairly quickly, with the embankment material forming a flow-slide event.

Visually, embankment toe bulging and associated foundation spreading was observed. Monitoring systems were in place, including crest wire-line extensometers and EDM prisms in front of the toe. Natural foundation EDM prism stations showed accelerating horizontal- and vertical-heave displacement components four days prior to more rapid embankment crest translations. The wire-line extensometers in the area indicated that average crest displacement rates were approximately 0.5 to 0.6 m/d prior to the rapid crest displacement. Placement operations were halted prior to rapid crest translation due to accelerating displacements recognized with the monitoring program.

The interpreted causes for the instability include general foundation yielding, which initiated due to rapid crest advance in the northeast placement area over a steep foundation with a weak soil foundation at the toe. Relatively high rains for four previous days prior to the event may have played a role. Also, embankment placement of 10,000 tonnes/d on the western placement platform (slope height over 145 m) may have contributed to overstressing the foundation soils. Finally, relatively fine-grained and wet overburden was suspected to be involved in the failure plane. Wet and fine types of mine overburden have generated static collapse hazards elsewhere (Dawson and Morgenstern 1995).

Experiences. Key geotechnical and operational experiences through the early observational period of Carstenzweide embankment placement resulted in the following findings:

1. It was decided to advance the Carstenzweide embankment with staged construction in lifts that ascend from the valley floor.
2. Visual monitoring of overburden placement activities should always be supplemented with aggressive use of crest monitoring. Poor (soft or deep) foundations increase the need for foundation monitoring systems.
3. Experience was gained with the concept of crest advance rate as a guide to assisting embankment construction management (recognizing that crest advance rate alone is not capable of explaining all the intricacies of Mother Nature).
4. Geotechnical indications suggested that sandy, nonsegregating overburden types (−2-mm sand contents between 20% and 40%) should be handled carefully (strategies, such as zoning, mixing, and low lift construction could be evaluated when possible).
5. Embankment displacement interaction was experienced between multiple placement platforms over a common, soft soil foundation.
6. Back-analysis confirmation of feasibility foundation strengths was provided for the "loose surficial" soil package.
7. A lower bound estimate of embankment shear strength was provided (C = 1.0 tonne/m^2, phi = 31.2° to 33.5°).

36.7 SUMMARY AND CONCLUSIONS

Foundations for the Carstenzweide overburden embankment should perform well under static conditions for controlled, ascending staged construction. A large seismic event may be capable of inducing some ground motions in the embankment or foundation toe area. Based on the technical work completed to date, permanent mine infrastructure has been restricted for distances within 275 m of the embankment toe in respect of the liquefaction run-out safety zone. Plans to provide for partial foundation dewatering, at least, are also being evaluated.

Uncertainty in the natural environment produces a degree of uncertainty in any geotechnical stability analysis (regardless of how well geotechnical engineers think that a particular site is understood). One must first wrestle with all the inherent ambiguity in geology, hydrology, and other natural events such as earthquakes. Then there is the uncertainty in investigating (e.g., mapping and drilling), sampling, testing, and predicting the spatial location and reoccurrence of these parameters. Finally, all parameters must be physically modeled in the environment of a simplified stability analysis. The observational engineering method in the framework of a mine-and-monitor approach provides a basic operational and geotechnical tool for managing the risks in the mining environment while measuring engineering performance over time.

Basic application of geology, hydrology, and geotechnical and mine engineering, tempered by the engineering observational method, is capable of providing relatively safe operational mining conditions as well as valuable design tools for the efficient development of a meaningful and economically optimized mine plan. Since experience with geological and geotechnical ground conditions increases with time and given the dynamics of daily operations versus short-to-long-term mine planning concerns, it is important to understand that mine design is a very iterative process, which includes investigation, prediction, operation to minimize risk, observation (monitoring), and adjustments over time as required.

Geotechnical monitoring of overburden embankments historically has been extremely effective at the Grasberg operation. Embankment construction has been controlled, while protecting personnel and equipment in a natural setting that is considered to be very physically challenging to mining activities.

36.8 REFERENCES

American Society For Testing and Materials (ASTM). 1988. *1988 Annual Book of Standards*. eds. R.A. Storer et al., Vol 04.08 *Soil and Rock; Building Stones; Geotextiles*. PCN: 01-040888-38. Easton, Maryland: ASTM.

Dawson, R.F., and N.R. Morgenstern. 1995. Liquefaction Flowslides in Rocky Mountain Coal Mine Waste Dumps, Phase 3. *Final Report to CANMET*. SSC File No 23440-3-9135/01=XSG. University of Alberta.

Glass, C.E. 1981. Influence of Earthquakes on Non-Impounding Mine Waste Embankments. In *Design of Non-Impounding Mine Waste Dumps*. ed. M.K. McCarter. New York: AIME.

Head, K.H. 1986. *Manual of Soil Laboratory Testing*, Vol. 3. New York: John Wiley.

Holtz, R.D., and W.D. Kovaks. 1981. *An Introduction to Geotechnical Engineering*. eds. N.M. Newmark and W.J. Hall. Englewood Cliffs, New Jersey: Prentice-Hall.

Hynes, M.E. and A.G. Franklin. 1989. Overburden correction for blow counts in gravels, wind and seismic effects. In *Proceedings 20th Joint Meeting of the U.S–Japan Cooperative Program in Natural Resources Panel on Wind and Seismic Effects*, NIST SP 760. U.S. Department of Commerce.

Julhendra. 12 June 1997. *Bali Failure Report*. Internal Memorandum. PT Freeport Indonesia (Internal Company Memorandum describing the Carstenzweide crest displacement of 16 January 1997).

Ladd, R.S., J.R. Dobry, P. Dutko, F.Y. Yokel, and R.M. Chung. 1989. Pore-water pressure buildup in clean sands because of cyclic straining. *Geotechnical Testing Journal*. 12:1:77. ASTM.

Leps, T.M. 1970. Review of shearing strength of rockfill. *J. Soil Mechanics and Foundations Division, Proceedings of the American Society of Civil Engineers*. 96(SM4):1159–1171.

Marcuson, W.F., M.E. Hynes, and A.G. Franklin. 1990. Evaluation and use of residual strength in seismic safety analysis of embankments. *Earthquake Spectra*.

McCarter, M.K. 1981. Monitoring stability of waste rock dumps. In *Design of Non-Impounding Mine Waste Dumps*. New York: AIME.

Ministry of Energy, Mines and Petroleum Resources (MEMPR). 1991. Mined Rock and Overburden Piles—Interim Report. In *Operating and Monitoring Manual—Interim Guidelines*. Prepared for the British Columbia Mine Dump Committee on Mined Rock and Overburden Piles.

Patwardhan, W., K. Sadigh, E. Idriss, and Young. 1978. Attenuation of Strong Motion Effect of Site Conditions, Transmission Path Characterisitics, and Focal Depths. submitted to the *Bull. Seismological Soc. America*.

Peck, R.B. 1969. Advantages and limitations of the observational method in applied soil mechanics, 9th Rankine Lecture. *Geotechnique*. 19:2:171.

Perloff, W.H. 1975. Pressure distribution and settlement. In *Foundation Engineering Handbook*, eds. H.F. Winterkorn and H.Y. Fang. Chapter 4. New York: Van Nostrand Reinhold.

Seed, H.B., and I.M Idriss. 1982. *Ground Motions and Soil Liquefaction During Earthquakes*. Monograph series. Berkeley, California: Earthquake Engineering Research Institute.

Seed, H.B., F.I. Makdisi, and P. de Alba. 1978. Performance of earth dams during earthquakes. *J. Geotechnical Engineering*. 104:7:967.

Seed, H.B., K. Tokimatsu, L.F. Harder, and R.M. Chung. 1984. *The Influence of SPT Procedures in Soil Liquefaction Resistance Evaluations*. Report No. UBC/EERC-84/15. Berkeley, California: University of California Earthquake Engineering Research Center.

Seed, R.B., and L.F. Harder. 1990. SPT-based analysis of cyclic pore pressure generation and undrained residual strength. *H.B. Seed Memorial Symposium*. Berkeley, California: University of California.

Sykora, D.W., J.P. Koester, R.E. Wahl, and M.E. Hynes. 1992. Post-Earthquake Slope Stability of Two Dams with Liquefied Gravel Foundations. In *Procedures Stability and Performance of Slopes and Embankments II*. Berkley, California: ASCE.

Wang, W. 1979. *Some Findings in Soil Liquefaction*. Beijing, China: Water Conservancy and Hydroelectric Power Scientific Research Institute.

Wright, S.G. 1986. *University of Texas Analysis of Slopes—Version 2*. (UTEXAS2). A computer program for slope stability calculations. Geotechnical Engineering Software GS 86-1. Austin, Texas: The University of Texas at Austin Civil Engineering Dept.

Zavodni, Z.M., J.D. Tygesen, and S.C. Pereus. 1984. *Open Pit Mine Rock Dump Geotechnical Evaluation. International Conference on Case Histories in Geotechnical Engineering*, ed. S. Prakash. Vol. 4. Rolla, St. Louis: University of Missouri.

CHAPTER 37

Construction and Operation of a Major Mined-Rock Disposal Facility at Elkview Coal Corporation, British Columbia

Brent K. Zeitz*

37.1 INTRODUCTION

Elkview Coal Corporation is a large-scale open-pit mining operation situated in the District of Sparwood, British Columbia (Figure 37.1). The coal mining history in the Kootenay Region of the province extends over 100 years. Large-scale surface mining in the region began in 1969 at the Elkview property.

Several owners have exploited reserves on the mine site. Elkview Coal Corporation is a wholly owned subsidiary of Teck Corporation of Vancouver, British Columbia. Teck purchased the property after the former owners declared bankruptcy in 1992. Elkview has a projected sales rate of more than 3.5 million clean metric tonnes of coal for the year 2000.

Waste rock removal involves drilling, blasting, loading, and hauling waste rock to uncover up to 26 coal seams. Currently, mining is focused in the Elk Pit. Overburden handling involves around 8 m^3 of spoil (20.0 t) per tonne of product coal.

With such a large quantity of overburden, rock haulage consumes a large proportion of the effort in the mining sequence. Efficiency of the mining operation dictated the needs to create a mined-rock disposal site centrally located and at an elevation close to that of the mining bench. The Bodie mined-rock pile (Bodie) halves the haulage distance to the alternative mined-rock pile in use at the time of Teck's acquisition of the Elkview property. The proposal to develop Bodie was made to the government of British Columbia in 1993. This started a large program of public information meetings and comprehensive reviews, and it ended with approval to construct Bodie in early 1996.

Bodie is situated on the southwest-facing slope of Michel Valley. This valley connects further to the east with Crowsnest Pass. The continuity of low valleys made this location ideal for the construction of infrastructure. Highway 3 and the southern route of the Canadian Pacific Railway are located in this valley. Numerous public utilities are also located in the valley, including a gas pipeline, telecommunications lines, and hydroelectric lines for two provincial utility companies. Figure 37.2 is a photo of the Michel Valley.

Due to the severity and unacceptable consequences of failure of a waste-rock structure above this public and private infrastructure, extensive consultation regarding the embankment construction was made. Golder Associates of Burnaby, British Columbia, was retained to consult on the construction techniques for the facility. After four years of work, Bodie was permitted on February 26, 1996.

As a result of the planning process, several areas requiring attention were identified. In order for the mined-rock pile to develop in an orderly fashion, foundation treatment, rollout protection, instrumentation, water management, orderly dumping plans, operations auditing, and reclamation had to occur.

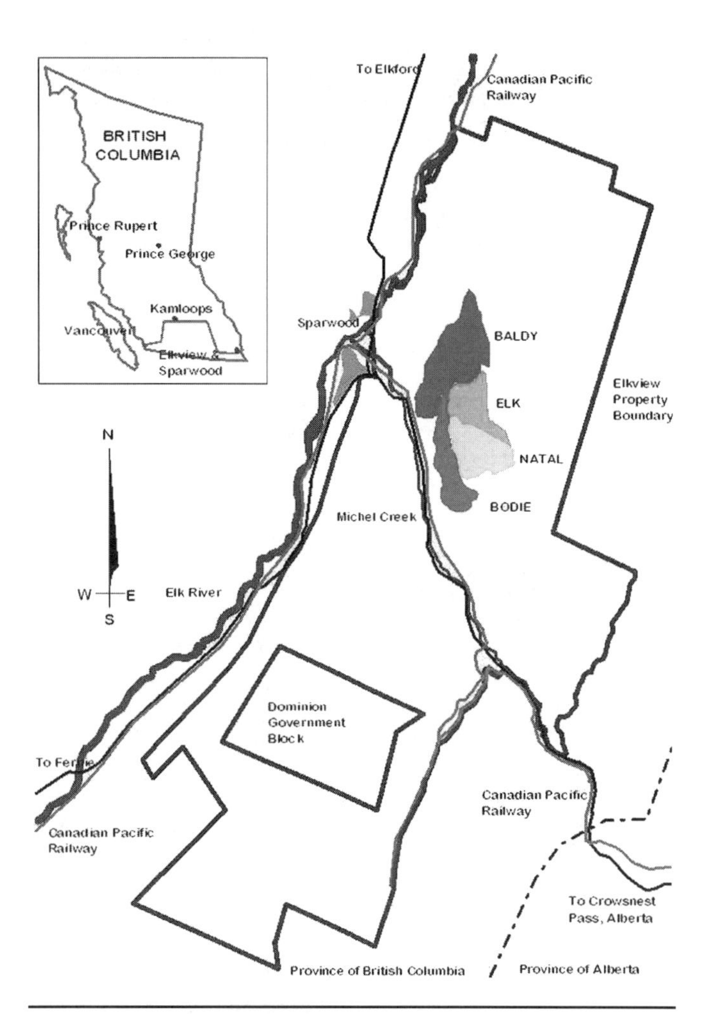

FIGURE 37.1 Location map

37.2 CONSTRUCTION PLAN

37.2.1 Foundation Treatment

In anticipation of regulatory approval, foundation treatment under the footprint of Bodie was planned to commence in March 1996. Embankment construction was suggested to take place in a bottom-up order. The lowest elevation of the structure (1,370 m elevation) was to be placed on a prepared foundation, with subsequent dumping "stacked" on the previous lift. Soils and weak

* Elkview Coal Corporation, Sparwood, British Columbia, Canada.

FIGURE 37.2 Michel Valley and Bodie embankment from Highway 3

foundation materials identified in four years of work were systematically identified and removed from the footprint to prepare for the first lift of material.

Bodie is nearly 2,500 m long and crosses three drainages. Each of these drainages was a steep valley trending to the Michel Valley in the west. Glaciation along the Michel Valley deposited large quantities of lacustrine clay and silt in the recesses along the valley walls. About 480,000 m^3 of clay and silt were removed in these drainages. This work involved an additional 260,000 m^3 of earthwork to establish access. Clay and silt were removed from the footprint to other areas on the mine site. In some instances, this material was used to reshape and reclaim other features on the mine site. Recovery of 80,000 m^3 of topsoil was also required as part of the foundation treatment. This soil material was retained for reclamation.

As dumping proceeds upslope onto undisturbed ground, soil recovery is completed to remove the weak layer and expose glacial till. Removal of the soil cover too far in advance could deteriorate the stiff glacial till foundation that is under a majority of the embankment footprint. Soils are used to cover the reclaimed ground following only a few hundred meters behind the advancing face. Timber is cut and sold if marketable. Otherwise, uneconomic timber is broken down and placed with soil on the rounded and resloped portions of the embankment face.

In May 1998, the initial 1,370-m lift was completed to establish the foundation of the upper dumping platforms. Additional lifts on the foundation are currently underway. The second pass (1,440 m elevation) is near completion, and the third lift (1,485 m elevation) follows behind.

37.2.2 Rollout Protection

Nearly the entire length of the embankment is protected by rollout protection berms. Berm construction began in the summer of 1995. The berms are about 3 m tall. Berming was needed to protect the infrastructure located below the embankment, which included the access road, power lines, a parking lot, and the security gate. To date, no rolling rocks have traveled far enough to reach this structure. Primary rollout protection, rock berms placed along the embankment toe, has stopped all rock movements on the face of the structure.

The rollout berms were designed to utilize old tires scrapped by the haulage fleet. A typical berm is two tires stacked on their sidewalls and covered with soil. Additional berm structures were placed along the towers carrying hydroelectric lines immediately below the structure.

37.3 INSTRUMENTATION AND MONITORING

Golder Associates recognized early in the predevelopment stages that large a number of instruments would be required to monitor the performance of the embankment. Personnel from Elkview and Agra Earth and Environmental implemented Golder's recommendations for inclinometers and piezometers before construction.

37.3.1 Inclinometers

Three 8.5-cm-outside-diameter vertical inclinometers were initially installed along the western (downslope) toe of the embankment. These tools range in depth from 25 to 80 m. The inclinometers are drilled through bedrock and the first major coal seam that occurs under the structure. Inclinometers were recommended to monitor the different stratigraphic features for shearing along their contacts.

Monitoring of these instruments occurs on a monthly basis, and to date there has been no deformation along their length. Measurements are compared to measurements at the toe of the hole. Cumulative deviations are calculated from an arbitrary zero. This zero point is established at the toe. Measurements are taken in ½-m intervals along two faces from the toe to the collar. Our torpedo-style tool measures both faces simultaneously. Inversion of the tool doubles both sets of angles and checks the precision of the measurements taken with respect to zenith. As of May 2000, no deformation in the foundation has been detected with these tools.

Two additional inclinometers were added after development of the 1,370-m platform. These tools were to measure a specific area. Other monitoring, including piezometric data and motion monitoring, dictated the need for the additional inclinometers to monitor for shearing in the rock slope and foundation.

37.3.2 Piezometers

Piezometers were installed to monitor the pore pressure under the embankment during construction. Pneumatic piezometers were placed specifically in the drainage basins along the topography. Standpipe piezometers could not be used in most areas of application, since they were needed under the structure. Forty-seven pneumatic piezometers were installed under the embankment in total. The pneumatic tools were placed in existing standpipe piezometers and in newly drilled Becker Hammer holes.

Installation of all the pneumatic piezometers started with the installation of a standpipe piezometer. Elkview used slotted 2.54-cm polyvinyl chloride piping, with a 1-m length at the target elevation.

The borehole around the slotted section of the PVC piping was packed with #9 environmental media (coarse sand). Installation proceeded as follows:

1. Three 0.25-m bentonite seals were made in the borehole, interspersed by 0.25 m of coarse sand.
2. The balance of the hole was refilled with drill cuttings if fine enough, or with pit run sands. In several instances, multiple installations or a "nest" were made. If there was a second or third piezometer for installation in the borehole, a bentonite seal was created below the target elevation, and the previously described installation took place. All the installation of the standpipe piezometers took place through the drill string of the Becker Hammer rig.
3. Pneumatic tools were mounted in the standpipe. The pneumatic tools used included a simple diaphragm in a housing, with two tubes to deliver and receive nitrogen gas from the measuring gauge. The diaphragm assemblies were lowered to the target elevation with the leads. The leads of the pneumatic tool were then strung to a central location outside of the Bodie footprint. Trenches used during the excavation of the foundation were utilized to bury

pneumatic piezometer leads to boreholes. Encasing them in 7.6-cm Schedule 80 PVC lines along their entire length provided additional protection of the leads.

Prefabricated sheds were used to house the measurement connections of the pneumatic tools.

The number of piezometers required under the structure was less than the 47 installed. This redundancy was important as dumping progressed. Although the collars of the drill holes were cemented after installation of the pneumatic tools and buried nearly three meters under the ground, there was a significant failure rate. Failure was attributed to damage as rock rolled down the face. Twenty-two of the original 47 pneumatic piezometers continue to operate in May 2000. These tools are monitored at least monthly. Additional readings are taken after significant rainfall and during spring runoff.

Observations of phreatic levels are compared to precipitation events and the advance of mined rock dumping. Occasionally, the piezometers have indicated increases in water pressure during periods of high runoff and during initial loading by overlying lifts. Generally, the increases are small in magnitude and duration.

37.3.3 Motion Monitoring

Elkview uses two different systems to monitor motion of the embankment. Wireline extensometers and remote monitoring are extensively used for day-to-day operation.

37.3.4 Wireline Extensometers

The simplest method in use to monitor motion on the embankments is by the use of wireline extensometers. Wirelines are set up and maintained by the Mine Operations Department. On an active crest, wirelines are established on 100-m centers or over cracks as they develop. A typical wireline setup will use six parts.

1. A tripod is established on the platform of the dump.
2. A pin is hammered into the crest of the dump.
3. A wire is strung from the pin, back to the tripod. This may or may not cross cracking on the dumping platform.
4. The wire is passed over a clothesline wheel, to hang in the center of the tripod.
5. A heavy weight is used to tension the wire, hanging in the center of the tripod.
6. Measurements are made on a meter stick suspended in the center of the tripod.

The shift supervisors and the mine dispatcher track the data. Measurements of the weight's displacement are taken and called into the dispatch office. After entering the displacements and time of measurement, the mine dispatcher can review the data and report the velocity of the crest.

On the Bodie dumping platforms, movement rates are calculated for each wireline every 4 hours. A danger criterion has been set for movement rates as follows: any movements that exceed 2.5 cm/hr require the closure of the dump crest in the affected area. This motion rate is quite conservative and half of the rate used on other mined rock piles at Elkview. Once a portion of the crest is closed, dumping may not resume until motion rates are less than 2.5 cm/hr for 12 consecutive hours. Close spacing of the wireline extensometers allows better definition of the affected area and reduces the length of crest affected by a closure.

Monitoring is reviewed on a daily basis by the geotechnical engineer to pinpoint areas that may be in distress. As a rule, the loading rate on dumping platforms (loads per meter of crest length) is reduced in areas where the wirelines exceed 1.5 cm/hr to avoid closure of the crest in the affected area.

Automated wirelines have recently been added to crest monitoring. These complement the manual monitoring done by the shift supervisors. The automated wirelines are part of the Modular Mining Systems Dispatch hardware. They radio in directly to the dispatcher. The mine dispatcher monitors the position of these wirelines as well.

37.3.5 Remote Monitoring

While wireline extensometers are a good tool for checking the motion of the crest relative to the platform of the dumping area, total embankment motion has become an important tool in management of the facility. Remote measurement from a static reference can detect larger scale instabilities in the structure.

Two systems are in use for total motion monitoring at Elkview. Both land-based surveying and global positioning system (GPS) surveying have been implemented in determining the availability of dumping platforms.

Land-Based Surveying. Permanent survey reflectors have been installed on the embankment face. These prisms are measured from a stationary point. Experimentation with these measurements has shown that total movement rates are often higher than those indicated by readings taken from wireline extensometers.

With each measurement, the position of the point is calculated. The changes in northing, easting, and elevation are calculated from the previous position. Squaring each of these components, summing them, and taking the square root calculates the overall change in displacement. Velocities are then calculated from this change in position and the time between measurements. Additionally, azimuth of motion and plunge are calculated.

Validation of the data involves the comparison of the two directions calculated, azimuth and plunge, with historic values. Due to atmospheric conditions and operator accuracy, calculated velocities can change rapidly, leading to premature closure of dumping platforms. The historical trend of the data, with respect to direction, can help to determine if there was some error present in the measurements taken.

GPS Surveying. GPS surveying was implemented during the summer of 1996 to determine total motion rates on Bodie. GPS was selected for the ease of application and cost effectiveness. Survey prisms can cost $200 per installation and are prone to disturbance and breakage by animals and operating activity on the embankment.

Survey stations are established on the slope using nails or 30-cm-long survey stakes. Similar to the land-based surveying procedure, the position of the point, the change in position, velocity, and direction of motion are calculated. GPS surveying has been difficult at times, due to variability of the satellite constellation. Times of low precision and inadequate satellite coverage have been experienced. These conditions often lead to error in the calculated position. Errors can be determined by comparison of the direction with historical trends in the motion.

Two distinct advantages of GPS surveying are as follows:

1. A walking inspection can also be completed during the process of measurement. Subtle changes in the terrain, cracks, and bulges are traversed with each set of measurements.
2. Surveys can be completed in periods of low visibility. Inclement weather and darkness are not a limitation for monitoring of motion.

Closure rates for total motion were suggested to be more conservative than the motion rates at the crest (2.5 cm/hr). A closure criterion is reached at a movement rate of 0.75 cm/hr. The more conservative rate was justified since these rates are indicative of larger-scale instabilities in the structure.

Confirmation of the data has been an issue with the application of these techniques. Both measurement techniques are used and are often compared to dispute the validity of suspect measurements. Additionally, maintenance of both techniques provides continuity of measurements when unavoidable conditions prohibit using one method or the other.

37.4 WATER MANAGEMENT

The three watercourses crossed by the embankment were maintained by the construction of segregated coarse-rock drains. Adequate rock drains were necessary for the overall stability of the structure. Poor drainage causes water pressure to build up under the embankment. Golder Associates recommended the construction of segregated rock drains with a minimum dumping height of 50 m. This height ensured that the coarsest rock rolled down to the drainage at the base of the rock slope. As experience was gained, this minimum height was adjusted to 30 m. Waste rock classification at the shovel was used to ensure material of appropriate size arrived for drain construction. Predominantly sandstone fragments greater than 0.3 m in diameter were used in the construction of the drains.

37.5 OPERATIONAL PLANS

37.5.1 Materials Classification

Five material classifications were determined to direct the placement of waste rock for Bodie. Simple classifications, ranging from high quality through poor quality, were based on rock type and particle size. Examples of rock quality are provided in Figure 37.3.

1. Poor quality–topsoil, clay till, and fragmented mudstone
2. Medium-poor quality–highly fragmented sandstone/siltstone/mudstone with fines
3. Medium quality–fragmented sandstone/siltstone with trace fines, particle size 0.1 m
4. Medium-high quality–sandstone and siltstone, particle size 0.2 to 0.3 m
5. High quality–largely sandstone, particle size 0.3 m

Material type was determined before loading the pattern from geophysical logging. A recent advent was the use of rock recognition technologies. Drill parameters such as pull-down force and bit rotation are used to fingerprint the various materials intercepted in the blasthole.

Geologists, geotechnical engineers, and technicians interpret the data generated from the geophysical logging and the rock recognition systems. Interpretation of this information and issuing rock-quality plans was critical in the construction of Bodie. The geotechnical engineer was responsible for assigning rock qualities at each shovel face and the appropriate location for each material type in the embankment.

The first lift of the embankment (1,370 m elevation) advanced a 100-m-wide rock spoil over the prepared foundation from north to south. This left an area to the east where lower-quality material would be stored between the high-quality rock on the outside shell of the structure and the original topography. High-quality material was also used to construct coarse rock drains to cover and protect the watercourses on the hillside.

The transfer of information from Engineering to Pit Operations was key to the safe management of the facility. Geotechnical Engineering performs this task twice weekly. Dumping plans are sketched with instructions regarding material type, loading rate, extensometer placement, and grade and direction control. Organizational meetings and written reports coincide with the distribution of the dumping plans.

37.5.2 Operations Auditing

Government permit requirements for Bodie dictated project auditing to ensure quality control. These audits have provided an opportunity for continuous improvement, providing feedback to Engineering and Pit Operations on the success or shortcomings of the project. Until the lowest platform (1,370 m elevation) was completed, quarterly external audits and monthly internal audits were performed. Internal audits involved the geotechnical engineer, the mine general foreman, and a representative of the Joint Occupational Health, Safety, and Environmental Committee. The

Poor

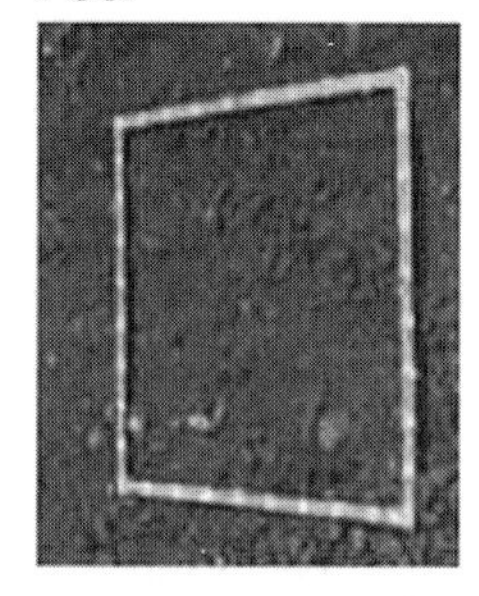

Medium-Poor

Medium

Medium-High

High

FIGURE 37.3 Rock quality

FIGURE 37.4 Elk herd grazing on reclaimed slopes of Bodie

external audits involved these same representatives from Elkview, the British Columbia Ministry of Energy and Mines, and independent consultants. Golder Associates, the principle consultant for the project, reviewed all the findings of the internal audits conducted between external audits.

Following the May 1998 external audit, the completion of the foundation treatment and development of the 1,370-m platform, all parties agreed that the frequency of external auditing could be adjusted to an annual frequency. Internal audits were also decreased in frequency, to a quarterly basis. Internal and external audits entailed the same elements:

- A review of the construction plan for the period and as-built status
- A thorough tour of the entire area of the embankment
- Interviews with operators in the pit and on Bodie to confirm their understanding of the rock quality at the workplace and required placement of material on the embankment
- Instrumentation review

The auditing process has been very successful and has been adapted for use on other projects at Elkview.

37.6 RECLAMATION

Bodie is being reclaimed as a matter of policy. Elkview Coal Corporation made a commitment to reclaim this area as soon as mining is complete. This commitment was made in recognition of the high wildlife values in the area.

Numerous elk, mule deer, and white-tail deer use Bodie and the surrounding area to graze in the winter. Figure 37.4 shows part of a resident elk herd on the revegetated lower slopes of Bodie. The combination of grass, dense forest cover, and southwestern exposure make the land highly important for ungulate species. The area is classified as Class One Winter Range.

Bodie is highly visible from the Michel Valley and the community of Sparwood. Accelerated reclamation reduces the visual impact of the development.

37.7 CONCLUSION

Bodie waste rock embankment has been an extremely important part of the development of Elkview Coal Corporation. Without Bodie, the mine would not have been viable over the long term; certainly mining would have taken different directions than the current mine plan.

Over 80,000,000 m^3 of material have been managed in this structure to date. Regular monitoring of the facility will continue during operation. This will ensure ongoing availability of the structure to service mining operations.

CHAPTER 38

Steepened Spoil Slopes at Bridger Coal Company

William L. Gerhard*

38.1 INTRODUCTION

The Jim Bridger Mine is operated by Bridger Coal Company and is located in Sweetwater County, 35 miles (56 km) northeast of Rock Springs, Wyoming (Figure 38.1). The general location of the mine site is shown in Figure 38.2.

Bridger Coal is a joint venture of the operator, Interwest Mining Co., and Idaho Energy Resources Co. The permit boundary encompasses in excess of 20,000 acres (81 km^2) and is nestled against the western side of the Continental Divide.

Production started in August 1974, and by the end of 1999 the mine had produced 146 million tons (132 million t). Annual production averages 5 to 7 million tons (4.54 to 6.35 million t) of run-of-mine coal. The Jim Bridger Power Plant, a 2,000-MW generating station, consumes coal on site, and the electricity is transmitted to the Pacific Northwest grid (Oregon, Idaho, Northern California, etc.).

Dragline stripping operations are scheduled around the clock, two 12-hr shifts/day, 7 days/week, 355 days/year (10 holidays are observed). A dragline crew works a 12-hr shift each day for four days and then has the next four days off. The remaining operations are scheduled for three 8-hr shifts/day, 5 days/week, 250 days/year. Maintenance, warehouse, drilling/blasting, and heavy equipment operators work the same schedule as the draglines.

The mine employs three draglines, one hydraulic shovel, three front-end loaders, and one hydraulic backhoe in four to five active production faces. The total number of employees at the mine is 364. Of this number, approximately 67 are in supervisory or administrative positions.

38.2 DEPOSIT DESCRIPTION AND GEOLOGY

The Jim Bridger Mine is located on the northeastern flank of the Rock Springs uplift, a broad, asymmetrical anticline feature about 60 miles (97 km) long and 35 miles (56 km) wide with a north-trending axis. This uplift separates the Green River Basin to the west and the Great Divide and Washakie Basins to the east.

Precambrian rocks on the apex of the uplift are estimated to be 17,000 ft (5,200 m) stratigraphically above the Precambrian rocks in the Green River and Washakie Basins. Normal faults with generally less than 100-ft (30-m) vertical displacement cut through the uplift. The dip of the strata on the flanks is generally 3° to 15°.

The coal seams occur in the Deadman Coal Zone of the Fort Union Formation, which is Paleocene in age. The zone is about 60 to 80 ft (18 to 24 m) above the contact with the underlying Lance Formation. The Fort Union Formation is about 1,500 ft (460 m) thick, and the Lance Formation is about 900 ft (270 m) thick. There are five coal seams exposed in the mine, they are designated D5 through D1 from top to bottom. Various combinations of these seams exist throughout the field forming single-seam, two-seam, three-seam, and four-seam areas. The seams generally dip 2° to 5° to the northeast.

FIGURE 38.1 Dragline on the spoil side

Sedimentary rocks of the Fort Union Formation represent depositional processes of a fine-grained, fluvial-flood-basin complex with extensive swampy conditions and minor lacustrine influence. Recognized depositional environments within the three-dimensional framework include poorly drained swamps, well-drained swamps, crevasse splays, and fine-grained channel sandstone deposits.

Overburden and interburden materials consist of interstratified sandstones, siltstones, and claystones, and minor shale and thin, discontinuous limestone stringers. These stratigraphic units exhibit a high degree of rock variability, both laterally and vertically.

The coal resource is classified as subbituminous. The energy content averages 9,400 Btu/lb (21,864 kJ/kg), with 18% moisture, 9.5% ash, and 0.59% sulfur. Coal leases are held by the federal government, the State of Wyoming, and the Union Pacific Railroad.

38.3 MINE DEVELOPMENT

Soil is removed from the areas prior to stripping operations. In the early years of the mine, reclaimed areas were not available for direct application of soil. Therefore, soil was placed in stockpiles at various locations. Direct-applied soil is hauled from the highwall side of the pit across or around the pit and distributed on reclaimed areas. Soil depths range from 0 to 60 in. (0 to 1,524 mm) with a field average of 15 in. (381 mm). A buffer of 300 ft (90 m) is maintained from the highwall for operational flexibility.

Once the soil is removed, the overburden material is drilled and blasted. A scraper or truck and front-end loader fleet generally

* Jim Bridger Mine, Rock Springs, Wyoming.

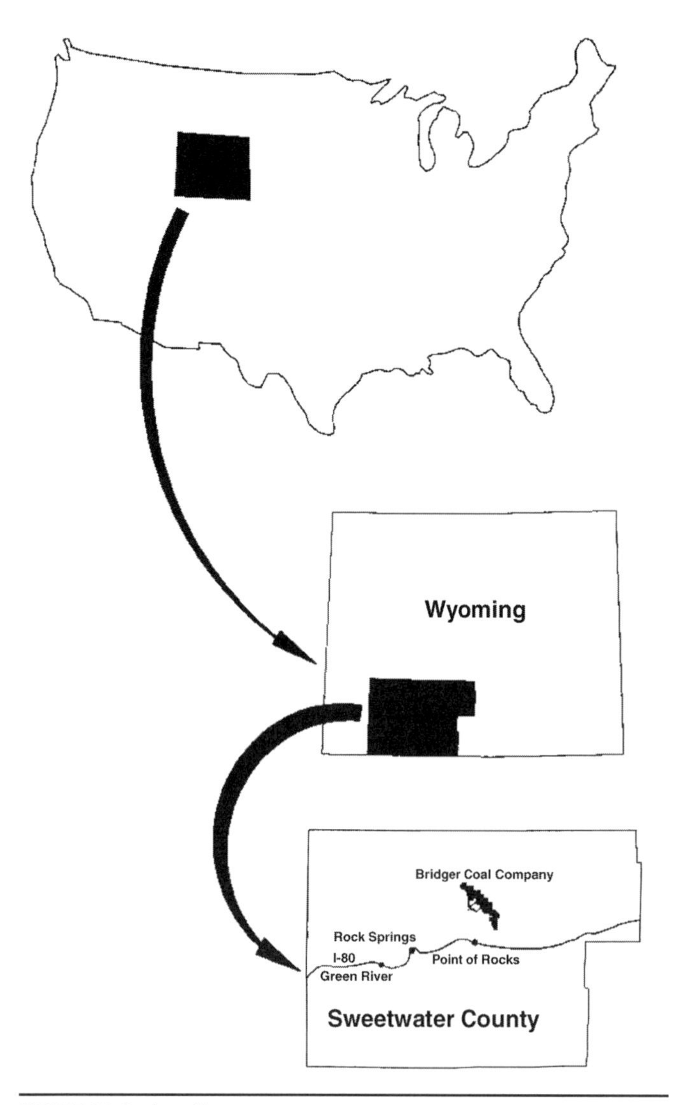

FIGURE 38.2 Location map

develop outcrop areas. The outcrop overburden material is placed to minimize out-of-pit spoil. After the outcrop is developed, the draglines remove overburden by simple side casting or extended bench methods up to approximately 150 ft (46 m) in depth. Depths greater than 150 ft (46 m) require a multiple-pass, spoil-side, dragline stripping method.

Highwall angles generally range from 50° to 90°, with an average of 70. Highwalls are presplit at 70, whenever possible, for highwall stability. The angle of repose for the spoil material is 36° to 38°.

General pit development is northwest to southeast along the strike of the coal beds. Each successive pit is farther down the dip of the coal seams. Pit widths vary greatly throughout the mine, but generally outcrop pits are 200 to 250 ft (61 to 76 m) in width. Where the overburden is less than 150 ft (46 m), the pit width varies from 150 to 200 ft (46 to 61 m). For overburden depths greater than 150 ft (46 m), pit widths ranges from 120 to 150 ft (37 to 46 m). An overall average pit width is 130 ft (39 m).

The active pit is approximately 13 miles (21 km) in length. Spread across this area are three draglines, a truck-and-shovel operation, and a scraper fleet operation. Therefore, as many as five faces are operative at any one time.

Pit access is available through a series of ramps or entries spaced at approximately 4,000-ft (1,220-m) intervals. The mine is located to the northeast of the power plant. The ramps start at the pit floor and advance perpendicular to the spoil side of the pit up a 5 to 8% grade to a haul-road network ending at one of three truck dump stations contiguous with an overland conveyor system directly linked to the power plant fuel-handling system.

Future mine development will consist of opening the remaining southern portions within the permit boundary, which will extend the pit approximately 6 miles (9.7 km). As the southern expansion is started, final reclamation will begin in the currently active area. The development philosophy minimizes the total area disturbed and allows portions of the mine to return to original livestock and wildlife land use as soon as possible.

38.4 UNIT OPERATIONS

38.4.1 Drilling and Blasting

Drilling is accomplished by four drills: two Drilltech D90K and C90KD drills (for overburden and parting), one Ingersall Rand DML 1XL (for overburden, parting, and coal), and one Schroeder Twin Mast (for coal).

Blast patterns are rectangular or square and vary with type of material and depth. An emulsion-product mix is used, which is a blend of 33% emulsion and 67% ANFO. In extremely wet areas, the product mix is raised to 50% emulsion and 50% ANFO. Powder factors average between 1.0 to 1.2 lb/yd^3 (0.60 to 0.72 kg/m^3) for cast blasting, for overburden, 0.7 to 0.9 lb/yd^3 (0.42 to 0.53kg/m^3), 0.8 lb/yd^3 (0.47 kg/m^3) for parting, and 0.4 lb/yd^3 (0.27 kg/m^3) for coal.

38.4.2 Overburden Removal

After outcrop development, draglines remove overburden by simple side casting or extended bench methods up to approximately 150 ft. (46 m) in depth. These methods are illustrated in Figure 38.3 (side casting method) and Figure 38.4 (extended bench method).

Depths greater than 150 ft (46 m) require a multiple pass spoil-side stripping method. The overburden is split into two lifts consisting of approximately 60% upper depth and 40% lower depth. The first or upper lift is drilled and stripped by simply side-casting the material into the empty pit (Figures 38.5 and 38.6).

Next the dragline maneuvers from the upper lift elevation down a ramp (cut out of the upper lift) to the lower lift elevation. Prior to cutting the ramp with the dragline, the lower material is drilled and blasted. The dragline strips a key cut that establishes the lower portion of the highwall and places that material behind the upper lift spoil.

Finally, the entire upper lift and lower lift key-cut spoils are leveled, and the dragline maneuvers onto the spoil-side pad. From this position, the dragline strips the remaining lower lift material and spoils the material in its final position.

Since mining operations occur in multiple-seam areas, supplemental stripping operations usually remove thin, 1- to 16-ft- (0.3- to 4.8-m-) thick interburden strata. In areas where the parting strata are greater than 16 ft (4.8 m) thick, the multiple-pass dragline operation described above is altered to allow the dragline to remove interburden materials.

Supplemental stripping operations consist of a truck-and-shovel fleet or a scraper fleet. These operations provide added flexibility to the mine by selectively handling toxic materials, performing advanced benching of the highwall for the dragline, or removing interburden materials in multiple-seam areas and opening pit ramps and entry ways.

38.4.3 Coal Production

Once the overburden or interburden has been removed and the top surface of the coal seam is exposed, the coal is cleaned by rubber-tired dozers, blades, or scrapers. The coal is then drilled using a twin-mast auger drill employing 5¼-in. (133-mm) bits. The material is lightly blasted to facilitate loading operations and then loaded by front-end loaders, a hydraulic shovel, or a hydraulic backhoe.

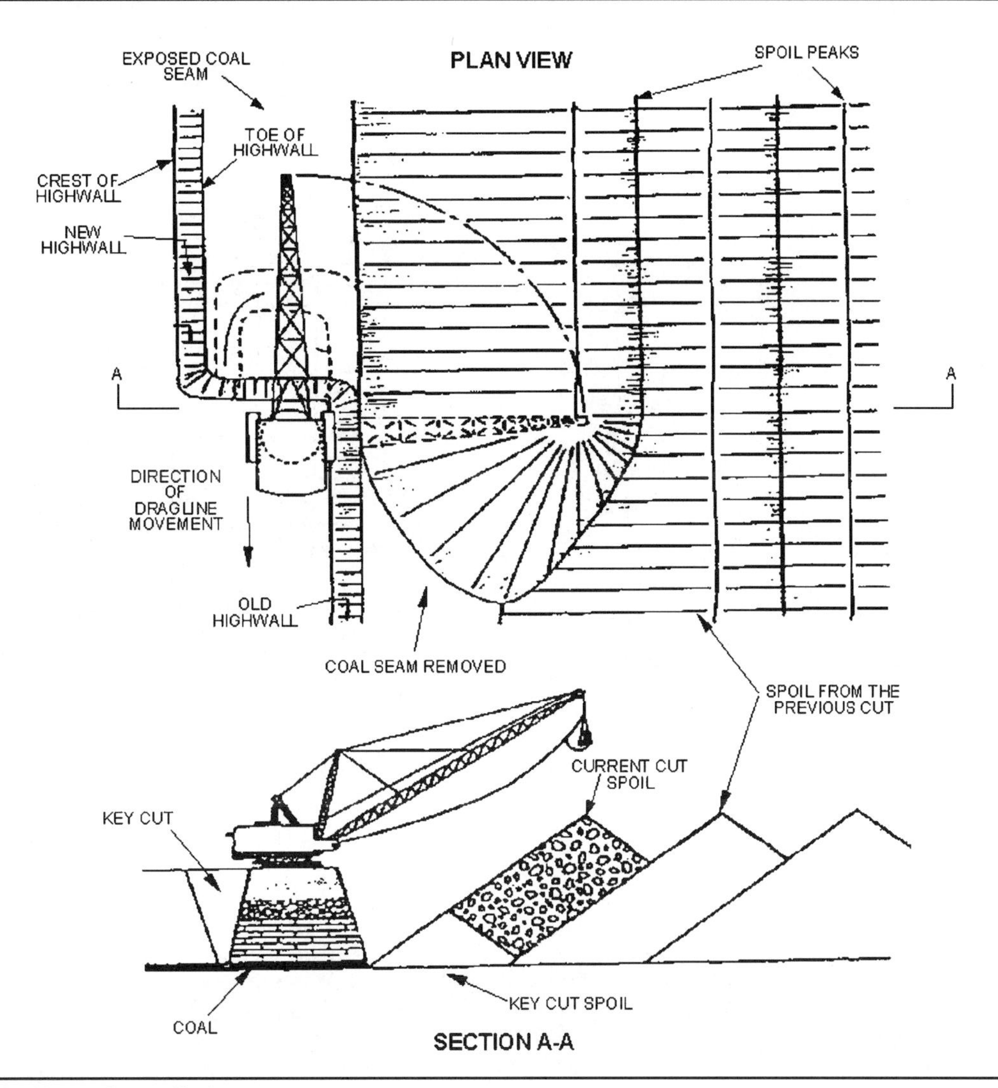

FIGURE 38.3 Simple side casting method

38.4.4 Haulage

Prior to 1989, coal was hauled 11 miles (17.7 km) from the mine to the Jim Bridger Power Plant by a fleet of thirteen 120-ton (109-t) bottom-dump trucks. Due to the length of haul, a 42-in. (1,067-mm) conveyor system was installed in 1989 to reduce the distance for truck haulage. Five 90-ton (82-t) end-dump trucks replaced the bottom-dump trucks. The trucks transfer coal a distance of approximately 4 to 12 miles (6.4 to 19.2 km) round trip from the pits to three truck dump sites feeding the conveyor system. Each truck dump site uses a feeder-breaker to size the coal to minus 6 in. (147 mm) prior to loading the belt.

The conveyor system consists of a 13,000-ft (3,962-m) main belt, an 8,000-ft (2,438-m) northern wing, and a 14,000-ft (4,268-m) southern wing. The belt runs at 825 ft/min (244 m/s) and conveys coal at a rate of 1,500 tph (1,361 t/h). Coal is discharged at the power plant's transfer building.

38.5 SPOIL STEEPENING

A spoil stability evaluation for steepened spoil slopes at the Jim Bridger Mine, Sweetwater County, Wyoming, was completed in June 1989. The study was undertaken by Bridger Coal Company utilizing the services of Seegmiller International (Seegmiller 1989). The modification of spoil placement determined to be economically beneficial to Bridger is demonstrated in Figure 38.7. Specifically, it is desired to have a steeper lower spoil slope to reduce spoil-side pullback rehandle.

The spoil piles are placed by dragline operations with some of the lower spoil zones being emplaced by blast casting (Figure 38.8). The spoil piles of interest for steepening are, generally, about three months old. Following any potential steepening, the spoil would have to remain stable for approximately 12 months. After that time, spoil instability would not affect the mining operations. Eventually, further spoil emplacement would cover up the steepened spoil zone.

The spoil stability study included (1) an evaluation of spoil density as a function of spoil depth, (2) an evaluation of spoil shear strength as a function of depth, and (3) an analysis of the safety factor and probability of failure with various steepened slopes. In addition, guidelines for steepened slope angle usage in spoil and procedures for field-evaluating potentially steepened spoil slopes were developed.

38.5.1 Geotechnical Investigation and Analysis

Three borings were placed into the spoil piles at the mine at strategic locations. The drilling was accomplished by a CME-75 Drill Rig with 9-in. (220.5-mm) diameter hollow stem augers. Samples were collected by means of a standard penetration test utilizing a 2.5-in.- (61-mm-) diameter split-barrel sampler driven either 18 or 24 in. (46 or 61 mm) by means of 140-lb (63.5-kg) hammer falling 30 in. (76 mm). Using established methods and correlating the blow count data with depth, it was possible to estimate the wet density of the spoil encountered during the

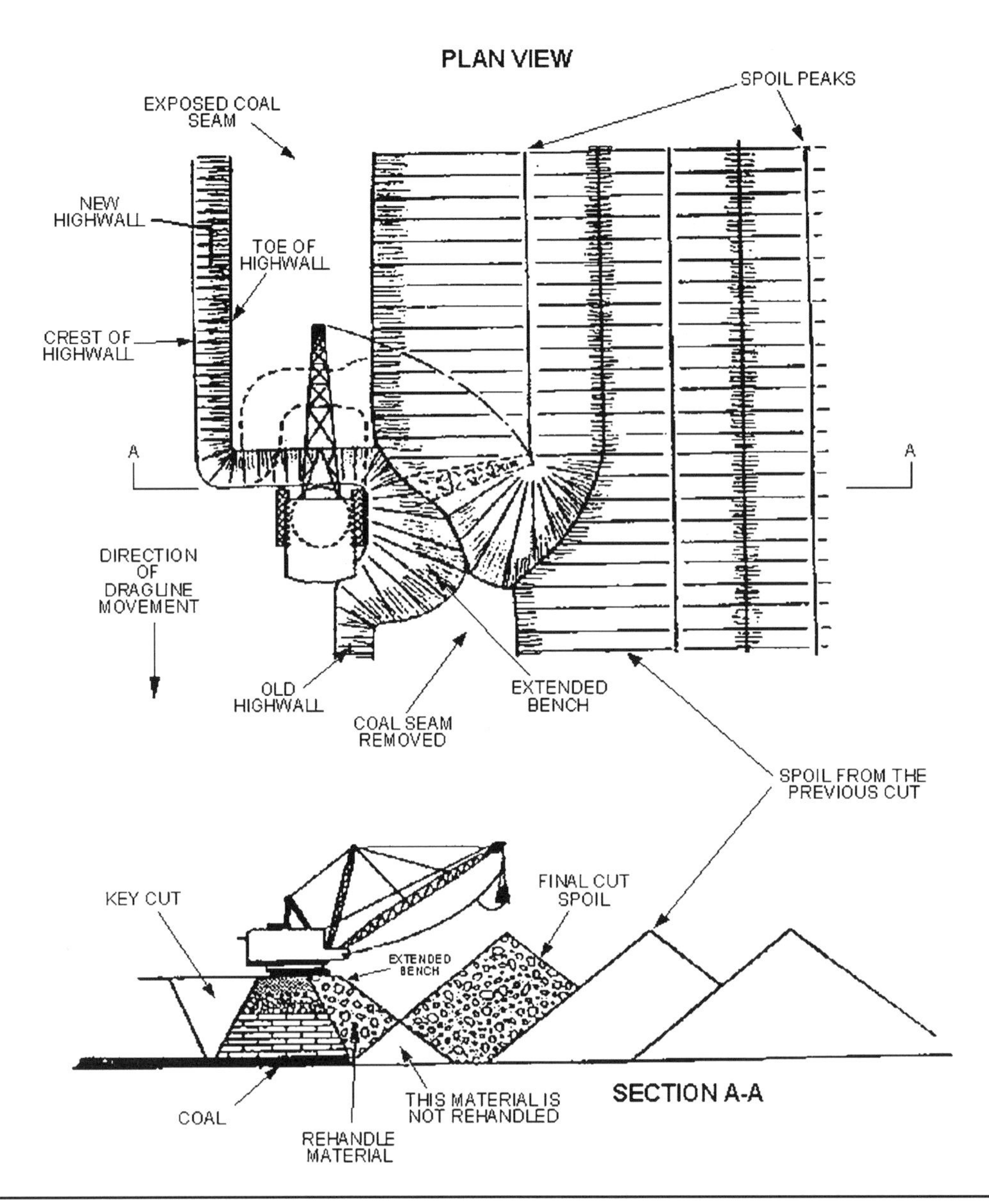

FIGURE 38.4 Extended bench method

penetration test. These data, combined with results of field density tests and laboratory direct shear tests, were used to generate the data presented as a relationship of wet density to depth.

Field density tests were conducted at six locations throughout the project to determine in-place density of the spoil-pile materials. Two methods were used to measure the in situ wet and dry densities. A Troxler 3411-B Nuclear Density Gage was the first device used. This instrument gives instantaneous results of wet density, dry density, and moisture content. As a means of correlating density as determined by the nuclear gage, six 12-in.- (30.5-mm-) sand cone tests were performed at the same locations as the nuclear gage tests. The tests were conducted in compliance with ASTM-D-1556, *Determination of In-Place Density by Means of the Sand Cone Device.* Considering the nature of the material, the way in which it was deposited, and the mineralogy, it was determined that the sand cone tests provided the most reliable information.

After reviewing the data collected during the field program, a series of mechanical analysis and plasticity tests were conducted on samples recovered from the sand cone tests. Based on a review of this information and the plots of the blow counts versus depth, it was decided to determine the gradation of the matrix material in five of the sand cone tests.

After a review of all available data, a decision was to fabricate direct shear samples on a composite of five samples collected from sand cone tests (Seegmiller 1989). This material was fabricated by combining equal amounts of −4 mesh material from each of the sand cone samples.

Because the matrix material would compact differently than the actual material containing rock greater than #4 mesh size, a corrected matrix density was used in the direct shear tests. An average rock density of 140 pcf (208 kg/m^3) with 1% water absorption was used in making this correction.

The first direct shear set was fabricated to a known wet density of 109 pcf (162 kg/m^3) before the normal loads were applied. Application of normal load compacted these samples to 135 to 137 pcf (201–204 kg/m^3) wet. This density was considered to be well in excess of the allowable weight for the matrix material, so a known weight of material was used in the test ring, and the maximum normal load was applied to determine the test envelope and the resulting wet density. This method was used for

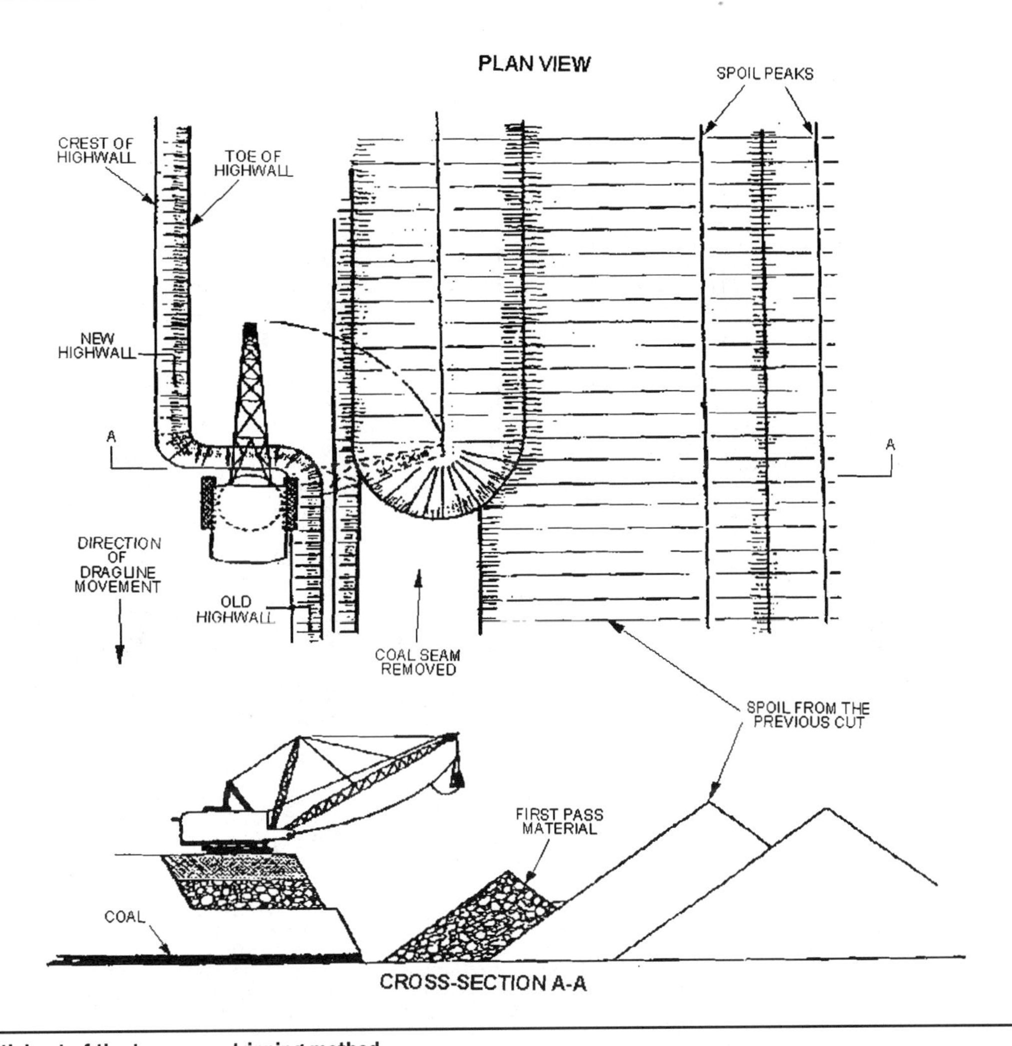

FIGURE 38.5 Initial cut of the two-pass stripping method

all other tests. Normal loads ranged from 0.711 to 11.55 ksf (34 to 553 Kpa).

Analysis of the test results on composite sample envelopes indicated that there was very little difference in strength regardless of the density or load applied. Two additional envelopes were run on the composite samples after adding 15.7% gravel to the original composite. The gravel added was –3/8 in. (–1 cm) to +8 mesh sieve material combined in equal portions from the sand cone samples.

38.5.2 Spoil Geometry

Four spoil profile models were selected for stability evaluation. The models were designated MODEL I, MODEL II, MODEL III, and MODEL IV and are shown in Figures 38.9 through 38.12, respectively. In each of the four models, the stability of the steepened slope angles was evaluated. The steepened slope angles were 40°, 45°, 50°, and 55°. For MODEL I and MODEL III, the angle of repose of 36° was also evaluated for comparison purposes. In total, some 18 individual spoil models were investigated for stability. MODEL I had a 90-ft- (27-m-) high upper zone, at a slope angle of 36. Its lower zone, which is also 90 ft (27 m) high, was successively steepened to 40°, 45°, 50°, and 55°. MODEL II is similar to MODEL I, except that an additional 40 ft (12 m) was included in the lower zone. The steepened zone would then become 130 ft (40 m) high, and 40°, 45°, 50°, and 55° slopes were evaluated. MODEL III is similar to MODEL I, except that the upper zone is 130 ft (40 m) high and the lower steepened zone is 140 ft (43 m) high. MODEL IV is similar to MODEL III, except that an additional 40 ft (12 m) zone is included in the lower steepened zone giving a total lower zone height of 180 ft (55 m). MODEL III and MODEL IV were evaluated using 40°, 45°, 50°, and 55° steepened slopes.

38.5.3 Spoil and Site Characterisitics

Discontinuities. In the classical sense, there are no geologic discontinuities known to exist in spoil materials, except for placement and angle-of-repose stratifications. These discontinuities may, in some respects, be similar to bedding planes in soil or rock. However, these stratification discontinuities may be irregular, as opposed to being planar, and they may vary in thickness, composition, and continuity. Consequently, they may in some cases represent the weakest element or a part of the weakest element in a spoil pile. The weakest element will be assumed to be the spoil matrix material located along the path of the highest shearing stress. The in situ rock under the lowermost spoil is assumed to be similar to other rock units in the highwall with respect to discontinuities. In essence, the base rock would probably have two sets of joint discontinuities that are steeply dipping and one relatively flat set of bedding discontinuities. These base rock discontinuities will be assumed to have shear strengths far in excess of the spoil mass. Consequently, any spoil mass instability will be assumed to occur above the base rock mass.

Groundwater and Surface Water. Atmospheric moisture amounts to about 7 in. (18 cm) per year and comes from winter snowfall and summer thunderstorms. Much of the snowfall sublimates and does not enter the spoil. The rainfall may form small

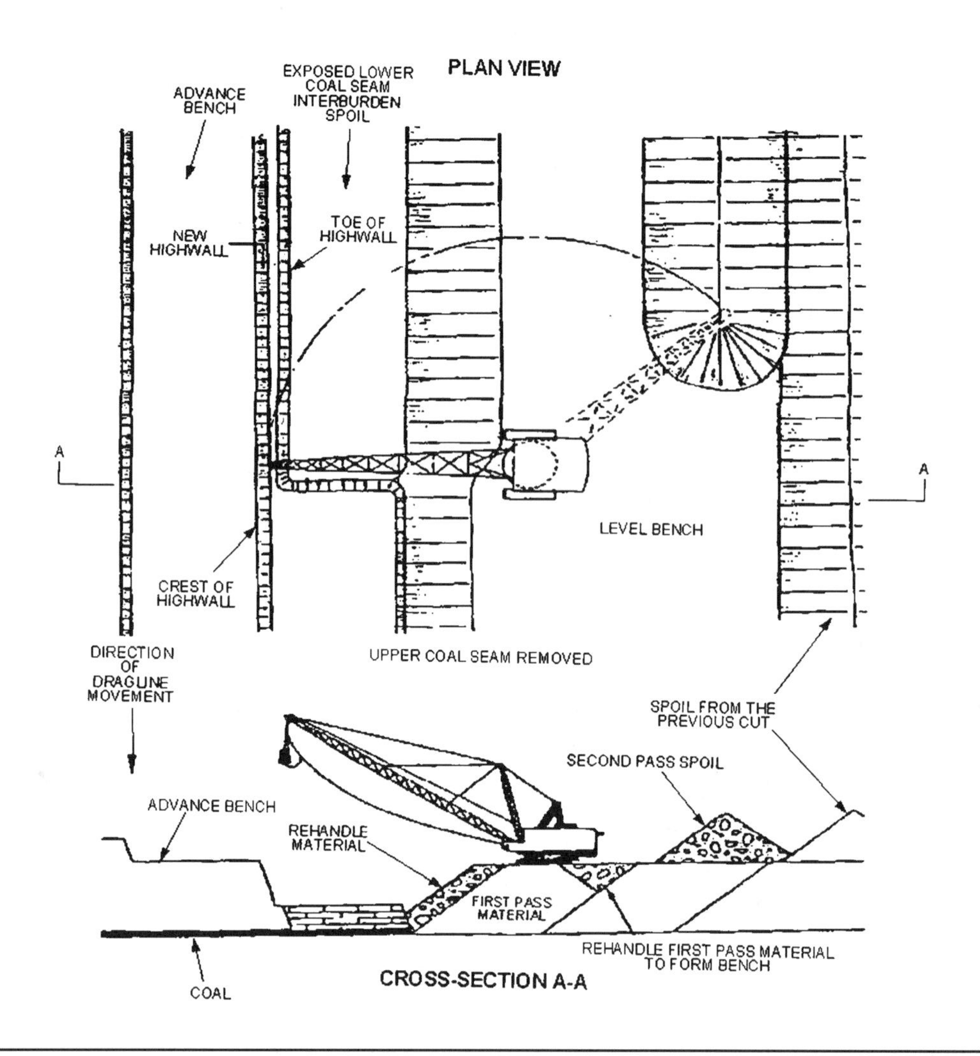

FIGURE 38.6 Second pass of the two-pass stripping method

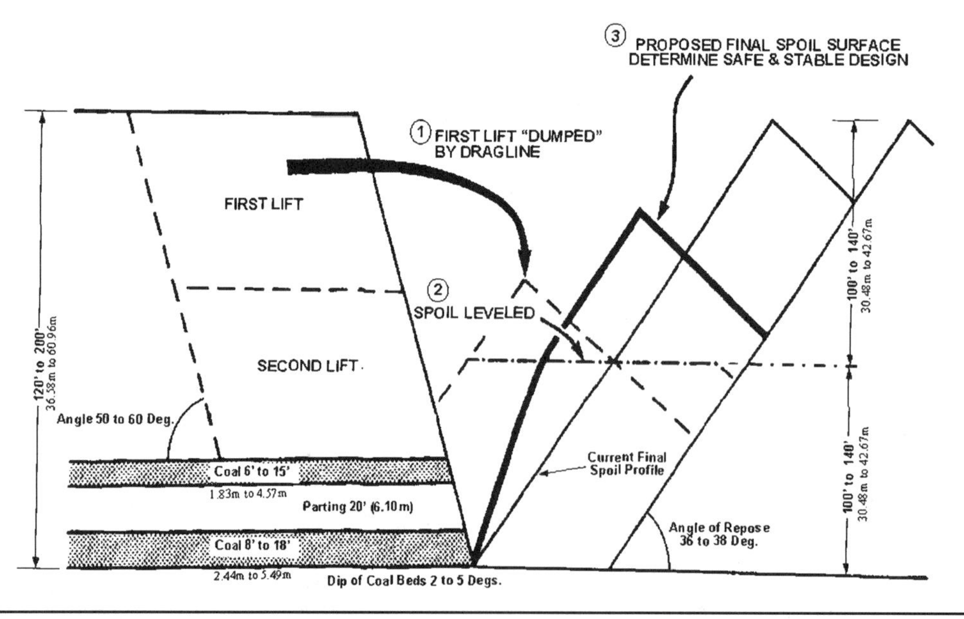

FIGURE 38.7 Spoil stability objectives

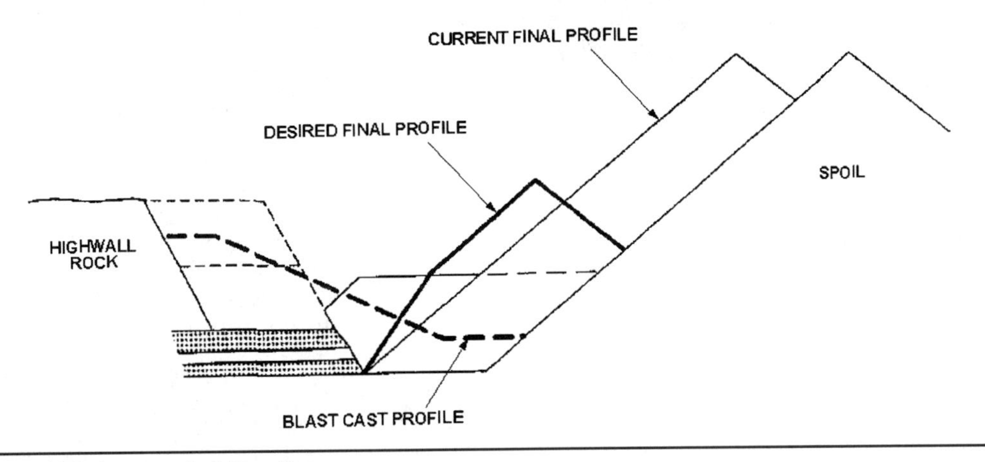

FIGURE 38.8 Conceptual blast cast material location

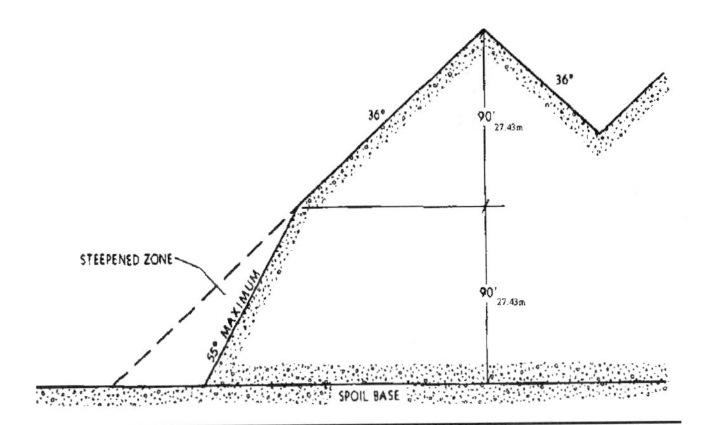

FIGURE 38.9 Spoil MODEL I cross section

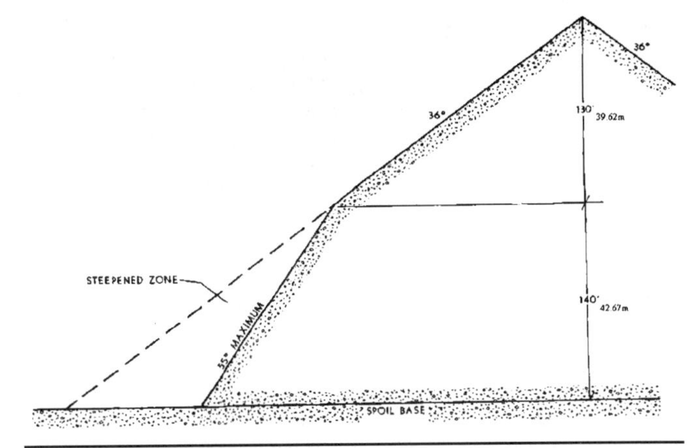

FIGURE 38.11 Spoil MODEL III cross section

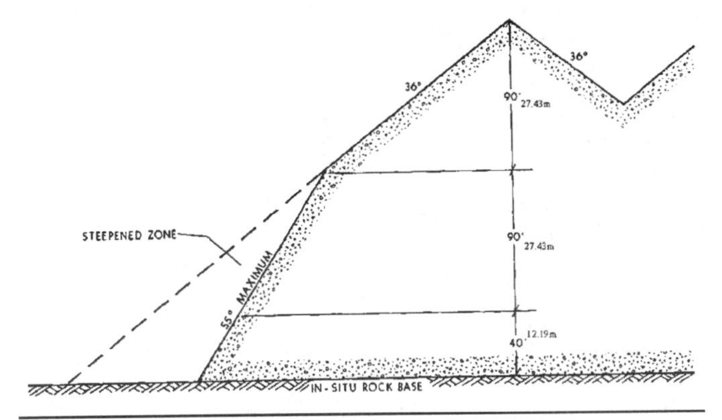

FIGURE 38.10 Spoil MODEL II cross section

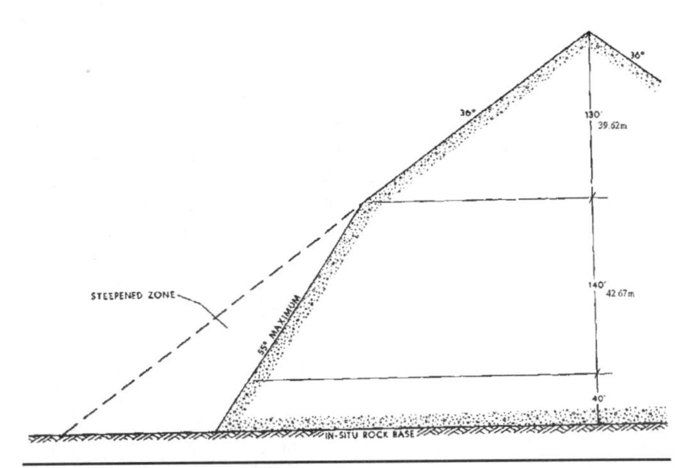

FIGURE 38.12 Spoil MODEL IV cross section

puddles in the valleys between spoil peaks and much of it may seep into the spoil. However, there is far too little water to create spoil groundwater with a true piezometric surface in most cases. The height of the piezometric surface in the spoil is thought to be on the order of 5 to 10 ft (1.52 to 3.05 m) or less. The extent of this small piezometric surface was unknown at the time of the study and therefore was not considered in the stability evaluation. Consequently, for the highwall stability evaluation, pore water was assumed to be completely absent from the spoil piles.

Field Density Evaluation. The field density data as determined by the sand cone and nuclear gage tests indicate a general increase in density with depth. However, one other factor could play a major role and that is spoil rebounding and expansion that occurs when upper spoil is removed and unloaded from lower spoil. The rebounding of lower spoil occurs due to the fact that stress and strain in spoil are proportional. When the upper spoil is removed, the stress on the lower spoil is lessened and the lower spoil rebounds upward. As the rebounding occurs, the spoil is less confined and probably expands and its density decreases. At all sample sites in the field that were deeper than 2 ft (0.61 m), the spoil had been removed by excavating equipment. Consequently, the true in situ density of spoil at a particular depth could not be accurately measured by the methods involving spoil removal to gain access. Engineering judgment was used to estimate the variation in spoil density.

In Situ Strength. Using the laboratory matrix shear strengths as a lower limit and rock fill strength obtained from the literature, a friction angle of 41 and cohesion of 200 psf (9.6 Kpa) was assumed for depths less than 25 ft (7.62 m). For depths greater than 25 ft (7.62 m), a friction angle of 35 and cohesion of 750 psf (35.9 Kpa) was used. These strength parameters represent only slight increases over the matrix strengths and are more reasonable.

Conservative Factors. At the Jim Bridger Mine, several factors exist that may actually give the spoil materials higher shear strengths than indicated by the spoil stability study. First, the lower zone in some of the spoil has been cast blasted into its present position. Cast blasting allows larger particle sizes to exist in the base of the spoil because the spoil is not handled by the dragline. Further, the blast cast material has a flatter or even nonexistent stratification, as compared to angle-of-repose stratifications that develop when the spoil is placed by dragline. It is assumed that the absence of these stratifications would give the spoil a higher shear strength. Second, there are certain portions of the lower spoil zones that are compacted by equipment travel before additional spoil is placed by the dragline. These compacted zones would have higher shear strengths than spoil zones that were constructed solely by the dragline. One additional factor is that within certain limits, the spoil continues to compact with time and, therefore, the shear strength probably also increases with time. Experience with other waste rock embankments indicates that after many years, compaction and solidification of materials may be so great that drilling and blasting may be required to remove materials (Seegmiller 1989). Consequently, the period of highest risk for instability should occur when the spoil pile is steepened or shortly thereafter. If the spoil does not become unstable at this time, gross failure is unlikely unless conditions change substantially.

38.5.4 Stability Analysis

Potential Failure Modes. For spoil embankments, there are a number of failure modes that may occur under various circumstances. These include edge slumps, deep-seated rotational shear, shallow flowslides, foundation spreading, and base translation. Edge slumps may occur near the crest or toe of an embankment, but they are only a very minor problem and should not be of concern for the Bridger Mine spoils. Deep-seated rotational failures are the major concern of the present study and they must be prevented in any steepened spoil slope. Shallow flowslides are generally the result of infiltration and loss of shear strength due to rain or snowmelt. Owing to the relatively dry climate, they should have a fairly low probability of occurrence. Even if they should occur, they are generally of very low volume and surface area and should represent only a minor problem at the mine site. Foundation spreading is usually initiated by excess pore water pressure on a weak, sloping foundation. While pore water could occur in some spoil foundations, most locations have fairly flat strata, which is much stronger than typical surface soils. Consequently, foundation spreading should not be a problem. Base translation can usually occur only on a sloping foundation surface. At the Bridger Mine, most foundations are much flatter than those upon which base translation is possible. In summary, the only failure mode of major concern for the Bridger Mine spoils is deep-seated rotational shear. Consequently, the spoil slope design is directed at their prevention.

Rotational Shear Methodology. The analysis procedure that was used involves the construction of spoil slope geotechnical models. These models were based on the geometry presented in MODEL I through MODEL IV, the applicable spoil shear strengths and densities, the applicable pore pressure conditions, and the applicable seismic conditions. These models were then evaluated using computer-aided methods to determine the sensitivity of the safety factor magnitude to steepened spoil slope angle using limiting equilibrium.

Limit-Equilibrium Method. This method is based on the definition of a safety factor that is the ratio of shearing strength, as determined by resistive forces, to the shearing or disruptive forces. At a safety factor of 1.00, the forces are exactly in balance. A safety factor less than 1.00 implies slope failure, and a safety factor greater than 1.00 indicates stability. When mean value material properties are used, the safety factor is defined as a deterministic safety factor. When the material properties are entered into the analyses as statistical populations, a probabilistic safety factor and an estimate of the probability of slope failure may be computed.

Computer Code. A limiting equilibrium/probabilistic code MCSLOPE program was used to perform the rotational shear analyses (Seegmiller 1989). This program is a derivative of the widely used STABL5 computer code developed at Purdue University. Numerous material types, groundwater scenarios, and slope geometries may be modeled and analyzed for stability using the MCSLOPE code. Both deterministic and probabilistic safety factors may be computed, as well as a probability of failure.

Safety Factor and Probability of Failure Magnitudes. The magnitude of acceptable safety factors generally ranges from 1.10 to 1.30, but it depends on a number of factors including the risk of failure, quality of the input data, applicable regulations, and the person doing the analysis. Safety factors may be static no earthquake loads) or dynamic earthquake loads). For the short-term spoil slopes that are considered here, there are no known applicable regulations. For long-term spoil slopes, which have been reclaimed, the Bridger spoils must have a minimum static safety factor of 1.30 to meet state regulatory requirements. The present input data are believed to be quite good and a realistic representation of the true spoil properties. However, the risk of failure is at least moderate because personnel and equipment may be working in the steepened toe area.

The selection of prudent short-term safety factors is not necessarily a simple task. The use of a probability of failure as a criterion for safety and risk is relatively new in geotechnical modeling and analysis. The specific probability magnitudes for judging safety and risk have not been established. Nevertheless, it is desired to have a low probability of failure and a high safety factor. Depending on the magnitude and variability of the input data, these two risk criteria can and will vary widely. It is entirely possible to have a static safety factor of 1.25 and a probability of failure of 25% with certain input data. Using input data specifying slightly lower strength magnitudes, but with much less variability in the data, the static safety factor could decrease to 1.20 and the probability of failure could drop to 5%. However, generally speaking, a high safety factor indicates a low probability of failure and a low safety factor indicates a higher probability of failure. Both criteria will be weighted equally, and preliminary slope steepening guidelines are based on both criteria. These guidelines specify a minimum short-term static safety factor on the order of 1.15 to 1.20 and a maximum probability of failure on the order of 5 to 10% (Seegmiller 1989).

38.5.5 Spoil Model Stability

- MODEL 1. This model is described by Figure 38.9. The 50° steepened slope has a fairly good static safety factor and a relatively low probability of failure and is illustrated in Figure 38.13. The 55° slope is not as stable as expected, but it still has a relatively good safety factor and acceptable probability of failure.
- MODEL II. The section presented in Figure 38.10 represents this model. Basically, the model is similar to MODEL I, but an additional 40 ft (12 m) has been cut at

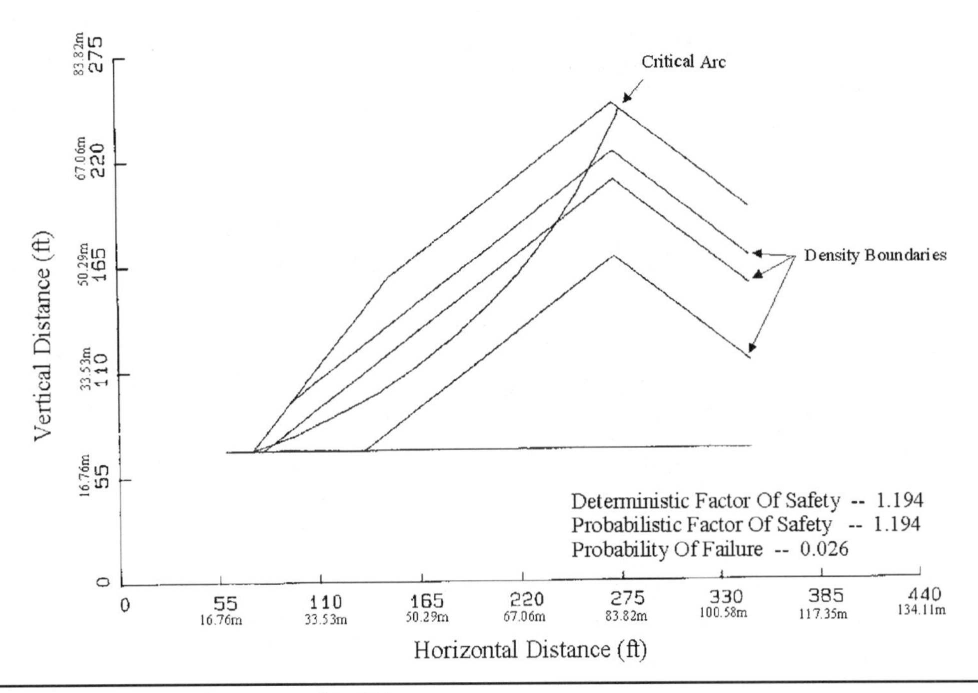

FIGURE 38.13 Spoil MODEL I—slope angle steepened to 50°

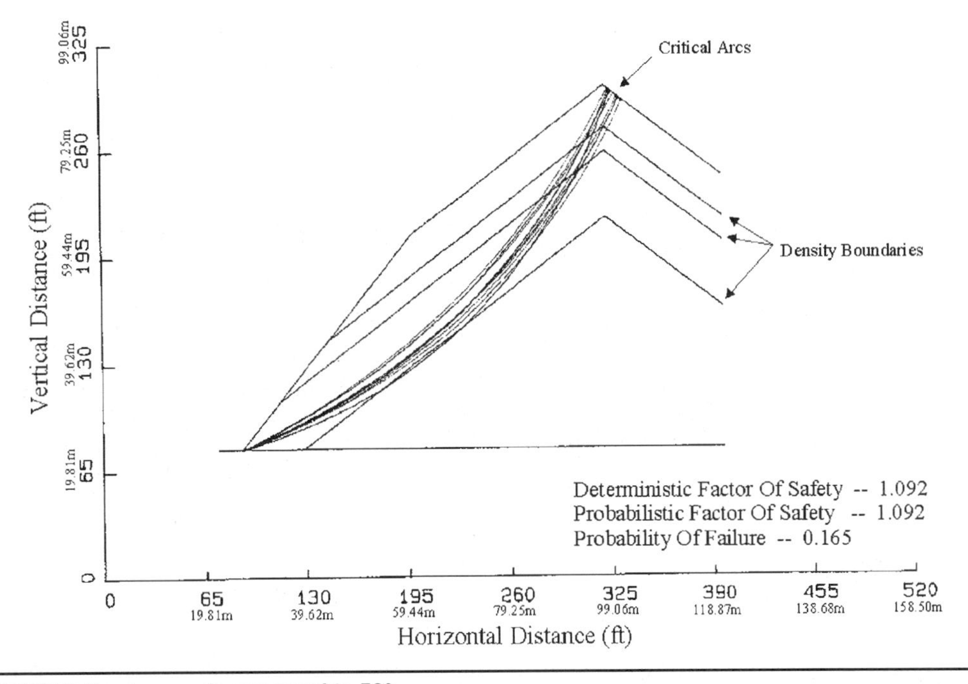

FIGURE 38.14 Spoil MODEL II—slope angle steepened to 50°

the slope toe. The 45° slope appears to have good stability, but the 50° slope, which is shown in Figure 38.14, may be too unstable as indicated by the low factor of safety and high probability of failure.

- MODEL III. This model has been defined previously in Figure 38.11. A summary of the slope stability with angle of repose slopes and with steeper slopes is given in Table 38.1. A slope angle of 45° appears to have good stability, but any slope as steep as 50° may be only marginally stable.
- MODEL IV. The section designated in Figure 38.6 represents this model. Basically, it is similar to MODEL III, but it has an additional 40-ft (12-m) cut at the toe. The 40° slope has good stability, but a 45° slope may only be marginally stable due to the low safety factor of 1.101. However, the probability of failure for the 45° slope is quite acceptable at 5.8%. (What are the values?) Consequently, a 45° slope may work well, but it should be used only after more is known about the field

TABLE 38.1 MODEL III stability summary

	Safety Factor (Static)		
Steepened Slope Angle	**Deterministic**	**Probabilistic**	**Probability of Failure**
36° (Angle of repose)	1.296	1.296	<0.1%
40°	1.226	1.224	0.9%
45°	1.151	1.151	5.6%
50°	1.084	1.085	9.2%
55°	1.025	1.024	39.1%

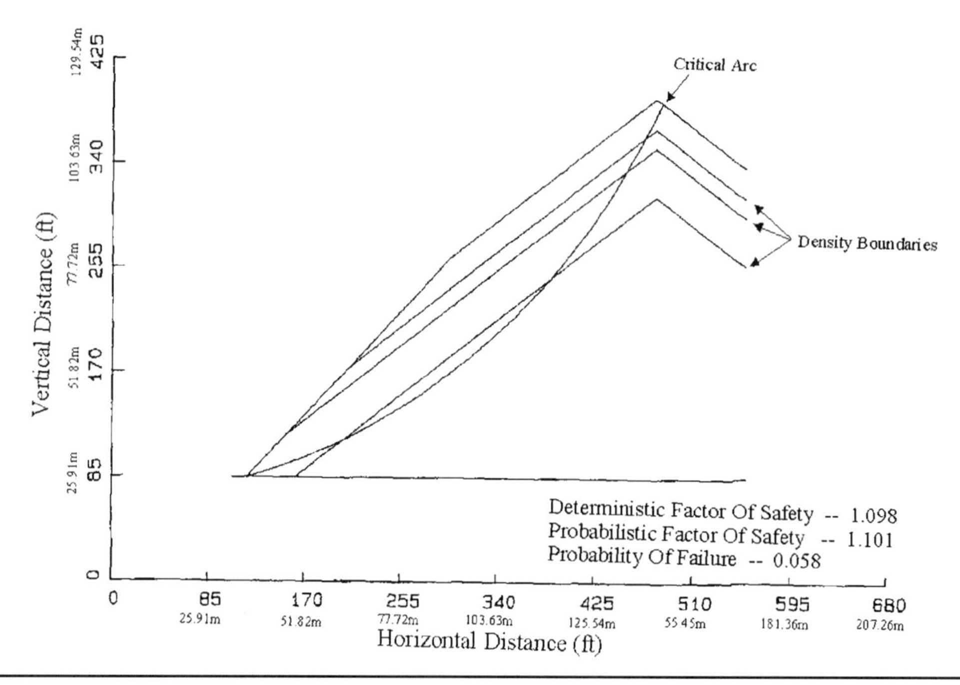

FIGURE 38.15 Spoil MODEL IV—slope angle steepened to 45°

performance of other acceptable slope models and a field trial using the MODEL IV configuration with a 40° slope. The 45° slope is shown in Figure 38.15.

38.6 SEISMIC EFFECTS

The potential of seismic effects on slope stability is relatively low. According to the US Army Corps of Engineers, the Jim Bridger Mine is in a zone 1. Such a zone is attributed to have a design seismic coefficient of 0.025, which is commonly the lowest coefficient used. It was found that dynamic analysis resulted in an average reduction of 0.045 safety factor units for MODEL I through MODEL IV. This reduction is similar to what may be expected by changing the slope angle about 1° or 2°. In other words, the dynamic safety factor of a 47° slope would be approximately equivalent to the static safety of a 49° slope. All in all, it is important to be aware of what effect of dynamic loading is on the stability of the various spoil configurations. These effects were found to be minor and not a serious concern for the spoil geometries considered.

38.7 SUMMARY

The choice of which slope angle to use in a particular spoil pile should be made by those with operating experience considering an acceptable degree of risk. The spoil models evaluated for the Bridger Mine provided guidance for selecting a workable spoil steepening program. Based on the preceding analyses, the Bridger mine has employed steepening at the toe of spoil embankments up to 45° with acceptable performance and a reduction of about 20% in the spoil that must be rehandled.

38.8 REFERENCES

Bricker, M.L. 1992. Monolithic Overburden, Horizontal Coal Seams: Bridger Mine. In *Mining Engineering Handbook*, p. 1407–1410. Littleton, Colorado: SME.

COAL AGE. 1982. Breadwinner mines western coal. October.

Seegmiller, B.L. 1989. Spoil Stability Evaluation–Steepened Spoil Slopes, Jim Bridger Mine, Sweetwater County, Wyoming. Professional report.

US Army Corps of Engineers. 1970. Engineering and Design, Stability of Earth and Rock-fill Dams, EM 1110-2-1902, Department of the Army, April (Revised 16 May 1983: Eng. Reg. 110-2-6).

CHAPTER 39

Design Objectives for Mine Waste Rock Disposal Facilities at Phosphate Mines in Southeastern Idaho

Brian Buck*

39.1 INTRODUCTION

Local environmental impacts due to the release of selenium from phosphate mining facilities in southeasten Idaho were discovered in 1996. Since then, the southeastern Idaho phosphate industry has been working with state and federal regulatory and land management agencies to identify causal factors for the selenium impacts and develop appropriate mitigative measures on a regional and site-specific basis. A great deal of locally derived information on this issue has been collected and documented by the Interagency/Phosphate Industry Selenium Working Group and added to the extensive existing literature on selenium in the environment.

39.2 ENVIRONMENTAL CHEMISTRY

Selenium has been known for some time to be an essential trace element in human nutrition, with a recommended daily allowance 0.070 mg/d for adult men (ATSDR 1989a). Studies in human nutrition have indicated that selenium may be important in the prevention of cancer, heart disease, and accelerated aging (EPA 1986). However, because selenium in soil and water can be concentrated in the food chain, animals exposed to chronic conditions of contaminated water and vegetation can accumulate harmful levels of selenium. Long-term consumption of high levels of selenium by livestock has been shown to have effects on the heart, kidney, and liver and to cause reproductive problems (EPA 1998a).

Selenium occurs in nature in four oxidation or valence states, including elemental selenium (0), selenide (−2), selenite (+4), and selenate (+6). The solubility and mobility of selenium are dependent on its valence and chemical environment. In soils and rocks, its mobility is affected by redox conditions, pH, hydrous iron and manganese oxide content, clay content, organic materials, and the presence of competing anions (NLM 2000).

Elemental selenium (0) is essentially insoluble, as are the selenide forms (-2) (NLM 2000). The selenide form is chemically immobile in the environment, and this form predominates in acidic environments and in environments with high organic carbon content (NLM 2000; EPA 1998b).

Weathering of selenium and/or selenide-containing rocks causes the anionic selenium to oxidize and combine with cations to form selenite (+4) or selenate (+6) compounds, the most common of which are sodium selenite (Na_2SeO_3), and sodium selenate (Na_2SeO_4) (ATSDR 1989b).

In neutral, well-drained soils and rocks, selenite is typically formed. It is soluble in water but is strongly adsorbed to soil minerals, organic material, and hydroxides of iron and manganese (NLM 2000). At pH levels below neutral and under mildly reducing conditions, selenites can be reduced to insoluble elemental selenium (NLM 2000). In acidic or neutral soils, selenium tends to remain in a relatively insoluble form, and the amount of biologically available selenium decreases (EPA 1998b).

In alkaline and oxidizing conditions, selenium can be further oxidized to selenate, a biologically active form. Selenates are not adsorbed in most soils or sediments and are quite mobile in water (NLM 2000).

In soils and aquatic sediments, selenium can also be biologically methylated and subsequently released as volatile dimethyl selenide and dimethyl diselenide (NLM 2000).

Based on the above description of the chemistry of selenium, the following chemical factors are recognized for the control of selenium and selenium compounds in the environment:

- Elemental and selenide forms of selenium are practically insoluble in water and are chemically immobile in the environment. They can be oxidized to selenite and selenate forms in oxidizing and alkaline conditions.
- Selenites predominate in neutral, well-drained conditions and are soluble, but they also are strongly adsorbed to iron and manganese oxides, clays, and organic materials. Selenite can also be chemically reduced to immobile forms in conditions that are acidic and reducing.
- Selenates predominate in oxidizing and alkaline conditions and are very mobile with minimal adsorption.
- Elemental and selenide forms of selenium should be handled to minimize oxidation and contact with water to minimize the generation of selenites and selenates and the dissolution of these compounds from the solid materials.

39.3 DESIGN OBJECTIVES

It is important that all mine waste rock disposal facility designs comply with applicable state and federal regulatory and land-use management requirements. In addition, it is necessary to use applicable best management practices that will provide both short- and long-term protection of the environment while still recognizing limitations imposed by safe and economic mine operations.

The conceptual objective of future designs for Idaho phosphate mine waste rock disposal facilities should be to isolate the seleniferous mine waste rock from weathering and natural transport influences in order to minimize the release of selenium to environmental media outside the waste rock disposal facility. These media include air, vegetative growth medium, surface water, ground water, vegetation, and fauna.

* JBR Environmental Consultants, Inc., Sandy, Utah.

39.4 CONCLUSIONS AND RECOMMENDATIONS

Based on the above conceptual objective, the following design objectives are proposed for future phosphate waste rock disposal facilities:

1. Selectively handle seleniferous mine overburden to control its final placement in the mining process.
2. Reduce the exposure of seleniferous mine overburden to weathering processes that may oxidize the selenium compounds in material.
3. Minimize infiltration of precipitation or run-on into seleniferous waste rock, which may cause dissolution of selenium compounds.
4. Minimize contact of seleniferous waste rock with ground water and surface water.
5. Minimize erosion and off-site sediment transport of seleniferous mine waste rock.
6. Prevent contact of seleniferous mine waste rock with roots of reclamation plant species and post-reclamation volunteer plant species.
7. Provide secondary, or backup, exposure protection methods whenever reasonably possible.

39.5 REFERENCES

Agency for Toxic Substances and Disease Registry (ATSDR). 1989a. *Toxicological profile for selenium.* Draft. Washington, DC: U.S. Department of Health and Human Services, Agency for Toxic Substances and Disease Registry.

Agency for Toxic Substances and Disease Registry (ATSDR). 1989b. Public Health Statement, Selenium. Washington, DC: U.S. Department of Health and Human Services, Agency for Toxic Substances and Disease Registry. http://atsdr.cdc.gov/ToxProfiles/phs8921.html. Site accessed Jan. 29, 2000.

National Library of Medicine (NLM). 2000. Hazardous Substances Data Bank. Washington, DC: Specialized Information Services, National Institutes of Health, U.S. Department of Health and Human Service. http://sis.nim.nih.gov/cgi-bin/sis/Tosnet. Site accessed Jan. 28, 2000.

U.S. Environmental Protection Agency (EPA). 1986. Drinking Water Criteria Document for Selenium. Final Draft. Washington, DC: U.S. EPA, Criteria and Standards Division, Office of Drinking Water.

U.S. Environmental Protection Agency (EPA). 1998a. Unified Air Toxics Website, Technology Transfer Network. Washington, DC: U.S. EPA, Office of Air Quality Planning and Standards. http://www.epa.gov/ttn/autw/hlthef/selenium.html. Site accessed Jan. 28, 2000.

U.S. Environmental Protection Agency (EPA). 1998b. Drinking Water and Health, Technical Factsheet on Selenium. Washington, DC: U.S. EPA, Office of Ground Water and Drinking Water. http://www.epa.gov/OGWDW/dwh/t-ioc/selenium.html. Site accessed Jan. 29, 2000.

SECTION 4

Tailings and Heap Leaching

CHAPTER 40

Tailings Dam Stability: Essential Ingredients for Success

Michael Davies,* Todd Martin,* and Peter Lighthall*

40.1 TAILINGS DAMS

Dams have been used in nonmining applications for water supply and/or flood control for thousands of years. More recently, dams have been developed for both hydroelectric power generation and the retention of industrial by-products such as mine tailings. Mine tailings dams, which really became recognized as "structures" near the beginning of the twentieth century, rival or, in many cases, exceed the scale of conventional water supply, flood control, or hydroelectric dams. Despite their size, and despite tailings impoundments representing some of the largest constructed structures, tailings dams have only gained recognition as "dams" in the last few decades.

Conventional dams continue to be constructed to greater heights with greater storage volumes. Conventional dams have had a generally good safety record, although catastrophic failures have occurred. However, the safety record of conventional dams has been steadily improving over the past 40 years to the point that the probability of a conventional dam failure in any given year is roughly 1 in 10,000. This safety trend does not exist for mine tailings dams, which appear to be failing at a rate at least 10 times that for conventional dams. Some make a different argument (Bruce, Louge, and Wichek 1997), implying that tailings dams are equally "safe" as conventional dams and that both are being built to at least the same "state-of-the-art" practice. This latter interpretation of the statistical database is common and worrisome, as it can lead to a complacent attitude. It also does not appear to account for the fact that tailings dams can undergo environmental failures while maintaining physical integrity–an issue not readily associated with conventional dams.

Recent trends and initiatives in tailings dam stewardship, spearheaded by the mining industry, are extremely positive and encouraging, though these initiatives tend to get ignored by a relatively biased news media. However, objective evaluation of the tailings dam failure database illustrates that many tailings dams are not being designed, constructed, and/or operated to adequate standards. Moreover, the safety record of tailings dams to date cannot be considered acceptable given the tremendous damage to the overall mining industry that every new failure provides.

40.2 TAILINGS DAM SLOPE FAILURES

To better illustrate the nature of tailings dam failures caused by slope instability, a few examples are given here.

40.2.1 Sullivan Mine, Canada—1991

Davies, Chin, and Dawson (1998) describe the static liquefaction event that occurred at the Active Iron Pond tailings impoundment at the Sullivan Mine in August 1991. The event resulted in a flowslide; fortunately, another tailings dyke contained the flow and no off-site impact was experienced. The dam had been built on a foundation of older tailings that were placed as beach below water (BBW) material. The failure occurred in this upstream-constructed facility due to the mobilization of shear stresses in the foundation tailings that exceeded the peak shear strengths. As the material strained, the pore pressures rose and drainage was impeded, leading to the liquefaction event. The downstream slope of the dyke was roughly 3H:1V, imposing stresses in excess of the collapse surface for the foundation tailings in an extension stress path and near to the collapse surface in compressive shear. The Sullivan tailings facility had been under the design and monitoring stewardship of a recognized consulting organization. This event served to demonstrate that "a well intentioned corporation employing apparently well-qualified consultants is not adequate insurance against serious incidents" (Morgenstern 1998).

Ironically, the 1991 event was similar in nature to a dyke failure that occurred at the same facility in 1948. Newspaper accounts from 1948 show the general public support for the mine and sympathy to the clean-up efforts. Had the 1991 incident progressed offsite, it is without doubt that the community and media response would have been dramatically different.

The passage of more than 40 years should not have been enough to induce the designers into TDA (tailings dam amnesia). As defined by Martin and Davies (2000), TDA refers to a state of tailings dam design or stewardship where lessons available at that very site are ignored in spite of ample available on-site information, visual evidence of previous event occurrence, and/or published accounts of incidents on a given project. Not succumbing to TDA is an essential ingredient to tailings dam stability.

40.2.2 Merriespruit, South Africa—1994

TDA struck again, but in a slightly different form, at the Harmony Mine adjacent to Merriespruit, South Africa. The Bafokeng tailings dam in South Africa, also a paddock upstream facility, failed in almost the same manner (involving static liquefaction) with a similar result, i.e., downstream fatalities. At Bafokeng in 1974, seepage/piping introduced the retrogressive liquefaction flowslide, whereas at Merriespruit, overtopping due to inadequate freeboard was ample trigger for liquefaction and slope failure once enough toe material eroded away.

The Merriespruit failure occurred on the evening of February 22, 1994. A massive failure of the north wall (see Figure 40.1) occurred following a heavy rainstorm. More than 600,000 m^3 of tailings and 90,000 m^3 of water were released. The slurry traveled about 2 km covering nearly 500,000 m^2. Given the downstream population, it is fortunate that only 17 people lost their lives in this tragedy.

40.2.3 Stava, Italy—1985

Perhaps the most tragic tailings dam failure to date occurred on July 19, 1985. A fluorite mine, located near Stava in northern Italy, had both of its tailings dams (Figure 40.2) fail suddenly and release approximately 240,000 m^3 of liquefied tailings. The remnant of the upper dam is shown in Figure 40.3. The liquefied

* AGRA Earth & Environmental Limited, An AMEC Company, Burnaby, British Columbia, Canada.

FIGURE 40.1 Merriespruit tailings dam failure, 1994

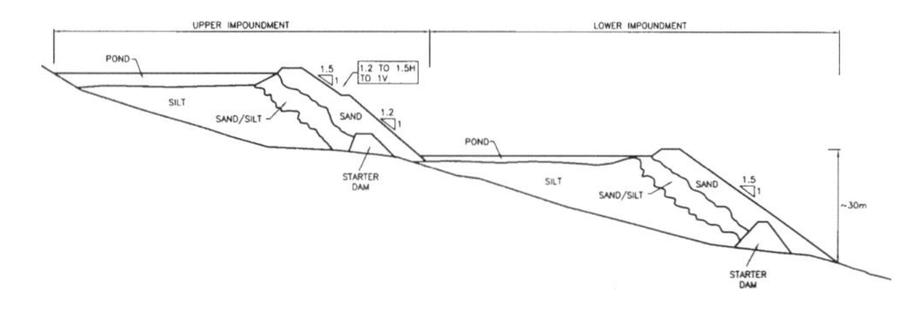

FIGURE 40.2 Schematic section of Stava tailings dams

FIGURE 40.3 Remnant of Stava upper tailings dam after failure

mass moved at speeds of up to 60 km/h, obliterating everything in its path for a stretch of some 4 km. The flowslide destroyed the village of Stava and also caused considerable damage at Tesero, at the junction of Stava Creek and the Avisio River 4 km from the mine.

The tailings dams were both nearly 25 m high, with one directly upstream of the other. The failure mechanism began with failure of the upper dam that, in turn, overtopped and failed the lower dam. The dams were upstream-constructed, with outer slopes from 1.2 to 1.5 horizontal to 1 vertical. Based on the likely state of the in situ tailings, the soil mechanics curiosity associated with this failure is that the dams could attain such a height prior to failure. These dams were in a precarious state, awaiting a trigger mechanism to initiate failure. For upstream tailings dams, unfortunately, such trigger mechanisms are numerous. There is no question that the design of these dams was not consistent with even the most elementary of engineering principles available at

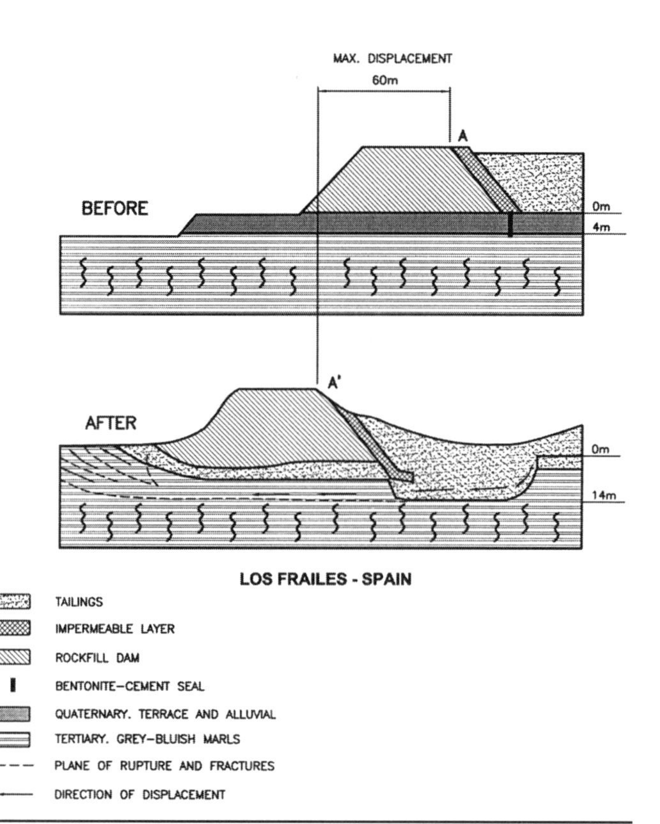

FIGURE 40.4 Los Frailes dam failure: schematic section

the time. There are a number of "rules" for upstream tailings dam engineering (Davies and Martin 2000) that were understood for many years prior to the Stava failure. Both Stava dams violated far more of these rules than they followed.

40.2.4 Los Frailes, Spain—1997

Possibly the most publicized tailings dam failure in history was the 1997 Los Frailes event in Spain. A shallow foundation failure involving a 60-m displacement of a portion of the dam (see Figure 40.4) led to release of more than 3×10^6 m^3 of process water and tailings from one of two adjacent ponds within an overall impoundment. For this failure, a lack of understanding of the prevailing foundation conditions was directly attributable to a design that was contraindicated by site conditions. That over-consolidated clay foundations represent a particularly troublesome embankment foundation condition due to their structure (laminated, frequently fissured) and brittle (strain-softening) nature has been well understood by the dam engineering community for several decades, yet this basic geotechnical knowledge was not applied at Los Frailes until the post-failure investigations. The lack of appreciation of the importance of geologic factors in the performance of the foundation for this and other tailings dam case histories demonstrates a need to treat each project with a thorough characterization of subsurface conditions.

40.3 WHY DO TAILINGS DAM INSTABILITIES OCCUR?

By objectively reviewing the instability database for tailings dams, it can be concluded that for the past 30 years there have been approximately two to five "major" tailings dam failure incidents per year. During no year were there less than two events (1970–1999, inclusive). If one assumes a worldwide inventory of 3,500 tailings dams (a tenuous extrapolation at best), then two to five failures per year equates to an annual probability of between 1 in 700 to 1 in 1,750. This rate of failure does not offer a favorable comparison with the 1 in 10,000 figure that appears representative for conventional dams. The comparison is even more unfavorable if less "spectacular" tailings dam failures are considered.

So why do failures of tailings dams continue to occur? The failures are not just of older facilities constructed without formal design, but include facilities designed and commissioned in the past 5 to 20 years–supposedly the "modern age" of tailings dam engineering.

The first step in evaluating the reasons for continued tailings dam failures comes from recognizing the uniqueness of mine tailings dams, which are discussed in detail by Szymanski (1999). These unique attributes include

- Tailings impoundments are among the largest synthetic structures, with several approaching 1×10^9 tonnes of stored slurried tailings.
- Tailings dams are built on a continuous basis by mine operators who are in business to extract wealth from the ground, not to build and operate dams.
- Tailings dams represent a cost to the mining process–they do not generate a revenue stream akin to a hydroelectric dam.

Mining companies typically do not have in-house geotechnical expertise, instead there is reliance on periodic design and perhaps construction monitoring from consulting engineers. Large-scale water supply and/or hydroelectric agencies more often than not have very capable dam designers and surveillance engineers/technicians in-house. Owners of large conventional dams also typically retain an independent board of eminent consultants to provide expert third-party review at appropriate project stages/intervals. This is not a typical practice in mining at this time, where all too often such boards are convened only to investigate the causes of failures. It would obviously be far more proactive to engage such boards to assist the owner and designer in preventing them.

Are the unique features of tailings dams the reason for the failure trends? A combination of factors, including a lack of input from appropriate external consultants and/or the reliance on, and trust in, third-party consultants without adequate review of their work, contributes to the failure trends. There are basic requirements for a designer working in tailings dam engineering and these requirements need to be followed.

Most, if not all, tailings impoundment instabilities fit into a very consistent set of trends. An understanding of the failure trends would make a large difference in reducing future failure frequencies. An improvement in the safety record of tailings dams would enhance public perception of the mining industry. If the mining industry collectively embraces the lessons from these trends, the current negative perceptions surrounding tailings dams can perhaps wane considerably as the safety record for tailings dams improves to the standard demanded by those who are so quick to criticize the industry.

40.4 LESSONS FROM PREVIOUS TAILINGS DAM INSTABILITIES

By combining published accounts of tailings dam instabilities with those available through reviews, industry contacts, and similar sources, several trends from the tailings dam instability database are evident, including:

- Active dams are more susceptible to failure–this trend may diminish over time if the current trend advocated by some to flood all tailings impoundments upon closure gains momentum.
- Upstream constructed dams = more incidents–this is not an entirely fair conclusion as the tailings dam population is dominated by upstream dams. However, upstream dams are more susceptible to liquefaction flow events and are solely responsible for all major static liquefaction events.
- Slope instability/earthquakes account for two-thirds of all upstream dam incidents.
- Seepage-related phenomena (e.g., piping due to poor filter design; this was evident in the Omai dam failure) is the main failure mode for nonupstream tailings dams, and, in failures such as Bafokeng (South Africa, 1974), can serve as the trigger for a slope failure.
- Earthquakes are of little consequence for most nonupstream dams.
- For inactive dams, overtopping is cited as the primary failure mode in nearly one-half of the incidents.
- TDA is a major contributor to the instability database.

The list of trends from the database can be expanded and has been presented in some form by many others. However, reviewers of the case histories seldom make the most important conclusion, i.e., there have been *no unexplained failure events*. If one becomes a student of tailings dam failure case histories, and all designers and regulators should indeed do just that, a single conclusion arises. Every failure was entirely predictable in hindsight. There are no unknown loading causes, no mysterious soil mechanics, no "acts of God," no "substantially different material behavior," and definitely no acceptable failures. In all of the cases over the past 30 years, the necessary knowledge existed to prevent the failure at either the design and/or operating stage. There is lack of design ability and poor stewardship (construction, operating, or closure), or a combination of the two, in every case history. If basic design and construction considerations are ignored, a tailings dam's candidacy as a potential failure case history is immediate.

40.5 WHY IS STABILITY IMPORTANT?

Tailings dam failures can have any or all of the following impacts:

- Extended production interruption
- Loss of life
- Environmental damage
- Damage to company and industry image
- Economic consequences to the company, and even the industry, as a whole
- Forcing higher standards for subsequent projects, perhaps to unnecessary and impractical levels
- Legal responsibility for company officers

For the mining company, the most tangible impact after ensuring public safety is the immediate and longer-term financial impact. The impact of the Los Frailes incident on the parent company's share price is illustrated in Figure 40.5. The public has high expectations for the mining industry in stewarding mine tailings. There are "fringe" groups that appear opposed to mining of any sort that have either not thought out their position with any real effort or advocate a return to a Paleolithic lifestyle. Given that society, at least implicitly by consumption patterns, places a high value on mined products, public perception of the industry should be commensurate with the value of the industry.

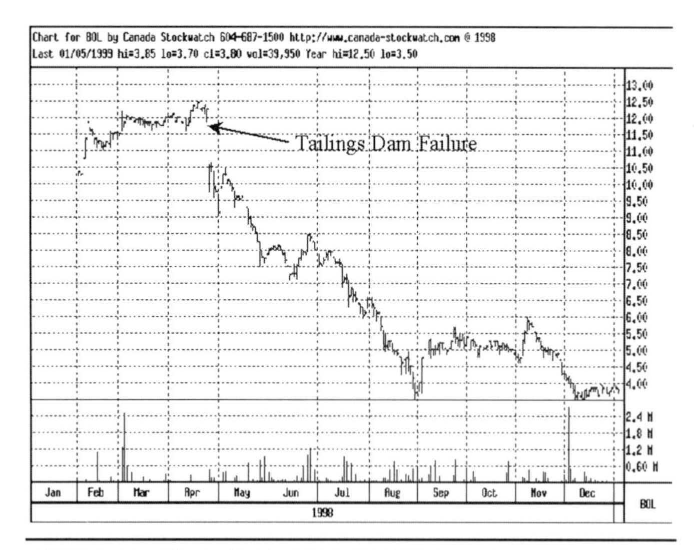

FIGURE 40.5 Effect of Los Frailes dam failure on share price

Tailings dams, particularly the well-publicized failure events, are lightning rods for public scrutiny of the industry. However, as noted below, this is not as new a public sentiment as many would believe.

> The strongest argument of the detractors of mining is that the fields are devastated by mining operations ... further, when the ores are washed, the water used poisons the brooks and streams, and either destroys the fish or drives them away ... thus it is said, it is clear to all that there is greater detriment from mining than the values of the metals which the mining produces.
>
> Agricola, 1556

And, unlike 450 years ago, the public now has instantaneous access to tailings dam events via the news media and the internet, from which Figure 40.4 was downloaded. Given the relatively constant frequency of tailings dam failures over the past 30 years, the public perception is that such events are on the rise due to the increase in publicity each successive event receives. The influence public sentiment can have on the viability of a proposed or existing mining project has never been higher. Public and some regulatory perception considerations are now largely driving project design decisions, as opposed to appropriate experiencc and tcchnical logic.

Failures of tailings dams tend to be viewed as events caused by the collective mining industry. It is naïve to assume that an individual corporation or regulatory jurisdiction is not affected by the dam failures of others. Whether the industry deserves the situation, each failure incident "raises the bar" with both the public and regulatory bodies for the next project.

40.6 INGREDIENT FOR SUCCESS NUMBER 1—NOMENCLATURE

In addition to the need to avoid TDA, which was mentioned earlier, the seemingly trivial question of what defines a tailings dam requires clarification. In spite of the limited number of geometrical options, it is surprising that there remains difficulty in naming these structures. A number of inventive, curious, and sometimes purely misleading names have been provided to tailings dams. The damage inaccurate naming can do by creating confusion among owners, regulators, and even designers is real (e.g., the latter start to believe if you call it something else, it will behave like something else!; case histories defend this assertion). Without any need for special cases, the following definitions, illustrated in Figure 40.6, should be adopted industry wide:

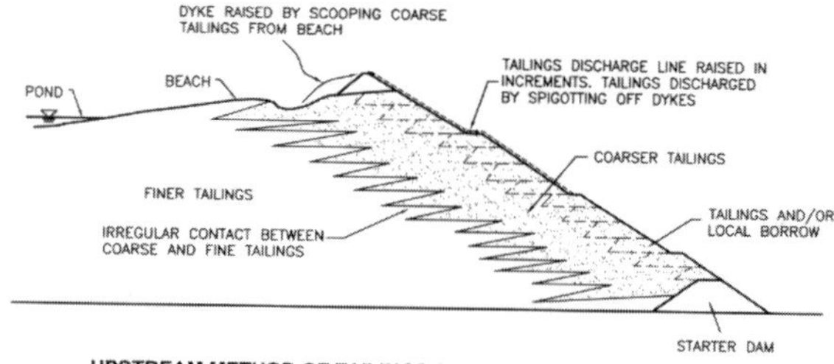

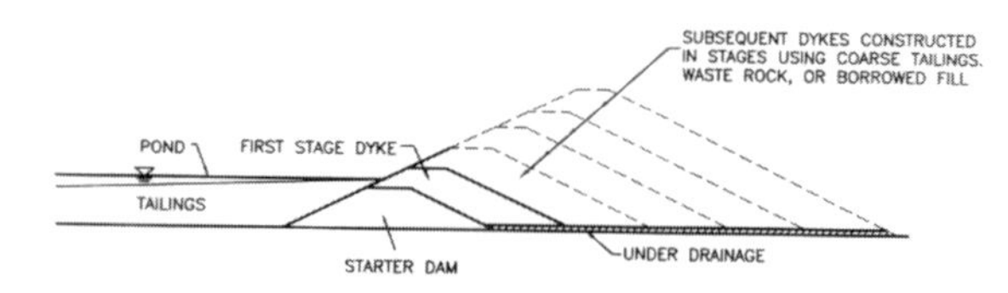

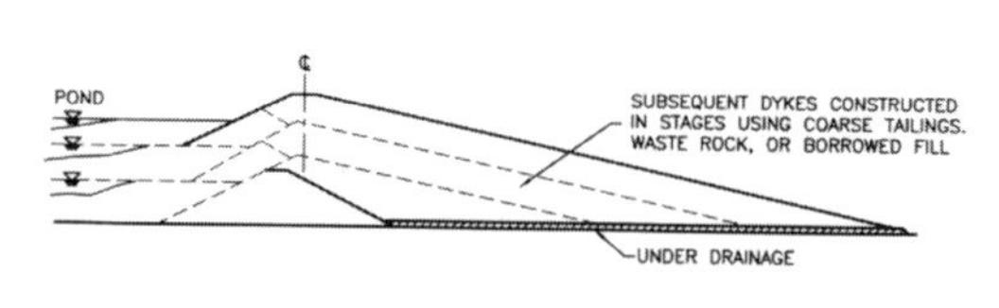

FIGURE 40.6 Tailings dam geometry definitions

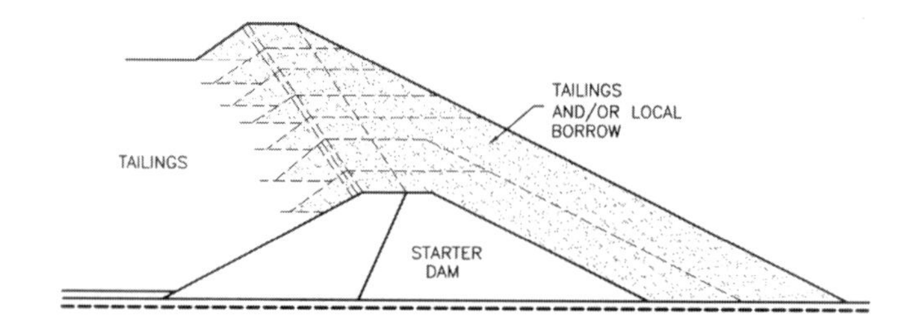

FIGURE 40.7 Hybrid geometry section: crest over tailings = upstream construction

- *Downstream Construction*—The dam crest during construction stages moves downstream from the crest of the initial raise (starter dam).
- *Centerline Construction*—The dam crest, upstream and downstream edges, remains fixed relative to upstream and downstream directions as the dam is sequentially raised.
- *Upstream Construction*—The dam crest, including the upstream edge of the crest, moves over beached tailings at any point in its sequential raising.

The above are simple and universal. There is absolutely no need to have any further manipulation to create misleading or convoluted terminology. Based on available literature and review of many projects, the greatest difficulty for many, it would appear, is in defining upstream dams. The best test for an upstream-constructed facility is whether a vertical line extended from the upstream crest ever touches beached tailings. If the answer is "yes," regardless of how much it touches, the facility is an upstream-constructed facility. For example, Figure 40.7 shows a hybrid geometry, involving elements of downstream and upstream construction. However, the crest has advanced onto tailings and therefore this is an upstream dam.

Based on available literature, there are at least 3,500 tailings dams worldwide. Although not a precise estimate, the authors suggest that more than 50% of these are upstream constructed. There have been fewer than 100 documented "significant" upstream failures, suggesting a failure rate of no more than 5 or 6%. Many upstream facilities, particularly those with wide beach-above-water (BAW) zones, have performed well during significant seismic events, as well as when subject to intense statistically infrequent rainfall events and in spite of questionable stewardship. In other words, if a set of conditions can be met, it appears as though upstream tailings dams can still be considered an option for modern tailings facilities. Centerline and downstream type dams, being generally more "robust" than upstream dams, have an even greater likelihood of success, but again only if certain basic requirements are satisfied.

40.7 INGREDIENT FOR SUCCESS NUMBER 2—UNDERSTANDING DESIGN REQUIREMENTS

From a "basics" perspective, the following is a handy checklist of issues to consider when designing a tailings dam:

- Foundation conditions
- Climatic conditions
- Tectonic setting
- Tailings properties
- Geotechnical considerations
- Geochemical considerations
- Operating criteria
- Monitoring and ongoing evaluation
- Closure plans
- Regulatory requirements

40.7.1 Foundation Conditions

For upstream dams, pervious foundations with a proven (by site investigation and design) resistance to liquefaction and/or extensive strain-softening under design static and seismic loads are preferred. Where foundation conditions can lead to rapid movements (e.g., centimeters per day or faster) and/or brittle strain-softening, upstream dam development is contraindicated. In these poorer foundation conditions, a more strain-independent dam is advisable.

Centerline and downstream-constructed tailings dams, while perceived as more robust than upstream dams, are no less dependent on the dam foundation for their stability, as the Los Frailes failure so dramatically demonstrated. Such dams are often placed on impervious foundations to minimize seepage losses. The potential for geologic defects, such as presheared surfaces, in such foundations must be recognized.

40.7.2 Climatic Conditions

Climate can dictate what type of tailings dam is most appropriate and what type is least appropriate. For example, upstream dams are far more likely to succeed in arid conditions (where long beaches can be truly subaerial) than in temperate conditions, although there are many successful examples of the latter. Tailings dam designs that are successful largely because of arid climates, such as in South Africa or Arizona, do not necessarily translate as sound concepts in wetter coastal climates, such as North America.

40.7.3 Tectonic Setting

Understanding the seismicity of the tailings dam site is essential. The importance is even greater when contemplating upstream construction. Seismic events have been responsible for roughly one quarter of upstream dam failures (conversely, there have been essentially no dramatic failures of other types of tailings dams due to seismic loading alone).

As noted by McLeod, Chambers, and Davies (1991), there are really no aseismic regions in the world. However, it is the regions that can provide mechanisms that create several significant cycles of peak ground accelerations greater than 0.1 g within the tailings that are of greatest concern. Analyses appropriately assessing the attenuation or amplification of seismic motions by the tailings dam are needed to determine seismic design parameters.

In designing for closure, the cumulative affect of multiple seismic events on a tailings dam over its long closure period should be considered (Szymanski 1999). Typical practice has been to assume that acceptable dam performance under a single maximum credible earthquake (MCE) event is an acceptable closure condition. However, what may actually govern is the cumulative effects of a number of smaller but more frequent events.

40.7.4 Tailings Properties

All tailings are not identical in mechanical behavior. It is essential that the gradation, fabric (including grain angularity), and bulk density be determined and understood. At the design stage, the likely pore pressure conditions need to be appropriately modeled and they need to be appropriately measured, plotted, and interpreted during operations (Stauffer and Obermeyer 1988; Martin 1998). Where pore pressure increases due to load additions exceed ratios greater than about 0.8, shear-induced pore pressures due to contractant response are a likely explanation and can present warnings of potential instability. A proper combination of pore pressures and bulk density is required to correctly estimate in situ stresses and, hence, in situ state. Ladd (1991), Carrier (1991), and Martin and McRoberts (1998) all touch on the nonconservatism of using incorrect values. From the authors' experience, too many facilities are being designed using a blanket specific gravity (S_G) = 2.67 to determine bulk density and hydrostatic conditions. Mineral tailings frequently have widely differing values of S_G. This complacency has been directly attributable to at least one failure where the designer's calculations showed the dam to be very safe.

40.7.5 Geotechnical Considerations

Transient loads leading to brittle response are the most important geotechnical consideration. Often, seismic loading is the most critical transient load.

Tailings facilities in several major parts of the world are subject to specific guidelines with respect to seismic design. For seismic liquefaction to occur, the number of significant cycles imparted by earthquakes must generate sufficient pore pressure rise to result in brittle behavior. The importance of defining the tectonic setting was discussed earlier. The resulting behavior of the tailings (and the dam foundation) in response to seismic loading depends on the initial state of the tailings (or the foundation soils) and the magnitude and nature of the seismic event.

Nonseismic transient loads can also lead to liquefaction. Static liquefaction is often a more difficult phenomenon to describe and/or anticipate than its seismic counterpart. However, it is a prime failure mode for upstream-constructed tailings dams. There is limited mention of static liquefaction as a phenomenon in available literature. As with seismic liquefaction, the most common way to address static liquefaction design issues is to use empirical relationships and/or laboratory testing. With laboratory testing, both compression and extension stress path triaxial data and direct simple shear data on the tailings are required to do a site-specific evaluation. These simplified stress paths are illustrated schematically in Figure 40.8. From this data, at least conceptually, the "collapse surface" can be approximately located within the lines of phase transformation (steady state).

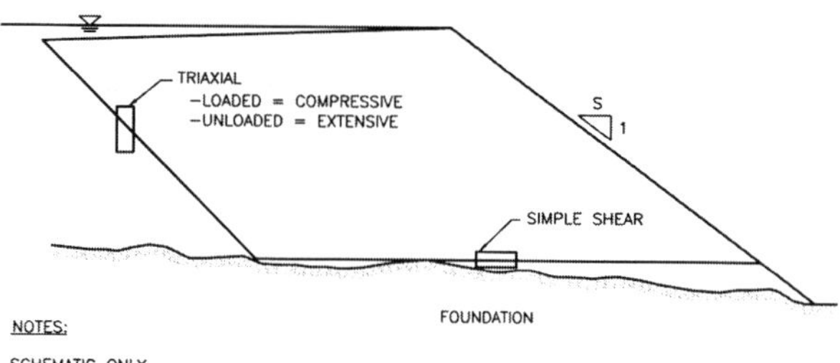

FIGURE 40.8 Simplified tailings impoundment stress states

The lines of phase transformation are identical in either compressive or extensive stress space, as shown in Figure 40.9. However, this isotropy is not evident with the collapse surface, which is steeper in compressive loading than in extension. This anisotropy is largely due to fabric/grain imbrication (almost always preferential to the horizontal plane) and increases with grain angularity and elongation. These two characteristics are common to ground mill tailings. The imbrication, due mainly to hydraulic deposition processes, results in elongated grains being aligned preferentially in horizontal to subhorizontal layers. Typically, this horizontal plane is normal to the maximum principle stress, resulting in additional cross-plane anisotropy in triaxial loading conditions.

For many tailings, a compressive loading slope of about 2H:1V to 3H:1V and an extensive loading slope of about 3.2H:1V to 3.9H:1V results; the implication being slopes flatter than these values have overall lower shear bias and hence lower risk of large-scale spontaneous/static liquefaction given equivalent in situ states. It is interesting to note from the instability database that no tailings dam with a slope flatter than 4 to 5H:1V has had a failure attributable to a "spontaneous" static liquefaction event (i.e., negligible trigger application to induce undrained behavior). On the other hand, there are several upstream-constructed tailings dams of overall or intermediate 2H:1V to 3H:1V slopes that have failed in this manner. Review of the Stava dams sections in Figure 40.2 in the context of Figure 40.9 shows just how precarious was the state of these dams. Kramer and Seed (1988) demonstrated in the laboratory that there is a marked increase in static liquefaction susceptibility with increase in principal effective stress ratio. This type of soil behavior has been observed by many other researchers and described in literature at least as far back as Bjerrum, Kringstad, and Kummeneje (1961).

Another geotechnical consideration, given full discussion in Carrier (1991) and Martin and McRoberts (1998), is the appropriate analytical framework for assessing slope stability. Effective stress analysis (ESA) is only appropriate *if* the pore pressures within the dam and foundations are known *accurately* at all times, including during shearing. For potential failure conditions, this involves estimation techniques, because the actual pore pressures at failure can only be measured once it is too late for the given dam. It is nearly impossible to monitor pore pressures accurately within an upstream dam undergoing raise-induced pore pressures and shear strains (e.g., due to foundation or internal dam deformations that are also inducing pore pressures). Further, there is little benefit in obtaining data on shear-induced pore pressures if the dam has to approach a state of failure to provide this data. The use of undrained strength analyses (USA), which implicitly includes consideration of shear-induced pore pressures via use of the undrained strength, should be included in the assessment of most tailings dams and *always* in assessing upstream tailings dams.

Another key geotechnical consideration is the potential for internal erosion. Incompatible filter materials must be avoided. The use of pipeworks (e.g., decants) in dams should also be discouraged, because these features rarely offer any economical or technical advantages and yet open the dam to an additional range of failure modes. Internal pipework should be used only if multiple lines of defense to internal erosion, especially filters, are present and the decant is not subject to potential rupture.

40.7.6 Geochemical Considerations

Although not a direct instability issue, geochemical issues are of paramount importance in selecting tailings dam type and operating criteria. The past 20 years have brought significant enlightenment to geochemical issues and manners with which to minimize/mitigate potential downstream impacts from stored mine tailings. From an environmental perspective, seepage that is contaminated by heavy metals can be just as severe a failure as a physical dam breach.

It is often necessary to maintain sulfide tailings ponds in a flooded condition to prevent oxidation upon closure. The alternative is typically "perpetual" collection and treatment. As discussed by Szymanski (1999) and Vick (1999), the environmental advantage gained by flooding sulfide tailings comes with an often unrecognized and undesirable dam safety side effect: flooded impoundments are more prone to failure than those that are closed with minimal impounded water. This fact appeared to have gone unrecognized in the initial "euphoria" over the efficacy of flooding in preventing mining's greatest environmental problem. Only recently are the dam safety implications of adopting this practice starting to be appreciated.

40.7.7 Operating Criteria

No amount of good design will overcome poor tailings impoundment stewardship. Poor stewardship typically comes from one of two sources:

- Operating criteria that are incompatible with the geotechnical design
- Facility operations that ignore the basic fundamentals of the design

Obviously, when both sources are present, the dam is provided with a much better opportunity to fail. In the first case, the rule is simple: If the design requirement is a necessary component of structural integrity, do not create an operating scenario that makes achieving the design requirement impossible. As an example, in 1996, an upstream dam in South America underwent a slope failure and subsequent flowslide due to excessive encroachment of the pond onto the beach. The designer's requirement of a wide BAW zone was combined with a pond size inadequate to allow settling of fines. This facility was therefore doomed to suffer either a geotechnical or operational failure. The unfortunate result, as shown in Figure 40.10, was a failure on both counts.

Another anecdotal example involves a facility where the design of a BAW zone was carried out using laboratory-derived beach slopes that, during regular tailings line flushing, were reduced to 0.5%, thereby causing pond encroachment and subsequently instability.

40.7.8 Use and Abuse of the Observational Method

Designers of tailings dams commonly invoke the observational method as an integral component of the construction and operation of a tailings facility. It is therefore worthwhile to review the use and abuse of this process.

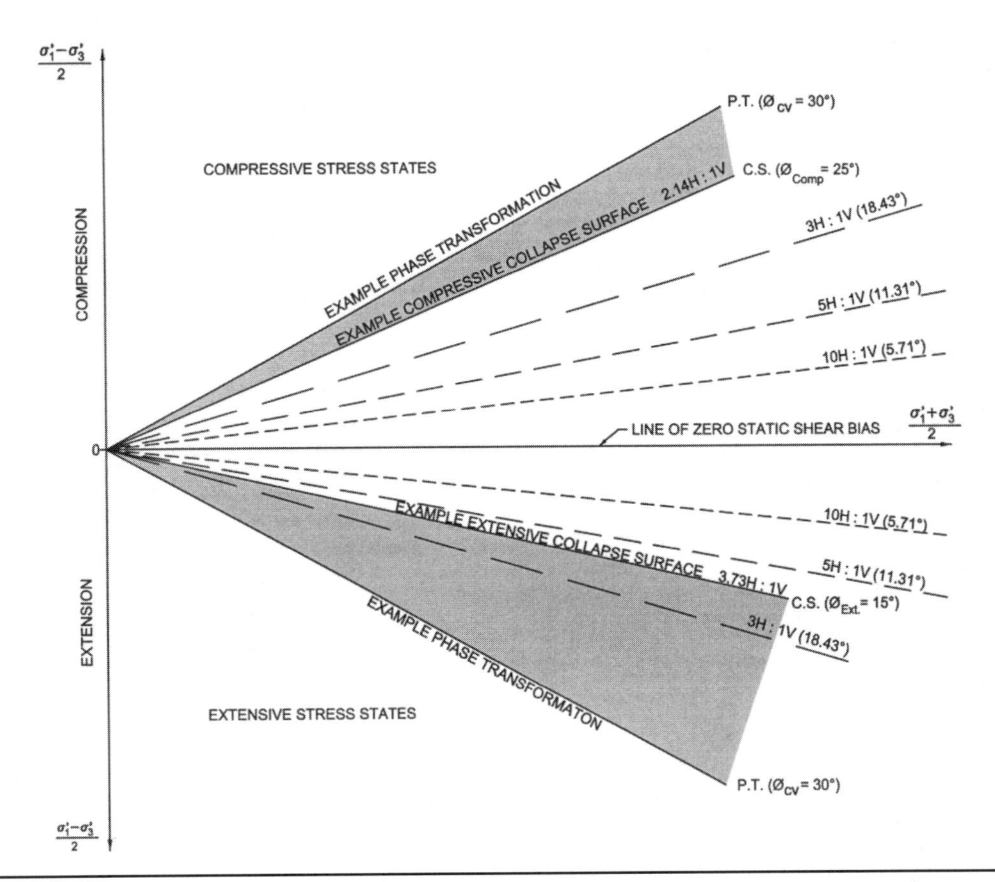

FIGURE 40.9 Tailings impoundment slopes in relation to example phase transformation and anisotropic collapse surfaces

FIGURE 40.10 Tailings dam failure in South America, 1996

The observational method in geotechnical engineering, as described by Peck (1969), requires proper execution of the following elements:

- Geological and geotechnical investigation sufficient to determine site conditions to suitable level of detail for design
- Assessment of the most probable conditions and the most unfavorable conceivable deviations from these conditions
- Design based on most probable conditions (with an appropriate degree of conservatism incorporated)
- Selection of parameters to be observed as construction proceeds, and prediction of values based on design analyses and assumptions
- Calculation of values of the same parameters that would correspond to the most unfavorable conditions conceivable
- Selection in advance of a course of action or design modification for every foreseeable significant deviation of the observed findings from those predicted in design
- Field measurement of parameters and evaluation of actual conditions relative to conservative conditions assumed for design
- Evaluate results and carry out design modifications as deemed necessary based on observed conditions

Even the safest dam cannot be without some finite risk. To manage this residual risk, dam engineering practice, in general, and tailings dam engineering practice, in particular, places considerable reliance on monitoring structure performance to confirm satisfactory performance and to confirm design assumptions, the two basic functions of tailings dam surveillance. Tailings dam construction, because it happens on a near-continuous basis, provides the opportunity (if not the necessity) to optimize design and construction due to the ability to monitor performance and adjust design and/or construction accordingly.

Failure to correctly apply each element results in following the "hope-for-the-best" method rather than the observational method. The most common abuses of the method involve one and, more typically, combinations of the following:

- Failure to recognize the most unfavorable conceivable deviations from assumed conditions and performance
- Failure to select in advance a course of action or design modification for every foreseeable deviation of observed conditions from assumed conditions
- Dependence on the method to detect potential failure modes that occur suddenly with few or no warning signs and/or with insufficient time to enact suitable countermeasures when (or if) such warning signs are detected

Finally, the method was never intended to "rescue" a substandard design that was based on inadequate site investigation and insufficient detail, forethought, or, in an alarming number of instances, general inability to properly design and construct tailings facilities. Nor was it intended to replace the appropriate degree of design conservatism required for tailings dams. These are the worst forms of abuse of the method, but ones seen all too frequently in tailings dam projects. Economic pressures and/or pressures in feasibility study stages to produce a positive outcome can result in a "don't worry about it now, we'll figure it out later as we go along" mentality. This approach almost always results in false economics.

40.7.9 Surveillance and Evaluation

Surveillance of tailings dams represents the underpinning of the use of the observational method in maintaining a stable condition. Any tailings dam, by its very nature, is a "work in progress." Conditions are constantly changing, and, in contrast to conventional dams, the steady-state condition of the impoundment is not established until into the closure phase (Szymanski 1999). Detailed monitoring and review of the facility is required throughout its life (which includes the closure period) to check that its performance and condition are satisfactory. It is essential that designers and operators of tailings facilities work as partners throughout the design, construction, operating, and closure phases to ensure that the expertise of each is appropriately applied throughout. It is also essential that monitoring and review requirements be specified as part of the design process and that the owner/operator is made aware of the importance of these facets of the design.

An effective tailings dam surveillance program should include all of the following elements:

- Determination of the data that needs to be gathered to confirm satisfactory performance
- Design of a surveillance program that will obtain the required data, with a suitable level of redundancy (repeatability)
- Training of personnel responsible for collection of the data as to how to obtain and manage the data and to provide awareness as to warning signs
- Provision of the installations and resources required (e.g., instrumentation, time, personnel, access) to permit responsible personnel to carry out the program
- Maintenance of the installations (e.g., piezometers, inclinometers, seepage weirs) required to carry out the program; a corollary is establishing a level of "respect" for installations by all on-site personnel to avoid, for example, damage to critical installations by vehicle traffic
- Appropriate presentation and communication of the data (e.g., plot of piezometer level versus Time, including trigger levels) to allow for quick and effective interpretation and response
- Prior definition of warning signs and response plans to be enacted in the event that warning signs are noted
- Clear and documented lines of communication and responsibility in terms of data gathering, plotting, reporting, and action
- Regular tailings dam designer review and input in terms of data requirements, interpretation, and response
- Appropriate documentation and database management practices to avoid the TDA malady
- Reviewing and updating the surveillance program, involving the dam designer, no less frequently than once per year (less frequent review is likely during closure phase)
- Periodic auditing of the program by an expert independent of the design team
- Documentation of the surveillance program requirements, procedures, and lines of responsibility, typically in the form of an operations manual

40.7.10 Closure Plans

Modern-day facilities should be designed and operated with closure in mind. For example, if a wide BAW zone is required for perpetual structural integrity under static/seismic loading, consider this prior to selecting any dam geometry for, as an example, sulfide-rich tailings that will require a closure water cover due to oxidation concerns (see Geochemical Considerations). It is less palatable to rehabilitate a dam at mine closure once the revenue stream is gone than to choose the appropriate dam geometry for closure conditions at the outset. Similarly, and as discussed previously, consider the dam safety implications of the permanent submergence scenario carefully before selecting it.

40.7.11 Regulatory Requirements

Regulators worldwide are becoming increasingly educated about tailings dam design and stewardship requirements. However, upstream dams remain unpopular with many regulators, even when the entire technical checklist points to success with the method. Improved communication (e.g., not renaming upstream dams) and disclosure of all issues on a technical checklist can go a long way in demonstrating necessary partnership with a regulator. Of course, there are jurisdictions in which upstream construction is more difficult to gain approval and the costs of such a process, including potential nonpermitting, need to be weighed against another storage geometry at a probable higher unit storage cost.

40.8 INGREDIENT FOR SUCCESS NUMBER 3—STEWARDSHIP

40.8.1 Industry Organizations

Tailings dam stewardship may be defined as the direction and implementation of all design (conceptual through detailed), construction, operations, inspection, surveillance, review, and managerial aspects (corporate policies, training, roles and responsibilities, documentation and reporting, etc.) involved in seeing a tailings facility through from conceptual design through closure. No single issue can be said to be more important than another.

Over the past several years, initiatives have been taken on a variety of fronts to improve the stewardship and therefore the safety of tailings impoundments. In addition, design practice for tailings dams continues to evolve and improve. Stewardship is of equal significance to technical design aspects because design errors can be detected and, in some instances, prevented from manifesting themselves as failures under proper stewardship. However, no stewardship will be sufficiently robust to cover all potential design flaws, and no design is sufficiently robust for the most negligent stewardship.

The initiatives discussed have been spearheaded by regulatory agencies, the United Nations Environment Programme (UNEP), the Canadian Dam Association (CDA), the Mining Association of Canada (MAC), and, most importantly, individual mining companies. The importance of the mining industry and individual mining companies leading these efforts cannot be overstated. If the industry cannot properly manage its tailings facilities, someone else (e.g., regulators) may impose proper management or their version of it. More failures will serve to strengthen the hand of mining's detractors. Tailings dam design consultants can take a significant role in assisting mining companies in developing and implementing proper stewardship programs.

Mining Association of Canada. The Mining Association of Canada (MAC) has recently published a document titled *A Guide to the Management of Tailings Facilities* (MAC 1998). This document, reviewed by industry and consultants, is very valuable, providing a framework of management principles, policies, and objectives and checklists for implementing the framework through the life cycle of a tailings facility. It is general in nature but demonstrates the need for individual mining companies and mining operations to establish their own specific programs to meet their own specific needs.

Canadian Dam Association. The Canadian Dam Association (CDA) recently updated its dam safety guidelines (CDA 1999). The update focused on incorporating elements specific to the safety of tailings dams. There are recommendations regarding the following issues that are very relevant to tailings dam stewardship:

- Responsibility for dam safety
- Scope and frequency of dam safety reviews, which includes a review of operations and maintenance and dam surveillance program
- Operation, maintenance, and surveillance, including need for an operations manual
- Emergency preparedness, including elements of an emergency preparedness plan (EPP)

ICOLD and Related Organizations. The International Committee on Large Dams (ICOLD) and related organizations have published materials with regard to tailings dams. Bulletin 74 (ICOLD 1989) presents guidelines for tailings dam safety. Bulletin 104 (ICOLD 1996) specifically addresses monitoring of tailings dams. Other ICOLD publications pertaining to tailings dams are as follows:

Bulletin 97. 1994. Tailings Dams–Design of Drainage–Review and Recommendations

Bulletin 98. 1998b. Tailings Dams and Seismicity–Review and Recommendations

Bulletin 101. 1995a. Tailings Dams–Transport, Placement and Decantation

Bulletin 103. 1996b. Tailings Dams and Environment–Review and Recommendations

Bulletin 106. 1996. A Guide to Tailings Dams and Impoundments–Design, Construction, Use and Rehabilitation

The U.S. Committee on Large Dams (USCOLD) published a compendium of tailings dam incidents (USCOLD 1994). This was probably the first attempt to catalogue and assess published information on tailings dam incidents. The document categorizes the various incidents in terms of technical causation, but there is no discussion of the extent to which inadequate stewardship played a role in the various incidents. The value of this document is that it discusses failure modes for different dam types, which is of benefit in scoping out requirements for a dam safety program, particularly dam surveillance.

UNEP/ICME. The United Nations Environment Programme, Industry and Environment (UNEP) and the International Council on Metals and the Environment (ICME) have been active in recent years in sponsoring of seminars and publishing of case studies (ICME-UNEP 1997, 1998) related to tailings management. Many of the topics covered directly address stewardship issues. Mining companies provided most of the contributions to these publications, thereby making these forums an excellent way to disseminate knowledge and experience to the international mining community. Key topics covered that relate to stewardship of tailings dams include

- Corporate policies and procedures regarding stewardship of tailings facilities
- Evolving regulatory climates and trends
- Definitions of roles and responsibilities
- Application of risk assessment techniques
- Environmental management systems
- Emergency preparedness and response
- Education and training

40.8.2 Mining Company Initiatives

The most encouraging trend in terms of tailings dam stewardship is that it is the mining industry, in general, and individual mining companies, in particular, that are leading the way in improving the state of practice and, equally as important, in sharing and publishing information. The following sections discuss some examples of proactive stewardship policies and practices being followed by mining companies.

Corporate Policies and Management Issues. Several major Canadian-based mining companies have established corporate policies and procedures to ensure that all personnel involved in stewardship of tailings facilities, from the corporate level to the operators, clearly understand their roles and responsibilities (Siwik 1997). Such an understanding and enforcement (performance measurement) of those roles and responsibilities is vitally important. The authors have reviewed many tailings disposal facilities where there was no such understanding and no "ownership" of key stewardship functions. A number of companies have also established policies with respect to degrees of training and competency required for the various roles involved in tailings facility stewardship (Siwik 1997; Brehaut 1997; Maltby 1997). This is extremely important, especially for tailings dam operators who have the most frequent exposure to the facility and usually have frontline responsibility for dam surveillance. It is essential that these personnel understand what to look for and why, what constitutes unfavorable conditions, and what to do about them.

Auditing of Tailings Facilities. Brehaut (1997) describes how, in an internal evaluation of its management systems, Placer Dome recognized that its tailings management systems were a priority for enhancement and that tailings management was an issue of great import at a corporate level. Placer Dome subsequently embarked on developing of guidelines for the design, construction, operation, and closure phases of tailings management systems. Placer Dome also determined that the application of risk assessment techniques was an essential next step in the review and enhancement of its tailings dam stewardship policies and procedures.

However, no sooner had this process been initiated than the Marcopper incident occurred in the Phillipines, involving release of about 2 million tons of tailings into a local river system. As is widely known within the industry, but not appreciated without, Placer Dome's response to the incident was exemplary. Brehaut (1997) indicated that the total cost to the company was estimated to be $43 million after insurance and tax recovery. The total cost to present far exceeds that value. Spurred on by the Marcopper incident, Placer Dome quickly initiated formal risk assessments of the tailings facilities at all of its operations. In many instances, these risk assessments were carried out and/or facilitated by geotechnical consultants who were not the engineers of record for the various facilities audited. This was an effective application of third-party review. The findings of these risk assessments (Brehaut 1997) revealed that any design deficiencies identified were of minor significance and the greatest weakness was related to management aspects of the stewardship of the facilities.

Other mining companies have implemented similar risk assessment programs for many of their tailings facilities.

Geotechnical Review Boards. Syncrude, Kennecott Utah Copper, Inco, and other mining companies retain a board of eminent geotechnical consultants to provide independent review and advice in terms of the design, operation, and management of their respective tailings facilities. Such review boards are independent of the design engineers, be they consultants or geotechnical personnel the mining company has on staff. Review boards are now considered to be standard for dam stewardship practice for owners of major water dams.

McKenna (1998) describes how Syncrude Canada Ltd., a large oilsands company in northern Alberta, Canada, has benefited from its geotechnical review board over the last 25 years, summarizing these benefits as follows:

- The board provides expert assistance in terms of assessing and managing risk.
- The board ensures that all of the bases are covered (i.e., posing the question "Has anything been missed?").
- Review board members bring to Syncrude a vast and varied practical experience and expertise.
- The board reassures senior management that an acceptable balance between risk taking and conservatism is maintained in an operation where the consequences of failure are extreme.
- Independent review by preeminent specialists gains the trust of regulators and the public and facilitates regulatory processes.
- Design engineers benefit through in-depth review of their work by preeminent specialists.

Another very important benefit afforded by a geotechnical review board is the continuity it provides. For example, over the 25-year period during which Syncrude has maintained a review board, there has likely been considerable turnover in staff and consultants. However, the current members of Syncrude's review board have been on the board, more or less continuously, since the board first formed in 1972. Such continuity represents a good defense against the development of TDA.

In summary, a review board can provide an objective view as to the potential, consequences, and cost of a potential failure and help the owner ensure that decisions on design alternatives are not based solely on minimizing capital and operating costs.

Information Database. McCann (1998) describes systems that Inco has implemented to develop an information matrix for maintaining records, in an easily retrievable manner. The records pertain to the design, construction, operation, and monitoring of its Copper Cliff tailings facilities, in use since the 1930s. For such a facility, given the inevitable turnover of operations and management personnel and design consultants, a good database is essential to maintaining continuity and preventing TDA.

After carrying out more than 30 risk assessments of tailings dams worldwide, the authors have reviewed the entire available "tailings library" at many mines. It is almost a certainty that any operation over 10 years of age will demonstrate TDA and will repeat costly studies, ignore essential design criteria, or unknowingly reinvent a tailings management plan without appreciating the "forgotten" earlier information. Maintaining the same consulting organization does not seem to stem the onset of TDA unless the mine itself is an active partner in tailings dam stewardship. The authors' review work also shows that tailings dams that have had past "incidents" are often well placed to have them reoccur.

40.8.3 The Role of Tailings Dam Designers

Tailings dam design consultants play an essential role in promoting good stewardship of tailings facilities, in addition to the obvious technical role of providing safe, cost-effective, practical, and enduring designs. Consultants obviously have common cause with the mining industry in this regard, and the following sections discuss a number of ways consultants have, are, and should be contributing to this effort.

Publications and Participation in Conferences. Tailings dam consultants work on a variety of projects, in many countries, and for many clients. By so doing, they amass a variety of experiences from which the mining industry and other consultants can and should benefit. Conferences such as the annual Tailings and Mine Waste Series in Fort Collins, Colorado, provide a forum for exchange of ideas and sharing of experiences. More emphasis, however, needs to be placed on stewardship issues rather than purely technical topics that typically dominate such conferences.

Consulting engineers specializing in tailings dams have also contributed by publishing entire books on the subject. The first such book with widespread distribution was written by Steve Vick, titled *Planning, Design, and Analysis of Tailings Dams* (Vick 1983). It is an excellent treatise recommended for all persons responsible for any aspect of tailings management. More recently, Maciej Szymanski published a book titled *Evaluation of Safety of Tailings Dams* (Szymanski 1999). This book provides a detailed discussion of elements of a comprehensive safety program specifically tailored to tailings dams. Issues related to good stewardship are discussed throughout.

The Design Product. The design product provided by tailings dam engineers to the owner is, all too often, rich in discussion of the finer points of soil mechanics and computer science but deficient in terms of detailed guidance for operation and surveillance. The condition of a tailings facility is governed by how it is operated and constructed, not necessarily by how it was analyzed and designed. Likewise, its safety is better judged based on surveillance rather than design analyses in the appendix of a design report. The design report must therefore include operational and surveillance requirements. Ideally, an operations manual or, at least, most elements of one should be provided with the design, otherwise the design is not complete.

Risk Assessments. Consultants are more frequently applying techniques that can be broadly categorized as "risk assessments" to various facets of tailings management, most notably in reviews of existing facilities. Mining companies are also applying such techniques in the stewardship of their tailings facilities. A risk assessment, by the authors' definition, provides answers to the following questions:

- What can go wrong?
- How likely is it that it will happen?
- If it does happen, what are the consequences?

- What can/should be done to reduce the likelihood and/or consequences of this potential occurrence?
- The authors have made frequent use of failure modes and effects analysis (FMEA) in workshop settings that include mine management and operating personnel. FMEA and most other qualitative risk assessment methods are nothing more than organized judgment, commonsense with a fancy name. Risk assessment techniques can be used to audit any number of technical and managerial aspects of tailings dam stewardship.

As an example, the FMEA technique, carried out in a workshop setting, is particularly effective in scoping out requirements for dam surveillance and ties in well to the correct application of the observational method that is so fundamental to tailings dam design and safety. The FMEA process captures the key elements of comprehensive dam surveillance, including

- Identification of potential failure modes
- Identification of warning signs for failure modes
- Consideration of how quickly failure could occur and how potential problems can be detected well in advance of their developing into incidents
- Consideration of the significance of temporal trends as opposed to single measurements/observations
- Allowance for "green light" (safe) versus "yellow light" (caution) versus "red light" (stop) limits/criteria to be established

The workshop format, involving personnel responsible for dam surveillance as well as management personnel, provides the following:

- Forum for interchange of ideas and concerns
- Technical, operational, environmental, and management input
- Transfer of essential knowledge from the designers to "frontline" personnel, and vice versa
- Development of a team approach to dam surveillance
- Buy-in from responsible parties

The FMEA process provides the following:

- A structured, repeatable, and documented process
- Assessment of current surveillance practices in terms of scope, frequency, reporting and interpretation, response to unusual conditions, and resources available versus resources required
- Identification of aspects requiring improvement
- Justification for allocation of resources to dam surveillance
- An action plan that evolves directly from the process
- Avoidance of "overconfusion" that quantitative risk assessments often bring (e.g., quantification of issues not readily quantified; the "black box" syndrome)

Training Seminars. Tailings dam designers are increasingly being called upon to provide on-site training to operators. Training facilitates transfer of key knowledge from the design engineer to the operators, who represent the designer's "eyes and ears." Such seminars are as beneficial to the designer as the operators, if not more so, providing the designer with feedback on the constructability of the designs and the practicality of the operating requirements imposed by the designs. Further, it is the authors' experience that designers of tailings dams can learn more of practical benefit to themselves and their clients from a day on site with experienced operators than a week of reading technical papers. Tailings dam operators truly appreciate such seminars, which acknowledge their importance to the operation and facilitate their buy-in, understanding, and commitment to good stewardship practices.

Designer Expertise and Perspective. As noted earlier, tailings dams and impoundments are unique structures in the engineering world. Design consultants should have appropriate educational and practical experience directly applicable to tailings dam design. A proven construction and operations history for the designer(s) projects is of extreme importance. There a limited number of qualified designers and organizations (more than one designer in house), and owners should share information with one another as much as possible regarding design consultants. It is an unfortunate fact that "all designers are not born equal." Independent peer reviews or review boards are ways mining companies can protect themselves from questionable designs. Design consultants should proactively encourage their clients to adopt these protective measures, which, as discussed previously, are also very much to the design consultants benefit.

Designers themselves need to maintain an appropriate balance between the real world (i.e., case history experience, both good and bad) and theory (e.g., laboratory testing and viscoplasticity theory), with the real world always being the most important. Before tailings dam engineering became a formal discipline in the late 1960s, tailings dams were designed and constructed based almost entirely on experience (trial and error), without the benefit of soil mechanics theory. Today, the pendulum may have swung too far in the opposite direction, with many tailings dams being designed "in the laboratory" and in computer simulations. Terzaghi is reputed to have said that nature has no contract with mathematics; she has even less of an obligation to laboratory test procedures and results. Peck (1980) pointed out that theory can inhibit judgment if used without discrimination and without critical evaluation.

General Regulatory Trends. Regulatory agencies typically do not "prescribe" stewardship practices to the industry, apart from some basic requirements (in some jurisdictions) such as requiring a dam surveillance program, annual reports, operations manuals, and so on. It is in terms of environmental issues (water quality, for example) that regulations are, necessarily, prescriptive. This is as it should be, because mining companies themselves, supported by their design consultants as appropriate, are best qualified to design stewardship programs appropriate to their particular facilities. An attempt by regulators to impose a uniform "code of stewardship" would be unsuccessful because each mining company and each tailings facility has their own unique requirements, resources, and constraints. Much of stewardship relates to the corporate and mine-specific organizations and personnel, aspects that cannot be regulated.

Regulatory Trends in Developing Countries. An interesting regulatory trend is that codes and standards with respect to tailings disposal are becoming "harmonized" internationally. Many developing countries, Bolivia being one example, have only recently enacted regulations covering tailings disposal. These regulations are largely modeled after World Bank guidelines and regulations in North American jurisdictions. Unfortunately, there appears to be a trend among developing countries to make their regulations technically prescriptive, with some (e.g., Peru) actually presenting a very elementary course in soil mechanics, describing methods of stability analysis and liquefaction potential screening methods. This may be due, in part, to the comparative lack of skill and expertise in tailings dam engineering on the part of the regulators and, in part, to their understandable desire to have tailings facilities in their respective countries conform to "international" (i.e., ICOLD) standards, which really do not exist in any tangible, readily referenced form. This development of prescriptive regulations is a misguided trend, because it can provide a false sense of security to the regulators and, worse, the mining companies. Many case histories of failures likely involved facilities that were in conformance with all regulations, except the most important of all (the dam failed).

Reliance on regulations that meet "international standards" and are full of prescriptive design criteria and elementary soil mechanics do nothing to address actual operating practices and stewardship issues. As discussed previously, even the most robust design can fail if not stewarded properly. Since regulations generally do not address stewardship issues, it is incumbent on the mining industry to take the lead in this regard. The mining industry and its consultants could be of great assistance to regulators in developing countries by providing training and workshops for regulatory personnel responsible for tailings facilities. It is in the mining industry's best interest to achieve a condition of "coregulation," whereby mining companies regulate themselves to a greater degree than do the regulatory agencies.

Regulatory Trends in Developed Jurisdictions. In more developed jurisdictions (e.g., the authors' home jurisdiction of British Columbia, Canada), regulators employ geotechnical engineers experienced in tailings management. Here, regulations are not prescriptive, neither in the technical sense nor the stewardship sense, because regulators have the expertise to assess the design of each tailings facility and the manner in which it is operated on a case-by-case basis. Regulations in British Columbia do require that an operations manual exist for each tailings facility, but the contents of those manuals are the responsibility of the owner. Similarly, a dam surveillance plan is also required (typically included in the operations manual). However, the details of tailings dam stewardship are left to the individual operations. An annual review report, prepared by a qualified geotechnical engineer, is also required.

In British Columbia, the regulators have actually assisted the industry by publishing of a document titled *Tailings Dam Inspection Manual*. This document is not a set of regulations but rather is intended to assist regulatory and mine personnel in inspecting tailings dams. British Columbia regulators are currently working with regulators in Peru to achieve a "knowledge transfer" to Peruvian regulators. This too is a welcome trend. The mining community is now international and includes mining companies, its consultants, and regulators. The more these three bodies can assist one another in facilitating good stewardship practices, the better off everyone will be.

40.9 CONCLUSIONS

The stability of tailings dams is inextricably linked to the design concepts and stewardship practices applied to their management and operation, from the concept stage through to closure. Positive and pro-active trends in stewardship of tailings dams, initiated primarily by the mining industry as a whole and by individual mining companies, provide the basis for considerable optimism that the trend of decreasing tailings dam failures will continue. Tailings dam design consultants can and should be of considerable assistance to the industry in this regard, placing as much emphasis and effort on stewardship issues as on technical issues in their practice. The goal of the industry, its consultants, and regulators must ultimately be to experience fewer failures than their colleagues dealing with conventional dams and to demonstrate and publicize mining industry successes to regulators and the public.

To provide some final "essential ingredients" for tailings dam stability, some minimum expectations for each of the four main participants in the tailings dam life cycle are suggested.

Owners–Only retain design assistance from reputable designers with track records that can be verified. Have submitted designs checked by independent professionals. Give serious consideration to retaining third-party review as part of a periodic audit process. During operations, retain a qualified person/firm charged with tailings dam stewardship and provide that individual with the authority and resources to retain professional assistance as deemed necessary. For older operations, be diligent in assessing the history of the operation–look for forgotten incidents involving tailings dam management.

Operators–Make certain you have an operations manual to guide you in the stewardship of your facility. If one does not exist, demand one. Maintain contact with the designer. Become educated on the surveillance requirements for your facility–understand not just what to look for but why it is important. Finally, do not allow a design to be imposed on you without your active participation. Object loudly when you are given a design that is not practical for operation.

Designers–Do not work out of your area of competence and/or experience. This includes not using "off the shelf" designs that may have been successful for you in the past but are possibly inappropriate for the climatic/tectonic/foundation conditions for the project at hand. Welcome independent review–do not view such as an attack on your design and/or competency but a benefit to you as much as your client. Do not impose a design that is theoretically satisfying but cannot be constructed and operated in the real world upon the operator. Include the operator as an active partner in the design process.

Regulators–Establish/maintain a database on all tailings dams, operating and otherwise, within your jurisdiction. Maintain candid assessments of the performance records of owners and designers and share such details with other regulators as appropriate. Facilitate developments where the owner presents an independently reviewed design that is consistent with standard design criteria. Work to repeal regulations that are incompatible with common sense.

All Participants–Do not fall victim to TDA. Lessons that are forgotten will be repeated. The knowledge required to prevent tailings dam instabilities exists–it is essential to seek it out and use it.

40.10 REFERENCES

Agricola, G. 1556. *De Re Metallica*. 1st edition.

Bjerrum, L., S. Kringstad, and O. Kummeneje. 1961. The shear strength of fine sand. *Proceedings of the Fifth International Conference on Soil Mechanics and Foundation Engineering,* Paris, France, pp. 29–37.

Brehaut, H. 1997. Tailings management–a Placer Dome perspective. *Proceedings of the International Workshop on Managing the Risks of Tailings Disposal*. ICME-UNEP, Stockholm, pp. 35–46.

Bruce. I., C. Louge, and L. Wichek. 1997. Trends in tailings dam safety. *Proceedings to CIM Annual General Meeting*, Vancouver.

Canadian Dam Association (CDA). 1999. *Dam Safety Guidelines*.

Carrier, W.D. 1991. Stability of tailings dams. *XV Ciclo di Conferenze di Geotecnica di Torino*, Italy, November.

Davies, M.P., Dawson, B.G., and Chin, B.G. 1998. Static liquefaction slump of mine tailings–a case history. *Proceedings, 51st Canadian Geotechnical Conference*, Edmonton.

Davies, M.P., and T.E. Martin. 2000. Upstream constructed tailings dams–A review of the basics. In *Proceedings of Tailings and Mine Waste '00*, Fort Collins, Colorado. Balkema Publishers, pp. 3–15.

International Committee on Large Dams (ICOLD). 1989. *Tailings Dam Safety Guidelines*. Bulletin 74.

International Committee on Large Dams (ICOLD). 1994. *Tailings Dams–Design of Drainage–Review and Recommendations.* Bulletin 97.

International Committee on Large Dams (ICOLD). 1995a. *Dam Failures Statistical Analysis*. Bulletin 99.

International Committee on Large Dams (ICOLD). 1995b. *Tailings Dams–Transport, Placement and Decantation*. Bulletin 101.

International Committee on Large Dams (ICOLD). 1996a. *Tailings Dams and Environment–Review and Recommendations.* Bulletin 103.

International Committee on Large Dams (ICOLD). 1996b. *A Guide to Tailings Dams and Impoundments–Design, Construction, Use and Rehabilitation*. Bulletin 106.

International Committee on Large Dams (ICOLD). 1998. *Tailings Dams and Seismicity–Review and Recommendations*. Bulletin 98.

International Council on Metals and the Environment (ICME) and United Nations Environment Programme (UNEP). 1997. *Proceedings of the International Workshop on Managing the Risks of Tailings Disposal*. Stockholm.

International Council on Metals and the Environment (ICME) and United Nations Environment Programme (UNEP). 1998. *Case Studies on Tailings Management*.

Kramer, S.L., and H.B. Seed. 1988. Initiation of soil liquefaction and static loading conditions. *ASCE Jour. Geotech. Engrg.* 114:4:412–430.

Ladd, C.C. 1991. Stability evaluation during staged construction. The 22nd Karl Terzaghi Lecture, Boston, 1986. *Jour. of Geotech. Eng., ASCE*, Vol. 117, pp. 537–615.

Maltby, J. 1997. Operational control. *Proceedings of the International Workshop on Managing the Risks of Tailings Disposal*. ICME-UNEP, Stockholm, pp. 161–169.

Martin, T.E., and M.P. Davies. 2000. Trends in the stewardship of tailings dams. In *Proceedings of Tailings and Mine Waste '00*, Fort Collins, Colorado. Balkema Publishers, pp. 393–407.

Martin, T.E. 1998. Characterization of pore pressure conditions in upstream tailings dams. *Proceedings of Tailings and Mine Waste '99*, Fort Collins, Colorado. pp. 303–314.

Martin, T.E., and E.C. McRoberts. 1998. Some considerations in the stability analysis of upstream tailings dams. *Proceedings of Tailings and Mine Waste '99*, Fort Collins, Colorado. pp. 287–302.

McCann, M. 1998. Sustaining the corporate memory at Inco's Copper Cliff operations. *Case Studies on Tailings Management*, UNEP-ICME, pp. 55–56.

McKenna, G. 1998. Celebrating 25 years: Syncrude's geotechnical review board. *Geotechnical News*, Vol. 16(3), September, pp. 34–41.

McLeod, H.N., R.W. Chambers, and M.P. Davies. 1991. Seismic design of hydraulic fill tailings structures. *Proceedings IX Panamerican Conference on Soil Mechanics and Foundation Engineering*, Viña del mar, Chile, pp. 1063–1081.

Mining Association of Canada. 1998. *A Guide to the Management of Tailings Facilities*.

Morgenstern, N.R. 1998. Geotechnics and mine waste management–An update. In *Proceedings of ICME/UNEP Workshop on Risk Assessment and Contingency Planning in the Management of Mine Tailings*, Buenos Aires, pp. 172–175.

Peck, R.B. 1969. Advantages and limitations of the observational method in applied soil mechanics. 9th Rankine Lecture. *Geotechnique*, 19:2:171–187.

Peck, R.B. 1980. Where has all the judgement gone? The 5th Laurits Bjerrum Memorial Lecture. *Canadian Geotechnical Journal*. 17:4:584–590.

Siwik, 1997. Tailings management: Roles and responsibilities. *Proceedings of the International Workshop on Managing the Risks of Tailings Disposal*. ICME-UNEP, Stockholm, pp. 143–158.

Stauffer, P.A., and J.R. Obermeyer. 1988. Pore water pressure conditions in tailings dams. *Hydraulic Fill Structures*, ASCE Spec. Publication No. 21, pp. 924–939.

Szymanski, M.B. 1999. *Evaluation of Safety of Tailings Dams*. BiTech.

UNEP. 1996. United Nations Environment Programme–Tailings Dam Incidents 1980–1996. A report by Mining Journal Research Services, United Kingdom.

United Nations Environment Programme. 1996. Environmental and safety incidents concerning tailings dams at mines: Results of a survey for the years 1980–1996. *Industry and Environment. Paris*, 1996, 129 pages.

U.S. Environmental Protection Agency. 1997. *Damage Cases and Environmental Releases from Mines and Mineral Processing Sites*. Office of Solid Waste, Washington, D.C., 231 pages.

USCOLD. 1994. *Tailings Dam Incidents*. A report prepared by the USCOLD Committee on Tailings Dams, November.

Vick, S.G. 1983. *Planning, Design and Analysis of Tailings Dams*. New York: John Wiley & Sons.

Vick, S.G. 1999. Tailings dam safety–implications for the dam safety community, keynote address. *Proceedings CDA Conference*, Sudbury, Ontario, pp. 1–12.

CHAPTER 41

A Database of Quantitative Risks in Tailing Management

Dr. F. Oboni* and Dr. I. Bruce†

41.1 INTRODUCTION

Recent releases of tailing effluents and solids from containment facilities around the world, including Merriespruitt in 1994, Omai in 1995, Marcopper in 1996, Los Frailes in 1999, and the recent events in Romania in 2000, have heightened awareness for mine owners and mine operators that risks associated with tailing containment must be fully addressed during all phases of a facility life. For companies owning or managing operations scattered on several continents, proper management and allocation of adequate resources dictates the need for relative risk evaluation and comparison. Indeed, attentive management of single or multiple mine sites requires objective quantitative risk assessments (QRA), detailed enough to allow comparisons, yet simple enough to be implemented by mine personnel on a regular basis.

Mine personnel can be trained during a two- to three-day workshop before applying the simplified QRA process. Typically, after the introductory course-workshop, mine personnel compile a site-specific QRA that can then be reviewed by the trainers, who in turn may ask for more detailed information on certain areas. Generally, after two to three iterations, the QRA is considered complete and ready to be passed on to the company risk managers.

Risk assessment methods such as failure modes and effects analysis (FMEA), a qualitative methodology where likelihood and consequences are evaluated purely in terms of verbal descriptions, are successful in the preliminary identification of areas of concern at one particular project. However, an FMEA does not allow objective quantitative comparisons between mines, and thus an FMEA does not offer mining corporations owning several properties the ability to compare properties and allocate resources among sites, an exercise that becomes particularly important during periods of shrinking human and financial resources. In fact, the use of qualitative descriptors, which may not be entirely consistent between mines, can be misleading and can deliver confusing results that are difficult to apply.

A method for QRA has been developed (Oboni et al. 1997, 1998; Oboni and Oldendorff 1998; Bruce and Oboni 2000) precisely to respond to the requirements of multioperation owners and has been already successfully applied to mines located in North and South America, Europe, and South Africa as described below.

41.2 QUANTITATIVE RISK ASSESSMENT

The term risk can be generally defined as the combination of the likelihood (p_H)of a specified hazard H being realized and the consequences of the event (harm and/or damage) (C_H). In many instances, the combination takes the form of a multiplication, thus leading to the risk $R_H = p_H \times C_H$. This definition is in compliance with recent standard definitions (IUGS 1997). Thus, QRAs require a numeric estimation of the probability of an event occurring coupled with the numeric assessment of the cost of damages

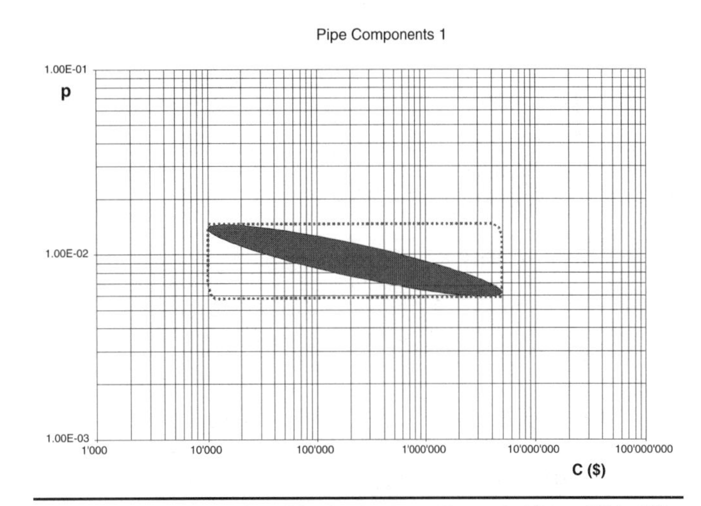

FIGURE 41.1 "Risk bubble" in the log-log p,C graph. The solid bubble corresponds to a scenario where low-cost, high-probability events coexist with higher-cost, lower-probability events (negative correlation), as in car accidents, for example.

or consequences of the event occurring. Both of these quantities are subject to uncertainties leading to various degrees of uncertainty in the definition of risk.

Once numeric values of risk are defined on an annual or lifetime basis, it is possible to compare the capital and operating costs of various mitigation techniques also on an annual basis or over the lifespan of the operation. The most cost-effective mitigation can then be defined by studying not only capital and operating costs but also by assessing the uncertainty in the assessments and the residual risks.

Since communicating the results of a QRA may constitute a challenge, in particular when addressing nontechnical people, a graphical display of risk helps people to grasp the meaning of the various terms is recommended. In this display, a risk scenario can be depicted as a "risk bubble" on a simple diagram of a p,C plan, as shown on Figure 41.1. The dimensions of the "bubble" graphically portray the uncertainty in the evaluation of p and C.

The height of the bubble represents the low to high estimate of the probability p, whereas the length of the bubble corresponds to the low to high estimate of the cost of consequences C. When a scenario has a negative correlation between p and C, as for example in the case of car accidents, where low-consequence low-speed impacts have higher probability than major crashes, the risk bubble slopes towards the right, as shown in Figure 41.1 for the solid bubble. If such a correlation does not exist, then the bubble is horizontal (dotted bubble in Figure 41.1). Log scales are used on

* Oboni Associates Inc., Vancouver, British Columbia, Canada.
† BGC Engineering Inc., Vancouver, British Columbia, Canada.

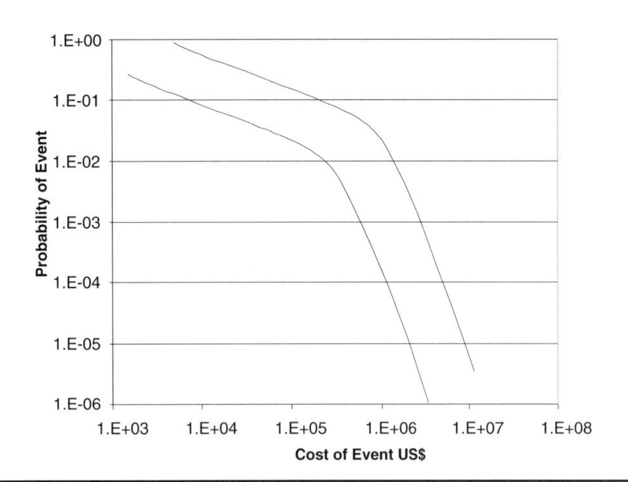

FIGURE 41.2 Risk tolerability curves (min, max) defining the tolerability band (uncertainty) in the p,C plan

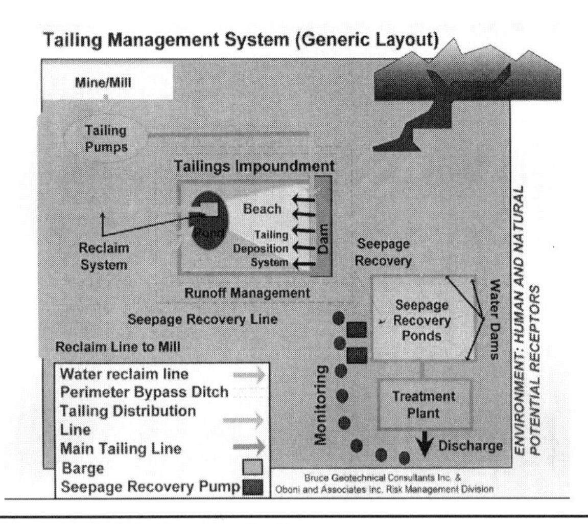

FIGURE 41.3 General layout of tailings system

both axes to allow graphical comparison between probabilities and costs that may be separated by orders of magnitude.

On the same p,C graph, it is possible to define curves of acceptable risk in consultation with the owner. Generally a band of two curves is defined in order to portray uncertainty on the owner's part related to the level of acceptability. Figure 41.2 depicts the p,C graph with an acceptability band defined by a mining client.

There are seven simple steps required to undertake a QRA. An example of how to undertake a QRA for part of a tailings containment facility is described below.

41.2.1 Step 1—Define the System

In order to fully assess the risks of a tailing containment facility, the "tailings system" is divided into different components (elements or links) starting at the tailing or slurry pump box in the mill and following through to final discharge of the effluent to the environment (Oboni et al. 1997), as shown in Figure 41.3. Components include but are not necessarily limited to

- The mill area where tailings are collected then pumped to the containment facility (pumpbox)
- Tailing transport, deposition and reclaim pipelines, and spill control systems
- Tailing embankment including primary dam and any secondary dams
- Tailing deposition beaches
- Reclaim facilities
- Ditches including diversion ditches and other stormwater controls
- Groundwater monitoring and collection systems
- All ponds associated with tailings including polishing ponds

41.2.2 Step 2—Identify the Hazards

An effective QRA requires proper identification of hazards or potential failure modes. Many of the hazards related to tailing containment facilities are unique to the mining industry (Oboni et al. 1998). For example, mines, and in particular tailing containment systems, are constructed over long periods of time by a changing work force and usually under changing design criteria. Tailings systems are complex and include humanmade components such as dams, pipelines, and ponds interacting with natural components such as slopes, seismically active faults, precipitation, and runoff. In addition, previous operations can have an impact on newly built components. Further adding to the complexity, tailing containment methods are process specific and therefore unique to each mine.

41.2.3 Step 3—Identify Hazards and Elemental Failure Modes

Failure modes that are attributable directly to single external causes are identified as "elemental failure modes" (Oboni et al. 1997, 1998). Elemental failure modes cannot be subdivided further. Examples of elemental failure modes of a pipeline are

- A pipe bursting as a result of mechanical failure
- A pipe bursting as a result of a traffic impact
- A pipe bursting as a result of over-pressurizing due to freezing or sanding

Lists of elemental failure modes are provided as a checklist as well as a "scenario builder" subapplication in the CASSANDRA database. These tools are constantly updated. The initial version of the dam elemental failure modes checklist was derived primarily from historic tailing dam incidents (USCOLD 1994).

41.2.4 Step 4—Combine Elemental Failure Modes into Compound Failure Modes

Any one of several elemental failure modes can affect a given element or link. By mathematically combining the probability of occurrence of each of the elemental failure modes, the compound failure modes can be assessed. An example of a compound failure mode is a pipeline rupturing on top of a dam. The rupture may intervene because of an elemental failure mode, but the final event, i.e., a major failure of the dam, is the result of a domino effect. Thus, such a compound failure mode encompasses the rupture of the pipeline, then the creation of an erosion channel, eventually followed by sloughing, local instability, and finally a major instability.

41.2.5 Step 5—Assess the Probability of Occurrence

The subjective probability of occurrence is defined initially based on the concept of historical occurrence. This is then modified by expert judgment following interview or group sessions with mine staff. The probability of occurrence of each of the elemental failure modes is assessed numerically and combined using simplified event trees.

41.2.6 Step 6—Estimate the Consequences

The mine staff are relied upon to estimate the range of costs that might be associated with each of the identified events. The costs are generally assigned a range of magnitude costs (Bruce and Oboni 2000).

41.2.7 Step 7—Present the Results

Once all the necessary evaluations have been performed, the results can be depicted as shown in Figure 41.4 and defined numerically in Table 41.1.

TABLE 41.1 Risks defined graphically in Figure 41.3

Number	Components/Links	Hazards	Probab.	Cost (US$)	P·C
1	Purification tanks	Explosion and fire	1.0E–01	290,000	29,000
2	Hafnium powder	Fire	2.5E–03	4,500,000	11,250
6	Staff transportation	Accident	1.5E–02	250,000	3,750
12	Water plant	Operational disruption	1.5E–03	1,800,000	2,700
7	XYZ pumping line	Spill, bursting or leak	4.5E–03	400,000	1,800
8	Leachate sump	Spill and/or filtration	2.5E–03	600,000	1,500
11	Products	Dishonest client	1.5E–03	1,000,000	1,500
4	Shipping system	Spill	1.5E–04	10,000,000	1,500
10	Sulphuric acid	Overstock	1.5E–03	700,000	1,050
9	Water pumping pipe	Drop	1.5E–03	600,000	900
3	General warehouse	Fire	1.5E–04	6,000,000	900
15	Roasting furnace	Falling of container dome	1.5E–04	3,820,000	573
14	Gas-gas heat exchangers	Damage	1.5E–04	2,450,000	368
16	Transformer	12 MVA failure	1.0E–04	2,450,000	245
13	Boiler	Overpressure	1.5E–04	1,566,000	235
5	Sulphuric acid	Accident of tank truck	1.5E–05	1,000,000	15
18	Boiler	Explosion of container	1.5E–05	7,900,000	119
17	Sulphuric acid storage tank	Drop of tanks	1.5E–05	2,000,000	30

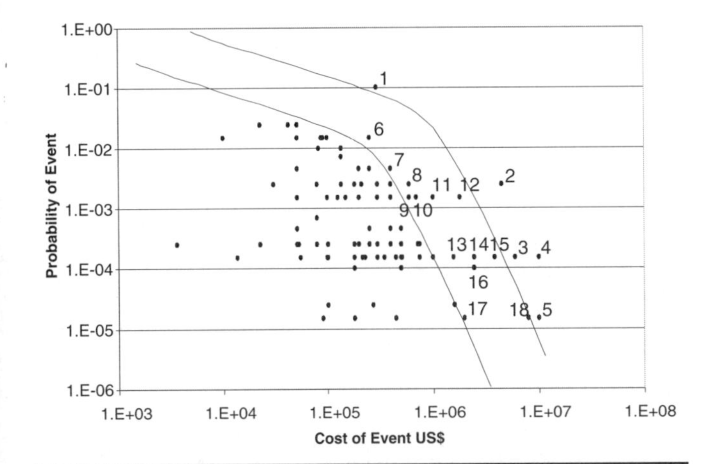

FIGURE 41.4 Risks for various components, elemental and compound, shown in the p,C graph together with the acceptibility curves. For readability purposes only, the centroids of the "bubbles" have been shown. The risks 1 to 18 are either in clear intolerability area (1,2,3,4,5) or in the "uncertainty area."

In addition to the "risk bubble" diagram, other displays are possible. An example from an industrial gas company can be used to describe the advantages of the QRA in terms of its capability to allow quantitative comparison between various risks in graphic format. In the example depicted in Figure 41.5, pipeline risks (monetary and in terms of casualties from leaks degenerating into explosions) are evaluated for a gas distribution system. Figure 41.5 shows the results of the comparison of the risks related to three types of gas pipes in four different environments (open fields to dense industrial environment). Both the risks in terms of physical damages (dollar risk) and casualties are depicted in the histogram, where the various combinations of gas pipes and environments are sorted by casualty risk. Such a diagram can be used to define, for example, staggered inspection schedules for the different pipes in different environments. Such a staggering is suggested in the picture by the four families of monitoring frequency F1 (minimal) to F4 (maximal). The thick vertical lines correspond to the proposed boundaries of the four families.

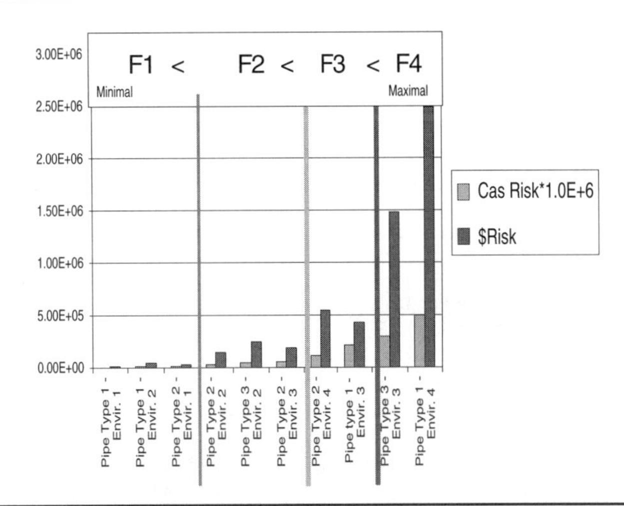

FIGURE 41.5 Use of QRA for comparison of risks related to different gas pipes. In the diagram, both physical damages (dollar risk) and casualty risks are depicted. Risks are ranked in terms of casualties and four families of monitoring (with different frequencies) are proposed based on these results.

41.3 DATABASE

In order to fulfill the needs of multioperations owners, including

- comprehensive quantitative comparisons between mines,
- benchmarking operations with respect to a worldwide population, and
- comparison of the risks associated with the mining industry with those of other industries, the CASSANDRA database of projects has been compiled.

The database and its related application (Figure 41.6) have been structured in such a way as to allow queries, reports, and cross-reporting by hazard, affected elements, and consequences.

The format used is a relational database structured around numerous tables. The architecture of the database, constituting the core of the application, has matured over the years and allows the application to be used on almost any type of facility, operation, or plant without modification.

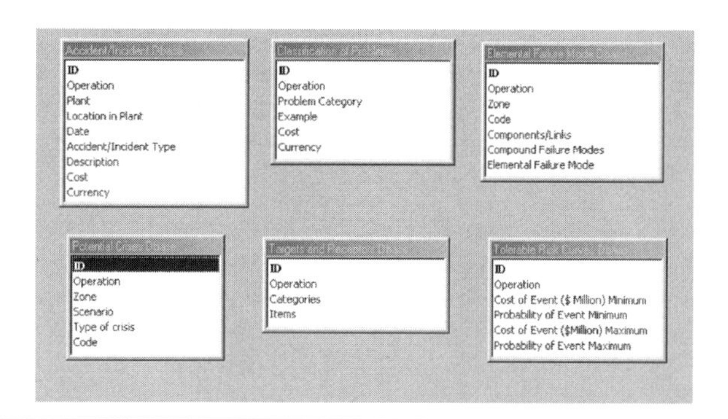

FIGURE 41.6 Example of CASSANDRA tables

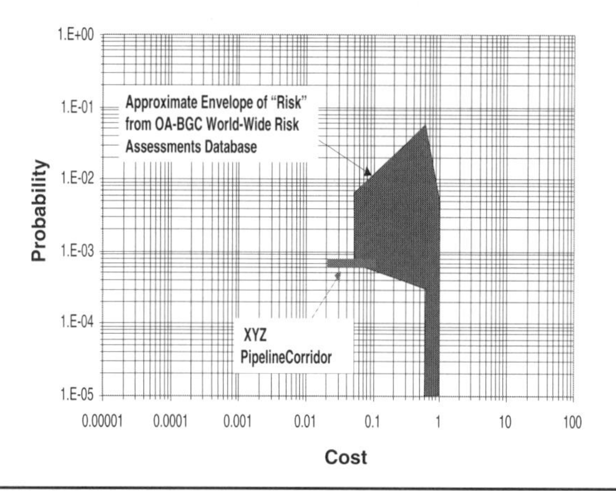

FIGURE 41.7 Comparison of the global risks of XYZ Pipeline Corridor and "the world." The comparison is favorable since the probability is well in the range of probabilities and the cost of consequence is in the low end. Before declaring that the assessment is good it would be recommended to check both p,C.

The database allows the storage of raw data as well as event trees, fault trees, and root cause analyses trees. The latter is mainly used for human error postaccident analyses. The database is designed to allow separate evaluations for physical damages (dollar value) and casualties (number of expected victims) of a potential accident.

When applied to a single mine, the system can be linked to a GIS. In the case of multiple operation comparison, the application runs independently from the GIS.

Figures 41.7 and 41.8 are examples of how comparisons can be made between the risks for a specific operation and the rest of the "world database." Figure 41.7 depicts the global risk of the XYZ tailing pipeline corridor and how it compares to the "world" of similar components stored in CASSANDRA. As it can be seen, the probability of occurrence of XYZ seems to be well in the range, whereas the cost of consequences lies at the low end of all the other estimates.

Figure 41.7 depicts cases related to tailings dams and shows two separate envelopes for the "world": the one on the left is related to overtopping and slope stability, whereas the one on the right is related to dam breach. Two examples of comparison are given in Figure 41.8: one for the overtopping of XYZ Creek Dam, the other for ZXY Dam breach. Given the substantial difference between these two examples and the "rest of the world," consideration might be given to confirming the evaluations performed by the mine personnel and more detailed studies may prove to be necessary.

By using such comparisons a multisite owner can define the areas of highest relative exposure within his system(s) or compare the "health" of his system (or systems) to the "world."

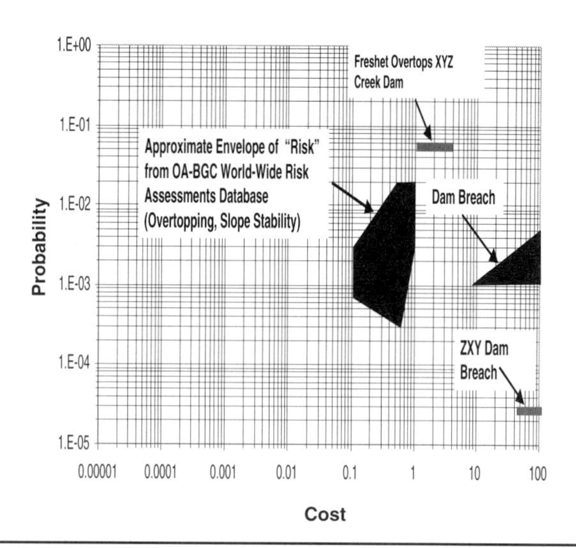

FIGURE 41.8 "World" envelopes for tailings dams overtop, slope stability, and dam breach. The overtopping of the XYZ Creek Dam seems to have both p,C higher than "world," whereas ZXY Dam breach seems to have a lower probability than the "world." Both will need to be checked before final approval.

41.4 CONCLUSIONS

Use of a QRA provides management with a decision making tool as part of a risk management program. A database application named CASSANDRA allows management and also field personnel to assess the risks, evaluate and benchmark proposed modifications, and understand how the modifications can affect the risk, and it thereby allows them to compare directly and immediately how capital expenditures can be directed to develop sustainable and rational risk reduction programs.

By comparing the results of any specific mine to the QRA generated database for mining and other related projects, owners can immediately determine their vulnerability on a worldwide basis both within the mining industry or in many instances with other industries. As the database expands and becomes more comprehensive the applicability of the results becomes more universal.

Although a QRA requires some additional efforts over those required to undertake a simpler qualitative assessment such as an FMEA, the benefits of having a quantitative risk assessment which can assist with risk management is patently obvious.

41.5 REFERENCES

Bruce, I.G., and F. Oboni. 2000. Tailings management using quantitative risk assessment. *Proceedings of Tailings Dams 2000, Association of State Dam Safety Officials, US Committee on Large Dams, March 28–30, 2000, Las Vegas, Nevada*. 449.

IUGS Working Group on Landslides, Committee on Risk Assessment, Quantitative Risk Assessment for Slopes and Landslides—The State of the Art. 1997. *Proceedings of the Meeting in Honolulu*. Rotterdam: Balkema.

Mehling, P., F. Oboni, and I. Bruce. 2000. Tailings: To flood or not to flood (presentation). *Tailings Dams 2000, Association of State Dam Safety Officials, US Committee on Large Dams, March 28–30, 2000, Las Vegas, Nevada.*

Oboni, F., I. Bruce, M. Aziz, and K. Ferguson. 1997. A risk management approach for tailing systems. *Golden Jubilee Conference, Canadian Geotechnical Society, Ottawa, Canada.*

Oboni, F., I. Bruce, M. Aziz, and K. Ferguson. 1998. A risk management approach for tailing systems. *Second International Conference on Environmental Management, Wollongong, Australia.*

Oboni, F., G. Oldendorff. 1998. Uncertainties, risk and decision making: risk management and crisis management integrated approaches. *Second International Conference on Environmental Management, Wollongong, Australia, 1998.*

USCOLD. 1994. United States Committee on Large Dams, Tailings Dam Incidents. November 1994.

CHAPTER 42

Management and Operational Background to Three Tailings Dam Failures in South Africa

Geoffrey Blight*

42.1 INTRODUCTION

During the past 120 years, the scale of mining in South Africa has grown to the extent that mining accidents, and particularly underground accidents, now claim scores of lives each year. The problems and scale of tailings disposal have grown commensurately with the size of the mining industry. Although there have been many tailings dam failures in South Africa, as well as the occurrence of extensive environmental damage, there have been few lives lost and only two accidents in which loss of life has exceeded 10.

All three of the failures referred to in the title were of ring-dyke tailings dams built by upstream methods. Four of the most basic requirements for the safety and stability of this type of dam are

- A moderate rate of rise so that each successive deposited layer of tailings can drain and then consolidate under its accumulating overburden
- Moderate side slope angles consistent with shear stability
- A minimum quantity of water retained in the basin of the impoundment
- The decant pool to be located as far as possible from the ring-dyke

The last two requirements keep the phreatic surface in the slopes as low as possible.

In all three of the cases to be described, one or more of these requirements were ignored.

Reports and papers dealing with geotechnical failures, including those of tailings dams, mainly concentrate on the technical aspects of failure. However, there are usually shortcomings in management and operation that are even more important, as they very often result in, or are the cause of, the technical shortcomings. This chapter will review the management and operational factors that contributed to tailings dam failures at Bafokeng in 1974, Saaiplaas in 1993, and Merriespruit in 1994. It will analyse and identify those factors that resulted from shortcomings in

- Technical knowledge
- Attention to design
- Operational methods and techniques
- Supervision, monitoring, and management

There is an additional very important factor that certainly played a part in all three failures, namely profit motive. The key question that must be asked is, "How can these shortcomings best be overcome in the future?" The mining industry has demonstrated that it cannot regulate itself. We also know that government regulation is usually inefficient and ineffective (especially in developing countries). Also, selfish shareholder profit motive cannot be resisted by mine management and can and does result in dangerous work practices (Voynick 1978).

42.2 BAFOKENG MINE, NOVEMBER 1974

42.2.1 Tailings Dam Flow Failure

On the morning of November 11, 1974, the southwestern wall of the No. 1 Tailings Dam at Bafokeng Platinum Mine, near the town of Rustenburg, South Africa, failed with disastrous results. Before the failure, the dam (85 ha in area and 20 m high) contained about 17×10^6 m^3 of tailings. Approximately 3×10^6 m^3 of liquefied tailings slurry flowed through the breach in the wall, engulfed a vertical shaft of the mine, and flowed into and on down the valley of the Kwa-Leragane River.

The slurry demolished or damaged many surface structures at the shaft, and the flood also carried away with it vehicles and items of equipment waiting to be taken underground. A large quantity of slurry flowed down the shaft, trapping some workers underground and tearing loose certain shaft equipment. Twelve died underground in the disaster. Miraculously, no one was killed on surface.

42.2.2 Events Leading to Failure at Bafokeng

The events leading to the failure of the No. 1 Tailings Dam at Bafokeng have been well reported by Midgley (1979) as follows:

During the early morning of Monday, November 11, 1974, heavy rain fell in the Rustenburg area. Rain gauges in the vicinity differed widely, but it was estimated that at the Bafokeng Mine 77 mm of rain fell between 02h00 and 04h00 on that day. At the time of the accident, as may be seen from Figure 42.1, the impoundment of Tailings Dam No. 1 was divided into two portions by a diagonal wall extending from the southeastern to the northeastern side. The area of the impoundment behind the dividing wall was about 73 ha and that of the triangular downstream portion about 11 ha. The peripheries of the ponds on top of the impoundment, as on the day before the failure, are indicated in Figure 42.1. The approximate water surface areas in the upper and lower compartments were 20 ha and 4.5 ha, respectively.

Adjacent to Dam No. 1 was Dam No. 2, its crest at a substantially lower elevation. Workers had cut a track (at B on Figure 42.1) through the dividing wall between dams 1 and 2, near the eastern corner, to bring in a front-end loader, reportedly for the purpose of pushing up barriers of dry tailings to drive the water back from the crest of the dam.

At about 08h45, workers noticed water overtopping the diagonal dividing wall, which soon breached, releasing water from the upper pool to the smaller lower one in which the water was judged to have been standing about 2 m lower. There was a sudden rise of level in the lower pool and water started to flow

* University of the Witwatersrand, Johannesburg, South Africa.

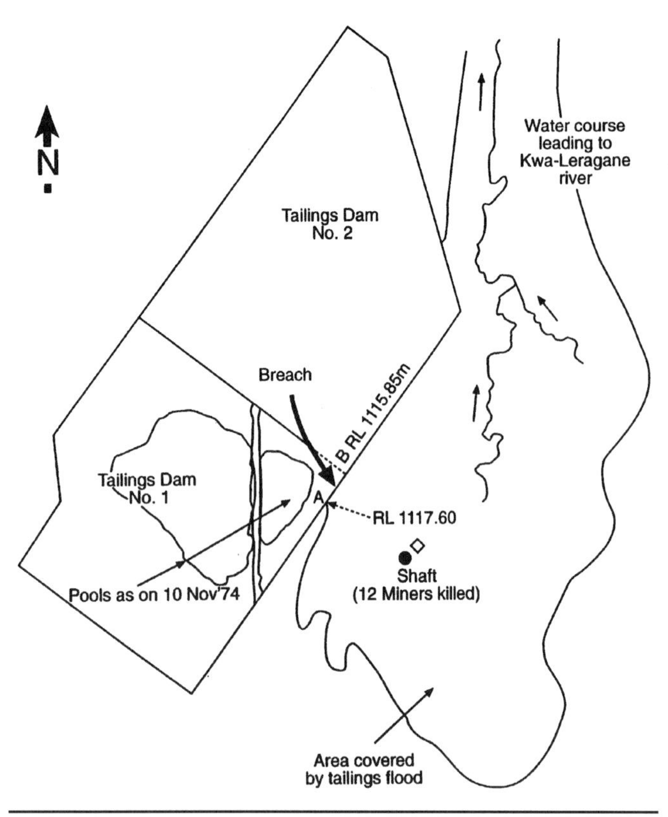

FIGURE 42.1 Plan of Bafokeng tailings dams showing position of breach, course of flow failure, and extent of pools prior to failure

through the cutting B from Dam No. 1 into Dam No. 2. Flow was not so strong, however, as to cause substantial erosional downcutting. The elevation of the crest of the overflow (B) section was subsequently measured as RL 1115.85 m compared with RL 1117.60 m at point A (the estimated elevation of the lowest part of the outer crest of the tailings dam).

From these elevations, Midgley concluded that the failure could not have been initiated by overtopping.

The process of failure appears to have been as follows:

1. At about 10h15, a leak appeared in the southeastern slope of the dam, appearing as a jet of water emerging about two thirds of the way up the slope and falling on the downstream slope.
1. The hole from which water was emerging quickly grew and the jet became a large stream pouring down the slope.
2. Above the stream of water two or more cracks developed in the slope, extending upward at fairly steep angles toward the crest, to form a wedge with its apex down.
3. Blocks of material collapsed out of this wedge, which fell as solid masses into the now rapidly increasing flood water and tailings.
4. The resulting gulley widened rapidly to about 130 m as three million cubic metres of liquefied tailings poured out. After spreading over the fairly level ground surface adjacent to the breach (and destroying the shaft and winder house), the escaping slurry found its way into a normally dry watercourse and thence into the Kwa-Leragane River.

These stages of the failure are illustrated in Figure 42.2, an elevation of the wall of the dam at the breach.

At a distance of 4 km from the breach in the dam, the flood of slurry had spread to a width of 0.8 km and was 10 m deep. The flood continued down the Kwa-Leragane River into the Elands River, and an estimated 2×10^6 m^3 of tailings eventually flowed

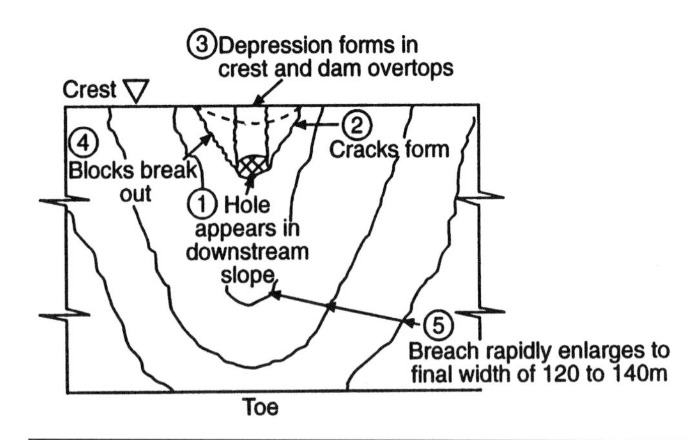

FIGURE 42.2 Stages of failure of Bafokeng tailings dam

into the reservoir of the Vaalkop (water-retaining) dam 45 km downstream of Bafokeng.

It should not be assumed, on the basis of the happenings at Bafokeng, that every failure of a tailings dam will result in a flowslide. In fact, flowslides such as that which occurred at Bafokeng (and 20 years later at Saaiplaas and Merriespruit) probably represent the exception rather than the rule. The mobility of tailings involved in a failure depends on the state of consolidation. The platinum tailings at Bafokeng are coarse and have a silt and clay tail. The dam was built by subaerial spigotting from the perimeter toward the center of the impoundment. Hence, the dyke would have been loose, though consolidated, but the impounded fines probably were not completely consolidated. It also appears that the mobility of the tailings is very dependent on the condition of the ground surface over which the material is moving. If the ground surface is already covered by a sheet of flowing stormwater, the viscosity of the tailings at the interface between the tailings and the ground surface will be much reduced, and this will facilitate the flow of the tailings. This is probably why the tailings from Bafokeng, diluted by water in the river channel, flowed so far down the Kwa-Leragane-Elands river valleys. The problem of whether or not the failure of the tailings dam will result in a flowslide has been examined in some detail by Blight (1997a).

42.2.3 Possible Causes of Failure at Bafokeng

It appears at first sight that the dam did not fail by conventional overtopping. Eyewitness accounts all point to a failure by erosion. However, a satisfactory explanation of how the initial hole formed in the wall was never reached. The following quotes from Jennings (1979) raise what are believed to be serious possibilities:

1. "The slimes dam foreman reported that he was on the dam when the dividing wall between the upper and lower ponds failed but that nevertheless he continued for some time thereafter to work with his front-end loader in the area of the crest of the dam where the breach subsequently occurred."
2. "... The foreman on the dam was not in a position to see water jetting from the wall. To him the failure would have appeared to be due to a deepening of a gulley which appeared in a place where he had been trying to push the water back with his front-end loader."

If the foreman was "trying to push the water back," it would appear that the crest of the dam must have been in imminent danger of overtopping, even though the quoted crest levels indicate a freeboard of 1.75 m or so.

A vibrating machine, even a light, wheeled front-end loader, working on a loose saturated sand could possibly cause the sand to liquefy at least locally. The gulley reported by the foreman and later the hole that appeared in the wall could have resulted from

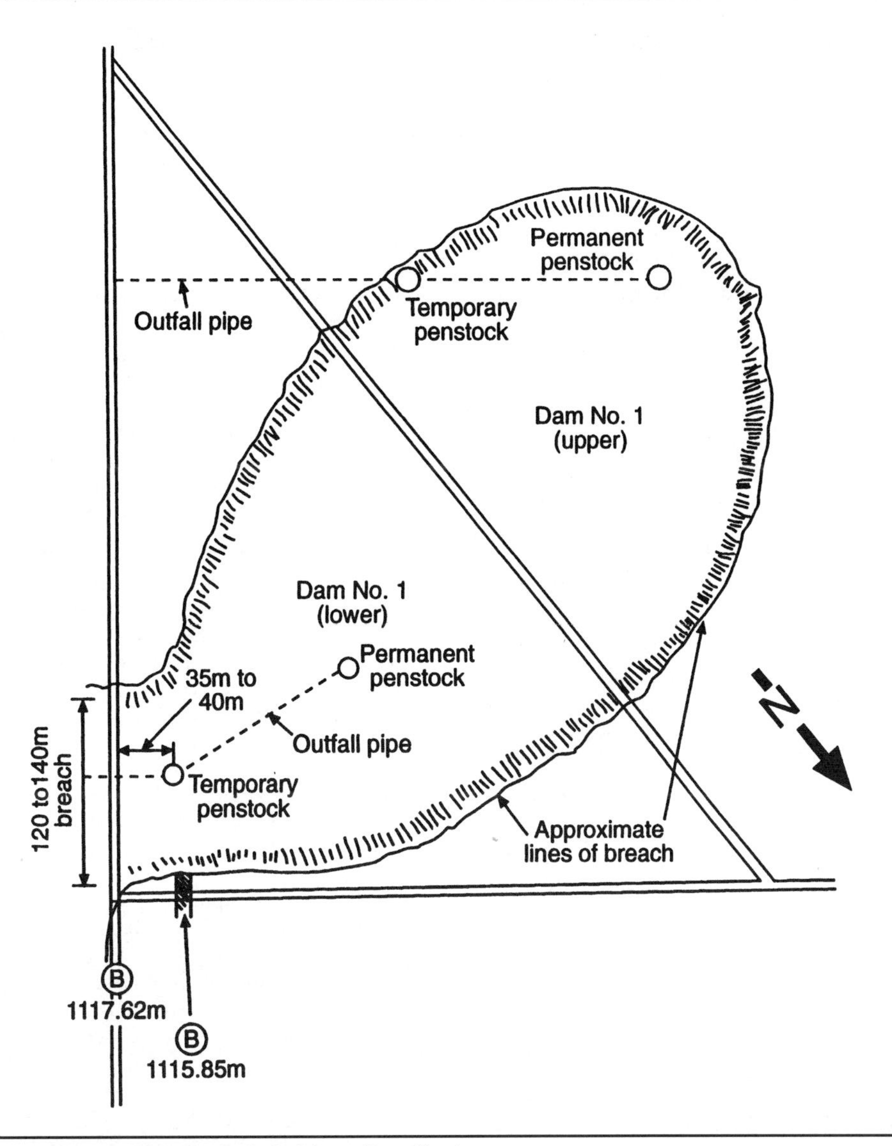

FIGURE 42.3 Plan showing positions of penstocks in Bafokeng Dam No. 1

this liquefaction. Loss of material through the hole would then have resulted in a deepening depression forming in the crest, followed by overtopping and breaching of the wall. Hence the foreman's desperate attempt to save the dam may have caused its destruction.

Only one of the papers that describes the failure even mentions the outlet penstocks. Midgley (1979) referred to attempts to lower the water level on the dam by lowering the crests of the decant penstocks, but gave no further details. Aerial photographs of the failure scar show no signs of any penstock shafts or towers. However, Figure 42.3 (taken from one of the original reports on the failure) shows that the dam was equipped with four penstock towers and two outfall pipes, all of which disappeared during the course of the failure. The temporary penstock in the triangular portion of the impoundment was less than 40 m from the crest of the wall at the breach. If this penstock was operational and if, in their zeal to lower the water level, too high a water head was allowed over the penstock crest, the penstock could have started to flow full. The shaft of the tower was built of stacked, precast concrete rings. It has long been known (e.g., US Bureau of Reclamation 1960) that if the head of water over the crest of a shaft spillway exceeds 0.25 times the shaft diameter, the shaft will tend to flow full and the flow may surge. The resulting pressure fluctuations have been known to dislodge rings and destroy similar penstock shafts. The vibrations from the surging could also have been the primary cause of liquefaction of the saturated sand, which resulted in the breach. Alternatively, if the penstock rings were dislodged, a cavity would have been eroded around the penstock shaft, and that cavity could have rapidly enlarged until it emerged through the downstream slope.

In summary, the dam must have failed by overtopping that was triggered by a form of erosion failure caused by either

- Localized liquefaction caused by vibrations set up by the front-end loader
- Dislodgement of rings from the temporary penstock shaft

Both of these possible causes could have resulted from overzealous and ignorant action by the contractor's staff in their attempts to save the dam.

42.2.4 Operational and Management Failings at Bafokeng

The inquest into the deaths of the 12 miners found that the failure was an Act of God and therefore that no one was to blame for the Bafokeng disaster! In almost every case of failure of an engineering structure, some blame attaches to the designers, constructors, supervisors, or operators of the structure. In the case of Bafokeng, at a distance in time of 26 years, the failings that resulted in the disaster are not very clear:

- One of the basic rules of tailings dam operation is that the pool should be kept as small as possible and be located as far as possible from the outer walls. Figure 42.1 shows that the pool was dangerously large prior to the failure, and the lower pool was only about 30 m from the wall that breached.
- The dam was not equipped with piezometers, so the position of the phreatic surface was not known.

For whatever reason, someone was responsible for the decisions not to install piezometers and to hold an excessive amount of water in the impoundment.

- It also appears that the operation of the dam was not subjected to regular examinations and supervision by a knowledgeable geotechnical engineer.

There was thus a lack of concern for and caution about the stability of the dam. Some blame for this must attach to the management of the mine.

To summarize the management failures,

- There was complacency about the state of safety of the dam.
- There was a lack of knowledge and understanding of how tailings dams function and what affects their safety. The decision to store large volumes of water on top of the dam could have arisen from this ignorance.
- There were no regular inspections or assessments of the dam's state of safety. In this regard, the lack of piezometers was a major shortcoming.
- There was obviously no emergency action plan that the operators of the dam were aware of. They were poorly trained and nonplussed when the emergency confronted them.

42.2.5 The Aftermath of the Bafokeng Failure

Engineering has a history of advancing via a series of disasters, and tailings dam engineering is no exception. For example, mine tailings disposal was largely unregulated in Britain until the Aberfan disaster occurred (Report of the Tribunal 1967). Prior to the failure of the Bafokeng dam, there was some rudimentary guidance available on tailings dam construction in South Africa, in the form of the Chamber of Mines Code of Practice for the Construction of Tailings Dams (Circular No. 92/68 of 1968). This was based on research carried out under the direction of J.E. Jennings and published by the Chamber of Mines in June 1959 as *An Investigation into the Stability of Slimes Dams with Particular Reference to the Nature of Their Foundations*. After the Bafokeng disaster, the Chamber of Mines commissioned the preparation by the present author of a comprehensive *Handbook of Guidelines for Environmental Protection—The Design, Operation and Closure of Metalliferous and Coal Residue Deposits*. This technical guideline to tailings dam design and operation appeared in 1979 and was voluntarily adopted by the whole gold mining industry. It was revised in 1983 and was again in the process of revision in 1993 and 1994 when the tailings dam failures at Saaiplaas and Merriespruit occurred. During the inquest on the Merriespruit failure, it emerged that the responsible staff on the mine were quite unaware of the existence of this publication, as were the site staff of the contractor who was operating the tailings dam. This made it quite clear that the mining industry in South Africa was not able to regulate itself and, in the interests of safety, regulation and coercion by the State were necessary.

Thus it appears that the lessons that should have been learned from the Bafokeng failure had largely been forgotten 20 years later when the next two incidents, one resulting in many fatalities, occurred.

42.3 SAAIPLAAS, MARCH 1993

42.3.1 The Tailings Dam Failure

The Saaiplaas No. 5A Dam is a ring-dyke gold tailings dam located near the town of Virginia in the Free State Province, built by upstream paddocking with subaerial deposition from several points off a ring-feed main into the impoundment. In March 1993, the dam was 28 m high with an average slope angle of 21°. At 12h15 on March 18, 1993, a rotational slide occurred in the western side of the impoundment wall. This was followed on March 19 by a second similar failure, separate but immediately adjacent to the first. At 07h30 on March 22, these two failures were followed by a third failure on the southern side of the impoundment wall. Figure 42.4 shows the layout of No. 5 Dam and the location of the three failures. Fortunately, no one was injured, possibly because very little water was ponded around the penstock and the tailings did not travel very far. Failure C was the most mobile of the three failures, but fortunately it took place into the return water pond, which dammed the tailings flow.

42.3.2 Background to the Failures

No. 5A Dam had been an old tailings deposit that was taken back for reprocessing in the late 1970s. In 1981, the 28 ha dam was recommissioned. At that time, Saaiplaas Gold Mine was considered to be on the brink of closure and the recommissioning was intended only to accommodate tailings for the few remaining months of the mine's operation. To save costs, deposition recommenced with no underdrains under the dyke, and no piezometers were installed to monitor the phreatic surface. To further save costs, the decant penstock was placed close to one corner of the rectangular impoundment instead of near the geometric center where it would have kept the pool further from the wall, thus depressing the phreatic surface.

Two of the required operating conditions for the recommissioned dam were that the rate of rise be limited to 1.5 m/yr and that the height be limited to 20 m. A 1.5 m/yr rate of rise should have resulted in a well-consolidated tailings deposit with adequate slope safety at the design overall slope angle of 26°.

The mine, however, did not close. Dam 5A continued to be operated, and by the end of 1993, was 28 m high and had risen at an average rate of 1.8 m/yr. Over the previous year, however, the rate of rise had averaged 2.57 m/yr, and in February 1993 it was 2.83 m/yr. For a number of years prior to the failure, the contractors who operated Dam 5A on behalf of the mine had pleaded for the installation of piezometers and had also insisted on the need to move the penstock away from the retaining dyke. The plea for piezometers fell on deaf ears, but the request to relocate the penstock was eventually acceded to, but only because the contractor had repeatedly drawn attention to the visibly wet condition of the southern slope of the dam. A further request to step back the wall, forming a berm and thus flattening the overall slope of the dam, was also refused.

Clearly, the profit motive was dominant in the minds of mine management who, in order to save costs, refused to entertain any warnings that the dam was unsafe.

There was no overtopping involved in these failures. The high rate of rise had resulted in an insufficiently consolidated wall, the shear strength of the tailings was deficient, and the wall failed in shear.

42.3.3 Effect of the Saaiplaas Failures

Because the failures did not result in injury or death, were confined to the boundaries of mine property, and damaged only mine property, they were successfully hushed up and not reported in local newspapers, let alone radio or television. Hence, these major failures went virtually unnoticed, and no lessons were learned from the incident by the gold mining industry. Nor were any alarms raised. Less than a year later, the failure at

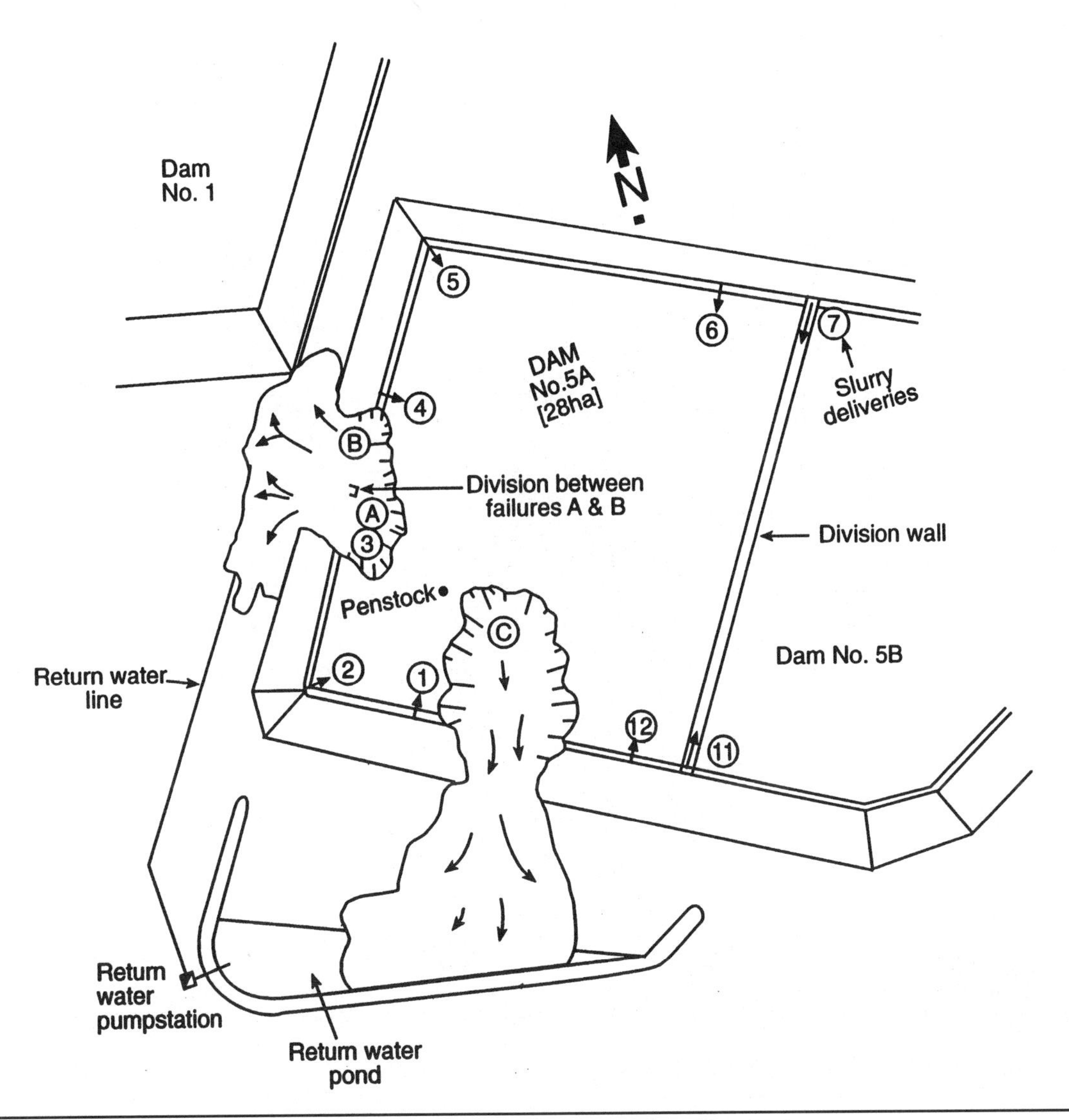

FIGURE 42.4 Plan of Saaiplaas Dam 5A showing locations of failures

Saaiplaas was followed by the tragedy at Merriespruit, barely 5 km away.

The failure could have been avoided had mine management had any regard for safety at all, but such regard was submerged by the all-important motive of profitable operation. However, the contractor at Saaiplaas Dam 5A also operated the Merriespruit Dam, and the management at Merriespruit was well aware of the Saaiplaas failures.

42.4 MERRIESPRUIT, FEBRUARY 1994

42.4.1 The Tailings Dam Failure

On the night of February 22, 1994, a 31-m-high tailings dam upslope of Merriespruit, a suburb of the town of Virginia and only about 5 km from Saaiplaas, failed with disastrous consequences. The dam breached a few hours after 30 to 55 mm of rain fell in approximately 30 minutes during a late afternoon thunderstorm. The failure resulted in some 600,000 m^3 of liquid tailings flowing through the town, causing the death of 17 people, widespread devastation of property, and environmental damage. The tailings flowed for a distance of about 4 km, with some material reaching a small tributary of the Sand River, which runs on the northern boundary of Virginia.

The wave of tailings was about 2.5 m high when it reached the first row of brick-built single-storey houses, about 300 m downslope of the dam. Some of these houses were swept off their foundation slabs, while walls were ripped off and roofs collapsed on others. The mudflow continued down the main street narrowly missing a home for the aged and lapping at the wall of a mine hostel housing several hundred sleeping mine workers. It was miraculous that only 17 people died.

From eyewitness reports, a strong stream of water had entered the top end of the town at dusk (about 17h00). The stream flowed down streets and through gardens. One person reported seeing water cascading over the top of the tailings dam wall. The mining company and contractor were informed, but when their representatives arrived on site it was already dark. An employee of the contractor rushed to the stacked ring penstock decant tower inlets and found water lapping the top ring but not overflowing. He removed several rings from the two penstock inlets. A second employee was below the dam near the overtopped area and saw blocks of tailings toppling from a recently constructed tailings buttress against the north wall of the dam. There was an attempt to raise an alarm, but before the inhabitants could be warned a loud bang was heard, followed by the wave of tailings that engulfed the town. (Technical aspects of the failure were examined by Wagener et al. 1997.)

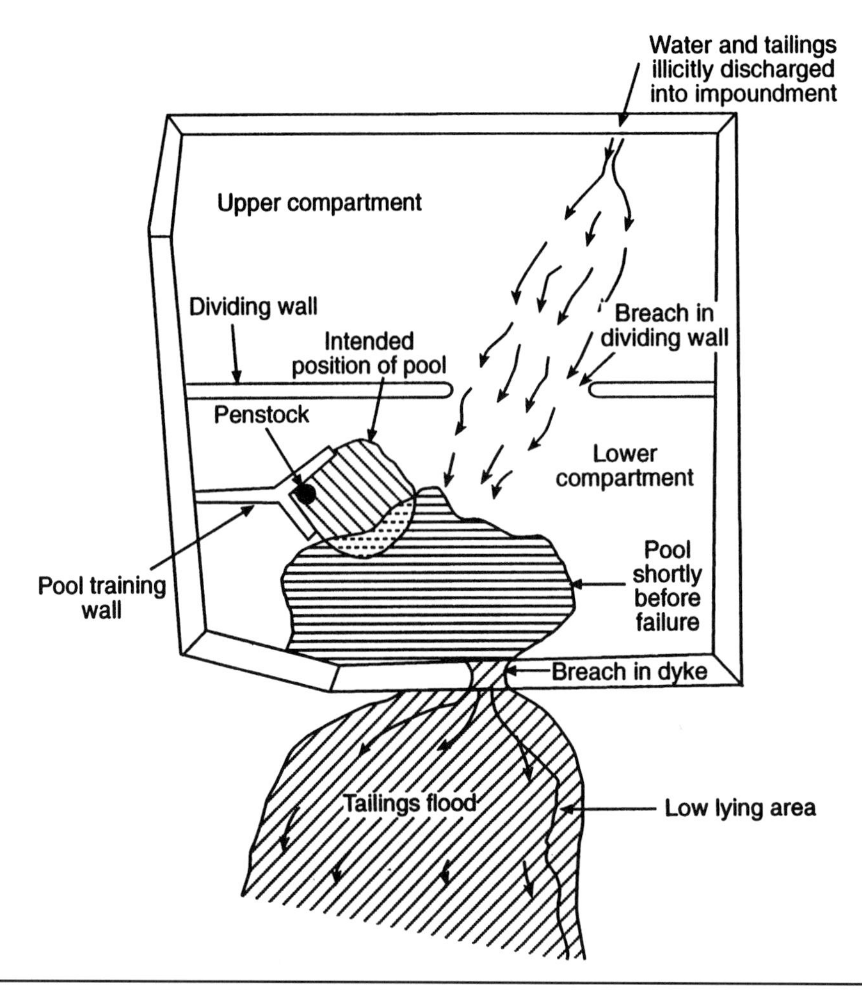

FIGURE 42.5 Plan of Merriespruit dam showing positions of pool at time of failure, intended position of pool, breach in dyke, and path of tailings flood

42.4.2 Condition of the Merriespruit Dam Prior to Failure

The northern wall of the Merriespruit tailings dam, only 300 m upslope of the village, had been showing distress for a number of years in the form of seepage and sloughing near the toe. Two years before the failure, a drained tailings buttress was constructed along a 90-m length to stabilize the wall. Continued sloughing resulted in a decision by the mine and the contractor to discontinue tailings disposition in the impoundment about 15 months prior to the disaster. According to the contractor, the freeboard on the dam was at least an acceptable 1.0 m at the time when the decision was made to retire the impoundment.

Sloughing at the toe continued after deposition had supposedly stopped. Shortly before the disaster, small slips occurred on the lower slope of the dam immediately above the buttress. The slips were repaired by dozing in dry tailings from the surface of the wall adjacent to the slips.

After the instruction to stop operating the impoundment, raising of the outer wall ceased. However, excess water was still illicitly delivered to and stored in the impoundment, and this water must often have contained tailings. The stored water was decanted through the penstock when required by the plant, but the tailings slowly accumulated and filled the available freeboard. At the time of failure, the freeboard was a mere 0.3 m. To a large extent, this situation arose because, for reasons of cost saving, there was no return-water pond in which excess water could be stored.

Merriespruit Dam is covered by the orbits of a Landsat satellite that transits the area every 16 days. The nearest date for which an image was available prior to the failure was February 1, 1994, because clouds masked the site on the next two passes. Several points emerged from a study of a series of the satellite photographs.

- The unauthorized deposition of water and tailings had not only reduced the freeboard to a dangerous and illegal level but had also pushed the pool away from the penstock causing it to be located next to the northern wall just prior to the failure, as shown in Figure 42.5.
- Wet conditions could be seen below some parts of the wall that eventually breached the north wall.
- Free water lay impounded near the north wall for a considerable period of time. Even in October 1993, there was a large pool of water very close to the north wall and well away from the penstock inlet.
- By superimposing the outlines of the impoundment and the water surface on the contour plan, investigators were able to estimate the water level on February 1, 1994, to have been about 0.45 m below the crest of the wall.

42.4.3 Most Likely Mode of Failure

The Merriespruit Dam was equipped with piezometers at several sections and these had been measured a week before the failure.

The overall shear stability of the tailings dam was shown by analysis to have been sufficient to rule out overall shear failure resulting from a high phreatic surface as a primary cause of the disaster. However, it is evident from the sloughing failures that

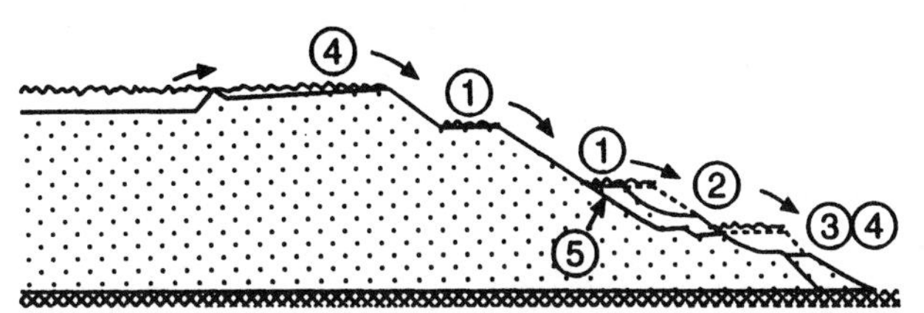

Critical section of north wall during early stages of failure

1. Berms overtop after thunderstorm.
2. Loose tailings infill to earlier failures on lower slope erodes.
3. Tailings buttress starts to fail.
4. Pool commences overtopping and erodes slopes and tailings buttress.
5. Unstable lower slope fails and failed material is washed away.

(a)

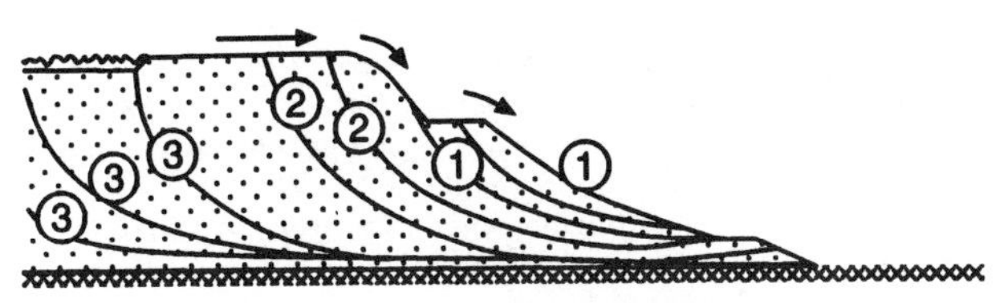

Critical section of north wall during failure

1. Lower slopes fail and are washed away.
2. Domino effect of local slope failures which are washed or flow away.
3. Major slope failures with massive flow of liquid tailings engulfing town.

(b)

FIGURE 42.6 Most likely development of flow failure at Merriespruit

occurred on the lower slope just above the tailings buttress prior to the disaster that local stability was not acceptable. As mentioned earlier, the localized failures had been "stabilized" by dozing dry tailings over them.

The most likely chain of events leading to the failure is shown diagrammatically in Figures 42.6a and 42.6b. Rain falling in the impoundment concentrated in the pool, which had moved away from the penstock toward the north wall. The pool already contained some 40,000 to 50,000 m^3 of water pumped from the plant. The impoundment had insufficient capacity to contain the additional inflow of about 50,000 m^3 that had resulted from the rainstorm and started overtopping where the pool touched the lowest point of the north wall. The spilling water, together with what had accumulated on the northern berms of the dam, eroded the loose tailings that had been pushed into earlier sloughs on the lower slope. This led to small slip failures; the sloughing tailings were removed by the cascading water thus preventing the buttressing effect of any material that would otherwise have accumulated at the toes of the unstable slopes. This resulted in a series of slip failures retrogressively cutting into the slope, ultimately leading to a massive overall slope failure that breached the wall and released the tailings flow.

This description not only accords with eyewitness accounts but also agrees with the results of the technical investigation and assessment carried out after the failure.

42.4.4 Management Failures at Merriespruit

The inquest judge laid the blame for the disaster at the doors of the contractor, the mine, and certain of the contractor's and mine's employees. Failings of these parties that were illuminated at the inquest were as follows:

- There was no review process for the operation of the impoundment that involved an independent reviewer. The mine's and contractor's familiarity with the problems of the impoundment resulted in complacency about their seriousness.
- The only involvement of a trained geotechnical engineer in the problems of the impoundment was that of an employee of the contractor, who became involved occasionally, only by request, and whose roles and responsibilities were ill defined.
- There were regular meetings between the mine and the contractor. However, decisions were poorly recorded, which led to confusion about responsibilities and agreed actions.
- The contractor's office at the mine did not keep the head office adequately informed of happenings at the impoundment. The head office was ignorant of problems and potential problems at the site and could thus not take corrective action.
- The contractor's local office was aware that water was being stored in the impoundment by the mine, but it took no action and did not inform either head office or seek the advice of its geotechnical engineer.
- Although the contractor had operated the impoundment since its inception, the contractor had never been requested to upgrade the facilities of the impoundment and so bring it in line with acceptable practice, as spelled

out in the industry guideline. (Chamber of Mines of South Africa 1979, 1983). Thus, the impoundment continued to be operated without a return-water pond. This necessitated storing water in the impoundment.

- Remedial measures taken to restore the stability of the northern wall were ad hoc and not the result of an adequate geotechnical investigation and design.

42.5 PREVENTING SIMILAR OCCURRENCES IN THE FUTURE

The Saaiplaas and Merriespruit disasters have provided a further impetus for both the mining industry and the State to take a more responsible and serious approach to the disposal of tailings. A fundamental reassessment of the philosophies for design, management, and operation of tailings dams was initiated in South Africa following the Merriespruit disaster. The mining industry has carried out independent self-audits and now (at least officially) views tailings dams and associated safety and environmental problems far more seriously than in the past.

The similarities of the circumstances, events leading to the failures, and the inadequate emergency actions of the operators showed that the lessons that should have been learned from the Bafokeng failure, and even the occurrence of the failure itself, had been forgotten. Twenty years later, there was a different generation of operators and officials who once again had become complacent and ignorant of the dangers of tailings dam failure.

In 1995, work started to draft an obligatory national Code of Practice for the design, operation, management, rehabilitation, and closure of tailings dams. This code was completed in 1998, but has not yet been made mandatory because organized labor considers it to be too discretionary and insufficiently based on hard technology. It appears at present (year 2000) that eventually the so-called Chamber of Mines Guidelines, which provide the hard technology basis, and the Code of Practice will be adopted as joint mandatory documents.

However, the most important problem of all remains unsolved. How will successive generations of tailings dam owners, tailings dam engineers, and tailings dam operators be kept continually aware that tailings dams are dangerous structures requiring continuous vigilance? How will they be kept constantly alert to the signs of distress that may portend an imminent failure? How will we prevent another major failure some years from now? We have all the answers, but we must learn continually to ask the right questions.

42.6 ACKNOWLEDGMENT

This chapter has been developed from an earlier document (Blight 1997b), which considered the failures at Bafokeng and Merriespruit in more detail, but omitted the Saaiplaas failure.

42.7 REFERENCES

Blight, G.E. 1997a. Destructive mudflows as a consequence of tailings dyke failures. *Geotech. Engng*. 125:9.

Blight, G.E. 1997b. Insights into tailings dam failure—technical and management factors. *Proceedings, International Workshop on Managing the Risks of Tailings Disposal,* Stockholm, Sweden. 17–34.

Chamber of Mines of South Africa. 1979; revised 1983 and 1995. *Handbook of Guidelines for Environmental Protection, v. 1. The Design, Operation and Closure of Metalliferous and Coal Residue Deposits.* Johannesburg: The Chamber. 3rd edition published 1996.

Jennings, J E. 1979. The failure of a slimes dam at Bafokeng. Mechanisms of failure and associated design considerations. *The Civil Engineer in South Africa.* 21:6:135.

Midgley, D.C. 1979. Hydrological aspects and a barrier to further escape of slimes. *The Civil Engineer in South Africa.* 21:6:151.

Report of the Tribunal Appointed to Inquire into the Disaster at Aberfan. 1967. HMSO, London, United Kingdom.

U.S. Bureau of Reclamation. 1960. *Design of Small Dams*. Washington, D.C.: U.S. Government Printing Office.

Voynick, S.M. 1978. *The Making of a Hardrock Miner*, Kearney, Neb.: Morris Publishing. ISBN 0-8310-7116-8.

Wagener, F., K. Strydom, H. Craig, and G. Blight. 1977. The tailings dam flow failure at Merriespruit, South Africa—causes and consequences, *Tailings and Mine Waste. 1997*, Rotterdam: Balkema. 657–666.

CHAPTER 43

Tailings Basin Water Management

Donald E. Welch*

43.1 INTRODUCTION

A tailings basin water balance is merely a mathematical tool that adds inflows and subtracts losses in a system. All water balance models use the same basic elements, shown on Figure 43.1. Experience has proven that a simple model, along with sound engineering judgment, can adequately predict performance over any range of anticipated conditions. This chapter introduces WATBAL, one such model that has been successfully used (Welch and Firlotte 1989; Welch et al. 1992, 1995). It is a simple spreadsheet accounting of water flows in both wet and dry climates. A simple definition of a wet climate is a climate in which annual accumulation of water has to be discharged to the environment to keep a tailings pond in balance on an annual basis.

An important objective of tailings basin water management is to ensure the safety of dams and to protect the environment. Problems arise when

- There is a prolonged period of greater-than-average precipitation
- Extreme flood events occur
- Discharge slurry density or recirculation decreases
- Water quality deteriorates
- Waterways and treatment plants are undersized
- A basin is not operated as designed

The effluent has to safely flow through a basin in a never-ending stream either to the environment or recycled to the mill. Costs and environmental effects are directly related to the quantity of water that has to be managed. In a wet climate, this can be minimized by

- Keeping the watershed as small as possible to minimize the inflow from runoff
- Providing large ponds to promote evaporation
- Maintaining a high slurry density to minimize the inflow of process water
- Maximizing recirculation to the mill to minimize the use of fresh make-up water

In dry climates, total containment (zero discharge) may be easy to attain while evaporation and seepage losses have to be minimized to conserve water.

A tailings basin does not have a single, unique water balance because the input data can vary from day to day, month to month, and even yearly. Precipitation and evaporation are variables and often are not well defined. Evaporation is a function of pond area. Seepage may only be an estimate. Recirculation to the mill is frequently a moving target, the discharge slurry density can vary, and even the milling rate can change.

43.2 REQUIREMENTS OF AN EFFECTIVE WATER BALANCE MODEL

A water balance model is only a predictive tool. Care must be taken not to build into a model sophistication that is not warranted for the degree of accuracy that is required to predict flows over a wide range of operating and hydrological conditions.

A tailings basin water balance should be

- Simple to use with easily recognizable input data
- Transparent (easy to understand, scrutinize, and criticize)

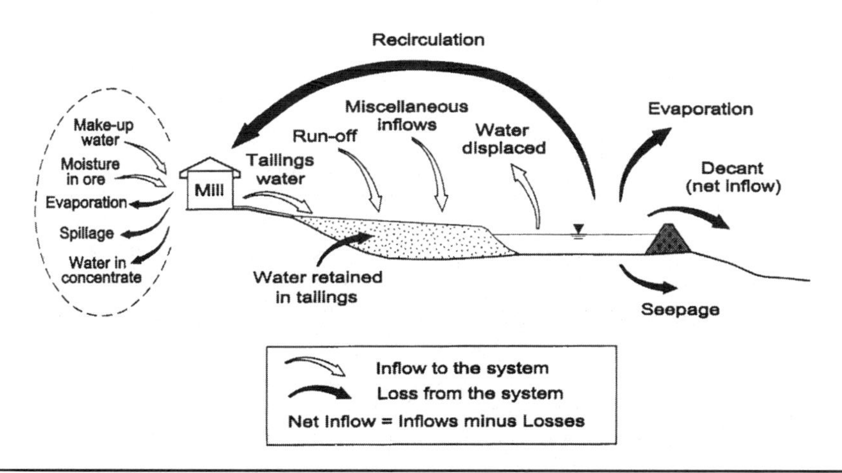

FIGURE 43.1 Elements of a tailings basin water balance

* Golder Associates, Mississauga, Ontario, Canada.

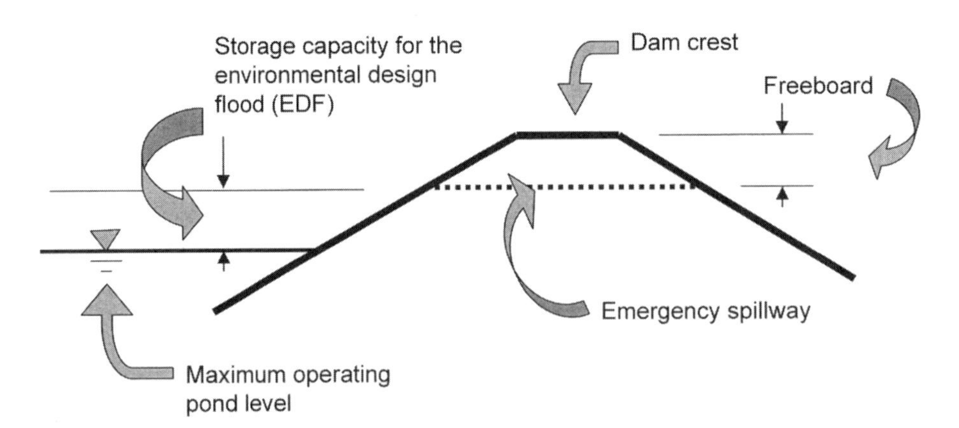

FIGURE 43.2 Management of storm flows

- Able to easily vary the input data to model changes in operating and climatic conditions
- Able to carry out sensitivity analyses
- Capable of being used by mine personnel and regulators during the life of a facility

43.3 PERIOD OF WATER BALANCE ACCOUNTING

A year covers one complete hydrological cycle; therefore, it is logical to at least summarize a water balance on an annual basis. An annual balance may be appropriate only when seasonal changes do not have a large effect on the outcome such as in a dry climate. In wet or cold climates, where seasonal changes do have a large effect, a monthly balance is required. The climatological data are usually conveniently summarized in a monthly format.

43.4 MANAGEMENT OF STORM FLOWS

A tailings basin has to be designed for a range of annual precipitation that statistically includes storm events and it must also be able to cope with extreme runoff events. One approach is to use two design floods: an environmental design flood (EDF) and a dam design flood (DDF). The EDF is a storm with a finite return period of somewhere between 10 to 100 years, possibly in conjunction with snowmelt or an antecedent wet period. It is the largest design runoff event that does not result in an unscheduled discharge of water to the environment. The EDF is retained between the maximum operating water level and the invert of the emergency spillway. Routed floods in excess of the EDF are allowed to spill over the emergency spillway. This concept is shown in Figure 43.2. If an unscheduled discharge cannot be tolerated, then the EDF and the DDF have to be the same.

Every tailings facility should have an emergency spillway because it is preferable to have an unscheduled spill of water over a spillway in a big storm than to have a dam overtop and lose tailings. Under no circumstances should a dam be allowed to overtop. If a basin does not have an emergency spillway, it must be able to store the probable maximum flood with a significant factor of safety.

43.5 WATBAL LOGIC

The elements of a water balance (Figure 43.1) are shown schematically on the logic diagram in Figure 43.3. The flows shown on these figures are listed below. The column numbers correspond to those on the WATBAL output column headings on Table 43.1.

Inflows

- Tailings slurry transport water (Column 2)
- Miscellaneous inflows such as mine water (Column 3)
- Runoff from precipitation (Column 4)
- Summation of the inflows (Column 5)

Losses

- Water retained in the tailings pore spaces (Column 6)
- Seepage (Column 7)
- Evaporation from ponds and wetted tailings surface (Column 8)
- Recirculation to the mill including slurry water that leaves the property in the concentrate pipeline (Column 9)
- Concentrate slurry water (Column 9a)
- Summation of the losses (Column 10)

WATBAL also includes

Accumulation

- Net inflow–Column 5 minus Column 10 (Column 11)
- Water displaced–if the pond volume has to be reduced on an annual basis to make room for tailings (Column 12)
- Total change–sum of Columns 11 and 12 (Column 13)
- Decant strategy (Column 14)
- Net change (Column 15)
- Accumulated pond volume (Column 16)

The net annual inflow (Column 11) is a significant number in the balance. It is the difference between the inflows and the losses. It is the volume of water that accumulates on an annual basis and has to be discharged to the environment each year if the pond is to remain in balance. If the pond volume has to be reduced to store tailings, then that volume (Column 12) has to be added to the net inflow to give the total to be decanted in Column 3.

The tailings water (Column 2) is computed from the percent solids (by weight) in the tailings slurry by solving for V_w in the following formula:

$$S = \left(\frac{W_s}{W_s + W_w}\right)100 \quad \text{where,} \quad V_w = \left(\frac{\frac{100W_s}{S} - W_s}{\rho_w}\right)$$

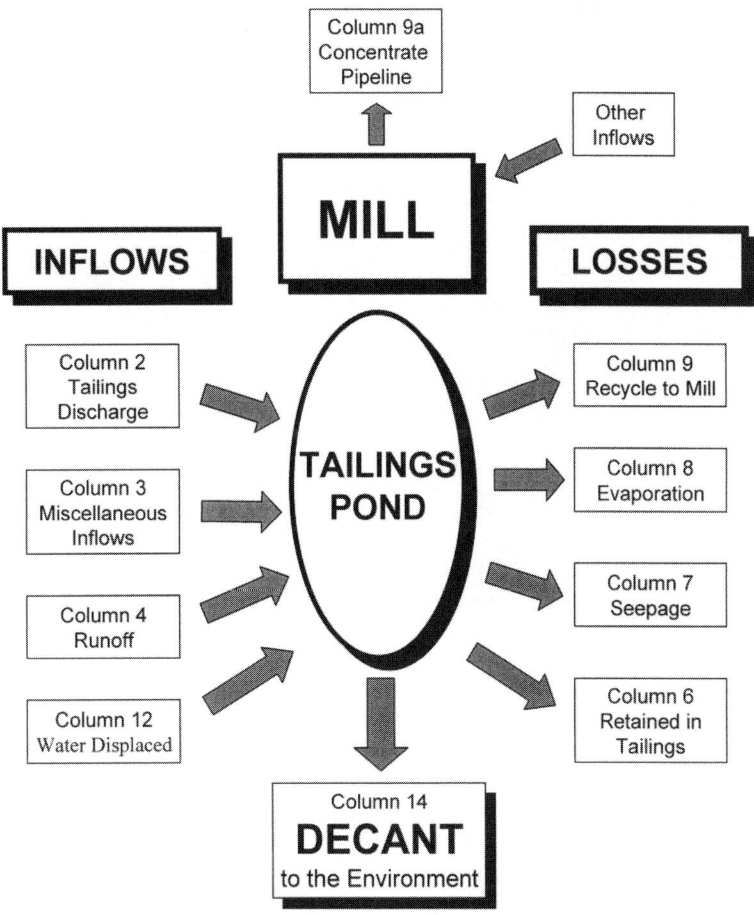

NOTES:

1. Column numbers refer to the output columns on the WATBAL spreadsheet.
2. Recycle = tailings discharge + concentrate slurry – other inflows into the mill
3. Decant (net inflow) = inflows – losses

FIGURE 43.3 WATBAL logic

Where:

Ws = dry weight of tailings solids in the slurry
W_w = weight of water in the slurry ($\rho_w \times V_w$)
V_w = volume of water in the slurry
ρ_w = density of water (unity in the metric system)
S = slurry density (percent solids by weight)

The tailings water is the total mill water requirements except for that which is lost as spillage and evaporation in the mill and any water that leaves the property with the product or concentrate. The first two are generally negligible and the latter is accounted for in the balance.

Miscellaneous inflows, such as mine water (Column 3), are sometimes put into the tailings pond. It may be advantageous to handle miscellaneous inflows separately; however, a single point of discharge from a property is attractive from a regulatory point of view.

If water is being discharged from pond to pond, as might be the case from a sedimentation pond to, say, a conditioning or polishing pond, then a separate balance is required for each pond, with the inflow into the receiving pond being entered into WATBAL as a miscellaneous inflow.

Surface runoff is computed in Column 4. Runoff is a function of watershed size and a runoff factor. Runoff can be assessed from three sources

- Precipitation records
- Streamflow measurements at the site or prorated from a nearby watershed
- Published regional specific runoff

In two separate versions of WATBAL, runoff can be assessed from either precipitation records or streamflow measurements. The streamflow version of the WATBAL works the same as the precipitation version except that the data used to compute runoff is entered either as "streamflow" or "specific runoff." Specific runoff, expressed as an average annual rate in L/s/km^2, is seldom used for a small watershed. In a large watershed, the annual runoff might be about 50% of the total precipitation. In a small watershed, such as a tailings basin without the benefit of lakes and extensive forest cover, the runoff could be much greater at, say, 65 to 75%. In dry climates, the specific runoff might be very low (10 to 20%) and possibly even zero in a desert environment.

Streamflow statistics seldom have a long enough period of measurement in a remote region. However, even cursory streamflow measurements are useful to assess the monthly runoff distribution, particularly in snow environments where runoff does not mirror precipitation.

WATBAL accounts for runoff from two different parts of a watershed: the natural land that surrounds the tailings and also the tailings and ponds. A runoff factor is used to compute the runoff

TABLE 43.1 WATBAL printout—precipitation version

Example - Dry Northern Climate

TABLE 1A
INPUT DATA

April 2000 000-0000

Precipitation Version

		UNITS	VALUE	Jan	Feb	Mar	Apr	May	Jun	Jul	Aug	Sep	Oct	Nov	Dec	Total
PROCESS WATER	Starting month	no.	1													
	Tailings production	t/day	1935													706275
	Solids (by weight) in discharge	%	45													
	Miscellaneous inflows	m^3/mo.	0													0
RUNOFF	Average precipitation	mm/mo.		17	15	17	12	17	21	42	49	29	32	25	24	300.0
	Change in precipitation	%	0													
	Total precipitation	mm/mo.		17	15	17	12	17	21	42	49	29	32	25	24	300.0
	Area of natural land in basin	ha	350													
	Runoff factor	%	70													
	Area of tailings and ponds	ha	170													
	Runoff factor	%	100													
	Monthly runoff (% of accumulation)	%		0	0	0	0	100	100	100	100	100	100	100	0	
DISPLACED	**IF TAILS DISPLACE POND**															
	Tailings submerged (% of total)	%														
	Deposited dry density	t/m^3	1.35													
LOSSES	Water retained in tailings (dry wt basis)	%	37													
	Estimated seepage losses	m^3/mo.	2000													24000
	Average Evaporation	mm/mo.		0	0	0	0	0	160	154	107	49	0	0	0	470
	Change in evaporation	%	0													
	Total evaporation	mm/mo.		0	0	0	0	0	160	154	107	49	0	0	0	470
	Concentrate production	t/day	60													21900
	Concentrate slurry density (by wt.)	%	40													
	Area of ponds and wetted tailings	ha	50													
	Recirculation to mill (% of process water)	%	90													
DECANT	Decant strategy (% of net inflow)	% / mo.		0	0	0	0	0	16.67	16.67	16.67	16.67	16.67	16.65		100
	Initial water volume in ponds	m^3	400000													

TABLE 1B
OUTPUT COMPUTATIONS

	INFLOWS (m^3mo.)				LOSSES (m^3mo.)						ACCUMULATION (m^3mo.)					(m^3)
								Recycle								
	Tailings Water	Misc. Inflows	Tailings Basin Runoff	Total	Retained in Tailings	Seepage	Pond Evap.	Total (includes co slurry)	Concent. Slurry	Total	Net Inflow	Water Displaced	Change	Decant	Net Change	Accum. Volume
1	2	3	4	5	6	7	8	9	9a	10	11	12	13	14	15	16
INITIAL																400000
Jan	73315	0	0	73315	22194	2000	0	65984	2790	90178	-16863	0	-16863	0	-16863	383137
Feb	66220	0	0	66220	20047	2000	0	59598	2520	81645	-15425	0	-15425	0	-15425	367712
Mar	73315	0	0	73315	22194	2000	0	65984	2790	90178	-16863	0	-16863	0	-16863	350850
Apr	70950	0	0	70950	21479	2000	0	63855	2700	87334	-16384	0	-16384	0	-16384	334466
May	73315	0	423300	496615	22194	2000	0	65984	2790	90178	406437	0	406437	0	406437	740903
Jun	70950	0	87150	158100	21479	2000	80000	63855	2700	167334	-9234	0	-9234	135194	-144427	596476
Jul	73315	0	174300	247615	22194	2000	77000	65984	2790	167178	80437	0	80437	135194	-54757	541719
Aug	73315	0	203350	276665	22194	2000	53500	65984	2790	143678	132987	0	132987	135194	-2207	539512
Sep	70950	0	120350	191300	21479	2000	24500	63855	2700	111834	79467	0	79467	135194	-55727	483785
Oct	73315	0	132800	206115	22194	2000	0	65984	2790	90178	115937	0	115937	135194	-19257	464528
Nov	70950	0	103750	174700	21479	2000	0	63855	2700	87334	87367	0	87367	135032	-47665	416863
Dec	73315	0	0	73315	22194	2000	0	65984	2790	90178	-16863	0	-16863	0	-16863	400000
TOTAL	863225	0	1245000	2108225	261322	24000	235000	776903	32850	1297224	811001	0	811001	811001	0	

from natural land as a percentage of precipitation. All of the precipitation that falls on the tailings surface and pond is assumed to enter the ponds, and evaporation is then accounted for separately in WATBAL from the pond and wetted tailings surface.

WATBAL can account for winter snow by entering a runoff distribution each month as a percentage of the total accumulated to date. For example, if there is no runoff in January, February, and March and 100% runoff in April, then the total accumulation for the four months will enter the inflow side of the water balance in April.

For all runoff inputs, the effect of changes can be easily evaluated by entering a plus or minus "% change" in the appropriate row.

Column 5 is the summation of the inflows.

Water retained in the pore spaces of the tailings (Column 6) is entered into WATBAL as a "water content" in the traditional soil mechanics sense. If the tailings are saturated then the water content can be simply calculated as follows:

$$w = \left(\frac{\rho_w}{\rho_d} - \frac{1}{G_s}\right)100$$

$$w = \left(\frac{e}{G_S}\right)100$$

Where:

w = saturated water content (percent—wt. water/dry wt. tails)
ρ_d = dry density of tailings
G_s = specific gravity of tailings particles
ρ_w = density of water (unity in the metric system)
e = void ratio (vol. of voids / vol. of solids

The dry density (or corresponding void ratio) that is chosen to calculate the saturated water content can affect a water balance. Density is a function of specific gravity. It also depends on gradation and the content of clay, gypsum, precipitates, ice inclusions, the degree of consolidation, and the distance from the discharge point. For design, an average value has to be used. A few general guidelines are given below.

- If the tailings are relatively coarse grained and do not contain clay minerals, then consolidation after the initial liquid/solids separation will be relatively minor and occur quickly. The typical average dry densities of such a tailings deposit might range between 1.3 and 1.4 t/m^3 (81 to 87 lb/ft^3) for tailings with a specific gravity of 2.7.
- If the tailings are finely ground or contain clay, gypsum, or other precipitates, then consolidation may take place more slowly and a lower density should be used.
- In extremely dry climates, there can be high evaporative losses on exposed beaches. For this case, a lower dry density should be used. This might typically be in the 1.0 t/m^3 range (62.4 lb/ft^3). It doesn't matter to the balance whether the water is lost as evaporation or permanently kept in the pore spaces.
- In cold regions, process water can become entrapped as ice. It may take years to thaw; in a permafrost environment, it may never thaw. This can be modeled in WATBAL by decreasing the dry density, which effectively increases the water content. If all the discharge water at a typical slurry density of 55% solids is entrapped as ice, then the effective dry density could be in the 0.85 t/m^3 (53 lb/ft^3) range.
- If a portion of a tailings mass is partially drained, then the average water content will be less than the saturated water content. The water content, in this case, has to be arbitrarily reduced by estimation.

Unrecoverable seepage (Column 7) may not be accurately known unless a detailed hydrogeological assessment has been carried out. In valley basins, where the groundwater is mounded in the hillsides above the tailings deposit, seepage from the basin does not usually represent a significant volume with respect to the balance, although it can be an environmental problem. However, in very dry climates, seepage can be a significant component of a water balance.

Evapotranspiration is implicitly taken into account in the runoff factor applied to the precipitation falling on the natural land surrounding the tailings. Lake evaporation (typically 70% of pan evaporation) is therefore applied to the pond and wetted surface of the tailings (Column 8). Evaporation data is normally obtained from the same source as precipitation and also in a monthly format. The effect of changes in evaporation can be easily evaluated by entering a plus or minus "% change" in the appropriate row.

Recycled water (Column 9) is entered into the model as a percentage of the tailings discharge water. In WATBAL, it also includes the water that leaves the property in the concentrate pipeline (summarized separately in Column 9a). The recycled water equals the tailings slurry discharge water plus the water that leaves the property in the concentrate pipeline less make-up water and other inflows into the mill. The rate of recirculation is readily available from designers and mill operators. Depending on the circumstances, a recirculation rate as high as 80 to 90% may be realistic, but 100% is seldom achievable.

In WATBAL, there has to be tailings discharge to calculate recirculation. However, a balance can be done for a pond, such as a polishing pond, that is not receiving tailings but from which water is being recycled by merely subtracting the recycled water from the inflow from the pond above.

Column 10 is the summation of the losses.

As discussed above the "Net Inflow" (Column 11) is the excess water that accumulates in the pond. It is the difference between Columns 5 and 10. It is the volume that has to be discharged to the environment if the pond volume is to remain in balance on an annual basis. The net inflow can be negative in some months.

Water displaced by tailings is calculated in Column 12. If an existing body of water has to be displaced to make room for tailings, then the displaced water has to be added to the net annual inflow if the basin is to remain in balance. It would, of course, be preferable to remove the water before start-up. The water retained in the tailings pores has already been accounted for in the balance, and therefore the volume displaced must equal the total space occupied by the deposited tailings slurry (solids plus water). This can be calculated by

$$\text{Water displaced} = \frac{\text{dry weight of submerged tailings}(W_s)}{\text{dry density of deposited tailings}(\rho_d)}$$

The tailings-displacing water does not have to be the total production.

Column 13 is the total change. It includes the net inflow (Column 11) plus any water that has to be displaced to make room for tailings.

The water that has to be decanted to the environment can be controlled by design and can vary by month and even be zero in the winter months if discharge cannot be permitted. Many mines are permitted to discharge only in certain seasons of the year, typically when receiver flows are high or when natural processes have provided optimal polishing. The amount of water decanted each month is entered into the program as a percentage of the total net inflow in Column 14.

Column 15 gives the net monthly change and Column 16 the accumulated volume in the pond at the end of each month.

43.6 SENSITIVITY ANALYSES

Development and operating costs can be minimized by optimizing the significant input flows such as

- Discharge slurry density
- Watershed diversion
- Precipitation and evaporation
- Recirculation to the mill
- Miscellaneous inflows such as mine water
- Water retained in the tailings (pore water)
- Seepage
- Pumping existing water from a basin prior to start-up

For example, if the slurry density is increased from 30 to 50% solids, then the tailings transport water is reduced by more than half. It is easy to evaluate the effect of diverting part of a watershed that may be an economical trade-off against managing and treating a larger volume of water. Evaporation is a function of pond area. A higher recirculation rate reduces the need to introduce clean make-up water in the mill.

Based on the results of a sensitivity analysis, parameters can be chosen to simulate best-case (optimistic) and worst-case (pessimistic) scenarios to establish the range of operating and hydrological conditions that a tailings facility might have to contend with. The worst-case parameters depend on an acceptable level of risk.

A water balance should also be reassessed as the basin changes shape over time. In a dry climate, this may have to be every year.

43.7 EXAMPLE OF A WATBAL APPLICATION

43.7.1 The Base Case

The first step is to establish a realistic set of "base case" design parameters that represent average operating and hydrological conditions. Then, based on a sensitivity analysis, the "best-case" and "worst-case" scenarios can be chosen that will set the boundary conditions on which to base the design.

As an example, let us assume a small mine operating in a relatively dry, severe, northern climate. A reasonable set of base case parameters for such a mill and tailings basin could be as follows:

Tailings production	
▪ design ore tonnage	2,100 t/d (metric)
▪ mill availability	95%
▪ concentrate production (3% of ore)	60 t/day
▪ nominal tailings tons (planned average annual over 365 days)	1,935 t/day
Operating period	12 months/year
Discharge tailings slurry density	45% solids by weight
Concentrate pipeline slurry density	40%
Miscellaneous inflows	zero m^3/m
Precipitation (average annual)	300 mm/y (monthly distribution required)
Total watershed area	5.2 km^2
Area of natural land	3.5 km^2
Area of tailings and ponds	1.7 km^2
Runoff factors (% of precipitation)	
▪ natural land surrounding the tailings	70% of precipitation
▪ tailings and ponds	100% of precipitation
Runoff period	7 months (May to November)
Specific gravity of tailings solids	2.70
Assumed void ratio of deposited tailings	1.0 (volume of voids/ volume of solids)
Dry density of tailings (calculated using assumed void ratio)	1.35 t/m^3 (no ice inclusions)
Saturated water content of tails (wt. water/dry wt. tails)	37% (assume tailings remain saturated)
Estimated basin seepage	2,000 m^3/month
Lake evaporation from ponds and wetted tailings surface	470 mm/yr. (monthly distribution required)
Area of pond and wetting tailings beach	0.5 km^2
Recirculation to the mill	90% of discharge water
% of tails displacing water	None
Decant period	6 months (June to November)
Starting month (beginning of an annual cycle)	December
Initial pond volume (end of Nov.)	400,000 m^3 (dead storage)

The WATBAL results are shown in Table 43.1 (precipitation version). The input data are shown in the upper part of Table 43.1. For most items, the data can either be entered as an average in the VALUE column or distributed with a variable distribution in the monthly columns. The VALUE column will take precedence.

The annual totals, from the total line on the table, are summarized below.

Inflows	**M m^3/year**
Tailings water	0.86
Miscellaneous inflows	0
Runoff	1.25
TOTAL	2.11
Losses	
Retained in tailings	0.26
Seepage	0.02
Evaporation	0.24
Recirculation (including concentrate slurry)	0.78
TOTAL	1.30
NET ANNUAL INFLOW	0.81
Water displaced by tails	0
TOTAL TO BE DECANTED	0.81
MAX. POND VOLUME	0.74 M m^3 (May)

The net annual inflow of 0.81 M m^3/y is the quantity of water that has to be discharged to the environment each year if the pond is to remain in balance. Even with a high recirculation rate to the mill and full saturation of the tailings, there is still a considerable amount of water that has to be discharged. Total containment (zero discharge) would be difficult, if not impossible, to attain, even with the relatively low precipitation used in the example.

The sensitivity of the balance is easily investigated by varying the significant input parameters such as precipitation, slurry density, recirculation, and a scenario where mine water is put into the basin as a miscellaneous inflow. The results are shown on Table 43.2. The analyses are carried out by varying the input parameter being investigated and keeping all of the others constant at the base case.

All the parameters investigated have a significant effect on the balance. If the annual precipitation increases by 25%, then

TABLE 43.2 Results of sensitivity analysis

Parameter	Range	Net Annual Flow (Column 11) M m^3/yr	% Change From Base Case
Precipitation	+50%	1.43	+77
	+25%	1.12	+38
	+10%	0.94	+16
	* Average	0.81 *	0
	-10%	0.69	-15
	-25%	0.50	-38
	-50%	0.19	-77
Slurry density (% Solids)	20	1.01	+25
	30	0.89	+10
	* 45	0.81 *	0
	50	0.80	-1
Recirculation (%)	0	1.59	+96
	30	1.33	+64
	60	1.07	+32
	* 90	0.81 *	0
	100	0.72	-11
Mine water (m^3/month)	* 0	0.81 *	0
	15,000	0.99	+22
	30,000	1.17	+44
	45,000	1.35	+67
	75,000	1.71	+111
	90,000	1.89	+133

* Denotes base case conditions.
NOTE:
The sensitivity analyses are carried out by varying the parameter being investigated and keeping all the other parameters constant at the base case.

the net annual inflow increases by 38%. The return period for such precipitation could be 10 to 15 years.

If the slurry density decreases from 45% solids to 20% solids, then the net annual inflow would increase by 25% even if the recirculation rate remains high at 90%. If it is 100%, there would be no change.

A drop in the recirculation rate to the mill has a major effect. A drop to zero increases the net inflow by 96%.

Putting the mine water into the tailings pond has a major effect on the balance. At a modest rate of 30,000 m^3/month (183 US gpm), the net annual inflow would increase by 44% from the base case, and at 90,000 m^3/month the increase would be 133%. The consequences of putting miscellaneous flows into a tailings pond should be considered at the design stage rather than after the mill has started up.

43.7.2 Best-Case and Worst-Case Scenarios

A combination of adverse conditions occurring simultaneously could have a disastrous effect on a tailings-basin water balance. It is not unreasonable to assume that the annual precipitation could vary by at least plus or minus 25% during the operating life of a mine. Recirculation to the mill and the discharge slurry density can drop as a result of problems in the mill. Also, it is not inconceivable that unplanned miscellaneous inflows such as mine water could be discharged to the tailings pond. A reasonable set of best and worst case parameters could arguably be as shown below.

Parameter	Best Case	Base Case	Worst Case
Slurry density (%)	60	45	25
Miscellaneous inflows – Mine water (m^3/month)	0	0	30,000
Precipitation	–25%	average	+25%
Recirculation (%)	100	90	30

TABLE 43.3 Best- and worst-case water balance scenarios

	Annual Flows: Reasonable Best-Case Scenario (M m^3/yr)	Average Base Case (M m^3/yr)	Reasonable Worst-Case Scenario (M m^3/yr)
Inflows			
Tailings water	0.47	0.86	2.12
Mine water	0	0	0.36
Run-off	0.93	1.25	1.56
Total	**1.40**	**2.11**	**4.04**
Losses			
Retained in tailings	0.26	0.26	0.26
Seepage	0.02	0.02	0.02
Evaporation	0.24	0.24	0.24
Recirculation	0.47	0.78	0.64
Total	**0.99**	**1.30**	**1.16**
Net annual flow	**0.41**	**0.81**	**2.88**
Max. pond volume (M m^3)	0.60	0.74	1.57
Input variables			
Tailings production (t/day)	1,935	1,935	1,935
Slurry density (% solids)	60	45	25
Mine water (m^3/month)	0	0	30,000
Runoff (%)	-25%	avg.	+25%
Recirculation (%)	100	90	30

The results are tabulated in Table 43.3. For the worst-case scenario, the net annual inflow increases by 256% (a factor of 3.5) over the base case; for the best case, it is approximately half. This clearly demonstrates the wide range over which the tailings basin would have to operate in order to prevent an unscheduled discharge to the environment and for the system to stay in balance on an annual basis.

Also, the pond size is an important consideration. For the worst-case scenario, the maximum pond capacity (in May) would have to be at least two times as large as the best case.

43.8 WATER BALANCE MODELING IN A DRY CLIMATE

A dry climate water balance is more difficult to model realistically than a wet climate balance because seepage (infiltration) and evaporation are large and difficult to predict, whereas seasonal variations may have little effect. The changing shape of a basin is important because as the basin increases in size, larger wetted areas become available for evaporation and infiltration. For these conditions, an annual balance for each year of operation is normally required. Flows can also be conveniently expressed in cubic meters of water lost per ton of tailings produced or ton of ore milled.

Frequently in a dry climate the tailings may have a high clay content. This causes a low initial deposited density and long consolidation times to reach ultimate density. The water released during consolidation can be relatively large, as is illustrated in the following example.

Discharge slurry density (solids by wt.)	48%
Specific gravity of tailings solids	2.7
Volume of slurry water (1/0.48 – 1)	1.08 m^3/t of tails (100%)
Initial liquid/solids separation	
Dry density (assume)	1.05 t/m^3
Pore water retained (saturated) (1/1.05 – 1/2.70)	0.58 m^3/t of tails (54%)
After consolidation	
Dry density (assume)	1.35 t/m^3
Pore water retained (saturated) (1/1.35 – 1/2.70)	0.37 m^3/t of tails (34%)
Water released by consolidation (0.58 – 0.37)	0.21 m^3/t of tails (19%)

In the above simple example, the water released by consolidation is 19% of the total discharge water, but it is very difficult to capture because most of it will be lost as evaporation and seepage.

Desiccation also occurs on inactive beaches. Desiccation has the same effect as consolidation except that water is removed by evaporation instead of being squeezed out by pressure. This can be modeled in a water balance by simply using a lower density to calculate the water retained in the pore spaces. The model understands that this is water that is permanently lost, and whether the loss is due to evaporation or to water permanently retained in the pore spaces doesn't really matter to the balance.

In a dry climate seepage is difficult to predict. It can consist of several components:

- Saturation of unsaturated soils beneath a basin
- Seepage from ponds in contact with natural soils, the area of which can change dramatically as a basin develops
- Seepage from the ponds on top of the tailings (depending on the hydraulic conductivity of the tailings this could be quite small)
- Pore water squeezed vertically downwards by consolidation may be lost as seepage
- Water may be permanently lost in infiltrating desiccated inactive beaches when they are rewetted and then permanently lost as evaporation

43.9 DISCUSSION

The WATBAL spreadsheet discussed in this chapter is Release 5.0 (WATBAL© Copyright 1989–2000, Golder Associates Ltd.). The program can be used free of charge. It is merely a mathematical tool that adds and subtracts inflows and losses to a system. No special skills are required to run such a model, but sound engineering judgment is required to ensure that the input data are appropriate for the intended application. The results are only as good as the numbers that are put into such a model. A hydrologist should advise on precipitation, evaporation, runoff factors, and the routing of storm flows.

In WATBAL, the input data are entered in easily recognizable terms that facilitate changes and enable sensitivity analyses to be carried out. In this respect, the program is transparent. It can also be used as an operating tool during the life of mine. An important conclusion that can be reached from the example is that a tailings basin must be able to operate over a wide range of operating and hydrological conditions. There is no such thing as a unique water balance for a tailings basin.

43.10 ACKNOWLEDGMENTS

As the modeling evolved over time, several engineers in the Golder organization contributed to the development of the logic and WATBAL spreadsheet, including John Gilby and Mike Ankenmann from Mississauga; Leon Botham, who is now in Saskatoon; Jim Johnson in Denver; Rick Firlotte from Montreal; and Mike Wilton in Vancouver. We hope that refinements will continue in the future.

43.11 REFERENCES

Welch, D.E., and F.W. Firlotte. 1989. Tailings management in the gold mining industry. *Proceedings Vol. 14, International Symposium on Tailings Effluent Management. Halifax, Nova Scotia, Aug. 1989.* The Metallurgical Society of the Canadian Institute of Mining and Metallurgy.

Welch, D.E., L.C. Botham, and D. Bronkhorst. 1992. Tailings basin water management, a simple effective water balance. *Environmental Management for Mining, 1992 Saskatchewan Conference.* Saskatchewan Mining Association.

Welch, D.E., L.C. Botham, and J.M. Johnson. 1995. Prediction of tailings effluent flows. *Tailings and Mine Waste '95 and Summitville Forum Conference.* Colorado State University, Fort Collins, Colorado, USA. January 17–20, 1995.

CHAPTER 44

The Gold Ridge Mine Tailings Storage Facility: An Australian Case History

Mike Gowan* and Glen Fergus*

44.1 INTRODUCTION

Most tailings storage facilities (TSFs) built and operated in Australia are in areas of relatively arid climatic conditions and low seismic risk. The same is not true of the mines located in the island nations north and east of Australia. This chapter describes the design and operation of the first mine in Melanesia to have a TSF designed to contain the life of mine tailings output.

The 100,000 oz/annum Gold Ridge Mine is the first mine developed in the Solomon Islands and represents a major milestone in the country's development.

Commissioned in the third quarter of 1998, it is 25 km southeast of the national capital Honiara, on Guadalcanal Island, Solomon Islands. From the outset, Ross Mining has been committed to contain the tailings on the mine site in a stable and rehabilitated structure. The implementation of best practice for the design of the TSF has been key to the high level of acceptance and support received from local landowners, the general community and the government of the Solomon Islands.

The mine, which has 20 million tonnes of ore and a projected life in excess of 10 years, currently operates two relatively small open pits on the lower northern slopes of Mt. Popomanaseu, a 2,300-m peak near the center of the island. A conventional CIP plant is located below the pits at an elevation of about 460 mRL. From there, the tailings slurry is piped 8 km down the mine access road to a TSF in the lower northern foothills, at a lowest elevation of 20 mRL.

44.2 OVERVIEW

The TSF is in a small, incised valley with an appropriately restricted catchment area. It includes

- Tailings storage (TS) which has a main retaining dam wall at the bottom of the valley that is 50 m high with a 1,000 m crest length and a smaller saddle dam wall in the southeast
- A return water dam (RWD) at the head of the valley, with a 30-m-high dam wall with a 200-m crest length

Both the TS and RWD walls are zoned earthfill embankments. The TS is being constructed to its final height in five stages, with a final height sufficient to contain 20 million tonnes of mine tailings at an average settled dry density of 1 t/m^3.

The tailings slurry is open-end discharged from the heads of the various valleys on the southern side of the TS, and spigoted off the TS dam wall. This deposition plan is operated so as to contain the supernatant pond in the head of the eastern valley, remote from the TS dam wall. Floating pumps are used to decant the supernatant water into the RWD.

44.3 DESIGN STANDARDS

Ross Mining required that best practice design and operating standards followed for the TSF, and throughout the design, receive third-party review, which was implemented by Ross Mining and by the Solomon Island government.

The design standards included the tailings dam guidelines published by ICOLD (1995, 1996a, 1996b) and ANCOLD (1986, 1998, 1999). In addition, there was a requirement that the TS comply with the water storage requirements set out in the Queensland Government *Guidelines on the Prevention of Water Pollution From Cyanide Use in Gold Ore Processing* (1990). These guidelines provide sufficient rainfall runoff capacity for gold mining operations in tropical regions where extended wet seasons are normal and cyclonic rainfall events are common.

44.4 SITE DESCRIPTION

The north coast of Guadalcanal is formed by a broad coastal outwash plain fringing the main central range of the island. The plain is crossed by frequent, braided, sandy streams draining from the range and is largely developed for oil palm, cocoa, and copra plantations. The lower-foothill terrain comprises uplifted and dissected alluvial outwash deposits, with a dense, often subparallel pattern of incised valleys separating long, low, steep-sided ridges (sometimes flat-topped). Locally, the ridge crests support patches of tall tropical grassland, with dense tropical rainforest in the valleys, heavily disturbed by intensive logging.

The TS site is in the incised valley of a small creek east of the mine access road, just beyond where it crosses the Tinahulu River and begins to climb away from the coastal plain. The valley is well suited to tailings storage, having only a small catchment, with a broad storage area downstream enclosed by moderately steep ridges. The small creek previously draining the valley was intermittent in the dry season.

The symmetrical V-shaped valley at the RWD is about 50 m deep, sloping at 20° to 25°. Downstream of the RWD, the valley opens into a broad basin, which will be filled by tailings in the TS. The TS dam wall occupies the alluvial flat-bottomed lower portion of the valley.

Guadalcanal is in the rear arc of the Solomon Islands' double chain of the "Pacific Ring of Fire," which is one of the most tectonically active areas in the southwest Pacific. A detailed study assessed the seismic risk in this area to be extreme, with magnitude 8+ earthquakes predicted to occur within 100 km of the site, with an average recurrence interval of about 20 years.

44.5 GROUND CONDITIONS AT THE TAILINGS STORAGE DAM WALL

At the site selected for the TS dam wall, the valley is about 150 to 200 m wide, with steep-sided abutment ridges to the northwest

* Golder Associates Pty Ltd, Brisbane, Australia.

and southeast. The valley floor is infilled with ~10 to 20 m of mostly clayey recent alluvium, overlying siltstone or sandstone basement material. The depth of infill appears to increase downstream.

44.5.1 Ground Conditions

The clayey alluvium is generally stiff to very stiff, with slickensided fissuring in the upper part grading to firm and intact below a depth of ~3 m. This generally clayey alluvium appears to be interspersed with occasional lenses or layers of more sandy/gravelly material. The interface between the clayey alluvium and the underlying weathered basement material is commonly indistinct, reflecting the general textural similarity of the materials. It may be best interpreted as corresponding to the change from firm to stiff/very stiff consistency. Boreholes drilled within the TS upstream of the TS dam wall encountered similar alluvial materials of gradually decreasing depth, tapering to ~8 m depth at ~500 m from the dam wall.

44.5.2 Groundwater

Measured groundwater levels within the valley infill alluvium generally range from 0.5 to 2 m depth. Pump-out testing of two boreholes indicated that are some of the more permeable sandy zones within the alluvium were of limited lateral extent.

Three boreholes drilled 21.5 m and 30 m through the alluvial infill into the basement material encountered artesian groundwater with a fairly steady flow of ~13 L/min over several months.

44.5.3 Borrow Material

All borrow material for construction of the dam walls will be located within the TS and RWD areas. The basement materials are suitable for select fill almost everywhere on the site, but clay borrow is only available in limited areas. Borrow sources that have been opened below the final storage level within the TS basin will be covered by a 500-mm compacted clay layer upon completion wherever nonclay subgrade has been exposed.

44.6 DAM WALLS

The design of the dam wall for the RWD follows standard water dam design practice, whereas the design for the TS dam walls differ somewhat, taking into account

- Function—they are to retain tailings and not a depth of water
- Tailings disposal—beaching away from the walls
- Site conditions—wall height and limitations set by staged construction

44.6.1 Return Water Dam

The RWD has a 30-m-high water-retaining dam wall and a storage capacity of 500,000 m^3. The overall average downstream slope is 2.5:1 (H:V), and the slope of the upstream face is 2.5:1. The dam wall has a sloping clay core on the upstream face, joined into a cutoff trench excavated to a nominal depth of 3 m. The clay core is protected by a filter drain, and seepage collected in the filter drain will be conveyed to the downstream toe of the RWD dam wall by discrete outlet drains. A layer of riprap covers the upper portion of the upstream face, to control wave-induced erosion.

The downstream face has been built at a slope of 2.2:1, with 4-m-wide berms for every 10-m rise. The berms, required to control runoff down the face, have been designed to collect this runoff and deliver it to the runoff control system, which drains down the outer edges of the dam wall. The downstream face has been covered with a 300-mm layer of topsoil and vegetated with local grasses and vines.

TABLE 44.1 Proposed construction staging of the TS dam wall

	Height (m)	Approximate Storage Volume (m^3)	TS Life at 2 Mtpa and 1 t/m^3
Stage 1A	10	1,000,000	0.5
Stage 1B	18	3,000,000	1.5
Stage 2	29	7,500,000	3.5
Stage 3	36.5	11,000,000	5.5
Stage 4	42.5	16,000,000	8
Stage 5	50	20,000,000	10

Whilst the RWD is relatively high, its hazard rating for both uncontrolled releases/seepage and embankment failure is classified as low, because

- All uncontrolled releases/seepage will be contained within the TS.
- Any water flows resulting from an embankment failure would be contained by the TS.

The RWD wall will be demolished at the end of the mine life, with its embankment fill used for rehabilitation cover material for the TS.

44.6.2 TS Dam Wall

The TS dam wall is being constructed in stages to a maximum height of about 50 m.

The proposed stage construction of this wall is shown in Table 44.1.

This staged construction is required to

- Provide for strengthening of the alluvial foundation soils through progressive consolidation
- Allow the mine to start production on time
- Spread the construction cost over the mine life, avoiding a high upfront capital cost

44.6.3 Dam Wall

A conventional water dam design was used for stage 1, with a wide central clay core, supported on each side by compacted selected fill and underlain by a compacted clay-filled cutoff trench. Due to the low initial strength of the foundation clays, berms were required to ensure the stability. Subsequent wall raising is being carried out in the downstream direction, incorporating wide berms for stability and foundation consolidation and with the clay core raised on the sloping upstream face of the wall (similar to the RWD).

A filter drain has been incorporated on the downstream side of the clay core to intercept any seepage through the clay core, which will be drained through outlet drains to a collection sump at the downstream toe of the dam. A full-width downstream blanket drain has also been provided beneath the downstream portion of stage 1 and 2 embankments to intercept any foundation seepage bypassing the cutoff trench. Discrete surface drains have been used in place of the blanket drain beneath the stage 3 raising footprint. These drains are required to prevent the buildup of pore water pressures within the general fill material downstream of the clay core, which would be detrimental to the stability of the TS wall.

To accelerate consolidation and strengthen of foundation clays, 6,000 wick drains were installed at a 1.2-m spacing beneath the stages 1 and 2 embankments' downstream footprint to a depth of between 10 and 13 m. The blanket drain also functions to collect water expelled from these wicks by the foundation consolidation. Water from these drains flows to a downstream collection sump, from where it is pumped back into the TS.

The slope of the upstream face of the TS dam wall starts at 2.5:1 for stage 1, and this slope has been increased to 2:1 for stages 2 to 5. Upstream stability at this steeper wall face slope is provided by the buttressing action of the tailings retained against the upstream face. The final downstream face of the dam wall is permanent and has been engineered for long-term performance. It will have an overall average slope of 3:1 (H:V), with 4-m-wide runoff control berms at every 10-m rise.

A layer of riprap, to be placed at facility closure on the upper face of the final stage wall, over the area that will be exposed above the tailings level, will provide long-term erosion protection for the wall. Provision is made at each tailings spigot point on the wall to prevent scour, using slotted delivery pipes.

44.6.4 Performance Monitoring

The rate of consolidation and strength gain of the foundation clays is critical to the stability of the TS dam wall. A monitoring system has been installed to monitor foundation performance in response to the staged construction of the dam wall. The system includes the following:

- Six water-filled settlement profile tubes instrumented with up to six low-range, high-accuracy, vibrating wire piezometers
- Thirteen vibrating wire piezometers to record pore pressures within the foundation in both wick-drained and unwicked areas
- V-notch flow-monitoring weirs on the seepage collection sumps

The performance-monitoring instrumentation records have been reviewed prior to each stage rise and, to date, have met the design requirements.

44.6.5 Hazard Rating

The hazard rating for the TS dam wall has been assessed as "high" for both uncontrolled releases or seepage and embankment failure. In view of the high hazard rating of the TS dam wall, the investigation and design process has been exhaustive and thorough, notwithstanding the considerable logistical difficulties of the mine location.

44.6.6 Saddle Dam

The TS requires a saddle dam wall to plug a low section on the southern rim of the basin. This dam will be approximately 20 m high on the centerline, but the batters to the downstream and upstream toes will be significantly higher due to the narrowness of the ridge at this point. The dam wall crest length will be approximately 250 m.

A central clay core design has been adopted, with details somewhat similar to stage 1 of the main TS dam wall, except

- A narrower tapering clay core in view of the lower final head to be resisted
- No wide berms because of the improved foundation conditions
- No downstream drainage blanket because of the improved foundation conditions and lower height

The saddle dam wall is to be built in one stage during 2000, coinciding with the construction of stage 3 of the TS dam wall.

44.7 STABILITY ANALYSES

The face slopes for the TS dam and the RWD walls were selected to ensure their stability under a range of possible failure conditions, while rehabilitation requirements limited the downstream slope. The stability of the dam walls has been considered for the following conditions:

TABLE 44.2 Lower quartile triaxial strength parameters

Failure Condition	Test Conditions	Angle of Friction Degree	Cohesion (kPa)
End of Construction	S_{uu}	$\varphi_u = 12$	$c_u = 75$
	Unconsolidated undrained	$\varphi_u = 12$	$c_u = 75$
Earthquake	S_{cu}	$\varphi_u = 12$	$c_u = 90$
	Consolidated undrained	$\varphi_u = 12$	$c_u = 90$
		$\varphi_u = 12$	$c_u = 90$
Long Term	S_{cup}	$\varphi' = 30$	$c' = 0$
	Consolidated undrained,	$\varphi' = 27$	$c' = 0$
	with pore pressure measurement	$\varphi' = 27$	$c' = 5$

TABLE 44.3 Minimum acceptable factors of safety

Failure Condition	Minimum Factor of Safety
End of construction	1.25
Rapid drawdown	1.50
Long-term/steady-state seepage	1.50
Earthquake	1.00

- End of construction–applies at the completion of each construction stage (undrained shear strength parameters)
- Long-term conditions with steady-state phreatic conditions (effective stress, shear strength parameters)
- Rapid drawdown–applicable for the RWD only
- Earthquake loading (consolidated undrained shear strength parameters)

The dam wall stability analyses were carried out using the SLOPE/W computer program.

44.7.1 Strength Paramaters

The shear strengths listed in Table 44.2 were determined in the laboratory for the compacted selected fill, clay core, and significant underlying foundation materials, respectively.

44.7.2 Factors of Safety

The factors of safety listed in Table 44.3 were adopted in determining the acceptable stability of the TS dam wall and RWD.

The results of the stability analyses showed that both the TS dam and the RWD walls will be acceptably stable for the various critical failure conditions investigated.

44.7.3 End of Construction Condition

The lower quartile strength parameters taken from the S_{uu} test results of the laboratory testing program were used to analyze the stability of the dam walls at the critical end-of-construction time for each stage. The calculated factors of safety listed in Table 44.4 were obtained for the most critical slips in each stage.

Downstream, noncircular, translational failures through the foundation soils were shown to be most critical for the TS dam wall, and berms are required on the downstream toe of the TS dam wall in order to achieve the required factor of safety.

A thorough sensitivity analysis was carried out to model the consolidation strengthening of the foundation soils due to the progressive increase in surcharge as each stage of the TS dam wall is constructed. This required calculation of estimated consolidated strengths for specific zones of the dam wall foundation relative to its geometry. Design details, such as berm height and length, batter slope and consolidated strength parameters in both the embankment and foundation materials, were varied during this analysis to achieve the required factors of safety. The final

TABLE 44.4 End-of-construction stability analysis

Dam Wall	Period	Dam Face	Factor of Safety
TS	Stage 1	Upstream	1.20
		Downstream	1.38
	Stage 2	Upstream	1.30
		Downstream	1.32
	Stage 3	Upstream	1.42
		Downstream	1.34
	Stage 4	Upstream	1.60
		Downstream	1.33
	Stage 5	Upstream	1.43
		Downstream	1.25
RWD	Complete	Upstream	1.85
		Downstream	1.63

design was based on the following criteria and took into account the calculated level of consolidation and associated strength parameters at the end of construction of each stage:

- In zones with wick drains, the foundation clay would be 100% consolidated under the surcharge of the previous stage and 50% consolidated under the surcharge of the newly completed dam wall stage.
- In zones without wick drains, the foundation clay would be 100% consolidated under the surcharge of the previous stage but would have 0% consolidation under the surcharge of the newly completed dam wall stage.
- The embankment material of all previous stages would have increased in strength to the consolidated lower quartile parameters determined in laboratory testing (from c = 75 kPa, $\varphi = 12°$ to c = 90 kPa, $\varphi = 12°$).

44.7.4 Long-Term Stability Under Steady-State Phreatic Conditions

The long-term stability of the dams was analyzed using lower quartile effective stress parameters based on the results of S_{cup} triaxial tests. The failure condition assumed, as a worst-case condition, that the TS and the RWD are filled to their maximum supply level and that steady-state seepage is established through the clay core and into the filter drain. A summary of the long-term, steady-state seepage analyses and the resulting factors of safety for the most critical failure surfaces in each dam are given in Table 44.5.

44.7.5 Rapid Drawdown in the RWD

The analysis of a rapid drawdown failure condition was considered relevant to the upstream face of the RWD, as this is the only dam that would be pumped down over a geotechnically short period of time. Previous analyses for sloping core dams have demonstrated that they are virtually immune to drawdown failure, because of the B_{bar} response effect, which approximately cancels the pore pressures within the sloping upstream core, due to the unloading on storage drawdown (provided $B_{bar} \approx 1.0$, as it is for saturated clay). No further analysis was deemed necessary.

44.7.6 Stability Under Earthquake Loading

The stability of the TS dam wall and the RWD under earthquake loading is critical to the long-term success of this project, and was analyzed using the Makdisi and Seed (1978) approach. This method is based on the classical Newmark (1965) sliding block method but modified to allow for the dynamic response of the dam walls. This is a widely used, simplified deformation analysis appropriate for dams not susceptible to liquefaction or strain weakening in the embankment or their foundations.

TABLE 44.5 Summary of long-term, steady-state seepage analyses

Dam Wall		Failure Mode	Factor of Safety
TS	Downstream	Circular	2.22
		Noncircular	2.28
RWD	Upstream		3.56
	Downstream		1.66

TABLE 44.6 Adopted earthquake parameters

Parameters	Unit	Embankment Fill	Foundation Soil
Bulk density	kN/m^3	18	18
Compression wave velocity	m/s	400	400
Shear wave velocity	m/s	200	200

It was considered that the loose sandy/gravelly material within the alluvial foundation materials was not sufficiently extensive to seriously compromise earthquake stability, particularly given the presence of positive pattern drainage and reinforcement provided by the foundation wick drains.

The analysis method determined the likely permanent deformations along critical slip surfaces at various heights up the embankments. Critical slopes with bases tangential at $y/h = 0.25$, 0.5, and 1.0 were considered, where y is the vertical distance from the dam crest to the base of the slope and h is the embankment height, with the full height slip consistently producing the highest deformations.

The amplification of the horizontal ground acceleration to obtain a maximum induced crest acceleration in the earthquake event was estimated using the simplified methods proposed by Sarma (1981) and by Makdisi and Seed (1978). Sarma's approach incorporates both the embankment and foundation geometry and uses estimations of the dynamic properties of the materials. These parameters were adopted based on experience and our present knowledge of the materials encountered on site. Alternatively, the Makdisi and Seed approach also involves an iterative method to determine the final dynamic properties based on their shear strain dependence. This approach, however, is restricted to the embankment only, ignoring the influence of a deep soil foundation. By considering the base ground acceleration from the highest earthquake considered, and by combining the results of both methods, consistent maximum crest acceleration were obtained (Table 44.6).

The general Makdisi and Seed analysis method for permanent deformations also requires determination of the yield accelerations k_y for the sliding masses considered, i.e., the magnitude of the horizontal seismic accelerations that would be required to produce a slip failure (factor of safety = 1.0) at that particular elevation.

These yield accelerations (Table 44.7) were determined using the computer program SLOPE/W. The material parameters used in these analyses were based on the results of a series of consolidated undrained triaxial tests on representative samples. These tests were structured toward modeling a quick undrained failure condition after a certain degree of consolidation had taken place. Lower quartile strength parameters of $\varphi_u = 12°$ and $c_u = 90$ kPa (with $\gamma_b = 18\ kN/m^3$) were adopted for the embankment fill materials, and fully consolidated strength parameters were used for the clay beneath the embankment.

The results of the yield analysis, based on the laboratory testing results, indicate that very flat slips with very high yield accelerations would be required to initiate failure. This is due to the value of the cohesion adopted for the consolidated undrained state and also to the flat overall downstream dam wall slopes adopted.

Based on the dynamic response of the dam, calculated from the Sarma analysis and further correlations by Makdisi and Seed,

TABLE 44.7 Yield acceleration

Failure Type	y/h	Factor of Safety	k_y (g)
Circular	0.25	1.0	0.82
	0.5	1.0	0.43
	1.0	1.0	0.23
Noncircular	0.25	1.0	>0.8
	0.5	1.0	>0.5
	1.0	1.0	>0.3

TABLE 44.8 Permanent deformations

Approach	Maximum Crest Acceleration*	Permanent Deformation (mm)
Makdisi and Seed	0.25 g	15
Sarma	0.5 g	2

* Based on a maximum ground acceleration of 0.51 g from magnitude M = 8.5 earthquake at an epicentral distance of F = 60 km.

TABLE 44.9 Seismic acceleration

Ground Acceleration (g)	Magnitude (M)	Epicenter Distance (F) km
1.0	9.0	51
1.0	8.5	42
1.0	8.0	34

values of the maximum seismic acceleration k_{max} were obtained at the base levels of the slips investigated. These were amplified from assumed ground accelerations induced from particular earthquake magnitudes.

The maximum ground acceleration determined from the highest magnitude earthquake considered in the seismic risk assessment was 0.5 g from a magnitude M = 8.5 earthquake at an epicentre distance of F = 60 km. The permanent deformations calculated for this ground acceleration are summarized in Table 44.8.

A back-analysis showed that a ground acceleration of 1.0 g would be required to produce a 1.0-m permanent but tolerable level of displacement as a result of a full dam wall height slip failure. Following the earthquake attenuation equations used in the seismic risk analysis, this acceleration corresponds to that listed in Table 44.9.

Under all conditions analyzed, the factor of safety for the TS dam wall and the RWD would be in excess of the criteria that are appropriate to their size and importance and has shown that they would perform adequately under an earthquake equivalent to the expected maximum credible earthquake.

44.8 CONSTRUCTION MATERIALS

The dam walls are being constructed as zoned embankments, requiring selection and placement of a range of materials. Careful selection and controlled placement has been required to ensure dam wall stability and integrity. The main zones having an affect on the stability analysis are described below.

44.8.1 Zone 1—Clay

Approximately 300,000 m^3 of low-permeability clay is required for construction of the cutoff trench and the clay core for the two dam walls. Sources of suitable clay within the TS and RWD areas have been identified during the geotechnical investigations.

44.8.2 Zone 2—Sand Filter

Sand filter material has been sourced from pits in the Ngalimbiu River 5 to 6 km to the northwest, with screening required to remove oversize from the river sand.

44.8.3 Zone 3—Selected Fill

Ultimately, approximately 3,000,000 m^3 of selected fill will be required for the three dam walls, borrowed from the basin and the ridges within the area of TS and RWD.

The stability of the dam walls immediately after construction is dependent on the compacted fill materials achieving a minimum shear strength of 75 kPa. The specification allows for compaction at slightly higher-than-normal moisture contents, the use of wet comparative density testing, and the use of in situ strength testing, instead of the normal dry density test methods. This is because most soils in the region have moisture contents often significantly above the optimum moisture content. In practice, the required strength has been consistently achieved.

Strict control of the material compaction has been maintained, and no new material is placed until the already-placed and compacted material has been tested, approved, and signed-off on.

44.9 SEEPAGE CONTROL

Limited seepage is expected from TS, as both the tailings and natural valley clay layers have low permeabilities in the order of 10^{-7} to 10^{-8} m/sec. Seepage flows through the TS dam wall will be monitored to check on the stability of the dam wall, but flows will be limited by the seepage interception systems included in the design.

44.9.1 Seepage Under the Dam Walls

Seepage under the TS and RWD walls is limited by clay-filled cutoff trenches, which are founded in low-permeability materials.

44.9.2 Seepage from the Tailings

Care has been taken during the construction of the TS dam wall to retain the integrity of the natural clay layers lining its basin. Where borrow sources are located below the final fill level of the dams, a 500-mm layer of clay has been replaced where a nonclay subgrade has been exposed.

44.9.3 Seepage Monitoring

A system of seepage-monitoring boreholes was installed downstream of the dam wall and around the periphery of the tailing storage area during stage 1 of TS dam wall construction.

44.10 PERFORMANCE

The mine has been operating for nearly 18 months. To date, the two dam walls have performed to expectation.

44.10.1 TS Dam Wall

Settlement records indicate substantial (up to in excess of 1 m) foundation settlement in response to the embankment load. It appears that much of this occurred rapidly as the embankment was constructed. This confirms that the intended foundation consolidation strengthening is proceeding. The records from piezometers beneath the constructed embankment show some response to the embankment load, followed by rapid dissipation. This indicates that the foundation drainage works (wick drains) are functioning as intended and is consistent with the rapid foundation settlement observed.

Two piezometers upstream of the cutoff trench show slower but still significant dissipation of embankment-induced pore pressures, confirming that wick drains did not need be installed in this area. The pressure at these two piezometers has subsequently risen in response to the rising tailings level in the storage. This rise is more marked than for the other piezometers downstream of the cutoff, which indicates that the cutoff is working to limit seepage.

There is no evidence of seepage through the TS dam wall and the seepage performance of the TS floor is currently not determinable. This is because water is still being expelled from the foundation soils to the drainage system by settlement and the tailings beach has not yet built up against the embankment and blankets the foundation.

44.10.2 Return Water Dam

Performance of the RWD to date is satisfactory, although slight (<0.01 L/s) seepage was apparent in late November 1998. Seepage was issuing from near the lowermost outlet drain with the storage level. However, there are indications that a natural seep may have been present in this area prior to construction.

44.11 REFERENCES

ICOLD. 1995. Bulletin 98, *Tailings Dams and Seismicity*.

ICOLD. 1996a. Bulletin 104, *Monitoring of Tailings Dams*.

ICOLD. 1996b. Bulletin 106, *A Guide to Tailings Dams and Impoundments*.

ANCOLD. 1986. *Guidelines on Design Floods for Dams*.

ANCOLD. 1998. *Guidelines on Design of Dams for Earthquakes*.

ANCOLD. 1999. *Guidelines on Tailings Dam Design, Construction, and Operation*.

Queensland Government. 1990. *Guidelines on the Prevention of Water Pollution From Cyanide Use in Gold Ore Processing*. Department of Environment and Heritage, January.

Australian National Committee of Large Dams. 1996. Interim Guidelines for Design of Dams for Earthquake. Section 6/3/3: Makdisi and Seed Analysis.

Makdisi, F.I., and H.B. Seed. 1978. Simplified Procedure for Estimating Dam and Embankment Earthquake Induced Deformations. *Jour. of Geotechnical. Eng. Div., ASCE*, 104:GT7:569–867.

Newmark, N.M. 1965. Effects of earthquakes on dams and embankments. *Geotechnique*, 15:2:139–160.

Sarma, S.K. 1981. A simplified method for the embankment resistant design of earth dams, dams and earthquake. TTL, London.

CHAPTER 45

Stability Analysis of a Waste Rock Dump of Great Height Founded over a Tailings Impoundment in a High Seismicity Area

Ramón Verdugo,* Carlos Andrade,† Sergio Barrera,† and José Lara†

45.1 INTRODUCTION

The stability analysis of a large waste rock dump from mining operations is presented in this article. As shown in Figure 45.1, the waste rock dump designed completely covers a tailings impoundment made up by a basin of saturated tailings, classified as very low plasticity silt, and a main dam made up of fine sands of medium compaction. The waste rock dump is 250 m in height and the tailings impoundment is approximately 40 m thick at the highest section. Due to space and operational restrictions, the waste rock dump will have side slopes of 1(V):1.5(H) and its compaction will be achieved from the transit of heavy trucks during the constructive phase. As can be seen, it is a very particular situation associated with a great mass of soils over a sort of "skate" consisting of soft material potentially liquefiable. The latter is especially of interest considering that the project area is of very high seismicity.

The geotechnical characterization of the materials and the stability analysis obtained from the traditional and dynamic methodologies are presented in this chapter, which concludes that both come up with very similar results. It is important to note that this project was, in the part regarding the pseudostatic analysis, described in detail in the article by Jamett, Andrade, and Barrera (1995). Nevertheless, with the objective that this chapter be all-inclusive in its content, both calculating methodologies (pseudostatic and dynamic) are presented.

45.2 SEISMIC BEHAVIOR OF TAILINGS DAMS

Chile has been known for mining throughout its history, with copper mining being that which had more influence in the economic development of the nation. The Chilean copper production is in the order of 3 million tons/yr, which places Chile among the top producers of this metal in the world. The copper mines in Chile have, as an average, a mineral grade in the order of 1%; this means that each ton of rock mined and processed generates 10 kg of copper and 990 kg of waste material. This results in the generation of significant volumes of waste, known as tailings. These need to be handled in a safe and economical manner making necessary the construction of containments with their corresponding dams or retention walls, and thus originating the so-called tailings dams. Structurally speaking, in a traditional tailings impoundment like the one where the waste rock dump is placed, two elements may be differentiated: the fine saturated tailings deposited in the basin (also called slimes) and one or more dams on the perimeter. A very attractive alternative due to costs consists of using the same tailings material for the construction of the dams but selecting the coarser fraction that classifies as sand. The lateral support necessary to contain the fine material deposited in the basin is provided in this manner.

The finer tailings placed in the basin, basically silt, are deposited with very low densities due to the initial condition of saturation. The coarser tailings (classified as fine silty sands) that compose the dams are compacted by some mechanical means to reach the compatible density necessary for their stability.

On the other hand, it is necessary to point out that due to the high seismic activity in Chile, those impoundments not designed adequately have failed catastrophically; this has been due to the occurrence of liquefaction both of the impounded slimes as well as the sands placed on the dams.

Table 45.1 incorporates the available information regarding seismic failures on tailings impoundments. It can be seen that the amount of dam failures is not negligible and, therefore, the engineering design of these structures is completely necessary. In the current project, a tailings impoundment is the foundation for a waste rock dump of great height, so the seismic stability analysis is very important.

45.3 UNDRAINED ULTIMATE RESISTANCE

To better visualize the use of the undrained ultimate resistance in the pseudostatic analysis, experimental results obtained from triaxial tests on the standard Toyoura sand (Verdugo 1992; Ishihara 1993) are presented. This material is classified as uniform fine sand, with D_{50} = 0.17 mm, Cu = 2, Gs = 2.65, e_{max} = 0.977, ye_{min} = 0.597.

Figure 45.2 shows the results of a series of tests with a void ratio of e = 0.735 and effective confining stresses of 0.1, 1, 2, and 3 MPa. For the density and range of stresses used, it is observed that the soil response is dilative with a marked drop in pore pressure, which is more significant as the level of initial confining stress decreases. Moreover, it is important to note that all the samples in this series developed the same ultimate resistance at large deformations, indicating that the undrained ultimate resistance is only a function of density and independent of the level of initial confining stress.

Figure 45.3 shows the results obtained from another series of tests perfomed with a void ratio after consolidation, e = 0.908 and effective confining stress of 0.1, 1, and 2 MPa. For the density and range of stress used, the response is mainly contractive, noticing for the sample tested at 1 MPa of confinement pressure a sharp drop of resistance until the ultimate resistance is reached for 2 MPa of stress. The results of this series also ratify that the undrained ultimate resistance is independent of the level of initial effective stress and that it only depends on the void ratio.

* IDIEM, Universidad de Chile, Santiago, Chile.
† Geotécnica Consultores S.A., Santiago, Chile.

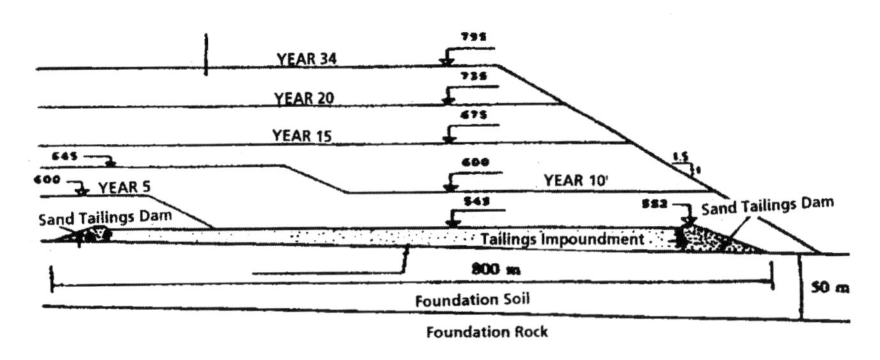

FIGURE 45.1 Diagram of the waste rock dump over the tailings impoundment

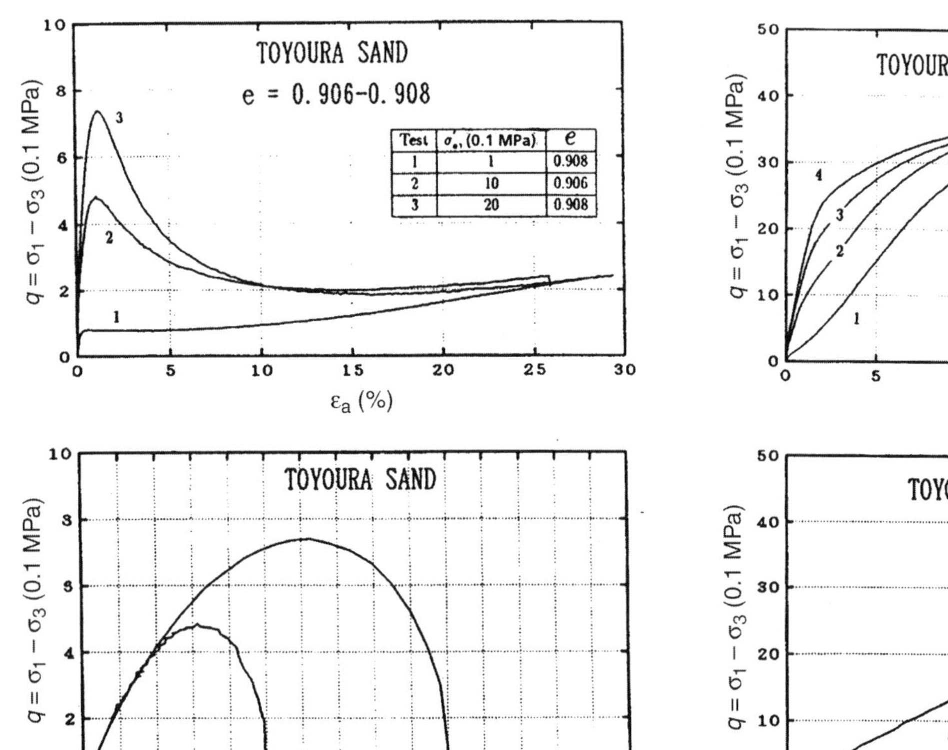

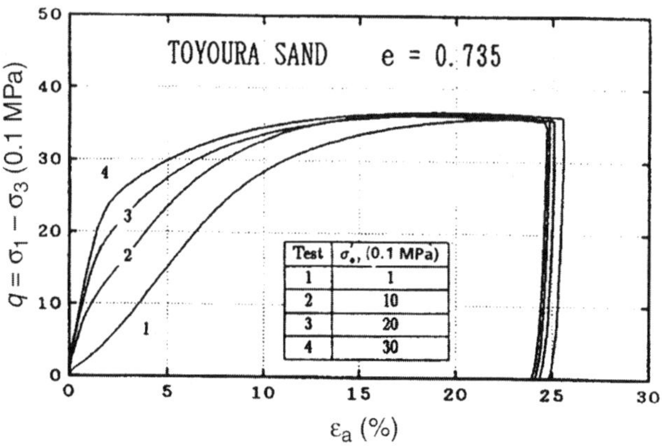

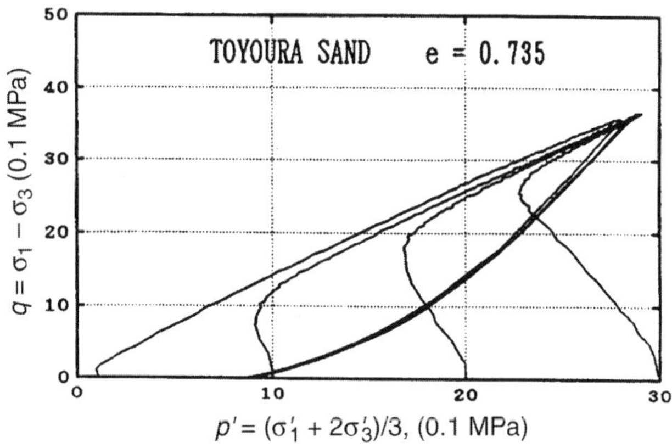

FIGURE 45.2 Undrained response, e = 0.735

FIGURE 45.3 Undrained response, e = 0.908

TABLE 45.1 Seismic failure of Chilean tailings impoundments

Impoundment	Year	V_1	V_2
Barahona	1928	27.0	4.0
El cobre viejo	1965	4.25	1.9
Cerro negro 3	1965	4.98	0.086
La Patagua	1965	—	0.036
Los Maquis	1965	0.042	0.021
Bellavista	1965	0.448	0.071
Cerro Negro 4	1985	2.0	0.130
Veta del Agua	1985	0.7	0.280

NOTES:
(*): Estimated value
V_1: Impounded volume (million m^3)
V_2: Emptied volume (million m^3)

The useful conceptual framework for viewing the soil response, drained and undrained, is the one called "steady state of deformation," or ultimate state (Castro 1969; Casagrande 1975; Poulos 1981). This state is characterized by the continuous deformation of the soil at constant shear stress, constant normal effective stress, without change in volume and at constant velocity.

With the objective of illustrating the cyclic response, and its relationship with the monotomic soil response, experimental results obtained in contractive and dilative samples, loaded cyclically and monotonically, are shown in Figures 45.4 and 45.5. It is observed that the monotonic response acts as a border over the cyclic response. Moreover, it is interesting to observe that the cyclic loading does not reduce the ultimate resistance as may be thought; therefore, the conclusion is that ultimate resistance is not affected by the stress history. This experimental evidence is very important for the seismic stability analysis of saturated noncohesive soils, because it implies that it is possible to use the ultimate undrained

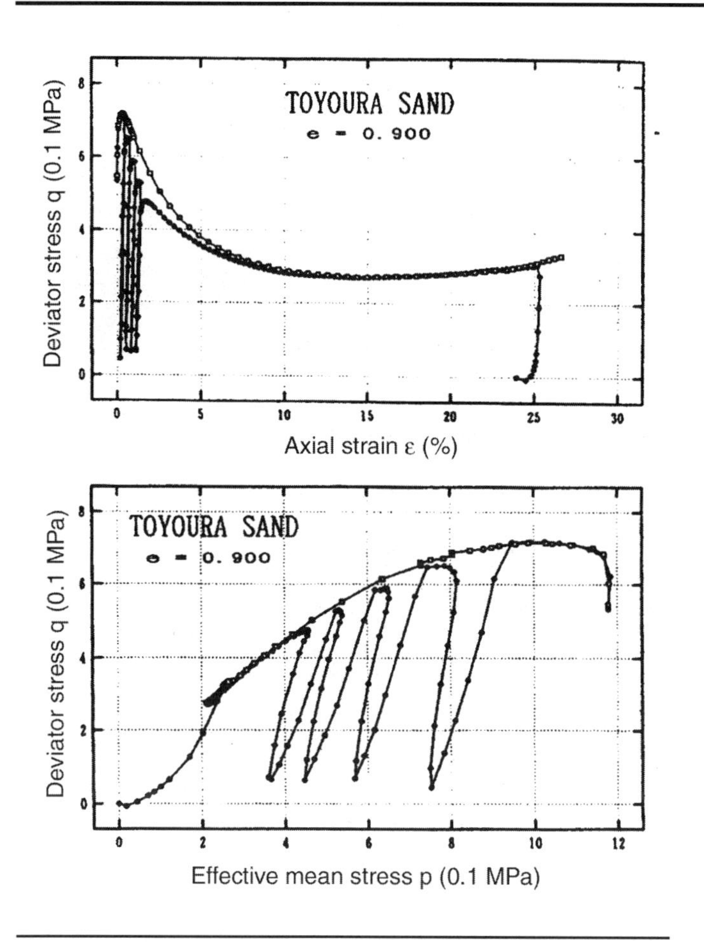

FIGURE 45.4 Cyclic and constant undrained contracting response

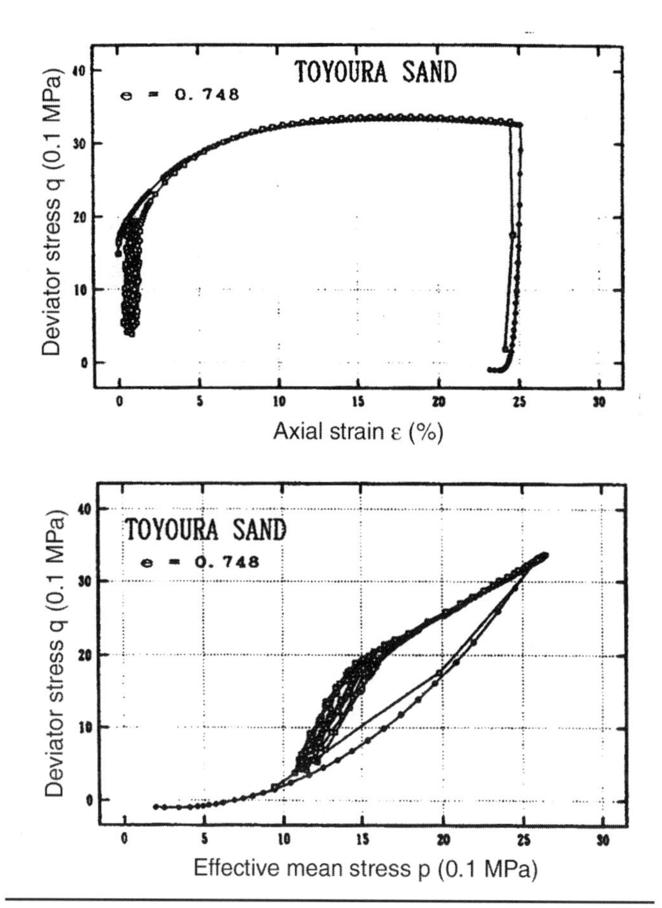

FIGURE 45.5 Cyclic and constant undrained expanding response

resistance, a parameter associated with static loads, as the seismic strength of the material.

The analysis based on steady state has been mainly developed by Casagrande and others (Castro 1969; Casagrande 1975; Poulos 1981; Poulos et al. 1985).

45.4 GEOTECHNICAL CONDITIONS OF THE PROJECT

A geotechnical investigation was carried out to characterize the different materials involved. The results obtained are summarized below.

45.4.1 Foundation

The foundation soil corresponds to an alluvium composed of sandy gravel with silty fines of medium to high permeability in a dense state. The alluvial layer has a maximum thickness estimated at 50 m, under which the bedrock is found. The geotechnical properties of this material are presented in Table 45.2.

45.4.2 Waste Rock Dump Material or Mine Waste Rock

The material that comprises the waste rock dump corresponds to a coarse granular soil called mine waste rock. The estimation of both the angle of internal friction and cohesion was based on the studies done by Leps (1970), Valenzuela (1989), and Seed et al. (1985). The shear resistance parameters of the waste rock deposit were separated according the different ranges of confining stress, which are presented in Table 45.2.

45.4.3 Cycloned Tailings Sand

The cycloned tailings sands make up the dam of the tailings impoundment over which the waste rock dump will be constructed. These sands have low moisture content, a fines content between 20 and 40%, and a medium compaction. According to the USCS system, this material corresponds to a silty sand (SM). The estimated geotechnical characteristics of this material are summarized in Table 45.2.

45.4.4 Fines in the Basin or Slimes

An important effort to characterize the fine tailings material was carried out; this was done mainly because of the strong incidence of this material on the global stability of the waste rock dump. This material corresponds to nonplastic silts (A) and sandy silts (B), ML, according to the USCS system. The observation of samples from Shelby tubes indicates that they are located in an alternating stratigraphy of small thickness lenses throughout the entire height of the deposit. Properties of this material are shown in Table 45.2.

Standard penetration tests (SPT) were done in the basin. The values measured were corrected considering the efficiency of the equipment and standardized to an effective stress of 100 KPa. The results indicate that the impoundment may be characterized in two sectors:

- From 0 to 15 m of depth: $1 < (N_1)_{60} < 3$ blows/ft
- Under 15 m of depth: $5 < (N_1)_{60} < 11$ blows/ft

Additionally, a series of eight triaxial tests CIU, six samples on fine slimes (A) and two on coarse slimes (B), were done on recompacted samples, using the subcompaction method proposed by Ladd (1978). Even though in the field the maximum level of vertical effective stress as in the order of 6 MPa, the series of triaxials reached a consolidation pressure of 2.4 MPa, corresponding to the capacity of the available equipment. The results indicate a contractive-dilative behavior for the range of stress used. A typical result is shown in Figure 45.6. From these triaxial tests the undrained residual shear strength of the fine tailings

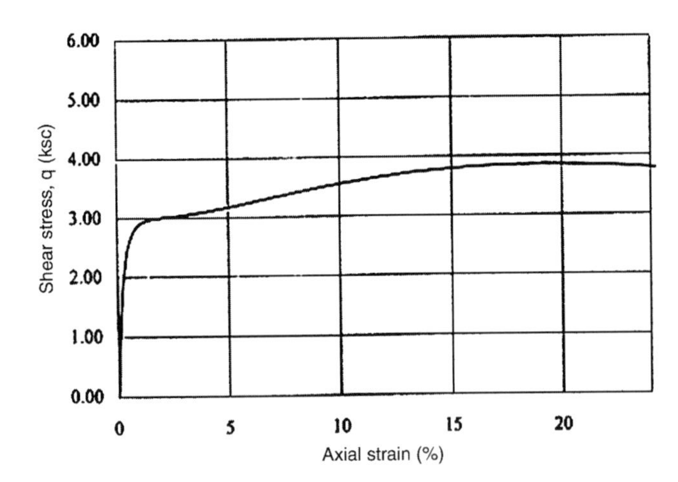

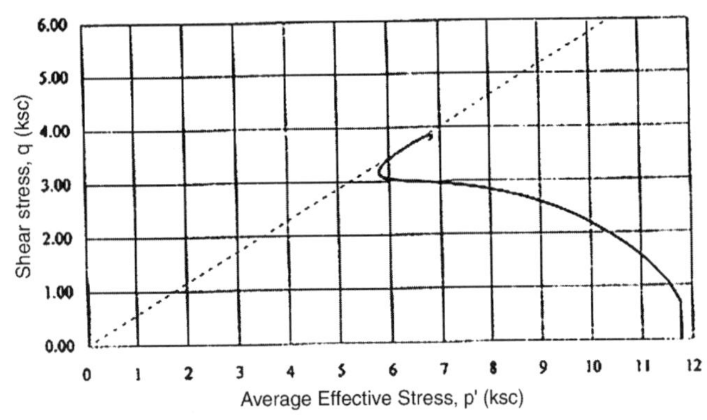

FIGURE 45.6 Fine tailings undrained response

TABLE 45.2 Properties of the material involved

	W_{nat} (%)	γ_{dry} (KN/m^3)	G_s	ϕ (°)	C (KPa)
Foundation soil	4.0	19.6	2.80	38.0	19.6
Waste rock, thickness < 30 m	4.0	22.1	3.00	45.5	0.0
Waste rock, thickness < 80 m	4.0	22.1	3.00	41.5	30.4
Waste rock, thickness > 80 m	4.0	22.1	3.00	37.0	123.6
Tailings sand	4.0	17.5	3.15	36.0	4.5
Fine tailings A	33	13.2	3.00	34.0	0.0
Fine tailings B	22	15.7	3.05	34.5	0.0

was evaluated; the results are shown in Figure 45.7. An approximately lineal relationship is observed between the effective confining stress and the undrained ultimate resistance, which may be expressed as $Su = 0.36\sigma_o'$.

Alternatively, using the correlations proposed by Seed and coworkers, investigators estimated the undrained ultimate resistance of the slimes from the penetration index SPT. The results obtained are included in Figure 45.7, from which the undrained resistance can be expressed as $Su = 0.05\sigma_o'$. Being on the safe side, the resistance given by this last expression was used. It is also possible to rewrite the latest expression as: $Su = tg\ 3°\sigma_o'$. This allows for the mathematic representation of the undrained ultimate resistance of the fine tailings by the parameters to a cohesion of zero and an equivalent internal friction angle of $\phi = 3°$.

45.5 CONVENTIONAL SEISMIC STABILITY ANALYSIS

First, a stability analysis using traditional methods was done for abandonment conditions. Figure 45.8 shows the failure surface obtained for which there was a factor of safety of 1. It is interesting

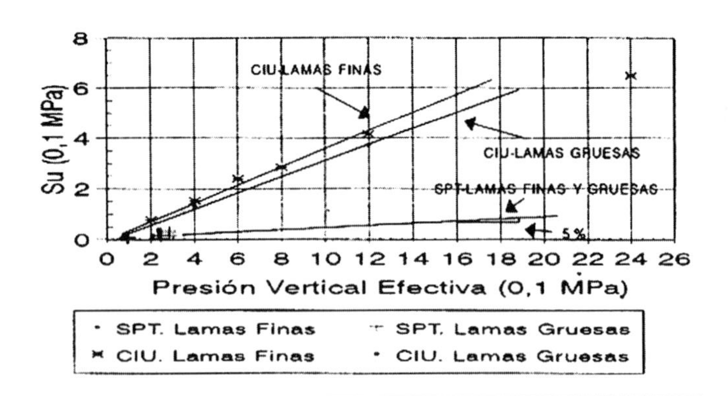

FIGURE 45.7 Fine tailings undrained ultimate resistance

TABLE 45.3 Elastic properties of the materials used

Material	G_{max}	ν
Foundation Soil	$9200 \times (\sigma_v * \sigma_h)^{0.24}$	0.3
Waste Rock	$6000 \times (\sigma_v * \sigma_h)^{0.24}$	0.25
Sands	$2000 \times (\sigma_v * \sigma_h)^{0.24}$	0.25

NOTE: σ_v and σ_h in t/m^2.

to note that in this analysis the resistance parameters for the previous indicated different materials were reduced, using additional factors of safety, with values according to the level of uncertainty with which these parameters were obtained. The factors of safety used were 1.1 for the waste rock, 1.3 for the sands, and 1.5 for the tailings in the basin (Jamett et al. 1995).

The evaluation of the permanent displacements from an earthquake of magnitude of 8.5 in the Richter scale was done using the method of Newmark (1965) and the software program SHAKE. The results obtained indicated a permanent displacement of nearly 3 m. This value corresponds to 1% of the height of the waste rock dump, which was considered an acceptable value and indicative that the waste rock dump is stable against an earthquake of great magnitude (Jamett et al. 1995).

45.6 DYNAMIC STABILITY ANALYSIS

Given the magnitude of the project and the unusual situation of using a tailings impoundment as the foundation for a waste rock dump of great height also, a seismic stability study was carried out using a dynamic analysis. This analysis was done by ITASCA-CHILE using the computer code FLAC 2D, modeling the response in effective stress of the fine tailings.

The resistance parameters used for the materials involved correspond to those already presented. The elastic properties of the materials used are those presented in Table 45.3.

The most important task of the numerical modeling corresponds to the estimate of increase in pore pressure and the eventual liquefaction of the fine tailings in the basin. The modeling adopted was the one proposed by Martin et al. (1975), in which the increment of pore pressure is evaluated from the tendency in increase in volumetric deformation given by the following expression:

$$\Delta\varepsilon_v = C_1(\gamma - C_2\varepsilon_v) + \frac{C_3\varepsilon_v^2}{\gamma + C_4\varepsilon_v} \qquad \text{EQ. 45.1}$$

Where:

ε_v = volumetric deformation
γ = amplitude of the shear deformation
C_i = coefficients of the model $_{(i = 1,2,3,4)}$

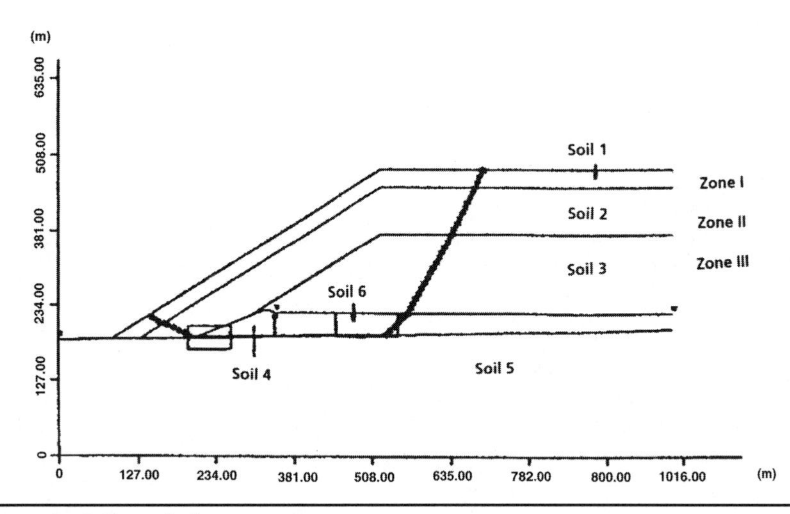

FIGURE 45.8 Failure surface pseudo-static analysis

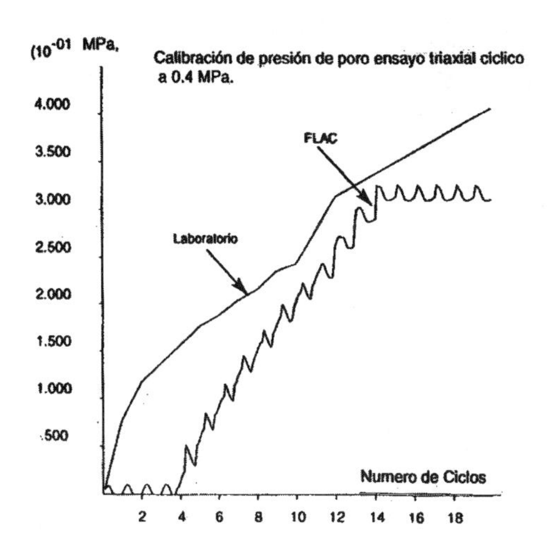

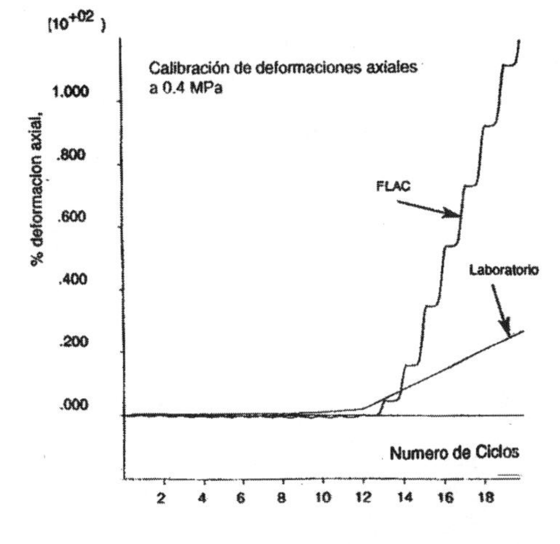

FIGURE 45.9 Model and testing comparison

The increase of cyclic pore pressure is evaluated from $_V$ times a constant of the material. A series of cyclic triaxial tests at different levels of confining stresses were done for the application of this model, finding the coefficients C_1 to C_4 of the material through the use of the code FLAC for the axial symmetry condition of the tests. The Ci constants were obtained directly from the undrained tests matching the obtained pore pressures. The coefficients for different

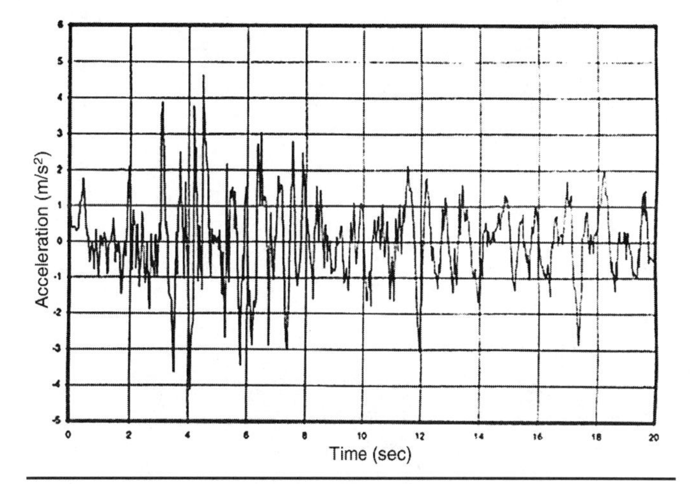

FIGURE 45.10 Design earthquake

TABLE 45.4 Coefficients C_i from the pore pressure model.

Coefficient	0.4 MPa	1.2 MPa	2.3 MPa
C_1	1.0600	7.7600	8.1533
C_2	0.8300	0.1500	0.0775
C_3	1.0560	3.5850	2.1287
C_4	1.1990	3.0020	4.2216

confining stresses are summarized in Table 45.4. A comparison between the testing and theoretical results is presented in Figure 45.9, in term of number of cycles as a function of pore pressure and axial deformation. It can be seen that the theoretical values are conservative in regards to the testing results since they predict a larger cyclic deformation for a same number of cycles.

The earthquake used for this analysis has a duration of 20 sec, a maximum acceleration of 0.47 g, and a dominant frequency content between 1 to 3 Hz. The acceleration record is shown in Figure 45.10.

In regards to the mesh for the finite elements analysis, Kuhlemeyer and Lysmer (1973) indicate that for the transmission of waves through a model like FLAC, the spatial size of the elements should be less than a tenth to an eight of the wave length associated with the highest frequency of the incidence wave. This led to the dimensioning of the model areas to be less than or equal to 3 m. This meant that the model in its entirety has approximately 29000 nodes.

The results of pore pressures in the slimes (tailings in the basin) and their displacements that the slimes do not reach a pore pressure associated with liquefaction and that the maximum horizontal deformation are less than 1.5 m, a value somewhat lower than with the one obtained applying the traditional methodology.

45.7 CONCLUSIONS

The main conclusion from this article is that the displacements calculated according to the traditional methodology using steady-state concepts, equilibrium limit, and displacements according to the Newmark model were very close to those obtained following a dynamic analysis including variations on pore pressure by seismic loading. Thus, for important projects, it is recommended the application of both methodologies, pseudo-static and dynamic analysis, is used. Both procedures should converge to similar results providing a reliable solution.

45.8 SPECIAL THANKS

The authors would like to thank Geotécnica Consultores S.A. for the access to the project data. Special thanks are given to the engineer, Carolina Vergara, for her help on the preparation of this manuscript.

45.9 REFERENCES

Casagrande, A. 1975. Liquefaction and cyclic deformation of sands—A Critical Review. *5th Pan American Conference on Soil Mechanics and Foundation Engineering.*

Castro, G. 1969. Liquefaction of sands. Harvard Soil Mechanics Series No. 81. Cambridge, Mass.: Harvard University.

Ishihara, K. 1993. Liquefaction and flow failure during earthquakes, The 33rd Rankine Lecture. *Geotechnique.* 43:3:51.

Jamett, R., C. Andrade, and S. Barrera. 1995. *Stability of a Tailings Impoundment with Overload.* X CPMSIC, p. 1474–1485, México.

Kuhlemeyer, R.L., and J. Lysmer. 1973. Finite element method accuracy for wave propagation problems. *Technical Notes.* May:421.

Ladd, R.S. 1978. Preparing test specimens using undercompaction. GTJ. 1:1:16.

Leps, T.M. 1970. Review of shearing strength of rockfill. *JSMFE, ASCE.* 96:SM4.

Martin, G., Finn, L. and Seed, H. 1975. Fundamentals of liquefaction under cyclic loading. *JGED, ASCE.* May:423.

Newmark, N.M. 1965. Effect of earthquakes on dams and embankments. *Geotechnique*, Institute of Civil Engineers, London. 15:2:139.

Poulos, S. 1981. The steady state of deformation. *JGED, ASCE.* 107:GT5:553.

Poulos, S.J., G. Castro, and W. France. 1985. Liquefaction evaluation procedure. *JGED, ASCE.* 3:6:772.

Seed, H.B., R.B. Seed, S.S. Lai, and B. Khamenehpour. 1985. Seismic Design of Concrete Face Rockfill Dams, Concrete Face Rockfill Dams—Design, Construction, and Performance. New York: ASCE. 459–478.

Valenzuela, L. 1989. Waste rock deposits stability. Memorias del 2nd Congreso Chileno de Ingeniería Geotécnica. 2:IV-41–IV-68. Santiago.

Verdugo, R. 1992. *Characterization of Sandy Soil Behavior Under Large Deformation*. Ph.D. diss. Departamento de Ing. Civil, Universidad de Tokio, Japón.

CHAPTER 46

Stability Issues Related to Tailing Storage and Heap Leach Facilities

Donald R. East* and Julio E. Valera*

46.1 INTRODUCTION

Slope stability is an extremely important consideration in the design and construction of tailing storage facilities and heap leach piles. Economics dictate the steepest possible side slopes consistent with acceptable factors of safety for most mining projects, because steeper side slopes allow a greater volume of material (either tailing or ore) to be placed within a given facility footprint. The task of the designer therefore is to minimize the ongoing construction costs of these facilities by maximizing the volume of material that the facilities can contain without unduly risking failure.

Earthen structures constructed during the life of a mining operation have some unique stability issues when compared to conventional civil engineering structures. Mining operations usually lend themselves to construction rates staged to match the mining and milling production rates. This provides the advantage that the observational approach to the design of each raise can be adopted, but it can also lead to discontinuous quality of design and construction and the loss of construction experience when staged over a long mine life.

The ever-increasing trend to construct higher heaps has tested the capability of geotechnical laboratories to measure shear strengths at correspondingly higher shear strengths.

The nonlinear nature of the failure envelope for a soil/liner interface has to be recognized, particularly for clayey soils as are typically used for construction of composite liner systems for heap leach pads.

46.2 TAILING STORAGE FACILITIES

Tailing storage facilities provide the possibility of raising the embankments, apart from the conventional downstream method of water dam construction, by centerline, modified centerline, and upstream methods. These methods of construction rely to varying extents on the strength and stability of the deposited tailings, and consequently a thorough geotechnical assessment of the tailings should be conducted prior to each construction phase.

Historically, it has been a challenge to obtain good-quality samples of tailing from the relatively soft and saturated deposits that exist in most tailing facilities. The use of in situ testing techniques that have proven to be reliable indicators for assessing the shear strength and permeability of most tailing deposits, without some of the disadvantages associated with more traditional sampling and laboratory testing techniques, will be described.

46.2.1 Tailing Facility Stability

The methods of analysis of slope stability of earthen and rockfill structures are generally well understood within the civil and geotechnical engineering profession, and there are a number of excellent software applications available to the designer. The key to any good geotechnical design, however, is not the mechanical performance of the analysis, but an in-depth understanding of the probable shape of the potential failure surface and the selection of appropriate input parameters.

Providing the correct strength parameters for soft tailings, for example, is not an easy task. Tailing deposits are very difficult to sample for laboratory testing without creating some sample disturbance that would then place in question the validity of the results of the tests. In situ testing, on the other hand, has developed in sophistication over the last 20 years to the point where sufficient research and case histories exist that demonstrate the validity of these methods. In particular, the electric piezocone (CPT), self-boring pressure meter, and field vane shear tests, used in combination with a limited amount of conventional drilling to provide positive identification of the material types, have become the current standard for evaluation of tailings deposits and replace earlier, less sensitive, tests such as the standard penetration test (SPT).

The CPT is equipped with precision strain gauges and pore-pressure transducers and can yield accurate and repeatable measurements of tip resistance, sleeve friction, and dynamic pore pressure when penetrating at a constant rate. The CPT is sufficiently sensitive to respond to changes in layers less than about 0.1 m (0.3 ft) thick within the deposit, and the continuous, electronically recorded data of these three parameters provide a very powerful tool for interpreting strength, material type, and pore-water conditions throughout the full depth probed. A typical CPT probe, which can exceed 60 m (200 ft), can usually be completed in less than one day. The time required to complete a similar conventional boring for SPTs cannot come close to achieving this efficiency. A typical CPT profile in tailings is shown in Figure 46.1.

In mine tailings, each individual layer can be characterized on the basis of tip resistance, sleeve friction, and pore-water pressure. The sandy layers can be readily differentiated from the silty and finer layers, and dense layers can be distinguished from loose layers. The stratigraphy of the deposit is therefore defined adequately for design purposes by the CPT, and only limited amounts of additional laboratory work are normally conducted to confirm these data.

Dynamic pore pressures are developed during penetration and shearing of the tailings by the piezocone. The difference between the dynamic pore pressure and the normal hydrostatic pressure is referred to as the excess pore pressure, which may be either positive or negative. If positive excess pore pressures are generated, the material exists in a loose, contractive state, and there is at least substantial evidence that pore-water pressures could increase during earthquake shaking.

* Knight Piésold, Denver, Colorado.

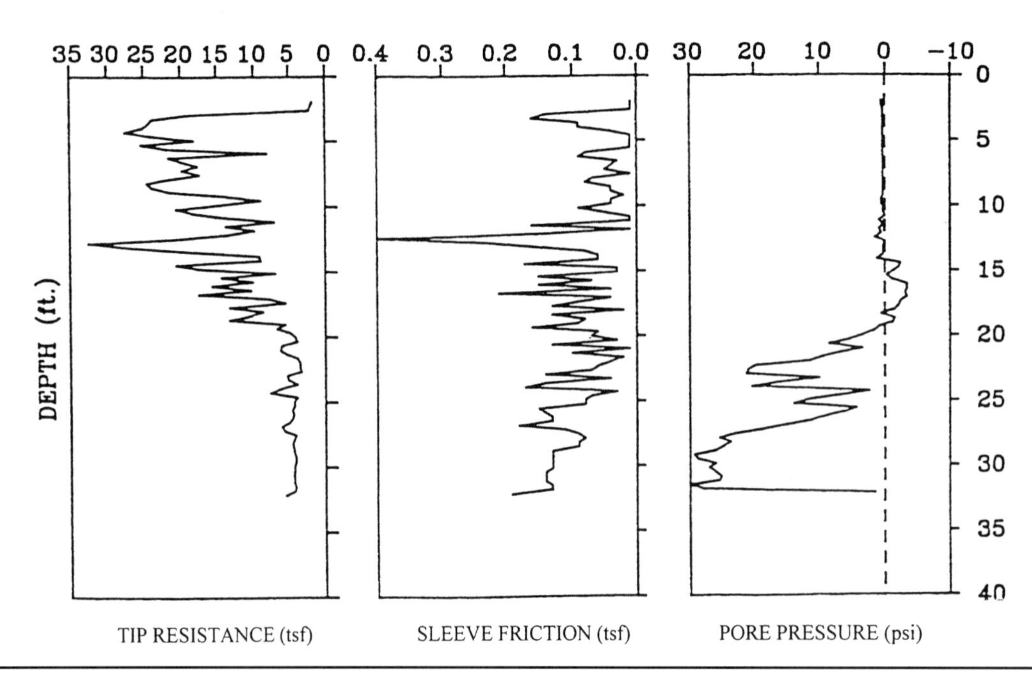

FIGURE 46.1 Typical plot of results measured by piezocone

The location of a phreatic surface, or phreatic surfaces in the case of perched water tables, within an embankment can have a significant effect on the stability of that structure. The pore-pressure measurement aspect of the CPT is extremely useful in evaluating not only the location of this water level (or multiple water levels) but also the permeability of the deposit at that point by means of a pore-pressure dissipation test. A pore-pressure dissipation test, which measures the decay of excess pore pressure with time when the probe is stopped, is relatively quick and simple to conduct. A relationship between normal hydrostatic and in situ static pore pressures can then be established. A comparison between the two provides information on the presence and direction of fluid flow or seepage within the tailing deposit.

Pore-pressure data from an operating tailing facility that was constructed with a full underdrainage system is shown in Figure 46.2. The phreatic surface was estimated to be approximately 5 m (15 ft) below the surface, but pore-pressure dissipation tests indicated that the in situ static pore pressures were considerably less than hydrostatic. This confirmed that water was draining vertically downward out of the tailing mass and that the underdrain was functioning properly. Such reduced pore pressures indicate that the effective confining stress on the tailing is higher than it would have been if the tailing had been undrained. Higher effective confining stresses lead to higher in situ densities, which is of importance in terms of the strength of the tailing, liquefaction, and the stability of the tailing embankment.

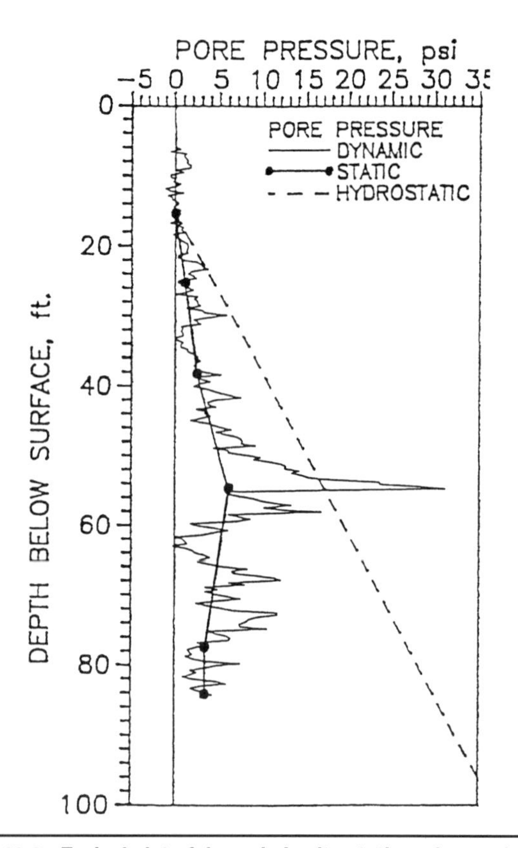

FIGURE 46.2 Typical plot of dynamic in situ static and normal hydrostatic pore pressures from CPT probing in drained tailings

46.2.2 Liquefaction of Tailings

Under strong earthquake shaking, certain loose, saturated materials may experience a state of momentary suspension. During this time, the effective interparticle stresses are minimized and the tailings are said to liquefy. Rather than acting as a frictional material, the tailings exhibit minimal undrained shear strength, and only very small driving forces may be required to cause unacceptable deformation.

The primary method used for liquefaction assessment is based on the original research of Seed and De Alba (1986), which was established using case history data and SPT values in clean sandy soils. They also presented a method by which CPT data could be used for assessing the liquefaction potential of natural soils. Since then, additional data obtained from CPT assessments in tailings and other materials have provided more refined correlations between SPT and CPT data (Ulrich and Hughes 1994; Robinson and Fear 1997), which has allowed for more refined evaluation of the liquefaction potential of tailing deposits.

46.2.3 Field Shear Vane

The field shear vane has been used for many years as a practical and economical means of measuring the in situ undrained

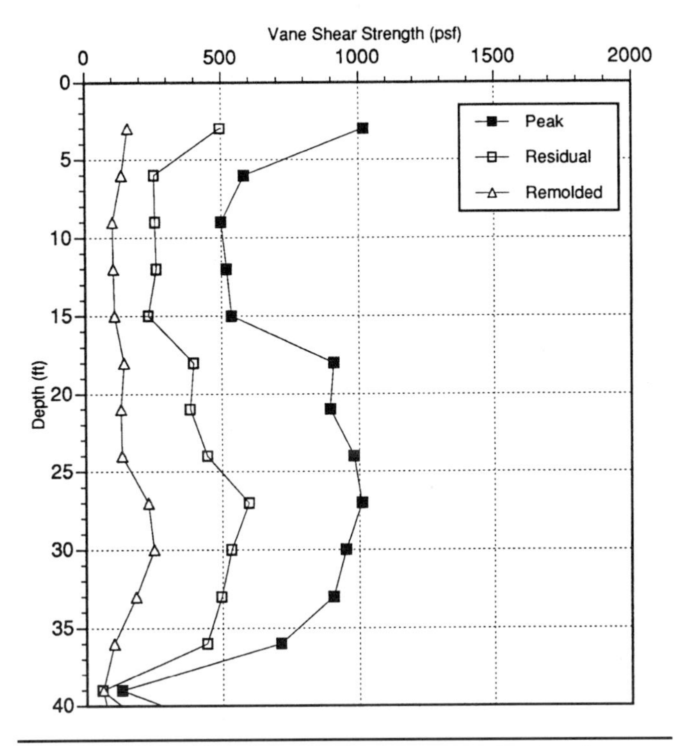

FIGURE 46.3 Summary of mean vane shear strength values from seven different locations

strength of cohesive soils. Only in recent years has the use of this equipment been tested for determining the undrained shear strength of tailings materials. The remolded undrained shear strength of a tailings deposit corresponds to its minimum strength at its in situ void ratio. The peak strength of tailings may be reduced to this residual strength through the introduction of any undrained loading, which causes sufficient strain as to exceed the strain at peak shear strength. The test is therefore of direct interest in cases where liquefaction of tailings is of concern.

Work undertaken by the authors used an Acker shear vane measuring 5 cm (2 in.) across with a load cell immediately behind the vane that virtually eliminated the usual problem of rod friction. The vane was advanced through the tailings by pushing from the surface with minimal disturbance from a backhoe. The tests were completed at a sufficiently high rate of rotation to limit pore-pressure drainage, and thereby to represent rapid or undrained loading conditions, whenever possible.

The results of field vane shear tests completed in a gold tailings facility indicate the typical scatter of peak strengths as would be expected from a highly layered deposit (Figure 46.3). The effects of layering become less recognizable for the remolded values due to the fact that the initial structure of the tailings has been destroyed. It should be noted that the remolded strengths may be as low as 20% of the peak strength. For postearthquake conditions, this remolded residual strength has to be taken into account when analyzing the stability of tailings facilities.

46.3 HEAP LEACH FACILITIES

The stability assessment of a heap leach facility comprises a significant portion of the geotechnical analysis pertaining to the heap. Each aspect examined during the geotechnical investigation is represented in the stability analysis, including the seismicity of the area, foundation characteristics, ore type and characteristics (gradation, permeability, strength), liner/geomembrane interface (clay liner material and geomembrane type), and location of the phreatic surface within the heap. Other parameters, such as pad topography, heap geometry, loading sequence and leachate application schedule, also impact the overall stability of the slope.

As heap heights continue to rise, and sites with good foundation conditions become scarcer, the role of the stability assessment in the overall design of a heap leach facility becomes of greater importance. In order that an adequate stability analysis can be conducted, it is critical that each aspect of the heap design be properly evaluated. This includes a seepage analysis to establish the flow regime throughout the heap and the height of the phreatic surface above the base of the heap; stability analysis to establish the static factor of safety of the downstream face of the heap; and analysis to estimate the magnitude of permanent slope deformation that could take place during the occurrence of the design earthquake event. Adequate stability of the heap will assure that the geomembrane liner will not rupture and impair the environmental performance of the facility.

In conjunction with the various aspects of designing a heap leach facility, it is conventional to conduct limit equilibrium slope-stability analyses to provide guidance on the overall heap layout and material selection, such as geomembrane type. In order to conduct a proper slope-stability analysis, which includes the examination of potential wedge-type failures with the failure plane passing along the soil/liner interface, it is vital that direct shear tests of the compacted soil/geosynthetic liner interface be carried out to assess the appropriate shear strength relationship to be adopted for the analysis.

Based on the results of extensive, high-quality direct shear testing conducted on various soil types and geomembranes, a number of key issues developed that deserve special consideration. These considerations are discussed in the following sections.

46.3.1 Curvature of the Failure Envelope

As additional reserves of low-grade ores are identified, it has become common practice to load existing heaps to greater and greater heights. This practice has illuminated a phenomenon that is not always otherwise apparent. As evidenced in other materials, the degree of curvature in a failure envelope for a soil/liner interface is often more pronounced when data are available for a wide range of stresses. Typical curvature of a failure envelope is shown in Figure 46.4. The failure envelope shown in this figure was established for a compacted silty clay (CH, PI = 67, 96% fines) tested against a 60-mil textured high-density polyethylene geomembrane. Such curvature of the failure envelope has been observed for a variety of soil and geomembrane combinations. Typically, however, this curvature is less pronounced for granular, nonplastic soils.

46.3.2 Extrapolation of Data

For a heap with an average total density of 1.8 t/m^3 (110 pounds per cubic foot, pcf), the maximum normal load for conventional test equipment corresponds to a heap height of about 40 m (130 ft). This is well below the heights to which heaps are currently being designed. In order to overcome this deficiency, a rigid insert was built to reduce the sample dimensions to 0.2 m × 0.2 m (8 in. × 8 in.). Given this reduced sample size, the maximum normal stress is increased by a factor of 2.25 to 1,500 kPa (32,400 psf)—equivalent to a heap height of nearly 100 m (300 ft), based on an average total density of 1.8 t/m^3 (110 pcf).

One of the greatest dangers of extrapolating laboratory shear test data is in establishing a formula through linear regression. While selection of, say, a power function may appear to be appropriate for the stress range represented by the test data, there is no warranty that such a function will represent the failure envelope at higher stress ranges; perhaps a hyperbolic function would be more appropriate. In many cases, this 0.2-m (8-in.) insert for the shear box device alleviates the requirement of extrapolating the failure envelope.

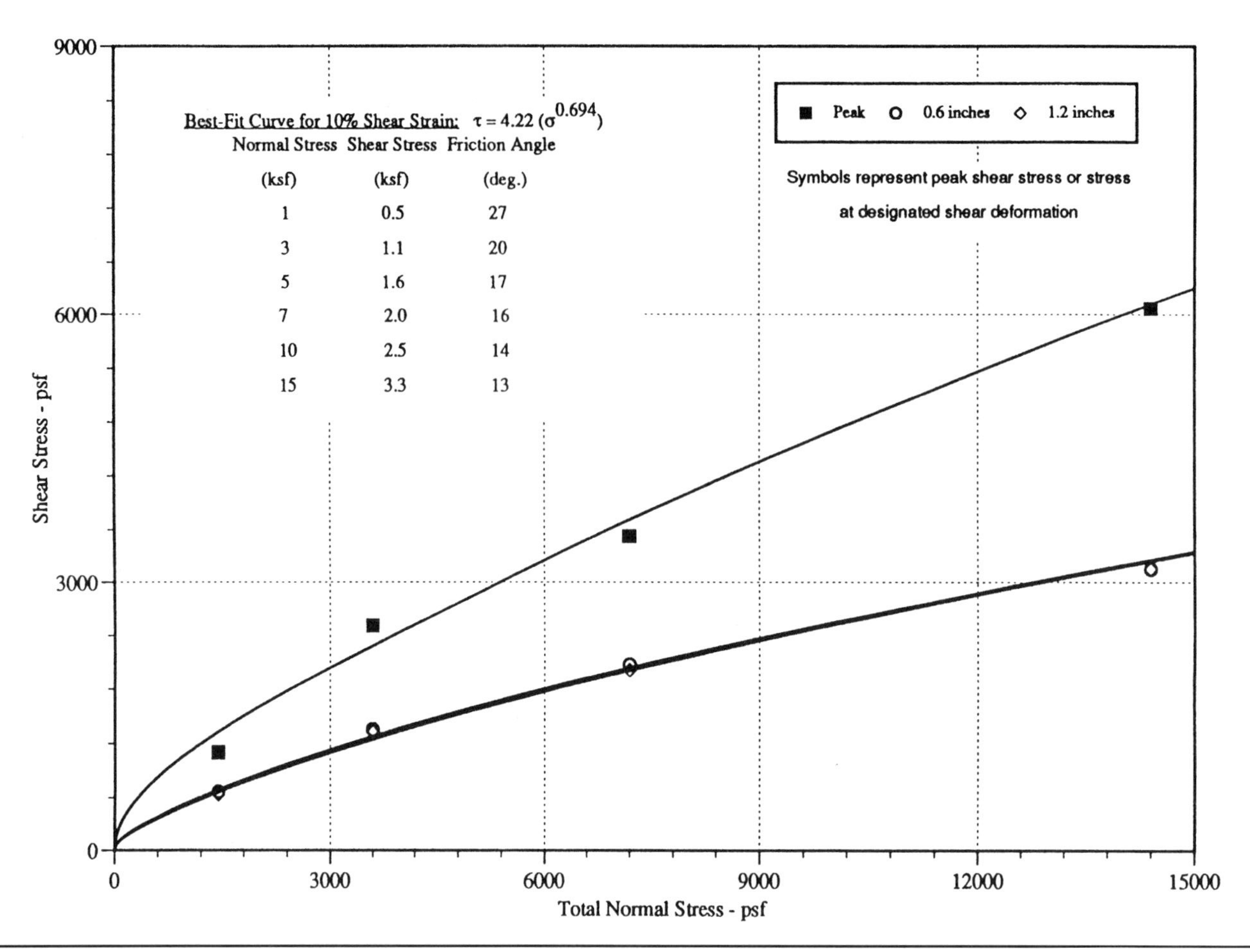

FIGURE 46.4 Interface test results for a clay liner and a 1.5-mm (60-mil) textured HDPE geomembrane (0.3 m [12 in.] shear box)

Figure 46.5 represents test results conducted on a clayey sand with gravel (LL = 43, PI = 20, 46% fines, 23% gravel) tested against a 1.5-mm (60-mil) smooth flexible polyethylene geomembrane. Tests conducted using both the 0.2-m and 0.3-m (8-in. and 12-in.) shear boxes were observed to yield essentially the same results for normal stresses less than 690 kPa (14,400 psf). However, notice that extrapolation of the failure envelope for data obtained using the larger box (in this case using a power function developed through linear regression analysis) underpredicts the shear strength at higher levels of normal stress. The opposite effect has also been observed where extrapolation of the failure envelope overpredicts the shear strength at higher levels of normal stress. Thus, extrapolation of a curved failure envelope should be avoided whenever possible.

46.3.3 Failure Criteria: Peak Versus Postpeak Strength

The peak interface shearing resistance of a clay liner/geomembrane is often mobilized at relatively low shear deformations and may be followed by a rapid loss of shear strength. An example of this is shown in Figure 46.6.

It can be seen from the shear stress-shear deformation plots presented in Figure 46.6 that the peak shearing resistance is mobilized at shear deformations of about 5 mm (0.2 in.) or less for all tests. The results for the tests conducted using a smooth geomembrame show a more or less constant strength at greater shear deformations once the peak value is reached. On the other hand, the tests conducted using a textured geomembrane indicate a significant reduction in strength with increasing shear deformations after the peak value is reached. It should be noted that, depending on the geomembrane/clay liner combination tested, results for smooth geomebranes may show a reduction in interface strength after the peak value is reached. As illustrated in Figure 46.4, the shear strength envelope for the tests shown in Figure 46.5 suggests that the interface shear strength at peak shearing resistance is approximately twice that available at a postpeak condition corresponding to shear deformations equal to or greater than 15 mm (0.6 in.).

Use of a peak shearing resistance in a slope stability analysis for interfaces such as shown in Figure 46.4 may not be conservative, as minor strains induced in the interface during installation and initial loading may be sufficient to bring about a postpeak strength condition. As suggested by Mitchell et al. (1990), Seed and Boulanger (1991), and Eigenbrod and Locker (1987), it is evident from test results such as those shown in Figure 46.4 that a postpeak strength should be adopted for use in a slope stability analysis.

46.3.4 Effects on Slope Stability

In order to illustrate the effects of considering a nonlinear interface failure envelope, slope stability analyses were carried out on an actual heap configuration using both linear and nonlinear envelopes. The analyses were conducted using the computer program XSTABL. The modeled heap is illustrated in Figure 46.7. The results of the direct shear liner interface testing are shown in Figure 46.8. These results were obtained for a clay liner tested against a 1.5-mm (60-mil) textured flexible polyethylene geomembrane at normal stresses up to about 1,725 kPa (36,000 psf).

Also shown in Figure 46.8 is a conventional linear interpretation of the nonlinear failure envelope. For the linear envelope, the test results obtained using the 0.2-m (8-in.) insert were not

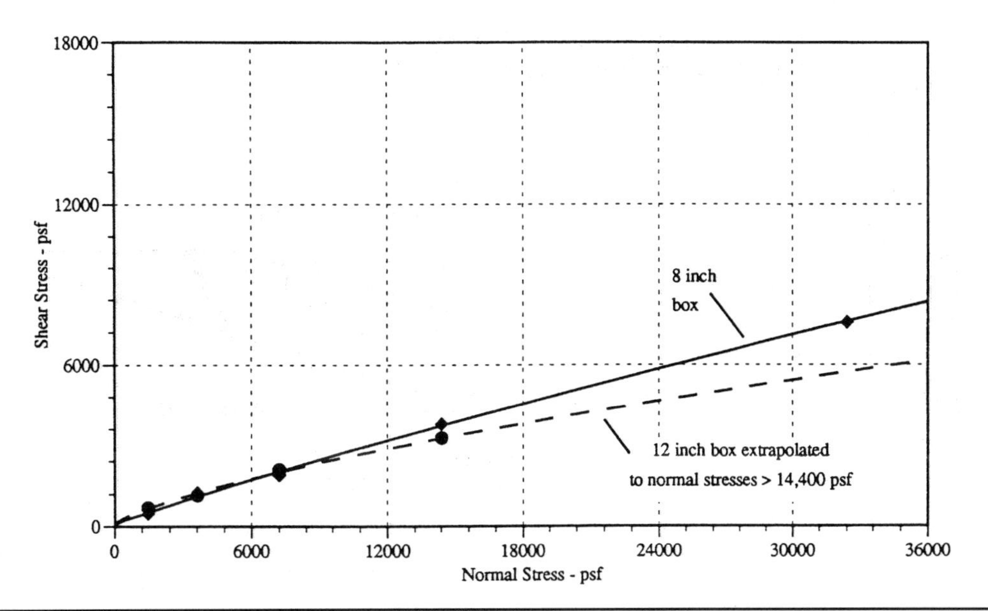

FIGURE 46.5 Comparison of interface test results using 0.2-m and 0.3-m (8-in. and 12-in.) shear boxes—clayey sand versus 1.5-mm (60-mil) smooth flexible polyethylene geomembrane

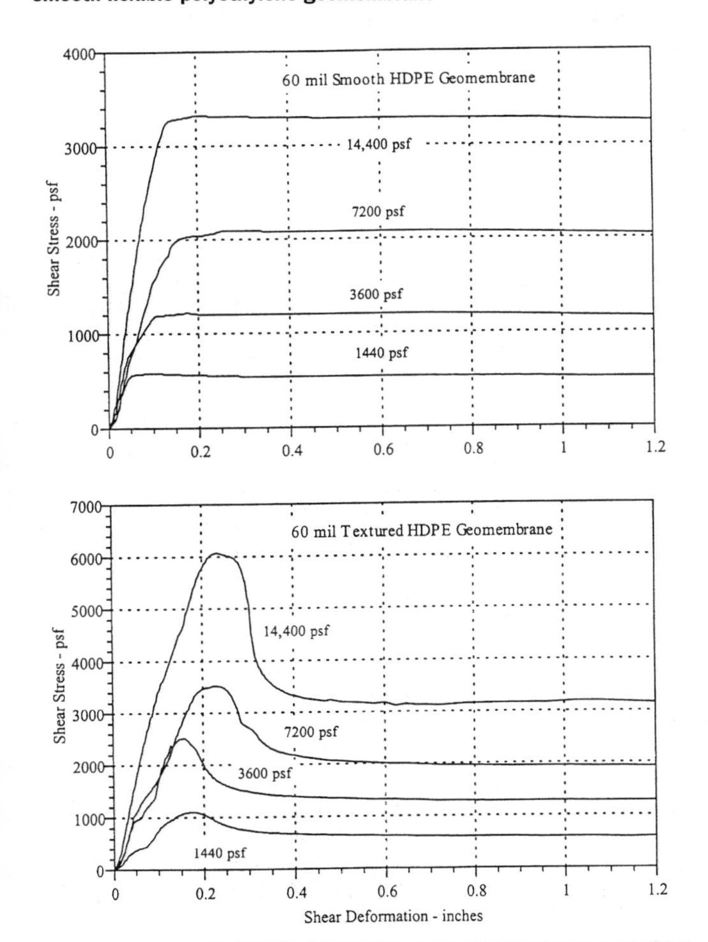

FIGURE 46.6 Liner/geomembrane interface tests—sandy clay liner versus 1.5-mm (60-mil) HDPE geomembrane

used to establish the linear interpretation since conventionally, the insert would not be available, and a straight line would be extrapolated from the available results.

Factors of safety obtained for the two analyses are summarized at the top of Figure 46.7. Use of a nonlinear failure envelope results in a factor of safety of 1.0, whereas the linear analysis gives a factor of safety of 1.2. If, as is conventionally assumed for nonwater-retaining heaps, the desired factor of safety is 1.3, neither of the analyses fully satisfies the requirements. This difference in the factor of safety can be explained by examination of the failure envelopes plotted in Figure 46.8. For normal stresses less than about 720 kPa (15,000 psf) both analyses should give about the same results. However, at higher normal stresses, corresponding to the ultimate heap height, the available shear strength is less than that estimated by using the interpreted straight line failure envelope. Thus, it can be seen that for very high heaps, as are presently being constructed throughout the world, use of a linear failure envelope in assessing the stability of the heap is not necessarily conservative.

46.4 RECOMMENDATIONS

Based on an extensive amount of in situ testing of tailings material and laboratory testing of soil/liner interfaces, the following recommendations are provided.

1. The piezometric cone can provide reliable detailed information on the in situ shear strength of tailings within the usually highly layered nature of the deposits. The pore-pressure distribution, both static and dynamic, can also be obtained within the full profile of the test. These tests have been completed to depths in excess of 60 m (200 ft).
2. The field shear vane has been shown to be useful for determining the underdrained and residual shear strength of tailings.

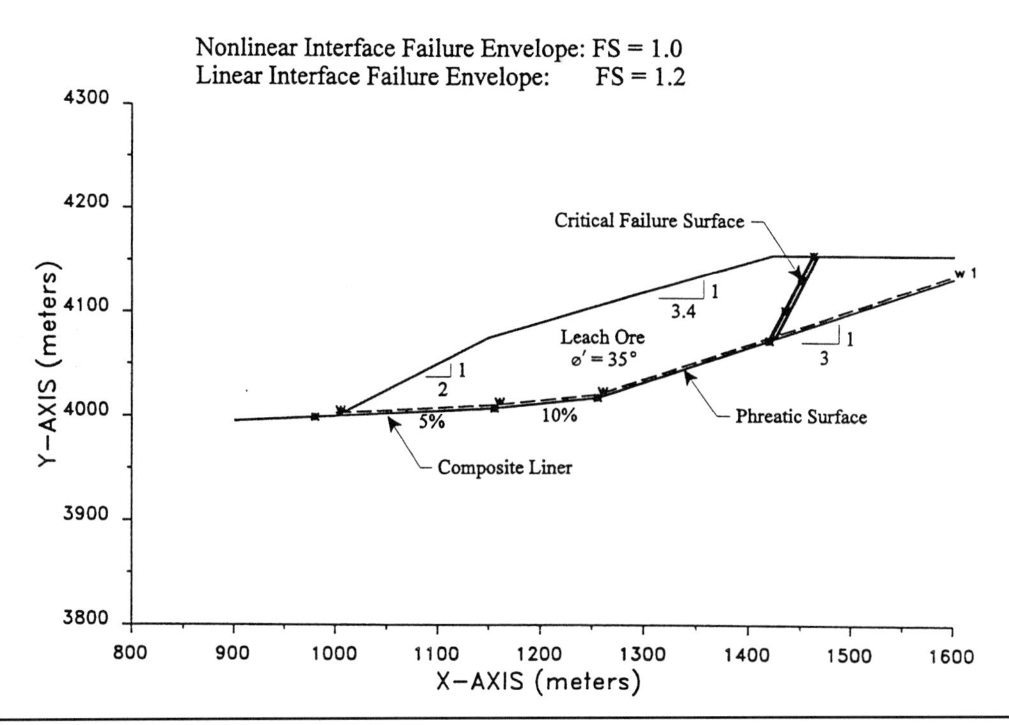

FIGURE 46.7 Heap leach stability

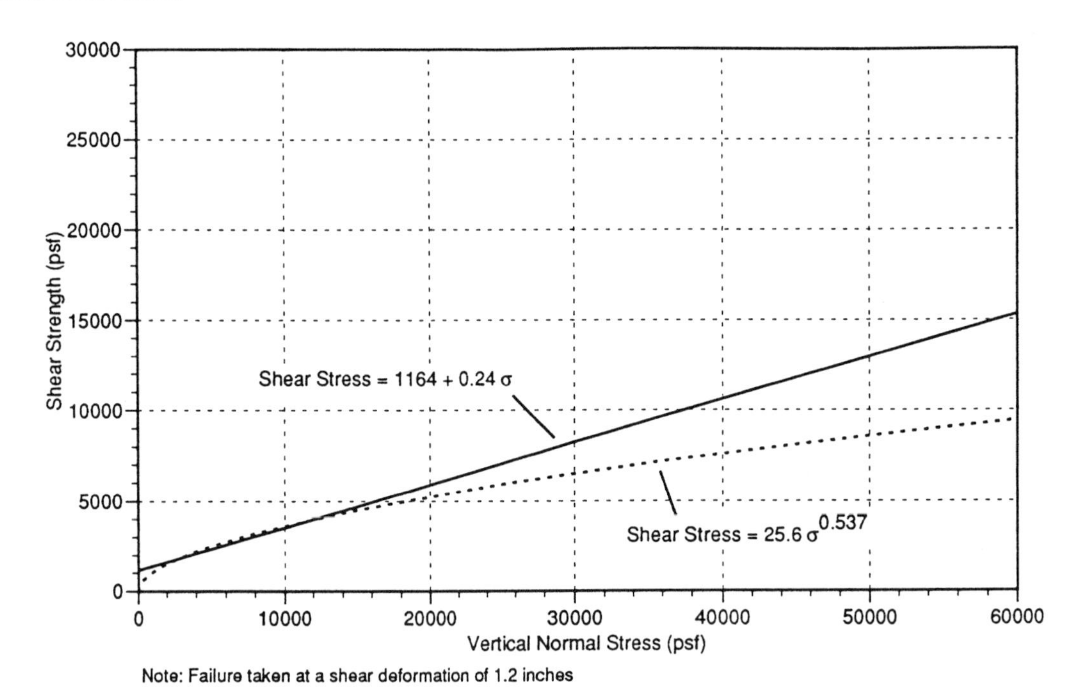

FIGURE 46.8 Nonlinear versus constant liner-interface strength envelope (clay liner versus 1.5-mm [60-mil] flexible polyethylene textured liner)

3. The interface strength of a clay liner/geomembrane depends on a number of factors. These include the soil characteristics, geomembrane type (smooth versus textured), field compaction effort and moisture content, and applied normal stress.
4. The failure envelope for a clay liner geomembrane interface is generally nonlinear and the nonlinearity becomes more pronounced at higher normal stresses. Since heaps are now being designed to greater and greater heights, it is important to include the nonlinearity of the failure envelope in stability analyses. Existing stability computer programs such as XSTABL and SLOPE/W can readily incorporate nonlinear failure envelopes.
5. Limited available data for rockfill materials indicates that their failure envelope is nonlinear. At the present time a constant friction angle corresponding to the angle of repose of the ore is used in stability analyses of heaps. It is recommended that a nonlinear failure envelope be used for the ore when analyzing the stability of very high heaps.

6. Review of shear strain-shear stress plots for various types of interface tests indicates that in many cases, the peak stress occurs at relatively small deformations and is followed by a significant reduction in strength with increasing deformation to a residual strength value. It is strongly recommended against use of the peak shear strength in stability analyses. Instead, a postpeak shear strength should be used.

It is also recommended that site-specific liner interface testing be conducted for all heap pad designs using the appropriate clay liner material and geomembrane types. In addition, the range of normal stresses used in the testing should correspond at least to the maximum heap height being designed whenever possible.

46.5 REFERENCES

Barton, N., and B. Kjaernsli. 1981. Shear strength of rockfill. ASCE Geotechnical Journal. 107: GT7:873.

Eigenbrod, K.D., and J.G. Locker. 1987. Determination of friction values for the design of side slopes lined or protected with geosynthetics. Canadian Geotechnical Journal. 24:509.

Gilbert, R.B., S.G. Wright, and E. Liedtke. 1998. Uncertainty in back analysis of slopes: Kettleman Hills case history. ASCE Journal of Geotechnical and Geoenvironmental Engineering. 124:12:1167.

Indrartna, B., D. Ionescu, and H.D. Christie. 1998. Shear behavior of railway ballast based on large-scale triaxial tests. ASCE Journal of Geotechnical and Geoenvironmental Engineering. 124:5:439.

Leps, T.M. 1970. Review of shearing strength of rockfill. ASCE Journal of Soil Mechanics and Foundation Div. 96:SM4:1159.

Maksimovic, M. 1996. A family of nonlinear failure envelopes for non-cemented soils and rock discontinuities. Electronic Journal of Geotechnical Engineering. Oklahoma State University.

Mitchell, J.K. 1993. Fundamentals of Soil Behavior, 2nd ed., New York: Wiley Interscience.

Mitchell, J.K., R.B. Seed, and H.B. Seed. 1990. Kettleman Hills waste landfill slope failure. I: liner-system properties. ASCE Journal of Geotechnical Engineering. 116:4:647.

Robertson, P.K., and C.E. Fear. 1997. Soil liquefaction and its evaluation based on SPT and CPT. Proceedings, Liquefaction Workshop (Draft), Salt Lake City, NCEER.

Seed, R.B., and P. De Alba. 1986. Use of SPT and CPT tests for evaluating the liquefaction potential of sands. ASCE Geotechnical Special Publication No. 6.

Seed, R.B., and R.W. Boulange. 1991. Smooth HPDE-clay liner interface shear strengths: compaction effects. ASCE Journal of Geotechnical Engineering, 117:4:686.

Sharma, S. 1998. XSTABL Version 5.202. Moscow, Idaho: Interface Software Designs, Inc.

Stark, T.D., and A.R. Poeppel. 1994. Landfill liner interface strengths from torsional-ring-shear tests. ASCE Journal of Geotechnical and Geoenvironmental Engineering. 120:3:597.

Ulrich, B.F. 1993. Consolidation study of an active tailings pond. Third International Conference on Case Histories in Geotechnical Engineering. St. Louis, Mo. 329–336.

Ulrich, B.F., and D.R. East. 1995. Experience with in situ tests in mine tailings for reclamation and facility expansion. Tailings and Mine Waste Symposium, Fort Collins, Colo.

Ulrich, B.F., and J.M.O. Hughes. 1994. SPT/CPT correlations for mine tailings. Proceedings of First International Conference on Tailings and Mine Waste, Fort Collins, Colo.

Valera, J.E., and B.F. Ulrich. 2000. Geomembrane/soil interface strength relationship for heap leach facility design. Tailings and Mine Waste Symposium, Fort Collins, Colo. 193–01.

. . . .
CHAPTER 47

Case Study: Stability Analysis of the Cresson Valley Leach Facility (Cripple Creek and Victor Gold Mining Company)

John F. Lupo* and Terry Mandziak†

47.1 INTRODUCTION

The Cripple Creek and Victor Gold Mining Company (CC&V) operates the Cresson Project located in Teller County, Colorado. The site location is shown in Figure 47.1. The Cresson Project is at an elevation of approximately 2,900 m above mean sea level (a.m.s.l.) and includes open pit mines, crushing facilities, a lined valley leach facility (VLF), and an adsorption/desorption/recovery (ADR) plant. Gold-bearing ore is produced from the mines, crushed, and placed within the VLF, where a weak, cyanide solution is applied. Gold is recovered from the pregnant-solution in the ADR plant through a carbon-adsorption process.

Since 1994, the VLF has been expanded in phases to accommodate increased ore production and reserves. The VLF expansions have included increased footprint of the facility as well as increased crushed ore height. The current VLF design consists of a lined footprint of more than 1.5 million m^2 and a maximum crushed ore height of approximately 180 m.

A major design component for expansion of the VLF is the stability of the ore slopes and liner foundation, which are critical to the operational, engineering, and environmental aspects of this project. This chapter presents the design approach, methodology, and results for the slope-stability analysis of the phase IV expansion of the Cresson Project VLF.

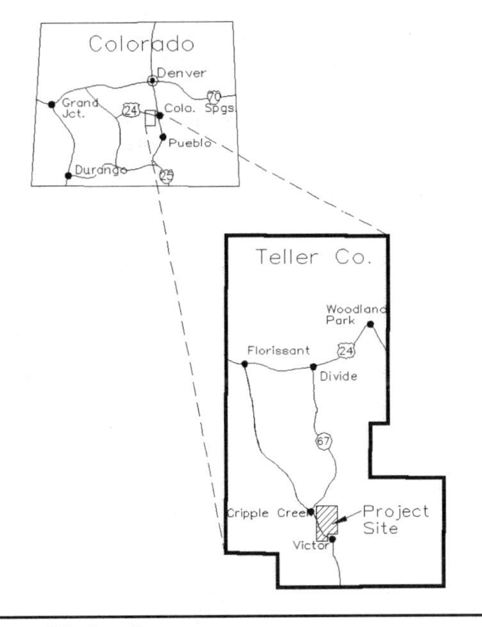

FIGURE 47.1 Site location map

47.2 VALLEY LEACH FACILITY DESIGN OVERVIEW

The Cresson VLF is situated within a steeply to gently sloping valley downslope of the mines and crusher facility. The Phase IV expansion has a design capacity of approximately 235 million tonnes of crushed ore, which will be placed within a 1.5-million m^2 geomembrane-lined facility. The design includes three internal pregnant-solution storage areas (two of which already exist), for a total internal storage of more than 550,000 m^3 of pregnant solution.

Two different liner systems are employed in the VLF design. A schematic of the liner systems used is shown in Figure 47.2. For the ore storage area (the area outside of the internal pregnant-solution storage areas), the liner system is a composite system consisting of an 80-mil geomembrane liner, which is placed over a compacted, low-permeability clay soil liner. The geomembrane liner materials used in the ore storage areas are either very low density polyethylene (VLDPE) or linear low density polyethylene (LLDPE), depending on the VLF phase.

Within the pregnant-solution storage areas, higher hydraulic heads will be present on the liner system. In these areas, the liner system consists of upper and lower 100-mil geomembrane liners. In between the liners is a solution-collection gravel, which is used to detect and collect any seepage through the upper geomembrane liner. The lower geomembrane liner is placed over a compacted, low-permeability clay soil liner.

Before crushed ore is placed on the ore or pregnant-solution storage areas, a drain cover layer is placed on top of the geomembrane liner for protection. The crushed ore is then placed on the VLF by end-dump from CAT 777 haul trucks. A plan view of the designed ultimate VLF configuration is shown in Figure 47.3.

47.3 DESIGN STABILITY CONSIDERATIONS

Stability of the VLF (both ore slopes and foundation) is critical to its overall performance. Failure of the ore slopes and/or the foundation could overstress the liner system, thereby compromising the hydraulic containment of the facility. The design of the VLF must include a thorough evaluation of stability to ensure short- and long-term performance.

In general, factors that can influence slope stability of leaching facilities include

- Ore material properties (shear strength, weight, permeability)
- Shear strength of the ore–geomembrane interface and shear strength of the clay soil–geomembrane interface at the anticipated normal stress

* Golder Associates Inc., Lakewood, Colorado.
† Cripple Creek & Victor Gold Mining Company, Victor, Colorado.

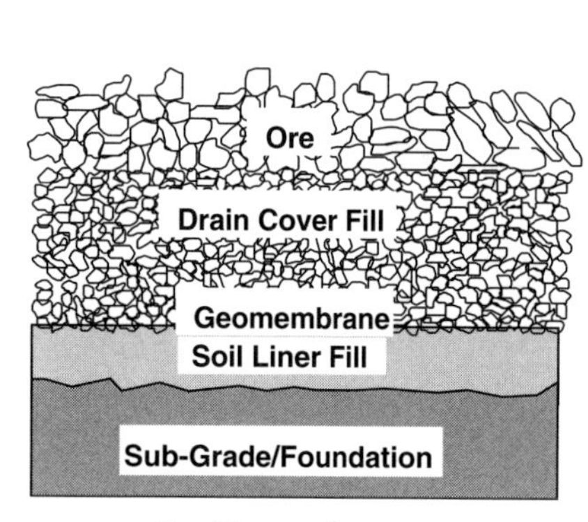

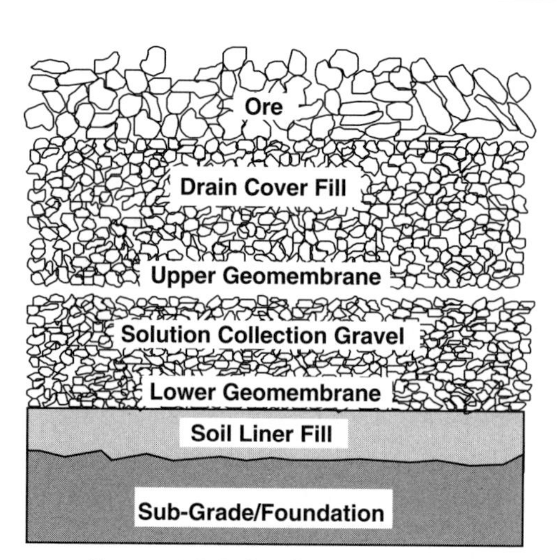

FIGURE 47.2 VLF liner systems

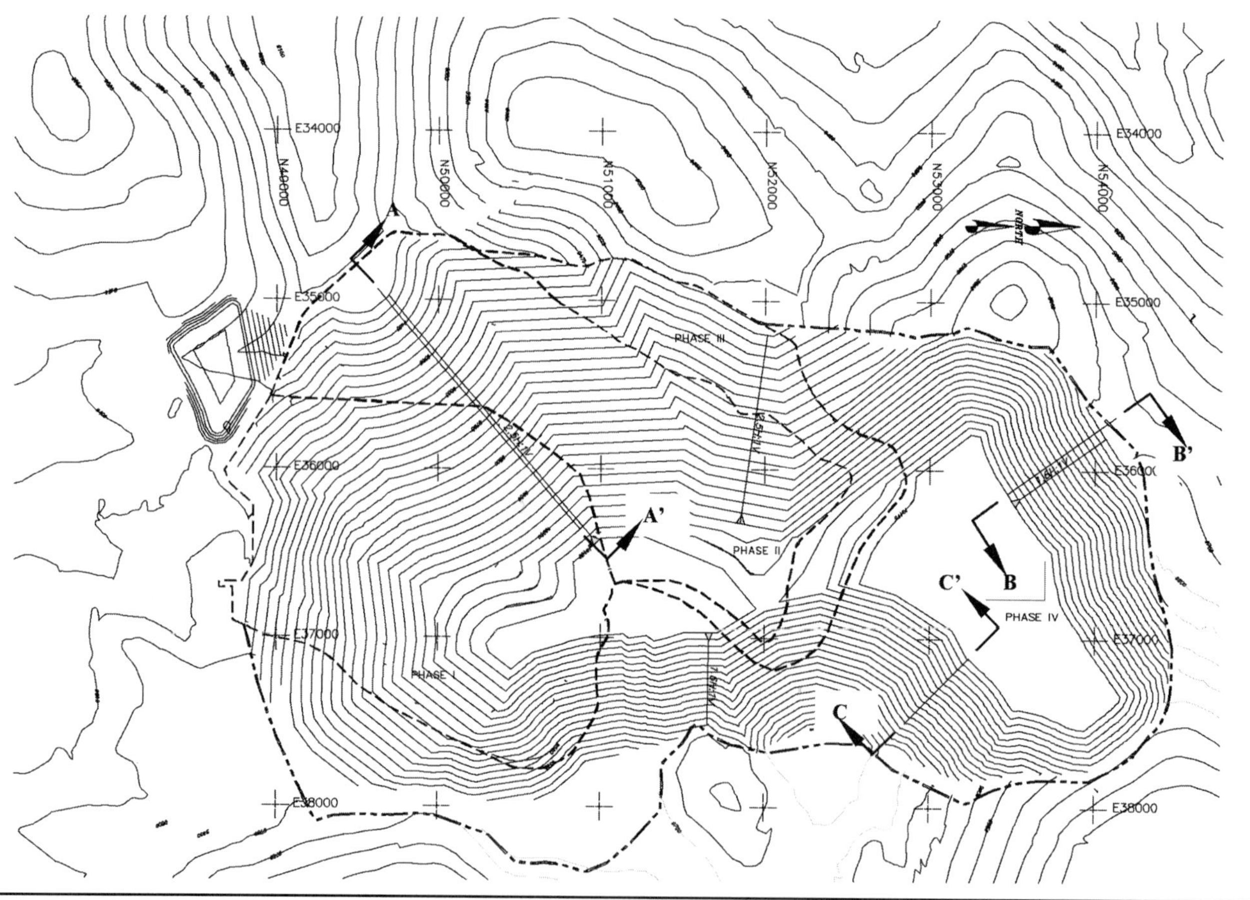

FIGURE 47.3 VLF plan map

- Shear strength of the foundation materials
- Slope of the foundation
- Ore side-slope angle
- Loading conditions (static and earthquake)
- Method of ore placement
- Location of phreatic (groundwater) surface beneath liner
- Location of phreatic surface (hydraulic solution head) above liner

Experience has shown that the modes of slope failure at leaching facilities (heap, dump, and valley) can include both shallow and deep failures. Shallow slope failures generally range from simple raveling of the ore slope to infinite slope failures involving one or more ore lifts. For the most part, shallow slope

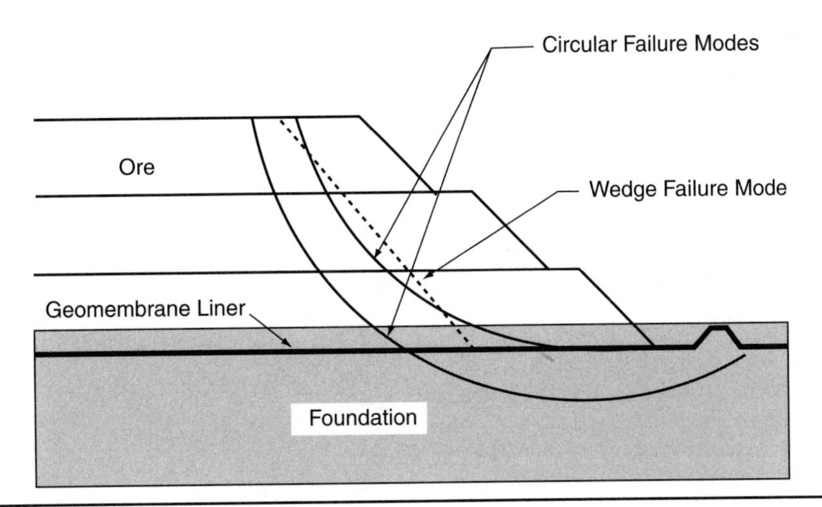

FIGURE 47.4 Deep failure modes

failures do not impact the liner system and are not considered an important design aspect for the purpose of this chapter.

Deep slope failure modes are illustrated in Figure 47.4 and may include

- Circular and noncircular failures contained within the ore material
- Wedge failures through the ore and along the ore–liner interface
- Circular and noncircular failures through the ore and into the foundation material(s)

Deep failure modes in leach facilities have the potential to damage the liner systems and must be evaluated as part of the design. It is also important that the design consider all potential modes of failure.

47.4 CRESSON PROJECT VALLEY LEACH FACILITY STABILITY ANALYSIS

This section summarizes the stability analysis conducted for the phase IV VLF. The general methodology for conducting the VLF slope-stability analyses included

- Identifying the location of critical stability sections
- Developing representative cross sections (two-dimensional) for each critical stability section
- Selecting the method for analysis (numerical, analytical, empirical)
- Identifying the different material types present in the cross section and assigning the appropriate material properties
- Performing site-specific laboratory testing to characterize the materials
- Identifying the boundary conditions and loading for each section
- Performing stability analysis evaluation

The critical stability sections for the VLF were determined by identifying areas within the facility with the steepest foundation slopes, the least amount of ore material at the toe (resisting force), and the greatest amount of ore material at the crest (driving force). Figure 47.3 shows the locations of the critical sections. Section A-A' was chosen because it has a downsloping foundation, the maximum ore height, and a relatively small buttress at the phase II toe berm.

Section B-B' was chosen because this section of the VLF will be supported with a structural berm on a downsloping foundation. This section also includes the maximum ore thickness. Foundation stability will be an important issue for the performance of this area of the VLF section.

Section C-C' was chosen as being representative of several areas within the VLF. This section is located perpendicular to grade, resulting in a flat foundation grade. Section C-C' can be used to test the stability of the VLF under maximum ore thickness and ore slope with a flat foundation. Other areas within the VLF have upsloping foundations and do not need to be evaluated for stability. If section C-C' is stable with a flat foundation, then areas with upsloping foundations will also be stable or have higher factors of safety.

47.4.1 Stability Analysis Method

To evaluate the stability of the VLF, a deterministic limit-equilibrium approach was selected. Limit-equilibrium methods analyze slope stability by evaluating moment and force equilibrium within the slope. This is typically done by dividing the slope into vertical slices. The moment and force components for each slice are then evaluated for the boundary and loading conditions. The stability of the slope can then be evaluated by summing the force and moments. If the summation of the force and moments exceed the available strength, the slope will fail. Typically, a factor of safety can be defined for the slope by simply comparing the shear stress to the shear strength. A static factor of safety of 1.0 indicates the shearing stresses are equal to the available shear strength. A static factor of safety less than 1.0 indicates slope instability, resulting in slope movement or deformation, while a factor of safety greater than 1.0 is indicative of a stable slope.

There are numerous limit-equilibrium methods that can be used to evaluate slope stability. A summary of the methods is presented in Table 47.1. For the VLF analysis, Janbu's simplified method (force equilibrium) was used to analyze the slopes and generate a range of failure surfaces. The critical failure surface (i.e., lowest factor of safety) was identified and then reevaluated using Spencer's method (force-moment equilibrium) to calculate the factor of safety on the critical surface. This approach was utilized because the Janbu method can quickly iterate a variety of failure surfaces and identify which failure surface has the lowest factor of safety. The critical surface used is then input into Spencer's method, as this approach is mathematically more rigorous, efficient, and less prone to numerical problems than other approaches and the stress boundary conditions are more realistic.

TABLE 47.1 Limit-equilibrium methods

Limit-Equilibrium Method	Characteristics
Method of Slices (Fellenius 1927)	Method addresses circular failure surfaces only; satisfies moment equilibrium only; does not satisfy force equilibrium
Bishops Modified Method (Bishop 1955)	Method addresses circular failure surfaces only; satisfies vertical force equilibrium only
Janbu's Generalized Method of Slices (Janbu 1968)	Method can be applied to any shaped surface; satisfies all conditions of equilibrium
Spencer's Method (Spencer 1967)	Method can be applied to any shaped surface; satisfies all conditions of equilibrium, but assumes side forces are parallel
Morgenstern and Price Method (Morgenstern and Price 1955)	Applies to any shaped surface; satisfies all conditions of equilibrium, but side forces can be variables
Force Equilibrium (U.S. Corps of Engineers 1970)	Applies to any shaped surface, but only satisfies force equilibrium

To simulate earthquake loading, a pseudostatic approach was used. Under this approach, seismic loading is simulated by incorporating a constant horizontal force. The horizontal force is calculated from an applied acceleration. A peak ground acceleration (PGA) value is either estimated for a project site from local references or determined from a survey of seismic activity in the area. Because the PGA value is sustained for such a short time, it is usually reduced by two-thirds (Newmark 1965; Corps of Engineers 1984) to yield a design acceleration that is used in the stability analysis. For the Cresson Project VLF, a PGA of 0.14 g (which was also used as the design acceleration) was used to simulate earthquake loading. This acceleration was determined from a site-specific seismicity study. The stability analysis is then repeated using the design acceleration and a new critical surface is calculated. A psuedo-static factor of safety greater than 1.0 indicates the slope will be stable during an earthquake event. A pseudo-static factor of safety less than 1.0 indicates the slope may move or deform as a result of the earthquake. To determine the amount of movement under earthquake loading, a separate deformation or yield analysis would be performed. For geomembrane-lined facililities, deformation of the lined foundation is generally considered unacceptable, while movement of the ore slope that does not impact the liner is acceptable.

47.4.2 Material Properties

The selection of materials used to construct any facility is not only important from an availability, costing, and constructability perspective but also equally important from a stability perspective. Generally, but not always, the shear strengths of soil materials will exceed that of soil/geosynthetic interface shear strengths. Consequently, the introduction of any synthetic materials in the design needs to be properly accounted for in the stability analysis. For example, the shear strength of a geotextile/geomembrane interface is lower than that of a soil/geotextile interface, and the shear strength of a soil/linear low-density polyethelene (LLDPE) interface may be higher than a soil/high-density polyethelene (HDPE) interface. The use of a textured geomembrane may increase the shear strength of an interface, as compared to a smooth geomembrane. A synthetic drainage layer between two geomembrane liners is easier to install as a leak-collection system, but a granular drainage layer may be better suited from an interface perspective. Careful consideration of the normal loads also needs to be given to ensure that the geomembrane is not installed in a loading condition that could damage the texturing. In addition, the interface shear strength may not vary linearly with normal stress. Testing must be conducted to fully define the interface shear strength of the chosen geomembrane. Therefore, laboratory testing on site-specific materials under a range of design conditions is crucial in the design of any containment facility.

TABLE 47.2 Material shear strengths

Material	Effective Friction Angle (degrees)	Effective Cohesion (kPa)
Foundation bedrock	40	0
Foundation native soil	26	0
Cresson crushed ore	40	0
Structural fill material	38	0
80-mil textured LLDPE/clay soil interface (phase III and IV liner system)	18	0
80-mil smooth VLDPE/clay soil interface (phase I liner system)	24	100
80-mil textured LLDPE/amended clay soil interface (phase II liner system)	15	1,600

The primary materials present in the critical slope-stability cross sections for the Cresson Project VLF include the following:

- Crushed Cresson ore
- Geomembrane liners (for phase I, II, III, and IV)
- Foundation bedrock
- Foundation native soils
- Structural fill

The methodology used to develop the material properties and shear-strength parameters for the stability analysis is presented in the following sections. A summary of the shear strength parameters for each material in the stability sections is presented in Table 47.2.

Cresson Ore. For the stability analysis, laboratory tests were conducted to determine the physical properties (bulk and dry unit weights) and shear strength of the crushed ore. The Cresson ore shear strength was determined by conducting a series of direct shear tests. These tests were conducted on representative samples of crushed ore. The laboratory test results indicate a curvilinear failure envelope for the ore, with internal friction angles ranging between 45° (under low-normal stress) and 36° (under high-normal stress) in the stress range expected along critical stability failure surfaces in the VLF. Weighting the internal friction angle to the normal stress levels expected along critical failure surfaces results in an internal friction angle of 38° to 40°. These test results agree well with published shear-strength data for rockfill materials (Leps 1970), where the internal friction angles range between 39° and 50° for similar-sized materials under similar normal stresses. Based on the results of the shear tests conducted on the Cresson ore and published data, an internal friction angle of 40° was chosen for the stability analyses. The in-place bulk density of the Cresson ore was determined to be 17.2 kN/m^3.

Geomembrane Liner Interface Shear Strength. Interface shear strength testing was performed on the liner system materials used for the VLF (phases I, I, II, and IV). Liner interface shear-strength tests were conducted in the laboratory using a standard shear box configuration. A representative section of the liner system for each VLF phase was constructed in the shear box, loaded under a specific normal stress, and sheared at a constant rate.

The liner system section was tested under several normal stresses, thereby allowing a shear strength versus normal stress curve for the liner configuration to be generated. This relationship was then used as input into the stability analyses.

Interface shear-strength tests were conducted on the primary liner systems for the VLF (phases I through IV) using different normal stress values ranging between 340 and 3,400 kPa. The maximum design normal stress for the VLF is 3,100 kPa, which is equivalent to an ore thickness of 180 meters with an ore bulk unit weight of 17.2 kN/m^3.

Additional evaluations were as follows:

- For the phase I ore storage area, the liner system consists of a 0.3-m-thick layer of compacted clay soil overlain by an 80-mil, smooth, VLDPE geomembrane liner overlain by a drain cover layer. The interface test results indicate that a linear shear strength defined with an angle of internal friction of 24° and 4.7-kPa cohesion can be used for the phase I liner system. The results of these tests also showed that the primary failure plane occurred between the liner and the clay soil.
- For the phase II liner system, the ore storage area liner system consists of a 0.3-m-thick layer of compacted bentonite-amended clay soil overlain by an 80-mil, single-sided, textured, LLDPE geomembrane liner overlain by a drain cover layer. The test results indicate that the primary failure plane occurs between the textured liner and the bentonite-amended clay soil. The shear strength can be defined as a linear failure envelope with an angle of internal friction of 15° and 74.6-kPa cohesion.
- The liner system for the phase II pregnant-solution storage area consists of an upper and lower, 100-mil, LLDPE geomembrane liners. In between the liners is solution-collection gravel, which is used to detect and collect any seepage through the upper geomembrane liner. The lower geomembrane liner is placed over a compacted, low-permeability clay-soil liner. Shear tests were conducted on a section consisting of solution-collection gravel/100-mil LLDPE textured geomembrane/compacted clay soil. The interface test results for this configuration were similar to those for the ore storage area, so both liner systems were modeled using the same shear strength properties.
- The phase III and phase IV liner systems for the ore storage liner area consist of a 0.3-m-thick layer of compacted clay soil overlain by an 80-mil, single-sided, textured, LLDPE geomembrane liner. The test results indicated an interface shear strength defined as a linear envelope with an angle of internal friction of 18 and zero cohesion.

Foundation Material. Foundation conditions within the VLF were evaluated prior to the design by conducting field investigation consisting of excavation of test pits and drilling test holes within the footprint of the VLF.

Test pits were excavated into the foundation to a nominal depth of 3.0 m. The test pits encountered approximately 0.3 m of growth medium, underlain by colluvial and residual soils and weathered bedrock. The soils encountered in the test pits generally consisted of silty sands classified as SM under the Unified Soil Classification System (USCS). The thickness of the soil ranged from 0 to more than 9 m. The underlying bedrock consists of weathered granodiorite, breccia, and phonolite. Groundwater was not encountered in any of the test pits.

Geotechnical holes were also drilled into the foundation to a nominal depth of 23.0 m to evaluate the subsurface conditions. The geotechnical holes were drilled using various methods, including augering, Becker hammer, and air-track. The results of the geotechnical drilling program indicate that the thickness of the weathered bedrock ranges between 0.6 and more than 2.0 m, with the quality of the bedrock increasing significantly with depth. Standard penetration tests (SPTs) were conducted in the colluvial/residual soil layer. The results of the SPTs, combined with Becker hammer drilling, indicate that the subsurface soils are firm, with blow counts in the low to mid 20s.

Groundwater was not encountered in any of the geotechnical holes. Historic mining records indicate that several deep drainage tunnels were excavated beneath and adjacent to the VLF area. Therefore, groundwater will not be an issue for the VLF stability.

The shear strength for the foundation bedrock and soils can be estimated from the test pit and geotechnical drilling data. The colluvial/residual soil shear strength can be conservatively estimated from the SPT blow count data using the empirical relationships presented by the Corps of Engineers (NAVFAC 1982). Using a nominal blow count of 22, the estimated friction angle for the native soils is 26° with zero cohesion.

The shear-strength parameters for the foundation bedrock were estimated using unconfined and point load test data supplied by the mine. The shear strength for the bedrock was evaluated using the Hoek–Brown (Hoek et al. 1998) criteria. The results of the evaluation indicated a shear strength of the bedrock with a friction angle of 45° and cohesion of 1,800 kPa. To account for weathering, the shear strength was defined with a friction angle of 40° and zero cohesion.

The structural fill used in constructing the VLF berms consists of coarse rock (minus 0.6-m-diameter) and soil. Based on the typical fill gradation and composition, the structural fill was conservatively estimated to have a shear strength defined by a friction angle of 38° and zero cohesion.

Piezometric Surface. A phreatic surface within the ore material can impact the stability of the facility, just like in any earthen slope. As a result, the permeability of the ore should be evaluated to establish what type of head, if any, would be generated on the composite liner. For this project, ore percolation testing was performed in which solution was applied to the ore at a rate of 10 times the solution application rate under loads corresponding to 3,100 kPa. Based on this testing, as well as a review of the ore gradation, the ore was determined to be free draining. While the piezometric surface within the VLF is dynamic and changes with the leach cycle and ore placement, the VLF was designed for a maximum hydraulic head of 0.63 m. All of the stability analyses included a phreatic surface that was 0.6 m above the liner in the ore storage areas. In the pregnant-solution storage areas, the maximum hydraulic head is assumed to be present, which was at the crest elevation of the toe berm.

If the ore at the Cresson Project was not free draining or if agglomerated ore was used, perched phreatic surfaces could be generated in the ore during leaching. Laboratory testing can be performed that can identify the ore percolation rate as a function of normal load, which can be used to identify the maximum ore depth before which ponding would occur. Two options would then be available. The first would be to revise the ore depth to reduce the potential for generating excess pore water pressures within the ore. The second option would be to lay back the ore

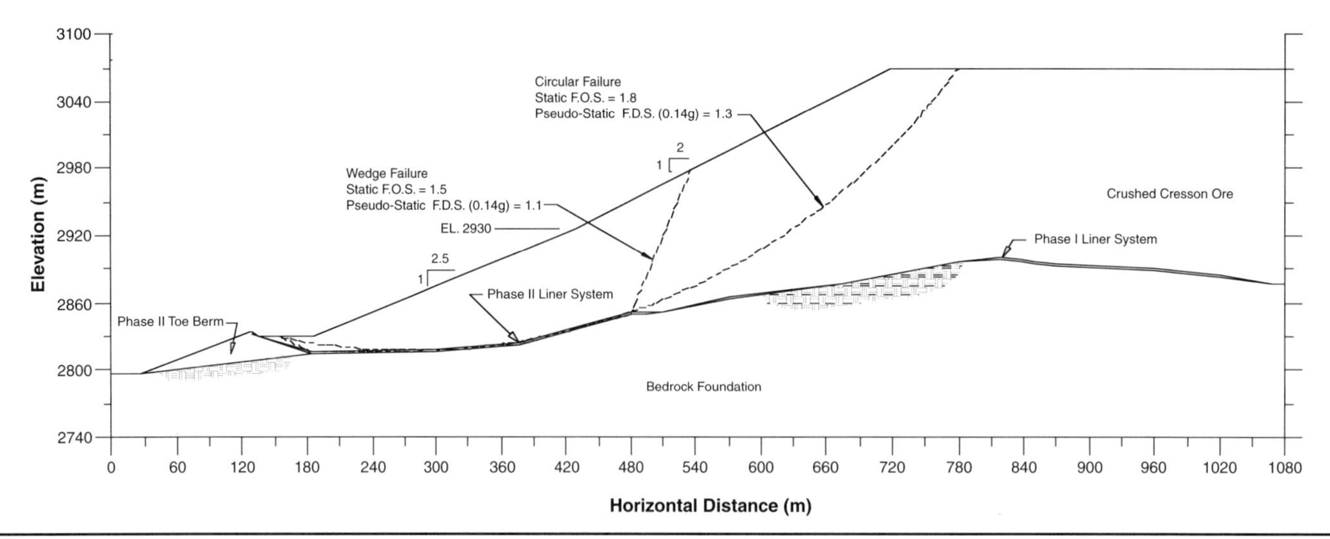

FIGURE 47.5 Stability section A-A'

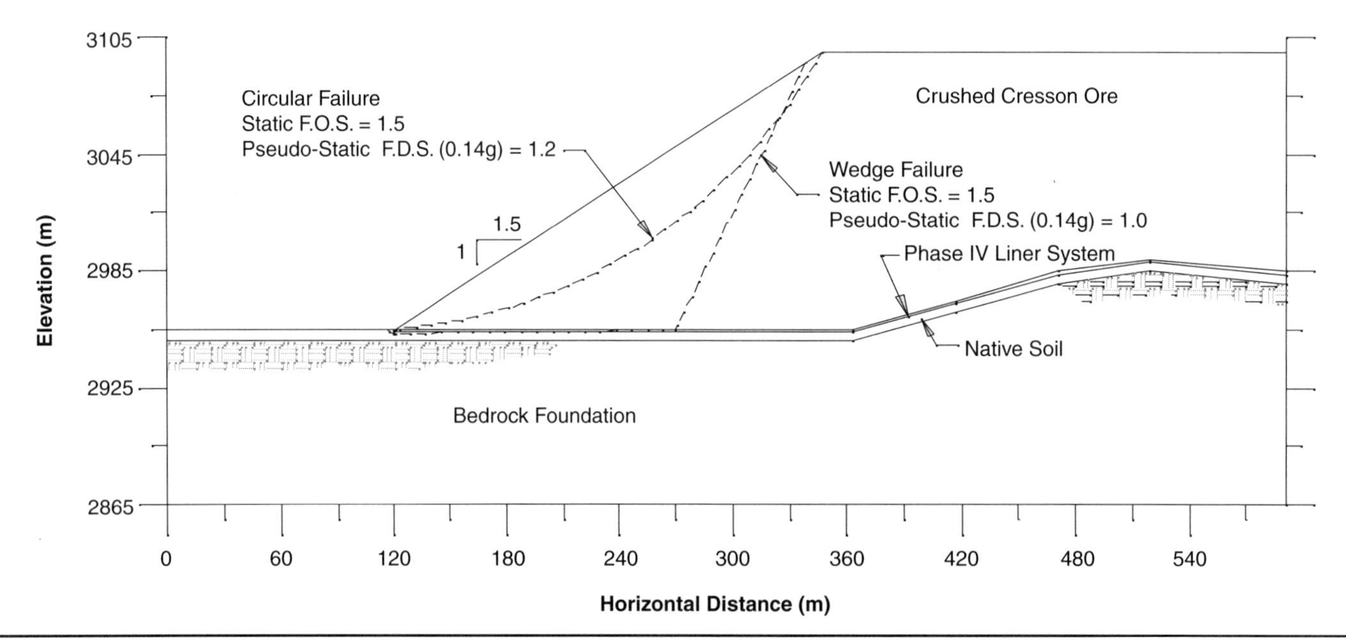

FIGURE 47.6 Stability section B-B'

slopes to account for the reduced factor of safety values calculated for the increased-head scenario. Either option would result in reduced ore capacity.

47.5 VALLEY LEACH FIELD STABILITY RESULTS

Stability analyses were performed on the critical sections shown on Figure 47.3. The stability analyses considered both circular and wedge failure modes within the VLF and through the foundation. The cross sections showing the stability results are presented in Figures 47.5 through 47.7, and a summary of the stability analyses is presented in Table 47.3. As shown, the stability results for the critical VLF sections indicate that the static factor of safety values are 1.5 or greater, and pseudo-static factor of safety values are 1.0 or greater for design acceleration of 0.14 g. These results indicate that the design slopes, liner system configuration, ore depth, and foundation for the phase IV VLF will be stable during the operational period of the mine life.

When the VLF is decommissioned, the ore slopes will be regraded to be flatter than the operational slopes, thereby ensuring long-term post-closure stability.

47.6 CONCLUSIONS

As part of the phase IV VLF design, an analysis was conducted to evaluate the stability of the ore side slopes and foundation. The analysis used site-specific data on the ore and foundation materials to develop shear-strength parameters. Laboratory testing was also used to determine the failure mode and shear strength of the geomembrane liner interfaces. These data were used in a limit-equilibrium approach to assess the stability of the VLF for deep failure modes.

The results of the VLF slope-stability analyses indicate that the designed ore side-slopes ore-depth configuration will be stable under static and earthquake-loading conditions.

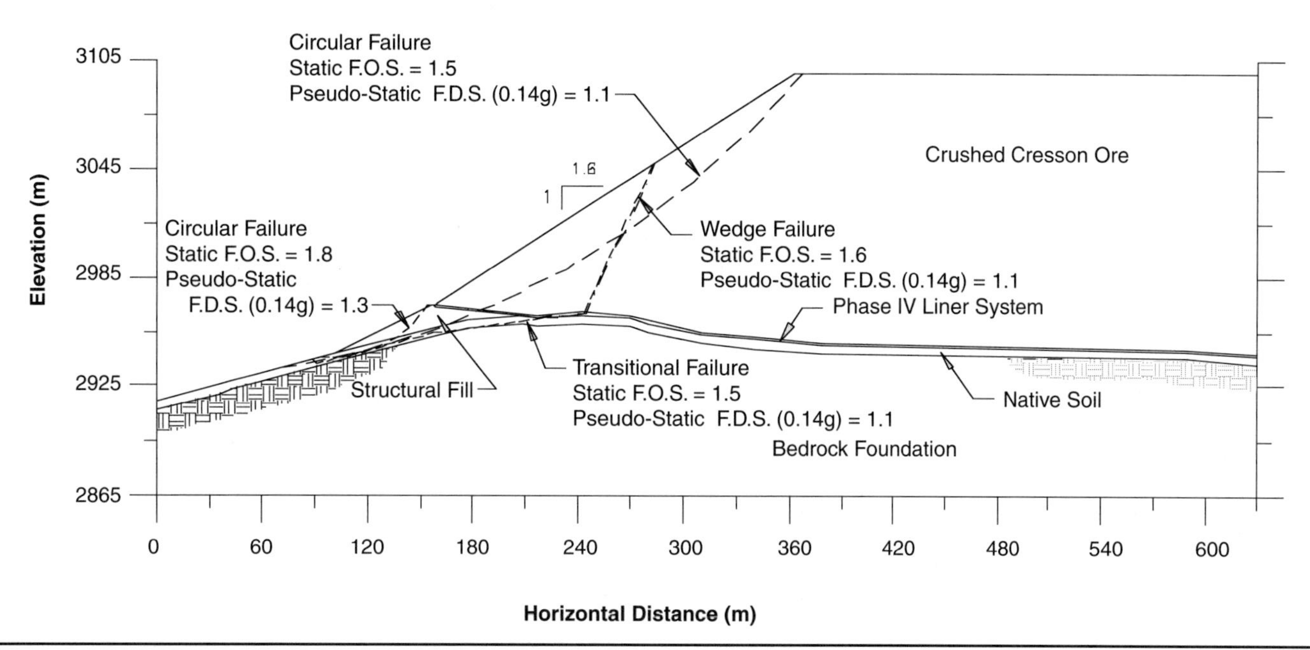

FIGURE 47.7 Stability section C-C'

TABLE 47.3 VLF stability results

Section	Static Factor of Safety	Pseudo-Static Factor of Safety 0.14 g
A-A' wedge failure mode along geomembrane/clay soil interface	1.5	1.1
A-A' circular failure mode	1.8	1.3
B-B' wedge failure mode along geomembrane/clay soil interface	1.6	1.1
B-B' wedge failure mode through foundation soil	1.5	1.1
B-B' circular failure mode	1.5	1.1
B-B' circular failure mode through saddle berm	1.8	1.3
C-C' wedge failure mode along geomembrane/clay soil interface	1.5	1.0
C-C' circular failure mode	1.5	1.2

47.7 ACKNOWLEDGMENTS

The authors thank CC&V and AngloGold North America for permission to publish the data and results of this chapter for the VLF phase IV design.

47.8 REFERENCES

Bishop, A.W. 1955. The use of the slip circle in the stability analysis of slopes. *Geotechnique,* London, 5(1), 7–17.

Hoek, E., Kaiser, P.K., and Bawden, W.F. 1998. *Support of Underground Excavations in Hard Rock,* Balkema.

Janbu, N. 1968. Slope stability computations. Soil Mech. and found. Engrg. Rep., The Technical University of Norway, Trondheim, Norway.

Janbu, N. 1954. Stability analysis of slopes with dimensionless parameters, Harvard Soil Mechanics Series, No. 46.

Leps, T.M. 1970. Review of shearing strength of rockfill. J. Soil Mechanics and Foundations Division, ASCE, 96, SM4.

Morgenstern, N.R., and Price, V.E. 1965. The analysis of the stability of general slip surfaces, *Geotechnique,* London, 15(1), 79–93.

NAVFAC. 1982. NAVFAC DM-7.1 Soil Mechanics, Naval Facilities Engineering Command, U.S. Navy.

Newmark, N.M. 1965. Effects of earthquakes on dams and embankments, *Geotechnique,* London, 15(2), 139–159.

Fellenius, W. 1927. Erdstatische Berechnungen mit Reibung und Kolhasion, Ernst, Berlin (in German).

Spencer, E. 1967. A method of analysis of the stability of embankments assuming parallel interslice forces. *Geotechnique,* London, 17(1), 11–26.

U.S. Army Corps of Engineers. 1970. Engineering and design–stability of earth and rock fill dams. Engr. Manual, EM 1110-2-1902, Dept. of the Army, Corps of Engrs.

U.S. Army Corps of Engineers. 1984. Rationalizing the seismic coefficient method. Misc. Paper GL-84-13, Dept. of the Army, Corps of Engrs.

CHAPTER 48

Radomiro Tomic Secondary Heap Leach Facility

Carlos Andrade,* Edgar Bard O,* Hugo Garrido,† and José Campaña*

48.1 DESCRIPTION

Radomiro Tomic is a copper deposit owned by Codelco-Chile. It is in the arid Atacama Desert in northern Chile, 15 km north of the Chuquicamata Mine, at an elevation of 3,000 m above sea level. The region is seismically active. Ore reserves have been estimated at 938 million tons, with an average of 0.59% total Cu. The copper oxide ore is mainly atacamite and chrysocolla.

Radomiro Tomic is an open-pit mine that has been in operation since 1997. The expected life of the project is 25 years. The SX/EW plant has the capacity to process 100,000 tons of ore/day, to produce 150,000 tons of Cu cathodes/yr. A plant expansion to 250,000 tpy of Cu cathodes is currently being developed.

The ore is blasted and crushed to minus 2 in. through a two-stage crushing system. After agglomeration and curing with water and sulfuric acid, the ore is stacked on an 8-m-high primary heap leach pad. After 45 days, the ore is removed with a bucket-wheel excavator and transferred by a conveyor and spreader to a secondary leach area. The secondary leach facility consists of a pie-shaped dump with a radial length of 3 km, covering a total surface area of 5.5 million m^2. The site is predominantly flat with 1 to 7% slopes.

Site preparation includes placing a double-textured high-density polyethylene (HDPE) geomembrane and installing 100-mm-diameter drain pipes, 5 m apart at their centers, supplemented with basal drains of selected material. Stacking the secondary leach ore takes place directly onto the liner from a spreader, producing a normal resting angle of about 37°. The dump height is currently about 60 m and the proposed final height is 110 m.

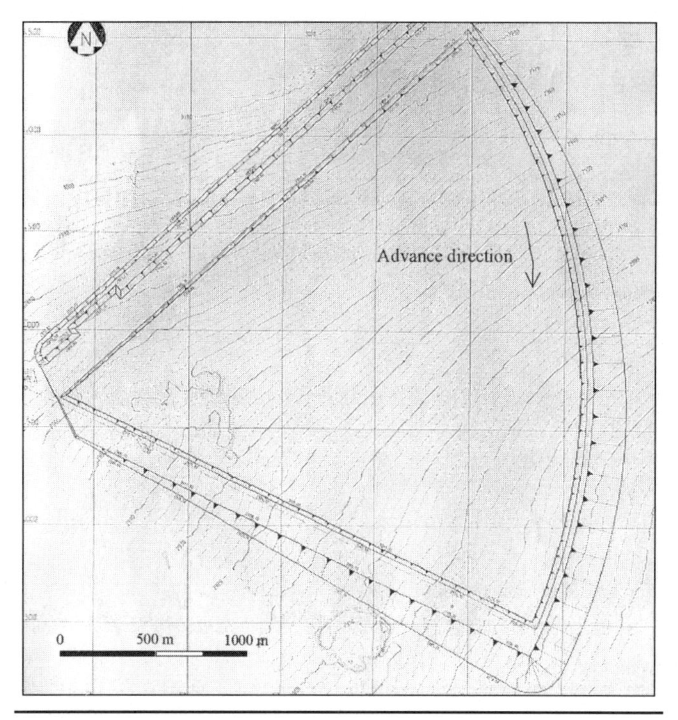

FIGURE 48.1 Site layout and foundation topography of the secondary heap leach facility, final scheme year 2018

48.2 STACKING THE SECONDARY HEAP LEACH

Topographic contours of the foundation and the outline of the secondary heap leach facility once completed are shown in Figure 48.1. A 3-km-long conveyor with a 60-m-long boom conveyor is used to distribute the ore along the top of the heap to the spreader. The height of the boom is adjustable up to a maximum of about 25 m from the surface of the previously deposited material.

A low cast and a high cast are constructed with the spreader system. A scheme of the stacking procedure as well as typical dimensions between the conveyor and spreader crests of low and high casts are presented in Figure 48.2. The low cast, presently at a height of 60 m, is deposited in a downslope direction on the pad, while the high cast is deposited on top of the low-cast material, which is ripped before placement of the high cast.

48.3 SOILS FOUNDATION AND ORE CHARACTERISTICS

Field investigations and laboratory tests were performed on representative foundation materials, ores, and geomembrane samples in order to characterize these materials, estimate index and mechanical parameters, and evaluate the shear strength of the ore/geomembrane/foundation soil interface. Grain size distribution curves for the foundation materials and ores are shown in Figure 48.3. A summary of their physical characteristics is presented in Table 48.1.

48.3.1 Soils Foundation

From a geological/geotechnical point of view, the foundation soil of the dump is homogeneous. The materials present at the surface are quaternary colluvials that correspond to a saline strata highly cemented with gravel lenses of 3 in. maximum size (caliche) and a soil cover of loose, silty-sandy soil (SM–ML) (chusca). These surficial materials present a total thickness of about 4 m, with an average permeability of $k = 10^{-3}$ cm/s. They are above a thick deposit of gravel of torrential fluvial deposition of Pleistocene age, known as RT Gravel. This deposit has an estimated thickness of between 130 and 220 m and has a permeability of the same order as the surface soils.

Bearing in mind technical, economic, and environmental aspects, it was determined, based on these factors, that the secondary leaching of the ore would only be feasible if an impermeable geomembrane were used at the base of the dump to collect leaching solutions.

* Arcadis Geotécnica Consultores S.A, Santiago, Chile.
† Corporación Nacional del Cobre de Chile, Codelco, Santiago, Chile.

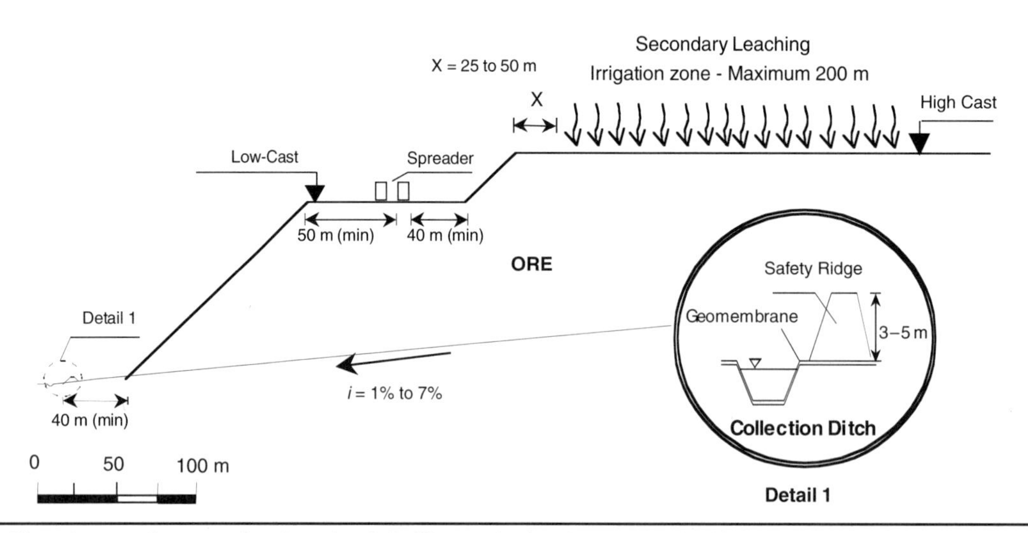

FIGURE 48.2 Stacking of ore on the secondary heap leach facility results in a low cast and a high cast

TABLE 48.1 Physical characteristics of the ore and soil foundation

Characteristic	Soils Foundation	Class 1 Ore	Class 2 Ore
Percent gravel	21	60 – 65	40 – 65
Percent sand	56	25	25 – 35
Percent minus # 200 sieve ASTM (by washing)	23	<10	10 – 25
Liquid limit	12 – 18	20 – 31	35 – 40
Plasticity index	NP – 4	7 – 13	10 – 22
USCS classification	SM-SM(ML)	GP-GC	GC
Saturated permeability (cm/s)			
▪ 0 to 30 m overburden		1×10^{-2}	1×10^{-2}
▪ 30 to 60 m overburden		1×10^{-2}	1×10^{-3}
▪ 60 to 90 m overburden		1×10^{-2}	1×10^{-4}
Moisture content when moved to secondary leach area (%)		8 – 10	10 – 12

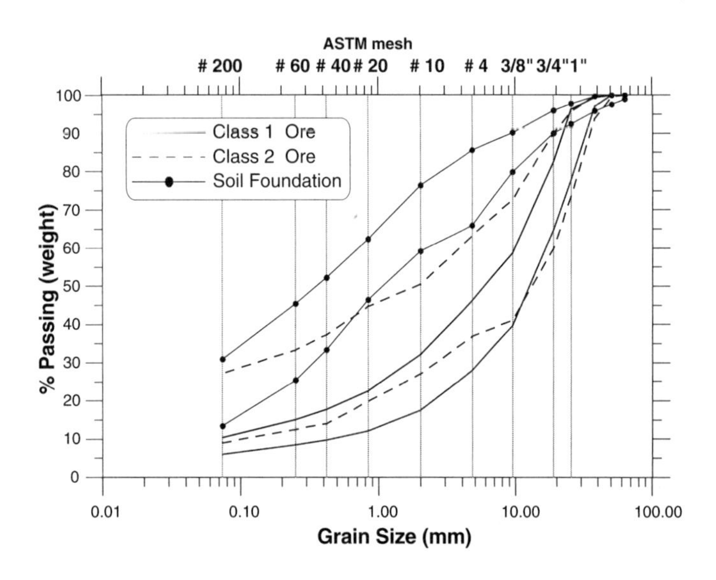

FIGURE 48.3 Grain size distribution, soil foundation, and ores

48.3.2 Ore Characteristics

A geological evaluation by Radomiro Tomic indicates that two major ore zones are present in the pit: the upper oxide and the lower oxide. Also, five mineralization types have been identified: atacamite, clays with copper, chrysocolla, copper wad, and others. A block model of the orebody was developed to obtain the distribution of ore types with depth. The model shows that the upper oxide zone contains 31% clay with copper mineralization, mainly montmorrilonite and kaolinite; the lower oxide zone contains only 11% of this type of mineralization.

According to mine development, from a geotechnical point of view, two classes of ore have been identified: class 1 and class 2. The class 1, or normal, ore consists mostly of atacamite and chrysocolla. Class 2 ore is scarce and its presence is very punctual. This material consists of clay with copper minerals and presents a greater fine content that is more plastic. Also, the moisture content of the class 2 ore is higher when it is deposited at the secondary leach facility. Grain size distribution curves of these materials are shown in Figure 48.3 and a summary of their physical characteristics is included in Table 48.1.

Infiltration rates were determined for class 1 and class 2 ores using water as the infiltrating fluid and imposing different hydraulic gradients on representative samples of these materials. For this purpose, both ore types were placed in molds with rigid walls at an initial dry density on the order of 16 kN/m^3. They were later saturated and subjected to axial loads of 0.6, 1.2, and 2.4 MPa. The evolution of the permeability coefficient of the

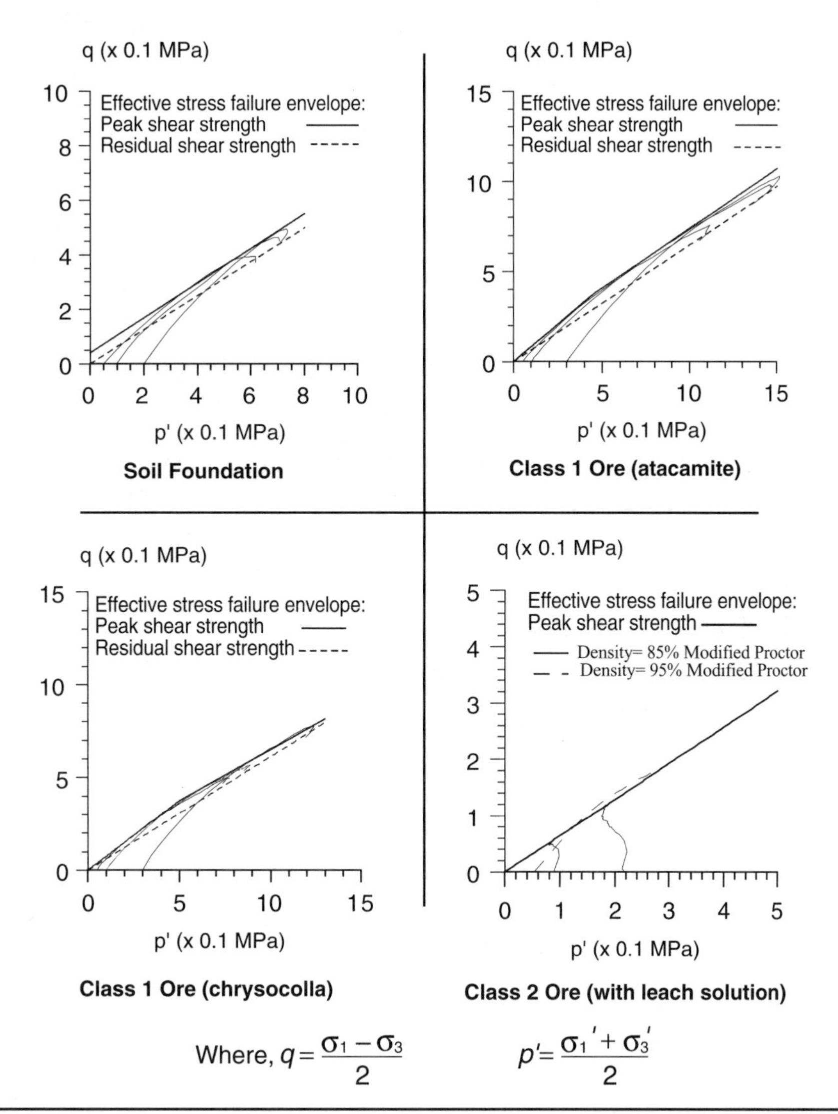

FIGURE 48.4 Results CIU triaxial test

materials under axial loads was determined simulating the effect of the load generated by the dump itself.

The saturated permeability of the class 1 ore ($k = 1 \times 10^{-2}$ cm/s) does not decrease with increased overburden stress, although the saturated permeability of the class 2 ore decreases with overburden loading to about $k = 1 \times 10^{-4}$ cm/s at an equivalent overburden height of more than 60 m.

In addition, in situ infiltration tests were performed on the more plastic ore deposited on the dump. The results confirm that the permeability of this material tends to decrease with an increase in height of the overburden and varies, in general, between 10^{-1} cm/s to 10^{-3} cm/s, for a height of less than 50 m of stacked material.

48.3.3 Shear Strength—Foundation Soils, Ore and Interface

Shear Strength of Soils. CIU triaxial tests were performed on both ores, and the results are presented in Table 48.2 and detailed in Figure 48.4. These tests were used to determine the evolution of the strength parameters with an increase in height of dump. It is important to note that the same acid leach solution as that used in the leaching process was used as fluid in the tests performed on the more plastic ore.

The Table 48.3 shows the adopted mechanical properties for class 1 ore for stability analysis purposes.

Interface. A series of large-scale direct shear tests reproducing field conditions were programmed to evaluate the strength properties of the foundation soil geomembrane-ore interface. These tests were done in a 60 × 60 cm box, with typical normal stresses of 0.3, 0.6, and 1.0 MPa.

Provided by four different manufacturers, 18 types of geomembrane were tested. Among the samples were double-textured HDPE, double-textured LLDPE, double-textured VFPE, VLDPE, and polypropylene. Additionally, geomembrane/geotextile composites were tested, with the geotextile welded on or placed next to one or two faces of the geomembrane.

The criteria for determining the best geomembrane or geocomposite were very strict. Indeed, preliminary stability analysis on the dump determined that the interface friction angle would need to be greater than 25°. Once the test was over, it was necessary that samples subjected to the shear test did not show any puncturing or loss of significant texture.

Table 48.4 is a summary of the shear test results based on the type of geomembrane or geocomposite. The large-scale direct shear test is shown in Figures 48.5.

TABLE 48.2 Summary results CIU triaxial test

Materials	Peak Shear Strength		Residual Shear Strength	
	c′ (kPa)	ϕ′ (°)	c′ (kPa)	ϕ′ (°)
Soils foundation (Chusca)	50	38	0	38
Class 1 ore				
▪ Atacamite			0, if p′<0.5 MPa	56, if p′<0.5 MPa
			10, if p′>0.5 MPa	41.5, if p′>0.5 MPa
▪ Chrysocolla			0, if p′<0.5 Mpa	48, if p′<0.5 MPa
			11, if p′>0.5 MPa	38, if p′>0.5 MPa
Class 2 ore			0	41

TABLE 48.3 Mechanical properties in the body of the heap leach for class 1 ore, (consider 50% atacamite and 50% chrysocolla)

Depth (m)	Cohesion, c′ (kPa)	ϕ′ (°)
z = 0 to z = 20	0	38
z = 20 to z = 40	10	37.8
z = 40 to z = 60	20	37.8
z = 60 to z = 80	40	37.6

TABLE 48.4 Summary results of geomembrane characterization

Test	Type	e (mm)	Strength Parameters: Peak c′ (MPa)	Peak ϕ′ (°)	Residual c′ (MPa)	Residual ϕ′ (°)	$\sigma'_{n\,max}$ (MPa)
1	1	2	0	32	0	26	0.95
2	2	1	0.076	27	0.137	23.6	0.95
3	3	1	a)		b)		1.2
4	4	1	0	31	0	18	1.0
5	5	2	0	34	0	30	0.9
6	1	1.5	0	29	0	25	1.1
7	5	1.5	0	33	0	28	1.0
8	1	1.5	0	33	0	25	1.05
9	1	1.5	0	33	0	30	1.05
10	1	1.5	c)		d)		0.94
11	6	1.5	0	23	0	15	1.08
12	1	2	e)		f)		1.17
13	7	2	g)		h)		0.95
14	7	1.5	i)		j)		1.1
15	8	1.5	0	30	0	25	1.1
16	8	2	0	27	0	25	1.05
17	9	2	0	28	0	23	1.15
18	9	1.5	k)		l)		1.11

Where: *e* = thickness, σ'_n = normal stress and,

Symbol	Description
Type 1	HDPE double textured
Type 2	Polypropylene HTS-1161
Type 3	VLDP + geotextile 270 g/m^2
Type 4	VLDP + geotextile 450 g/m^2
Type 5	LLDPE double textured
Type 6	HDPE textured + geotextile 150 g/m^2
Type 7	HDPE double textured (1 white size)
Type 8	VFPE double textured (Ultraflex)
Type 9	HDPE GSE HDT

Symbol	σ'_n (MPa)	c (MPa)	ϕ (°)
a	<0.58	0	34
	>0.58	0.17	22
b	<0.58	0	34
	>0.58	0.17	22
c	<0.3	0	35
	>0.3	0.12	18
d	<0.3	0	32
	>0.3	0.12	15
e	<0.45	0	35
	>0.45	0.17	18
f	<0.45	0	25
	>0.45	0.08	16
g	<0.4	0	29
	>0.4	0.035	25
h	<0.4	0	25
	>0.4	0.058	18
i	<0.4	0	35
	>0.4	0.12	22
j	<0.4	0	28
	>0.4	0.11	15
k	<0.5	0	37
	>0.5	0.24	15
l	<0.5	0	32
	>0.5	0.22	12

View of Large-Scale Direct Shear Equipment

Detail of Test Preparation

FIGURE 48.5 Large-scale direct shear test

48.4 BASAL DRAINAGE SYSTEM

The phreatic water level generated by leaching activities must be maintained at levels compatible with the stability of the advancing front of the dump. Therefore, the flow produced by those activities must be evacuated by a basal drainage system. The possible presence, even in isolated form, of ore with a lower permeability led to consideration of a basal drainage carpet for total heights of more than 60 m.

The basal drainage system should be comprised of corrugated and perforated HDPE pipes, 4 in. in diameter, 5 m apart, covered with a 1-m layer of class 1 ore with a low fine content. Intercepting drains, 5 m wide, of selected material will be placed every 25 m. This carpet will keep the phreatic level to a maximum of 1 m at the advancing front of the dump, strengthening the safety of the dump.

The height of the phreatic level within the dump was determined using rational formulas commonly accepted for drainage applications. The design criterion that only 25% of the drainage pipes be in operating condition was used. These analyses were complemented with bidimensional flow modeling, using the finite element software SEEP/W, considering the irrigation in a steady flow state.

To maintain a freatic level compatible with the dump's global stability at the advancing front, the lengths of the irrigation areas were adjusted. An analysis determined that it was necessary to restrict the high cast to a maximum irrigation length of 200 m, measured perpendicularly to the slope. Moreover, it is feasible to evacuate the amount of flow into the dump with the anticipated pipes.

A drainage carpet will also do the following:

- Assure the continuous presence of drainage material underneath the deposited ore
- Increase the drainage system's capacity, since the drainage carpet constitutes a preferential medium for the flow of the leaching solution
- Allow for better structural behavior of the pipes, providing them with a continuous support medium over the base geomembrane, and improve the confinement, which leads to a decrease in vertical deflection and thus an increase in the overall efficiency of the system
- Assure, at the same time, the correct placement and continuity of the pipes by preventing their disconnection due to potential relative movements originated by the deposition of the ore
- Maintain the phreatic level of the advancing front of the dump to within 1 m
- Protect the base geomembrane against eventual landslides

48.4.1 Radial Intercepting Ditch

Construction of radial intercepting ditches beneath the body of the dump are being considered. The objective is to intercept the leaching solutions, transport them to the perimeter ditch, and convey them to the process plant.

The geometrical design of the excavations was established based on the minimum ditch dimensions for arching development over the pipe. This phenomenon greatly reduces the vertical stress on the pipes and therefore reduces the potential vertical deflections and the possibility of a pipe failure.

To verify the efficiency of the proposed design, the radial collecting ditch number 1 was modeled using the finite difference software Fast Lagrangian Analysis of Continua (FLAC). The results obtained with the adopted model indicate arching development over the pipe, with deflection on the order of 2.5% for an overload equivalent to a dump of 120 m in height. This deflection value is acceptable for this type of pipe. Figure 48.6 is the radial collecting ditch number 1 during construction and FLAC software outputs used in the design.

48.5 DUMP STABILITY

The stability analyses were performed using limit-equilibrium methods, such as the simplified Bishop Method and Block Method, which allow for the determination of safety factors associated with different potential slope failure surfaces. In these analytical methods, the seismic loads are considered using seismic coefficients that allow for a pseudo-static analysis.

For the pseudo-static analysis of the dump, a horizontal seismic coefficient of 0.12 g derived from a seismic risk analysis was chosen. Safety factors of 1.2 for the static case and 1.0 for seismic conditions during operations have been considered acceptable. Nevertheless, slightly lower factors of safety do not necessarily result in a dump failure but rather have variable deformations associated with them, which must be verified in terms of the tolerance of structures and equipment involved.

The evolution of the degree of compaction of the deposited ores has been estimated using large odometer tests performed both on class 1 and class 2 ore. Figure 48.7 presents the density variation of both materials versus the height of the dump.

48.5.1 Stability Analysis Results

The stability analyses were performed considering potential sliding block-type failure surfaces, which are generated when the interface ore-geomembrane passes through the vertex of the slope's crest and therefore spawns potential sliding blocks. This surface failure corresponds to a translational failure of a block along the geomembrane/ore interface and results are critical.

Using the values obtained from stability analyses, the chart presented in Figure 48.8 was created. It is easy to observe the

Radial Ditch 1 and Collection Pipe

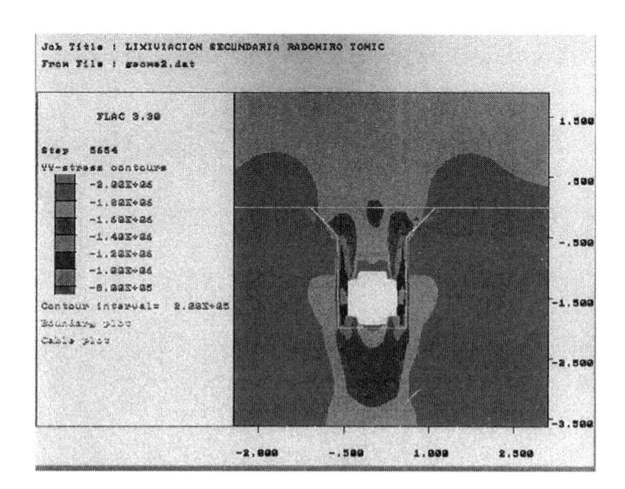

Output σ_{yy}—stress contours

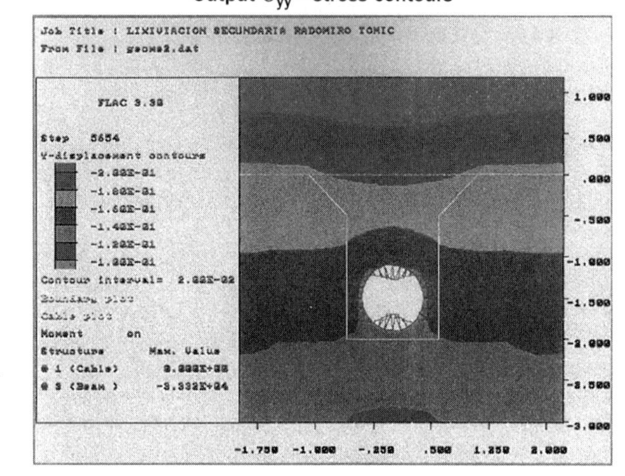

Output Y—displacement contours
Example Output FLAC Software

FIGURE 48.6 Ditch and collection pipe 1

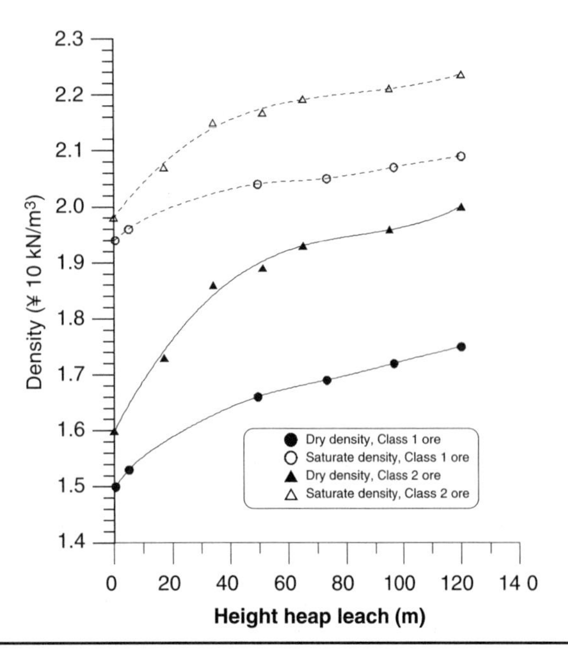

FIGURE 48.7 Variation of the density class 1 and class 2 ore

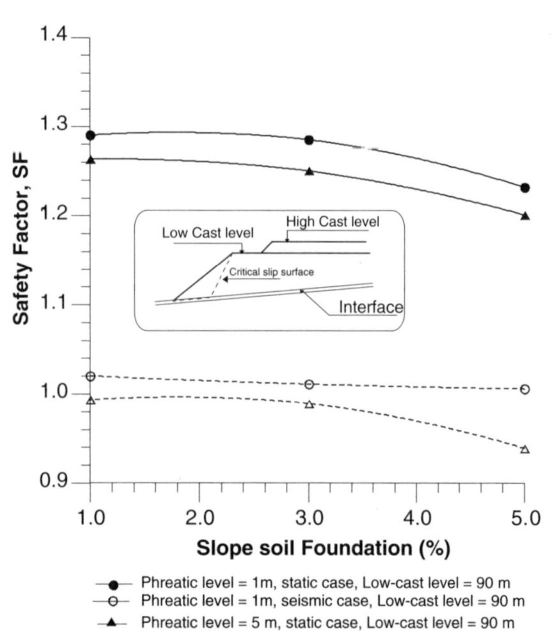

FIGURE 48.8 Variation of safety factor

Bucket-Wheel Excavator in Primary Heap Leach

View Secondary Heap Leach—Sector 1

View Stacking of Secondary Heap Leach

View Ore Conveyor and Spreader

FIGURE 48.9 Illustrative photographs

evolution of the static and seismic safety factors for a translational failure against the height of the low cast, the slope of the natural ground, and the phreatic level within the dump. The following conclusions were drawn after analyzing the results:

- The adopted stability criteria for operation process (FS_{static} >1.2 and $FS_{seismic} \approx 1$) where FS = factor of safety are verified for the different critical situations analyzed by maintaining the phreatic level at 1 m in the advancing front or at the toe of the final slopes considered.
- There isn't significant change in the safety factor values, both static and pseudo-static, for natural ground slopes of less than 3%. Higher slopes lead to a decrease in the safety factor of the dump.

The maximum height of the phreatic level compatible with the global stability of the dump is 4 m. Indeed, for a natural ground slope of 4%, the corresponding seismic factor of safety is less than 1, which does not comply with the adopted design criteria. This situation shows the sensitivity of the stability to minor variations of the adopted parameters. Therefore, the freatic level will be permanently monitored either in the advancing front or the final slope, and appropriate measures (reduction of the irrigated area, reduction of the irrigation rate, or proceed with an intermittent irrigation) must be taken to quickly reduce the level.

48.6 CONSTRUCTION STAGE

The deposition area of the ore has been divided into sectors. Currently, deposition of the first sector has been completed. The second sector has been started, with the current height of the dump approximately 75 m at the high-cast level. The boundary between sectors 1 and 2 corresponds to radial ditch number 1 (see Figure 48.7).

An industrial-scale leaching test has been programmed to evaluate the global behavior of the dump during the leaching stage from a geotechnical and metallurgy perspective. A 75-m × 75-m area has been conditioned for this purpose. The area has been under continuous irrigation since February 2000, at a rate of 8 l/h/m^2. This will continue for six to eight months. Six boreholes have been excavated and fitted with Casagrande-type piezometers to measure the phreatic level within the dump. Additionally, in situ geophysical tests will verify the evolution of the permeability, density, and moisture content within the dump and then compare them to the estimates previously established.

The operation of radial collecting ditch number 1 has been verified as optimum during the industrial-scale leaching test. Currently, the geotechnical data collected at the time of the industrial-scale leaching test is being analyzed.

Photos of the ore dump are shown in Figures 48.9 and 48.10.

48.7 ACKNOWLEDGMENT

The authors gratefully acknowledge Codelco-Chile, División Radomiro Tomic, for giving permission to publish this chapter.

48.8 REFERENCES

ADS Specification Manual (Leachate Collection Pipe Testing).

Arcadis Geotécnica, "Characterization of Interfaces," REP-1798-GE-01, Rev. 0, August 1999.

Arcadis Geotécnica, "Permeability Class 1 and Class 2 Ore," REP-1798-GE-02, Rev.0, September 1999.

Detail Stacking in Advancing Front

Leaching Test Area:
Casagrande-Type Piezometer for Phreatic Level Control

View Construction Stage of the Basal Drain—Sector 2 Secondary Heap Leach

FIGURE 48.10 Illustrative photographs

Arcadis Geotécnica, "Stability and Drains Secondary Heap Leach," REP-1798-GE-03, Rev. 0, February 2000.

Arcadis Geotécnica, "Mechanical Properties and Characterization of Interfaces Secondary Heap Leach," REP-1632-GE-02, Rev. 0, April 1998.

"Diseño y Construcción de Sistemas de Geomembranas en Proyectos de Ingeniería." Universidad Católica - Vector Engineering, Mayo 5,6 - 1994.

Geotécnica Consultores S.A., "Geotechnical Report Stability Secondary Heap Leach," REP-1521-GE-01, Rev. 1, 1997.

Geotécnica Consultores S.A., "Mechanical Properties and Characterization of Interfaces Secondary Heap Leach," REP-1521-GE-02, Rev. 0, April 1998.

Geotécnica Consultores S.A., "Report Geologic-Geotechnical for the Disposition In Place of Secondary Heap Leach - Project Radomiro Tomic," Rev. 0, April 1997.

Geotécnica Consultores S.A, "Geotechnical Report Primary Heap Leach," REP-866-001, Rev. 0, May 1993.

Geotécnica Consultores S.A. and Golder Associates, "Short course on Geotechnical and Environmental Aspects in the Design of Heap Leach Facilities," May 1996.

van Zyl, Dirk, "Technical Note Field Visit and Review Design Analysis Criteria, Secondary Heap Leach Radomiro Tomic Project," August 1999.

van Zyl, Hutchison, Kiel. "Introduction to Evaluation, Design and Operation of Precious Metal Heap Leaching Projects."

CHAPTER 49

Influence of the Direction of Ore Placement on the Stability of Ore Heaps on Geomembrane-Lined Pads

M.E. Smith* and J.P. Giroud†

49.1 INTRODUCTION

Based on experience, it is believed in the mining industry that ore heaps resting on geomembrane-lined graded pads are more likely to be unstable during stacking if the ore is stacked in the down-gradient direction than in the up-gradient direction. This belief is sometimes questioned because, to the best of the authors' knowledge, no demonstration is available. This chapter describes such a demonstration, which confirms the validity of the belief.

49.2 HEAP LEACH PAD DESCRIPTION AND OPERATION

A typical heap leach pad is shown in Figure 49.1. The pad has a gradient to collect the solution that has leached through the ore. In reality, the pad gradient is rarely uniform: It may increase or decrease locally to accommodate the natural ground topography while minimizing earthwork costs. Typically, an overall gradient is specified (e.g., 2%), with a maximum value (e.g., 5%), accepted in localized areas. In addition, the gradient must be positive to ensure proper collection of the solution (i.e., at any location, the gradient must be at least zero). The requirement for a positive gradient is well understood, and every effort is generally made in the field to avoid negative gradients, which would result in ponding of the solution. In contrast, the requirement for a maximum gradient is not always understood, and, as a result, pads are sometimes constructed with a gradient that, at some locations, is significantly greater than the specified maximum gradient (e.g., the pad gradient locally may be 10%). It will be shown that high pad gradients may have a detrimental impact on slope stability, especially when associated with stacking of the ore in the down-gradient direction.

The geomembrane used to line the pad has interface shear strengths with the overlying material (ore or geotextile) and the underlying material (soil or geotextile) that are smaller than the internal shear strength of the ore. As a result, potential slip surfaces for an ore slide run through the ore and then along the upper or lower face of the geomembrane.

The ore is stacked in a very loose state, paying particular attention to avoiding any undue compactive effort to keep the ore as permeable as possible. Therefore, even when it contains fines, ore is generally cohesionless or possessing a relatively low cohesion. As a result, if the ore is dumped or pushed, it will form a slope with the angle of repose, e.g., 35°, or 1 vertical:1.4 horizontal (1V:1.4H). Indeed, in heap leach pads, each lift of ore is typically stacked at the angle of repose. Stacking of the ore can proceed in the down-gradient direction (Figure 49.2a), in the up-gradient direction (Figure 49.2b), or in a direction parallel with the pad contours (not shown in the figures).

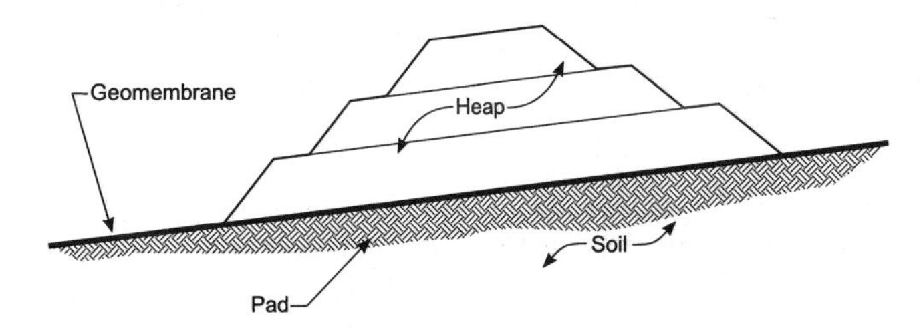

FIGURE 49.1 Schematic cross section of a typical ore heap leach pad (pad gradient and ore slopes are exaggerated)

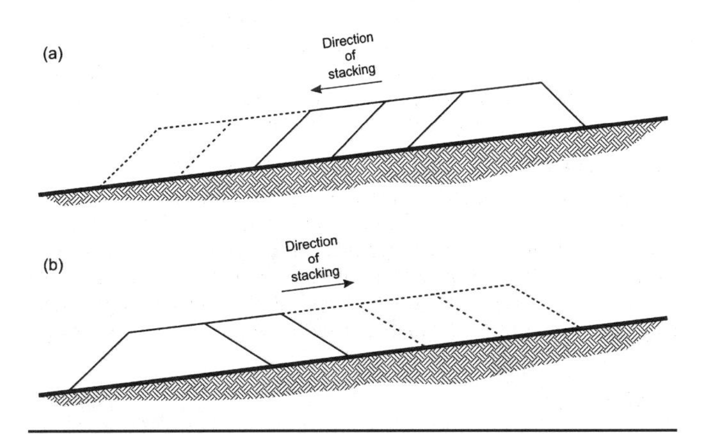

FIGURE 49.2 Stacking of ore: (a) down-gradient stacking; (b) up-gradient stacking (pad gradient and ore slopes are exaggerated)

A heap is generally constructed in successive lifts, and each lift is irrigated with a leaching solution (typically alkalis for precious metals and acids for base metals) before the subsequent lift is placed. Typical lift heights are 4 to 10 m. At each lift, a bench reduces the average overall slope. Further, as the loose ore is leached, there is a degree of hydraulic compaction, and the additional weight of the subsequent lift further compacts the lower lift(s); the resulting increased density often increases the shear strength of the ore in the lower lift(s). Finally, the importance of equipment loading relative to ore heap weight is greater on the first lift than at subsequent stages when the heap comprises several lifts. Therefore, for the above reasons (decreased average slope, increased shear strength, and lesser relative importance of equipment loading) a heap comprising several lifts with benches is often more stable than the first lift. This statement is valid only: (i) if the cohesion of the ore is zero or small; (ii) for ores that

* Vector Engineering, Inc., Grass Valley, California.
† GeoSyntec Consultants, Boca Raton, Florida.

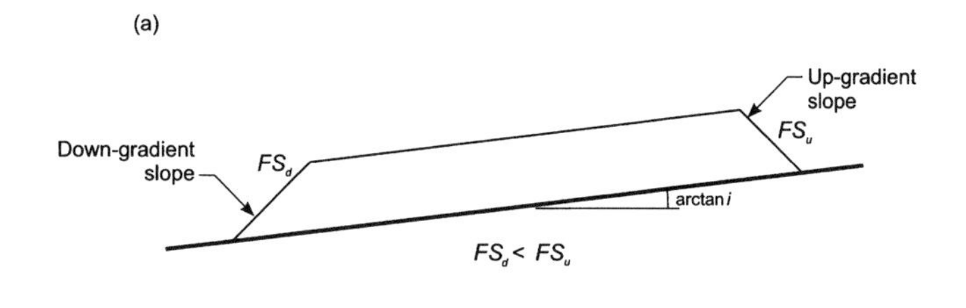

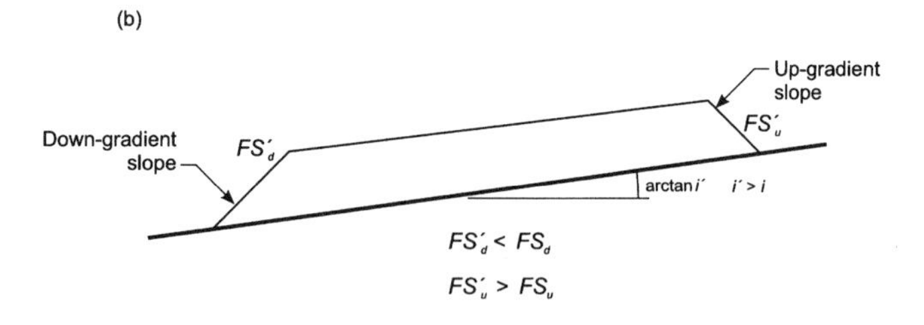

FIGURE 49.3 Influence of pad gradient on factor of safety: (a) pad with uniform gradient *i*; (b) pad with uniform gradient *i'* (four ore slopes are equal)

become more dense after initial placement; (iii) for failure surfaces not limited within the top lift if there is more than one lift; and (iv) for failure surfaces not involving the foundation of the pad (i.e., for failure surfaces entirely within the heap and the pad). These, in fact, are the most common conditions. In particular, the last two conditions are met if the failure surface is located along the geomembrane, which is likely, due to the low interface shear strength associated with the geomembrane.

49.3 BASIC STABILITY CONSIDERATIONS

In this section, an ideal geomembrane-lined pad with a uniform gradient *i* is considered. If an ore heap with identical slopes on the up-gradient and down-gradient sides (hereafter referred to as the up-gradient slope and down-gradient slope, respectively) is constructed on this pad, the factor of safety of the down-gradient slope is smaller than the factor of safety of the up-gradient slope (see Figure 49.3a):

$$FS_d < FS_u \qquad \text{EQ. 49.1}$$

Where:

FS_d = factor of safety of the down-gradient slope when the pad gradient is *i*

FS_u = factor of safety of the up-gradient slope when the pad gradient is *i*

Equation 49.1 is an obvious relationship that can easily be verified by elementary slope stability calculations for the case in which the slip surface runs along the geomembrane.

If the same ore heap with the same slopes were constructed on a pad lined with the same geomembrane having a greater gradient ($i' > i$), the following relationships would exist between the factors of safety (Figure 49.3b):

$$FS'_d < FS_d \qquad \text{EQ. 49.2}$$

$$FS'_u > FS_u \qquad \text{EQ. 49.3}$$

Where:

FS'_d = factor of safety of the down-gradient slope when the pad gradient is *i'*

FS'_u = factor of safety of the up-gradient slope when the pad gradient is *i'*

Equations 49.2 and 49.3 are obvious relationships that can easily be verified by elementary slope stability calculations for the case in which the slip surface runs along the geomembrane.

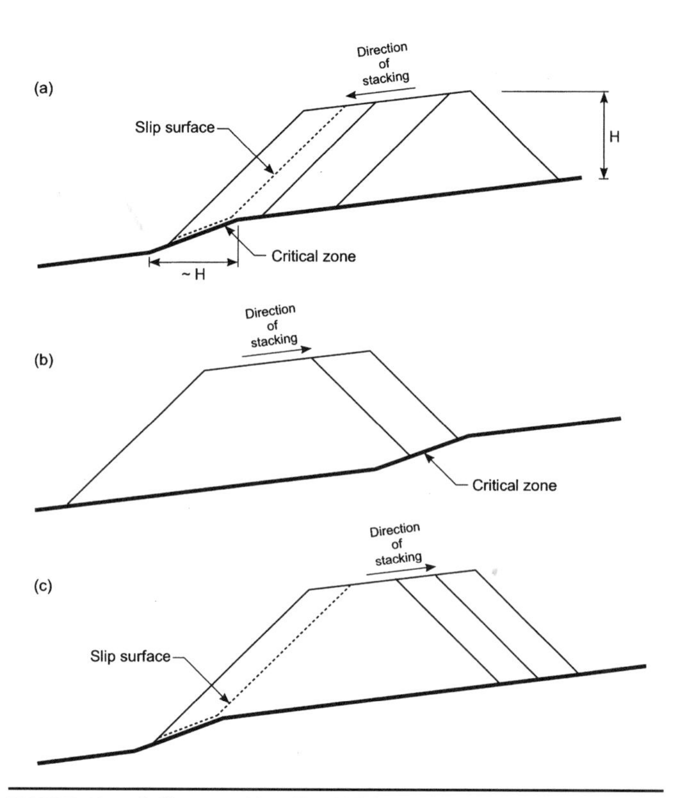

FIGURE 49.4 Influence of localized high pad gradient on ore stability during construction: (a) down-gradient stacking; (b) up-gradient stacking; (c) up-gradient stacking with a critical zone at the down-gradient edge of the pad (The bold solid lines represent the geomembrane liner; the solid lines represent successive stages of stacking; the dashed lines represent the slip surface.)

According to Eq. 49.2 and 49.3, increasing the pad gradient decreases the stability of a down-gradient slope and increases the stability of an up-gradient slope. Equations 49.1 to 49.3 constitute the basis for the demonstration presented in the next section.

As a practical matter, many heaps are designed with average overall up-gradient slopes steeper than average overall down-gradient slopes by using narrower benches on the up-gradient slopes, such that the resulting design factors of safety are approximately equal for the up-gradient and down-gradient slopes.

49.4 STABILITY DURING STACKING

A geomembrane-lined pad with a gradient that varies locally is considered. The gradient is everywhere equal to or greater than a minimum value, such as zero. It is assumed that there are "critical zones" in the pad that are sufficiently large and where the gradient is sufficiently high that the factor of safety of a down-gradient slope constructed in such a zone would be less than 1.0. In other words, a down-gradient slope in a critical zone would slide. To have a significant impact on ore stability, a critical zone must be large; it is assumed that, to have a significant impact on ore stability, a critical zone should have an extent greater than or equal to the height of the first lift of ore. The same concept applies to the total heap height, but the analysis is somewhat different and is beyond the scope of this discussion.

If stacking is done in the down-gradient direction (Figure 49.4a), an ore slide will occur when the ore being stacked reaches a critical zone. In contrast, if stacking is done in the up-gradient direction (Figure 49.4b), the stability of the ore slope will in fact increase (based on Eq. 49.3) upon reaching a critical zone. An ore slide will occur in the case of up-gradient stacking only if a critical zone is located at the down-gradient edge of the pad (Figure 49.4c), in which case an ore slide would occur in the down-gradient direction.

Ignoring factors other than pad gradient, the probability for a slide to occur is equal to the probability for a critical zone to exist (which is dependent only on pad topography and is, therefore, independent of the direction of stacking) multiplied by the probability for such a zone to be located in an area where it impacts stability. The latter is 1.0 in the case of stacking in the down-gradient direction and H/L in the case of stacking in the up-gradient direction, if it is assumed that the distance from the edge of the pad where a critical zone may cause a slide in the down-gradient direction is approximately equal to the height H of the ore lift (L being the distance between the down-gradient and up-gradient edges of the pad if the pad slopes in a single direction, or the distance between the edge of the pad under consideration and the top of the pad if the pad slopes in two directions). Therefore, the probability for a slide to occur is L/H times greater if stacking is done in the down-gradient direction than if stacking is done in the up-gradient direction. Clearly there is a far greater risk of ore sliding during stacking if stacking is done in the down-gradient direction.

It follows from the above that, if the pad gradient is specified not to exceed a certain critical gradient (divided by a safety factor), no ore slide can theoretically occur when stacking is done in the up-gradient direction. It is very easy and not costly to take this measure, thereby theoretically eliminating the risk of ore instability.

It should, however, be noted that parameters other than the stacking direction (down-gradient or up-gradient) have an influence on ore stability, such as ore characteristics and water content, which may vary significantly from one point to another and from time to time. Such variations may decrease the gradient required for a zone to become critical. As there are always uncertainties in such parameters, it is not possible to totally eliminate the risk of failure in the case of the up-gradient stacking, but at least this risk can be greatly minimized.

The model considered above is essentially two-dimensional (i.e., only cross sections are considered). As a result, only two directions are involved in the demonstrations, the down-gradient direction and the up-gradient direction. In reality, heap leach pads are three-dimensional, and all directions may have an impact on ore slope stability. However, the discussions above show that the only relevant characteristic of a given stacking direction regarding ore stability is whether it is down-gradient or up-gradient. Therefore, the conclusions derived from the discussions are applicable to actual three-dimensional situations.

In the discussions above, the driving forces likely to cause ore instability were not mentioned. Typical driving forces are the self-weight of the ore and the load (including vertical and horizontal components) applied by the stacking equipment. The equipment load often plays an important role, because very large equipment is often used over a relatively thin lift of ore. It is important to note that the conclusions drawn from these discussions are applicable regardless of the type of driving force.

49.5 LONG-TERM STABILITY

The discussions above clearly show that stability during stacking is affected by the stacking direction. It is also possible that the stacking direction (down-gradient or up-gradient) affects the long-term stability of the ore heap, although this hypothesis is not supported by clear field evidence or experience in the mining industry.

The hypothesis that the stacking direction affects long-term ore heap stability is based on the following considerations:

- An ore heap that is completed or at an advanced stage of stacking contains many internal planes sloping at the angle of repose of the ore (Figures 49.5 and 49.6). These internal planes are the temporary ore slopes at each construction phase.

FIGURE 49.5 Lateral view of an ore lift showing parallel planes sloping at the angle of repose

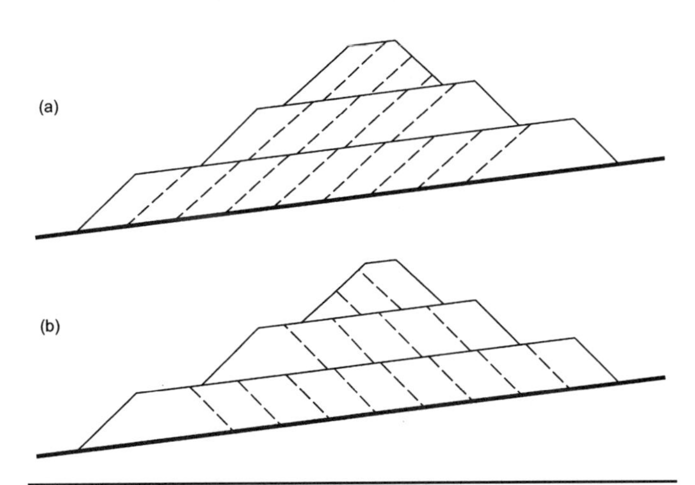

FIGURE 49.6 Internal planes in an ore heap: (a) heap stacked in the down-gradient direction; (b) heap stacked in the up-gradient direction

- It is possible, but not certain, that the ore shear strength along these internal planes is less than that along any other surface through the ore. This could be because there is less interlocking between ore particles along these planes than there is within the ore mass. Another factor could be the concentration of finer particles along these planes, as the larger stones tend to roll to the base of the heap (Figure 49.7). While there is no documented direct evidence that segregation creates weak planes, it is well recognized in the mining industry that this segregation leads to non-uniform flow of solutions within the heap and, in extreme cases, creates zones of instability due to perched water and erosion of the side slopes.
- Since ore heap instability is more likely to take place in the down-gradient than in the up-gradient direction, as discussed earlier, low shear-strength internal planes will be more likely to cause instability if they are sloping in the down-gradient direction than in the up-gradient direction.
- In an ore heap stacked in the down-gradient direction, the internal planes are all sloping in the down-gradient direction (Figure 49.6a), whereas in an ore heap stacked in the up-gradient direction, the internal planes are all sloping in the up-gradient direction (Figure 49.6b).

FIGURE 49.7 Ore segregation

- Based on the above rationale, ore heaps stacked in the down-gradient direction are more likely to be unstable in the long term than ore heaps stacked in the up-gradient direction. However, it should be noted that the rationale above is based on the assumption that low shear strength internal planes result from stacking phases, an assumption that has not been evaluated experimentally.

There is another mechanism through which the stacking direction may impact the long-term stability of the ore heap. Localized failures that occur during stacking, if they are not serious enough to require liner repair, may result in displacements such that only post-peak (even only residual) shear strength remains available for future potential slides along the same slip surface. Since localized failures during stacking and failure of the complete ore heap at a later stage are likely to have portions of slip surface in common along the geomembrane liner, the factor of safety for long-term stability of the ore heap is decreased if local failure occurs during stacking. Since the probability of localized failures during stacking is greater with down-gradient stacking than with up-gradient stacking, it follows that the probability for decreasing the factor of safety for long-term stability of the ore heap (through the post-peak shear strength mechanism) is greater in the case of down-gradient stacking than in the case of up-gradient stacking.

49.6 CONCLUSIONS

It has been demonstrated that, when the pad gradient is not rigorously uniform (which is generally the case), the probability of an ore slide during stacking of ore on a leach pad is much greater if the ore is stacked in the down-gradient direction than in the up-gradient direction. This is consistent with experience in the mining industry. It has also been shown that, with simple precautions, it is possible to minimize and perhaps eliminate the risk of an ore slide during construction in the case of stacking in the up-gradient direction.

It has also been shown that there is a possibility that an ore heap may be more likely to exhibit long-term instability if it has been stacked in the down-gradient direction than if it has been stacked in the up-gradient direction. However, this possibility has not been evaluated experimentally. Also, it should be noted that, in a number of cases, ore instability is less likely to occur in the long term than during stacking of the first lift for the reasons discussed here.

Furthermore, it should be noted that, in some cases, ore heap instability is more likely to occur during ore placement than in the long term because the first lift of ore has a steeper slope than the average overall slope, including benches between lifts.

Index